MATHEMATICS
FOR
DEGREE STUDENTS

B Sc First Year

MATHEMATICS FOR DEGREE STUDENTS

B Sc First Year

As per UGC Model Curriculum

- Algebra and Trigonometry
- Calculus (Differential & Integral) and Ordinary Differential Equations
- Vector Analysis and Geometry

Dr. P.K. MITTAL

M Sc, PhD

Formerly Head of Mathematics Department
Govt. Post Graduate College
Rishikesh (Uttarakhand)
(Reviser of Prof. Shanti Narayan's Mathematics Books)

S Chand And Company Limited

(An ISO 9001 : 2008 Company)
Ram Nagar, New Delhi – 110 055

S Chand And Company Limited
(An ISO 9001:2008 Company)
Head Office: 7361, RAM NAGAR, NEW DELHI - 110 055
Phone: 23672080-81-82, 9899107446, 9911310888 Fax: 91-11-23677446
www.schandpublishing.com; e-mail: helpdesk@schandpublishing.com

Branches

Ahmedabad	:	Ph: 27541965, 27542369, ahmedabad@schandpublishing.com
Bengaluru	:	Ph: 22268048, 22354008, bangalore@schandpublishing.com
Bhopal	:	Ph: 4274723, 4209587; bhopal@schandpublishing.com
Chandigarh	:	Ph: 2625356, 2625546, chandigarh@schandpublishing.com
Chennai	:	Ph: 28410027, 28410058, chennai@schandpublishing.com
Coimbatore	:	Ph: 2323620, 4217136, coimbatore@schandpublishing.com (Marketing Office)
Cuttack	:	Ph: 2332580, 2332581, cuttack@schandpublishing.com
Dehradun	:	Ph: 2711101, 2710861, dehradun@schandpublishing.com
Guwahati	:	Ph: 2738811, 2735640, guwahati@schandpublishing.com
Hyderabad	:	Ph: 27550194, 27550195, hyderabad@schandpublishing.com
Jaipur	:	Ph: 2219175, 2219176, jaipur@schandpublishing.com
Jalandhar	:	Ph: 2401630, 5000630, jalandhar@schandpublishing.com
Kochi	:	Ph: 2378740, 2378207-08, cochin@schandpublishing.com
Kolkata	:	Ph: 22367459, 22373914, kolkata@schandpublishing.com
Lucknow	:	Ph: 4026791, 4065646, lucknow@schandpublishing.com
Mumbai	:	Ph: 22690881, 22610885, mumbai@schandpublishing.com
Nagpur	:	Ph: 6451311, 2720523, 2777666, nagpur@schandpublishing.com
Patna	:	Ph: 2300489, 2302100, patna@schandpublishing.com
Pune	:	Ph: 64017298, pune@schandpublishing.com
Raipur	:	Ph: 2443142, raipur@schandpublishing.com (Marketing Office)
Ranchi	:	Ph: 2361178, ranchi@schandpublishing.com
Siliguri	:	Ph: 2520750, siliguri@schandpublishing.com (Marketing Office)
Visakhapatnam	:	Ph: 2782609, visakhapatnam@schandpublishing.com (Marketing Office)

© 2010, P K Mittal

All rights reserved. No part of this publication may be reproduced or copied in any material form (including photocopying or storing it in any medium in form of graphics, electronic or mechanical means and whether or not transient or incidental to some other use of this publication) without written permission of the copyright owner. Any breach of this will entail legal action and prosecution without further notice.

Jurisdiction: All disputes with respect to this publication shall be subject to the jurisdiction of the Courts, Tribunals and Forums of New Delhi, India only.

First Edition 2010
Reprints 2012, 2013, 2014 (Twice), 2016
Reprint 2017

ISBN : 978-81-219-3240-0

PRINTED IN INDIA

By Vikas Publishing House Pvt. Ltd., Plot 20/4, Site-IV, Industrial Area Sahibabad, Ghaziabad-201010 and Published by S Chand And Company Limited, 7361, Ram Nagar, New Delhi-110 055.

PREFACE

I feel pleasure in presenting this book to the teachers and students of B.Sc.–I year of all the Indian Universities. The book has been prepared keeping in view the syllabi prepared by different universities on the basis of UGC Model Curriculum. The present book has evolved out of my experience of more than three decades of classroom teaching at B.Sc. level.

The following have been added to make the book student-friendly :

- Illustrative Solved Examples.
- Uptodate Questions added of different important Indian Universities.
- Objective Type Questions given as per the requirement of the topic.
- The Questions that have been provided in the Exercise are as per latest pattern of Examination.
- All the important subjects like Algebra, Trigonometry, Differential Calculus, Integral Calculus, Differential Equations, Vector Analysis and Geometry given in the single volume book.

I have tried my level best to make the book useful and fruitful for B.Sc. I year students. I acknowledge my sincere thanks to S. Chand & Company Ltd. for giving me an opportunity to prepare this book.

I will look forward to receive valuable and useful suggestions from our users for further improvement in the book.

Dr. P.K. Mittal

RISHIKESH
(Uttarakhand)

UGC MODEL SYLLABUS

BA/B Sc (HONOURS) PART-I MATHEMATICS

BMH 101 (a & b) ALGEBRA AND TRIGONOMETRY

ALGEBRA *(Duration : Two Semesters/One Year)*

Mappings : Equivalence relations and partitions. Congruence modulo n.

Symmetric. Skew symmetric. Hermitian and skew Hermitian matrices. Elementary operations on matrices. Inverse of a matrix. Linear independence of row and column matrices. Row rank, column rank and rank of a matrix. Equivalence of column and row ranks. Eigenvalues, eigenvectors and the characteristic equation of a matrix. Cayley Hamilton theorem and its use in finding inverse of a matrix. Applications of matrices to a system of linear (both homogeneous and non-homogeneous) equations. Theorems on consistency of a system of linear equations.

Relations between the roots and coefficients of general polynomial equation in one variable. Transformation of equations. Descarte's rule of signs. Solution of cubic equations (Cardon method). Biquadratic equations.

Definition of a group with examples and simple properties. Subgroups. Generation of groups. Cyclic groups. Coset decomposition. Lagrange's theorem and its consequences. Fermat's and Euler's theorems. Homomorphism and Isomorphism. Normal subgroups. Quotient groups. The fundamental theorem of homomorphism. Permutation groups. Even and odd permutations. The alternating groups An. Cayley's theorem. Introduction to rings, subrings, integral domains and fields. Characteristic of a ring.

TRIGONOMETRY

De Movre's theorem and its applications. Direct and inverse circular and hyperbolic functions. Logarithm of a complex quantity. Expansion of trigonometrical functions. Gregory's series. Summation of series.

BMH 102 (a & b) CALCULUS

DIFFERENTIAL CALCULUS *(Duration : Two Semesters/One Year)*

ε–δ definition of the limit of a function. Basic properties of limits. Continuous functions and classification of discontinuities. Differentiability. Successive differentiation. Leibnritz theorem. Maclaurin and Taylor series expansions. Asymptotes. Curvature. Tests for concavity and convexity. Points of inflexion. Multiple points. Tracing of curves in Cartesian and polar coordinates.

INTEGRAL CALCULUS

Integration of irrational algebraic functions and transcendental functions. Reduction formulae. Definite integrals. Quadrature. Rectification. Volumes and surfaces of solids of revolution.

ORDINARY DIFFERENTIAL EQUATIONS

Degree and order of a differential equation. Equations of first order and first degree. Equations in which the variables are separable. Homogeneous equations. Linear equations and equations reducible to the linear form. Exact differential equations. First order higher degree equations solvable for x, y, p. Clairaut's form and singular solutions. Geometrical meaning of a differential equation. Orthogonal trajectories. Linear differential equations with constant coefficients. Homogeneous linear ordinary differential equations.

Linear differential equations of second order. Transformation of the equation by changing–the dependent variable/the independent variable. Method of variation of parameters.

Ordinary simultaneous differential equations.

BMH 103 (a & b) VECTOR ANALYSIS AND GEOMETRY

VECTOR ANALYSIS *(Duration : Two Semesters/One Year)*

Scalar and vector product of three vectors. Product of four vectors. Reciprocal Vectors. Vector differentiation. Gradient, divergence and curl. Vector integration. Theorems of Gauss, Green, Stokes and problems based on these.

GEOMETRY

General equation of second degree. Tracing of conics. System of conics. Confocal conics. Polar equation of a conic.

Plane, The Straight line and the plane. Sphere, Cone Cylinder.

Central conicoids. Paraboloids. Plane Sections of Conicoids. Generating lines. Confocal Conicoids. Reduction of Second degree equations.

CONTENTS

BMH 101 (a & b) ALGEBRA AND TRIGONOMETRY

ALGEBRA — 1–211

1. Functions and Relations — 3 – 27
2. Congruence of Integers — 28 – 35
3. Some Special Types of Matrices — 36 – 43
4. Elementary Operations and Inverse of a Matrix — 44 – 51
5. Linear Dependence of Vectors — 52 – 59
6. Rank of a Matrix — 60 – 68
7. Linear Equations — 69 – 78
8. Characteristic Roots and Vectors — 79 – 90
9. Theory of Equations — 91 – 118
10. Group — 119 – 156
11. Permutations — 157 – 167
12. Homomorphism and Isomorphism — 168 – 179
13. Normal Sub-Groups — 180 – 189
14. Rings and Subrings — 190 – 201
15. Integral Domain and Field — 202 – 211

TRIGONOMETRY — 213–283

1. De Moivre's Theorem and Deductions — 215 – 229
2. Hyperbolic Functions — 230 – 239
3. Inverse Hyperbolic Functions — 240 – 246
4. Logarithms of Complex Quantities — 247 – 253
5. Gregory's Series — 254 – 259
6. Trigonometrical Expansions — 260 – 266
7. Summation of Series — 267 – 283

BMH 102 (a & b) CALCULUS

DIFFERENTIAL CALCULUS — 285–386

1. Limits and Continuity — 287 – 298
2. Differentiability — 299 – 308
3. Successive Differentiation — 309 – 321
4. Expansion of Functions — 322 – 327

5. Asymptotes — 328 – 340
6. Curvature — 341 – 363
7. Concavity, Convexity and Singular Points — 364 – 376
8. Curve Tracing — 377 – 386

INTEGRAL CALCULUS — 387–501

1. Integration of Irrational Algebraic Functions — 389 – 406
2. Integration of Transcendental Functions — 407 – 426
3. Reduction Formulae — 427 – 453
4. Definite Integrals — 454 – 464
5. Area of Curves (Quadrature) — 465 – 478
6. Lengths of Curves (Rectification) — 479 – 487
7. Volume and Surface of Solid and Revolution — 488 – 501

ORDINARY DIFFERENTIAL EQUATIONS — 503–621

1. Differential Equations : An Introduction — 505 – 508
2. Differential Equations of First Order and First Degree — 509 – 533
3. Differential Equations of First Order but Not of First Degree — 534 – 550
4. Geometrical Interpretation and Orthogonal Trajections — 551 – 558
5. Linear Differential Equations with Constant Coefficients — 559 – 583
6. Homogeneous Linear Differential Equations — 584 – 591
7. Linear Differential Equations of Second Order with Variable Coefficients — 592 – 611
8. Simultaneous Ordinary Differential Equations — 612 – 621

BMH 103 (a & b) VECTOR ANALYSIS AND GEOMETRY

VECTOR ANALYSIS — 623–696

1. Multiple Products — 625 – 634
2. Differential of Vectors — 635 – 642
3. Differential Operators — 643 – 670
4. Integration of Vectors — 671 – 684
5. Gaus's, Green's and Stoke's Theorem — 685 – 696

GEOMETRY — 697–1023

[(a) Two-Dimensional]

1. General Equation of Second Degree and Tracing of Conics — 699 – 723
2. System of Conics : Confocal Conics — 724 – 743
3. Polar Equations — 744 – 770

(x)

[(b) Three-Dimensional]

1. Systems of Coordinates — 773 – 790
2. The Plane — 791 – 808
3. The Straight Line — 809 – 834
4. The Sphere — 835 – 860
5. Cones, Cylinders — 861 – 892
6. Coincide — 893 – 928
7. Plane Section of Conicoids — 929 – 952
8. Generating Lines of Conicoids — 953 – 978
9. General Equation of the Second Degree — 979 – 1008
10. Confocal Conicoids — 1009 – 1023

BMH 101 (a & b) Algebra & Trigonometry
- Algebra
- Trigonometry

ALGEBRA

Mappings. Equivalence relations and partitions. Congruence modulo n.

Symmetric, Skew symmetric, Hermitian and skew Hermitian matrices. Elementary operations on matrices, inverse of a matrix. Linear independence of row and column matrices. Row rank, column rank and rank of a matrix. Equivalence of column and row ranks. Eigen values, eigen vectors and the characteritic equation of a matrix. Cayley Hamilton theorem and its use in finding inverse of a matrix. Applications of matrices to a system of linear (both homogeneous and non-homogeneous) equations. Theorems on consistency of a system of linear equations.

Relation between the roots and coefficients of general polynomial equation in one variable. Transformation of equations. Descartes' rule of signs. Solution of cubic equations (Cardon method). BIquadratic equations.

Definition of a group with examples and simple properties. Subgroups. Generation of groups. Cyclic groups. Coset decomposition . Lagrange's theorem and its consequences. Fermat's and Euler's theorems. Homomorphism and isomorphism. Permutation groups. Even and odd permutations. The alternating groups. Cayley's Theorem. Introduction to rings, sub rings, integral domains and fields. Characteristic of a ring.

1
Functions and Relations

1.1 MAPPING OR FUNCTION

The concept of mapping of one set into another is of great importance in mathematics. It is not a new concept of any of us as we have been considering mapping from the beginning of our mathematical training. For example, plotting of the relation $y = x^2$ is nothing but to study the particular mapping which takes every real number into its square. The following discussion will make the concept of mapping clear.

Suppose A and B are any two non-empty sets. Let $A = \{a, b, c, d\}$, $B = \{x, y, z\}$. Suppose by some rule or other, we assign to each element A unique element of B. Suppose a is associated to x, b is associated to y, c is associated to x and b is associated to z. The set of such assignments is called a *'function'* or *'mapping'* from A to B. If we denote this set by f than we write

$$f : A \to B$$

which is read as ``$f$ *is a function of $A$ to $B$*'' or ``f *is a mapping from A to B*''.

A mapping from one set into another may be defined as given below :

Definition : *Let A and B be two given sets. Suppose there exists a rule denoted by f, which associated to each member of A, a unique member of B. Then f is called a function or a mapping of A to B. The mapping f to A to B is denoted by* $f : A \to B$ *or by* $A \xrightarrow{f} B$.

Further, if $a \in A$, then the element in B which is assigned to a is called the f image of a or the value of the function f for a and is denoted by $f(a)$.

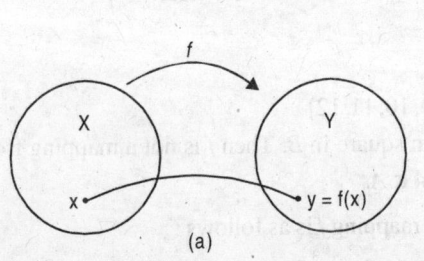 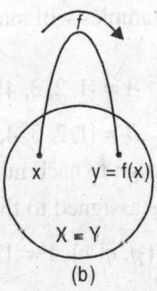

(a)   (b)

Following diagrams [fig. (a), (b)] will help in understanding the idea of a function.

Now, we think of a definition which serves to make the concept of a mapping precise and thus formally, we define mapping as follows :

Definition : *If A and B are non-empty sets than a mapping from A to B is a subset C of $A \times B$ such that for every $a \in A$ there is a unique $b \in B$ such that the ordered pair (a, b) is in C.*

For all practical purposes the following notion of mapping will be used hereafter and thus we arrive at another form of definition.

Definition : *A mapping $f : A \to B$, is a rule which associates any element $a \in A$ with some element $b \in B$, the rule being associate (or map) $a \in A$ with $b \in B$ if $(a, b) \in C, C$ being the subset of $A \times B$.*

Note : The rule f should posses the characteristics that there may be some element of the set B which are not associated to any element of the set A but each element of the set A must be associated to one and only one element of the set B. Two or more elements of the set A may be associated to the same element of the set B but associated of one element of A to more than one element in B is not permissible.

Domain, Co-domain and Range of Functions : Let f be a mapping of A into B. Then A is called the *'domain'* of the function f and B the *'co-domain'* of the function f. It is evident from the definition that each element of B need not appear as the image of an element in A. We define the *'range'* of f to consist of all those elements in B which appear as f image of atleast one element in A. There can be more than one elements of A which have the same image in B. The image set $f[A]$ is called the range of f.

Functions Defined as Sets of Ordered Pairs : Let A and B are any two non-empty sets, then a mapping f of A to B is subset f of $A \times B$ satisfying the following conditions :

(i) for each $a \in A, (a, b) \in f$, for some $b \in B$.

(ii) if $(a, b) \in f$ and $(a, b') \in f$, then $b = b'$.

The first condition ensures that we have a rule that assigns to each element $a \in A$ some element $b \in B$. Thus each element in A will have image. The second condition guarantees that the image is unique. Accordingly, f is a function from A to B.

Note : If $f : A \to B$, it is important to distinguish between a function f and the value $f(x)$ of f for any element x. While f is a subset of $A \times B$, $f(x)$ is an element of the set B.

Operator : *If the domain and co-domain of a function f are both the same set, say*

$$f : A \to A$$

the f is called an operator or transformation on A.

Following examples will make the notion of a function more clear;

Examples :

(i) Let $\quad A = \{1, 2, 3, 4\}$

and $\quad B = \{1, 2, 3, 4, 5, 6, 7, 8, 9, 10, 11, 12\}$

Now, let f assign to each number in A in square in B. Then f is not a mapping from A to B since no number of B is assigned to the element $4 \in A$.

(ii) Let $X = \{p, q, r\}, Y = \{1, 2, 3\}$ and mapping f is as follows :

$$f(p) = 3, \ f(q) = 1, \ f(r) = 2,$$

then f-image of X is $\{3, 1, 2\}$ or in other words f-image of X is Y.

(iii) Let f be a mapping of N into N such that

$$f(1) = 3; \ f(2) = 5; \ f(3) = 7; \ldots$$

We can express this mapping by the functional notion as

$$f : N \to N \ ; \ \text{defined by } f(x) = 2x + 1, \ \forall \ x \in N$$

(iv) Let A be the set of countries in the world and B be the set of capital cities then (i) every country has a capital assigned to it, (ii) no country will have two capitals. Thus, f is a function from the set A to B and the image of India under f is New Delhi, *i.e.,* f (India) = New Delhi. Similarly, f (Nepal) = Kathmandu.

Functions and Relations

Here, domain of f is the list of countries in the world and the co-domain is the list of capital cities of these countries.

Diagrammatic Representation of Function : Sometimes a function may be represented by a diagram as will be obvious from the following example :

Let $\quad A = \{a, b, c, d\}$ and $B = \{u, x, y, z\}$.

Let $f : A \to B$ defined by the correspondance $f(a) = y$, $f(b) = x$, $f(c) = z$ and $f(d) = y$.

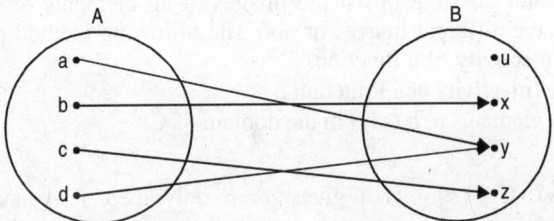

We represen the two sets A and B by the points inside the circles. The mapping $f : A \to B$ is represented by means of arrows joining the points which represent the elements of A to points representing the corresponding elements of B. By the definition of function, it is obvious that.

(i) every point in A is joined to some point in B by an arrow.
(ii) a point in A cannot be joined to two or moer distinct points in B.
(iii) two or more points in A may be joined to the same point in B.
(iv) there may be some points in B which are not joined to any point in A.

Equal Functions : *Two functions f and g are said to be equal iff :*
(i) *the domain of f = domain of g*
(ii) *the co-domain of f = the co-domain of g and*
(iii) *f(x) = g(x) for every x belonging to their common domain.*

Example : Let $A = \{1, 2, 3\}$, $B = \{2, 3, 4\}$ and f, g- and h be three sub-sets of $A \times B$ as given below :

$$f = \{(1,2), (2,3), (3,4)\}, g = \{(1, 2), (1, 3), (2, 3), (3, 4)\}, h = \{(1, 3), (2, 4)\}$$

Then f is a function from A to B but g and h are not functions from A to B, because $1 \in A$ has two images 2 and 3 in B and h is not a function from A to B because $3 \in A$ has no image in B.

1.1.2 KINDS OF FUNCTIONS

If $f : A \to B$ is a function, then f associates all elements of set A to elements in set B such that an element of set A is associated to a unique element of set B. Following these two conditions we may associate :

(i) different elements of set A to different elements of set B, or
(ii) more than one element of set A may be associated to the same element of set B, or
(iii) there may be some elements in B which do not have their pre-images in A or
(iv) all elements in B may have their pre-images in A.

Corresponding to each of these possibilities we define a type of a function as given below :

One-One Function (Injection) : *A function $f : A \to B$ is said to be one-one function or an injection if different elements of A have different images in B.*

Thus, $f : A \to B$ is one-one $\Leftrightarrow a \neq b$

Examples :
(i) A function which assoicates to each country in the world, its capital, is one-one because different countries have their different capital.

(ii) Let $X = \{1, 2, 3, 4\}, Y = \{1, 4, 9, 16\}$ and $f : X \to Y$ s.t. $f(x) = x^2 \; \forall \; x \in X$, then f is one-one mapping of X into Y, as no two distinct elements of X have the same f-image in Y.

Let $f : A \to B$ be a function such that A is an infinite set and we wish to check the injectivity of f. In such a case it is not possible to list the images of all elements of set A to see whether different elements of A have different images or not. The following method provides a systematic procedures to check the injectivity of a function.

Method of Check the Injectivity of a Function :
(i) Take two arbitrary elements a, b (say) in the domain of f.
(ii) Put $f(a) = f(b)$
(iii) Solve $f(a) = f(b)$. If $f(a) = f(b)$ gives $a = b$ only, then $f : A \to B$ is a one-one function (or an injection) otherwise not.

Note : Let $f : A \to B$ and let $a, b \in A$. Then $a = b \Rightarrow f(a) = f(b)$ is always true from the definition. But, $f(a) = f(b) \Rightarrow a = b$ is true only when f is injective.

Many-one Function : *A function $f : A \to B$ is said to be a many-one function of two or more elements of set A have the same images is B.*

Thus, $f : A \to B$ is a many one function if there exists, $x, y \in A$ such that $x \neq y$ but $f(x) = f(y)$.

In other words, $f : A \to B$ is a many one function if it is not a one-one function :

Examples :
(i) $A = \{-1, 1, -2, 2\}$ and $B = \{1, 4, 9, 16\}$. Consider $f : A \to B$ s.t. $f(x) = x^2$. Then $f(-1) = 1, f(1) = 1, f(-2) = 4, f(2) = 4$.

Clearly 1 and –1 have the same image. Similarly, 2 and –2 also have the same imaeg. So, f is a many one function.

(ii) Consider a function $f : Z \to Z$ given by $f(x) = |x|$.

Then f is a many-one function because for every $a \in Z, a \neq 0, a \neq -a$ but $f(a) = f(-a) [\because |a| = |-a|]$.

Onto Function (Surjection) : *A function $f : A \to B$ is said to be an onto function or a surjection if every element of B is the f-image of some element of A, i.e., if $f(A) = B$ or range of f is the co-domain of f.*

Thus, $f : A \to B$ is a surjection iff for each $b \in B, \exists \; a \in A$ such that $f(a) = b$.

Into Function : *A function of $f : A \to B$ is an onto function if there exists an element in B being no pre-image in A.*

In other words, $f : A \to B$ is into function if it is not an onto function.

Functions and Relations

Examples:

(i) Let $A = \{-1, 1, 2, -2\}$, $B = \{1, 4\}$ and $f : A \to B$ be a function defined by $f(x) = x^2$. The f is onto because $f(A) = \{f(-1), f(1), f(2), f(-2)\}$.

$$= \{1, 4\} = B.$$

(ii) A function $f : N \to N$ defined by $f(x) = 2x$ is not an onto function, beacuse $f(N) = \{2, 4, 6, ...\} \neq N$ (co-domain).

Method for Checking the Surjectivity of a Function: Let $f : A \to B$ be the given function.

(i) Choose an arbitrary element b in B.

(ii) Put $f(a) = b$.

(iii) Solve the equation $f(a) = b$ for a and obtain a in terms of b. Let $a = g(b)$.

(iv) If for all values $b \in B$, the values of a obtained from $a = g(b)$ are in A, then f is onto.

If there are some $b \in B$ for which a given by $a = g(b)$, is not in b. Then f is not onto.

Bijection (One-One Onto Function): *A function* $f : A \to B$ *is a bijection if it is one-one as well as onto.*

In other words, a function $f : A \to B$ *is a bijection if:*

(i) *it is one-one, i.e., f(x) = f(y)* $\Rightarrow$ *x = y for all x, y* $\in A$.

(ii) *it is onto, i.e., for all* $y \in B$, *there exists* $x \in A$ *such that* $f(x) = y$.

Examples: Let A set of even integers and B be the set of odd integers, then the mapping $f : A \to B$ given by

$$f(x) = x + 1, \forall\, x \in A$$

is one-one onto.

One-One into Mapping: *Any mapping which is one-one as well as into is called one-one into mapping.*

Example: Let X be the set of integers and Y the set of all even integers, then the mapping $f : X \to Y$, s.t. $f(x) = 2x$, $x \in X$ is an into mapping which is also one-one.

Many-one into Mapping: *A function which is many one as well as into is called many-one into mapping.*

Example: Let $X = \{a, b, c, d\}$, $Y = \{3, 4, 5\}$ and $f(a) = 3$, $f(b) = 4$, $f(c) = 4$, $f(d) = 5$, then it is many-one onto mapping.

Diagrammatic Representation of Different Types of Mappings

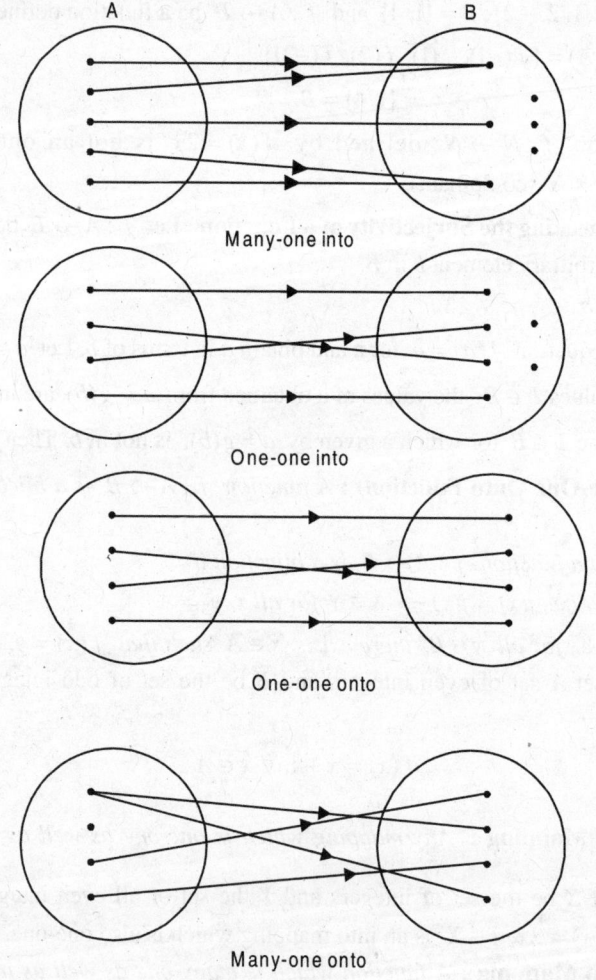

EXAMPLES

1. Let $A = \{-2, -1, 0, 1, 2\}$. Let the function $f : A \to R$ is defined by $f(x) = x^2 + 1$. Find the range of f.

Solution. The range of f consists of those elements of R which appear as f-images of different elements of A. So, we calculate the f-images of each elements of A.

$$f(-2) = (-2)^2 + 1 = 4 + 1 = 5$$
$$f(-1) = (-1)^2 + 1 = 1 + 1 = 2$$
$$f(0) = (0)^2 + 1 = 0 + 1 = 1$$
$$f(1) = (1)^2 + 1 = 1 + 1 = 2$$
$$f(2) = (2)^2 + 1 = 4 + 1 = 5$$

The range of f is the set $\{5, 2, 1, 2, 5\}$ *i.e.*, the set $\{1, 2, 5\}$.

Functions and Relations

2. *Find the domain for which the function* $f(x) = 2x^2 - 1$ *and* $g(x) = 1 - 3x$ *are equal.*

Solution. We have $f(x) = g(x) \Rightarrow 2x^2 - 1 = 1 - 3x$

$\Rightarrow \quad 2x^2 + 3x - 2 = 0 \quad \Rightarrow (x+2)(2x-1) = 0$

$\Rightarrow \quad x = -2, \dfrac{1}{2}$

Thus, $f(x)$ and $g(x)$ are equal on the set $\left\{-2, \dfrac{1}{2}\right\}$.

3. *Show that the mapping* $f : I \to I$ *defined by* $f(x) = x^2$, $x \in I$ *where I is the set of positive integers, is one-one into.*

Solution. $f(x) = x^2$ means that the function f is such that f-images of x is x^2. Domain of the mapping is $\{1, 2, 3, ...\}$ and the range is $\{1, 4, 9, ...\}$.

Thus, f-images is the subset of its domain, $\{f(x)\} \subset I$. It is mapping of 1 into 1. Here two different elements of domain necessarily correspond to different elements of the range so that it is one-one mapping. Hence, it is one-one into mapping.

4. *Show that the function* $f : R \to R$ *defined by* $f(x) = 3x^3 + 5$ *for all* $x \in R$ *is a bijection.*

Solution. Injectivity : Let x, y be any two elements of R (domain).

Then $f(x) = f(y) \Rightarrow 3x^3 + 5 = 3y^3 + 5 \Rightarrow x^3 = y^3 \Rightarrow x = y$

Thus, $f(x) = f(y) \Rightarrow x = y$ for all, $x, y \in R$. So, f is an injective map.

Surjectivity : Let y be an arbitrary elements of R (co-domain). Then

$$f(x) = y \Rightarrow 3x^3 + 5 = y \Rightarrow x^3 = \dfrac{y-5}{3}$$

$\Rightarrow \quad x = \left(\dfrac{y-5}{3}\right)^{1/3}.$

Thus, we find that for all $y \in R$ (co-domain) there exists $x = \left(\dfrac{y-5}{3}\right)^{1/3} \in R$ (domain) such that

$$f(x) = f\left[\left(\dfrac{y-5}{3}\right)^{1/3}\right] = 3\left[\left(\dfrac{y-5}{3}\right)^{1/3}\right]^3 + 5 = y - 5 + 5 = y$$

This shows that every element in the co-domain has its pre-image in the domain. So, f is a surjection.

Hence, f is a bijection.

1.2 INCLUSION MAP

If $X \subset Y$, the function $f : X \to Y$, defined by $f(x) = x$ for each $x \in X$, is called the inclusion map.

Example : Let $X = \{-3, -2, -1, 0, 1, 2, 3\}$ and $Y = \{... -4, -3, -2, -1, 0, 1, 2, 3, 4, ...\}$ and $f(-3) = -3$, $f(-2) = -2$, $f(-1) = -1$, $f(0) = 0$, $f(1) = 1$, $f(2) = 2$, $f(3) = 3$, then it is an inclusion map of X to Y because $X \subset Y$ and $f(x) = x \ \forall \ x \in X$.

1.2.1 Identity Map or identity Function

Let X be any set and the function $f : X \to X$ be defined by the formula $f(x) = x \ \forall \ x \in X$, i.e., each element of X is mapped on itself, then f is called the identity map or the identity function (transformation) on X.

We denote this function by I_x usually. Thus, if I_x denotes the identity mapping on a set X, we have

$$I_x(x) = x, \ \forall \ x \in X.$$

Example : Let $A = \{a, b, c, d\}$. Then $f = \{(a, a), (b, b), (c, c), (d, d)\}$ is an identity mapping of A.

Identity mapping is always one-one onto.

1.2.2 Cardinally Equivalent Sets

Let there be a mapping $f : X \to Y$ which is one-one and onto then the two sets X and Y are said to be *cardinally equivalent* or *equinumerous*. The fact is denoted by writing X-Y.

Note : If any X is equivalent to N, the set of natural numbers then X is said to be *denumerable* set.

(i) Let $X = \{1, 2, 3\}$, $Y = \{1, 4, 9\}$ then $f : X \to Y$ defined by $f(x) = x^2 \ \forall \ x \in X$ is one-one and onto mapping, therefore, the set X and Y are cardinally equivalent, *i.e.*, $X \sim Y$.

(ii) The set $A = \left\{1, \dfrac{1}{2}, \dfrac{1}{3}, ..., \dfrac{1}{n}, ...\right\}$ is denumerable set as it can be put in one-to-one correspondence with the set of natural numbers.

1.3 INVERSE IMAGE OF AN ELEMENT

Let f be a function defined from the set X to the set Y then the inverse image of an element $b \in Y$ under f denoted by $f^{-1}(b)$ to be read as f-image b and

$$f^{-1}(b) = \{x : x \in X \text{ and } f(x) = b\}$$

i.e., $f^{-1}(b)$ is the set of those elements in X which have b as their f-image.

Example : Let $f : R \to R$ where $f(x) = x^2$ and R is the set of real numbers, then $f^{-1}(9) = \{3, -3\}$ for 9 is the f-image of both 3 and -3.

Inverse Image of a Subset : Let f be a function defined from the set X to the set Y and B be a subset of Y, *i.e.*, $B \subset Y$ then the inverse of B under f is given by $f^{-1}(B) = \{x : x \in X \text{ and } f(x) \in B\}$.

Obviously,

$$f^{-1}(B) \in Y \text{ and } f^{-1}[f(x)] = X$$

Note : If $b \in Y$, $\{b\} \subset Y$ and $f^{-1}(b) = f^{-1}\{(b)\}$.

Functions and Relations

Example : Let $f : X \to Y$, where $X = \{1, 2, 3, 4, 5\}$ and $Y = \{1, 2, 4, 6, 9, 15, 16, 25, 30\}$ and let $f(x) = x^2 \; \forall \; x \in X$, let $B_1 = \{2, 4, 9, 15, 16\}$ and $B_2 = \{2, 6\}$. Then $f^{-1}(B_1) = \{2, 3, 4\}$ and $f^{-1}(B_2) = \phi$.

1.3.1 Inverse Mapping (Or Function)

Let X and Y be two sets and let $f : X \to Y$ be a function. If we allow a rule in which elements of Y are associated to their pre-images, then we find that under such a rule there may be some elements in Y which are not associated to elements in A. This happens when f is not an onto map. Therefore, all elements in Y will be associated to some elements in X if f is an *onto* map. Also, if it is a many-one function then under the said rule an element in Y may be associated to more than one element in X. Therefore, an element in Y will be associated to a unique element in X if f is a *injective* map.

It follows from the above discussion that if $f : X \to Y$ is a bijection, we can define a new function from Y to X which associates each elemnent $y \in Y$ to its pre-image $f^{-1}(y) \in X$. Such a function is known as the inverse of function f and is denoted by f^{-1}.

Definition : *Let $f : X \to Y$ be a bijection. Then a function $g : Y \to X$ which associates each element $y \in Y$. to a unique element $x \in X$ such that $f(x) = y$ is called the inverse of f*, i.e.,

$$f(x) = y \Leftrightarrow g(y) = x.$$

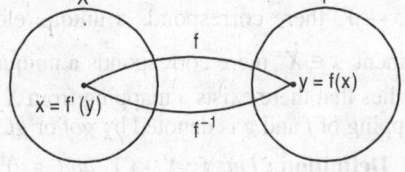

The inverse of f is generally denoted by f^{-1}.

Thus, if $f : X \to Y$ is a bijection, then $f^{-1} : Y \to X$ is such that

$$f(x) = y \Leftrightarrow f^{-1}(y) = x$$

Example : If $A = \{1, 2, 3, 4\}$, $B = \{2, 4, 6, 8\}$ and $f : A \to B$ is given by $f(x) = 2x$, then

$$f(1) = 2, f(2) = 4, f(3) = 6 \text{ and } f(4) = 8.$$

So, $\qquad f = \{(1, 2), (2, 4), (3, 6), (4, 8)\}$

Which is clearly a bijection.

$\therefore \qquad f^{-1} = \{(2, 1), (4, 2), (6, 3), (8, 4)\}$

Method to Find the Inverse of a Bijection : Let $f : X \to Y$ be a bijection. To find the inverse of f we proceed as follows :

(i) Put $f(x) = y$ where $y \in Y$ and $x \in X$.

(ii) Solve $f(x) = y$ to obtain x in terms of y.

(iii) In the relation obtained in step (ii) replace x by $f^{-1}(y)$ to obtain the inverse of f.

Some Results on Inversible Functions

Theorem : *Let X and Y be two non-empty sets. Let Y be a one-one mapping of X onto Y. Then f^{-1} is also one-one onto i.e., inverse of an invertible mapping is invertible.*

Now, $f^{-1}(y_1) = f^{-1}(y_2) \Rightarrow x_1 = x_2$
$\Rightarrow f(x_1) = f(x_2) \Rightarrow y_1 = y_2$

This, shows that f^{-1} is one-one.

Now, let x be an arbitrary element of X. Then by definition of f there exists a unique element $y \in Y$ such that $f(x) = y$ or $x = f^{-1}(y)$. Thus, for each $x \in X$, there exists a unique element $y \in Y$ such that $f^{-1}(y) = x$. So, f^{-1} is onto.

Hence, f^{-1} is one-one and onto and therefore invertible.

1.4 PRODUCT OF MAPPINGS OR COMPOSITE OF FUNCTIONS

The idea of a function of a function in mathematics has an analogue in abstract mathematics and it is known as a composite function. Let X, Y, Z be three sets and f be a function defined from X to Y and g be a function defined from Y to Z.

i.e., $\qquad f : X \to Y$ and $g : Y \to Z$.

But $f : X \to Y$ we mean that to every element $x \in X$, there corresponds a unique element $f(x) \in Y$. Since the domain of g is Y, so by the function $g : Y \to Z$ we mean that to every element $f(x) \in Y$ there corresponds a unique element of $g[f(x)] \in Z$. Thus, we notice that to every element $x \in X$ there corresponds a unique element $g[f(x)] \in Z$ under the mappins f and g. This implies that there exists a mapping from X to Z. This mapping is called the composite or the product mapping of f and g is denoted by gof or gf.

Definition : *Let* $f : X \to Y$ *and* $g : Y \to Z$; *then the composite of the function f and g denoted by gof or gf is mapping gof :* $X \to Z$ *s.t.* $(gof)(x) = g[f(x)]$, $\forall \ x \in X$.

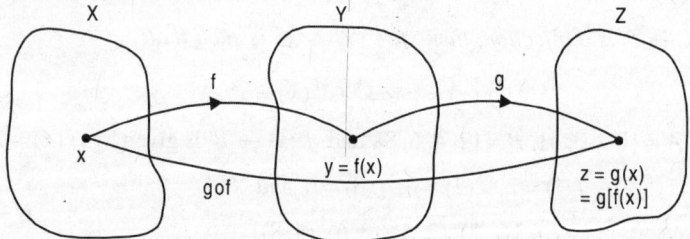

Example : Let $f : R \to R$ be given by the relation $f(x) = \sin x \ \forall \ x \in R$ and the mapping $g : R \to R$ be given by the relation

$$g(x) = x^2, \forall \ x \in R,$$

then the composite function $(gof) = R \to R$ is given by relation.

$$(gof)(x) = g[f(x)] = g[\sin x] = (\sin x)^2 = \sin^2 x$$

Also, $\qquad (fog)(x) = f[g(x)] = f[x^2]$

$$= \sin x^2 \ \forall \ x \in R.$$

Note : (i) If $f : X \to X$ and $g : X \to X$, then both the composite of functions gof and fog are defined. But in general $gof \neq fog$.

Functions and Relations

(ii) It may be noted that the product *(gof)* is defined only when range $(f) \subset$ dom (g). When *gof* is defined then it is not necessary that *(fog)* is also defined. However, *(fog)* is defined when, range $(g) \subset$ dom (f).

Some Results on Composite Mappings :

Theorem 1. *The product of any function with the identity function is the function itself.*

Proof : Let $f : X \to Y$ and let us denote I_x and I_y the identity functions on X and Y respectively. Then we must show that

$$I_y of = f \text{ and } foI_x = f$$

Since $\qquad f : X \to Y$ and $I_y : Y \to Y$, so $I_y of : X \to Y$

Now, let x be an arbitrary element of X and let $f(x) = y$.

Then, $\qquad (I_y of)(x) = I_y [f(x)]$

$$= I_y(y) = y = f(x)$$

$\therefore \qquad I_y of = f.$

Again, since $I_x = X \to X$ and $f : X \to Y$, so $foI_x = X \to Y$.

Now, for an arbitrary $x \in X$, we have $(foI_x)(x) = f[I_x(x)] = f(x)$.

$\therefore \qquad fof^{-1} = I_y.$

Theorem 3. *Composite of functions is associative.*

Proof : Let X, Y, Z, S be the four non-empty sets.

Let $f : X \to Y, g : Y \to Z$ and $h : Z \to S$. Then we have to show that (hog) $of = ho$ (gof).

Let x be an arbitrary element of x. Then

$$[(hog) of](x) = (hog)[f(x)]$$

$$= h[g(f(x))] = h[(gof)(x)]$$

$$= [ho(gof)](x)$$

Hence, $\qquad (hog) of = ho (gof).$

Theorem 4. *Let X, Y and Z be any non-empty sets and let f and g be one-one mappings of X onto Y and Y onto Z respectively so that f and g are both invertible. Then (gof) is also invertible and*

$$(gof)^{-1} = f^{-1} og^{-1}.$$

Proof : In order to show that *(gof)* is invertible, we must show that, it is one-one and onto.

Now, $\qquad (gof)(x_1) = (gof)(x_2)$

$\Rightarrow \qquad g[f(x_1)] = g[f(x_2)]$

$\Rightarrow \qquad f(x_1) = f(x_2) \qquad\qquad\qquad\qquad [\because g \text{ is one-one}]$

$\Rightarrow \qquad x_1 = x_2 \qquad\qquad\qquad\qquad\qquad [\because f \text{ is one-one}]$

Hence, *(gof)* is one-one.

Now, in order to show that *(gof)* is onto, let z be an arbitrary element of Z. Then g being onto $\exists$ an element of in Y such that $g(y) = z$. Also, f being onto, corresponding to the element $y \in Y$ $\exists$ an element x in X such that $f(x) = y$.

$\therefore \qquad (gof)(x) = g[f(x)] = g(y) = z$

Thus, corresponding to the element $z \in Z, \exists$ such that $(gof)(x) = z$. Consequently, (gof) is onto.

Thus, (gof) is one-one onto and hence invertible.

Now $\qquad (gof)(x) = z \Rightarrow (gof)^{-1}(z) = x \qquad$ ...(i)

Also, $\qquad (f^{-1}og^{-1})(z) = f^{-1}[g^{-1}(z)]$

$\qquad\qquad\qquad\qquad = f^{-1}(y) \qquad [\because g(y) = z \Rightarrow y = g^{-1}(z)]$

$\qquad\qquad\qquad\qquad = x \qquad [\because f(x) = y \Rightarrow x = f^{-1}(y)]$

Thus, $\qquad (f^{-1}og^{-1})(z) = x \qquad$ ...(ii)

Hence, from (i) and (ii) we have

$$(gof)^{-1} = (f^{-1}og^{-1})$$

EXAMPLES

1. *Let $R \to R$ defined by $f(x) = ax + b$, where $a, b, x \in R$ and $a \neq 0$. Prove that f is invertible.*

Solution. f is one-one for

$\qquad\qquad x_1, x_2 \in R, f(x_1) = f(x_2)$

$\Rightarrow \qquad\qquad ax_1 + b = ax_2 + b \Rightarrow ax_1 = ax_2$

$\Rightarrow \qquad\qquad x_1 = x_2$

f is onto :

Let $y \in R$ such that

$\qquad\qquad y = f(x) \Rightarrow y = ax + b$

$\Rightarrow \qquad\qquad ax = y - b$ and $a \neq 0 \in R$

$\Rightarrow \qquad\qquad x = \dfrac{1}{a}(y - b) \in R$

$\therefore$ Given $y \in R, \exists$ some $x = \dfrac{1}{a}(y - b) \in R$ s.t. $f(x) = y$.

$\therefore \quad f : R \to R$ is both one-one and onto hence f is invertible.

2. *Let f be a function defined from the set X to the set Y and let A, B be the subsets of Y, then*

(i) $f^{-1}(A \cup B) = f^{-1}(A) \cup f^{-1}(B)$

(ii) $f^{-1}(A \cap B) = f^{-1}(A) \cap f^{-1}(B)$

Solution. (a) Let x be an element of $f^{-1}(A \cup B)$, then

$\qquad\qquad x \in f^{-1}(A \cup B)$

$\Rightarrow \qquad\qquad f(x) \in A \cup B$

$\Rightarrow \qquad\qquad f(x) \in A$ or $f(x) \in B$

$\Rightarrow \qquad\qquad x \in f^{-1}(A)$ or $x \in f^{-1}(B)$

$\Rightarrow \qquad\qquad x \in \{f^{-1}(A) \cup f^{-1}(B)\}$

$\therefore \qquad\qquad f^{-1}(A \cup B) \subset \{f^{-1}(A) \cup f^{-1}(B)\} \qquad$ ...(1)

Functions and Relations

Again, let y be an element of $f^{-1}(A) \cup f^{-1}(B)$, then

$$y \in f^{-1}(A) \cup f^{-1}(B)$$

$\Rightarrow$ $\quad y \in f^{-1}(A)$ or $y \in f^{-1}(B)$

$\Rightarrow$ $\quad f(y) \in A$ or $f(y) \in B$

$\Rightarrow$ $\quad f(y) \in A \cup B$

$\Rightarrow$ $\quad y \in f^{-1}(A \cup B)$

Therefore, $\quad f^{-1}(A) \cup f^{-1}(B) \subset f^{-1}(A \cup B)$ ...(2)

(1) and (2), imply.

$$f^{-1}(A \cup B) = f^{-1}(A) \cup f^{-1}(B)$$

Similarly, 2nd result can be proved.

3. If $f : R \to R : f(x) = |x|$, then prove that $fof = f$.

Solution. $\quad (fof)(x) = f[f(x)] = f(|x|)$

$\qquad = \|x\| = |x| = f(x)$

$\therefore \qquad fof = f$

EXERCISES

1. Define a function as a set of ordered pairs.
2. Let $A = \{-2, -1, 0, 1, 2\}$ and $f : A \to Z$ be a function defined by $f(x) = x^2 - 2x - 3x$. Find (a) range of f, i.e., $f(A)$, (b) pre-images of 6, –3 and 5.
3. If a function $f : R \to R$ be defined by

$$f(x) = \begin{cases} 3x - 2, & x < 0 \\ 1, & x = 0 \\ 4x + 1, & x > 0 \end{cases}$$

   find $f(1), f(-1), f(0), f(2)$.

4. Let $f : R \to R$ and $g : C \to C$ be two functions defined as $f(x) = x^2$ and $g(x) = x^2$. Are they equal functions?
5. Classify the following functions as injection, surjection or bijection:

   (i) $f : R \to R, f(x) = |x|$
   (ii) $f : Z \to Z, f(x) = x^2 + x$
   (iii) $f : Z \to Z, f(x) = x + 5$
   (iv) $f : R \to R, f(x) = \sin x$
   (v) $f : R \to x, f(x) = x^3 - x$
   (vi) $f : R \to R, f(x) = \sin^2 x + \cos^2 x$
   (vii) $f : R \to R, f(x) = x^3 + 1$
   (viii) $f : Q - \{3\} \to Q, f(x) = \dfrac{2x + 3}{x - 3}$
   (ix) $f : Q \to Q, f(x) = x^3 + 1$
   (x) $f : R \to R, f(x) = 5x^3 + 4$

6. Show that the function $f : N \to N$ given by $f(n) = n - (-1)^n$ for all $n \in N$ is one-one and onto.
7. Prove that the function $f : N \to N$, defined by $f(x) = x^2 + x + 1$ is one-one but not onto.
8. Let $f = \{(3, 1), (9, 3), (12, 4)\}$ and $g = \{(1, 3), (3, 3), (4, 9), (5, 9)\}$. Show that fog and gof are both defined. Also find fog and gof.

9. Let $A = \{a, b, c\}$, $B = \{u, v, w\}$ and let f and g be two functions from A to B and from B to A respectively defined as :
$$f = \{(a, v), (b, u), (c, w)\}$$
$$g = \{(u, b), (v, a), (w, c)\}$$
Show that f and g both are bijections and find fog and gof.

10. Find $fog\,(2)$ and $gof\,(1)$ when : $f : R \to R$, $f(x) = x^2 + 8$ and $g : R \to R$, $g(x) = 3x^3 + 1$.

11. Let R^+ be the set of all non-negative real numbers. If $R^+ \to R^+$ and $g : R^+ \to R^+$ are defined as $f(x) = x^2$ and $g(x) = +\sqrt{x}$. Find fog and gof. Are they equal functions ?

12. Let $A = \{x \in R \mid -1 \le x \le 1\}$ and $f : A \to A$, $g : A \to A$ be two functions defined by $f(x) = x^2$ and $g(x) = \sin\left(\dfrac{\pi x}{2}\right)$. Show that g^{-1} exists but f^{-1} does not exist. Also find g^{-1}.

Answers

2. (a) $f(A) = \{-4, -3, 0, 5\}$ (b) $\phi, \{0, 2,\}, -2$
3. $f(1) = 5, f(-1) = -5, f(0) = 1, f(2) = 9$
4. No, since domain of $f \ne$ domain of g.
5. (i) Neither an injection nor a bijection.
   (ii) Neither injective nor surjective.
   (iii) Bijective,   (iv) Neither injective nor surjective,
   (v) Surjective but not injective,   (vi) Neither injective nor surjective
   (vii) Bijective   (viii) injective but not surjective
   (ix) Injective,   (x) Bijective.
8. $gof = \{(3, 3), (9, 3), (12, 9)\}$, $fog = \{(1, 1), (3, 1), (4, 3), (5, 3)\}$
9. $fog = \{(u, u), (v, v), (w, w)\}$, $gof = \{(a, a), (b, b), (c, c)\}$
10. $fog\,(2) = 623$, $gof\,(1) = 2188$
11. $fog\,(x) = x$, $gof\,(x) = x$
12. $g^{-1}(x) = \dfrac{2}{\pi} \sin^{-1} x$

1.5 RELATIONS

The word relation is not unfamiliar even to a child. He is aware of his various relations with family members. As he grows up he comes across the other types of relations also, *e.g.*, relation between money and its purchasing power, relation of education with school etc. The description of relation in above language is very vague. Here we shall try to define and describe relation in the precise language of mathematics. For this we shall use the concept of ordered pairs and Cartesian product.

Definition : *Let A and B be two sets. A relation from A to B is a subset of $A \times B$. Symbolically, R is a relation from A to B iff $R \subset A \times B$.*

If (x, y) be a member of a relation set R, we express it by writing $x\,R\,y$ and say that 'x is in the relation R to y'. Thus,

$$(x, y) \in R \iff xRy$$

Functions and Relations

For example, if $A = \{2, 3, 5, 6\}$ and R means 'divides' then

$2R2, 2R6, 3R3, 3R6, 5R5, 6R6$ and as such relation set

$$R = \{(2, 2), (2, 6), (3, 3), (3, 6), (5, 5), (6, 6)\}.$$

Furthermore if I be the set of all integers, the statement "x is less than y where $x, y \in I$" determines a relation in I. If we denote this relation by R, then we may describe the set R in the set builder notation as given below.

$$R = \{(x, y) : x, y \in I, x < y\}$$

Binary Relation : A relation R between pairs of element of a set A is called a *binary relation*. In the discussions here after the word relation will mean binary relation.

Let A be any set and R be the set of all those pairs (x, y) of $A \times A$ for which $x = y$. Then the relation R is called the equality relation in the set A. If is also called the *diagonal relation of A* and is denoted by Δ. Thus,

$$\Delta = \{(x, y) : x \in A\}$$

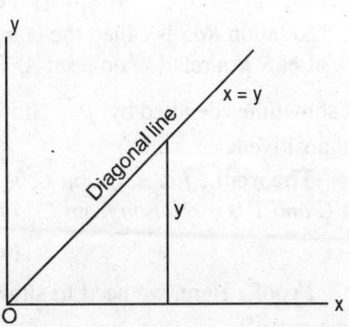

1.5.1 Domain and Range of a Relation

Let R be a relation from A to B. Then the set of all first co-ordinates of the members of the relation set R is called the *domain* of R and the set of all second members of R is called the *range* of R. Thus,

Domain $R = \{x : (x, y) \in R\}$

Range $R = \{y : (x, y) \in R\}$

For example, $A = \{1, 2, 3, 4\}, B = \{3, 4, 5\}$, Then

$A \times B = \{(1, 3), (1, 4), (1, 5), (2, 3), (2, 4)$

$(2, 5), (3, 3), (3, 4), (3, 5), (4, 3), (4, 4), (4, 5)\}$

Now, if R stands for 'less than', then relation set

$R = \{(1, 3), (1, 4), (1, 5), (2, 3), (2, 4), (2, 5), (3, 4), (3, 5), (4, 5)\}$

Domain R = set of first co-ordinates of ordered pairs in the relation set = $\{1, 2, 3, 4\}$.

Range R = set of second co-ordinates of ordered pairs in the relation set = $\{3, 4, 5\}$.

1.5.2 TOTAL NUMBER OF RELATIONS

Let A and B be two non-empty finite sets consisting of m and n elements, respectively. Then $A \times B$ consists of mn ordered pairs. So, total number of subsets of $A \times B$ is 2^{mn}. Since each subset of $A \times B$ defines a relation from A to B, so total number of relations from A to B is 2^{mn}. Among these 2^{mn} relations the void relation ϕ and the universal relation $A \times B$ are trivial erlations from A to B.

Relation on a Set : Let A be a non-void set. Then, a relation of $A \times A$, is called relation on set A.

Inverse Relation : Let A, B be two sets and let R be a relation from a set A to a set B. Then the inverse of R, denoted by R^{-1}, is a relation from B to A and is defined by

$$R^{-1} = \{(b, a) : (a, b) \in R\}$$

Clearly, $(a, b) \in R \Leftrightarrow (b, a) \in R^{-1}$

Also Dom (R) = Range (R^{-1}) and

Range (R) = Dom (R^{-1})

1.5.3 Composition of Relations

Let A, B and C be sets, and let R be a relation from A to B and let S be a relation from B to C. Obviously, R is a subset of $A \times B$ and S is a subset of $B \times C$. Then R and S give rise to a relation from A to C denoted by RoS and defined by

$$a\,(RoS)\,c \text{ if for some } b \in B \text{ we have } a\,R\,b \text{ and } b\,S\,c.$$

Symbolically :

$$RoS = \{(a, c) : \forall\, b \in B \text{ for which } (a, b) \in R \text{ and } (b, c) \in S\}$$

Relation RoS is called the *composition of R and S*. It is sometimes denoted as RS.

Let R is a relation on a set A. Then RoR, the composition of R with itself is always defined, and is sometimes denoted by R^2. Similarly, $R^3 = R^2 oR = RoRoR$ and so on. Thus, R^n is defined for all positive n.

Theorem : *Let A, B and C be sets. Suppose R is a relation from A to B, S is a relation from B to C and T is a relation from C to D. Then show that*

$$(RoS)\,oT = Ro\,(SoT)$$

Proof : Here, we need to show that each ordered pair in $(RoS)\,oT$ belongs to $Ro\,(SoT)$ and vice-versa.

Let (a, d) belongs to $(RoS)\,oT$.

Then, there exists ac in C such that $(a, c) \in RoS$ and $(c, d) \in T$.

Since $(a, c) \in RoS$, there exists ab in B such that $(a, b) \in R$ and $(b, c) \in S$.

Since $(b, c) \in S$ and $(c, d) \in T$

we have $(b, d) \in SoT$.

Again, since $(a, b) \in R$ and $(b, d) \in SoT$,

we have $(a, d) \in Ro\,(SoT)$

Therefore, $(RoS)\,oT \subset Ro\,(SoT)$ ...(1)

Similarly, $Ro\,(SoT) \subset (RoS)\,oT$ ...(2)

From (1) and (2), we get

$$Ro\,(SoT) = (RoS)\,oT$$

1.6 TYPES OF RELATIONS

Void Relation : Let A be as set. Then $\phi \subset A \times A$ and so it is a relation on A. This relation is called the *void or empty relation* on A.

Universal Relation : Let A be a set. Then $A \times A \subset A \times A$ and so it is a relation on A. This relation is called the *universal relation* on A.

Note : Void and the universal relations on a set A are respectively the smallest and the largest relations on A.

Identity Relation : Let A be a set. Then the relation $I_A = \{(a, a) : a \in A\}$ on A is called the *identity relation* on A.

Obviously, the relation I_A on A is called the identity relation if every element of A is related to itself only.

Reflexive Relation : *A relation R on a set A is said to be reflexive if every element of A is related to itself.* Thus, R is reflexive $\Leftrightarrow (a, a) \in R$ for all $a \in A$.

Functions and Relations

A relation R on a set A is not reflexive if there exists an element $a \in A$ such that $(a, a) \notin R$.

Example:

(i) The universal relation on a non-void set A is reflexive.

(ii) The relation R on N defined by $(x, y) \in R \Leftrightarrow x \geq y$ is a reflexive relation on N, because every natural number is greater than or equal to itself.

Symmetric Relation : *A relation R on a set A is said to be a symmetric relation iff $(a, b) \in R \Rightarrow (b, a) \in R$ for all $a, b \in A$.*

i.e., $\qquad a\,R\,b \Rightarrow b\,R\,a$ for all $a, b \Rightarrow A$.

Example:

(i) The identity and the universal relation on a non-void set are symmetric relations.

(ii) Let L be the set of all lines in a plane and let R be a relation defined on L by the rule $(x, y) \in R \Leftrightarrow x$ is perpendicular to y. Then R is a symmetric relation on L, because $L_1 \perp L_2 \Rightarrow L_2 \perp L_1$, i.e., $(L_1, L_2) \in R \Rightarrow (L_2, L_1) \in R$.

Note : A relation R on set A is not a symmetric relation if there are at least two elements $a, b \in A$ such that $(a, b) \in A$ but $(b, a) \notin R$.

Transitive Relation : *Let A be any set. A relation R on A is said to be a transitive relation iff $(a, b) \in R$ and $(b, c) \in \Rightarrow (a, c) \in R$ for all $a, b, c \in R$.*

i.e., $\qquad a\,R\,b$ and $b\,R\,c \Rightarrow a\,R\,c$ for all $a, b, c \in A$.

Example:

(i) The relation R on the set N of all natural numbers defined by $(x, y) \in R \Leftrightarrow x$ divides y, for all $x, y \in N$ is transitive.

(ii) On the set N of natural numbers, the relation R defined by $x\,R\,y \Rightarrow x$ is less than y is transitive, because for any $x, y, z \in N$.

$$x < y \text{ and } y < z \Rightarrow x < z \Rightarrow$$
$$x\,R\,y \text{ and } y\,R\,z \Rightarrow x\,R\,z.$$

Antisymmetric Relation : *Let A be any set. A relation R on set A is said to be an antisymmetric relation iff*

$$(a, b) \in R \text{ and } (b, a) \in R \Rightarrow a = b \text{ for all } a, b \in A.$$

Examples:

(i) The identity relation on a set A is an antisymmetric relation.

(ii) Let R be a relation on the set N of natural numbers defined by

$$x\,R\,y \Leftrightarrow \text{`}x \text{ divides } y\text{' for all } x, y \in N.$$

This relation is an anti-symmetric relation on N. Since for any two numbers, $a, b \in N$.

$$a|b \text{ and } b|a \Rightarrow a = b, \text{ i.e., } a\,R\,b$$
and $\qquad b\,R\,a \Rightarrow a = b$.

Note : If $(a, b) \in R$ and $(b, a) \notin R$, then also R is an anti-symmetric relation.

Equivalence Relation : *A relation R on a set A is said to be an equivalence relation on A iff.*

(i) it is reflexive, i.e., $(a, a) \in R$ for all $a \in A$.

(ii) it is symmetric, i.e., $(a, b) \in R \Rightarrow (b, a) \in R$ for all $a, b \in A$.

(iii) it is transitive, i.e., $(a, b) \in R$ and $(b, c) \in R \Rightarrow (a, c) \in R$ for all $a, b, c \in A$.

1.6.1 Some Theorems on Equivalence Relations

Theorem 1. *If R and S are two equivalence relations on a set A than $R \cap S$ is also an equivalence relation on A.*

or

The intersection of two equivalence relations on a set is an equivalence relation on the set.

Proof : It is given that R and S are relations on set A.

$\therefore \quad R \subset A \times A$ and $S \subset A \times A \Rightarrow R \cap S \subset A \times A \Rightarrow R \cap S$ is also a relation on A.,

Now, we shall show that it is an equivalence relation on A.

We observe the following properties :

Reflexivity : Let a be an arbitrary element of A. Then.

$$a \in A \Rightarrow (a, a) \in S \text{ and } (a, a) \in S \qquad [\because R \text{ and } S \text{ are reflexive.}]$$
$$\Rightarrow \qquad (a, a) \in R \cap S$$

Thus, $(a, a) \in R \cap S$ for all $a \in A$, so $R \cap S$ is a reflexive relation on A.

Symmetry : Let $a, b \in A$ such that $(a, b) \in R \cap S$. Then

$$(a, b) \in R \cap S \Rightarrow (a, b) \in R \text{ and } (a, b) \in S$$
$$\Rightarrow \qquad (b, a) \in R \text{ and } (b, a) \in S \qquad [\because R \text{ and } S \text{ are symmetric}]$$
$$\Rightarrow \qquad (b, a) \in R \cap S.$$

Thus, $(a, b) \in R \cap S \Rightarrow (b, a) \in R \cap S$ for all $(a, b) \in R \cap S$. So, $R \cap S$ is symmetric an A.

Transitivity : Let $a, b, c \in A$ such that $(a, b) \in R \cap S$ and $(b, c) \in R \cap S$. Then $(a, b) \in R \cap S$.

$$\Rightarrow \qquad \{(a, b) \in R \text{ and } (a, b) \in S\} \text{ and } \{(b, c) \in R \text{ and } (b, c) \in S)\}$$
$$\Rightarrow \qquad \{(a, b) \in R \text{ and } (b, c) \in R\} \text{ amd } \{(a, b) \in S \text{ and } (b, c) \in S\}$$
$$\Rightarrow \qquad (a, c) \in R \text{ and } (a, c) \in S$$

$(\because R$ and S are transitive, so, $(a, b) \in R$ and $(b, c) \in R \Rightarrow (a, c) \in R)$

$$\Rightarrow \qquad (a, c) \in R \cap S$$

Thus, $(a, b) \in R \cap S$ and $(b, c) \in R \cap S$.

$$\Rightarrow \qquad (a, c) \in R \cap S. \text{ So, } R \cap S \text{ is transitive on } A.$$

Here, $R \cap S$ is an equivalence relation on A.

Theorem II. *The union of two equivalence relations on a set is not necessarily an equivalence relation on the set.*

Proof : Let $A = \{a, b, c\}$ and let R and S be two relations on A, given by

$$R = \{(a, a), (b, b), (c, c), (a, b), (b, a)\}$$

and $\qquad S = \{(a, a), (b, b), (c, c), (b, c), (c, b)\}$

It can be easily that each one of R and S is an equivalence relation on A. But $R \cup S$ is not transitive, because $(a, b) \in R \cup S$ and $(b, c) \in R \cup S$ but $(a, c) \notin R \cup S$. Hence, $R \cup S$ is not an equivalence relation on A.

Theorem III : *If R is an equivalence relation on a set A, then R^{-1} is also an equivalence relation on A.*

Or

The inverse of an equivalence relation is an equivalence relation.

Proof : Since R is a relation on A. So, $R \subset A \times A \Rightarrow R^{-1} \subset A \times A \Rightarrow R^{-1}$ is also a relation on A.

Functions and Relations

Now, we shal show that R^{-1} is an equivalence relation on A.

Reflexivity : Let a be an arbitrary element of A. Then

$$a \in A \Rightarrow (a, a) \in R \qquad (\because R \text{ is reflexive})$$

$$\Rightarrow (a, a) \in R^{-1}$$

Thus, $(a, a) \in R^{-1}$ for all $a \in A$. So, R^{-1} is reflexive on A.

Symmetry : Let $(a, b) \in R^{-1}$. Then

$$(a, b) \in R^{-1} \Rightarrow (b, a) \in R \qquad (\text{By def. of } R^{-1})$$

$$\Rightarrow (a, b) \in R \qquad (\because R \text{ is symmetric})$$

$$\Rightarrow (b, a) \in R^{-1} \qquad (\text{By def. of } R^{-1})$$

Thus, $(a, b) \in R^{-1} \Rightarrow (b, a) \in R^{-1}$ for all $a, b \in A$. So, R^{-1} is symmetric on A.

Transitivity : Let $(a, b) \in R^{-1}$ and $(b, c) \in R^{-1}$. Then $(a, b) \in R^{-1}$ and $(b, c) \in R^{-1}$.

$$\Rightarrow (c, b) \in R \text{ and } (b, a) \in R \Rightarrow (c, a) \in R \qquad (\because R \text{ is transitive})$$

$$\Rightarrow (a, c) \in R^{-1} \qquad (\text{By def. of } R^{-1})$$

Thus, $(a, b) \in R^{-1}$ and $(b, c) \in R^{-1}$

$$\Rightarrow (a, c) \in R^{-1} \text{ for all } a, b, c \in A. \text{ So, } R^{-1} \text{ is transitive on } A.$$

Hence, R^{-1} is an equivalence relation on A.

1.6.2 Closure Properties

Let A is a set. Consider the collection of all relations on A. Let P be a property of such relations such as being symmetric or being transitive. A relation with property P will be called a P-relation. P will be called R-closable if P satisfies the following two conditions.

(i) There is a P-relation S containing R.
(ii) The intersection of P-relations is a P-relation.

P-closure of an arbitrary relation R on A, written P(R), is a P-relation such that

$$R \subseteq P(R) \subseteq S$$

Reflexive, Symmetric and Transitive Closures : Let $\Delta_A = \{(a, a) : a \in A\}$ is the equality relation on A. Let R be a relation on set A, then

(i) $R \cup R^{-1}$ is symemtric closure of R.

(ii) $R \cup \Delta_A$ is reflexive closure of R.

(iii) $R^* = \bigcup_{i=1}^{\infty} R^i$ is the transitive closure of R.

1.7 EQUIVALENCE CLASSES

Let A be any non-empty set and let R be an equivalence relation in A. Further let a be an arbitray element of A for which $a\,R\,b$, is known as equivalence class of a. We shall denote this equivalence class by $[a]$ or $\bar{a}$. Evidently *an equivalence class is a subset of the given set, such that any two elements on it are equivalent to each other.*

Some authors denote the equivalence class of a determined by a relation R in A by the symbol $\dfrac{a}{R}$. The set of all equivalence classes will be denoted by $\dfrac{A}{R}$ and read as "A modulo R". This is called the quotient set of A by R.

Thus the set of all disjoint equivalence classes defined by an equivalence relation R over a set A is called the Quotient set of A relative to R and also denoted by $\overline{A}$.

Example : Let A be the set of all triangles in a plane and let R be an equivalence relation in A defined by "x is congruent to y", $x, y \in A$. When $x \in A$ we shall mean by the equivalence class $[x]$ the set of all triangles of A congruent to the triangle x.

Similarly, when $y \in A$ we shall mean by the equivalence class $[y]$ the set of all triangles of A congruent to the triangle y.

Properties of Equivalence Classes

Let A be non-empty set and let R be an equivalence relation in A. Let x and y be arbitrary elements in A. Then

(i) $x \in [x]$,

(ii) If $y \in [x]$, then $[y] = [x]$

(iii) $[x] = [y] \Leftrightarrow (x, y) \in R$ i.e., iff $x R y$.

(iv) Either $[x] = [y]$ or $[x] \cap [y] = \phi$.

i.e., two equivalence classes are either disjoint or identical.

1.8 PARTITION OF A SET

Let X be a non-empty set. A set $P = \{A, B, C\}$ of non-empty subsets of X will be called a partition of X if

(i) $A \cup B \cup C ... = X$, i.e., the set X is the union of the sets in P and

(ii) The intersection of every pair of distinct subsets of $X \in P$ is the null set i.e., if A and $B \in P$ then either $A = B$ or $A \cap B = \phi$.

Example : Consider the set $X = \{1, 2, ..., 9, 10\}$ and its subsets $B_1 = \{1, 3\}$, $B_2 = \{7, 8, 10\}$, $B_3 = \{2, 5, 6\}$, $B_4 = \{4, 9\}$.

The set $P = \{B_1, B_2, B_3, B_4\}$ is such that

(i) B_1, B_2, B_3, B_4 are all non-empty subsets of X.

(ii) $B_1 \cup B_2 \cup B_3 \cup B_4 = X$.

(iii) For any set B_i, $B_i \cap B_j = \phi$.

Hence, the set $\{B_1, B_2, B_3, B_4\}$ is a partition of X.

Theorem 1. *An equivalence relation defined in a set decomposes the set into disjoint equivalence classes.*

Proof : Let an equivalence relation R be defined in a set S. Let $a \in S$ and T be a subset of S consisting of all those elements which are equivalent to a, i.e.,

$$T = \{x : x \in S \text{ and } x R a\}$$

Then $a \in T$, for $a R a$ (R is reflexive). Any two elements of T are equivalent to each other, for $x, y \in T$, then $x R a$ and $y R a$.

Again $\quad\quad\quad x R a, y R a \Rightarrow x R a, a R y \quad\quad\quad$ (R is symmetric)

$\Rightarrow \quad\quad\quad\quad\quad\quad\quad x R y \quad\quad\quad\quad\quad\quad\quad$ (R is transitive)

Functions and Relations

Thus, T is an equivalence class.

Let T_1 be another equivalence class, i.e.,

$$T_1 = \{x : x \in S \text{ and } x \, R \, b\}$$

where b is not equivalent to a. Then the class T and T_1 must be disjoint. For if they have a common elements s, $s \, R \, a$ and $s \, R \, b$, so that $b \, R \, a$ which is contrary to our hypothesis.

The set S can now be decomposed into equivalence classes $T, T_1, T_2, \ldots$ such that every element of S belongs to one of these classes. Since these classe are mutually disjoint, we obtain the required partition of S.

Theroem 2. *If R is an equivalence relation in a non-empty set X, then the quotient set X/R is a partition of x.*

Proof : Each $x \in X$ must belongs to some equivalence class. Also, the equivalence classes are pairwise disjoint, for if $z \in \dfrac{x}{R} \cap \dfrac{y}{R}$ then $x \, R \, z, y \, R \, z$. Since and $z \, R \, y \Rightarrow x \, R \, y$, it follows that $\dfrac{x}{R}$ and $\dfrac{y}{R}$ must be identical. Hence, two equivalence classes are either disjoint or identical. The set of equivalence classes is therefore a *partition*. Further, if x, y be any two members of the some set of this partition, they stand in a relation R to each other, showing that the partition induces the relation R.

Converse : If C be a partition of X, then the induced relation is an equivalence relation whose set of equivalence classes is X/C.

1.9 PARTIAL ORDER RELATIONS

A relation $\leq$ on a set A is called a *partial order relation iff* it is *reflexive, antisymmetric and transitive* and in this case the set A is called a *partial ordered set* and is denoted by the symbol $(A, \leq)$.

Remark : The relation in N defined by '$x < y$' is a partial order and is called *natural order or usual order in N*.

Comparable Elements : Two elements x, y in a partially ordered set $(A, \leq)$ are said to be *comparable* if either $x \leq y$ or $y \leq x$.

Uncomparable Elements : Two elements x, y in a partially ordered set $(A, \leq)$ are said to *uncompoarable* if $x \leq y$ and $y \leq x$ both do not hold.

Example : Let $A = \{1, 2, 3, 4, 5, 8, 16\}$. Define a relation $<$ in A by requiring $x < y$ iff $2x = y$.

The relation $<$ is A is a partial order. Here the elements 1, 3 are not comparable. So, we write $1 \not< 3$.

Similarly, $1 \not< 5, 5 \not< 1, 3 \not< 8, 8 \not< 3$.

The elements 4, 8 are comparable and so write $4 < 8$. Similarly, $8 < 16, 2 < 4$ etc.

1.9.1 Totally Ordered Set

Let (A, R) be a partially ordered set and let $B \subset A$ be arbitrary. Then the partiald order in A induces a partial order R, in B such that

$$aR'b : a, b \in B \Rightarrow aRb.$$

Then we say that (B, R') is a subset of a partially ordered set (A, R). Generally, the subset of a partially ordered set $(A, \leq)$ is denoted by $(B, \leq)$. *It is clear that any subset of a partially ordered set is a partially ordered set and similar is the case for a linearly ordered set.*

1.9.3 Upper and Lower Bounds

Let $(X, \leq)$ be a partially ordered set and let $(A, \leq)$ be a subset of $(X, \leq)$.

An element $a \in X$ is said to be an *upper bound of A if* $x \leq a \ \forall \ x \in A$.

An upper bound b of A is said to be a *least bound or supremum of A*. If $b \leq a$ for every upper bound a of A.

In brief, we write *l.u.b.* in place of 'least upper bound'. Similarly, we write *sup.* for the word supremum. The supremum of A is denoted by the symbol sup. (A).

An element $a \in X$ is called *lower bound of A* if

$$a \leq x \ \forall \ x \in A.$$

A lower bound b of A is called *greater lower bound or infimum of A*, if $a \leq b$ for every lower bound a of A.

In short, greatest lower bound is written as *g.l.b.*
Similarly, the infimum of A denoted by the symbol inf. (A).

Examples:

(i) Let $A = \{x \in Q; 2 < x^2 < 5\}$ be a set in the usual order. Then A contains an infinite number of lower bounds and upper bounds, but sup. (A) and inf. (A) do not exist.

(ii) Let $X = \{1, 2, 3, 4, 5, 6\}$ be a set in the natural order. Then $A = \{2, 3, 4\}$ is a subset of the partially order set X.

Lower bounds of A are 2, 1.
Upper bounds of A are 4, 5, 6.

$\therefore \qquad \sup(A) = 4, \inf.(A) = 2$

ILLUSTRATIVE EXAMPLES

1. *Prove that the relation defined by 'is perpendicular to' in the set of straight lines in a plane is symmetric but neither reflexive nor transitive.*

Solution. Let R stands for "is perpendicular to". Let L be the set of straight lines in a plane.

1. Reflexivity: Since no straight line is perpendicular to itself, hence relation is not reflexive, *i.e.*,

$$a \not R \ a \text{ for } a \in L$$

2. Symmetry: If straight line a is $\perp$ to straight line b then b is also $\perp$ to a, *i.e.*,

$$a \ R \ b \Rightarrow b \ R \ a \text{ for } a, b \in L.$$

Hence relation is symmetric.

3. Transitivity: If for $a, b, c \in L$, $a \perp b$ and $b \perp c$ (it may be parallel).

Hence, $a \ R \ b, b \ R \ c$ does not implies $a \ R \ c \ \forall \ a, b, c \in L$.
Hence relation is not transitive.

2. *Prove that a relation R on a set A is symmetric iff* $R = R^{-1}$.

Solution. First, let R be a symmetric relation on set A. Then we have to prove that $R = R^{-1}$. In order to prove this we have to prove that $R \subset R^{-1}$ and $R^{-1} \subset R$.

Now, $\qquad (a, b) \in R \Rightarrow (b, a) \in R \qquad\qquad (\because R \text{ is symmetric})$

Functions and Relations

$\Rightarrow \qquad\qquad\qquad (a, b) \in R^{-1}$ (by def. of inverse relation)

Thus, $(a, b) \in R \Rightarrow (a, b) \in R^{-1}$ for all $a, b \in A$.

So, $\qquad\qquad\qquad R \subset R^{-1}$ ...(i)

Now, let (x, y) be an arbitrary elements of Then

$\qquad\qquad (x, y) \in R^{-1} \Rightarrow (y, x) \in R$ (by def. of inverse relation)

$\Rightarrow \qquad\qquad (x, y) \in R$ ($\because R$ is symmetric)

Thus, $\qquad (x, y) \in R^{-1} \Rightarrow (x, y) \in R$ for all $x, y \in A$.

So, $\qquad\qquad\qquad R^{-1} \subset R$ ...(ii)

Thus, from (i) and (ii), we get $R = R^{-1}$.

Conversely, let R be a relation on set A such that $R = R^{-1}$. Then we have to prove that R is a symmetric relation on set A.

Let $(a, b) \in R$. Then

$\qquad\qquad (a, b) \in R \Rightarrow (b, a) \in R^{-1}$ (by def. of inverse relation)

$\Rightarrow \qquad\qquad (b, a) \in R,$ ($\because R = R^{-1}$)

Thus, $(a, b) \in R \Rightarrow (b, a) \in R$ for all $a, b \in A$.

So, R is a symmetric relation on A.

Hence, R is symmetric iff $R = R^{-1}$.

Congruence Modulo m : Let m be an arbitrary but fixed integer. Two integers a and b are said to be congruence modulo m if $a - b$ is divisible by m and we write $a \equiv b \pmod{m}$.

Thus, $a \equiv b \pmod{m} \Leftrightarrow a - b$ is divisible by m.

For example, $18 \equiv 3 \pmod 5$ because $18 - 3 = 15$ which is divisible by 5. But $25 \neq 2 \pmod 4$ because 4 is not a divisor of $25 - 2 = 23$.

3. Prove that the relation 'congruence modulo m' on the set Z of all integers is an equivalence relation.

Solution. (i) Reflexivity : Let a be an arbitrary integer. Then

$\qquad a - a = 0 = 0 \times m \Rightarrow a = a$ is divisible by m.

$\Rightarrow \qquad\qquad\qquad a \equiv a \pmod m$

Thus, $a \equiv a \pmod m$ for all $a \in Z$. So, ``congruence modulo m" if reflexive.

(ii) Symmetry : Let $a, b \in Z$ such that $a \equiv b \pmod m$.

Then, $\qquad a \equiv b \pmod m \Rightarrow a - b$ is divisible by m

$\Rightarrow \qquad\qquad a - b = \lambda m$ for $\lambda \in Z$.

$\Rightarrow \qquad\qquad b - a = (-\lambda) m \Rightarrow b = a$ is divisible by m.

$\Rightarrow \qquad\qquad b \equiv a \pmod m$

So, ``congruence modulo m'' is symmetric on Z.

(iii) Transitivity : Let $a, b, c \in Z$ such that

$\qquad a \equiv b \pmod m$ and $b \equiv c \pmod m$. Then

$a \equiv b \pmod{m} \Rightarrow a - b$ is divisible by m.

$\Rightarrow \qquad a - b = \lambda_1 m$ for some $\lambda_1 \in Z$

$b \equiv c \pmod{m} \Rightarrow b - c$ is divisible by m.

$\Rightarrow \qquad b - c = \lambda_2 m$ for some $\lambda_2 \in Z$

$\therefore \qquad (a - b) + (b - c) = \lambda_1 m + \lambda_2 m = (\lambda_1 + \lambda_2) m$

Thus, $a \equiv b \pmod{m}$ and $b \equiv c \pmod{m}$

$\Rightarrow \qquad a \equiv c \pmod{m}$

So, `congruence modulo m' is transitive on Z.

Hence, `congruence modulo m' is an equivalence relation on Z.

4. *Consider a set* $A = \{p, q, r\}$ *and the relation R on A defined by*

$$R = \{(p, p), (p, q), (q, r), (r, r)\} \quad \text{find}$$

(a) *reflexive (R); (b) symmetric (R); and (c) transitive (R).*

Solution. (a) The reflexive closure on R is obtained by adding all diagonal pairs of $A \times A$ to R which are not currently in R. Hence, reflexive $(R) = R \cup \{(p, q)\}$.

$$= \{(p, p), (p, q), (q, q), (q, r), (r, r)\}$$

(b) The symmetric closure of R is obtained by adding all the pairs in R^{-1} to R which are not currently in R. Hence,

Symmetric $\qquad (R) = R \cup \{(q, p), (r, q)\}$

$$= \{(p, p), (p, q), (q, p), (q, r), (r, q), (r, r)\}$$

(c) The transitive closure on R, since A has three elements, is obtained by taking the union of R with $R^2 = RoR$ and $R^3 = RoRoR$. Now,

$$R^2 = RoR = \{(p, p), (p, q), (p, r), (q, r), (r, r)\}$$

$$R^3 = RoRoR = \{(p, p), (p, q), (p, r), (q, r), (r, r)\}$$

Hence, transitive $(R) = R \cup R^2 \cup R^3$

$$= \{(p, p), (p, q), (p, r), (q, r), (r, r)\}$$

5. *Show that in the set N of all natural numbers, the relation R defined by aRb if a divides to is a partial order relation.*

Solution. (i) We have $\forall a \in N$, a is a divisor of a, i.e., aRa. Therefore, R is reflexive.

(ii) Again, if a is a divisor of b then b can not be a divisor of a unless $a = b$. Thus, aRb and $bRa \Rightarrow a = b$. Therefore, R is anti-symmetric.

(iii) Finally, a is a divisor of b and b is a divisor of c implies a is a divisor of c. Therefore, R is transitive.

Since R is reflexive, anti-symmetric and transitive, therefore, R is a partial order relation.

6. *If A be a family of sets, and R be the relation in A defined by ``X is a subset of Y''. Then show that R is a partial order relation.*

Solution. Since every set is a subset of itself therefore, $\forall X \in A$ we have $X R X$, i.e., R is reflexive.

Again R is anti-symmetric since

$$A \subset B \text{ and } B \subset A \Rightarrow A = B.$$

Functions and Relations

Finally, $A \subset B$ and $B \subset C \Rightarrow A \subset C$.

Therefore, R is transitive.

Since R is reflexive, anti-symmetric and transitive, therefore, R is a partial order relation.

EXERCISES

1. Define a relation. When a relation R on a set A is known as symmetric, reflexive, transitive and anti-symmetric ? Give an example for each.
2. Show that in the set of all real numbers, the relation `greater than' is transitive but not reflexive.
3. Show that the relation R in the set of natural numbers N defined by $a\,R\,b$ if a divides b, is reflexive and transitive but not symmetric.'
4. Give an example of the relation which is :
   (a) reflexive, symmetric but not transitive
   (b) symmetric, transitive but not reflexive.
   (c) symmetric but neither transitive nor reflexive.
   (d) neither symmetric nor reflexive, nor transitive.
5. If R and S are equivalence relations on a set A, prove that $R \cap S$ is an equivalence relation in A.
6. Which of the following relations in the set of real numbers are equivalence relations.
   (a) $a\,R\,b$ if $|a|=|b|$,
   (b) $a\,R\,b$ if $|a| \geq |b|$,
   (c) $a\,R\,b$ if $a - b \geq 0$?
7. S is the set of real numbers $a\,R\,b$, if $a = \pm b$, determine whether R is an equivalence relation ?
8. Prove that if a relation R is transitive then its inverse relation R^{-1} is also transitive.
9. Find all the partitions of $\{a, b, c\}$.
10. If $x = \{1, 2\}$ and C (a partition of x) = $\{(1), (2)\}$ find what equivalence relation is induced by C.

ANSWERS

6. (a) an equivalence   (b) not an equivalence   (c) not an equivalence
9. $\{(a), (b), (c)\}, \{(a), (b, c)\}, \{\{b\},\{a, c\}\}$

   $\{\{c\},\{a, b\}\}, \{\{a\}, \{b\}, \{c\}\}$.

2
Congruence of Integers

2.1 CONGRUENCE OF INTEGERS

(Relations of ``congruence modulo m'' in the set of integers)

Definition : *Let m be any fixed positive integer, i.e., $m > 0$. Then an integer a is said to be congruent to another integer b modulo m if $m \mid (a - b)$, i.e., if m is a divisor of $a - b$.*

Symbolically, we write
$$a \equiv b \pmod{m}.$$
It will be read as ``a is congruent to b modulo m''.

Thus $a \equiv b \pmod{m}$ if $a - b = km$ for some integer k.

For example,

$13 \equiv 1 \pmod{12}$, because $13 - 1 = 12$ which is multiple of 12.

$27 \equiv 3 \pmod{12}$, because $27 - 3 = 24 = 2 \cdot 12$

$25 \equiv 0 \pmod{5}$, because $25 - 0 = 5 \cdot 5$

$3m + 2 \equiv 2 \pmod{m}$, because $3m + 2 - 2 = 3m$.

Remark : It is clear that if some number is to be expressed as congruence mod m, then that number should be divided by m and then remainder is the desired number.

Theorem 1. *If a and b are two integers, then $a \equiv b \pmod{m}$ if and only if a and b have the same remainder when divided by m.*

Proof : Necessary Condition : Let a and b have the numbers r_1 and r_2 respectively when divided by m. Then for some integers q_1 and q_2, we have
$$a = q_1 m + r_1, \; 0 \le r_1 < m$$
and
$$b = q_2 m + r_2, \; 0 \le r_2 < m.$$

Now, suppose that $a \equiv b \pmod{m}$. Then to prove that
$$r_1 = r_2.$$
We have
$$a - b = m(q_1 - q_2) + r_1 - r_2$$
$$\therefore \quad r_1 - r_2 = (a - b) + m(q_2 - q_1).$$

Now $a \equiv b \pmod{m} \Rightarrow m \mid (a - b)$. Also, $m \mid m(q_2 - q_1)$.

$\therefore \quad m \mid \{(a - b) + m(q_2 - q_1)\}$

$\Rightarrow \quad m \mid (r_1 - r_2)$

$\Rightarrow \quad r_1 - r_2 = 0 \qquad [\because 0 \le (r_1 - r_2) < m]$

$\Rightarrow \quad r_1 = r_2.$

Congruence of Integers

Sufficient condition : Let $r_1 = r_2$. Then to prove that
$$a \equiv b \pmod{m}$$
we have
$$a - b = m(q_1 - q_2) + r_1 - r_2$$
$$= m(q_1 - q_2) \text{ if } r_1 = r_2$$
$$\therefore \quad m \mid (a - b)$$
$$\Rightarrow \quad a \equiv b \pmod{m}.$$

Theorem 2. *If a is any integer, then $a \equiv r \pmod{m}$ where r is the remainder obtained on dividing a by m.*

Proof : Let r be the remainder obtained on dividing a by m. Then for some integer q, we have
$$a = mq + r$$
$$\Rightarrow \quad a - r = mq$$
$$\Rightarrow \quad m \mid (a - r) \Rightarrow a \equiv r \pmod{m}.$$

Theroem 3. *Let d be the GCD of c and m, i.e., $(c, m) = d$ and let $m = m_1 d$. Then prove that $ca \equiv cb \pmod{m} \Rightarrow a \equiv b \pmod{m_1}$. Also prove the converse, i.e., if $a \equiv b \pmod{m_1}$, Then $ca \equiv cb \pmod{m}$.*

Proof : We have $(c, m) = d$ and $m = m_1 d$. Let $c = c_1 d$. Since d is the greatest common divisor of c and m, therefore, the greatest common divisor of c_1 and m_1 msut be 1, i.e, $(c_1, m_1) = 1$, because c_1 and m_1 are the integers obtained on dividing c and m respectively by their greatest common divisor d. Now
$$ca \equiv cb \pmod{m}$$
$$\Rightarrow \quad m \mid (ca - cb) \Rightarrow m \mid c(a - b)$$
$$\Rightarrow \quad m_1 d \mid c_1 d (a - b) \qquad [\because m = m_1 d \text{ and } c = c_1 d]$$
$$\Rightarrow \quad m_1 \mid c_1 (a - b) \qquad [\because d \neq 0]$$
$$\Rightarrow \quad m_1 \mid (a - b) \qquad [\because m_1 \text{ and } c_1 \text{ are relatively prime}]$$
$$\Rightarrow \quad a \equiv b \pmod{m_1}.$$

Converse : We have $a \equiv b \pmod{m_1}$
$$\Rightarrow \quad m_1 \mid (a - b) \Rightarrow m_1 \mid c_1 (a - b)$$
$$\Rightarrow \quad m_1 d \mid c_1 d (a - b)$$
$$\Rightarrow \quad m \mid c (a - b)$$
$$\Rightarrow \quad m \mid (ca - cb)$$
$$\Rightarrow \quad ca \equiv cb \pmod{m}.$$

Theroem 4. *If $ca \equiv cb \pmod{m}$ and $(c, m) = 1$, then $a \equiv b \pmod{m}$.*

Proof : We have $(c, m) = 1$, i.e., c and m are relatively prime.
We have $\quad ca \equiv cb \pmod{m}$
$$\Rightarrow \quad m \mid (ca - cb)$$
$$\Rightarrow \quad m \mid c (a - b)$$
$$\Rightarrow \quad m \mid (a - b) \qquad [\because (c, m) = 1]$$
$$\Rightarrow \quad a \equiv b \pmod{m}.$$

2.3 RESIDUE CLASSES OR CONGRUENCE CLASSES

If m is a fixed positive integer, then ``congruence modulo m'' is an equivalence relation in the set of integers, consequently it will partition I into equivalence classes. These equivalence classes are called residue classes modulo m or congruence classes moduo m.

The set of all residue classes of integers modulo m is represented by I_m. It is called the set of integrs modulo m. Some authors denote this set by J_m or by $I/(m)$.

If $a \in I$, then the residue class $[a] \in I_m$ is given by

$$[a] = \{x : x \in I \text{ and } x \equiv a \pmod{m}\}.$$

Similarly, if $b\ I$, then the residue class $[b] \in I_m$ is given by

$$[b] : \{y : y \in I \text{ and } y \equiv b \pmod{m}, i.e., m \mid (y - b)\}.$$

We know that two equivalent classes are either disjoint or identical. Therefore, if $[a] \in I_m$ and $[b] \in I_m$, then either $[a] = [b]$ or $[a] \cap [b] = \phi$.

Also $[a] = [b]$ if and only if $a \equiv b \pmod{m}$, i.e., if and only if $m \mid (a-b)$. Thus $[a] = [a+m] = [a+2m] = ...$ so on. Similarly, $[1] = [1+m] = [1+2m]$ and so on. Also, $[1] = [1+m] = [1+2m]$ and so on. Also $[0] = [m] = [2m] = [-m]$ and so on.

The residue classes modulo 3, *i.e.*, the elements of the set I_3 are

$$[0] = \{..., -9, -6, -3, 0, 3, 6, 9, ...\}$$
$$[1] = \{..., -8, -5, -2, 1, 4, 7, 10, ...\}$$
$$[2] = \{..., -7, -4, -1, 2, 5, 8, 11, ...\}$$

Basic Properties : The basic properties of the residue classes modulo m are :

(i) If a and b are elements of the same residue class $[s]$, then

$$a \equiv b \pmod{m}.$$

(ii) If $[s]$ and $[t]$ are two distinct residue classes with $a \in [s]$ and $b \in [t]$, then $a \equiv b \pmod{m}$.

Theorem : *The set I_m of all residue classes of integers modulo m contains exactly m distinct elements.*

Proof : We have $I_m = \{[0], [1], [2], ..., [m-1]\}$

First we shall show that the m residue classes $[0], [1], ..., [m-1]$ are all distinct.

Let $0 \leq i < m, 0 \leq j < m$ and $j > i$.

Then $\qquad\qquad\qquad [i] = [j]$

$\Rightarrow \quad i \equiv j \pmod{m} \Rightarrow i - j$ is divisible by m

$\Rightarrow \quad j - i$ is divisible by m.

But according to our assumption $j - i$ is a positive integer less than m. So, it can not be divisible by m. Therefore, $[i] \neq [j]$ and thus $[0], [1], ..., [m-1]$.

By division algorithm, we have

$\qquad\qquad\qquad a = km + r, \qquad\qquad\qquad$ where $k, r \in I$ and $0 \leq r < m$.

$\Rightarrow \quad a - r = km$

$\Rightarrow \quad (a - r)$ is divisible by m

$\Rightarrow \quad a \equiv r \pmod{m}$

$\Rightarrow \quad [a] = [r]$.

Since $0 \leq r \leq m-1$, therefore, the residue class $[a] = [r]$ is one of the residue classes $[0], [1], ..., [m-1]$.

Hence the set I_m has m distinct elements.

Congruence of Integers

2.3 ALGEBRA OF RESIDUE CLASSES

Addition of residue classes : If $[a], [b] \in \mathbf{I_m}$, then we define
$$[a] + [b] = [a + b].$$
Here '+' on L.H.S. stands for addition of residue classes, and '+' on the R.H.S. stands for addition of integers.

Since $a, b \in \mathbf{I} \Rightarrow a + b \in \mathbf{I}$, therefore $[a + b]$ is also a residue class, i.e., $[a + b] \in \mathbf{I_m}$. Now we know that $[a] = [a + m] = [a + 2m]$, and so on. Thus a residue class can be represented in several ways. Therefore, we must know that addition of residue classes is well defined, i.e., it is independent of the representation of any residue class. For this we are to show that if $[a] = [c]$ and $[b] = [d]$, then $[a] + [b] = [c] + [d]$.

We have $\qquad [a] = [c] \Rightarrow a - c$ is divisible by m.
Also $\qquad [b] = [d] \Rightarrow b - d$ is divisible by m.

Now $m \mid (a - c)$ and $m \mid (b - d) \Rightarrow m \mid \{(a - c) + (b - d)\}$

$\Rightarrow \qquad m \mid \{(a + b) - (c + d)\}$
$\Rightarrow \qquad a + b \equiv c + d \pmod{m}$
$\Rightarrow \qquad [a + b] = [c + d]$
$\Rightarrow \qquad [a] + [b] = [c] + [d]$

Thus, $[a] = [c]$ and $[b] = [d]$
$\Rightarrow \qquad [a] + [b] = [c] + [d]$

Hence, addition of residue classes is well defined.

Multiplication of residue classes.

If $[a], [b] \in \mathbf{I_m}$ then we define
$$[a][b] = [ab].$$

Since, $a, b \in \mathbf{I} \Rightarrow ab \in \mathbf{I}$, therefore, $[ab]$ is also a residue class, i.e., $[ab] \in \mathbf{I_m}$. But we must show that multiplication of residue classes is well defined. For this we are to show that if $[a] = [c]$ and $[b] - [d]$, then $[a][b] = [c][d]$.

We have $\qquad [a] = [c] \Rightarrow a \equiv c \pmod{m}$
$\qquad \Rightarrow a - c$ is divisible by m.
$\qquad \Rightarrow b(a - c)$ is divisible by m.

Also $\quad [b] = [d] \quad \Rightarrow b - d$ is divisible by m
$\qquad \Rightarrow c(b - d)$ is divisible by m

$\therefore \quad [a] = [c]$ and $[b] = [d] \Rightarrow \{b(a - c) + c(b - d)\}$ is divisible by m
$\qquad \Rightarrow ab - cd$ is divisible by m
$\qquad \Rightarrow ab \equiv cd \pmod{m}$
$\qquad \Rightarrow [ab] = [cd]$
$\qquad \Rightarrow [a][b] = [c][d]$

Hence multiplication of residue classes is well defined.

2.3.1 Linear Congruences

Consider the congruence

$$ax \equiv b \pmod{m}$$

In this a, b, m are fixed integers with $m > 0$ and x is an unknown integer. This equation is called a *linear congruence*.

By a solution of equation (i) we mean an integer $x = x_1$ such that $ax_1 \equiv b \pmod{m}$, i.e., $m \mid (ax_1 - b)$.

Theorem : *If x_1 is solution of $ax \equiv b \pmod{m}$, then*

$$ax_1 \equiv b \pmod{m}.$$

Now if $x_2 \equiv x_1 \pmod{m}$, then

$$ax_2 \equiv ax_1 \pmod{m}$$

Since the relation `congruent modulo m' is transitive, therefore,

$ax_2 \equiv ax_1 \pmod{m}$ and $ax_1 \equiv b \pmod{m}$

$\Rightarrow \qquad ax_2 \equiv b \pmod{m}.$

$\therefore \quad x_2$ is a solution of $ax \equiv b \pmod{m}$.

2.3.2 Incongruent Solutions

Suppose x_1 and x_2 are two solutions of the congruence $ax \equiv b \pmod{m}$. If $x_1 \not\equiv x_2 \pmod{m}$, then x_1 and x_2 are called incongruent solutions of $ax \equiv b \pmod{m}$.

For example consider the congruence $6x \equiv 2 \pmod{4}$. We see that 1 and 3 are its solutions. Since $1 \not\equiv 3 \pmod 4$, therefore, 1 and 3 are incongruent solution of $6x \equiv 2 \pmod 4$. On the other hand 5 is also a solution of $6x \equiv 2 \pmod 4$. But $5 \equiv 1 \pmod 4$. Therefore, 5 and 1 are congruent solutions of this linear congruence :

2.3.3 Existence of Solutions of $ax \equiv b \pmod{m}$

Theorem : *The congruence $ax \equiv b \pmod{m}$ has a solution if and only if the greatest common divisor of a and m, i.e., (a, m) divides b.*

Proof : Let $d = (a, m)$.

Let $ax \equiv b \pmod{m}$ has a solution of $x = x_1$.

Then $\qquad ax_1 \equiv b \pmod{m}$

$\Rightarrow \qquad m \mid (ax_1 - b)$

$\Rightarrow \qquad ax_1 - b = mk$ for some integer k

$\Rightarrow \qquad b = ax_1 - mk.$

Now $d \mid a$ and $d \mid m$. Let $a = a_1 d$ and $m = m_1 d$ where a_1 and m_1 are some integers. Then

$$b = a_1 dx_1 - m_1 dk = d(a_1 x_1 - m_1 k)$$

$\therefore \ d \mid b$. This shows that if a solution exists, then it is necessary that $d \mid b$.

Converse : Suppose that $d \mid b$. Then $b = b_1 d$ where b_1 is some integer. Since $d = (a, m)$, therefore, by Euclidean algorithm there exists integers u and v such that

$$d = ua + vm$$

$\Rightarrow \qquad db_1 = uab_1 + vmb_1 \qquad\qquad$ [multiplying by b_1]

$\Rightarrow \qquad b = a(ub_1) + (vb_1)m \qquad\qquad$ [$\because db_1 = b$]

Congruence of Integers

$\Rightarrow \quad a(ub_1) - b = -(vb_1)m$

$\Rightarrow \quad m \mid \{a(ub_1) - b\}$

$\Rightarrow \quad a(ub_1) \equiv b \pmod{m}$

$\Rightarrow \quad x = ub_1$ is a solution of $ax \equiv b \pmod{m}$.

Thus, if $d \mid b$, then it is sufficient to say that $ax \equiv b \pmod{m}$ has a solution.

2.4 NUMBER OF INCONGRUENT SOLUTIONS

Theorem : If $d = (a, m)$ divides b, then the congruence $ax \equiv b \pmod{m}$ has exactly d incongruent solutions which can be expressed in the form $x_0 + rm_1$ for $r = 0, 1, 2, ..., d-1$, where x_0 is an arbitrary solution and $m = dm_1$.

Proof : Let $b = b_1 d$, $a = a_1 d$, $m = m_1 d$ where b_1, a_1, m_1 are some integers. Since d is the greatest common divisor of a and m, therefore, $(a_1, m_1) = 1$.

Since $d \mid b$, therefore, $ax \equiv b \pmod{m}$ possesses a solution. Let x_0 be an arbitrary solution of this congruence. Let x_1 be any other solution of this congruence. We know that $x = y$ is a solution of $ax \equiv b \pmod{m}$ if and only if y is a solution of $a_1 x \equiv b_1 \pmod{m_1}$. Therefore, x_0 and x_1 are also solutions of $a_1 x \equiv b_1 \pmod{m_1}$. But $a_1 x \equiv b_1 \pmod{m_1}$ has a unique incongruent solution modulo m_1 because $(a_1, m_1) = 1$. Therefore,

$$x_1 \equiv x_0 \pmod{m_1}$$

$\Rightarrow \quad m_1 \mid (x_1 - x_0)$

$\Rightarrow \quad x_1 - x_0 = rm_1$ for some integer r

$\Rightarrow \quad x_1 = x_0 + rm_1$.

Hence, every solution x_1 of the congruence $ax \equiv b \pmod{m}$ can be expressed in the form $x_1 = x_0 + rm_1$ for some integer r.

Further, for every integer r, we have $x_0 + rm_1 \equiv x_0 \pmod{m_1}$. Therefore, for every integer r, $x_0 + rm_1$ is a solution of $a_1 x \equiv b_1 \pmod{m_1}$ and so also of $1x \equiv b \pmod{m}$. Hence all the solutions of $ax \equiv b \pmod{m}$ are found among the set of integers $x_0 + rm_1$.

Let us consider the following d integers in this set :

$$x_0, x_0 + m_1, x_0 + 2m_1, ..., x_0 + (d-1)m_1.$$

Now no two of these d integers are congruent to each other modulo m. For if $0 \le i < d$ and $0 \le j < d$, then

$$x_0 + im_1 \equiv x_0 + jm_1 \pmod{m}$$

$\Rightarrow \quad m \mid \{(x_0 + im_1) - (x_0 + jm_1)\}$

$\Rightarrow \quad m \mid (i - j)m_1$

$\Rightarrow \quad m_1 d \mid (i - j)m_1$ $\quad [\because m = m_1 d]$

$\Rightarrow \quad d \mid (i - j)$

$\Rightarrow \quad i - j = 0$ $\quad [\because 0 \le i - j < d]$

$\Rightarrow \quad i = j$

$\Rightarrow \quad x_0 + im_1 = x_0 + jm_1$.

Hence any two of these d integers are incongruent $\pmod{m}$.

Further if r is any integer, then we shall show that $x_0 + rm_1$ is congruent to one of the d integers given above.

Applying division algorithm for the integers r and d, we have

Then
$$x_0 + rm_1 = x_0 + (ds + q) m_1$$
$$= x_0 + dsm_1 + qm_1$$
$$= (x_0 + qm_1) + dsm_1$$

Now $(x_0 + rm_1) - (x_0 + qm_1) = (x_0 + qm_1) + dsm_1 - (x_0 + qm_1)$
$$= dsm_1 = sm.$$

$\therefore \quad m \mid \{(x_0 + rm_1) - (x_0 + qm_1)\}$

$\Rightarrow \quad x_0 + rm_1 \equiv x_0 + qm_1 \pmod{m}$

Since $0 \le q < d$, therefore, $x_0 + qm_1$ is one of the above d integers.

Hence $ax \equiv b \pmod{m}$ has exactly d incongruent solutions modulo m.

ILLUSTRATIVE EXAMPLES

1. *Find the least positive integer to which the integer 107, 429 and 107×429 are congruent modulo 9.*

Solution. Since $107 = 11 \cdot 9 + 8$ we have
$$107 \equiv 8 \pmod{9}$$

Also $\quad 429 = 9 \cdot 47 + 6 \quad\quad \therefore\ 429 \equiv 6 \pmod{9}$

and therefore, $\quad 107 \times 429 \equiv 8 \times 6 \pmod{9}$
$$\equiv 48 = 9 \cdot 5 + 3$$
$$\equiv 3 \pmod{9}.$$

2. *Show that the congruence $x + 50 \equiv 39 \pmod{7}$ possesses a solution.*

Solution. Since $\quad 50 = 7 \cdot 7 + 1 \quad\quad \therefore\ 50 \equiv 1 \pmod{7}$

Again since $\quad 39 = 7 \cdot 5 + 4 \quad\quad \therefore\ 39 \equiv 4 \pmod{7}$

$\therefore \quad x + 50 \equiv 39 \pmod{7}$

$\Rightarrow \quad x + 1 \equiv 4 \pmod{7}$

$\Rightarrow \quad x \equiv 3 \pmod{7}.$ \hfill (on adding -1 on both sides.)

3. *Show that $15x \equiv 12 \pmod{5}$ does not possesses a solution.*

Solution. Since greatest common divisor of 15 and 5 is 5 and 12 is not divisble by 5 ; hence it does not possess a solution.

4. *Find the least positive incongruent solutions of $52x \equiv 28 \pmod{20}$.*

Solution. We have $(52, 20) = 4$ and $4 \mid 28$. Therefore, $52x \equiv 28 \pmod{20}$ has 4 incongruent solutions.

Now $\quad 52x \equiv 28 \pmod{20}$ is equivalent to
$$13 \cdot 4x \equiv 7 \cdot 4 \bmod (5 \cdot 4)$$

or $\quad\quad 13x \equiv 7 \pmod{5}$

or $\quad\quad 3x \equiv 7 \pmod{5}$ \hfill $[\because 10x \equiv 0 \pmod{5}]$

Congruence of Integers

or	$3x \equiv 2 \pmod 5$	$[\because 7 \equiv 2 \pmod 5]$
or	$3x \equiv 12 \pmod 5$	$[\because 0 \equiv 10 \pmod 5]$
or	$x \equiv 4 \pmod 5$	$[\because (3, 5) = 1]$

showing that $x = 4$ is a solution of $13x \equiv 7 \pmod 5$.

Now 4 is the least positive integer satisfying $13x \equiv 7 \pmod 5$ since 4 is the least positive integer in the residue class $[4] \in I_5$. Therefore, the four required solutions of $52x \equiv 28 \pmod{20}$ are 4, $4 + 1 \cdot 5, 4 + 2 \cdot 5, 4 + 3 \cdot 5$ i.e., 4, 9, 14, 19.

EXERCISE

1. Show that :
    (a) $91 \equiv 1 \pmod 9$;
    (b) $23 \cdot 215 \equiv 3 \pmod 7$;
    (c) $7.6.5.4 \equiv 0 \pmod 6$;
    (d) $109 \cdot 431 \equiv 9 \pmod{11}$;
2. Show that the congruence $a + 58 \equiv 53 \pmod 9$ possesses a solution and find it.
3. Show that $25x \equiv 12 \pmod{10}$ does not possesses a solution.
4. Find the least positive incongruent solutions of :
    (a) $13x \equiv 9 \pmod{25}$
    (b) $259x \equiv 5 \pmod{11}$
5. Find the least positive incongruent solutions of :
    (a) $35x \equiv 14 \pmod{21}$
    (b) $222x \equiv 12 \pmod{18}$
    (c) $x + 50 \equiv 39 \pmod 7$
6. Find the least positive incongruent solutions of :
    (a) $2x + 1 \equiv 4 \pmod 5$
    (b) $2x + 1 \equiv 4 \pmod{10}$
    (c) $51x \equiv 32 \pmod 7$
    (d) $7x \equiv 5 \pmod{256}$
    (e) $104x \equiv 16 \pmod{296}$
    (f) $45x \equiv 24 \pmod{348}$.

Answers

4. (a) 18 ; (b) 10 ;
5. (a) 1, 4, 7, 10, 13, 16, 19 ; (b) 2, 5, 8, 11, 14, 17 ; (c) 3 ;
6. (a) 4, (b) no solution, (c) 2, (d) 147, (e) 3; (f) 16.

3

Some Special Types of Matrices

3.1 TRANSPOSE OF A MATRIX

The matrix of order $n \times m$ obtained by interchanging the rows and columns of a matrix A of order $m \times n$ is called the **transpose of the matrix** A or **the transposed matrix** of A and is denoted by A'.

If $A = [a_{ij}]$ be a matrix of order $m \times n$, then the matrix $B = [b_{ji}]$ of order $n \times m$, such that $b_{ij} = a_{ji}$ is known as transposed matrix of A or the transpose of the matrix A and is denoted by A'.

For example : If
$$A = \begin{bmatrix} 2 & 3 & 4 \\ 1 & 5 & 6 \end{bmatrix}$$

then
$$A' = \begin{bmatrix} 2 & 1 \\ 3 & 5 \\ 4 & 6 \end{bmatrix}$$

3.1.1. Some Theorems on Transposed Matrix

Theorem 1. *If A be any matrix, then $(A')' = A$.*

Proof. Let $A = [a_{ij}]$ be an $m \times n$ matrix. Then A', the transpose of A is an $n \times m$ matrix, and $(A')'$, the transpose of A' is an $m \times n$ matrix.

Now, (i, j) th element of $(A')' = (j, i)$ th element of $A' = (i, j)$ the element of A.

Hence the matrices $(A')'$ and A are comparable and their corresponding elements are equal each to each. Therfore, $(A')' = A$.

Theorem 2. *The transpose of the sum of two matrices is the sum of their transposes, i.e., $(A + B)' = A' + B'$, A and B being comparable.*

Proof. Let $A = [a_{ij}]$, $B = [b_{ij}]$ be two matrices of order $m \times n$ each. Obviously $A + B$ exists and is of order $m \times n$. Consequently $(A + B)'$ is of order $n \times m$. Now A' and B' both are order $n \times m$ and therefore $A' + B'$ also exists and is of order $n \times m$. Thus the matrices $(A + B)'$ and $A' + B'$ being of the same order.

Now, (i, j) th element of $(A + B)'$

$= (j, i)$ the element of $(A + B)$

$= a_{ji} + b_{ji}$

$= (j, i)$ th element of $A + (j, i)$ the element of B,

$= (i, j)$ th element of $(A' + B')$

Hence, we have,

(i) order of $(A + B)' = $ order of $A' + B'$

Some Special Types of Matrices

and (ii) (i, j)th element of $(A + B)' = (i, j)$th element of $(A' + B')$

∴ $(A + B)' = A' + B'$.

Theorem 3. *If A and B two matrices such that they are conformable for multiplication AB then $(A B)' = B'A'$ (Reversal Law)*

Proof. Let $A = [a_{ij}]_{m \times n}$, $B = [b_{jk}]_{n \times p}$ be two matrices. Clearly product AB is possible and it is $m \times p$ matrix. Evidently $(A B)'$ will be a $(p \times m)$ matrix.

Now, $A' = [a'_{ji}]_{n \times m}$, $B' = [b'_{kj}]_{p \times n}$,

where $a'_{ji} = a_{ij}, b'_{kj} = b_{jk}$, Hence $B'A'$ is possible and it is a matrix of order $p \times m$.

Also (k, i)th element of $(A B)'$

$$= (i, k) \text{ th element of } AB$$

$$= \sum_{j=1}^{n} a_{ij} b_{jk}$$

$$= \sum_{j=1}^{n} a'_{ji} b'_{ki}$$

$$= \sum_{j=1}^{n} b'_{kj} a'_{ji}$$

$$= (k, i) \text{ th element of } B'A' \text{ for all } i, k.$$

Hence,

(i) Matrices $(A B)'$ and $(B'A')$ are of same order, and

(ii) Corresponding elements of the two matrices are equal.

Hence $(A B)' = B'A'$.

3.2. SYMMETRIC MATRIX

A square matrix A such that $A' = A$ is called symmetric matrix, *i.e.,* matrix $[a_{ij}]$ is symmetric provided $a_{ij} = a_{ji}$ for all values of i and j.

For example :

$$A = \begin{bmatrix} a & h & g \\ h & b & f \\ g & f & c \end{bmatrix} \text{ is a symmetric matrix.}$$

Skew-symmetric matrix

A square matrix A such that $A' = -A$ is called skew-symmetric, *i.e.,* the matrix $[a_{ij}]$ is skew-symmetric provided $a_{ij} = -a_{ji}$, for all values of i and j.

For example :

$$A = \begin{bmatrix} 0 & 1 & -3 \\ -1 & 0 & 5 \\ 3 & -5 & 0 \end{bmatrix} \text{ is a skew-symmetric matrix.}$$

Remark. In a skew-symmetric matrix, we have $a_{ij} = -a_{ji}$. For diagonal elements $a_{ii} = -a_{ii}$, *i.e.,* $2a_{ii} = 0$, or $a_{ii} = 0$.

Thus every diagonal element of a skew-symmetric matrix is zero.

EXAMPLES

1. *Show that every matrix can be uniquely expressed as the sum of a symmetric and a skew-symmetric matrix.*

Solution. Let A be any square matrix.
Now, we have

$$A = \frac{1}{2}A + \frac{1}{2}A = \frac{1}{2}(A + A') + \frac{1}{2}(A - A') \qquad \ldots(i)$$

Now, $\qquad (A + A')' = A' + (A')' = A' + A = A + A'$
(since matrix addition is commutative)
and $\qquad (A - A')' = A' - (A')' = A' - A = -(A - A')$

Hence $(A + A')$ is symmetric and $(A - A')$ is skew-symmetric.

Consequently $\frac{1}{2}(A + A') = P$ (say) is a symmetric matrix and $\frac{1}{2}(A - A') = Q$ (say) is a skew-symmetric matrix.

Hence $\qquad A = P + Q.$

Thus, any square matrix, can be expressed as the sum of a symmetric and a skew symmetric matrix.

Uniqueness. To show that this representation is unique, let us suppose that another representation $A = R + S$ is possible, where R is symmetric and S is skew-symmetric, *i.e.,* $R = R'$ and $S = -S'$.

Now $\qquad A' = (R + S)' = R' + S' = R - S$
Also, $\qquad A + A' = (R + S) + (R - S) = 2R$
and $\qquad A - A' = (R + S) - (R - S) = 2S$
or $\qquad R = \frac{1}{2}(A + A')$ and $S = \frac{1}{2}(A - A')$

Hence $\qquad A = \frac{1}{2}(A + A') + \frac{1}{2}(A - A'),$ is a unique representation.

2. *If A is a skew-symmetric matrix, then show that $AA' = A'A$ and A^2 is symmetric.*

Solution. We have, A is a skew-symmetric matrix, hence

$$A' = -A \qquad \ldots(i)$$

Pre-multiplying both sides by A, we get

$$AA' = -AA = -A^2 \qquad \ldots(ii)$$

Post-multiplying both sides of (i) by A, we get

$$A'A = -AA = -A^2 \qquad \ldots(iii)$$

From (ii) and (iii), we have

$$AA' = A'A.$$

We also know that AA' and $A'A$ are symetric matrices. Hence from (ii) and (iii) we get $-A^2$ is also a symmetric matrix.

3. *Prove that positive odd integral powers of a skew-symmetric matrix are skew symmetric but positive even integral powers are symmetric.*

Solution. Let **A** be a skew symmetric matrix. Then

$$A' = -A$$

Some Special Types of Matrices

Now, we have
$$(A^p)' = (A . A . A ... p \text{ times})'$$
$$= (A' . A' . A' ... p \text{ times}) \quad \text{(by reversal law)}$$
$$= [(-A)(-A)(-A) ... p \text{ times}]$$
$$= (-1)^p A^p.$$

Now when p is even integer, then
$$(A^p)' = (-1)^p A^p = A^p.$$

Hence A^p is a symmetric matrix.

But, if p is an odd integer, then $(A^p)' = (-1)^p A^p = -A^p$,

i.e., A^p is a skew-symmetric matrix.

EXERCISES (A)

1. If $A = \begin{bmatrix} 1 & 2 & 0 \\ 3 & -1 & 0 \end{bmatrix}$, find AA' and $A'A$ and show that AA' and $A'A$ are symmetric but $AA' \neq A'A$.

2. If $A = \begin{bmatrix} 1 & -1 & 0 \\ 2 & 1 & 3 \\ 4 & 1 & 8 \end{bmatrix}$ and $B = \begin{bmatrix} 4 & 1 & 0 \\ 2 & -3 & 1 \\ 1 & 1 & -1 \end{bmatrix}$ then verify: $(AB)' = B'A'$.

3. Show that A^2 is symmetric, if either A is symmetric or A is skew-symmetric.
4. If A and B are symmetric (skew-symmetric), then show that $A + B$ is symmetric (skew-symmetric).
5. Show that all positive integral powers of a symmetric matrix are symmetric.
6. If A is symmetric (skew-symmetric), show that $B'AB$ is symmetric (skew-symmetric).
7. If A, B are symmetric, show that $AB + BA$ is symmetric and $AB - BA$ is skew-symmetric.
8. If A be any square matrix, show that $A + A'$ is symmetric and $A - A'$ is skew-symmetric.

ANSWERS

1. $\begin{bmatrix} 5 & 1 \\ 1 & 26 \end{bmatrix}$; $\begin{bmatrix} 10 & -1 & 12 \\ -1 & 5 & -4 \\ 12 & -4 & 16 \end{bmatrix}$

3.3 CONJUGATE OF A MATRIX

The matrix obtained from any given matrix A by replacing its elements by the corresponding conjugate complex numbers is called the **conjugate of A** and is denoted by $\overline{A}$.

For example; if
$$A = \begin{bmatrix} 1-i & i \\ 3-i & 2 \end{bmatrix}, \text{ then } \overline{A} = \begin{bmatrix} 1+i & -i \\ 3+i & 2 \end{bmatrix}.$$

3.3.1. Conjugate Transpose of a Matrix

The conjugate of the transpose of a matrix A is called the conjugate transpose of A and is denoted by A^*.

For example, if
$$A = \begin{bmatrix} 1-i & i \\ 3-i & 2 \end{bmatrix}, \quad \text{then} \quad A^* = \begin{bmatrix} 1+i & 3+i \\ -i & 2 \end{bmatrix}.$$

3.3.2. Hermitian Matrix

A square matrix is said to be **Hermitian matrix if its conjugate transpose is equal to the matrix itself.**

Thus any square matrix $A = [a_{ij}]$ is Hermitian, if $A^* = A$, i.e., iff $a_{ij} = \overline{a}_{ji}$ for all i and j.

For example, the matrix

$$A = \begin{bmatrix} 2 & a+ib & c+id \\ a-ib & 3 & x+iy \\ c-id & x-iy & 4 \end{bmatrix} \text{ is Hermitian.}$$

Skew Hermitian Matrix

A suquare matrix $A = [a_{ij}]$ is said to be **skew-hermitian** if $A^* = -A$, i.e., iff $a_{ij} = -\overline{a}_{ji}$, for all i and j.

For example, the matrix

$$A = \begin{bmatrix} 2i & a+ib & x \\ -a+ib & -i & c+id \\ -x & -c+id & 0 \end{bmatrix} \text{ is skew-hermitian.}$$

3.4. SOME THEOREMS

Theorem 1. *If A and B are any two matrices conformable for addition, then*
$$(A+B)^* = A^* + B^*.$$

Proof. We have, $(A+B)^* = \overline{(A+B)'} = (\overline{A+B})'$ $\quad [\because \overline{A+B} = \overline{(A+B)}]$
$$= (\overline{A})' + (\overline{B})'$$
$$= A^* + B^*$$

Theorem 2. *If A and B are any two matrices conformable for multiplication, then* $(AB)^* = B^* A^*$

(Reversal Law for Transpose conjugate)

Proof. We have
$$(AB)^* = (\overline{AB})' = (\overline{B}'\,\overline{A}') = (\overline{B}')(\overline{A}')$$
$$= B^* A^*.$$

Generalising the above rule, we get
$$(A\,B\,C\,...\,KL)^* = L^* K^* ... C^* B^* A^*.$$

Theorem 3. *The diagonal elements of a Hermitian matrix are necessarily real.*

Proof. Let $A = [a_{ij}]$ be a hermitian matrix of order $n \times n$. Then we have $a_{ij} = \overline{a}_{ji}$ for i, j such that $1 \le i \le n, 1 \le j \le n$.

For diagonal elements, we have
$$a_{ii} = \overline{a}_{ii}$$

Now let $a_{ii} = \alpha + i\beta$, α and β being real. Then
$$\overline{a}_{ii} = \alpha - i\beta$$
$$\therefore \quad \alpha + i\beta = \alpha - i\beta$$
or $\quad 2i\beta = 0$ or $\beta = 0$.

$\therefore \quad a_{ii} = \alpha$, which is purely real.

Some Special Types of Matrices

Note. similarly we can prove that the diagonal elements of a skew hermitian matrix are either purely imaginary or zero.

Theorem 4. *Every square matrix can be uniquely expressed as the sum of a hermitian and a skew-hermitian matrix.*

Proof. Let A be any square matrix then we can write

$$A = \frac{1}{2}A + \frac{1}{2}A = \frac{1}{2}(A + A^*) + \frac{1}{2}(A - A^*)$$
$$= P + Q \quad \text{(say)}$$

where $\quad P = \frac{1}{2}(A + A^*) \text{ and } Q = \frac{1}{2}(A - A^*)$

Now, $\quad P^* = \frac{1}{2}(A + A^*)^* = \frac{1}{2}[A^* + (A^*)^*]$

$\quad = \frac{1}{2}(A^* + A) = \frac{1}{2}(A + A^*) = P \quad\quad\quad [\because (A^*)^* = A]$

and $\quad Q^* = \frac{1}{2}(A - A^*)^* = \frac{1}{2}[A^* - (A^*)^*]$

$\quad = \frac{1}{2}[A^* - A] = -\frac{1}{2}(A - A^*) = -Q.$

Hence P is hermitian matrix and Q is skew-hermitian matrix.

To prove that this representation is unique, let $A = R + S$ be another representation of A, where R is hermitian and S is skew-hermitian. Hence $R^* = R$ and $S^* = -S$.

Now, $\quad\quad A^* = (R + S)^* = R^* + S^* = R - S$

Hence $\quad\quad A + A^* = 2R \text{ i.e., } R = \frac{1}{2}(A + A^*) = P$

and $\quad\quad A - A^* = 2S \text{ i.e., } S = \frac{1}{2}(A - A^*) = Q.$

Hence $\quad\quad A = R + S = P + Q.$

Thus the representation is unique.

Theorem 5. *Every square matrix can be uniquely expressed as $P + iQ$, where P and Q are hermitian.*

Proof. Let A be a square matrix. Then we can write

$$A = \frac{1}{2}(A + A^*) + \frac{i}{2i}(A - A^*)$$
$$= P + iQ \quad\quad\quad\quad\quad\quad\quad\quad\quad\quad\quad\quad \ldots(i)$$

where $\quad P = \frac{1}{2}(A + A^*) \text{ and } Q = \frac{1}{2i}(A - A^*)$

Now, $\quad P^* = \left[\frac{1}{2}(A + A^*)\right]^* = \frac{1}{2}[A^* + (A^*)^*] = \frac{1}{2}(A^* + A) = \frac{1}{2}(A + A^*) = P$

and $\quad Q^* = \left[\frac{1}{2i}(A - A^*)\right]^* = \left(\frac{1}{2i}\right)[A^* - (A^*)^*]$

$\quad = -\frac{1}{2i}(A^* - A) = \frac{1}{2i}(A - A^*) = Q$

Hence $P^* = P$ and $Q^* = Q$, i.e., P and Q are hermitian.

Uniqueness. To prove the uniqueness of the representation, let us supose that another representation of the form
$$A = R + iS \qquad \text{...(ii)}$$
is possible, where R and S are both hermitian, i.e., $R^* = R$ and $S^* = S$.

Then
$$A^* = (R + iS)^* = R^* + (iS)^*$$
$$= R^* - iS^*$$
$$= R - iS \qquad \text{...(iii)}$$

from (ii) and (iii), we get
$$R = \frac{1}{2}(A + A^*) = P \quad \text{and} \quad S = \frac{1}{2i}(A - A^*) = Q$$

Hence the representation is unique.

EXAMPLES

1. *Prove that the matrix* $A = \begin{bmatrix} 1 & 1-i & 2 \\ 1+i & 3 & i \\ 2 & -i & 0 \end{bmatrix}$ *is hermitian. If k is a complex number, verfy if kA is hermitian ?*

Solution. We have
$$A' = \begin{bmatrix} 1 & 1+i & 2 \\ 1-i & 3 & -i \\ 2 & i & 0 \end{bmatrix} \quad \text{and} \quad A^* = \begin{bmatrix} 1 & 1-i & 2 \\ 1+i & 3 & i \\ 2 & -i & 0 \end{bmatrix} = A$$

Since $A^* = A$ Hence A is hermitian.

Now for second part, we have
$$(kA)^* = \overline{(kA)'} = \overline{k}\,(\overline{A})' = \overline{k}\,A^* \neq k\,A^*$$

because $\overline{k} \neq k$ as k is a complex number. Hence kA is not hermitian.

2. *If A is hermitian (skew-hermitian), show that B^*AB is hermitian (skew-hermitian).*

Solution. If A is hermitian then
$$A^* = A$$
Now,
$$(B^*AB)^* = (AB)^*(B^*)^*$$
$$= B^*A^*B \qquad \text{(By reversal law)}$$
$$= B^*AB.$$

Hence B^*AB is also hermitian.

2nd part. When A is skew-hermitian we have $A^* = -A$. Then
$$(B^*AB)^* = (AB)^*(B^*)^*$$
$$= B^*A^*B = B^*(-A)B$$
$$= -(B^*AB)$$

Hence B^*AB is also skew-hermitain.

3. *If A and B are hermitian show that AB is hermitian, iff A and B commute.*

Solution. If A and B are hermitian
then $A^* = A$ and $B^* = B$.

Now, $\qquad (AB)^* = B^*A^* = BA$

Some Special Types of Matrices

$= AB$ only iff A and B commute.

Hence AB is hermitian only if A and B commute.

4. *If A and B are hermitian matrices then show that BAB is also hermitian.*

Solution. If A and B are hermitian matrices, then we have

$$A^* = A \text{ and } B^* = B.$$

Now, we have $(BAB)^* = (AB)^* B^* = B^* A^* B^*$ (By reversal law)

$$= BAB \text{ (given)}.$$

Hence BAB is a hermitian matrix.

EXERCISES

1. Prove that $\begin{bmatrix} 3 & 7-4i & -2+5i \\ 7+4i & -2 & 3+i \\ -2-5i & 3-i & 4 \end{bmatrix}$ is a hermitian matrix.

2. Show that if A and B are hermitian (skew-hermitian) so is also $A + B$.

3. If A is a square matrix, show that $A + A^*$ is hermitian and $A - A^*$ is skew hermitian.

4. If A is a hermitian matrix, examine whether kA and ikA are hermitian matrices, where k is a real number.

5. If A is any matrix, show that AA^* and A^*A are both hermitian.

6. If A and B are hermitian, show that $AB + BA$ is hermitian and $AB - BA$ is skew-hermitian.

4
Elementary Operations and Inverse of a Matrix

4.1 ELEMENTARY OPERATIONS

The following three operations are said to be **elementary operations** on a given matrix.
(i) The interchange of two rows (or columns).
(ii) The multiplication of a row (or column) by a non-zero scalar.
(iii) The addition of a multiple of one row (or column) to another row (or column).

These operations are known as **E-operations** also. An elementary operation is said to be **row operation** or a **column operation** according as it is performed on rows or columns.

Notations : The following notations may be adopted to express the elementary row or column operations.

(i) The interchange of ith and jth rows will be denoted by writing $R_i \leftrightarrow R_j$ and that of column by $C_i \leftrightarrow C_j$.

(ii) The multiplication of ith row by a number $p\ (\neq 0)$ will be denoted by $R_i \to p\,R_i$ and similarly the multiplication of ith column by a number $p\ (\neq 0)$ by $C_i \to pC_i$.

(iii) The addition of p times the jth row to the ith row will be denoted by writing $R_i \to R_i + p\,R_j$ and the addition of p times the jth column to the ith column by $C_i \to C_i + p\,C_j$.

The elementary operation which can undo a give elementary operation is said to be the **inverse** of the given elementary operation.

4.2 ELEMENTARY MATRICES

A matrix obtained from a unit marix by a single elementary operation is known as an **elementary matrix** or an **E-matrix**.

Types of Elementary Matrices

1. E_{ij} denotes the elementary matrix obtained by interchanging the ith and jth rows (or columns) of an identity matrix.
2. $E_{i(k)}$ denotes the elementary matrix obtained by multiplying the ith row (or column) of a unit matrix by k.
3. $E_{ij(k)}$ denotes the elementary matrix obtained by adding to the elements of the ith row (or column) of a unit matrix, k times the correspondign elements of the jth row (or column).
4. $E'_{ij(k)}$ denotes the transpose of $E_{ij(k)}$ and can be obtained by adding to the elements of ith row (or column) of the unit matrix, k times the corresponding elements of the jth row (or column).

4.2.1. Some Theorems on Elementary Operations

Theorem 1. *An elementary row operation on the product of two matrices is equivalent to the same elementary row operation on the pre-factor.*

Elementary Operations and Inverse of a Matrix

Proof. Let $A = [a_{ij}]$ and $B = [b_{ij}]$ be $m \times n$ and $n \times p$ matrices respectively. Let C be the matrix obtained from A by an E–row operation, D the matrix obtained from AB by the same E–row operation, then we have to show that $CB = D$.

Case I. Let C, D be the matrices obtained from A and AB respectively by the E-row operation $R_i \leftrightarrow R_j$.

Since A is of order $m \times n$, hence C will also be of order $m \times n$. Also, since AB is of order $m \times p$, hence, D is of order $m \times p$. Hence CB and D are the same ordered matrices.

Since C differs from A only in the ith and the jth rows, hence, it follows that CB differs from AB in the ith and the jth rows only. Now,

(i, k) the element of $D = (j, k)$ th element of AB

$$= \sum_{i=1}^{n} a_{ji} \cdot b_{ik}$$

$$= \sum_{i=1}^{n} \{(i, j) \text{th element of } C\}.$$

$\{(j, k) \text{ th element of } B\}$

$= (i, k)$ th element of CB.

Hence, the ith row of D and CB are identical. Similarly, the jth rows of D and CB are identical. Hence $CB = D$.

Case II. Let C, D be the matrices obtained from A and AB respectively by the E–row operation $R_i \to \lambda R_j$ ($\lambda \neq 0$).

As we have discussed above in case I, the matrices CB and D are of the same order, and CB and D differ from AD and from each other, in the ith row only.

Now, (i, k) th element of $D = \lambda \cdot \{(i, k) \text{ the element of } AB\}$

$$= \lambda \sum_{j=i}^{n} a_{ij} \cdot b_{jk}$$

$$= \sum_{i=1}^{n} (\lambda \, a_{ij}) \, b_{jk}$$

$$= \sum_{i=1}^{n} \{(i, j) \text{ the element of } C\}$$

$.(j, k)$ th element of $B\}$

$= (i, k)$ th element of CB.

Hence the ith rows of D and CB are identical. Hence

$$CB = D$$

Case III. Let C, D be the matrices obtained from A and AB respectively by the E–row operation $R_i \to R_i + \lambda R_j$. Now,

$$= \sum_{j=1}^{n} (a_{ij} \, b_{jk}) + \lambda \cdot \sum_{j=1}^{n} a_{ij} \cdot b_{jk}$$

$$= \sum_{j=1}^{n} (a_{ij} + \lambda \, a_{ij}) \, b_{jk}$$

$$= \sum_{j=1}^{n} \{(i, j) \text{th element of } C\} \cdot \{(j, k) \text{th element of } B\}$$

$$= (i, k) \text{th element of } CB.$$

Hence, $CB = D$.

Thus, if C, D be the matrices obtained from A and AB respectively by the same E–row operation of any type, then

$$CB = D.$$

Theorem 2. *An elementary operation on the product of two matrices is equivalent to the same column operation on the post-factor.*

Proof. Let $C(M)$ denotes the matrix obtained from a matrix M by an elementary column operation and $C^*(M)$ the matrix obtained from M' (the transpose of M) by the corresponding elementary row-operation. Then the matrix $C(M) = $ transpose of $C^*(M')$.

Let A, B be two matrices of order $m \times n$ and $n \times p$ respectively and $C(B), C(AB)$ respectively denote the matrices obtained from the matrices B and AB by the same elementary column operation.

Then,

$$C(AB) = [C^*(AB)']'$$
$$= [C^*(B' A')]' \qquad \text{[by reversal law of Transposes]}$$
$$= [C^*(B') A')]' \qquad \text{[by theorem I]}$$
$$= (A')' [C^*(B')]' \qquad \text{[by reversal law of transposes]}$$
$$= AC(B).$$

Theorem 3. *Each elementary row (column) operation on a matrix can be effected by pre– (post–) multiplication by the corresponding elementary matrix of an appropriate order.*

Proof. Let A be any $m \times n$ matrix. Let $r(A)$ denotes the matrix obtained from A by an E-row operation and $r(I_m A)$ denote matrix obtained from $(I_m A)$ by the same E–row operation, then

$$r(A) = r(I_m A) \qquad [\because A = I_m A]$$
$$= r(I_m) \cdot A$$

Where $r(I_m)$ is the matrix obtained from I_m by the same E–row oberation as for $r(A)$.

This shows that $r(A)$ can be obtained from A by pre-multiplying it by the elementary matrix $r(I_m)$. Similarly, we can prove the theorem for E–column operations.

Theorem 4. *The inverse of an elementary matrix is an elementary matrix.*

Proof. Let I_n be an identity matrix of order n and let E_{ij} denotes the matrix obtained by interchanging the ith and jth rows of I_n. Obviously, the interchanging of ith and jth rows of E_{ij} will give back the original matrix I_n. This operation can be effected by pre-multiplying E_{ij} with E_{ij}.

Thus, $$E_{ij} \cdot E_{ij} = I_n.$$

This shows that E_{ij} is its own inverse i.e.,

$$(E_{ij})^{-1} = E_{ij}.$$

Now, let $E_i(\lambda)$ denotes the matrix obtained on multiplying the ith row of I_n by λ. $E_i(\lambda)$

Elementary Operations and Inverse of a Matrix

can be transformed into the unit matrix I_n on multiplying the ith row of $E_i(\lambda)$ by $1/\lambda$. This operation can be effected by pre-multiplying $E_i(\lambda)$ with $E_i(1/\lambda)$. Hence

$$E_i(1/\lambda) \cdot E_i(\lambda) = I_n$$

Hence $[E_i(\lambda)]^{-1} = E_i(1/\lambda)$

Similarly, we can show that

$$[E_{ij}(\lambda)]^{-1} = E_{ij}(-\lambda).$$

4.3 EQUIVALENT MATRICES

Two matrices of the same order are said to be **equivalent**, if it is possible to obtain one matrix from the other by a finite chain of the application of elementary operations.

Thus, if a matrix B is obtained from a matrix A by application of E–operations, then A is said to be **equivalent** to B and is expressed as $A \sim B$ symbolically.

4.3.1. Theorems on Equivalent Matrices

Theorem 1. *If A and B are equivalent matrices than there exist non-singular matrices R and C such that $B = R A C$.*

Proof. As the matrices A and B are equivalent, hence the matrix B can be obtained from A, by the application of a chain of elementary row and column operations. Also, elementary transforatmions on A can be effected by pre-or post-multiplication of A by elementary matrices of appropriate orders. Hence

$$(R_n \cdot R_{n-1} \ldots R_2 R_1) A (C_1 C_2 \ldots C_{m-1} C_m) = B$$

Hence $B = R A C$

where $R = R_n \cdot R_{n-1} \ldots R_2 R_1$ and $C = C_1 C_2 \ldots C_{m-1} C_m$ being the product of elementary matrices are non-singular matrices.

Cor. We have

$$R A C = B$$

$\therefore \qquad R^{-1}(R A C) = A^{-1} B$

or $\qquad (R^{-1}R)(AC) C^{-1} = A^{-1} B C^{-1}$

or $\qquad I A (C C^{-1}) = A^{-1} B C^{-1}$

or $\qquad A I = A^{-1} B C^{-1}$

or $\qquad A = A^{-1} B C^{-1}$

Theorem 2. *If a sequence of elementary operations on non-singular matrix A transforms it into a unit matrix I then the same sequence transforms I to A^{-1}.*

Proof. We have that $S A = I$, where S is the product of the elementary matrices, *i.e.*,

$$(E_n \cdot E_{n-1} \ldots E_2 E_1) A = I$$

where E_i denotes the elementary matrices.

or $\qquad (E_n \cdot E_{n-1} \ldots E_2 E_1) A A^{-1} = I A^{-1}$

or $\qquad (E_n \cdot E_{n-1} \ldots E_2 E_1) I = A^{-1}.$

4.4 COMPUTATION OF THE INVERSE OF A MATRIX BY ELEMENTARY OPERATIONS

We know that
$$IA = A$$
where I is the unit matrix of the same order as A. Now, by performing elementary row operations on the right hand side of the above identity, we can reduce it to the matrix I. By applying all these operations to the pre-factor I on the left hand side of the identity, the matrix I reduces to same matrix, say B, such that $BA = I$. The matrix B is then the required inverse of A.

Such a reduction would not be possible by any elementary operation what so ever if A is not invertible.

EXAMPLES

1. *Compute the following elementary matrices of order 4, E_{23}, $E_2(4)$, $E_{31}(-2)$, $E'_{34}(-2)$.*

Solution. The identity (or unit) matrix of order four is given by

$$I_4 = \begin{bmatrix} 1 & 0 & 0 & 0 \\ 0 & 1 & 0 & 0 \\ 0 & 0 & 1 & 0 \\ 0 & 0 & 0 & 1 \end{bmatrix}$$

(i) $E_{23} = \begin{bmatrix} 1 & 0 & 0 & 0 \\ 0 & 0 & 1 & 0 \\ 0 & 1 & 0 & 0 \\ 0 & 0 & 0 & 1 \end{bmatrix}$ [Interchanging R_2 and R_3 or C_2 and C_3]

(ii) $E_2(4) = \begin{bmatrix} 1 & 0 & 0 & 0 \\ 0 & 4 & 0 & 0 \\ 0 & 0 & 1 & 0 \\ 0 & 0 & 0 & 1 \end{bmatrix}$ [Replacing R_2 by $4R_2$ or C_2 by $4C_2$]

(iii) $E_{34}(-2) = \begin{bmatrix} 1 & 0 & 0 & 0 \\ 0 & 1 & 0 & 0 \\ 0 & 0 & 1 & -2 \\ 0 & 0 & 0 & 1 \end{bmatrix}$ [Replacing R_3 by $R_3 - 2R_4$]

(iv) $E'_{34}(-2) = \begin{bmatrix} 1 & 0 & 0 & 0 \\ 0 & 1 & 0 & 0 \\ 0 & 0 & 1 & 0 \\ 0 & 0 & -2 & 1 \end{bmatrix}$ [Replacing C_3 by $C_3 - 2C_4$]

[Here students should note that $E'_{34}(-2)$ is nothing but the transpose matrix of $E_{34}(-2)$].

2. *Find the inverse of the matrix*

$$\begin{bmatrix} 2 & 0 & -1 \\ 5 & 1 & 0 \\ 0 & 1 & 3 \end{bmatrix}$$

$$\begin{bmatrix} \cos\theta & -\sin\theta \\ \sin\theta & \cos\theta \end{bmatrix} = \begin{bmatrix} 1 & \tan\theta/2 \\ -\tan\theta/2 & 1 \end{bmatrix}$$

$$= \begin{bmatrix} \cos\theta + \sin\theta \tan\dfrac{\theta}{2} & \cos\theta \tan\dfrac{\theta}{2} - \sin\theta \\ \sin\theta - \cos\theta \tan\dfrac{\theta}{2} & \sin\theta \tan\dfrac{\theta}{2} + \cos\theta \end{bmatrix}$$

$$= \begin{bmatrix} \dfrac{\cos\theta\cos\dfrac{\theta}{2} + \sin\theta\sin\dfrac{\theta}{2}}{\cos\theta/2} & \dfrac{\cos\theta\sin\dfrac{\theta}{2} - \sin\theta\cos\dfrac{\theta}{2}}{\cos\theta/2} \\ \dfrac{\sin\theta\cos\dfrac{\theta}{2} - \cos\theta\sin\dfrac{\theta}{2}}{\cos\theta/2} & \dfrac{\sin\theta\sin\dfrac{\theta}{2} + \cos\theta\cos\theta/2}{\cos\theta/2} \end{bmatrix}$$

$$= \begin{bmatrix} 1 & -\tan\theta/2 \\ \tan\theta/2 & 1 \end{bmatrix}$$

If $\begin{bmatrix} 1 & \tan\theta/2 \\ -\tan\theta/2 & 1 \end{bmatrix} = A$ (say), then we can write

$$\begin{bmatrix} \cos\theta & -\sin\theta \\ \sin\theta & \cos\theta \end{bmatrix} A = \begin{bmatrix} 1 & -\tan\theta/2 \\ \tan\theta/2 & 1 \end{bmatrix}$$

Post multiplying both sides by A^{-1}, we have

$$\begin{bmatrix} \cos\theta & -\sin\theta \\ \sin\theta & \cos\theta \end{bmatrix} AA^{-1} = \begin{bmatrix} 1 & -\tan\theta/2 \\ \tan\theta/2 & 1 \end{bmatrix} A^{-1}$$

or $\begin{bmatrix} \cos\theta & -\sin\theta \\ \sin\theta & \cos\theta \end{bmatrix} I = \begin{bmatrix} 1 & -\tan\theta/2 \\ \tan\theta/2 & 1 \end{bmatrix} \times \begin{bmatrix} 1 & \tan\theta/2 \\ -\tan\theta/2 & 1 \end{bmatrix}^{-1}$

$\therefore \begin{bmatrix} \cos\theta & -\sin\theta \\ \sin\theta & \cos\theta \end{bmatrix} = \begin{bmatrix} 1 & -\tan\theta/2 \\ \tan\theta/2 & 1 \end{bmatrix} \times \begin{bmatrix} 1 & \tan\theta/2 \\ -\tan\theta/2 & 1 \end{bmatrix}^{-1}$

EXERCISES

1. Apply the row operations $R_4(-3)$ and $R_{21}(4)$ to the matrix
$$\begin{bmatrix} 4 & -1 & 2 & 3 \\ -1 & 8 & -3 & -4 \\ 2 & 3 & 4 & -1 \\ -3 & -4 & -1 & 8 \end{bmatrix}$$

2. Apply the column operation $C_3(4)$ and $C_{12}(-3)$ to the matrix
$$\begin{bmatrix} 0 & 1 & 2 & 3 & 4 \\ 1 & 2 & 3 & 4 & 0 \\ 3 & 4 & 0 & 1 & 2 \\ 2 & 0 & 1 & 3 & 4 \end{bmatrix}$$

Elementary Operations and Inverse of a Matrix

Solution. Consider the identity

$$\begin{bmatrix} 1 & 0 & 0 \\ 0 & 1 & 0 \\ 0 & 0 & 1 \end{bmatrix} \begin{bmatrix} 2 & 0 & -1 \\ 5 & 1 & 0 \\ 0 & 1 & 3 \end{bmatrix} = \begin{bmatrix} 2 & 0 & -1 \\ 5 & 1 & 0 \\ 0 & 1 & 3 \end{bmatrix}$$

By performing the elementary raw operation $R_2 \to R_2 - \dfrac{5}{2} R_1$ on the right hand side as well as on the per-factor on the left, we get

$$\begin{bmatrix} 1 & 0 & 0 \\ -\dfrac{5}{2} & 1 & 0 \\ 0 & 0 & 1 \end{bmatrix} \begin{bmatrix} 2 & 0 & -1 \\ 5 & 1 & 0 \\ 0 & 1 & 3 \end{bmatrix} = \begin{bmatrix} 2 & 0 & -1 \\ 0 & 1 & 5/2 \\ 0 & 1 & 3 \end{bmatrix}$$

Replacing R_3 by $R_3 - R_2$ in the matrix on the right as well as in the pre-factor on the left, we get

$$\begin{bmatrix} 1 & 0 & 0 \\ -\dfrac{5}{2} & 1 & 0 \\ \dfrac{5}{2} & -1 & 1 \end{bmatrix} \begin{bmatrix} 2 & 0 & -1 \\ 5 & 1 & 0 \\ 0 & 1 & 3 \end{bmatrix} = \begin{bmatrix} 2 & 0 & -1 \\ 0 & 1 & 5/2 \\ 0 & 1 & 1/2 \end{bmatrix}$$

Replacing R_1 by $R_1 + 2R_2$, R_2 by $R_2 - 5R_3$ in the matrix on the right as well as in the pre-factor on the left, we have

$$\begin{bmatrix} 6 & -2 & 2 \\ -15 & 6 & -2 \\ 5/2 & -1 & -1 \end{bmatrix} \begin{bmatrix} 2 & 0 & -1 \\ 5 & 1 & 0 \\ 0 & 1 & 3 \end{bmatrix} = \begin{bmatrix} 2 & 0 & 0 \\ 0 & 1 & 0 \\ 0 & 0 & 1/2 \end{bmatrix}$$

Performing $R_1 \to \dfrac{1}{2} R_1$ and $R_3 \to 2R_3$ on the matrix on the right as well as on the pre-factor on the left, we get

$$\begin{bmatrix} 3 & -1 & 1 \\ -15 & 6 & -5 \\ 5 & -2 & -2 \end{bmatrix} \begin{bmatrix} 2 & 0 & -1 \\ 5 & 1 & 0 \\ 0 & 1 & 3 \end{bmatrix} = \begin{bmatrix} 1 & 0 & 0 \\ 0 & 1 & 0 \\ 0 & 0 & 1 \end{bmatrix}$$

Hence the required inverse is

$$\begin{bmatrix} 3 & -1 & 1 \\ -15 & 6 & -5 \\ 5 & -2 & -2 \end{bmatrix}.$$

3. *Show that*

$$\begin{bmatrix} \cos\theta & -\sin\theta \\ \sin\theta & \cos\theta \end{bmatrix} = \begin{bmatrix} 1 & -\tan\theta/2 \\ \tan\theta/2 & 1 \end{bmatrix} \begin{bmatrix} 1 & \tan\dfrac{\theta}{2} \\ -\tan\dfrac{\theta}{2} & 1 \end{bmatrix}^{-1}$$

Solution. We have

Elementary Operations and Inverse of a Matrix

3. With the help of elementary operations find inverse of following matrices

(a) $\begin{bmatrix} 1 & 2 & -2 \\ -1 & 3 & 0 \\ 0 & -2 & 1 \end{bmatrix}$

(b) $\begin{bmatrix} 1 & 2 & 3 & 1 \\ 1 & 3 & 3 & 2 \\ 2 & 4 & 3 & 3 \\ 1 & 1 & 1 & 1 \end{bmatrix}$

ANSWERS

1. $\begin{bmatrix} 4 & -1 & 2 & 3 \\ -1 & 8 & -3 & -4 \\ 2 & 3 & 4 & -1 \\ 9 & 12 & 3 & -24 \end{bmatrix}$ and $\begin{bmatrix} 4 & -1 & 2 & 3 \\ 15 & 4 & 5 & 8 \\ 2 & 3 & 4 & -1 \\ -3 & -4 & -1 & 8 \end{bmatrix}$

2. $\begin{bmatrix} 0 & 1 & 8 & 3 & 4 \\ 1 & 2 & 12 & 1 & 2 \\ 3 & 4 & 0 & 1 & 2 \\ 2 & 0 & 4 & 3 & 4 \end{bmatrix}$ and $\begin{bmatrix} 3 & 1 & 2 & 3 & 4 \\ -5 & 2 & 3 & 4 & 0 \\ -9 & 4 & 0 & 1 & 2 \\ 2 & 0 & 1 & 3 & 4 \end{bmatrix}$

3. (a) $\begin{bmatrix} 3 & 2 & 6 \\ 1 & 1 & 2 \\ 2 & 2 & 5 \end{bmatrix}$

(b) $\begin{bmatrix} 1 & -2 & 1 & 0 \\ 1 & -2 & 2 & -3 \\ 0 & 1 & -1 & 1 \\ 2 & 3 & -2 & 3 \end{bmatrix}$

☐☐☐

5
Linear Dependence of Vectors

5.1 ORDERED SET OF NUMBERS

Two objects a, b can presented in two different ways, (i) first a then b, (ii) first b then a. This presentation may be represented as (a, b) or (b, a) depending which order we wish to indicate (a, b) is called as n **ordered pair.** The word ordered means that the order of writing the two obejcts a and b is important, *i.e.,* (a, b) is to be considered as a different pair than (b, a) unless $a = b$.

In the ordered pair (a, b), we call a the **first member** and b the **second member.**

In coordinate geometry the ordered pair (x, y) of real numbers is used to denote a point in a plane. Similarly the ordered triple (x, y, z) of real numbers is used to denote a point in three dimensional space. Some physical entities such as velocity, acceleration, force etc. can also be represented as ordered triples of real numbers, the numbers being the resolved parts along the three rectangular coordinate axes. Accordingly, the set of ordered numbers can be interpreted in several ways in relation to different applications.

5.2 VECTORS

any ordered n-tuple of numbers is called an n-vector.

By an ordered n-tuple we mean a set consisting of n numbers in which the place of each number is fixed. If $x_1, x_2, ..., x_n$ be any n numbers, then the ordered n-tuple $X = (x_1, x_2, ..., x_n)$ is called an n-vector. The ordered triad (x_1, x_2, x_3) is called a 3-vector. Similarly $(1, 0, 1, -1)$ and $(2, 3, -4, 5)$ are 4-vectors. The n numbers $x_1, x_2, ..., x_n$ are called components of the n-vector $X = (x_1, x_2, ..., x_n)$. A vector may be written either as a **row vector** or as a **column vector.** If $\mathbf{A}$ be a matrix of the type $m \times n$, then each row of $\mathbf{A}$ will be an m-vector. A vector whose components are all zero is called a **zero vector** and will be denoted by $\mathbf{O}$.

If k be any number and X be any vector, then relative to the vector X, k is called *a scalar*.

5.2.1 Algebra of Vectors

Since an n-vector is nothing but a row matrix or a column matrix, therefore an algebra of vector can be developed in the same manner as the algebra of matrices.

Equality of two Vectors. Two n-vectors X and Y where $X = (x_1, x_2, ..., x_n)$ and $Y = (y_1, y_2, ..., y_n)$ are said to be **equal** if and only if their corresponding components are equal, *i.e.,* if $x_i = y_i$, for all $i = 1, 2, ..., n$. For example if $X = (1, 3, 6)$ and $Y = (1, 3, 6)$, then $X = Y$. But if $X = (1, 3, 6)$ and $Y = (3, 6, 1)$, then $X \ne Y$.

Addition of two Vectors. If $X = (x_1, x_2, ..., x_n)$ and $Y = (y_1, y_2, ..., y_n)$ then $X + Y = (x_1 + y_1, x_2 + y_2, ..., x_n + y_n)$.

Thus $X + Y$ is an n-vector whose components are the sums of corresponding components of X and Y.

If $X = (1, 2, 4)$ and $Y = (2, -3, 7)$ then $X + Y = (1 + 2, 2 - 3, 4 + 7) = (3, -1, 11)$.

Linear Dependence of Vectors

Multiplication of a Vector by a Scalar

If k be any number and $X = (x_1, x_2, ..., x_n)$, then by definition
$$kX = (kx_1, kx_2, ..., kx_n),$$
The vector kX is called the scalar multiple of the vector X by the scalar k.

If $X = (2, 4, -7)$, then $3X = (6, 12, -21)$ and $-2X = (-4, -8, 14)$.

Properties of addition and scalar multiplication of Vectors

If X, Y, Z be any three n-vectors and p, q be any two numbers, then
(i) $X + Y = Y + X$
(ii) $X + (Y + Z) = (X + Y) + Z$
(iii) $p(X + Y) = pX + pY$
(iv) $(p + q)X = pX + qX$
(v) $p(qX) = (pq)X$.

5.3 LINEAR DEPENDENCE AND LINEAR INDEPNDENCE OF VECTORS

(a) Linear dependent set of vectors

A set of r n-vectors $X_1, X_2, ..., X_r$ is said to be linearly dependent if there exists r scalars $k_1, k_2, ..., k_r$, not all zero, such that
$$k_1 X_1 + k_2 X_2 + ... + k_r X_r = 0,$$
where 0 denotes the n-vector whose components are all zero.

(b) Linearly independent set of vectors

A set of r n-vectors $X_1, X_2, ..., X_r$ is said to be linearly indepndent if every relation of the type
$$k_1 X_1 + k_2 X_2 + ... + k_r X_r = 0,$$
implies $k_1 = k_2 = ... = k_r = 0$.

5.3.1. Linear Combination of Vectors

A vector X which can be expressed in the form
$$X = k_1 X_1 + k_2 X_2 + ... + k_r X_r,$$
is said to be a linear combination of the set of vectors
$$X_1, X_2, ... X_r,$$
where $k_1, k_2, ..., k_r$ are any numbers.

The following two results are quite obvious :
(i) If a set of vectors in linearly dependnet, then at least one members of the set can be expressed as a linear combination of the remaining members.
(ii) If a set of vectors is linearly independent then no member of the set can be expressed as a linear combination of the remaining membes.

EXAMPLES

1. Show that the vectors $X_1 = [1, 2, 4]$, $X_2 = [3, 6, 12]$ are linearly dependent.

Solution. Let us assume that
$$k_1 X_2 + k_2 X_2 = 0$$
i.e., $$k_1 [1, 2, 4] + k_2 [3, 6, 12] = 0$$

i.e., $\quad [k_1 + 3k_2, 2k_1 + 6k_2, 4k_1 + 12k_2] = [0, 0, 0]$

From the definition of equality, we have

$$k_1 + 3k_2 = 0, \quad 2k_1 + 6k_2 = 0$$

or $\quad k_1 + 3k_2 = 0 \quad$ and $\quad 4k_1 + 12k_2 = 0,$

or $\quad k_1 + 3k_2 = 0$

This relation gives a number of non-zero values satisfying it. Hence the given vectors are linearly dependent.

2. *If* $X_1 = [3, 1, -4]$, $X_2 = [2, 2, -3]$, $X_3 = [0, -4, 1]$, *show that*

(i) the vectors X_1 *and* X_2 *are linearly independent over the field of rational numbers;*

(ii) the vectors X_1, X_2 *and* X_3 *are linearly dependent over the field of rational numbers.*

Solution. (i) Let us assume that

$$k_1 X_1 + k_2 X_2 = 0.$$

i.e., $\quad k_1 [3, 1, -4] + k_2 [2, 2, -3] = [0, 0, 0]$

or $\quad [3k_1 + 2k_2, k_1 + 2k_2, -4k_1 - 3k_2] = [0, 0, 0]$

From the definition of equality of vectors, this implies that

$$3k_1 + 2k_2 = 0,$$
$$k_1 + 2k_2 = 0,$$
$$-4k_1 + 2k_2 = 0,$$

Solving these linear equations, we find that $k_1 = 0$ and $k_2 = 0$. Thus $k_1 X_1 + k_2 X_2 = 0$ implies that $k_1 = 0, k_2 = 0$. Hence the vectors X_1 and X_2 are linearly independent.

(ii) We have

$$k_1 X_1 + k_2 X_2 + k_3 X_3 = [3k_1 + 2k_2, k_1 + 2k_2 - 4k_3, 4k_1 - 3k_2 - k_3]$$

If $\quad k_1 X_1 + k_2 X_2 + k_3 X_3 = [0, 0, 0],$

then $\quad 3k_1 + 2k_2 = 0$

$$k_1 + 2k_2 - 4k_3 = 0$$
$$4k_1 - 3k_2 - k_3 = 0.$$

From these relations, we see that $\quad k_1 = -\dfrac{2}{3} k_2 = -2k_3 \neq 0.$

This shows that scalars k_1, k_2, k_3 (not all zero) exist such that $k_1 X_1 + k_2 X_2 + k_3 X_3 = 0$. Hence the vectors X_1, X_2 and X_3 are linearly dependent.

5.4 SOME BASIC THEOREMS

Theorem 1. *The set* $\{X_1, X_2, ..., X_m\}$ *of n-vectors is a linearly dependent set over F if at least one of the vectors of the set is the zero vector.*

Proof. Let us suppose that $X_1 = 0$. Then since $k_1 X_1 = 0$ for any $k_1 \neq 0$ in F, and $0 X_i = 0$, we have

$$k_1 X_1 + 0 X_2 + ... + 0 \cdot X_m = 0$$

In this relation $k_1 \neq 0$. Hence by definition the set $\{X_1, X_2, ..., X_m\}$ is linearly dependent.

Theorem 2. *If the n-vectors* $X_1, X_2, ..., X_m,$ *are linearly independent over F, then none of them can be the zero vector.*

Linear Dependence of Vectors

Proof. For if $X_1 = 0$ (say), then
$$k_1 X_1 + 0.X_2 + ... + 0.X_m = 0 \text{ for any } k_1 \neq 0 \text{ if F.}$$
Hence the vectors $X_1, X_2, ..., X_m$ are not linearly independent, a contradiction. Accordingly, $X_1 \neq 0$ and since X_1 is arbitrary hence the theorem.

Theorem 3. *The set $\{X\}$, consisting of the one vector X, is linearly independent if and only if $X = 0$.*

Proof. Let $X \neq 0$. Then $kX \neq 0$ implies that $k = 0$. Hence the set $\{X\}$ is linearly independent. Again, if the set $\{X\}$ be linearly independent, then by Theorem 2, $X \neq 0$.

Theorem 4. *If the set $\{X_1, X_2, ..., X_m\}$ of n-vectors is linearly independent over F, then any non-empty subset of this set is linearly independent.*

or

Any non-empty subset of a linearly independent set of vectors is linearly independent.

Proof. Let us consider the subset $\{X_1, X_2, ..., X_p\}$ where $1 \leq p < m$. Now, let
$$k_1 X_1 + k_2 X_2 + ... + k_p X_p = 0$$
Then,
$$k_1 X_1 + k_2 X_2 + ... + k_p X_p + 0.X_{p+1} + ... 0.X_m = 0.$$
Since the set $\{X_1, X_2, ..., X_m\}$ is linearly independent, it follows that we must have $k_1 = 0, k_2 = 0, ..., k_p = 0$.

Thus $k_1 X_1 + k_2 X_2 + ... + k_p X_p = 0$ only if all the k's are equal to zero. Hence the set $\{X_1, X_2, ..., X_p\}$ is linearly independent.

EXAMPLES

1. *Show that the vectors $X_1 = [1, 2, 3]$, $X_2 = [2, -2, 0]$ form a linearly independent set.*

Solution. Consider the matrix
$$A = \begin{bmatrix} 1 & 2 & 3 \\ 2 & -2 & 0 \end{bmatrix}$$

The minor $\begin{vmatrix} 1 & 2 \\ 2 & 2 \end{vmatrix}$ of A is not equal to zero. Therfore rank $A = 2$.

$\therefore$ rank of $A = 2 =$ the maximum number of linearly independent rows of A. Hence the vectors $[1, 2, 3]$ and $[2, -2, 0]$ are linearly independent.

2. *Show, using a matrix, that the vectors*
$$X_1 = [1, 2, -3, 4], X_2 = [3, -1, 2, 1], X_3 = [1, -5, 8, -7]$$
are linearly dependent. Determine a maximum subset of linearly independent vectors and express the other as linear combination of these.

Solution. Consider the matrix
$$A = \begin{bmatrix} 1 & 2 & -3 & 4 \\ 3 & -1 & 2 & 1 \\ 1 & -5 & 8 & -7 \end{bmatrix} \qquad ...(1)$$

Now by performing the operation $R_{21} (-2)$, we have

$$A \sim \begin{bmatrix} 1 & 2 & -3 & 4 \\ 1 & -5 & 8 & -7 \\ 1 & -5 & 8 & -7 \end{bmatrix} \qquad \ldots(2)$$

By $R_{32}(-1)$, $A \sim \begin{bmatrix} 1 & 2 & -3 & 4 \\ 1 & -5 & 8 & -7 \\ 0 & 0 & 0 & 0 \end{bmatrix}$

This shows that Rank (A) = 2, i.e., < 3,

Hence the given vectors are linearly deepndent, and since rank (A) = 2, there are two linearly independent vectors, say X_1 and X_2.

:From (1) and (2) it is clear that by the operation $R_{21}(-2)$, the 2nd and 3rd rows of the matrix A become identical, i.e., on multiplying the first row by -2 and adding it to the second row we obtain the third row. Recalling that A is the matrix with the given vectors as rows, it follows that

$$X_3 = -2X_1 + X_2.$$

EXERCISES

1. If $X_1 = [1, 0, 0]$, $X_2 = [0, 1, 0]$, $X_3 = [0, 0, 1]$, compute the vectors:
   (i) $2X_1 + X_2 - X_3$, (ii) $X_1 - 2X_2 + 3X_3$.
2. Examine the linear dependence or indepndence of the following set of vectors:
   (a) $[1, -1, 1], [2, 1, 1], [2, 0, 2]$
   (b) $[1, 0, -1], [2, 1, 3], [-1, 0, 0], [1, 0, 1]$
   (c) $[1, 2, 1, 2], [0, 1, 2, 0], [1, 4, 3, 2]$.
3. Show, using a matrix, that the set of vectors
   $X_1 = [2, 3, 1, -1]$, $X_2 = [2, 3, 1, -2]$, $X_3 = [4, 6, 2, -3]$
   is linearly dependent. Determine a maximum subset of linearly independent vectors and express the others as a linear combination of these.

ANSWERS

1. (i) $[2, 1, -1]$, (ii) $[1, -2, 3]$.
2. (a) linearly dependent, (b) linearly depndent, (c) linearly independent.
3. $X_3 = X_1 + X_2$.

5.5 THE n-VECTOR SPACE

The set of all n-vectors of a field F is called the n-vector space over F. It is usually denoted by $V_n(F)$ or simply by V_n.

5.5.1. Sub-space of an n-vectror space V_n

A non-empty set, S, of vectors of V_n is called a vector subspace of V_n. If $\mathbf{a} = (a_1, a_2, a_3)$ is any non-zero vector of V_3, then the set S of vector $k\mathbf{a}$ is a subspace of V_3, where k is a variable scalar which can take any value.

Vector sub-space spanned by a given system of vectors: A vector space which arises as a set of all linear combinations of any given set of vectors, is said to be spanned by the given set of vectors.

Basis and dimension of a subspace: A set of vector $a_1, a_2, a_3, \ldots, a_k$ belonging to the subspace S is said to be a basis of S, if

(i) The subspace S is spanned by the set $\mathbf{a}_1, \mathbf{a}_2, \ldots, \mathbf{a}_k$ and

Linear Dependence of Vectors

(ii) The vectors $a_1, a_2, ..., a_k$ are linearly independent. The number of members in any basis of a subspace is called the dimension of the subspace.

5.5.2. Row rank of a Matrix

Let $A = [a_{ij}]$ be any $m \times n$ matrix. Each of the m rows of A consists of n elements. Therefore the row vectors of A are n-vectors. These rows vectors of A will span a sub-space R of V_n. This sub-space R is called the **row space** of the matrix A. The dimension r of R is called the **row rank** of A. in other words the row rank of a matrix A is equal to the maximum number of linearly independent rows of A.

Left nullity of a matrix : Suppose X is an m-vector written in the form of a row vector. Then the matrix product XA is defined. The subspace S of V_m generated by the row vectors X belonging to V_m such that $XA = O$ is called the *row null* space of the matrix A. The dimension s of S is called the left nullity or *row nullity* of the matrix A.

Sum of the row rank and the row nullity of a matrix is equal to the number of rows, i.e.,

$$r + s = m.$$

5.5.3. Column rank of a matrix

Let $A = [a_{ij}]$ be any $m \times n$ matrix. Each of the n columns of A consists of m elements. Therfore the column vectors of A are m-vectors. These column vectors of A are m-vectors. These column vectors of A will space a subspace C of V_m. This subspace C is called the *column space* of the matrix A. The dimension c of C is called the column rank of A. In other words the column rank of a matrix A is equal to the maximum number of linearly independent columns of A.

Right nullity of a matrix : Suppose Y is an n-vector written in the form of a column vector. Then the matrix product AY is defined. The subspace T of V_n generated by the column vectors Y belonging to V_n such that $AY = O$ is called the column null space of the matrix A. The dimension t of T is called the right nullity or column nullity of the matrix A. Again, $c + t = n$.

5.6. Invariance of row rank under E-row operations

Theorem : *Row equivalent matrices have the same row rank.*

Proof : Let A be any given $m \times n$ matrix. Let B be a matrix row equivalent to A. Since B is obtainable from A by a finite chain of E-row operations and every E-row operation is equivalent to pre-multiplication by the corresponding E-matrix, there exists E-matrices $E_1, E_2, ..., E_k$ each of the type $m \times n$ such that

$$B = (E_k E_{k-1} ... E_2 E_1) A,$$

i.e., $B = PA,$

where $P = E_k E_{k-1} E_2 E_1$ is a non-singular matrix of the type $m \times m$.

Let us write

$$B = PA = \begin{bmatrix} p_{11} & p_{12} & \cdots & p_{1m} \\ p_{21} & p_{22} & \cdots & p_{2m} \\ \cdots & \cdots & \cdots & \cdots \\ \cdots & \cdots & \cdots & \cdots \\ p_{m1} & p_{m2} & \cdots & p_{mm} \end{bmatrix} \begin{bmatrix} R_1 \\ R_2 \\ \cdots \\ \cdots \\ R_m \end{bmatrix} \qquad ...(i)$$

where the matrix A has been expressed as a matrix of its row sub-matrices $R_1, R_2, ..., R_m$.

From the product of the matrices on R.H.S. of (i), we observe that the rows of the matrix B are

$$p_{11}R_1 + p_{12}R_2 + ... + p_{1m}R_m,$$
$$p_{21}R_1 + p_{22}R_2 + ... + p_{2m}R_m,$$
$$.................................$$
$$.................................$$
$$p_{m1}R_1 + p_{m2}R_2 + ... + p_{mm}R_m.$$

Thus we see that the rows of B are all linear combinations of the rows $R_1, R_2, ..., R_m$ of A. Therefore every member of the row space of B is also a member of the row space of A.

Similarly by writing $A = P^{-1}B$ and giving the same reasoning we can prove that every member of the row space of A is also a member of the row space of B. Therefore the row spaces of A and B are identical.

Thus we see that elementary row operations do not after the row space of a matrix. Hence the row rank of a matrix remains invariant under E-row transformations.

Note : From the above theorem we also conclude that premultiplication by a non-singular matrix does not alter the row rank of a matrix.

5.6.1. Invariance of column rank under E-column operations

Theorem : *Column equivalent matrices have the same column rank.*

or

Post-multiplication by a non-singular matrix does not alter the column rank of matrix.

Proof : Proceeding in the same way as in § 5.6 we can show the post multiplication with a non-singular matrix does not alter the column space and therefore the column rank of a matrix.

Note : Since every n-rowed E-matrix is obtainable from I_n by a single E-operation (row or column operation as may be desired), therefore the row rank and column rank of an E-matrix are each equal to n.

5.6.2. Invaraince of a column rank under E-row operations.

Theorem : *Row equivalent matrices have the same column rank.*

Let A be any given $m \times n$ matrix and let B be a matrix row equivalent to A. Then there exists a non-singular matrix P such that $B = PA$.

For every column vector X such that $AX = 0$, we have

$$BX = (PA) X = P(AX) = P0 = 0.$$

Since, $B = PA$, therefore $A = P^{-1}B$.

Therfore for every vector X such that $BX = 0$, we have

$$AX = (P^{-1}B) X = P^{-1}(BX) = P^{-1}0 = 0.$$

Thus we see that the matrices A and B have the same right nullities and consequently their column ranks are equal.

Similarly we can prove that column equivalent matrices have the same row rank.

5.6.3. Theorem

If r, be the row rank of an $m \times n$ matrix A then there exsits a non-singular matrix, P such that

$$PA = \begin{bmatrix} K \\ O \end{bmatrix},$$

where K is an $r \times n$ matrix consisting of a set of r linearly independent rows of A.

Linear Dependence of Vectors

Proof. If the row rank r of A is zero, we have nothing to prove. Therefore let us assume that $r > 0$. The matrix A has then r linearly independent rows. By elementary row operations on A we can bring these linearly independent rows in the first r places. Since the last $m - r$ rows are now linearly combinations of the first r rows, they can be made zero by E-row operations without altering the first r rows.

Thus we see that the matrix A is row equivalent to a matrix B such that $B = \begin{bmatrix} K \\ O \end{bmatrix}$, where K is an $r \times n$ matrix consisting of a set of r linearly independent rows of A.

Since every elementary row operation is equivalent to premultiplication by the corresponding E-matrix and the product of E-matrices is a non-singular matrix, therefore there exists a non-singular matrix P such that

$$PA = \begin{bmatrix} K \\ O \end{bmatrix}.$$

Similarly considering column transformations instead of row transformations, we can show that if c be the column rank of a matrix A, then there exists a non-singular matrix R such that

$$AR = [L \quad O],$$

where L is an $m \times c$ matrix consisting of c, linearly independent columns of A.

6
Rank of a Matrix

6.1 MINOR OF A MATRIX

Let A be a $m \times n$ matrix. If we retain any r rows and r columns of A, we shall have a square sub-matrix of order r. **The determinant of the square sub-matrix of order r is called a minor of A of order r.** From a given matrix we can form square sub-matrices of order 1, 2, 3, ... m if m is less than n or of order 1, 2, 3, ..., n if n is less than m.

For example, if the matrix is 3×4, naturally we can have square sub-matrices of order 1, 2, 3. Let $A = \begin{bmatrix} a_{11} & a_{12} & a_{13} & a_{14} \\ a_{21} & a_{22} & a_{23} & a_{24} \\ a_{31} & a_{32} & a_{33} & a_{34} \end{bmatrix}$.

Minors of A of order 1.

Each element of A is a minor of order one.

Minor of A of order 2.

Retain any two rows and two columns of A and the determinants of the square sub-matrices of order 2 thus formed are called minors of order 2, *i.e.*,

$\begin{vmatrix} a_{11} & a_{12} \\ a_{21} & a_{22} \end{vmatrix}$, $\begin{vmatrix} a_{11} & a_{13} \\ a_{21} & a_{23} \end{vmatrix}$, $\begin{vmatrix} a_{11} & a_{14} \\ a_{21} & a_{24} \end{vmatrix}$,

$\begin{vmatrix} a_{11} & a_{14} \\ a_{21} & a_{24} \end{vmatrix}$, $\begin{vmatrix} a_{21} & a_{22} \\ a_{31} & a_{32} \end{vmatrix}$ etc. are all minors of order 2 of A.

Minors of A of order 3.

Retain any three rows and any three columns of A and the determinant of square sub-matrices of order 3 thus formed are called the minors of A of order 3. Thus,

$\begin{vmatrix} a_{11} & a_{12} & a_{13} \\ a_{21} & a_{22} & a_{23} \\ a_{31} & a_{32} & a_{33} \end{vmatrix}$, $\begin{vmatrix} a_{11} & a_{13} & a_{14} \\ a_{21} & a_{23} & a_{24} \\ a_{31} & a_{33} & a_{34} \end{vmatrix}$,

$\begin{vmatrix} a_{11} & a_{12} & a_{14} \\ a_{21} & a_{22} & a_{24} \\ a_{31} & a_{32} & a_{34} \end{vmatrix}$ and $\begin{vmatrix} a_{12} & a_{13} & a_{14} \\ a_{22} & a_{23} & a_{24} \\ a_{32} & a_{33} & a_{34} \end{vmatrix}$

are the minors of A of order 3.

6.2 RANK OF A MATRIX

The rank of a given matrix A is said to be r if
(a) Every minor of A of order $r+1$ is zero.
(b) There is at least one minor of A of order r which does not vanish.

Rank of a Matrix

If a minor of A is zero the corresponding **sub-matrix is singular** and if a minor of A is not zero then corresponding **sub-matrix is non-singular.**

Hence we can also say that the rank of matrix A is said to be r if

(a) Every square sub-matrix of order $r+1$ is singular.
(b) There is at least one square sub-matrix of order r which is non-singular.

The rank r of matrix A is written as $\rho(A) = r$.

6.2.1 Nullity of a Matrix

If A is a square matrix of order r then $n - \rho(A)$ is called the **nullity of the matrix A** and is denoted by $N(A)$. Thus a non-singular square matrix of order n has rank equal to n and the nullity of such a matrix is equal to zero.

EXAMPLES

1. *Find the rank of matrix A, where*

$$A = \begin{bmatrix} 6 & 1 & 3 & 8 \\ 4 & 2 & 6 & -1 \\ 10 & 3 & 9 & 7 \\ 16 & 4 & 12 & 15 \end{bmatrix}$$

Solution. The given matrix is 4×4 and we can have minors of order 1, 2, 3, 4.

Minor of order 4 is $\begin{vmatrix} 6 & 1 & 3 & 8 \\ 4 & 2 & 6 & -1 \\ 10 & 3 & 9 & 7 \\ 16 & 4 & 12 & 15 \end{vmatrix}$

$= \begin{vmatrix} 6 & 1 & 3 & 8 \\ 4 & 2 & 6 & -1 \\ 10 & 3 & 9 & 7 \\ 6 & 1 & 3 & 8 \end{vmatrix}$ by $R_4 - R_3$

$= 3 \begin{vmatrix} 6 & 1 & 1 & 8 \\ 4 & 2 & 2 & -1 \\ 10 & 3 & 3 & 7 \\ 6 & 1 & 1 & 8 \end{vmatrix} = 0$ as C_1 and C_2 are identical.

Hence, $\rho(A) < 4$.

Minor of order 3.

$= \begin{vmatrix} 6 & 1 & 3 \\ 4 & 2 & 6 \\ 10 & 3 & 9 \end{vmatrix} = 3 \begin{vmatrix} 6 & 1 & 1 \\ 4 & 2 & 2 \\ 10 & 3 & 3 \end{vmatrix} = 0$

since C_2 and C_3 are identical.

Similarly, we can show that all minors of order 3 are zero. Hence, $\rho(A) < 3$. But one of the minor of order 2 is

$$\begin{vmatrix} 6 & 1 \\ 4 & 2 \end{vmatrix} = 12 - 4 = 8 \neq 0.$$

Hence, $\rho(A) = 2$.

2. *Show that the rank of the transpose of a matrix is the same as that of the original matrix.*

Solution. We know that the value of a determinant remains unchanged if rows be changed into columns and columns into rows.

Now, let $\rho(A) = r$, then there is atleast one minor of A of order $(r+1)$ is zero and hence every minor of A' of order $(r+1)$ is zero.

Hence, we have $\rho(A') = r$.

3. *Prove that the points $(x_1, y_1), (x_2, y_2), (x_3, y_3)$ are collinear if and only if the rank of the matrix.*

$$A = \begin{bmatrix} x_1 & y_1 & 1 \\ x_1 & y_2 & 1 \\ x_3 & y_3 & 1 \end{bmatrix} \text{ is less than 3.}$$

Solution. Case I. If the rank of the matrix A is less than 3 then every minor of order 3 should vanish.

$$\therefore \quad \begin{vmatrix} x_1 & y_1 & 1 \\ x_2 & y_2 & 1 \\ x_3 & y_3 & 1 \end{vmatrix} = 0 \qquad \ldots(i)$$

But this is the condition for the three points to be collinear.

Case II. Let the three points be collinear so that we have condition (i) showing that all the minors of order 3 of matrix A vanish. Thus the rank of A must be less than 3.

EXERCISE

Find the rank of the following matrices :

1. $\begin{bmatrix} 1 & 1 & 1 & -1 \\ 1 & 2 & 3 & 4 \\ 3 & 4 & 5 & 2 \end{bmatrix}$ [Ans. 2]

2. $\begin{bmatrix} 2 & 1 & -1 \\ 0 & 3 & -2 \\ 2 & 4 & -3 \end{bmatrix}$ [Ans. 2]

3. Prove that for a $m \times n$ matrix, whose every element is unity, the rank is 1.

4. If $A = \begin{bmatrix} 1 & 1 & -1 \\ 2 & -3 & 4 \\ 3 & -2 & 3 \end{bmatrix}; B = \begin{bmatrix} -1 & -2 & -1 \\ 6 & 12 & 6 \\ 5 & 10 & 5 \end{bmatrix}$ then show that $\rho(AB) \neq \rho(BA)$, where ρ denotes its rank.

5. Under what condition the rank of the following matrix A is 3 ?

$$A = \begin{bmatrix} 2 & 4 & 2 \\ 3 & 1 & 2 \\ 1 & 0 & x \end{bmatrix} \qquad \left[\text{Ans. } x \neq \frac{3}{5}\right]$$

6.3 NORMAL FORM

Every $m \times n$ matrix A of rank r can be reduced to any of the form

$$\begin{bmatrix} I_r & 0 \\ 0 & 0 \end{bmatrix}, [I_r, 0], \begin{bmatrix} I_r \\ 0 \end{bmatrix}, [I_r]$$

Each one of these four forms is called **normal form** or **canonical form** of the given matrix A.

Rank of a Matrix

Procedure: To obtain the normal form of the matrix A, the elementary transformations are applied in the following manner:

(i) Interchange rows (or columns) to obtain a non-zero element in the first row and the first column of the given matrix.

(ii) Divide the first row by this element if it is not zero.

(iii) Obtain zeros in the remainder of the first column by subtracting appropriate multiples of the first row from other rows.

(iv) Obtain zeros in the remainder of the first row by subtracting appropriate multiples of the first column from the other columns.

(v) Repeat the above four steps starting with element in the second row and second column.

(vi) Continue the process down the **main diagonal** either until the end of the diagonal is reached or until all the remaining elements of the matrix are zero.

Note 1. As the rank of the matrix is not altered due to elementary transformations, the rank of the normal form will be same as the rank of a given matrix A.

Note 2. In evaluation of the rank of a matrix by method of elementary transformations, if certain rows and columns are reduced to zeros entirely, we can remove them without affecting the rank of the matrix. This method is known as **Sweep Out** or **Pivotal method**.

6.3.1 Some Theorems on Rank

Theorem 1. *Elementary transformations do not alter the rank of matrix, i.e., equivalent matrices have the same rank.*

Proof: Let A be a matrix of order $m \times n$, i.e.,

$$A = \begin{bmatrix} a_{11} & a_{12} & \cdots & a_{1n} \\ a_{21} & a_{22} & \cdots & a_{2n} \\ \cdots & \cdots & \cdots & \cdots \\ \cdots & \cdots & \cdots & \cdots \\ a_{m1} & a_{m2} & \cdots & a_{mn} \end{bmatrix}$$

Let a sub-matrix A_r of order r belongs to the first r rows of the matrix A. Then from the properties of the determinants, following three conditions are always satisfied:

(i) If two rows are interchanged then any determinant $|A|$, either remains unaltered or changes into $-|A|$.

(ii) If one row of the determinant is multiplied by a non-zero scalar k, then the determinant $|A_r|$ becomes $k|A_r|$.

(iii) If any row is changed by adding to it another row, then any determinant $|A_r|$ remains unaltered.

Similar statements are true for elementary transformations on columns.

Now, let the matrix B be equivalent to the matrix A, i.e.,

$$B \sim A.$$

From (i), (ii) and (iii), it is obvious that if all the determinants of order r in A are zero, then all the determinants of the same order r in B will also be zero.

Hence, $\qquad$ rank $(B) \leq$ rank (A)

But since $A \sim B$, it implies that rank $(A) \leq$ rank (B).

hence, we have rank $(A) =$ rank (B).

Theorem 2. *If A and B are equivalent matrices, there exist non-singular matrices C and D such that $B = C A D$.*

Proof: Since A and B are elementary matrices, B is obtained from A by applying to A a sequence of elementary row and column transformations. But we know that elementary row transformations can be accomplished by pre-multiplying A by elementary matrices of appropriate order and elementary column transformation can be accomplished by post-multiplying A by elementary matrices of appropriate order. Hence

$$C_1 C_2 \ldots C_r A \cdot D_1 D_2 \ldots D_s = B$$

or
$$C A D = B$$

where $C = C_1 \cdot C_2 \cdot C_3 \ldots C_r$,

and $D = D_1 \cdot D_2 \cdot D_3 \ldots D_s$

Since elementary matrices are non-singular, C and D are non-singular.

Theorem 3. *The rank of a product of two matrices cannot exceed the rank of either matrix,* i.e.,

$$\rho(AB) \leq \rho(A) \text{ and } \rho(AB) \leq \rho(B)$$

Proof : Let the rank of matrices AB, A and B be r, r_1 and r_2 respectively.

Let there exists a non-singular matrix P such that

$$\rho(P A B) = \rho(AB) = r \text{ and } PA = \begin{bmatrix} G \\ \ldots \\ O \end{bmatrix} \qquad \ldots(i)$$

where G is a matrix of rank r_1 with r_1 rows.

Post-multiplying (i) by B, we have

$$P A B = \begin{bmatrix} G \\ \ldots \\ O \end{bmatrix} B \qquad \ldots(ii)$$

$\therefore \qquad \rho(AB) = \rho(PAB) \, \rho\left(\begin{bmatrix} G \\ O \end{bmatrix} B\right)$

Since G has r_1 non-zero rows, $\begin{bmatrix} G \\ O \end{bmatrix} B$ cannot have more than r_1 non-zero rows.

Consequently, $\qquad \rho\left(\begin{bmatrix} G \\ O \end{bmatrix} B\right) \leq r_1$

i.e., $\qquad$ rank of $(AB) \leq r_1$

i.e., $\qquad$ rank of $(AB) \leq $ rank of A.

Again $\qquad \rho(AB) = \rho(AB)'$

$\qquad\qquad\qquad = \rho(B'A')$

$\qquad\qquad\qquad \leq \rho(B') = \rho(B)$

$\therefore \qquad \rho(AB) \leq \rho(B)$

Theorem 4. *The rank of a matrix does not alter by pre-(post-) multiplication with any non-singular matrix.*

Proof : We know that A can be reduced to the normal form $\begin{bmatrix} I_r & 0 \\ 0 & 0 \end{bmatrix}$ by chain of elementary row and column operations on A. Also, we know that each row operation is equivalent to pre-multiplication of A by corresponding elementary matrix and each column operation is equivalent to post-multiplication of A by corresponding elementary matrices. If these matrices be $P_1, P_2, \ldots, P_r, Q_1, Q_2, \ldots, Q_s$ then $P_1 . P_2 \ldots P_{r-1} P_r A Q_1 . Q_2 \ldots Q_s = \begin{bmatrix} I_r & 0 \\ 0 & 0 \end{bmatrix}$.

Rank of a Matrix

or
$$PAQ = \begin{bmatrix} I_r & 0 \\ 0 & 0 \end{bmatrix}$$

where $\qquad P = P_1 \cdot P_2 \ldots P_r$

and $\qquad Q = Q_1 \cdot Q_2 \ldots Q_s$

Since each of the elementary matrices $P_1, P_2, \ldots P_r$ and $Q_1, Q_2, \ldots, Q_s$ is non-singular, therefore, their product P and Q are also non-singular matrices.

EXAMPLES

1. *Reduce the following matrix A to its normal form and find its rank*

$$A = \begin{bmatrix} 1 & 3 & 4 & 5 \\ 1 & 2 & 6 & 7 \\ 1 & 5 & 0 & 1 \end{bmatrix}$$

Solution. We have

$$A = \begin{bmatrix} 1 & 3 & 4 & 5 \\ 1 & 2 & 6 & 7 \\ 1 & 5 & 0 & 1 \end{bmatrix}$$

$$\sim \begin{bmatrix} 1 & 3 & 4 & 5 \\ 0 & -1 & 2 & 2 \\ 0 & 2 & -4 & -4 \end{bmatrix},$$

Replacing R_2 by $R_2 - R_1$ and R_3 by $R_3 - R_1$

$$\sim \begin{bmatrix} 1 & 0 & 0 & 0 \\ 0 & -1 & 2 & 2 \\ 0 & 2 & -4 & -4 \end{bmatrix},$$

Replacing C_2 by $C_2 - 3C_1$, C_3 by $C_3 - 4C_1$ and C_4 by $C_4 - 5C_1$.

$$\sim \begin{bmatrix} 1 & 0 & 0 & 0 \\ 0 & 1 & -2 & -2 \\ 0 & 2 & -4 & -4 \end{bmatrix}, \qquad \text{Replacing } R_2 \text{ by } (-1)R_2.$$

$$\sim \begin{bmatrix} 1 & 0 & 0 & 0 \\ 0 & 1 & -2 & -2 \\ 0 & 0 & 0 & 0 \end{bmatrix}, \qquad \text{Replacing } R_3 \text{ by } R_3 - 2R_1$$

$$\sim \begin{bmatrix} 1 & 0 & 0 & 0 \\ 0 & 1 & 0 & 0 \\ 0 & 0 & 0 & 0 \end{bmatrix}, \qquad \text{Replacing } C_3 \text{ by } C_3 + 2C_2 \text{ and } C_4 \text{ by } C_4 + 2C_2$$

$$\sim \begin{bmatrix} I_2 & 0 \\ 0 & 0 \end{bmatrix},$$

Hence, $\rho(A) = 2$.

2. Find non-singular matrices P and Q such that PAQ is in the normal form, where

$$A = \begin{bmatrix} 1 & 2 & 3 \\ 3 & 2 & 1 \\ 1 & 3 & 2 \\ 2 & 1 & 3 \end{bmatrix}$$

Solution. Since A has four rows, we shall find elementary matrix by applying row operations on a unit matrix I_4 to get pre-factor P. Also to get post-factor Q, we shall find elementary matrix by applying column operations on a unit matrix I_3 of order three because A has three columns. Thus to get PAQ we write $A = I_4 A I_3$.

or
$$\begin{bmatrix} 1 & 2 & 3 \\ 3 & 2 & 1 \\ 1 & 3 & 2 \\ 2 & 1 & 3 \end{bmatrix} = \begin{bmatrix} 1 & 0 & 0 & 0 \\ 0 & 1 & 0 & 0 \\ 0 & 0 & 1 & 0 \\ 0 & 0 & 0 & 1 \end{bmatrix} A \begin{bmatrix} 1 & 0 & 0 \\ 0 & 1 & 0 \\ 0 & 0 & 1 \end{bmatrix}$$

Applying $R_2 \to R_2 - 3R_1, R_3 \to R_3 - R_1, R_4 \to R_4 - 2R_1$ we get,

$$\begin{bmatrix} 1 & 2 & 3 \\ 0 & -4 & -8 \\ 0 & 1 & -1 \\ 0 & -3 & -3 \end{bmatrix} = \begin{bmatrix} 1 & 0 & 0 & 0 \\ -3 & 1 & 0 & 0 \\ -1 & 0 & 1 & 0 \\ -2 & 0 & 0 & 1 \end{bmatrix} A \begin{bmatrix} 1 & 0 & 0 \\ 0 & 1 & 0 \\ 0 & 0 & 1 \end{bmatrix}$$

Applying $C_2 \to C_2 - 2C_1, C_3 \to C_3 - 3C_1$, we have

$$\begin{bmatrix} 1 & 2 & 3 \\ 0 & -4 & -8 \\ 0 & 1 & -1 \\ 0 & -3 & -3 \end{bmatrix} = \begin{bmatrix} 1 & 0 & 0 & 0 \\ -3 & 1 & 0 & 0 \\ -1 & 0 & 1 & 0 \\ -2 & 0 & 0 & 1 \end{bmatrix} A \begin{bmatrix} 1 & -2 & -3 \\ 0 & 1 & 0 \\ 0 & 0 & 1 \end{bmatrix}$$

Applying $R_2 \to -\dfrac{1}{4} R_2$

$$\begin{bmatrix} 1 & 0 & 0 \\ 0 & 1 & 2 \\ 0 & 1 & -1 \\ 0 & -3 & -3 \end{bmatrix} = \begin{bmatrix} 1 & 0 & 0 & 0 \\ 3/4 & -1/4 & 0 & 0 \\ -1 & 0 & 1 & 0 \\ -2 & 0 & 0 & 1 \end{bmatrix} A \begin{bmatrix} 1 & -2 & -3 \\ 0 & 1 & 0 \\ 0 & 0 & 1 \end{bmatrix}$$

Applying $R_3 \to R_3 - R_2, R_4 \to R_4 + 3R_2$

$$\begin{bmatrix} 1 & 0 & 0 \\ 0 & 1 & 2 \\ 0 & 0 & -3 \\ 0 & 0 & 3 \end{bmatrix} = \begin{bmatrix} 1 & 0 & 0 & 0 \\ 3/4 & -1/4 & 0 & 0 \\ -7/4 & 1/4 & 1 & 0 \\ 1/4 & -3/4 & 0 & 1 \end{bmatrix} A \begin{bmatrix} 1 & -2 & -3 \\ 0 & 1 & 0 \\ 0 & 0 & 1 \end{bmatrix}$$

Applying $C_3 \to C_3 - 2C_2$

$$\begin{bmatrix} 1 & 0 & 0 \\ 0 & 1 & 2 \\ 0 & 0 & -3 \\ 0 & 0 & 3 \end{bmatrix} = \begin{bmatrix} 1 & 0 & 0 & 0 \\ 3/4 & -1/4 & 0 & 0 \\ -7/4 & 1/4 & 1 & 0 \\ 1/4 & -3/4 & 0 & 1 \end{bmatrix} A \begin{bmatrix} 1 & -2 & 1 \\ 0 & 1 & -2 \\ 0 & 0 & 1 \end{bmatrix}$$

Rank of a Matrix

Changing R_3 by $-\dfrac{1}{3}R_3$

$$\begin{bmatrix} 1 & 0 & 0 \\ 0 & 1 & 0 \\ 0 & 0 & 1 \\ 0 & 0 & 3 \end{bmatrix} = \begin{bmatrix} 1 & 0 & 0 & 0 \\ 3/4 & -1/4 & 0 & 0 \\ -7/12 & -1/12 & -1/3 & 0 \\ 1/4 & -3/4 & 0 & 1 \end{bmatrix} A \begin{bmatrix} 1 & -2 & 1 \\ 0 & 1 & -2 \\ 0 & 0 & 1 \end{bmatrix}$$

Applying $R_4 \to R_4 - 3R_3$

$$\begin{bmatrix} 1 & 0 & 0 \\ 0 & 1 & 0 \\ 0 & 0 & 1 \\ 0 & 0 & 0 \end{bmatrix} = \begin{bmatrix} 1 & 0 & 0 & 0 \\ 3/4 & -1/4 & 0 & 0 \\ 7/12 & -1/12 & -1/3 & 0 \\ -3/2 & -1/2 & -1 & 1 \end{bmatrix} A \begin{bmatrix} 1 & -2 & 1 \\ 0 & 1 & -2 \\ 0 & 0 & 1 \end{bmatrix}$$

This is the required normal form, where

$$P = \begin{bmatrix} 1 & 0 & 0 & 0 \\ 3/4 & -1/4 & 0 & 0 \\ 7/12 & -1/12 & -1/3 & 0 \\ -3/2 & -1/2 & 1 & 1 \end{bmatrix} \text{ and } Q = \begin{bmatrix} 0 & -2 & 1 \\ 0 & 1 & -2 \\ 0 & 0 & 1 \end{bmatrix}$$

3. Show that rank $(AA') = $ rank (A).

Solution. We know that

$$\text{rank } (A) = \text{rank } (A') \quad \text{....(i)}$$

and $\qquad$ rank $(AA') \leq$ rank (A) $\qquad$ ...(ii)

Let $B = AA'$, then rank $(B) = $ rank $(AA') \leq $ rank (A) by (ii)

or $\qquad$ rank $(B) \leq $ rank (A) $\qquad$ ...(iii)

$B = AA'$, gives

$$A^{-1}B = A^{-1}(AA') = (A^{-1}A)A' = IA' = A'$$

$\therefore \qquad$ rank $(A) = $ rank $(A') = $ rank $(A^{-1}B) \leq $ rank (B)

i.e., $\qquad$ rank $(A) \leq $ rank (B)

By (ii) and (iv), we get

$$\text{rank } (A) = \text{rank } (B) = \text{rank } (AA')$$

4. Show that the rank of a skew-symmetrix cannot be one.

Solution. Let $A = \begin{bmatrix} 0 & a & b \\ -a & 0 & c \\ -b & -c & 0 \end{bmatrix}$ be a skew-symmetric matrix of order 3. If each of a, b, c is zero then $A = 0$, i.e., null matrix which is not the case. Hence at least one of the numbers a, b, c is non0-zero. Now let $a \neq 0, b = 0$ and $c = 0$, then second order minor $\begin{vmatrix} 0 & a \\ -a & 0 \end{vmatrix}$ is non-zero. Hence rank $(A) = 2$. Hence, $\rho(A) > 1$.

EXERCISE

1. Find the rank of following matrices by elementary transformations:

   (a) $\begin{bmatrix} 1 & 2 & 3 & 2 \\ 2 & 3 & 5 & 1 \\ 2 & 3 & 4 & 5 \end{bmatrix}$
   (b) $\begin{bmatrix} 2 & 3 & -1 & -1 \\ 1 & -1 & -2 & -4 \\ 3 & 1 & 3 & -2 \\ 6 & 3 & 0 & -7 \end{bmatrix}$

   (c) $\begin{bmatrix} 3 & -2 & 0 & -1 \\ 0 & 2 & 2 & 1 \\ 1 & -2 & -3 & 2 \\ 0 & 1 & 2 & 1 \end{bmatrix}$

2. Reduce the following matrices to their normal forms and find their ranks.

   (a) $\begin{bmatrix} 1 & 2 & 0 & -1 \\ 3 & 4 & 1 & 2 \\ -2 & 3 & 2 & 5 \end{bmatrix}$
   (b) $\begin{bmatrix} 2 & 3 & -1 & -1 \\ 1 & -1 & -2 & -4 \\ 3 & 1 & 3 & -2 \\ 6 & 3 & 0 & -7 \end{bmatrix}$

   (c) $\begin{bmatrix} 0 & 1 & 2 & -2 \\ 4 & 0 & 2 & 6 \\ 2 & 1 & 3 & 1 \end{bmatrix}$

3. For the following matrices find non-singular matrices P and Q, so that PAQ is in the normal form:

   (a) $\begin{bmatrix} 1 & 1 & 2 \\ 1 & 2 & 3 \\ 0 & -1 & -1 \end{bmatrix}$
   (b) $\begin{bmatrix} 1 & -1 & 2 & -1 \\ 4 & 2 & -1 & 2 \\ 2 & 2 & -2 & 0 \end{bmatrix}$

4. Find the rank of matrix A by reducing into the normal form

   $$A = \begin{bmatrix} 2 & -2 & 0 & 6 \\ 4 & 2 & 0 & 2 \\ 1 & -1 & 0 & 3 \\ 1 & -2 & 1 & 2 \end{bmatrix}$$

   Also prove that

   (i) rank (A) = rank (A^*)
   (ii) rank (AA^*) = rank (A)

Answers

3. (a) $P = \begin{bmatrix} 1 & 0 & 0 \\ -1 & 1 & 0 \\ -1 & 1 & 1 \end{bmatrix}$, $Q = \begin{bmatrix} 1 & -1 & -1 \\ 0 & 1 & -1 \\ 0 & 0 & 1 \end{bmatrix}$

   (b) $P = \begin{bmatrix} 1 & 0 & 0 \\ -2/3 & 1/6 & 0 \\ -1/3 & 1/3 & -1/2 \end{bmatrix}$, $Q = \begin{bmatrix} 1 & 1 & 0 & -1/2 \\ 0 & 1 & -1 & 3/2 \\ 0 & 0 & 0 & 1 \\ 0 & 0 & 1 & 0 \end{bmatrix}$

4. 3

7
Linear Equations

7.1 LINEAR EQUATIONS

Consider a system of m linear equations in n unknowns given as below

$$a_{11}x_1 + a_{12}x_2 + \ldots + a_{1n}x_n = b_1$$
$$a_{21}x_1 + a_{22}x_2 + \ldots + a_{2n}x_n = b_2$$
$$\ldots \ldots \ldots \ldots \ldots \ldots \ldots$$
$$a_{m1}x_1 + a_{m2}x_2 + \ldots + a_{mn}x_n = b_m$$

...(i)

All the a's and b's in the above are constants and $x_1, x_2, \ldots, x_n$ are the n unknowns to be determined.

The above equations can be written in matrix form as follows:

$$\begin{bmatrix} a_{11} & a_{12} & \ldots & a_{1n} \\ a_{21} & a_{22} & \ldots & a_{2n} \\ \ldots & \ldots & \ldots & \ldots \\ \ldots & \ldots & \ldots & \ldots \\ a_{m1} & a_{m2} & \ldots & a_{mn} \end{bmatrix} \begin{bmatrix} x_1 \\ x_2 \\ \vdots \\ \vdots \\ x_m \end{bmatrix} = \begin{bmatrix} b_1 \\ b_2 \\ \vdots \\ \vdots \\ b_m \end{bmatrix}$$

or
$$A X = B,$$

Where A is $m \times n$ matrix and x is $n \times 1$ matrix so that $A X$ is conformabale for multiplication and AX is $m \times 1$ matrix and B is also $m \times 1$ matrix.

The matrix

$$A = \begin{bmatrix} a_{11} & a_{12} & \ldots & a_{1n} \\ a_{21} & a_{22} & \ldots & a_{2n} \\ \ldots & \ldots & \ldots & \ldots \\ \ldots & \ldots & \ldots & \ldots \\ a_{m1} & a_{m2} & \ldots & a_{mn} \end{bmatrix}$$

is called coefficient matrix of the given system of equations.

The matrix

$$A = \begin{bmatrix} a_{11} & a_{12} & \ldots & a_{1n} & b_1 \\ a_{21} & a_{22} & \ldots & a_{2n} & b_2 \\ \ldots & \ldots & \ldots & \ldots & \ldots \\ a_{m1} & a_{m2} & \ldots & a_{mn} & b_n \end{bmatrix}$$

of type $m \times (n+1)$ is called the **Augmented matrix.**

Any set of values of the unknowns $x_1, x_2, x_3, \ldots x_n$ which satisfy the equations (i) is called the **solution of the system.**

When the system of equations has one or more solutions, the system is called **consistent.** If it has no solution, the system of equations is called **inconsistent.** A consistent system has either one solution or infinitely many solutions.

7.2 SOLUTION OF A SYSTEM OF n NON-HOMOGENEOUS LINEAR EQUATIONS IN n UNKNOWNS

Let the system of equations be given by

$$a_{11}x_1 + a_{12}x_2 + \ldots + a_{1n}x_n = b_1 \qquad \ldots(i)$$
$$a_{21}x_1 + a_{22}x_2 + \ldots + a_{2n}x_n = b_2 \qquad \ldots(ii)$$
$$\ldots \ldots \ldots \ldots \ldots \ldots$$
$$a_{n1}x_1 + a_{n2}x_2 + \ldots + a_{nn}x_n = b_n \qquad \ldots(n)$$

The number of equations being the same as the number of unknowns. These can be written as

$$\begin{bmatrix} a_{11} & a_{12} & \ldots & a_{1n} \\ a_{21} & a_{22} & \ldots & a_{2n} \\ \ldots & \ldots & \ldots & \ldots \\ a_{n1} & a_{n2} & \ldots & a_{nn} \end{bmatrix} \begin{bmatrix} x_1 \\ x_2 \\ \vdots \\ \vdots \\ x_n \end{bmatrix} = \begin{bmatrix} b_1 \\ b_2 \\ \vdots \\ \vdots \\ b_n \end{bmatrix}$$

or $\qquad AX = B$

Ist Method. Cramer's Rule

If in the set of equations (i), (ii) ... (n) the value of the determinant of the coefficients is not zero, i.e., $|A| \neq 0$, the set of equations is said to be **regular** and the system has unique solution given by

$$\frac{x_1}{|A_1|} = \frac{x_2}{|A_2|} = \frac{x_3}{|A_3|} = \ldots = \frac{x_n}{|A_n|} = \frac{1}{|A|}$$

Where A_i denotes the determinant obtained on replacing the ith column of $|A|$ by the column of b's.

This method is called **Cramer's rule.**

Note. If $|A| = 0$, the set of equations is said to be **singular** and if $|A| \neq 0$, it is said to be **non-singular.** In case $|A| = 0$, Cramer's rule fails in solving the linear equations.

2nd Method. Solution by the help of Inverse

The given equations in matrix form is

$$AX = B \qquad \ldots(A)$$

Now if A is non-singular then A^{-1} exists and $A^{-1}A = I$.

Multiplying both sides of (A) by A^{-1} we get,

$$A^{-1}AX = A^{-1}B \qquad \text{or} \qquad IX = A^{-1}B$$

or $\qquad X = A^{-1}B$

This will give us the unique solution of given equations provided A is non-singular, i.e., its determinant $|A| \neq 0$.

Uniqueness. Solution (B) of equations is unique. If it be not unique let X_1 and X_2 be two solutions of $AX = B$, so that

$$AX_1 = B \qquad \text{and} \qquad AX_2 = B$$

$\therefore \qquad AX_1 = AX_2$

or $\qquad A^{-1}(AX_1) = A^{-1}(AX_2) \qquad \text{or} \qquad (A^{-1}A)X_1 = (A^{-1}A)X_2$

or $\quad I X_1 = I X_2 \quad$ or $\quad X_1 = X_2$.
Hence the solution is unique.

7.3 SOLUTION OF m NON-HOMOGENEOUS LINEAR EQUIATIONS IN n VARIABLES

Let the given equations be

$$a_{11}x_1 + a_{12}x_2 + \ldots + a_{1n}x_n = b_1$$
$$a_{21}x_1 + a_{22}x_2 + \ldots + a_{2n}x_n = b_2$$
$$\ldots \ldots \ldots \ldots \ldots \ldots \ldots \ldots$$
$$a_{m1}x_1 + a_{m2}x_2 + \ldots + a_{mn}x_n = b_n$$

...(i)

or

$$\begin{bmatrix} a_{11} & a_{12} & \ldots & a_{1n} \\ a_{21} & a_{22} & \ldots & a_{2n} \\ \ldots & \ldots & \ldots & \ldots \\ a_{m1} & a_{m2} & \ldots & a_{mn} \end{bmatrix} \begin{bmatrix} x_1 \\ x_2 \\ \vdots \\ x_n \end{bmatrix} = \begin{bmatrix} b_1 \\ b_2 \\ \vdots \\ b_m \end{bmatrix}$$

or $\quad A X = B \quad$ ...(ii)

Where coefficient matrix

$$A = \begin{bmatrix} a_{11} & a_{12} & \ldots & a_{1n} \\ a_{21} & a_{22} & \ldots & a_{2n} \\ \ldots & \ldots & \ldots & \ldots \\ a_{m1} & a_{m2} & \ldots & a_{mn} \end{bmatrix}$$

and augmented matrix

$$C = \begin{bmatrix} a_{11} & a_{12} & \ldots & a_{1n} & b_1 \\ a_{21} & a_{22} & \ldots & a_{2n} & b_2 \\ \ldots & \ldots & \ldots & \ldots & \ldots \\ a_{m1} & a_{m2} & \ldots & a_{mn} & b_m \end{bmatrix}$$

1st Case. When $m = n$. We have discussed the solution in 7.2. When coefficient matrix A is non-singular (or the rank of matrix A is n), the solution is unique and could be found by Cramer's rule or writing $A X = B$ as $X = A^{-1}B$.

2nd Case. If $m \neq n$. The system of equation $A X = B$ will be consistent, *i.e.*, they have a solution if and only if **the coefficient matrix A and augmented matrix C have the same rank.** If rank A < rank C then the system of equations $A X = B$ are said to be inconsistent. (see 7.4).

Nature of Solution. When the equations are consistent *i.e.*, rank (A) = rank $(C) = r$, say, then we shall assign arbitrary values to $n - r$ variables and the remaining r variables shall be uniquely determined in terms of these arbitrary chosen values and the system is said to have infinite number of solutions. When rank (A) = rank $(C) = r = n$, then $n - r = n - n = 0$ then none of the variables is to be assigned arbitrary values and hence in this case there will be unique solution as discussed in 7.2.

7.4 SOME THEOREMS ON CONSISTENCY OF EQUATIONS

Theorem 1. *If rank (A) < rank (C), the equation $A X = B$ are inconsistent.*

Proof. Let the given system of equations be consistent. Let rank $(C) = r >$ rank (A). Obviously, there will be atleast one non-zero determinant Δ_r (say) of order r in the matrix C. This Δ_r must contain the last column of C, otherwise it would be contained in A and this will be contrary to our hypothesis that rank $(A) <$ rank $(C) = r$.

Now, without any loss of generality, let us suppose that Δ_r lies in the top right hand corner of C. Since rank $(A) < r$, first r rows of A are linearly dependent and hence the r linear functions $f_1, f_2, ..., f_r$ forming the first r rows of A are linearly dependent, so that we can find constants $k_1, k_2, ..., k_r$ not all zero, such that

$$k_1 f_1 + k_2 f_2 + ... + k_r f_r = 0$$

$\therefore \quad k_1 (F_1 - b_1) + k_2 (F_2 - b_2) + ... + k (F_r - b_r) = 0$

or $\quad k_1 F_1 + k_2 F_2 + ... + k_r F_r = k_1 b_1 + k_2 b_2 + ... + k_r b_r = k$ \hfill (say)

Obviously $k \neq 0$, for otherwise the linear functions $F_1, F_2, ..., F_r$ forming the first r rows of C would be linearly dependent, which is not the case as Δ_r, the top right hand corner determinant of order r in C is not zero.

But the first r equations are

$$F_1 = F_2 = ... = F_r = 0$$

So that $k = 0$.

Hence there is a contradiction, so that rank $(A) <$ rank (C) and then the equations are inconsistent.

Theorem 2. *If rank $(A) =$ rank (C) then the equations $AX = B$ are consitent.*

Proof. Since A is sub-matrix of the augmented matrix C, hence rank $(A) \leq$ rank (C).

Now, by applying the elementary row operations to the augmented matrix C, we reduce the matrix C to the row reduced echelon matrix R (say), i.e.,

$$R = \begin{bmatrix} 1 & 0 & 0 & ... & ... & ... & \beta_1 \\ 0 & 1 & 0 & ... & ... & ... & \beta_2 \\ 0 & 0 & 0 & ... & ... & ... & \beta_3 \\ ... & ... & ... & ... & ... & ... & ... \\ 0 & 0 & 0 & ... & ... & ... & \beta_m \end{bmatrix}$$

If at least one of $\beta_{r+1} ... \beta_m$ is not zero, then the number of non-zero rows in the coefficient matrix is not equal to the number of non-zero rows in the augmented matrix.

Hence coefficient matrix A and the augmented matrix C do not have the same rank, i.e.,

rank $(A) <$ rank (C),

Which shows that at least one of equations $\beta_j = 0, j = r+1, ..., m$ is false. So the system is inconsistent.

If all $\beta_{r+1} = \beta_{r+2} = ... = \beta_m = 0$, then

rank $(A) <$ rank (C),

Which shows that the system has one or more solutions.

Again if rank $(A) =$ rank $(C) = r = n$, then the system of equations has unique solution.

Also, if rank $(A) =$ rank $(C) < n$, then the system has infinite solution.

Remark. From the above discussions, we have

(i) If rank $(A) <$ rank (C), the equations $AX = B$ are inconsistent and have no solution.

(ii) If rank $(A) =$ rank (C) i.e., $r = n$, then the system of equations have unique solution.

(iii) **If rank $(A) =$ rank $(C) < n$, then the system has infinite solutions.**

7.5 HOMOGENEOUS LINEAR EQUATIONS

The system of equations $AX = B$ having m linear equations in n unknowns is called the

Linear Equations

system of **homogeneous Linear equations** if $B = 0$. Thus the above equation reduces to
$$AX = 0$$
For such a system of equations, the ranks of coefficient matrix A and its augmented matrix C are always equal. Hence a system of homogeneous linear equations is always consistent. In fact $X = 0$, i.e., $x_1 = x_2 = x_3 = \ldots = x_n = 0$ is always a solution. This solution is called trivial solution.

If A is a non-singular matrix of order n, rank $(A) = n$ and can be solved by Cramer's rule to give the unique solution $x_1 = x_2 = x_3 = \ldots = x_n = 0$, i.e., the system has only the trivial solution.

But, if rank $(A) = r < n$, then A can be reduced to a matrix which has r non-zero rows and $(n - r)$ zero rows. In this case the system is consistent and has infinite solutions.

EXAMPLES

1. *Solve the following equations with the help of matrices.*
$$x + 2y + 3z = 14$$
$$3x + y + 2z = 11$$
$$2x + 3y + z = 11$$

Solution. Here, we have
$$A = \begin{bmatrix} 1 & 2 & 3 \\ 3 & 1 & 2 \\ 2 & 3 & 1 \end{bmatrix}$$

and Augmented matrix $C = \begin{bmatrix} 1 & 2 & 3 & : & 14 \\ 3 & 1 & 2 & : & 11 \\ 2 & 3 & 1 & : & 11 \end{bmatrix}$

Replacing R_2 by $R_2 - 3R_1$ and R_3 by $R_3 - 2R_1$, we get
$$C = \begin{bmatrix} 1 & 2 & 3 & : & 14 \\ 0 & -5 & -7 & : & -31 \\ 0 & -1 & -5 & : & -17 \end{bmatrix}$$

By $R_2 \to (-1) R_2$ and $R_3 \to (-1) R_3$
$$C \sim \begin{bmatrix} 1 & 2 & 3 & : & 14 \\ 0 & 5 & 7 & : & 31 \\ 0 & 1 & 5 & : & 17 \end{bmatrix}$$

By interchanging R_2 and R_3,
$$C \sim \begin{bmatrix} 1 & 2 & 3 & : & 14 \\ 0 & 1 & 5 & : & 17 \\ 0 & 5 & 7 & : & 31 \end{bmatrix}$$

By replacing R_3 by $R_3 - 5R_2$, we get
$$C \sim \begin{bmatrix} 1 & 2 & 3 & : & 14 \\ 0 & 1 & 5 & : & 17 \\ 0 & 0 & -18 & : & -54 \end{bmatrix}$$

By $R_3 \to \left(-\dfrac{1}{18}\right) R_3$

$$C \sim \begin{bmatrix} 1 & 2 & 3 & : & 14 \\ 0 & 1 & 5 & : & 17 \\ 0 & 0 & 1 & : & 3 \end{bmatrix} \qquad \text{...(i)}$$

This show that rank (A) = rank $(B) = 3$ = number of variables. Thus the given equations are consistent and have a unique solution. From (i) the equivalent system of equations is

$$x + 2y + 3z = 14$$
$$y + 5z = 17$$
$$z = 3$$

Hence, we get $x = 1$, $y = 2$, $z = 3$.

2. *Investigate for what values of λ, be the equations*

$$x + y + z = 6$$
$$x + 2y + 3z = 10$$
$$x + 2y + \lambda z = \mu$$

have (i) no solution, (ii) unique solution (iii) an infinite number of solutions.

Solution. We have

$$A = \begin{bmatrix} 1 & 1 & 1 \\ 1 & 2 & 3 \\ 1 & 2 & \lambda \end{bmatrix}, C = \begin{bmatrix} 1 & 1 & 1 & : & 6 \\ 1 & 2 & 3 & : & 10 \\ 1 & 2 & \lambda & : & \mu \end{bmatrix}$$

By applying $R_2 \to R_2 - R_1$, $R_3 \to R_3 - R_1$, on augmented matrix C, we get

$$C = \begin{bmatrix} 1 & 1 & 1 & : & 6 \\ 0 & 1 & 2 & : & 4 \\ 0 & 1 & \lambda - 1 & : & \mu - 6 \end{bmatrix}$$

By applying $R_3 \to R_3 - R_2$,

$$C \sim \begin{bmatrix} 1 & 1 & 8 & : & 6 \\ 0 & 1 & 2 & : & 4 \\ 0 & 0 & \lambda - 3 & : & \mu - 10 \end{bmatrix}$$

Now, we have

(i) For no solution, we have rank $(A) \neq$ rank (C). Hence if $\lambda = 3$ and $\mu \neq 10$, then rank $(A) = 2$ and rank $(B) = 3$ which satisfies the above condition. Thus, for no solution the values of λ and μ are given by $\lambda = 3$, $\mu \neq 10$.

(ii) For unique solution, we have $|A| \neq 0$; hence rank (A) = rank $(B) = 3$ = number of variables, which is possible only if $\lambda \neq 3$, whatever may be the value of μ. Thus there is a unique solution when $\lambda \neq 3$ for all values of μ.

(iii) For an infinite number of solutions, we have rank (A) = rank $(B) < 3$. Now we have if $\lambda = 3$ and $\mu = 10$, then rank (A) = rank $(B) = 2 <$ number of variables. Thus, for values $\lambda = 3$ and $\mu = 10$ the equations are consistent and there is an infinite number of solutions.

3. *Show that the equations*

$$x + 2y - z = 3,$$
$$3x - y + 2z = 1,$$
$$2x - 2y + 3z = 2,$$

Linear Equations

are consistent and solve them.

Solution. The given system of equations is equivalent to the single matrix equation

$$AX = \begin{bmatrix} 1 & 2 & -1 \\ 3 & -1 & 2 \\ 2 & -2 & 3 \\ 1 & -1 & 1 \end{bmatrix} \begin{bmatrix} x \\ y \\ z \end{bmatrix} = \begin{bmatrix} 3 \\ 1 \\ 2 \\ -2 \end{bmatrix} = B$$

The augmented matrix is

$$[A \ B] = \begin{bmatrix} 1 & 2 & -1 & : & 3 \\ 3 & -1 & 2 & : & 1 \\ 2 & -2 & 3 & : & 2 \\ 1 & -1 & 1 & : & -1 \end{bmatrix}$$

$$\sim \begin{bmatrix} 1 & 2 & -1 & : & 3 \\ 0 & -7 & 5 & : & -8 \\ 0 & -6 & 5 & : & -4 \\ 0 & -3 & 2 & : & -4 \end{bmatrix}$$

replacing R_2 by $R_2 - 3R_1$, R_3 by $R_3 - 2R_1$ and R_4 by $R_4 - R_1$

$$\sim \begin{bmatrix} 1 & 2 & -1 & : & 3 \\ 0 & -1 & 0 & : & -4 \\ 0 & -6 & 5 & : & -4 \\ 0 & -3 & 2 & : & -4 \end{bmatrix}, \quad \text{replacing } R_2 \text{ by } R_2 - R_3$$

$$\sim \begin{bmatrix} 1 & 2 & -1 & : & 3 \\ 0 & -1 & 0 & : & -4 \\ 0 & 0 & 5 & : & 20 \\ 0 & 0 & 2 & : & 8 \end{bmatrix}, \quad \text{Replacing } R_3 \text{ by } R_3 - 6R_2 \text{ and } R_4 \text{ by } R_4 - 3R_2$$

$$\sim \begin{bmatrix} 1 & 2 & -1 & : & 3 \\ 0 & -1 & 0 & : & -4 \\ 0 & 0 & 1 & : & 4 \\ 0 & 0 & 1 & : & 4 \end{bmatrix} \quad \text{multiplying } R_3 \text{ by } \frac{1}{5} \text{ and } R_4 \text{ by } \frac{1}{2}$$

$$\sim \begin{bmatrix} 1 & 2 & -1 & : & 3 \\ 0 & -1 & 0 & : & -3 \\ 0 & 0 & 1 & : & 4 \\ 0 & 0 & 0 & : & 0 \end{bmatrix}, \quad \text{replacing } R_4 \text{ by } R_4 - R_3$$

Thus the matrix [A B] has been reduced to Echelon form. It is clear that rank [A B] = number of non-zero rows in the Echelon form = 3.

Also, we have

$$A = \begin{bmatrix} 1 & 2 & -1 \\ 0 & -1 & 0 \\ 0 & 0 & 1 \\ 0 & 0 & 0 \end{bmatrix}$$

Clearly, rank $A = 3$. Since rank $[A\ B] = $ rank A, hence the given equations are consistent. since ranks $A = 3 = $ the number of unknowns, therefore the given equations have unique solution. The given equation are equivalent to $x + 2y - z = 3, -y = -4, z = 4$.

∴ we have $x = -1, y = 4, z = 4$.

4. *Investigate for what values of λ, μ the simultaneous equations*

$$x + y + z = 6, x + 2y + 3z = 10, x + 2y + \lambda z = \mu$$

have (i) no solution, (ii) a unique solution, (iii) an infinite number of solutions.

Solution. The matrix form of the given system of equations is

$$AX = \begin{bmatrix} 1 & 1 & 1 \\ 1 & 2 & 3 \\ 1 & 2 & \lambda \end{bmatrix} \begin{bmatrix} x \\ y \\ z \end{bmatrix} = \begin{bmatrix} 6 \\ 10 \\ \mu \end{bmatrix} = B$$

The augmented matrix

$$[A\ B] = \begin{bmatrix} 1 & 1 & 1 & : & 6 \\ 1 & 2 & 3 & : & 10 \\ 1 & 2 & \lambda & : & \mu \end{bmatrix}$$

$$\sim \begin{bmatrix} 1 & 1 & 1 & : & 6 \\ 0 & 1 & 2 & : & 4 \\ 0 & 1 & \lambda - 1 & : & \mu - 6 \end{bmatrix},$$

replacing R_2 by $R_2 - R_1$ and R_3 by $R_3 - R_1$ respectively

$$\sim \begin{bmatrix} 1 & 1 & 1 & : & 6 \\ 0 & 1 & 2 & : & 4 \\ 0 & 0 & \lambda - 3 & : & \mu - 10 \end{bmatrix},$$ replacing R_3 by $R_3 - R_2$

Case I : If $\lambda \neq 3$, we have rank $[A\ B] = 3 = $ rank A

So in this case the given system of equations is consistent since rank $A = $ the number of unknowns, therefore the given system of equation possesses a unique solution.

Thus if $\lambda \neq 3$, the given system of equations possesses a **unique solution** for any value of μ.

Case II : If $\lambda = 3$ and $\mu \neq 10$, we have rank $[A\ B] = 3$ and rank $A = 2$. Thus, rank $[A\ B] \neq $ rank A and so the given system of equations is inconsistant, *i.e.,* possesses **no solution.**

Case III : If $\lambda = 3$ and $\mu = 10$, we have rank $[A\ B] = 2 = $ rank A

So in this case the given system of equations is again consistent. Since rank $A < $ the number of unknowns, therefore in this case the given system to equations possesses an **infinite number of solutions.**

5. *Solve completely the following system of equations.*

$$2w + 3x - y - z = 0$$
$$4w - 6x - 2y + 2z = 0$$
$$-6w + 12x + 3y - 4z = 0$$

Solution. We have

$$A = \begin{bmatrix} 2 & 3 & -1 & -1 \\ 4 & -6 & -2 & 2 \\ -6 & 12 & 3 & -4 \end{bmatrix}$$

Linear Equations

Applying $R_2 \to R_2 - 2R_1, R_3 \to R_3 + 3R_1$,

$$A \sim \begin{bmatrix} 2 & 3 & -1 & -1 \\ 0 & -12 & 0 & 4 \\ 0 & 21 & 0 & -7 \end{bmatrix} \quad \text{By } R_2 \to \left(-\frac{1}{4}\right)R_2, R_3 \to \left(\frac{1}{7}\right)R_3,$$

$$A \sim \begin{bmatrix} 2 & 3 & -1 & -1 \\ 0 & 3 & 0 & -1 \\ 0 & 3 & 0 & -1 \end{bmatrix} \quad \text{By } R_3 \to R_3 - R_2$$

$$A \sim \begin{bmatrix} 2 & 3 & -1 & -1 \\ 0 & 3 & 0 & -1 \\ 0 & 0 & 0 & 0 \end{bmatrix} \quad \text{...(i)}$$

∴ rank $(A) = 2 < n$ ($n = 4$, number of variables).

Hence the equations are consistent and have an infinite number of solutions. Such equivalent system of equation is

$$2w + 3x - y - z = 0$$
$$3x - z = 0$$

On taking $z = k_1$, $y = k_2$, the above equations give $x = \frac{1}{3}k_1$ and $w = \frac{1}{2}k_2$.

Hence the complete solution is

$$x = \frac{1}{3}k_1, \, y = k_2, \, z = k_1, \, w = \frac{1}{2}k_2.$$

EXERCISES

1. Solve the following equations by using matrix method :
   (a) $x + y + z = 29$
       $2x + 5y + 7z = 52$
       $2x + y - z = 10$
   (b) $x + 2y + z = 2$
       $2x + 4y + 3z = 3$
       $3x + 6y + 5z = 4$
   (c) $x + 2y + 3z = 4$
       $2x + 2y + 8z = 7$
       $x - y + 9z = 1$

2. Show that the following equations are consistent and solve them :
   (a) $x + y + z = 3$
       $x + 2y + 3z = 4$
       $x + 4y + 9z = 6$
   (b) $y + 2z = a$
       $x + 2y + 3z = b$
       $3x + y + z = c$

3. Find the value of λ, so that the following equations may have a solution and solve them completely in each case
   $$x + y + z = 1$$
   $$x + 2y + 4z = \lambda$$
   $$x + 4y + 10z = \lambda^2$$

4. Solve the following equations by matrix method
   $$\lambda x - 2y - 2z - 1 = 0$$
   $$4x + 2\lambda y - z - 2 = 0$$

$$6x + 6y + \lambda z - 3 = 0$$

Considering specially the case when $\lambda = 2$.

5. Investigate for what valuer of λ, μ the simultaneous equations $x + 2y + z = 8$; $2x + y + 3z = 13$; $3x + 4y - \lambda z = \mu$, have (i) no solution, (ii) a unique solution; infinitely many solutions.

6. Find the solution of the system of equations
$$x_1 + x_2 + x_3 + x_4 = 1$$
$$2x_1 - x_2 + x_3 - 2x_4 = 2$$
$$3x_1 + 2x_2 - x_3 - x_4 = 3$$

7. Show that the following system of equations has only a trivial solution :
$$x + 2y + 3z = 0$$
$$3x + 4y + 4z = 0$$
$$7x - 10y + 12z = 0$$

ANSWERS

1. (a) $x = 1, y = 3, z = 5$ (b) $x = 1, y = 1, z = -1$;
   (c) $x = 2 - 7k, y = 1 + 2k, z = k$ where k arbitrary

2. (a) $x = 2, y = 1, z = 0$;
   (b) $x = \frac{1}{2}(a - b + c), y = -4a + 3b - c, z = \frac{1}{2}(5a - 3b + c)$

3. $x = 2z + 1, y = -3z$

4. $y = \frac{1}{2} - k, y = k, z = 0$ where k is arbitrary.

5. (i) $\lambda = -\frac{11}{3}, \mu \neq 22$, (ii) $\lambda \neq -\frac{11}{3}; \mu \neq 22$
   (iii) $3\lambda + 11 = 0, \mu = 22$

6. (a) $x_1 = -(11 + 9k)/11, x_2 = -12k/11, x_3 = -8k/11, x_4 = k$, where k is arbitrary.

8
Characteristic Roots and Vectors

8.1 CHARACTERISTIC VALUE PROBLEM

Let A be a square matrix, such that $A = [a_{ij}]_{n \times n}$. Corresponding to matrix A to determine the scalars and non-zero vectors $X = [x_1, x_2, ..., x_n]$ which satisfy the equation $AX = \lambda X$ is known as the **Characteritic value problem.** Now,

$$AX = \lambda X = (\lambda I) X$$

or
$$(A - \lambda I) X = 0$$

where I is a unit matrix or order n.

In matrix form, we have the characteristic value problem as

$$\begin{bmatrix} a_{11} & a_{12} & ... & a_{1n} \\ a_{21} & a_{22} & ... & a_{2n} \\ ... & ... & ... & ... \\ ... & ... & ... & ... \\ a_{n1} & a_{n2} & ... & a_{nn} \end{bmatrix} \begin{bmatrix} x_1 \\ x_2 \\ ... \\ ... \\ x_n \end{bmatrix} = \begin{bmatrix} \lambda & 0 & - & - & 0 \\ 0 & \lambda & - & - & 0 \\ - & - & - & - & - \\ - & - & - & - & - \\ 0 & 0 & - & - & \lambda \end{bmatrix}$$

On subtracting the R.H.S. from both sides and simplifying the above equation we get

$$\begin{bmatrix} a_{11} - \lambda & a_{12} & ... & a_{1n} \\ a_{21} & a_{22} - \lambda & ... & a_{2n} \\ ... & ... & ... & ... \\ a_{n1} & a_{n2} & ... & a_{nn} - \lambda \end{bmatrix} \begin{bmatrix} x_1 \\ x_2 \\ ... \\ ... \\ x_n \end{bmatrix} = \begin{bmatrix} 0 \\ 0 \\ ... \\ ... \\ 1 \end{bmatrix}$$

This equation represents a system of n homogeneous linear equations

$$(a_{11} - \lambda) x_1 + a_{12} x_2 + ... + a_{1n} x_n = 0$$
$$a_{21} x_1 + (a_{22} - \lambda) x_2 + ... + a_{2n} x_n = 0$$
$$... \quad ... \quad ... \quad ... \quad ... \quad ... \quad ...$$
$$a_{n1} x_1 + a_{n2} x_2 + ... + (a_{nn} - \lambda) x_n = 0$$

The coefficient matrix of the above system is $(A - \lambda I)$. Since X is a non zero vector, i.e., all x's are not zero and it must have a non-trivial solution. The condition for this is the rank of coefficient matrix $(A - \lambda I)$ is less than number of variables n.

But this will be possible if and only if the coefficient matrix is singular, i.e.,

$$|A - \lambda I| = 0.$$

Characteristic Matrix : If $A = [a_{ij}]$ is a square matrix of order n then the matrix $(A - \lambda I)$ is called the **characteristic matrix** of A.

Characteritic polynomial : The determinant of a characteristic matrix $(A - \lambda I)$ of a square matrix A is called the **characteritic polynomial** of A and is generally denoted by $\phi(\lambda)$.

Characteristic Equtaion : The equation $|(A - \lambda I)| = 0$ is called the **characteristic equation** or **secular equation** of the square matrix A.

Characteristic roots : The roots of the characteristic equation $|A - \lambda I| = 0$ of a square matrix A are called **characteristic roots or latent roots or proper roots or eigen values** of a square matrix A. The set of eigen values of A is called the **spectrum** of A.

If the matrix is of order n, then the degree of the characteristic equation is n and consequently there exist n roots, (not necessarily distinct) of the matrix A.

Characteristic vectors : Corresponding to a characteristic root λ of a square matrix A, if there exists a non-zero vector X such that $(A - \lambda I) X = 0$, then X is called the **characteritic vector or latent vector or proper vector or eigen vector or invariant vector** corresponding to the characteristic root λ.

8.2 SOME FUNDAMENTAL THEOREMS

Theorem 1. *Corresponding to a characteristic vector X of a square matrix A, there exists one and only one characteristic root where as corresponding to a characteristic root there exists more than one characteristic vectors.*

Proof. Let us assume that there exist two distinct characteristic roots λ_1 and λ_2 corresponding to a given characteristic vector X of a square matrix A. Then, we have

$$A X = \lambda_1 X, \; A X = \lambda_2 X$$

On subtracting, we get

$$(\lambda_1 - \lambda_2) X = 0$$

as $\quad \lambda_1 - \lambda_2 \neq 0$, hence $X = 0$

This is a contradiction that X is a non-zero vector. Hence corresponding to a characteristic vector X there is only one characteristic root of the square matrix A.

Again, if λ be the characteristic root of A, then corresponding characteristic vector X will be given by

$$A X = \lambda X$$

Let k be any non-zero scalar, then

$$k(A X) = k(\lambda X)$$

i.e., $\quad A(k X) = \lambda (k X).$

Thus, $k X$ is also a characteristic vector A corresponding to the same characteristic root λ.

Theorem 2. *The product of the characteristic roots of a square matrix of order **n** is equal to the determinant of the matrix.*

Proof. Let $A = [a_{ij}]$ be a given square matrix. Let $\lambda_1, \lambda_2, ..., \lambda_n$ be the characeristic roots of A. If $\phi(\lambda)$ is the characteristic function, then

$$\phi(\lambda) = |A - \lambda I|$$

$$= \begin{vmatrix} a_{11} - \lambda & a_{12} & ... & a_{1n} \\ a_{21} & a_{22} - \lambda & ... & a_{2n} \\ ... & ... & ... & ... \\ a_{n1} & a_{n2} & ... & a_{nn} - \lambda \end{vmatrix}$$

$$= (-1)^n [\lambda^n + p_1 \lambda^{n-1} + p_2 \lambda^{n-2} + ... + p_n] \qquad ...(i)$$

$$= (-1)^n (\lambda - \lambda_1)(\lambda - \lambda_2)...(\lambda - \lambda_n) \qquad ...(ii)$$

On putting $\lambda = 0$, we have

Characteristic Roots and Vectors

$$\phi(0) = |A| = \lambda_1 \cdot \lambda_2 \cdot \lambda_3 \ldots \lambda_n = (-1)^n p_n \qquad \ldots(iii)$$

Hence the product characteristic roots of a square matrix is equal to the determinant of the matrix.

Theorem 3. *For a square matrix A, λ is a characteristic root, if and only if there exist a non-zero vector X such that $AX = \lambda X$.*

or

The equation $AX = \lambda X$ has a non-trivial solution X if λ is latent root of A.

or

The scalar λ is a charactersitic root of the matrix A if and only if the matrix $(A - \lambda I)$ is singular.

Proof. Let λ be a characteristic root of the square matrix A. Then by definition λ must satisfy the characteristic equation of A,

i.e., $\qquad |A - \lambda I| = 0$

This implies that the matrix $A - \lambda I$ must be singular. Hence, if $(A - \lambda I)$ is singular, then λ is a characteristic root of a matrix. Conversely, if $|A - \lambda I| = 0$ then for some non-zero vector X, we have

$$(A - \lambda I) X = 0$$

or, $\qquad AX = \lambda X$

which show that λ is a characteristic root of the square matrix A.

Theorem 4. *The characteristic roots of a Hermitian matrix are all real*

Proof. Let λ be the characteristic roots of a Hermitian matrix A. Then there exists a non-zero characteristic vector X such that

$$AX = \lambda X \qquad \ldots(i)$$

By pre-multiplying both sides of (i) by X^*, we get

$$X^*(AX) = X^*(\lambda X) \qquad \ldots(ii)$$

By taking transpose-conjugate of (ii), we get

$$[X^*(AX)]^* = (\lambda X^* X)^*$$

or $\qquad X^* A^* (X^*)^* = X^* (X^*)^* \lambda^* \qquad$ (by reversal law)

or $\qquad X^* A^* X = X^* X \bar{\lambda} \qquad$ (since $\lambda^* = \bar{\lambda}$)

But A is a hermitian matrix, hence

$$A^* = A$$

Hence the above equation becomes

$$X^* A X = \bar{\lambda} X^* X \qquad \ldots(iii)$$

From (ii) and (iii), we have

$$X^*(\lambda X) = \bar{\lambda} X^* X$$

or $\qquad (\lambda - \bar{\lambda}) X^* X = 0 \qquad \ldots(iv)$

Since X is a non-zero characteristic vector, $X^* X \neq 0$. Hence from (iv), we have

$$\lambda - \bar{\lambda} = 0 \qquad \text{or} \qquad \lambda = \bar{\lambda}$$

Let $\lambda = \alpha + i\beta$,

$$\alpha + i\beta = \alpha - i\beta \quad \text{or} \quad \beta = 0$$

Hence the charactersitic roots of a Hermitian matrix are all real.

Theroem 5. *The characteristic roots of a real symmetric matrix are all real.*

Proof. Let A be a real symmetric matrix,

So that $$A' = A \text{ and } \bar{A} = A$$

Hence $A^* = (\bar{A})$, i.e., A is a Hermitian matrix. Hence by Theorem 4 above, the latent roots of A are all real.

Theorem 6. *The characteristic roots of a skew-Hermitian matrix are either all zero or purely imaginary.*

Proof. Let λ be a latent root relative to the non-zero characteristic vector X of A, then
$$AX = \lambda X \qquad \ldots(i)$$
Since A is skew-Hermitian matrix, hence
$$A^* = -A \qquad \ldots(ii)$$
On multiplying both sides of (i) by i, we get
$$i(AX) = i(\lambda X)$$
i.e., $$(iA)X = i\lambda(X) \qquad \ldots(iii)$$
The relation (iii) shows that $i\lambda$ is a latent root of iA.

Also, $$(iA)^* = \bar{i}A^* = (-i)A^* = -i(-A)$$
$\therefore$ $$(iA)^* = iA$$

This shows that iA is a Hermitian matrix. Hence its latent root $i\lambda$ must be real. This implies that either $\lambda = 0$ or λ is purely imaginary number.

Corollary. *The characteristic roots of a real skew-symmetric matrix are either all zero or purely imaginary.*

Proof. Let A be a real skew-symmetric matrix. Hence
$$A' = -A \text{ and } \bar{A} = A$$
Therefore, we have
$$A^* = (\bar{A})' = -A$$
Hence A is skew-Hermitian.

Hence by Theorem 6 above, the latent roots of a real skew-symmetric matrix are either all zero or purely imaginary.

Theorem 7. *The characteristic roots of an orthogonal matrix are of unit modulus.*

Proof. Let A be an orthogonal matrix. Hence
$$A'A = AA' = I$$
Also, let λ be a latent root of the matrix A and X its corresponding latent vector, we have
$$AX = \lambda X \qquad \ldots(i)$$
On taking transpose of both sides of (i), we get
$$(AX)' = (\lambda X)' \qquad \ldots(ii)$$
On multiplying (i) and (ii), we get
$$(AX)'(AX) = (\lambda X)'(\lambda X)$$
or $$(X'A')(AX) = \lambda^2 X'X$$
or $$X'(A'A)X = \lambda^2 X'X$$
or $$X'X = \lambda^2 X'X$$
or $$(1-\lambda^2)X'X = 0$$
or $$(1-\lambda^2)X'X = 0$$
Since X is a latent vector, $X \neq 0$ and consequently $XX' \neq 0$

Characteristic Roots and Vectors

Hence, we get
$$1 - \lambda^2 = 0 \quad \text{or} \quad \lambda = \pm 1$$
$$\therefore \quad |\lambda| = 1$$

Theorem 8. *The characteristic roots of a unitary matrix are of unit modulus.*

Proof. Let A be a unitary matrix. Hence
$$A^* A = I \qquad \ldots(i)$$

Let λ be a charactersitic root of the matrix A and X its latent vector, hence
$$AX = \lambda X \qquad \ldots(ii)$$

On taking transposed conjugate of (ii) we get
$$(AX)^* = (\lambda X)^*$$
or
$$X^* A^* = \bar{\lambda} X^* \qquad \ldots(iii)$$

On multiplying (ii) and (iii), we get
$$(X^* A^*)(AX) = (\bar{\lambda} X^*)(\lambda X)$$
or
$$X^* (A^* A) \lambda = (\lambda \bar{\lambda})(X^* X)$$
or
$$X^* X = (\lambda \bar{\lambda})(X^* X)$$
or
$$(1 - \lambda \bar{\lambda}) X^* X = 0 \qquad \ldots(iv)$$

Since X is a latent vector, hence $X \neq 0$ and consequently, $X^* X \neq 0$.
Hence from (iv), we get
$$1 - \lambda \bar{\lambda} = 0 \quad \text{or} \quad \lambda \bar{\lambda} = 1$$
$$\therefore \quad |\lambda|^2 = 1 \quad \text{or} \quad |\lambda| = 1$$

Hence the latent roots of a unitary matrix are of unit modulus.

EXAMPLES

1. *Find the characteristic roots and associated characteristic vectors for the matrix*
$$A = \begin{bmatrix} 8 & -6 & 2 \\ -6 & 7 & -4 \\ 2 & -4 & 3 \end{bmatrix}$$

Solution. We know that the characteristic equation is $|A - \lambda I| = 0$, i.e.,
$$\begin{vmatrix} 8-\lambda & -6 & 2 \\ -6 & 7-\lambda & -4 \\ 2 & -4 & 3-\lambda \end{vmatrix} = 0$$

or $\quad \{(8-\lambda)(7-\lambda)(3-\lambda) - 16\} + 6\{(3-\lambda)(-6) + 8\} + 2\{24 - 2(7-\lambda)\} = 0$

or $\quad -\lambda^3 + 8\lambda^2 - 45\lambda = 0$

or $\quad \lambda(\lambda^2 - 3)(\lambda - 15) = 0$

$\therefore \quad \lambda = 0, 3, 15.$

Hence the characteristic roots are $\lambda_1 = 0, \lambda_2 = 3, \lambda_3 = 15$. The characteristic vector associated with is $\lambda_1 = 0$ is given by

$$\begin{bmatrix} 8 & -6 & 2 \\ -6 & 7 & -4 \\ 2 & -4 & 3 \end{bmatrix} \begin{bmatrix} x_1 \\ x_2 \\ x_3 \end{bmatrix} = \begin{bmatrix} 0 \\ 0 \\ 0 \end{bmatrix}$$

This gives
$$8x_1 - 6x_2 + 2x_3 = 0$$
$$-6x_1 + 7x_2 - 4x_3 = 0$$
$$2x_1 - 4x_2 + 3x_3 = 0$$
On solving these equations, we get
$$\frac{x_1}{1} = \frac{x_2}{2} = \frac{x_3}{2} = k_1 \quad \text{(say)}$$
Hence the required characteristic vector corresponding to the characteristic root $\lambda_1 = 0$, is
$$X = \begin{bmatrix} x_1 \\ x_2 \\ x_3 \end{bmatrix} = \begin{bmatrix} k_1 \\ 2k_1 \\ 2k_1 \end{bmatrix}$$
The characteristic vector corresponding to the root $\lambda_2 = 3$ is given by
$$\begin{bmatrix} 8-3 & -6 & 2 \\ -6 & 7-3 & -4 \\ 2 & -4 & 3-3 \end{bmatrix} \begin{bmatrix} x_1 \\ x_2 \\ x_3 \end{bmatrix} = \begin{bmatrix} 0 \\ 0 \\ 0 \end{bmatrix}$$

or
$$\begin{bmatrix} 5 & -6 & 2 \\ -6 & 4 & -4 \\ 2 & -4 & 0 \end{bmatrix} \begin{bmatrix} x_1 \\ x_2 \\ x_3 \end{bmatrix} = \begin{bmatrix} 0 \\ 0 \\ 0 \end{bmatrix}$$

This gives
$$5x_1 - 6x_2 + 2x_3 = 0$$
$$6x_1 + 4x_2 - 4x_3 = 0$$
$$2x_1 - 4x_2 = 0$$
On solving these equation, we get
$$\frac{x_1}{2} = \frac{x_2}{1} = \frac{x_3}{-2} = k_2 \quad \text{(say) } k_2 \neq 0$$

Thus $x = \begin{bmatrix} x_1 \\ x_2 \\ x_3 \end{bmatrix} = \begin{bmatrix} 2k_2 \\ k_2 \\ -2k_2 \end{bmatrix}$ is the required characteristic vector for $\lambda = 3$.

Similarly, for $\lambda = 15$, the characteristic vector will be
$$\begin{bmatrix} 8-15 & -6 & 2 \\ -6 & 7-15 & -4 \\ 2 & -4 & 3-15 \end{bmatrix} \begin{bmatrix} x_1 \\ x_2 \\ x_3 \end{bmatrix} = \begin{bmatrix} 0 \\ 0 \\ 0 \end{bmatrix}$$

or
$$\begin{bmatrix} -7 & -6 & 2 \\ -6 & -8 & -4 \\ 2 & -4 & -12 \end{bmatrix} \begin{bmatrix} x_1 \\ x_2 \\ x_3 \end{bmatrix} = \begin{bmatrix} 0 \\ 0 \\ 0 \end{bmatrix}$$

which give
$$7x_1 + 6x_2 - 2x_3 = 0$$
$$3x_1 + 4x_2 + 2x_3 = 0$$
$$x_1 - 2x_2 - 6x_3 = 0$$
On solving these, we get

Characteristic Roots and Vectors

$$\frac{x_1}{2} = \frac{x_2}{-2} = \frac{x_3}{1} = k_3 \text{ (say)}, k_3 \neq 0$$

Hence, latet vector corresponding to the latent root, $\lambda_3 = 15$ will be

$$X = \begin{bmatrix} x_1 \\ x_2 \\ x_3 \end{bmatrix} = \begin{bmatrix} 2k_3 \\ -2k_3 \\ k_3 \end{bmatrix}$$

2. *If A be a square matrix, show that the latent roots of the matrix A are the same as those of its transpose A'.*

Solution. The characteristic equation of the square matrix A is given by

$$|A - \lambda I| = 0$$

Similarly the characteristic equation of the matrix A' is $(A' - \lambda I) = 0$

Now, we have to prove that the characteristic roots of $|A - \lambda I| = 0$ and $(A' - \lambda I) = 0$ are identical.

Since interchange of row and column does not change the value of the determinant, hence we have

$$|A - \lambda I| = |A' - \lambda I|$$

Hence the roots of the equations $|A - \lambda I| = 0$ and $|A' - \lambda I| = 0$ are same.

EXERCISES (A)

1. Find the characteristic roots and vectors of the matrix.

$$A = \begin{bmatrix} -2 & -1 \\ 5 & 4 \end{bmatrix},$$

2. If $A = \begin{bmatrix} 2 & -1 & 1 \\ -1 & 2 & -1 \\ 1 & -1 & 2 \end{bmatrix}$ find the characteristics roots and the associated characteristic vectors for matrix A.

3. Determine the characteristic roots and the corresponding characteristic vectors of the matrix

$$A = \begin{bmatrix} 6 & -2 & 2 \\ -2 & 3 & -1 \\ 2 & -1 & 3 \end{bmatrix}$$

4. Find the eigenvalues and corresponding eigen-vectors of the matrix

$$A = \begin{bmatrix} 3 & 1 & 4 \\ 0 & 2 & 6 \\ 0 & 0 & 5 \end{bmatrix}$$

5. Find the latent roots and latent vectors for the matrix

$$A = \begin{bmatrix} 1 & 2 & 0 \\ 2 & -1 & 0 \\ 0 & 0 & 1 \end{bmatrix}$$

ANSWERS

1. $-1, 3$; corresponding to $\lambda = -1$, $X_1 = \begin{bmatrix} 1 \\ -1 \end{bmatrix}$ and for $\lambda = 3$, $X_2 = \begin{bmatrix} 1 \\ -5 \end{bmatrix}$.

2. $\lambda = 1, 1, 4$; for $\lambda = 1$, $X_1 = \begin{bmatrix} 0 \\ 0 \\ 0 \end{bmatrix}$, for $\lambda = 4$, $X_2 = \begin{bmatrix} 2k_2 \\ -k_4 \\ k_4 \end{bmatrix}$

3. λ 2, 2, 8; for $\lambda = 2$, $X_1 = \begin{bmatrix} k_2 + k_3 \\ 2k_2 \\ -2k_3 \end{bmatrix}$ and for $\lambda = 8$, $X_2 = \begin{bmatrix} 2k_4 \\ -k_4 \\ k_4 \end{bmatrix}$

4. $\lambda = 5, 2, 3$; for $\lambda = 5$, $X = \begin{bmatrix} 3 \\ 2 \\ 1 \end{bmatrix}$ and for $\lambda = 2, 3$ these can not be calculated.

5. $\lambda = 1, \sqrt{5}, -\sqrt{5}$; for $\lambda = 1$, $X_1 = \begin{bmatrix} 0 \\ 0 \\ 0 \end{bmatrix}$, for $\lambda = \sqrt{5}$, $X_2 = \begin{bmatrix} 2k \\ (1-\sqrt{5})k \\ 0 \end{bmatrix}$ and for

$\lambda = -\sqrt{5}$ $X_3 = \begin{bmatrix} 2k \\ (1+\sqrt{5})k \\ 0 \end{bmatrix}$.

8.3 CAYLEY-HAMILTON THEOREM

Statement. *A square matrix A satisfies its own characteristic equation.*
The characteristic equation of matrix A is given by

$$|A - \lambda I| = \phi(\lambda) = \begin{vmatrix} a_{11} - \lambda & a_{12} & a_{1n} \\ a_{21} & a_{22} - \lambda & a_{2n} \\ a_{n1} & a_{n2} & a_{nn} - \lambda \end{vmatrix}$$ ...(i)

$$= (-1)^n \{\lambda^n + p_1 \lambda^{n-1} + p_2 \lambda^{n-2} + ... + p_n\} = 0$$

or $\quad \lambda^n + p_1 \lambda^{n-1} + p_2 \lambda^{n-2} + ... + p_n = 0$ ...(ii)

then the theorem states that the matrix A satisfies the equation (ii), *i.e.*,

$$A^n + p_1 A^{n-1} + p_2 A^{n-2} + ... + p_n I = 0$$

or $\quad \phi(A) = 0$

In other words if λ be replaced by matrix A in the characteristic equation then it becomes a zero matrix.

Proof. The elements of matrix $A - \lambda I$ are of at the most first degree in λ and hence when we replace them by their co-factors then they will be at most of $(n-1)$ degree in λ. Hence adj $(A - \lambda I)$ may be expressed as a matirx polynomial in λ given by

$$\text{adj } (A - \lambda I) = B_0 \lambda^{n-1} + B_1 \lambda^{n-2} + ... + B_{n-1}$$

where $B_0, B_1, ..., B_{n-1}$ are all $(n-1) \times (n-1)$ matrices whose elements are functions of a_{ij}^s of the elements of matrix A.

Now, we have

$$A \cdot (\text{adj } A) = |A| I$$

∴ $\quad (A - \lambda I) \text{ adj } (A - \lambda I) = |A - \lambda I| I$

Characteristic Roots and Vectors

or
$$(A - \lambda I)(B_0 \lambda^{n-1} + B_1 \lambda^{n-2} + ... + B_{n-1})$$
$$= (-1)^n [\lambda^n + p_1 \lambda^{n-1} + ... + p_n] I$$

Comparing the coefficients of like powers of λ in both sides, we get

$$-B_0 = (-1)^n I$$
$$AB_0 - B_1 = (-1)^n p_1 I$$
$$AB_1 - B_2 = (-1)^n p_2 I$$
$$... ... ...$$
$$... ... ...$$
$$AB_{n-1} = (-1)^n p_n I$$

Pre-multiplying the above equations by $A^n, A^{n-1}, ... I$ respectively and adding, we get

$$0 = (-1)^n [A^n + p_1 A^{n-1} + p_2 A^{n-2} + ... + p_n I]$$

or
$$A^n + p_1 A^{n-1} + p_2 A^{n-2} + ... + p_n I = 0$$

8.3.1 Computation of the Inverse by Cayley-Hamilton Theorem

Let A be a non-singular square matrix of order n. Hence by Cayley Hamilton theorem, we have

$$p_0 A^n + p_1 A^{n-1} + p_2 A^{n-2} + ... + p_{n-1} A + p_n I = 0 \qquad ...(i)$$

On multiplying by A^{-1}, we get

$$p_0 A^{n-1} + p_1 A^{n-2} + p_2 A^{n-3} + ... + p_{n-1} I + p_n A^{-1} = O \cdot A^{-1}$$

or
$$p_0 A^{n-1} + p_1 A^{n-2} + p_2 A^{n-3} + ... + p_{n-1} I + p_n A^{-1} = 0$$

or
$$p_n A^{-1} = -(p_0 A^{n-1} + p_1 A^{n-2} + ... + p_{n-1} I)$$

$$\therefore \quad A^{-1} = -\frac{1}{p_n} [p_0 A^{n-1} + p_1 A^{n-2} + ... + p_{n-1} I]$$

Thus, inverse of a matrix A can be evaluated by putting the values of $A^{n-1}, A^{n-2}, ...$ etc.

EXAMPLES

1. *Verify Cayley-Hamilton's theorem for the matrix.*

$$A = \begin{bmatrix} 0 & 0 & 1 \\ 3 & 1 & 0 \\ -2 & 1 & 4 \end{bmatrix}$$

Solution. We have

$$|A - \lambda I| = \begin{vmatrix} 0-\lambda & 0 & 1 \\ 3 & 1-\lambda & 0 \\ -2 & 1 & 4-\lambda \end{vmatrix}$$

$$= 5 - 6\lambda + 5\lambda^2 - \lambda^3$$

Hence, the characteristic equation of A is

$$\lambda^3 - 5\lambda^2 + 6\lambda - 5 = 0 \qquad ...(i)$$

Now, $A^2 = \begin{bmatrix} 0 & 0 & 1 \\ 3 & 1 & 0 \\ -2 & 1 & 4 \end{bmatrix} \begin{bmatrix} 0 & 0 & 1 \\ 3 & 1 & 0 \\ -2 & 1 & 4 \end{bmatrix} = \begin{bmatrix} -2 & 1 & 4 \\ 3 & 1 & 3 \\ -5 & 5 & 14 \end{bmatrix}$

and $A^3 = A^2 \cdot A = \begin{bmatrix} -2 & 1 & 4 \\ 3 & 1 & 3 \\ -5 & 5 & 14 \end{bmatrix} \begin{bmatrix} 0 & 0 & 1 \\ 3 & 1 & 0 \\ -2 & 1 & 4 \end{bmatrix} = \begin{bmatrix} -5 & 5 & 14 \\ -3 & 4 & 15 \\ -13 & 19 & 51 \end{bmatrix}$

$\therefore \quad A^3 - 5A^2 + 6A - 5I = \begin{bmatrix} -5 & 5 & 14 \\ -3 & 4 & 15 \\ -13 & 19 & 51 \end{bmatrix}$

$-5 \begin{bmatrix} -2 & 1 & 4 \\ 3 & 1 & 3 \\ -5 & 5 & 14 \end{bmatrix} + 6 \begin{bmatrix} 0 & 0 & 1 \\ 3 & 1 & 0 \\ -2 & 1 & 4 \end{bmatrix} - 5 \begin{bmatrix} 1 & 0 & 0 \\ 0 & 1 & 0 \\ 0 & 0 & 1 \end{bmatrix}$

$= \begin{bmatrix} 0 & 0 & 0 \\ 0 & 0 & 0 \\ 0 & 0 & 0 \end{bmatrix} = O$

Where O is the null matrix.
Hence the matrix A satisfies its own characteristic equation.

2. *Verify that the matrix* $A = \begin{bmatrix} 2 & -1 & 1 \\ -1 & 2 & -1 \\ 1 & -1 & 2 \end{bmatrix}$ *satisfies Cayley-Hamilton theorem and hence find* A^{-1}.

Solution. The characteristic equation of the matrix A is

$$|A - \lambda I| = 0$$

i.e., $\begin{vmatrix} 2-\lambda & -1 & 1 \\ -1 & 2-\lambda & -1 \\ 1 & -1 & 2-\lambda \end{vmatrix} = 0$

or $(2-\lambda)\{(2-\lambda)^2 - 1\} + 1\{-1(2-\lambda) + 1\} + 1 \cdot \{1 - (2-\lambda)\} = 0$

or $(6 - 11\lambda + 6\lambda^2 - \lambda^3) - 2 + \lambda + 1 + (1 - 2 + \lambda) = 0$

or $-\lambda^3 + 6\lambda^2 - 9\lambda + 4 = 0$

or $\lambda^3 - 6\lambda^2 + 9\lambda - 4 = 0$ ...(i)

To verify that the matrix A satisfies its own characteristic equation, we must establish that

$$A^3 - 6A^2 + 9A - 4I = 0$$

Now, we have

$A^2 = A \cdot A = \begin{bmatrix} 2 & -1 & 1 \\ -1 & 2 & -1 \\ 1 & -1 & 2 \end{bmatrix} \begin{bmatrix} 2 & -1 & 1 \\ -1 & 2 & -1 \\ 1 & -1 & 2 \end{bmatrix}$

Characteristic Roots and Vectors

$$= \begin{bmatrix} 6 & -5 & 5 \\ -5 & 6 & -5 \\ 5 & -5 & 6 \end{bmatrix}$$

and
$$A^3 = A^2 \cdot A = \begin{bmatrix} 6 & -5 & 5 \\ -5 & 6 & -5 \\ 5 & -5 & 6 \end{bmatrix} \begin{bmatrix} 2 & -1 & 1 \\ -1 & 2 & -1 \\ 1 & -1 & 2 \end{bmatrix}$$

$$= \begin{bmatrix} 22 & -21 & 21 \\ -21 & 22 & -21 \\ 21 & -21 & 22 \end{bmatrix}$$

Now, $A^3 - 6A^2 + 9A - 4I = \begin{bmatrix} 22 & -21 & 21 \\ -21 & 22 & -21 \\ 21 & -21 & 22 \end{bmatrix}$

$$-6\begin{bmatrix} 6 & -5 & 5 \\ -5 & 6 & -5 \\ 5 & -5 & 6 \end{bmatrix} + 9 \begin{bmatrix} 2 & -1 & 1 \\ -1 & 2 & -1 \\ 1 & -1 & 2 \end{bmatrix} - 4 \begin{bmatrix} 1 & 0 & 0 \\ 0 & 1 & 0 \\ 0 & 0 & 1 \end{bmatrix}$$

$$= \begin{bmatrix} 22-36+18-4 & -21+30-9+0 & 21-30+9+0 \\ -21+30-9+0 & 22-36+18-4 & -21+30-9+0 \\ 21-30+9+0 & -21+30-9+0 & 22-36+18-4 \end{bmatrix}$$

$$= \begin{bmatrix} 0 & 0 & 0 \\ 0 & 0 & 0 \\ 0 & 0 & 0 \end{bmatrix}$$

Hence the given matrix satisfies Cayley-Hamilton theorem.

Computation of A^{-1}.

We have $\qquad A^3 - 6A^2 + 9A - 4I = 0$

or $\qquad 4I = (A^3 - 6A^2 + 9A)$

On pre-multiplying by A^{-1}, we get

$$A^{-1}I = 1/4 \, (A^{-1}A^3 - 6A^{-1}A^2 + 9A^{-1}A)$$

$A^{-1} = 1/4 \, (A^2 - 6A + 9I)$

On putting the values of A^2 and A, we get

$$A^{-1} = \frac{1}{4}\begin{bmatrix} 6 & -5 & 5 \\ -5 & 6 & -5 \\ 5 & -5 & 6 \end{bmatrix} - \frac{1}{4}\begin{bmatrix} -12 & 6 & -6 \\ 6 & -12 & 6 \\ -6 & 6 & -12 \end{bmatrix} + \frac{1}{4}\begin{bmatrix} 9 & 0 & 0 \\ 0 & 9 & 0 \\ 0 & 0 & 9 \end{bmatrix}$$

$$= \frac{1}{4}\begin{bmatrix} 3 & 1 & -1 \\ 1 & 3 & 1 \\ -1 & 1 & 3 \end{bmatrix} = \begin{bmatrix} \frac{3}{4} & \frac{1}{4} & -\frac{1}{4} \\ \frac{1}{4} & \frac{3}{4} & \frac{1}{4} \\ -\frac{1}{4} & \frac{1}{4} & \frac{3}{4} \end{bmatrix}$$

EXERCISES

1. Show that the matrix $A = \begin{bmatrix} 1 & 2 \\ 1 & 1 \end{bmatrix}$ satisfies Cayley Hamilton Theorem.

2. (a) Show that the matrix $A = \begin{bmatrix} 2 & 3 & 1 \\ 1 & 3 & 1 \\ 1 & 2 & 2 \end{bmatrix}$ satisfies Cayley-Hamilton theorem.

   (b) Verify that the following matrix satisfies its characteristic equation :
   $$A = \begin{bmatrix} 1 & 2 & 1 \\ -1 & 0 & 3 \\ 3 & -1 & 1 \end{bmatrix}$$

3. Use Cayley-Hamilton theorem to find the inverse of the following matrix.
   $$\begin{bmatrix} 1 & 2 & 3 \\ 1 & 3 & 5 \\ 1 & 5 & 12 \end{bmatrix},$$

4. Show that $A = \begin{bmatrix} 1 & 2 & 2 \\ 2 & 1 & 2 \\ 2 & 2 & 1 \end{bmatrix}$ satisfies its characteristic equation and deduce the inverse of matrix A.

ANSWERS

3. $\dfrac{1}{3} \begin{bmatrix} 11 & -9 & 1 \\ -7 & 9 & -2 \\ 2 & -3 & 1 \end{bmatrix};$

4. $\begin{bmatrix} 5/4 & 0 & 0 \\ -3/4 & 1 & 0 \\ 1/2 & -1/4 & 1/4 \end{bmatrix}$

9
Theory of Equations

9.1 SOME DEFINITIONS

(i) **Equation** : If for some values of x, two polynomials become equal, then such a relation is called an equation and the values of x which makes these two polynomials equal are called the roots of this equation. For example $x^3 + 11x = 6x^2 + 6$ is an equation, since the two polynomials $x^3 + 11x$ and $6x^2 + 6$ become equal for $x = 1, 2$ or 3. Hence, 1, 2, 3 are the roots of the equation $x^3 + 11x = 6x^2 + 6$ or $x^3 - 6x^2 + 11x - 6 = 0$.

Thus $a_0 x^n + a_1 x^{n-1} + a_2 x^{n-2} + \ldots + a_{n-1} x + a_n = 0$ is called an equation. If this equation contains all the powers of x from n to 0 then it is called complete otherwise incomplete i.e., while $7x^4 - 9x^3 + 3x^2 - 2x + 8 = 0$ is a complete equation, $3x^5 - 2x^2 + 4 = 0$ is an incomplete equation.

An incomplete equation can be made complete by supplying the missing terms with zero coefficients. The above incomplete equation can be written in complete form as
$$3x^5 + 0 \cdot x^4 + 0 \cdot x^3 - 2x^2 + 0 \cdot x + 4 = 0$$

(ii) **Algebraic equation** : If the coefficient of various powers of the variable x (say) in an equation are algebraic then it is said to be algebraic e.g., $a_0 x^n + a_1 x^{n-1} + \ldots + a_{n-1} x + a_n = 0$.

(iii) **Numerical equation** : The equation of the type $5x^3 + 6x^2 + 7x + 9 = 0$, where the coefficient of various powers of x are numerical, is called numerical equation.

(iv) **Roots of an equation** : The value of the variable x which when substituted in the equation reduces it to an identity is called a root of the equation.

(v) **Solution of an equation** : By the solution of an equation we mean the determination of all the roots of that equation.

(vi) **Degree of an equation** : By the degree of an equation we mean the highest power of x occuring in the equation.

For example the equation $x^5 + 5x^4 + 6x^3 + 7x^2 + 8x + 9 = 0$ is an equation of 5th degree.

An equation of **first degree** is called **linear.** The equations of second, third, fourth, fifth and sixth degree are called **quadratic, cubic, biquadratic, quintic** and **sextic** respectively.

An equation of nth degree is called **quantic** and symbolically written as $f_n(x) = 0$, where $f_n(x)$ is a polynomial of degree n.

(vii) **Theory of equations** : The branch of mathematics which deals with solution of equations (algebraical and numerical of any degree) is called theory of equations.

9.2 THEOREMS CONCERNING ROOTS OF AN EQUATION

Theorem I. *If $f_n(x)$ is divisible by $(x - \alpha)$, then α shall be a root of the equation $f_n(x) = 0$.*

Conversely, *if α be a root of the equation $f_n(x) = 0$, then $f_n(x)$ shall be divisible by $(x - \alpha)$.*

If $f_n(x)$ is exactly divisible by $(x - \alpha)$ then the remainder R should be zero and from (i) we get $f_n(x) = (x - \alpha) Q$.

This shows that $f_n(x) = 0$ when $x = \alpha$ i.e., α is root of the equation $f_n(x) = 0$.

Conversely, if α is a root of the equation $f_n(x) = 0$, then we have

$$f_n(\alpha) = 0 \qquad \text{...(ii)}$$

Also from (i) we have $\qquad f_n(\alpha) = (\alpha - \alpha) Q + R \qquad$ (iii)

$\therefore$ From (ii) and (iii) we get $R = 0$ i.e., $f_n(x)$ is exactly divisible by $(x - \alpha)$.

Theorem II. *Every equation of the nth degree has n roots and no more.*

Let $\qquad f_n(x) \equiv a_0 x^n + a_1 x^{n-1} + a_2 x^{n-2} + \ldots + a_{n-1} x + a_n = 0 \qquad$ ...(i)

be given equation.

We assume the proposition that every equation has a root (real or imaginary).

Let α_1 be such a root of equation (i).

Then $f_n(x)$ is exactly divisible $(x - \alpha_1)$, see theorem I above, so that

$$f_n(x) = (x - \alpha_1) f_{n-1}(x) \qquad \text{...(ii)}$$

where $f_{n-1}(x)$ is a function of $(n-1)$ th degree in x.

Again the equation $f_{n-1}(x) = 0$ must have a root (real or imaginary) and let α_2 be such a root. Then $f_{n-1}(x)$ is exactly divisible by $(x - \alpha_2)$, so that

$$f_{n-1}(x) = (x - \alpha_2) f_{n-2}(x) \qquad \text{...(iii)}$$

where $f_{n-2}(x)$ is a function of $(n - 2)$ th degree in x.

$\therefore$ From (ii) and (iii) we have

$$f_n(x) = (x - \alpha_1)(x - \alpha_2) f_{n-2}(x) \qquad \text{...(iv)}$$

Similarly if α_3 be a root (real or imaginary) of the equation $f_{n-2}(x) = 0$, then $f_{n-2}(x)$ is exactly divisible by $(x - \alpha_3)$, so that

$f_{n-2}(x) = (x - \alpha_3) f_{n-3}(x)$, where $f_{n-3}(x)$ is a function of $(n - 3)$ th degree in x.

And from (iv), we have

$$f_n(x) = (x - \alpha_1)(x - \alpha_2)(x - \alpha_3) f_{n-3}(x)$$

Reasoning in this manner we find that

$$f_n(x) \equiv (x - \alpha_1)(x - \alpha_2)(x - \alpha_3) \ldots (x - \alpha_n) Q \qquad \text{...(v)}$$

As both $f_n(x)$ and $(x - \alpha_1)(x - \alpha_2) \ldots (x - \alpha_n)$ are of degree n in x, therefore Q must be a constant and equal to a_0, the coefficient of x^n in $f_n(x)$ as given by (i).

Theory of Equations

Now the right hand side of the identity (v) vanishes when $x = \alpha_1, \alpha_2, ..., \alpha_n$. Hence the equation $f_n(x) = 0$ has n roots, since if x has any value other than $\alpha_1, \alpha_2, ..., \alpha_n$ then no factor in the right hande side of (v) vanishes and as such $f_n(x)$ does not vanish. Thus the given equation (i) of nth degree in x viz. $f_n(x) = 0$ can not have more than n roots.

Note : It is not necessary that $\alpha_1, \alpha_2, \alpha_3, ..., \alpha_n$ are all different.

Theorem III. *In an equation with real coefficients, complex roots occur in conjugate pairs.*

Let $f_n(x) = 0$ be an eqaution with real coefficients and let $a + ib$ be complex root of this equation, where a and b are real quantities and $b \neq 0$.

Now we are to prove that $a - ib$ is also a root of the equation $f_n(x) = 0$.

Let the polynomial $f_n(x)$ be divided by $\{(x-a)^2 + b^2\}$ i.e.,

$$\{(x-a)^2 + i^2 b^2\} \text{ i.e., } \{(x-a+ib)(x-a-ib)\}$$

Let Q be the quotient and $(Rx + R')$ be the remainder, if any.

Then $f_n(x) \equiv \{(x-a)^2 + b^2\} Q + (Rx + R')$

or $\quad f_n(x) \equiv \{(x-a+ib)(x-a-ib)\} Q + (Rx + R')$ ...(i)

Putting $x = a + ib$ we find that $x - a - ib$ vanishes and also $f_n(x)$ vanishes as $a + ib$ is a root of the equation $f_n(x) = 0$.

∴ From (i), we have

$$R(a+ib) + R' = 0 \quad \text{or} \quad (Ra + R') + iRb = 0$$

Equating real and imaginary parts on both sides of this relation we have $\quad Ra + R' = 0$ ...(ii) $\quad$ and $\quad Rb = 0$ ...(iii)

∵ $b \neq 0$, so we have from (iii) $R = 0$ and ∴ from $R' = 0$

∴ from (i), we have $f_n(x) = \{(x-a+ib)(x-a-b)\} Q$

∴ $f_n(x)$ vanishes when $x = a - ib$ i.e., $a - ib$ is a root of the equation $f_n(x) = 0$.

Cor. *Every equation of odd degree has atleast one real root.*

In Theorem III above, we have proved that imaginary roots occur in conjugate pairs and the products of factors corresponding to such conjugate pairs of roots give rise to quadratic factors of the type $\{(x-a)^2 + b^2\}$ of the equation $f_n(x) = 0$, which being of odd degree must have at least one linear factor and consequently $f_n(x) = 0$ must have at least one real root.

Theorem IV. *In an equation with rational coefficients, irrational roots of the form $a \pm \sqrt{b}$ occur in pairs, b being not a perfect sqaure or zero.*

Let $f_n(x) = 0$ be an eqaution with rational coefficients and let $a + \sqrt{b}$ be a root of this equation, where a and b are rational and b is not a perfect square or zero.

We are to prove that $a - \sqrt{b}$ is also a root of $f_n(x) = 0$.

Let the polynomial $f_n(x)$ be divided by $\{(x-a)^2 - b\}$

i.e., $\quad \{(x-a)^2 - (\sqrt{b})^2\}$ i.e., $\{(x-a+\sqrt{b})(x-a-\sqrt{b})\}$

Let Q be the quotient and $(Rx + R')$ be the remainder, if any.

Then $\quad f_n(x) \equiv \{(x-a+\sqrt{b})(x-a-\sqrt{b})\} Q + (Rx+R')$ ...(i)

Putting $x = a+\sqrt{b}$ we find that $x-a-\sqrt{b}$ vanishes and also $f_n(x)$ vanishes as $a+\sqrt{b}$ is a root of the equation $f_n(x) = 0$.

∴ from (i) we have $R(a+\sqrt{b}) + R' = 0$ or $(Ra + R') + R\sqrt{b} = 0$ ...(ii)

Equating rational and irrational parts on both sides of this relation we have

$\quad\quad\quad\quad Ra + R' = 0 \quad$ and $\quad R\sqrt{b} = 0$ ...(iii)

∵ $b \neq 0$, so we have from (iii) $R = 0$ and ∴ from (ii) $R' = 0$

∴ from (i) we have $f_n(x) = \{(x-a+\sqrt{b})(x-a-\sqrt{b})\} Q$

∴ $f_n(x)$ vanishes when $x = a - \sqrt{b}$ i.e., $a - \sqrt{b}$ is a root of the equation $f_n(x) = 0$.

9.3 TO FORM AN EQUATION WITH GIVEN ROOTS

We know that if α is a root of the equation $f_n(x) = 0$, then $f_n(x)$ is exactly divisible by $(x - \alpha)$.

Now let $\alpha_1, \alpha_2, \alpha_3 ...$ be the given roots, then the required equation is $a_0(x - \alpha_1)(x - \alpha_2)(x - \alpha_3) ... = 0$, where a_0 is constant.

To obtain the equation in the usual form the factors on the left hand side of the above equation are multiplied together and the terms are arranged in descending powers of x.

9.4 DESCARTE'S RULE OF SIGNS

The numbers of the positive roots of the equation $f(x) = 0$ cannot exceed the number of changes of sign (from + to – or form – to +) in the terms occurring in $f(x)$.

To illustrate what is meant by the number of changes of sign, consider the following equations:

$$x^3 - 2x^2 + 3x - 7 = 0,$$
$$x^3 + 2x^2 + 3x - 7 = 0,$$
$$x^3 - 2x^2 + 3x + 7 = 0.$$

In the first equation the first term has the sign +, the next –, the next after that + and the last one –. Writing these signs consecutively, we have + – + – . Counting the changes of sign, we see that there are 3 changes of sign. Similarly, the second equation has only one change of sign, and the last one two changes of sign.

Consider the case of any equation taken at random. Suppose the signs of the terms are as follows :

$$+ + + - + - - - - + + - -$$

Multiply the equation by $x - \alpha$ where α is any positive number. The signs of the terms in the multiplication will be as shown in the following scheme.

$$+ + + - + - - - - + + - -$$
$$+ -$$
$$\overline{+ + + - + - - - - + + - -}$$
$$- - - + - + + + + - - + +$$
$$\overline{+ \pm \pm - + - \mp \mp \mp + \pm - \mp +} \quad ...(1)$$

In the sum, certain terms are certainly positive, being the sum of two positive terms; others are similarly certainly negative; but in the case of those terms which are obtained by adding two terms of diffeent signs (one positive and the other negative), we cannot say for certain whether the sum

Theory of Equations

is positive or negative; the sign will depend upon the numerical values of the terms. The sign has, therefore, been indicated by $\pm$ or $\mp$, showing that it may be positive or negative. Also, if the given equation contains n signs, the sum (1) will contain $n+1$ signs.

Now if we deliberately assign to the ambiguous signs $\pm$ such signs (+ or −) as will give the least number of changes of sign in (1), there cannot be less changes of sign between the first n signs in (1) than the number of changes of sign in the given equation. Moreover, there is an extra sign in (1) and that cetainly involves one more change of sign.

Thus multiplication in $f(x)$ by $x - \alpha$, where α is positive, has increased the number of changes of sign by at least one.

Suppose now that a polynomial is formed by multiplying out the factors which give negative and imaginary roots. Then, if $\alpha, \beta, \gamma, \ldots$ are the positive roots, successive multiplication by $x - \alpha, x - \beta, \ldots$ will each increase the number of changes of sign by at least one. Hence in the complete equation there will be at least as many changes of sign as it has positive roots.

Corollary : *The number of negative roots of $f(x) = 0$ cannot exceed the number of changes of sign in $f(-x)$.*

For, if
$$f(x) = (x - \alpha)(x - \beta)(x - \gamma)\ldots, \text{ then}$$
$$f(-x) = (-1)^n (x + \alpha)(x + \beta)(x + \gamma)\ldots$$

Therefore, the roots of $f(-x) = 0$ are $-\alpha, -\beta, \ldots, i.e.,$ are numerically equal to the roots of $f(x) = 0$, but of the opposite sign. consequently the number of negative roots of $f(x) = 0$ is the same as the number of positive roots of $f(-x) = 0$.

If the number of positive and negative roots of an equation of degree n is found by Descarte's rule to be not more than n', where $n' > n$, we can infer at once that at least $n - n'$ of the roots of $f(x) = 0$ are imaginary.

EXAMPLE

Solve the equation $x^4 + 2x^2 - 22x + 7 = 0$, having given that one of its roots is $2 + \sqrt{3}$.

Sol. Since irrational roots of the form $a + \sqrt{b}$ occur in pairs therefore if $2 + \sqrt{3}$ is a given root then $2 - \sqrt{3}$ will also be a root of the given equation

The product of the factors corresponding to these roots is
$$\{x - (2+\sqrt{3})\}\{x - (2-\sqrt{3})\} = \{(x-2) - \sqrt{3}\}\{(x-2) + \sqrt{3}\}$$
$$= (x-2)^2 - 3 = x^2 - 4x + 1.$$

Dividing $x^4 + 2x^3 - 16x^2 - 22x + 7$ by $x^2 - 4x + 1$ we get $x^2 + 6x + 7$ which when equated to zero gives other roots of the given equation as $\frac{1}{2}[-6 \pm \sqrt{(36-28)}] = -3 \pm \sqrt{2}$.

EXERCISES

1. Form an equation whose roots are 1, 2 and 3. [**Ans.** $x^3 - 6x^2 + 11x - 6 = 0$]

2. Find the equation of fourth degree with rational coefficients one of whose roots is $\sqrt{2} + i\sqrt{3}$.

    [**Ans.** $x^4 + 2x^2 + 25 = 0$]

3. Two roots of the equation $x^4 - 6x^3 + 8x^2 - 30x + 25 = 0$ are $\alpha + \beta i$ and $\beta + \alpha i$, where α and β are real. Solve it completely. [**Ans.** $\alpha + \beta i, \beta + \alpha i$]

9.5 RELATION BETWEEN THE ROOTS AND COEFFICIENTS

We already know that if α and β be the roots of the quadratic equation $ax^2 + bx + c = 0$ then

Sum of the roots $= \alpha + \beta = -b/a$ and

Product of the roots $= \alpha\beta = c/a$

Here we shall find out above type of relations between the roots and coefficients when the equation is of nth degree.

Let $\alpha_1, \alpha_2, \alpha_3, ..., \alpha_n$ be the n roots of equation

$$a_0 x^n + a_1 x^{n-1} + a_2 x^{n-2} + ... + a_{n-1} x + a_n = 0 \qquad ...(i)$$

Then evidently we have

$$a_0 x^n + a_1 x^{n-1} + ... + a_{n-1} x + a_n \equiv a_0 (x - \alpha_1)(x - \alpha_2)...(x - \alpha_n)$$

Dividing both sides of this identity by a_0, we get

$$x^n + \frac{a_1}{a_0} x^{n-1} + \frac{a_2}{a_0} x^{n-2} + ... + \frac{a_{n-1}}{a_0} x + \frac{a_n}{a_0} \equiv (x - \alpha_1)(x - \alpha_2)...(x - \alpha_n)$$

$$\equiv x^n - (\alpha_1 + \alpha_2 + \alpha_3 + ...) x^{n-1} + (\alpha_1\alpha_2 + \alpha_1\alpha_3 + ... + \alpha_2\alpha_3 + ...) x^{n-2}$$

$$- (\alpha_1\alpha_2\alpha_3 + \alpha_1\alpha_2\alpha_4 + ...) x^{n-3} + ... + (-1)^n \alpha_1\alpha_2 ... \alpha_n$$

i.e., $$x^n + \frac{a_1}{a_0} x^{n-1} + \frac{a_2}{a_0} x^{n-2} + ... + \frac{a_{n-1}}{a_0} x + \frac{a_n}{a_0} \equiv x^n - (\Sigma \alpha_1) x^{n-1}$$

$$+ (\Sigma \alpha_1 \alpha_2) x^{n-2} - (\Sigma \alpha_1 \alpha_2 \alpha_3) x^{n-3} + ... + (-1)^n (\alpha_1 \alpha_2 ... \alpha_n)$$

Comparing the coefficients of like powers of x on both sides we have

$$\Sigma \alpha_1 = -\frac{a_1}{a_0} = (-1) \frac{a_1}{a_0}; \quad \Sigma \alpha_1 \alpha_2 = \frac{a_2}{a_0} = (-1)^2 \frac{a_2}{a_0};$$

$$\Sigma \alpha_1 \alpha_2 \alpha_3 = -\frac{a_3}{a_0} = (-1)^3 \frac{a_3}{a_0}; \quad \Sigma \alpha_1 \alpha_2 \alpha_3 \alpha_4 = \frac{a_4}{a_0} = (-1)^4 \frac{a_4}{a_0}; ... ...$$

$$(-1)^n \alpha_1 \alpha_2 \alpha_3 ... \alpha_n = \frac{a_n}{a_0} \quad \text{or} \quad \alpha_1 \alpha_2 ... \alpha_n = (-1)^n \frac{a_n}{a_0}; \quad \because (-1)^{2n} = 1$$

Here $\Sigma \alpha_1$ stands for the sum of the roots; $\Sigma \alpha_1 \alpha_2$ stand for the sum of the product of the roots taking two at a time; etc.

Working Rule : In a complete equation, sum of the product of the roots taken r at a time

$$= (-1)^r \frac{\text{coefficient of } (r+1)\text{th term}}{\text{coefficient of 1st term}},$$

If the equation is not complete, then we should first of all make the equation complete by introducing zero as the coefficient of missing powers of x.

Deductions :

(I) If in an equation a_0 (i.e., the first coefficient) is positive and its all roots be positive then

$$\Sigma \alpha_1 = -\frac{a_1}{a_0} = \text{positive i.e., } a_1 \text{ is negative,}$$

Theory of Equations

$$\Sigma\alpha_1\alpha_2 = \frac{a_2}{a_0} = \text{positive} \; (\because \text{ all roots are negative}) \; i.e., a_2 \text{ is positive}$$

$$\Sigma\alpha_1\alpha_2\alpha_3 = -\frac{a_3}{a_0} = \text{negative } i.e., a_3 \text{ is positive and so on.}$$

i.e., all the coefficients of the equation are positive.

Cubic Equation : Let α, β, γ be the roots of the equation

$$a_0 x^3 + a_1 x^2 + a_2 x + a_3 = 0$$

Then
$$\Sigma\alpha = \alpha + \beta + \gamma = -a_1/a_0$$
$$\Sigma\alpha\beta = \alpha\beta + \alpha\gamma + \beta\gamma = a_2/a_0$$

Biquadratic Equation : Let $\alpha, \beta, \gamma, \delta$ be the roots of the equation $a_0 x^4 + a_1 x^3 + a_2 x^2 + a_3 x + a_4 = 0$

Then $\Sigma\alpha = \alpha + \beta + \gamma + \delta = -a_1/a_0$

$\Sigma\alpha\beta = \alpha\beta + \alpha\gamma + \alpha\delta + \beta\gamma + \beta\delta + \gamma\delta = a_2/a_0$ (six terms)

or $\qquad (\alpha + \beta)(\gamma + \delta) + \alpha\beta + \gamma\delta = a_2/a_0$

$\Sigma\alpha\beta\gamma = \alpha\beta\gamma + \alpha\beta\delta + \alpha\gamma\delta + \beta\gamma\delta = -a_3/a_0$ (four terms)

EXAMPLES

1. *Show that the condition, that the cubic equation* $x^3 + px^2 + qx + r = 0$, *should have two roots* α, β *connected by the relation* $\alpha\beta + 1 = 0$ *is* $1 + q + pr + r^2 = 0$.

Sol. If γ is the third root of the given cubic

then $\qquad \alpha\beta\gamma = -r$

or $\qquad (-1)\gamma = -r$ since $\alpha\beta = -1 \;\therefore\; \gamma = r.$

But γ is root of the given equation

$\therefore \qquad \gamma^3 + p\gamma^2 + q\gamma + r = 0$

Putting $\gamma = r$ we get $r^3 + pr^2 + qr + r = 0$

or $\qquad r^2 + pr + q + 1 = 0$

2. *Find the equation that the roots of the equation* $x^3 - px^2 + qx - r = 0$ *may be in arithmetical progression and hence solve* $x^3 - 12x^2 + 39x = 28.$

Sol. Let the roots of the equation be $a - d, a, a + d$

Then sum of the roots $= (a - d) + a + (a + d) = p$

or $\qquad 3a = p$ ...(i)

But a is a root of the given equation, so we have

$$a^3 - pa^2 + qa - r = 0$$

or $\qquad (1/3\, p)^2 - p(1/3\, p)^2 + q(1/3\, p) - r = 0,$ $\qquad$ from (i) $a = \frac{1}{3}p$

or $\qquad 2p^3 - 9pq + 27r = 0$ is the required condition.

Now let $a - d, a$ and $a + d$ be the roots of $x^3 - 12x^2 + 39x - 28 = 0$

∴ $\Sigma\alpha = (a-d) + a + (a+d) = 12$

or $3a = 12$

or $a = 4$ ...(ii)

$\Sigma\alpha\beta = (a-d)a + (a-d)(a+d) + a(a+d) = 39$

or $3a^2 - d^2 = 39$

or $48 - d^2 = 39$, $\because a = 4$ from (ii)

or $d^2 = 9$

or $d = \pm 3$

∴ The roots are $a-d, a, a+d$, i.e., $1, 4, 7$.

3. *Find the condition that the roots of the equations $x^3 - px^2 + qx - r = 0$ be in geometrical progression.*

Sol. Let the roots of the equation be $\dfrac{\alpha}{\rho}, \alpha, \alpha\rho$

Then product of the roots $= \dfrac{\alpha}{\rho} \cdot \alpha \cdot \alpha\rho = r$

or $\alpha^3 = r$ ...(i)

Also α is a root of the given equation, so we have

$\alpha^3 - p\alpha^2 + q\alpha - r = 0$ or $r - p\alpha^2 + q\alpha - r = 0$, from (i)

or $p\alpha^2 = q\alpha$ or $p\alpha = q$ or $p^3\alpha^3 = q^3$

or $p^2 r = q^3$ [from (i)], which is the required condition.

EXERCISES

1. Solve the equation $x^3 - 3x^2 + 4 = 0$, two of its roots being equal.   [Ans. $2, 2, -1$]
2. Find the condition which must be satisfied by the coefficients of the equation $x^3 - px^2 + qx - r = 0$, when the sum of its two roots is zero.   [Ans. $pq = r$]
3. Solve the equation $x^3 - 13x^2 + 15x + 189 = 0$ having given that one rots exceeds the other by 2.   [Ans. $9, 7, -3$]
4. The cubic $2x^3 - 9x^2 + 12x + \lambda = 0$ has two equal roots. Find λ and solve the equation completely.   [Ans. $2, 2, 1/2$ when $\lambda = -4$, $1, 1, 5/2$ when $\lambda = -5$]
5. Solve the equation $x^3 - 6x^2 + 3x + 10 = 0$, the roots being in arithmetical progression.   [Ans. $-1, 2, 5$]
6. Solve the equation $x^4 + 15x^3 + 70x^2 + 120x + 64 = 0$ whose roots are in G.P.   [Ans. $-1, -2, -4, -8$]

9.6 SYMMETRIC FUNCTION

Symmetric functions of the roots of an equation are those functions which remain unchanged when any two roots are interchanged.

Theory of Equations

Symmetric functions of the roots are usually represented by writting Σ (sigma) before one of the terms of the functions. For example $\Sigma \alpha\beta$ represents $\alpha\beta + \beta\gamma + \gamma\alpha$, where α, β, γ are the roots of any cubic equation.

Order of a symemtric function : The highest degree in which any root occurs in the function is called the order of a symmetric function of the roots of an equation. For example the order of the symmetric function $\Sigma\alpha\beta^2\gamma^4$ is 4.

Weight of the symmetric function : The sum of the degrees of all the roots occurring in any term of the function is called the weight of the symmetric function. For example the weight of the symmetric function $\Sigma\alpha\beta^2\gamma^4$ is $1 + 2 + 4 = 7$.

EXAMPLE

If α, β, γ be the roots of the cubic $x^3 + px^2 + qx + r = 0$, calculate the values of the following symmetric functions :

(a) $\Sigma\alpha^2$;   (b) $\Sigma\alpha^2\beta$;   (c) $\Sigma\alpha^3$;

(d) $\Sigma\alpha^2\beta^2$;   (e) $\Sigma\alpha^4$ *and*   (f) $\Sigma\dfrac{1}{\alpha}$.

Sol. $\because$ α, β, γ are the roots of the given cubic

$\therefore$ we have

$\Sigma\alpha = -p$ ...(i)

$\Sigma\alpha\beta = q$ ...(ii)

and $\alpha\beta\gamma = -r$ ...(iii)

(a) We know $(\Sigma\alpha)^2 = (\alpha + \beta + \gamma)^2 = \Sigma\alpha^2 + 2\Sigma\alpha\beta$

or $(-p)^2 = \Sigma\alpha^2 + 2q$, from (i) and (ii)

or $\Sigma\alpha^2 = p^2 - 2q$

(b) We know $\Sigma\alpha \cdot \Sigma\alpha\beta = (\alpha + \beta + \gamma)(\alpha\beta + \beta\gamma + \gamma\alpha)$

$= \Sigma\alpha^2\beta + 3\alpha\beta\gamma$

or $(-p)(q) = \Sigma\alpha^2\beta + 3(-r)$, from (i), (ii) and (iii)

or $\Sigma\alpha^2\beta = 3r - pq$

(c) We know $\Sigma\alpha \cdot \Sigma\alpha^2 = (\alpha + \beta + \gamma)(\alpha^2 + \beta^2 + \gamma^2)$

$= \Sigma\alpha^3 + \Sigma\alpha^2\beta$

or $(-p)(p^2 - 2q) = \Sigma\alpha^3 + (3r - pq)$,

substituting the values of $\Sigma\alpha, \Sigma\alpha^2$ and $\Sigma\alpha^2\beta$.

or $\Sigma\alpha^3 = -p(p^2 - 2q) + pq - 3r = 3pq - 3r - p^3$

(d) We know $(\Sigma\alpha\beta)^2 = (\alpha\beta + \beta\gamma + \gamma\alpha)^2$

$= \alpha^2\beta^2 + \beta^2\gamma^2 + \gamma^2\alpha^2 + 2\alpha\beta^2\gamma + 2\alpha^2\beta\gamma + 2\alpha\beta\gamma^2$

$= \Sigma\alpha^2\beta^2 + 2\alpha\beta\gamma(\alpha + \beta + \gamma)$

or $(q)^2 = \Sigma\alpha^2\beta^2 + 2(-r) \cdot (-p)$, from (i), (ii) and (iii)

or
$$\Sigma a^2\beta^2 = q^2 - 2pr$$

(e) $(\Sigma \alpha^2)^2 = (\alpha + \beta + \gamma)^2 = \Sigma \alpha^4 + 2 \Sigma \alpha^2 \beta^2$

or $(p^2 - 2q)^2 = \Sigma \alpha^4 + 2(q^2 - 2pr)$, substituting the values of $\Sigma \alpha^2$ and $\Sigma \alpha^2 \beta^2$ from parts (a), (d) above.

or
$$\Sigma \alpha^4 = (p^2 - 2q)^2 - 2(q^2 - 2pr) = p^4 + 4q^2 - 4p^2 q - 2q^2 + 4pr$$
$$= p^4 + 2q^2 + 4pr - 4p^4 q$$

(f) $\Sigma \dfrac{1}{\alpha} = \dfrac{1}{\alpha} + \dfrac{1}{\beta} + \dfrac{1}{\gamma} = \dfrac{\Sigma \alpha \beta}{\alpha \beta \gamma} = \dfrac{q}{-r}$

EXERCISES

If α, β, γ be the roots of the cubic $x^2 + px^2 + qx + r = 0$, calculate the values of the following symmetric functions:

1. $\Sigma \alpha^2 \beta \gamma$;
2. $\Sigma \alpha^3 \beta$;
3. $\Sigma \alpha^3 \beta^2$;
4. $\Sigma \alpha^3 \beta^3$;
5. $\Sigma \dfrac{1}{\alpha^2}$;
6. $\Sigma \dfrac{\beta^2 + \gamma^2}{\beta + \gamma}$;
7. $\Sigma \dfrac{\alpha^2 + \beta \gamma}{\beta + \gamma}$;
8. $\left(\dfrac{1}{\alpha^2} - \dfrac{1}{\beta^2}\right)\left(\dfrac{1}{\beta^2} - \dfrac{1}{\gamma^2}\right)\left(\dfrac{1}{\gamma^2} - \dfrac{1}{\alpha \beta}\right)$

ANSWERS

1. pr;
2. $p^2 q - 2q^2 - pr$;
3. $2p^2 r - pq^2 + rq$;
4. $q^3 - 3pqr + 3r^2$;
5. $\dfrac{q^2 - 2pr}{r^2}$;
6. $\dfrac{2p^2 q - 2q^2 - 4pr}{r - pq}$;
7. $\dfrac{p^4 - 3p^2 q + q^2 + 5pr}{r - pq}$;
8. $\dfrac{q^3 - rp^3}{r^4}$.

9.7 TRANSFORMATION OF EQUATIONS

In order to solve a given equation in an easier way, we have to transform it into another, whose roots bear a certain assigned relations with the roots of the given equation. After solving the transformed equation we can find the roots of the given equation with the help of the given relation between the roots of the two equations. Here, we shall discuss the methods of elementary transformations of equations.

9.7.1 To transform an equation into another, whose roots are equal in magnitude but opposite in sign to those of the given equation.

Let the given equation be
$$a_0 x^n + a_1 x^{n-1} + a_2 x^{n-2} + \ldots + a_{n-1} x + a_n = 0 \qquad \ldots(i)$$

Let its roots be $\alpha_1, \alpha_2, \alpha_3, \ldots, \alpha_n$. Then we can write
$$a_0 x^n + a_1 x^{n-1} + a_2 x^{n-2} + \ldots + a_{n-1} x + a_n \equiv a_0 (x - \alpha_1)(x - \alpha_2) \ldots (x - \alpha_n)$$

Substituting $-x$ for x in this identity and simplifying we get
$$a_0 x^n - a_1 x^{n-1} + a_2 x^{-1} - \ldots - (-1)^n a_n \equiv a_n (x + \alpha_1)(x + \alpha_2) \ldots (x + \alpha_n) \qquad \ldots(ii)$$

Theory of Equations

The R.H.S. of (ii) vanishes when x has any one of the values $-\alpha_1, -\alpha_2, ..., -\alpha_n$

∴ From (ii) the required equation is

$$a_0 x^n - a_1 x^{n-1} + a_2 x^{n-2} - ... - (-1)^n a_n = 0$$

Working Rule : If the equation be complete (if not, it should be made so) then this transformation is affected by changing the signs of second, fourth, sixth terms and so on.

Example : *Transform the equation $x^5 + 4x^4 - 3x^2 - 2x^2 - x + 5 = 0$ into another whose roots shall be equal in magnitude but opposite in sign to those of this equation.*

Sol. Changing x into $-x$ in the given equation, we get the required equation as

$$(-x)^5 + 4(-x)^4 - 3(-x)^3 - 2(-x)^2 - (-x) + 5 = 0$$

or $\quad x^5 - 4x^4 - 3x^3 + 2x^2 + x - 5 = 0$

9.7.2 To transform an equation into another whose roots are m times the roots of the given eqaution.

Let the given equation be

$$a_0 x^n + a_1 x^{n-2} + a_2 x^{n-2} + ... + a_{n-1} x + a_n = 0 \qquad ...(1)$$

If x be a root of the given equation and y that of the transformed equation, then $y = mx$ or $x = y/m$.

Hence the transformed equation is obtained by putting y/m for x in (i) and the required equation is

$$a_0 (y/m)^n + a_1 (y/m)^{n-1} + a_2 (y/m)^{n-2} + ... + a_{n-1}(y/m) + a_n = 0$$

multiplying each term by m^n, we get

$$a_0 y^n + a_1 m y^{n-1} + a_2 m^2 y^{n-2} + ... + a_{n-1} m^{n-1} y + a_n m^n = 0,$$

which is the required equation.

Working Rule : If the given equation be complete, (if not, it should be made so), then its transformation is affected by multiplying the successive terms beginning with second by

$$m, m^2, m^3, ...; m^{n-1}, m^n$$

Example. *Form an equation whose rotos shall be -6 times the roots of the equation $x^3 - 4x^2 + \frac{1}{2} x - \frac{1}{9} = 0$.*

Sol. Let $y = -6x$ or $x = -\frac{1}{6} y$

Substituting $x = -\frac{1}{6} y$ in the given equation, the required equation is

$$\left(-\frac{1}{6} y\right)^3 - 4\left(-\frac{1}{6} y\right)^2 + \frac{1}{2}\left(-\frac{1}{6} y\right) - \frac{1}{9} = 0$$

or $\qquad -\frac{1}{216} y^3 - \frac{1}{9} y^2 - \frac{1}{12} y - \frac{1}{9} = 0$

or $\qquad y^3 + 24 y^2 + 18 y + 24 = 0$

9.7.3. To transform an equation into another whose roots are the reciprocals of the roots of the given equation

Let the given equation be

$$a_0 x^n + a_1 x^{n-1} + a_2 x^{n-2} + \ldots + a_{n-1} x + a_n = 0 \quad \ldots(i)$$

If x be a root of the given equation and y that of the transformed equation then $y = 1/x$ or $x = 1/y$.

Hence the transformed equation is obtained by putting $1/y$ for x in (i) and the required equation is

$$a_0 (1/y)^n + a_1 (1/y)^{n-1} + a_2 (1/y)^{n-2} + \ldots + a_{n-1}(1/y) + a_n = 0$$

or $a_n y^n + a_{n-1} y^{n-1} + \ldots + a_2 y^2 + a_1 y + a_0 = 0$, multiplying each term by y^n.

Working Rule : If the given equation be complete (if not, it should be made so) then this transformation is affected by taking the coefficients in the reverse order i.e., last coefficeints to be the first, last but one to be the second and so on.

Example : *The roots of the equation $81x^3 - 18x^2 - 36x + 8 = 0$ are in H.P. Transform it into another with integral coefficients and unity for the coefficient of the first term, so that the terms of the transformed equation may be in A.P.*

Sol. The roots of the given equation are in H.P. and those of the required equation are in A.P., therefore the roots of the required equation must be reciprocal of those of the given equation.

Let y be the roots of the transformed equation and x that of the given equation, then $y = 1/x$ or $x = 1/y$.

Substituting $1/y$ for x in the given equation, we get

$$81 (1/y)^3 - 18 (1/y)^2 - 36 (1/y) + 8 = 0$$

or $$8y^3 - 36y^2 - 18y + 81 = 0$$

or $$y^3 - \frac{9}{2} y^2 - \frac{90}{4} y + \frac{81}{8} = 0, \quad \text{dividing each term by 8.}$$

Multiplying the roots of this equation by m, we get

$$y^3 - \frac{9}{2} m y^2 - \frac{9}{4} m^2 y + \frac{81}{8} m^3 = 0$$

The least value of m for which the fractions will disappear is 2.

Therefore substituting 2 for m we have the required equation as $y^3 - 9y^2 - 9y + 81 = 0$.

9.8 RECIPROCAL EQUATIONS

The equation which remains unchanged when the variable x is replaced by $1/x$ are called reciprocal equation.

For example (i) $6x^6 + 25x^5 + 31x^4 + 31x^2 + 25x + 6 = 0$

(ii) $x^5 - 5x^4 + 6x^3 - 6x^2 + 5x - 1 = 0$

In (i) the coefficient of terms from the beginning and end are equal and of same sign (such equations are known as reciprocal equations of 1st class), where as in (ii) they are equal but of opposite signs. Such eqautions are known as the reciprocal equations of 2nd class.

Theory of Equations

Note 1. If α be a root of the reciprocal equation, then $1/\alpha$ is also a root. Hence the roots of reciprocal equations occur in pairs $\alpha, 1/\alpha; \beta, 1/\beta; \gamma, 1/\gamma$ and so on.

Note 2. If the equation be of an odd degree, then one of the roots must be its reciprocal *i.e.*, one of its roots must be either -1 or $+1$. If the coefficient of terms from the beginning and end have same sign, *i.e.*, is of 1st class -1 is a root and if they have opposite signs, *i.e.*, is of 2nd class $+1$ is a root.

Note 3. If the equation be of an even degree and the coefficient of terms equidistant from the begining and the end be equal but of opposite signs, then $x^2 - 1$ will always be a factor of the given equation.

For example : $\qquad 6x^6 - 25x^5 + 31x^4 - 31x^2 + 25x - 6 = 0$

or $\qquad 6(x^6 - 1) - 25x(x^4 - 1) + 31x^2(x^2 - 1) = 0$

or $\qquad (x^2 - 1)[6(x^4 + x^2 + 1) - 25x(x^2 + 1) + 31x^2] = 0$

or $\qquad (x^2 - 1)(6x^4 - 25x^3 + 37x^2 - 25x + 6) = 0$

Here we observe that $(x^2 - 1)$ is a factor of the given equation and also that $6x^4 - 25x^3 + 37x^2 - 25x + 6 = 0$ is equation of even degree where the coefficients of the terms equidistant from the beginning and the end are equal and of the same sign. This is called the 'Standard Form' to which all reciprocal equations can be reduced.

9.8.1. To find the condition that the general equation

$$p_0 x^n + p_1 x^{n-1} + p_2 x^{n-2} + \ldots + p_{n-1} x + p_n = 0$$

may be a reciprocal equation.

The given equation is $p_0 x^n + p_1 x^{n-1} + \ldots + p_{n-1} x + p_n = 0$ ...(i)

If (i) represents a reciprocal equation, then it should remain unaltered when x is changed into $1/x$.

Changing x into $1/x$ from (i) after simplifying we have

$$p_n x^n + p_{n-1} x^{n-1} + \ldots + p_1 x + p_0 = 0 \qquad \ldots(ii)$$

Comparing the coefficients of (i) and (ii) we have

$$\frac{p_0}{p_n} = \frac{p_1}{p_{n-1}} = \frac{p_2}{p_{n-2}} = \ldots = \frac{p_r}{p_{n-r}} = \ldots = \frac{p_{n-1}}{p_1} = \frac{p_n}{p_0}$$

From these we get $p_n^2 = p_0^2$, or $p_n = \pm p_0$

If $p_n = p_0$, then $p_{n-1} = p_1, p_{n-2} = p_2$, etc. the coefficient of the terms equidistant from the beginning and the end are equal and of the same sign. If $p_n = -p_0$ then $p_{n-1} = -p_1; p_{n-2} = -p_2$ etc. *i.e.*, the coefficients of the terms equidistant from the beginning and the end are equal but of opposite signs. In this case we have $p_n = -p_{n-r}$ and therefore if the equation be of an even degree say $n = 2m$, then we have $p_m = -p_{2m-m}$ *i.e.*, $p_m = -p_m$ *i.e.*, $p_m = 0$ the middle term is missing.

Example : *Solve the reciprocal equation*
$$x^4 - 10x^3 + 25x^2 - 10x + 1 = 0$$

Sol. The given reciprocal equation is in the standard form ($\because$ the given reciprocal equation is of even degree and the coefficients of terms equidistant from the beginning and the end are equal of the same sign).

Dividing each term of the given equation by x^2.

we get $\qquad x^2 - 10x + 26 - \dfrac{10}{x} + \dfrac{1}{x^2} = 0$

or $\qquad \left(x^2 + \dfrac{1}{x^2}\right) - 10\left(x + \dfrac{1}{x}\right) + 26 = 0$ ...(i)

Let $x + \dfrac{1}{x} = y$ then we get $x^2 + \dfrac{1}{x^2} + 2 = y^2$, squaring both sides

or $\qquad x^2 + 1/x^2 = y^2 - 2$

$\therefore$ From (i) we have $(y^2 - 2) - 10(y) + 26 = 0$

or $\qquad y^2 - 10y + 24 = 0$ or $y^2 - 6y - 4y + 24 = 0$

or $\qquad (y - 6)(y - 4) = 0$ or $y = 6, 4$

If $y = 6$, then $x + (1/x) = 6$, $\because y = x + (1/x)$

or $\qquad x^2 - 6x + 1 = 0 \qquad$ or $\qquad x = \dfrac{1}{2}[6 \pm \sqrt{(36-4)}] = 3 \pm 2\sqrt{2}$

If $y = 4$, then $x + (1/x) = 4$ or $x^2 - 4x + 1 = 0$

or $\qquad x = \dfrac{1}{2}[4 \pm \sqrt{(16-4)}] = 2 \pm \sqrt{3}$.

$\therefore$ The roots of the given equation are $3 \pm 2\sqrt{2}, 2 \pm \sqrt{3}$.

EXERCISES

1. Transform the equation $x^7 - 7x^6 + 3x^4 + 4x^2 - 3x - 2 = 0$ into another whose roots shall be equal in magnitude but opposite in sign to those of this equation.

    [**Ans.** $x^7 + 7x^6 + 3x^4 - 4x^2 - 3x + 2 = 0$]

2. Transform the equation $x^4 - 3x^3 + x^2 - x + 5 = 0$ into another whose roots shall be reciprocals of the roots of this equation. [**Ans.** $5x^4 - x^3 + x^2 - 3x + 1 = 0$]

3. Reduce the following reciprocal equations to the standard form
    (a) $7x^5 - 5x^4 - 2x^3 - 2x^3 - 5x + 7 = 0$ (b) $7x^5 - 5x^4 - 2x^3 + 2x^2 + 5x - 7 = 0$

    [**Ans.** (a) $7x^4 + 12x^3 + 10x^2 - 12x + 7 = 0$ (b) $7x^4 + 2x^3 + 2x + 7 = 0$]

Theory of Equations

9.9. TO INCREASE OR DIMINISH THE ROOTS OF A GIVEN EQUATION BY AN ASSIGNED QUANTITY h (SAY).

Let the given equation be

$$f(x) \equiv a_0 x^n + a_1 x^{n-1} + a_2 x^{n-2} + \ldots + a_{n-1} x + a_n = 0 \qquad \ldots(i)$$

If y be a root of the transformed equation, then we have $y = x - h$ (provided roots are decreased by h)

or $\qquad x = y + h \qquad \ldots(ii)$

[Note. If the roots are to be increased by h, then take $y = x + h$ and proceed].

Substituting $y + h$ for x in (i) we get

$$f(y+h) = 0 \text{ i.e., } a_0(y+h)^n + a_1(y+h)^{n-1} + \ldots + a_{n-1}(y+h) + a_n = 0,$$

which can be written as

$$A_0 y^n + A_1 y^{n-1} + A_2 y^{n-2} + \ldots + A_{n-1} y + A_n = 0 \qquad \ldots(iii)$$

or $\qquad A_0 (x-h)^n + A_1 (x-h)^{n-1} + \ldots + A_{n-1}(x-h) + A_n = 0, \qquad \ldots(iv)$

since $y = x - h$.

Evidently L.H.S. of (i) and (iv) are identical, therefore if $f(x)$ be divided by $(x-h)$, the remainder is A_n and the quotient is $A_0(x-h)^{n-2} + A_1(x-h)^{n-3} + \ldots + A_{n-2}$. Proceeding in this way we can find all the coefficient $A_n / n - 1, A_{n-2}, \ldots, A_2, A_1$ and the last coefficient A_0 is evidently equal to a_0.

Working Rule : First of all we should make the equation $f(x) = 0$ complete if it is not already so. Then in order to find the coefficients of the transformed equation, we should divide the given complete equation by $(x - h)$, the resulting quotient again by $(x - h)$ and so on. The successive remainders obtained in the above procedure of division are the successive coefficients begining from the end, the first coefficient being the same as that of the given equation.

Example : *If α, β, γ are the roots of the equation $8x^3 - 4x^2 + 6x - 1 = 0$, find the equation whose roots are $\alpha + \frac{1}{2}, \beta + \frac{1}{2}, \gamma + \frac{1}{2}$.*

Sol. We are to find the equation whose roots are those of the given equation increased by $\frac{1}{2}$.

Let the transformed equation be

$$A_0 y^3 + A_1 y^2 + A_2 y + A_3 = 0$$

where $y = x + \frac{1}{2}$.

Then $A_0 =$ the first coeff. of the given equation $= 8$ and the other coefficient A_1, A_2, A_3 are calculated by synthetic division as follows.

	8	−4	6	−1
−1/2		−4	4	−5
	8	−8	10	−6 = A_3
−1/2		−4	6	
	8	−12	16 = A_2	
−1/2		−4		
	8	−16 = A_1		

Hence from (i) the required equation is

$$8y^3 - 16y^2 + 16y - 6 = 0 \text{ or } 4y^3 - 8y^2 + 8y - 3 = 0$$

EXERCISES

1. Find the equation whose roots are the roots of $x^4 - 5x^3 + 7x^2 - 17x + 11 = 0$ each diminished by 4.

   [**Ans.** $y^4 + 11y^3 + 43y^2 + 55y - 9 = 0$]

2. Diminish the root of the equation $2x^5 - x^3 + 10x - 8 = 0$ by 5.

   [**Ans.** $2x^5 + 50x^4 + 499x^3 + 2485x^2 + 6185x + 6167 = 0$]

9.10 TO REMOVE A PARTICULAR TERM FROM A GIVEN EQAUTION

Let the equation be

$$f(x) \equiv a_0 x^n + a_1 x^{n-1} + a_2 x^{n-2} + \ldots + a_{n-1} x + a_n = 0 \qquad \ldots(i)$$

Let the roots of (i) be diminished by h by putting $y = x - h$ or $x = y + h$ in (i) and the transformed equation is

$$a_0(y+h)^n + a_1(y+h)^{n-1} + \ldots + a_{n-1}(y+h) + a_n = 0$$

or
$$a_0 y^n + (na_0 h + a_1) y^{n-1} + [{}^nC_2 a_0 h^2 + (n-1) a_1 h + a_2] y^{n-2}$$
$$+ [{}^nC_3 a_0 h^3 + {}^{n-1}C_2 a_1 h^2 + (n-2) a_2 h + a_3] y^{n-3} + \ldots$$
$$\ldots + [a_0 h^3 + a_1 h^{n-1} + \ldots + a_{n-1} h + a_n] = 0 \qquad \ldots(iii)$$

Now if we wish to remove the second term of the transformed equation, then the coefficients of y^{n-1} in (ii) should be zero i.e., $n a_0 h + a_1 = 0$ or $h = -a_1 / na_0$.

∴ To remove the second term of the transformed equation we should diminish the roots of the given equation by $-a_1 / na_0$.

i.e., $-\dfrac{\text{Second coefficient}}{(\text{degree of the equation})(\text{first coeff})}$.

Theory of Equations   107

In this way if we wish to remove any term of (ii), then we equate to zero the coefficient of that particular term of (ii). The value of h obtained from it will give the quantity by which the roots of the given equation should be diminished in order to remove any particular term.

Thus if we wish to remove third term of the transformed equation by h, where h is given by $^nC_2 a_0 h^2 + (n-1) a_1 h + a_2 = 0$ (which is obtained by equating to zero the coefficient of y^{n-2} in (ii) above) and so on.

Working Rule : First of all diminish the roots of the given equation by h. Then equate to zero the coefficient of the term which is to be removed from the transformed equation, this will give the value of h in the transformed equation.

1. *Solve the equation* $x^4 + 20x^3 + 143x^2 + 430x + 462 = 0$ *by removing its second term.*

Sol. Let the roots of the given equation be diminished by h, so that $y = x - h$ or $x = y + h$.

Substituting $x = y + h$ in the given equation, we get

$$(y+h)^4 + 20(y+h)^3 + 143(y+h)^2 + 430(y+h) + 462 = 0$$

$$y^4 + (4h + 20) y^3 + ... = 0$$

As we have to remvoe the second term, so equating the coefficient of the second term in (i) to zero, we get $4h + 20 = 0$ or $h = -5$.

∴ We have to diminish the roots of the given equation by -5 *i.e.*, increase the roots by 5.

∴ Let the transformed equation be

$$A_0 y^4 + A_1 y^3 + A_2 y^2 + A_3 y + A_4 = 0. \qquad ...(iii)$$

where $A_0 = 1$ and A_1, A_2, etc. are calculated as follows :

	1	20	143	430	462
-5		-5	-75	-340	-450
	1	15	68	90	12 = A_4
-5		-5	-50	-90	
	1	10	18	0 = A_3	
-5		-5	-25		
	1	5	-7 = A_2		
-5		-5			
	1	0 = A_1			

∴ From (iii) the transformed equation is $y^4 - 7y^2 + 12 = 0$

or $\qquad (y^2 - 4)(y^2 - 3)$ or $y^2 = 3, 4$ or $y = \pm \sqrt{3}, \pm 2$.

Hence the roots of the given equation are given by $x = y + h$, where $h = -5$.

i.e., $\pm \sqrt{3} - 5, \pm 2 - 5$ *i.e.*, $\pm \sqrt{3} - 5, -3, -7$.

9.11 TRANSFORMATION IN GENERAL

Let the given equation be $f(x) = 0$.

Let y be a root of the transformed equation. Let x and y be connected by a given relation $\phi(x, y) = 0$.

Then to obtain the transformed equation, we should eliminate x between (i) and (ii).

Example : If α, β, γ are the roots of the cubic $x^3 + px^2 + px + r = 0$ form the equation whose roots are

(a) $\beta\gamma + (1/\alpha), \gamma\alpha + (1/\beta), \alpha\beta + (1/\gamma)$.

(b) $\dfrac{\alpha}{\beta+\gamma-\alpha}, \dfrac{\beta}{\gamma+\alpha-\beta}, \dfrac{\gamma}{\alpha+\beta-\gamma}$

(c) $\dfrac{\alpha^2 - \beta\gamma}{\alpha}, \dfrac{\beta^2 - \gamma\alpha}{\beta}, \dfrac{\gamma^2 - \alpha\beta}{\gamma}$

Sol. Here we get

$$\Sigma\alpha = -p, \Sigma\alpha\beta = q, \alpha\beta\gamma = -r$$

(a) Let $y = \beta\gamma + \dfrac{1}{\alpha} = \dfrac{\alpha\beta\gamma + 1}{\alpha} = \dfrac{-r+1}{\alpha}, \because \alpha\beta\gamma = -r$

or $\qquad y = (-r+1)/x$, replacing α by x.

or $\qquad x = (1-r)/y$.

Substituing this value of x in the given cubic, we have the required equation as

$$\left\{\dfrac{(1-r)}{y}\right\}^3 + p\left\{\dfrac{(1-r)}{y}\right\}^2 + p\left\{\dfrac{(1-r)}{y}\right\} + r = 0$$

or $\qquad ry^3 + q(1-r)y^2 + p(1-r^2)y + (1-r)^3 = 0$.

(b) Let $\qquad y = \dfrac{\alpha}{\beta+\gamma-\alpha} = \dfrac{\alpha}{(\alpha+\beta+\gamma) - 2\alpha}$

$\qquad\qquad\qquad = \dfrac{\alpha}{-p-2\alpha},$ $\qquad\qquad \because \alpha+\beta+\gamma = \Sigma\alpha = -p$

or $\qquad y = \dfrac{x}{-p-2x},$ $\qquad$ replacing α by x.

or $\qquad -py - 2xy = x$ or $x = -py/(1+2y)$

Substituting this value of x in the given cubic, we have the required equation as

$$\left\{\dfrac{-py}{(1+2y)}\right\}^3 + p\left\{\dfrac{-py}{(1+2y)}\right\}^2 + q\left\{\dfrac{-py}{(1+2y)}\right\} + r = 0$$

or $\qquad -p^3y^3 + p^3(1+2y)y^2 - pq(1+2y)^2 y + r(1+2y)^3 = 0$

or $\qquad (p^3 - 4pq + 8r)y^3 + (p^3 - 4py + 12r)y^2 + (6r - pq)y + r = 0$

(c) Let $\qquad y = \dfrac{\alpha^2 - \beta\gamma}{\alpha} = \alpha - \dfrac{\alpha\beta\gamma}{\alpha^2} = \alpha + \dfrac{r}{\alpha^2},$ $\qquad$ since $\alpha\beta\gamma = -r$

or $\qquad y = x + r/x^2$, replacing α by x

or $\qquad x^3 - yx^2 + r = 0$

Also the given cubic is $\qquad x^3 + px^2 + qx + c = 0$ ...(i)

Subtracting (i) from (ii) we get

$$(p+y)x^2 + qx = 0 \quad \text{or} \quad x = -\dfrac{q}{p+y}$$

Theory of Equations

Substituting this value of x in (ii), we have the required equation as

$$\left\{-\frac{q}{p+y}\right\}^3 + p\left\{-\frac{q}{p+y}\right\}^2 + q\left\{-\frac{q}{p+y}\right\} + r = 0$$

or $\qquad -q^3 + pq^2(p+y) - q^2(p+y)^2 + r(p+y)^3 = 0$

or $\qquad ry^3(3pr - q^2)y^2 + (-pq^2 + 3p^2r)y + (rp^3 - q^3) = 0$

EXERCISE

1. If α, β, γ are the roots of the equation $x^3 - px^2 + qx - r = 0$, find the equation whose roots are $(\alpha + \beta), (\beta + \gamma), (\gamma + \alpha)$. [**Ans.** $y^3 - 2py^2 + (p^2 + q)y - (pq - r) = 0.$]

2. If the roots of the equation $x^3 - 6x^2 + 11x - 6 = 0$ be α, β, γ find the equation whose roots are $\beta^2 + \gamma^2 + \alpha^2, \alpha^2 + \beta^2$. [**Ans.** $y^3 - 28y^2 + 245y - 650 = 0$]

3. If α, β, γ be the roots of the cubic $x^3 - px^2 + qx - r = 0$ then form the equation whose roots are $\beta\gamma + 1/\alpha, \gamma\alpha + 1/\beta, \alpha\beta + 1/\gamma$. [**Ans.** $ry^3 - q(1+r)y^2 + p(1+r)^2 y$]

4. If α, β, γ are the roots of the cubic $x^3 + qx + r = 0$, find the equation whose roots are :

   (a) $\beta^2\gamma^2, \gamma^2\alpha^2, \alpha^2\beta^2$

   (b) $\dfrac{\beta^2 + \gamma^2}{\alpha^2}, \dfrac{\gamma^2 + \alpha^2}{\beta^2}, \dfrac{\alpha^2 + \beta^2}{\gamma^2}$ and

   (c) $\beta^2 + \beta\gamma + \gamma^2, \gamma^2 + \gamma\alpha + \alpha^2, \alpha^2 + \alpha\beta + \beta^2$

   [**Ans.** (a) $y^3 - q^2 y^2 - 2qr^2 y - r^4 = 0$

   (b) $r^2 y^3 + (3r^2 + 2q^3)y^2 + (3r^2 - 4q^2)y - (r^2 + 2q^3) = 0$, (c) $(y+q)^3 = 0$]

9.12 CARDON'S METHOD OF SOLVING THE CUBIC EQUATION

Let the cubic equation be $a_0 x^3 + 3a_1 x^2 + 3a_2 x + a_3 = 0$ ...(i)

Removing the second term and multiplying the roots by a_0 this eqaution (i) can be reduced to the form

$$z^3 + 3Hz + G = 0, \qquad ...(ii)$$

where $z = a_0 x + a_1$, G and H have their usual meanings, i.e.,

$$G = a_0^2 a_2 - 3a_0 a_1 a_2 + 2a_1^3 \quad \text{and} \quad H = a_0 a_2 - a_1^2$$

Let us assume that $\qquad z = u^{1/3} + v^{1/3} \qquad$ ...(iii)

be a solution of (ii).

Cubing both sides of (iii), we get $z^3 = (u^{1/3} + v^{1/3})^3$

or $\qquad z^3 = u + v + 3u^{1/3}(u^{1/3} + v^{1/3}) = u + v + 3u^{1/3} v^{1/3} z$

or $\qquad z^3 - 3u^{1/3} v^{1/3} z - (u+v) = 0 \qquad$ ...(iv)

Comparing the coefficients of (ii) and (iv) we get

$$u^{1/3} v^{1/3} = -H \quad \text{or} \quad uv = -H^3 \quad \text{and} \quad u+v = -G \qquad ...(v)$$

Hence u and v are the roots of the quadratic

$$t^2 + Gt - H^3 = 0,$$

which gives $t = \frac{1}{2}[-G \pm \sqrt{(G^2 + 4H^3)}]$

$\therefore \quad u = \frac{1}{2}[G + \sqrt{(G^2 + 4H^3)}] \qquad$ and $\qquad v = [-G - \sqrt{(G^2 + 4H^3)}] \qquad$ ...(vii)

Now taking cube roots we shall get three values each for cube root of u and cube root of v viz. $u^{1/3}, \omega u^{1/3}, \omega^2 u^{1/3}$ and $v^{1/3}, \omega v^{1/3}, \omega^2 v^{1/3}$. Combining these in all possible ways we shall have nine values of the expression $u^{1/3} + v^{1/3}$ which is a root of the cubic (ii). But there should be only three roots of a cubic, consequently while combining the values of $u^{1/3}$ and $v^{1/3}$ we should make use of $u^{1/3} v^{1/3} = -H$ and thus shall have only three values of $u^{1/3} + v^{1/3}$ i.e., z.

$\therefore$ If z_1, z_2, z_3 be the three roots of the cubic (ii), then

$$z_1 = u^{1/3} + v^{1/3} = u^{1/3} - \frac{H}{u^{1/3}}, \quad \because u^{1/3} \cdot v^{1/3} = -H \quad \text{or} \quad v^{1/3} = -H/u^{1/3}$$

$$z_2 = \omega u^{1/3} - \frac{H}{\omega u^{1/3}} = \omega u^{1/3} - \frac{\omega^2 H}{\omega^3 u^{1/3}} = \omega u^{1/3} - \frac{\omega^2 H}{u^{1/3}} = \omega u^{1/3} + \omega^2 v^{1/3}, \quad \because \omega^3 = 1$$

and $z_3 = \omega^2 u^{1/3} - \dfrac{H}{\omega^2 u^{1/3}} = \omega^2 u^{1/3} - \dfrac{\omega H}{\omega^3 u^{1/3}} = \omega^2 u^{1/3} + \dfrac{\omega H}{u^{1/3}} = \omega^2 u^{1/3} + \omega v^{1/3}$

The corresponding values of x can be deduced from $z = a_0 x + a_1$.

9.12.1. Application of Cardon's method to Numerical Equations

If $G^2 + 4H^3 < 0$, all the roots of the cubic, given by (i) of last article, are real but from result (vii) of the last article we find that Cardon's solution gives them in an imaginary form as u and v are both imaginary. Also there is no arithmetical process for finding the cube roots of complex numbers and hence Cardon's Method fails to give the solution of a cubic when all its roots are real and different. This case is generally known as the **irreducible case of Cardon's Solution.** However in such a case we can use De Moivre's Theorem of Trigonometry to extract the cube roots of complex numbers.

If $G^2 + 4H^3 \geq 0$ i.e., when the cubic has two imaginary or two equal roots the values of u and v from result (vii) of last article are real. And then by some suitable arithmetical process we can find the cubic roots of real quantities i.e., we can find the values of $u^{1/3}$ and $v^{1/3}$.

9.12.3 Trigonometrical Solutions

When the roots of the given cubic are all real, we use the following method :

Let $\qquad -G = p, G^2 + 4H^3 = -q^2$, so that

$$u = \frac{1}{2}p + \frac{1}{2}qi \qquad \text{and} \qquad v = \frac{1}{2}p - \frac{1}{2}qi$$

$\therefore \qquad z = u^{1/3} + v^{1/3} = \left(\dfrac{1}{2}p + \dfrac{1}{2}qi\right)^{1/3} + \left(\dfrac{1}{2}p - \dfrac{1}{2}qi\right)^{1/3} \qquad$ ...(i)

Let $\dfrac{1}{2} p = r \cos\theta$ and $\dfrac{1}{2} q = r \sin\theta$, so that

$$r^2 = \left(\frac{1}{2}p\right)^2 + \left(\frac{1}{2}q\right)^2 = \frac{1}{4}(p^2 + q^2) = \frac{1}{4}(G^2 - G^2 - 4H^3) = -H^3$$

Theory of Equations

and
$$\tan\theta = \frac{\frac{1}{2}q}{\frac{1}{2}p} = \frac{q}{p} = -\sqrt{[-(G^2 + 4H^3)]}/G$$

∴ From (i) we get

$$z = (r\cos\theta + ir\sin\theta)^{1/3} + (r\cos\theta - ir\sin\theta)^{1/3}$$

$$= r^{1/3}\left[\left\{\cos\left(\frac{2n\pi+\theta}{3}\right) + i\sin\left(\frac{2n\pi+\theta}{3}\right)\right\} + \left\{\cos\left(\frac{2n\pi+\theta}{3}\right) - i\sin\left(\frac{2n\pi+\theta}{3}\right)\right\}\right]$$

$$= 2r^{1/3}\cos\frac{1}{3}(2n\pi+\theta), \text{ where } n = 0, 1, 2.$$

∴ The roots of the equation

$$2r^{1/3}\cos\frac{1}{3}(\theta),\ 2r^{1/3}\cos\frac{1}{3}(2\pi+\theta) \quad \text{and} \quad 2r^{1/3}\cos\frac{1}{3}(4\pi+\theta)$$

or $\quad 2r^{1/3}\cos\frac{1}{3}\theta,\ 2r^{1/3}\cos\frac{1}{3}(2\pi+\theta)$ and $2r^{1/3}\cos\frac{1}{3}(2\pi-\theta),$

since $\quad \cos\frac{1}{3}(4\pi+\theta) = \cos\left\{2\pi - \frac{1}{3}(2\pi-\theta)\right\} = \cos\frac{1}{3}(2\pi-\theta)$

or $\quad 2r^{1/3}\cos\frac{1}{3}\theta,\ 2r^{1/3}\cos\frac{1}{3}(2\pi\pm\theta)$

or $\quad 2(-H)^{1/2}\cos\frac{1}{3}\theta,\ 2(-H)^{1/2}\cos\frac{1}{3}(2\pi\pm\theta),$

since $\quad r^2 = -H^3$ or $r^{1/3} = (-H)^{1/2}$

EXAMPLES

1. *Solve* $x^3 + x^2 - 16x + 20 = 0$ *by Cardan's Method.*

Sol. Comparing the given cubic with

$$a_0 x^3 + 3a_1 x^2 + 3a_2 x + a_3 = 0$$

we find $\quad a_0 = 1,\ a_1 = \frac{1}{3},\ a_2 = -\frac{16}{3}$ and $a_3 = 20$

∴ $\quad H = a_0 a_2 - a_1^2 = 1\left(-\frac{16}{3}\right) - \left(\frac{1}{3}\right)^2 = -\frac{49}{9}$

and $G = a_0^2 a_2 - 3a_0 a_1 a_2 + 2a_1^3 = (1)(20) - 3(1)\left(\frac{1}{3}\right)\left(-\frac{16}{3}\right) + 2\left(\frac{1}{3}\right)^3 = \frac{686}{27}$

∴ For the given cubic, the standad cubic is $z^3 + 3Hz + G = 0$

or $\quad z^3 - \frac{1}{3}(49)z + \frac{1}{27}(686) = 0,$ ...(i)

where $\quad z = a_0 x + a_1 = x + \frac{1}{3}$ or $x = z - \frac{1}{3}$ ...(ii)

Let $\quad z = u^{1/3} + v^{1/3},$ ...(iii)

then the t-quadratic is $t^2 + \frac{1}{27}(686)t + \left(\frac{49}{9}\right)^3 = 0$

or $\qquad 27t^2 + 686t + \{(49)^3/27\} = 0$

or $\qquad t = \dfrac{1}{54}[-686 \pm \sqrt{\{(686)^2 - 4(49)^3\}}] = -\dfrac{343}{27},\qquad$ on simplifying.

or $\qquad t = \left(-\dfrac{7}{3}\right)^3$

$\therefore\qquad u^{1/3} = -\dfrac{7}{3}\qquad$ and $\qquad v^{1/3} = -\dfrac{7}{3}$

Hence from (iii) we get $z_1 = u^{1/3} + v^{1/3} = -\dfrac{7}{3} - \dfrac{7}{3} = -\dfrac{14}{3}$;

$\qquad z_2 = \omega u^{1/3} + \omega^2 v^{1/3} = -\dfrac{7}{3}(\omega + \omega^2) = \dfrac{7}{3} \qquad \because 1 + \omega + \omega^2 = 0$

and $\qquad z_3 = \omega^2 u^{1/3} + \omega v^{1/3} = -\dfrac{7}{3}(\omega^2 + \omega) = \dfrac{7}{3}$

$\therefore$ From (ii) we get $x = z - \dfrac{1}{3} = -\dfrac{14}{3} - \dfrac{1}{3}; \dfrac{7}{3} - \dfrac{1}{3}; \dfrac{7}{3} - \dfrac{1}{3}$

$\qquad = -5, 2, 2.$

2. *Prove that roots of $x^3 - 3x + 1 = 0$ are $2\cos(2\pi/9); 2\cos(8\pi/9)$ and $2\cos(14\pi/9).$*

Sol. Here $\qquad G = 1, H = 1$

$\therefore\quad G^2 + 4H^3 = 1^2 + 4(-1)^3 = -3$ *i.e.*, $G^2 + 4H^3 < 0$, hence all the roots of the given cubic are real.

Let $\qquad x = u^{1/3} + v^{1/3} \qquad\qquad\qquad\qquad\qquad\qquad$ ...(i)

Then the '*t*-quadratic' viz', $t^2 + Gt - H^3 = 0$ reduces to $t^2 + t + 1 = 0$

or $\qquad t = \dfrac{1}{2}[-1 \pm \sqrt{(1-4)}] = -\dfrac{1}{2} \pm i\dfrac{\sqrt{3}}{2}$

$\therefore \qquad u = -\dfrac{1}{2} + i\dfrac{\sqrt{3}}{2}; v = -\dfrac{1}{2} - \dfrac{i\sqrt{3}}{2} \qquad\qquad\qquad$ ...(ii)

Let $\qquad -\dfrac{1}{2} + i\dfrac{\sqrt{3}}{2} = r(\cos\theta + i\sin\theta) \qquad\qquad\qquad$ ...(iii)

Then $r\cos\theta = -\dfrac{1}{2}$ and $r\sin\theta = \dfrac{\sqrt{3}}{2}$, comparing real and imaginary parts.

$\therefore$ Squaring and adding we have $r^2 = \dfrac{1}{4} + \dfrac{3}{4} = 1$ or $r = 1$

and $\therefore \qquad \cos\theta = -\dfrac{1}{2} \qquad$ and $\qquad \sin\theta = \dfrac{\sqrt{3}}{2}, \qquad\qquad$ which gives $\theta = \dfrac{2}{3}\pi$

$\therefore$ from (ii) and (iii) we have $u = -\dfrac{1}{2} + i\dfrac{\sqrt{3}}{2} = \cos\dfrac{2\pi}{3} + i\sin\dfrac{2\pi}{3}$

and $\qquad v = -\dfrac{1}{2} - i\dfrac{\sqrt{3}}{2} = \cos\dfrac{2\pi}{3} - i\sin\dfrac{2\pi}{3}$

$\therefore\qquad u^{1/3} = \left(\cos\dfrac{2}{3}\pi + i\sin\dfrac{2}{3}\pi\right)^{1/3} = \left[\cos\left(2n\pi + \dfrac{2}{3}\pi\right) + i\sin\left(2n\pi + \dfrac{2}{3}\pi\right)\right]^{1/3},$

Theory of Equations

where $n = 0, 1, 2$

or
$$u^{1/3} = \cos\frac{1}{3}\left(2n\pi + \frac{2}{3}\pi\right) + i\sin\frac{1}{3}\left(2n\pi + \frac{2}{3}\pi\right),$$ where $n = 0, 1, 2$

and similarly
$$v^{1/3} = \cos\frac{1}{3}\left(2n\pi + \frac{2}{3}\pi\right) - i\sin\frac{1}{3}\left(2n\pi + \frac{2}{3}\pi\right)$$ where $n = 0, 1, 2$

∴ from (i)
$$x = u^{1/3} = 2\cos\frac{1}{3}\left(2n\pi + \frac{2}{3}\pi\right),$$ where $n = 0, 1, 2$

$$= 2\cos\frac{2}{9}\pi, 2\cos\frac{8}{9}\pi, 2\cos\frac{14}{9}\pi.$$

EXERCISES

Solve following cubic equations by carbon's method :

1. $x^3 - 6x - 9 = 0$ [Ans. $3, \frac{1}{2}(-3 \pm i\sqrt{3})$]
2. $x^3 - 15x^2 - 33x + 847 = 0$ [Ans. $-7, 11, 11$]
3. $x^3 - 15x^2 - 357x + 5491 = 0$ [Ans. $-19, 17, 17$]
4. $x^3 - 21x - 344 = 0$ [Ans. $8, -4 \pm 3i\sqrt{3}$]
5. $28x^3 - 9x^2 + 1 = 0$ [Ans. $-\frac{1}{4}, \frac{1}{7}(2 \pm i\sqrt{3})$]

9.13 TRANSFORMATION OF THE BIQUADRATIC

To reduce the biquadratic

$$f(x) \equiv a_0 x^4 + 4a_1 x^3 + 6a_2 x^2 + 4a_3 x + a_4 = 0 \qquad \ldots(1)$$

to the form
$$z^4 + 6Hz^2 + 4Gz + (a_0^2 I - 3H^2) = 0.$$

Here we have to remove the second term from (1). To effect the required transformation. Let the roots of (1) be diminished by h, so that we put $y = x - h$, or $x = y + h$ and the resulting Eqn. is

$$f(y+h) \equiv a_0(y+h)^4 + 4a_1(y+h)^3 + 6a_2(y+h)^2 + 4a_3(y+h) + a_4 = 0$$
$$= a_0 y^4 + 4(a_0 h + a_1) y^3 + 6(a_0 h^2 + 2a_1 h + a_2) y^2$$
$$+ 4(a_0 h^3 + 3a_1 h^2 + 3a_2 h + a_3) y$$
$$+ a_0 h^4 + 4a_1 h^3 + 6a_2 h^2 + 4a_3 h + a_4 = 0. \qquad \ldots(2)$$

Therefore in order to remove the second term, we must have

$$4(a_0 h + a_1) = 0; \text{ giving } h = \frac{a_1}{a_0}.$$

Substituing this in (2), the transformed equation is

$$a_0 y^4 + 6\left(\frac{a_1^2}{a_0} - \frac{2a_1^2}{a_0} + a_2\right) y^2 + 4\left(-\frac{a_1^3}{a_0^2} + 3\frac{a_1^3}{a_0^2} - 3\frac{a_1 a_2}{a_0} + a_3\right) y$$

$$+ \frac{a_1^4}{a_0^3} - 4\frac{a_1^4}{a_0^3} + 6\frac{a_1^2 a_3}{a_0^2} - 4\frac{a_1 a_3}{a_0} + a_4 = 0$$

or $\quad a_0 y^4 + \dfrac{6}{a_0}(a_0 a_2 - a_1^2) y^2 + \dfrac{4}{a_0^2}(a_0^2 a_3 - 3a_0 a_1 a_2 + 2a_1^3) y$
$$+ \dfrac{1}{a_0^3}(a_0^3 a_4 - 4a_0^2 a_1 a_3 + 6a_0 a_1^2 a_2 - 3a_1^4) = 0$$

Now let
$$H = a_0 a_2 - a_1^2,$$
$$G = a_0^2 a_3 - 3a_0 a_1 a_2 + 2a_1^3;$$
$$I = a_0 a_4 - 4a_1 a_3 + 3a_1^4.$$

$\therefore \quad a_0^3 a_4 - 4a_0^2 a_1 a_3 + 6a_0 a_1^2 a_2 - 3a_1^4$
$$= a_0^2(a_0 a_4 - 4a_1 a_3 + 3a_2^2) - 3(a_0^2 a_2^2 - 3a_0 a_1^2 + a_1^4)$$
$$= a_0^2(a_0 a_4 - 4a_1 a_3 + 3a_2^2) - 3(a_0 a_2 - a_1^2)^2$$
$$= a_0^2 I - 3H^2.$$

Hence the transformed equation is
$$a_0 y^4 + \dfrac{6H}{a_0} y^2 + \dfrac{4G}{a_0^2} y + \dfrac{a_0^2 I - 3H^2}{a_0^3} = 0$$

or $\quad y^4 + \dfrac{6H}{a_0^2} y^2 + \dfrac{4G}{a_0^3} y + \dfrac{a_0^2 I - 3H^2}{a_0^4} = 0 \qquad \ldots(3)$

Multiplying the roots of (3) by a_0, it reduces to
$$z^4 + 6Hz^2 + 4Gz + (a_0^2 I - 3H^2) = 0 \qquad \ldots(4)$$

The variables x, y, z are connected by the relations
$$z = a_0 y = a_0\left(x + \dfrac{a_1}{a_0}\right) = a_0 x + a_1.$$

Thus if α, β, γ are the roots of the original biquadratic (1), those of the transformed equation (3) and (4) are respectively.

$$\alpha + \dfrac{a_1}{a_0}, \beta + \dfrac{a_1}{a_0}, \gamma + \dfrac{a_1}{a_0}, \delta + \dfrac{a_1}{a_0}$$

and $\quad a_0 \alpha + a_1, a_0 \beta + a_1, a_0 \gamma + a_1, a_0 \delta + a_1$

Now from the given equation, we have
$$\alpha + \beta + \gamma + \delta = -\dfrac{4a_1}{a_0}; \quad \therefore \quad \dfrac{a_1}{a_0} = -\dfrac{1}{4}(\alpha + \beta + \gamma + \delta)$$

Hence the roots of the equation (4) are
$$a_0 \alpha - \dfrac{1}{4} a_0(\alpha + \beta + \gamma + \delta); \; a_0 \beta - \dfrac{1}{4} a_0(\alpha + \beta + \gamma + \delta); \ldots \text{etc.}$$

i.e., $\quad \dfrac{1}{4} a_0 (3\alpha - \beta - \gamma - \delta); \dfrac{1}{4} a_0 (3\beta - \gamma - \delta - \alpha); \ldots$ etc.

Note. The expression
$$a_0 a_2 a_4 + 2a_1 a_2 a_3 - a_0 a_3^2 - a_1^2 a_4 - a_2^3$$
is denoted by the letter J. It may be easily seen that
$$G^2 + 4H^3 = a_0^2(HI - a_0 J) \qquad \ldots(A)$$

Theory of Equations

9.13.1. Resolution of the biquadratic into quadratic factors or solving a biquadratic by expressing it as the difference of two squares.

Ferrari's Method of Solving a Biquadratic Equation.

Let the given biquadratic be

$$f(x) = a_0 x^4 + 4a_1 x^3 + 6a_2 x^2 + 4a_3 x + a_4 = 0. \qquad \ldots(1)$$

Also we assume that

$$a_0 f(x) \equiv (a_0 x^2 + 2a_1 x + a_2 + 2a_0\theta)^2 - (2Mx + N)^2 \qquad \ldots(2)$$

Comparing the coefficients of like powers of x from both sides, we have

$$M^2 = a_1^2 - a_0 a_2 + a_0^2 \theta \qquad \ldots(i)$$

$$MN = a_1 a_2 - a_0 a_3 + 2a_0 a_1 \theta \qquad \ldots(ii)$$

and $\qquad N^2 = (a_2 + 2a_0\theta)^2 - a_0 a_4 \qquad \ldots(iii)$

Eliminating M and N between (i), (ii), we get

$$(a_1^2 - a_0 a_2 + a_0^2 \theta)\{(a_2 + 2a_0\theta)^2 - a_0 a_4\} = (a_1 a_2 - a_0 a_3 + 2a_0 a_1 \theta)^2$$

which on simplification reduces to

$$4a_0^3 \theta^3 - (a_0 a_4 - 4a_1 a_3 + 3a_2^2) 3a_0\theta + a_0 a_2 a_4 + 2a_1 c_2 a_3 - a_0 a_3^2 - a_4 a_1^2 - a_2^3 = 0$$

i.e., $\qquad 4a_0^3 \theta^3 - a_0 I\theta + J = 0. \qquad \ldots(3)$

Equation (3) is called the **Reducing Cubic** of the given biquadratic. Since imaginary roots occur in pairs, a real root of (3) always exists. Finding the real value of θ (by solving the above cubic) the corresponding values of M and N can be determined. Then from (2) the biquadratic breaks into the quadratics

$$a_0 x^2 + 2(a_1 - M)x + a_2 + 2a_0\theta - N = 0 \qquad \ldots(4)$$

and $\qquad a_0 x^2 + 2(a_1 + M)x + a_2 + 2a_0\theta + N = 0. \qquad \ldots(5)$

Example : *Solve the equation*

$$x^4 - 2x^3 - 5x^2 + 10x - 3 = 0$$

by Ferrari's method.

Sol. Let $\qquad f(x) \equiv x^4 - 2x^3 - 5x^2 + 10x - 3 = 0 \qquad \ldots(1)$

Also we assume that

$$f(x) \equiv (x^2 - x + \lambda)^2 - (Mx + N)^2 \qquad \ldots(2)$$

Comparing the coefficients of like powers of x from both sides, we have

$$M^2 = 2\lambda + 6,$$

$$MN = -\lambda - 5,$$

and $\qquad N^2 = \lambda^2 + 3.$

Eliminating M and N between these relations, we get

$$(2\lambda + 6)(\lambda^2 + 3) = (\lambda + 5)^2$$

i.e., $\qquad 2\lambda^3 + 5\lambda^2 - 4\lambda - 7 = 0$

which gives $\lambda = -1$. Therefore $M = 2$, and $N = -2$.

Hence the given biquadratic reduces to

or
$$(x^2 - x - 1)^2 - (2x - 2)^3 = 0$$
$$(x^2 + x - 3)(x^2 - 3x + 1) = 0$$
∴
$$x^2 + x - 3 = 0 \text{ and } x^2 - 3x + 1 = 0$$

Hence
$$x = \frac{-1 \pm \sqrt{(1+12)}}{2} = \frac{-1 \pm \sqrt{(13)}}{2}$$

and
$$x = \frac{3 \pm \sqrt{(9-4)}}{2} = \frac{3 \pm \sqrt{(5)}}{2}.$$

EXERCISES

Solve the following biquadratic by Ferrari's Method :

1. $x^4 - 3x^3 + x^2 - 2 = 0.$ [Ans. $\frac{1}{2}(-1 \pm \sqrt{3}), \frac{1}{2}(-1 \pm \sqrt{3}i)$]

2. $x^4 + 4x^3 + 12x^2 - 8x + 95 = 0$ [Ans. $= \pm \sqrt{(10)}$ i.e., $1 \pm 2i$]

3. $x^4 + 12x - 5 = 0.$ [Ans. $-1 \pm \sqrt{2}, 1 \pm 2i$]

9.14 RESOLUTION OF THE BIQUADRATIC INTO QUADRATIC FACTORS

Descarte's method of solving a Biquadratic Equation.

Let the biquadratic be
$$a_0 x^4 + 4a_1 x^3 + 6a_2 x^2 + 4a_3 x + a_4 = 0 \qquad ...(1)$$

This equation transform to
$$f(z) \equiv z^4 + 6Hz^2 + 4Gz + (a_0^2 I - 3H^2) = 0 \qquad ...(2)$$

where $z = a_0 x + a_1$. Now we assume that
$$f(z) \equiv (z^2 + kz + m)(z^2 - kz - n) \qquad ...(3)$$

where k, m, n are unknown constants. Comparing the coefficients of like powers of z from both sides, we get

$$m + n - k^2 = H \qquad ...(i)$$
$$k(n - m) = 4G \qquad ...(ii)$$

and
$$mn = a_0^2 I - 3H^2. \qquad ...(iii)$$

Solving (i) and (ii), for m and n, we get $2m = 6H + k^2 - 4G/k$

and
$$2n = 6H + k^2 + 4G/k$$

Substituting these values in (iii), obtain
$$(6H + k^2 - 4G/k)(6H + k^2 + 4G/k) = 4mn = 4(a_0^2 I - 3H^2)$$

i.e.,
$$k^6 + 12Hk^4 - 4(a_0^2 I - 12H^2)k^2 - 16G^2 = 0,$$

which is a cubic in k^2 and it always has one real root. The value of k^2 obtained from this equation give the values of m and n. Then solving the two quadratics

$$z^2 + kz + m = 0 \quad \text{and} \quad z^2 - kx + n = 0$$

we get the roots of the transformed equation in z and hence of the original biquadratic where $z = a_0 x + a_1$.

Theory of Equations

EXAMPLES

1. *Solve the biquadratic $x^4 - 5x^2 - 6x - 5 = 0$ by Descarte's method.*

Sol. Let $f(x) \equiv (x^2 + kx + m)(x^2 - kx + n)$.

Comparing the coefficients of like powers of x, we get

$$m + n - k^2 = -5 \qquad \text{...(i)}$$
$$k(n - m) = -6 \qquad \text{...(ii)}$$
and $$mn = -5 \qquad \text{...(iii)}$$

But $$(m+n)^2 = (n-m)^2 + 4mn$$

Therefore $$(k^2 - 5)^2 = (-6/k)^2 + 4(-5)$$

i.e., $$k^6 - 10k^4 + 45k^2 - 36 = 0$$

This equation gives $k = 1$. Therefore

$$m + n = -4 \quad \text{and} \quad n - m = -6$$

These give $m = 1$ and $n = -5$. Hence

$$f(x) \equiv (x^2 + x + 1)(x^2 - x - 5) = 0$$

∴ solving $x^2 + x + 1 = 0$ and $x^2 - x - 5 = 0$, we get

$$x = \frac{-1 \pm i\sqrt{3}}{2} \quad \text{and} \quad \frac{1 \pm \sqrt{21}}{2}.$$

2. *Solve the bi-quadratic*

$$x^4 - 4x^3 + 9x^2 - 12x + 18 = 0$$

by resolving it into quadratic factors.

Sol. Given equation is

$$x^4 - 4x^3 + 9x^2 - 12x + 18 = 0 \qquad \text{...(i)}$$

First remove the second term by diminishing the roots by

$$h = -\left(-\frac{4}{4 \cdot 1}\right) = 1$$

```
1 | 1   -4    9   -12   18
  |     1   -3    6    -6
  |----------------------------
    1   -3    6   -6  | 12
        1   -2    4
    1   -2    4  | -2
        1   -1
    1   -1  | 3
        1
    1  | 0
    1
```

∴ The transformed equation is

$$y^4 + 3y^2 - 2y + 12 = 0, \qquad \text{where } y = x - 1 \text{ ...(ii)}$$

Let
$$y^4 + 3y^2 - 2y + 12 \equiv (y^2 + ky + m)(y^2 - ky + n)$$
comparing the coefficients of x^2, x and constant terms from both sides, we get
$$m + n - k^2 = 3 \qquad \text{...(iii)}$$
$$k(n - m) = -2 \qquad \text{...(iv)}$$
But
$$(m+n)^2 = (n-m)^2 + 4mn$$
$$\therefore \quad (3+k^2)^2 = \left(-\frac{2}{k}\right)^2 + 48$$
or
$$k^6 + 6k^4 - 39k^2 - 4 = 0$$
which is a cubic in k^2. Let us find a value of k^2 by inspection method.

Put $k^2 = 1$, $\quad 1 + 6 - 39 - 4 \neq 0$

$k^2 = 4$, $\quad 64 + 96 - 156 - 4 = 0$

$\therefore k = \pm 2$. Taking $k = 2$
$$m + n = 7$$
$$m - n = 1$$
which gives $m = 4, n = 3$ satisfying (v).

$\therefore$ The two quadratic equations are
$$y^2 + 2y + 4 = 0 \qquad \text{...(vi)}$$
and
$$y^2 - 2y + 3 = 0 \qquad \text{...(vii)}$$

Solving (vi) $\quad y = \dfrac{-2 + \sqrt{(4-16)}}{2} = -1 \pm i\sqrt{3}$

Solving (vii) $\quad y = \dfrac{2 \pm \sqrt{(4-12)}}{2} = 1 \pm i\sqrt{2}$

Hence roots of (ii) are
$$-1 \pm i\sqrt{3} \quad \text{and} \quad 1 \pm i\sqrt{2}.$$
The roots of given equation (i) are obtained by adding 1 to the roots of (ii)

$\therefore$ Roots of (i) are
$$\pm i\sqrt{3}, 2 \pm i\sqrt{2}.$$

EXERCISES

Solve the following equations by Descarte's method :

1. $x^4 + 12x - 5 = 0.$ $\qquad$ [Ans. $-1 \pm \sqrt{2}, 1 \pm 2i$]

2. $x^4 - 10x^2 - 20x - 16 = 0.$ $\qquad$ [Ans. $4, -2, -1 \pm i$]

3. $x^4 - 3x^2 - 42x - 40 = 0.$ $\qquad$ $\left[\text{Ans. } -1, 4; \dfrac{-3 \pm i\sqrt{31}}{2}\right]$

4. $x^4 - 8x^2 - 24x + 7 = 0.$ $\qquad$ [Ans. $2 \pm \sqrt{3}; -2 \pm i\sqrt{3}$]

5. $x^4 - 4x^3 + 5x + 2 = 0.$ $\qquad$ $\left[\text{Ans. } \dfrac{1 \pm \sqrt{5}}{2}; \dfrac{3 \pm \sqrt{17}}{2}\right]$

□□□

10
Group

The theory of groups, an important part in present day Mathematics, started early in ninteenth century in connection with the solutions of algebraic equations, Originally a group was the set of all permutations of the roots of an algebraic equation which has the property that combination of any two of these permutations again belongs to the set. Later the idea was generalized to the concept of an abstract group. An abstract group is essentially the study of a set with an operation defined on it. Group theory has many useful applications both within and outside mathematics. Groups arise in a number of apparently unconnected subjects. In fact they appear in crystallography and quantum mechanics, in geometry and topology, in analysis and algebra and even in biology.

10.1 GROUP

Definition

An algebraic structure (G, o) where G is a non-empty set with a operation 'o' defined on it is said to be a group, if the operation satisfies the following axioms (called group axioms).

(G_1) **Closure axiom.** G is closed under the operation o, i.e., $a \, o \, b \in G$, for all $a, b, \in G$.

(G_2) **Associative axiom.** The binary operation o is associative i.e.,
$$(a \, o \, b) \, o \, c = a \, o \, (b \, o \, c) \; \forall \, a, b, c \in G.$$

(G_3) **Identity axiom.** There exists an element $e \in G$ such that
$$e \, o \, a = a \, o \, e = a \; \forall \, a \in G.$$
The element e is called the identity of 'o' in G.

(G_4) **Inverse axiom.** Each element of G possesses inverse, i.e., for each element $a \in G$, there exists an element $b \in G$ such that
$$a \, o \, b = b \, o \, a = e$$
The element b is then called the inverse of a with respect to 'o' and we write $b = a^{-1}$. Thus a^{-1} is an element of G such that
$$a^{-1} \, o \, a = a \, o \, a^{-1} = e.$$

10.1.1 Abelian Group or Commutative Group Definition

A group (G, o) is said to be abelian or commutative if the composition 'o' is commutative, i.e., if,
$$a \, o \, b = b \, o \, a \; \forall \, a, b \in G.$$
A group which is not abelian is called non-abelian.

Examples : (i) The structures $(N, +)$ and $(N, \times)$ are not groups i.e., the set of natural numbers considered with the addition composition or the multiplication composition, does not form a group. For, the postulate (G_3) and (G_4) in the former case, and (G_4) in the latter case, are not satisfied.

(ii) The structure $(Z, +)$ is a group, i.e., the set of integers with the addition composition is a group. This is so because addition in numbers is associative, the additive identity 0 belongs to Z, and the inverse of every element a, viz., $-a$ belong to Z. This is known as *additive group of integers*.

The structure $(Z, \times)$, i.e., the set of integers with the multiplication composition does not form a group, as the axiom (G_4) is not satisfied.

(iii) The structures $(Q, +), (R, +), (C, +)$ are all groups, i.e., the sets of rational numbes, real numbers, complex numbers, each with the additive composition, form a group.

But the same sets with the multiplication composition do not form a group, for the multiplicative inverse of the number zero does not exist in any of them.

(iv) The structure $(Q_0, \times)$ is group, where Q_0 is the set of non-zero rational numbers. This is so because the operation is associative, the multiplicative identity 1 belongs to Q_0 and the multiplicative inverse of every element a in the set is $1/a$, which also belongs to Q_0. This is known as the *multiplicative group of non-zero rationals.*

Obviously $(R_0, \times)$ and $(C_0, \times)$ are groups, where R_0 and C_0 are respectively the sets of non-zero real numbers and non-zero complex numbers.

(v) The structure $(Q^+, \times)$ is a group, where Q^+ is the set of positive rational numbers. It can easily be seen that all the postulates of a group are satisfied.

Similarly, the structure $(R^+, \times)$ is a group, where R^+ is the set of positive real numbers.

(vi) The groups in (ii), (iii), (iv) and (v) above are all *abelian groups,* since addition and multiplication are both commutative operations in numbers.

10.1.2 Finite and Infinite Groups

If a grouip contains a finite number of distinct elements, it is called *finite group* otherwise an *infinite group.*

In other words, a group (G, o) is said to be finite or infinite according as the underlying set G is finite or infinite.

10.1.3 Order of a Group

The number of elements in a finite group is called the *order* of the grouip. An infinite group is said to be of infinite order.

Note : It should be noted that the smallest group for a given composition is the set $\{e\}$ consisting of the identity element e alone.

10.2 SOME MORE DEFINITIONS

Quasi Group : A non-empty set G together with a binary operation o is called as **quasi group** or **groupoid.**

Loop : A quasi group with an identity element is called a **loop.**

Semi-group : A quasi group (G, o) is called a semi-group if o is associative in G, i.e., if

$$(a \, o \, b) \, o \, c = a \, o \, (b \, o \, c) \, \forall \, a, b, c \in G$$

Alternatively : A non-empty set G together a binary operation o is called a semi-group if o is associative in G.

Monoid : A semi-group with an identity element is called a **monoid.**

Quaternion Group : Let $G = \{\pm 1, \pm i, \pm j, \pm k\}$. Define a binary operator of multiplication as

$$i^2 = j^2 = k^2 = -1, ij = k, ki = ik = j, jk = -kj = i.$$

It can be easily be shown that G is a non-abelian group for this operation. This group is called Quaternion group and its order is 8.

EXAMPLES

1. *Show that the set of all even integers (including zero) with additive property is an abelian group.*

Group

Solution. The set of all even integers (including zero) is
$$I = \{0, \pm 2, \pm 4, \pm 6 \ldots\}$$
Now we will discuss the group axioms one by one :

(G_1) The sum of two even integers is always an even integer, therefore **closure axiom** is satisfied.

(G_2) The addition is associative for even integers, hence **associative axiom** is satisfied.

(G_3) $0 \in I$, which is an additive identity in I, hence **identity axiom** is satisfied.

(G_4) Inverse of an even integer a is the even integer $-a$ in the set, so **axiom of inverse** is satisfied.

(G_5) Commutative law is also satisfied for addition of even integers. Hence the set forms an abelian group.

2. *If S be the set of all real numbers of the form $(m + n\sqrt{2})$ where $m, n \in Q$, set of rational numbers, prove that S is a multiplicative or additive group, m, n not vanishing simultaneously.*

Solution. Let us first prove the question for multiplication as composition.

(G_1) Suppose $m_1 + n_1\sqrt{2}, m_2 + n_2\sqrt{2} \in S$, so that $m_1, n_1, m_2, n_2 \in R$.
Now
$$(m_1 + n_1\sqrt{2}) \cdot (m_2 + n_2\sqrt{2}) = (m_1 m_2 + 2n_1 n_2) + (m_1 n_2 + m_2 n_1)\sqrt{2}$$
which belongs to S because $(m_1 m_2 + 2n_1 n_2), (m_1 n_2 + m_2 n_1) \in Q$.
Hence **closure axiom** is satisfied.

(G_2) The elements of S are all real numbers and the multiplication of real numbers is associative. Hence **associative axiom** is satisfied.

(G_2) $1 = (1 + 0 \cdot \sqrt{2}) \in S$ is an **identity element** to multiplication.

(G_4) If $(m + n\sqrt{2}) \in S$ then
$$(m + n\sqrt{2})^{-1} = \frac{1}{m + n\sqrt{2}} = \frac{m - n\sqrt{2}}{m^2 - 2n^2}$$
$$= \left(\frac{m}{m^2 - 2n^2}\right) + \left(\frac{-n}{m^2 - 2n^2}\right)\sqrt{2}$$
$$= \alpha + \beta\sqrt{2} \in S$$
where $\alpha = \dfrac{m}{m^2 - 2n^2}$, $\beta = \dfrac{-n}{m^2 - 2n^2} \in Q, \forall\, m, n \in Q$.

Hence the **inverse axiom** is satisfied.

Therefore $(S, .)$ is a group. Similarly, we can show that $(S, +)$ is also a group.

3. *Prove that the set of matrices*
$$A_\alpha = \begin{bmatrix} \cos\alpha & -\sin\alpha \\ \sin\alpha & \cos\alpha \end{bmatrix}$$
where α is a real number, forms a group under multiplication.

Solution.

(G_1) **Closure Property.** Let $A_\alpha, A_\beta \in G$. Then
$$A_\alpha A_\beta = \begin{bmatrix} \cos\alpha & -\sin\alpha \\ \sin\alpha & \cos\alpha \end{bmatrix} \begin{bmatrix} \cos\beta & -\sin\beta \\ \sin\beta & \cos\beta \end{bmatrix}$$

$$= \begin{bmatrix} \cos(\alpha+\beta) & -\sin(\alpha+\beta) \\ \sin(\alpha+\beta) & \cos(\alpha+\beta) \end{bmatrix}$$

$$= A_{\alpha+\beta} \in G$$

(as $\alpha+\beta$ is a real number whenever α and β are real).

(G_2) **Associativity.** Let A_α, A_β and A_γ belong to G. Then

$$(A_\alpha \cdot A_\beta) A_\gamma = A_{\alpha+\beta} \cdot A_\gamma = A_{(\alpha+\beta)+\gamma}$$
$$= A_{\alpha+(\beta+\gamma)} = A_\alpha \cdot (A_\beta \cdot A_\gamma).$$

(G_3) **Identity axiom.** Since 0 is a real number, therefore

$$A_0 = \begin{bmatrix} \cos 0 & -\sin 0 \\ \sin 0 & \cos 0 \end{bmatrix} = \begin{bmatrix} 1 & 0 \\ 0 & 1 \end{bmatrix} \in G$$

Now $A_\alpha \cdot A_0 = A_{\alpha+0} = A_\alpha$

Thus $A_0 = I_2$ is the identity element.

(G_4) **Inverse Axiom.** Let $A_\alpha \in G$, α is a real number. Then $A_{-\alpha} \in G$, since $-\alpha$ is also a real number, and we have

$$A_\alpha \cdot A_{-\alpha} = A_{\{\alpha+(-\alpha)\}} = A_0 = A_{-\alpha} \cdot A_\alpha.$$

Hence $A_{-\alpha}$ is the inverse of A_α.

Thus G is a group under matrix multiplication.

4. *Prove that the set Q of all rational numbers other than 1 with the operation defined by $a \circ b = a + b - ab$ constitutes an abelian group.*

Solution. (G_1) **Closure Property.** Let $a, b \in Q$ so that both a and b are rational numbers other than 1.

∴ $a \circ b = a + b - ab$ is also a rational number other than 1 because if it is equal to 1,

i.e., if $a + b - ab = 1$ then $a + b - ab - 1 = 0$

or $(a-1)(b-1) = 0.$

This would mean that $a = 1$ or $b = 1$ which is not true. Hence $a \circ b \in Q$, i.e., the closure property is staisfied.

(G_2) **Associative Law.** Let $a, b, c \in Q$, then

$$(a \circ b) \circ c = (a+b-ab) \circ c$$
$$= (a+b-ab) + c - (a+b-ab)c$$
$$= a+b+c-ab-ac-bc+abc.$$

Also, $a \circ (b \circ c) = a \circ (b+c-bc)$
$$= a + (b+c-bc) - a(b+c-bc)$$
$$= a+b+c-ab-ac-bc+abc.$$

Thus $(a \circ b) \circ c = a \circ (b \circ c)$ and hence associative law is satisfied.

(G_3) **Existence of Identity.** If e be the identity then

$$a \circ e = a \Rightarrow a + e - ae = a$$

⇒ $e(1-a) = 0$ $e = 0$ as $a \neq 1$

Group

and in the case $a \, o \, e = a \, o \, 0 = a + 0 - a \cdot 0 = a$
Hence 0 is the identity.

(G_4) **Existence of Inverse.**

$$a \, o \, b = a + b - ab = 0 \text{ the identity.}$$

$\Rightarrow \quad \dfrac{a}{a-1} = b$ is the inverse of a

Also, since $a - 1 \neq 0$

$\therefore \quad \dfrac{a}{a-1}$ is a ratioal number other than 1 and as such belongs to Q. Therefore inverse exists

and belongs to the set.

(G_5) **Commutative Law.**

$$a \, o \, b = a + b - a \, b = b + a - b \, a = b \, o \, a$$

Hence commutative.
Therefore, Q is an abelian group.

EXERCISES

1. Show that the set of all odd integers with addition as operation is not a group.
2. Show that the set I of integers with the binary operation o defined as

$$a \, o \, b = a - b \, (a, b \in I)$$

   is not a group.
3. Show that the set

$$G = \{....., -4m, -3m, -2m, -m, 0, m, 2m, 3m, 4m, ...\}$$

   of multiples of integers by a fixed integer m is a group with respect to addition.
4. Verify that the totality of all positive rationals form a group under the composition defined by

$$a \, o \, b = ab/2$$

5. Show that the set of all numbers $\cos \theta + i \sin \theta$ forms an infinite abelian group with respect to ordinary multiplication; where θ runs over all a rational numbers.
6. Prove that the four matrices

$$\begin{bmatrix} 1 & 0 \\ 0 & 1 \end{bmatrix}, \begin{bmatrix} -1 & 0 \\ 0 & 1 \end{bmatrix}, \begin{bmatrix} 1 & 0 \\ 0 & -1 \end{bmatrix}, \begin{bmatrix} -1 & 0 \\ 0 & -1 \end{bmatrix}$$

   form a multiplicative group.
7. Whether the set I of all integers with the operation o defined by $a \, o \, b = a + b + 1$, form a group ?

10.3 COMPOSITION (OR OPERATION) TABLE

A binary operation in a finite set can completely be described by means of a table. This table is known as *composition table*. The composition table helps us to verify most of the properties satisfied by the binary operations.

This table can be formed as follows :

(i) Write the elements of the set (which are finite in number) in a row as well as in a column.

(ii) Write the element associated to the ordered pair (a_i, a_j) at the intersection of the row headed by a_i and the column headed by a_j. Thus

(*i*th entry on the left) . (*j*th entry on the top)
  = (entry on the *i*th row and *j*th column intersect).

For example, the composition table for the group $\{(0, 1, 2, 3, 4)\}$ for the operation of addition is given below :

+	0	1	2	3	4
0	0	1	2	3	4
1	1	2	3	4	5
2	2	3	4	5	6
3	3	4	5	6	7
4	4	5	6	7	8

In the above example, the first element of the first row in the body of the table, 0 is obtained by adding the first element 0 of head row and the first element 0 of the head column. Similarly the third element of 4th row (5) is obtained by adding the third element 2 of the head row and the fourth element 3 of the head column and so on.

An operation represented by the composition table wil be binary, if every entry of the composition table belongs to the given set. It is to be noted that composition table contains all possible combinations of two elements of the set with respect to the operation.

Note :
(i) If should be noted that the elements of the set should be written in the same order both in top border and left border of the table, while preparing the composition table.
(ii) Generally a table which define a binary operation '.' on a set is called *multiplication table,* when the operation is '+' the table is called an *addition table.*

10.3.1 Group Tables

The composition tables are useful in examining the following axioms in the manner explained below :

1. Closure Property : If all the elements of the table belong to the set G (say) then G is closed under the composition o (say). If any of the elements of the table does not belong to the set, the set is not closed.

2. Existence of Identity : The element (in the vertical column) to the left of the row identical to the top row (border row) is called an identity element in the G with respect to operation 'o'.

3. Existence of Inverse : If we mark the identity elements in the table then the element at the top of the column passing through the identity element is the inverse of the element in the extreme left of the row passing through the identity element and vice versa.

4. Commutativity : If the table is such that the entries in every row coincide with the corresponding entries in the corresponding column *i.e.,* the composition table is symmetrical about the principal or main diagonal, the composition is said to have satisfied the commutative axiom otherwise it is not commutative.

The process will be more clear with the help of following illustrative examples.

EXAMPLES

1. *Prove that the set of cube rootso f unity is an abelian finite grouip with respect to multiplication.*

Solution. The set of cube roots of unity is, $G = \{(1, \omega, \omega^2)\}$. Let us form the composition table as given below :

Group

·	1	ω	ω^2
1	1	ω	ω^2
ω	ω	ω^2	$\omega^3 = 1$
ω^2	ω^2	$\omega^3 = 1$	$\omega^4 = \omega$

(G_1) **Closure axiom.** Since each element obtained in the table is a unique element of the given set G, multiplication is a binary operation. Thus the closure axiom is satisfied.

(G_2) **Associative axiom.** The elements of G are all complex numbers and we known that multiplication of complex numbers is always associative. Hence associative axiom is also satisfied.

(G_3) **Identity axiom.** Since row 1 of the table is identical with the top border row of elements of the set, 1 (the element of the extreme left of this row) is the identity element in G.

(G_4) **Inverse axiom.** The inverse of 1, ω, ω^2 are 1, ω^2 and ω respectively.

(G_5) **Commutative axiom.** Multiplication is commutative in G because the elements equi-distant with the main diagonal are equal each to each.

The number of elements in G is 3. Hence $(G,.)$ is a finite group of order 3.

2. *Prove that the set $\{1, -1, i, -i\}$ is an abelian multiplicative finite group of order 4.*

Solution. Let $G = \{1, -1, i, -i\}$ The following will be the composition table for $(G,.)$

·	1	-1	i	$-i$
1	1	-1	i	$-i$
-1	-1	1	$-i$	i
i	i	$-i$	$-i$	1
$-i$	$-i$	i	1	-1

(G_1) **Closure axiom.** Since all the entries in the composition table are elements of the set G, the set G is closed under the operation multiplication. Hence closure axiom is satisfied.

(G_2) **Associative Axiom.** Multiplication for complex numbers is always associative.

(G_3) **Identity axiom.** Row 1 of the table is identical with that at the top border, hence the element 1 in the extreme **left column heading** row 1 is the identity element.

(G_4) **Inverse axiom.** Inverse of 1 is 1. Inverse of -1 is -1. Inverse of i is $-i$ and of $-i$ is i. Hence inverse axiom is satisfied in G.

(G_5) **Commutative axiom.** Since in the table the 1st row is identical with 1st column, 2nd row is identical with 2nd column, 3rd row is identical with the 3rd column and 4th row is identical with 4th column, hence the multiplication in G is commutative.

The number of elements in G is 4. Hence G is an abelian finite group of order 4 with respect to multiplication.

EXERCISES

1. Show that the set $G = \{1\}$ forms a group with respect to multiplication.
2. Show that the set $A = \{a, b, c\}$ forms a finite abelian group of order three under the operation denoted multiplicatively and defined by the following table :

•	a	b	c
a	a	b	c
b	b	c	a
c	c	a	b

10.4 GENERAL PROPERTIES OF GROUPS

Theorem 1. *The identity element of a group is unique.*

Proof : Let us suppose e and e' are two identity elements of group G, with respect to operation o.

Then $e\, o\, e' = e$ if e' is identity.

and $e\, o\, e' = e'$ if e is identity.

But $e\, o\, e'$ is unique element of G, therefore,

$$e\, o\, e' = e \text{ and } e\, o\, e' = e' \Rightarrow e = e'$$

Hence the identity element in a group is unique

Theorem 2. *The inverse of each element of a group is unique, i.e., in a group G with operation o for every $a \in G$, there is only one element a^{-1} such that $a^{-1} o\, a = a\, o\, a^{-1} = e$, e being the identity.*

Proof. Let a be any element of a group G and let e be the identity element. Suppose there exist a^{-1} and a' two inverses of a in G then

$$a^{-1} o\, a = e = a\, o\, a^{-1}$$

and $$a'\, o\, a = e = a\, o\, a'$$

Now, we have

$$a^{-1} o\, (a\, o\, a') = a^{-1} o\, e \qquad \text{(since } a\, o\, a' = e)$$
$$= a^{-1} \qquad (\because e \text{ is identity})$$

Also, $$(a^{-1} o\, a) o\, a' = e\, o\, a' \qquad (\because a^{-1} o\, a = e)$$
$$= a' \qquad (\because e \text{ is identity})$$

But $a^{-1} o\, (a\, o\, a') = (a^{-1} o\, a) o\, a'$ as in a group composition is associative

$\therefore \qquad a^{-1} = a'$.

Theorem 3. *If the inverse of a is a^{-1} then the inverse of a^{-1} is a, i.e., $(a^{-1})^{-1} = a$.*

Proof. If e is the identity element, we have $a^{-1} o\, a = e$ (by definition of inverse)

$\Rightarrow \quad (a^{-1})^{-1} o\, (a^{-1} o\, a) = (a^{-1})^{-1} o\, e \qquad [\because a^{-1} \in G \Rightarrow (a^{-1})^{-1} \in G]$

$\Rightarrow \quad [(a^{-1})^{-1} o\, a^{-1}] o\, a = (a^{-1})^{-1}$

$\qquad\qquad\qquad\qquad [\therefore$ Composition in G is associative and e is identity element]

$\Rightarrow \quad e\, o\, a = (a^{-1})^{-1}$ since $a^{-1} \in G$

$\Rightarrow \quad a = (a^{-1})^{-1}$

$\Rightarrow \quad (a^{-1})^{-1} = a$.

Group

Theorem 4. *The inverse of the product of two elements of a group G is the product of the inverse taken in the reverse order i.e.,*
$$(a \, o \, b)^{-1} = b^{-1} \, o \, a^{-1} \; \forall \, a, b \in G$$

Proof. Let us suppose a and b are any two elements of G. If a^{-1} and b^{-1} are inverses of a and b respectively, then

$$a^{-1} o\, a = e = a \, o \, a^{-1} \quad \text{(} e \text{ being the identity element)}$$

and $\quad b^{-1} o\, b = e = b \, o \, b^{-1}$

Now, $\quad (a \, o \, b) \, o \, (b^{-1} \, o \, a^{-1}) = [(a \, o \, b) \, o \, b^{-1}] \, o \, a^{-1}$ (by associativitiy)

$\hspace{4.5cm} = [a \, o \, (b \, o \, b^{-1})] \, o \, a^{-1}$ (by associativity)

$\hspace{4.5cm} = (a \, o \, e) \, o \, a^{-1}$ $\quad [\because b \, o \, b^{-1} = e]$

$\hspace{4.5cm} = a \, o \, a^{-1}$ $\quad [\because a \, o \, e = a]$

$\hspace{4.5cm} = e$ $\quad [\because a \, o \, a^{-1} = e]$

Also $(b^{-1} \, o \, a^{-1}) \, o \, (a \, o \, b) = b^{-1} \, o \, [a^{-1} \, o \, (a \, o \, b)]$ (by associativity)

$\hspace{4.5cm} = b^{-1} o \, [(a^{-1} \, o \, a) \, o \, b]$

$\hspace{4.5cm} = b^{-1} o \, (e \, o \, b)$ $\quad [\because a^{-1} o\, a = e]$

$\hspace{4.5cm} = b^{-1} o \, b$ $\quad [\because e \, o \, b = b]$

$\hspace{4.5cm} = e.$

Hence, we have
$$(b^{-1} o\, a^{-1}) \, o \, (a \, o \, b) = e = (a \, o \, b) \, o \, (b^{-1} o\, a^{-1})$$

Therefore, by definition of inverse, we have
$$(a \, o \, b)^{-1} = b^{-1} o\, a^{-1}$$

This theorem can be generalised as :

if $a, b, c, ..., k, l, m \in G$, then
$$(a \, o \, b \, o \, c \, ... \, k \, o \, l \, o \, m)^{-1} = m^{-1} \, o \, l^{-1} \, o \, k^{-1} \, o \, ... \, c^{-1} \, o \, b^{-1} \, o \, a^{-1}$$

Theorem 5. *Cancellation laws hold good in a group, i.e., if a, b, c, are any elemnets of G, then*

$\hspace{2cm} a \, o \, b = a \, o \, c \Rightarrow b = c \hspace{2cm}$ (left cancellation law)

and $\hspace{1cm} b \, o \, a = c \, o \, a \Rightarrow b = c \hspace{2cm}$ (right cancellation law)

Proof. Let $a \in G$. Then

$$a \in G \Rightarrow G \text{ such that } a^{-1} o\, a = e = a \, o \, a^{-1},$$

where e is the identity element.

Now, let us assume that
$$a \, o \, b = a \, o \, c$$

then $\quad a \, o \, b = a \, o \, c \Rightarrow a^{-1} o \, (a \, o \, b) = a^{-1}(a \, o \, c)$

$\Rightarrow \quad (a^{-1} o\, a) \, o \, b = (a^{-1} o\, a) \, o \, c \hspace{2cm}$ (by associative law)

$\Rightarrow \quad e \, o \, b = e \, o \, c \hspace{4cm} (\because a^{-1} o\, a = e)$

$\Rightarrow \quad b = c.$

Similarly, $\quad b \, o \, a = c \, o \, a$

$\Rightarrow \quad (b\,o\,a)\,o\,a^{-1} = (c\,o\,a)\,o\,a^{-1}$

$\Rightarrow \quad b\,o\,(a\,o\,a^{-1}) = c\,o\,(a\,o\,a^{-1})$

$\Rightarrow \quad b\,o\,e = c\,o\,e$

$\Rightarrow \quad b = c.$

Theorem 6. *If G is a group with binary operation o and if a and b are any elements of G, then the linear equations*

$$a\,o\,x = b \text{ and } y\,o\,a = b$$

have unique solutions in G.

Proof. Now $\quad a \in G \Rightarrow a^{-1} \in G,$

and $\quad a^{-1} \in G, b \in G \Rightarrow a^{-1}o\,b \in G.$

Substituting $a^{-1}o\,b$ for x in the equation $a\,o\,x = b$, we obtain

$$a\,o\,(a^{-1}o\,b) = b$$

$\Rightarrow \quad (a\,o\,a^{-1})\,o\,b = b$

$\Rightarrow \quad e\,o\,b = b$

$\Rightarrow \quad b = b \qquad\qquad\qquad\qquad [\because e \text{ is the identity}]$

Thus $x = a^{-1}o\,b$ is a solution of the equation $a\,o\,x = b.$

To show that the solution is unique let us suppose that the equation $a\,o\,x = b$ has two solutions given by

$$x = x_1 \text{ and } x = x_2$$

Then $a\,o\,x_1 = b$ and $a\,o\,x_2 = b$

$\Rightarrow \quad a\,o\,x_1 = a\,o\,x_2 = b$

$\Rightarrow \quad x_1 = x_2$

$\Rightarrow \quad x_1 = x_2 \qquad\qquad\qquad\qquad$ (by left cancellation law)

In a similar manner, we can prove that the equation

$$y\,o\,a = b$$

has the unique solution

$$y = b\,o\,a^{-1}.$$

Theorem 7. *If corresponding to any element $a \in G$; there is an element which satisfies one of the conditions*

$$a + 0_a = a \text{ or } 0_a + a = a$$

then it is necessary that $0_a = 0$, where 0 is the identity element of the group.

Proof. Since 0 is the identity element,
We have

$$a + 0 = a \qquad\qquad\qquad\qquad\qquad\qquad ...(i)$$

Also, it is given that

$$a + 0_a = a \qquad\qquad\qquad\qquad\qquad\qquad ...(ii)$$

Group

Hence, from (i) and (ii)
$$a + 0_a = a + 0$$
or $\quad\quad 0_a = 0 \quad\quad$ (by left cancellation law)

Again, we have
$$0 + a = a \quad\quad\quad\quad ...(iii)$$
and $\quad\quad 0_a + a = a \quad$ (given) $\quad\quad ...(iv)$

Hence, from (iii) and (iv), we get
$$0_a + a = 0 + a$$
so that $0_a = 0$ (by right cancellation law.)

Theorem 8. *A finite non-empty semigroup satisfying the cancellation laws is a grouip.*

Alternately. A finite set G with a binary composiiton denoted multiplicatively is group if the composition is associative and the right and left cancellation laws hold in G.

Proof. Suppose, a set G with a binary composition (.), consisting on n distinct elements
$$a_1, a_2, ..., a_n.$$

Also, suppose that left and right cancellation laws hold good in G and the operation is associative in G.

To prove that G is a group, we must show that the equations
$$ax = b, \; ya = b; \; a, b \in G$$
have unique solutions in G.

If a is any arbitrary element of G, then forming the products
$$aa_1, aa_2, ..., aa_n \quad\quad\quad ...(1)$$
All these elements are distinct.

For if possible let
$$aa_r = aa_s; \; r \neq s \text{ and } r, s = 1, 2, ..., n$$
By left cancellation law this gives
$$a_r = a_s \text{ where } r \neq s.$$
This is a contradiction for
$$r \neq s \Rightarrow a_r \neq a_s$$

Hence, the elements (1) are n distinct elements of G placed in some different order.

Thus, if $a, b \Rightarrow G$, then $\exists$ a unique element $c \in G$ such that $ac = b$. More precisely the equation
$$ax = b; \; a, b \in G$$
has a unique solution in G.

Similarly, by forming the products
$$a_1 a, a_2 a, ..., a_n a$$
we can show that the equation
$$ya = b; \; a, b \in G$$
has a unique solution in G.

Hence, G is a group.

10.5 SOME DEFINITIONS

(i) If for every $a \in G$, there exists an element e in G, such that $eoa = a$, then e is called the **left identity**.

(ii) If for every $a \in G$, there exists an element e in G such that $aoe = a$, then e is called the **right identity**.

(iii) If for an $a \in G$ there exists an element a^{-1} in G such that $a^{-1} o\, a = e$, then a^{-1} is called the **left inverse** of a.

(iv) If for an $a \in G$ there eixsts an element a^{-1} in G such that $a\, o\, a^{-1} = e$, then a^{-1} is called the **right inverse** of a.

Theorem 1. *The left identity is also the right identity i.e.,*
$$eoa = a = aoe \; \forall \; a \in G$$

Proof. If a^{-1} be the left inverse of a,

then $\qquad a^{-1} o\, (a\, o\, e) = (a^{-1} \, o\, a)\, o\, e \qquad$ (By associative law)

or $\qquad a^{-1} o\, (a\, o\, e) = e\, o\, e \qquad$ (by definition of left inverse)

$\qquad\qquad\qquad\qquad = e$

$\qquad\qquad\qquad\qquad = a^{-1} o\, a$

Thus $\qquad a^{-1} o\, (a\, o\, e) = a^{-1} o\, a$

$\therefore \qquad\qquad aoe = a \qquad$ (by left cancellation law)

Hence e is also the right identity element.

Theorem 2. *The left inverse of an element is also its right inverse, i.e.,*
$$a^{-1} o\, a = e = a\, o\, a^{-1}$$

Proof. Now,

$\qquad a^{-1} o\, (a\, o\, a^{-1}) = (a^{-1} o\, a)\, o\, a^{-1} \qquad$ (by associative law)

$\qquad\qquad\qquad\qquad = e\, o\, a^{-1}$

$\qquad\qquad\qquad\qquad = a^{-1} o\, e \qquad$ (by theorem 1)

Thus $\qquad a^{-1} o\, (a\, o\, a^{-1}) = a^{-1} o\, e$

$\therefore \qquad\qquad a\, o\, a^{-1} = e \qquad$ (by left cancellation law)

Hence a^{-1} is also the right inverse of a.

10.6 AN ALTERNATIVE DEFINITION FOR A GROUP

A set G with a binary composition denoted multiplicatively is a group if
(i) the composition is associative.

(ii) for every pair of elements $a, b \in G$, the equations $a\, x = b$ and $ya = b$ have unique solutions in G.

Proof. Binary operation implies that the set G, under consideration is closed under the operation. Now to prove that set G is a group, we have to show that the left identity exists and each element of G possesses left inverse with respect to the operation under consideration.

It is given that for every pair of elements $a, b \in G$ the equation $ya = b$ has a solution in G. Therefore, if $a \in G$, then taking $b = a$, we observe that there exists an element, say $e \in G$ such that

Group

$$ea = a \quad \ldots(i)$$

Now, let us suppose that b is any arbitrary element of G.
Therefore, there exists $x \in G$ such that

$$ax = b \quad \ldots(ii)$$

Thus $\qquad b = ax \Rightarrow eb = e(ax)$

$\Rightarrow \qquad eb = (ea)x \qquad$ (associative law)

$\Rightarrow \qquad eb = ax \qquad$ (from (i))

$\Rightarrow \qquad eb = b \qquad$ (from (ii))

Therefore $\exists\, e \in G$ such that

$$eb = b \;\forall\; b \in G$$

∴ e is the last identity.

Now, let $b = e \in G$ be an element.

∴ $\qquad ya = b \Rightarrow ya = e$

$\Rightarrow$ y is inverse of a in G.

Let $y = a^{-1}$ such that

$$a^{-1}a = e \text{ then } a^{-1} \in G \text{ as } ya = e \text{ has got solutio in } G.$$

Thus a^{-1} is the left inverse of a in G. Therefore each element of G possesses left inverse. Hence G is a group for the given composition if the postulates (i) and (ii) are satisfied.

EXAMPELS

1. *Show that if every element of a group G is its own inverse, then G is abelian.*

Solution. Let a and b be any two elements of G, then ab is also an element of G. (Closure law)

Therefore, $(ab)^{-1} = ab$ (because it is given that every element in G is its own inverse)

Hence, $\qquad (ab)^{-1} = ab \Rightarrow b^{-1}a^{-1} = ba$

$\Rightarrow \qquad ba = ab \qquad [\because a^{-1} = a, b^{-1} = b]$

Hence, we have

$$ab = ba \;\forall\; a, b \in G$$

∴ G is an abelian group.

2. *Show that a finite set G with an associative binary operation is a group if the right and left cancellation laws hold in G. Show further that validity of the cancellation laws does not characterise infinite groups but charactertise finite groups only.*

Solution. Let $G = \{a_1, a_2, a_3, \ldots, a_n\}$ where $a_1, a_2, \ldots, a_n$ are distinct. Let us consider an element a of G then $a_1a, a_2a, a_3a, \ldots a_na \in G$, the set G being closed under the binary operation and these elements are all distinct, because, otherwise $a_ia = a_ja$, for $a_i, a_j \in G$ $(i \neq j) \Rightarrow a_i = a_j$ (by the right cancellation law).

Thus $a_1a, a_2a, a_3a, \ldots, a_na$ are nothing but some permutation of elements $a_1, a_2, a_3, \ldots, a_n$ of G.

Therefore, if b is an element of G, then it must be one of the elements $a_1a, a_2a, \ldots a_na$.

Hence $a_ia = b$, where $a, b, a_i \in G$.

Thus we arrive at the conclusion that the equation $ya = b$ $(a, b \in G)$ has a unique solution in G. Similarly by forming the products $aa_1, aa_2, aa_3, ..., aa_n$ and by using the left cancellation law,

We can show that the equation $ax = b$ $(a, b \in G)$ has a solution in G.

Hence G is a finite group of order n.

However, an infinite set will not necessarily form a group even if the binary composition in a set is associative and both the cancellation laws hold good in it. The following example will illustrate this fact.

The set N of all natural numbers 1, 2, 3, 4, ... is not a group for multiplication even though N is closed under multiplication, multiplication is associative and both the cancellation laws hold good for multiplication in N.

EXERCISES

1. Prove that for every element a in a group G, $a^2 = e$, where e is the identity, then G is an abelian group.
2. Show that if a, b are any two elements of a group G, then $(ab)^2 = a^2 b^2$, if and only if G is abelian.
3. If G is a finite group, show that for each $a \in G$, there exists a positive integer n such that $a^n = e$.
4. If G is a group of even order, prove that it has an element $a \neq e$, satisfying $a^2 = e$.

10.7 MODULO SYSTEM

It is of common experience that railway time-table is fixed with the provision of 24 hours in a day and night. When we say that a particular train is arriving at 15 hours, it implies that the train will arrive at 3 p.m. according to our watch. Thus all the timing starting from 12 to 23 hours correspond to one of 0, 1, 2 ... 11 O'clock as indicated in watches. In other words all integers from 12 to 23 are equivalent to one or the other of integers 0, 1, 2, 3, ... 11 with modulo 12. In saying like this the integers in question are divided into 12 classes.

In the manner described above the integer could be divided into 2 classes, or 5 classes or m (m being a positive integer) classes and then we would have written mod 2 or mod 5 or mod m. This system of representing integers is called **modulo system.**

Addition modulo m

We shall now define a new type of addition known as "addition modulo m" and written as $a +_m b$ where a and b are any integers and m is a fixed positive integer.

By definition, we have

$$a +_m b = r, \quad 0 \leq r < m$$

where r is the least non-negative remainder when $a + b$, i.e., the ordinary sum of a and b, is divided by m.

For example, $5 +_6 3 = 2$, since $5 + 3 = 8 = 1(6) + 2$, i.e., 2 is the least non-negative remainder when $5 + 3$ is divided by 6. Similarly, $5 +_7 2 = 0, 4 +_3 2 = 0; 3 +_3 1 = 1, 15 +_5 7 = 2$.

Thus to find $a +_m b$, we add a and b in the ordinary way and then from the sum, we remove integral multiples of m in such a way that the remainder r is either 0 or a positive integer less than m.

When a and b are two integers such that $a - b$ is divisible by a fixed positive integer m, then we write

$$a = b \pmod{m}$$

which is read as "a is congurent to b modulo m."

Group

Thus $a = b \pmod{m}$ if $a - b$ is divisible by m. For example $13 = 3 \pmod 5$ since $13 - 3 = 10$ is divisibly by 5, $5 = 5 \pmod 5$, $16 = 4 \pmod 6$; $-20 = 4 \pmod 6$.

Multiplication modulo p

We shall now define a new type of multiplication known as "multiplication modulo p" and written as $a \times_p b$ where a and b are any integers and p is a fixed positive integer.

By definition we have

$$a \times_p b = r, 0 \leq r \leq p,$$

where r is the least non-negative remainder when ab, i.e., the ordinary product of a and b, is divided by p. For example $4 \times_7 2 = 1$, since $4 \times 2 = 8 = 1(7) + 1$.

It can be easily shown that if $a = b \pmod p$, then $a \times_p c = b \times_p c$

10.7.1 Additive Group of Integers Modulo m

The set $G = \{0, 1, 2, \ldots m - 1\}$ of first m non-negative integers is a group. The composition being addition reduced mudulo m.

Closure Property. We have by definition of addition modulo m,

$$a +_m b = r$$

where r is the least non-negative remainder when the ordinary sum $a + b$ is divided by m. Obviously $0 \leq r \leq m - 1$. Therefore for all $a, b \in G$ we have $a +_m b \in G$ and thus G is closed with respect to the composition addition modulo m.

Associative Property. Let a, b, c be any arbitrary elements in G.. Then

$$(a + b) +_m c = (a +_m b) +_m c \qquad [\because b +_m c = b + c \pmod m]$$

= least non-negative remainder when $a + (b + c)$ is divided by m

= least non-negative remainder when $(a + b) + c$ is divided by m, since

$$a + (b + c) = (a + b) + c$$

$$= (a + b) +_m c \qquad \text{[by definition of } +_m \text{]}$$

$$= (a +_m b) +_m c \qquad [\because a + b = a +_m b \pmod m]$$

$\therefore$ '$+_m$' is an associative composition.

Existence of Identity Element : We have $0 \in G$. Also, if a is any element of G, then $0 +_m a = a = a +_m 0$. Therefore 0 is the identity element.

Existence of Inverse. The inverse of 0 is 0 itself. If $r \in G$ and $r \neq 0$, then $m - r \in G$. Also $(m - r) +_m r = 0 = r +_m (m - r)$. Therefore $(m - r)$ is the inverse of r.

Commutative Property. The composition '$+_m$' is commutative also. Since

$$a +_m b = \text{least-non-negative remainder when } a + b \text{ is divided by } m$$

$$= \text{least non-negative remainder when } b + a \text{ is divided by } m$$

$$= b +_m a.$$

The set G contains m elements.

Hence $(G, +_m)$ is a finite abelian group of order m.

10.7.2 Multiplicative Group of Integers Modulo p where p is Prime

The set G of $(p - 1)$ integers $1, 2, 3, \ldots p - 1$, p being prime, is finite abelian group of order $p - 1$ the composition being multiplication modulo p.

Let $G = \{1, 2, 3, \ldots p-1\}$ where p is prime.

Closure Property. Let a and b be any elements of G. Then $1 \leq a \leq p-1, 1 \leq b \leq p-1$. Now by definition, $a \times_p b = r$ where r is the least non-negative remainder when the ordinary product $a\,b$ is divided by p. Since p is prime, therefore $a\,b$ is not exactly divisible by p. Therefore r can not be zero and we shall have $1 \leq r \leq p-1$. Thus $a \times_p b \in G \ \forall \ a, b \in G$. Hence the closure axiom is satsified.

Associative Law. Let a, b, c be any arbitrary elements of G.

Then $\qquad a \times (_p b \times _p c) = a \times _p (bc) \qquad [\because b \times _p c = bc \pmod{p}]$

$\qquad$ = least non-negative remainder when $a\,(bc)$ is divided by p

$\qquad$ = least non-negative remainder when $ab\,(c)$ is divided by p

$\qquad = (ab) \times _p c$

$\qquad = (a \times _p b) \times _p c \qquad [\because ab = a \times _p b \pmod{p}]$

$\therefore$ '$\times_p$' is an associative composition.

Existence of left identity. We have $1 \in G$. Also if a is any element of G, then $1 \times _p a = a$. Therefore 1 is the left identity.

Existence of left inverse. Let s be any member of G.

Then $\qquad 1 < s < p-1$.

Let us consider the following $(p-1)$ products :

$\qquad 1 \times _p s, 2 \times _p s, 3 \times _p s, \ldots (p-1) \times _p s.$

All these are elements of G. Also no two of these can be equal as shown below :

Let i and j be two unequal integers such that

$\qquad 1 \leq i \leq p-1, 1 \leq j \leq p-1$ and $i > j$.

Then $\qquad i \times _p s = j \times _p s$

$\Rightarrow \quad is$ and js leave the same least non-negative remainder when divided by p

$\Rightarrow \quad is - js$ is divisibly by p.

$\Rightarrow \quad (i-j)\,s$ is divisible by p.

Since $1 \leq (i-j) < p-1; 1 \leq s \leq p-1$ and p is prime, therefore $(i-j)\,s$ cannot be divided by p.

$\therefore \qquad i \times _p s \neq j \times _p s.$

Thus $1 \times _p s, 2 \times _p s, \ldots (p-1) \times _p s$ are $(p-1)$ distinct elements of the set G. Therefore one of these elements must be equal to 1.

Let then $s' \times _p s = 1$. Then s' is the left inverse of s.

Commutative Law. The composition 'X_p' is commutative, since

$\qquad a \times _p b$ = least no-negative remainder when ab is divisible by p

$\qquad$ = least non-negative remainder when ba is divided by p

$\qquad = b \times _p a$

$\therefore \quad (G, \times_p)$ is a finite abelian group of order $p-1$.

Theorem 1. *The residue classes modulo form a finite group with respect to addition of residue classes.*

Proof. Let G be the set of residue classes (mod m), then
$$G = \{\{0\}, \{1\}, \ldots \{r_1\}, \ldots \{r_2\}, \ldots \{m-1\}\}$$
or $$G = \{0, 1, 2, \ldots r_1, \ldots r_2 \ldots m-1 \,(\mathrm{mod}\ m)\}$$

Closure axiom. $\{r_1\} + \{r_2\} = \{r_1 + r_2\} = \{r\} \in G$,

where r is the least positive integer obtained as remainder when $r_1 + r_2$ is divided by $m\ (0 \le r < m)$.

Thus, the closure axiom is satisfied.

Associative axiom. The addition is associative.

Identity axiom. $\{0\} \in \{G\}$ and $\{0\} + \{r\} = \{r\}$. Hence the identity for addition is $\{0\}$.

Inverse axiom. Since $\{m-r\} + \{r\} = \{m\} = \{0\}$, the additive inverse of the element $\{r\}$ is $\{m-r\}$.

Hence G is a finite group with respect to addition modulo m.

Theorem 2. *The set of non-zero residue classes modulo p, where p is a prime, forms a group with respect to multiplication of residue classes.*

Proof. Let $I = \{\ldots, -3, -2, -1, 0, 1, 2, 3, \ldots\}$ be the set of integers. Let $a \in I$, then $\{a\}$ is residue class modulo p of I, if $\{a\} = \{x : x \in I$ and $x - a$ is divisible by $p\}$.

If $p\,|\,a$ then $\{a\} = \{0\}$ which is called the zero residue class. Let G be the set of non-zero residue classes mod p (p being prime) then
$$G = \{1, 2, 3, \ldots, (p-1)\}$$

Closure axiom. Let $r_1, r_2 \in G$ then
$$r_1 \cdot r_2 = r\ (\mathrm{mod}\ p)$$
where r is the least non-negative integer such that $0 \le r \le p-1$ obtained after dividing r_1, r_2 by p.

Also, since p is prime r_1, r_2 is not divisible by p. Hence r cannot be zero.

Hence, $$r_1 \cdot r_2 = r = G.$$

Thus closure axiom is satisfied.

Associative axiom. Multiplication of residue classes is associative.

Existence of Identity. $1 \in G$ and $a \cdot 1 = 1 \cdot a = a\ \forall\ a \in G$.

Therefore 1 is the identity element in G with respect to multiplication.

Existence of Inverse. Let $s \in G$ then $1 \le s \le p-1$. Let us consider following $(p-1)$ elements.
$$1 \cdot s, 2 \cdot s, 3 \cdot s \ldots, (p-1) \cdot s.$$

All these elements are elements of G because the closure law is true. All these elements are distinct as otherwise if
$$i \cdot s = j \cdot s \text{ for } i \ne j \text{ and } i, j \in G$$
then $$i \cdot s = j \cdot s \Rightarrow i \cdot s - j \cdot s \text{ is divisible by } p$$
$\Rightarrow \quad (i-j) \cdot s$ is divisibly by p
$\Rightarrow \quad (i-j)$ is divisibly by p $\quad [\because 1 \le s < p-1]$
$\Rightarrow \quad i = j$ which is contrary to our assumption that $i \ne j$,

Therefore above $(p-1)$ elements are the same as the elements of G. Hence some one of them should be 1 also. Let $s' \cdot s = 1$ where $1 \le s' \le p-1$. Hence s' is inverse of s. Hence inverse axiom is also satisfied.

∴ G is a grouip under multiplication mod p.
Note: Since $r.s = s.r \ \forall \ r, s \in G$.

G is finite abelian group of order $(p-1)$.

EXAMPLES

1. *Prove that the set $G = \{0, 1, 2, 3, 4\}$ is a finite abelian group of order 5 with respect to addition modulo 5.*

Solution. Let us prepare a composition table as given below:

+ (mod 5)	0	1	2	3	4
0	0	1	2	3	4
1	1	2	3	4	0
2	2	3	4	0	1
3	3	4	0	1	2
4	4	0	1	2	3

Closure Property. All the entries in the composition table are elements of the set G. Hence G is closed under addition modulo 5.

Associative Property. Addition modulo 5 is always associative.

Identity. $0 \in G$ is the identity element.

Inverse. It is clear from composition table.

Element   —   0   1   2   3   4
Inverse   —   0   4   3   2   1

∴ Inverse exists for every element of G.

Commutative law. The composition is commutative as the corresponding rows and columns in the table are identical.

The number of elements in G are 5.

Hence $\{G, +_5\}$ is a finite abelian group of order 5.

2. *Prove that the set $G = \{1, 2, 3, 4, 5, 6\}$ is a finite abelian group of order 6 with respect to multiplication modulo 7.*

Solution. Let us prepare the following composition table:

$\times_7$	1	2	3	4	5	6
1	1	2	3	4	5	6
2	2	4	6	1	3	5
3	3	6	2	5	1	4
4	4	1	5	2	6	3
5	5	3	1	6	4	2
6	6	5	4	3	2	1

Closure Property. All the entries in the table are elements of G. Therefore G is closed with respect to multiplication modulo 7.

Group

Associative Property. Multiplication modulo 7 is always associative.

Identity. Since first row of the table is identical to the row of elements of G in the horizontal border, the element to the left of first row in vertical border is identity element, *i.e.,* 1 is identity element in G with respect to multiplication modulo 7.

Inverse. From the table it is obvious that inverses of 1, 2, 3, 4, 5, 6 are 1, 4, 5, 2, 3 and 6 respectively. Hence inverse of each element in G exists.

Commutative Property. The composition is commutative because the elements equidistant from principal diagonal are equal each to each.

The set G has 6 elements. Hence $(G, \times_7)$ is a finite abelian group of order 6.

EXERCISES

1. Show that the residue classes modulo 5 form a group with respect to addition.
2. Show that the residue classes modulo 6 with respect to multiplication do not form a group.
3. Show that set of residue classes modulo m is an abelian group of order m with respect to addition of residue classes.

10.8 INTEGRAL POWERS OF AN ELEMENT

Suppose G is a group and the composition has been denoted multiplicatively. Let $a \in G$. Then by closure property $a, aa, aaa, aaaa$, etc., are all elements of G. Since the composition in G obeys general associative law, therefore $a\, a\, a \ldots a$ to n factors is independent of the manner in which the factors may be grouped.

If n is a positive integer, we define $a^n = a\, a\, a \ldots a$ to n factors. Obviously $a^n \in G$.

If e is the identity element of the group G, then we define

$$a^0 = e.$$

If n is a positive integer then $-n$ is a negative integer. Now we define $a^{-n} = (a^n)^{-1}$ where $(a^n)^{-1}$ is the inverse of a^n in G. Thus $a^{-n} \in G$.

Thus we have defined for all integral values of n positive, zero or negative.

Integral multiples of an element of a group.

If in a group G the composition has been denoted additively, then in place of using the word integral powers of an element of a group we use the word integral multiples of an element of a group. The difference is only of notation otherwise the meaning is the same. Thus in this case if n is a positive integer we write na in place of a^n and we define $n\, a = a + a + \ldots + a$ upto a terms.

In place of a^0 we write $0a$. Thus we define $0a = e$ where e is the identity of G.

If n is a positive integer, then in place of a^{-n} we write $(-n)\, a$. Thus we define $(-n)\, a = -(n\, a)$ denotes the inverse of na in G.

In multiplicative notation the following laws of indices can be easily proved :

$$a^m a^n = a^{m+n},$$

and
$$(a^m)^n = a^{mn},$$

$\forall\, a \in G$ and $\forall\, m; n \in I$ where I is the set of integers.

In additive notation the following laws of multiples can be easily proved :

$$ma + na = (m + n)\, a,$$
$$n\,(m\, a) = (n\, m)\, a,$$
$\forall\, a \in G$ and $\forall\, m, \in I.$

10.8.1 Order of an Element of a Group

Definition. *If G is a group and $a \in G$, the order (or period) of a is the least positive integer n such that*

$$a^n = e.$$

If there exists no such integer, we say that a is of infinite order or of zero order.

We shall use the notation $o(a)$ for the order of a.

Note that the only element of order one in a group is the identity element e.

Important note. If there exists a positive integer m such that $a^m = e$, then the order of a is definitely finite. Also we must have $o(a) \le m$. When $a^m = e$, then the question of order of a being greater than m does not arise. At the most it can be equal to m. If m itself is the least positive integer such that $a^m = e$; then we will have

$$o(a) = m$$

EXAMPLE

Find the order of each element of the multiplicative group G, where $G = \{1, -1, i, -i\}$.

Solution. Since 1 is the identity element, its order is 1.

Now $\qquad (-1)^1 = -1, (-1)^2 = (-1)(-1) = 1$

Hence order of -1 is 2.

$$i^1 = i, i^2 = -1, i^3 = -i, i^4 = 1$$

Therefore order of i is 4.

Similarly, $\qquad (-i)^1 = -i, (-i)^2 = i^2 = -1$

$$(-i)^3 = +i, (-i)^4 = 1.$$

Hence order of $-i$ is 4.

10.8.2 Some Theorems

Theorem 1. *The order of every element of a finite group is finite.*

Proof. Let G be a finite group and let $a \in G$. We consider all positive integral powers of a, i.e.,

$$a, a^2, a^3, a^4, \ldots$$

Every one of these powers must be an element of G. But G is of finite order. Hence these elements can not all be different. We may, therefore, suppose that

$$a^s = a^r, s > r.$$

Now, $\qquad a^s = a^r \Rightarrow a^s . a^{-r} = a^r . a^{-r}.$

$$\Rightarrow a^{s-r} = a^0$$

$$\Rightarrow a^{s-r} = e$$

$$\Rightarrow a^t = e \qquad \text{(putting } s = r = t\text{)}.$$

Since $s > r$, t is a positive integer.

Hence there exists a positive integer t such that $a^t = e$.

Now, we know that every set of positive integers has a least number. It follows that the set of all those positive integers such that $a^t = e$ has a least member, say m. Thus there exists a least positive integers m such that $a^m = e$, showing that the order of every element of a finite group is finite.

Theorem 2. *The order of an element of a group is the same as that of its inverse a^{-1}.*

Proof. Let n and m be the order of a and a^{-1} respectively.

Then $\qquad a^n = e$ and $(a^{-1})^m = e$.

Now, $\qquad a^n = e \Rightarrow (a^n)^{-1} = e^{-1}$

$\qquad\qquad \Rightarrow (a^{-1})^n = e$

$\qquad\qquad \Rightarrow o(a^{-1}) \leq n \Rightarrow m \leq n$.

Also $\qquad o(a^{-1}) = m \Rightarrow (a^{-1})^m = e$

$\qquad\qquad \Rightarrow (a^m)^{-1} = e \Rightarrow a^m = e \qquad\qquad [\because b^{-1} = e \Rightarrow b = c]$

$\qquad\qquad \Rightarrow o(a) \leq m \Rightarrow n \leq m$.

Now, $\quad m \leq n$ and $n \leq m \Rightarrow m = n$.

If the order of a is infinite, then the order of a^{-1} cannot be finite. Because $o(a^{-1}) = m \Rightarrow o(a) \leq m \Rightarrow o(a)$ is finite. Therefore if the order of a is infinite, then the order of a^{-1} must also be infinite.

Theorem 3. *The order of any integral power of an element a can not exceed the order of a.*

Proof. Let a^k be any integral power of a. Let $o(a) = n$.

Now, $\qquad o(a) = n \Rightarrow a^n = e \qquad\qquad$ (identity element)

$\qquad\qquad \Rightarrow (a^n)^k = e^k$

$\qquad\qquad \Rightarrow a^{nk} = e \Rightarrow (a^k)^n = e$

$\qquad\qquad \Rightarrow o(a^k) \leq n$.

Theorem 4. *If the element a of a group G is of order n, then $a^m = e$ iff n is a divisor of m.*

Proof. Since m must be greater than n,

let $\qquad m = nq + r$, where $0 \leq r < n$.

Now $\qquad a^m = a^{nq+r} = (a^n)^q \cdot a^r = a^r$.

since $\qquad a^n = e$. Therefore $a^r = e$.

But this is not possible for $0 < r < n$, since n is the least integer for which $a^n = e$.

It follows that $r = 0$, i.e., $m = nq$.

Conversely, if $m = nq$, then $a^m = a^{nq} = (a^n)^q = e^q = e$. More generally, if m is an integer (not necessarily positive), then also $m = nq$, where q is an integer. In either case we say that m is a multiple of n, or that n is divisor of m.

Theorem 5. *The order of the elements a and $x^{-1}ax$ are the same where a, x are any two elements of a group.*

Proof. Let n and m be the orders of a and $x^{-1}ax$ respectively

Now $\qquad (x^{-1}ax)^2 = (x^{-1}ax)(x^{-1}ax)$

$\qquad\qquad = x^{-1}a(xx^{-1})ax$

$\qquad\qquad = x^{-1}(ae)ax = x^{-1} a a x = x^{-1}a^2 x$.

In general, we get
$$(x^{-1}ax)^n = x^{-1}a^n x = x^{-1}ex \qquad [\because o(a) = n \Rightarrow a^n = e]$$
$$= x^{-1}x = e.$$

$\therefore \qquad o(x^{-1}a x) \leq n \Rightarrow m \leq n.$

Again $\qquad o(x^{-1}a x) \leq m \Rightarrow (x^{-1}a x)^m = e$

$$\Rightarrow x^{-1}a^m x = e$$
$$\Rightarrow x^{-1}a^m x = x^{-1}x$$
$$\Rightarrow a^m x = x \qquad \text{(by left cancellation law)}$$
$$\Rightarrow a^m x = ex$$
$$\Rightarrow a^m = e \qquad \text{(by right cancellation law)}$$
$$\Rightarrow o(a) \leq m \Rightarrow n \leq m.$$

Finally $m \leq n, n \leq m \Rightarrow m = n.$

Cor. *Order of a b is the same as that of ba where a and b are any elements of a group.*

Proof. We have $\qquad a^{-1}(a b) a = (a^{-1}a)(b a)$
$$= e(b a) = (b a)$$

Thus $\qquad b a = a^{-1}(a b) a$

$\Rightarrow$ order of $b a =$ order of $a^{-1}(a b) a$

$\Rightarrow$ order of $\Rightarrow ba =$ order $ab \qquad [\because o(x^{-1}a x) = o(a)]$

Theorem 6. *If a is an element of order n and p is prime to n, then a^p is also of order n.*

Proof. Let m be the order of a^p.

Now, $\qquad o(a) = n \Rightarrow a^n = e \Rightarrow (a^n)^p = e^p = e$
$$\Rightarrow (a^p)^n = e \Rightarrow (a^n)^p = e^p = e$$
$$\Rightarrow m \leq n.$$

Since p, n are relative primes, there exists integers x and y such that
$$px + ny = 1$$

$\therefore \qquad a = a^1 = a^{px+ny} = a^{px} \cdot a^{ny} = a^{px}(a^n)^y$
$$= a^{px} \cdot e^y = a^{px} \cdot e = a^{px} = (a^p)^x.$$

Now, $\qquad a^m = [(a^p)^x]^m = (a^p)^{mx} = [(a^p)^m]^x$
$$= e^x \qquad [\because o(a^p) = m \Rightarrow (a^p)^m = e]$$
$$= e$$

$\therefore \qquad o(a) \leq m \Rightarrow n \leq m.$

Finally $m \leq n$ and $n \leq m \Rightarrow m = n.$

EXAMPLES

1. *Find the order of each element in the following multiplicative group G.*
$$G = \{a, a^2, a^3, a^4, a^5, a^6 = e\}$$

Solution. The order of a is a 6 as $a^6 = e$

Group

The order of a^2 is 3 as $(a^2)^3 = a^6 = e$

The order of a^3 is 2 as $\quad (a^3)^2 = a^6 = e$

The order of a^4 is 3 as $\quad (a^4)^3 = a^{12} = (a^6)^2$
$$= e^2 = e.$$

The order of a^5 is 6 as $\quad (a^5)^6 = a^{30} = (a^6)^5$
$$= e^5 = e$$

and the order of a^6 is 1 as $\quad (a^6)^1 = a^6 = e.$

2. *If the elements a, b of a group commute and $o(a) = m, o(b) = n$, where $(m, n) = 1$, prove that $o(ab) = mn$.*

Solution. We are given that the two elements a and b of a group commute and $o(a) = m, o(b) = n$ with $(m, n) = 1$

We have to prove that $\qquad o(ab) = mn$

Let $\qquad o(ab) = k.$

Since a and b commute, we have
$$(ab)^{mn} = a^{mn}b^{mn} = (a^m)^n (b^n)^m = e^n e^m$$
$$= ee = e$$

Hence $k \mid mn$ [Theorem 4, §10.8.2]

Again $\quad (ab)^k = a^k b^k$ i.e., $e = a^k b^k$ $\qquad$ [$\because o(ab) = k$ so that $(ab)^k = e$]

But $\qquad a^k b^k = e \Rightarrow a^k = (b^k)^{-1}$
$$\Rightarrow o(a^k) = o\{(b^k)^{-1}\}$$
$$\Rightarrow o(a^k) = o(b^k) \qquad \text{[Theorem 4, §10.8.2]}$$

Now, by Theorem 4, §12.8.2, we have $o(a^k) \mid o(a)$ and $o(b^k) \mid o(b)$, i.e., $o(a^k) \mid m$ and $o(b^k) \mid n$.

Since $o(a^k) = o(b^k)$, we see that either $o(a^k)$ or $o(b^k)$ divides both m and n and so it divides their H.C.F. $(m, n) = 1$. In other words, $o(a^k) = o(b^k) = 1$, so that $a^k = e$ and $b^k = e$. Hence, we have $m \mid k$ and $n \mid k$. Together with $(m, n) = 1$, this gives $mn \mid k$ and combining this with $k \mid mn$, we obtain $k = mn$, i.e., $o(ab) = mn$.

3. *If G is a group such that $(ab)^m = a^m b^m$ for three consecutive integers m for all $a, b, \in G$, show that G is abelian.*

Solution. Let a, b be any two elements of G. Suppose $m, m+1, m+2$ are three consecutive integers such that
$$(ab)^m = a^m b^m, (ab)^{m+1} = a^{m+1} b^{m+1},$$
and $\qquad (ab)^{m+2} = a^{m+2} b^{m+2}$

We have $\qquad (ab)^{m+2} = (ab)^{m+1} (ab)$
$$\Rightarrow aa^{m+1}b^{m+1}b = aa^m b^m bab$$
$$\Rightarrow a^{m+1} b^{m+1} = a^m b^m (ba) \qquad \text{[by left and right cancellation laws]}$$

$$\Rightarrow (ab)^{m+1} = (ab)^m (ba)$$
$$\Rightarrow (ab)^m (ab) = (ab)^m (ba)$$
$$\Rightarrow ab = ba \qquad \text{[by left cancellation law]}$$
$$\Rightarrow G \text{ is abelian}.$$

EXERCISES

1. Find the order of the element 1 of the additive group I whose elements are all integers.
2. $G = \{[0], [1], [2], [3]\}$ is the additive group of residue classes modulo 4. Find the order of each element. [**Ans.** $o([0]) = 1, o[1] = 4, o([2]) = 2, o([3]) = 4$]
3. Show that if a, b are any two elements of a group G, then $(ab)^2 = a^2 b^2$ if and only if G is abelian.

10.9 CYCLIC GROUP

Definition. *A group G is called cyclic if, for some $a \in G$, every element $x \in G$ is of the form a^n, where n is some integer. The element a is then called a generator of G.*

There may be more than one generators of a cyclic group. If G is a cyclic group generated by a, then we shall write $G = \{a\}$ or $G = (a)$. The elements of G will be of the form ..., $a^{-3}, a^{-2}, a^{-1}, a^0 (= e), a, a^2, a^3, ...$ of course they are not necessarily all distinct.

Examples:

(i) The multiplicative group $\{1, \omega, \omega^2\}$ is cyclic. The generators are ω and ω^2.

(ii) The multiplicative group $G = \{1, -1, i, -i\}$ is cyclic. We can write $G = \{i, i^2, i^3, i^4\}$. The generators are i and $-i$.

(iii) The multiplicative group of n n^{th} roots of unity is cyclic, a generator being $e^{2\pi i/n}$.

10.9.1 Properties of Cyclic Groups

Theorem 1. *Every cyclic group is abelian.*

Proof. Let a be a generator of a cyclic group G and let $a^r, a^s \in G$ for any $r, s \in I$ then

$$a^r . a^s = a^{r+s} = a^{s+r} \qquad (\because r+s = s+r \text{ for } r, s \in I)$$
$$= a^s . a^r$$

Thus the operation is commutative and hence the cyclic group G is abelian.

Note. For the addition composition the above proof could have been written as

$$a^r a^s = ra + sa = sa + ra \qquad \text{(addition of integers is commutative)}.$$
$$= a^s + a^r$$

Thus the operation + is commutative in G.

Theorem 2. *The order of a cyclic group is same as the order of its generator.*
Proof. Let the order of a generator a of a cyclic group be n, then

$$a^n = e \text{ while } a^s \neq e \text{ for } 0 < s < n$$

When $s > n$, $s = nq + r, 0 \leq r < n$ (say), we observe that

$$a^s = a^{nq+r} = (a^n)^q . a^r = e^q . a^r$$
$$= e . a^r = a^r.$$

Thus there are exactly n elements in the group by a^r, where $0 \leq r < n$. Therefore there are n and only n distinct elements in the cyclic group, *i.e.*, the order of the group is n.

Group

Theorem 3. *The generator of a cyclic group of order n are all the elements a^p, p being prime to n and $0 < p < n$.*

Proof. We know that
$$(a^p)^n = (a^n)^p = e, \text{ therefore the order of } a^p \text{ is } n.$$

Also $(a^p)^s = a^{ps} \neq e$ if $0 < s < n$,
because n does not divide p, nor does it divide s, therefore it does not divide ps.

Now, let
$$ps = nq + r, 0 < r < n$$
and
$$(a^p)^s = a^{ps} = a^{nq+r}$$
$$= (a^n)^q \cdot a^r = e^q \cdot a^r$$
$$= e \cdot a^r = a^r \neq 0 \qquad \text{(since } 0 < r < n\text{).}$$

Thus, a^p is a generator of the group.

EXAMPLES

1. *How many elements of the group $G = \{a, a^2, a^3, a^4, a^5, a^6 = e\}$ can be used as generators of the group?*

Solution. From the definition of the generator of a group, it is evident that a is a generator of the group G.

Also, by the property of cyclic group, a^p is a generator of G, provided p is prime to n, i.e., p is prime to 6 ($\because$ Here $n = 6$). Therefore, here $p = 5$.

Hence a and a^5 can be used as generators of G.

2. *Prove that the set of nth roots of unity is a finite abelian cyclic group with multiplicative composition.*

Solution. We know that
$$1 = \cos 0 + i \sin 0$$
$$\therefore \quad 1^{1/n} = (\cos 2r\pi + i \sin 2r\pi)^{1/n}$$
$$= \left(\cos \frac{2r\pi}{n} + i \sin \frac{2r\pi}{n} \right)$$

where $r = 0, 1, 2, \ldots, (n-1)$
$$= e^{2r\pi i/n}$$

Let G denotes the set of nth roots of unity, then
$$G = \{e^{2r\pi i/n} : r = 0, 1, 2, \ldots, (n-1)\}$$

Now, let us verify the group axioms:

(i) Closure axiom. Let
$$e^{2\pi r_1 i/n}, e^{2\pi r_2 i/n} \in G$$
then
$$e^{2\pi r_1 i/n} \cdot e^{2\pi r_2 i/n} = e^{2\pi(r_1 + r_2)i/n} \in G$$
$$r_1 + r_2 \leq n - 1.$$

If $r_1 + r_2 > n - 1$, let us suppose $r_1 + r_2 = n + k$ where $k \leq n - 2$ then
$$e^{2\pi r_1 i/n} \cdot e^{2\pi r_1 i/n} = e^{2\pi(r_1 + r_2)i/n}$$
$$= e^{2\pi(n+k)i/n} = e^{2\pi i} \cdot e^{2\pi/n}$$
$$= 1 \cdot e^{2\pi ki/n} = e^{2\pi ki/n} \in G.$$

Hence the set G is closed under multiplication.

(ii) Associative axiom. The multiplication for complex number is always associative.

(iii) Identity axiom. $\cos 0 + i \sin 0 = 1 = e^{2\pi i 0/n}$ is an identity element is G.

(iv) Inverse axiom. The inverse of $e^{2\pi r i/n}$ is $e^{2\pi(n-r)i/n}$

because $e^{2\pi r i/n} e^{2\pi(n-1)i/n} = e^{2\pi r i/n} e^{2\pi r i} \cdot e^{-2\pi r i/n}$

$$= 1 \cdot e^{(2\pi r i/n - 2\pi r i/n)} = e^0 = 1.$$

(v) Commutative axiom. Since multiplication of complex numbers is commutative, G is a multiplicative abelian group.

Also, the number of elements in G in n, i.e., finite and all the elements are generated by $e^{2\pi i/n}$. Hence G is a finite abelian cyclic group.

3. *Show that there exist only two generators of any infinite cyclic group.*

Solution. Let G be any cyclic group whose generator is a, i.e., $G = \{a\}$.

Then as any integral power of a can be expressed as some integral power of a^{-1} (i.e., the inverse of a) and vice versa.

So a^{-1} is also a generator of G. Thus we have a and a^{-1} as the two generators of G. If possible let a^p be another generator of G besides these two, then $a = (a^p)^m$, for some $m \in I$, the set of integers.

$\Rightarrow aa^{-1} = (a^p)^m a^{-1} \Rightarrow e = a^{pm-1}$, $\qquad (\because a a^{-1} = e)$,

$\Rightarrow a^0 = a^{pm-1}$

$\Rightarrow$ either $p = m = 1$ or $p = m - 1$

$\Rightarrow a$ and a^{-1} are only two generators of G.

EXERCISES

1. How many elements of the cyclic group of order 8 can be used as generators of the group ?

[Ans. a, a^3, a^5, a^7]

2. Show that $\{1, -1, i, -i\}$ is a cyclic group under multiplication. Also find its generator.

[Ans. $i, -i$]

3. If ω be the imaginary cube root of unity, show that the set $\{1, \omega, \omega^2\}$ is a cyclic group of order 3 with respect to multiplication.

4. Find all the generators of the cyclic group $(\{1, 2, 3, 4, 5, 6\}, \times_7)$.

5. (i) $G = \{a, a^2, a^3, ..., a^8 = e\}$

   How many generators are there ?

   (ii) How many generators are there of a cyclic group of order 10 ?

6. Prove that any group of order 3 is cyclic.

10.10 SUBGROUPS

Let G be a group and H any subset of G. Let a, b be any two elements of H. Now a, b being members of G, the product ab belongs to G, but it may or may not belong to H. If, however, ab belongs to H, we say that H is **stable** for the composition in G and that the composition in G has **induced** the composition in H. If H is itself a group for the induced composition, then we say that H is a **subgroup** of G.

Definition. *A non-empty subset H of a group G is said to be a subgroup of G if the composition in G induces a composition in H and if H is a group for the induced composition.*

The two sub-groups (i) consisting of the identity element alone, and (ii) the group G itself are always present in a group G.

Group

There are, however, **trivial** subgroups. A sub-group other than these two is known as **proper sub-group**.

A **complex** is any subset of a group, whether it is a sub-group or not.

It is easy to prove that
(i) the identity of a sub-group is the same as that of the group.
(ii) the inverse of any element of a sub-group is the same as the inverse of the element regarded as a member of the group.
(iii) the order of any element of a sub-group is the same as that of the element regarded as a member of the group.

Examples:
(i) The additive group of integers is a sub-group of the additive group of rational numbers.
(ii) The multiplicative group of positive rational nuimbers is a sub-group of the multiplicative group of non-zero real numbers.
(iii) The multiplicative group $\{1, -1\}$ is a sub-group of the multiplicative group $\{1, -1, i, -i\}$.

10.10.1 Necessary and Sufficient Condition

The necessary and sufficient conditions for a subset of a group to be a sub-group are stated in the following two theorems.

Theorem 1. *A subset H of a group G is a sub-group iff*

(i) $(a \in H, b \in H) \Rightarrow a \, o \, b \in H$.

and (ii) $a \in H \Rightarrow a^{-1} \in H$.

Proof. Suppose H is a sub-group of G then H must be closed with respect to composition o in G, i.e., $a \in H, b \in H \Rightarrow a \, o \, b \in H$.

Let $a \in H$ and a^{-1} be the inverse of a in G. As H itself is a group, each element of H will possess inverse in it, i.e., $a \in H \Rightarrow a^{-1} \in H$.

Thus the condition is *necessary*. Now let us examine the *sufficiency* of the condition.

(i) **Closure axiom.** $a \in H, b \in H \Rightarrow a \, o \, b \in H$. Hence closure axiom is satisfied with respect to the operation o.

(ii) **Associativity.** Since the elements of H are also the elements of G, the composition is associative in H also.

(iii) **Existence of identity.** The identity of the subgroup is the same as the identity of the group because,

$$a \in H \Rightarrow a^{-1} \in H \qquad \text{[given condition (ii)]}$$

Hence $\qquad a \, o \, a^{-1} \in H \qquad \text{[given condition (i)]}$

i.e., $\qquad e \in H$

∴ The identity e is an element of H.

(iv) **Existence of inverse.** Since $a \in H \Rightarrow a^{-1} \in H, \forall a \in H$.

Therefore each element of H possesses inverse.

Thus H itself is a group for the composition in G. Hence H is a sub-group.

Theorem 2. *A necessary and sufficient condition for a non-empty subset H of a grouip G to be a sub-group is that* $a \in H, b \in H \Rightarrow a \, o \, b^{-1} \in H$ *where* b^{-1} *is the inverse of b in G*.

Proof. *The condition is necessary*. Suppose H is a sub group of G and let $a \in H, b \in H$.

Now each element of H must possess inverse because H itself is a group.

$$b \in H \Rightarrow b^{-1} \in H.$$

Also H is closed under the composition o in G. Therfore

$$a \in H, b^{-1} \in H \Rightarrow a o b^{-1} \in H.$$

[*The condition is sufficient.*] As $a \in H, b \in H \Rightarrow aob^{-1} \in H$, then we have to prove that H is a sub group.

(i) **Closure property.** Let $a, b \in H$ then $b \in H \Rightarrow b^{-1} \in H$ (as shown above)
Therefore by the given condition
$$a \in H, b^{-1} \in H \Rightarrow a\, o\, (b^{-1})^{-1} \in H.$$
$$\Rightarrow aob \in H.$$
Thus H is a closed with respect to the composition o in G.

(ii) **Associative property.** Since the elements of H are also the elements of G, the composition is asosciative in H.

(iii) **Existence of Identity.** Since
$$a \in H, a^{-1} \in H \Rightarrow aoa^{-1} \in H \Rightarrow e \in H.$$
Thus the identity element belongs to H.

(iv) **Existence of Inverse.** Let $a \in H$ then
$$e \in H, a \in H \Rightarrow eoa^{-1} \in H \Rightarrow a^{-1} \in H.$$
Thus each element of H possesses inverse.
Hence H itself is a group for the composition o in group G.

10.10.2 Properties of Subgroups

Theorem 1. *The intersection of two sub-groups of a group G is a sub-grouip of G.*

Proof. Let H_1 and H_2 be any two sub-groups of G.

Then $H_1 \cap H_2 \neq \phi$ (because at least the identity element e in common in both H_1 and H_2.)

Now to prove that $H_1 \cap H_2$ is a subgroup of G

It is sufficient to show that
$$a \in H_1 \cap H_2, b \in H_1 \cap H_2 \Rightarrow aob^{-1} \in H_1 \cap H_2, (o \text{ being composition in } G).$$
Since $a \in H_1 \cap H_2 \Rightarrow a \in H_1$ and $a \in H_2$
and $b \in H_1 \cap H_2 \Rightarrow b \in H_1$ and $b \in H_2$ and
H_1, H_2 are sub-groups of G, we see that
$$a \in H_1, b \in H_1 \Rightarrow aob^{-1} \in H_1 \text{ and similarly}$$
$$a \in H_2, b \in H_2 \Rightarrow aob^{-1} \in H_2$$
Thus, $aob^{-1} \in H_1, aob^{-1} \in H_2 \Rightarrow aob^{-1} \in H_1 \cap H_2$

Hence $a \in H_1 \cap H_2, b \in H_1 \cap H_2 \Rightarrow aob^{-1} H_1 \cap H_2$, which establishes that $H_1 \cap H_2$ is a sub-group of G.

Theorem 2. *The union of two sub-groups is not necessarily a sub group.*

Proof. For example, let G be the additive group of integers, and let
$$H_1 = \{0, \pm 2, \pm 4, \pm 6, ...\}$$
$$H_2 = \{0, \pm 3, \pm 6, \pm 9, ...\}$$
Then H_1, H_2 are sub-groups of G, but
$$H_1 \cup H_2 = \{0, \pm 2, \pm 3, \pm 4, \pm 6, ...\},$$

which is not a group. It is evident that the closure property is not satisfied. For, $2+3=5$, which does not belong to $H_1 \cup H_2$.

The set $\qquad H_1 \cap H_2 = \{0, \pm 6, \pm 12, ...\}$
which is certainly a group.
Hence the union of two subgrouips is not necessarily a group.

Theorem 3. *The union of two sub-groups is a sub-group if and only if one is contained in the other.*

Proof. Let H_1 and H_2 be two sub-groups of a group G.

(i) Let $\qquad H_1 \subset H_2$ or $H_2 \subset H_1$.
Then $\qquad H_1 \cup H_2 = H_2$ or H_1.
But H_1, H_2 are sub-groups so that $H_1 \cup H_2$ is also a sub-group.

(ii) Next suppose $H_1 \cup H_2$ is a sub-group.
To prove that $H_1 \subset H_2$ or $H_2 \subset H_1$, Assume if possible that
$$H_1 \not\subset H_2, H_2 \not\subset H_1$$

Now, $\qquad H_1 \not\subset H_2 \Rightarrow \exists s \in H_1$ and $s \notin H_2$ ...(i)
and $\qquad H_2 \subset H_1 \Rightarrow \exists t \in H_2$ and $t \notin H_1$ ...(ii)
From (i) and (ii), it follows that
$$s \in H_1 \cup H_2, t \in H_1 \cup H_2$$
since $H_1 \cup H_2$ is a sub-group, we see that
$st = k$ (say) is also an element of $H_1 \cup H_2$.
But $\qquad st = k \in H_1 \cup H_2 \Rightarrow st = k \in H_1$ or H_2
Suppose $st = k \in H_1$
then $\qquad t = s^{-1}k \in H_1 \qquad\qquad [\because H_1$ is a subgroup, $s^{-1} \in H_1]$
This contradicts (ii). Hence either $H_1 \subset H_2$ or $H_2 \subset H_1$

10.10.3 Sub-Group of Cyclic Groups

Theorem 1. *Every sub-group of a cyclic group is cyclic.*

Proof. Let $G = \{a\}$ be a cyclic group generated by a. Let H be a sub-group of G. Now every element of G, hence also of H, has the form a^s, s being an integer. Let m be the smallest possible integer such that $a^m \in H$. We claim that $H = \{a^m\}$. For this it is sufficient to show that $a^s \in H$, then $s = mh$ for then $a^s = (a^m)^h$. Now, if m does not divides, then there exist integers q and r such that
$$s = mq + r, 0 \leq r < m.$$
Then $\qquad a^s = a^{mq+r} = a^{mq} a^r \qquad$ or $\qquad a^r = a^s \cdot (a^{mq})^{-1}$ ...(i)
Since $a^m \in H$, it follows that $a^{mq} \in H$ and hence its inverse $(a^{mq})^{-1} \in H$.

But $a^s \in H$ by supposition. Then from (i) follows that $a^r \in H$ contrary to the choice of m since m was assumed to be least positive integer such that $a^m \in H$.

Therefore $\quad r = 0$ and so $s = mq$. But then $\qquad a^s = (a^m)^q$

Thus every element a^s of H is of the form $(a^m)^q$. Hence $H = \{a^m\}$.

Theorem 2. *Every sub-group of an infinite cyclic group is infinite.*

Proof. Let $G = \{a\}$ be infinite cyclic group. Let H be a sub-group of G. Then by the preceding theorem, $H = \{a^m\}$ where m is the least positive integer such that $a^m \in H$. Now suppose, if possible, that H is finite.

This implies that $(a^m)^s = e$ for some $s > 0$.

It follows that a is of finite order and this in turn implies taht G is finite, contrary to the hypothesis. Hence H must be infinite cyclic sub-group of G.

EXAMPLES

1. *If $S = \{0, 2, 4\}$ then show that $(S, +_6)$ a sub-group of the group $(I_+, +_6)$.*

Solution. We know $I_6 = \{0, 1, 2, 3, 4, 5\}$

$\therefore$ $S \leq I_6$ and is non-empty.

Also, for the system $(S, +_6)$ the composition table is

$+_6$	0	2	4
0	0	2	4
2	2	4	0
4	4	0	2

From the above table we conclude that

(i) **The closure property** is satisfied since each entry in the table is an element of S.

(ii) $0 +_6 (2 +_6 4) = 0 +_6 0,$ $\quad \because \ 2 +_6 4 = 0$ from the table

$\qquad = 0,$ from the table

and $(0 +_6 2) +_6 4 = 2 +_6 4$ $\quad \because \ 0 +_2 2 = 2$ from the table

$\qquad = 0$ from the table

Hence we find $0 +_6 (2 +_6 4) = (0 +_6 2) +_6 4$

Similarly, we can show that

$$a +_6 (b +_6 c) = (a +_6 b) +_6 c, \forall \, a, b, c \in S.$$

Hence the **associative property** is satisfied.

(iii) From the table it is evident that

$$a +_6 0 = 0 = 0 +_6 a, \forall \, a, \in S.$$

i.e., 0 is the **identity element** is S for the operation.

(iv) From the table we observe that the inverse of 0, 2 and 4 are 0, 4 and 2 respectively.

Thus all the group postulates are satisfied and so the set S forms a group under addition modulo 6.

Hence $(S, +_6)$ is a sub-group of the group $(I_+, +_6)$.

2. *Find all the sub-groups of a cyclic group of order 1 2.*

Solution. We know that the integral divisors of 12 are 1, 2, 3, 4, 6, 12. Now, there exists one and only one sub-group of each of these orders.

Let a be the generator of the group and m be a divisor of 12. Then there exists one and only one element G whose order is m, *i.e.,* $a^{12/m}$.

$\therefore$ All the elements of order 1, 2, 3, 4, 5, 6, 12 will give sub-groups.

$\therefore$ $(a^{12}) = \{e\}, (a^6), (a^4), (a^3), (a^2), (a)$ are the required sub-groups.

Group

3. *(i) Can an abelian group have a non-abelian sub-group ?*
(ii) Can a non-abelian group have an abelian sub-group ?
(iii) Can a non-abelian group have a non-abelian sub-group ?

Solution. (i) Every sub-group of an abelian group is abelian. If G is an abelian group and H is a sub-group of G, then the operation on H is commutative because it is already commutative in G and H is a subset of G. Hence an abelian group have a non-abelian sub-group.

(ii) A non-abelian group can have an abelian sub-group. For example, the symmetric group P_3 of permutations of degree 3 is non-abelian while its sub-group A_3 is abelian.

(iii) A non-abelian group can have a non-abelian sub-group. For example P_4 is a non-abelian group and its sub group A_4 is also non-abelian.

EXERCISES

1. Show that integral multiples of 5 form a sub-group of the additive group of all integers including 0.
2. Show that the set E of even integers is a sub-group of the additive group of all integers.
3. Show that the set $S = \{1, -1, i, -i\}$ is a sub-group of the multiplicative group of complex numbers.
4. Let G be the multiplicative group of all positive real numbers and R the additive group of all real numbers. Is G a sub-group of R ? [Ans. No.]

10.11 ALGEBRA OF COMPLEXES OF A GROUP

Let us consider the set of all complexes of a group G, which is nothing but power set of G. Let it be denoted by $P(G)$. Now, we define three binary compositions in $P(G)$. The two compositions namely union and intersection of sets are familiar ones. Here we define the multiplication of complexes.

Multiplication of Complexes

Let H and K be two complexes of a group G whose composition has been denoted multiplicatively, then the product of H and K denoted by HK is defined as

$$HK = \{hk : h \in H, \text{ and } k \in K\}.$$

In other words HK is the set of all possible products of elements of H with those of K.

It is evident that $hk \in HK$

$\Rightarrow h \in H, k \in K$
$\Rightarrow h, k \in G$ ($\because H \subset G, K \subset G$)
$\Rightarrow hk \in G$ (Closure law in (i))
$\therefore HK \subset G$

Thus the product of two complexes is also a complex of the group.

Multiplication of Complexes is associative

Let H, K and L be three coplexes of a group G whose composition is denoted multiplicatively. Then

$HK = \{hk : h \in H, k \in H\}$
$\therefore \quad (HK) L = \{(hk) l : h \in H, k \in K, l \in L\}$
$\qquad = \{h (kl) : h \in H, k \in K, l \in L\}$

$\qquad\qquad$ [$\because (hk) l = h(kl)$, multiplication in G being associative]

Also $\quad H(KL) = \{h(kl) : h \in H, k \in K, l \in L\}$
$\therefore \quad (HK) L = H(KL).$

Inverse of a complex in a group

Let H be any complex of G and let us define

$$H^{-1} = \{h^{-1} : h \in H\}$$

then H^{-1} is the complex of G consisting of the inverse of the elements of H. This H^{-1} is called the inverse of complex H.

Theorem 1. *If H and K are any two complexes of a group G, then $(HK)^{-1} = K^{-1}H^{-1}$ (reversal law).*

Proof. Let $x \in (HK)^{-1}$ then

$$x \in (HK)^{-1} \Rightarrow x = (hk)^{-1} \text{ where } h \in H, k \in K$$

$$\Rightarrow x = k^{-1}h^{-1}$$

$$\Rightarrow x \in K^{-1}H^{-1}$$

[Since $k^{-1} \in K^{-1}$ and $h^{-1} \in H^{-1}$ $\therefore k^{-1}h^{-1} \in K^{-1}H^{-1}$]

$$\therefore (HK)^{-1} \subseteq K^{-1}H^{-1} \qquad \text{...(i)}$$

Again, let $y \in K^{-1}H^{-1}$ then

$$y \in K^{-1}H^{-1} \Rightarrow y = k^{-1}h^{-1}, k \in K, h \in H$$

$$\Rightarrow y = (hk)^{-1} \qquad [\because (hk)^{-1} = k^{-1}h^{-1}]$$

$$\Rightarrow y \in (HK)^{-1} \qquad [\because (HK)^{-1} = \{(hk)^{-1} : h \in H, k \in K\}] \qquad \text{...(ii)}$$

$$\therefore K^{-1}H^{-1} \subseteq (HK)^{-1}$$

$\therefore$ From (i) and (ii)

$$(HK)^{-1} = K^{-1}H^{-1}.$$

Theorem 2. *If H is any subgroup of G then $H^{-1} = H$. Also show that the converse is not true.*

Proof. Let $h^{-1} \in H^{-1}$ then $h \in H$.
Since H is a sub-group of G.

$$h \in H \Rightarrow h^{-1} \in H \qquad \text{(Inverse axiom)]}$$

Therfore, $\quad h^{-1} \in H^{-1} \Rightarrow h^{-1} \in H$

$\therefore \quad H^{-1} \subseteq H.$ \qquad ...(i)

Again $\quad h \in H \Rightarrow h^{-1} \in H$

$$\Rightarrow (h^{-1})^{-1} \in H^{-1} \Rightarrow h \in H^{-1}$$

$\therefore \quad H \subseteq H^{-1}$ \qquad ...(ii)

Hence from (i) and (ii), $H^{-1} = H$.

10.12 COSETS

Definition. *If G is a group, H is a sub-group and a any element in G, then the set*

$$\{h\,a : h \in H\}$$

is called the right coset generated by a and H and is denoted by Ha.
Similarly, the set

$$\{a\,h : h \in H\}$$

is the left coset denote by aH.

Group

If the group operation is 'addition', we define the right coset of H in G by
$$H + a = \{h + a : h \in H\}$$
Similarly left coset in an additive group shall be written as
$$a + H.$$
It must be noticed that cosets are not necessarily sub-groups of G. They are only special types of complexes which are sometimes called **residue classes modulo sub-group.**

In general $aH \neq Ha$. In the case of an abelian group each right coset coincides with the corresponding left coset.

EXAMPLES

1. *If H, K are two sub-groups of a group G, then HK is a sub-group of G iff $HK = KH$.*

Solution. (i) Condition is necessary. Let $HK = KH$ then we have to prove that HK is a sub-group of G. For this we will prove $(HK)(HK)^{-1} = HK$.

Now, $(HK)(HK)^{-1} = (HK)(K^{-1}H^{-1})$ (reversal law)
$= H(KK^{-1})H^{-1}$ (associativity)
$= (HK)H^{-1}$ $[\because K$ is a sub-group $\therefore KK^{-1} = K]$
$= (KH)H^{-1}$ $[\because HK = KH]$
$= KH$ $[\because H$ is a sub-group $\therefore HH^{-1} = H]$

$\therefore \quad HK = KH \Rightarrow HK$ is a sub-group.

(ii) Condition is sufficient. Suppose that HK is a sub-group.

Then $(HK)^{-1} = HK \Rightarrow K^{-1}H^{-1} = HK$
$\Rightarrow KH = HK$ $[\because H$ is a subgroup $\Rightarrow H^{-1} = H$ and K is a subgroup $\Rightarrow K^{-1} = K]$

Hence the result.

2. $G = \{a, a^2, a^3, a^4 = 1\}, o(G) = 4, H = \{1, a^2\}$ *is a sub-group of G. Find all the cosets of H in G and prove that G is equal to union of all these cosets and also establish that any two cosets are either disjoint or identical.*

Solution. We have
$1.H = \{1, a^2\} = H$
$a.H = \{a, a^3\},$
$a^2.H = \{a^2, a^4\} = \{a^2, 1\} = H$ $[\because a^4 = 1]$
$a^3.H = \{a^3, a^5\} = \{a^3, a\} = \{a, a^3\} = a.H$ $[\because a^4 = 1]$

Thus there are only two **distinct** cosets namely H and aH which are disjoint. Also H is identical with a^2H and aH is identical with a^3H.

Again $H \cup aH = \{1, a^2\} \cup \{a, a^3\}$
$= \{1, a, a^2, a^3\} = G.$

10.12.1 Properties of Cosets

Theorem 1. *If $h \in H$, then the right (or left) coset Hh (or hH) of H is identical with H, and conversely.*

Proof. Let h' be an arbitrary element of H so that $hh' \in hH$. Again, since H is a sub-group, we have
$$h \in H, h' \in H \Rightarrow hh' \in H$$
Thus every element of hH is also an element of H.

Hence
$$hH \subset H \qquad \ldots(i)$$

Again $\qquad h' = (hh^{-1}) h' = h(h^{-1}h') \in hH \qquad [\because h^{-1} \in H, h' \in H \Rightarrow h^{-1}h' \in H]$

This shows that every element of H is also an element of hH. Hence
$$H \subset hH \qquad \ldots(ii)$$

From (i) and (ii) it follows that $\quad hH = H$

Similarly, we can show that $\quad Hh = H$

Conversely, $\qquad Hh = H \Rightarrow eH \in H \Rightarrow h \in H$ and similarly $hH \Rightarrow H \Rightarrow h \in H$.

Theorem 2. *Any two right (or left) cosets of H are either disjoint or identical.*

Proof. Let H be a sub-group of a group G and let aH and bH be two left cosets. Suppose these cosets are not disjoint. Then they possess an element, say c, in common. Then c may be written as $c = ah$, and also as $c = ah'$, where h and h' are in H.

Therefore, $\qquad ah = bh' \quad$ or $\quad a = bh'h^{-1}$

Since H is a sub-group, $h'h^{-1} \in H$.

Let $\qquad h'h^{-1} = h''$

Then $\qquad a = bh''$

Hence $\qquad aH = (bh'') H$
$$= b(h''H) = bH \qquad \text{[Theorem 1]}$$

Therefore the two left cosets are identical if they are not disjoint. Thus either $aH \cap bH = \phi$ or $aH = bH$.

A similar result can be shown to hold for right cosets.

Theorem 3. *If H is finite the number of elements in a right (or left) coset of H is equal to order of H.*

Proof. The mapping $f : H \to Ha$, defined by $f(h_i) = h_i a$, is one-one onto.

It is one-one since
$$f(h_i) = f(h_j) \Rightarrow h_i a = h_j a \Rightarrow h_i = h_j,$$
(from the right cancellation law) it is onto, since an element belonging to Ha is the f-image of h belonging to H.

It follows that the number of elements in a right coset of H is the same as that in H.

Similarly the number of elements in a left coset of H is the same as that in H.

Theorem 4. *If $Ha = Hb$, where $a, b \in G$, then $ab^{-1} \in H$, and conversely if $aH = bH$, then $a^{-1}b \in H$.*

Proof. If $Ha = Hb$, then a belonging to Ha, is equal to some element $h_i b$ in Hb, i.e., $a = h_i b$, or $ab^{-1} = h_i \in H$

Conversely, if $ab^{-1} \in H$, then $ab^{-1} = h_i$ i.e., $a = h_i b$.

Group

Therefore, $Ha = H(h_i b) = (Hh_i)b$

$\qquad = Hb.$

Similarly we can slow that $\qquad aH = bH$ iff $a^{-1}b \in H.$

10.12.2 Coset Decomposition

Let H be a sub-group of G. We know that no right coset of H in G is empty and any two right cosets of H in G are either disjoint or identical.

The union of all right cosets of H in G is equal to G. Hence the set of all right cosets of H in G gives a partition of G.

This partition is called *right coset decomposition of G*. The procedure to obtain distinct members of this partition is given below :

H itself is a right coset. Now suppose $a \in G$ and $a \notin H$ then Ha will be another distinct right coset. Again let b be another such element that $b \in G$ and $b \notin H$ and also $b \notin Ha$, then Hb will be another distinct right coset. Proceeding in this way all distinct right cosets of H in G will be obtained.

Thus $G = H \cup Ha \cup Hb \cup Hc \ldots$, where a, b, c are elements of G so chosen that all right cosets are distinct.

In the same way *left coset decomposition* of G can be obtained.

10.12.3 Relation of Congruence Modulo a Subgroup h is a Group g

Let H be a sub-group of a group G. If the element a of G belongs to the right coset Hb, *i.e.*, if $a \in Hb$, *i.e.*, if $ab^{-1} \in H$ then it is said that a is congruent to b modulo H.

Definition. *Let H be a sub-group of a group G. For $a, b \in G$ we say that a is congruent to b mod H if and only if $ab^{-1} \in H$.*

Symbolically, it can be expressed as $a \equiv b \pmod{H}$ iff $ab^{-1} \in H$.

Theorem 5. *The relation of congruency in a group G defined by $a \equiv b \pmod{H}$ iff $ab^{-1} \in H$ is an equivalence realtion.*

Proof. (i) Reflexivity. Let $a \in G$ then $aa^{-1} = e \in H$ because H is a sub-group of G.

Hence $a \equiv b \pmod{H} \ \forall \ a \in G$.
The relation is reflexive.

(ii) Symmetry. $a \equiv b \pmod{H}$

$\qquad \Rightarrow ab^{-1} \in H$

$\qquad \Rightarrow (ab^{-1})^{-1} \in H \qquad\qquad$ [H is a sub-group of G]

$\qquad \Rightarrow ba^{-1} \in H$

$\qquad \Rightarrow b \equiv a \pmod{H}$

Hence the relation is symmetric.

(iii) Transitivity.

$\qquad a \equiv b \pmod{H}$ and $b \equiv c \pmod{H}$

$\qquad \Rightarrow ab^{-1} \in H$ and $bc^{-1} \in H$

$\qquad \Rightarrow (ab^{-1})(bc^{-1}) \in H \qquad\qquad$ (Closure property)

$\qquad \Rightarrow a(b^{-1}b)c^{-1} \in H$

$$\Rightarrow ac^{-1} \in H \qquad [\because bb^{-1} = c].$$
$$\Rightarrow a \equiv c \pmod{H}.$$

Hence the relation is transitive.

Thus the relation congruence mod H is an equivalence relation in G.

10.13 LAGRANGE'S THEOREM

The order of a subgroup of a finite group is a divisor of the order of the grouip.

Proof. Let H be any sub-group of order m of a finite group G of order n. Let us consider the left coset decomposition of G relative to H.

We will first show that each coset aH consists of m different elements.

Let
$$H = \{h_1, h_2, ..., h_m\}$$

Then $ah_1, ah_2, ..., ah_m$, are the members of aH, all distinct.

For, we have
$$ah_i = ah_j \Rightarrow h_i = h_j$$

by cancellation law in G

Since G is a finite group, the number of distinct left cosets will also be finite, say k. Hence the total number of elements of all cosets is km which is equal to the total number of elements of G. Hence
$$n = mk.$$

This show that m, the order of H, is a divisor of n, the order of the group G.

We also find that the index k is also a divisor of the order of the group.

Corollary 1. *If G is of finite order n, then the order of any $a \in G$ divides the order of G and in particular $a^n = e$.*

Proof. Let a be of order m so that m is the least positive integer such that
$$a^m = e.$$

Then it is easy to verify that the elements
$$a, a^2, a^3, ..., a^{m-1}, a^m, = e$$

of G are all distinct and form a sub-group.

Since this sub-group is of order m, it follows that m, the order of a, is a divisor of the order of the group.

We may write $n = mk$, where k is a positive integer.

Then
$$a^n = a^{mk} = (a^m)^k = e.$$

Corollary 2. *A finite group of prime order has no proper sub-groups.*

Proof. Let the order of the group G be a prime number p.

Since p is prime, its only divisors are 1 and p.

Therefore the only sub-groups of G are $\{e\}$ and G, i.e., the group G has no proper sub-groups.

Corollary 3. *Every group of prime order is cyclic.*

Proof. Let G be a group of prime order p, and let $a \neq e \in G$. Since the order of a is a divisor of p, it is either 1 or p.

But $o(a) \neq 1$, since $a \neq e$.

Therefore, $o(a) = p$, and the cyclic subgroup of G generated by a is also of order p.

It follows that G is identical with the cyclic sub-group generated by a, i.e., G is cyclic.

Corollary 4. *Every finite group of composite order possesses proper subgroups.*

Proof. Let G be a finite group of order $n = pq$, where $p \neq 1, q \neq 1$,

If G be cyclic, and a be its generator, the group generated by a^p (or a^q) is a sub-group of G of order q (or p).

If G be not cyclic, and $a \neq e \in G$, the order of a is a proper divisor of n. Therefore the cyclic group generated by a is a proper sub-group of G.

Corollary 5. Fermat's Theorem. *If p is a prime number which does not divide the integer a, then $a^{p-1} \equiv 1 \pmod{p}$.*

Proof. The multiplicative groups of non-zero residue classes modulo p is
$$G = (\{1, 2, 3, ..., p-1\} \times_p),$$
which is of order $p-1$, with the identity 1.

Since a is not divisible by p, it is congruent to one of the numbers $1, 2, 3, ..., p-1$, modulo p.

It follows from Cor. 1, that $a^{p-1} \equiv 1 \pmod{p}$.

Corollary 6. *Euler's Function.*

Definition. The Euler's function, $\phi(n)$ represents number of elements in the set A_n where
$$A_n = \{m \in N : 1 \leq m < n \text{ s.t. } (m, n) = 1\}$$
$(m, n) = 1$ means m and n are relatively prime.

Euler's Theorem : *If a and m are positive integers relatively prime, then*
$$a^{\phi(m)} \equiv 1 \pmod{m}, \qquad \text{where } \phi(m)$$
represents Euler's functions and $m > 1$.

Proof. For any integer x, let $[x]$ denotes the residue class of the set of integers mod m. Let
$$G = \{[a] : a \text{ is an integer relative to } m\}$$

Then we know that with respect to multiplication of residue classes G is a group of order $\phi(m)$. The identity element of this group is the residue class $[1]$. We have
$$[a] \in G \Rightarrow [a]^{o(G)} = [1]$$
$$\Rightarrow [a]^{\phi(m)} = [1]$$
$$\Rightarrow [a][a][a]... \text{ upto } \phi(m) = [1]$$
$$\Rightarrow [a\, a\, a\, ... \text{ upto } \phi(m) \text{ times}] = [1]$$
$$\Rightarrow [a]^{\phi(m)} = 1 \text{ [Note that } [a][b] = [ab]]$$
$$\Rightarrow [a]^{\phi(m)} \equiv 1 \pmod{m}$$

EXAMPLES

1. *Show that two right cosets Ha, Hb are distinct if and only if the two left cossets $a^{-1}H, b^{-1}H$ are distinct.*

Solution. Suppose Ha and Hb are distinct. We have to prove that $a^{-1}H, b^{-1}H$ are distinct.

If $a^{-1}H, b^{-1}H$ are not distinct, then they are equal, *i.e.,*
$$a^{-1}H = b^{-1}H.$$
Now $\quad a^{-1}H = b^{-1}H \Rightarrow a^{-1} \in b^{-1}H \Rightarrow ba^{-1} \in H.$
$$\Rightarrow (ba^{-1})^{-1} \in H \Rightarrow ab^{-1} \in H$$
$$\Rightarrow a \in Hb \Rightarrow Ha = Hb.$$
But this is a contradiction, since Ha and Hb are distinct.

It follows that $a^{-1}H, b^{-1}H$ are not equal, *i.e.,* they are distinct. The converse follows in the same manner.

2. *Use Lagrange's theorem to prove that a finite group can not be expressed as the union of two of its proper sub-groups.*

Solution. Let G be a finite group of order n. Suppose G is the union of two of its proper sub-groups H and K.

Since $e \in$ both H and K and $G = H \cup K$, therefore atleast one of H and K (say, H) must contain more than half the elements of G. Let $o(H) = p$. Then $n/2 < p < n$, since H is a proper sub-groups of G.

Since $n/2 < p < n$ therefore p cannot be a divisor of n. This contradicts Lagrange's theorem which states that the order of each sub-group of a finite group is a divisor of the order of the group.

Hence our initial assumption is wrong and so a finite group cannot be expressed as the union of two of its proper sub-groups.

3. *Prove that the sets H and H^x are of the same order.*

Solution. We define H^x as follows :

$$H^x = x^{-1} H x$$

Thus the elements of H^x are of the form $x^{-1}hx$, for some x and all $h \in H$.

Then $h \Rightarrow x^{-1}hx$

Conversely $xhx^{-1} = x(x^{-1}hx)x^{-1}$

$$= (xx^{-1}) h (xx^{-1})$$

$$= h$$

Hence the correspondence between H and H^x is one-one.

Hence H and H^x have the same number of elements *i.e.,* their order is the same.

EXERCISES

1. If G is a cyclic group of order n, generated by a then show that for any divisor d of n, there exists a unique sub-group of G of order d.
2. Prove that there is a one-one correspondence between the set of left cosets of H and the set of right cosets of H.
3. If H be a sub-group of a group G, and $a \in G$, then show that

$$(Ha)^{-1} = a^{-1}H.$$

4. By using Lagrange's Theorem for finite groups, prove that $\{0, 1, 2, 3\}$ is not a sub-group of $(Z_9, +_9)$.
5. Let H be a subgroup of G. For $a, b \in G$, prove that

$$Ha = Hb \Leftrightarrow a^{-1}H = b^{-1}H$$

11
Permutations

11.1 PERMUTATIONS

Definitions. *Suppose S is a finite set having n distinct elements. Then a one-one mapping of S onto itself is called a permutation of degree n.*

The number of elements in the finite set S is known as the degree of permutation.

Symbol for a permutation. Let $S = \{a_1, a_2, a_3, ..., a_n\}$ be a finite set having n distinct elements. If $f : S \to S$ is one-one onto mapping, then f is permutation of degree n.

Let $f(a_1) = b_1, f(a_2) = b_2, f(a_3) = b_3, ... f(a_n) = b_n$, where $\{b_1, b_2, ..., b_n\}$ = $\{a_1, a_2, a_3, ..., a_n\}$, i.e., $b_1, b_2, ..., b_n$ is one arrangement of the n elements $a_1, a_2, ..., a_n$.

It is customary to write a permutation in a two line symbol. In this notation we write

$$f = \begin{pmatrix} a_1 & a_2 & a_3 & \cdots & a_n \\ b_1 & b_2 & b_3 & \cdots & b_n \end{pmatrix}$$

i.e., each element in the second row is the f-image of the element of the first row lying directly above it.

If $S = \{1, 2, 3, 4\}$ be a finite set of order four then

$$S = \begin{pmatrix} 1 & 2 & 3 & 4 \\ 2 & 4 & 1 & 3 \end{pmatrix}, g = \begin{pmatrix} 1 & 2 & 3 & 4 \\ 1 & 3 & 2 & 4 \end{pmatrix}, h = \begin{pmatrix} 1 & 2 & 3 & 4 \\ 1 & 2 & 4 & 3 \end{pmatrix}$$

etc. are all permutations of degree four, here in the permutation f the element 1, 2, 3, 4, have been replaced respectively by the elements 2, 4, 1, 3. Thus $f(1) = 2, f(2) = 4, f(3) = 1, f(4) = 3$.

Similarly $g(1) = 1, g(2) = 3, g(3) = 2, g(4) = 4$
and $h(1) = 1, h(2) = 2, h(3) = 4, h(4) = 3$.

Thus for a permutation f on S, we just put the elements of S in one row in any order we like and below each element of this row we put down its image under f, g or h to obtain another row of elements of S.

Equality of two permutations

Two permutations f and g of degree n are said to be equal if we have $f(a) = g(a) \ \forall \ a \in S$.

Example : If $f = \begin{pmatrix} 1 & 2 & 3 & 4 \\ 2 & 3 & 4 & 1 \end{pmatrix}$, and $g = \begin{pmatrix} 2 & 4 & 1 & 3 \\ 3 & 1 & 2 & 4 \end{pmatrix}$

are two permutations of degree 4, then we have $f = g$. Here we see that both f and g replace 1 by 2, 2 by 3 by 4 and 4 by 1.

If $f = \begin{pmatrix} a_1 & a_2 & a_3 & \cdots & a_n \\ b_1 & b_2 & b_3 & \cdots & b_n \end{pmatrix}$ is a permutation of degre n, we can write it in several ways.

The interchange of columns wil not change the permutation. Thus, we can write

If $f = \begin{pmatrix} a_2 & a_3 & a_1 & \cdots & a_n \\ b_2 & b_3 & b_1 & \cdots & b_n \end{pmatrix} = \begin{pmatrix} a_n & a_1 & \cdots & a_{n-1} \\ b_n & b_1 & \cdots & b_{n-1} \end{pmatrix}$

$$= \begin{pmatrix} a_{n-1} & a_n & a_{n-2} & \cdots & a_1 \\ b_{n-1} & b_n & b_{n-2} & \cdots & b_1 \end{pmatrix}$$

Therefore, if f and g are two permutations of the same elements of degree n, then it is always possible to write g in such a way that the first row of g coincide with the second row of f.

For example, if

$$f = \begin{pmatrix} a_1 & a_2 & a_3 & a_4 \\ a_2 & a_4 & a_1 & a_3 \end{pmatrix} \text{ and } g = \begin{pmatrix} a_1 & a_2 & a_3 & a_4 \\ a_4 & a_3 & a_2 & a_1 \end{pmatrix}$$

the by interchanging the columns of g we can write

$$g = \begin{pmatrix} a_2 & a_4 & a_1 & a_3 \\ a_3 & a_1 & a_4 & a_2 \end{pmatrix}$$

Total number of distinct permutations of degree n. If S is a finite set having n distinct elements, then we shall have $n!$ distinct arrangements of the elements of S. Therefore there will be $n!$ distinct permutations of degree n. If P_n be the set consisting of all permutations of degree n then the set P_n will have $n!$ distinct elements. This set P_n is called the **symmetric set of permutations** of degree n. Sometimes it is also denoted by S_n. Thus $P_n = (f : f$ is a permutation of degree $n)$.

The set P_3 of all permutations of degree 3 will have 3! i.e., 6 elements. Obviously

$$P_3 = \left\{ \begin{pmatrix} 1 & 2 & 3 \\ 1 & 2 & 3 \end{pmatrix}, \begin{pmatrix} 1 & 2 & 3 \\ 3 & 1 & 2 \end{pmatrix}, \begin{pmatrix} 1 & 2 & 3 \\ 3 & 1 & 2 \end{pmatrix}, \begin{pmatrix} 1 & 2 & 3 \\ 3 & 2 & 1 \end{pmatrix}, \begin{pmatrix} 1 & 2 & 3 \\ 3 & 2 & 1 \end{pmatrix}, \begin{pmatrix} 1 & 2 & 3 \\ 1 & 3 & 2 \end{pmatrix}, \begin{pmatrix} 1 & 2 & 3 \\ 2 & 1 & 3 \end{pmatrix} \right\}.$$

Identity Permutation. *If I is a permutation fo degree n such that I replaces each element by the element iself, I is called the identity permutation of degree n.*

Thus
$$I = \begin{pmatrix} 1 & 2 & 3 & \cdots & n \\ 1 & 2 & 3 & \cdots & n \end{pmatrix}$$

or
$$\begin{pmatrix} a_1 & a_2 & a_3 & \cdots & a_n \\ a_1 & a_2 & a_3 & \cdots & a_n \end{pmatrix}$$

or
$$\begin{pmatrix} b_1 & b_2 & b_3 & \cdots & b_n \\ b_1 & b_2 & b_3 & \cdots & b_n \end{pmatrix}$$

is the identity permutation of degree n.

Product or Composite of two permutations. *The product or composite of two permutations f and g of degree n denoted by fg, is obtained by first carrying out the operation defined by f then by f then by g.*

Let us suppose P_n is the set of all permutations of degree n. Let

$$f = \begin{pmatrix} a_1 & a_2 & a_3 & \cdots & a_n \\ b_1 & b_2 & b_3 & \cdots & b_n \end{pmatrix} \text{ and}$$

$$g = \begin{pmatrix} b_1 & b_2 & b_3 & \cdots & b_n \\ c_1 & c_2 & c_3 & \cdots & c_n \end{pmatrix}$$

be any two elements of P_n.

Here the permutation g has been written in such a way that the first row of g coincides with the second row of f. If the product of the permutations f and g is denoted multiplicatively, i.e., by fg, then by definition

$$fg = \begin{pmatrix} a_1 & a_2 & a_3 & \cdots & a_n \\ c_1 & c_2 & c_3 & \cdots & c_n \end{pmatrix}$$

Permutations

For, f replaces a_1 by b_1 and then g replaces b_1 by c_1 so that $f g$ replaces a_1 by c_1. Similarly fg replaces a_2 by c_2, a_3 by c_3, ..., a_n by c_n.

Obviously, fg is also a permutation of degree n. Thus the product of two permutations of degree n is also a permutation of degree n. Therefore $f\, g \in P_n \; \forall \; f, g \in P_n$.

Inverse of permutations. If f be a permutation of degree n, defined on a finite set S consisting of a n distinct elements, then by definition f is a one-one mapping of S onto itself. Since f is one-one onto, it is invertible. Let f^{-1} be the inverse of map f then f^{-1} will be one-one onto map of S onto itself. Thus, f^{-1} is also a permutation of degree n on S. This f^{-1} is known as the inverse of the permutations f.

Thus, if

$$f = \begin{pmatrix} a_1 & a_2 & a_3 & \cdots & a_n \\ b_1 & b_2 & b_3 & \cdots & b_n \end{pmatrix}$$

then

$$f^{-1} = \begin{pmatrix} b_1 & b_2 & b_3 & \cdots & b_n \\ a_1 & a_2 & a_3 & \cdots & a_n \end{pmatrix}$$

Note : Evidently f^{-1} is obtained by interchanging the rows of f because $f(a_1) = b_1 \Rightarrow f^{-1}(b_1) = a_1$ etc.

EXAMPLES

1. If $f = \begin{pmatrix} 1 & 2 & 3 \\ 1 & 3 & 2 \end{pmatrix}$ and $g = \begin{pmatrix} 1 & 2 & 3 \\ 2 & 3 & 1 \end{pmatrix}$ be two permutations of degree 3. Find fg and gf and show that $fg \ne gf$.

Solution.

$$\mathbf{fg} = \begin{pmatrix} 1 & 2 & 3 \\ 1 & 3 & 2 \end{pmatrix}\begin{pmatrix} 1 & 2 & 3 \\ 2 & 3 & 1 \end{pmatrix}$$

$$= \begin{pmatrix} 1 & 2 & 3 \\ 1 & 3 & 2 \end{pmatrix}\begin{pmatrix} 1 & 3 & 2 \\ 2 & 1 & 3 \end{pmatrix} = \begin{pmatrix} 1 & 2 & 3 \\ 2 & 1 & 3 \end{pmatrix}$$

and

$$\mathbf{gf} = \begin{pmatrix} 1 & 2 & 3 \\ 2 & 3 & 1 \end{pmatrix}\begin{pmatrix} 1 & 2 & 3 \\ 1 & 3 & 2 \end{pmatrix}$$

$$= \begin{pmatrix} 1 & 2 & 3 \\ 2 & 3 & 1 \end{pmatrix}\begin{pmatrix} 2 & 3 & 1 \\ 3 & 2 & 1 \end{pmatrix}$$

$$= \begin{pmatrix} 1 & 2 & 3 \\ 3 & 2 & 1 \end{pmatrix}$$

Obviously, $\mathbf{fg} \ne \mathbf{gf}$.

2. If $f = \begin{pmatrix} a_1 & a_2 & a_3 & \cdots & a_n \\ b_1 & b_2 & b_3 & \cdots & b_n \end{pmatrix}$ then prove that $ff^{-1} = I$ where I is the identity permutation of degree n.

Solution.

$$ff^{-1} = \begin{pmatrix} a_1 & a_2 & a_3 & \cdots & a_n \\ b_1 & b_2 & b_3 & \cdots & b_n \end{pmatrix}\begin{pmatrix} b_1 & b_2 & b_3 & \cdots & b_n \\ a_1 & a_2 & a_3 & \cdots & a_n \end{pmatrix}$$

$$\left[f^{-1} = \begin{pmatrix} b_1 & b_2 & b_3 & \cdots & b_n \\ a_1 & a_2 & a_3 & \cdots & a_n \end{pmatrix} \right]$$

$$= \begin{pmatrix} a_1 & a_2 & a_3 & \cdots & a_n \\ a_1 & a_2 & a_3 & \cdots & a_n \end{pmatrix} = I.$$

$\Rightarrow \quad ff^{-1} = I$

11.2 GROUP OF PERMUTATIONS

The set P_n of all permutations on n symbols is a finite group of order n! with respect to composite of mappings as the operation. For $n \leq 2$, this group is abelian and for $n > 2$ it is always non-abelian.

Let $S = [a_1, a_2, a_3, \ldots, a_n]$ be a finite set having n distinct elements. Thus there are $n!$ permutations possible on S. If P_n denotes the set of all permutations of degree n then multiplication of permutations on P_n satisfies the following axioms.

Closure axiom. Let $f, g \in P_n$ then each of them is one-one mapping of S onto itself and, therefore, their composite mapping $(g \circ f)$ is a one-one mapping of S onto itself. Thus $(g \circ f)$ is a permutation of degree n on S, i.e.,

$$f, g \in P_n \Rightarrow f\ g \in P_n.$$

This shows that P_n is closed under multiplication.

Associative axiom. Since the product of two permutations on a set S is nothing but the product of two one-one onto mappings on S and the product of mapping being associative, the product of permutations also obeys the associative law. Hence

$$(f\ g)\ h = f\ (g\ h) \text{ for } f, g, h \in P_n.$$

For example let

$$f = \begin{pmatrix} a_1 & a_2 & \cdots & a_n \\ b_1 & b_2 & \cdots & b_n \end{pmatrix}.$$

$$g = \begin{pmatrix} b_1 & b_2 & \cdots & b_n \\ c_1 & c_2 & \cdots & c_n \end{pmatrix}, h = \begin{pmatrix} c_1 & c_2 & \cdots & c_n \\ d_1 & d_2 & \cdots & d_n \end{pmatrix}$$

where $b_1, b_2, \ldots, b_n; c_1, c_2, \ldots, c_n$ and $d_1, d_2, \ldots, d_n$ are simply different arrangements of the same n elements $a_1, a_2, \ldots, a_n$.

Now, $\quad f\ g = \begin{pmatrix} a_1 & a_2 & \cdots & a_n \\ c_1 & c_2 & \cdots & c_n \end{pmatrix},$

$$(f\ g)\ h = \begin{pmatrix} a_1 & a_2 & \cdots & a_n \\ c_1 & c_2 & \cdots & c_n \end{pmatrix} \begin{pmatrix} c_1 & c_2 & \cdots & c_n \\ d_1 & d_2 & \cdots & d_n \end{pmatrix}$$

$$= \begin{pmatrix} a_1 & a_2 & \cdots & a_n \\ d_1 & d_2 & \cdots & d_n \end{pmatrix}$$

Also, $\quad g\ h = \begin{pmatrix} b_1 & b_2 & \cdots & b_n \\ c_1 & c_2 & \cdots & c_n \end{pmatrix} \begin{pmatrix} c_1 & c_2 & \cdots & c_n \\ d_1 & d_2 & \cdots & d_n \end{pmatrix}$

$$= \begin{pmatrix} b_1 & b_2 & \cdots & b_n \\ d_1 & d_2 & \cdots & d_n \end{pmatrix}$$

Permutations

$$\therefore \quad f(g\,h) = \begin{pmatrix} a_1 & a_2 & \cdots & a_n \\ b_1 & b_2 & \cdots & b_n \end{pmatrix} \begin{pmatrix} b_1 & b_2 & \cdots & b_n \\ d_1 & d_2 & \cdots & d_n \end{pmatrix}$$

$$= \begin{pmatrix} a_1 & a_2 & \cdots & a_n \\ d_1 & d_2 & \cdots & d_n \end{pmatrix} = (f\,g)\,h.$$

Hence $\quad f(gh) = (fg)\,h$

Identity axiom. Identity permutation $I \in P_n$ is identity of multiplication in P_n because $If = fI = f \ \forall \ f \in P_n$.

Inverse axiom. Let $f \in P_n$ then f is one-one onto mapping, hence it is invertible. Hence f^{-1}, the inverse mapping of f is also one-one and onto. Consequently, f^{-1} is also a permutation in P_n.

$$\therefore \quad f^{-1}f = f\,f^{-1} = I.$$

For example let $\quad f = \begin{pmatrix} a_1 & a_2 & \cdots & a_n \\ b_1 & b_2 & \cdots & b_n \end{pmatrix}$ be any element of P_n then

$$f^{-1} = \begin{pmatrix} b_1 & b_2 & \cdots & b_n \\ a_1 & a_2 & \cdots & a_n \end{pmatrix}$$

Now, $\quad f^{-1}f = \begin{pmatrix} b_1 & b_2 & \cdots & b_n \\ a_1 & a_2 & \cdots & a_n \end{pmatrix} \begin{pmatrix} a_1 & a_2 & \cdots & a_n \\ b_1 & b_2 & \cdots & b_n \end{pmatrix}$

$$= \begin{pmatrix} b_1 & b_2 & \cdots & b_n \\ b_1 & b_2 & \cdots & b_n \end{pmatrix} = I$$

Similarly $\quad ff^{-1} = \begin{pmatrix} a_1 & a_2 & \cdots & a_n \\ b_1 & b_2 & \cdots & b_n \end{pmatrix} \begin{pmatrix} b_1 & b_2 & \cdots & b_n \\ a_1 & a_2 & \cdots & a_n \end{pmatrix} = \begin{pmatrix} a_1 & a_2 & \cdots & a_n \\ a_1 & a_2 & \cdots & a_n \end{pmatrix} = I$

Hence $\quad f^{-1}f = ff^{-1} = I$

Therefore f^{-1} is the multiplicative inverse of f.

Thus the symmetric set P_n of all permutations of degree n defined on a finite set forms a finite group of order n! with respect to the composite of permutations as the composition.

Commutative axiom. If we consider the the symmetric group (P_1, o) of permutations of degree 1 with respect to permutations product o, then it consists of a single permutations namely the identity permutation I. Since $I \, o \, I = I$, (P_1, o) is an abelian group. If we conisder the symmetric group (P_2, o) of all permutations of degree 2, *i.e.*, the group of all permutations defined on a set of two elements (a_1, a_2), then

$$P_2 \left\{ \begin{pmatrix} a_1 & a_2 \\ a_1 & a_2 \end{pmatrix}, \begin{pmatrix} a_1 & a_2 \\ a_2 & a_1 \end{pmatrix} \right\}$$

Now $\quad \begin{pmatrix} a_1 & a_2 \\ a_1 & a_2 \end{pmatrix} o \begin{pmatrix} a_1 & a_2 \\ a_2 & a_1 \end{pmatrix} = \begin{pmatrix} a_1 & a_2 \\ a_2 & a_1 \end{pmatrix}$

and $\quad \begin{pmatrix} a_1 & a_2 \\ a_2 & a_1 \end{pmatrix} o \begin{pmatrix} a_1 & a_2 \\ a_1 & a_2 \end{pmatrix} = \begin{pmatrix} a_1 & a_2 \\ a_2 & a_1 \end{pmatrix} \begin{pmatrix} a_1 & a_2 \\ a_2 & a_1 \end{pmatrix} = \begin{pmatrix} a_1 & a_2 \\ a_2 & a_1 \end{pmatrix}$

Therefore operation having commutative (P_2, o) is abelian group of order 2.

But when $n > 2$ then permutation product is not necessarily commutative. Hence (P_n, o) then is not necessarily an abelian group.

EXAMPLES

Show that the four permutations $I, (a\ b), (c\ d), (a\ b)(c\ d)$ on four symbols a, b, c, d from a finite abelian group with respect to the permutation multiplication.

Solution. Let $I = f_1, (ab) = f_2, (cd) = f_3, (ab)(cd) = f_4$. To prepare the composition table, we observe that f_2 and f_3 are transpositions. Therefore $f_2 f_2 = f_1, f_3 f_3 = f_1$. Also f_2 and f_3 are disjoint cycles.

Therefore, $\qquad f_2 f_3 = f_3 f_2 = f_4.$

Further $\qquad f_2 f_4 = (a\ b)(a\ b)(c\ d)$
$\qquad\qquad = I(c\ d) = (c\ d) = f_3$

Similarly, $\qquad f_3 f_4 = (c\ d)(a\ b)(c\ d)$
$\qquad\qquad = (c\ d)(c\ d)(a\ b) = I(a\ b)$
$\qquad\qquad = (a\ b) = f_2 \qquad\qquad [\because (a\ b)(c\ d) = (c\ d)(a\ b)]$

Also, $\qquad f_4 f_4 = (a\ b)(c\ d)(a\ b)(c\ d)$
$\qquad\qquad = (a\ b)(a\ b)(c\ d)(c\ d)$
$\qquad\qquad = (I)(I) = I = f_1.$

Similarly, making all other calculations, the composition table is

Production of Permutations	f_1	f_2	f_3	f_4
f_1	f_1	f_2	f_3	f_4
f_2	f_2	f_1	f_4	f_3
f_3	f_3	f_4	f_1	f_2
f_4	f_4	f_3	f_2	f_1

From the table it is clear that

(i) All the entries in the composition table are elements of the given set. Therefore, the **closure axiom** is astisfied.

(ii) f_1 is the **identity** element.

(iii) Each element possesses **inverse**. In fact $f_1^{-1} = f_1, f_2^{-1} = f_2, f_3^{-1} = f_3, f_4^{-1} = f_4$.

(iv) The composition is **commutative.**

(v) The multiplication of permutations is an **associative** composition. Hence, the given set is a finite abelian group of order 4 with respect to the permutation multiplication.

11.3 ORBIT OF PERMUTATIONS

Let f be a permutation on a set S. If a relation $\sim$ is defined on S such that $a \sim b \Leftrightarrow f^{(n)}(a) = b$ for some integral $n\ \forall\ a, b \in S$, we observe that the relation is

(i) **Reflexive,** because $a \sim a \Leftrightarrow f^{(o)}(a)$

Permutations

$\Rightarrow I(a) = a \; \forall \; a \in S.$

(ii) **Symmetric**, because $a \sim b \Rightarrow f^{(n)}(a) = b$ for some integer n

$\Rightarrow a = f^{-(n)}(b)$

$\Rightarrow b \sim a$ for $a, b \in S.$

(iii) **Transitive**, because $a \sim b$ and $b \sim c$

$\Rightarrow f^{(n)}(a) = b, f^{(m)}(b) = c$ for some integers n and m.

$\Rightarrow f^m(f^n(a)) = f^{(m)}(b) = c$

$\Rightarrow f^{m+n}(a) = c$ for some integer $(m+n)$

$\Rightarrow a \sim c.$

Thus above defined relation $\sim$ is an equivalence relation on S and hence partitions it into mutually disjoint classes. Each equivalence class determined by the above relation is called an orbit of f.

11.4 CYCLIC PERMUTATIONS

A permutation of the type

$\begin{pmatrix} a_1 & a_2 & a_3 & \cdots & a_{n-1} & a_n \\ a_2 & a_3 & a_4 & \cdots & a_n & a_1 \end{pmatrix}$ is called a cyclic permutation or a cycle. It is usually

denoted by the symbol $(a_1, a_2, ..., a_n)$.

Thus if f is a permutation of degree n on a set S having n distinct elements and if it is possible to arrange some of the elements (say m in number) of the set S in a row such that the f-image of each element in this row is the following it and the f-image of the last element in the row is the first element and the remaining $(n-m)$ elements of the set S remain invariant under f, then f is called a cyclic permutation or a cycle of length m.

The number of objects permuted by the cycle is called the **length of cycle.**

Thus by the cycle of length one we mean a permutation in which the image of each element remains unchanged under a permutation f. Consequently, the cycle permutation of length one is the identity permutation.

Oner row symbol. One row symbol is used to denote a cyclic permutation. In this notation the element of S are arranged in such a way that the image of each element in this row is the element which follows it and that of the last element is the first element. Those elements of X which remain invariant need not be written in the row.

Let $f = \begin{pmatrix} 1 & 2 & 3 & 4 & 5 & 6 \\ 2 & 4 & 1 & 3 & 5 & 6 \end{pmatrix}$ be a cyclic permutation.

Since ther elements 1, 2, 3, 4 are such that $f(1) = 2, f(2) = 4, f(4) = 3$ and two remaining elements 5 and 6 remain invariant under f, f is a cycle of length 4 or a 4-cycle and can be expressed as $f(1\,2\,4\,3)$.

Transposition. *A cycle of length two is called a transposition.* Thus the cycle (1, 3) is a transposition. It is a 2-cycle such that the image of 1 is 3 and image of 3 is 1 and the remaining missing **elements are** invariant.

Disjoint cycles. Two cycles are said to be disjoint if when expressed in one row notations, they have no element in **common.**

11.5 SOME THEOREMS

Theorem 1. *The product of disjoint cycles is commutative.*

Proof. Let f and g be any two disjoint cycles, *i.e.*, there is no element common in two when they are expressed in one row notation. Therefore, the elements permuted by f are invariant under g and vice-versa.

Hence $f \circ g = g \circ f$; i.e., the product of disjoint cycles is commutative.

Theorem 2. *Every permutation can be expressed as a composite of disjoint cycles.*

Proof. Let the given permutation f be denoted by the usual two row symbol with in a bracket. Let a be any element in second row exactly beneath a, i.e., $f(a) = b$. Similarly, let $f(b) = c$. Continuing this process, an element 1 may be found in the upper row such that its f-image is a. Then $(a, b, c, \ldots l)$ is one circular permutation. If there are additional elements a', b' etc., in the original permutation f, follow the above process to obtain **another** cycle $(a', b', c', \ldots, l')$. Even now if, some element or elements are left in original permutation this procedure can be repeated to the extent that all elements of f are exhausted. In this way the original permutation can be put as the product of disjoint cycles.

Theorem 3. *Every permutation can be expressed as a product of transpositions.*

Proof. To prove the above result, we shall first show that every cycle can be expressed as a composite of transpositions. Let us consider a cycle $(a_1, a_2, \ldots, a_n)$ then

$$(a_1, a_2, \ldots, a_n) = (a_1\ a_n)(a_1\ a_{n-1}), \ldots, (a_1, a_2).$$

We have already proved that every permutation can be expressed as a composition of disjoint cycles. Therefore in the light of the two results stated above every permutation can be expressed as a product of transpositions.

EXAMPLE

Find $(a_1\ a_2)(a_1\ a_3)$

Solution.
$$(a_1\ a_2)(a_1\ a_3) = \begin{pmatrix} a_1 & a_2 & a_3 \\ a_2 & a_1 & a_3 \end{pmatrix} \begin{pmatrix} a_1 & a_2 & a_3 \\ a_3 & a_2 & a_1 \end{pmatrix}$$

$$= \begin{pmatrix} a_1 & a_2 & a_3 \\ a_2 & a_1 & a_3 \end{pmatrix} \begin{pmatrix} a_2 & a_1 & a_3 \\ a_2 & a_3 & a_1 \end{pmatrix}$$

$$= \begin{pmatrix} a_1 & a_2 & a_3 \\ a_2 & a_3 & a_1 \end{pmatrix} = (a_1\ a_2\ a_3)$$

11.6 EVEN AND ODD PERMUTATIONS

A permutation is said to be an even permutation if it can be expressed as a prdouct of an even number of transpositions, otherwise, it is said to be an odd permutation.

Theorem 1. *A permutation can not be both even and odd, i.e., if a permutation f is expressed as a product of transpositions then the number of transpositions is either always even or always odd.*

Proof. Let us consider the polynomail A in distinct symbol $x_1, x_2, \ldots, x_n$. It is defined as the product of $1/2\, n(n-1)$ factors of the form $x_i - x_j$ where $i < j$.

Thus
$$A = \prod_{i \leq j = 1}^{n} (x_i - x_j)$$

$$= (x_1 - x_2)(x_1 - x_3)(x_1 - x_4) \ldots (x_1 - x_n)$$
$$(x_2 - x_3)(x_2 - x_4) \ldots (x_2 - x_n)$$
$$(x_3 - x_4) \ldots (x_3 - x_n) \ldots (x_{n-1} - x_n)$$

Consider now any permutation P on n symbols 1, 2, 3, ..., n. By AP we mean the polynomial

Permutations

obtained by permutation the subscripts $1, 2, ..., n$ of the x_i as prescribed by P.

For example, taking $n = 4$, we have
$$A = (x_1 - x_2)(x_1 - x_3)(x_1 - x_4)(x_2 - x_3)(x_2 - x_4)(x_3 - x_4)$$
and if P (1 3 4 2), then
$$AP = (x_3 - x_1)(x_3 - x_4)(x_3 - x_2)(x_1 - x_4)(x_1 - x_2)(x_4 - x_3)$$
In particular if $P = (1, 2)$ we have
$$AP = (x_2 - x_1)(x_2 - x_3)(x_2 - x_4)(x_1 - x_3)(x_1 - x_4)(x_3 - x_4)$$
$$= -A.$$
This shows that the effect of a transposition on A is to change the sign of A.

In general a transposition $(i, j) \, i < j$ has the following effects on A:

(i) Any factor which involves neither the suffix i nor j remains unchanged.

(ii) The single factor $(x_i - x_j)$ changes its sign.

(iii) The remaining factors which involve either the suffix i or j but not both can be grouped into pairs of products, $\pm(x_m - x_i)(x_m - x_j)$ where $m \neq i$ or j and such a product remains unaltered when x_i and x_j are interchanged.

Hence the net effect of transposition (i, j) on A is to change its sign, i.e., A operated upon transposition (i, j) gives $-A$.

Now the permutation P considered as a product of s transpositions when operated upon A gives $(-1)^s A$ so that $AP = (-1)^s A$ and considered as a product of t transpositions gives $(-1)^t A$ so that $AP = (-1)^t A$.

Hence $\qquad (-1)^s A = (-1)^t A$

or $\qquad (-1)^s = (-1)^t$

Now, this equation will hold only if s and t are either both even or both odd. Hence the theorem.

Theorem 2. *Of the $n!$ permutations on n symbols, $\frac{1}{2}n!$ are even permutations and $\frac{1}{2}n!$ are odd permutations.*

Proof. Let the even permutations be $e_1, e_2, ..., e_m$ and the odd permutations be $O_1, O_2, ..., O_k$.

Then $\quad m + k = n!$

Now let t be any transposition. Since t is evidently an odd permutation, we see that $te_1, te_2, ..., te_m$ are odd permutations and that $tO_1, tO_2, ..., tO_k$ are even permutations. Since an odd permutation is never an even permutation, we have
$$te_i \neq tO_j$$
for any $i = 1, 2, ..., m; j = 1, 2, ..., k$. Furthermore, if
$$te_i = te_j,$$
then $e_i = e_j$ by cancellation law.

Similarly $\qquad tO_i \neq tO_j$ if $i \neq j$

It follows that all the m even permutations must appear in the list $tO_1, tO_2, ..., tO_k$, which are all distinct so that their number is m.

Similarly, all of the k odd permutations must be in the list.

$te_1, te_2, ..., te_m$.

which are all distinct as shown above and their number is k.

Note 1. A cycle containing an odd number of symbols is an even permutation, whereas a cycle containing an even number of symbols is an odd permutation, since a permutation on n symbols can be expressed as a product of $(n-1)$ transpositions.

Note 2. Inverse of an even permutation is an even permutation and the inverse of an odd permutation is an odd permutation.

Note 3. Product of two permutations is an even permutation if either both the permutations are even or both are odd and the product is an odd permutation if one permutation is odd and the other even.

EXAMPLES

1. *Write the following permutations as the product of disjoint cycles.*

(a) $f = \begin{pmatrix} 1 & 2 & 3 & 4 & 5 & 6 & 7 & 8 & 9 \\ 2 & 3 & 4 & 5 & 1 & 6 & 7 & 9 & 8 \end{pmatrix}$

(b) $g = \begin{pmatrix} 1 & 2 & 3 & 4 & 5 & 6 \\ 6 & 5 & 4 & 3 & 1 & 2 \end{pmatrix}$

Solution. We have

$$(f) = (6)\,(7)\,(1\ 2\ 3\ 4\ 5)\,(8\ 9)$$

or $f = (1\ 2\ 3\ 4\ 5)\,(8\ 9)$, omitting cycles of length I as they represent identity permutation.

(b) We have $g = (1\ 6\ 2\ 5)\,(3\ 4)$.

2. *Determine which of the following are even permutations :*

(a) $f = (1\ 2\ 3)\,(1\ 2)$

(b) $g = (1\ 2\ 3\ 4\ 5)\,(1\ 2\ 3)\,(4\ 5)$

(c) $h = (1\ 2)\,(1\ 3)\,(1\ 4)\,(2\ 5)$.

Solution. (a) We can write $f = (1\ 2)\,(1\ 3)\,(1\ 2)$. The number of transpositions is 3, *i.e.,* odd, Hence f is odd permutations.

(b) We have $g = (1\ 2)\,(1\ 3)\,(1\ 4)\,(1\ 5)\,(1\ 2)\,(1\ 3)\,(4\ 5)$

The number of transpositions is 7, *i.e.,* odd.

Hence g is an odd permutation.

(c) $h = (1\ 2)\,(1\ 3)\,(1\ 4)\,(2\ 5)$.

The number of transpositions is 4, *i.e.,* even. Hence h is an even permutation.

3. *Show that the set of six transformations $f_1, f_2, f_3, f_4, f_5, f_6$, on the set of complex number defined by*

$$f_1(z) = z,\ f_2(z) = 1/z,\ f_3(z) = 1-z$$

$$f_4(z) = \frac{z}{z-1},\ f_5(z) = \frac{1}{1-z},\ f_6(z) = \frac{z-1}{z}$$

forms a finite non-abelian group of order 6 with respect to the composite composition.

Solution. First of all we prepare a composition table. Clearly identity element for the composition is the identity mapping f_1, calculations can be made in the following manner :

$$(f_2 \circ f_2)(z) = f_2(f_2(z)) = f_2(1/z) = z$$

$$= f_1(z) \text{ so that } f_2 \circ f_2 = f_1$$

$$(f_2 \circ f_3)(z) = f_2(f_3(z)) = f_2(1-z) = 1/1-z$$

$$= f_5(z),$$

Permutations

so that $\quad f_2 \circ f_3 = f_5$.

$$(f_2 \circ f_4)(z) = f_2(f_4(z)) = f_2\left(\frac{z}{z-1}\right) = \frac{z-1}{z}$$

$$= f_6(z)$$

so that $\quad (f_2 \circ f_4) = f_6$.

$$(f_2 \circ f_5)(z) = f_2\left(\frac{1}{1-z}\right) = 1 - z = f_3(z)$$

$$(f_2 \circ f_6)(z) = f_2\left(\frac{z-1}{z}\right) = \frac{z}{z-1} = f_4(z)$$

This completes the second row of the table. Similarly other entries can be filled up. These are left as an exercise for the students. We thus get the following composition table.

$\circ$	f_1	f_2	f_3	f_4	f_5	f_6
f_1	f_1	f_2	f_3	f_4	f_5	f_6
f_2	f_2	f_1	f_5	f_6	f_3	f_4
f_3	f_3	f_6	f_1	f_5	f_4	f_2
f_4	f_4	f_5	f_6	f_1	f_2	f_3
f_5	f_5	f_4	f_2	f_3	f_6	f_1
f_6	f_6	f_3	f_4	f_2	f_1	f_5

EXERCISES

1. Write down all permutations on three symbols a, b, c. Which of these permutations are even?
   [**Ans.** 1 (identity permutation), $(a\ b), (b\ c), (a\ b\ c), (a\ c\ b)$; even permutations are $I, (a\ b\ c), (a\ c\ b)$]

2. Define a permutation. If $A = \begin{pmatrix} 1 & 2 & 3 \\ 2 & 3 & 1 \end{pmatrix}$ and $B = \begin{pmatrix} 1 & 2 & 3 \\ 3 & 1 & 2 \end{pmatrix}$, find AB and BA.
   [**Ans.** Both AB and BA are identity permutations]

3. Find the inverse of each of the following permutations :

   (i) $\begin{pmatrix} 1 & 2 & 3 & 4 \\ 1 & 3 & 4 & 2 \end{pmatrix}$ (ii) $\begin{pmatrix} 1 & 2 & 3 & 4 \\ 3 & 4 & 1 & 2 \end{pmatrix}$ (iii) $\begin{pmatrix} 1 & 2 & 3 & 4 & 5 \\ 2 & 3 & 1 & 5 & 4 \end{pmatrix}$

   $\left[\textbf{Ans.} \text{ (i) } \begin{pmatrix} 1 & 2 & 3 & 4 \\ 1 & 4 & 2 & 3 \end{pmatrix} \text{ (ii) } \begin{pmatrix} 1 & 2 & 3 & 4 \\ 3 & 4 & 1 & 2 \end{pmatrix} \text{ (iii) } \begin{pmatrix} 1 & 2 & 3 & 4 & 5 \\ 3 & 1 & 2 & 5 & 4 \end{pmatrix}\right]$

4. Decompose the following permutations into transpositions :

   (i) $\begin{pmatrix} 1 & 2 & 3 & 4 & 5 & 6 & 7 \\ 6 & 5 & 2 & 4 & 3 & 1 & 7 \end{pmatrix}$ (ii) $\begin{pmatrix} 1 & 2 & 3 & 4 & 5 & 6 & 7 & 8 \\ 3 & 1 & 4 & 7 & 2 & 5 & 8 & 6 \end{pmatrix}$

   [**Ans.** (i) $(1\ 6)(2\ 5)(2\ 3)$; (ii) $(1\ 3)(1\ 4)(1\ 7)(1\ 8)(1\ 6)(1\ 5)(1\ 2)$]

12
Homomorphism and Isomorphism

Just as the idea of congruence plays an important role in geometry, the idea which is common to all aspects of abstract algebra is the notion of isomorphism. Two congruent figures are considered to be geometrically equivalent in the sense that geometrical properties possessed by one are also possessed by the congruent figure. In abstract algebra congruence is replaced by the term 'isomorphism' which results by setting up one to one correspondence between two algebraic system of the same type, which preserves structure. With particular reference to groups, this leads us to the identification of groups on different sets and with different types of binary operations. We say that two isomorphic groups are algebraically equivalent in the sense that they possess the same algebraic properties. We make this more precise and illustrative in what follows, and make a beginning with a more general term called **Homomorphism**.

12.1 HOMOMORPHISM

By homomorphism we mean a mapping from one algebraic system to a like algebraic system which preserves structure.

Definition. Let G and G' be any two groups with binary operation 'o' and 'o'' respectively. Then a mapping $f : G \to G'$ said to be a homomorphism if for all $a, b \in G$, $f(aob) = f(a) o' f(b)$.

A homomorphism f, which at the same time is also onto is said to be an **epimorphism**.

A homomorphism f which is also one-one is called a **monomorphism**.

A group G' is called a homomorphic image of a group G, if there exists a homomorphism f of g onto G'.

A homomorphism of a group G into itself is called an **endomorphism**.

Examples:

(i) Let G be any group under binary operation o. If $f(x) = x$ for every $x \in G$, then $f : G \to G'$ is a homomorphism because
$$f(xy) = xy = f(x) f(y).$$

(ii) Let G be the group of integers under addition, let G' be the group of integers under addition modulo n. If $f : G \to G'$ be defined by f = remainder of x on division by n, then this is a homomorphism.

(iii) Let g be any group under addition. If $f(x) = e \, \forall \, x \in G$ then the mapping $f : G \to G'$, is a homomorphism because for all $x, y \in G$, $f(x + y) = e$ and $f(x) + f(y) = e + e = e$, so that
$$f(x + y) = f(x) + f(y)$$

(iv) Let G be the group of integers under addition and let $G' = G$. If for all $x \in G$, $f(x) = 2x$ then f is a homomorphism because
$$f(x + y) = 2(x + y) = 2x + 2y = f(x) + f(y)$$

12.2 KERNEL OF HOMOMORPHISM

Definition. If f is a homomorphism of a group G into a group G', then the set K of all those elements of G which are mapped by f onto the identity e' of G' is called the kernel of the homomorphism f.

Theorem. Let G and G' be any two groups and let e and e' be their respective identities. If f is a homomorphism of G into G', then

Homomorphism and Isomorphism

(1) $f(e) = e'$,

(2) $f(x^{-1}) = [f(x)]^{-1}$ for all $[x \in G.]$

(3) K is a normal subgroup of G.

Proof. (1) We know that for $x \in G$, $f(x) \in G'$,

$$f(x).e' = f(x) = f(xe) = f(x).f(e)$$

and therefore by using left cancellation law, we have

$$e' = f(e) \text{ or } f(e) = e'$$

(2) Since for any $x \in G$, $xx^{-1} = e$, we get

$$f(x).f(x^{-1}) = f(xx^{-1}) = f(e) = e'$$

Similarly $\quad x^{-1}x = e$, gives $f(x^{-1}).f(x) = e'$

Hence by definition of $[f(x)]^{-1}$ in G' we obtain the result.

$$f(x^{-1}) = [f(x)]^{-1}$$

(3) Since $f(e) = e'$, $e \in K$. This shows that $K \neq \phi$

Now let $\quad a, b \in K, x \in G, a \in K, b \in K$.

$\Rightarrow f(a) = e', f(b) = e' \quad \Rightarrow f(a) = e', f(b^{-1}) = [f(b)]^{-1} = e'$

$\Rightarrow f(ab^{-1}) = f(a)[f(b)]^{-1} = e'.e' = e' \quad \Rightarrow ab^{-1} \in K.$

This establish that K is a sub-group of G.

Now, to show that it is also normal we prove the following.

$$f(x^{-1}ax) = f(x^{-1})f(a)f(x)$$
$$= [f(x)]^{-1} f(a) f(x)$$
$$= [f(x)]^{-1} e' f(x) = [f(x)]^{-1} f(x) = e'$$

∴ $\quad x^{-1}ax \in K.$

Hence the result.

Theorem 2. *Let f be a homomorphism of a group G onto a group G' with kernel K and a be a given element of G such that $f(a) = a' \in G$. Then the set of all those elements of G which have their image a' in G' is the coset Ka of K in G.*

or

The set of all inverse image of a' under f is the right coset Ka, where a is any particular inverse image of a' in G.

$$K = \{x \in G : f(x) = e'\}.$$

Proof. Let us denote by A the set of all those elements of G which have their image a' in G'.

∴ $\quad A = \{x \in G : f(x) = a'\}$

We have to prove that $A = Ka$

Let $y \in Ka \Rightarrow y = ka \quad$ where $k \in K$ i.e., $f(k) = e'$

∴ $\quad f(y) = f(ka) = f(k) f(a) = e' f(a) = f(a) = a'$.

Since $f(y) = a'$, it follows that $y \in A$.

∴ $\quad y \in Ka \Rightarrow y \in A$ ∴ $Ka \subset A$

Again, let $z \in A \Rightarrow f(z) = a'$

Now $f(za^{-1}) = f(z) f(a)^{-1} = f(z)[f(a)]^{-1}$
$= a'(a')^{-1} \quad \because f(a) = a' \quad$ given

$\therefore \quad f(za^{-1}) = e' \Rightarrow za^{-1} \in K$ by def. of K.

$\Rightarrow \quad (za^{-1}) a \in Ka \Rightarrow z \in Ka.$

$\therefore \quad z \in a \Rightarrow z \in Ka. \quad \therefore A \subset Ka.$

Hence from (1) & (2) we prove that $A = Ka$.

Theorem 3. *Every quotient group $G \mid H$ of a given group G is a homomorphism image of the group. Also prove that kernel of f is H.*

Let G be a given group and H be any normal sub group of G so that $G \mid H$ is a quotient group whose elements are of the type $Ha, a \in G$.

We have to show that $G \mid H$ is homomorphic image of G. For this we will show that mapping f of G onto $G \mid H$ such that
$$f(ab) = f(a) f(b) \; \forall \; a, b \in G$$

Consider that mapping $f: G \to G \mid H$ defined as
$$f(ab) = Ha, a \in G, Ha \in G \mid H.$$

The mapping is onto. Corresponding to each element $Ha, (a \in G)$ of $G \mid H \; \exists \, a \in G$ such that $f(a) = Ha$.

Now if $a, b \in G$ than $f(a) = Ha, f(b) = Hb.$
$$f(ab) = H(ab) = (Ha)(Hb) = f(a) f(b).$$

Hence f is a bomomorphism of G onto $G \mid H$,

Kernel of f. $f: G \to G \mid H.$

By definition kernel of f will be the sub set K of G whose f image is the identity of $G \mid H$ which is H.
$$K = \{a \in G : f(a) = H\}$$

We have to prove that $K = H$.

Let $a \in K \Rightarrow f(a) = H$ the identity of $G \mid H$.

By definition of f given above $f(a) = Ha$

$\therefore \quad Ha = H \Rightarrow a \in H$

$\therefore \quad a \in K \Rightarrow a \in H \; \therefore \; K \subseteq H$ ...(1)

Again let $h \in H$ then $Hh = H$ by def. of cosets.

Now $f(h) = Hh = H$ by def. of f

$\therefore \quad h \in K$

Hence $\quad h \in H \Rightarrow h \in K \; \therefore H \subseteq K$ ...(2)

$\therefore$ from (1) and (2) we get $K = H$

i.e., kernel of f is H where f is a bomomorphism of G onto $G \mid H$.

EXAMPLES

1. *If $G = \{1, -1, i, -i\}$ which forms a group under multiplication and $I = $ the group of all integers under addition. Prove that the mapping f from I onto G such that $f(x) = i^n \forall \, n \in I$ is a homomorphism.*

Homomorphism and Isomorphism

Solution. Since $f(x) = i^n$, $f(m) = i^m$, for all $m, n, \in I$.

$$f(m+n) = i^{m+n} = i^m, i^n = f(m) f(n).$$

Hence f is a homomorphism.

2. Let G be the group of all ordered pairs (a, b) of real numbers with the binary operation denoted additively and defined by

$$(a, b) + (c, d) = (a + c, b + d).$$

Further let G' be the additive-group of all real numbers. Then the mapping

$$f : G \to G'$$

defined by

$$f(a, b) = a \ \forall \ (a, b) \in G$$

is a homomorphism of G onto G'.

Solution. It can be easily proved that G is a group with respect to the given binary operation. The ordered pair $(0, 0)$ is the identity element and the ordered pair $(-a, -b)$ is the inverse (a, b).

Let (a, b) and (c, d) be any two elements of G.
They by definition of f, we have

$$f(a, b) = a, \ f(c, d) = c.$$

Now
$$f[(a, b) + (c, d)] = f[a + c, b + d]$$
$$= a + c$$
$$= f(a, b) + f(c, d)$$

Also, obviously f is onto G', Therefore f is a homomorphism of G onto G'.

EXERCISES

1. Let $G = \{a, a^2, a^3, ..., a^{12} = e\}$ be a cyclic group under multiplication. Let $G' = \{a^2, a^4, a^6, ..., a^{12}\}$ be its subgroup then prove that mapping $a^n \to a^{2n}$ is a homomorphism G onto G'.

2. G is the group of real numbers under addition and $G' = G$, prove that $f : G \to G'$ defined by $f(x) = 12x, \ \forall \ x \in G$ is a homomorphism.

3. Show that the mapping $f : C \to R$ such that $f(x + iy) = x$ is a homomorphism of the additive group C of complex number onto the additive group R of real numbers. Find the kernel of f. [**Ans.** Kernel consists of all complex numbers whose real part is zero.]

12.3 ISOMORPHISM

Definition. *Let G, G' be two groups with binary operations o and o' respectively. If there exists a one-one onto mapping $f : G \to G'$ such that*

$$f(a \ o \ b) = f(a) \ o' f(b), \text{ for all } a, b \in G,$$

then the group G is said to be isomorphic to the group G', and the mapping f is said to be an isomorphism. If G is isomorphism to G', we write $G \cong G'$ or $G \equiv G'$.

In other words, a group G is isomorphic to the group G' if there exists one-one mapping of G to G' such that the *image of the product of two elements is the product of the images of the elements,* with respect to the compositions in the respective groups.

The last condition may also be stated as follows :

If $ab = c$, where $a, b, c \in G$,

and $f(a) = a'$, $f(b) = b'$, $f(c) = c'$,

then $a'b' = c'$, where $a', b', c' \in G'$

12.4 PROPERTIES OF ISOMORPHISM

Theorem 1. *If isomorphism exists between two groups, then the identities correspond, i.e., if $G \to G'$ is an isomorphism and e, e' are respectively the identities in G, G' then $f(e) = e'$.*

Proof. Let e be the identity in G, and let $f(e) = e'$. Then we will prove that e' is the identity in G'.

Let $\quad\quad\quad\quad x' \in G'$

Since f is one-one onto, there is an $x \in G$ such that $f(x) = x'$ since f is an isomorphism. Thus
$$x' = f(x) = f(e) f(x) = e'x'$$
for every $x' \in G$, if follows that e' is the identity in G'.

Theorem 2. *If isomorphism exists between two groups, then the inverses correspond, i.e., if $f : G \to G'$ is an isomorphism and $f(a) = a'$ where $a \in G, a' \in G'$ then $f(a^{-1}) = a'^{-1} = [f(a)]^{-1}$.*

Proof. Let $f(a) = a'$ where $a \in G, a' \in G'$. We will prove that the inverse of a' is $f(a^{-1})$

Since $aa^{-1} = e$ the identity G, we have
$$f(a) \cdot f(a^{-1}) = f(a \cdot a^{-1})$$
$$= f(e) = e', \text{ the identity in } G'$$
i.e., $\quad\quad\quad\quad a' \cdot f(a^{-1}) = e'$

if follows that $a'^{-1} = f(a^{-1})$.

Theorem 3. *In an isomorphism the order of an element is preserved, i.e., if $f : G \to G'$ is an isomorphism, and the order of a is n, then the order of $f(a)$ is also n.*

Proof. As $f(a) = a'$ then we have
$$f(a^2) = f(a.a) = f(a) \cdot f(a) = a' \cdot a' = a'^2$$
and in general,
$$f(a^n) = a'^n$$
But $\quad\quad\quad f(a^n) = f(e) = e',$ $\quad\quad\quad\quad\quad\quad\quad\quad$ (by Theorem 1)

Therefore $a'^n = e'$.

Also $\quad\quad\quad a'^m \ne e'$ for $m < n$ i.e., $o(a') = n$

It follows that the order of an element in G, if finite, is equal to the order of its image in G'.

If the order of a is infinite, we can similarly show that the order of a' cannot be finite,

Theorem 4. *The relation of isomorphism in the set of groups is an equivalence relation.*

Proof. Reflexivitiy. It can be easily shown that every group G is isomorphic to itself, and isomorphism being the identity mapping defined by
$$f : G \to G : f(x) = x \ \forall \ x \in G.$$
The mapping is evidently one-one onto,

Moreover, we have
$$f(xy) = xy, \text{ by definition of } f$$

Homomorphism and Isomorphism

$$= f(x) f(y).$$

Thus f is composition preserving as well.

Hence $G \cong G.$

Symmetry. Let a group G is isomorphic to another group G' i.e.,

$$G \cong G'.$$

and let f be an isomorphic mapping of G onto G'. Since f is a one-one onto mapping, it is invertible, *i.e.*, f^{-1} exists and is also one-one onto. We shall show that f^{-1} is an isomorphic mapping of G' onto G.

Let $x', y' \in G'$

Then there exists elements $x, y \in G$ such that

$$f^{-1}(x') = x, f^{-1}(y') = y \qquad \qquad \text{...(i)}$$

so that $\qquad x' = f(x)$ and $y' = f(y) \qquad \qquad \text{..(ii)}$

Also since f is an isomorphism of G onto G', we have

$$f(xy) = f(x) f(y) \; \forall \; x, y \in G \qquad \qquad \text{...(iii)}$$

Now $\qquad f^{-1}(x'y') = f^{-1}[f(x) f(y)] \qquad$ by (ii)

$\qquad \qquad \qquad = (f^{-1} o f)(xy) \qquad$ by (iii)

$\qquad \qquad \qquad = xy$ since $f^{-1} o f$ is the identity mapping

$\qquad \qquad \qquad = f^{-1}(x') f^{-1}(y') \qquad$ by (i)

Hence $G' \cong G.$

Transitivity. Let G, G', G'' be three groups such that $G \cong G'$ and $G' \cong G''$. Let f and ϕ be the respective isomorphic mappings.

Since f, ϕ are one-one onto, $\phi o f$ is also one-one onto.

Moreover $\qquad [\phi o f](x \; y) = \phi \{f(xy)\} x, y \in G$

$\qquad \qquad \qquad = \phi \{f(x) f(y)\}, \; \because \; f$ is an isomorphism

$\qquad \qquad \qquad = [(\phi o f)(x)][(\phi o f)(y)].$

Hence f is an isomorphic mapping of G onto G'', so that

$$G \cong G''.$$

Thus the theorem is completely established.

12.5 ISOMORPHISM OF CYCLIC GROUPS

Theorem 1. *Cyclic groups of same order are isomorphic.*

Proof. Let G and G' be two cyclic groups of order n, which are generated by a and b respectively. Then

$$G = \{a, a^2, a^3, ..., a^n = e\} \quad \text{and}$$

$$G' = \{b, b^2, b^3, ..., b^n = e'\}$$

The mapping $f : G \to G'$ defined by $f(a^r) = b^r$ is isomorphism. For,

$f(a^r . a^s) = f(a^{r+s}) = b^{r+s} = b^r . b^s$

$$= f(a^r) . f(a^s)$$

Therefore the groups are isomorphic.

Theorem 2. *An infinite cyclic group is isomorphic to the additive group of integers.*

Proof. Let G be an infinite cyclic group generated by a. Then
$$G = \{..., a^{-2}, a^{-1}, a^0 = e, a^1, a^2, a^3 ...\}$$
$$= \{a^r, r \text{ is an integer}\}$$

The mapping $f : G \to Z$ defined by $f(a^r) = r$ is an isomorphism. For it is one-one onto and further
$$f(a^r . a^s) = f(a^{r+s}) = r + s$$
$$= f(a^r) + f(a^s)$$
If follows that G is isomorphic to Z.

Theorem 3. *A cyclic group of order n is isomorphic to the additive group of residue classes modulo n.*

Proof. Let G be the cyclic group of order n, generated by a. Then
$$G = \{a, a^2, a^3, ... a^{n-1}, a^n = e\}$$
Let G' be the additive group of residue classes (mod n), i.e.,
$$G' = \{[1], [2], ... [n] = [0]\}$$
The mapping $f : G \to G'$ defined by $f(a^r) = [r]$ is an isomorphism.
For, it is one-one onto, and further,
$$f(a^r . a^s) = f(a^{r+s}) = [r + s] = [r] + [s]$$
$$= f(a^r) + f(a^s)$$
It follows that G is isomorphic to G'.

Theorem 4. *A subgroup of the infinite cyclic group is isomorphic to the additive group of integral multiples of an integer.*

Proof. Let $G = \{..., a^{-2}, a^{-1}, a^0 = e, a^1, a^2, a^3, ...\}$ and let H be a subgroup of G, given by
$$H = \{..., a^{-2m}, a^{-m}, a^0 = e, a^m, a^{2m}, ...\}$$
$$= \{(a^m)^n : n \in Z\}.$$
Then H is isomorphic to the additive group H', given by
$$H' = \{0, \pm m, \pm 2m, \pm 3m, ...\}$$
$$= \{nm : n \in z\}$$
For, the mapping $f : H \to H'$ defined by $f(a^{nm}) = nm$ is one-one onto and if $r, s \in z$ then
$$f(a^{rm}, a^{sm}) = f(a^{(r+s)m}) = (r + s)m$$
$$= rm + sm = f(a^{rm}) + f(a^{sm})$$
It will be observed that H is itself an infinite cyclic group, and as such it is isomorphic to G. Thus a *subgroup of an infinite cyclic group is isomorphic to the group itself*.

12.6 CAYLEY'S THEOREM

Every finite group is isomorphic to a permutation group.

Proof. Let G be a finite group of order n. If $a \in G$, then $\forall x \in G \Rightarrow ax \in G$. Now consider a function from G into G defined by
$$f_a(x) = ax \;\forall\; x \in G$$
For, $\qquad x, y \in G, f_a(x) = f_a(y) \Rightarrow ax = ay$
$$\Rightarrow x = y \qquad\qquad\qquad\qquad \text{(by left cancellation law)}$$

Homomorphism and Isomorphism

Therefore, function f_a is one-one.

The function f_a is also onto because if x is any element of G then there exists an element f_a such that

$$f_a(a^{-1}x) = a(a^{-1}x) = (aa^{-1})x = ex = x.$$

Thus f_a is one-one from G onto G. Therefore f_a is a permutation on G. Let G' denote the set of all such one-to-one functions defined on G corresponding to every element of G, i.e.,

$$G' = \{f_2 : a \in G\}$$

Now, we show that G' is a group with respect to the product of functions.

(i) Closure axiom. Let $f_a, f_b \in G'$, where $a, b \in G$, then

$$(f_a \circ f_b)x = f_a[f_b(x)] = f_a(bx) = a(bx)$$
$$= (ab)x = f_{ab}(x) \; \forall \; x \in G$$

Hence $\qquad f_a \circ f_b = f_{ab}$ ...(i)

Since $ab \in G$, therefore $f_{ab} \in G'$ and thus G' is closed under the product of functions.

(ii) Associative axiom. Let $f_a, f_b, f_c \in G'$ where $a, b, c \in G$

Then
$$f_a \circ (f_b \circ f_c) = f_a \circ f_{bc} \qquad \text{[From (i)]}$$
$$= f_a(bc) \qquad \text{[From (i)]}$$
$$= (ab)c \qquad \text{[by associative law in } G\text{]}$$
$$= f_{ab} \circ f_c \qquad \text{[From (i)]}$$
$$= (f_a f_b) \circ f_c \qquad \text{[From (i)]}$$

∴ Product of functions is associative in G'.

(iii) Identity axiom. If e the identity element in G, then f_e is the identity of G' because $\forall \; f_x \in G'$ we have $f_e \circ f_x = f_{ex} = f_x$ and $f_x \circ f_e = f_{xe} = f_x$.

(iv) Inverse axiom. If a^{-1} is the inverse of a in G, then f_a^{-1} is the inverse of f_a in G' because

$$f_a^{-1} \circ f_a = f_{aa}^{-1} = f_e \text{ and } f_a \circ f_a^{-1} = f_{aa}^{-1} = f_e.$$

Hence G' is a group with respect to composite of functions denoted by the symbol o.
Now consider the function g from G into G' defined by

$$g(a) = f_a \; \forall \; a \in G$$

g is one-one because for $a, b \in G$

$$g(a) = g(b) \Rightarrow f_a = f_b \Rightarrow f_a(x) = f_b(x)$$
$$\Rightarrow ax = bx \Rightarrow a = b \; \forall \; x \in G$$

g is onto because if $f_a \in G'$ then for $a \in G$, we have

$$g(a) = f_a$$

Here g preserves composition in G and G' because if $ab \in G$ then

$$g(ab) = f_{ab} \qquad \text{[by definition of } g\text{]}$$
$$= f_a \circ f_b \qquad \text{[from (i)]}$$
$$= g(a) \circ g(b)$$

∴ $\qquad G \cong G'$

EXAMPLES

1. *Show that the multiplicative group G consisting of the three cube roots of unity $1, \omega, \omega^2$ is isomorphic to the group G' residue classes (mod 3) under addition of residue classes (mod 3).*

Solution. Let us construct the composition tables of two structures G, G' as given below:

	1	ω	ω^2
1	1	ω	ω^2
ω	ω	ω^2	1
ω^2	ω^2	1	ω

+ (mod 3)	[0]	[1]	[2]
[0]	[0]	[1]	[2]
[1]	[1]	[2]	[0]
[2]	[2]	[0]	[1]

From these tables it is evident that if $1, \omega, \omega^2$ are repalced by $\{0\}, \{1\} \{2\}$ respectively in the composition table for G we get the composition table for G'. This leads to the fact that mapping f of G onto G' defined by $f(1) = \{0\}, f(\omega) = \{1\}, f(\omega^2) = \{2\}$ is an isomorphism.

Also $\quad f(\omega, \omega^2) = f(1) = \{0\} = \{1\} + \{2\}$

$\qquad\qquad = f(\omega) + f(\omega^2)$

2. *Let G be the additive group of real numbers and G', the multiplicative group of all positive real numbes. Show that the following mappings are isomorphic.*

(i) $f : G \to G' : f(x) = e^x, x \in G$

(ii) $g : G' \to G : g(x) = \log x, x \in G'$.

Solution. (i) Since $x \neq y \Rightarrow e^x \neq e^y \ \forall \ x, y \in G$, the mapping f is one-one.

The mapping f is also onto since to every positive real number $x \in G'$, there is a real number $\log_e x \in G$ such that

$$f(\log_e x) = e^{\log_e x} = x$$

Moreover $\quad \forall \ x, y \in G$, we have

$\qquad f(x+y) = e^{x+y} \qquad$ [by definition of f]

$\qquad\qquad\quad = e^x . e^y$

$\qquad\qquad\quad = f(x) f(y) \qquad$ [by definition of f]

This show that f preserves compositions in G and G'.

Hence f is an isomorphicm of G and G'

(ii) The mapping g is one-one since any two different real positive numbers have two different logarithms.

It is onto since every real number has a unique anti-logarithm as a positive real number, *i.e.*,

$\qquad a \neq b \Rightarrow \log a \neq \log b \ \forall \ a, b \in G'$

and $\qquad \forall \ x \in G$ we have $x = \log_e e^x$

$\qquad\qquad = g(e^x)$, where $e^x \in G'$.

Also, $\quad \forall \ x, y \in G'$, we have

Homomorphism and Isomorphism

$$g(xy) = \log(xy) \qquad \text{[by definition of } g\text{]}$$
$$= \log x + \log y$$
$$= g(x) + g(y) \qquad \text{[by definition of } g\text{]}$$

Hence g is an isomorphic mapping of G' onto G.

3. *Find the regular permutation group isomorphic to the group.*

$$G = [\{f_1, f_2, f_3, f_4\}, o],$$

where $f_1(z) = z$, $f_2(z) = -z$, $f_3(z) = 1/z$ and $f_4(z) = -1/z$.

Solution. We have already set up a composition table for G. We reproduce it here for convenience.

o	f_1	f_2	f_3	f_4
f_1	f_1	f_2	f_3	f_4
f_2	f_2	f_1	f_4	f_3
f_3	f_3	f_4	f_1	f_2
f_4	f_4	f_3	f_2	f_1

Then by Cayley's theorem the regular permutation group G' consists of the four permutations P_1, P_2, P_3, P_4, given as below

$$P_1 = \begin{pmatrix} f_1 & f_2 & f_3 & f_4 \\ f_1 o f_1 & f_1 o f_2 & f_1 o f_3 & f_1 o f_4 \end{pmatrix} = \begin{pmatrix} f_1 & f_2 & f_3 & f_4 \\ f_1 & f_2 & f_3 & f_4 \end{pmatrix}$$

$$= I$$

$$P_2 = \begin{pmatrix} f_1 & f_2 & f_3 & f_4 \\ f_2 o f_1 & f_2 o f_2 & f_2 o f_3 & f_2 o f_4 \end{pmatrix} = \begin{pmatrix} f_1 & f_2 & f_3 & f_4 \\ f_2 & f_1 & f_4 & f_3 \end{pmatrix}$$

$$= (f_1 f_2)(f_3 f_4)$$

$$P_3 = \begin{pmatrix} f_1 & f_2 & f_3 & f_4 \\ f_3 o f_1 & f_3 o f_2 & f_3 o f_3 & f_3 o f_4 \end{pmatrix} = \begin{pmatrix} f_1 & f_2 & f_3 & f_4 \\ f_3 & f_4 & f_1 & f_2 \end{pmatrix}$$

$$= (f_1 f_3)(f_2 f_4)$$

$$P_4 = \begin{pmatrix} f_1 & f_2 & f_3 & f_4 \\ f_4 o f_1 & f_4 o f_2 & f_4 o f_3 & f_4 o f_4 \end{pmatrix} = \begin{pmatrix} f_1 & f_3 & f_3 & f_4 \\ f_4 & f_3 & f_2 & f_1 \end{pmatrix}$$

$$= (f_1 f_4)(f_2 f_3).$$

EXERCISES

1. Show that the multiplicative group $(1, -1, i, -i)$ is isomorphic to the permutation group
2. Show that the group $G = [\{1, -1\}]$ is isomorphic to the group
$$G' = [\{f_1, f_2\}, o]$$
where $f_1 : R \to R : f_1(x) = x$, $f_2 : R \to R : f_2(x) = -x$
3. Prove that the group $G = [\{1, -1, i, -i\}]$ is isomorphic to the group $G' = [\{0, 1, 2, 3\}, +_4]$ where $+_4$ stands for addition modulo 4.
4. Prove that the symmetric group S_2 of degree 2, is isomorphic to the multiplicative group $(\{1, -1\}, \times)$ and also to the additive group of residue classes modulo 2.

12.7 FUNDAMENTAL THEOREM ON HOMOMORPHISM OF GROUPS OR FIRST THEOEM ON ISOMORPHISM

Every homomorphic image of a group G is isomorphic to some quotient group of G.

Proof. Let G' be the homomorphic image of a group G and f be the corresponding hommorphism. Then f is a homomorphism of G onto G'. Let K be the kernel of this homomorphism. Then K is a normal sub-group of G. We shall prove that
$$G \mid K \cong G'.$$
If $a \in G$, then $Ka \in G/K$ and $f(a) \in G'$. Consider the mapping
$$\phi : G/K \to G' \text{ such that } \phi(Ka) = f(a) \; \forall \; a \in G.$$

First we shall show that the mapping ϕ is well-defined i.e., if $a, b \in G$ and $Ka = Kb$, then $\phi(Ka) = \phi(Kb)$.

We have
$$Ka = Kb$$
$$\Rightarrow ab^{-1} \in K$$
$$\Rightarrow f(ab^{-1}) = e' \quad \text{(identity of } G')$$
$$\Rightarrow f(a) f(b^{-1}) = e'$$
$$\Rightarrow f(a) [f(b)]^{-1} = e'$$
$$\Rightarrow f(a) [f(b)]^{-1} f(b) = e' f(b)$$
$$\Rightarrow f(a) e' = f(b)$$
$$\Rightarrow f(a) = f(b)$$
$$\Rightarrow \phi(Ka) = \phi(Kb).$$

$\therefore \quad \phi$ is well-defined.

ϕ is one one. We have
$$\phi(Ka) = \phi(Kb)$$
$$\Rightarrow f(a) = f(b)$$
$$\Rightarrow f(a) [f(b)]^{-1} = f(b) [f(b)]^{-1}$$
$$\Rightarrow f(a) f(b^{-1}) = e'$$

Homomorphism and Isomorphism

$$\Rightarrow f(ab^{-1}) = e'$$
$$\Rightarrow ab^{-1} \in K \qquad [\because K \text{ is kernel}]$$
$$\Rightarrow Ka = Kb.$$

$\therefore \quad \phi$ is one-one.

ϕ **is onto** G'. Let y be any element of G'. Then $y = f(a)$ for some $y = f(a)$ because f is onto G'. Now $Ka \in G/K$ and we have

$$\phi(Ka) = f(a) = y. \qquad \therefore \quad \phi \text{ is onto } G'.$$

Finally we have $\phi[(Ka)(Kb)] = \phi(Kab)$
$$= f(ab)$$
$$= f(a) f(b)$$
$$= \phi(Ka) \phi(Kb),$$

ϕ is an isomorphism of G/K onto G'.

Hence $G/K \cong G'$.

❏❏❏

13

Normal Sub-Groups

Let G be an abelian group, the composition in G being denoted multiplicatively. Let H be any sub-group of G. If x is an element of G, then Hx is a right coset of H in G and xH is a left coset of H in G. Also G is abelian, therefore we must have $Hx = xH \; \forall \; x \in G$. However, it is possible that G is not abelian, yet it possesses a sub-group H such that $HX = xH \; \forall \; x \in G$. Such sub-group of G come under the category of *normal sub-groups* and these are very important.

13.1 NORMAL SUB-GROUPS

Definition. *A sub-group N of a group G is said to be a normal sub-group of G if for every $x \in G$ and for every $n \in N$, $x n x^{-1} \in N$.*

From this definition we can immediately conclude that N is a normal sub-group of G if and only if

$$xNx^{-1} \subset N \; \forall \; x \in G.$$

Theorem 1. *A sub-group N of a group G is normal if and only if*

$$xNx^{-1} = N \; \forall \; x \in G.$$

Proof. Let $xNx^{-1} = N \; \forall \; x \in G$

Then $x N x^{-1} \subset N \; \forall \; x \in G$. Therefore N is a normal sub-group of G.

Conversely, Let N be a normal sub-group of G.

Then $$xNx^{-1} \subset N \; \forall \; x \in G \qquad \ldots(i)$$

Also $x \in G^{-1} \Rightarrow x^{-1} \in G$. Therefore we have

$$x^{-1} N (x^{-1})^{-1} \subset N \; \forall \; x \in G$$
$$\Rightarrow x^{-1} N x \subset N \; \forall \; x \in G$$
$$\Rightarrow x (x^{-1} N x) x^{-1} \subset x N x^{-1} \; \forall \; x \in G$$
$$\Rightarrow N \subset x N x^{-1} \text{ for all } x \in G$$

From (i) and (ii), we conclude that

$$xNx^{-1} = N \text{ for all } x \in G.$$

Theorem 2. *A sub-group N of a group G is a normal sub-group of G if and only if each left coset of N in G is a right coset of N in G.*

Proof. Let N be a normal sub-group of G.

Then $$xNx^{-1} = N \; \forall \; x \in G$$
$$\Rightarrow (x N x^{-1}) x = N x \text{ for all } x \in G$$
$$\Rightarrow xN = Nx \text{ for all } x \in G$$
$$\Rightarrow \text{ each left coset } xN \text{ is the right coset } Nx.$$

Conversely, let each left coset of N in G be a right coset of N in G. It means that if x is any element of G, then the left coset xN is also a right coset. Now $e \in N$ therefore $xe = x \in xN$. So x must also belong to that right coset which is equal to the left coset xN. But x is an element of the

Normal Sub-Groups

right coset Nx and two right cosets are either disjoint or identical. Therefore Nx is the unique right coset which is equal to the left coset xN. Therefore, we have

$$xN = Nx \ \forall \ x \in G$$
$$\Rightarrow xNx^{-1} = Nxx^{-1} \ \forall \ x \in G$$
$$\Rightarrow xNx^{-1} = N \ \forall \ x \in G$$
$$\Rightarrow N \text{ is a normal sub-group of } G.$$

Theorem 3. *A sub-group N of a group G is a normal sub-group of G if and only if the product of two right cosets of N in G is again a right coset of N in G.*

Proof. Let N be a normal sub-group of a group G. Let a, b be any two elements of G. Then Na and Nb are two right cosets of N in G. We have

$$(Na)(Nb) = N(aN)b$$
$$= (Na)b \qquad [\because N \text{ is normal} \Rightarrow Na = aN]$$
$$= N N ab$$
$$= N a b$$

Since $a \in G, b \in G \Rightarrow ab \in G$, therefore $N a b$ is also a right coset of N in G. Thus the product of the right cosets Na and Nb is the right coset Nab.

Conversely, let N be a sub-group of G such that the product of two right cosets of N in G is again a right coset of N in G. Let x be any element of G. Then $x^{-1} \in G$. Therefore Nx and Nx^{-1} are two right cosets of N in G. Consequently, by hypothesis $NxNx^{-1}$ is also a right coset of N in G. Since $e \in N$, therefore $e x e x^{-1} = e$ is an element of the right coset $NxNx^{-1}$. But N itself is a right coset of N in G and $e \in N$. Also if two right cosets have one element common, then they must be identical. Therefore we must have

$$NxNx^{-1} = N \ \forall \ x \in N$$
$$\Rightarrow n_1 x \, nx^{-1} \in N \ \forall \ x \in G \text{ and } n_1, n \in N$$
$$\Rightarrow n_1^{-1}(n_1 x \, nx^{-1}) \in n_1^{-1} N \ \forall \ x \in G \text{ and } \forall \ n_1, n \in n \in N$$
$$\Rightarrow x \, nx^{-1} \in N \ \forall \ n_1, n \in N$$
$$\Rightarrow x \, nx^{-1} \in N \ \forall \ x \in G \text{ and } \forall \ n \in N \quad [\because n_1^{-1}N = N \text{ as } n_1^{-1}N \text{ since } n_1 \in N]$$
$$\Rightarrow N \text{ is a normal sub-group of } G.$$

Theorem 4. *The intersection of two normal sub-groups of a group is a normal sub-group.*

Proof. Let H and K be two normal sub-groups of a group G. Since H and K are sub-groups of G, therefore $H \cap K$ is also a sub-group of G. Now to prove that $H \cap K$ is a normal sub-group of G.

Let x be any element of G and n be any element of $H \cap K$.

We have $n \in H \cap K \Rightarrow n \in H, N \in K$.

Since H is a normal sub-group of G, therefore

$$x \in G, n \in H \Rightarrow x n x^{-1} \in H$$

Similarly, $x n x^{-1} \in K$.

Now, $xnx^{-1} \in H, xnx^{-1} \in K$

$$\Rightarrow x n x^{-1} \in H \cap K.$$

Thus we have $x \in G, n \in G, n \in H \cap K$

$$\Rightarrow x n x^{-1} \in H \cap K$$
$$\Rightarrow H \cap K \text{ is a normal sub-group of } G.$$

13.2 CENTRE OF A GROUP

Definition. *The set Z of all those elements of a group G which commute with every element of G is called the centre of G. Symbolically*
$$Z = [z \in G : zx = xz \Rightarrow x \in G]$$

Theorem. *The centre Z of a group G is a normal sub-group of G.*

Proof. We have $Z = [z \in G : zx = xz \ \forall \ x \in G]$
First we shall prove that Z is a sub-group of G.
Let $z_1, z_2 \in Z$. Then
$$z_1 x = x z_1 \text{ and } z_2 x = x z_2 \text{ for all } x \in G.$$
We have $z_2 x = x z_2 \ \forall \ x \in G$
$$\Rightarrow z_2^{-1}(z_2 x) z_2^{-1} = z_2^{-1}(x z_2) z_2^{-1}$$
$$\Rightarrow x z_2^{-1} = z_2^{-1} \ \forall \ x \in G$$
$$\Rightarrow z_2^{-1} \in Z$$
Now $(z_1 z_2^{-1}) x = z_1 (z_2^{-1} x) = z_1 (x z_2^{-1})$
$$= (z_1 x) z_2^{-1} = (x z_1) z_2^{-1}$$
$$= x (z_1 z_2^{-1})$$
$\therefore \quad z_1 z_2^{-1} \in Z.$

Thus $z_1, z_2 \in Z \Rightarrow z_1 z_2^{-1} \in Z.$
$\therefore$ Z is a sub-group of G.
Now, we shall show that Z is a normal sub-group of G. Let $x \in G$ and $z \in Z$. Then
$$xzx^{-1} = (xz) x^{-1} = (zx) x^{-1}$$
$$= z \in Z.$$
Thus $x \in G, z \in Z, x z x^{-1} \in Z.$
$\therefore$ Z is a normal sub-group of G.

EXAMPLES

1. *Prove that every sub-group of an abelian group is normal.*
Solution. Let H be a sub-group of an abelian group G. Let a be any element of G and h that of H.
Then we have
$$aha^{-1} = aa^{-1}h, \qquad \because \ ha^{-1} = a^{-1}h, \text{ as } G \text{ is abelian.}$$
$$= eh, \qquad \because \ eh = h$$
$$\in H$$

Therefore $a \in G, h \in H \Rightarrow aha^{-1} a \in H$ i.e., H is normal in G.

2. *If N and M are normal subgroups then show that NM is also a normal sub-group of G.*
Solution. We can easily prove that NM is a sub-group of G.
Now let a be any element of G and nm that of NM.
Then $n \in N$ and $m \in M$ and $a (nm) a^{-1} = (ana^{-1}) (ama^{-1})$

Normal Sub-Groups

$\in NM, \because ana^{-1} \in N, ama^{-1} \in M$ as N

$\therefore$ NM is normal sub-group of G.

3. *Suppose that N and M are two normal sub-groups of G such that* $N \cap M = \{e\}$. *Show that every element of N commutes with every element of M.*

Solution. Let x be any element of N and y be any element of M. Then we have to prove that $xy = yx$.

Let us consider the element $xyx^{-1}y^{-1}$.

Since N is normal, $yx^{-1}y^{-1} \in N$,

Also $x \in N$, so that

$$xyx^{-1}y^{-1} \in N.$$

Furthermore M is normal, $xyx^{-1} \in M$

Also $y^{-1} \in M$.

So that $xyx^{-1}y^{-1} \in M$. Thus we see that

$$xyx^{-1} \in N \text{ and } xyx^{-1}y^{-1} \in M.$$

This implies that $xyx^{-1}y^{-1} \in N \cap M$.

But $N \cap M = \{e\}$, so we have

$$xyx^{-1}y^{-1} = e, \text{ i.e., } xy = yx.$$

EXERCISE

1. If H be a sub-group of G and N be a normal sub-group of G, then prove that $H \cap N$ is a normal sub-group of H.
2. Let H be a sub-group of a group G. Let for $x \in G$, $xHx^{-1} = \{xhx^{-1} : h \in H\}$. Prove that xHx^{-1} is a sub-group of G.
3. If a cyclic sub-group N of G is normal in G, then prove that every sub-group of N is normal in G.
4. If an abelian group G is simple, then show that the order of G is prime.

13.3 QUOTIENT GROUPS

Definition. *If G is a group and N is a normal sub-group of G, then the set* $G | N$ *of all cosets of N in G is a group with respect to multiplication of cosets. It is called the quotient group or factor group of G by N.*

The identity element of the quotient group $G | N$ is N.

Theorem. *The set of all cosets of a normal sub-group is a group with respect to multiplication of complexes as the composition.*

Proof. Let N be a normal sub-group of a group G. Since N is normal in G, therefore each right coset will be equal to the corresponding left coset.

Thus there is no distinction between right and left cosets and we shall call them simply as cosets. Let $G | N$ be the collection of all cosets of N in G, i.e., let

$$G | N = \{Na : a \in G\}.$$

Closure property. Let $a, b \in G$.

Then $(Na)(Nb) = N(aN)b$

$= N(Na)b$ [$\because N$ is normal]

$= NNab$

$$= N\,a\,b$$

Since $ab \in G$, therefore $N\,a\,b$ is also a coset of N in G. So $Nab \in G|N$. Thus $G|N$ is closed with respect to coset multiplication.

Associativity. Let $a, b, c \in G$. Then $Na, Nb, Nc \in G|N$.

We have $Na\,[(Na)(Nc)] = Na\,(Nbc)$

$$= Na\,(bc)$$
$$= N\,(ab)\,c \qquad\qquad [\because a\,(bc) \neq (ab)\,c]$$
$$= (Nab)\,Nc$$
$$= [(Na)(Nb)]\,Nc.$$

Thus the product in $G|N$ satisfies the associative law.

Existence of Identity. We have $N = Ne \in G|N$. Also if Na is any element of $G|N$, then

$$N\,(Na) = (Ne)(Na) = Nea = Na$$

and similarly $\qquad (Na)\,N = (Na)(Ne) = Nae = Na$

Therefore the coset N is the identity element.

Existence of Inverse. Let $Na \in G|N$, then $Na^{-1} \in G|N$.

We have $(Na)(Na^{-1}) = Naa^{-1} = Ne = N$.

and $\qquad\qquad (Na^{-1})(Na) = Na^{-1}a = Ne = N$.

$\therefore$ The coset Na^{-1} is the inverse of Na. Thus each element of $G|N$ possesses inverse. Hence $G|N$ is a group with respect to product of cosets.

EXAMPLES

1. *If H be a normal subgroup of a finite group G, then prove that*

$$o\,(G|H) = \frac{o\,(G)}{o\,(H)}$$

Solution. $o\,(G|H) =$ number of distinct right (or left) cosets of H in G, as $G|H$ is the collection of all right (or left) cosets of H in G.

$$= \frac{\text{number of distinct elements in } G}{\text{number of distinct elemnets in } H}$$

$$= \frac{o\,(G)}{o\,(H)}, \text{ by Largrange's theorem.}$$

2. *Show that every quotient group of a cyclic group is cyclic but not conversely.*

Solution. Let H be a sub-group of a cyclic group G. Then H is also cyclic.

$\because$ every cyclic group is abelian.

$\therefore$ H is a normal sub-group is G.

Let a be a generator of G and a^n be any element of G, where n is some integer.

Then Ha^n is any element of $G|H$.

Also, it can be proved easily that $(Ha)^n = Ha^n$, for every integer n.

$\therefore$ $G|H$ is cyclic and its generator is Ha.

Its converse is not ture. For example, if P_3 and A_3 be the symmetric and alternating groups on the three symbols a, b, c then the quotient group $P_3 | A_3$ is cyclic, whereas P_3 is not.

Normal Sub-Groups

3. *Show that every quoteint group of an abelian group is abelian but its converse is not true.*
Solution. Let H be a sub-group of an abelian group G, so H is a normal sub-group of G.
Let $a, b \in G$ be arbitrary, then Ha, Hb are any two elements of the quotient group $G|H$.
Then we have

$$(Ha).(Hb) = Hab = Hba \quad \because ab = ba \text{ if } G \text{ is abelian.}$$
$$= (Hb).(Ha)$$

$\therefore \quad G|H$ is abelian.

Its converse is not true. For example if P_3 and A_3 be the symmetric and alternating groups on the three symbols a, b, c then the quotient group $P_3|A_3$ being of order 2 is abelian whereas P_3 is not.

13.4 RELATION OF CONJUGACY IN A GROUP

Conjugate Element. *If $a, b \in G$, then b is said to be a conjugate of a in G if there exists an element $x \in G$ such that $b = x^{-1}ax$.*

Symbolically, we shall write $a \sim b$ for this and shall refer to this relation as **conjugacy**.

Thus $b \sim a \Leftrightarrow b = x^{-1}ax$ for some $x \in G$.

Example. In P_3, since $(2, 3) = (1\,2\,3)^{-1} (1\,2) (1\,2\,3), (2\,3)$ is the conjugate of $(1\,2)$ by $(1\,2\,3)$.

Theorem. *Conjugacy is an equivalence relation in a group.*
Proof.

(i) Reflexivity. Let $a \in G$, then $a = e^{-1}ae$, hence $a \sim a \; \forall \; a \in G$ i.e., the relation of conjugacy is reflexive.

(ii) Symmetry. Let $a \sim b$, so there exists an $x \in G$ such that

$$a = x^{-1}bx, a, b \in G$$

Now $\quad a \sim b \Rightarrow a = x^{-1}bx \quad \Rightarrow xa = x(x^{-1}bx)$

$\Rightarrow xax^{-1} = (xx^{-1})b(xx^{-1}) \quad \Rightarrow b = xax^{-1}$

$\Rightarrow b = (x^{-1})^{-1}ax^{-1}, x^{-1} \in G \quad \Rightarrow b \sim a$

Thus $a \sim b \Rightarrow b \sim a$
Henc relation is symmetric.

(ii) Transitivity. Let there exists two elements $x, y \in G$ such that $a = x^{-1}bx$, and $b = y^{-1}cy$ for $a, b, c \in G$.

Hence $\quad a \sim b, b \sim c$

$\Rightarrow a = x^{-1}bx$ and $b = y^{-1}cy \quad \Rightarrow a = x^{-1}(y^{-1}cy)x$

$= a = (x^{-1}y^{-1})c(yx) \quad \Rightarrow a = (yx)^{-1}c(yx)$

Where $yx \in G, G$ being the group.

Therefore, $a \sim b, b \sim c \Rightarrow a \sim c$.
Hence relation is transitive.
Thus conjugacy is equivalence relation on G.

Conjugate classes. For $a \in G, C(a) = \{x : x \in G \text{ and } a \sim x\}$, $C(a)$ the equivalence class of a in G under conjugacy relation is usually called the *conjugate class* of a G. It consists of the set of all distinct elements of the type $y^{-1}ay$ as y ranges over G.

13.5 NORMALIZER OR CENTRALIZER OF AN ELEMENT OF A GROUP

If $a \in G$, then $N(a)$, the normalizer of a in group G, is the set
$N(a) = \{x \in G : xa = ax\}$.

Note. $N(a)$ consists of precisely those elements in G which commute with a. $N(a)$ is the symbol used for normalizer.

Remark 1. Since $ex = xe \ \forall \ x \in G$, e being the identity element, $N(a) = G$.
Thus the normalizer of the identity element of a group coincides with the group.

Remark. If we consider an element n of an abelian group G, then $ax = xa \ \forall \ x \in G$ and therefore $N(a) = G$ i.e., the *normalizer of every element of an abelian group concides with the group.*

13.6 SOME THEOREMS

Theorem 1. *The centre of a group is always a normal subgroup of the group.*
Proof. Let Z be the centre of G so that
$$Z = \{z \in G : zx = xz \ \forall \ x \in G\}$$
Let $z_1, z_2 \in Z$ then
$$z_1, z_2 \in Z \Rightarrow z_1 x = xz_1 \text{ and } z_2 x = xz_2 \ \forall \ x \in G.$$

Now $\qquad z_2 x = xz_2 \ \forall \ x \in G \Rightarrow z_2^{-1} z_2 x z_2^{-1} = z_2^{-1} x z_2 z_2^{-1} \ \forall \ x \in G$

$\qquad\qquad\qquad \Rightarrow z x_2^{-1} = z_2^{-1} x \ \forall \ x \in G.$

But $\qquad (z_1 z_2^{-1}) x = z_1 (z_2^{-1} x)$ \hfill (associativity)

$\qquad\qquad\qquad = z_1 (xz_2^{-1})$ \hfill $[\because z_2^{-1} x = xz_2^{-1}]$

$\qquad\qquad\qquad = (z_1 x) z_2^{-1}$ \hfill (associativity)

$\qquad\qquad\qquad = (xz_1) z_2^{-1}$ \hfill $[\because z_1 x = xz_1]$

$\qquad\qquad\qquad = x (z_1 z_2^{-1}) \ \forall \ x \in G$

$\therefore \quad z_1 z_2^{-1} \in Z.$

Hence $z_1 \in Z, z_2 \in z \Rightarrow z_1 z_2^{-1} \in Z$.
and therefore Z is a subgroup of G.
Again let $x \in G$ and $z \in Z$ then

$\qquad\qquad xzx^{-1} = (xz) x^{-1}$ \hfill (associativity)

$\qquad\qquad\qquad = (zx) x^{-1}$ \hfill $[\because xz = xz]$

$\qquad\qquad\qquad = z(xx^{-1})$ \hfill (associativity)

$\therefore \quad xzx^{-1} \in Z \ \forall \ x \in G \text{ and } z \in Z.$

Hence Z is a normal in G. Thus we establish that z is a normal subgroup of G.

Theorem 2. *The normalizer of any element of a group is always a subgroup of the same.*
Proof. Let G be any group of $N(a)$ the normalizer of $a (\in G)$ in G. Let $x, y \in N(a)$ then
$$x \in N(a) \text{ and } z \in N(a) \Rightarrow ax = xz \text{ and } ay = ya.$$
$$\Rightarrow y^{-1} a = a y^{-1}$$

and therefore $\qquad a (xy^{-1}) = (ax) y^{-1} = (xa) y^{-1} = x (ay^{-1})$

$\qquad\qquad\qquad\qquad = x (y^{-1} a) = (xy^{-1}) a$

Normal Sub-Groups

which shows that $xy^{-1} \in N(a)$.

Hence $N(a)$ subgroup of G.

NOTE. The normalizer of an element of a group G is not necessarily a normal subgroup. Symmetric group P_3 of all permutations of degree 3 defined on three symbols is an example of the same.

Theorem 3. *The number of elements conjugate to an element a of a finite group G is the index of the normalizer of a in G.*

Proof. Since the index in G of the normalizer $N(a)$ of a is finite. G has a right decomposition

$$G = Nx_1 \cup Nx_2 \cup ... \cup Nx_m$$

where one-one correspondence

$$Nx_i \leftrightarrow x_i^{-1} a x_i$$

exists.

Hence the number of the conjugates of a must be $m = (G : N(a))$.

13.7 NORMALIZER OR CENTRALIZER OF A SUBGROUP

If $a \in g$, then the set $= \{x \in H : xa = ax\}$, where H is a subgroup of G, is called the centralizer of a in G and is denoted by $C(a)$ or $N(H)$.

Note. If $G_1 = G$, then $C(a)$ is said to be centre of G and the elements in $C(a)$ are called the central elements of G.

13.8 THEOREMS

Theorem 1. *The normalizer $N(H)$ of a subgroup H of a group G is a subgroup of G.*

Proof. $N(H) \neq \phi$ because if e is identity in G then

$$eH = He = H \text{ and therefore } e \in (H).$$

Let $a, b \in N(H)$ then

$$aH = Ha \text{ and } bH = Hb.$$

But $\qquad bH = Hb \Rightarrow b^{-1}bHb^{-1} = b^{-1}Hbb^{-1}$

$$\Rightarrow Hb^{-1} = b^{-1}H.$$

Also $\qquad (ab^{-1})H = a(b^{-1}H) = a(Hb^{-1})$

$$= (aH)b^{-1} = (Ha)b^{-1} = H(ab^{-1})$$

$\therefore \quad ab^{-1} \in N(H)$.

Hence $N(H)$ is a subgroup of G.

Theorem 2. *If H is a subgroup of a group G there is one-one correspondence between the right cosets of the normalizer $N(H)$ in G and the conjugates of H.*

Proof. If C_H be the set of all conjugates of H then

$$C_H = \{x^{-1}Hx : x \in G\}.$$

Let S be the set of all right cosets of $N(H)$ in G then

$$S = \{N(H)x : x \in G\}.$$

Let f be a mapping from S to C_H such that

$$f(N(H)x) = x^{-1}(Hx) \ \forall \ x \in G$$

then $\qquad N(H)x = N(H)y$

$$\Rightarrow xy^{-1} \in N(H) \Rightarrow xy^{-1}H = Hxy^{-1}$$
$$\Rightarrow x^{-1}xy^{-1}H = x^{-1}Hxy^{-1}$$
$$\Rightarrow y^{-1}H = x^{-1}Hxy^{-1}$$
$$\Rightarrow y^{-1}Hy = H^{-1}Hxy^{-1}y$$
$$\Rightarrow y^{-1}Hy = x^{-1}Hx$$
$$\Rightarrow f(N(H)y) = f(N(H)x).$$

Hence f is well defined. It is one-one also because
$$f(N(H)x) = f(N(H)y)$$
$$\Rightarrow x^{-1}Hx = y^{-1}Hy$$
$$\Rightarrow xx^{-1}Hxy^{-1} = xy^{-1}Hyy^{-1}$$
$$\Rightarrow Hxy^{-1} = xy^{-1}H \Rightarrow xy^{-1} \in N(H)$$
$$\Rightarrow N(H)x = N(H)y$$

Now let $x^{-1}Hx \in C_H$ then $x \in G$ and therefore $N(H)x$ is an element of S such that
$$f(N(H)x) = x^{-1}Hx$$

Hence f is onto.

Thus there is a one-one correspondence between the right cosets of $N(H)$ in G and the conjugates of H.

13.9 COMMUTATOR SUBGROUP OF A GROUP

If G is a group and x, y are any two elements of G then the element $xyx^{-1}y^{-1}$ is called the **commutator** of the ordered pair (x, y).

The commutator of ordered pair (x, y) is the inverse of ordered pair (x, y) because

Commutator of $(x, y) = xyx^{-1}y^{-1}$ and commutator of $(y, x) = yxy^{-1}x^{-1}$ and $(yxy^{-1}x^{-1})^{-1} = xyx^{-1}y^{-1}$.

The commutator of an ordered pair is identity if and only if the elements commute because if $x, y \in G$ then
$$xyx^{-1}y^{-1} = e \Leftrightarrow (xyx^{-1}y^{-1})^{-1} = e^{-1} = e$$
$$\Leftrightarrow yxy^{-1}x^{-1} = e$$
$$\Leftrightarrow yxy^{-1}x^{-1}x = ex$$
$$\Leftrightarrow yxy^{-1} = x \Leftrightarrow yxy^{-1}y = xy$$
$$\Leftrightarrow yx = xy.$$

The complex formed of the commutators of all the pair of elements of a group G is not necessarily a group of G because prodcut of commutators is not necessarily a commutator. The *subgroup generated by the complexes consisting of the commutators of all the ordered pairs of elements of the group is called the commutator subgroup of the group.*

Note. The commutator subgroup of a group G is the smallest subgroup of G containing the complex formed by the commutators of the pairs of elements of G.

Theorem. *If G' be a commutator subgroup of a group G, then (i) G' is normal in G, (ii) $G|G'$ is abelian, and (iii) a quotient group G/H is abelian iff contains the commutator subgroup G' of G.*

Normal Sub-Groups

Proof. Let $K = \{xyx^{-1}y^{-1} : x, y \in G\}$. Then the commutator subgroup G' of a grouip G is the smallest subgroup of G containing K.

(i) Let a be an arbitrary element of G and c be any element of G'.

$\therefore \qquad aca^{-1} = ce^{-1}aca^{-1} = c[c^{-1}aca^{-1}]$

$\qquad \qquad = c[c^{-1}a(c^{-1})^{-1}a^{-1}]. \qquad \qquad \ldots(\text{i})$

Also $\qquad c \in G' \Rightarrow c^{-1} \in G' \Rightarrow c^{-1} \subset G$.

$\therefore \quad c^{-1} \in G, a \in G \Rightarrow c^{-1}a(c^{-1})^{-1}a^{-1}$ is a commutator of $(c^{-1}a)$.

$\Rightarrow c^{-1}a(c^{-1})^{-1}a^{-1} \in K, \Rightarrow c^{-1}a(c^{-1})^{-1}a^{-1} \in G'$. $\qquad$ [as $K \subset G$]

But $c \in G', c^{-1}a(c^{-1})^{-1}a \in G' \Rightarrow c(c^{-1}a(c^{-1})^{-1}a^{-1}) \in G'$ $\qquad$ (by closure law)

$\Rightarrow aca^{-1} \in G'$ $\qquad$ [Using (1)]

i.e. $\qquad aca^{-1} \in G' \,\forall\, a \in G$ and $\forall\, c \in G'$.

Therefore, G' is normal in G.

(ii) Let $G'x$ and $G'y$ be may two elements of $G|G'$.

Therefore $x, y \in G$ and the comutator of (x, y) is $xyx^{-1}y^{-1}$ which consequently belongs to K.

Since $\qquad xyx^{-1}y^{-1} \in K \Rightarrow xy^{-1}y^{-1} \in G'$ $\qquad$ [$\because K \subseteq GG'$]

$\qquad \Rightarrow (xy)(yx)^{-1} \in G'$ $\qquad$ [$\because (yx)^{-1} = x^{-1}y^{-1}$]

$\qquad \Rightarrow (G'xy) = (G'xy)$

$\qquad \Rightarrow (G'x) = (G'y) = (G^{-1}y)(G'x)$.

Hence $G|G'$ is abelian.

(iii) Let Hz and Hy be any two elements of the quotient group $G|H$ so that $x, y\, G$.

Since $G|H$ is abelian, $(Hx)(Hy) = Hy(Hx)$.

But $(Hx)(Hy) = (Hy)(Hx) \Rightarrow Hxy = Hyx \Rightarrow (xy)(yx)^{-1} \in H$

$\qquad \Rightarrow xyx^{-1}y^{-1} \in H$

i.e., commutator of (x, y) is in H. Hence $K \subseteq H$.

H is a subgroup of G and $K \subseteq H$. Therefore by definition G' is smallest subgroup of G containing K, and as such $G' \subseteq H$.

Conversely let $G' \subseteq H$ then $K \subseteq H$ as $K \subseteq G'$

But $K \subseteq H \Rightarrow \forall\, x, y \in G$, commutator of $(x, y) \in H$

$\qquad \Rightarrow xyx^{-1}y^{-1} \in H\, \forall\, x, y \in G \Rightarrow Hxy = Hyx\, \forall\, x, y \in G$

$\qquad \Rightarrow (Hx)(Hy) = (Hy)(Hx)\, \forall\, x, y\, G$.

Hence $G|H$ is abelian.

13.10 MAXIMAL NORMAL SUBGROUPS OF GROUP

A normal subgroup H of a group G is called the maximal normal subgroup of G, if there exists no proper normal subgroup of G which contains H properly.

Remark. If H is a maximal normal subgroup of a group G and K is a normal subgroup of G such that $H \subset K \subset G$ then $K = G$.

❏❏❏

14
Rings and Subrings

10.1 RING

The concept of a group has its origin in the set of mappings or permutations, of a set onto itself. So far we have considered sets with one binary operation only. But rings are the outcome of the motivation which arises from the fact that integers follow a definite pattern with respect to the addition and multiplication. Thus we now aim at studying rings which are algebraic systems with two suitably restricted and related binary operations.

Definition. *An algebraic structure* $(R, +.)$, *where R is a non-empty set and + and . are two defined operations in R, is called a ring if for all a, b, c in R, the following axioms are satisfied :*

$R_1 . (R, +)$ *is an abelian group, i.e.,*

$(R_{11})\; a + b \in R$ (Closure law for addition)

$(R_{12})\; (a + b) + c = a + (b + c)$ (associative las for addition)

$(R_{13})\; R$ has an identity, to be denoted by 0, with respect to addition,

    i.e., $a + 0 = a \;\forall\; a \in R$ (Existence of additive identity)

(R_{14}) There exists an additive inverse for every element in R,

    i.e., there exists an element a in R such that

$a + (-a) = 0 \;\forall\; a \in R$ (Existence of additive inverse)

$(R_{15})\; a + b = b + a$ (Commutative law for addition)

$R_2 . (R, .)$ *is a semigroup, i.e.,*

$(R_{21})\; a . b \in R$ (Closure law for multiplication)

$(R_{22})\; (a . b) . c = a . (b . c)$ (associative law for multiplication)

R_3 . *Multiplication is left as well as right distributive over addition, i.e.,*

$a . (b + c) = a.b + a.c$

and $(b + c) . a = b.a + c.a$

14.2 ELEMENTARY PROPERTIES OF A RING

Theorem. *If R is a ring, then for all* $a, b \in R$.

(a) $a.0 = 0.a = 0$

(b) $a(-b) = (-a)b = -(ab)$

(c) $(-a)(-b) = ab$

Proof. (a) We know that

$a.0 = a(0 + 0) = a.0 + a.0 \;\forall\; a \in R$ (using distributive law)

Since R is a group under addition, applying right cancellation law,

$a.0 = a.0 + a.0 \Rightarrow 0 + a.0 = a.0 + a.0 \Rightarrow a.0 = 0$

Similarly, $0.a = (0 + 0)a = 0.a + 0.a$ (using distributive law)

$\therefore$ $0 + 0.a = 0.a + 0.a$ $(\because\; 0 + 0.a = 0.a)$

Rings and Subrings

Applying right cancellation law for addition, we get
$$0 = 0.a \text{ i.e., } 0.a = 0$$
Thus $\quad a.0 = 0.a = 0.$

(b) To prove that $a(-b) = -ab$ we would show that
$$ab + a(-b) = 0$$
We know that $\quad a[b+(-b)] = a.0 \qquad\qquad [\because b+(-b)=0]$
$$= 0$$
(with the virtue of result (a) above)

or $\qquad\qquad ab + a(-b) = 0 \qquad\qquad$ (by distributive law)

$\therefore \qquad\qquad a(-b) = -(ab).$

Similarly, to show $(-a)b = -ab$, we must show that
$$ab + (-a)b = 0$$
But $\qquad ab + (-a)b = [a + (-a)]b = 0.b = 0$

$\therefore \qquad -(a)b = -(ab)$

Hence the result.

(c) Actually to prove $(-a)(-b) = ab$ is a special case of foregoing article. However its proof is given as under :
$$(-a)(-b) = -[a(-b)] \qquad\qquad \text{[by result } b]$$
$$= -[-(ab)] \qquad\qquad [\because a(-b) = -ab]$$
$$= ab$$

because $-(-x) = x$ is a consequence of the fact that in a group inverse of the inverse of an element is element itself.

14.3 SOME SPECIAL TYPES OF RINGS

1. Commutative rings.

Definition. *A ring R is said to be commutative, if the multiplication composition in R is commutative, i.e.,*
$$ab = ba \;\forall\; a, b \in R.$$

2. Rings with unity elements

Definition. *A ring R is said to be a ring with unity element if R has a multiplicative identity, i.e., if there exists an element R denoted by 1, such that*
$$1.a = a.1 = a \;\forall\; a \in R.$$

The ring of all $n \times n$ matrices with elements as integers (rational, real or complex numbers) is a ring with unity. The unity matrix

$$I_n = \begin{bmatrix} 1 & 0 & 0 & \ldots & 0 \\ 0 & 1 & 0 & \ldots & 0 \\ 0 & 0 & 1 & \ldots & 0 \\ 0 & 0 & 0 & \ldots & 1 \end{bmatrix}$$

is the unity element of the ring.

3. Rings with or without zero divisors

While dealing with an arbitrary ring R, we may find elements a and b in R neither of which is zero, and their product may be zero. We call such elements **divisors of zero** or **zero divisors.**

Definition. *A ring element $a (\neq 0)$ is called a divisor of zero if there exsits an element $b (\neq 0)$ in the ring such that either*

$$ab = 0 \quad \text{or} \quad ba = 0$$

We also say that a *ring R is without zero divisors if the product of no two non-zero elements is zero, i.e., if*

$$ab = 0 \Rightarrow \text{either } a = 0 \text{ or } b = 0 \text{ or both } a = 0 \text{ and } b = 0.$$

14.4 CANCELLATION LAWS IN A RING

We say that *cancellation laws hold in a ring R if*

$$ab = ac \, (a \neq 0) \Rightarrow b = c$$

and $ba = ca \, (a \neq 0) \Rightarrow b = c$ where a, b, c are in R.

Thus in a ring with zero divisors, it is impossible to define a cancellation law.

Theorem. *A ring has no divisor of zero if and only if the cancellation laws hold in R.*

Proof. Suppose that R has no zero divisors. Let a, b, c be any three elements of R such that $a \neq 0, ab = ac$.

Now, $\qquad ab = ac \Rightarrow ab - ac = 0$

$\qquad\qquad\qquad \Rightarrow a(b - c) = 0$

$\qquad\qquad\qquad \Rightarrow b - c = 0 \qquad\qquad (\because R \text{ is without zero divisors and } a \neq 0)$

$\qquad\qquad\qquad \Rightarrow b = c$.

Thus the left cancellation law holds in R. Similarly, it can be shown that right cancellation law also holds in R.

Conversely, suppose that the cancellation laws hold in R.

Let $a, b \in R$ and if possible let $ab = 0$ with $a \neq 0, b \neq 0$ then $ab = a.0 \qquad (\because a.0 = 0)$

Since $a \neq 0, ab = a.0 \Rightarrow b = 0$ (by left cancellation law)

Hence we get a contradiction to our assumption that $b \neq 0$ and therefore the theorem is established.

Division Ring

A ring is called a division ring if its non-zero elements form a group under multiplication.

Pseudo Ring

A non-empty set R with binary operations '+' and '.' satisfying all the postulates of a ring except right and left distributive laws, a called a pseudo ring if

$$(a + b).(c + d) = a.c + a.d + b.c + b.d \text{ for all } a, b, c, d \in R.$$

EXAMPLES

1. *I is the set of all integers (positive, negative and zero); . is the ordinary multiplication and + is the ordinary addition of integers. Then show that I is a commutative ring with unit element.*

Solution. For the set I we find that :

(i) Closure Property. We know the sum and product of two integers is also an integer

i.e., $\qquad\qquad a + b \in I, a \bullet b \in I \, \forall \, a, b \in I$

Thus the set of integers is closed with respect to addition and multiplication.

(ii) $\qquad\qquad a + b = b + a \, \forall \, a, b \in I$

i.e., addition of integers is commutative.

(iii) Associative Laws for addition and multiplication.

$\qquad\qquad (a + b) + c = a + (b + c) \text{ and } (a.b).c = a.(b.c) \, \forall \, a, b, c \in I$

i.e., the addition and multiplication of integers is associative.

(iv) $\qquad a(b + c) = a.b + a.c \text{ and } (b + c).a = b.a + c.a \, \forall \, a, b, c \in I.$

Rings and Subrings

i.e., the multiplication is both left and right distributive with respect to addition.

(v) There exists an element 0 in **I**, such that
$$a + 0 = 0 + a = a \ \forall \ a \in \mathbf{I}$$
(vi) There exists an element $-a$ in **I** such that
$$a + (-a) = 0 = (-a) + a, \ \forall \ a \in \mathbf{I}$$

Thus all the postulates are satisfied and hence $(\mathbf{I}, +, .)$ is a ring.

Also $\qquad a.b = b.a \ \forall \ a \in \mathbf{I}$

∴ **I** is a commutative ring.

Also $\qquad 1.a = a.1 = a \ \forall \ a \in \mathbf{I}.$

∴ $(\mathbf{I}, +, .)$ is a ring with unit element.

2. *A Gaussian integer is a complex number $a + ib$, where a and b are integers. Show that the set $J(i)$ of Gaussian integers forms a ring under ordinary addition and multiplication of complex numbers.*

Solution. Let $a_1 + ib_1$ and $a_2 + ib_2$ be any two elements of $J(i)$ then
$$(a_1 + ib_1) + (a_2 + ib_2) = (a_1 + a_2) + i(b_1 + b_2)$$
$$= A + iB \text{ (say)}$$
and $\qquad (a_1 + ib_1).(a_2 + ib_2) = (a_1 a_2 - b_1 b_2) + i(a_1 b_2 + b_1 a_2)$
$$= C + iD \text{ (say)}$$

These are Gaussian integers and therefore $J(i)$ is closed under addition as well as multiplication of complex numbers.

Addition and multiplication are both associative and commutative compositions for complex numbers.

Also, multiplication distributes with respect to addition.

$0 \ (= 0 + 0i) \in J(i)$ is the additive identity.

The additive inverse of $a + ib \in J(i)$ is
$$(-a) + (-b)i \in J(i) \text{ as}$$
$$(a + ib) + (-a) + (-b)i$$
$$= (a - a) + (b - b)i$$
$$= 0 + 0i = 0.$$

The Gasussian integer $1 + 0.i$ is the multiplicative identity.

Therefore, the set of Gasussian integers is a commutative ring with unity.

3. *Prove that the set I of integers is a commutative ring in which 'addition' and 'multiplication' are defined below indicated by $\oplus$ and $\odot$ respectively :*
$$a \oplus b = a + b - 1; \ a \odot b = a + b - ab, \text{ where } a, b \in \mathbf{I}.$$

Solution. Here we have

R_1 : (i) The set **I** is closed with respect to the composition $\oplus$, since
$$a \in \mathbf{I}, b \in \mathbf{I} \Rightarrow a \oplus b = a + b - 1 \in \mathbf{I}$$

(ii) $\forall \ a, b, c \in \mathbf{I}$, we have
$$(a \oplus b) \oplus c = (a + b - 1) \oplus c, \qquad \qquad \because \quad a \oplus b = a + b - 1$$
$$= (a + b - 1) + c - 1$$
$$= a + b + c - 2$$

And $\qquad a \oplus (b \oplus c) = a \oplus (b + c - 1), \qquad \qquad \because \quad b \oplus c = b + c - 1$

$$= a + (b + c - 1) - 1 = a + b + c - 2$$

∴ We have $(a \oplus b) \oplus c = a \oplus (b \oplus c)$ i.e., the set **I** obeys the associative law for the composition $\oplus$.

(iii) Let e be the additive identity of the set **I** under the given composition $\oplus$. Then $a \oplus e = a \ \forall \ a \in \mathbf{I}$

$$\Rightarrow a + e - 1 = a, \qquad \because \quad a \oplus b = a + b - 1$$
$$\Rightarrow e = 1$$

Hence the additive identity (or zero of the ring) exists and is 1.

(iv) Let $b \in \mathbf{I}$ the additive inverse of $a \in \mathbf{I}$, then we must have $a \oplus b = e$ the additive identy

$$\Rightarrow a + b - 1 = 1, \because e = 1 \text{ and } a \oplus b = a + b - 1$$
$$\Rightarrow b = 2 - a$$

Hence the additive inverse of each element $a \in \mathbf{I}$ exists and is equal to $2 - a$.

(v) Here we have $a \oplus b = a + b - 1, b \oplus a = b + a - 1$

∴ $a \oplus b = b \oplus a$ i.e., the commutative law holds for the operation $\oplus$ in **I**.

R_2 : Here $a \odot b = a + b - ab \in \mathbf{I}$ if $a, b \in \mathbf{I}$

i.e., the set **I** is closed under the operation $\odot$

R_3 : $\forall \ a, b, c \in \mathbf{I}$, we have

$$(a \odot b) \odot c = (a + b - ab) \odot c, \qquad \because \quad a \odot b = a + b - ab$$
$$= (a + b - ab) + c - (a + b - ab) c$$
$$= a + b + c - ab - ac - bc + abc$$

and $\qquad a \odot (b \odot c) = a \odot (b + c - bc), \qquad \because \quad b \odot c = b + c - bc$
$$= a + (b + c - bc) - a (b + c - bc)$$
$$= a + b + c - bc - ab - ac + abc$$

∴ $(a \odot b) \odot c = a \odot (b \odot c)$ i.e., the set **I** obeys the associative law for the composition $\odot$.

R_4 : $\forall \ a, b, c \in \mathbf{I}$, we have

$$a \odot (b \oplus c) = a \odot (b + c - 1), \qquad \because \quad a \oplus b = a + b - 1$$
$$= a + (b + c - 1) - a (b + c - 1), \qquad \because \quad a \odot b = a + b - ab$$
$$= a + b + c - 1 - ab - ac + a$$
$$= (a + b - ab) + (a + c - ac) - 1,$$

by associative law of integers.

$$= (a \odot b) + (a \odot c) - 1 = (a \odot b) \oplus (a \odot c)$$

Also $\qquad (b \oplus c) \odot a = (b + c - 1) \odot a, \qquad \because \quad a \oplus b = a + b - 1$
$$= (b + c - 1) + a - (b + c - 1) a, \qquad \because \quad a \odot b = a + b - ab$$
$$= b + c - 1 + a - ba - ca + a$$
$$= (b + a - ba) + (c + a - ca) - 1,$$

by associative law of integers.

$$= (b \odot a) + (c \odot a) - 1 = (b \odot a) \oplus (c \odot a)$$

Hence the operation $\odot$ is left and right distributive over the operation $\oplus$.

R_5 : Here $a \odot b = a + b - ab, b \odot a = b + a - ba \ \forall \ a, b \in \mathbf{I}$

Rings and Subrings

∴ $a \odot b = b \odot a$, as addition and multiplication of integers is commutative.

i.e., the commutative law is satisfied for the composition $\odot$.

Hence all the ring postulates are satisfied and so $(\mathbf{I}, \oplus, \odot)$ is a commutative ring.

[Moreover we find that
$$a \odot 0 = a + 0 - a \cdot 0,$$
i.e., $\qquad a \odot 0 = a \ \forall \ a \in \mathbf{I}$
$\qquad\qquad\qquad\qquad\qquad\qquad\qquad\qquad\qquad$ ∵ $a \odot b = a + b - ab$

Similarly $\qquad 0 \odot a = 0 + a - 0.a = a \ \forall \ a \in \mathbf{I}$

∴ Identity element for the composition $\odot$ exists and is 0.

Hence $(\mathbf{I}, \oplus, \odot)$ is a ring with unity.

4. *If R is a ring with unity element 1, then prove that*

$\qquad$ (α) $(-1)a = -a = a \cdot (-1), \ \forall \ a \in R$;

$\qquad$ (β) $(-1) \cdot (-1) = 1$.

Solution. We know if 1 be the unity element in the ring R, then
$$a \cdot 1 = 1 \cdot a = a \ \forall \ a \in R \qquad\qquad ...(i)$$

(α) Now $\quad [1 + (-1)] \cdot a = 1 \cdot a + (-1) \cdot a, \ \forall \ a \in R$

or $\qquad 0.a = a + (-1) \cdot a, \qquad$ from (i)

or $\qquad 0 = a + (-1) \cdot a$

or $\qquad (-1) \cdot a = -a$

Again from (i) we get $a = a \cdot 1$

or $\quad a + a \cdot (-1) = a \cdot 1 + a \cdot (-1)$, adding $a \cdot (-1)$ on the right of both sides

or $\qquad a + a \cdot (-1) = a \cdot [1 + (-1)] = a \cdot [0] = 0$

or $\qquad a \cdot (-1) = -a$

(β) From part (α) we have $a \cdot (-1) = -a$

Replacing a by (-1) in this result we get $(-1) \cdot (-1) = -(-1)$

i.e., $(-1) \cdot (-1)$ = additive inverse of (-1), where -1 is itself the additive inverse of 1 in R.

$\qquad$ = 1 Hence proved.

5. *If in a ring left identity is also right identity then prove that both are equal.*

Solution. Let e_1 and e_2 be left and right identities respectively and $e_1, e_2 \in R$

Now, taking e_1 as left identity and $e_2 \in R$,
$$e_1 e_2 = e_2. \qquad\qquad ...(i)$$

Again, taking e_2 as right identity and $e_1 \in R$
$$e_1 e_2 = e_1. \qquad\qquad ...(ii)$$

Hence, from (i) and (ii),
$$e_1 e_2 = e_2 = e_1$$

i.e., two identities are equal.

EXERCISES

1. If E is the set of even integers under ordinary operations of addition and multiplication, then show that E is a commutative ring but has no unity element.
2. If R is the set of real numbers under ordinary addition and multiplication of real numbers, then show that R is a commutative ring with unit element.

3. Prove that the set of all matrices of the form $\begin{bmatrix} 0 & a \\ 0 & b \end{bmatrix}$, a and b being real numbers with matrix addition and matrix multiplication is a ring.
4. Prove that the set $\{0, 1, 2, 3, 4\}$ is a commutative ring with respect to addition and multiplication modulo 5.
5. Prove that the set C of complex numbers is a ring with respect to ordinary addition and multiplication. Prove also that it is commutative.

14.5 SUBRINGS

Definition : *A non-empty subset S of a ring R is called a* **subring** *of R if S is itself a ring with respect to the operations of addition and multiplication in R.*

From the above definition it is quite apparent that a subset S of R cannot possibly be a subring of R unless S is closed under the operations of addition and multiplication on R since, otherwise, we would have no operations on the set S.

In all examples below the operations are usual addition and multiplication unless otherwise stated.

Illustrations

(1) The set **I** is a subrirng of **R**, the ring of real numbers.

(2) The set **E** of *even integers* is a subring of the ring **I** (set of integers). However, the set of all *odd integers* cannot be a subring of **I** since this set is not closed under addition.

(3) The set of all those integers which are multiples of a given integer, say m, that is,
$$\{..., -2m, -m, 0, m, 2m, ...\}$$
is a subring of the ring **I**.

(4) The set of all rational numbers is a subring of the ring of real numbers.

(5) The set of all real numbers is a subring of the ring of all complex numbers.

(6) The set of complex numbers $\{m + ni; m, n \in \mathbf{I}\}$ is a subring of **C**.

(7) Let R be a ring with unity e, then the set $N = \{ne ; n \in \mathbf{I}\}$ is a subring of R.

(8) Every ring R has two trivial subrings : R itself and the set consisting of the zero element alone. These are called *improper* subrinhgs of R.

(9) The set of all $n \times n$ matrices over the rational numbers is a subring of $n \times n$ matrices over the real numbers under addition and multiplication of matrices.

Subrings are characterised by the following theroems :

14.6 SOME THEOREMS ON SUBRINGS

Theorem 1. *A non-empty subset S of a ring R, is a subring of R, if and only iff,*

(1) $a - b \in S$, for all $a, b \in S$;

(2) $ab \in S$, for all $a, b \in S$.

Proof. *Necessary condition.* Let $(S, +, .)$ be a sub-ring of $(R, +, .)$ and let $a, b \in S$.

Since S is a group with respect to addition, so
$$b \in S \Rightarrow -b \in S$$
Also as S is closed with respect to addition, so we have
$$a \in S, b \in S \Rightarrow a \in S, -b \in S$$
$$\Rightarrow a + (-b) \in S$$
$$\Rightarrow a - b \in S$$

Also we know that a non-empty stable subset S of a ring $(R, +, .)$ is a sub-ring of R if S is a ring for the induced operations.

Hence S is closed under multiplication, so
$$a \in S, b \in S \Rightarrow a.b \in S.$$

Rings and Subrings

Sufficient condition. Let the conditions (i) and (iii) given above hold.

Then we have $\quad a \in S, a \in S \Rightarrow a - a = 0 \in S$

Again $\quad 0 \in S, a \in S \Rightarrow 0 - a = -a \in S.$

$\therefore$ If $b \in S$, then $\quad -b \in S.$

Therefore $\quad a \in S, -b \in S \Rightarrow a - (-b) = a + b \in S$

Also since associativity and commutivity of addition hold in R so these hold in S. $\therefore$ $(S, +)$ is an abelian group.

The remaining postulates also hold in S since they hold in R.

Corollary. *The necessary and sufficient conditions for a non-emtpy subset S of a ring R to be a sub-ring of R are that*

(i) $S + (-S) = S$, (ii) $SS \subseteq S$

Proof. *Necessary condition.* Let S be a sub-ring of R, then S is a sub-group of the additive group of R.

Let $\quad a + (-b) \in S + (-S) \Rightarrow a \in S, -b \in -S$

$\quad\quad\quad\quad\quad\quad\quad\quad \Rightarrow a \in S, b \in S.$

$\quad\quad\quad\quad\quad\quad\quad\quad \Rightarrow a - b \in S, \quad\quad\quad\quad\quad\quad \because S$ is a sub-group

$\therefore \quad\quad\quad\quad S + (-S) \subseteq S \quad\quad\quad\quad\quad\quad\quad\quad\quad\quad\quad\quad\quad\quad$ ...(i)

Now if a is any element of S, then $a = a + 0$

Also as S is a sub-group so $0 \in S$ or $0 \in -S$

$\therefore \quad\quad\quad\quad 0 + 0 \in S + (-S)$

$\therefore \quad\quad\quad\quad S \subseteq S + (-S) \quad\quad\quad\quad\quad\quad\quad\quad\quad\quad\quad\quad\quad\quad$ ...(ii)

$\therefore$ From (i) and (ii) we get $S + (-S) = S$

Again S is closed with respect to multiplication

$\therefore \quad\quad\quad\quad a \in S, b \in S \Rightarrow a, b \in S$

And ab is an arbitrary element of SS

Hence $\quad\quad\quad\quad SS \subseteq S$

Sufficient Condition. Let S be a non-empty sub-set of R such that $SS \subseteq S$ and $S + (-S) = S$

Now $\quad\quad\quad\quad SS \subseteq S \Rightarrow ab \in S, \forall a, b \in S$

$\quad\quad\quad\quad\quad\quad\quad\quad \Rightarrow a + (-b) \in S,$ if $a, b \in S$

$\Rightarrow S$ is a sub-group of the additive group of R

Hence S is a sub-ring of R.

Theorem II. *The intersection of two subrings is a subring.*

Proof. Let R_1 and R_2 be two subrings of a ring $(R, +, .)$.

Firstly $0 \in R_1$ and $0 \in R_2$

Since R_1 and R_2 are subtrings.

Hence $\quad\quad\quad\quad 0 \in R_1 \cap R_2$

$\therefore \quad\quad\quad\quad R_1 \cap R_2 \not\in \phi$

Let $\quad\quad\quad\quad a, b \in R_1 \cap R_2$

Now $a \in R_1 \cap R_2 \Rightarrow a \in R_1$ and $a \in R_2$

$\quad\quad b \in R_1 \cap R_2 \Rightarrow b \in R_1$ and $b \in R_2$

Since R_1 and R_2 are sub-rings, we have

$\quad\quad\quad\quad a, b \in R_1 \Rightarrow a - b \in R_1$ and $ab \in R_1$

Also
$$a, b \in R_2 \Rightarrow a - b \in R_2 \text{ and } ab \in R_2$$
$$a - b \in R_1, a - b \in R_2 \Rightarrow a - b \in R_1 \cap R_2$$
$$ab \in R_1, ab \in R_2 \Rightarrow ab \in R_1 \cap R_2$$

Thus
$$a, b \in R_1 \cap R_2 \Rightarrow a - b \in R_1 \cap R_1 \text{ and } ab \in R_1 \cap R_2$$

Hence $R_1 \cap R_2$ is a sub-ring of the ring R.

Theorem III. *Intersection of the arbitrary collection of subrings in a subring.*

Proof. Let, $\{A\alpha : \alpha \in A\}$ is an arbitrary collection of the subrings of a ring $(R, +, .)$ where Λ is a index set in such a way that $\forall \alpha \in \Lambda$, $A\alpha$ is a subring of the ring $(R, +, .)$,

Now, let $A = \cap A\alpha = \{x \in R : X \in A\alpha, \forall \alpha \in A\}$

which is the intersection of the collection of the subring of R, then we are to prove that A is the subring of R.

Now, clearly $A \neq \phi$ because at least zero element of R is in $A\alpha$, $\forall \alpha \in \Lambda$.

Now, let $a, b, \in A$ are any two arbitrary elements, then
$$a \in \bigcap_{\alpha \in \Lambda} A\alpha \Rightarrow a \in A\alpha, \forall \alpha \in \Lambda$$

and
$$b \in \bigcap_{\alpha \in \Lambda} A\alpha \Rightarrow A\alpha, \forall \alpha \in \Lambda$$

But $\forall \alpha \in \Lambda$, $A\alpha$ is a subring of the ring R.

Hence, $a, b \in A\alpha \Rightarrow a - b \in A\alpha$ and $ab \in A\alpha, \forall \alpha \in \Lambda$

$\Rightarrow$ $a - b \in \bigcap_{\alpha \in \Lambda} A\alpha$ and $ab \in \bigcap_{\alpha \in \Lambda} A\alpha$

Now, $a, b \in \bigcap_{\alpha \in \Lambda} A\alpha \Rightarrow a - b \in \bigcap_{\alpha \in \Lambda} A\alpha$ and $ab \in \bigcap_{\alpha \in \Lambda} A\alpha$

$\therefore \bigcap_{\alpha \in \Lambda} A\alpha$ is a subring of ring R.

14.7 SMALLEST SUBRING

Let $(R, +, .)$ be a ring, M is a subset of R. Again, let S, be a subring such that $M \subseteq S$ and if T, is a ring of R which includes M, then $S \subseteq T$; then S is subring of R generated by the subset M. In short, S including M is the smallest subring of R, then S is set subring generated by M. Also, we write
$$S = \{M\}.$$

14.8 HOMOMORPHISM OF RINGS

Definition : *Let R, R' be rings. If $f : R \to R'$ is a map such that for all values of $a, b \in R$,*
$$f(a + b) = f(a) + f(b),$$
$$f(a . b) = f(a) . f(b),$$
*then f is called a **homomorphism** of the ring R into the ring R'.*

Isomorphism of Rings

Let R, R' be two rings. if $f : R \to R'$ is a mapping such that for all values of $a, b \in R$.

Then f is said to be isomorphism of the ring R into the ring R', if f satisfies following :

(i) f is one-one mapping such that if $f(a) = f(b)$ then $a = b$
(ii) f is on-to
(iii) f is homomorphism such that $f(a + b) = f(a) + f(b)$, $f(a . b) = f(a) . f(b)$.

Rings and Subrings

EXAMPLES

1. *Show that the set of matrices* $\begin{bmatrix} a & b \\ 0 & c \end{bmatrix}$ *is the sub-ring of a ring of* 2×2 *matrices.*

Solution. Let M_2 be the set of matrices of order 2×2. Then we can prove that M_2 is a ring.

Let S be a subset of M_2 let the elements of S be matrices of the type $\begin{bmatrix} a & b \\ 0 & c \end{bmatrix}$.

Let $A = \begin{bmatrix} a_1 & b_1 \\ 0 & c_1 \end{bmatrix}$ and $B = \begin{bmatrix} a_2 & b_2 \\ 0 & c_2 \end{bmatrix}$ be any two elements of S

Then $A - B = \begin{bmatrix} a_1 - a_2 & b_1 - b_2 \\ 0 & c_1 - c_2 \end{bmatrix} \in S$

And $A \cdot B = \begin{bmatrix} a_1 & b_1 \\ 0 & c_1 \end{bmatrix} \cdot \begin{bmatrix} a_2 & b_2 \\ 0 & c_2 \end{bmatrix}$

$= \begin{bmatrix} a_1.a_2 + b_1.0 & a_1.b_2 + b_1.c_2 \\ 0.a_2 + c_1.0 & 0.b_2 + c_1.c_2 \end{bmatrix} = \begin{bmatrix} a_1 a_2 & a_1 b_2 + b_1 c_2 \\ 0 & c_1 c_2 \end{bmatrix} \in S$

$\therefore \quad A \in S, B \in S \Rightarrow A - B \in S \text{ and } A.B \in S$

2. *Show that the set* $\{..., -3k, -2k, -k, 0, k, 2k, 3k, ...\}$, *where k is any fixed integer, is the sub-ring of the ring of integers.*

Solution. Let R be the ring of the integers and let $S = \{...,-2k, -k, 0, k, 2k, ...\}$, be any subset of R, where k is a fixed integer.

Let $a = mk$ and $b = nk$ be any two elements of S, then m and n are some integers.

Now $\quad a - b = mk - nk = (m-n)k$

or $\quad a - b \in S$, as $m - n$ is some integer

and $\quad a.b \in (mk).(nk) = (mnk)k$

or $\quad a.b \in S$, as mnk is also some integer, m, n, k being integers.

$\therefore \quad a \in S, b \in S \Rightarrow a - b \in S \text{ and } a.b \in S.$

Hence S is a sub-ring of R.

EXERCISES

1. Show that the set S of all elements of ring M_2 (set of all matrices of order 2×2 over integers) of the form $\begin{bmatrix} x & 0 \\ y & z \end{bmatrix}$, where $x, y, z \in I$ is a sub-ring of M_2.

2. Show that the set of matrices $\begin{bmatrix} 0 & x \\ 0 & y \end{bmatrix}$, where x, y are real numbers, is the sub-ring of 2×2 matrices.

3. Let R be a ring and let a be a fixed element of R. If $I_\alpha = \{x \in R : ax = 0\}$, then show that I_α is a sub-ring of R.

14.9 CHARACTERISTIC OF A RING

Let $(R, +, .)$ be a ring with 0 as its zero element. If there eixsts a positive integer n such that

$$n.a + a + a + a + ... \text{ to } n \text{ terms} = 0 \ \forall \ a \in R,$$

then such smallest positive integer n is called the characteristic of the ring $(R, +, .)$.

If no such positive integer n exists, then the ring $(R, +, .)$ is said to be of characteristic zero or infinite.

Example : *The rings of integers of rational numbers of real numbers are of characteristic zero as there exists no positive integer n for which $n . a = 0, \forall a \in$ the set of integers of rational numbers or real numbers.*

Theorem. *The characteristic of a ring with unity is zero or $n > 0$ according as the unity element 'e' regarded as a member of the additive group of the ring has the order or n:*

Proof. Since e is the unity element of the ring R (say).

so $o(e) = o \Rightarrow$ characteristic of ring R is 0

Let $o(e) = n =$ a finite number so that n is the least positive integer such that $ne = 0$.

Also as $$ea = a = ae \, \forall \, a \in R$$
$\therefore$ $$na = n(ea)$$
$$= (ne)a = 0 \cdot a, \qquad \because ne = 0$$
$$= 0$$

$\therefore$ n is the least positive integer such that $na = 0$

i.e., characteristic of the ring R is n, by definition.

Kernel of a Ring Homomorphism

Definition. *If f is a homomorphism of a ring R into a ring R' then the set S of all those elements of R which are mapped on the additive identity 0 of R' is called Kernel of the homomorphism.*

In other words Kernel of f is defined as the set
$$S = \{r \in R \mid f(r) = 0\}$$

Theorem 1. If $f : R \to R'$ be a ring homomorphism then

(i) $f(0) = 0$ (ii) $f(-a) = -f(a)$ for all $a \in R$

Proof. (i) We have $$f(x) = f(x + 0)$$
$$= f(x) + f(0) \text{ for every } x \in R$$

As $f(x) \in R'$ we must have
$$f(x) = f(x) + 0$$
then $$f(x) + f(0) = f(x) + 0$$

Hence by cancellation law (additive) we have $f(0) = 0$.

(ii) Now $\qquad a + (-a) = 0$

$\Rightarrow \qquad f[a + (-a)] = f(0) = 0 \qquad$ [by (i)]

$\Rightarrow \qquad f(a) + f(-a) = 0$

$\Rightarrow \qquad f(-a) = -f(a)$

Theorem 2. *The Kernel of a homomorphism from a ring R to a ring R' is as ideal of R.*

Proof. Let $f : R \to R'$ be a homomorphism

As $f(0) = 0', 0 \in \ker f$

$\therefore$ $\ker f$ is non empty

Now $\qquad x, y \in \ker f$

$\Rightarrow \qquad f(x) = 0, f(y) = 0$

Thus $\qquad f(x - y) = f[x + (-y)] = f(x) + f(-y)$
$$= f(x) + [-f(y)]$$

Rings and Subrings

$$= f(x) - f(y) = 0$$

$$\Rightarrow \quad x - y \in \ker f$$

Again $r \in R$, $x \in \ker f \Rightarrow f(xr) = f(x) f(r)$ similarly $f(rx) = 0 \Rightarrow xr \in \ker f$ and $rx \in \ker f$. Hence $\ker f$ is an ideal of R.

Theorem 3. *Iff* $f : R \to R'$ *be a homomorphism and* A *be an ideal of* R *then* $f(A)$ *is an ideal of* $f(R)$.

Proof. We know that

$$f(A) = \{x \in R' : a \in A \text{ such that } x = f(a)\}$$

Obviously $\quad 0 \in f(A)$ as $0 = f(0)$

Therefore $f(A)$ is not empty

Let $x, y \in f(A)$ then

$$x = f(a), y = f(b) \text{ for some } a, b \in A$$

$\therefore \qquad x - y \in f(A)$

or $\qquad a - b \in A$

If $r_1 \in f(R)$, then there exists $r \in R$ such that $r_1 = f(r)$

$\therefore \qquad r_1 x = f(r) f(a) = f(ra)$

$\Rightarrow \qquad r_1 x \in f(A) \qquad\qquad\qquad\qquad\qquad$ as $ra \in A$

Similarly $\qquad xr_1 \in f(A)$

Hence $f(A)$ is an ideal of $f(R)$.

Ideals

Definition. *A non empty subset* S *of a ring* R *is said to be an ideal (or two sided ideal) of* R *if.*
(i) S *is a subgroup of* R *under addition.*
(ii) For every $r \in S$ *and* $r \in R$ *both* sr *and* rs *are in* S.

It is clear that ideal S is necessarily a subring of ring R because condition (2) asserts that both Sr and $rS \in S$, where r is specially restricted to remain in R.

Unit ideal : The ring R itself is an ideal of R and is called the unit ideal.
Zero ideal : The subring $\{0\}$ of R is also an ideal of R and is called the zero ideal.
Improper ideal : The unit ideal and the zero ideal are called improper ideals of R the rest are proper ideals.

EXAMPLE

Find the characteristic of the ring $(I_4, +_4, \cdot_4)$ *of integers modulo 4.*

Solution. Here we have $I_4 = \{0, 1, 2, 3\}$ and we get

$$0 +_4 0 +_4 0 +_4 0 = 0$$
$$1 +_1 1 +_4 1 +_4 1 = 4 = 0 \pmod{4}$$
$$2 +_4 2 +_4 2 +_4 2 = 8 = 0 \pmod{4}$$
$$3 +_4 3 +_4 3 +_4 3 = 12 = 0 \pmod{4}$$

i.e., $\qquad 4a = 0, \forall a \in I_4$

Hence the given ring has characteristic 4.

15
Integral Domain and Field

15.1 INTEGRAL DOMAIN

A commutative ring with unit element having at least two elements, and no divisors of zero is called an integral domain.

From the above definition it is obvious that a ring is an integral domain if (i) it is commutative, (ii) it possesses an unit element, (iii) it has atleast two elements and (iv) it is without zero divisors.

The System $(D, +, .)$ is an integral domain if the following postulates are satisfied :

D_1 : The system $(D, +)$ is an abelian group, so we have the following properties :
(i) Closure property.
$$a \in D, b \in D \Rightarrow a + b \in D, \forall a, b \in D.$$
(ii) Associativity of addition.
$$(a + b) + c = a + (b + c), \forall a, b, c \in D.$$
(iii) Existence of zero (or additive identity)
$$a + 0 = a = 0 + a, \forall a \in D.$$
(iv) Existence of additive inverse (or negative)
$$a + (-a) = 0 = (-a) + a, \forall a \in D.$$
(v) Commutative of addition.
$$a + b = b + a, \forall a, b \in D$$

D_2 : The system $(D, .)$ is an abelian semi-group with unity, so we have the following properties :
(i) Closure property.
$$a \in D, b \in D \Rightarrow a . b \in D \; \forall a, b \in D.$$
(ii) Associativity of multiplication.
$$(a . b) . c = a . (b . c), \forall a, b, c \in D.$$
(iii) Existence of unity.
$$a . 1 = 1 . a, \forall a \in D.$$
(iv) Commutativity of mutliplication.
$$a . b = b . a, \forall a, b \in D.$$

D_3 : **Multiplication composition is right and left distributive with respect to addition :**
$$a . (b + c) = a . b + a . c \text{ and } (b + c) . a = b . a + c . a, \forall a, b, c \in D.$$

D_4 : If the product of two elements is zero, then one of them at least is zero.
i.e., $\quad a . b = 0 \Rightarrow a = 0 \quad \text{or} \quad b = 0, \forall a, b \in D.$
i.e., product of non-zero elements is non-zero.

Examples. The rings of integers real numbers, rational numbers and complex numbers are integral domains, their zero element is 0 and unit element is 1. But the ring of even integers with zero though it does not have zero divisors is not an integral domain as it does not contain the unity element.

Also it can be seen that the ring $(\{0, 1, 2, 3, 4\}, +_5, \cdot_5)$ is a finite integral domain.

Integral Domain and Field

15.1.1 Euclidean Rings

An integral domain R is said to be a Euclidean ring if for every $a \neq 0$ in R there is defined a non-negative integer, to be denoted by $d(a)$, such that :

(i) for all $a, b \in R$, both non-zero, $d(a) \leq d(ab)$,

(ii) for any $a, b \in R$, both non-zero, there exists $q, r \in R$ such that $a = qb + r$ when either $r = 0$ or $d(r) < d(b)$.

EXAMPLES

1. *Prove that the ring of integers is an integral domain.*

Solution. It can easily be proved that the set of integers is a commutative ring with unit element. Also this ring does not possess zero divisors because if a, b are any two integers (*i.e.*, elements of this ring) such that $a \cdot b = 0$ then either a or b or both must be zero.

The number of elements of the set of integers is more than two.

Hence, according to the definition of integral domain the ring of integers is an integral domain.

2. *Show that the set E of even integers is not an integral domain with respect to addition and multiplication.*

Solution. We can easily prove that the set E of even integers is a commutative ring with respect to ordinary addition and multiplication but with no unit element (*i.e.*, unity).

Also this ring $(E, +, .)$ does not possess zero divisors because if a, b are any two even integers (*i.e.*, elements of E) such that $a \cdot b = 0$ then either a or b or both must be zero.

Hence according to the definition of integral domain the ring $(E, +, .)$ is not an integral domain.

3. *Prove that the ring of complex numbers C is an integral domain.*

Solution. Let $\qquad J(i) = \{a + bi : a, b \in I\}.$

It is easy to prove that $J(i)$ is a commutative ring with unity.

The zero element $0 + 0 \cdot i$ and unit element $1 + 0 \cdot i$.

Also this ring is free from zero-divisors because the product of two non-zero complex numbers can not be zero. Hence $J(i)$ is an integral domain.

EXERCISES

1. Does the set O of odd integers forms an integral domain with respect to addition and multiplication. [Ans. No]
2. Prove that the set of the form $a + b\sqrt{5}$, where a, b are integers is a integral domain.
3. If $a, b \in \mathbf{Q}$, the set of rational numbers, examine whether numbers of teh form $a + b\sqrt{3}$ constitue an integral domain.
4. Prove that the ring of integers with zero is not an integral domain.
5. Verify that the set S of residue classes modulo 6 is a ring. Is the ring an integral domain. [Ans. Yes]
6. The set of natural numbers does not form an integral domain. Explain why ?

15.2 FIELD

Definition I. If every element $a \neq 0$ of an integral domain has a multiplicative inverse a^{-1} in the integral domain, then it (the integral domain) is called *a field* and is gernerally denoted by F.

Definition II. A ring F whose non-zero elements form an abelain, multiplicative group is known as a *field*.

Definition III. A system $(F, +, .)$ having atleast two elements (*i.e.*, the additive and multiplicative identities) is called a field F, if

F_1 : the system is an abelian group with respect to addition.

F_2 : the distributive laws are satisfied,

and F_3 : the subset of non-zero elements of F is an abelian multiplicative group.

Definition IV. A ring $(R, +, .)$ with at least two elements is called a field, if it is a commutative ring with unity and is such that its each non-zero element a (*i.e.*, $a \neq 0$) possesses multiplicative inverse.

A field is generally written as $(F, +, .)$ or simply as F.

Postulates for Field :

The system $(F, +, .)$ is a field if the following postulates are satisfied :

F_1 : The system $(F, +, .)$ is an abelian group, so we have the following properties :

(i) Closure property.
$$a \in F, b \in F \Rightarrow a+b \in F, \forall a, b \in F.$$

(ii) Associativity of addition
$$(a+b)+c = a+(b+c), \forall a, b, c \in F.$$

(iii) Existence of Additive Identity.
$$a+0 = a = 0+a, \forall a \in F$$

(iv) Existence of Additive Inverse :
$$a+(-a) = 0 = (-a)+a, \forall a \in F$$

(v) Commutativity of Addition.
$$a+b = b+a, \forall a, b \in F$$

F_2 : Multiplication composition is left and right distributive with respect to addition :
$$a.(b+c) = a.b + a.c \text{ and } (b+c).a = b.a + c.a, \forall a, b, c \in F.$$

F_3 : The subset of non-zero elements of F forms an abelian multiplicative group and so we have the following properties :

(i) Closure Property.
$$a \in F, b \in F \Rightarrow a.b \in F \; \forall a, b, \in F$$

(ii) Associativity of multiplication
$$(a.b).c = a.(b.c), \forall a, b, c \in F.$$

(iii) Existence of Multiplicative Identity.
$$a.1 = a = 1.a, \forall a \in F$$

(iv) Existence of Multiplicative Inverse.
$$a.a^{-1} = 1 = a^{-1}.a, \forall a, b \in F$$

(v) Commutativity of Multiplication.
$$a.b = b.a, \forall a, b \in F.$$

Examples of Field. The rings of real numbers, rational numbers and complex numbers are fields as each one of them is a commutative ring with unity and each non-zero element of each of the above rings possesses multiplicative inverse.

Integral Domain and Field

15.3 COMPARISON OF RING, INTEGRAL DOMAIN AND FIELD

S. No.	Ring (R, +, .)	Integral Domain (D, +, .)	Field (F, +, .)
(i)	(R, +) is an abelian group.	(D, +) is an abelian group.	(F, +) is an abelian group.
(ii)	(.) is associative.	(.) is associative.	(.) is associative.
(iii)	Distributive laws are satisfied.	Distributive laws are satisfied.	Distributive laws are satisfied.
(iv)	......	(.) is commutative.	(.) is commutative.
(v)	......	Unity belongs to D	Unity belongs to F.
(vi)	......	....	Multiplicative inverse of each non-zero element of F exists and belongs to F
(vii)	may or may not possess proper zero divisors	does not possess proper zero divisors	does not possess proper zero divisors

From the above table, it is clear that the only difference between an integral domain and field is that every element of a field possesses a multiplicative inverse whereas in an integral domain it is not so.

An important property of a field F.
If $a, b \in F$, then
$$a + b \in F, a - b \in F, a \cdot b \in F, a \div b \text{ or } ab^{-1} \in F$$

15.4 SUB-FIELD

Definition. *A subset E (containing more than one element) of a field F is called a sub-field of F, if E is a field with respect to addition and multiplication in F.*

This definition may be restated as follows :
(1) $a + b \in E$, for all $a, b \in E$ and $a \in E \Rightarrow -a \in e$; also $0 \in E$.

(2) $a \cdot b \in E$, for all $a, b \in E, a \neq 0$ and $a \in E \Rightarrow a^{-1} \in E$ and $1 \in E$.

Illustrations :
(1) The field of rational numbers is a subfield of the field of complex numbers.
(2) The field of real numbers is a sub-field of the field of complex numbers.

15.5 PRIME FIELD

Definition. *A field is called a prime field if it contains no proper sub-field.*
Illustrations :
(1) The set Q of rational numbers is a prime sub-field of the field F of real numbers, if F is of characteristic zero.
(2) The set of residue classes modulo a prime p is also a sub-field of F, is of characteristic p.

15.6 SOME THEOREMS

Theorem 1. *The multiplicative inverse of a non-zero element of a field is unique.*

Proof. Let there be two multiplicative inverse a^{-1} and a' for a non-zero element $a \in F$. Let 1 be the unity of the field F.

$\therefore \qquad aa^{-1} = 1$ and $a.a' = 1$ so that $a.a^{-1} = a.a'$

Since $F - \{0\}$ is a multiplicative group, applying left cancellation law, we get $a^{-1} = a'$.

Theorem 2. *A field is necessary an integral domain.*

Proof. Since a field is a commutative ring with unity, therefore, in order to show that every field is an integral domain we only need proving that a field is without zero divisors.

Let F be any field and let, $a, b \in F$ with $a \neq 0$ such that $ab = 1$. Let 1 be the unity of F. Since $a \neq 0$, a^{-1} exists in F and therefore,

$$ab = 0 \Rightarrow a^{-1}(ab) = a$$
$$\Rightarrow \qquad (a^{-1}a)b = 0$$
$$\Rightarrow \qquad 1.b = 0 \qquad\qquad (\because a^{-1}a = 1)$$
$$\Rightarrow \qquad b = 0 \qquad\qquad (\because 1.b = 0)$$

Similarly if $b \neq 0$ then it can be shown that
$$ab = 0 \Rightarrow a = 0$$

Thus $ab = 0 \Rightarrow a = 0$ or $b = 0$.

Hence, a field is necessarily an integral domain.

Corollary : Since integral domain has no zero divisors and field is necessarily an integral domain, therefore, field has no zero divisor.

Theorem 3. *If a, b are any two elements of a field F and $a \neq 0$ there exists unique element x such that $a.x = b$.*

Proof. Let 1 be unity of $F = a^{-1}$ the inverse of a in F then
$$a.(a^{-1}b) = (aa^{-1}).b = 1.b = b$$
$\therefore \qquad ax = b \Rightarrow a.x = a.(a^{-1}b)$
$\Rightarrow \qquad x = a^{-1}b \quad \text{(by left cancellation)}$

Thus $x = a^{-1}b \in F$.

Now, suppose there are two such elements x_1, x_2 (say) then

Theroem 4. *Every finite integral domain is a field.*

or

A finite commutative ring with no zero divisor is a field.

Proof. Let D be an integral domain with finite number of distinct elements $a_1, a_2, \ldots a_n$.

i.e., $\qquad D = \{a_1, a_2, \ldots a_n\}$

Let a_i be any fixed element of D. Now multiplying each element of D by a_i then we get same set in some different order.

Which is $a_i a_1, a_i a_2, a_i a_3, \ldots, a_i a_n$.

Let $\qquad a_i a_k = a_i a_j$

$\Rightarrow \qquad a_k = a_j \qquad\qquad \text{(by left cancellation law)}$

Integral Domain and Field

which is a contradiction

$$a_k \neq a_j$$

Therefore $a_i a_k \neq a_i a_j$.

Hence all elements of the form $a_j a_k$ are distinct.

Now one element of these elements is unity.

Therefore we must have $a_i a_k = 1$ for some k such that $1 \leq k \leq n$ since a_i is an arbitrary non-zero element. Therefore, every element of the type a_i has a multiplicative inverse and therefore D is a field.

For example : *Ring of integers* $(I, +, .)$ *is an integral domain but it is not a field. The inverse elements is the ring of integers are* 1 *and* -1.

Note : For a field $(F, +, .)$, identity and zero are two different elements means 1 and 0.

Now, let $a \in F$ is a non-zero element, then

$$a^{-1} = 0 \Rightarrow a \cdot a^{-1} = 0$$
$$\Rightarrow 1 = 0$$
$$\Rightarrow a \cdot 1 = a \cdot 0$$
$$\Rightarrow a = 0$$

which is a contradiction. Now, because a field does not contain zero-divisor, hence

$$1 = a^{-1} \, a \neq 0.$$

Theorem 5. *A skew field is without zero divisor.*

Proof. Let $(D, +, .)$ is a skew field, then D is a ring with unity element 1 and each non-zero element of D has a multiplicative inverse. Let $a, b \in D$, when $a \neq 0$ has arbitrary element such that

$$ab = 0.$$

Since $\qquad 0 \neq a \in D \Rightarrow \exists \, a^{-1} \in D$

$\therefore \qquad ab = 0 \Rightarrow a^{-1}(ab) = a^{-1} 0$

$\Rightarrow \qquad (a^{-1} a) b = 0,$ \hfill (By associative law)

$\Rightarrow \qquad 1 \cdot b = 0$ \hfill $[\because a^{-1} a = 1]$

$\Rightarrow \qquad b = 0$

Similarly, let $ab = 0$, when $b \neq 0$.

Since $\qquad 0 \neq b \in D \Rightarrow \exists \, b^{-1} \in D.$

$\therefore \qquad ab = 0 \Rightarrow (ab) b^{-1} = 0 b^{-1}$

$\Rightarrow \qquad a(bb^{-1}) = 0$

$\Rightarrow \qquad a1 = 0$

$\Rightarrow \qquad a = 0.$

Therefore, a skew field is without zero divisors.

Theorem 6. *The necessary and sufficient condition for non-empty subset F' of a field $(F, +, .)$ to be a sub-field is*

(i) $a \in F', b \in F' \Rightarrow a - b \in F'.$

(ii) $a \in F', b \neq 0 \, F' \Rightarrow ab^{-1} \in F'.$

Proof. Let F' be a non-empty subset of the field $(F, +, .)$ such that $(F', +, .)$ is a sub-field of $(F, +, .)$. Hence, $(F', +, .)$ is a sub-field of $(F, +, .)$

$\Rightarrow \qquad (F', +, .)$ **and** $(F, +, .)$ both are fields and $F' \subseteq F$.

$\Rightarrow \qquad (F', +)$ is additive sub-group of $(F, +)$

$\Rightarrow \qquad a - b \in F', \forall a, b \in F'$

or $\qquad a, b \in F' \Rightarrow a - b \in F'.$

Hence, condition (i) is satisfied.

Again, $(F', +, .)$ is a field.

$\Rightarrow$ non-zero elements of F' make a multiplicative group.

Hence, $a, b \in F'$ such that, $b \neq 0$

$\Rightarrow \qquad a, b^{-1} \in F'.$

$\Rightarrow \qquad ab^{-1} \in F'$

Hence, condition (ii) is satisfied.

Conversely : Let F' be a non-empty subset such that conditions (i) and (ii) are satisfied.

Now, to prove that $(F', +, .)$ is a sub-field of $(F, +, .)$ we shall show that $(F', +, .)$ is a field.

(1) It is clear from the condition (i) that $(F', +)$ is a sub-group of the abelian group $(F, +)$. Hence, $(F', +)$ is itself an abelian group.

(2) $\qquad a, b, c \in F' \Rightarrow a, b, c \in F$. Also, $(F, +, .)$ is a field

$\Rightarrow \qquad (ab) c = a (bc)$ and

$a (b + c) = ab + ac, (b + c) a = ba + ca$

$\Rightarrow$ associative law for multiplication and distributive law of addition (or multiplication) are true in F'.

From condition (ii),

$a \in F', a \neq 0 \Rightarrow aa^{-1} \in F'$

$\Rightarrow \qquad 1 \in F'$

$\Rightarrow$ unity element is included in F'.

Again, from condition (ii),

$1 \in F', a \in F'$ such that $a \neq 0$

$\Rightarrow \qquad 1a^{-1} \in F'$

Hence, inverse of each non-zero element is in F'. Now,

$a, b \in F' \neq ab \in F$

$\Rightarrow \qquad ab = ba.$

Hence, non-zero elements of F' make a commutative group.

Hence, $(F', +, .)$ is a field.

EXAMPLES

1. *If the operations be addition and multiplication (mod p), prove that the set $\{0, 1, 2, ..., p - 1\}$ (mod p),*

where p is a prime, is a field.

Solution. Let this set be denoted by $I / (p)$. We have already shown that $I(p)$ is a

Integral Domain and Field

commutative ring with unity [1]. Let $[r]$ be any non-zero element of $I/(p)$. In order to show that it is a field we need to show that there exists an element $[x]$ of $I/(p)$ such that
$$[r].[x] = [1].$$
Now, $[r] \neq [0] \Rightarrow r \not\equiv 0 \pmod{p}$
$\Rightarrow$ r is not divisible by p
$\Rightarrow$ r and p are relatively prime
i.e., there exists integers x, y such that
$$rx + py = 1$$
$\Rightarrow \qquad rx \equiv 1 \pmod{p}$ as $py \equiv 0 \pmod{p}$.
Therefore, $[r].[x] = [1]$.

2. *Is the set I of all integers a field with respect to ordinary addition and multiplication?*
Solution. We can prove that the set I of all integers is commutative ring with unity.
Also we find that its each element a, where $a \neq 0$, does not possess multiplicative inverse in I
e.g., the multiplicative inverse of 2 is $\frac{1}{2} \notin I$ [as $2.\frac{1}{2} = 1$, the multiplicative identity in I]

Hence by definition the given set **I** is not a field with respect to ordinary addition and multiplication.

3. *Show that the set M of all matrices of the form* $\begin{bmatrix} x & y \\ -y & x \end{bmatrix}$, *where x, y are real numbers, is a field with respect to matrix addition and matrix multiplication.*

Solution. R_1 : **(i) Closure property.**

Let $\qquad A = \begin{bmatrix} x & y \\ -y & x \end{bmatrix}$ and $B = \begin{bmatrix} u & v \\ -v & u \end{bmatrix}$

be any two members of the set M.

Then $\qquad A + B = \begin{bmatrix} x+u & y+v \\ -y-v & x+u \end{bmatrix} = \begin{bmatrix} x+u & y+v \\ -(y+v) & x+u \end{bmatrix} \in M$,

since x, y, u, v are real numbers, so $x+u, y+v$ are also real numbers.

(ii) Associative Law : Since matrix addition is associative,
so $\qquad (A+B)+C = A+(B+C), \forall A, B, C \in M$

(iii) Existence of Identity element.

There exists the matrix $O = \begin{bmatrix} 0 & 0 \\ -0 & 0 \end{bmatrix} = \begin{bmatrix} 0 & 0 \\ 0 & 0 \end{bmatrix} \in M$,

such that $\qquad O + A = A = A + O, \forall A \in M$
i.e., O is the additive identity element.

(iii) Existence of addition inverse.

The matrix $\begin{bmatrix} -x & -y \\ y & -x \end{bmatrix}$ is the additive inverse of the matrix A and it belongs to M, since

$$\begin{bmatrix} -x & -y \\ y & -x \end{bmatrix} + \begin{bmatrix} x & y \\ -y & x \end{bmatrix} = \begin{bmatrix} 0 & 0 \\ 0 & 0 \end{bmatrix} = \begin{bmatrix} x & y \\ -y & x \end{bmatrix} + \begin{bmatrix} -x & -y \\ y & -x \end{bmatrix}$$

(v) Commutative Law : Since matrix addition is commutative, so
$$A + B = B + A, \forall A, B \in M$$

$R_2:$
$$A.B = \begin{bmatrix} x & y \\ -y & x \end{bmatrix} . \begin{bmatrix} u & v \\ -v & u \end{bmatrix}$$
$$= \begin{bmatrix} xu - yv & xv - yu \\ -yu - xv & -yv + xv \end{bmatrix}$$
$$= \begin{bmatrix} xu - yv & xv - yu \\ -(xv + yu) & xu - yv \end{bmatrix}$$
$$\in M$$

i.e., the set M is closed with respect to multiplication composition.

$R_3:$ Since matrix multiplication is associative, so
$$(A.B).C = A.(B.C), \forall A, B, C \in M$$

$R_4:$ As the matrix multiplication is distributive with respect to addition, so $A.(B+C) = A.B + A.C$ and $(B+C).A = B.A + C.A \; \forall A, B, C \in M$

$R_5:$
$$B.A = \begin{bmatrix} u & v \\ -v & u \end{bmatrix} . \begin{bmatrix} x & y \\ -y & x \end{bmatrix}$$
$$= \begin{bmatrix} ux - vy & uy + vx \\ -vx - uy & -vy + ux \end{bmatrix}$$
$$= A.B, \text{ as proved in } R_2 \text{ above}$$

i.e., $B.A = A.B, \forall A, B \in M$

Hence all the ring postulates are satisfied and so the given set M is a commutative ring with unity.

Also
$$\begin{bmatrix} 1 & 0 \\ 0 & 1 \end{bmatrix} \begin{bmatrix} x & y \\ -y & x \end{bmatrix} = \begin{bmatrix} 1.x + 0.(-y) & 1.y + 0.x \\ 0.x + 1.(-y) & 0.y + 1.x \end{bmatrix}$$
$$= \begin{bmatrix} x & y \\ -y & x \end{bmatrix}$$

Similarly we can show that $\begin{bmatrix} x & y \\ -y & x \end{bmatrix} . \begin{bmatrix} 1 & 0 \\ 0 & 1 \end{bmatrix} = \begin{bmatrix} x & y \\ -y & x \end{bmatrix}$

$\therefore \begin{bmatrix} 1 & 0 \\ 0 & 1 \end{bmatrix}$ or $\begin{bmatrix} 1 & 0 \\ -0 & 1 \end{bmatrix}$ is the multiplicative identity.

Again if $x^2 + y^2 \neq 0$, then

$$\begin{bmatrix} x & y \\ -y & x \end{bmatrix} . \begin{bmatrix} \dfrac{x}{x^2+y^2} & \dfrac{-y}{x^2+y^2} \\ \dfrac{y}{x^2+y^2} & \dfrac{x}{x^2+y^2} \end{bmatrix} = \begin{bmatrix} \dfrac{x^2+y^2}{x^2+y^2} & \dfrac{-xy+xy}{x^2+y^2} \\ \dfrac{-yx+xy}{x^2+y^2} & \dfrac{y^2+x^2}{x^2+y^2} \end{bmatrix}$$

$$= \begin{bmatrix} 1 & 0 \\ 0 & 1 \end{bmatrix}$$

i.e., the multiplicative identity.

$\therefore$ The multiplicative inverse of $A \neq O$, where $A \in M$, exists.

Hence by definition the given set M is a field.

Integral Domain and Field

4. If $a, b, c, d \in F$ and $b \neq 0, d \neq 0$, prove that

$$\frac{a}{b} = \frac{c}{d} \leftrightarrow ad = bc.$$

Solution. We have $\dfrac{a}{b} = \dfrac{c}{d}$

$\Rightarrow \qquad ab^{-1} = cd^{-1}$

$\Rightarrow \qquad (ab^{-1})(bd) = (cd^{-1})(bd)$

$\Rightarrow \qquad (b^{-1}b)(ad) = (cb)(d^{-1}d)$

$\Rightarrow \qquad 1 \cdot (ad) = (cb) \cdot 1$

$\Rightarrow \qquad ad = cb = bc.$

Again, $ad = bc \Rightarrow add^{-1} = bcd^{-1}$

$\Rightarrow \qquad a = bcd^{-1}$

$\Rightarrow \qquad b^{-1}a = b^{-1}(bcd^{-1})$

$\Rightarrow \qquad b^{-1}a = (b^{-1}b)(cd^{-1})$

$\Rightarrow \qquad b^{-1}a = cd^{-1}$

$\Rightarrow \qquad \dfrac{a}{b} = \dfrac{c}{d}$

Hence, $\dfrac{a}{b} = \dfrac{c}{d} \leftrightarrow ad = bc.$

EXERCISES

1. Show that the set Q of rational numbers is a field with respect to ordinary addition and multiplication.
2. Prove that the set of all even integers forms a commutative ring but not a field.
3. Show that the set J of Gaussian integers does not form a field with respect to ordinary addition and multiplication of complex numbers.
4. Show that the set R of residue classes modulo m, where m is a positive prime integer is a field with respect to addition and multiplication of residue classes (mod m).
5. Verify that the set of residue classes modulo 7 is a field.

TRIGONOMETRY

De Moivre's theorem and its applications. Direct and inverse circular and hyperbolic functions. Logarithm of a complex quantity. Expansion of trigonometrical functions. Gregory's series. Summation of series.

1
De Moivre's Theorem and Deductions

1.1 DE-MOIVRE'S THEOREM

Whatever be the value of n, integral or fractional, positive or negative, the value or one of the values of $(\cos\theta + i\sin\theta)^n$ *is* $(\cos n\theta + i\sin n\theta)$.

Case I. Let n be a positive integer
By simple multiplication, we have

$$(\cos\alpha + i\sin\alpha)(\cos\beta + i\sin\beta)$$
$$= \cos\alpha\cos\beta - \sin\alpha\sin\beta + i(\sin\alpha\cos\beta + \cos\alpha\sin\beta)$$
$$= \cos(\alpha+\beta) + i\sin(\alpha+\beta)$$

Multiplying on both sides by $(\cos\gamma + i\sin\gamma)$, we have

$$(\cos\alpha + i\sin\alpha)(\cos\beta + i\sin\beta)(\cos\gamma + i\sin\gamma)$$
$$= \{\cos(\alpha+\beta) + i\sin(\alpha+\beta)\}(\cos\gamma + i\sin\gamma)$$
$$= \cos(\alpha+\beta+\gamma) + i\sin(\alpha+\beta+\gamma).$$

By proceeding in this way, we obtain the product of any number of factors of the form $(\cos\alpha + i\sin\alpha)$.

Thus if there are n factors of such form, we have

$$(\cos\alpha + i\sin\alpha)(\cos\beta + i\sin\beta)(\cos\gamma + i\sin\gamma) \times (\cos\delta + i\sin\delta)\ldots \text{ to } n \text{ factors}$$
$$= \cos(\alpha+\beta+\gamma+\delta+\ldots \text{ to } n \text{ terms}) + i\sin(\alpha+\beta+\gamma+\delta+\ldots \text{ to } n \text{ terms})$$

Now putting on both sides,

$$\alpha = \beta = \gamma = \delta \ldots = \theta \text{ (each)}.$$

we have $\quad(\cos\theta + i\sin\theta)^n = \cos n\theta + i\sin n\theta$

This proves De Moivre's Theorem when n is positive integer.

Case II. Let n be a negative integer.
Suppose $n = -m$, when m is a positive integer. Then

$$(\cos\theta + i\sin\theta)^n = (\cos\theta + i\sin\theta)^{-m}$$
$$= \frac{1}{(\cos\theta + i\sin\theta)^m}$$
$$= \frac{1}{(\cos m\theta + i\sin m\theta)} \qquad \text{(by case I)}$$
$$= \frac{\cos m\theta - i\sin m\theta}{(\cos m\theta + i\sin m\theta)(\cos m\theta - i\sin m\theta)}$$

[on multiplying the numerator and denominator by $(\cos m\theta - i\sin m\theta)$]

$$= \frac{\cos m\theta - i \sin m\theta}{\cos^2 m\theta + \sin^2 m\theta}$$

$$= \cos m\theta - i \sin m\theta$$

$$= \cos(-m\theta) + i \sin(-m\theta)$$

$$= \cos n\theta + i \sin n\theta$$

This proves De Moivre's Theorem when n is a negative integer.

Thus, we see from the above two cases that, when n is an integer, positive or negative.

$$(\cos \theta + i \sin \theta)^n = \cos n\theta + i \sin n\theta$$

Extracting nth root of both sides, it is easy to follow that $(\cos \theta + i \sin \theta)$ is one of the value of

$$(\cos n\theta + i \sin n\theta)^{1/n},$$

where n is any integer.

Case III. Let n be a fraction, positive, or negative.

Suppose, $n = \dfrac{p}{q}$, where q is a positive integer and q is an integer, positive or negative. Then

$$(\cos \theta + i \sin \theta)^n = (\cos \theta + i \sin \theta)^{p/q}$$

$$= [(\cos \theta + i \sin \theta)^p]^{1/q}$$

$$= [\cos p\theta + i \sin p\theta]^{1/q} \qquad \text{by case I or II.}$$

and by what has just been proved in the last few lines of the case II, one of the values of $(\cos p\theta + i \sin p\theta)^{1/q}$ is

$$\left(\cos \frac{p\theta}{q} + i \sin \frac{p\theta}{q}\right),$$

that is to say, $(\cos n\theta + i \sin n\theta)$ is one of the values of $(\cos \theta + i \sin \theta)^n$, when n is a fraction.

Thus, De Moivre's Theorem is completely established.

Corollary : For all values of n, integral or fractional positive or negative,

$$(\cos \theta - i \sin \theta)^n = \cos n\theta - i \sin n\theta$$

ILLUSTRATIVE EXAMPLES

1. *Show that if n is a positive interger, then*

$$\left(\frac{1 + \sin \phi + i \cos \phi}{1 + \sin \phi - i \cos \phi}\right)^n = \cos\left(\frac{n\pi}{2} - n\phi\right) - i \sin\left(\frac{n\pi}{2} - n\phi\right)$$

Solution. $1 + \sin \phi = 1 + \cos\left(\dfrac{\pi}{2} - \phi\right) = 2\cos^2\left(\dfrac{\pi}{4} - \dfrac{\phi}{2}\right)$

and $\cos \phi = \sin\left(\dfrac{\pi}{2} - \phi\right)$

$$= 2 \cos\left(\frac{\pi}{4} - \frac{\phi}{2}\right) \sin\left(\frac{\pi}{4} - \frac{\phi}{2}\right)$$

Therefore, $1 + \sin \phi \pm i \cos \phi$

De Moivre's Theorem and Deductions

$$= 2\cos\left(\frac{\pi}{4} - \frac{\phi}{2}\right)\left[\cos\left(\frac{\pi}{4} - \frac{\phi}{2}\right) \pm i\sin\left(\frac{\pi}{4} - \frac{\phi}{2}\right)\right]$$

$$\therefore \left(\frac{1 + \sin\phi + i\cos\phi}{1 + \sin\phi - i\cos\phi}\right)^n = \left[\frac{\cos\left(\frac{\pi}{4} - \frac{\phi}{2}\right) + i\sin\left(\frac{\pi}{4} - \frac{\phi}{2}\right)}{\cos\left(\frac{\pi}{4} - \frac{\phi}{2}\right) - i\sin\left(\frac{\pi}{4} - \frac{\phi}{2}\right)}\right]^n$$

$$= \left[\cos\left(\frac{\pi}{2} - \phi\right) + i\sin\left(\frac{\pi}{2} - \phi\right)\right]^n$$

$$= \cos n\left(\frac{\pi}{2} - \phi\right) + i\sin n\left(\frac{\pi}{2} - \phi\right)$$

2. *If* $x + \dfrac{1}{x} = 2\cos\theta$, *and* $y + \dfrac{1}{y} = 2\cos\phi$, *prove that*

$$\sqrt{(x^m y^n)} + \frac{1}{\sqrt{(x^m y^n)}} = 2\cos\frac{1}{2}(m\theta + n\phi).$$

Solution. $\because$ $\quad x + \dfrac{1}{x} = 2\cos\theta,$

$\therefore \quad x^2 - 2x\cos\theta + 1 = 0$

$\therefore \quad x = \dfrac{2\cos\theta \pm \sqrt{(4\cos^2\theta - 4)}}{2} = \cos\theta \pm i\sin\theta$

Similarly, $y = \cos\phi \pm i\sin\phi$, then

$$x^{m/2} = (\cos\theta \pm i\sin\theta)^{m/2} = \cos\frac{m\theta}{2} \pm i\sin\frac{m\theta}{2}$$

and $\quad y^{n/2} = (\cos\phi \pm i\sin\phi)^{n/2} = \cos\dfrac{n\phi}{2} \pm i\sin\dfrac{n\phi}{2}$

Therefore, $\sqrt{(x^m y^n)} = \left(\cos\dfrac{m\theta}{2} \pm i\sin\dfrac{m\theta}{2}\right)\left(\cos\dfrac{n\phi}{2} \pm i\sin\dfrac{n\phi}{2}\right)$

$$= \cos\frac{m\theta + n\phi}{2} \pm i\sin\frac{m\theta + n\phi}{2} \qquad \ldots(i)$$

$\therefore \quad \dfrac{1}{\sqrt{(x^m y^n)}} = \left[\cos\dfrac{m\theta + n\phi}{2} \pm i\sin\dfrac{m\theta + n\phi}{2}\right]^{-1}$

$$= \cos\frac{m\theta + n\Phi}{2} \mp i\sin\frac{m\theta + n\Phi}{2} \qquad \ldots(ii)$$

Adding (i) and (ii)

$$\sqrt{(x^m y^n)} + \frac{1}{\sqrt{(x^m y^n)}} = 2\cos\left(\frac{m\theta + n\phi}{2}\right)$$

$$= 2\cos\frac{1}{2}(m\theta + n\phi)$$

3. *If* $\cos \alpha + \cos \beta + \cos \gamma = \sin \alpha + \sin \beta + \sin \gamma = 0$. *Prove that*
$$\cos 3\alpha + \cos 3\beta + \cos 3\gamma = 3 \cos (\alpha + \beta + \gamma)$$
and
$$\sin 3\alpha + \sin 3\beta + \sin 3\gamma = 3 \sin (\alpha + \beta + \gamma)$$

Solution. To prove it, we have to remember the following algebraical identity :
If $a + b + c = 0$, then we have the identity
$$a^3 + b^3 + c^3 = 3abc$$

Let
$$a = \cos \alpha + i \sin \alpha,$$
$$b = \cos \beta + i \sin \beta,$$
$$c = \cos \gamma + i \sin \gamma,$$

so that $a + b + c = (\cos \alpha + \cos \beta + \cos \gamma) + i (\sin \alpha + \sin \beta + \sin \gamma) = 0$.
Now substituting values of a, b, c in the above algebraical identity, we have
$$(\cos \alpha + i \sin \alpha)^3 + (\cos \beta + i \sin \beta)^3 + (\cos \gamma + i \sin \gamma)^3$$
$$= 3 (\cos \alpha + i \sin \alpha)(\cos \beta + i \sin \beta)(\cos \gamma + i \sin \gamma)$$

i.e.,
$$(\cos 3\alpha + i \sin 3\alpha) + (\cos 3\beta + i \sin 3\beta) + (\cos 3\gamma + i \sin 3\gamma)$$
$$= 3 [\cos (\alpha + \beta + \gamma) + i \sin (\alpha + \beta + \gamma)]$$

or
$$(\cos 3\alpha + \cos 3\beta + \cos 3\gamma) + i (\sin 3\alpha + \sin 3\beta + \sin 3\gamma)$$
$$= 3 [\cos (\alpha + \beta + \gamma) + i \sin (\alpha + \beta + \gamma)].$$

Equating real and imaginary parts, we have
$$\cos 3\alpha + \cos 3\beta + \cos 3\gamma = 3 \cos (\alpha + \beta + \gamma)$$
$$\sin 3\alpha + \sin 3\beta + \sin 3\gamma = 3 \sin (\alpha + \beta + \gamma)$$

Thus the required results are proved.

EXERCISE

1. If a denotes $\cos 2\alpha + i \sin 2\alpha$, with similar expressions for b, c, d prove that

(i) $a + b = 2 \cos (\alpha - \beta) [\cos (\alpha + \beta) + i \sin (\alpha + \beta)]$

(ii) $a - b = 2i \sin (\alpha - \beta) [\cos (\alpha + \beta) + i \sin (\alpha + \beta)]$

(iii) $\dfrac{a - b}{a + b} = i \tan (\alpha - \beta)$

(iv) $ab + cd = 2 \cos (\alpha + \beta - \gamma - \delta) [\cos (\alpha + \beta + \gamma + \delta) + i \sin (\alpha + \beta + \gamma + \delta)]$

(v) $(a + b)(c + d) = 4 \cos (\alpha - \beta) \cos (\gamma - \delta)$
$$\times \{\cos (\alpha + \beta + \gamma + \delta) + i \sin (\alpha + \beta + \gamma + \delta)\}$$

(vi) $\dfrac{(b + c)(c + a)(a + b)}{abc}$ is real.

2. If $x_r = \cos \left(\dfrac{\pi}{2^r}\right) + i \sin \left(\dfrac{\pi}{2^r}\right)$, prove that
$$x_1 x_2 x_3 \ldots ad \ inf. = -1.$$

3. Prove that
$$(a + ib)^{m/n} + (a - ib)^{m/n} = 2(a^2 + b^2)^{m/2n} \cos \left(\dfrac{m}{n} \tan^{-1} \dfrac{b}{a}\right).$$

4. Prove that $(1 + i)^n + (1 - i)^n = 2^{(1/2)n + 1} \cos \dfrac{n\pi}{4}$ if n be a positive integer.

De Moivre's Theorem and Deductions

5. Show that

$$[(\cos\theta + \cos\Phi) + i(\sin\theta + \sin\Phi)]^n + [(\cos\theta + \cos\Phi) - i(\sin\theta + \sin\Phi)]^n$$

$$= 2^{n+1} \cos^n \frac{\theta - \Phi}{2} \cos n \frac{(\theta + \Phi)}{2}$$

6. If $2\cos\theta = x + \dfrac{1}{x}$, $2\cos\alpha = y + \dfrac{1}{y}$, prove that

(i) $x^m y^n + \dfrac{1}{x^m y^n} = 2\cos(m\theta + n\phi)$

(ii) $\dfrac{x^m}{y^n} + \dfrac{y^n}{x^m} = 2\cos(m\theta - n\phi)$.

1.2 THE q-ROOTS OF $(\cos\theta + i\sin\theta)^{p/q}$

By De Moivre's theorem, we know that when n is fractional, $\cos n\theta + i\sin n\theta$ is one of the values of $(\cos\theta + i\sin\theta)^n$. But we should have n values of the roots. To determine remaining values, we shall proceed as follows :

Obviously, if θ of the expression $\cos\theta + i\sin\theta$ is replaced by $\theta + 2r\pi$, where r is any integer or zero, its value remains unchanged. Hence,

$$(\cos\theta + i\sin\theta)^n = \{\cos(\theta + 2r\pi) + i\sin(\theta + 2r\pi)\}^n$$

where $r = 0, 1, 2, 3, ...$

Taking $n = p/q$, where p and q are integers prime to each other, we get

$$(\cos\theta + i\sin\theta)^{p/q} = \{\cos(\theta + 2r\pi) + i\sin(\theta + 2r\pi)\}^{p/q}$$

$$= \left\{\cos\frac{p}{q}(\theta + 2r\pi) + i\sin\frac{p}{q}(\theta + 2r\pi)\right\} \qquad ...(i)$$

Giving to r the values $0, 1, 2, ..., q-1$, we get the different values of L.H.S. as

$$\cos\left(\frac{p}{q}\theta\right) + i\sin\left(\frac{p}{q}\theta\right),$$

$$\cos\left\{\frac{p}{q}(\theta + 2\pi)\right\} + i\sin\left\{\frac{p}{q}(\theta + 2\pi),\right\}$$

....

....

$$\cos\left[\frac{p}{q}\{\theta + 2(q-2)\pi\}\right] + i\sin\left[\frac{p}{q}\{\theta + 2(q-2)\pi\}\right],$$

and $\quad \cos\left[\dfrac{p}{q}\{\theta + 2(q-1)\pi\}\right] + i\sin\left[\dfrac{p}{q}\{\theta + 2(q-1)\pi\}\right]$

It may be noted that no two values of the above set can be equal, for the angles involved are neither zero nor they differ by a multiple of 2π.

Hence by giving r the values $0, 1, 2, ..., (q-1)$ in the R.H.S. of (i) we obtain q and only q distinct values of $(\cos\theta + i\sin\theta)^{p/q}$.

Note : To find the roots of any real or complex number, put the given quantity in the polar form $r(\cos\theta + i\sin\theta)$ and follow the method given above.

EXAMPLES

1. *Prove that n, nth roots of unity from a series in G.P.*
Solution. We have
$$1 = \cos 0 + i \sin 0 = \cos 2r\pi + i \sin 2r\pi$$

$\therefore \quad (1)^{1/n} = (\cos 2r\pi + i \sin 2r\pi)^{1/n}$

$$= \left(\cos \frac{2r\pi}{n} + i \sin \frac{2r\pi}{n}\right)$$

Giving r the values $0, 1, 2, ..., (n-1)$, the n, nth roots of unity will be

$$(\cos 0 + i \sin 0), \left(\cos \frac{2\pi}{n} + i \sin \frac{2\pi}{n}\right), \left(\cos \frac{4\pi}{n} + i \sin \frac{4\pi}{n}\right) ...$$

and
$$\left[\cos \frac{2(n-1)\pi}{n} + i \sin \frac{2(n-1)\pi}{n}\right]$$

i.e.,
$$1, \left(\cos \frac{2\pi}{n} + i \sin \frac{2\pi}{n}\right), \left(\cos \frac{4\pi}{n} + i \sin \frac{4\pi}{n}\right) ....$$

and
$$\left[\cos \frac{2(n-1)\pi}{n} + i \sin \frac{2(n-1)\pi}{n}\right]$$

Let
$$\cos \frac{2\pi}{n} + i \sin \frac{2\pi}{n} = x$$

then
$$x^2 = \left(\cos \frac{2\pi}{n} + i \sin \frac{2\pi}{n}\right)^2 = \left(\cos \frac{4\pi}{n} + i \sin \frac{4\pi}{n}\right)$$

$$x^3 = \left(\cos \frac{2\pi}{n} + i \sin \frac{2\pi}{n}\right)^3 = \left(\cos \frac{6\pi}{n} + i \sin \frac{6\pi}{n}\right)$$

....
....

$$x^{n-1} = \left(\cos \frac{2\pi}{n} + i \sin \frac{2\pi}{n}\right)^{n-1}$$

$$= \left[\cos \frac{2(n-1)\pi}{n} + i \sin \frac{2(n-1)\pi}{n}\right]$$

Hence, the n, nth roots of unity are $1, x, x^2, x^3, ..., x^{n-1}$, where $x = \cos \frac{2\pi}{n} + i \sin \frac{2\pi}{n}$.

2. *Solve the equation $x^7 + 1 = 0$.*
Solution. We have,

$$x^7 + 1 = 0 \quad \text{or} \quad x^7 = -1$$

or
$$x = (-1)^{1/7} = (\cos s\pi + i \sin \pi)^{1/7}$$

$$= \{\cos (2r\pi + \pi) + i \sin (2r\pi + \pi)\}^{1/7}$$

$$= \left[\cos \frac{(2r\pi + \pi)}{7} + i \sin \left(\frac{2r\pi + \pi}{7}\right)\right]$$

Giving r the values $0, 1, 2, 3, 4, 5$ and 6 we get the required roots are

De Moivre's Theorem and Deductions

$$\left(\cos\frac{\pi}{7}+i\sin\frac{\pi}{7}\right), \left(\cos\frac{3\pi}{7}+i\sin\frac{3\pi}{7}\right),$$

$$\left(\cos\frac{5\pi}{7}+i\sin\frac{5\pi}{7}\right), \left(\cos\frac{7\pi}{7}+i\sin\frac{7\pi}{7}\right),$$

$$\left(\cos\frac{9\pi}{7}+i\sin\frac{9\pi}{7}\right), \left(\cos\frac{11\pi}{7}+i\sin\frac{11\pi}{7}\right)$$

and $\left(\cos\frac{13\pi}{7}+i\sin\frac{13\pi}{7}\right)$ ...(i)

Out of these values

$$\cos\frac{7\pi}{7}+i\sin\frac{7\pi}{7}=\cos\pi+i\sin\pi=-1$$

$$\cos\frac{9\pi}{7}+i\sin\frac{9\pi}{7}=\cos\left(2\pi-\frac{5\pi}{7}\right)+i\sin\left(2\pi-\frac{5\pi}{7}\right)$$

$$=\left(\cos\frac{5\pi}{7}-i\sin\frac{5\pi}{7}\right)$$

$$\cos\frac{13\pi}{7}+i\sin\frac{13\pi}{7}=\cos\left(2\pi-\frac{\pi}{7}\right)+i\sin\left(2\pi-\frac{\pi}{7}\right)$$

$$=\left(\cos\frac{\pi}{7}-i\sin\frac{\pi}{7}\right)$$

Hence from (i) the required values of $(-1)^{1/7}$ of the roots of the given equations are

$$-1, \left(\cos\frac{\pi}{7}\pm i\sin\frac{\pi}{7}\right), \left(\cos\frac{3\pi}{7}\pm i\sin\frac{3\pi}{7}\right), \left(\cos\frac{5\pi}{7}\pm i\sin\frac{5\pi}{7}\right)$$

or $\quad -1, \left(\cos\frac{r\pi}{7}\pm i\sin\frac{r\pi}{7}\right), \quad$ where $r=1,3,5.$

3. *Solve equation* $x^4-x^3+x^2-x+1=0.$
Solution. We have

$$x^5+1=(x+1)(x^4-x^3+x^2-x+1)=0$$

or $\quad x^5=-1 \quad$ or $\quad x=(-1)^{1/5}$

or $\quad x=(\cos\pi+i\sin\pi)^{1/5}$

$$=[\cos(2r\pi+\pi)+i\sin(2r\pi+\pi)]^{1/5}$$

$$=\left[\cos\left(\frac{2r\pi+\pi}{5}\right)+i\sin\left(\frac{2r\pi+\pi}{5}\right)\right]$$

Giving r the values 0, 1, 2, 3 and 4 we get the roots of equation $x^5+1=0$ as

$$\left(\cos\frac{\pi}{5}+i\sin\frac{\pi}{5}\right), \left(\cos\frac{3\pi}{5}+i\sin\frac{3\pi}{5}\right),$$

$$(\cos\pi+i\sin\pi), \left(\cos\frac{7\pi}{5}+i\sin\frac{7\pi}{5}\right) \text{ and } \left(\cos\frac{9\pi}{5}+i\sin\frac{9\pi}{5}\right)$$

or
$$\left(\cos\frac{\pi}{5} \pm \sin\frac{\pi}{5}\right), \left(\cos\frac{3\pi}{5} \pm i\sin\frac{3\pi}{5}\right) \text{ and } -1.$$

But the root $x = -1$ corresponds to the factor $x+1=0$, hecne the remaining roots are those of the given equation. Hene, the roots of the given equation are
$$\left(\cos\frac{\pi}{5} \pm i\sin\frac{\pi}{5}\right) \text{ and } \left(\cos\frac{3\pi}{5} \pm i\sin\frac{3\pi}{5}\right)$$

EXERCISES

1. Find all the values of $(-1)^{1/3}$.
2. Find all the values of $(8i)^{1/3}$.
3. Find all the values of $(32)^{1/5}$.
4. Find all the values of $(1+i)^{1/3}$.
5. Solve the equation $x^{12} - 1 = 0$ and find which of its roots satisfy the equation $x^4 + x^2 + 1 = 0$.
6. Solve the equation $x^9 - x^5 + x^4 - 1 = 0$.
7. Solve the equation $x^6 + x^5 + x^4 + x^3 + x^2 + x + 1 = 0$.
8. Solve the equation $x^{16} - 47x^8 + 1 = 0$.
9. If $x = \cos\theta + i\sin\theta$ and $\sqrt{(1-c^2)} = nc - 1$, show that
$$1 + c\cos\theta = \frac{c}{2n}(1+nx)\left(1+\frac{n}{x}\right).$$
10. Find the cube roots of $(1 - \cos\phi - i\sin\phi)$, where ϕ is real.

Answers

1. $\frac{1}{2}(1 \pm i\sqrt{3})$ and -1.
2. $\pm\sqrt{3} + i$ and $-2i$.
3. $2, 2\left[\cos\frac{2\pi}{5} \pm i\sin\frac{2\pi}{5}\right]$ and $2\left[\cos\frac{4\pi}{5} \pm i\sin\frac{4\pi}{5}\right]$.
4. $2^{-1/3}(-1+i)$ and $2^{1/6}\left[\cos\frac{r\pi}{12} + i\sin\frac{r\pi}{12}\right]$ where $r = 1, 17$.
5. $\pm 1, \pm i, \left(\pm\cos\frac{\pi}{6} \pm i\sin\frac{\pi}{6}\right)$ and $\left(\pm\cos\frac{\pi}{5} \pm i\sin\frac{\pi}{3}\right)$.
6. $\pm 1, \pm i, (\cos\pi + i\sin\pi), \left(\cos\frac{\pi}{5} \pm i\sin\frac{\pi}{5}\right)$ and $\left(\cos\frac{3\pi}{5} \pm i\sin\frac{3\pi}{5}\right)$.
7. $\left(\cos\frac{2r\pi}{7} \pm i\sin\frac{2r\pi}{7}\right)$, where $r = 1, 2, 3$.
8. $\left(\frac{\sqrt{5} \pm 1}{2}\right)\left(\cos\frac{r\pi}{4} \pm i\sin\frac{r\pi}{4}\right)$, where $r = 0, 1, 2$ and 3.

De Moivre's Theorem and Deductions

1.3 EXPANSION OF $\sin n\theta$ AND $\cos n\theta$

From De moivre's Theorem, we know that

$$(\cos\theta + i\sin\theta)^n = \cos n\theta + i\sin n\theta \qquad \ldots(i)$$

Also if n be positive integer, then by Binomial Theorem, we have

$$(\cos\theta + i\sin\theta)^n = \cos^n\theta + {}^nC_1 \cos^{n-1}\theta\,(i\sin\theta)$$
$$+ {}^nC_2 \cos^{n-2}\theta\,(i\sin\theta)^2 + {}^nC_3 \cos^{n-3}\theta\,(i\sin\theta)^3 + \ldots$$

$$= \cos^n\theta + i\cdot{}^nC_1 \cos^{n-1}\theta\sin\theta$$
$$- {}^nC_2 \cos^{n-2}\theta\sin^2\theta - i\,{}^nC_3 \cos^{n-3}\theta\sin^3\theta + \ldots$$

$$= (\cos^n\theta - {}^nC_2 \cos^{n-2}\theta\sin^2\theta + {}^nC_4 \cos^{n-4}\theta\sin^4\theta - \ldots$$
$$+ i\,({}^nC_1 \cos^{n-1}\theta\sin\theta - {}^nC_3 \cos^{n-3}\theta\sin^3\theta + \ldots) \qquad \ldots(ii)$$

find (i) and (ii), we get

$$\cos n\theta + i\sin n\theta = (\cos^n\theta - {}^nC_2 \cos^{n-2}\theta\sin^2\theta + {}^nC_4 \cos^{n-4}\theta\sin^4\theta - \ldots)$$
$$+ i\,({}^nC_1 \cos^{n-1}\theta\sin\theta - {}^nC_3 \cos^{n-3}\theta\sin^3\theta + \ldots)$$

Equating real and imaginary parts on both sides, we have

$$\cos n\theta = \cos^n\theta - {}^nC_2 \cos^{n-2}\theta\sin^2\theta + {}^nC_4 \cos^{n-4}\theta\sin^4\theta - \ldots$$

and

$$\sin n\theta = {}^nC_1 \cos^{n-1}\theta\sin\theta - {}^nC_3 \cos^{n-3}\theta\sin^3\theta + {}^nC_5 \cos^{n-5}\theta\sin^5\theta - \ldots$$

i.e.,

$$\cos n\theta = \cos^n\theta - \frac{n(n-1)}{2!}\cos^{n-2}\theta\sin^2\theta$$
$$+ \frac{(n-1)(n-2)(n-3)}{4!}\cos^{n-3}\theta\sin^3\theta$$

and

$$\sin n\theta = n\cos^{n-1}\theta\sin\theta - \frac{n(n-1)(n-2)}{3!}\cos^{n-3}\theta\sin^3\theta$$
$$+ \frac{n(n-1)(n-2)(n-3)(n-4)}{5!}\cos^{n-5}\theta\sin^5\theta - \ldots$$

Note that in each of the series, the terms are altenately positive and negative and each continues till a factor in the numerator vanishes.

1.4 EXPANSION OF $\tan n\theta$

From § 1.3, we get

$$\tan n\theta = \frac{\sin n\theta}{\cos n\theta}$$

$$= \frac{n\cos^{n-1}\theta\sin\theta - \dfrac{n(n-1)(n-2)}{3!}\cos^{n-3}\theta\sin^3\theta + \ldots}{\cos^n\theta - \dfrac{n(n-1)}{2!}\cos^{n-2}\theta\sin^2\theta + \ldots}$$

Dividing the numerator and denominators of R.H.S. by $\cos^n\theta$, we get

$$\tan n\theta = \frac{n\tan\theta - \dfrac{n(n-1)(n-2)}{3!}\tan^3\theta + \ldots}{1 - \dfrac{n(n-1)}{2!}\tan^2\theta + \dfrac{n(n-1)(n-2)(n-3)}{4!}\tan^4\theta - \ldots}$$

EXAMPLES

1. *Prove that*

$$\sin n\theta \cos^n \theta = n \tan \theta - \frac{n(n+1)(n+2)}{1.2.3} \tan^3 \theta + \ldots$$

and

$$\cos n\theta \cos^n \theta = 1 - \frac{n(n+1)}{1.2} \tan^2 \theta + \frac{n(n+1)(n+2)(n+3)}{1.2.3.4} \tan^4 \theta - \ldots$$

Solution. We have by De Moivre's Theorem

$$(\cos n\theta + i \sin \theta)^{-n} = \cos n\theta - i \sin n\theta$$

Also with the help of Binomial theorem, we have

$$(\cos \theta + i \sin \theta)^{-n} = [\cos \theta (1 + i \sin \theta)]^{-n}$$

$$= \sec^n \theta \left[1 - n(i \tan \theta) + \frac{(-n)(-n-1)}{2!}(i \tan \theta)^2 + \frac{(-n)(-n-1)(-n-2)}{3!} (i \tan \theta)^3 + \frac{(-n)(-n-1)(-n-2)(-n-3)}{4!} (i \tan \theta)^4 + \ldots \right]$$

$$= \sec^n \theta \left[1 - ni \tan \theta - \frac{n(n+1)}{1.2} \tan^2 \theta + \frac{n(n+1)(n+2)}{1.2.3} i \tan^3 \theta + \frac{n(n+1)(n+2)(n+3)}{1.2.3.4} \tan^4 \theta - \ldots \right]$$

from (i) and (ii), we get

$$\cos n\theta - i \sin n\theta = \sec^n \theta \left[\left\{ 1 - ni \tan \theta - \frac{n(n+1)}{1.2} \tan^2 \theta + \frac{n(n+1)(n+2)(n+3)}{1.2.3.4} \tan^4 \theta - \ldots \right\} - i \left\{ n \tan \theta - \frac{n(n+1)(n+2)}{1.2.3} \tan^3 \theta + \ldots \right\} \right]$$

Equating real and imaginary parts on both sides, we get

$$\cos n\theta = \sec^n \theta \left[1 - \frac{n(n+1)}{1.2} \tan^2 \theta + \frac{n(n+1)(n+2)(n+3)}{1.2.3.4} \tan^4 \theta - \ldots \right]$$

and

$$\sin n\theta = \sec^n \theta \left[n \tan \theta - \frac{n(n+1)(n+2)}{1.2.3} \tan^3 \theta + \ldots \right]$$

i.e.,

$$\cos n\theta \cos^n \theta = 1 - \frac{n(n+1)}{1.2} \tan^2 \theta + \frac{n(n+1)(n+2)(n+3)}{1.2.3.4} \tan^4 \theta - \ldots$$

and

$$\cos^n \theta \sin n\theta = n \tan \theta - \frac{n(n+1)(n+2)}{1.2.3} \tan^3 \theta + \ldots$$

2. *Expand* $\tan 5\theta$.

Solution. We have

$$\tan n\theta = \frac{n \tan \theta - \frac{n(n-1)(n-2)}{1.2.3} \tan^3 \theta + \ldots}{1 - \frac{n(n-1)}{1.2} \tan^2 \theta + \frac{n(n-1)(n-2)(n-3)}{1.2.3.4} \tan^4 \theta \ldots}$$

De Moivre's Theorem and Deductions

$$\therefore \quad \tan 5\theta = \frac{5\tan\theta - \frac{5.4.3}{1.2.3}\tan^3\theta + \frac{5.4.3.2.1}{1.2.3.4.5}\tan^5\theta}{1 - \frac{5.4}{1.2}\tan^2\theta + \frac{5.4.3.2}{1.2.3.4}\tan^4\theta}$$

$$= \frac{5\tan\theta - 10\tan^3\theta + \tan^5\theta}{1 - 10\tan^2\theta + 5\tan^4\theta}$$

EXERCISE

1. Expand:
   (a) $\cos 7\theta$, (b) $\sin 7\theta$

   [**Ans.** (a) $\cos^7\theta - 21\cos^5\theta\sin^2\theta + 35\cos^3\theta\sin^4\theta - 7\cos\theta\sin^6\theta$;

   (b) $7\cos^6\theta\sin\theta - 35\cos^4\theta\sin^3\theta + 21\cos^2\theta\sin^5\theta - \sin^7\theta$.]

2. Expand $\sin 8\theta$.

   [**Ans.** $8\cos^7\theta\sin\theta - 56\cos^5\theta\sin^3\theta + 56\cos^3\theta\sin^5\theta - 8\cos\theta\sin^7\theta$.]

3. When n is even prove that the last terms in the expansion of $\sin n\theta$ and $\cos n\theta$ are respectively.
   $$n(-1)^{(n-2)/2}\cos\theta\sin^{n-1}\theta \text{ and } (-1)^{n/2}\sin^n\theta.$$

4. Expand $\tan 8\theta$. $\left[\text{**Ans.** } \dfrac{8\tan\theta - 56\tan^3\theta + 56\tan^5\theta - 8\tan^7\theta}{1 - 28\tan^2\theta + 70\tan^4\theta - 28\tan^6\theta + \tan^8\theta}\right]$

1.5 EXPANSION OF $\tan(\alpha + \beta + \gamma + \delta + ...)$

We know that
$$(\cos\alpha + i\sin\alpha)(\cos\beta + i\sin\beta)(\cos\gamma + i\sin\gamma)(\cos\delta + i\sin\delta)...$$
$$= \cos(\alpha + \beta + \gamma + \delta + ...) + i\sin(\alpha + \beta + \gamma + \delta + ...) \quad ...(i)$$

Now, we have
$$\cos\alpha + i\sin\beta = \cos\alpha(1 + i\tan\alpha),$$
Similarly, $\quad \cos\beta + i\sin\beta = \cos\beta(1 + i\tan\beta),$
$$\cos\gamma + i\sin\gamma = \cos\gamma(1 + i\tan\gamma), ...$$

Substituting these values in (i), we get
$$\cos(\alpha + \beta + \gamma + \delta + ...) + i\sin(\alpha + \beta + \gamma + \delta + ...)$$
$$= \cos\alpha(1 + i\tan\alpha)\cos\beta(1 + i\tan\beta)\cos\gamma(1 + i\tan\gamma)\cos\delta(1 + i\tan\delta)...$$
$$= \cos\alpha\cos\beta\cos\gamma\cos\delta...(1 + i\tan\alpha)(1 + i\tan\beta)(1 + i\tan\gamma)(1 + i\tan\delta)...$$
$$= \cos\alpha\cos\beta\cos\gamma\cos\delta...[1 + i(\tan\alpha + \tan\beta) + \tan\gamma + \tan\delta + ...)$$
$$+ i^2(\tan\alpha\tan\beta + \tan\beta\tan\gamma + ...)$$
$$+ i^3(\tan\alpha\tan\beta\tan\gamma + \tan\beta\tan\gamma\tan\delta + ...) + ...]$$
$$= \cos\alpha\cos\beta\cos\gamma\cos\delta...[1 + is_1 + i^2s_2 + i^3s_3 + ...] \quad ...(iv)$$

where $s_1 = \Sigma \tan \alpha$, $s_2 = \Sigma \tan \alpha \tan \beta$,

$s_3 = \Sigma \tan \alpha \tan \beta \tan \gamma$ ....

Hence, from (ii)

$$\cos(\alpha + \beta + \gamma + \delta + ...) + i \sin(\alpha + \beta + \gamma + \delta + ...)$$
$$= \cos \alpha \cos \beta \cos \gamma \cos \delta ... [1 + is_1 - s_2 - is_3 + s_4 + ...]$$

Equating real and imaginary parts, we get

$$\cos(\alpha + \beta + \gamma + \delta + ...) = \cos \alpha \cos \beta \cos \gamma \cos \delta ... (1 - s_2 + s_4 - ...)$$

and $\sin(\alpha + \beta + \gamma + \delta + ...) = \cos \alpha \cos \beta \cos \gamma \cos \delta ... (s_1 - s_3 + s_5 ...)$

Dividing, we get

$$\tan(\alpha + \beta + \gamma + \delta + ...) = \frac{s_1 - s_3 + s_5 ...}{1 - s_2 + s_4 - ...}$$

1.6 EXPANSION OF $\sin \alpha$ AND $\cos \alpha$ IN SERIES OF POWER OF α.

(i) Expansion of $\sin \alpha$

From § 1.3, we have

$$\sin n\theta = n \cos^{n-1}\theta \sin\theta - \frac{n(n-1)(n-2)}{3!} \cos^{n-3}\sin^3\theta + ...$$

Let $n\theta = \alpha$ or $n = \alpha/\theta$, we get

$$\sin \alpha = \frac{\alpha}{\theta} \cos^{n-1}\theta \sin\theta - \frac{\alpha\left(\frac{\alpha}{\theta} - 1\right)\left(\frac{\alpha}{\theta} - 2\right)}{3!} \cos^{n-3}\theta \sin^3\theta + ...$$

$$= \frac{\alpha}{\theta} \cos^{n-1}\theta \sin\theta - \frac{\alpha(\alpha - \theta)(\alpha - 2\theta)}{\theta^3 \cdot 3!} \cos^{n-3}\theta \sin^3\theta + ...$$

or $\sin \alpha = \alpha \cos^{n-1}\theta \left(\frac{\sin\theta}{\theta}\right) - \frac{\alpha(\alpha - \theta)(\alpha - 2\theta)}{3!} \cos^{n-3}\theta \left(\frac{\sin\theta}{\theta}\right)^3 + ...$

Let n is indefinitely large and α remaining constant, then we have $\frac{\sin\theta}{\theta}$ and $\cos\theta$ in the limit, when θ is very small are equal to unity and hence, we have

$$\sin \alpha = \alpha - \frac{\alpha^3}{3!} + \frac{\alpha^5}{5!} ... \text{ ad. inf.}$$

(ii) Expansion of $\cos \alpha$

We again have from § 1.3,

$$\cos n\theta = \cos^n\theta + \frac{n(n-1)}{2!} \cos^{n-2}\theta \sin^2\theta + \frac{n(n-1)(n-2)(n-3)}{4} \cos^{n-4}\theta \sin^4\theta ...$$

Putting $n\theta = \alpha$ or $n = \frac{\alpha}{\theta}$, we get

$$\cos \alpha = \cos^n\theta - \frac{\frac{\alpha}{\theta}\left(\frac{\alpha}{\theta} - 1\right)}{2!} \cos^{n-2}\theta \sin^2\theta$$

De Moivre's Theorem and Deductions 227

$$+ \frac{\frac{\alpha}{\theta}\left(\frac{\alpha}{\theta}-1\right)\left(\frac{\alpha}{\theta}-2\right)\left(\frac{\alpha}{\theta}-3\right)}{4!} \cos^{n-4}\theta \sin^4\theta + \ldots$$

$$= \cos^n\theta - \frac{\alpha(\alpha-\theta)}{2!}\cos^{n-2}\theta\left(\frac{\sin\theta}{\theta}\right)^2$$

$$+ \frac{\alpha(\alpha-\theta)(\alpha-2\theta)(\alpha-3\theta)}{4!}\cos^{n-4}\theta\left(\frac{\sin\theta}{\theta}\right)^4 - \ldots$$

Again taking n infinitely large and $\lim \theta \to 0$, we get

$$\cos\alpha = 1 - \frac{\alpha^2}{2!} + \frac{\alpha^4}{4!} - \ldots \text{ ad. inf.}$$

1.7 EXPANSION OF TAN α IN POWERS OF α

We have

$$\tan\alpha = \frac{\sin\alpha}{\cos\alpha} = \frac{\alpha - \frac{\alpha^3}{2!} + \frac{\alpha^5}{5!} - \ldots}{1 - \frac{\alpha^2}{2!} + \frac{\alpha^4}{4!} - \ldots}$$

$$= \left(\alpha - \frac{\alpha^3}{6} + \frac{\alpha^4}{120} - \ldots\right)\left(1 - \frac{\alpha^2}{2} + \frac{\alpha^4}{24} - \ldots\right)^{-1}$$

$$= \left(\alpha - \frac{\alpha^3}{6\cdot} + \frac{\alpha^5}{120} - \ldots\right)\left[1 - \left(\frac{\alpha^2}{2} - \frac{\alpha^4}{24} + \ldots\right)\right]^{-1}$$

$$= \left(\alpha - \frac{\alpha^3}{6} + \frac{\alpha^5}{120} - \ldots\right)\left[1 + \left(\frac{\alpha^2}{2} - \frac{\alpha^4}{24} + \ldots\right) + \left(\frac{\alpha^2}{2} - \frac{\alpha^4}{24} + \ldots\right)^2 + \ldots\right]$$

$$= \left(\alpha - \frac{\alpha^3}{6} + \frac{\alpha^5}{120} \ldots\right)\left(1 + \frac{\alpha^2}{2} - \frac{\alpha^4}{24} + \frac{\alpha^4}{24} + \ldots\right)$$

$$= \left(\alpha - \frac{\alpha^3}{6} + \frac{\alpha^5}{120} \ldots\right)\left(1 + \frac{\alpha^2}{2} + \frac{5\alpha^4}{24} + \ldots\right)$$

$$= \alpha + \frac{\alpha^3}{2} + \frac{5\alpha^5}{24} + \ldots - \frac{\alpha^3}{6} - \frac{\alpha^5}{12} - \ldots + \frac{\alpha^5}{120} + \ldots$$

$$= \alpha + \alpha^3\left(\frac{1}{2} - \frac{1}{6}\right) + \alpha^5\left(\frac{5}{24} - \frac{1}{12} + \frac{1}{12}\right) + \ldots$$

or $$\tan\alpha = \alpha + \frac{\alpha^2}{3} + \frac{2}{15}\alpha^5 + \ldots \text{ ad. inf.}$$

EXAMPLES

1. If α, β, γ be the roots of the equation $x^3 + px^2 + qx + p = 0$, prove that $\tan^{-1}\alpha + \tan^{-1}\beta + \tan^{-1}\gamma = n\pi$ radians except in one particular case.

Solution. If α, β, γ be the roots of given equation, then
$$\alpha + \beta + \gamma = -p,$$
$$\alpha\beta + \beta\gamma + \gamma\alpha = q \text{ and}$$
$$\alpha\beta\gamma = -p$$

Also, if $\alpha = \tan\theta_1, \beta = \tan\theta_2$ and $\gamma = \tan\theta_3$,

then
$$\tan(\theta_1 + \theta_2 + \theta_3) = \frac{s_1 - s_3}{1 - s_2}$$

$$= \frac{(\tan\theta_1 + \tan\theta_2 + \tan\theta_3) - \tan\theta_1 \tan\theta_2 \tan\theta_3}{1 - (\tan\theta_1 \tan\theta_2 + \tan\theta_2 \tan\theta_3 + \tan\theta_1 \tan\theta_3)}$$

$$= \frac{(\alpha + \beta + \gamma) - \alpha\beta\gamma}{1 - (\alpha\beta + \beta\gamma + \gamma\alpha)}$$

$$= \frac{(-p) - (-p)}{1 - q}$$

$= 0$, except when $q = 1$.

or $\qquad \theta_1 + \theta_2 + \theta_3 = n\pi$

or $\qquad \tan^{-1}\alpha + \tan^{-1}\beta + \tan^{-1}\gamma = n\pi$, except when $q = 1$.

2. If $\dfrac{\sin\theta}{\theta} = \dfrac{2165}{2166}$, prove that θ is the number of radians of $3°$ nearly.

Solution. Since $\dfrac{2165}{2166}$ is nearly equal to 1, hence $\dfrac{\sin\theta}{\theta}$ is also equal to and consequently θ is very small.

Now,
$$\frac{\sin\theta}{\theta} = \frac{2165}{2166}$$

or
$$\frac{\theta - \dfrac{\theta^3}{3!} + \dfrac{\theta^5}{5!} \cdots}{\theta} = \frac{2165}{2166}$$

or $\quad 1 - \dfrac{\theta^2}{3!} = \dfrac{2165}{2166}$, neglecting higher powers of θ as it is very small.

or
$$\frac{\theta^2}{6} = 1 - \frac{2165}{2166} = \frac{1}{2166}$$

or
$$\theta^2 = \frac{6}{2166} = \frac{1}{361}$$

or
$$\theta = \frac{1}{19} \text{ radians.}$$

$$= \frac{1}{19} \times \frac{180}{\pi} \text{ degrees}$$

$$= \frac{180 \times 7}{19 \times 22} \text{ degrees} = 3° \text{ nearly.}$$

De Moivre's Theorem and Deductions

3. *Prove that*

$$\frac{1}{6}\sin^3\theta = \frac{\theta^3}{3!} - (1+3^2)\frac{\theta^5}{5!} + (1+3^2+3^4)\frac{\theta^7}{7!} - \ldots$$

Solution. We have $\sin 3\theta = 3\sin\theta - 4\sin^3\theta$

or $\sin^3\theta = \frac{1}{4}(3\sin\theta - \sin 3\theta)$

$$= \frac{1}{4}\left[3\left\{\theta - \frac{\theta^3}{3!} + \frac{\theta^5}{5!} - \frac{\theta^7}{7!} + \ldots\right\} - \left\{(3\theta) - \frac{(3\theta)^3}{3!} + \frac{(3\theta)^5}{5!} - \frac{(3\theta)^7}{7!} + \ldots\right\}\right]$$

$$= \frac{1}{4}\left[(3\theta - 3\theta) - \frac{\theta^3}{3!}(3 - 3^3) + \frac{\theta^5}{5!}(3 - 3^5) - \frac{\theta^7}{7!}(3 - 3^7) + \ldots\right]$$

$$= \frac{3}{4}\left[\frac{\theta^3}{3!}(3^2 - 1) - \frac{\theta^5}{5!}(3^4 - 1) + \frac{\theta^7}{7!}(3^6 - 1) - \ldots\right]$$

$$= \frac{3}{4}(3^2 - 1)\left[\frac{\theta^3}{3!} - \frac{\theta^5}{5!}(3^2 + 1) + \frac{\theta^7}{7!}(3^4 + 3^3 + 1) - \ldots\right]$$

or $\sin^3\theta = \frac{3}{4}\times 8\left[\frac{\theta^3}{3!} - \frac{\theta^5}{5!}(3^2 + 1) + \frac{\theta^7}{7!}(3^4 + 3^2 + 1) - \ldots\right]$

or $\frac{1}{6}\sin^3\theta = \frac{\theta^3}{3!} - \frac{\theta^5}{5!}(3^2 + 1) + \frac{\theta^7}{7!}(3^4 + 3^2 + 1) - \ldots$

EXERCISE

1. If α, β, γ and δ be the roots of the equation
$$x^4 - x^3\sin 2\theta + x^2\cos 2\theta - x\cos\theta - \sin\theta = 0,$$
prove that $\tan^{-1}\alpha + \tan^{-1}\beta + \tan^{-1}\gamma + \tan^{-1}\delta = n\pi + \frac{1}{2}\pi - \theta$.

2. Prove that the equation $\sin 3\theta = a\sin\theta + b\cos\theta + c$ has six roots and the sum of the six value of θ, which satisfy it, is equal to an odd multiple of π radians.

3. Prove that the equation $ah\sec\theta - bk\csc\theta = a^2 - b^2$, has four roots and that the sum of the values of θ, which satisfy it, is equal to an odd multiple of π radians.

4. If $\frac{\sin\theta}{\theta} = \frac{5045}{5046}$, prove that θ is equal to $1°\ 58'$ nearly.

❑❑❑

2
Hyperbolic Functions

2.1 DEFINITIONS

When x is real, then we know that

$$e^x = 1 + x + \frac{x^2}{2!} + \frac{x^3}{3!} + \dots \qquad \dots(i)$$

$$\sin x = x - \frac{x^3}{3!} + \frac{x^5}{5!} - \frac{x^7}{7!} + \dots \qquad \dots(ii)$$

$$\cos x = 1 - \frac{x^2}{2!} + \frac{x^4}{4!} - \frac{x^6}{6!} + \dots \qquad \dots(iii)$$

Let us define that when x is complex then e^x, $\sin x$ and $\cos x$ are only short forms of denoting series (i), (ii) and (iii) respectively.

2.2 EULER'S EXPONENTIAL VALUES

From § 2.1, we have

$$e^{i\theta} = 1 + i\theta + \frac{(i\theta)^2}{2!} + \frac{(i\theta)^3}{3!} + \frac{(i\theta)^4}{4!} + \frac{(i\theta)^5}{5!} + \dots$$

$$= \left(1 - \frac{\theta^2}{2!} + \frac{\theta^4}{4!} - \dots\right) + i\left(\theta - \frac{\theta^3}{3!} + \frac{\theta^5}{5!} - \dots\right)$$

$$= \cos\theta + i\sin\theta \qquad \dots(i)$$

[From eqns. (ii) and (iii) of § 2.1]

Also,
$$e^{-i\theta} = 1 - i\theta + \frac{(-i\theta)^2}{2!} + \frac{(-i\theta)^3}{3!} + \frac{(-i\theta)^4}{4!} + \dots$$

$$= \left(1 - \frac{\theta^2}{2!} + \frac{\theta^4}{4!} - \dots\right) - i\left(\theta - \frac{\theta^3}{3!} + \frac{\theta^5}{5!} - \dots\right)$$

$$= \cos\theta - i\sin\theta \qquad \dots(ii)$$

From (i) and (ii) it follows that (On addition and subtraction)

$$\cos\theta = \frac{e^{i\theta} + e^{-i\theta}}{2} \quad \text{and} \quad \sin\theta = \frac{e^{i\theta} - e^{-i\theta}}{2i}$$

whatever θ may be, real or complex. These formulae are known as **Euler Exponential Values** for $\cos\theta$ and $\sin\theta$ respectivley.

Hyperbolic Functions

From above, we have

$$\tan\theta = \frac{\sin\theta}{\cos\theta} = \frac{e^{i\theta} - e^{-i\theta}}{i(e^{i\theta} + e^{-i\theta})} \quad \text{and} \quad \cot\theta = \frac{\cos\theta}{\sin\theta} = \frac{i(e^{i\theta} + e^{-i\theta})}{(e^{i\theta} - e^{-i\theta})}$$

2.3 PERIODICITY OF e^z

Let $z = x + iy$, then

$$e^z = e^{x+iy} = e^x \cdot e^{iy}$$
$$= e^x [\cos y + i \sin y]$$
$$= e^x [\cos(2n\pi + y) \, i \sin(2n\pi + y)]$$
$$= e^x \cdot e^{i(2n\pi + y)}$$
$$= e^{(x+iy) + i \cdot 2n\pi}$$
$$= e^{z + 2n\pi i}$$

which shows that e^z is periodic function of period $2\pi i$.

2.4 HYPERBOLIC FUNCTIONS

In analogy with the definitions of circular sines, cosines and tangents we have the definitions of hyperbolic sines, cosines and tangents.

The hyperbolic sines, cosines, tangents, cotangents, secants and cosecants of x are denoted by sinh x, cosh x, tanh x, coth x, sech x, and cosech x respectively.

The expressions

$$\frac{1}{2}(e^x + e^{-x}) \quad \text{and} \quad \frac{1}{2}(e^x - e^{-x})$$

where the exponentials have their principal values, are defined as cosh x and sinh x.

Similarly, we have

$$\tanh x = \frac{\sinh x}{\cosh x} = \frac{e^x - e^{-x}}{e^x + e^{-x}}$$

$$\coth x = \frac{\cosh x}{\sinh x} = \frac{e^x + e^{-x}}{e^x - e^{-x}}$$

$$\operatorname{sech} x = \frac{1}{\cosh x} = \frac{2}{e^x + e^{-x}}$$

and

$$\operatorname{cosech} x = \frac{1}{\sinh x} = \frac{2}{e^x - e^{-x}}$$

2.5 RELATIONS BETWEEN HYPERBOLIC AND CIRCULAR FUNCTIONS

(i) We have $\cos x = \dfrac{e^{ix} + e^{-ix}}{2}$

Put $x = iy$, $\cos iy = \dfrac{e^{i^2 y} + e^{-i^2 y}}{2} = \dfrac{e^{-y} + e^{y}}{2} = \cosh y$

∴ **cos iy = cos hy**

(ii) $$\sin x = \frac{e^{ix} - e^{-ix}}{2i}$$ put $x = iy$,

$$\sin iy = \frac{e^{i^2 y} - e^{-i^2 y}}{2i} = \frac{i(e^{-y} - e^{y})}{2i^2}$$

$$= i\left(\frac{e^y - e^{-y}}{2}\right) = i \sinh y$$

∴ **sin $iy = i$ sinh y** ...(ii)

(iii) $$\tan iy = \frac{\sin iy}{\cos iy} = \frac{i \sinh y}{\cosh y} = i \tanh y$$

∴ **tan $iy = i$ tanh y** ...(iii)

Similarly, we have

$$\cot iy = \frac{1}{\tan iy} = \frac{1}{i \tanh y} = -i \coth y$$

$$\sec iy = \frac{1}{\cos iy} = \frac{1}{\cosh y} = \operatorname{sech} y$$

$$\operatorname{cosec} iy = \frac{1}{\sinh y} = \frac{1}{i \sinh y} = -i \operatorname{cosech} y$$

2.6 FORMULAE FOR HYPERBOLIC FUNCTIONS

(i) $\cosh^2 x - \sinh^2 x = 1$

L.H.S. $$= \left(\frac{e^x + e^{-x}}{2}\right)^2 - \left(\frac{e^x - e^{-x}}{2}\right)^2$$

$$= \frac{e^{2x} + e^{-2x} + 2 - e^{2x} - e^{-2x} + 2}{4} = \frac{4}{4} = 1$$

$= $ R.H.S.

(ii) $\cosh^2 x + \sinh^2 x = \cosh 2x$

L.H.S. $$= \left(\frac{e^x + e^{-x}}{2}\right)^2 + \left(\frac{e^x - e^{-x}}{2}\right)^2$$

$$= \frac{e^{2x} + e^{-2x} + 2 + e^{2x} + e^{-2x} - 2}{4}$$

$$= \frac{e^{2x} + e^{-2x}}{2} = \cosh 2x = \text{R.H.S.}$$

(iii) By adding and subtracting relations (i) and (ii), we get

$$\cosh 2x = 2 \cosh^2 x - 1 = 1 + 2 \sinh^2 x$$

(iv) $\sinh 2x = 2 \sinh x \cosh x$

R.H.S. $$= 2 \cdot \frac{e^x - e^{-x}}{2} \cdot \frac{e^x + e^{-x}}{2} = \frac{e^{2x} - e^{-2x}}{2}$$

$= \sinh 2x = $ R.H.S.

Hyperbolic Functions

(v) From relation (i), we have
$$\cosh^2 x - \sinh^2 x = 1$$

Dividing both sides by $\cosh^2 x$, we get
$$1 - \tanh^2 x = \operatorname{sech}^2 x$$

Again, dividing both sides of relation (i) by $\sinh^2 x$, we get
$$\coth^2 x - 1 = \operatorname{cosech}^2 x$$

(vi) $\sinh(x \pm y) = \sinh x \cosh y \pm \cosh x \sinh y$

we have
$$\sinh(x \pm y) = \frac{1}{i} \sin i(x \pm y)$$
$$= \frac{1}{i}[\sin ix \cos iy \pm \cos ix \sin iy]$$
$$= \frac{1}{i}[i \sinh x \cosh y \pm i \cosh x \sinh y]$$
$$= \sinh x \cosh y \pm \cosh x \sinh y$$

(vii) $\cosh(x \pm y) = \cosh x \cosh y \pm \sinh x \sinh y$

we have
$$\cosh(x \pm y) = \cos i(x \pm y)$$
$$= \cos ix \cos iy \mp \sin ix \sin iy$$
$$= \cosh x \cosh y \mp i^2 \sinh x \sinh y$$
$$= \cosh x \cosh y \pm \sinh x \sinh y$$

(viii) $\tanh(x \pm y) = \dfrac{\tanh x \pm \tanh y}{1 \pm \tanh x \tanh y}$

we have
$$\tanh(x \pm y) = \frac{1}{i} \tan i(x \pm y)$$
$$= \frac{1}{i} \frac{\tan ix \pm \tan iy}{1 \pm \tan ix \tan iy}$$
$$= \frac{1}{i} \frac{i \tanh x \pm i \tanh y}{1 \pm i^2 \tanh x \tanh y}$$
$$= \frac{\tanh x \pm \tanh y}{1 \pm \tanh x \tanh y}$$

Similarly, we can show that
$$\coth(x \pm y) = \frac{\coth x \coth y \pm 1}{\cot y \pm \coth x}$$

2.7 EXPANSIONS FOR $\sinh x$ AND $\cosh x$

We have
$$\sinh x = \frac{1}{2}(e^x - e^{-x})$$
$$= \frac{1}{2}\left[\left(1 + x + \frac{x^2}{2!} + \frac{x^2}{2!} + \ldots\right) - \left(1 - x + \frac{x^2}{2!} - \frac{x^3}{3!} + \ldots\right)\right]$$

$$= x + \frac{x^3}{3!} + \frac{x^5}{5!} + \frac{x^7}{7!} + ...$$

Again,
$$\cosh x = \frac{1}{2}(e^x + e^{-x})$$

$$= \frac{1}{2}\left[\left(1 + x + \frac{x^2}{2!} + \frac{x^3}{3!} + ...\right) + \left(1 - x + \frac{x^2}{2!} - \frac{x^3}{3!} + ...\right)\right]$$

$$= 1 + \frac{x^2}{2!} + \frac{x^4}{4!} + \frac{x^6}{6!} + ...$$

2.8 PERIODS OF HYPERBOLIC FUNCTIONS

We have
$$\sinh(x + 2n\pi i) = \frac{1}{2}\{e^{(x+2n\pi i)} + e^{-(x+2n\pi i)}\}$$

$$= \frac{1}{2}\{e^{2n\pi i} \cdot e^x - e^{-2n\pi i} \cdot e^{-x}\}$$

But
$$e^{2n\pi i} = \cos 2n\pi + i \sin 2n\pi = 1$$

and
$$e^{-2n\pi i} = \cos 2n\pi - i \sin 2n\pi = 1$$

∴
$$\sinh(x + 2n\pi i) = \frac{1}{2}(e^x - e^{-x}) = \sinh x$$

Similarly,
$$\cosh(x + 2n\pi i) = \frac{1}{2}\{e^{(x+2n\pi i)} + e^{-(x+2n\pi i)}\}$$

$$= \frac{1}{2}(e^x - e^{-x}) = \cosh x$$

Hence it is clear that the **period of both hyperbolic sine and cosine is imaginary and is equal to $2\pi i$.**

Again
$$\tanh(x + \pi i) = \frac{e^{x+\pi i} - e^{-(x+\pi i)}}{e^{(x+\pi i)} + e^{-(x+\pi i)}}$$

But,
$$e^{\pi i} = \cos \pi + i \sin \pi = -1$$

and
$$e^{-\pi i} = \cos \pi - i \sin \pi = -1$$

Hence,
$$\tanh(x + \pi i) = \frac{e^x - e^{-x}}{e^x + e^{-x}} = \tanh x$$

Hence period of $\tanh x$ is πi. Similarly, we can show that period of $\coth x$ is also πi.

2.9 SEPARATION INTO REAL AND IMAGINARY PARTS

By separation of a given complex function, we mean to put it in the form $x + iy$, where x is called the real part and y the imaginary part.

When we have to separate a fraction where numerator and denominator are both complex functions, then we shall always make the denominator real, by multiplying the given fraction above and below by the complex conjugate of the denominator. Thus,

$$\frac{f_1(\alpha + i\beta)}{f_2(\alpha + i\beta)} = \frac{f_1(\alpha + i\beta) f_2(\alpha - i\beta)}{f_2(\alpha + i\beta) f_2(\alpha - i\beta)}$$

Hyperbolic Functions

We shall find that the denominator becomes real. (since $f_2 = (\alpha + i\beta)$ and $f_2 = (\alpha - i\beta)$ are complex conjugates of each other and product of any complex quantity with its conjugate is alwayus real).

EXAMPLES

1. *Separate into real and imaginary parts:*
   (a) $\cot(\alpha + i\beta)$ (b) $\sinh(\alpha + i\beta)$ (c) $\text{sech}(\alpha + i\beta)$

Solution.

(a) $\cot(\alpha + i\beta) = \dfrac{\cos(\alpha + i\beta)}{\sin(\alpha + i\beta)} = \dfrac{2\cos(\alpha + i\beta)}{2\sin(\alpha + i\beta)} \times \dfrac{\sin(\alpha - i\beta)}{\sin(\alpha - i\beta)}$

$= \dfrac{\sin 2\alpha - \sin(2\alpha\beta)}{\cos(2i\beta) - \cos 2\alpha} = \dfrac{\sin 2\alpha - i \sinh 2\beta}{\cosh 2\beta - \cos 2\alpha}$

(b) $\sinh(\alpha + i\beta) = \sinh\alpha \cosh(i\beta) + \cosh\alpha \sinh(i\beta)$

$= \sinh\alpha \cos\beta + \cosh\alpha (i \sin\beta)$

$= \sinh\alpha \cos\beta + i \cosh\alpha \sin\beta$

(c) $\text{sech}(\alpha + i\beta) = \dfrac{1}{\cosh(\alpha + i\beta)}$

$= \dfrac{1}{\cosh\{i(\alpha + i\beta)\}} = \dfrac{1}{\cos(i\alpha - \beta)}$

$= \dfrac{1}{\cos(i\alpha - \beta)} \times \dfrac{\cos(-i\alpha - \beta)}{\cos(-i\alpha - \beta)}$

$= \dfrac{2\cos(i\alpha + \beta)}{2\cos(i\alpha - \beta)\cos(i\alpha + \beta)}$

$= \dfrac{2(\cos i\alpha \cos\beta - \sin i\alpha \sin\beta)}{\cos(2i\alpha) + \cos(-2\beta)}$

$= \dfrac{2(\cosh\alpha \cos\beta - i\sinh\alpha \sin\beta)}{\cosh 2\alpha + \cos 2\beta}$

$= \dfrac{2\cosh\alpha \cos\beta}{\cosh 2\alpha + \cos 2\beta} - \dfrac{2i\sinh\alpha \sin\beta}{\cosh 2\alpha + \cos 2\beta}$

2. *If* $\tan(\theta + i\phi) = \cos\alpha + i\sin\alpha$, *show that*

(a) $\theta = \dfrac{n\pi}{2} + \dfrac{\pi}{4}$

(b) $e^{2\phi} = \tan\left(\dfrac{\pi}{4} + \dfrac{\alpha}{2}\right)$ *or* $\phi = \dfrac{1}{2}\log\tan\left(\dfrac{\pi}{4} + \dfrac{\alpha}{2}\right)$

Solution. We have

$\tan(\theta + i\phi) = \cos\alpha + i\sin\alpha$

so that $\tan(\theta - i\phi) = \cos\alpha - i\sin\alpha$

Hence, $\tan 2\theta = \tan[(\theta + i\phi) + (\theta - i\phi)]$

$= \dfrac{\tan(\theta + i\phi) + \tan(\theta - i\phi)}{1 - \tan(\theta + i\phi)\tan(\theta - i\phi)}$

$$= \frac{2\cos\alpha}{1-(\cos^2\alpha+\sin^2\alpha)}$$

$$= \frac{2\cos\alpha}{0} = \infty$$

$\therefore \qquad \tan 2\theta = \tan\dfrac{\pi}{2}$

Hence $\qquad 2\theta = n\pi + \dfrac{\pi}{2} \quad \text{or} \quad \theta = \dfrac{n\pi}{2} + \dfrac{\pi}{4}$

Again, $\quad \tan(2i\phi) = \tan\{(\theta+i\phi)-(\theta-i\phi)\}$

$$= \frac{\tan(\theta+i\phi)-\tan(\theta-i\phi)}{1+\tan(\theta+i\phi)\tan(\theta-i\phi)}$$

$$= \frac{2i\sin\alpha}{1+(\cos^2\alpha+\sin^2\alpha)} = \frac{2i\sin\alpha}{2}$$

$\Rightarrow \qquad \tan 2i\theta = i\sin\alpha$

$\therefore \qquad i\tanh 2\phi = i\sin\alpha \quad \Rightarrow \quad \tanh 2\phi = \sin\alpha$

i.e., $\qquad \dfrac{e^{2\phi}-e^{-2\phi}}{e^{2\phi}+e^{-2\phi}} = \dfrac{\sin\alpha}{1}$

cross-multiplying, we have

$$e^{2\phi}-e^{-2\phi} = \sin\alpha \cdot e^{2\phi} + \sin\alpha\, e^{-2\phi}$$

or $\qquad (1-\sin\alpha)e^{2\phi} = (1+\sin\alpha)e^{-2\phi}$

or $\qquad e^{4\phi} = \dfrac{1+\sin\alpha}{1-\sin\alpha} = \dfrac{1-\cos\left(\dfrac{\pi}{2}+\alpha\right)}{1+\cos\left(\dfrac{\pi}{2}+\alpha\right)}$

$$= \frac{2\sin^2\left(\dfrac{\pi}{4}+\dfrac{\alpha}{2}\right)}{2\cos^2\left(\dfrac{\pi}{4}+\dfrac{\alpha}{2}\right)} = \tan^2\left(\dfrac{\pi}{4}+\dfrac{\alpha}{2}\right)$$

$\therefore \qquad e^{2\phi} = \tan\left(\dfrac{\pi}{4}+\dfrac{\phi}{2}\right) \quad \Rightarrow \quad 2\theta = \log\tan\left(\dfrac{\pi}{\gamma}+\dfrac{\alpha}{2}\right)$

or $\qquad 2\theta = \log\tan\left(\dfrac{\pi}{\gamma}+\dfrac{\alpha}{2}\right)$

3. *If* $\tan(\theta+i\phi) = \tan\alpha + i\sec\alpha$, *prove that*

(a) $e^{2\phi} = \pm\cot\dfrac{\alpha}{2}$ \qquad (b) $2\theta = n\pi + \dfrac{\pi}{2}+\alpha$

Solution. Given that $\quad \tan(\theta+i\phi) = \tan\alpha + i\sec\alpha \qquad$ ...(i)

Hence, $\qquad \tan(\theta-i\phi) = \tan\alpha - i\sec\alpha \qquad$ ...(ii)

Hyperbolic Functions

(a)
$$\tan(2i\phi) = \tan[(\theta + i\phi) - (\theta - i\phi)]$$
$$= \frac{\tan(\theta + i\phi) - \tan(\theta - i\phi)}{1 + \tan(\theta + i\phi)\tan(\theta - i\phi)}$$
$$= \frac{\tan\alpha + i\sec\alpha - (\tan\alpha - i\sec\alpha)}{1 + (\tan\alpha + i\sec\alpha)(\tan\alpha - i\sec\alpha)}$$

or
$$i\tanh 2\phi = \frac{2i\sec\alpha}{1 + \tan^2\alpha + \sec^2\alpha}$$

or
$$\tanh 2\phi = \frac{2\sec\alpha}{2\sec^2\alpha} = \frac{1}{\sec\alpha} = \frac{\cos\alpha}{1}$$

or
$$\frac{e^{2\phi} - e^{-2\phi}}{e^{2\phi} + e^{-2\phi}} = \frac{\cos\alpha}{1}$$

cross-multiplying, we get
$$e^{2\phi} - e^{-2\phi} = e^{2\phi}\cos\alpha + e^{-2\phi}\cdot\cos\alpha$$

or
$$e^{2\phi}(1 - \cos\alpha) = e^{-2\phi}(1 + \cos\alpha)$$

or
$$e^{4\phi} = \frac{1 + \cos\alpha}{1 - \cos\alpha} = \frac{2\cos^2\frac{\alpha}{2}}{2\sin^2\frac{\alpha}{2}}$$

or
$$e^{2\phi} = \pm\cot\frac{\alpha}{2}$$

(b)
$$\tan 2\theta = \tan[(\theta + i\phi) + (\theta - i\phi)]$$
$$= \frac{\tan(\theta + i\phi) + \tan(\theta - i\phi)}{1 - \tan(\theta + i\phi)\tan(\theta - I\phi)}$$
$$= \frac{\tan\alpha + i\sec\alpha + \tan\alpha - i\sec\alpha}{1 - (\tan\alpha + i\sec\alpha)(\tan\alpha - i\sec\alpha)}$$
$$= \frac{2\tan\alpha}{1 - (\tan^2\alpha + \sec^2\alpha)}$$
$$= \frac{2\tan\alpha}{1 - \tan^2\alpha - (1 + \tan^2\alpha)}$$
$$= \frac{2\tan\alpha}{-2\tan^2\alpha} = -\frac{1}{\tan\alpha}$$
$$= -\cot\alpha$$

or
$$\tan 2\theta = -\cot\alpha = \tan\left(\frac{\pi}{2} + \alpha\right)$$

or
$$2\theta = n\pi + \left(\frac{\pi}{2} + \alpha\right)$$

4. If $\cos(\theta + i\phi) \cos(\alpha + i\beta) = 1$, show that

$$\tanh^2 \phi \cosh^2 \beta = \sin^2 \alpha$$

and $\qquad \tanh^2 \beta \cosh^2 \phi = \sin^2 \theta$

Solution. From the given relation, we have

$$\cos(\theta + i\phi) = \sec(\alpha + i\beta)$$

$\therefore \qquad \cos(\theta - i\phi) = \sec(\alpha - i\beta)$

Also, $\qquad \sin(\theta + i\phi) = \sqrt{[1 - \cos^2(\theta + i\phi)]}$

$$= \sqrt{[1 - \sec^2(\alpha + i\beta)]},$$

$$= \sqrt{[-\tan^2(\alpha + i\beta)]}$$

so that $\qquad \sin(\theta + i\phi) = i \tan(\alpha + i\beta)$

and $\qquad \sin(\theta - i\phi) = -i \tan(\alpha - i\beta)$

$\therefore \qquad \cos 2i\phi = \cos[(\theta + i\phi) - (\theta - i\phi)]$

$$= \cos(\theta + i\phi) \cos(\theta - i\phi) + \sin(\theta + i\phi) \sin(\theta - i\phi)$$

$$= \sec(\alpha + i\beta) \sec(\alpha - i\beta) - i^2 \tan(\alpha + i\beta) \tan(\alpha - i\beta)$$

$$= \frac{1}{\cos(\alpha + i\beta) \cos(\alpha - i\beta)} + \frac{\sin(\alpha + i\beta) \sin(\alpha - i\beta)}{\cos(\alpha + i\beta) \cos(\alpha - i\beta)}$$

i.e., $\qquad \dfrac{\cosh 2\phi}{1} = \dfrac{1 + \sin(\alpha + i\beta) \sin(\alpha - i\beta)}{\cos(\alpha + i\beta) \cos(\alpha - i\beta)}$

By componendo and dividendo, we have

$$\frac{\cos 2\phi - 1}{\cosh \phi + 1} = \frac{1 + \sin(\alpha + i\beta)\sin(\alpha - i\beta) - \cos(\alpha + i\beta)\cos(\alpha - i\beta)}{1 + \sin(\alpha + i\beta)\sin(\alpha - i\beta) + \cos(\alpha + i\beta)\cos(\alpha - i\beta)}$$

or $\qquad \dfrac{2 \sinh^2 \theta}{2 \cosh^2 \phi} = \dfrac{1 - \cos(\alpha + i\beta + \alpha - i\beta)}{1 + \cos(\alpha + i\beta - \alpha + i\beta)}$

or $\qquad \tanh^2 \phi = \dfrac{1 - \cos 2\alpha}{1 + \cosh 2\beta} = \dfrac{2 \sin^2 \alpha}{2 \cosh^2 \beta}$

Hence, $\qquad \tanh^2 \phi \cosh^2 \beta = \sin^2 \alpha$

By interchanging θ by α and ϕ by β, we will at once get the second result.

EXERCISE

1. Separate into real and imaginary parts:
   (a) $\operatorname{cosec}(\alpha + i\beta)$,
   (b) $\cosh(\alpha + i\beta)$,
   (c) $\sec(\alpha + i\beta)$,
   (d) $\tanh(\alpha + i\beta)$

[Ans. (a) $\dfrac{2 \sin \alpha \cosh \beta}{\cosh 2\beta - \cos 2\alpha} - \dfrac{2i \cos \alpha \sinh \beta}{\cosh 2\beta - \cos 2\alpha}$

(b) $\cosh \alpha \cos \beta + i \sinh \alpha \sin \beta$,

(c) $\dfrac{2 \cos \alpha \cosh \beta}{\cos 2\alpha + \cosh 2\beta} + i \cdot \dfrac{2 \sin \alpha \sinh \beta}{\cos 2\alpha + \cosh 2\beta}$]

2. Separate the following into real and imaginary parts :

(a) $\dfrac{e^{i\beta}}{1 - ce^{i\alpha}}$, (b) $e^{\sin(x+iy)}$, (c) $\exp(\sin i\theta)$

[Ans. (a) $\dfrac{\{\cos\beta - c\cos(\beta - \alpha)\} + i\{\sin\beta - c\sin(\beta - \alpha)\}}{1 - 2c\cos\alpha + c^2}$

(b) $e^{\sin x \cosh y}[\cos(\cos x \sinh y) + i\sin(\cos x \sinh y)]$

(c) $\cos(\sinh\theta) + i\sin(\sinh\theta)$.]

3. Separate into real and imaginary parts :

(a) $(1 + c\cos\alpha + ic\sin\alpha)^{1/2}$, (b) $\dfrac{\sin\theta i}{x + iy}$

(c) $m\exp\left(\dfrac{x - a + iy}{x + a + iy}\right)$

[Ans. (a) $(1 + c^2 + 2c\cos\alpha)^{1/4}\left(\cos\dfrac{\theta}{2} + i\sin\dfrac{\theta}{2}\right)$, where $\tan\theta = \dfrac{c\sin\alpha}{1 + c\cos\alpha}$

(b) $(y + ix)\sin h\,\theta/(x^2 + y^2)$;

(c) $e^p(\cos q + i\sin q)$, where $p = \dfrac{x^2 + y^2 - a^2}{(x + a)^2 + y^2}$ and $q = \dfrac{2ay}{(x + a)^2 + y^2}$.]

4. Prove that $\tan\dfrac{u + iv}{2} = \dfrac{\sin u + i\sinh v}{\cos u + \cosh v}$.

5. If $\cos(x + iy) = \cos\alpha + i\sin\alpha$, show that
$\cosh 2y + \cos 2x = 2$

6. If $\sin(\theta + i\phi) = \tan\alpha + i\sec\alpha$, prove that
$\cos 2\theta \cosh 2\phi = 3$

7. If $\cos(\theta + i\phi) = \cos\alpha + i\sin\alpha$, prove that

(i) $\sin\alpha = \pm \sin^2\theta$, (ii) $\sin\alpha = \pm\sinh^2\phi$.

8. If $\cosh(\alpha + i\beta) = x + iy$ prove that

$\dfrac{x^2}{\cosh^2\alpha} + \dfrac{y^2}{\sinh^2\alpha} = 1$ and $\dfrac{x^2}{\cos^2\beta} - \dfrac{y^2}{\sin^2\beta} = 1$

9. If $\sin(\theta + i\phi) = \cos\alpha + i\sin\alpha$ prove that

(a) $\cos^2\theta = \pm\sin\alpha$ (b) $\sinh^2\phi = \pm\sin\alpha$

10. If $\tan(\theta + i\phi) = \sin(x + iy)$, then
$\cot h\,y \sin h\,2\phi = \cot x \sin 2\theta$

3
Inverse Hyperbolic Functions

3.1 INVERSE CIRCULAR FUNCTIONS OF COMPLEX QUANTITIES

If $\sin(x+iy) = u+iv$, then $x+iy$ is said to be the sine inverse of $u+iv$, and we denote it as

$$x+iy = \sin^{-1}(u+iv)$$

We know that $\sin(x+iy) = \sin[n\pi + (-1)^n(x+iy)]$. Hence, general value of the inverse sine of $u+iv$ will be

$$n\pi + (-1)^n(x+iy).$$

Hence, $\sin^{-1}(u+iv)$ is a many valued function. **The general value** of $\sin^{-1}(u+iv)$ is denoted by $\sin^{-1}(u+iv)$. Therefore,

$$\sin^{-1}(u+iv) = n\pi + (-1)^n(x+iy)$$

$$= n\pi + (-1)^n \sin^{-1}(u+iv) \qquad \ldots(i)$$

The **Principal Value** of $\sin^{-1}(u+iv)$ is that value of $n\pi + (-1)^n(x+iy)$ whose real part lies between $\dfrac{-\pi}{2}$ and $\dfrac{\pi}{2}$. The principal value of $\sin^{-1}(u+iv)$ is denoted by $\sin^{-1}(u+iv)$.

Similarly, if $\cos(x+iy) = u+iv$, then the inverse cosine of $u+iv$ will be $x+iy$ and we may write it as

$$x+iy = \tan^{-1}(u+iv)$$

Also, since $\tan(x+iy) = \tan\{n\pi + (x+iy)\}$

we have $\tan^{-1}(u+iv) = n\pi + (x+iy)$

$$= n\pi + \tan^{-1}(u+iv)$$

The principal value of this many valued functions is that in which the real part lies between $-\pi/2$ and $\pi/2$.

3.2 INVERSE HYPERBOLIC FUNCTIONS

If $y = \sinh x$, then x is said to be inverse hyperbolic sine of y and is written as

$$x = \sinh^{-1} x$$

The other inverse hyperbolic functions, $\cosh^{-1} y$, $\tanh^{-1} y$, $\operatorname{cosech}^{-1} y$, $\operatorname{sech}^{-1} y$ and $\coth^{-1} y$ are defined similarly.

These inverse hyperbolic functions can be expressed as logarithmic functions, as discussed below:

Inverse Hyperbolic Functions

(i) If $\sinh y = x$ then $y = \sinh^{-1} x = \log\{x + \sqrt{(x^2+1)}\}$

By definition,

$$\sinh y = x \quad \text{or} \quad \frac{e^y - e^{-y}}{2} = x$$

or $\quad e^y - \dfrac{1}{e^y} = 2x \quad$ or $\quad (e^y)^2 - 2xe^y - 1 = 0$

$\therefore \quad e^y = \dfrac{2x \pm \sqrt{(4x^2+4)}}{2} = \{x \pm \sqrt{(x^2+1)}\}$

The negative sign is rejected and it has been agreed that

$$e^y = \{x + \sqrt{(x^2+1)}\}$$

Taking log we get

$$y = \sinh^{-1} x = \log_e \{x + \sqrt{(x^2+1)}\}$$

(ii) $\cosh y = x$, then $y = \cosh^{-1} x = \log\{x + \sqrt{\{x^2-1\}}\}$

By definition,

$$\cosh y = x \quad \therefore \quad \frac{e^y + e^{-y}}{2} = x$$

or $\quad (e^y)^2 - 2xe^y + 1 = 0$

$\therefore \quad e^y = \dfrac{2x \pm \sqrt{4x^2 - 4}}{2} = x \pm \sqrt{(x^2-1)}$

The negative sign is again rejected and it has been accepted that

$$e^y = \{x + \sqrt{(x^2-1)}\}$$

Taking log we get

$$y = \cosh^{-1} x = \log_e \{x + \sqrt{(x^2-1)}\}$$

(iii) If $\tanh y = x$, then $y = \tanh^{-1} x = \dfrac{1}{2} \log \dfrac{1+x}{1-x}$

By definition, $\quad \dfrac{e^y - e^{-y}}{e^y + e^{-y}} = \dfrac{x}{1}$

By cross-multiplication, we have

$$e^y - e^{-y} = xe^y + xe^{-y}$$

or $\quad (1-x)e^y = (1+x)e^{-y}$

or $\quad e^{2y} = \dfrac{1+x}{1-y}$

Taking log of both sides, we get

$$2y = \log_e \left(\frac{1+x}{1-x}\right)$$

or $\quad y = \tanh^{-1} x = \dfrac{1}{2} \log_e \left(\dfrac{1+x}{1-x}\right)$

3.3 RELATION BETWEEN INVERSE HYPERBOLIC AND INVERSDE CIRCULAR FUNCTIONS

If $y = \sinh x$, then $x = \sinh^{-1} y$ ...(i)

and $iy = i \sinh x = \sin ix$ $\therefore$ $ix = \sinh^{-1} iy$ ...(ii)

Equating the two values of x from (i) and (ii), we get

$$\sinh^{-1} y = \frac{1}{i} \sin^{-1} (iy) = -i \sin^{-1} (iy)$$

Similarly, $\cosh^{-1} y = -i \cos^{-1} (iy)$

and $\tanh^{-1} y = -i \tan^{-1} (iy)$

3.4 INVERSE HYPERBOLIC FUNCTIONS OF COMPLEX QUANTITIES

If $\sin h (x + iy) = u + iv$, then $x + iy$ is said to be inverse hyperbolic sine of $u + iv$ and we write it as

$$x + iy = \sinh^{-1} (u + iv)$$

Since $\sinh (x + iy) = \sinh \{n\pi + (-1)^n (x + iy)\}$

We get $u + iv = \sinh \{n\pi i + (-1)^n (x + iy)\}$

This shows that the general value of the inverse hyperbolic sine of $u + iv$ is

$$\{n\pi i + (-1)^n (x + iy)\}$$

or we write it as

$$\sinh^{-1} (u + iv) = n\pi i + (-1)^n (x + iy)$$

$$= n\pi i + (-1)^{-1} \sinh^{-1} (u + iv)$$

The principal value of $\sinh^{-1} (u + iv) = n\pi i + (-1)^n (x + iy)$ is the value of $n\pi i + (-1)^n (x + iy)$, whose imaginary part lies between $-\frac{1}{2} i\pi$ and $\frac{1}{2} i\pi$.

Similarly, inverse hyperbolic cosine and tangent of $(u + iv)$ may be defined.

EXAMPLES

1. *Express* $\cos^{-1} (x + iy)$ *and* $\sin^{-1} (x + iy)$ *in the form* $A + iB$.

Solution. We have

$$\cos^{-1} (x + iy) = A + iB$$

or $x + iy = \cos (A + iB)$

$= \cos A \cos (iB) - \sin A \sin (iB)$

$= \cos A \cosh B - i \sin A \sinh B$

Equating real and imaginary parts on both sides, we get

$x = \cos A \cosh B$ ...(i)

$y = -\sin A \sinh B$ ...(ii)

From (i) and (ii) we have

$$\frac{x^2}{\cos^2 A} - \frac{y^2}{\sin^2 A} = \cosh^2 B - \sinh^2 B = 1$$

Inverse Hyperbolic Functions

or $\qquad x^2 \sin^2 A - y^2 \cos^2 A = \sin^2 A \cos^2 A$

or $\qquad x^2 \sin^2 A + y^2 (1 - \sin^2 A) = \sin^2 A (1 - \sin^2 A)$

or $\qquad \sin^4 A + (x^2 + y^2 - 1) \sin^2 A - y^2 = 0$

$\therefore \qquad \sin^2 A = \dfrac{-(x^2 + y^2 - 1) \pm \sqrt{\{(x^2 + y^2 - 1)^2 + 4y^2\}}}{2}$

$\qquad = \dfrac{-(x^2 + y^2 - 1) + \sqrt{\{(x^2 + y^2 - 1)^2 + 4y^2\}}}{2}$

negative sign is rejected as $\sin^2 A$ is positive

or $\qquad \sin A = \pm \dfrac{\sqrt{\sqrt{\{(x^2 + y^2 - 1)^2 + 4y^2\}} - (x^2 + y^2 - 1)}}{2}$

or $\qquad A = \pm \sin^{-1} \left[\dfrac{\sqrt{\{(x^2 + y^2 - 1)^2 + 4y^2\}} - (x^2 + y^2 - 1)}{2} \right]^{1/2}$

Again from (i) and (ii)

$\qquad \dfrac{x^2}{\cosh^2 B} + \dfrac{y^2}{\sinh^2 B} = \cos^2 A + \sin^2 A = 1$

or $\qquad x^2 \sinh^2 B + y^2 \cosh^2 B = \cosh^2 B \sinh^2 B$

or $\qquad x^2 \sinh^2 B + y^2 (1 + \sinh^2 B) = (1 + \sinh^2 B) \sinh^2 B$

or $\qquad \sinh^4 B + (1 - x - y^2) \sinh^2 B - y^2 = 0$

$\therefore \qquad \sinh^2 B = \dfrac{-(1 - x^2 - y^2) + \sqrt{\{(1 - x^2 - y^2)^2 + 4y^2\}}}{2}$

negative sign is rejected as $\sinh^2 B$ is positive

or $\qquad \sinh B = \pm \left[\dfrac{\sqrt{\{(1 - x^2 - y^2)^2 + 4y^2\}} - (1 - x^2 - y^2)}{2} \right]^{1/2}$

or $\qquad B = \pm \sinh^{-1} \left[\dfrac{\sqrt{\{(1 - x^2 - y^2)^2 + 4y^2\}} - (1 - x^2 - y^2)^{1/2}}{2} \right]$

The general value of $\cos^{-1}(x + iy)$ is $\cos^{-1}(x + iy)$ and is given by

$\qquad \cos^{-1}(x + iy) = 2n\pi + \cos^{-1}(x + iy)$

$\qquad = 2n\pi + A + iB$

Where A and B are given by (iii) and (iv) respectively.

Again, $\qquad \sin\left\{\dfrac{1}{2}\pi - (A + iB)\right\} = \cos(A + iB)$

$\qquad\qquad\qquad\qquad\qquad = x + iy$

or $$\frac{1}{2}\pi - (A + iB) = \sin^{-1}(x + iy)$$

or $$\sin^{-1}(x + iy) = \frac{\pi}{2} - (A + iB)$$

$$= \frac{\pi}{2} - \cos^{-1}(x + iy)$$

Putting the value of $\cos^{-1}(x + iy)$ from above we have the required value of $\sin^{-1}(x + iy)$.

2. *Prove that* $\tanh^{-1} x = \sinh^{-1} \dfrac{x}{\sqrt{(1 - x^2)}}$.

Solution. Let $\tanh^{-1} x = \theta$, then $\tanh \theta = x$ ....(1)

$$\therefore \quad \sinh^{-1}\frac{x}{\sqrt{(1-x^2)}} = \sinh^{-1}\left[\frac{\tanh\theta}{\sqrt{(1-\tanh^2\theta)}}\right]$$

$$= \sinh^{-1}\left[\frac{\sinh\theta/\cosh\theta}{\sqrt{\left(\dfrac{\cosh^2\theta - \sinh^2\theta}{\cosh^2\theta}\right)}}\right]$$

$$= \sinh^{-1}[\sinh\theta], \qquad \because \cosh^2\theta - \sinh^2\theta = 1$$

$$= \theta = \tanh^{-1} x.$$

3. *Separate* $\tan^{-1}(\alpha + i\beta)$ *into real and imaginary parts.*

Solution. Let $\tan^{-1}(\alpha + i\beta) = x + iy$ ...(i)

so that $\tan^{-1}(\alpha - i\beta) = x - iy$ ....(ii)

Adding (i) and (ii), we get

$$2x = \tan^{-1}(\alpha + i\beta) + \tan^{-1}(\alpha - i\beta)$$

$$= \tan^{-1}\frac{(\alpha + i\beta) + (\alpha - i\beta)}{1 - (\alpha + i\beta)(\alpha - i\beta)}$$

$$= \tan^{-1}\frac{2\alpha}{1 - (\alpha^2 + \beta^2)}$$

$$\therefore \quad x = \frac{1}{2}\tan^{-1}\frac{2\alpha}{1 - \alpha^2 - \beta^2}$$

Again, subtracting (ii) from (i), we get

$$2iy = \tan^{-1}(\alpha + i\beta) - \tan^{-1}(\alpha - i\beta)$$

$$= \tan^{-1}\frac{(\alpha + i\beta) - (\alpha - i\beta)}{1 + (\alpha + i\beta) - (\alpha - i\beta)}$$

$$= \tan^{-1}\frac{2i\beta}{(1 + \alpha^2 + \beta^2)}$$

Inverse Hyperbolic Functions

$$= i \tanh^{-1} \frac{2\beta}{(1+\alpha^2+\beta^2)}$$

$$\therefore \quad y = \frac{1}{2} \tanh^{-1} \frac{2\beta}{(1+\alpha^2+\beta^2)}$$

Hence, $\tan^{-1}(\alpha + i\beta) = \frac{1}{2} \tan^{-1} \frac{2\alpha}{1-\alpha^2-\beta^2} + \frac{i}{2} \tanh^{-1} \frac{2\beta}{1+\alpha^2+\beta^2}$

Remark : Considering the many valuedness of $\tan^{-1}(\alpha + i\beta)$, we get the general value as

$$\tan^{-1}(\alpha + i\beta) = n\pi + \tan^{-1}(\alpha + i\beta)$$

$$= n\pi + \frac{1}{2} \tan^{-1} \frac{2\alpha}{1-\alpha^2-\beta^2} + \frac{i}{2} \tanh^{-1} \frac{2\beta}{1+\alpha^2+\beta^2}$$

4. *Show that*

$$\tan^{-1}\left(i \frac{x-a}{x+a}\right) = -\frac{i}{2} \log \frac{a}{x}$$

Solution. Let, $\tan^{-1}\left(i \frac{x-a}{x+a}\right) = y$

Then $i \frac{x-a}{x+a} = \tan y$

or $\frac{x-a}{x+a} = \frac{1}{i} \tan y = -i \tan y$ $\quad \left(\because \frac{1}{i} = -i\right)$

or $\frac{x-a}{x+a} = \frac{-i \sin y}{\cos y}$

$\Rightarrow \frac{(x+a)-(x-a)}{(x+a)+(x-a)} = \frac{\cos y + i \sin y}{\cos y - i \sin y}$ (by componendo and dividendo)

$\Rightarrow \frac{2a}{2x} = \frac{e^{iy}}{e^{-iy}} \Rightarrow \frac{a}{x} = e^{2iy}$

Hence $2iy = \log_e \left(\frac{a}{x}\right)$

or $y = \frac{1}{2i} \log_e \frac{a}{x} = -\frac{i}{2} \log \left(\frac{a}{x}\right)$

EXERCISE

1. Prove that

(a) $\coth^{-1} \frac{2}{y} = \sinh^{-1} \frac{y}{\sqrt{(4-y^2)}}$

(b) $\tanh^{-1} x + \tanh^{-1} y = \tanh^{-1}\left(\frac{x+y}{1+xy}\right)$

2. Prove that

$$\sin^{-1}(ix) = 2n\pi - i\log[\sqrt{(1+x^2)} - x].$$

and hence deduce the value of $\sin^{-1}(i)$. [**Ans.** $2n\pi - i\log(\sqrt{2}-1)$]

3. Separate into real and imaginary parts :

(a) $\tan^{-1}(\cos\theta + i\sin\theta)$ (b) $\cos^{-1}(\cos\theta + i\sin\theta)$

[**Ans.** (a) $\pm\dfrac{\pi}{4} + \dfrac{i}{4}\log\dfrac{1+\sin\theta}{1-\sin\theta}$, according as $\cos\theta >$ or < 0.

(b) $\sin^{-1}(\sqrt{\sin\theta}) + i\log\{\sqrt{(1+\sin\theta)} - \sqrt{\sin\theta}\}$]

4. (a) If $\cosh^{-1}(x+iy) = \alpha + i\beta$, find the values of α and β.

(b) If $\tanh^{-1}(x-iy) = \alpha + i\beta$, find α and β.

[**Ans.** (a) $\alpha = \pm\sinh^{-1}\left[\dfrac{\sqrt{\{(1-x^2-y^2)^2 + 4y^2\}} - (1-x^2-y^2)}{2}\right]^{1/2}$

$\beta = +\sin^{-1}\left[\dfrac{\sqrt{\{(x^2+y^2-1)^2+4y^2\}} - (x^2+y^2-1)}{2}\right]^{1/2}$

(b) $\alpha = \dfrac{1}{2}\tanh^{-1}\left(\dfrac{2x}{1+x^2+y^2}\right)$, $\beta = \dfrac{1}{2}\tan^{-1}\left(\dfrac{2y}{1-y^2-x^2}\right)$]

5. Prove that

$$\cot^{-1}\dfrac{3-2i}{3+2i} = \dfrac{\pi}{4} + \dfrac{i}{2}\log 5$$

6. Prove that

$$\tan^{-1}(\sinh x) = \dfrac{1}{i}\log\tan\left(\dfrac{\pi}{4} + \dfrac{ix}{2}\right)$$

7. Prove that

$$\tan^{-1}(e^{i\theta}) = n\pi + \dfrac{\pi}{4} - \dfrac{i}{2}\log\tan\left(\dfrac{\pi}{4} - \dfrac{\theta}{2}\right).$$

4
Logarithms of Complex Quantities

4.1 LOGARITHM OF COMPLEX QUANTITIES

If $e^{x+iy} = u+iv$, then $x+iy$ is called *a* logarithm of $u+iv$ to the base e and is written as
$$x + iy = \log_e (u + iv)$$
We know that $e^{2n\pi i} = \cos 2n\pi + i \sin 2n\pi = 1$, wher n is zero or any integer, hence
$$e^{x+iy} \cdot e^{2n\pi i} = e^{x+i(y+2n\pi)} = u + iv$$
$$\therefore \quad \log_e (u+iv) = x + i(2n\pi + y)$$

This shows that the logarithm of a complex quantity is a many valued function. The **principal value** of $\log_e (u+iv)$ is that for which $n = 0$. The general value of logarithm of $u + iv$ is denoted by $\text{Log}_e (u+iv)$ and the principal value by $\log_e (u+iv)$. Thus,
$$\mathbf{Log}_e\ (u + iv) = 2n\pi i + \log_e\ (u + iv)$$

Note : Since the logarithm of a complex quantity has been defined in the same way as the logarithm of a real quantity, hence following results will also be true.
$$\log_e mn = \log_e m + \log_e n$$
$$\log_e (m/n) = \log_e m - \log_e n$$
$$\log_e m^n = n \log_e m,\ \text{etc.}$$
where m and n are complex quantities.

4.1.1 To seperate $\log_e (x+iy)$ and $\text{Log}_e (x+iy)$ into real and imaginary parts

Let
$$x + iy = r(\cos\theta + i\sin\theta)$$
Then
$$r = \sqrt{(x^2 + y^2)} \quad \text{and} \quad \tan\theta = \frac{y}{x}$$
Hence
$$\log_e (x+iy) = \log [r(\cos\theta + i\sin\theta)]$$
$$= \log(re^{i\theta})$$
$$= \log r + \log e^{i\theta}$$
$$= \log \sqrt{(x^2+y^2)} + i\tan^{-1}\left(\frac{y}{x}\right)$$
$$= \frac{1}{2}\log(x^2+y^2) + i\tan^{-1}\frac{y}{x}$$
$$\therefore \quad \log_e (x+iy) = \frac{1}{2}\log(x^2+y^2) + i\tan^{-1}\frac{y}{x}$$

And the general value is given by adding $2n\pi i$ as
$$\mathbf{Log}_e\ (x+iy) = \frac{1}{2}\log(x^2+y^2) + i\left(2n\pi + \tan^{-1}\frac{y}{x}\right)$$

4.1.2 Some Important Results

(a) Logarithm of a real negative quantity.

We have $\qquad -x = x \times (-1) = x(\cos \pi + i \sin \pi)$

or $\qquad -x = xe^{\pi i}$

$\therefore \qquad \log(-x) = \log x + \log e^{\pi i} = \log x + i\pi,$

and hence $\quad \log(-x) = 2n\pi i + \log(-x)$

$\qquad\qquad\qquad = 2n\pi i + \log x + \pi i$

$\qquad\qquad\qquad = (2n+1)\pi i + \log x$

(b) Logarithm of a purely imaginary quantity.

We have,

$$xi = x\left(\cos\frac{\pi}{2} + i\sin\frac{\pi}{2}\right) = xe^{\pi i/2},$$

$\therefore \qquad \log(xi) = \log xe^{i\pi/2} = \log x + \frac{\pi}{2}i$

Hence $\qquad \log(xi) = 2n\pi i + \log(xi)$

$$= 2n\pi i + \log x + \frac{\pi}{2}i$$

$$= \log x + i\left(2n + \frac{1}{2}\right)\pi.$$

EXAMPLES

1. *Seperate $\log \sin(x+iy)$ into real and imaginary parts.*

Solution. We have,

$\qquad \sin(x+iy) = \sin x \cos(iy) + \cos x \sin iy$

$\qquad\qquad\qquad = \sin x \cosh y + i \cos x \sinh y$

$\qquad\qquad\qquad = r(\cos\theta + i\sin\theta)$

Equating real and imaginary parts, we get

$\qquad\qquad r\cos\theta = \sin x \cosh y$

and $\qquad\qquad r\sin\theta = \cos x \sinh y$

Which give

$\qquad r^2 = \sin^2 x \cosh^2 y + \cos^2 x \sinh^2 y$

$\qquad\quad = \dfrac{1}{4}[(1-\cos 2x)(\cosh 2y + 1) + (1+\cos 2x)(\cosh 2y - 1)]$

$\qquad\quad = \dfrac{1}{4}[2(\cosh 2y - \cos 2x)]$

or $\qquad r^2 = \dfrac{1}{2}(\cosh 2y - \cos 2x)$ ...(iii)

and $\qquad \tan\theta = \dfrac{\cos x \sinh y}{\sin x \cosh y} = \cot x \tanh y$ ...(iv)

Now, from (i)

$\qquad \sin(x+iy) = r(\cos\theta + i\sin\theta)$

Logarithms of Complex Quantities 249

$$= re^{i\theta}$$

$\therefore \quad \log \sin(x+iy) = \log re^{i\theta} \cdot e^{2n\pi i}$

$$= \log r + i(2n\pi + \theta)$$

$$= \log \sqrt{\left\{\frac{1}{2}(\cosh 2y - \cos 2x)\right\}} + i[2n\pi + \tan^{-1}(\cot x \tanh y)]$$

[from (iii) and (iv)]

2. *If* $\log \log \log (\alpha + i\beta) = p + iq$, *then prove that*

(a) $e^{e^p \cos q} \cdot \cos(e^p \sin q) = \frac{1}{2}\log(\alpha^2 + \beta^2)$

(b) $e^{e^p \cos q} \cdot \sin(e^p \sin q) = \tan^{-1}\left(\frac{\beta}{\alpha}\right)$

Solution. Let $\quad x + iy = \log(\alpha + i\beta)$ ....(i)

$$= \frac{1}{2}\log(\alpha^2 + \beta^2) + i\tan^{-1}\left(\frac{\beta}{\alpha}\right)$$

Equating real and imaginary parts on both sides, we get

$$x = \frac{1}{2}\log(\alpha^2 + \beta^2) \qquad \qquad ...(ii)$$

and $\quad y = \tan^{-1}\left(\frac{\beta}{\alpha}\right)$ ...(iii)

Now, we are given that

$$\log \log \log(\alpha + i\beta) = p + iq$$

or $\qquad \log \log(x + iy) = p + iq,$ from (i)

$$\log(x + iy) = e^{p+iq} = e^p \cdot e^{iq}$$

$$= e^p(\cos q + i \sin q)$$

or $\quad x + iy = e^{e^p(\cos q + i \sin q)} = e^{e^p \cos q} \cdot e^{ie^p \sin q}$

or $\quad x + iy = e^{e^p \cos q}[\cos(e^p \sin q) + i \sin(e^p \sin q)]$

Equating real and imaginary parts on both sides, we get

$$x = e^{e^p \cos q} \cos(e^p \sin q)$$

and $\quad y = e^{e^p \cos q} \sin(e^p \sin q),$

with the help of (ii), and (iii), putting the values of x and y,

(a) $\frac{1}{2}\log(\alpha^2 + \beta^2) = e^{e^p \cos q} \cos(e^p \sin q)$

(b) $\tan^{-1}\left(\frac{\beta}{\alpha}\right) = e^{e^p \cos q} \sin(e^p \sin q)$

3. *Prove that*

$$\tan\left\{i \log \frac{a - ib}{a + ib}\right\} = \frac{2ab}{a^2 - b^2}$$

Solution. Let $\quad a = r\cos\theta$ and $b = r\sin\theta$

then $r = \sqrt{(a^2 + b^2)}$ and $\theta = \tan^{-1}\left(\dfrac{b}{a}\right)$ ...(i)

$\therefore \quad \log \dfrac{a - bi}{a + bi} = \log\left[\dfrac{r(\cos\theta - i\sin\theta)}{r(\cos\theta + i\sin\theta)}\right]$

$= \log\left[\dfrac{\cos\theta - i\sin\theta}{\cos\theta + i\sin\theta}\right]$

$= \log \dfrac{e^{-i\theta}}{e^{i\theta}} = \log(e^{-2i\theta})$

$= -2i\theta$

or $\quad i\log\left(\dfrac{a - ib}{a + ib}\right) = i(-2\theta i) = 2\theta$

$= 2\tan^{-1}\left(\dfrac{b}{a}\right),\quad$ from (i)

$= \tan^{-1}\left\{\dfrac{2b/a}{1 - (b^2/a^2)}\right\}$

$= \tan^{-1}\left[\dfrac{2ab}{a^2 - b^2}\right]$

or $\quad \tan\left[i\log\dfrac{a - ib}{a + ib}\right] = \dfrac{2ab}{a^2 - b^2}$

EXERCISE

Prove that:

1. (a) $\log(-1) = \pi i$ \quad (b) $\log i = \dfrac{1}{2}\pi i$

2. $\log(1 + i) = \dfrac{1}{2}\log_e 2 + i\left(2n\pi + \dfrac{\pi}{4}\right)$

3. (a) $\log_e \dfrac{a + bi}{a - bi} = 2i\tan^{-1}\left(\dfrac{b}{a}\right)$ \quad (b) $i\log\dfrac{x - i}{x + i} = \pi - 2\tan^{-1} x$

4. If $\log_e \log_e (x + iy) = p + iq$, then show that
$$y = x\tan\left\{\tan q \log_e \sqrt{(x^2 + y^2)}\right\}$$

5. Prove that
$$\log\log\sin(x + iy) = \dfrac{1}{2}\log\{p^2 + q^2\} + i\tan^{-1}\left\{\dfrac{p}{q}\right\},$$
where $p = \dfrac{1}{2}\log\left\{\dfrac{1}{2}(\cosh 2y - \cos 2x)\right\}$ and $q = \tan^{-1}(\cot x \tanh y)$

Logarithms of Complex Quantities

4.2 VALUE OF a^z WHERE a AND z ARE COMPLEX QUANTITIES

When a and z are both real, then we define
$$a^z = e^{z \log_e a}$$
Similarly, when a and z are complex numbers, we define
$$a^z = e^{z \log_e a}$$
but $\log_e a = 2n\pi i + \log_e a$

$\therefore \qquad a^z = a^{z(2n\pi i + \log_e a)}$

This shows that a^z is a many valued function. The principal value of a^z is obtained by putting $n = 0$.

4.3 SEPERATION OF $(a+ib)^{x+iy}$ INTO REAL AND IMAGINARY PARTS

$$(a+ib)^{(x+iy)} = e^{(x+iy)\log_e(a+ib)}$$
$$= e^{(x+iy)[2n\pi i + \log_e(a+ib)]}$$

But
$$\log_e(a+ib) = \frac{1}{2}\log_e(a^2+b^2) + i\tan^{-1}\left(\frac{b}{a}\right)$$
$$= A + iB, \text{ say}$$

$\therefore \quad (a+ib)^{x+iy} = e^{(x+iy)\{2n\pi i + A + iB\}}$

$= e^{Ax - 2n\pi y - By + i(2n\pi x + xB + yA)}$

$= e^{xA - y(2n\pi + B)} \cdot e^{i\{2n\pi x + Bx + Ay\}}$

$= e^{xA - y(2n\pi + B)} \cdot e^{i\{(2n\pi + B)x + Ay\}}$

$= e^{xA - y(2n\pi + B)} [\cos\{(2n\pi + B)x + Ay\} + i\sin\{(2n\pi + B)x + Ay\}]$

where $A = \frac{1}{2}\log_e(a^2+b^2)$ and $B = \tan^{-1}\frac{b}{a}$.

EXAMPLES

1. *Prove that*
$$i^i = e^{-(4n+1)\pi/2}$$

Hence show that the values of i^i form a geometric progression whose common ratio is $e^{-2\pi}$.

Solution. We have
$$i^i = e^{i\log_e i}$$
$$= e^{i(2n\pi i + \log_e i)}$$
$$= e^{i\left(2n\pi i + \frac{1}{2}\pi i\right)}, \qquad \text{as } i = e^{i\pi/2}$$
$$= e^{-(2n+1/2)\pi}$$

Putting $n = 0, 1, 2, \ldots$, the various values of i^i are $e^{-\frac{1}{2}\pi}, e^{-\frac{5}{2}\pi}, e^{-\frac{9}{2}\pi}, \ldots$

This is a geometric progression whose common ratio is $e^{-2\pi}$.

2. If $i^{\alpha + i\beta} = \alpha + i\beta$, prove that
$$\alpha^2 + \beta^2 = e^{-(4n+1)\pi\beta}$$

Solution.
$$\alpha + i\beta = i^{\alpha + i\beta} = e^{(\alpha + i\beta)\log i}$$
$$= e^{(\alpha + i\beta)\{2n\pi i + \log i\}}$$
$$= e^{(\alpha + i\beta)\left\{2n\pi i + \log\left(\cos\frac{\pi}{2} + i\sin\frac{\pi}{2}\right)\right\}}$$
$$= e^{(\alpha + i\beta)\{2n\pi i + \log e^{i\pi/2}\}}$$
$$= e^{i\alpha\left(\frac{\pi}{2} + 2n\pi\right) - \beta\left(\frac{\pi}{2} + 2n\pi\right)}$$
$$= e^{-\beta\pi\left(2n + \frac{1}{2}\right)} \times e^{\pi\alpha\left(2n + \frac{1}{2}\right)i}$$
$$= e^{-\frac{1}{2}\pi\beta(4n+1)}\left[\cos\left\{\frac{1}{2}\pi\alpha(4n+1)\right\} + i\sin\left\{\frac{1}{2}\pi\alpha(4n+1)\right\}\right]$$

Equating real and imaginary parts on both sides we get,
$$\alpha = e^{-\frac{1}{2}\pi\beta(4n+1)} \cdot \cos\left\{\frac{1}{2}\pi\alpha(4n+1)\right\}$$
and
$$\beta = e^{-\frac{1}{2}\pi\beta(4n+1)} \cdot \sin\left\{\frac{1}{2}\pi\alpha(4n+1)\right\}$$

squaring and adding, we get
$$\alpha^2 + \beta^2 = e^{-\pi\beta(4n+1)} \times \left[\cos^2\left\{\frac{1}{2}\pi\alpha(4n+1)\right\} + \sin^2\left\{\frac{1}{2}\pi\alpha(4n+1)\right\}\right]$$
$$= e^{-\pi\beta(4n+1)}$$

3. *If all the values of* $(1 + i\tan\alpha)^{1 + i\tan\beta}$ *are real, prove that one of them is* $(\sec\alpha)^{\sec^2\beta}$.

Solution. Let $x = (1 + i\tan\alpha)^{1 + i\tan\beta}$
$$= e^{\log(1 + i\tan\alpha)^{(1 + i\tan\beta)}} = e^{(1 + i\tan\beta)\log(1 + i\tan\alpha)}$$
$$= e^{(1 + i\tan\beta)\left[\frac{1}{2}\log(1 + \tan^2\alpha) + i\tan^{-1}(\tan\alpha)\right]}$$
$$= e^{(1 + i\tan\beta)\left[\frac{1}{2}\log(\sec^2\alpha) + i\alpha\right]} = e^{(1 + i\tan\beta)[\log\sec\alpha + i\alpha]}$$
$$= e^{(\log\sec\alpha - \alpha\tan\beta) + i(\alpha + \tan\beta\log\sec\alpha)}$$

or $= e^{(\log\sec\alpha - \alpha\tan\beta)}[\cos(\alpha + \tan\beta\log\sec\alpha) + i\sin(\alpha + \tan\beta\log\sec\alpha)]$ ...(i)

If x has real values only, then imaginary component is zero.
Hence $\sin(\alpha + \tan\beta\log\sec\alpha) = 0$
∴ $\cos(\alpha + \tan\beta\log\sec\alpha) = 1$ and $(\alpha + \tan\beta\log\sec\alpha) = 0$
or $\alpha = -\tan\beta\log\sec\alpha$
substituting these values in (i), we have

Logarithms of Complex Quantities

$$x = e^{(\log \sec \alpha - \alpha \tan \beta)} [1 + i0] = e^{[\log \sec \alpha - (-\tan \beta \log \sec \alpha) \tan \beta]}$$

substituting the value of α from (ii)

$$x = e^{(1 + \tan^2 \beta) \log \sec \alpha} = e^{\sec^2 \beta \log \sec \alpha} = e^{\log (\sec \alpha)^{\sec^2 \beta}}$$

$$= (\sec \alpha)^{\sec^2 \beta}$$

EXERCISE

1. Prove that $i^a = \cos\left(2n + \dfrac{1}{2}\right) \pi a + i \sin\left(2n + \dfrac{1}{2}\right) \pi a$.

2. If $(i^i)^i = (\cos \theta - i \sin \theta)$, show that $\theta = \pi (4n + 1)/2$.

3. Prove that $(-i)^{-i} = e^{(4n-1)\pi/2}$.

4. If $\dfrac{(1+i)^{p+qi}}{(1-i)^{p-qi}} = \alpha + i\beta$, prove that one of the values of $\tan^{-1}\left(\dfrac{\beta}{\alpha}\right)$ is $\dfrac{p\pi}{2} + q \log_e 2$.

5. If $(a + ib)^p = m^{x+iy}$, prove that one of the values of $\left(\dfrac{y}{x}\right)$ is

$$\dfrac{2 \tan^{-1}\left(\dfrac{b}{a}\right)}{\log_e (a^2 + b^2)}$$

6. If $a^{\alpha + i\beta} = (x + iy)^{p+iq}$, principal values only being considered, prove that

$$\alpha = \dfrac{p}{2} \log_a (x^2 + y^2) - q \tan^{-1} \dfrac{y}{x} \log_a e$$

7. If $(a + ib)^{c+id}$ is wholly real and principal values only are considered, prove that

$$(a + ib)^{c+id} = (a^2 + b^2)^{(c^2 + d^2)/2c}.$$

◻◻◻

5

Gregory's Series

5.1 IF $\dfrac{-\pi}{4} \le \theta \le \dfrac{\pi}{4}$, **THEN** $\theta = \tan \theta - \dfrac{1}{3} \tan^3 \theta + \dfrac{1}{5} \tan^5 \theta - \ldots$

We have
$$1 + i \tan \theta = 1 + i \dfrac{\sin \theta}{\cos \theta}$$
$$= (\cos \theta + i \sin \theta) \sec \theta$$
$$= e^{i\theta} \sec \theta$$

Taking logarithms, we get
$$\log(1 + i \tan \theta) = \log \sec \theta + i\theta \qquad \ldots(i)$$

Since $-\dfrac{\pi}{4} \le \theta \le \dfrac{\pi}{4}$, $\tan \theta$ can not be numerically greater than unity, hence the modulus of $i \tan \theta$ cannot be greater than unity. Hence by logarithmic theorem, we get

$$\log(1 + i \tan \theta) = i \tan \theta - \dfrac{1}{2}(i \tan \theta)^2 + \dfrac{1}{3}(i \tan \theta)^3 - \dfrac{1}{4}(i \tan \theta)^4 + \ldots$$
$$= i \tan \theta + \dfrac{1}{2} \tan^2 \theta - \dfrac{1}{3} i \tan^3 \theta - \dfrac{1}{4} \tan^4 \theta + \ldots \qquad \ldots(ii)$$

From (i) and (ii), we get
$$\log \sec \theta + i\theta = i \tan \theta + \dfrac{1}{2} \tan^2 \theta - \dfrac{1}{3} i \tan^3 \theta - \dfrac{1}{4} \tan^4 \theta + \ldots$$
$$= \left(\dfrac{1}{2} \tan^2 \theta - \dfrac{1}{4} \tan^4 \theta + \ldots \right) + i \left(\tan \theta - \dfrac{1}{3} \tan^3 \theta + \dfrac{1}{5} \tan^5 \theta - \ldots \right) \qquad \ldots(iii)$$

Equating the imaginary parts, we obtain
$$\theta = \tan \theta - \dfrac{1}{3} \tan^3 \theta + \dfrac{1}{5} \tan^5 \theta \qquad \ldots(A)$$

This is known as **Gregory' sSeries.**

Putting $\tan \theta = x$, i.e., $\theta = \tan^{-1} x$, we obtain another form of Gregory's series, as

$$\tan^{-1} x = x - \dfrac{1}{3} x^3 + \dfrac{1}{5} x^5 - \ldots \qquad \ldots(B)$$

where $|x| \le 1$,

Again, putting $\theta = \dfrac{\pi}{4}$ in (A), we get

Gregory's Series

$$\frac{\pi}{4} = 1 - \frac{1}{3} + \frac{1}{5} - \frac{1}{7} + \ldots$$

Also equating the real parts on the two sides of (iii), we get

$$\log \sec \theta = \frac{1}{2} \tan^2 \theta - \frac{1}{4} \tan^4 \theta + \frac{1}{6} \tan^6 \theta \ldots$$

5.2 GENERALISED FORM OF GREGORY'S SERIES

If $n\pi - \frac{\pi}{4} \leq \theta \leq n\pi + \frac{\pi}{4}$, then $\theta - n\pi = \tan \theta - \frac{1}{3} \tan^3 \theta + \frac{1}{5} \tan^5 \theta \ldots \infty$.

Put $\theta - n\pi = \phi$

Hence, ϕ must lie in the interval $-\frac{\pi}{4}$ to $\frac{\pi}{4}$. Therefore, by Gregory's series.

$$\phi = \tan \phi - \frac{1}{3} \tan^3 \phi + \frac{1}{5} \tan^5 \phi - \ldots \infty$$

Now, $\phi = n\pi + \phi$;

$\therefore \qquad \tan \theta = \tan (n\pi + \phi) = \tan \phi$

$\therefore \qquad \theta - n\pi = \tan \theta - \frac{1}{3} \tan^3 \theta + \frac{1}{5} \tan^3 \phi - \ldots \infty$

Suppose θ lies between $\frac{7\pi}{4}$ and $\frac{9\pi}{4}$, i.e., θ lies between $2\pi - \frac{\pi}{4}$ and $2\pi + \frac{1}{4}\pi$,

$\therefore \qquad n = 2.$

Hence, by above,

$$\theta - 2\pi = \tan \theta - \frac{1}{3} \tan^3 \theta + \frac{1}{5} \tan^5 \theta - \ldots \infty$$

Similarly, if θ lies between $\frac{-11\pi}{4}$ and $\frac{-13\pi}{4}$, i.e., θ lies between $-3\pi + \frac{\pi}{4}$ and $-3\pi - \frac{\pi}{4}$,

$\therefore \qquad n = -3,$

hence, we have

$$\theta + 3\pi = \tan \theta - \frac{1}{3} \tan^3 \theta + \frac{1}{5} \tan 5\theta \ldots \infty$$

5.3 APPLICATIONS OF GREGORY'S SERIES

With the help of Gregory's series we can find out the value of π to any degree of accuracy.
(a) We have from Gregory's series.

$$\tan^{-1} x = x - \frac{x^3}{3} + \frac{x^5}{5} - \frac{x^7}{7} + \frac{x^9}{9} - \ldots \infty$$

By putting $x = 1$, we get

$$\frac{\pi}{4} = 1 - \frac{1}{3} + \frac{1}{5} - \frac{1}{7} + \frac{1}{9} - \ldots \infty$$

We may evaluate value of π from above series.
(b) Euler's Series.

$$\tan^{-1} \frac{1}{2} + \tan^{-1} \frac{1}{3} = \frac{\pi}{4}$$

We have $\tan^{-1}\dfrac{1}{2} + \tan^{-1}\dfrac{1}{3} = \tan^{-1}\dfrac{\dfrac{1}{2}+\dfrac{1}{3}}{1-\dfrac{1}{2}\cdot\dfrac{1}{3}} = \tan^{-1}1 = \dfrac{\pi}{4}.$

Expanding by Gregory's series, we have

$$\dfrac{\pi}{4} = \left\{\dfrac{1}{2} - \dfrac{1}{3}\cdot\dfrac{1}{2^3} + \dfrac{1}{5}\cdot\dfrac{1}{2^5} - \ldots\right\} + \left\{\dfrac{1}{3} - \dfrac{1}{3}\cdot\dfrac{1}{3^3} + \dfrac{1}{5}\cdot\dfrac{1}{3^5} - \ldots\right\}$$

$$= \left(\dfrac{1}{2}+\dfrac{1}{3}\right) - \dfrac{1}{3}\left(\dfrac{1}{2^3}+\dfrac{1}{3^3}\right) + \dfrac{1}{5}\left(\dfrac{1}{2^5}+\dfrac{1}{3^5}\right) - \ldots$$

(c) Machin's Series.

$$\dfrac{\pi}{4} = 4\tan^{-1}\dfrac{1}{5} - \tan^{-1}\dfrac{1}{239}.$$

We have $2\tan^{-1}\dfrac{1}{5} = \tan^{-1}\dfrac{2/5}{1-\dfrac{1}{25}} = \tan^{-1}\dfrac{5}{12}$

$$4\tan^{-1}\dfrac{1}{5} = 2\tan^{-1}\dfrac{5}{12} = \tan^{-1}\dfrac{2\cdot\dfrac{5}{12}}{1-\dfrac{25}{144}} = \tan^{-1}\dfrac{120}{119}$$

$\therefore\quad \tan^{-1}\dfrac{120}{119} - \tan^{-1}\dfrac{1}{239} = \tan^{-1}\dfrac{\dfrac{120}{119}-\dfrac{1}{2}}{1+\dfrac{120}{119}\cdot\dfrac{1}{239}}$

$$= \tan^{-1}\dfrac{120\times 239 - 119}{119\times 239 + 120} = \tan^{-1}1 = \dfrac{\pi}{4}$$

$\therefore\quad \dfrac{\pi}{4} = 4\tan^{-1}\dfrac{1}{5} - \tan^{-1}\dfrac{1}{239}$

$$= 4\left\{\dfrac{1}{5} - \dfrac{1}{3}\cdot\dfrac{1}{5^3} + \dfrac{1}{5}\cdot\dfrac{1}{5^2} - \ldots\infty\right\} - \left\{\dfrac{1}{239} - \dfrac{1}{3}\cdot\dfrac{1}{(239)^3} + \dfrac{1}{5}\cdot\dfrac{1}{(239)^5} - \ldots\infty\right\}$$

(d) Rutherford's Series.

$$4\tan^{-1}\dfrac{1}{5} - \tan^{-1}\dfrac{1}{70} + \tan^{-1}\dfrac{1}{99} = \dfrac{\pi}{4}.$$

L.H.S. $= 4\tan^{-1}\dfrac{1}{5} - \left[\tan^{-1}\dfrac{\dfrac{1}{70}-\dfrac{1}{99}}{1+\dfrac{1}{70}\cdot\dfrac{1}{99}}\right]$

$= 4\tan^{-1}\dfrac{1}{5} - \tan^{-1}\dfrac{1}{239} = \dfrac{\pi}{4}$, by Machin's series.

$\therefore\quad \dfrac{\pi}{4} = 4\left[\dfrac{1}{5} - \dfrac{1}{3}\cdot\dfrac{1}{5^3} + \dfrac{1}{5}\cdot\dfrac{1}{5^5} - \ldots\infty\right] - \left\{\dfrac{1}{70} - \dfrac{1}{3}\cdot\dfrac{1}{(70)^3} + \dfrac{1}{5}\cdot\dfrac{1}{(70)^5} - \ldots\infty\right\}$

$$+ \left\{\dfrac{1}{99} - \dfrac{1}{3}\cdot\dfrac{1}{(99)^3} + \dfrac{1}{5}\cdot\dfrac{1}{(99)^5} - \ldots\infty\right\}$$

Gregory's Series

EXAMPLES

1. Show that

$$\frac{\pi}{4} = \left(\frac{2}{3} + \frac{1}{7}\right) - \frac{1}{3}\left(\frac{2}{3^3} + \frac{1}{7^3}\right) + \frac{1}{5}\left(\frac{2}{3^5} + \frac{1}{7^5}\right) \ldots$$

Solution. We have

$$\text{R.H.S.} = 2\left\{\frac{1}{3} - \frac{1}{3}\cdot\frac{1}{3^3} + \frac{1}{5}\cdot\frac{1}{3^5} - \ldots\right\} + \left\{\frac{1}{7} - \frac{1}{3}\cdot\frac{1}{7^3} + \frac{1}{5}\cdot\frac{1}{7^5} - \ldots\right\}$$

$$= 2\tan^{-1}\frac{1}{3} + \tan^{-1}\frac{1}{7}$$

$$= \tan^{-1}\frac{\frac{2}{3}}{1 - \frac{1}{9}} + \tan^{-1}\frac{1}{7}$$

$$= \tan^{-1}\frac{3}{4} + \tan^{-1}\frac{1}{7} = \tan^{-1}\frac{\frac{3}{4} + \frac{1}{7}}{1 - \frac{3}{4}\cdot\frac{1}{7}}$$

$$= \tan^{-1}\frac{25}{25} = \tan^{-1} 1 = \frac{\pi}{4}.$$

2. Prove that

$$\frac{\tan^{-1} x}{x} + \frac{\tan^{-1} y}{y} + \frac{\tan^{-1} z}{z} = 3\left\{1 - \frac{1}{7} + \frac{1}{13} - \frac{1}{19} + \frac{1}{25} - \ldots\right\}$$

where x, y, z are cube roots of unity.

Solution. We know that cube roots of unity are $1, \omega,$ and ω^2, where $\omega = \frac{-1 + i\sqrt{3}}{2}$.

Hence, $x = 1, y = \omega$ and $z = \omega^2$.
Now we have

$$\text{L.H.S.} = \tan^{-1} 1 + \frac{\tan^{-1}\omega}{\omega} + \frac{\tan^{-1}\omega^2}{\omega^2}$$

$$= \left(1 - \frac{1}{3} + \frac{1}{5} - \ldots\right) + \frac{1}{\omega}\left(\omega - \frac{1}{3}\omega^2 + \frac{1}{5}\omega^5 - \ldots\right)$$

$$+ \frac{1}{\omega^2}\left(\omega^2 - \frac{1}{3}\omega^6 + \frac{1}{5}\omega^{10} - \frac{1}{7}\omega^{14} + \ldots\right)$$

$$= \left(1 - \frac{1}{3} + \frac{1}{5} - \frac{1}{7} + \ldots\right) + \frac{1}{\omega}\left(\omega - \frac{1}{3}\omega^2 + \frac{1}{5}\omega^2 - \frac{1}{7}\omega + \ldots\right)$$

$$+ \frac{1}{\omega^2}\left(\omega^2 - \frac{1}{3} + \frac{1}{5}\omega - \frac{1}{7}\omega^2 + \ldots\right), \text{ as } \omega^3 = \omega^6 = \ldots = 1$$

$$= (1 + 1 + 1) - \frac{1}{3}\left(1 + \frac{1}{\omega} + \frac{1}{\omega^2}\right) + \frac{1}{5}\left(1 + \omega + \frac{1}{\omega}\right) - \frac{1}{7}(1 + 1 + 1) + \ldots$$

$$= 3 - \frac{1}{3}\frac{1+\omega+\omega^2}{\omega^2} + \frac{1}{5}\frac{(1+\omega+\omega^2)}{\omega} - \frac{3}{7} + \ldots$$

$$= 3 - \frac{3}{7} + \frac{3}{13} - \ldots, \text{ as } 1+\omega+\omega^2 = 0$$

$$= 3\left(1 - \frac{1}{7} + \frac{1}{13} - \frac{1}{19} + \ldots\right) = \text{R.H.S.}$$

3. *Prove that*

$$\log(\tan^{-1} x) - \log x = -\frac{1}{3}x^2 + \frac{13}{90}x^4 - \ldots \infty$$

where x lies between 0 and 1.

Solution. $\log(\tan^{-1} x) - \log x = \log \dfrac{\tan^{-1} x}{x}$

$$= \log\left\{\frac{x - \dfrac{x^3}{3} + \dfrac{x^5}{5} - \ldots}{x}\right\}$$

$$= \log\left\{1 - \left(\frac{x^2}{3} - \frac{x^4}{5} - \ldots\right)\right\}$$

$$= \log(1-k), \qquad \text{where } k = \frac{x^2}{3} - \frac{x^4}{5} \ldots \text{ and is } < 1.$$

$$= -k - \frac{k^2}{2} \ldots = -\left(\frac{x^2}{3} - \frac{x^4}{5} \ldots\right) - \frac{1}{2}\left(\frac{x^2}{3} - \frac{x^4}{5} - \ldots\right)$$

$$= -\frac{1}{3}x^2 + x^4\left(\frac{1}{5} - \frac{1}{18}\right) - \ldots$$

$$= -\frac{1}{3}x^2 + \frac{13}{90}x^4 - \ldots$$

EXERCISE

1. Prove that

$$\frac{\pi}{8} = \frac{1}{1.3} + \frac{1}{5.7} + \frac{1}{9.11} + \ldots \text{ ad. inf.}$$

2. Prove that

$$\pi = 2\sqrt{3}\left(1 - \frac{1}{3^2} + \frac{1}{5.3^2} - \frac{1}{7.3^2} + \ldots \text{ ad.inf.}\right)$$

3. Prove that

$$1 - 2\left\{\frac{1}{3.5} + \frac{1}{7.9} + \frac{1}{11.13} + \ldots \text{ ad inf.}\right\} = \frac{\pi}{4}$$

Gregory's Series

4. If $x > 0$, show that

$$\tan^{-1} x = \frac{\pi}{4} + \frac{x-1}{x+1} - \frac{1}{3}\left(\frac{x-1}{x+1}\right)^3 + \frac{1}{5}\left(\frac{x-1}{x+1}\right)^5 - \ldots \text{ad.inf}.$$

5. If θ lies between 0 and $\frac{\pi}{2}$, prove that

$$\tan^{-1}\frac{(1-\cos\theta)}{(1+\cos\theta)} = \tan^2\frac{\theta}{2} - \frac{1}{3}\tan^6\frac{\theta}{2} + \frac{1}{5}\tan^{10}\frac{\theta}{2} - \ldots \text{ad.inf}.$$

6
Trigonometrical Expansions

6.1 EXPANSION OF $\log(1 - 2x\cos\theta + x^2)$ IN A SERIES OF COSINES OF MULTIPLAS OF θ

We have
$$1 - 2x\cos\theta + x^2 = 1 - x(e^{i\theta} + e^{-i\theta}) + x^2$$
$$= (1 - xe^{i\theta})(1 - xe^{-i\theta})$$

$\therefore \quad \log(1 - 2x\cos\theta + x^2) = \log(1 - xe^{i\theta}) + \log(1 - xe^{-i\theta})$

$$= -\left\{ xe^{i\theta} + \frac{1}{2}x^2 e^{2i\theta} + \frac{1}{3}x^3 e^{3i\theta} + \ldots \right\}$$

$$- \left\{ xe^{-i\theta} + \frac{1}{2}x^2 e^{-2i\theta} + \frac{1}{3}x^3 e^{-3i\theta} + \ldots \right\}$$

or $\quad \log(1 - 2x\cos\theta + x^2) = -2\left\{ x\cos\theta + \frac{1}{2}x^2\cos 2\theta + \frac{1}{3}x^3\cos 3\theta + \ldots \right.$

Obviously, the above expansion will be valid only when moduli of $xe^{i\theta}$ and $xe^{-i\theta}$ will both be less than unity. But $|xe^{i\theta}| = |xe^{-i\theta}| = x$. Henec, the above expansion will be valid when $x < 1$.

6.2 IN $\sin x = n \sin(\alpha + x)$, EXPAND x IN A SERIES OF ASCENDING POWERS OF n, WHERE n IS LESS THAN UNITY

We are given that
$$\sin x = n \sin(\alpha + x)$$

i.e., $\quad \dfrac{e^{ix} - e^{-ix}}{2i} = n \cdot \dfrac{e^{i(\alpha+x)} - e^{-i(\alpha+x)}}{2i}$

or $\quad e^{ix} - e^{-ix} = n\{e^{i(\alpha+x)} - e^{-i(\alpha+x)}\}$

or $\quad e^{2ix} - 1 = n\{e^{2ix} \cdot e^{i\alpha} - e^{-i\alpha}\}$

On multiplying both sides by e^{ix}

or $\quad e^{2ix}(1 - ne^{i\alpha}) = 1 - ne^{-i\alpha}$

or $\quad e^{2ix} = \dfrac{1 - ne^{-i\alpha}}{1 - ne^{i\alpha}}$

Taking logarithms of both sides, we get

$2ix = \log(1 - ne^{-i\alpha}) - \log(1 - ne^{i\alpha})$

$$= ne^{-i\alpha} - \frac{1}{2}n^2 e^{-2i\alpha} - \frac{1}{3}n^3 e^{-3i\alpha} - \ldots + ne^{i\alpha} + \frac{1}{2}n^2 e^{2i\alpha} + \frac{1}{3}n^2 e^{3i\alpha} + \ldots$$

Trigonometrical Expansions

$$= n(e^{i\alpha} - e^{-i\alpha}) + \frac{1}{2}n^2(e^{2i\alpha} - e^{-2i\alpha}) + \ldots \text{ad.inf.}$$

$\therefore \quad x = \dfrac{n(e^{i\alpha} - e^{-i\alpha})}{2i} + \dfrac{1}{2}n^2 \cdot \dfrac{(e^{2i\alpha} - e^{-2i\alpha})}{2i} + \ldots \text{ad.inf.}$

$\therefore \quad x = n\sin\alpha + \dfrac{1}{2}n^2 \sin 2\alpha + \dfrac{1}{3}n^3 \sin 3\alpha + \ldots \text{ad. inf.}$

6.3 IF $\tan x = n \tan y$, FIND A SERIES FOR x.

We are given that

$$\frac{e^{ix} - e^{-ix}}{e^{ix} + e^{-ix}} = n \cdot \frac{e^{iy} - e^{-iy}}{e^{iy} + e^{-iy}}$$

or
$$\frac{e^{2ix} - 1}{e^{2ix} + 1} = n \cdot \frac{(e^{2iy} - 1)}{(e^{2iy} + 1)}$$

Cross-multiplying, we get

$$e^{2ix}(e^{2iy} + 1) - e^{2iy} - 1 = e^{2ix}(ne^{2iy} - n) + ne^{2iy} - n$$

or
$$e^{2ix} = \frac{e^{2iy}(1+n) + (1-n)}{(1+n) + (1-n)e^{2iy}}$$

$$= \frac{e^{2iy}(1 + me^{-2iy})}{1 + me^{2iy}}, \qquad \text{where } m = \frac{1-n}{1+n}$$

Taking logarithms of both sides, we get

$$2ix = 2iy + \log(1 + me^{-2iy}) - \log(1 + me^{2iy})$$

$$= 2iy + me^{-2iy} - \frac{1}{2}m^2 e^{-4iy} + \frac{1}{3}m^3 e^{-6iy} - \ldots$$

$$- \left\{ me^{2iy} - \frac{1}{2}m^2 e^{4iy} + \frac{1}{3}m^3 e^{6iy} - \ldots \right\}$$

$$= 2iy - m(e^{2iy} - e^{-2iy}) + \frac{1}{2}m^2(a^{4iy} - e^{-4iy}) - \ldots$$

Hence $\quad x = y - m\sin 2y + \dfrac{1}{2}m^2 \sin 4y - \dfrac{1}{3}m^3 \sin 6y + \ldots \text{ad. inf.}$

6.4 EXPANSION OF $e^{ax} \cos bx$ IN A SERIES OF ASCENDING POWERS OF x.

We have

$$e^{ax} \cos bx = e^{ax} \cdot \frac{e^{ibx} + e^{-ibx}}{2} = \frac{1}{2}\{e^{(a+ib)x} + e^{(a-ib)x}\}$$

$$= \frac{1}{2}\left[\left\{1 + (a+ib)x + \frac{(a+ib)^2}{2!}x^2 + \ldots\right\} + \left\{1 + (a-ib)x + \frac{(a-ib)^2}{2!}x^2 + \ldots\right\} \right]$$

$$= \frac{1}{2}[2 + x\{(a+ib) + (a-ib)\} + \frac{x^2}{2!}\{(a+ib)^2 + (a-ib)^2\}$$

$$+ \ldots + \frac{x^n}{n!}\{(a+ib)^n + (a-ib)^n\} + \ldots]$$

Let $a = r \cos \alpha$ and $b = r \sin \alpha$, then the coefficient of x^n in the above expansion is

$$= \frac{1}{2} \cdot \frac{1}{n!} \{r^n (\cos \alpha + i \sin \alpha)^n + r^n (\cos \alpha - i \sin \alpha)^n\}$$

$$= \frac{r^n}{2 \cdot n!} \{(\cos n\alpha + i \sin n\alpha) + (\cos n\alpha - i \sin n\alpha)\} \quad \text{by De Moivre's thoerem}$$

$$= \frac{r^n \cos n\alpha}{n!} = \frac{(a^2 + b^2)^{n/2}}{n!} \sin\left(n \tan^{-1} \frac{b}{a}\right)$$

Hence $\quad e^{ax} \sin bx = \sum_{n=1}^{\infty} \left\{ \frac{x^n}{n!} (a^2 + b^2)^{n/2} \sin\left(n \tan^{-1} \frac{b}{a}\right) \right\}$

6.6 EXPANSION OF $\dfrac{\sin n\theta}{\sin \theta}$ IN A SERIES OF DESCENDING POWERS OF $\cos \theta$.

We have

$$\frac{x \sin \theta}{1 - 2x \cos \theta + x^2} = \frac{x (e^{i\theta} - e^{-i\theta})}{2i (1 - xe^{i\theta})(1 - xe^{-i\theta})}$$

$$= \frac{1}{2i} \left\{ \frac{1}{1 - xe^{i\theta}} - \frac{1}{1 - xe^{-i\theta}} \right\}$$

$$= \frac{1}{2i} \{(1 - xe^{i\theta})^{-1} - (1 - xe^{-i\theta})^{-1}\}$$

$$= \frac{1}{2i} [(1 + xe^{i\theta} + x^2 e^{2i\theta} + x^3 e^{3i\theta} + \ldots) - (1 + xe^{-i\theta} + x^2 e^{-2i\theta} + x^3 e^{-3i\theta} + \ldots)]$$

$$= \frac{x (e^{i\theta} - e^{-i\theta})}{2i} + \frac{x^2 (e^{2i\theta} - e^{-2i\theta})}{2i} + \ldots$$

$$= x \sin \theta + x^2 \sin 2\theta + x^3 \sin 3\theta + \ldots \text{ ad. inf.}$$

$$\therefore \quad (1 - 2x \cos \theta + x^2)^{-1} = \sum_{n=1}^{\infty} x^{n-1} \frac{\sin n\theta}{\sin \theta}$$

This shows that $\dfrac{\sin n\theta}{\sin \theta}$ is equal to the coefficient of x^{n-1} in the expansion of $(1 - 2x \cos \theta + x^2)^{-1}$.

Again, $\quad (1 - 2x \cos \theta + x^2)^{-1}$

$$= \{1 - x(2 \cos \theta - x)\}^{-1}$$

$$= 1 + x(2 \cos \theta - x) + x^2 (2 \cos \theta - x)^2 + \ldots + x^{n-1} (2 \cos \theta - x)^{n-1} + \ldots$$

Now, coefficient of x^{n-1}

in $\quad x^{n-1} (2 \cos \theta - x)^{n-1} \quad$ is $\quad (2 \cos \theta)^{n-1}$

in $\quad x^{n-2} (2 \cos \theta - x)^{n-2} \quad$ is $\quad -(n-2)(2 \cos \theta)^{n-3}$

in $\quad x^{n-3} (2 \cos \theta - x)^{n-3} \quad$ is $\quad \dfrac{(n-3)(n-4)}{2!} (2 \cos \theta)^{n-5}$

Trigonometrical Expansions

and so on. Hence,
$$\frac{\sin n\theta}{\sin \theta} = (2\cos\theta)^{n-1} - (n-2)(2\cos\theta)^{n-3} + \frac{(n-3)(n-4)}{2!}(2\cos\theta)^{n-5} - \ldots$$

The general term
$$= (-1)^r \frac{n(n-r-1)(n-r-2)\ldots(n-2r)}{r!}(2\cos\theta)^{n-2r-1}$$

and the last term will be
$$(-1)^{(n/2)-1}(2\cos\theta) \quad \text{or} \quad (-1)^{(n-1)/2}$$
according as n is even or odd.

6.7 EXPANSION OF $\dfrac{1+x\cos\theta}{1+2x\cos\theta+x^2}$ IN A SERIES OF COSINES OF MULTIPLES OF θ, WHERE $x<1$.

We have
$$\frac{1+x\cos\theta}{1+2x\cos\theta+x^2} = \frac{1+\frac{1}{2}x(e^{i\theta}+e^{-i\theta})}{1+x(e^{i\theta}+e^{-i\theta})+x^2}$$

$$= \frac{1}{2}\frac{(1+xe^{i\theta})+(1+xe^{-i\theta})}{(1+xe^{i\theta})(1+xe^{-i\theta})}$$

$$= \frac{1}{2}\left\{\frac{1}{(1+xe^{-i\theta})}+\frac{1}{(1+xe^{i\theta})}\right\}$$

$$= \frac{1}{2}[(1+xe^{-i\theta})^{-1}+(1+xe^{i\theta})^{-1}]$$

$$= \frac{1}{2}\{(1-xe^{-i\theta}+x^2e^{-2i\theta}-x^3e^{-3i\theta}+\ldots)(1-xe^{i\theta}+x^2e^{2i\theta}-x^3e^{3i\theta}+\ldots)\}$$

$$= 1 - x\frac{e^{i\theta}+e^{-i\theta}}{2} + x^2\frac{e^{2i\theta}+e^{-2i\theta}}{2} - \ldots$$

$$= 1 - x\cos\theta + x^2\cos 2\theta - x^3\cos 3\theta + \ldots$$

The condition of validity is $x<1$.

6.8 EXPANSION OF $\dfrac{\cos\theta - x\cos(\theta-\phi)}{1-2x\cos\phi+x^2}$, WHERE $x<1$.

We have
$$\frac{\cos\theta - x\cos(\theta-\phi)}{1-2x\cos\phi+x^2} = \frac{\frac{1}{2}(e^{i\theta}+e^{-i\theta}) - \frac{1}{2}x\{e^{i(\theta-\phi)}+e^{-i(\theta-\phi)}\}}{1-x(e^{i\phi}+e^{-i\phi})+x^2}$$

$$= \frac{1}{2}\frac{e^{i\theta}(1-xe^{-i\phi})+e^{-i\theta}(1-xe^{i\phi})}{(1-xe^{-i\phi})(1-xe^{i\phi})}$$

$$= \frac{1}{2}\left[\frac{e^{i\theta}}{(1-xe^{i\phi})}+\frac{e^{-i\theta}}{(1-xe^{-i\phi})}\right]$$

$$= \frac{1}{2}[e^{i\theta}(1-xe^{i\phi})^{-1} + e^{-i\theta}(1-xe^{-i\phi})^{-1}]$$

$$= \frac{1}{2}[e^{i\theta}(1+xe^{i\phi}+x^2e^{2i\phi}+x^3e^{3i\phi}+...) + e^{-i\theta}(1+xe^{-i\phi}+x^2e^{-2i\phi}+x^3e^{-3i\phi}+...)]$$

$$= \frac{1}{2}(e^{i\theta}+e^{-i\theta}) + \frac{1}{2}x\{e^{i(\theta+\phi)}+e^{-i(\theta+\phi)}\} + \frac{1}{2}x^2\{e^{i(\theta+2\phi)}+e^{-i(\theta+2\phi)}\} + ... \text{ad. inf.}$$

$$= \cos\theta + x\cos(\theta+\phi) + x^2\cos(\theta+2\phi) + ... \text{ad. inf.}$$

EXAMPLES

1. *If $x < 1$, then prove that*

$$\frac{1-x^2}{1-2x\cos\theta+x^2} = 1 + 2x\cos\theta + 2x^2\cos 2\theta + ... + 2x^n\cos n\theta + ... \text{ad. inf.}$$

Solution. We have

$$\frac{1-x^2}{1-2x\cos\theta+x^2} = -1 + \frac{2-2x\cos\theta}{(1-2x\cos\theta+x^2)}$$

$$= -1 + \frac{2-x(e^{i\theta}+e^{-i\theta})}{1-x(e^{i\theta}+e^{-i\theta})+x^2}$$

$$= -1 + \frac{(1-xe^{-i\theta})+(1-xe^{i\theta})}{(1-xe^{i\theta})(1-xe^{-i\theta})}$$

$$= -1 + (1-xe^{i\theta})^{-1} + (1-xe^{i\theta})^{-1}$$

$$= -1 + \{1 + xe^{-i\theta} + x^2e^{-2i\theta} + x^3e^{-3i\theta} + ...\}$$

$$+ \{1 + xe^{i\theta} + x^2e^{2i\theta} + x^3e^{3i\theta} + ...\}$$

$$= 1 + x(e^{i\theta}+e^{-i\theta}) + x^2(e^{2i\theta}+e^{-2i\theta}) + x^3(e^{3i\theta}+e^{-3i\theta}) + ...$$

$$= 1 + 2x\cos\theta + 2x^2\cos 2\theta + 2x^3\cos 3\theta + ... \text{ad. inf.}$$

2. *In any triangle where $a < b$ prove that*

$$\log c = \log a - \frac{b}{a}\cos C - \frac{1}{2}\frac{b^2}{a^2}\cos 2C - \frac{1}{3}\frac{b^3}{a^3}\cos 3C - ...$$

Solution. We know that in any triangle

$$c^2 = a^2 + b^2 - 2ab\cos C$$

$$= a^2 + b^2 - 2ab \cdot \frac{1}{2}(e^{iC}+e^{-iC})$$

$$= a^2\left[1 - \frac{b}{a}(e^{iC}+e^{-iC}) + \frac{b^2}{a^2}\right]$$

or

$$c^2 = a^2\left(1 - \frac{b}{a}e^{iC}\right)\left(1 - \frac{b}{a}e^{-iC}\right)$$

Taking log of both sides, we get

$$2\log c = 2\log a + \log\left(1 - \frac{b}{a}e^{iC}\right) + \log\left(1 - \frac{b}{a}e^{-iC}\right)$$

Trigonometrical Expansions

$$= 2\log a + \left[-\frac{b}{a}e^{iC} - \frac{b^2}{2a^2}e^{2iC} - \frac{b^3}{3a^3}e^{3iC} - \ldots\right]$$

$$+ \left[-\frac{b}{a}e^{-iC} - \frac{b^2}{2a^2}e^{-2iC} - \frac{b^3}{3a^3}e^{-3iC} - \ldots\right]$$

$$= 2\log a - \frac{b}{a}(e^{iC} + e^{-iC}) - \frac{b^2}{2a^2}(e^{2iC} + e^{-2iC}) - \frac{b^3}{3a^2}(e^{3iC} + e^{-3iC}) - \ldots$$

or $\quad 2\log c = 2\log a - \dfrac{2b}{a}\cos C - \dfrac{b^2}{2a^2}\cdot 2\cos 2C - \dfrac{b^3}{3a^3}\cdot 2\cos 3C - \ldots$

or $\quad \log c = \log a - \dfrac{b}{a}\cos C - \dfrac{b^2}{2a^2}\cos 2C - \dfrac{b^2}{3a^3}\cos 3C - \ldots$

3. *If* $\tan(\theta + \phi)\cos 2\alpha = \tan\phi$, *prove that*

$$\theta = \tan^2\alpha \sin 2\phi + \frac{1}{2}\tan^4\alpha \sin 4\phi + \frac{1}{3}\tan^6\sin 6\phi + \ldots ad.\,inf.$$

Solution. Given that
$$\tan(\theta + \phi)\cos 2\alpha = \tan\phi$$

or $\quad \cos 2\alpha = \dfrac{\tan\phi}{\tan(\theta + \phi)} = \dfrac{\sin\phi\cos(\theta + \phi)}{\cos\phi\sin(\theta + \phi)}$

or $\quad \dfrac{\cos 2\alpha}{1} = \dfrac{\sin\phi\cos(\theta + \phi)}{\cos\phi\sin(\theta + \phi)}$

Applying componendo and dividendo, we get

$$\frac{1 + \cos 2\alpha}{1 - \cos 2\alpha} = \frac{\cos\phi\sin(\theta + \phi) + \sin\phi\cos(\theta + \phi)}{\cos\phi\sin(\theta + \phi) - \sin\phi\cos(\theta + \phi)}$$

or $\quad \dfrac{2\cos^2\alpha}{2\sin^2\alpha} = \dfrac{\sin(\theta + 2\phi)}{\sin\theta}$

or $\quad \sin\theta = \tan^2\alpha \sin(\theta + 2\phi)$,

which is of the form $\sin x = n\sin(x + \alpha)$,

where $x = \theta$, $n = \tan^2\alpha$ and $\alpha = 2\phi$,

then $\quad x = n\sin\alpha + \dfrac{n^2}{2}\sin 2\alpha + \dfrac{n^3}{3}\sin 3\alpha + \ldots ad.\,inf.$

Putting $x = \theta$, $n = \tan^2\alpha$ and $\alpha = 2\phi$, then

$$\theta = \tan^2\alpha \sin 2\phi + \frac{1}{2}\tan^4\alpha \sin 4\phi + \frac{1}{3}\tan^6\alpha \sin 6\phi + \ldots ad.\,inf.$$

EXERCISE

1. Prove that
$$\frac{2x \cos \theta}{1 - 2x \sin \theta + x^2} = 2x \cos \theta + 2x^2 \sin 2\theta - 2x^3 \cos 3\theta - 2x^4 \sin 4\theta + \ldots \text{ad. inf.}$$

2. Prove that
$$\frac{\sin \theta}{1 - \sin \alpha \cos \theta} = 2 \operatorname{cosec} \left[\tan \frac{\alpha}{2} \sin \theta + \tan^2 \frac{\alpha}{2} \sin 2\theta + \ldots \right]$$

3. Prove that
$$\log \frac{a^2}{a^2 \cos^2 \theta + b^2 \sin^2 \theta} = 4 \left[c \sin \theta - \frac{1}{2} c^2 \sin^2 2\theta + \frac{1}{3} c^3 \sin^3 3\theta - \ldots \right]$$
where $c = \frac{(a-b)}{(a+b)}$.

4. Prove that, if θ be an angle whose cosine is positive
$$\log \cos \theta = -\log 2 + \cos 2\theta - \frac{1}{2} \cos 4\theta + \frac{1}{3} \cos 6\theta - \ldots \text{ ad. inf.}$$

5. In any triangle when $a < c$, show that
$$\frac{\cos nA}{b^n} = \frac{1}{c^n} \left\{ 1 + \frac{na}{c} \cos B + \frac{n(n+1)}{1.2} \frac{a^2}{c^2} \cos 2B + \ldots \right\}$$
and
$$\frac{\sin nA}{b^n} = \frac{n}{c^n} \left\{ \frac{a}{c} \sin B + \frac{(n+1) a^2}{1.2 c^2} \sin 2B + \ldots \right\}$$

6. If $\cot \phi = \frac{1}{x} + \cot \theta$, prove that
$$\phi = \frac{x}{\sin \theta} \cdot \sin \theta - \frac{1}{2} \cdot \frac{x^2}{\sin^2 \theta} \cdot \sin 2\theta + \frac{1}{3} \cdot \frac{x^3}{\sin^3 \theta} \cdot \sin 3\theta - \ldots$$

7. If $\tan 2\theta = \sin \alpha \tan 2\beta$, show that
$$\theta = \sin \alpha \tan \beta + \frac{1}{3} \sin 3\alpha \tan^3 \beta + \frac{1}{5} \sin 5\alpha \tan^5 \beta + \ldots$$

7
Summation of Series

7.1 SUMMATION OF SERIES

Here we shall give some important methods of summing up trigonometric series. For this we shall utilize the results of the previous chapters. The series, may be finite or infinite.

7.2 C+ IS METHOD

This is the most effective method of summing a trigonometric series. In this method the given series is connected with some standard series. This is effected with the help of complex quantities. These standard series may be of the following forms :
  (i) Series in geometric progression.
  (ii) Binomial series or which can be reduced to it.
  (iii) Exponential series or the allied series.
  (iv) Logarithmic series or the Gregory's series.

Suppose we are to find out the sum of a series in terms of cosines of multiple angles. Then we shall denote this series containing cosines of multiple angles by C and write down another corresponding series in terms of sines of multiple angles by replacing cosine by sine every where and denoting this new series by S. (In case this series contains sines of multiple angles, then we shall denote this series by S and will write down another corresponding series in terms of cosines of multiple angles by replacing sine by cosine everywhere). When C and S are so chosen, we shall multiply each term of the series S by i and add to the corresponding term of C series and thus we shall have a new series C + iS. In this series we shall replace each term of the form $(\cos \theta + i \sin \theta)$ by $e^{i\theta}$ and we shall have a series in any one of the above four forms, whose sum can easily be calculated. After obtaining the sum of C + iS series and by equating real and imaginary parts on both sides we can find out the values of C and S which give the sum of C series and S series respectively.

7.3 SUMMATION DEPENDING UPON GEOMETRIC PROGRESSION (ANGLES IN ARITHMETICAL PROGRESSION)

To find the sum of a series of sines or cosines of angles in arithmetical progression.
Let n angles in A.P. be

$$\alpha, \alpha + \beta, \alpha + 2\beta, ..., \{\alpha + (n-1)\beta\}$$

Then the sine series will be

$$S = \sin \alpha + \sin (\alpha + \beta) + \sin (\alpha + 2\beta) + ... + \sin \{\alpha + (n-1)\beta\}$$

Then the cosine series will be denoted as

$$C = \cos \alpha + \cos (\alpha + \beta) + \cos (\alpha + 2\beta) + ... + \cos \{\alpha + (n-1)\beta\}$$

Multiplying the sine series by i, $(i^2 = -1)$ and adding it to cosine series, we have

$$C + iS = (\cos \alpha + i \sin \alpha) + \{\cos (\alpha + i\beta) + i \sin (\alpha + \beta)\}$$
$$+ \{\cos (\alpha + 2\beta) + i \sin (\alpha + \beta) + ... \text{ to } n \text{ terms.}\}$$
$$= e^{i\alpha} + e^{i(\alpha + \beta)} + e^{i(\alpha + 2\beta)} + ... \text{ to } n \text{ terms.}$$
$$= e^{i\alpha} (1 + e^{i\beta} + e^{2i\beta} + ... \text{ to } n \text{ terms})$$

Which is a geometric series whose common ratio is $e^{i\beta}$.

$$\therefore \quad C + iS = \frac{e^{i\alpha}(1 - e^{ni\beta})}{(1 - e^{i\beta})}$$

$$= \frac{e^{i\alpha}(1 - e^{ni\beta})}{e^{i\beta/2}(e^{-i\beta/2} - e^{i\beta/2})} = -\frac{e^{i(\alpha - \beta/2)}(1 - e^{ni\beta})}{2i \sin \beta/2}$$

$$= \frac{i[e^{i(\alpha - \beta/2)} - e^{-i(\alpha - \beta/2 + n\beta)}]}{2 \sin \beta/2}$$

$$= \frac{i\left[\cos\left(\alpha - \frac{\beta}{2}\right) + i \sin\left(\alpha - \frac{\beta}{2}\right) - \cos\left(\alpha - \frac{1}{2}\beta + n\beta\right) - i \sin\left(\alpha - \frac{1}{2}\beta + n\beta\right)\right]}{2 \sin \frac{\beta}{2}}$$

$$= \frac{i \cos\left(\alpha - \frac{\beta}{2}\right) - \sin\left(\alpha - \frac{\beta}{2}\right) - i \cos\left(\alpha - \frac{\beta}{2} + n\beta\right) + \sin\left(\alpha - \frac{\beta}{2} + n\beta\right)}{2 \sin \frac{\beta}{2}}$$

Equating real and imaginary parts, we get

$$C = \frac{\sin\left(\alpha - \frac{\beta}{2} + n\beta\right) - \sin\left(\alpha - \frac{\beta}{2}\right)}{2 \sin \frac{\beta}{2}}$$

$$= \frac{2 \cos\left(\alpha - \frac{\beta}{2} + \frac{n\beta}{2}\right) \sin \frac{n\beta}{2}}{2 \sin \frac{\beta}{2}}$$

$$= \frac{\cos\left\{\alpha + (n-1)\frac{\beta}{2}\right\} \sin n \frac{\beta}{2}}{\sin \frac{\beta}{2}}$$

and

$$S = \frac{\cos\left(\alpha - \frac{\beta}{2}\right) - \cos\left(\alpha - \frac{\beta}{2} + n\beta\right)}{2 \sin \frac{\beta}{2}}$$

$$= \frac{2 \sin\left(\alpha - \frac{\beta}{2} + \frac{n\beta}{2}\right) \sin \frac{n\beta}{2}}{2 \sin \frac{\beta}{2}}$$

Thus, we have

$$\sin \alpha + \sin(\alpha + \beta) + \ldots + \sin\{\alpha + (n-1)\beta\}$$

$$= \sin\left\{\frac{\text{first angle} + \text{last angle}}{2}\right\} \frac{\sin\left(\frac{n \times \text{diff.}}{2}\right)}{\sin\left(\frac{n \times \text{diff.}}{2}\right)}$$

Summation of Series

and
$$\cos\alpha + \cos(\alpha + \beta) + \cos(\alpha + 2\beta) + \ldots + \cos\{\alpha + (n-1)\beta\}$$

$$= \cos\left\{\frac{\text{first angle} + \text{last angle}}{2}\right\} \frac{\sin\left(\frac{n \times \text{diff.}}{2}\right)}{\sin\left(\frac{\text{diff.}}{2}\right)}$$

$$= \frac{\sin\left\{\alpha + (n-1)\frac{\beta}{2}\right\} \sin n\frac{\beta}{2}}{\sin\frac{\beta}{2}}$$

EXAMPLES

1. *Sum the following series to n terms and to infinity*

$$1 + a\cos\theta + a^2\cos 2\theta + a^3\cos 3\theta + \ldots$$

where a is less than unity.

Solution. Let

$$C = 1 + a\cos\theta + a^2\cos 2\theta + \ldots + a^{n-1}\cos(n-1)\theta$$

and
$$S = a\sin\theta + a^2\sin 2\theta + \ldots + a^{n-1}\sin(n-1)\theta$$

$\therefore$
$$C + iS = 1 + a(\cos\theta + i\sin\theta) + a^2(\cos 2\theta + i\sin 2\theta)$$
$$+ \ldots + a^{n-1}\{\cos(n-1)\theta + i\sin(n-1)\theta\}$$

$$= 1 + ae^{i\theta} + a^2 e^{2i\theta} + \ldots + a^{n-1}e^{i(n-1)\theta}$$

This is a G.P. whose common ratio is $ae^{i\theta}$. Hence, by summing the G.P., we obtain

$$C + iS = \frac{1 \cdot (1 - a^n e^{in\theta})}{1 - ae^{i\theta}}, \qquad \text{since } a < 1$$

$$= \frac{(1 - a^n e^{in\theta})(1 - ae^{-i\theta})}{(1 - ae^{i\theta})(1 - ae^{-i\theta})}$$

$$= \frac{1 - ae^{-i\theta} - a^n e^{in\theta} + a^{n+1} e^{i(n-1)\theta}}{1 - a(e^{i\theta} + e^{-i\theta}) + a^2 e^{i\theta} \cdot e^{-i\theta}}$$

$$= \frac{1 - ae^{-i\theta} - a^n e^{in\theta} + a^{n+1} e^{i(n-1)\theta}}{1 - 2a\cos\theta + a^2}$$

$$= \frac{1 - a(\cos\theta - i\sin\theta) - a^n(\cos n\theta + i\sin\theta)}{1 - 2a\cos\theta + a^2}$$
$$\phantom{=\frac{1}{1}} \frac{+ a^{n+1}\{\cos(n-1)\theta + i\sin(n-1)\theta\}}{1 - 2a\cos\theta + a^2}$$

Equating the real parts, we obtain

$$C = \frac{1 - a\cos\theta - a^n\cos n\theta + a^{n+1}\cos(n-1)\theta}{1 - 2a\cos\theta + a^2}$$

When n tends to infinity, a^n and a^{n+1} both tend to zero, as $a < 1$. Hence, sum of the required series upto infinity is

$$C_\infty = \frac{1 - a\cos\theta}{1 - 2a\cos\theta + a^2}$$

Again equating the imaginary parts from the two sides, of (i) we obtain

$$S = \frac{a\sin\theta - a^n \sin n\theta + a^{n+1} \sin(n-1)\theta}{1 - 2a\cos\theta + a^2}$$

and $$S_\infty = \frac{a\sin\theta}{1 - 2a\cos\theta + a^2}.$$

2. *If S_n be the sum of n terms of the series*

$$\sin x + \sin 2x + \sin 3x + \dots$$

prove that $\lim_{n\to\infty} \dfrac{S_1 + S_2 + S_3 + \dots + S_n}{n} = \dfrac{1}{2}\cot\dfrac{x}{2}.$

Solution. From the above method, we have

$$S_n = \sin\frac{(n+1)x}{2} \cdot \frac{\sin\frac{n}{2}x}{\sin\frac{x}{2}} \qquad \text{(by 7.3)}$$

$$= \operatorname{cosec}\frac{x}{2} \sin\frac{n+1}{2}x \sin\frac{nx}{2}$$

$$= \frac{1}{2}\operatorname{cosec}\frac{x}{2}\left\{\cos\frac{x}{2} - \cos\frac{2n+1}{2}x\right\}$$

Putting $n = 1, 2, 3, \dots n$, this gives

$$S_1 + S_2 + S_3 + \dots + S_n$$

$$= \frac{1}{2}\operatorname{cosec}\frac{x}{2}\left[\left(\cos\frac{x}{2} - \cos\frac{3}{2}x\right) + \left(\cos\frac{x}{2} - \cos\frac{5x}{2}\right) + \dots + \left(\cos\frac{x}{2} - \cos\frac{(2n+1)x}{2}\right)\right]$$

$$= \frac{1}{2}\operatorname{cosec}\frac{x}{2}\left[n\cos\frac{1}{2}x - \left(\cos\frac{3}{2}x + \cos\frac{5}{2}x + \dots + \cos\frac{1}{2}(2n+1)x\right)\right]$$

$$= \frac{1}{2\sin\frac{x}{2}}\left\{n\cos\frac{x}{2} - \cos\frac{\left(\frac{3}{2}x + \frac{1}{2}(2n+1)x\right)}{2}\cdot\frac{\sin\frac{n}{2}x}{\sin\frac{x}{2}}\right\}$$

$$= \frac{1}{2\sin\frac{x}{2}}\left\{n\cos\frac{x}{2} - \frac{\cos(n+2)x}{2}\cdot\frac{\sin\frac{n}{2}x}{\sin\frac{x}{2}}\right\}$$

$$\therefore \quad \lim_{n\to\infty} \frac{S_1 + S_2 + S_2 + \dots + S_n}{n}$$

$$= \lim_{n\to\infty} \frac{1}{2\sin\frac{x}{2}}\left\{\cos\frac{x}{2} - \frac{1}{n}\cos\frac{(n+2)x}{2}\cdot\frac{\sin\frac{n}{2}x}{\sin\frac{x}{2}}\right\}$$

Summation of Series

$$= \frac{\cos\frac{x}{2}}{2\sin\frac{x}{2}} = \frac{1}{2}\cos\frac{x}{2}.$$

because as $n \to \infty$, $\frac{1}{n} \to 0$ and $\sin\frac{1}{2}nx$, $\cos\frac{1}{2}(n+2)x$ remain finite.

EXERCISE

1. Find the sum to n terms of the series
   (a) $\sin\alpha - \sin(\alpha+\beta) + \sin(\alpha+2\beta) - ...$
   (b) $\cos\alpha - \cos(\alpha+\beta) + \cos(\alpha+2\beta) - ...$

   [Ans. (a) $\dfrac{\sin\left\{\alpha + \frac{1}{2}(n-1)(\pi+\beta)\right\}\sin\frac{1}{2}n(\pi+\beta)}{\cos\frac{\beta}{2}}$

   (b) $\cos\left\{\alpha + \frac{1}{2}(n-1)(\pi+\beta)\right\} \cdot \dfrac{\sin\frac{1}{2}n(\pi+\beta)}{\cos\frac{\beta}{2}}$]

2. Sum the series $\cos\alpha \cdot \cos\alpha + \cos^2\alpha\cos 2\alpha + \cos^3\alpha\cos 3\alpha + ...$ ad. inf.

   [**Ans.** 0, provided α does not equal a multiple of π.]

3. Sum the series $\sin\alpha + c\sin(\alpha+\beta) + c^2\sin(\alpha+2\beta) + ...$ to n terms and to infinity where c is less tha unity.

   [**Ans.** $\dfrac{\sin\alpha - c\sin(\alpha-\beta) - c^n\sin(\alpha+n\beta) + c^{n+1}\sin\{\alpha+(n-1)\beta\}}{1 - 2c\cos\beta + c^2}$]

4. Sum the series $\sin^2\alpha + \sin^2(\alpha+\beta) + \sin^2(\alpha+2\beta) + ...$ to n terms.

   [**Ans.** $\frac{1}{2}n - \frac{1}{2}\cos\{2\alpha + (n-1)\beta\}\sin n\beta \operatorname{cosec}\beta$]

5. Sum the following series

   $\cos^2\alpha + \cos^2\left(\alpha + \frac{\pi}{2}\right) + \cos^2\left(\alpha + \frac{2\pi}{2}\right) + ...$ to n terms.

   [**Ans.** $\frac{1}{2}n + \frac{1}{2}\cos\left\{2\alpha + \frac{1}{n}(n-1)\pi\right\}\sin\frac{1}{2}n\pi$]

7.3.1 Summation depending on Arithmetico-Geometric Series

EXAMPLE

1. *Find the sum of the series*

   $$3\sin\alpha + 5\sin 2\alpha + 7\sin 3\alpha + ... \text{ to } n \text{ terms}$$

Solution. Let

$$S = 3\sin\alpha + 5\sin 2\alpha + 7\sin 3\alpha + ... \text{ to } n \text{ terms}$$

and

$$C = 3\cos\alpha + 5\cos 2\alpha + 7\cos 3\alpha + ... \text{ to } n \text{ terms}$$

$\therefore$

$$C + iS = 3e^{i\alpha} + 5e^{2i\alpha} + 7e^{3i\alpha} + ... + (2n+1)e^{ni\alpha}$$

This is an arithmetico-geometric series whose common ratio is $e^{i\alpha}$. Multiplying both sides by $e^{i\alpha}$, we get

$$(C + iS)\, e^{i\alpha} = 3e^{2i\alpha} + 5e^{3i\alpha} + \ldots + (2n - 1)\, e^{ni\alpha} + (2n + 1)\, e^{i(n+1)\alpha}$$

Subtracting, we get

$$(C + iS)(1 - e^{i\alpha}) = 3e^{i\alpha} + 2e^{2i\alpha} + 2e^{3i\alpha} + \ldots + 2e^{ni\alpha} - (2n + 1)\, e^{i(n+1)\alpha}$$

$$= e^{i\alpha} + 2\{e^{i\alpha} + e^{2i\alpha} + \ldots \text{to } n \text{ terms}\} - (2n + 1)\, e^{i(n+1)\alpha}$$

$$= e^{i\alpha} + \frac{2e^{i\alpha}(1 - e^{ni\alpha})}{1 - e^{i\alpha}} - (2n + 1)\, e^{i(n+1)\alpha}$$

$$= e^{i\alpha} - (2n + 1)\, e^{i(n+1)\alpha} - \frac{2(1 - e^{in\alpha})}{1 - e^{-i\alpha}}$$

$$\therefore \quad C + iS = \frac{e^{i\alpha} - (2n + 1)\, e^{i(n+1)\alpha}}{1 - e^{i\alpha}} - \frac{2(1 - e^{in\alpha})}{(1 - e^{i\alpha})(1 - e^{-i\alpha})}$$

$$= \frac{\{e^{i\alpha} - (2n + 1)\, e^{i(n+1)\alpha}\}(1 - e^{-i\alpha})}{(1 + e^{i\alpha})(1 - e^{-i\alpha})} - \frac{2(1 - e^{in\alpha})}{(1 - e^{i\alpha})(1 - e^{-i\alpha})}$$

$$= \frac{e^{i\alpha} - 1 - (2n + 1)\, e^{i(n+1)\alpha} + (2n + 1)\, e^{in\alpha} - 2(1 - e^{in\alpha})}{1 - (e^{i\alpha} + e^{-i\alpha}) + 1}$$

$$\Rightarrow \quad C + iS = \frac{e^{i\alpha} - (2n + 1)\, e^{i(n+1)\alpha} + (2n + 3)\, e^{in\alpha} - 3}{2(1 - \cos \alpha)}$$

Equating the imaginary parts, we get

$$S = \frac{\sin \alpha - (2n + 1) \sin (n + 1)\alpha + (2n + 3) \sin n\alpha}{2(1 - \cos \alpha)}$$

EXERCISE

Find the sum of the following series :

1. $\cos \alpha + 2 \cos 2\alpha + 3 \cos 3\alpha + \ldots + n \cos n\alpha$ $\quad\left[\text{Ans. } \dfrac{(n+1) \cos n\alpha - n \cos (n+1)\alpha - 1}{2(1 - \cos \alpha)}\right]$

2. $1 + \dfrac{2}{2} \cos \theta + \dfrac{3}{2^2} \cos 2\theta + \dfrac{4}{2^3} \cos 3\theta + \ldots \text{ad. inf.}$ $\quad\left[\text{Ans. } \dfrac{4(4 - 4 \cos \theta + \cos 2\theta)}{(5 - 4 \cos \theta)^2}\right]$

7.4 SUMMATION DEPENDING UPON BINOMIAL SERIES

To sum the series depending on binomial series, following useful expansions are worth remembering,

$$(1 + x)^n = 1 + nx + \frac{n(n-1)}{2!} x^2 + \frac{n(n-1)(n-2)}{3!} x^3 + \ldots \text{ to } (n+1) \text{ terms}$$

$$(1 - x)^{-n} = 1 + nx + \frac{n(n+1)}{2!} x^2 + \frac{n(n+1)(n+2)}{3!} x^3 + \ldots \text{ad. inf.}$$

$$(1 + x)^{1/2} = 1 + \frac{1}{2} x - \frac{1}{2.4} x^2 + \frac{1.3}{2.4.6} x^3 - \ldots$$

Summation of Series

$$(1-x)^{-1/2} = 1 + \frac{1}{2}x + \frac{1.3}{2.4}x^2 + \frac{1.3.5}{2.4.6}x^3 + \ldots$$

$$(1+x)^{1/3} = 1 + \frac{1}{3}x - \frac{1.2}{3.6}x^2 + \frac{1.2.5}{3.6.9}x^3 - \ldots$$

$$(1-x)^{-1/3} = 1 + \frac{1}{3}x + \frac{1.4}{3.6}x^2 + \frac{1.4.7}{3.6.9}x^3 + \ldots$$

Examples

1. *Show the series*

$$\sin\alpha + n\sin(\alpha+\beta) + \frac{n(n-1)}{2!}\sin(\alpha+2\beta) + \ldots \text{ to } (n+1) \text{ terms}$$

Solution. Let $S = \sin\alpha + n\sin(\alpha+\beta) + \frac{n(n-1)}{2!}\sin(\alpha+2\beta) + \ldots$ to $(n+1)$ terms

and $\quad C = \cos\alpha + n\cos(\alpha+\beta) + \frac{n(n-1)}{2!}\cos(\alpha+2\beta) + \ldots$ to $(n+1)$ terms

Then $\quad C + iS = (\cos\alpha + i\sin\alpha) + n\{\cos(\alpha+\beta) + i\sin(\alpha+\beta)\}$

$$+ \frac{n(n-1)}{2!}\{\cos(\alpha+2\beta) + i\sin(\alpha+2\beta) + \ldots \text{ to } (n+1) \text{ terms}\}$$

$$= e^{i\alpha} + ne^{i(\alpha+\beta)} + \frac{n(n-1)}{2!}e^{i(\alpha+2\beta)} + \ldots \text{ to } (n+1) \text{ terms}$$

$$= e^{i\alpha}\left[1 + e^{i\beta} + \frac{n(n-1)}{2!}e^{2i\beta} + \ldots \text{ to } (n+1) \text{ terms}\right]$$

$$= e^{i\alpha}(1 + e^{i\beta})^n$$

$$= (\cos\alpha + i\sin\alpha)[1 + \cos\beta + i\sin\beta]^n$$

$$= (\cos\alpha + i\sin\alpha)\left[2\cos^2\frac{\beta}{2} + 2i\sin\frac{\beta}{2}\cos\frac{\beta}{2}\right]^n$$

$$= (\cos\alpha + i\sin\alpha) \cdot 2^n \cos^n\frac{\beta}{2}\left(\cos\frac{\beta}{2} + i\sin\frac{\beta}{2}\right)^n$$

$$= 2^n \cos^n\frac{\beta}{2}\left[\cos\left(\alpha + \frac{n\beta}{2}\right) + i\sin\left(\alpha + \frac{n\beta}{2}\right)\right]$$

Equating imaginary parts on both sides, we get

$$S = 2^n \cos^n\frac{\beta}{2}\sin\left(\alpha + \frac{n\beta}{2}\right)$$

2. *Sum the series*

$$1 + \frac{1}{2}\cos\alpha + \frac{1.3}{2.4}\cos 2\alpha + \frac{1.3.5}{2.4.6}\cos 3\alpha + \ldots ad.\inf.$$

Solution. Let $C = 1 + \frac{1}{2}\cos\alpha + \frac{1.3}{2.4}\cos 2\alpha + \frac{1.3.5}{2.4.6}\cos 3\alpha + \ldots$

and $\quad S = \frac{1}{2}\sin\alpha + \frac{1.3}{2.4}\sin 2\alpha + \frac{1.3.5}{2.4.6}\sin 3\alpha + \ldots$

$$\therefore\quad C + iS = \frac{1}{2}(\cos\alpha + i\sin\alpha) + \frac{1.3}{2.4}(\cos 2\alpha + i\sin 2\alpha) + \frac{1.3.5}{2.4.6}(\cos 3\alpha + i\sin 3\alpha) + \ldots$$

$$= 1 + \frac{1}{2}e^{i\alpha} + \frac{1.3}{2.4}e^{2i\alpha} + \frac{1.3.5}{2.4.6}e^{3i\alpha} + \ldots$$

$$= (1 - e^{i\alpha})^{-1/2} = (1 - \cos\alpha - i\sin\alpha)^{-1/2}$$

$$= \left[2\sin^2\frac{\alpha}{2} - i \cdot 2\sin\frac{\alpha}{2}\cos\frac{\alpha}{2}\right]^{-1/2}$$

$$= \left(2\sin\frac{\alpha}{2}\right)^{-1/2}\left(\sin\frac{\alpha}{2} - i\cos\frac{\alpha}{2}\right)^{-1/2}$$

$$= \left(2\sin\frac{\alpha}{2}\right)^{-1/2}\left[\cos\left(\frac{\pi}{2} - \frac{\alpha}{2}\right) - i\sin\left(\frac{\pi}{2} - \frac{\alpha}{2}\right)\right]^{-1/2}$$

$$= \left(2\sin\frac{\alpha}{2}\right)^{-1/2}\left[\cos\left(\frac{\pi}{4} - \frac{\alpha}{4}\right) - i\sin\left(\frac{\pi}{4} - \frac{\alpha}{4}\right)\right]$$

Equating real and imaginary parts on both sides, we have

$$C = \left(2\sin\frac{\alpha}{2}\right)^{-1/2}\cos\left(\frac{\pi}{4} - \frac{\alpha}{4}\right)$$

EXERCISE

Sum the following series :

1. $n\sin\alpha + \dfrac{n(n-1)}{1.2}\sin 2\alpha + \dfrac{n(n-1)(n-2)}{1.2.3}\sin 3\alpha + \ldots$ to n terms

$$\left[\textbf{Ans. } \left(2\cos\frac{\alpha}{2}\right)^n \sin\frac{1}{2}n\alpha\right]$$

2. $\sin\alpha + \dfrac{1}{2}\sin 3\alpha + \dfrac{1.3}{2.4}\sin 5\alpha + \ldots \text{ad. inf.}$

$$\left[\textbf{Ans. } (2\sin\alpha)^{-1/2}\sin\left(\frac{\pi}{4} + \frac{\alpha}{2}\right), \text{ except when } \alpha = n\pi\right]$$

3. $1 - \dfrac{1}{2}\cos\alpha + \dfrac{1.3}{2.4}\cos 2\alpha - \dfrac{1.3.5}{2.4.6}\cos 3\alpha + \ldots \text{ad. inf.}$ $\quad\left[\textbf{Ans. } \left(2\cos\dfrac{\alpha}{2}\right)^{-1/2}\cos\dfrac{\alpha}{4}\right]$

4. $n\sin\alpha + \dfrac{n(n+1)}{1.2}\sin 2\alpha + \dfrac{n(n+1)(n+2)}{1.2.3}\sin 3\alpha + \ldots \text{ad. inf.}$

$$\left[\textbf{Ans. } \left(2\sin\frac{\alpha}{2}\right)^{-n}\sin\frac{n}{2}(\pi - \alpha), \text{ if } n \text{ be} < 1.\right]$$

5. $1 + \dfrac{1}{3}c\cos\alpha + \dfrac{1.4}{3.6}c^2\cos 2\alpha + \ldots \text{ad. inf., if } c < 1.$

$$\left[\textbf{Ans. } \frac{\cos\frac{1}{3}\theta}{r^{1/3}}, \text{ when } r = \sqrt{(1 - 2c\cos\alpha + c^2)}, \alpha = \tan^{-1}\frac{c\sin\alpha}{1 - c\cos\alpha}\right]$$

Summation of Series

7.5 SUMMATION DEPENDING UPON EXPONENTIAL SERIES

To sum the trigonometric series depending upon exponential series, following formulae are worth remembering:

(i) $e^x = 1 + x + \dfrac{x^2}{2!} + \dfrac{x^3}{3!} + ..$ ad. inf.

(ii) $e^{-x} = 1 - x + \dfrac{x^2}{2!} - \dfrac{x^3}{3!} + ...$ ad. inf.

(iii) $\sin x = x - \dfrac{x^3}{3!} + \dfrac{x^5}{5!} + ..$ ad. inf.

(iv) $\sinh x = x + \dfrac{x^3}{3!} + \dfrac{x^5}{5!} + ...$ ad. inf.

(v) $\cosh x = 1 + \dfrac{x^2}{2!} + \dfrac{x^4}{4!} + ...$ ad. inf.

EXAMPLES

1. Sum the series:

$$\cos\theta + \sin\theta + \frac{\cos^2\theta \sin 2\theta}{2!} + \frac{\cos^3\theta \sin 3\theta}{3!} + ... \text{ad. inf.}$$

Solution. Let $S = \sin\theta\cos\theta + \dfrac{\sin 2\theta \cos^2\theta}{2!} + \dfrac{\sin 3\theta \cos^3\theta}{3!} + ...$ ad. inf.

and $C = 1 + \cos\theta\cos\theta + \dfrac{\cos 2\theta \cos^2\theta}{2!} + \dfrac{\cos 3\theta \cos^3\theta}{3!} + ...$ ad. inf.

$\therefore \quad C + iS = 1 + \cos\theta\, e^{i\theta} + \dfrac{\cos^2\theta\, e^{2i\theta}}{2!} + \dfrac{\cos^2\theta\, e^{3i\theta}}{3!} + ...$ ad. inf.

$$= e^{\cos\theta \cdot e^{i\theta}} = e^{\cos\theta(\cos\theta + i\sin\theta)}$$

$$= e^{\cos^2\theta + i\sin\theta\cos\theta} = e^{\cos^2\theta} \cdot e^{i\sin\theta\cos\theta}$$

$$= e^{\cos^2\theta} [\cos(\sin\theta\cos\theta) + i\sin(\sin\theta\cos\theta)]$$

Equating imaginary parts on both sides, we get

$$S = e^{\cos^2\theta} \sin(\sin\theta\cos\theta) = e^{\cos^2\theta} \sin\left(\frac{1}{2}\sin 2\theta\right)$$

2. Sum the series

$$1 + e^{\sin\alpha}\cos(\cos\alpha) + \frac{e^{2\sin\alpha}\cos(2\cos\alpha)}{2!} + ... \text{ad. inf.}$$

Solution. Let $C = 1 + e^{\sin\alpha}\cos(\cos\alpha) + \dfrac{e^{2\sin\alpha}}{2!}\cos(2\cos\alpha) + ...$ ad. inf.

and $S = e^{\sin\alpha}\sin(\cos\alpha) + \dfrac{e^{2\sin\alpha}}{2!}\sin(2\cos\alpha) + ...$ ad. inf.

$$\therefore \quad C + iS = 1 + e^{\sin\alpha} \, e^{i\cos\alpha} + \frac{e^{2\sin\alpha} \cdot e^{2i\sin\alpha}}{2!} + \ldots \text{ad. inf.}$$

$$= 1 + e^{(\sin\alpha + i\cos\alpha)} + \frac{e^{2(\sin\alpha + i\cos\alpha)}}{2!} + \frac{e^{3(\sin\alpha + i\cos\alpha)}}{3!} + \ldots \text{ad. inf.}$$

$$= e^{e^{(\sin\alpha + i\cos\alpha)}}$$

$$= e^{\{e^{\sin\alpha} \cdot e^{i\cos\alpha}\}} = e^{e^{\sin\alpha}[\cos(\cos\alpha) + i\sin(\cos\alpha)]}$$

$$= e^{[e^{\sin\alpha}\cos\alpha(\cos\alpha) + ie^{\sin\alpha}\sin\alpha(\cos\alpha)]}$$

$$= e^{e^{\sin\alpha}\cos(\cos\alpha)} \times e^{ie^{\sin\alpha}\sin(\cos\alpha)}$$

$$= e^{e^{\sin\alpha}\cos(\cos\alpha)} [\cos\{e^{\sin\alpha}\sin(\cos\alpha)\} + i\sin\{e^{\sin\alpha}\sin(\cos\alpha)\}]$$

Equating real parts on both sides, we get

$$C = e^{e^{\sin\alpha}\cos(\cos\alpha)} \cdot \cos\{e^{\sin\alpha}\sin(\cos\alpha)\}$$

3. *Find the sum of the series*

$$\frac{5\cos\theta}{1!} + \frac{7\cos 3\theta}{3!} + \frac{9\cos 5\theta}{5!} + \ldots \text{ad. inf.}$$

Solution. Let $\quad C = \dfrac{5\cos\theta}{1!} + \dfrac{7\cos 3\theta}{3!} + \dfrac{9\cos 5\theta}{5!} + \ldots \text{ad. inf.}$

and $\quad S = \dfrac{5\sin\theta}{1!} + \dfrac{7\sin 3\theta}{3!} + \dfrac{9\sin 5\theta}{5!} + \ldots \text{ad. inf.}$

$\therefore \quad C + iS = \dfrac{5e^{i\theta}}{1!} + \dfrac{7e^{3i\theta}}{3!} + \dfrac{9e^{5i\theta}}{5!} + \ldots \text{ad. inf.}$

$$= 4\left(\frac{e^{i\theta}}{1!} + \frac{e^{3i\theta}}{3!} + \frac{e^{5i\theta}}{5!} + \ldots \text{ad. inf.}\right) + \left(\frac{e^{i\theta}}{1!} + \frac{3e^{3i\theta}}{3!} + \frac{5e^{5i\theta}}{5!} + \ldots \text{ad. inf.}\right)$$

$$= 4\left(\frac{e^{i\theta}}{1!} + \frac{e^{3i\theta}}{3!} + \ldots\right) + e^{i\theta}\left(1 + \frac{e^{2i\theta}}{2!} + \frac{e^{4i\theta}}{4!} + \ldots\right)$$

$$= 4\sinh(e^{i\theta}) + e^{i\theta}\cosh(e^{i\theta})$$

$$= \frac{4}{i}\sin(ie^{i\theta}) + e^{i\theta}\cos(ie^{i\theta})$$

$$= -4i\sin\{i(\cos\theta + i\sin\theta)\} + e^{i\theta}\cos\{i(\cos\theta + i\sin\theta)\}$$

$$= -4i\sin(i\cos\theta - \sin\theta) + (\cos\theta + i\sin\theta)\cos\{i\cos\theta - \sin\theta\}$$

$$= -4i\,[i\sinh(\cos\theta)\cos(\sin\theta) - \cosh(\cos\theta)\sin(\sin\theta)] + (\cos\theta + i\sin\theta)$$
$$[\cosh(\cos\theta)\cos(\sin\theta) + i\sinh(\cos\theta)\sin(\sin\theta)]$$

Equating the real parts, we get

$$C = 4\sinh(\cos\theta)\cos(\sin\theta) + \cos\theta\cosh(\cos\theta)\cos(\sin\theta)$$
$$- \sin\theta\sinh(\cos\theta)\sin(\sin\theta)$$

Summation of Series

EXERCISE

Sum the following series :

1. $\sin \alpha - a \sin (\alpha + \beta) + \dfrac{a^2}{2!} \sin (\alpha + 2\beta) - \ldots \text{ad. inf.}$ [Ans. $e^{-a \cos \beta} \sin (\alpha - a \sin \beta)$]

2. $\sin \alpha - \dfrac{\sin 2\alpha}{2!} + \dfrac{\sin 3\alpha}{3!} - \ldots \text{ad. inf.}$ [Ans. $e^{-\cos \alpha} \sin (\sin \alpha)$]

3. (a) $\sin \alpha + a \sin (\alpha + \beta) + \dfrac{a^2}{2!} \sin (\alpha + 2\beta) + \ldots \text{ad. inf.}$

   (b) $\cos \alpha + a \cos (\alpha + \beta) + \dfrac{a^2}{2!} \cos (\alpha + 2\beta) + \ldots \text{ad. inf.}$

   [Ans. (a) $e^{a \cos \beta} \sin (\alpha + a \sin \beta)$, (b) $e^{a \cos \beta} \cos (\alpha + a \sin \beta)$]

4. (a) $\cos \alpha + \dfrac{\sin \alpha \cos 2\alpha}{1!} + \dfrac{\sin^2 \alpha \cos 3\alpha}{2!} + \ldots \text{ad. inf.}$

   (b) $\cos \alpha + \dfrac{\cos \alpha \cos 2\alpha}{1!} + \dfrac{\cos^2 \alpha \cos 3\theta}{3!} + \ldots \text{ad. inf.}$

   [Ans. (a) $e^{\sin \alpha \cos \alpha} \cos (\alpha + \sin^2 \alpha)$, (b) $e^{\cos^2 \alpha} \cos (\alpha + \sin \alpha \cos \alpha)$]

7.6 SUMMATION DEPENDING UPON LOGARITHMIC SERIES

To sum such series, following useful results are worth remembering :

(i) $\log (1 + x) = x - \dfrac{x^2}{2} + \dfrac{x^3}{3} - \dfrac{x^4}{4} + \ldots \text{ad. inf.}$

(ii) $\log (1 - x) = - x - \dfrac{x^2}{2} - \dfrac{x^3}{3} - \dfrac{x^4}{4} - \ldots \text{ad. inf.}$

(iii) $\log (1 + x) + \log (1 - x) = - 2 \left(\dfrac{x^2}{2} + \dfrac{x^4}{4} + \dfrac{x^6}{6} + \ldots \text{ad. inf.} \right)$

(iv) $\log (1 + x) - \log (1 - x) = 2 \left(x + \dfrac{x^3}{3} + \dfrac{x^5}{5} + \ldots \text{ad. inf.} \right)$

where modulus of $x \leq 1$, but $x \neq -1$.

EXAMPLES

1. If $\theta - \alpha = \tan^2 \dfrac{\phi}{2} \sin 2\theta - \dfrac{1}{2} \tan^4 \dfrac{\phi}{2} \sin 4\theta + \dfrac{1}{3} \tan^6 \dfrac{\phi}{2} \sin 6\theta - \ldots \text{ad. inf.}$

prove that $\tan \alpha = \tan \theta \cos \phi$.

Solution. We have

$$\theta - \alpha = \tan^2 \dfrac{\phi}{2} \sin 2\theta - \dfrac{1}{2} \tan^4 \dfrac{\phi}{2} \sin 4\theta + \dfrac{1}{3} \tan^6 \dfrac{\phi}{2} \sin 6\theta - \ldots \text{ad. inf.} \quad \ldots(i)$$

$$= S \text{ (say)}$$

and let

$$C = \tan^2 \dfrac{\phi}{2} \cos 2\theta - \dfrac{1}{2} \tan^4 \dfrac{\phi}{2} \cos 4\theta + \dfrac{1}{3} \tan^6 \dfrac{\phi}{2} \cos 6\theta - \ldots \text{ad. inf.}$$

$$\therefore\ C + iS = \tan^2\frac{\phi}{2}\cdot e^{2i\theta} - \frac{1}{2}\tan^4\frac{\phi}{2} e^{4i\theta} + \frac{1}{3}\tan^6\frac{\phi}{2} e^{6i\theta} - \ldots \text{ad. inf.}$$

$$= \log\left[1 + \tan^2\frac{\phi}{2}\cdot e^{2i\theta}\right]$$

$$= \log\left[1 + \tan^2\frac{\phi}{2}(\cos 2\theta + i\sin 2\theta)\right]$$

$$= \log\left[\left(1 + \tan^2\frac{\phi}{2}\cos 2\theta\right) + i\tan^2\frac{\phi}{2}\sin 2\theta\right]$$

$$= \frac{1}{2}\log\left\{\left(1 + \tan^2\frac{\phi}{2}\cos 2\theta\right)^2 + \left(\tan^2\frac{\phi}{2}\sin 2\theta\right)^2\right\} + i\tan^{-1}\left(\frac{\tan^2\frac{\phi}{2}\sin 2\theta}{1 + \tan^2\frac{\phi}{2}\cos 2\theta}\right)$$

Equating imaginary parts on both sides we get

$$S = \tan^{-1}\frac{\tan^2\frac{\phi}{2}\sin 2\theta}{1 + \tan^2\frac{\phi}{2}\cos 2\theta}$$

∴ from (1)

$$\theta - \alpha = \tan^{-1}\frac{\tan^2\frac{\phi}{2}\sin 2\theta}{1 + \tan^2\frac{\phi}{2}\cos 2\theta}$$

or

$$\tan(\theta - \alpha) = \frac{\tan^2\frac{\phi}{2}\sin 2\theta}{1 + \tan^2\frac{\phi}{2}\cos 2\theta}$$

or

$$\frac{\sin(\theta - \alpha)}{\cos(\theta - \alpha)} = \frac{\tan^2\frac{\phi}{2}\sin 2\theta}{1 + \tan^2\frac{\phi}{2}\cos 2\theta}$$

Cross multiplying and simplifying, we get

$$\sin(\theta - \alpha) = \tan^2\frac{\phi}{2}[\sin 2\theta \cos(\theta - \alpha) - \cos 2\theta \sin(\theta - \alpha)]$$

$$= \tan^2\frac{\phi}{2}\sin(2\theta - \theta + \alpha)$$

$$= \tan^2\frac{\phi}{2}\sin(\theta + \alpha)$$

or

$$\frac{\tan^2\frac{\phi}{2}}{1} = \frac{\sin(\theta - \alpha)}{\sin(\theta + \alpha)}$$

Applying componendo and dividendo, we get

$$\frac{1 - \tan^2\frac{\phi}{2}}{1 + \tan^2\frac{\phi}{2}} = \frac{\sin(\theta + \alpha) - \sin(\theta - \alpha)}{\sin(\theta + \alpha) + \sin(\theta - \alpha)}$$

Summation of Series

or
$$\cos\phi = \frac{2\cos\theta\sin\alpha}{2\sin\theta\cos\alpha}$$

or
$$\cos\phi = \frac{2\cos\theta\sin\alpha}{2\sin\theta\cos\alpha}$$

or
$$\cos\phi\tan\theta = \tan\alpha$$

2. *If a, b, c are the sides of a triangle, show that the sum of the series*

$$\log a - \frac{b}{a}\cos C - \frac{b^2}{2a^2}\cos 2C - \frac{1}{3}\frac{b^3}{a^3}\cos 3C - ad\ inf.\ is\ \log c.$$

Solution. We have

$$\log a - \frac{b}{a}\cos C + \frac{1}{2}\frac{b^2}{a^2}\cos 2C - \frac{1}{3}\frac{b^3}{3a^3}\cos 3C - ...\ ad.\ inf.$$

$$= \log a - C\ (\text{say}) \qquad ...(i)$$

Then
$$C = \frac{b}{a}\cos C + \frac{b^2}{2a^2}\cos 2C + \frac{b^3}{3a^3}\cos 3C + ...\ ad.\ inf.$$

and let
$$S = \frac{b}{a}\sin C + \frac{b^2}{2a^2}\sin 2C + \frac{b^3}{3a^3}\sin 3C + ...\ ad.\ inf.$$

Than
$$C + iS = \frac{b}{a}e^{iC} + \frac{1}{2}\frac{b^2}{a^2}e^{2iC} + \frac{1}{3}\frac{b^3}{a^3}e^{3iC} + ...\ ad.\ inf.$$

$$= -\log\left\{1 - \left(\frac{b}{a}\right)e^{iC}\right\}$$

$$= -\log\left[1 - \frac{b}{a}(\cos C + i\sin C)\right]$$

$$= -\log\left[\frac{a - b\cos C - ib\sin C}{a}\right]$$

$$= -\frac{1}{2}\log\{(a - b\cos C)^2 + (-b\sin C)^2\} - i\tan^{-1}\left(\frac{-ib\sin C}{a - b\cos C}\right) + \log a$$

$$= -\frac{1}{2}\log(a^2 + b^2 - 2ab\cos C) + \log a + i\tan^{-1}\left\{\frac{b\sin C}{(a - b\cos C)}\right\}$$

$$= \frac{1}{2}\log(c^2) + \log a + i\tan^{-1}\left(\frac{b\sin C}{a - b\cos C}\right)$$

$$\because\quad c^2 = a^2 + b^2 - 2ab\cos C$$

$$= -\log c + \log a + i\tan^{-1}\left[\frac{b\sin C}{a - b\cos C}\right]$$

Equating real parts on both sides, we get
$$C = -\log c + \log a$$

Hence, from (i) the sum of the given series
$$= \log a - (-\log c + \log a)$$
$$= \log c$$

EXERCISES

Sum the following series taking the value of c not greater than unity numerically :

1. (a) $c \cos \alpha - \dfrac{c^2}{2} \cos 2\alpha + \dfrac{c^3}{3} \cos 3\alpha - \ldots \text{ad. inf}.$

   (b) $c \cos \alpha - \dfrac{c^2}{2} \cos 2\alpha + \dfrac{c^3}{3} \cos 3\alpha - \ldots \text{ad. inf}.$

   [Ans. (a) $\dfrac{1}{2} \log (1 + 2c + \cos \alpha + c^2)$, (b) $\tan^{-1} \dfrac{c \sin \alpha}{1 + c \sin \alpha}$,

   except when (i) $c = 1$, $\alpha = (2n + 1)\pi$ when (ii) $c = -1$, $\alpha = 2n\pi$]

2. (a) $\cos \alpha - \dfrac{1}{2} \cos 2\alpha + \dfrac{1}{3} \cos 3\alpha - \ldots \text{ad. inf}.$

   (b) $\sin \alpha - \dfrac{1}{2} \sin 2\alpha + \dfrac{1}{3} \sin 3\alpha - \ldots \text{ad. inf}.$

   $\left[\text{Ans. (a) } \log\left(2 \cos \dfrac{\alpha}{2}\right), \text{ (b) } \dfrac{1}{2}\alpha, \text{ except when } \alpha = (2n + 1)\pi\right]$

3. $\cos \alpha \sin \alpha + \dfrac{1}{2} \cos^2 \alpha \sin 2\alpha + \dfrac{1}{3} \cos^3 \alpha \sin 3\alpha + \ldots \text{ad. inf}.$ $\left[\text{Ans. } \dfrac{1}{2}\pi - \alpha\right]$

4. $c \sin^2 \alpha - \dfrac{c^2}{2} \sin^2 2\alpha + \dfrac{c^3}{3} \sin^2 3\alpha + \ldots \text{ad. inf}.$

   [Ans. $\dfrac{1}{2} \log [(1 + c)/\sqrt{(1 + 2c \cos 2\alpha + c^2)}]$, except when

   (i) $c = 1$, $\alpha = \left(n + \dfrac{1}{2}\right)\pi$ and (ii) $c = -1$]

5. (a) $c \cos \alpha - \dfrac{c^2}{2} \cos (\alpha + \beta) + \dfrac{c^3}{3} \cos (\alpha + 2\beta) - \ldots \text{ad. inf}.$

   (b) $c \sin \alpha - \dfrac{c^2}{2} \sin (\alpha + \beta) + \dfrac{c^3}{3} \sin (\alpha + 2\beta) - \ldots \text{ad. inf}.$

   [Ans. (a) $\dfrac{1}{2} \cos ((\alpha - \beta) \log (1 + 2c \cos \beta + c^2) - \sin (\alpha - \beta) \tan^{-1} \left\{\dfrac{c \sin \beta}{(1 + c \cos \beta)}\right\}$

   (b) $\dfrac{1}{2} \sin (\alpha - \beta) \log (1 + 2c \cos \beta + c^2) + \cos (\alpha - \beta) \tan^{-1} \left\{\dfrac{c \sin \beta}{(1 + c \cos \beta)}\right\}$

   except when (i) $c = 1$, $\beta = (2n + 1)\pi$ and (ii) $c = -1$, $\beta = 2n\pi$]

7.8 SUMMATION DEPENDING UPON GREGORY'S SERIES

We know that, if $-1 \leq x \leq 1$, than by Gregory's series

$$\tan^{-1} x = x - \dfrac{x^3}{3} + \dfrac{x^5}{5} - \dfrac{x^7}{7} + \ldots \text{ad. inf.}$$

Summation of Series

EXAMPLES

1. *Sum the series*

$$e^\alpha \cos \beta - \frac{1}{3} e^{3\alpha} \cos 3\beta + \frac{1}{5} e^{5\alpha} \cos 5\beta - \ldots \text{ad. inf.}$$

Solution. Let $C = e^\alpha \cos \beta - \frac{1}{3} e^{3\alpha} \cos 3\beta + \frac{1}{5} e^{5\alpha} \cos 5\beta - \ldots \text{ad. inf.}$

and $\quad S = e^\alpha \sin \beta - \frac{1}{3} e^{3\alpha} \sin 3\beta + \frac{1}{5} e^{5\alpha} \cos 5\beta - \ldots \text{ad. inf.}$

$\therefore \quad C + iS = e^\alpha \cdot e^{i\beta} - \frac{1}{3} e^{3\alpha} \cdot e^{3i\beta} + \frac{1}{5} e^{5\alpha} \cdot e^{5i\beta} - \ldots \text{ad. inf.}$

$\quad = e^{(\alpha + i\beta)} - \frac{1}{3} e^{3(\alpha + i\beta)} + \frac{1}{5} e^{5(\alpha + i\beta)} - \ldots \text{ad. inf.}$

$\quad = \tan^{-1} [e^{\alpha + i\beta}] = \tan^{-1} [e^\alpha \cdot e^{i\beta}]$

or $\quad C + iS = \tan^{-1} [e^\alpha (\cos \beta + i \sin \beta)]$

$\therefore \quad C - iS = \tan^{-1} [e^\alpha (\cos \beta - i \sin \beta)]$

Adding

$2C = \tan^{-1} [e^\alpha (\cos \beta + i \sin \beta)] + \tan^{-1} [e^\alpha (\cos \beta - i \sin \beta)]$

$\quad = \tan^{-1} \left[\dfrac{e^\alpha (\cos \beta + i \sin \beta) + e^\alpha (\cos \beta - i \sin \beta)}{1 - e^{2\alpha} (\cos^2 \beta - i^2 \sin^2 \beta)} \right]$

$\quad = \tan^{-1} \left[\dfrac{2e^\alpha \cos \beta}{1 - e^{2\alpha}} \right] = \tan^{-1} \left[\dfrac{2e^\alpha \cos \beta}{e^\alpha (e^{-\alpha} - e^\alpha)} \right]$

$\quad = \tan^{-1} \left[\dfrac{2 \cos \beta}{-(e^\alpha - e^{-\alpha})} \right] = -\tan^{-1} \left[\dfrac{2 \cos \beta}{2 \sinh \alpha} \right]$

$\therefore \quad C = -\dfrac{1}{2} \tan^{-1} \left(\dfrac{\cos \beta}{\sinh \alpha} \right)$

2. *Sum the series*

$$\frac{2}{1 \cdot 3} \sin 2x - \frac{4}{3 \cdot 5} \sin 4x + \frac{6}{5 \cdot 7} \sin 6x + \ldots \quad 0 < x < \pi/2$$

Solution. The given series can be written as

$\dfrac{1}{2} \left[\left(1 + \dfrac{1}{3}\right) \sin 2x - \left(\dfrac{1}{3} + \dfrac{1}{5}\right) \sin 4x + \left(\dfrac{1}{5} + \dfrac{1}{7}\right) \sin 6x + \ldots \right]$

$= \dfrac{1}{2} \left[\left(\sin 2x - \dfrac{1}{3} \sin 4x + \dfrac{1}{5} \sin 6x \ldots \right) - \left\{ 1 - \dfrac{1}{3} \sin 2x + \dfrac{1}{5} \sin 4x - \dfrac{1}{7} \sin 6x + \ldots \right\} + 1 \right]$

$= \dfrac{1}{2} [S_1 - S_2 + 1] \quad \ldots(i)$

Let $\quad S_1 = \sin 2x - \dfrac{1}{3} \sin 4x + \dfrac{1}{5} \sin 4x - \ldots$

and $\quad C_1 = \cos 2x - \dfrac{1}{3} \cos 4x + \dfrac{1}{5} \cos 6x - \ldots$

$$\therefore \quad C_1 + iS_1 = e^{2ix} - \frac{1}{3}e^{4ix} + \frac{1}{5}e^{6ix} - \ldots$$

$$= e^{ix}\left[e^{ix} - \frac{1}{3}e^{3ix} + \frac{1}{5}e^{5ix} - \ldots\right]$$

$$= e^{ix} \tan^{-1}(e^{ix})$$

Let $\quad S_2 = 1 - \dfrac{1}{3}\sin 2x + \dfrac{1}{5}\sin 4x - \ldots$

and $\quad C_2 = -\dfrac{1}{3}\cos 2x + \dfrac{1}{5}\cos 4x - \ldots$

$$\therefore \quad C_2 + iS_2 = 1 - \frac{1}{3}e^{2ix} + \frac{1}{5}e^{4ix} - \ldots$$

$$= e^{-ix}\left[e^{ix} - \frac{1}{3}e^{3ix} + \frac{1}{5}e^{5ix} - \ldots\right]$$

$$= e^{-ix} \tan^{-1}(e^{ix})$$

$\therefore \quad S_1 - S_2$ is imaginary part of $(C_1 + iS_1) - (C_2 + iS_2)$ or of $e^{ix} \tan^{-1}(e^{ix}) - e^{-ix} (\tan^{-1} e^{ix})$ or of $(e^{ix} - e^{-ix}) \tan^{-1}(e^{ix})$.

or of $2i \sin x\, (p + iq)$ where $\tan^{-1}(e^{ix}) = p + iq$

$$\therefore \qquad S_1 - S_2 = 2p \sin x \qquad \ldots(ii)$$

where $\qquad p + iq = \tan^{-1}(e^{ix}) = \tan^{-1}(\cos x + i \sin x)$

$\therefore \qquad p - iq = \tan^{-1}(\cos x - i \sin x)$

Adding, we get

$$2p = \tan^{-1}(\cos x + i \sin x) + \tan^{-1}(\cos x - i \sin x)$$

$$= \tan^{-1} \frac{\cos x + i \sin x + \cos x - i \sin x}{1 - (\cos^2 x - i^2 \sin^2 x)}$$

$$= \tan^{-1} \infty = \frac{\pi}{2}$$

$$\therefore \qquad p = \frac{\pi}{4}$$

Hence, from (ii)

$$S_1 - S_2 = 2p \sin x = \frac{1}{2}\pi \sin x$$

$\therefore$ from (i) the sum of the series is

$$\frac{1}{2}[S_1 - S_2 + 1] = \frac{1}{2}\left[\frac{1}{2}\pi \sin x + 1\right]$$

Summation of Series

EXERCISES

Sum the following series:

1. $c \sin \alpha - \dfrac{c^3}{3} \sin 3\alpha + \dfrac{c^5}{5} \sin 5\alpha - \ldots$ ad. inf.

   and $c \cos \alpha - \dfrac{c^3}{3} \cos 3\alpha + \dfrac{c^5}{5} \cos 5\alpha - \ldots$ ad. inf.

   $\left[\text{Ans. } \dfrac{1}{2} \tanh^{-1} \left(\dfrac{2c \sin \alpha}{1+c^2}\right)\right]$

2. $\cos \alpha - \dfrac{1}{3} \cos 3\alpha + \dfrac{1}{5} \cos 5\alpha - \ldots$ ad. inf.

   $\left[\text{Ans. } \dfrac{\pi}{4}, \dfrac{-\pi}{4} \text{ or } 0 \text{ according as } \cos \alpha \text{ is positive, negative or zero.}\right]$

3. $c \sin \alpha - \dfrac{c^3}{3} \sin (\alpha + 2\beta) + \dfrac{c^5}{5} \sin (\alpha + 4\beta) - \ldots$ ad. inf.

   $\left[\text{Ans. } \dfrac{\sin(\alpha-\beta)}{2} \tan^{-1} \dfrac{2c \cos \beta}{1-c^2} + \dfrac{\cos(\alpha-\beta)}{2} \tanh^{-1} \dfrac{2c \sin \beta}{1+c^2}\right]$

4. $c \sin \alpha \sin \beta - \dfrac{c^3}{3} \sin 3\alpha \sin 3\beta + \dfrac{c^5}{5} \sin 5\alpha \sin 5\beta - \ldots$ ad. inf. where $0 < c < 1$.

   $\left[\text{Ans. } \dfrac{1}{4} \left\{ \tan^{-1} \dfrac{2c \cos(\alpha-\beta)}{1-c^2} - \tan^{-1} \dfrac{2c \cos(\alpha+\beta)}{1-c^2} \right\}\right]$

5. If $\sin \theta - \dfrac{1}{3} \sin 3\theta + \dfrac{1}{5} \sin 5\theta - \ldots = v$, show that $\tanh 2v = \sin \theta$.

6. If $u = \cos \alpha - \dfrac{1}{3} \cos 3\alpha + \dfrac{1}{5} \cos 5\alpha - \ldots$ ad. inf.

   and $v = \sin \alpha - \dfrac{1}{3} \sin 3\alpha + \dfrac{1}{5} \sin 5\alpha - \ldots$ ad. inf.

   prove that (i) $u = \dfrac{1}{4}\pi$ and (ii) $\cosh 2\theta = \sec \alpha$, $0 \leq \alpha \leq \dfrac{\pi}{2}$.

BMH 102 (a & b) Calculus

- **Differential Calculus**
- **Integral Calculus**
- **Ordinary Differential Equations**

DIFFERENTIAL CALCULUS

$\varepsilon - \delta$ definition of the limit of a function. Basic properties of limits. Continuous functions and classification of discontinuities. Differentiability. Successive differentiation. Leibnitz theorem. Maclaurin and Taylor series expansions. Asymptotes. Curvature. Tests for concavity and convexity. Points of inflexion. Multiple points. Tracing of curves in Cartesian and polar coordinates.

1
Limits and Continuity

1.1 LIMIT

We have a rough idea of what is meant when we say, 'The variable x approaches the constant a as a limit' which never means that x equals a, or that x is near a. In this statement we mean that x is in the neighbourhood of a. We first define sequence. If to each positive integer $1, 2, 3, ..n, ...$ there correspnds a definite real number x_n, then the numbers $x_1, x_2, ..., x_n, ...$ are said to form a sequence. Such a sequence is denoted by the symbol $\{x_n\}$. A sequence $\{x_n\}$ is said to converge to a number a, or to have limit a, if the numerical values of the diffrences

$$a - x_1, a - x_2, a - x_3, ... a - x_n$$

remain smaller than any preassigned arbitrarily small positive number however small.
The relation defined is written as
$$\text{Lim } x = a.$$

We use $|a - x_n|$ to denote the numerical value or positive value of $a - x_n$.

Thus if for any positive number ε howevery small, there can be found a positive number N such that $|a - x_n| < \varepsilon$ for every $n > N$, then the sequence $\{x_n\}$ has limit a. We can also say that $x \to a$ (x approaches a).

1.1.1 Limit of a Function

l is said to be the limit of $f(x)$ at $x = a$, if $|f(x) - l| < \varepsilon$ for every value of x, other than a, such that $0 < |x - a| < \varepsilon$, where $|x - a|$ means the absolute value of $x - a$ without any consideration of sign.

We conclude that the numerical value of $|f(x) - l|$ should be made as small as we please by making the numerical difference between x and a sufficiently small.

It is necessary and sufficient that corresponding to any positive number ε (not zero), a choice of δ is possible.

Now again $0 < |x - a| < \delta$ means that numerical value of $x - a$ lies between 0 and δ, i.e., difference is less than δ, where δ is a small positive number.

This can be shown as

(i) when $x > a$, then $x - a < \delta$, i.e., $x < a + \delta$

and (ii) when $x < a$, then $a - x < \delta$, i.e., $x > a - \delta$.

Thus this result can also be expressed as
$$a - \delta < x < a + \delta,$$

i.e., x contains values between $a - \delta$ and $a + \delta$.

The interval $(a - \delta, a + \delta)$ is called the neighbourhood of x.
The following example shall help us in understanding the meaning of limit:
Let us consider any function

$$f(x) = x^2 + x - 1 \text{ at } x = 3.$$

Here if 11 is the limit of $f(x)$ at $x = 3$, then $|f(x) - 11|$ shall be less than ε if

$$|x-3|<\delta.$$

Now $$|x^2+x-1-11|=|x^2+x-12|$$
$$=|x+4||x-3|.$$

Again let x lie between 2 and 4.

Then for all values of x between 2 and 4, $x+4$ is positive and is less than $4+4=8$.

So when $2<x<4$, we have
$$|f(x)-11|<8|x-3|,$$
i.e., $$|x-3|<\varepsilon/8.$$

Thus we see that there exists a positive number $\varepsilon/8$, such that
$$|f(x)-11|<\varepsilon, \qquad\qquad \text{where } x-3<\varepsilon/8.$$

Hence limit of $f(x)$ at $x=3$ is 11. Again we consider the function $f(x)=\sin x$ for every value of x.

Now we have to show that the limit exists at $x=a$.

Here if $\sin a$ is the limit of $f(x)$ at $x=a$, then $|f(x)-f(a)|$ should be less than ε, any arbitrarily assigned positive number.

Now
$$|f(x)-f(a)|=|\sin x - \sin a|$$
$$=\left|2\cos\frac{x+a}{2}\sin\frac{x-a}{2}\right|$$
$$=2\left|\cos\frac{x+a}{2}\right|\left|\sin\frac{x-a}{2}\right|.$$

Now $\left|\cos\dfrac{x+a}{2}\right|\leq 1$ for every value of x and a.

Also $\left|\sin\dfrac{x-a}{2}\right|\leq\left|\dfrac{x-a}{2}\right|.$

Thus we have
$$|\sin x - \sin a|\leq 2\cdot\left|\frac{x-a}{2}\right|=|x-a|.$$

Hence $(\sin x - \sin a)\leq \varepsilon$ when $(x-a)<\varepsilon$.

Hence there exists an interval $(a-\varepsilon, a+\varepsilon)$ around a, such that for every value of x in this interval.
$$|\sin x - \sin a|<\varepsilon.$$

1.2 DIFFERNCE BETWEEN THE LIMIT AND THE VALUE OF THE FUNCTION

We know that if l is the limit of $f(x)$ at $x=a$, then
$$|f(x)-l|<\varepsilon, \qquad\qquad\qquad ...(1)$$
for all values of x, other than a, such that $|x-a|<\delta$.

We observe that x does not take up the value $x=a$, we can also say that the relation (1) holds good for all values of x, other than a.

But, we obtain the value of the function, when we put a for x in the given function.

Note. The limit of $f(x)$ has the relationship with the value of the function only when the function is continuous; at $x=a$, the two are equal.

Limits and Continuity

For example, we consider the function $f(x)$ which is such that

$$\left.\begin{array}{l} f(x) = x^2, \quad \text{when } x \neq a, \\ f(x) = a^2 + 1 \text{ when } x = a. \end{array}\right\} \qquad \ldots(1)$$

The value of $f(x)$ for $x = a$ is $a^2 + 1$ while the limit of $f(x)$ for $x = a$ is $\underset{x \to a}{\text{Lim}}\, x^2 = a^2$.

Thus the limit and value of the function are not the same.
As another example, we consider the function

$$f(x) = \frac{x^2}{1+x^2} + \frac{x^2}{(1+x^2)^2} + \frac{x^2}{(1+x^2)^3} + \ldots$$

Thus value of the function $f(x)$ for $x = 0$ is

$$f(x) \text{ for } x = 0 = \frac{0}{1+0} + \frac{0}{1+0} + \frac{0}{1+0} + \ldots = 0.$$

Limit of $f(x)$. When $x \neq 0$,

$$f(x) = \frac{x^2}{1+x^2} + \frac{x^2}{(1+x^2)^2} + \frac{x^2}{(1+x^2)^3} + \ldots$$

$$= \frac{x^2}{1+x^2}\left[1 + \frac{1}{(1+x^2)} + \frac{1}{(1+x^2)^2} + \ldots\right]$$

$$= \frac{x^2}{1+x^2}\left[\frac{1}{1 - 1/(1+x^2)}\right] = 1 \left(\text{since } \frac{1}{1+x^2} < 1\right)$$

Thus if $x \neq 0$, $f(x) = 1$ whatever that value of x may be and therefore $\underset{x \to 0}{\text{Lim}}\, f(x)$ is equal to 1 and the value of $f(x) = 0$, which are different.

1.3 RIGHT HAND AND LEFT HAND LIMITS

First we shall discuss, what we mean by these limits. This we shall do by taking an example.

Example $y = f(x) = 3$ for $0 < x \leq 1$,

$y = 6$ for $1 < x \leq 2$,

$y = 9$ for $2 < x \leq 3$,

We observe here that $|y - 3|$ becomes and remains less than any arbitrarily small positive number which is equal to 0 for x approaching 1 through values less than 1 ('from the left').

But $|y - 3| = 3$ for x approaching 1 through values greater than 1 ('from the right') and therefore cannot become and remain less than any arbitrarily small positive number. Since it is to have 3 as a limit $|y - 3|$ must become less than any arbitrarily small positive number; it does not matter how x approaches 1. Thus the limit of y is not 3.

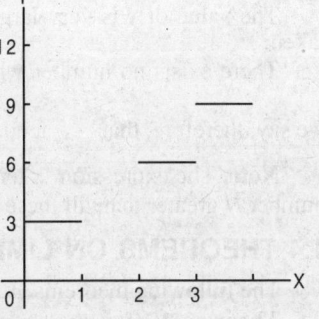

Therefore, we may say the left hand limit of y is 3,

i.e., $\underset{x \to 1^-}{\text{Lim}}\, y = 3.$

Here the symbol $x \to 1^-$ indicates that x approaches 1 through values less than 1.
Similarly we may have the right hand limit of y as 6,

i.e., $\underset{x \to 1^+}{\text{Lim}}\, y = 6.$

Hence the symbol $x \to 1^+$ indicates that x approaches 1 through values greater than 1. Now we define them in usual manner, as

A function $f(x)$ is said to tend to a limit l as x tends to a from above or from the right corresponding to any positive number arbitrarily small, there exists a positive number δ, such that $|f(x) - l| < \varepsilon$,

for all value of x for which
$$0 < x - a < \delta.$$

We write the above in symbolic form as
$$\underset{x \to a+0}{\text{Lim}} f(x) = l.$$

This is known as *Right hand limit*.

Similarly a function $f(x)$ is said to tend to a limit l as x tends to a from below or from the left if corresponding to any positive number ε, arbitrarily small, there exists a positive number δ, such that
$$|f(x) - l| < \varepsilon$$

for all values of x for which
$$a - \delta < x < a.$$

We write it in the following way.
$$\underset{x \to a-0}{\text{Lim}} f(x) = l,$$

which is known as *Left hand limit of $f(x)$*.

Limit of a Function

A function has a limit for $x \to a$ only if the left and right hand limits exist and are equal.

Thus
$$\underset{x \to a}{\text{Lim}} f(x) = l,$$

If and only if
$$\underset{x \to a-0}{\text{Lim}} f(x) = l = \underset{x \to a+0}{\text{Lim}} f(x),$$

Infinite Limits

We consider the function y which is equal to $\frac{1}{x^2}$ for every value of x.

The value of y is very large when x is very small y becomes larger as smaller values of x are taken.

There exists no number which is greater than any number even if we choose x small enough; we say, therefore, that $\frac{1}{x^2}$ tends to infinity (or the limit of $\frac{1}{x^2}$ is infinity) as x tends to zero.

Note. The expression 'x tends to infinity' means only x goes on increasing and there erxists no number N greater than all these values of x which are less.

1.4 THEOREMS ON LIMITS

The following theorems are useful while operating upon the limits of the functions.

Theorem 1. *If each one of a finite numbers of functions of x has a limit as x approaches a, then the limit of the sum of the functions is equal to the sum of their limits,*

i.e., $\underset{x \to a}{\text{Lim}} [f_1(x) + f_2(x)] = \underset{x \to a}{\text{Lim}} f_1(x) + \underset{x \to a}{\text{Lim}} f_2(x)$

Example. $\underset{x \to 1}{\text{Lim}} (x^3 + x^2) = \underset{x \to 1}{\text{Lim}} x^3 + \underset{x \to 1}{\text{Lim}} x^2 = 1 + 1 = 2.$

Limits and Continuity

Theorem 2. *If each one of a finite number of functions of x has a limit as x approaches a, then the limit of the product of the functions is equal to the product of their limits,*

$$\lim_{x \to a} [f_1(x) \cdot f_2(x)] = \lim_{x \to a} f_1(x) \cdot \lim_{x \to a} f_2(x).$$

Theorem 3. *If each of two functions of x has a limit as x approaches a, then the limit of the quotient of the functions is equal to quotient of their limits, provided that the limit of denominator is not zero.*

i.e., $\qquad \lim\limits_{x \to a} \dfrac{f_1(x)}{f_2(x)} = \dfrac{\lim\limits_{x \to a} f_1(x)}{\lim\limits_{x \to a} f_2(x)} \qquad$ provided $\lim\limits_{x \to a} f_2(x) \neq 0.$

For example, $\lim\limits_{x \to 2} \dfrac{x^2+4}{x^2+1} = \dfrac{\lim\limits_{x \to 2} x^2+4}{\lim\limits_{x \to 2} x^2-1} = \dfrac{8}{3}.$

EXAMPLES

1. *Discuss the limits for the following function :*

$\phi(x) = 0 \qquad\qquad$ when $x = 0$

$\phi(x) = \dfrac{1}{2} - x \qquad$ when $0 < x < \dfrac{1}{2}$

$\phi(x) = \dfrac{1}{2} \qquad\qquad$ when $x = \dfrac{1}{2}$

$\phi(x) = \dfrac{3}{2} - x \qquad$ when $\dfrac{1}{2} < x < 1$

$\phi(x) = 1 \qquad\qquad$ when $x = 1.$

Solution. When $x \to 0$ from the right,

$$\phi(x) = \frac{1}{2} - x \text{ tends to } \frac{1}{2}$$

while function is not defined for the values of x less than zero.

Hence only right ahnd limit exists. Again as $x \to \dfrac{1}{2}$ from the left,

$$\phi(x) = \frac{1}{2} - x \to 0.$$

Also as $x \to \dfrac{1}{2}$ from the right hand side.

$$\phi(x) = \frac{3}{2} - x \to 1.$$

Thus left hand limit is 0 and right hand limit is 1, thus no limit exsits.

Similarly we can discuss the limit of function $\phi(x)$ for $x = 1$ from the left hand side and right hand side.

2. *Evaluate the limit* $\lim\limits_{x \to \infty} \dfrac{\sin x}{x}.$

Solution. We are unable to take the quotient of the limits here since the numerator *i.e.*, $\sin x$ oscillated between -1 and 1 while the denominator increases without limit.

However since x is positive, we have $-1 \leq \sin x < 1$ and $-\dfrac{1}{x} \leq \dfrac{\sin x}{x} \leq \dfrac{1}{x}$; we observe then

$\dfrac{\sin x}{x}$ is contained between $-\dfrac{1}{x}$ and $\dfrac{1}{x}$.

But as x increases identifinitely both $-\dfrac{1}{x}$ and $\dfrac{1}{x}$ approach 0; therefore $\dfrac{\sin x}{x}$ must also approach 0.

Hence $0 = \underset{x \to \infty}{\text{Lim}} \left(-\dfrac{1}{x}\right) \leq \underset{x \to \infty}{\text{Lim}} \dfrac{\sin x}{x} \leq \text{Lim} \dfrac{1}{x} = 0$.

Hence $\underset{x \to \infty}{\text{Lim}} \dfrac{\sin x}{x} = 0$.

EXERCISES

1. Evaluate $\underset{x \to 2}{\text{Lim}} \dfrac{x^2 - 3x + 2}{x - 2}$. [Ans. 1]

2. Evaluate $\underset{x \to 5}{\text{Lim}} \dfrac{x^2 - 25}{x - 5}$. [Ans. 10]

3. Evaluate $\underset{x \to 0}{\text{Lim}} \dfrac{(1 + x)^n - 1}{x}$. [Ans. n]

4. (a) Evaluate $\underset{x \to 0}{\text{Lim}} \dfrac{e^x - 1}{x}$. [Ans. 1]

   (b) $\underset{x \to 0}{\text{Lim}} \dfrac{a^x - 1}{x}$. [Ans. $\log a$]

5. Evaluate $\underset{x \to 0}{\text{Lim}} (1 + x)^{1/x}$. [Ans. e]

6. Show that $\underset{x \to 0}{\text{Lim}} \dfrac{\sqrt{(a + x)} - \sqrt{(a)}}{x} = \dfrac{1}{2\sqrt{2}}$.

1.5 CONTINUITY

A function $f(x)$ is said to be continuous at a point $x = a$ if and only if the function is defined at the point and the limit of the function as $x \to a$ is equal to the value of the function for $x = a$. This may be expressed in symbols as follows :

$f(x)$ is continuous at $x = a$ if and only if $\underset{x \to a}{\text{Lim}} f(x) = f(a)$.

We can further say that $f(x)$ is continuous at a point $x = a$ provided $f(x)$ exists and $\underset{h \to 0}{\text{Lim}} f(a + h) = f(a)$ when the right-hand limit is considered.

We will also say that $f(x)$ is continuous at $x = a$, provided $f(a)$ exists and $\underset{h \to 0}{\text{Lim}} f(a - h) = f(a)$ when the left hand limit is taken.

Thus if $f(x)$ is continuous in the right hand neighbourhood of $x = a$ as well as in the left hand neighbourhood $x = a$, it is said to be continuous at $x = a$, and then following conditions must be satisfied.

(1) $f(x)$ must be defined in the neighbourhood of $x = a$ including at $x = a$.

(2) $f(a + 0) = f(a - 0) = f(a)$.

Limits and Continuity

Definition of the Continuity. If $f(x)$ is defined in the δ-neighbourhood of $x = a$ including the point $x = a$, it is said to be continuous there if given $\varepsilon > 0$ (an arbitrarily chosen small number) there exists a number δ, such that $|f(x) - f(a)| < \varepsilon$ for all values of x such that
$$|x - a| < \delta.$$

This obviously means that $f(x)$ lies between $f(a) - \varepsilon$ and $f(a) + \varepsilon$ for all values of x lying between $a - \delta$ and $a + \delta$.

1.5.1 Explanation by Graph

Let MN represent the graph of $f(x)$ and P be any point on it for which $x = a$ and $y = f(a)$ while M' and N' are

$(a - \delta, 0)$ and $(a + \delta, 0)$.

Also the lines parallel to x-axis are

$y = f(a) - \varepsilon$ and $f(a) + \varepsilon$

and MM' and NN' are $x = a - \delta, x = a + \delta$.

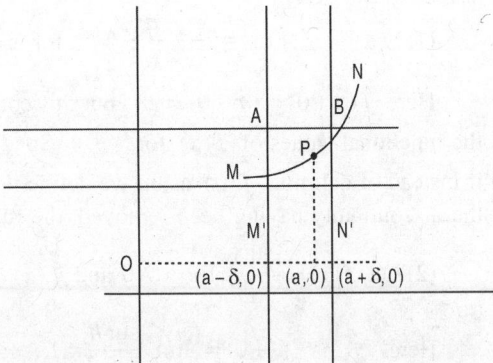

Now the relations $f(a) - \varepsilon$ and $f(a) + \varepsilon$ represent lines which are parallel to x-axis and lie on different sides of P and ε (an arbitrarily chosen positive number), measures the degree of closeness of the lines from each other.

The continuity of $f(x)$ at $x = a$, then required that we should be able to draw the set of lines $x = a - \delta$ and $x = a + \delta$, which are parallel to y-axis and which lie around a, such that every point of the graph of $f(x)$ between them lies also between the lines
$$y = f(a) - \varepsilon \text{ and } y = f(a) + \varepsilon.$$

1.6 DISCONTINUITY

A function which is not continuous at $x = a$ is said to be discontinuous at $x = a$.
A function may be discontinuous at $x = a$ because

(i) $f(x)$ is not defined at $x = a$ even though it is defined in the δ-neighbourhood of a,

(ii) $f(a) = f(a + 0) \neq f(a - 0)$

(iii) $f(a) = f(a - 0) \neq f(a + 0)$,

(iv) $f(a) \neq f(a - 0) \neq f(a + 0)$,

(v) $f(a)$ may exist but either $f(a + 0)$ or $f(a - 0)$ may not exist. The following graphs shall make it clear:

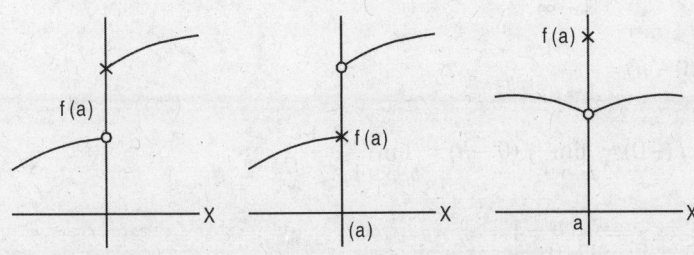

1.6.1 CLASSIFICATION OF DISCONTINUITIES OF A FUNCTION

(a) Removable discontinuity.

If, for a point $x = a$, $f(a+0)$ and $f(a-0)$ both exist, and are equal in value, but their common value is not equal to $f(a)$, then the function is said to have a *removable discontinuity* at $x = a$, for by defining $f(a)$ as the common value of $f(a+0)$ and $f(a-0)$, the function $f(x)$ can be made continuous at $x = a$, as it will then satisfy the conditions of continuity.

Examples.

(1) Let $\quad f(x) = \dfrac{x^2 - a^2}{x - a} \quad$ for $x \neq a$, $f(x) = 3a$ for $x = a$.

Here $f(a+0) = (a-0) = 2a$, but this common value of the two limits at $x = a$ is not equal to the functional values of $f(x)$ for $x = a$. So $f(x)$ has a *removable discontinuity* at $x = a$ because, if instead of defining $f(a)$ as $3a$ we define it as $2a$, we have $f(a+0) = f(a-0) = 2a$, and now the discontinuity having been removed, the function becomes continuous at $x = a$.

(2) Let $f(x) = \dfrac{\sin x}{x}$ for $x \neq 0$ and $f(x) = 0$ for $x = 0$. Consider $f(x)$ for continuity at $x = 0$.

Here, $\quad f(+0) = \lim\limits_{h \to 0} \dfrac{\sin h}{h} = 1$,

$$f(-0) = \lim_{h \to 0} \dfrac{\sin(-h)}{-h} = \lim_{h \to 0} \dfrac{\sin h}{h} = 1.$$

But $f(0) = 0$, so that $f(+0) = f(-0) \neq f(0)$. Hence the function has removable discontinuity at $x = 0$, as by defining $f(0)$ to be equal to 1 instead of 0, the discontinuity is removed.

(b) Discontinuity of the first kind.

If $f(a+0)$ and $f(a-0)$ both exist but are unequal, and $f(a)$ is equal to either or neither of these two limits, then $f(x)$ is said to have **discontinuity of the first kind** or **ordinary discontinuity** at $x = a$.

Examples.

(1) Show that $f(x) = \dfrac{1}{1 - e^{1/x}}$ has an ordinary discontinuity at $x = 0$.

Here, $\quad f(0 + h) = \dfrac{1}{1 - e^{1/h}}$

$\therefore \quad f(+0) = \lim\limits_{h \to 0} f(0 + h) = \lim\limits_{h \to 0} \dfrac{1}{1 - e^{1/h}}$

$$= \dfrac{1}{1 - \infty} = 0.$$

Again $\quad f(0 - h) = \dfrac{1}{1 - e^{-1/h}}$

$\therefore \quad f(-0) = \lim\limits_{h \to 0} f(0 - h) = \lim\limits_{h \to 0} \dfrac{1}{1 - e^{-1/h}}$

$$= \dfrac{1}{1 - 0} = 1.$$

Limits and Continuity

Thus the two limits $f(+0)$ and $f(-0)$ exist but are unequal; hence $f(x)$ has a discontinuity of the first kind at the point $x=0$.

(2) *Show that* $f(x) = \dfrac{2}{\pi} \lim\limits_{h \to 0} \tan^{-1} n/x$ *has an ordinary discontinuity at* $x=0$.

$$f(0+h) = f(h) = \frac{2}{\pi} \lim_{h \to 0} \tan^{-1} n/h$$

$$= \frac{2}{\pi} \cdot \frac{\pi}{2} = 1$$

$\therefore \qquad f(+0) = 1.$

$$f(0-h) = \frac{2}{\pi} \lim_{h \to 0} \tan^{-1}(-n/h).$$

$\therefore \qquad f(-0) = \dfrac{2}{\pi}\left(-\dfrac{\pi}{2}\right) = -1.$

So, $\qquad f(+0) \neq f(-0).$

$\therefore$ there is an ordinary discontinuity of $f(x)$ at $x=0$.

(c) **Discontinuity of the Second kind.**

If both of the limits $f(a+0), f(a-0)$ do not exist, the function is said to have a *discontinuity of the second kind* at $x=a$.

Example. Consider for continuity the function $f(x) = \sin \dfrac{1}{x}$ for $x \neq 0$, $f(0) = 0$.

$$\lim_{h \to 0} (0 \pm h) = \lim_{h \to 0} \sin \frac{1}{0 \pm h} = \lim_{h \to 0} \sin \frac{1}{\pm h}$$

Since $\sin \dfrac{1}{\pm h}$ does not tend to a definite number, nor does it tend to be infinite with a definite sign as $\pm h$ tends to zero, none of the two limits $f(+0)$ and $f(-0)$ exists at the point $x=0$. So $\sin \dfrac{1}{x}$ has a discontinuity of second kind at $x=0$.

(d) **Mixed discontinuity.** If only one of the two limits $f(a+0)$ and $f(a-0)$ exists and not the other, the function is said to hae *mixed discontinuity* at $x=a$.

Example. $f(x) = x$ for $x \leq 0$; $f(x) = \sin \dfrac{1}{x}$ for $x > 0$.

Here $\qquad f(+0) = \lim\limits_{h \to 0} f(0+h) = \lim\limits_{h \to 0} f(h)$

$$= \lim_{h \to 0} \sin \frac{1}{h'}$$

which does not exist; $f(-0) = \lim\limits_{h \to 0} f(0-h) = \lim\limits_{h \to 0} f(-h) = \lim\limits_{h \to 0} (-h) = 0.$

Thus, at the point $x=0$, the left hand limit exists but the right hand limit does not exist.

Hence $f(x)$ has mixed discontinuity at the point $x=0$.

(e) **Infinite discontinuity.** If one or more of the four limits $f(a+0), \overline{f(a+0)}, \overline{f(a-0)}, f(a-0)$ be indefinitely great, the point a is said to have **infinite discontinuity** at $x=a$.

(f) **Totally discontinuous functions.** A function is said to be totally discontinuous if it is discontinuous for every value of x.

1.6.2 Theorem on Continuity

If $f_1(x)$ and $f_2(x)$ are continuous at $x = a$, then $f_1(x) \pm f_2(x)$ and $f_1(x) f_2(x)$ are continuous at $x = a$. Also if $f_1(x)$ and $f_2(x)$ are continuous at $x = a$, $f_1(x)/f_2(x)$ is continuous at $x = a$ provided $f_2(a) \neq 0$.

Also if $z = g[f(x)]$ defines a composite function of g and f and $y = f(x)$ is continuous at $x = a$, then $g[f(x)]$ is also continuous at $x = a$ provided $g(y)$ is continuous at $y = f(a)$.

1.6.3 Continuity in an Interval

A function $f(x)$ is said to be continuous in an interval $a \leq x \leq b$ if it is continuous at every point of the interval.

EXAMPLES

1. *Given*

$$f(x) = 3x + \frac{3}{2} \qquad -\frac{3}{2} \leq x < 0$$

$$= -3x + \frac{3}{2} \qquad 0 \leq x < \frac{3}{2}$$

$$= -3x - \frac{3}{2} \qquad = -3x - \frac{3}{2}$$

Discuss the continuity at $x = -\frac{3}{2}, 0, \frac{3}{2}$.

Solution. (a) The function is not defined to the left of $x = -\frac{3}{2}$.

$$f\left(-\frac{3}{2}\right) = 3\left(-\frac{3}{2}\right) + \frac{3}{2} = -3.$$

Also $\qquad f\left(-\frac{3}{2} + h\right) = 3\left(-\frac{3}{2} + h\right) + \frac{3}{2} = -3 + 3h$

Hence $\underset{h \to 0}{\text{Lt}} f\left(-\frac{3}{2} + h\right) = -3,$

Therefore $f(x)$ is continuous in the right neighbourhood of $x = -\frac{3}{2}$.

(b) At $x = 0$, $\qquad f(0) = +\frac{3}{2},$

$$f(0 - h) = -3h + \frac{3}{2},$$

$$f(0 + h) = -3h + \frac{3}{2},$$

$$\underset{h \to 0}{\text{Lim}} f(0 - h) = \underset{h \to 0}{\text{Lim}} f(0 + h) = f(0) = \frac{3}{2}.$$

Hence $f(x)$ is continuous at $x = 0$.

Limits and Continuity

(c) $\quad x = \dfrac{3}{2}, f\left(\dfrac{3}{2}\right) = -6.$

$$f\left(\dfrac{3}{2}+h\right) = -6 - 3h, \ f\left(\dfrac{3}{2}-h\right) = -3 + 3h.$$

Now $\quad \underset{h \to 0}{\text{Lim}} f\left(\dfrac{3}{2}+h\right) = -6 \quad$ and $\quad \underset{h \to 0}{\text{Lim}} f\left(\dfrac{3}{2}-h\right) = -3.$

Here $\quad f(a+h) = f(a)$
but $\quad f(a-h) \neq f(a).$

Hence $f(x)$ is continuous in the right neighbourhood of $x = \dfrac{3}{2}$ and $f(x)$ is discontinuous in the left hand neighbourhood of $x = \dfrac{3}{2}$; consequently $f(x)$ is discontinuous at $x = \dfrac{3}{2}.$

2. Show that a polynomial is continuous for all real values of x.

Solution. Let $\phi(x) = a_0 x^n + a_1 x^{n-1} + \ldots + a_n$ ...(i)

Consider first a function $f(x) = x^n$, where n is a positive integer.

Now, $\quad f(a) = a^n$ and $\lim\limits_{x \to a} f(x) = \lim\limits_{x \to a} x^n$

$$= a^n, \forall \, a \in \mathbf{R}$$

Hence $\lim\limits_{x \to a} f(x) = f(a) \, \forall \, a \in \mathbf{R}$

i.e., $f(x)$ or x^n is continuous for every real value of x.

Let a_0 be any constant, then $a_0 x^n$ is continuous for every real value of x.

Now from (i), we have $\phi(x)$ as a polynomial of nth degree and is the sum of $(n+1)$ terms of the type $a_0 x^n$, each of which is continuous for every real value of x, hence $\phi(x)$ is continuous for all real values of x.

3. Examine the continuity of the following function $f(x)$ at $x = 0$:

$$f(x) = \dfrac{xe^{1/x}}{1 + e^{1/x}}, \ x \neq 0$$

Solution. We have $f(0) = 0$ ...(1)

Also, $\quad f(+0) = \lim\limits_{h \to 0} f(0+h)$

$$= \lim\limits_{h \to 0} \dfrac{(0+h) e^{1/(0+h)}}{1 + e^{1/(0+h)}}$$

$$= \lim\limits_{h \to 0} \dfrac{h e^{1/h}}{1 + e^{1/h}} = \lim\limits_{h \to 0} \dfrac{h}{e^{-1/h} + 1}$$

$$= \dfrac{0}{0+1} \quad \because e^{-1/h} \to 0 \text{ as } h \to 0$$

$$= 0$$

And $\quad f(-0) = \lim\limits_{h \to 0} f(0-h)$

$$= \lim_{h \to 0} \frac{(0-h) e^{1/(0-h)}}{1 + e^{1/(0-h)}} = \lim_{h \to 0} \frac{-h e^{-1/h}}{1 + e^{-1/h}}$$

$$= \frac{0}{1+0} = 0, \qquad \qquad \qquad \qquad \qquad \dots(3)$$

∴ From (1), (2) and (3) we get

$$f(+0) = f(-0) = f(0) = 0.$$

Hence, the given function is continuous at $x = 0$.

4. *Test the continuity of the following function :*

$$f(x) = \frac{1}{(x-a)} \operatorname{cosec}\left(\frac{1}{x-a}\right) \text{ when } x \ne 0$$

$$= 0, \text{ when } x = a.$$

Solution. We are given

$$f(a) = 0 \qquad \qquad \qquad \qquad \qquad \dots(i)$$

Also $\lim_{h \to 0} f(a+h) = \lim_{h \to 0} \frac{1}{a+h-a} \operatorname{cosec}\left(\frac{1}{a+h-a}\right)$

$$= \lim_{h \to 0} \frac{1}{h} \operatorname{cosec}\left(\frac{1}{h}\right) = \lim_{h \to 0} \left[\frac{1}{h \sin(1/h)}\right]$$

$$= \frac{1}{0'} \qquad \because \quad \lim_{h \to 0} h \sin\left(\frac{1}{h}\right) \to 0,$$

$$= \infty \qquad \qquad \qquad \qquad \qquad \dots(ii)$$

Also $\lim_{h \to 0} f(a-h) = \lim_{h \to 0} \frac{1}{a-h-a} \operatorname{cosec}\left(\frac{1}{a-h} - a\right)$

$$= \lim_{h \to 0} \frac{1}{h} \operatorname{cosec}\left(\frac{1}{h}\right) = \infty, \text{ as above} \qquad \dots(iii)$$

From (i), (ii) and (iii) we find that right and left hand limits of $f(x)$ as $x \to a$ though equal to each other are not equal to the value of $f(x)$ at $x = a$, i.e., $f(a)$.

Hence the given function $f(x)$ is discontinuous at $x = a$.

EXERCISE

1. The function y of x is defined as follows :

$y = 5x - 4$ when $0 < x \le 1$

$\quad = 4x^3 - 3x$ when $1 < x < 2.$

Examine whether or not it is continuous at $x = 1$.   **[Ans.** discontinuous**]**

2. Show that the function $f(x) = x^2$ when $x \ne 1$ when $x = 1$ is discontinuous at $x = 1$.

3. Prove that $\sin^2 x$ is continuous for every value of x.

4. Show that $f(x) = |x|$ is continuous at $x = 0$.

5. Show that $x = a, f(x)$ is continuous when

$$f(x) = (x - a) \sin \frac{1}{x-a} \quad x \ne a$$

and $f(a) = 0$.

6. Show that $f(x) = \sin(1/x)$ when $x \ne 0$ and $f(0) = 0$ at $x = 0$ is discontinuous.

2

Differentiability

2.1 CONCEPT OF THE DERIVATIVE

Here we shall consider differentiability of some continuous functions, which is the aim of differential calculus. In this chapter our aim is to find how the value of a function varies when the value of the function on which it depends varies. In short, *'if the variable y is a continuous function of the variable x, then what will be the change in y relative to x'*. Thus the main idea of differential calculus is to find the rate of change in the variable y which depends on another independent variable x.

Thus our aim is to find a function from the relation between two variables which depicts fully 'the idea of the rate of change'. This 'derived function is called the **derivative** of the 'original function' and we say that the original function has been **differentiated.**

2.2 DERIVATIVE

If $\lim_{h \to 0} \dfrac{f(a+h) - f(a)}{h}$ and $\lim_{h \to 0} \dfrac{f(a-h) - f(a)}{(-h)}$ both exist and are

(i) each equal to a finite number A, or

(ii) they are both infinite with the same sign, then number A or $+\infty$ or $-\infty$ as the case may be, is said to be the **Differential Coefficient** or the **Derivative** of $f(x)$ at the point $x = a$.

Then the function $f(x)$ is said to be **differentiable** or **derivable** at $x = a$.

Non-Differentiability

$f(x)$ is said to be **non-differentiable** at $x = a$ if either or both of the above two limits do not exist, or if they are different when they both exist.

2.2.1 Progressive and Regressive Derivatives or Differential Coefficeints

h being taken as positive, if the two limits

$$\lim_{h \to 0} \frac{f(a+h) - f(a)}{h} \quad \text{and} \quad \lim_{h \to 0} \frac{f(a-h) - f(a)}{-h}$$

exist but are different, then the former is the **Progressive** and the latter the **Regressive** Differential coefficient of $f(x)$ at $x = a$.

The progressive differential coefficient of $f(x)$ at $x = a$ is otherwise called the **Derivative of** $f(x)$ **on the right** at $x = a$ and is denoted by $Rf'(a)$, whereas the regressive differential coefficient is called the **Derivative of** $f(x)$ **on the left** at $x = a$ and is denoted by $Lf'(a)$. Thus

$$Rf'(a) = \lim_{h \to 0} \frac{f(a+h) - f(a)}{h};$$

$$Lf'(a) = \lim_{h \to 0} \frac{f(a-h) - f(a)}{-h}$$

where h is positive.

2.2.2 Upper and Lower Derivatives on the Right and on the Left

If $\lim\limits_{h \to 0} \dfrac{f(a+h) - f(a)}{h}$ (where h is positive) does not exist as a definite limit but assumes various values depending upon the manner in which h tends to the limit zero, the highest of these values is called the **Upper Derivative on the right** and is denoted by $D^+ f(a)$, while the lowest of the values is called **Lower Derivative on the right** and is denoted by $D_+ f(a)$.

If the $\lim\limits_{h \to 0} \dfrac{f(a-h) - f(a)}{-h}$ (where h is positive) does not exist as one definite limit but assumes various values, the highest of these is called the **Upper derivative on left** and is denoted by $D^- f(a)$, whereas the lowest of these values is called **Lower Derivative on the left** and is denoted by $D_- f(a)$.

2.3 A NECESSARY CONDITION FOR THE EXISTENCE OF A FINITE DERIVATIVE

Theorem. *Continuity is a necessary but not a sufficient condition for the existence of a finite derivative.*

Proof. Let f be differebtiable at $x = a$. Then $\lim\limits_{x \to a} \dfrac{f(x) - f(a)}{x - a}$ exists and equals $f'(a)$. Now, we can write

$$f(x) - f(a) = \dfrac{f(x) - f(a)}{x - a}(x - a) \text{ if } x \neq a.$$

Taking limits as $x \to a$, we get

$$\lim_{x \to a} [f(x) - f(a)] = \lim_{x \to a} \left\{ \dfrac{f(x) - f(a)}{(x - a)}(x - a) \right\}$$

$$= \lim_{x \to a} \dfrac{f(x) - f(a)}{x - a} \cdot \lim_{x \to a} (x - a)$$

$$= f'(a) \cdot 0 = 0$$

so that, $\lim\limits_{x \to a} f(x) = f(a).$

Hence f is continuous at a. Thus continuity is a necessary condition for differentiability but it is not a sufficient condition for the existence of a finite derivative. The following example illustrates the fact :

Let $\quad f(x) = x \sin\left(\dfrac{1}{x}\right), x \neq 0 \quad$ and $\quad f(a) = 0.$

This function is continuous at $x = 0$.

but not differentiable at $x = 0$, as

$$Rf'(0) = \lim_{h \to 0} \dfrac{f(0+h) - f(0)}{h}$$

$$= \lim_{h \to 0} \dfrac{f(h) - f(0)}{h}$$

$$= \lim_{h \to 0} \dfrac{h \sin(1/h) - 0}{h}$$

Differentiability

$$= \lim_{h \to 0} \sin \frac{1}{h}.$$

which does not exist. Similarly $Lf'(0)$ does not exist.

Thus $f(x)$ is not differentiable at $x = 0$, though it is continuous there.

2.4 MEANING OF THE SIGN OF DERIVATIVE

Let $y = f(x)$ be a function of x. Let δy be the increment in y corresponding to an increment δx in x.

If δy and δx be of the same sign then the function $f(x)$ or y is said to be an *increasing function at the point x*. In this case y increases as x increses and y decreases as x decreases.

On the other hand, if δy and δx are of opposite signs, then the function $f(x)$ or y is said to be *decreasing function at the point x*. In this case y decreases as x increases and y increases as x decreases.

If $f(x)$ is increasing function at the point x, then $f'(x) > 0$ and if $f(x)$ is decreasing function at the point x, then $f'(x) < 0$.

Also if $f'(x) > 0$ for all x in (a, b), then $f'(x)$ is an increasing function of x for each x in (a, b) and $f(b) > f(a)$.

2.4.1 Geometrical Meaning of Derivative

Let $P(x, y)$ and $Q(x + \delta x, y + \delta y)$ be two neighbouring points on a curve
$$y = f(x).$$
Then the equation of the line PQ is
$$Y - y = \frac{(y + \delta y) - y}{(x + \delta x) - x}(X - x),$$
where X and Y are the current coordinates.

or $\qquad Y - y = \dfrac{\delta y}{\delta x}(X - x) \qquad$ ...(i)

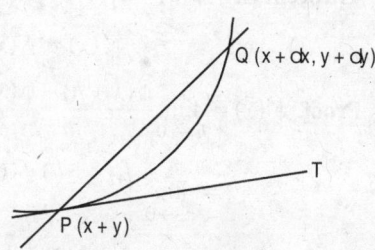

As Q tends to P (*i.e.*, Q approaches P) along the curve, the straight line PQ tends to a definite straight line PT called the tangent to the curve at P.

Now as $Q \to P$, $\delta y \to 0$, $\delta x \to 0$,

$$\frac{\delta y}{\delta x} \to \frac{dy}{dx}$$

and the line PQ tends to the tangent PT to the curve at P, so from (i) the equation of the tangent to the curve at P is

$$Y - y = \frac{dy}{dx}(X - x)$$

or, $\qquad Y = \left(\dfrac{dy}{dx}\right) X + \left(y - x \dfrac{dy}{dx}\right) \qquad$ ...(ii)

which is of the form
$$Y = mX + c, \qquad \text{...(iii)}$$

which represents a straight line whose gradient is m, *i.e.*, the line makes an angle, with the positive direction of X-axis, whose tangent is m.

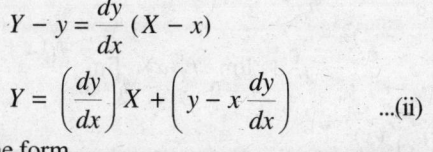

Hence, if the tangent PT makes an angle ψ with the positive direction of X-axis, then comparing (ii) and (iii), we get

$$\frac{dy}{dx} = \tan \psi,$$

i.e., the differential coefficient of $\frac{dy}{dx}$ at the point $P(x, y)$ of a curve $y = f(x)$ is the tangent of the angle ψ which the tangent to the curve at P makes with the positive direction of x-axis.

2.5 ALGEBRA OF DERIVATIVES

If $f_1(x), f_2(x)$ are derivable for every point x of a certain domain, then

Theorem 1. If $\phi(x) = f_1(x) \pm f_2(x)$, then $\phi'(x) = f_1'(x) \pm f_2'(x)$

Proof.
$$\phi'(x) = \lim_{h \to 0} \frac{\phi(x+h) - \phi(x)}{h}$$

$$= \lim_{h \to 0} \frac{[f_1(x+h) \pm f_2(x+h)] - [f_1(x) \pm f_2(x)]}{h}$$

$$= \lim_{h \to 0} \frac{[f_1(x+h) - f_1(x)] \pm [f_2(x+h) - f_2(x)]}{h}$$

$$= f_1'(x) \pm f_2'(x).$$

$\therefore \quad \phi'(x) = f_1'(x) \pm f_2'(x).$

Theorem II. If $\phi(x) = f_1(x) \cdot f_2(x)$; then

$$\phi'(x) = f_1(x) f_2'(x) + f_2(x) \cdot f_1'(x).$$

Proof. $\phi'(x) = \lim_{h \to 0} \frac{\phi(x+h) - \phi(x)}{h}$

$$= \lim_{h \to 0} \frac{f_1(x+h) f_2(x+h) - f_1(x) f_2(x)}{h}$$

$$= \lim_{h \to 0} \frac{f_1(x+h) f_2(x+h) - f_1(x+h) f_2(x) + f_1(x+h) f_2(x) - f_1(x) f_2(x)}{h}$$

$$= \lim_{h \to 0} \frac{f_1(x+h)\{f_2(x+h) - f_2(x)\} + f_2(x)\{f_1(x+h) - f_1(x)\}}{h}$$

$$= \lim_{h \to 0} f_1(x+h) \left\{ \frac{f_2(x+h) - f_2(x)}{h} \right\} + \lim_{h \to 0} f_2(x) \left\{ \frac{f_1(x+h) - f_1(x)}{h} \right\}$$

$$= \lim_{h \to 0} f_1(x+h) \cdot \lim_{h \to 0} \frac{f_2(x+h) - f_2(x)}{h}$$

$$\quad + \lim_{h \to 0} f_2(x) \cdot \lim_{h \to 0} \frac{f_1(x+h) - f_2(x)}{h}$$

$$= f_1(x) f_2'(x) + f_2(x) f_1'(x).$$

$\therefore \quad \phi'(x) = f_1(x) f_2'(x) + f_2(x) f_1'(x).$

Theorem III. If $\phi(x) = \dfrac{f_1(x)}{f_2(x)}$, and $f_2(x) \neq 0$, then

Differentiability

$$\phi'(x) = \frac{f_2(x) f_1'(x) - f_1(x) f_2'(x)}{\{f_2(x)\}^2}$$

Proof. We have,

$$\phi'(x) = \lim_{h \to 0} \frac{\phi(x+h) - \phi(x)}{h}$$

$$= \lim_{h \to 0} \frac{\left\{\dfrac{f_1(x+h)}{f_2(x+h)} - \dfrac{f_1(x)}{f_2(x)}\right\}}{h}$$

$$= \lim_{h \to 0} \frac{f_1(x+h) f_2(x) - f_1(x) f_2(x+h)}{h f_2(x) f_2(x+h)}$$

$$= \lim_{h \to 0} \frac{f_1(x+h) f_2(x) - f_1(x) f_2(x) + f_1(x) f_2(x) - f_1(x) f_2(x+h)}{h f_2(x) f_2(x+h)}$$

$$= \lim_{h \to 0} \frac{f_2(x)\{f_1(x+h) - f_1(x)\} - f_1(x)\{f_2(x+h) - f_2(x)\}}{h f_2(x) f_2(x+h)}$$

$$= \lim_{h \to 0} \frac{1}{f_2(x) f_2(x+h)} \left[\frac{f_2(x)\{f_1(x+h) - f_1(x)\}}{h} - \frac{f_1(x)\{f_2(x+h) - f_2(x)\}}{h} \right]$$

$$= \frac{1}{f_2(x) \cdot f_2(x)} [f_2(x) f_1'(x) - f_1(x) f_2'(x)]$$

$$= \frac{f_2(x) f_1'(x) - f_1(x) f_2'(x)}{\{f_2(x)\}^2}$$

Theorem IV. If $\phi(x) = \psi[f(x)]$, then $\phi'(x) = \psi'[f(x)] f'(x)$. **(Chain Rule)**

Proof. Let $f(x) = t$, then $f(x+h) = t + \delta t$,

Then
$$\phi'(x) = \lim_{h \to 0} \frac{\phi(x+h) - \phi(x)}{h}$$

$$= \lim_{h \to 0} \frac{\psi[f(x+h)] - \psi[f(x)]}{h}$$

$$= \lim_{h \to 0} \frac{\psi(t + \delta t) - \psi(t)}{\delta t} \times \frac{\delta t}{h}$$

$$= \lim_{h \to 0} \frac{\psi(t + \delta t) - \psi(t)}{\delta t} \times \frac{f(x+h) - f(x)}{h}$$

Now as $h \to 0$, $\delta t \to 0$.

$$\therefore \quad \phi'(x) = \lim_{\delta t \to 0} \frac{\psi(t + \delta t) - \psi(t)}{\delta t} \times \lim_{h \to 0} \frac{f(x+h) - f(x)}{h}$$

$$= \psi'(t) f'(x)$$

$$\therefore \quad \phi'(x) = \psi'[f(x)] f'(x).$$

2.6 DERIVATIVE OF THE INVERSE FUNCTION

Theorem. *If f be a continuous one-to-one function defined on an interval and let f be differentiable at x_0, with $f'(x_0) \neq 0$, then the inverse of the function f is differentiable at $f(x_0)$ and its derivative at $f(x_0)$ is $1/f'(x_0)$.*

Proof. Before going through the actual proof of the theorem we should note that if the domain of f be X and its range be Y, then the inverse function g of f usually denoted by f^{-1} is the function with domain Y and range X such that $f(x) = y \Leftrightarrow g(y) = x$. Again g exists if f is one-one.

Let $\quad y = f(x) \quad$ and $\quad y_0 = f(x_0)$.

Since f is differentiable at x_0, we have

$$\lim_{x \to x_0} \frac{f(x) - f(x_0)}{x - x_0} = f'(x_0)$$

or $\quad f(x) - f(x_0) = (x - x_0)[f'(x_0) + \lambda(x)] \quad$ ...(i)

where $\lambda(x) \to 0$ as $x \to x_0$. Further we have

$$g(y) - g(y_0) = (x - x_0), \text{ by definition of } g.$$

$$\therefore \quad \frac{g(y) - g(y_0)}{y - y_0} = \frac{x - x_0}{y - y_0} = \frac{x - x_0}{f(x) - f(x_0)}$$

$$= \frac{1}{f'(x_0) + \lambda(x)}, \quad \text{by (i)}$$

It can easily be seen that if $y \to y_0$, then $x \to x_0$.

In fact, f is continuous at x_0 implies that $g = f^{-1}$ is continuous as $f(x_0) = y_0$ and consequently

$$g(y) \to g(y_0) \text{ as } y \to y_0$$

i.e., $x \to x_0$ as $y \to y_0$, so that $\lambda(x) \to 0 \quad$ as $\quad y \to y_0$.

$$\therefore \quad \lim_{y \to y_0} \frac{g(y) - g(y_0)}{y - y_0} = \lim_{y \to y_0} \frac{1}{f'(x_0) + \lambda(x)}$$

$$= \frac{1}{f'(x_0)}$$

or $\quad g'(y_0) = \dfrac{1}{f'(x_0)} \quad$ or $\quad g'\{f(x_0)\} = \dfrac{1}{f'(x_0)}$.

2.7 INTERMEDIATE VALUE THEOREM FOR DERIVATIVES OR DARBOUX THEOREM

Theorem. *If f is finitely differentiable in a closed interval $[a, b]$ and $f'(a), f'(b)$ are of opposite signs, then there exists at least one point $c \in [a, b]$, such that $f'(c) = 0$.*

Proof. Let $f'(a) > 0$ and $f'(b) < 0$. Then there exists open intervals $(a, a+h)$ and $(b-h, b), h$ being positive such that

$$f(x) > f(a) \; \forall \; x \in (a, a+h) \quad \text{...(i)}$$
$$f(x) > f(b) \; \forall \; x \in (b-h, b) \quad \text{..(ii)}$$

Differentiability

Frther f being finitely differentiable, is continuous in (a, b) and hence it is bounded on (a, b) and attains its each bound at some point of (a, b).

Thus if M be the least upper bound of f in $[a, b]$, then there exists $c \in (a, b)$ such that $f(c) = M$. It is clear from (i) and (ii) that the upper bound is not attained at the end points a and b so that $c \in (a, b)$. We should prove that $f'(c) = 0$.

If $f'(c) > 0$, then there exists an interval $(c, c+h), h > 0$ such that $f(x) > f(c) = M \ \forall \ x \in (c, c+h)$, which is not possible since M is the least upper bound of the function $f(x)$ is (a, b).

If $f'(c) < 0$, then there exsits an interval $(c-h, c), h > 0$ such that $f(x) > f(c) = M \ \forall \ x \in (c-h, c)$, which is again not possible for the some reason as mentioned earlier.

Hence we conclude that $f'(c) = 0$.

Corollary 1. *If f is finitely differentiable on $[a,b]$ and $f'(a) \neq f'(b)$ and k is any number lying between $f'(a)$ and $f'(b)$, there exist at least one point $c \in (a, b)$ such that $f'(c) = k$. In other words $f'(x)$ takes every value intermeidate between $f'(a)$ and $f'(b)$.*

Proof. Let k be any real number lying between $f'(a)$ and $f'(b)$. Let us define a function ϕ as $\phi(x) = f(x) - kx$. Since f and kx both are finitely differentiable in (a, b), ϕ is also finitely differentiable is (a, b).

We have $\qquad \phi'(x) = f'(x) - k \ \forall \ x \in (a, b)$.

Hence $\qquad \phi'(a) = f'(a) - k$ and $\phi'(b) = f'(b) - k$.

Since k lies between $f'(a)$ and $f'(b)$, $\phi'(a)$ and $\phi'(b)$ are of opposite signs. Hence by the above theorem, there exist at least one point c of (a,b) such that $\phi'(c) = 0$

or $\qquad f'(c) - k = 0 \qquad$ or $\qquad f'(c) = k$.

Corollary 2. *If f is finitely differentiable on (a,b) and $f'(x) \neq 0$ for any $x \in (a, b)$ then $f'(x)$ retains the same sign, positive or negative in open interval (a, b), i.e., $f'(x)$ is either positive or negative for all values of $x \in (a, b)$.*

Proof. If possible, let x_1 and x_2 be two distinct elements of (a, b) and let $f'(x_1)$ and $f'(x_2)$ be of opposite signs. Then by the above theorem, there exists $c \in (a, b)$ such that $f'(c) = 0$ which contradicts the fact that $f'(x) \neq 0 \ \forall \ x \in (a, b)$. Hence $f'(x)$ must retain the same sign in (a, b).

Corollary 3. *If f is finitely differentiable on $I = [a, b]$, then the range $f'(I)$ of f' on I is either an interval or a singleton.*

Proof. Let $f'(I) = I$ and let p_1, p_2 be two distinct elements of J. Then ther exists two distinct element x_1, x_2 of I such that $f'(x_1) = p_1$ and $f'(x_2) = p_2$. Suppose that $x_1 < x_2$. Then
$$[x_1, x_2] \subset (a, b).$$

Let p be any real number lying between p_1 and p_2. Then, by the corrollary 1, there exists $c \in (x_1, x_2) \subset (a, b)$ such that $f'(c) = p$. Hence $p \in J$. This shows that every number lying between p_1 and p_2 belongs to J. Hence J is an interval.

If J does not contain at least two distinct elements, then obviously it is a singleten.

EXAMPLES

1. If $f(x) = x^2 \sin(1/x)$, for $x \neq 0$ and $f(0) = 0$, then show that $f(x)$ is continuous and differentiable everywhere and that $f'(0) = 0$. Also show that the function $f'(x)$ is discontinuous at $x = 0$.

Solution. We have,

$$f(0+0) = \lim_{h \to 0} f(0+h) = \lim_{h \to 0} (0+h)^2 \sin \frac{1}{0+h}$$

$$= \lim_{h \to 0} h^2 \sin \frac{1}{h} = 0;$$

$$f(0-0) = \lim_{h \to 0} f(0-h) = \lim_{h \to 0} f(-h)$$

$$= \lim_{h \to 0} (-h)^2 \sin(-1/h)$$

$$= -\lim_{h \to 0} h^2 \sin \frac{1}{h} = 0.$$

$\therefore \quad f(0+0) = f(0-0) = f(0)$, so the function is continuous at $x = 0$.

Now,

$$Rf'(0) = \lim_{h \to 0} \frac{f(0+h) - f(0)}{h}$$

$$= \lim_{h \to 0} \frac{f(h) - f(0)}{h}$$

$$= \lim_{h \to 0} \frac{h^2 \sin(1/h) - 0}{h}$$

$$= \lim_{h \to 0} h \sin(1/h) = 0;$$

and

$$Lf'(0) = \lim_{h \to 0} \frac{f(0-h) - f(0)}{-h}$$

$$= \lim_{h \to 0} \frac{(-h)^2 \sin(-1/h) - 0}{-h}$$

$$= \lim_{h \to 0} h \sin(1/h) = 0.$$

Thus $Rf'(0) = Lf'(0)$ implies that $f(x)$ is differentiable at $x = 0$ and $f'(0) = 0$.

For all other values of x, $f(x)$ is easily seen to be continuous and differentiable.

Now, $\qquad f'(x) = 2x \sin \frac{1}{x} - \cos \frac{1}{x}$ at $x \neq 0$ and $f'(0) = 0$.

$\therefore \qquad f'(0+0) = \lim_{h \to 0} f'(0+h)$

$$= \lim_{h \to 0} \left(2h \sin \frac{1}{h} - \cos \frac{1}{h} \right),$$

which does not exist.

Similarly it can be shown that $f'(0-0)$ does not exist.

Hence f' is discontinuous at the origin.

Differentiability

2. *Show that the function* $f(x) = |x| + |x-1|$ *is not differentiable at* $x=0$ *and* $x=1$.

Solution. We know that if $x < 0$, then $|x| = x$ and
$$|x-1| = |1-x| = 1-x;$$
and if $x > 1$, then $|x| = x$ and $|x-1| = x-1$.

$\therefore$ then function $f(x)$ is given by
$$f(x) = 1 - 2x, \quad \text{if} \quad x < 0$$
$$= 1, \quad \text{if} \quad 0 \le x \le 1$$
$$= 2x - 1, \quad \text{if} \quad x > 1.$$

At $x = 0$. We have

$$Rf'(0) = \lim_{h \to 0} \frac{f(0+h) - f(0)}{h}$$

$$= \lim_{h \to 0} \frac{f(h) - f(0)}{h} = \lim_{h \to 0} \frac{1-1}{h}, \text{ as } f(x) = 1 \text{ if } 0 \le x \le 1$$

$$= \lim_{h \to 0} 0 = 0,$$

and $\quad Lf'(0) = \lim_{h \to 0} \frac{f(0-h) - f(0)}{-h}$

$$= \lim_{h \to 0} \frac{f(-h) - f(0)}{-h} = \lim_{h \to 0} \frac{[1 - 2(-h)] - 1}{-h}$$

$$[\because f(x) = 1 - 2x, \text{ if } x < 0]$$

$$= \lim_{h \to 0} \frac{2h}{-h} = \lim_{h \to 0} -2 = -2.$$

$\therefore$ $Rf'(0) \ne Lf'(0)$, so the given function is not differentiable at $x = 0$.

At $x = 1$. We have

$$Rf'(1) = \lim_{h \to 0} \frac{f(1+h) - f(1)}{h}$$

$$= \lim_{h \to 0} \frac{[2(1+h) - 1] - 1}{h} = \lim_{h \to 0} \frac{2 + 2h - 1 - 1}{h}$$

$$= \lim_{h \to 0} \frac{2h}{h} = \lim_{h \to 0} 2 = 2,$$

and $\quad Lf'(1) = \lim_{h \to 0} \frac{f(1-h) - f(1)}{-h}$

$$= \lim_{h \to 0} \frac{1-1}{-h} = \lim_{h \to 0} 0 = 0$$

$\therefore$ $Rf'(1) \ne Lf'(1)$, so the given function $f(x)$ is not differentiable at $x = 1$.

EXERCISES

1. If $f(x) = \dfrac{x}{1+e^{1/x}}$, show that there is a discontinuity in $f'(x)$ as x passes through zero.

2. $f(x) = x^2 \sin 1/x$, $f(0) = 0$, show that the differential coefficient $f'(x)$ exists everywhere and is finite. At the point $x = 0$, $f'(x) = 0$. Also show that $f'(x)$ is discontinuous at $x = 0$.

3. Test the character of $f(x)$ as regards continuity and differentiability at $x = 1$, where
   $f(x) = x, 0 \le x \le 1.$
   $f(x) = 2 - x, 1 \le x,$ [**Ans.** Continuous and differentiable]

4. Prove that the function $f(x) = |x|$ is continuous at $x = 0$, but not differentiable at $x = 0$, where $|x|$ means the numerical value or the absolute value of x.

3
Successive Differentiation

3.1 SUCCESSIVE DIFFERENTIAL COEFFICIENTS

If y is a function of x, i.e., $f(x)$ then its derivaitve $\dfrac{dy}{dx}$ is denoted by $f'(x)$.

i.e., $$\dfrac{dy}{dx} = f'(x)$$

Differentaiting both sides with respect to x, we obtain

$$\dfrac{d}{dx}\left(\dfrac{dy}{dx}\right) = \dfrac{d^2y}{dx^2} = \dfrac{d}{dx}\{f'(x)\} = f''(x)$$

then $$\dfrac{d^2y}{dx^2} = f''(x)$$

In like manner $\dfrac{d}{dx}\left(\dfrac{d^2y}{dx^2}\right)$ is represented by $\dfrac{d^3y}{dx^3}$ and so on.

Hence $$\dfrac{d^3y}{dx^3} = f'''(x)$$

Similarly $$\dfrac{d^n y}{dx^n} = f^{n'}(x)$$

The expressions

$$\dfrac{dy}{dx}, \dfrac{d^2y}{dx^2}, \dfrac{d^3y}{dx^3}, \ldots, \dfrac{d^n y}{dx^n}$$

are called the first, second, third, ... nth differential coefficients of y regarded as a function of x. These functions are sometimes represented by

$$y', y'', y''', \ldots y^{(n)'}$$

Also either of the following notations can be used to denote the nth derivative of $y = f(x)$.

$$f_n(x), f^n(x), D_n f(x), D^n f(x), \dfrac{d^n}{dx^n} f(x), y^n(x), D_n y(x) \dfrac{d^n}{dx^n} y(x) \ldots$$

On some occasions, we shall denote nth derivative by y_n.

3.2 CALCULATIONS OF nth DERIVATIVE

(I) Let $y = x^m$, then

$$\dfrac{dy}{dx} = mx^{m-1}, \dfrac{d^2y}{dx^2} = m(m-1)x^{m-2}$$

etc., and in general

$$\dfrac{d^n y}{dx^n} = m(m-1)(m-2)\ldots(m-n+1)x^{m-n}$$

$$\therefore \quad D^n(x^m) = m(m-1)(m-2)\ldots(m-n+1)x^{m-n}$$

If n be a positive integer, then we have,

$$D^n(x^n) = 1.2.3\ldots n = n!$$

(II) Let $y = (ax+b)^m$,

then
$$\frac{dy}{dx} = m(ax+b)^{m-1} a$$

$$\frac{d^2 y}{dx^2} = m(m-1)(ax+b)^{m-2} a^2$$

....
....

$$\frac{d^n y}{dx^n} = m(m-1)(m-2)\ldots(m-n+1)(ax+b)^{m-n} a^n$$

$$\therefore \quad D^n(ax+b)^m = a^n m(m-1)(m-2)\ldots(m-n+1)(ax+b)^{m-n}$$

(III) Let $y = (ax+b)^{-1}$

then
$$y_1 = (-1) a (ax+b)^{-2}$$

$$y_2 = (-1)^2 a^2 (ax+b)^{-3} . 2$$

$$y_3 = (-1)^3 a^3 (ax+b)^{-4} . 2.3$$

....
....

$$y_n = (-1)^n a^n (ax+b)^{-n-1} . 2.3.4 \ldots n$$

$$\therefore \quad D^n (ax+b)^{-1} = (-1)^n a^n n! (ax+b)^{-n-1}$$

(IV) Let $y = \sin(ax+b)$

then
$$y_1 = a \cos(ax+b) = a \sin\left(ax+b+\frac{\pi}{2}\right)$$

$$y_2 = a^2 \cos\left(ax+b+\frac{\pi}{2}\right) = a^2 \sin\left(ax+b+\frac{2\pi}{2}\right)$$

$$y_3 = a^3 \cos\left(ax+b+\frac{2\pi}{2}\right) = a^3 \sin\left(ax+b+\frac{3\pi}{2}\right)$$

....
....

$$y_n = a^n \sin\left(ax+b+\frac{n\pi}{2}\right)$$

$$\therefore \quad D^n \sin(ax+b) = a^n \sin\left(ax+b+\frac{n\pi}{2}\right)$$

(V) Let $y = \cos(ax+b)$

then
$$y_1' = -a \sin(ax+b) = a \cos\left(ax+b+\frac{\pi}{2}\right)$$

$$y_2 = -a^2 \sin\left(ax+b+\frac{\pi}{2}\right) = a^2 \cos\left(ax+b+\frac{2\pi}{2}\right)$$

Successive Differentiation

$$y_3 = -a^3 \sin\left(ax + b + \frac{2\pi}{2}\right) = a^3 \cos\left(ax + b + \frac{3\pi}{2}\right)$$

....
....

$$y_n = a^n \cos\left(ax + b + \frac{n\pi}{2}\right)$$

$\therefore \quad D^n \cos(ax+b) = a^n \cos\left(ax + b + \frac{n\pi}{2}\right)$

(VI) Let $\quad y = e^{ax+b}$

then
$\quad y_1 = ae^{ax+b}$
$\quad y_2 = a^2 e^{ax+b}$
$\quad y_3 = a^3 e^{ax+b}$

....
....

$\quad y_n = a^n e^{ax+b}$

$\therefore \quad D^n e^{ax+b} = a^n e^{ax+b}$

(VII) Let $\quad y = a^{bx+c}$

then
$\quad y_1 = ba^{bx+c} \log_e a$
$\quad y_2 = b^2 a^{bx+c} (\log_e a)^2$
$\quad y_3 = b^3 a^{bx+c} (\log_e a)^3$

....
....

$\quad y_n = b^n a^{bx+c} (\log_e a)^n$

$\therefore \quad D^n (a^{bx+c}) = b^n (\log_e a)^n a^{bx+c}$

(VIII) Let $\quad y = \log(ax+b)$

then $\quad y_1 = \dfrac{a}{ax+b} = a \cdot (ax+b)^{-1}$ ...(1)

Now, we know that $D^n (ax+b)^{-1} = (-1)^n a^n n! (ax+b)^{-n-1}$ and then $D^{n-1}(ax+b)^{-1} = (-1)^{n-1} a^{n-1}!(ax+b)^{-n}$; differentiating eqn. (i) $(n-1)$ times more, we have

$$y_n = a \cdot (-1)^{n-1}(a^{n-1})(n-1)!(ax+b)^{-n}$$

$\therefore \quad D^n \{\log(ax+b)\} = \dfrac{a^n (-1)^{n-1}(n-1)!}{(ax+b)^n}$

(IX) Let $\quad y = e^{ax} \sin(bx+c)$

then $\quad y_1 = ae^{ax} \sin(bx+c) + be^{ax} \cos(bx+c)$

Let $\quad a = r\cos\phi$ and $b = r\sin\phi$; then

$$\phi = \tan^{-1}\frac{b}{a} \quad \text{and} \quad r = \sqrt{(a^2+b^2)}$$

$\therefore$ $\quad y_1 = e^{ax} [r \cos \phi \sin (bx+c) + r \sin \phi \cos (bx+c)]$

$\quad\quad = e^{ax} r \sin (bx + c + \phi)$

Similarly, $\quad y_2 = e^{ax} r^2 \sin (bx + c + 2\phi)$

$\quad\quad y_3 = e^{ax} r^3 \sin (bx + c + 3\phi)$

$\quad\quad \cdots\cdots\cdots\cdots\cdots\cdots\cdots$

$\quad\quad y_n = e^{ax} r^n \sin (bx + c + n\phi)$

$\therefore$ $\quad D^n e^{ax} \sin (bx + c) = e^{ax} (a^2 + b^2)^{n/2} \sin \left(bx + c + n \tan^{-1} \dfrac{b}{a} \right)$

(X) Let $\quad y = e^{ax} \cos (bx + c)$

then $\quad y_1 = a e^{ax} \cos (bx + c) - b e^{ax} \sin (bx + c)$

Let $\quad a = r \cos \phi \quad$ and $\quad b = r \sin \phi;$

then $\quad \phi = \tan^{-1} \dfrac{b}{a} \quad$ and $\quad r = \sqrt{(a^2 + b^2)}$

$\therefore$ $\quad y_1 = e^{ax} r [\cos \phi \cos (bx + c) - \sin \phi \sin (bx + c)]$

$\quad\quad = e^{ax} r \cos (bx + c + \phi)$

Similarly, $\quad y_2 = e^{ax} r^2 \cos (bx + c + 2\phi)$

$\quad\quad y_3 = e^{ax} r^3 \cos (bx + c + 3\phi)$

$\quad\quad \cdots\cdots\cdots\cdots\cdots\cdots\cdots$

$\quad\quad y_n = e^{ax} r^n \cos (bx + c + n\phi)$

$\therefore$ $\quad D^n e^{ax} \cos (bx + c) = e^{ax} (a^2 + b^2)^{n/2} \cos \left(bx + c + n \tan^{-1} \dfrac{b}{a} \right)$

EXAMPLES

1. Find y_n when $y = \dfrac{x}{x^2 + a^2}$

Solution. We have $\quad \dfrac{x}{x^2 + a^2} = \dfrac{x}{(x - ai)(x - ai)}$

$\quad\quad = \dfrac{1}{2} \dfrac{1}{(x - ai)} + \dfrac{1}{2} \dfrac{1}{(x + ai)}$

$\therefore$ $\quad y = \dfrac{1}{2} (x - ai)^{-1} + \dfrac{1}{2} (x + ai)^{-1}$

Differentiating n time, we obtain

$\quad y_n = \dfrac{1}{2} (-1)^n n! (x - ai)^{-n-1} + \dfrac{1}{2} (-1)^n n! (x + ai)^{-n-1}$

Now, we can make use of De Moivre's Theorem and can render the result free from 'i'

Let $\quad x = r \cos \theta \quad$ and $\quad a = r \sin \theta;$

then $\quad \theta = \tan^{-1} \dfrac{a}{x}, r = \sqrt{(a^2 + x^2)}$

Successive Differentiation

$$\therefore \quad y_n = \frac{1}{2}(-1)^n \, n! \, \frac{1}{r^{n+1}} [(\cos\theta - i\sin\theta)^{-n-1} + (\cos\theta + i\sin\theta)^{-n-1}]$$

$$= \frac{1}{2} \frac{(-1)^n \, n!}{r^{n+1}} [2\cos(n+1)\theta]$$

$$= (-1)^n \cdot \frac{n! \cos(n+1)\theta}{r^{n+1}}$$

where $r = \sqrt{(a^2 + x^2)}$ and $\theta = \tan^{-1}(a/x)$

2. If $y = (x^2 - 1)^n$, prove that $y_{2n} = (2n)!$

Solution. We have

$$y = (x^2 - 1)^n$$

$$= x^{2n} - n \cdot x^{2n-2} + \frac{n(n-1)}{2!} x^{2n-4} - \dots \qquad \text{(By Binomial Theorem)}$$

Differentaiting $2n$ times, we have

$$y_{2n} = (2n)! - 0 + 0 - \dots$$

$$= (2n)!$$

3. Find the nth differential coefficient of $\tan^{-1}\frac{x}{a}$.

Solution. We have, $\qquad y = \tan^{-1}\frac{x}{a}$

$$\therefore \quad y_1 = \frac{1}{1 + x^2/a^2} \cdot \frac{1}{a} = \frac{a}{a^2 + x^2}$$

$$= \frac{1}{2i} \left[\frac{1}{x - ai} - \frac{1}{x + ai} \right]$$

$$= \frac{1}{2i} [(x - ai)^{-1} - (x + ai^{-1})]$$

Now differentaiting $(n - 1)$ times more, we get

$$y_n = \frac{1}{2i} [(-1)^{n-1} (n-1)!(x - ai)^{-n} - (-1)^{n-1} (n-1)!(x + ai)^{-n}]$$

$$= \frac{1}{2i} (-1)^{n-1} (n-1)! [(x - ai)^{-n} + (x + ai)^{-n}]$$

Put $\qquad x = r\cos\theta$ and $a = r\sin\theta$

then $\qquad \theta = \tan^{-1}\frac{a}{x}$ and $r = \frac{a}{\sin\theta}$

therefore, $\quad y_n = \frac{1}{2i} (-1)^{n-1} (n-1)! \, r^{-n} [(\cos\theta - i\sin\theta)^{-n} - (\cos\theta + i\sin\theta)^{-n}]$

$$= \frac{1}{2i} (-1)^{n-1} (n-1)! \, r^{-n} \cdot 2i \sin n\theta$$

$$= (-1)^{n-1} (n-1)! \, a^{-n} (\sin\theta)^n \sin n\theta$$

where $\qquad \theta = \tan^{-1} a/x$

EXERCISES

Find nth differential coefficient of :

1. $\dfrac{x^4}{(x-1)(x-2)}$ $\left[\text{Ans. } (-1)^n \, n! \left[\dfrac{16}{(x-2)^{n+1}} - \dfrac{1}{(x-1)^{n+1}} \right], n > 2 \right]$

2. $\dfrac{x^2 + 4x + 1}{x^3 + 2x^2 - x - 2}$ $\left[\text{Ans. } n!(-1)^n \left[\dfrac{1}{(x-1)^{n+1}} + \dfrac{1}{(x+1)^{n+1}} - \dfrac{1}{(x+2)^{n+1}} \right] \right]$

3. $\dfrac{x}{(x-a)(x-b)(x-c)}$ $\left[\text{Ans. } (-1)^n \, n! \left[\dfrac{a}{(a-c)(a-b)(x-a)^{n+1}} + ... \right] \right]$

4. $\tan^{-1} \dfrac{1+x}{1-x}$ $\left[\text{Ans. } (-1)^{n-1}.(n-1)! \sin^n \theta \sin n\theta, \text{ where } \theta = \tan^{-1}\left(\dfrac{1}{x}\right) \right]$

5. $\sin^{-1} \dfrac{2x}{1+x^2}$ $\left[\text{Ans. } 2(-1)^{n-1}.(n-1)! \sin^n \alpha \sin n\alpha, \text{ where } \alpha = \tan^{-1}\left(\dfrac{1}{x}\right) \right]$

6. $\cos^{-1} \dfrac{1-x^2}{1+x^2}$ $\left[\text{Ans. } 2(-1)^{n-1}(n-1)! \sin^n \alpha \sin n\alpha, \text{ where } \alpha = \tan^{-1}\left(\dfrac{1}{x}\right) \right]$

7. $\cos x \cos 2x \cos 3x$

$\left[\text{Ans. } \dfrac{1}{4}\left[2^n \cos\left(2x + \dfrac{n\pi}{2}\right) + 4^n \cos\left(4x + \dfrac{n\pi}{2}\right) + 6^n \cos\left(6x + \dfrac{n\pi}{2}\right) \right] \right]$

8. $e^{ax} \cos x \sin^2 2x$

$\left[\text{Ans. } \dfrac{1}{4} e^{2x} \left[25^{n/2} \cos\left(x + n \tan^{-1} \dfrac{1}{2}\right) - 13^{n/2} \cos\left(3x + n \tan^{-1} \dfrac{3}{2}\right) \right.\right.$
$\left.\left. - 5^n \cos\left(5x + \dfrac{n\pi}{2}\right) \right] \right]$

9. If $y = \log(x^2 + a^2)$, prove that

$$y_n = \dfrac{2(-1)^{n-1}(n-1)! \sin^n \theta \cos n\theta}{a^n}$$

where $\theta = \cos^{-1}\left(\dfrac{x}{a}\right)$.

3.3 LEIBNITZ'S THEOREM

To find the nth differential coefficient of the product of two functions of x.

Let $y = uv$, where u and v are functions of x; then u_n and v_n denote nth derivative of u and v with regards to x. Then by Leibnitz's Theorem

$$D^n(uv) = D^n u \cdot v + {}^n C_1 \, D^{n-1} u \cdot Dv$$
$$+ {}^n C_2 \, D^{n-2} u \cdot D^2 v + ... + {}^n C_r D^{n-r} u \cdot D^r v + ... + u \cdot D^n v$$

Proof. We shall prove this theorem by method of mathematical induction.

We have $D(uv) = Du \cdot v + u \cdot Dv$

$D^2(uv) = D^2 u \cdot v + Du \cdot Dv + Du \cdot Dv + u \cdot D^2 v$

$= D^2 u \cdot v + 2 \, Du \cdot Dv + u \, D^2 v$

Successive Differentiation

Similarly, $\quad D^3(uv) = D^3u \cdot v + 3D^2u \cdot Dv + 3Du \cdot D^2v + u D^3v$

and $\quad D^4(uv) = D^4u \cdot v + 4D^3u \cdot Dv + 6D^2u \cdot D^2v + 4Du \cdot D^3v + u \cdot D^4v$

We observe that the coefficient are the same as those in expansion of $(a+b), (a+b)^2, (a+b)^3$ and $(a+b)^4$.

Suppose that the same law holds for nth differential coefficient and that

$$D^n(uv) = D^nu \cdot v + {}^nC_1 D^{n-1}u \cdot Dv + {}^nC_2 D^{n-2}u \cdot D^2v$$
$$+ \ldots + {}^nC_1 D^{n-r}u \cdot D^r v + \ldots + u D^n v$$

Differentaiting again, we obtain

$$D^{n+1}(uv) = [D^{n+1}u \cdot v + D^nu \cdot Dv] + {}^nC_1 [D^nu \cdot Dv + D^{n-1}D^2v]$$
$$+ {}^nC_2 [D^{n-1}u \cdot D^2u + D^{n-2}u \cdot D^3v]$$
$$+ \ldots + {}^nC_r [D^{n-r+1}u \cdot D^r v + D^{n-r} u D^{r+1}v]$$
$$+ \ldots + [Du \cdot D^n v + u D^{n+1}v]$$

Rearranging terms, we get

$$D^{n+1}(uv) = D^{n+1}u \cdot v + ({}^nC_1 + 1) D^n u \cdot Dv + ({}^nC_1 + {}^nC_2) D^{n-1}u \cdot D^2 v + \ldots$$
$$+ \left({}^nC_r + {}^nC_{r+1}\right) D^{n-r} u \cdot D^{r+1}v + \ldots + u D^{n+1}v$$

But $\quad {}^nC_r + {}^nC_{r+1} = \dfrac{n!}{r!(n-r)!} + \dfrac{n!}{(r+1)!(n-r-1)!}$

$$= \dfrac{n!(r+1+n-r)}{(r+1)!\{n-(r+1)\}!}$$

$$= {}^{n+1}C_{r+1}$$

Hence $D^{n+1}(uv) = D^{n+1} u \cdot v + {}^{n+1}C_1 D^n u \ldots Dv + {}^{n+1}C_2 D^{n-1}u \cdot D^2v$
$$+ \ldots + {}^{n+1}C_r D^{n-r} u \cdot D^{r+1}v + \ldots + u D^{n+1}v$$

Thus it can be seen that theorem which holds for a particular value of n, also holds for the next higher value of n, i.e., $(n+1)$.

This establishes the theorem.

EXAMPLES

1. If $y = x^2 e^x$, show that

$$\frac{d^n y}{dx^n} = \frac{n}{2}(n-1)\frac{d^2 y}{dx^2} - n(n-2)\frac{dy}{dx} + \frac{1}{2}(n-1)(n-2) y$$

Solution. We have $\dfrac{d^n y}{dx^n} = D^n (e^x \cdot x^2)$

$$= D^n e^x \cdot x^2 + n \cdot D^{n-1} e^x \cdot D x^2 + \frac{n(n-1)}{2!} D^{n-2} e^x D^2 x^2$$

$$= e^x \cdot x^2 + n e^x \cdot 2x + \frac{n(n-1)}{2!} e^x \cdot 2$$

$$= e^x [x^2 + 2nx + n(n-1)] \qquad \ldots(i)$$

Now, $y = x^2 e^x$,

$\therefore \quad \dfrac{dy}{dx} = e^x \cdot x^2 + e^x \cdot 2x = e^x(x^2 + 2x)$

$\dfrac{d^2 y}{dx^2} = e^x(x^2 + 2x) + e^x(2x + 2)$

Hence $\dfrac{n}{2}(n-1)\dfrac{d^2 y}{dx^2} - n(n-2)\dfrac{dy}{dx} + \dfrac{(n-1)(n-2)}{2} y$

$= \dfrac{n(n-1)}{2} e^x (x^2 + 4x + 2) - n(n-2) e^x (x^2 + 2x) + \dfrac{(n-1)(n-2)}{2} x^2 e^x$

$= e^x \left[\dfrac{n(n-1)}{2}(x^2 + 4x + 2) - n(n-2)(x^2 + 2x) + \dfrac{(n-1)(n-2)}{2} \right]$

$= e^x [x^2 + 2nx + n(n-1)]$ ...(ii)

The relations (i) and (ii) are the same.

2. If $I_n = \dfrac{d^n}{dx^n}(x^n \log x)$, prove that

$$I_n = nI_{n-1} + \{(n-1)!\}$$

Solution. We have

$I_n = \dfrac{d^n}{dx^n}(x^n \log x) = \dfrac{d^{n-1}}{dx^{n-1}}\left[\dfrac{d}{dx}(x^n \log x)\right]$

$= \dfrac{d^{n-1}}{dx^{n-1}}[nx^{n-1} \log x + x^{n-1}]$

$= n \dfrac{d^{n-1}}{dx^{n-1}}[x^{n-1} \log x] + \dfrac{d^{n-1}}{dx^{n-1}}(x^{n-1})$

$= nI_{n-1} + \{(n-1)!\}$

EXERCISES

1. Find the nth differential coefficient of $x^2 \tan^{-1} x$.

[**Ans.** $(-1)^{n-3}(n-3)! [(n-1)(n-2) x^2 \sin^n \theta \sin n\theta$
$- {}^nC_1 \cdot 2(n-2) x \sin^{n-1} \theta \sin(n-1)\theta + {}^nC_2 \cdot 2 \cdot \sin^{n-2} \theta \sin(n-2)\theta]$
where $\theta = \tan^{-1}(1/x)$

2. Find nth differential coefficient of $x^{n-1} \log x$. [**Ans.** $\dfrac{(n-1)!}{x}$]

[**Hint.** Let $y = x^{n-1} \log x$, then $\dfrac{dy}{dx} = x^{n-1} \cdot \dfrac{1}{x} + (n-1) x^{n-2} \log x$

or $xy_1 = x^{n-1} + (n-1) y$

Differentiate this $(n-1)$ times to obtain the required result.]

3. If $y = \dfrac{\log x}{x}$, prove that $y_n = \dfrac{(-1)^n \cdot n!}{x^{n+1}}\left[\log x - 1 - \dfrac{1}{2} - \dfrac{1}{3} - \ldots - \dfrac{1}{n}\right]$

Successive Differentiation

4. Prove that the nth derivative of $x^n(1-x)^n$ is

$$n!(1-x)^n \left[1 - \frac{n^2}{1^2} \cdot \frac{x}{(1-x)} + \frac{n^2(n-1)^2}{1^2 \cdot 2^2} \cdot \frac{x^2}{(1-x^2)} + \ldots \right]$$

5. If $x + y = 1$, prove that

$$\frac{d^n}{dx^n}(x^n y^n) = n![y^n - (^nC_1)^2 y^{n-1} x + (^nC_2)^2 y^{n-2} x^2 \ldots + (-1)^n x^n]$$

3.3.1 Use of Leibnitz's Theorem

EXAMPLES

1. If $y = [x + \sqrt{(x^2+1)}]^m$, prove that

$$(x^2+1) y_{n+2} + (2n+1) xy_{n+1} - (m^2 - n^2) y_n = 0$$

Solution. Given that $y = [x + \sqrt{(x^2+1)}]^m$

$$\therefore \quad y_1 = m[x + \sqrt{(x^2+1)}]^{m-1} \left[1 + \frac{x}{\sqrt{(x^2+1)}}\right]$$

$$= \frac{m}{\sqrt{(x^2+1)}}[x + \sqrt{(x^2+1)}]^m$$

or $\qquad \sqrt{(x^2+1)} \, y_1 = my$

or $\qquad (x^2+1) y_1^2 = m^2 y^2$

Differentiating again w.r.t. x, we get

$$2y_1 y_2 (1+x^2) + y_1^2 \cdot 2x = 2m^2 yy_1$$

or $\qquad (1+x^2) y_2 + xy_1 - m^2 y = 0$

Differentiating n times, we get

$$(1+x^2) y_{n+2} + ny_{n+1} \cdot 2x + \frac{n(n-1)}{2!} \cdot y_n \cdot 2 + xy_{n+1} + ny_n - m^2 y_n = 0$$

or $\qquad (x^2+1) y_{n+2} + (2n+1) xy_{n+1} - (m^2 - n^2) y_n = 0$

2. If $y = e^{a \sin^{-1} x}$, prove that

$$(1-x^2) y_{n+2} - (2n+1) xy_{n+1} - (n^2 + a^2) y_n = 0$$

Solution. We have $y = e^{a \sin^{-1} x}$

$$\therefore \quad y_1 = \frac{a}{\sqrt{(1-x^2)}} e^{a \sin^{-1} x}$$

or $\qquad \sqrt{(1-x^2)} \, y_1 = ay$

or $\qquad (1-x^2) y_1^2 = a^2 y^2$

then $\qquad (1-x^2) \cdot 2y_1 y_2 - 2xy_1^2 = 2a^2 yy_1$

or $\qquad (1-x^2) y_2 - xy_1 - a^2 y = 0$

Differentiating n times by Leibnitz's Theorem, we get

$$(1-x^2)\, y_{n+2} - 2nxy_{n+1} - \frac{2n(n-1)}{2!}\, y_n - xy_{n+1} - ny_n - a^2 y_n = 0$$

or $\qquad (1-x^2)\, y_{n+2} - (2n+1)\, xy_{n+1} - (n^2 + a^2)\, y_n = 0$

3. If $y^{1/m} + y^{-1/m} = 2x$, *prove that* $(x^2 - 1)\, y_{n+2} + (2n+1)\, xy_{n+1} + (n^2 - m^2)\, y_n = 0$, where y_n denotes the nth derivative of y.

Solution. We have

$$y^{1/m} + \frac{1}{y^{1/m}} = 2x$$

or $\qquad y^{2/m} - 2xy^{1/m} + 1 = 0$

or $\qquad y^{1/m} = \dfrac{2x \pm \sqrt{(4x^2 - 4)}}{2} = x \pm \sqrt{(x^2 - 1)}$

or $\qquad y = [x \pm \sqrt{(x^2 - 1)}]^m$

On differentiating, we get

$$y_1 = m[x \pm \sqrt{(x^2-1)}]^{m-1} \left[1 \pm \frac{2x}{2\sqrt{(x^2-1)}} \right]$$

$$= m[x \pm \sqrt{(x^2-1)}]^{m-1}\, \frac{[\sqrt{(x^2-1)} \pm x]}{\sqrt{(x^2-1)}}$$

or $\qquad \sqrt{(x^2-1)}\, y_1 = \pm m[x \pm \sqrt{x^2-1}]^{m-1}\, [x \pm \sqrt{x^2-1}]$

$\qquad\qquad = \pm m[x \pm \sqrt{x^2-1}]^m$

or $\qquad \sqrt{(x^2-1)}\, y_1 = \pm my$

or $\qquad (x^2 - 1)\, y_1^2 = m^2 y^2$

Differentiating both sides w.r.t. x, we get

$$y_1^2 \cdot 2x + 2y_1 y_2 (x^2 - 1) = m^2 \cdot 2yy_1$$

or $\qquad (x^2 - 1)\, y_2 + xy_1 - m^2 y = 0$

Now, differentaiting n times, we get

$$D^n[(x^2-1)\, y_2] + D^n[xy_1] - m^2 D^n(y) = 0$$

or $\left\{ (x^2-1)\, y_{n+2} + ny_{n+1}\, (2x) + \dfrac{n(n-1)}{2!}\, y_n\, (2) \right\} + \{xy_{n+1} + ny_n \cdot 1\} - m^2 y_n = 0$

or $\qquad (x^2 - 1)\, y_{n+2} + (2n+1)\, xy_{n+1} + (n^2 - m^2)\, y_n = 0$

EXERCISES

1. If $y = \sin^{-1} x$, prove that

$(1 - x^2)\, y_2 - xy_1 = 0$ and differentaite it n times with respect to x.

[**Ans.** $(1 - x^2)\, y_{n+2} - (2n+1)\, xy_{n+1} - n^2 y_n = 0$]

Successive Differentiation

2. If $y = \sin(m \sin^{-1} x)$, prove that $(1-x^2)y_2 - xy_1 + m^2 y = 0$ and deduce that
$(1-x^2)y_{n+2} - (2n+1)xy_{n+1} - (n^2 - m^2)y_n = 0$

3. If $y = e^{\tan^{-1} x}$, show that $(1+x^2)y_2 + (2x-1)y_1 = 0$, and
$(1+x^2)y_{n+2} - \{2(n+1)x - 1\}y_{n+1} + n(n+1)y_n = 0$

4. If $\cos^{-1}\dfrac{y}{b} = \log\left(\dfrac{x}{n}\right)^n$, prove that $x^2 y_{n+2} + (2n+1)xy_{n+1} + 2n^2 y_n = 0$

3.3.2 nth Differential Coefficeint When $x = 0$

EXAMPLES

1. Find $(y_n)_0$ if $y = [x + \sqrt{(x^2+1)}]^m$.

Solution. Given $\quad y = [x + \sqrt{(x^2+1)}]^m$

$\therefore \qquad y_1 = m[x + \sqrt{(x^2+1)}]^{m-1}\left[1 + \dfrac{x}{\sqrt{(x^2+1)}}\right]$

$\qquad y_1 \sqrt{(1+x^2)} = m[x + \sqrt{(x^2+1)}]^m$

or $\qquad y_1^2(1+x^2) = m^2 y^2$

Now differentiating it w.r.t. x, we obtain

$\qquad (x^2+1) \, 2 y_1 y_2 + y_1^2 \cdot 2x = 2m^2 yy_1$

$\therefore \qquad (x^2+1)y_2 + xy_1 - m^2 y = 0$

Differentiating n times, we obtain

$\qquad y_{n+2}(1+x)^2 + ny_{n+1} \cdot 2x + \dfrac{n(n-1)}{2!} y_n \cdot 2 + xy_{n+1} + ny_n - m^2 y_n = 0$

or $\qquad (1+x^2)y_{n+2} + (2n+1)xy_{n+1} - (m^2 - n^2)y_n = 0$

Now put $x = 0$

$\therefore \qquad (y_{n+2})_0 = (m^2 - n^2)(y_n)_0$

$[(y_n)_0$ means nth differential coefficient at $x = 0]$

Putting $(n-2)$ for n, we get

$\qquad (y_n)_0 = [m^2 - (n-2)^2](y_{n-2})_0$

$\qquad\qquad = [m^2 - (n-2)^2][m^2 - (n-4)^2](y_{n-4})_0$

Case I. When n is even, proceeding further,

$\qquad (y_n)_0 = [m^2 - (n-2)^2][m^2 - (n-4)^2] \cdot [m^2 - (n-6)^2] \dots (m^2 - 2^2)(y_2)_0 \qquad \dots(i)$

from $\qquad (1+x^2)y_2 + xy_1 - m^2 y = 0$

we have $\qquad (y_2)_0 = m^2(y)_0 = m^2 [x + \sqrt{(x^2+1)}]_{x=0}$

$\qquad\qquad = m^2$

Hence from (i)

$$(y_n)_0 = \{m^2 - (n-2)^2\}\{m^2 - (n-4)^2\}...(m^2 - 2^2)m^2$$

Case II. When n is odd, proceeding further,

$$(y_n)_0 = \{m^2 - (n-2)^2\}\{m^2 - (n-4)^4\}...(m^2 - 1^2)(y_1)_0$$
$$= \{m^2 - (n-2)^2\}\{m^2 - (n-4)^2\}...(m^2 - 1^2)m$$

because $\quad y_1 = \dfrac{m}{\sqrt{(1+x^2)}}[x + \sqrt{(x^2+1)}]^m$

$\therefore \quad (y_1)_0 = m.$

2. Find $(y_n)_0$, when $y = e^{m \sin^{-1} x}$.

Solution. Given $\quad y = e^{m \sin^{-1} x}$

$\therefore \quad y_1 = e^{m \sin^{-1} x} \cdot \dfrac{m}{\sqrt{(1-x^2)}}$

or $\quad (1-x^2) y_1^2 = m^2 y^2$

Differentiating again with respect to x, we get

$$2 y_1 y_2 (1-x^2) - 2x y_1^2 = 2m^2 y y_1$$

or $\quad y_2 (1-x^2) - x y_1 - m^2 y = 0.$

Differentiating again n times by Lebnitz's Theorem, we get

$$y_{n+2}(1-x^2) + n y_{n+1}(-2x) + \dfrac{n(n-1)}{2!} y_n(-2) - x y_{n+1} - n y_n - m^2 y_n = 0$$

or $\quad (1-x^2) y_{n+2} - (2n+1) x y_{n+1} - (n^2 + m^2) y_n = 0$

Put $x = 0$; then we obtain

$$(y_{n+2})_0 = (n^2 + m^2)(y_n)_0$$

Putting $(n-2)$ for n, we get

$$(y_n)_0 = [(n-2)^2 + m^2](y_{n-2})_0$$
$$= [m^2 + (n-2)^2][m^2 + (n-4)^2](y_{n-4})_0 \qquad ...(i)$$

Again $y = e^{m \sin^{-1} x}, (y)_0 = e^0 = 1$

from $(1-x^2) y_1^2 = m^2 y^2, (y_1)_0 = m,$

$$(1-x^2) y_2 - x y_1 = m^2 y, (y_2)_0 = m^2 \cdot 1$$
$$(y_n)_0 = [m^2 + (n-2)^2](y_{n-2})_0$$

Put $n = 3$.

$$(y_3)_0 = (m^2 + 1^2)(y_1)_0 = (m^2 + 1^2) \cdot m,$$
$$(y_4)_0 = (m^2 + 2^2)(y_2)_0 = (m^2 + 2^2) \cdot m^2,$$
$$(y_5)_0 = (m^2 + 3^2)(y_3)_0$$
$$= (m^2 + 3^2)(m^2 + 1^2)(y_1)_0$$
$$= (m^3 + 3^2)(m^2 + 1^2) \cdot m$$

Successive Differentiation

$$(y_6)_0 = (m^2 + 4^2)(y_4)_0$$
$$= (m^2 + 4^2)(m^2 + 2^2) m^2$$

Case I. From (i) when n is even
$$(y_n)_0 = [m^2 + (n-2)^2][m^2 + (n-4)^2](y_{n-4})_0$$
$$= [m^2 + (n-2)^2][m^2 + (n-4)^2] \ldots (m^2 + 4^2)(m^2 + 2^2) m^2$$

Case II. When n is odd
$$(y_n)_0 = [m^2 + (n-2)^2][m^2 + (n-4)^2](y_{n-4})_0$$
$$= [m^2 + (n-2)^2][m^2 + (n-4)^2] \ldots (m^2 + 3^2)(m^2 + 1^2) \cdot m$$

EXERCISES

1. If $y = \sin^{-1} x$, find $(y_n)_0$.

   [**Ans.** 0, when n is even and $1^2 \cdot 3^2 \ldots (n-4)^2 (n-2)^2$, when n is odd.]

2. Prove that the value when $x = 0$ of $D^n (\tan^{-1} x)$ is 0, $(n-1)!$ or $-(n-1)!$ according as n of the form $2p, 4p+1,$ or $4p+3$ respectively.

3. If $y = e^{m \cos^{-1} x}$, show that $(1 - x^2) y_{n+2} - (2n+1) xy_{n+1} - (n^2 + m^2) y_n = 0$. Hence evaluate $(y_n)_0$.

   [**Ans.** $-m(1^2 + m^2)(3^2 + m^2) \ldots \{(n-2)^2 + m^2\} e^{m\pi/2}$ when n is odd.;

   $m^2 (2^2 + m^2)(4^2 + m^2) \ldots \{(n-2)^2 + m^2\} e^{m\pi/2}$, when n is even.]

4. If $y = \log[x + \sqrt{(x^2 + a^2)}]$, prove that $(a^2 + x^2) y_2 + xy_1 = 0$. Differentaite this differential equation n times and prove that
$$\lim_{x \to 0} \frac{y_{n+2}}{y_n} = -\frac{n^2}{a^2}$$

❏❏❏

4

Expansion of Functions

4.1 MACLAURIN'S THEOREM

Let $f(x)$ be a function of x. Let this function be expanded in ascending powers of x and let the expansion be differentiable term by term any number of times.

Let $\quad f(x) = A_0 + A_1 x + A_2 x^2 + A_3 x^3 + A_4 x^4 + \ldots$

Now by successive differentiation, we get

$$f'(x) = A_1 + 2A_2 x + 3A_3 x^2 + 4A_4 x^3 + \ldots$$
$$f''(x) = 2.1\, A_2 + 3.2\, A_3 x + 4.3\, A_4 x^2 + \ldots$$
$$f'''(x) = 3.2.1\, A_3 + 4.3.2\, A_4 x + \ldots$$
$$f^n(x) = n(n-1)(n-2)\ldots 2.1\, A_n$$
$$+ (n+1)n(n-1)\ldots 3.2\, A_{n+1} . x + \ldots$$

Putting $x = 0$ in each term, we get

$$f(0) = A_0,\ f'(0) = A_1,\ f''(0) = 2!\, A_2$$
$$f'''(0) = 3!\, A_3, \ldots f^n(0) = n!\, A_n \ldots$$

with the help of these values of $A_0, A_1, A_2, A_3, \ldots$ we have

$$f(x) = f(0) + x f'(0) + \frac{x^2}{2!} f''(0) + \ldots + \frac{x^n}{n!} f^n(0)$$

This result is known as **Maclaurin's Theorem**.

EXAMPLES

1. *Expand $\cos x$ by Maclaurin's series.*

Solution. We have $\quad y = \cos x,\ y_1 = -\sin x = \cos\left(\dfrac{\pi}{2} + x\right)$

$$y_2 = -\sin\left(\frac{\pi}{2} + x\right) = \cos\left(x + \frac{2\pi}{2}\right)$$

then $\quad y_n = \cos\left(x + \dfrac{n\pi}{2}\right)$

Now put $x = 0$; then

$$(y)_0 = 1,\ (y_1)_0 = 0,\ (y_2)_0 = -1$$
$$(y_3)_0 = \cos\left(\frac{3\pi}{2}\right) = 0$$

In general $\quad (y_n)_0 = \cos\dfrac{n\pi}{2}$

Now, put $n = 4$; then $(y_4)_0 = \cos 2\pi = 1$

Expansion of Functions

Hence $(y_n)_0 = \cos \dfrac{n\pi}{2} = 0$, if n is odd

$\qquad\qquad\qquad = (-1)^{n/2}$ if n is even.

Then $\qquad f(x) = f(0) + xf'(0) + \dfrac{x^2}{2!} f''(0) + \ldots$

$$\cos x = 1 - \dfrac{x^2}{2!} + \dfrac{x^4}{4!} - \dfrac{x^6}{6!} + \ldots + (-1)^{n/2} \dfrac{x^n}{n!} + \ldots$$

when n is even.

2. Show if $y = \sin(m \sin^{-1} x)$, then $(1-x^2) \dfrac{d^2 y}{dx^2} - x \dfrac{dy}{dx} + m^2 y = 0$. Hence or otherwise expand $\sin m\theta$ in powers of $\sin \theta$.

Solution. We have $\qquad y = \sin(m \sin^{-1} x)$

then $\qquad\qquad y_1 = \cos(m \sin^{-1} x) \cdot \dfrac{m}{\sqrt{(1-x^2)}}$

or $\qquad\qquad (1-x^2) y_1^2 = m^2 (1 - y^2)$

or $\qquad\qquad 2 y_1 y_2 (1-x^2) + y_1^2 (-2x) = -2m^2 y y_1$

or $\qquad\qquad (1-x^2) y_2 - x y_1 + m^2 y = 0$

Differentiating n times, we have

$\qquad y_{n+2}(1-x^2) + n y_{n+1}(-2x) + \dfrac{n(n-1)}{2!} y_n(-2) - x y_{n+1} - n y_n + m^2 y_n = 0$

or $\qquad (1-x^2) y_{n+2} - (2n+1) x y_{n+1} + (m^2 - n^2) = 0$.

Now, put $x = 0$, then

$$(y_{n+2})_0 = (n^2 - m^2)(y_n)_0$$

Now, $(y)_0 = 0;\ (y_1)_0 = m,\ (y_2)_0 = 0$

$\qquad\qquad (y_3)_0 = (1^2 - m^2) \cdot m;\ (y_4)_0 = 0$

$\qquad\qquad (y_5)_0 = (3^2 - m^2)(1^2 - m^2) \cdot m$ etc.

hence by Maclaurin's Theorem, we get

$$\sin(m \sin^{-1} x) = mx + m(1^2 - m^2) \dfrac{x^3}{3!} + m(3^2 - m^2) \cdot \dfrac{x^5}{5!} + \ldots$$

Put $\qquad\qquad \sin^{-1} x = \theta;$ then $x = \sin \theta$

Hence $\qquad\qquad \sin m\theta = m \sin \theta + \dfrac{1}{3!} m(m^2 - 1) \sin^3 \theta$

$\qquad\qquad\qquad\qquad + m(m^2 - 1^2)(m^2 - 3^2) \sin^5 \theta + \ldots$

EXERCISES

1. (a) $e^{x \cos x}$, (b) $\log_e (1 + \sin x)$

   [Ans. (a) $1 + x + \dfrac{x^2}{2} - \dfrac{x^3}{3} - \dfrac{11x^4}{24} - \dfrac{x^5}{5} + ...$ (b) $x - \dfrac{x^2}{2} + \dfrac{x^3}{6} - \dfrac{x^4}{12} + \dfrac{x^5}{24} + ...$]

2. (a) $\log_e (1 + \tan x)$; (b) $e^x \log_e (1 + x)$; (c) $e^{a \cos^{-1} x}$

   [Ans. (a) $x - \dfrac{1}{2} x^2 + \dfrac{2}{3} x^3 - ...$ (b) $x + \dfrac{x^2}{2!} + \dfrac{2x^3}{3!} + \dfrac{9x^5}{5!} + ...$

   (c) $e^{a\pi/2} \left\{ 1 - ax + \dfrac{a^2 x^2}{2!} - a(1 + a^2) \dfrac{x^3}{3!} + (2^2 + a^2) \dfrac{a^2 x^4}{4!} + ... \right\}$]

3. $e^x \sec x$ or $\dfrac{e^x}{\cos x}$

   [Ans. $1 + x + x^2 + \dfrac{2x^3}{3!} + ...$]

4. Prove that $e^{ax} \cos bx = 1 + ax + \dfrac{a^2 - b^2}{2!} x^2 + \dfrac{a(a^2 - 3b^2)}{3!} x^3$

   $+ ... + \dfrac{(a^2 + b^2)^{n/2}}{n!} x^n \cos(n \tan^{-1} b/a) + ...$

5. Expand $e^{a \sin^{-1} x}$ in powers of x by Maclaurin's Theorem and hence obtain the value of e^θ.

   [Ans. $e^{a \sin^{-1} x} = 1 + a \cdot x + \dfrac{a^2}{2!} x^2 + \dfrac{x^3}{3!} (1^2 + a^2) a + ...$]

6. Prove that $e^x \cos x = 1 + x - \dfrac{2x^3}{3!} - \dfrac{2^2 x^4}{4!} - \dfrac{2^2 x^5}{5!} + \dfrac{2^3 x^7}{7!} + ...$

7. Prove that $\log_e \sec x = \dfrac{1}{2} x^2 + \dfrac{1}{12} x^4 + \dfrac{1}{45} x^6 + ...$

8. Prove that for all finite values of x,

   $e^x \sin x = x + x^2 + \dfrac{2}{3!} x^3 - \dfrac{2^2}{5!} x^5 - ... + \sin\left(\dfrac{n\pi}{4}\right) \dfrac{2^{n/2}}{n!} x^n + ...$

9. If $y^3 - 6xy - 8 = 0$, prove that $y = 2 + x - \dfrac{1}{2} \cdot \dfrac{x^3}{3!} + \dfrac{x^4}{4!} + ...$

4.2 TAYLOR'S THEOREM

Let $f(x + h)$ be a function of h (x being independent of h) which can be expanded in powers of h and the expansion be differentiable any number of times, then

$$f(x + h) = f(x) + hf'(x) + \dfrac{h^2}{2!} f'(x) + ... + \dfrac{h^n}{n!} f^n(x) . + ...$$

Let $f(x + h) = A_0 + A_1 h + A_2 h^2 + A_3 h^3 + A_4 h^4 + ...$ where $A_0, A_1, A_2, ...$ be functions of x alone which are to be determined.

Now by successive differentiation with respect to h, we get

$$f'(x + h) = A_1 + 2A_2 h + 3A_3 h^2 + 4A_4 h^3 + ...$$

Expansion of Functions

$$f''(x+h) = 2.1 A_2 + 3.2 A_3 h + 4.3 A_4 h^2 + ...$$
$$f'''(x+h) \; 3.2.1 A_3 + 4.3.2 A_4 h + ... \text{etc.}$$

Putting $h = 0$,
$$f(x) = A_0, f'(x) = A_1, f''(x) = 2! A_2, f'''(x) = 3! A_3, ...$$

Hence,
$$f(x+h) = f(x) + hf'(x) + \frac{h^2}{2!} f''(x) + \frac{h^3}{3!} f'''(x) + ... \frac{h^n}{n!} f^n(x) + ...$$

Other forms :

Putting $x = a$ in Taylor's Theorem.

$$f(a+h) = f(a) + hf'(a) + \frac{h^2}{2!} f''(a) + ... + \frac{h^2}{n!} f^n(a) + ... \qquad ...(i)$$

Now, putting $a = 0$ and $h = x$ in (i), we get

$$f(0) + f(0) + xf'(0) + \frac{x^2}{2!} f''(0) + ... + \frac{x^n}{n!} f^n(0) + ...$$

which is nothing but Maclaurin's Theorem.

Putting $h = x - a$ in (i), we get the expansion of $f(x)$ in powers of $(x-a)$ which is given below :

$$f(x) = f(a) + (x-a) f'(a) + \frac{(x-a)^2}{2!} f''(a) + \frac{(x-a)^3}{3!} f'''(a) + ...$$

$$+ \frac{(x-a)^n}{n!} f^n(a) + ...$$

EXAMPLES

1. *Exapnd* $\sin x$ *in powers of* $\left(x - \frac{\pi}{2}\right)$.

Solution. We have
$$f(x) = \sin x = \sin\left\{\frac{1}{2}\pi + \left(x - \frac{1}{2}\pi\right)\right\}$$

Now, by Taylor's Theorem, we have

$$f(x) = f(a) + (x-a) f'(a) + \frac{(x-a)^2}{2!} f''(a) + \frac{(x-a)^3}{3!} f'''(a) + ... \qquad ...(1)$$

Here, we have $f(x) = \sin x$ and $a = \frac{1}{2}\pi$, as we are to expand in powers of $\left(x - \frac{1}{2}\pi\right)$

Now, $f'(x) = \cos x, f''(x) = -\sin x, f'''(x) = -\cos x,$

$f^{iv}(x) = \sin x$, etc.

Putting $x = \frac{\pi}{2}$, we get

$$f\left(\frac{\pi}{2}\right) = \sin\frac{\pi}{2} = 1, f'\left(\frac{\pi}{2}\right) = \cos\frac{\pi}{2} = 0,$$

$$f''\left(\frac{\pi}{2}\right) = -\sin\frac{\pi}{2} = -1,\ f'''\left(\frac{\pi}{2}\right) = -\cos\frac{\pi}{2} = 0,$$

$$f^{iv}\left(\frac{\pi}{2}\right) = \sin\frac{\pi}{2} = 1,\ \text{etc.}$$

Now, in (1), putting $a = \dfrac{\pi}{2}$ and substituting these values, we get

$$\sin x = 1 + \left(x - \frac{\pi}{2}\right)(0) + \frac{\left(x - \frac{\pi}{2}\right)^2}{2!}(-1) + \frac{\left(x - \frac{1}{2}\pi\right)^3}{3!}(0) + \frac{\left(x - \frac{1}{2}\pi\right)^4}{4!}(1) + \ldots$$

$$= 1 - \frac{1}{2!}\left(x - \frac{\pi}{2}\right)^2 + \frac{1}{4!}\left(x - \frac{\pi}{2}\right)^4 - \ldots$$

2. Use Taylor's theorem to prove that

$$\tan^{-1}(x + h) = \tan^{-1} x + h \sin x \cdot \frac{\sin z}{1} - (h \sin z)^2 \cdot \frac{\sin 2z}{2} + (h \sin z)^3 \frac{\sin 3z}{3} \ldots$$

where $z = \cot^{-1} x$.

Solution. Let $f(x + h) = \tan^{-1}(x + h)$

$\therefore\quad f(x) = \tan^{-1} x$

then $\quad f^n(x) = (-1)^{n-1}(n-1)!\sin^n z \sin nz$

where $\quad z = \tan^{-1}\dfrac{1}{x} = \cot^{-1} x$

$\therefore\quad f'(x) = \sin z \sin z$

$f''(x) = -(1!)\sin^2 z \sin 2z,$

$f'''(x) = (2!)\sin^3 z \sin 3z,$ etc.

$\therefore\quad \tan^{-1}(x + h) = \tan^{-1} x + \dfrac{h}{1!}\sin z \cdot \sin z$

$$-\frac{h^2}{2!}(1!)\sin^2 z \sin 2z + \frac{h^3}{3!}\sin^3 z \sin 3z - \ldots$$

$$= \tan^{-1} x + h \sin z \cdot \frac{\sin z}{1} - (h \sin z)^2 + (h \sin z)^3 \cdot \frac{\sin 3z}{3} - \ldots$$

3. Prove that $f\left(\dfrac{x^2}{1+x}\right) = f(x) - \dfrac{x}{1+x} f'(x) + \dfrac{x^2}{(1+x)^2} \cdot \dfrac{f''(x)}{2!} + \ldots$

Solution. We can write $\quad f\left(\dfrac{x^2}{1+x}\right) = f\left(x - \dfrac{x}{1+x}\right)$

Also, by Taylor's Theorem $\quad f(x + h) = f(x) + hf'(x) + \dfrac{h^2}{2!}f''(x) + \ldots$

Expansion of Functions

Putting $h = -\left(\dfrac{x}{1+x}\right)$ in this expansion, we get

$$f\left(x - \dfrac{x}{1+x}\right) = f(x) - \dfrac{x}{1+x} f'(x) + \dfrac{x^2}{(1+x)^2} \cdot \dfrac{1}{2!} f''(x) + \dots$$

or

$$f\left(\dfrac{x^2}{1+x}\right) = f(x) - \left(\dfrac{x}{1+x}\right) f'(x) + \dfrac{x^2}{(1+x)^2} \dfrac{f''(x)}{2!} + \dots$$

EXERCISES

1. Prove that

$$\log_e \sin (x+h) = \log_e \sin x + h \cot x - \dfrac{h^2}{2} \operatorname{cosec}^2 x + \dfrac{h^3}{3} \operatorname{cosec}^2 x \cot x + \dots$$

2. Prove that $\dfrac{1}{x+h} = \dfrac{1}{x} - \dfrac{h}{x^2} + \dfrac{h^2}{x^3} - \dfrac{h^3}{x^4} + \dots$

3. Expand $\sin^{-1}(x+h)$ in powers of x as far as the terms in x^3.

[**Ans.** $\sin^{-1} h + x(1-h^2)^{-1/2} + \dfrac{x^2}{2!} h(1-h^2)^{-3/2} + \dfrac{x^3}{3!}\{(1-h^2)^{-5/2}(1+2h^2)\} + \dots$]

4. Expand $\tan^{-1} x$ in powers of $\left(x - \dfrac{\pi}{4}\right)$.

$$\left[\text{Ans. } \tan^{-1}\dfrac{\pi}{4} + \dfrac{\left(x - \dfrac{\pi}{4}\right)}{\left(1 + \dfrac{1}{16}\pi^2\right)} - \dfrac{\pi\left(x - \dfrac{\pi}{4}\right)^2}{4\left(1 + \dfrac{1}{16}\pi^2\right)^2} + \dots\right]$$

5. Prove that

$$f(mx) = f(x) + (m-1) x f'(x) + \dfrac{1}{2!}(m-1)^2 x^2 f''(x) + \dfrac{1}{3!}(m-1)^3 x^3 f'''(x) + \dots$$

$$\left[\text{Ans. } 1 - \dfrac{1}{2!}\left(x - \dfrac{\pi}{2}\right)^2 + \dfrac{1}{4!}\left(x - \dfrac{\pi}{2}\right)^4 - \dots\right]$$

❏❏❏

5
Asymptotes

5.1 INTRODUCTION

First we discuss the intersection of a straight line with a curve of nth degree.

Let us consider the straight line whose equation is $y = mx + c$ and substitute $(mx + c)$ instead of y in the equation of the curve; then the roots of the resulting equation in x represent the abscissae of the points which are the points of intersection of the straight line and curve.

As the equation of the curve is of nth degree, it follows that every straight line meets a curve of the nth degree in not more than n points, real or imaginary.

If two roots are equal, then two of the points of intersection become coincident and the straight line becomes a tangent.

Let us suppose the equation of the curve

$$y^n + a_1 y^{n-1} x + a_2 y^{n-2} x^2 + \ldots + a_n x^n = 0$$

This equation can be written in the form as given below :

$$x^n \left[\left(\frac{y}{x}\right)^n + a_1 \left(\frac{y}{x}\right)^{n-1} + a_2 \left(\frac{y}{x}\right)^{n-2} + \ldots + a_n \right]$$

i.e., $\quad x^n f_0(y/x) + x^{n-1} f_1(y/x) + x^{n-2} f_2(y/x) + \ldots = 0$

Substituting $m + \dfrac{c}{x}$ for $\dfrac{y}{x}$ in this it becomes

$$x^n f_0\left(m + \frac{c}{x}\right) + x^{n-1} f_1\left(m + \frac{c}{x}\right) + \ldots = 0$$

On expanding by Taylor's Theorem, we get

$$x^n f_0(m) + x^{n-1}[c f_0'(m) + f_1(m)]$$

$$+ x^{n-2}\left[\frac{1}{2!} c^2 f_0''(m) + c f_1'(m) + f_2(m)\right] + \ldots = 0 \quad \ldots(i)$$

The roots of the equation are worth noting :

1. Every straight line must intersect a curve of an odd degree in at least one real point; for every equation of an odd degree has one real root.
2. A tangent to a curve of nth degree can not meet it in more than $(n - 2)$ points besides its point of contact (since at the point of contact two roots coincide for the tangent).
3. Every tangent to a curve of an odd degree must meet it in one other real point besides its point of contact.
4. Every tangent to a curve for the third degree meets the curve in one and the other at real point.

5.2 ASYMPTOTE

An symptote is a tangent to a curve in the limiting position when its point of contact is situated at an infinite distance.

It may also be defined as follows :

Asymptotes

A line which lies in the finite portion of the plane, but meets a curve in at least two coincident points at infinity is called an asymptote of the curve. An asymptote is, therefore, a tangent to the given curve at infinity.

(i) No asymptotes to a curve of the nth degree can meet it in more than $(n-2)$ points distinct from that at infinity.

(ii) Each asymptote to a curve of the third degree intersects the curve in one point besides that at infinity.

5.2.1 Conditinos for the Existence of the Asymptote

If the curve is $y = f(x)$, then the tangent to the curve is

$$Y - y = \frac{dy}{dx}(X - x)$$

or

$$Y = X\frac{dy}{dx} + y - x\frac{dy}{dx}$$

This will be an asymptote (except the asymptotes that are parallel to the y-axis) if

$$\lim_{x \to \infty} \frac{dy}{dx} \text{ and } \lim_{x \to \infty}\left(y - x\frac{dy}{dx}\right)$$

both tend to finite limit say m and c

Then the asymptote becomes

$$y - mx - c = 0$$

Then

$$m = \lim_{x \to \infty}\left(\frac{y}{x} - \frac{c}{x}\right)$$

$$= \lim_{x \to \infty}\left(\frac{y}{x}\right)$$

Since

$$c = y - x\frac{dy}{dx},$$

then

$$\lim_{x \to \infty} \frac{c}{x} = \lim_{x \to \infty} \frac{y}{x} - \lim_{x \to \infty} \frac{dy}{dx}$$

i.e.,

$$\lim_{x \to \infty} \frac{y}{x} = \lim_{x \to \infty} \frac{dy}{dx}$$

Hence the m of an asymptote is equal to the limit of y/x as x tends to infinity.

5.2.2 Method of Determining the Asymptotes to a Curve of the nth Degree

If one of the points of intersection of $y = mx + c$ with the curve be at an infinite distance, one root of the equation (i) § 5.1 must be infinite, i.e.,

$$f_0(m) = 0 \qquad \qquad \text{...(ii)}$$

Again if two of the roots be infinite, we have in addition

$$cf_0'(m) + f_1(m) = 0 \qquad \qquad \text{...(iii)}$$

From these results, the value of m and c are determined; then the corresponding line

$$y = mx + c$$

meets the curve in two points in infinity and consequently is an asymptote.

Hence if m_1 be a root of the equation $f_0(m) = 0$, the line

$$y = m_1 x - \frac{f_1(m_1)}{f_0'(m_1)}$$

is in general an asymptote to a curve.

If $f_1(m) = 0$ and $f_0(m) = 0$ have a common root (m_1 say) the corresponding asymptote in general passes through the origin and is represented by the equation

$$y = m_1 x$$

In this case $x^n f_0(y/x)$ and $x^{n-1} f_1(y/x)$ evidently have a common factor.

To each root of $f_0(m) = 0$ corresponds an asymptote, or in other words every curve of the nth degree has in general n asymptotes, real or imaginary.

5.2.3 Working Rule for Finding the Asymptotes

1. Substitute $mx + c$ for y in the equation of curve,

$$x^n f_0(y/x) + x^{n-1} f_1(y/x) + \ldots = 0;$$

then

$$x^n f_0\left(m + \frac{c}{x}\right) + x^{n-1} f_1\left(m + \frac{c}{x}\right) + \ldots = 0$$

or

$$x^n \left\{ f_0(m) + \frac{c}{x} f_0'(m) + \frac{c^2}{2! x^2} f_0''(m) + \ldots \right\}$$

$$+ x^{n-1} \left\{ f_1(m) + \frac{c}{x} f'(m) + \ldots \right\} + \ldots = 0$$

2. Equate to zero the coefficients of the two highest powers of x.
3. Determine m and c from these equations.
4. Then $y = m_1 x + c_1,\ y = m_2 x + c_2, \ldots$ are the equations of the asymptote.

Second Method (Shorter Method)

1. Put $x = 1$ and $y = m$ in the highest degree terms of the equation to the curve.
2. Then equate highest degree terms of m to zero.
3. Solve for m; let $m_1, m_2, \ldots, m_n$ be roots.
4. Then equate next higher degree terms to zero.

   Then the values of $c_1, c_2, c_3, \ldots$ are obtained by substituting $m_1, m_2, m_3 \ldots$ in turn in the formula

   $$c = -\frac{f_1(m)}{f_0'(m)}$$

5. Then the asymptotes are

   $$y = m_1 x + c_1,\ y = m_2 x + c_2, \ldots$$

EXAMPLE

Find all the asymptotes of the curve

$$3x^3 + 2x^2 y - 7xy^2 + 2y^3 - 14xy + 7y^2 + 4x + 5y = 0.$$

Solution. First method. Substituting $y = mx + c$ in the given equation, we get

Asymptotes

$$3x^3 + 2x^2(mx+c) - 7x(mx+c)^2 + 2(mx+c)^3 - 14x(mx+c)$$
$$+ 7(mx+c)^2 + 4x + 5(mx+c) = 0$$

or $\quad x^3(3 + 2m - 7m^2 + 2m^3) + x^2(2c - 14mc + 6m^2c - 14m + 7m^2) + \ldots = 0$

Equating the coefficient of x^3 and x^2 seperately to zero, we get

$$2m^3 - 7m^2 + 2m + 3 = 0$$

or $\quad (m-1)(2m+1)(m-3) = 0$

or $\quad m = 1, 3, -\dfrac{1}{2}$

and then $\quad 2c - 14mc + 6m^2c - 14m + 7m^2 = 0$

or $\quad c = -\dfrac{7m^2 - 14m}{2 - 14m + 6m^2}$

Put $\quad m = 1, c = -\dfrac{7 - 14}{2 - 14 + 6} = -\dfrac{7}{6};$

$\quad m = 3, c = -\dfrac{7.9 - 14.3}{2 - 14.3 + 6.9} = -\dfrac{3}{2};$

$\quad m = -\dfrac{1}{2}, c = \dfrac{7\left(-\dfrac{1}{2}\right)^2 - 14\left(-\dfrac{1}{2}\right)}{2 - 14\left(-\dfrac{1}{2}\right) + 6\left(-\dfrac{1}{2}\right)^2}$

$\quad = -5/6.$

Hencethe asymptotes are

$$y = x - \dfrac{7}{6}; \ y = 3x - \dfrac{3}{2} \ \text{and} \ y = -\dfrac{1}{2}x - \dfrac{5}{6}.$$

Second method (Shorter) :

By putting $x = 1$ and $y = m$, we get

$$f_0(m) = 3 + 2m - 7m^2 + 2m^3$$
$$= (m-1)(2m+1)(m-3)$$

Now put $f_0(m) = 0$; then

$$(m-1)(2m+1)(m-3) = 0$$

$\therefore \quad m = 1, -\dfrac{1}{2}, 3$

and also $\quad f_1(m) = 7m^2 - 14m$

and $\quad f_0'(m) = 6m^2 - 14m + 2$

Then $\quad c = -\dfrac{7m^2 - 14m}{6m^2 - 14m + 2}$

When $\quad m = 1; c = -\dfrac{7-14}{6-14+2} = -\dfrac{7}{6}$

$$m = 3; c = \frac{7.9 - 14.3}{6.3^3 - 14.3 + 2} = -\frac{3}{2}$$

$$m = -\frac{1}{2}; c = \frac{7\left(-\frac{1}{2}\right)^2 - 14\left(-\frac{1}{2}\right)}{6\left(-\frac{1}{2}\right)^2 - 14\left(-\frac{1}{2}\right) + 12}$$

$$= -5/6 \text{ etc.}$$

EXERCISES

Find the asymptotes of the following curves :

1. $y^3 - x^2y - 2xy^2 + 2x^3 - 7xy + 3y^2 + 2x^2 + 2x + 2y + 1 = 0$

   [**Ans.** $y = x - 1, y + x + 2 = 0$ and $y = 2x$;]

2. $x^3 + 2x^2y - xy^2 - 2y^3 + 4y^2 + 2xy + y - 1 = 0$

   [**Ans.** $y = x + 1, y = -x + 1,$ and $y = -\frac{1}{2}x$]

3. $x^3 + 3x^2y - xy^2 - 3y^3 + x^2 - 2xy + 3y^2 + 4x + 5 = 0$

   [**Ans.** $y + \frac{1}{3}x + \frac{3}{4} = 0, y = x + \frac{1}{4}$ and $y + x = \frac{3}{2}$]

4. $x^3 + y^3 = 3axy$ [**Ans.** $y + x + a = 0$]

5.3 CASE OF REPEATED ROOTS (PARALLEL ASYMPTOTES)

If $f_0(m)$ has repeated roots $m = m_1$ say, then $f_0'(m_1) = 0$. If now $f_1(m_1) \neq 0, c$ becomes infinite.

The line $y = m_1x + c$ cuts the curve at two coincident points at infinity while itself has an intercept on y-axis. The straight line lies wholly at infinity and therefore is not called an asymptote.

But if for a repeated root $f_1(m_1) = 0$ also, the coefficient of x^{n-1} in the equation (i) of § 5.2 is identically zero and c can not be determined, this implies that there are infinite number of lines $y = m_1x + c$ which meet the curve in two coincident points at infinity. Out of these lines we now choose two lines given by

$$\frac{c^2}{2!} f_0''(m_1) + c f_1'(m_1) + f_2(m_1) = 0$$

If c_1, c_2 are two real roots of this equation, the asymptotes $y = m_1x + c_1, y = m_1x + c_2$ are parallel and cut the curve in three points at infinity.

5.3.1 Working Rule

(i) Substitute $mx + c$ for y in the given equation.
(ii) Equate to zero the coefficient of the highest power of x.
(iii) If one value of m derived from the equation in (ii) makes the coefficient of x^{n-1} identically zero, then put the coefficient of x^{n-2} equal to zero.
(iv) Then determine the values of c and substituite in $y = mx + c$.

Asymptotes

EXAMPLE

Find all the asymptotes of the curve $y^3 - x^2y + x^3 + x^2 - y^2 - 1 = 0$.

Solution. Substituting $y = mx + c$ in the given equation,

We have $(mx + c)^3 - x(mx + c)^2 - x^2(mx + c) + x^3 + x^2 - (mx + c)^2 - 1 = 0$

or $(m^3 - m^2 - m + 1)x^3 + (3m^2c - 2mc - c + 1 - m^2)x^2$

$$+ (3mc^2 - c^2 - 2mc)x + ... = 0 \qquad ...(1)$$

Equating the coefficient of highest power of x i.e., x^3 in (1) to zero, we get

$$m^3 - m^2 - m + 1 = 0 \text{ or } (m^2 - 1)(m - 1) = 0$$

or $m = 1, 1, -1$.

Equating the coefficient of x^2 in (1) to zero, we get

$$3m^2c - 2mc - c + 1 - m^2 = 0$$

or $$c = \frac{m^2 - 1}{(3m^2 - 2m - 1)} \qquad ...(2)$$

When $m = -1$, we get $c = 0$ and therefore the corresponding asymptote is $y = -x$ or $x + y = 0$.

When $m = 1$, from (2), we get $c = \frac{0}{0}$, which is indeterminate.

∴ Equating the coefficient of x in (1) to zero,

we have $3mc^2 - c^2 - 2mc = 0$

When $m = 1, 2c^2 - 2c = 0$

i.e., $2c(c - 1) = 0$ or $c = 0$ and 1.

∴ The corresponding asymptotes are $y = x$ and $y = x + 1$.

∴ The required asymptotes are

$$x + y = 0, x - y = 0 \text{ and } y = x + 1.$$

EXERCISES

Find the asymptotes of the following curves :

1. $x^3 + x^2y - xy^2 - y^3 + x^2 - y^2 = 2$ [**Ans.** $x \mp y = 0; x + y + 1 = 0;$]

2. $x^3 + 3x^2y - 4y^3 - x + y + 3 = 0$ [**Ans.** $y = -\frac{1}{2}x \pm \frac{1}{2}; y = x$]

3. $(x^2 - y^2)^2 - 4y^2 + y = 0$ [**Ans.** $y - x = \pm 1; y + x = \pm 1$]

4. $(y^2 - x^2)^2 - 2(x^2 + y^2) = 1$ [**Ans.** $y \pm x = \pm 1$]

5.4 ASYMPTOTES PARALLEL TO THE COORDINATE AXES

Let the equation of the curve be

$$(a_0 y^n + a_1 y^{n-1} x + a_2 y^{n-2} x^2 + ... + a_n x^n)$$

$$+ (b_1 y^{n-1} + b_2 y^{n-2} x + ... + b_{n-1} y x^{n-2} + b_n x^{n-1})$$

$$+ (c_2 y^{n-2} + c_3 y^{n-3} x + ... + c_n x^{n-2}) + ... = 0$$

Let it be arranged according to powers of x, we get

$$a_0 x^n + (a_{n-1} y + b_n) x^{n-1} + \ldots = 0$$

Then if $\quad a_0 = 0$ and $a_{n-1} y + b_n = 0$

or
$$y = -\frac{b_n}{a_{n-1}}$$

two of the roots of equation in x becomes infinite and consequently the line $(a_{n-1} y + b_n) = 0$ is an asymptote.

In other words, whenever the highest power of x is wanting in the equation of a curve, the coefficient of the next higher power equated to zero represents an asymptote parallel to the axis of x.

If $a_n = 0$ and $b_n = 0$, the axis of x is itself an asymptote.

If x^n and x^{n-1} be both wanting, the coefficient of x^{n-2} represents a pair of asymptotes real or imaginary, parallel to the axis of x, and so on.

We can state as follows :

Asymptote or asymptotes parallel to x-axis are determined by equating the coefficient of the highest power of x to zero, provided that the coefficient of the highest power of x is not constant.

Similarly, *the asymptote or asymptotes parallel to the axis of y can be determined by equating the coefficient of the highest power of y to zero, provided that the coefficient is not constant.*

EXAMPLES

1. *Find all the asymptotes to the curve*

$$\frac{a^2}{x^2} - \frac{b^2}{y^2} = 1$$

Solution. The given curve is

$$x^2 y^2 + b^2 x^2 - a^2 y^2 = 0$$

This equation is of fourth degree; it can not have more than four asymptotes.

Equating the coefficient of highest power of y (*i.e.*, of y^2) to zero, the asymptotes parallel to y-axis are given by

$$x^2 - a^2 = 0, \, x = \pm a,$$

which gives two real asymptotes.

Again equating to zero the coefficient of the highest power of x, *i.e.*, of x^2, the asymptotes parallel to x-axis are given by

$$y^2 + b^2 = 0$$

which gives the imaginary asymptotes.

Hence the only asymptotes are $x = \pm a$.

2. *Prove that the asymptotes of the curve* $x^2 y^2 = a^2(x^2 + y^2)$ *form a square of side* $2a$.

Solution. Equating to zero the coefficients of highest power of x and y, *i.e.*, of x^2 and y^2 in the given equation we find the asymptotes parallel to x and y axes of the given curve are

$$y^2 - a^2 = 0 \text{ and } x^2 - a^2 = 0 \text{ respectively.}$$

∴ The asymptotes of the curve are $x = \pm a$ and $y = \pm a$ which evidently form a square, and the distance between two parallel sides $x = -a$ and $x = +a$ is $a + a = 2a$.

∴ The side of above square is $2a$.

Asymptotes

EXERCISES

1. Show that the asymptotes of
$$x^2y^2 - a^2(x^2 + y^2) - a^3(x+y) + a^4 = 0$$
form a square, through two of whose angular points the curve passes.

2. Find the asymptotes parallel to axes of the curve
$$x^2(x-y)^2 + a^2(x^2 - y^2) - a^2 xy = 0.$$ [Ans. $x = \pm a$]

3. Show that one of the asymptote of
$$ay^3 - (a^2 + a)xy + a^2x^2y - 3y^2 + 3xy - 2x + 5 = 0$$
always touches a parabola, whatever be the value of a.

5.5 ASYMPTOTES OF THE CURVE

$$y = mx + c + \frac{A}{x} + \frac{B}{x^2} + \ldots$$

Obviously the part $\frac{A}{x} + \frac{B}{x^2} + \ldots$ is convergent for sufficiently large values of x.

The curve is $y = mx + c + \frac{A}{x} + \frac{B}{x^2} + \ldots$; then $\frac{dy}{dx} = m + 0 - \frac{A}{x^2} - \frac{2B}{x^3} - \ldots$

The equation of the tangent is

$$Y - y = \left(m - \frac{A}{x^2} - \frac{2B}{x^3} - \ldots\right)(X - x)$$

or $\quad Y = \left(m - \frac{A}{x^2} - \frac{2B}{x^3} - \ldots\right)X + c + \frac{2A}{x} + \frac{3B}{x^2} + \ldots$

But $x \to \infty$ for asymptote,
$$Y = mX + c.$$

Hence the required asymptote is $Y = mX + c$.

Example. *Find the asymptotes of the hyperbola* $\frac{x^2}{a^2} - \frac{y^2}{b^2} = 1$.

Solution. The given curve is $\frac{y^2}{b^2} = \frac{x^2}{a^2} - 1$

or $\quad y = \pm \frac{b}{a} x \left[1 - \frac{a^2}{x^2}\right]^{1/2} = \pm \frac{bx}{a}\left[1 - \frac{a^2}{2x^2} - \frac{a^4}{8x^4} - \ldots\right]$

When $x \to \infty$, we get $y = \pm b/a\, x$, which are the required asymptotes.

EXERCISES

Find the asymptotes of :

1. $y^3 = x^3 + ax^2$ [Ans. $y = x + \frac{1}{3}a$]

2. $y^3 = x^2(x - a)$ [Ans. $x = \frac{1}{3}a$]

3. $y = x\frac{(x^2 + a^2)}{x^2 - a^2}$ [Ans. $y = x$]

5.6 ASYMPTOTES BY INSPECTION

If any equation of the curve can be put in the form

$$\phi_n + \phi_{n-2} = 0.$$

Where ϕ_n can be factorized into n linear factors, then each of these factors equated to zero gives an asymptote, for it will cut the curve in two coincident points at infinity.

5.6.1 Alternative Method for Determining Asymptotes of Algebraic Curves

If $ax + by$ is a non-repeated factor of the nth degree terms of the equation to the curve, then the equation of the curve can be written as

$$(ax + by) F_{n-1} + P_{n-1} = 0 \qquad \ldots(i)$$

Where F_{n-1} contains only terms of degree $(n-1)$ and P_{n-1} contains terms of various degrees, none of which is of a degree higher than $(n-1)$,

Thus $ax + by + c = 0$ is an asymptote where c is to be determined, which is equal to

$$c = \lim_{x \to \infty} [-(ax + by)] \text{ and } (x, y) \text{ lies on (i)}$$

Then
$$ax + by = -\frac{P_{n-1}}{F_{n-1}}$$

Thus
$$c = \lim_{x \to \infty} \frac{P_{n-1}}{F_{n-1}}$$

EXAMPLES

1. *Find all the asymptotes of*

$$y^2(x^2 - a^2) = x^2(x^2 - 4a^2).$$

Solution. The given equation can be written as

$$x^2(y^2 - x^2) + 4a^2 x^2 - y^2 a^2 = 0$$

or $\qquad x^2(y - x)(y + x) + 4a^2 x^2 - y^2 a^2 = 0.$

Equating the coefficients of highest power of y, i.e., y^2 to zero, we have $x^2 - a^2 = 0$ or $x = \pm a$ as the asymptotes parallel to y-axis.

Also asymptote corresponding to the factor $(y - x)$ is

$$y - x = \lim_{x \to \infty} \frac{a^2 y^2 - 4a^2 x^2}{x^2(y + x)}$$

$$= \lim_{\substack{x \to \infty \\ y/x \to 1}} \frac{a^2(y^2/x^3) - 4a^2(1/x)}{(y/x) + 1} = 0$$

or $\qquad y - x = 0.$

And asymptote corresponding to the factor $(y + x)$ is

$$y + x = \lim_{\substack{x \to \infty \\ y/x \to 1}} \frac{a^2 y^2 - 4a^2 x^2}{x^2 cy - x} = 0, \text{ as above.}$$

∴ The required asymptotes are $x = \pm a$, $y = \pm x$.

2. Find the asymptotes of the curve
$$x^3 - 2x^2y + xy^2 + x^2 - xy + 2 = 0$$
Solution. The equation of the curve can be written as
$$x(x-y)^2 + x(x-y) + 2 = 0 \qquad \text{...(i)}$$
Let $x = k$ be an asymptote of the curve with the help of this eliminating x from equation (i), we get
$$k(k-y)^2 + k(k-y) + 2 = 0$$
Now equating to zero the coefficient of highest power of y in the above equation, we have
$$k = 0$$
Hence the asymptote parallel to y-axis is $x = 0$.

Again put $x - y = k_1$ to find the asymptote parallel to $x - y = 0$.

Therefore, $y = x - k_1$

Substituting this value of y in (i), we have
$$xk_1^2 + xk_1 + 2 = 0.$$
Equating the coefficient of highest power of x to zero, we get
$$k_1^2 + k_1 = 0, \text{ which gives } k_1 = 0, -1.$$
Hence the asymptotes parallel to $x - y = 0$ are $x - y = 0$ and $x - y + 1 = 0$.
Therefore asymptotes of given curve are
$$x = 0, x - y = 0 \text{ and } x - y + 1 = 0.$$

EXERCISES

Find all the asymptotes of :

1. $x(y-3)^3 = 4y(x-1)^3$ [Ans. $x = 0, y = 0, 2y + 4x = 15, 2y - 4x = 3$]
2. $(x+y)^2(x+2y+2) = x + 9y - 2$ [Ans. $x + y = \pm 2\sqrt{2}; x + 2y + 2 = 0$]
3. $(x-y)(x-2y)^2 - 4y(x-2y) - (8x+7y) = 0$ [Ans. $x - y + 4 = 0; x - 2y = 2 \pm 3\sqrt{3}$]
4. $x^2(x-y)^2 + a^2(x^2 - y^2) = a^2xy$ [Ans. $x = \pm a, x - y = \pm a$]

5.7 INTERSECTION OF THE CURVE AND ITS ASYMPTOTES

We have already discussed that a straight line cuts a curve of degree n in general in n points. As one of these points of intersection is kept fixed and another point of intersection is made to approach to it, and then the line tends to the tangent at the first point, hence a tangent (or asymptote) will in general, cut the curve in $(n-2)$ points. Therefore n asymptotes will cut the curve in $n(n-2)$ points.

Let the equation of the curve having asymptotes $F_n = 0$ be put in the form
$$F_n + F_{n-2} = 0.$$
Therefore $(F_n + F_{n-2}) - F_n = 0$ is a curve which passes through the intersection of $F_n + F_{n-2} = 0$ and $F_n = 0$, where
$$F_n + F_{n-2} = 0 \quad \text{(Curve)}$$
and
$$F_n = 0 \quad \text{(asymptotes)}$$

Thus a curve of degree $(n-2)$ or less can be made to pass through $n(n-2)$ points of intersection of a curve of degree n and its asymptotes.

EXAMPLES

1. *Show that asymptotes of the cubic*
$$x^3 - 2y^3 + xy(2x - y) + y(x - y) + 1 = 0$$
cut the curve again in three points which lie on the straight line $x - y + 1 = 0$.

Solution. The equation of the curve is
$$(x^3 - 2y^3 + 2x^2y - xy^2) + (xy - y^2) + 1 = 0 \qquad ..(i)$$
$$f_0(m) = 1 - 2m^3 + 2m - m^2 = 0$$

i.e., $\qquad m = 1, -1, -\dfrac{1}{2}$

and $\qquad f_0'(m) = -6m^2 + 2 - 2m$

and $\qquad f_1(m) = m - m^2$.

Now $\qquad c = -\dfrac{f_1(m)}{f_0'(m)} = -\dfrac{m - m^2}{2 - 2m - 6m^2}$

For $m = 1, c = 0$; $m = -1, c = -1$; $m = -\dfrac{1}{2}, c = \dfrac{1}{2}$

Hence the asymptotes are
$$(y - x)(y + x + 1)(x + 2y - 1) = 0$$

i.e., $\qquad x^3 - 2y^3 + 2x^2y - xy^2 + xy - y^2 - x + y = 0 \qquad ...(ii)$

The asymptotes cut the curve in
$$n(n-2), \quad i.e., \quad 3(3-2) = 3 \text{ points}.$$

Subtracting (ii) from (i), we observe that points of intersection of the curve and asymptotes lie on
$$x - y + 1 = 0$$

2. *Find the equation of the straight line on which lie the three points of intersection of the curve* $(x + a)y^2 = (y + b)x^2$ *and its asymptotes.*

Solution. The asymptotes parallel to x-axis and y-axis are $y + b = 0$; $x + a = 0$.

Now putting $y - x = k$ or $x = -k + y$, the curve is
$$(-k + y)y^2 + ay^2 = (y + b)(-k + y)^2$$
or $\qquad y^2(-k + a - b + 2k) + ... = 0$

If $y - x = k$ is asymptote, the coefficient of y^2 is zero,

i.e., $\qquad k = -a + b$.

Hence third asymptote is $y - x = b - a$.

The combined equation of asymptotes is
$$(x + a)(y + b)(-y + x + b - a) = 0$$
or $\qquad xy(y - x) + ay^2 - bx^2 + a^2y - b^2x + ab(a - b) = 0.$

Subtracting the combined equation to the asymptotes from the curves, we obtain
$$a^2y - b^2x + ab(a - b) = 0 \qquad ...(i)$$

Hence the points of intersection of the curve and asymptotes are $3(3 - 2) = 3$ and they lie on (i).

Asymptotes

EXERCISES

1. Show that the asymptotes of the curve $x^2y - xy^2 + xy + y^2 + x - y = 0$ cut the curve again in three points which lie on the curve $x + y = 0$.
2. Prove that the three points in which the asymptotes of the curve $x^3 + 2x^2y - xy^2 - 2y^3 + x^2 - y^2 - 2x - 3y = 0$ cut the curve, lie on a straight line which will pass through origin.
3. Find all the asymptotes of the curve
$$3x^3 + 2x^2y - 7xy^2 + 2y^3 - 14xy + 7y^2 + 4x + 5y = 0.$$
Show that the asymptotes meet the curve again in three points which lie on a straight line; and find the equation of line.

[**Ans.** $6(y - x) + 7 = 0, 2(y - 3x) + 3 = 0; 3(2y + x) + 5 = 0, 106y - 381x + 105 = 0$]

5.8 ASYMPTOTES IN POLAR CO-ORDIANTES

If a curve is referred to polar coordinates the directions of its points at an infinite distance from the origin can be in general determined by making $r = \infty$ or $u = \dfrac{1}{r} = 0$ in its equation and solving the resulting equation in θ.

The position of the asymptotes corresponding to any such value of θ is obtained by finding the length of the corresponding polar subtangent i.e., by finding the value of $\dfrac{d\theta}{du} = 0$ corresponding to $u = 0$.

If the equation of the curve, arranged in powers of r is
$$r^n f_0(\theta) + r^{n-1} f_1(\theta) + \ldots + r f_{n-1}(\theta) + f_n(\theta) = 0$$

or
$$u^n f_n(\theta) + u^{n-1} f_{n-1}(\theta) + \ldots + u f_1(\theta) + f_0(\theta) = 0 \qquad \ldots(i)$$

by substituting u for $1/r$.

Consequently the directions of the asymptotes are given by putting $u = 0$, i.e., by the equation $f_0(\theta) = 0$.

Again differentiating (i) with respect to θ, we obtain

$$nu^{n-1} \frac{du}{d\theta} f_1(\theta) + (n-1) u^{n-2} \frac{du}{d\theta} f_{n-1}(\theta) + u^{n-1} f'_{n-1}(\theta) + \ldots$$

$$+ \frac{du}{d\theta} f_1(\theta) + u f'_1(\theta) + f'_0(\theta) = 0$$

Then $\dfrac{du}{d\theta} f_1(\theta) + f'_0(\theta) = 0$

is the equation corresponding to $u = 0$ provided that $f_1(\theta), f_2(\theta), \ldots$ are not infinite for the values of θ.

If α is the root of the equation $f_0(\theta) = 0$, the curve has an asymptote making the angle α, whose perpendicular distance from the origin is represented by

$$-\frac{f_1(\alpha)}{f'_0(\alpha)}.$$

Then the equation of the corresponding asymptote is

$$r \sin(\alpha - \theta) - \frac{f_1(\alpha)}{f'_0(\alpha)} = 0$$

EXAMPLES

1. Find the asymptotes of the curve
$$r = a \sec \theta + b \tan \theta.$$

Solution. Here
$$u = \frac{1}{a \sec \theta + b \tan \theta} = \frac{\cos \theta}{a + b \sin \theta}$$

For $\theta = \frac{\pi}{2}$, $u = 0$,

and
$$\frac{du}{d\theta} = \frac{-\sin \theta (a + b \sin \theta) - \cos \theta (b \cos \theta)}{(a + b \sin \theta)^2}$$

$$\therefore \left(\frac{du}{d\theta}\right)_{\theta = \pi/2} = -\frac{1}{a+b}$$

As the corresponding polar subtangent is $(a+b)$ and hence the perpendicular to radius vector at a distance of $(a+b)$ from the origin is an asymptote to the curve.

Again $u = 0$ for $\theta = 3\pi/2$; the corresponding value of polar subtangent is $a - b$, giving another asymptote.

Hence the asymptotes are
$$r \sin\left(\theta - \frac{\pi}{2}\right) = a + b \quad \text{and} \quad r \sin\left(\theta - \frac{3\pi}{2}\right) = a - b.$$

2. Find asymptotes for $r\theta = a$.

Solution. We have
$$u = \frac{1}{r} = \frac{\theta}{a}$$

Now $u = 0$ for $\theta = 0$ and $\frac{du}{d\theta} = \frac{1}{a}$.

Hence the perpendicular distance $= a$; then the asymptote is
$$r \sin (\theta - 0) = a$$
or
$$r \sin \theta = a.$$

EXERCISES

Find the asymptotes of :

1. $r = a \operatorname{cosec} \theta + b$. [Ans. $r \sin \theta = a$]

2. $r = \dfrac{2}{1 + 2 \sin \theta}$ [Ans. $r \sin\left(\theta \pm \dfrac{\pi}{6}\right) = \dfrac{2}{\sqrt{3}}$;]

3. $r = 4 (\sec \theta + \tan \theta)$ [Ans. $r \cos \theta = 8$]

4. $r = \dfrac{2\theta}{\sin \theta}$ [Ans. $r \sin \theta = 2k\pi; k = \pm 1, \pm 2, ...$]

5. $r \sin 2\theta = a$ [Ans. $r \sin \theta = \pm \dfrac{1}{2} a; r \cos \theta = \pm \dfrac{1}{2} a$]

6. $r \sin n\theta = a$ [Ans. $r \sin\left(\theta - \dfrac{m\pi}{n}\right) = \dfrac{a}{n \cos m\pi}$]

7. $r \cos \theta = a \sin \theta$ [Ans. $r \cos \theta = \pm a$]

8. $r \theta \cos \theta = a \cos 2\theta$ [Ans. $r \sin \theta = a, r \cos \theta = \dfrac{2a}{(2m+1)\pi}$]

6
Curvature

6.1 INTRODUCTION

We consider here two curve *AB* and *CD* as shown in the figure. One i.e., *AB* bends more sharply than *CD*. We can express it in other words as one has a greater curvature than the other. A mathematical definition is given below.

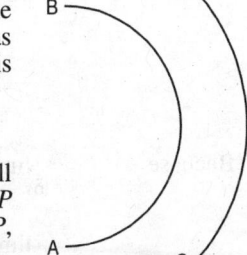

6.2 DEFINITIONS

Let *P* be a given point on a given curve, and *Q* another point in small neighbourhood of *P*. Let *N* be the point of intersection of normals at *P* and *Q* to the curve. If *N* tends to a definite position *C* as *Q* tends to *P*, then *C* is called the **centre of curvature** of the curve at the given point *P*.

The length *CP* is called the **radius of curvature** and is usually denoted by the Greek letter ρ.

The reciprocal of *CP*, i.e., $1/CP$ is called the **curvature** of the curve at the point *P*.

The circle with its centre *C* and radius *CP* is called the **circle of curvature** of the curve at *P*. Any chord of the circle of curvature, drawn through *P*, is called a **chord of curvature**.

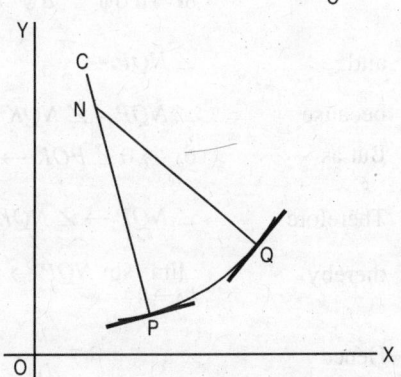

6.3 RADIUS OF CURVATURE

Let the length of arc of *AP* be *s*, and that of *Q* be $s + \delta s$.

Let the tangents at *P* and *Q* subtend the angle ψ and $\psi + \delta \psi$ respectively with the *x*-axis. The angle between the tangents at *P* and *Q* will be $\delta \psi$. This angle $\delta \psi$ is called the total curvature of the arc *PQ*.

Let the tangents at *P* and *Q* meet in *K* and normals at *N*. Join *PQ*.

Then the radius of curvature at *P*

$$= \rho = \lim_{\delta s \to 0} PN$$

In $\triangle PNQ$, $\quad \dfrac{PN}{\sin NQP} = \dfrac{\text{chord } PQ}{\sin PNQ}$

But $\angle PNQ = 180° - \angle PKQ$

(as $\angle NPK$ and $\angle NQK$ are right angles)

$= \angle TKT' = \delta \psi$.

Therefore $\quad \dfrac{PN}{\sin NQP} = \dfrac{\text{chord } PQ}{\sin \delta \psi}$

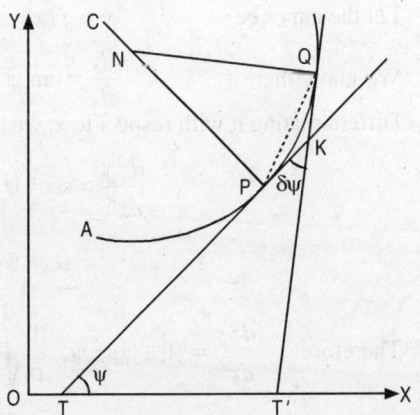

or
$$PN = \frac{\text{chord } PQ \cdot \sin NQP}{\sin \delta\psi}$$

$$= \frac{\text{chord } PQ}{\delta s} \cdot \frac{\delta s}{\delta \psi} \cdot \frac{\delta \psi}{\sin \delta\psi} \cdot \sin NQP$$

Hence,
$$\rho = \lim_{\substack{\delta s \to 0 \\ \delta \psi \to 0}} \frac{\text{chord } PQ}{\delta s} \cdot \frac{\delta s}{\delta \psi} \cdot \frac{\delta \psi}{\sin \delta\psi} \sin NQP$$

$$= \lim_{\delta \psi \to 0} \left(\lim_{\delta s \to 0} \frac{\text{chord } PQ}{\delta s} \right) \left(\frac{\delta s}{\delta \psi} \right) \frac{\delta \psi}{\delta \psi} \sin NQP$$

$$= \frac{ds}{d\psi}$$

Becuase
$$\lim_{\delta s \to 0} \frac{\text{chord } PQ}{\delta s} \to 1,$$

$$\lim_{\delta s \to 0} \frac{\delta s}{\delta \psi} \to \frac{ds}{d\psi}, \quad \lim_{\delta \psi \to 0} \frac{\delta \psi}{\sin \delta\psi} \to 1$$

and $\angle NQP \to \dfrac{\pi}{2}$

because $\angle NQP = \angle NQK - \angle PQK$

But as $\delta s \to 0 \angle PQK \to 0$

Therefore $\angle NQP \to \angle NQK$, i.e., $\dfrac{\pi}{2}$

thereby $\lim_{\delta s \to 0} \sin NQP \to 1$

Hence $\rho = \dfrac{ds}{d\psi}.$

6.3.1 Radius of Curvature for Cartesian Curves

Let the curve be $\quad y = f(x)$

We know that $\quad \dfrac{dy}{dx} = \tan \psi$

Differentiating it with respect to x, we have

$$\frac{d^2 y}{dx^2} = \sec^2 \psi \cdot \frac{d\psi}{dx}$$

$$= \sec^2 \psi \cdot \frac{d\psi}{ds} \cdot \frac{ds}{dx}$$

Therefore $\dfrac{d^2 y}{dx^2} = (1 + \tan^2 \psi) \cdot \dfrac{1}{\rho} \sqrt{\left\{1 + \left(\dfrac{dy}{dx}\right)^2\right\}}$

$$= \left\{1 + \left(\frac{dy}{dx}\right)^2\right\} \cdot \sqrt{\left\{1 + \left(\frac{dy}{dx}\right)^2\right\}} \frac{1}{\rho} \qquad \left(\text{as } \frac{dy}{dx} = \tan \psi\right)$$

Curvature

$$= \left\{1+\left(\frac{dy}{dx}\right)^2\right\}^{3/2} \cdot \frac{1}{\rho}$$

or
$$\rho = \frac{\left[1+\left(\frac{dy}{dx}\right)^2\right]^{3/2}}{\frac{d^2y}{dx^2}} = \frac{(1+y_1^2)^{3/2}}{y_2}$$

Note I. ρ denotes length; it is positive or negative according as $\frac{d^2y}{dx^2}$ is positive or negative. If this length becomes negative, the sign is left.

Note II. Since the value of ρ is independent of the choice of x-axis and y-axis, interchanging x and y we observe that

$$\rho = \frac{\left[1+\left(\frac{dx}{dy}\right)^2\right]^{3/2}}{\frac{d^2x}{dy^2}}$$

This formula is useful in case $\frac{dy}{dx}$ is infinite or $\frac{dx}{dy} = 0$.

EXAMPELS

1. *Prove that the radius of curvature at any point of the catenary $y = c \cosh \frac{x}{c}$ varies as the square of the ordinate.*

Solution. The given curve is $y = c \cosh \frac{x}{c}$.

Differentiating it with respect to x, we have

$$\frac{dy}{dx} = c \cdot \sinh \frac{x}{c} \cdot \frac{1}{c} = \sinh \frac{x}{c}$$

and
$$\frac{d^2y}{dx^2} = \cosh \frac{x}{c} \cdot \frac{1}{c}$$

$$\therefore \quad \rho = \frac{\left[1+\left(\frac{dy}{dx}\right)^2\right]^{3/2}}{\frac{d^2y}{dx^2}}$$

$$= \frac{c\left[1+\sinh^2 \frac{x}{c}\right]^{3/2}}{\cosh \frac{x}{c}}$$

$$= c \cosh^2 x/c = c \cdot \frac{y^2}{c^2} = \frac{y^2}{c}$$

Hence $\rho \propto (\text{ordinate})^2$.

2. *In the ellipse* $\dfrac{x^2}{a^2} + \dfrac{y^2}{b^2} = 1$, *show that the radius of curvature at an end of the major axis is equal to semi-latus rectum of the ellipse.*

Solution. The given curve is $\dfrac{x^2}{a^2} + \dfrac{y^2}{b^2} = 1$

Differentiating it with respect to x, we get

$$\dfrac{2x}{a^2} + \dfrac{2y}{b^2} \cdot \dfrac{dy}{dx} = 0$$

or $\dfrac{dy}{dx} = -\dfrac{b^2 x}{a^2 y}$

and $\dfrac{d^2 y}{dx^2} = -\dfrac{b^2}{a^2} \left[\dfrac{y - x\left(-\dfrac{b^2 x}{a^2 y}\right)}{y^2} \right]$

$$= -\dfrac{b^2}{a^4 y^3} \cdot (a^2 y^2 + b^2 x^2)$$

$$= -\dfrac{b^2}{a^4 y^3} \cdot a^2 b^2 \left(\dfrac{x^2}{a^2} + \dfrac{y^2}{b^2}\right)$$

$$= -\dfrac{b^4}{a^2 y^3}$$

Now, $\rho = \dfrac{\left[1 + \left(\dfrac{dy}{dx}\right)^2\right]^{3/2}}{d^2 y / dx^2}$

$$= -\dfrac{\left(1 + \dfrac{b^4 x^2}{a^4 y^2}\right)^{3/2}}{b^4 / a^2 y^3}$$

$$= -\dfrac{(a^4 y^2 + b^4 x^2)^{3/2}}{a^4 b^4}$$

Now, ρ at one end of major axis, *i.e.*, $(a, 0)$

$$= \dfrac{(a^4 \cdot 0 + b^4 \cdot a^2)^{3/2}}{a^4 b^4} = \dfrac{b^6 a^3}{a^4 b^4}$$

$$= \dfrac{b^2}{a} = \text{semi-latus rectum}.$$

3. *Find radius of curvature at any point of the curve*

$$x = ae^\theta (\sin\theta - \cos\theta),$$

$$y = ae^\theta (\sin\theta + \cos\theta)$$

Solution. From the given equation of curve, we have

Curvature

$$\frac{dx}{d\theta} = ae^\theta(\cos\theta + \sin\theta) + ae^\theta(\sin\theta - \cos\theta)$$

$$= 2ae^\theta \sin\theta$$

and $\quad \dfrac{dy}{d\theta} = 2ae^\theta \cos\theta$

$\therefore \quad \dfrac{dy}{dx} = \dfrac{dy/d\theta}{dx/d\theta} = \dfrac{2ae^\theta \cos\theta}{2ae^\theta \sin\theta} = \cot\theta$

$$\frac{d^2y}{dx^2} = \frac{d}{dx}\left(\frac{dy}{dx}\right) = \frac{d}{d\theta}(\cot\theta) \cdot \frac{d\theta}{dx}$$

$$= -\csc^2\theta \cdot \frac{1}{2ae^\theta \sin\theta} = -\frac{1}{2ae^\theta \sin^3\theta}$$

$\therefore \quad \rho = \dfrac{\left[1 + \left(\dfrac{dy}{dx}\right)^2\right]^{3/2}}{d^2y/dx^2}$

$$= \frac{(1 + \cot^2\theta)^{3/2}}{-\dfrac{1}{2ae^\theta \sin^3\theta}}$$

$$= -2ae^\theta \sin^3\theta \cdot \csc^3\theta$$

$$= 2ae^\theta \quad \text{(leaving negative sign as } \rho \text{ is a length).}$$

EXERCISES

1. Find the radius of curvature at the point (s, ψ) on the following curves :

   (i) $s = c \tan\psi$;   (ii) $s = 8a \sin^2 \psi/6$;

   (iii) $s = 4a \sin\psi$;   (iv) $s = c \log \tan\left(\dfrac{\pi}{4} + \dfrac{\psi}{2}\right)$

   (v) $s = c \sec^3 \psi$;   (vi) $s - a \log(\sec\psi \tan\psi) + a \tan\psi \sec\psi$;

   (vii) $x = a(e^{m\psi} - 1)$.

   [**Ans.** (i) $c \sec^2\psi$;   (ii) $\dfrac{4}{3} a \sin\psi/3$;   (iii) $4a \cos\psi$;   (iv) $c \sec\psi$;

   (v) $3c \sec^4\psi \tan\psi$;   (vi) $2a \sec^3\psi$;   (vii) $ame^{m\psi}$]

2. Find the radius of curvature of the following at (x, y):

   (i) $y^2 = 4ax$;   (ii) $xy = c^2$ (iii) $ay^2 = x^3$;
   (iv) $y = \log \sin x$ (v) $y = a \log \sec(x/a)$;
   (vi) $x^{2/3} + y^{2/3} = a^{2/3}$;   (vii) $a^2 y = x^3 - a^3$.

   [**Ans.** (i) $\dfrac{2}{\sqrt{a}}(x+a)^{3/2}$;   (ii) $\dfrac{(x^2+y^2)^{3/2}}{2c^2}$;   (iii) $\dfrac{\sqrt{x}(4a+9x)^{3/2}}{6a}$;

   (iv) $-\csc x$;   (v) $a \sec\dfrac{x}{a}$;   (vi) $3(axy)^{1/3}$;   (vii) $\dfrac{(a^4+9x^4)^{3/2}}{6xa^4}$]

3. Prove that for the ellipse $\dfrac{x^2}{a^2} + \dfrac{y^2}{b^2} = 1$, $\rho = \dfrac{a^2 b^2}{p^3}$, p being the perpendicular from the centre upon the tangent at (x, y).

4. If ρ_1, ρ_2 be the radii of curvature at the extremities of two conjugate diameters on an ellipse, prove that
$$(\rho_1^{2/3} + \rho_2^{2/3})\, a^{2/3} b^{2/3} = a^2 + b^2.$$

5. Show that for the curve $y = \dfrac{ax}{a+x}$
$$\left(\dfrac{2\rho}{a}\right)^{2/3} = \left(\dfrac{y}{x}\right)^2 + \left(\dfrac{x}{y}\right)^2.$$

6.3.2 Radius of Curvature (Pedal Equations)

From the figure
$$\psi = \theta + \phi$$
Differentiating both sides with respect to s, we obtain
$$\dfrac{d\psi}{ds} = \dfrac{d\theta}{ds} + \dfrac{d\phi}{ds}$$
$$= \dfrac{d\theta}{ds} + \dfrac{d\phi}{dr}\cdot\dfrac{dr}{ds}$$

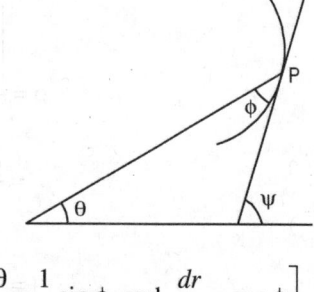

$\therefore \quad \dfrac{1}{\rho} = \dfrac{1}{r}\sin\phi + \cos\phi\cdot\dfrac{d\phi}{dr} \qquad \left[\because \dfrac{d\theta}{ds} = \dfrac{1}{r}\sin\phi \text{ and } \dfrac{dr}{ds} = \cos\phi\right]$

$\therefore \quad \dfrac{1}{\rho} = \dfrac{1}{r}\left(\sin\phi + r\cos\phi\cdot\dfrac{d\phi}{dr}\right)$

$\phantom{\therefore \quad \dfrac{1}{\rho}} = \dfrac{1}{r}\dfrac{d}{dr}(r\sin\phi)$

$\phantom{\therefore \quad \dfrac{1}{\rho}} = \dfrac{1}{r}\dfrac{dp}{dr}$

$\therefore \quad \rho = r\dfrac{dr}{dp}$

6.3.3 Radius of Curvature When Polar Equation of the Curve is Given

We know that
$$\dfrac{1}{p^2} = \dfrac{1}{r^2} + \dfrac{1}{r^4}\left(\dfrac{dr}{d\theta}\right)^2$$

Differentiating both sides of this equation with respect to t, we get
$$-\dfrac{2}{p^3}\dfrac{dp}{dr} = -\dfrac{2}{r^3} + \dfrac{2}{r^4}\left(\dfrac{dr}{d\theta}\right)\dfrac{d^2 r}{d\theta^2}\cdot\dfrac{d\theta}{dr} - \dfrac{4}{r^5}\left(\dfrac{dr}{d\theta}\right)^2$$

or
$$\dfrac{1}{p^3}\dfrac{dp}{dr} = \dfrac{1}{r^3} - \dfrac{1}{r^4}\dfrac{d^2 r}{d\theta^2} + \dfrac{2}{r^5}\left(\dfrac{dr}{d\theta}\right)^2$$

$$= \dfrac{r^2 - r\dfrac{d^2 r}{d\theta^2} + 2\left(\dfrac{dr}{d\theta}\right)^2}{r^5}$$

Curvature

or
$$r\frac{dr}{dp} = \frac{\dfrac{r^6}{p^3}}{r^2 - r\dfrac{d^2r}{d\theta^2} + 2\left(\dfrac{dr}{d\theta}\right)^2} \qquad ...(i)$$

But
$$\frac{1}{p^2} = \frac{1}{r^2} + \frac{1}{r^4}\left(\frac{dr}{d\theta}\right)^2$$

or
$$\frac{r^4}{p^2} = r^2 + \left(\frac{dr}{d\theta}\right)^2$$

or
$$\left(\frac{r^4}{p^2}\right)^{3/2} = \left[r^2 + \left(\frac{dr}{d\theta}\right)^2\right]^{3/2}$$

or
$$\frac{r^6}{p^3} = \left\{r^2 + \left(\frac{dr}{d\theta}\right)^2\right\}^{3/2}$$

and
$$\rho = r\frac{dr}{dp}$$

Hence from (i), we have

$$\rho = \frac{\left\{r^2 + \left(\dfrac{dr}{d\theta}\right)^2\right\}^{3/2}}{r^2 - r\dfrac{d^2r}{d\theta^2} + 2\left(\dfrac{dr}{d\theta}\right)^2}$$

Note 1. If $\dfrac{dr}{d\theta}$ and $\dfrac{d^2r}{d\theta^2}$ are denoted by r' and r'', then the above formula for radius of curvature ρ is given by

$$\rho = \frac{(r^2 - r'^2)^{3/2}}{r^2 - rr'' + 2r'^2}$$

Note 2. If we put $u = \dfrac{1}{r'}$ then $r = \dfrac{1}{u'}\dfrac{dr}{d\theta} = -\dfrac{1}{u^2}\dfrac{du}{d\theta}$ and

$$\frac{d^2r}{d\theta^2} = -\frac{2}{u^3}\left(\frac{du}{d\theta}\right)^2 - \frac{1}{u^2}\frac{d^2u}{d\theta^2}$$

then
$$\rho = \frac{\left\{r^2 + \left(\dfrac{dr}{d\theta}\right)^2\right\}^{3/2}}{r^2 - r\dfrac{d^2r}{d\theta^2} + 2\left(\dfrac{dr}{d\theta}\right)^2}$$

$$= \frac{\left\{\dfrac{1}{u^2} + \dfrac{1}{u^4}\left(\dfrac{du}{d\theta}\right)^2\right\}^{3/2}}{\dfrac{1}{u^2} - \dfrac{1}{u}\left\{-\dfrac{2}{u^3}\left(\dfrac{du}{d\theta}\right)^2 - \dfrac{1}{u^2}\dfrac{d^2u}{d\theta^2}\right\} + \dfrac{2}{u^4}\left(\dfrac{du}{d\theta}\right)^2}$$

On simplification, it gives

$$\rho = \frac{\left\{u^2 + \left(\dfrac{du}{d\theta}\right)^2\right\}^{3/2}}{u^3\left(u + \dfrac{d^2u}{d\theta^2}\right)}$$

Note 3. The students are advised to follow the following procedure as far as possible in the problems when the equation of the curve is given in polar coordinates :

(i) Find its pedal equation $p = f(r)$

(ii) Then apply $\rho = r \dfrac{dr}{dp}$.

6.3.4 Radius of Curvature When the Equation of the Curve is Given in p AND ψ or to Prove That

$$\rho = p + \frac{d^2p}{d\psi^2}$$

We have

$$\frac{dp}{d\psi} = \frac{dp}{dr} \cdot \frac{dr}{ds} \cdot \frac{ds}{d\psi}$$

$$= \frac{dp}{dr} \cdot \cos\phi \cdot \rho \qquad \left(\because \frac{dr}{ds} = \cos\phi\right)$$

$$= \frac{dp}{dr} \cos\phi \cdot r \frac{dr}{dp} \qquad \left(\because \rho = r \frac{dr}{dp}\right)$$

$$= r \cos\phi$$

Now again $p^2 + \left(\dfrac{dp}{d\psi}\right)^2 = r^2 \sin^2\phi + r^2 \cos^2\phi = r^2$.

Differentiating with respect to p, we obtain

$$2p + 2 \cdot \frac{dp}{d\psi} \cdot \frac{d^2p}{d\psi^2} \cdot \frac{d\psi}{dp} = 2r \frac{dr}{dp} \qquad \text{or} \qquad p + \frac{d^2p}{d\psi^2} = \rho$$

This is knows as tangential polar formula.

6.3.5 More Formulae for Radius of Curvature

(i) **When x and y are functions of s.**

$$\frac{ds}{dx} = \sqrt{1 + \left(\frac{dy}{dx}\right)^2} = \sqrt{(1 + \tan^2\psi)}$$

$$= \sec\psi$$

$$\therefore \qquad \cos\psi = \frac{dx}{ds}.$$

Differentiating with respect to s, we obtain

$$-\sin\psi \cdot \frac{d\psi}{ds} = \frac{d^2x}{ds^2} \qquad \text{or} \qquad -\sin\psi \cdot \frac{1}{\rho} = \frac{d^2x}{ds^2}$$

Curvature

$$\therefore \quad \rho = -\frac{\sin \psi}{d^2x/ds^2} \quad \text{while} \quad \sin \psi = dy/ds$$

Hence
$$\rho = -\frac{dy/ds}{\dfrac{d^2x}{ds^2}}$$

(ii) Similarly, we obtain
$$\rho = +\frac{dx/ds}{\dfrac{d^2y}{ds^2}}$$

by taking $\quad \sin \psi = \dfrac{dy}{ds}.$

(iii) Now again, we know
$$\rho = -\frac{dy/ds}{d^2x/ds^2}$$

$$= -\frac{\sin \psi}{d^2x/ds^2}$$

$$\therefore \quad \frac{\sin \psi}{\rho} = \frac{d^2x}{ds^2} \qquad \ldots(i)$$

and also $\quad \dfrac{\cos \psi}{\rho} = \dfrac{d^2y}{ds^2} \qquad \ldots(ii)$

Squaring and adding (i) and (ii),

we obtain
$$\frac{1}{\rho^2} = \left(\frac{d^2x}{ds^2}\right)^2 + \left(\frac{d^2y}{ds^2}\right)^2$$

EXAMPLES

1. *Show that for any curve*
$$\frac{r}{\rho} = \sin \phi \left(1 + \frac{d\phi}{d\theta}\right)$$

Solution. We have

R.H.S. $= \sin \phi \left(1 + \dfrac{d\phi}{d\theta}\right)$

$= r \dfrac{d\theta}{ds} \left(1 + \dfrac{d\phi}{d\theta}\right)$

$= r \left\{\dfrac{d\theta}{ds} + \dfrac{d\theta}{ds} \cdot \dfrac{d\phi}{d\theta}\right\}$

$= r \cdot \left\{\dfrac{d\theta}{ds} + \dfrac{d\phi}{ds}\right\}$

$= r \dfrac{d}{ds}(\theta + \phi) = r \dfrac{d\psi}{ds} = \dfrac{r}{\rho}.$

2. *Find the radius of curvature of the curve* $r = a(1 + \cos \theta).$

Solution. We have $\quad \dfrac{dr}{d\theta} = -a \sin \theta, \quad \dfrac{d^2r}{d\theta^2} = -a \cos \theta$

$$\therefore \quad \rho = \frac{\left\{r^2 + \left(\frac{dr}{d\theta}\right)^2\right\}^{3/2}}{r^2 + 2\left(\frac{dr}{d\theta}\right)^2 - r\frac{d^2r}{d\theta^2}}$$

$$= \frac{\{a^2(1+\cos\theta)^2 + a^2\sin^2\theta\}^{3/2}}{a^2(1+\cos\theta)^2 + 2a^2\sin^2\theta - a^2\cos\theta(1+\cos\theta)}$$

$$= \frac{4}{3}a\cos\frac{\theta}{2} = \frac{4}{3}a\frac{1}{2}\sqrt{4\cos^2\frac{\theta}{2}} = \frac{2a}{3}\sqrt{2(1+\cos\theta)}$$

$$= \frac{2a}{3}\sqrt{2\frac{r}{a}}$$

$$= \frac{2}{3}\sqrt{(2ar)}$$

Aliter. We have $\quad r = a(1+\cos\theta)$

$$\therefore \quad \frac{dr}{d\theta} = -a\sin\theta \text{ and}$$

$$\tan\phi = r\frac{d\theta}{dr} = \frac{a(1+\cos\theta)}{-a\sin\theta} = -\cot\frac{\theta}{2} = \tan\left(\frac{\pi}{2} + \frac{\theta}{2}\right)$$

$$\therefore \quad \phi = \frac{\pi}{2} + \frac{\theta}{2}$$

Now $\quad p = r\sin\phi$, i.e., $p = r\sin\left(\frac{\pi}{2} + \frac{\theta}{2}\right) = r\cos\frac{\theta}{2}$

or $\quad p^2 = r^2\cos^2\frac{\theta}{2} = \frac{1}{2}r^2(1+\cos\theta)$

$\therefore \quad 2ap^2 = r^3 \qquad$ (pedal equation)

or $\quad \sqrt{(2a)}\,p = r^{3/2}$

Differentiating with respect to r, we have

$$\sqrt{(2a)}\frac{dp}{dr} = \frac{3}{2}r^{1/2}$$

Hence $\quad \rho = r\frac{dr}{dp} = \frac{r \cdot 2\sqrt{(2a)}}{3\sqrt{r}}$

$$= \frac{2}{3}\sqrt{(2ar)}$$

3. Show that for any curve $r = f(\theta)$ the curvature is given by

$$\left(u + \frac{d^2u}{d\theta^2}\right)\sin^3\phi, \text{ where } u = 1/r$$

Solution. We have $\quad \rho = \dfrac{\left[u^2 + \left(\dfrac{du}{d\theta}\right)^2\right]^{3/2}}{u^3\left[u + \dfrac{d^2u}{d\theta^2}\right]}$

Curvature

$$\therefore \quad \frac{1}{\rho} = \frac{u^3 \left(u + \frac{d^2u}{d\theta^2}\right)}{\left[u^2 + \left(\frac{du}{d\theta}\right)^2\right]^{3/2}}$$

$$= \frac{u^3}{\left[u^2 + \left(\frac{du}{d\theta}\right)^2\right]^{3/2}} \left[u + \frac{d^2u}{d\theta^2}\right]$$

$$= \frac{1/r^3}{\left[\frac{1}{r^2} + \frac{1}{r^4}\left(\frac{dr}{d\theta}\right)^2\right]^{3/2}} \left[u + \frac{d^2u}{d\theta^2}\right]$$

$$= \frac{1/r^3}{1/p^3} \left[u + \frac{d^2u}{d\theta^2}\right] = \frac{p^3}{r^3}\left[u + \frac{d^2u}{d\theta^2}\right]$$

$$= \frac{r^3 \sin^3 \phi}{r^3} \left[u + \frac{d^2u}{d\theta^2}\right]$$

$$= \left[u + \frac{d^2u}{d\theta^2}\right] \sin^3 \phi$$

4. *Prove that for any curve*

$$\frac{1}{\rho} = \frac{d}{dx}\left(\frac{dy}{ds}\right)$$

Solution. We are given that

$$\frac{d}{dx}\left(\frac{dy}{ds}\right) = \frac{d}{dx}(\sin \psi)$$

$$= \frac{d}{d\psi}(\sin \psi) \cdot \frac{d\psi}{dx}$$

$$= \cos \psi \cdot \frac{d\psi}{ds} \cdot \frac{ds}{dx}$$

$$= \cos \psi \cdot \frac{d\psi}{ds} \cdot \sec \psi = \frac{d\psi}{ds} = \frac{1}{\rho}.$$

EXERCISES

1. If the pedal equation of an ellipse be $\frac{1}{p^2} = \frac{1}{a^2} + \frac{1}{b^2} - \frac{r^2}{a^2 b^2}$ find the value of ρ.

   [Ans. $(a^2 b^2)/\rho^3$**]**

2. Find the radius of curvature of $r = a \cos n\theta$ as a function of r; also show that at a point where $r = a$ its value is $\frac{a}{(1+n^2)}$.

3. Find the radius of curvature of the rectangular hyperbola $r^2 \cos 2\theta = a^2$.

   [Ans. r^3/a^2**]**

4. Find the radius of curvature at the point r for the curve $u^{-m} = a^m \sin m\theta$, where $u = 1/r$.

[Ans. $a^m / \{(m+1) r^{m-1}\}$]

5. Show for the curve $s = f(x)$, $\rho = \dfrac{\left(\dfrac{ds}{dx}\right)^2 \sqrt{\left\{\left(\dfrac{ds}{dx}\right)^2 - 1\right\}}}{\dfrac{d^2 s}{dx^2}}$

6. For any curve, prove that $\dfrac{d^2 r}{ds^2} = \dfrac{\sin^2 \phi}{r} - \dfrac{\sin \phi}{\rho}$.

6.4 RADIUS OF CURVATURE AT ORIGIN

We know that $\rho = \dfrac{\left\{1 + \left(\dfrac{dy}{dx}\right)^2\right\}^{3/2}}{\dfrac{d^2 y}{dx^2}}$ ...(i)

Now we may determine $\dfrac{dy}{dx}$ and $\dfrac{d^2 y}{dx^2}$ for the point $x = 0$ and $y = 0$, then substitute the values of $\dfrac{dy}{dx}$ and $\dfrac{d^2 y}{dx^2}$ in (i).

Aliter. We can exapnd $f(x)$ with the help of Maclaurin's Theorem as

$$f(x) = (y)_0 + x (y_1)_0 + \dfrac{x^2}{2!} + \dfrac{x^3}{3!} (y_3)_0 + \ldots \quad \text{...(i)}$$

Hence any $f(x)$ can also be written as

$$y = px + q \dfrac{x^2}{2!} + \ldots \quad \text{...(ii)}$$

The result (ii) may be obtained by expanding $f(x)$ in ascending powers of x by trigonometrical or algebraic methods.

Comparing (i) and (ii) we obtain

$$p = (y_1)_0 = \left(\dfrac{dy}{dx}\right)_{x=0, y=0}$$

$$q = (y_2)_0 = \left(\dfrac{d^2 y}{dx^2}\right)_{x=0, y=0}$$

Therefore ρ (at the origin) $= \left[\dfrac{\left\{1 + \left(\dfrac{dy}{dx}\right)^2\right\}^{3/2}}{\dfrac{d^2 y}{dx^2}}\right]_{x=0, y=0}$

Curvature

$$= \frac{(1+p^2)^{3/2}}{q}.$$

6.4.1 Radius of Curvature by Newton's Method

If the curve passes through the origin, i.e., $x = 0$, when $y = 0$. Also if x-axis is tangent to the curve at the origin, then at $x = 0$, $y = 0$, $\frac{dy}{dx} = 0$.

Then
$$y = (y)_0 + (y_1)_0 \frac{x}{1!} + (y_2)_0 \frac{x^2}{2!} + \ldots$$

or
$$y = 0 + 0.x + (y_2)_0 \cdot \frac{x^2}{2!} + \ldots$$

Dividing by $\frac{1}{2}x^2$ and taking the limit as $x \to 0$, we obtain

$$\lim_{x \to 0} \frac{2y}{x^2} = (y_2)_0 = q$$

Then ρ (at the origin) $= \frac{(1+p^2)^{3/2}}{q}$

$$= \frac{1}{q} \text{ as } p = 0$$

$$= \lim_{x \to 0} \frac{x^2}{2y}$$

Therefore when $\frac{dy}{dx} = 0$ and the curve passes through the origin, x-axis is a tangent to the curve at origin.

$$\therefore \quad \rho \text{ (at the origin)} = \lim_{\substack{x \to 0 \\ y \to 0}} \frac{x^2}{2y}$$

and also when y-axis is tangent to the curve at the origin, then

$$\rho \text{ (at the origin)} = \lim_{\substack{x \to 0 \\ y \to 0}} \frac{y^2}{2x}$$

6.4.2 Radius of Curvature at the Pole

We know $x = r \cos \theta$ and $y = r \sin \theta$ and at the pole (in case the initial line is tangent to the curve at the pole),

$$\rho = \lim_{x \to 0} \left(\frac{x^2}{2y}\right) = \lim_{x \to 0} \left[\frac{r^2 \cos^2 \theta}{2r \sin \theta}\right]$$

$$= \lim_{\theta \to 0} \frac{1}{2}\left[\frac{r}{\theta} \cdot \frac{\theta}{\sin \theta} \cdot \cos^2 \theta\right]$$

$$= \lim_{\theta \to 0} \frac{1}{2}\left(\frac{r}{\theta}\right) \text{ as } \frac{\theta}{\sin \theta} \to 1 \text{ and } \cos^2 \theta \to 1 \text{ as } \theta \to 0$$

Hence ρ (at pole) $= \lim_{\theta \to 0} \frac{1}{2}\left(\frac{r}{\theta}\right)$.

EXAMPLES

1. *Obtain the radii of curvature of the curve* $a(y^2 - x^2) = x^3$ *at the origin.*

Solution. The curve is $\quad ay^2 = x^3 + ax^2$

or
$$y = \sqrt{x^2 + \frac{x^3}{a}}$$

$$= \pm x \left(1 + \frac{x}{a}\right)^{1/2}$$

$$= \pm x \left(1 + \frac{1}{2}\frac{x}{a} - \frac{1}{8}\frac{x^2}{a^2} + \ldots\right)^{1/2}$$

$$= \pm \left(x + \frac{x^2}{2a} - \frac{1}{8a^2}x^3 + \ldots\right)$$

Hence
$$p = \left(\frac{dy}{dx}\right)_{at\ (0,0)} = \pm 1$$

and
$$q = \left(\frac{d^2y}{d^2x}\right)_{at\ (0,0)} = \pm \frac{1}{a}$$

Therefore
$$\rho = \frac{(1+p^2)^{3/2}}{q} = \frac{(2^{3/2})}{\pm 1/a} = \pm 2\sqrt{2}a$$

2. *Find* ρ *at the pole for the curve* $r = a \sin n\theta$.

Solution. The curve is $r = a \sin n\theta$. Here r and θ are 0. Hence initial line is tangent to the curve at origin.

we know,
$$\rho = \lim_{\theta \to 0} \frac{r}{2\theta}$$

$$= \lim_{\theta \to 0} \frac{a \sin n\theta}{2\theta}$$

$$= \lim_{\theta \to 0} \frac{\sin n\theta}{n\theta} \cdot \frac{na}{2}$$

$$= \frac{na}{2}.$$

3. *Apply Newton's formula to find the radius of curvature at the origin for the cycloid*
$$x = a(\theta + \sin\theta),\ y = a(1 - \cos\theta).$$

Solution. We have $\quad \dfrac{dx}{d\theta} = a(1 + \cos\theta),\ \dfrac{dy}{d\theta} = a \sin\theta$

$\therefore\quad \dfrac{dy}{dx} = \dfrac{dy/d\theta}{dx/d\theta} = \dfrac{a \sin\theta}{a(1 + \cos\theta)}$

$$= \tan\theta/2$$

$\therefore\quad \dfrac{dy}{dx} = 0,\ \text{when } \theta = 0$

Hence initial line is tangent to origin.

Curvature

Then
$$\rho = \lim_{\theta \to 0} \frac{x^2}{2y} = \lim_{\theta \to 0} \frac{a^2(\theta + \sin\theta)^2}{2a(1 - \cos\theta)}$$

$$= \lim_{\theta \to 0} \frac{a}{2} \left[\frac{2(\theta + \sin\theta)(1 + \cos\theta)}{\sin\theta} \right]$$

$$= \lim_{\theta \to 0} \frac{a}{2} \left[\frac{2.(1 + \cos\theta)^2 - \sin\theta(\theta + \sin\theta)}{\cos\theta} \right] \quad \text{[By Hospital's rule]}$$

$$= 4a.$$

EXERCISES

1. Find ρ at $(0, 0)$;
   (a) $y = xe^{-x}$, (b) $y = xe^{-x^2}$ [Ans. (a) $\sqrt{2}$, (b) $\frac{1}{e}$]

2. Find ρ at origin of the curve $y = x^4 - 4x^3 - 18x^2$. [Ans. $\frac{1}{36}$]

3. Find ρ at origin for the curve $2x^4 + 3y^4 + 4x^2y + xy - y^2 + 2x = 0$. [Ans. 1]

4. Find ρ at the origin for the curve $3x^3 + y^3 + 5y^2 + 3yx^2 + 2x = 0$. [Ans. $\frac{1}{5}$]

5. Find ρ at the origin for the curve $x^4 - y^4 + x^3 - y^3 + x^2 - y^2 + y = 0$. [Ans. $\frac{1}{2}$]

6. Find the radius of curvature of the cardioid $r = a(1 - \cos\theta)$ at the pole (origin). [Ans. 0]

6.5 CHORD OF CURVATURE

If there is any point P at the given curve and a circle having the radius of curvature ρ is drawn passing through P, then any chord PN is called the **chord of curvature**. If C be the centre of the circle then
$$PN = RP \cos R \ PN = 2\rho \cos RPN$$
$$= 2\rho \cos \alpha.$$

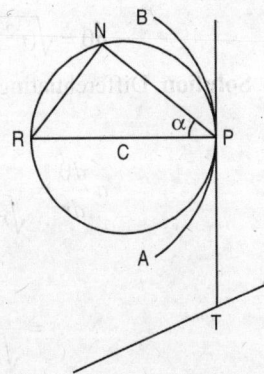

Different cases of chord of curvature

Case I. Chord of curvature through pole (origin).
In the figure PN is the required chord of curvature
$$PN = RP \cos RPN$$
$$= 2\rho \cos\left(\frac{\pi}{2} - \phi\right)$$
$$= 2\rho \sin\phi$$

Hence, **Chord of curvature through pole** $= 2\rho \sin\phi$.

Case II. Chord of curvature through the pole for the curve $p = f(r)$.

We know $p = r \sin\phi; \quad \therefore \quad \sin\phi = \frac{p}{r}$

Hence chord of curvature $= 2\rho \cdot \frac{p}{r}$

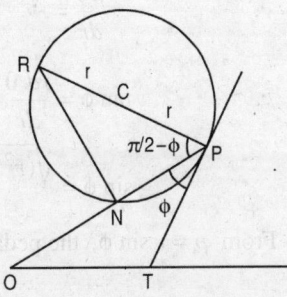

$$= 2 \cdot r \frac{dr}{dp} \cdot \frac{p}{r} = 2p \frac{dr}{dp}$$

$$= 2f(r) \frac{dr}{dp}, \text{ i.e., } \frac{2f(r)}{f'(r)}$$

Case III. Chord of curvature perpendicular to radius vector.
Here PN = chord of curvature perpendicular to radius vector
$$= RP \cos RPN$$
$$= 2\rho \cos \phi$$

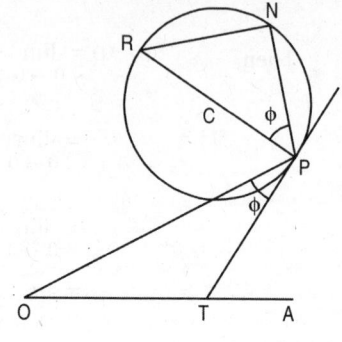

Case IV. Chord of curvature parallel to x-axis and y-axis
Chord of curvature parallel to x-axis
$$= PN = RP \cos RPN$$
$$= 2\rho \cos\left(\frac{\pi}{2} - \psi\right) = 2\rho \sin \psi$$
and chord of curvature parallel to y-axis
$$= PN' = RP \cos RPN' = 2\rho \cos \psi.$$

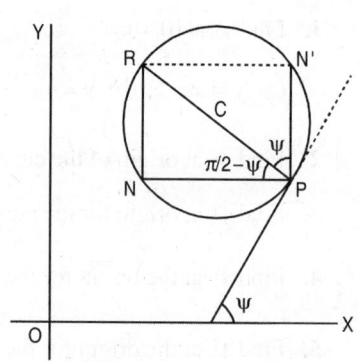

EXAMPLES

1. *Find the chord of curvature through the pole of the curve*
$$a\theta = \sqrt{(r^2 - a^2)} - \cos^{-1} \frac{a}{r}.$$

Solution. Differentiating the equation of given curve, w.r.t. r, we get

$$a \frac{d\theta}{dr} = \frac{r}{\sqrt{r^2 - a^2}} - \left(\frac{-1}{\sqrt{1 - \frac{a^2}{r^2}}}\right)\left(-\frac{a}{r^2}\right)$$

$$= \frac{r}{\sqrt{r^2 - a^2}} - \frac{a}{r\sqrt{r^2 - a^2}} = \frac{\sqrt{r^2 - a^2}}{r}$$

$$\therefore \quad \frac{d\theta}{dr} = \frac{\sqrt{(r^2 - a^2)}}{ar}$$

$$\therefore \quad \tan \phi = \frac{r \, d\theta}{dr} = \frac{\sqrt{(r^2 - a^2)}}{a}, \text{ so, } \cos \phi = \frac{a}{r},$$

$$\sin \phi = \frac{\sqrt{(r^2 - a^2)}}{r}$$

From $p = r \sin \phi$, the pedal equation of the given curve is

$$p = r \cdot \frac{\sqrt{(r^2 - a^2)}}{r} = \sqrt{(r^2 - a^2)}$$

Curvature

$$\frac{dp}{dr} = \frac{1}{2}(r^2 - a^2)^{-1/2} \cdot 2r = \frac{r}{\sqrt{r^2 - a^2}}$$

$$\therefore \quad \rho = r \frac{dr}{dp} = r \cdot \frac{\sqrt{r^2 - a^2}}{r} = \sqrt{(r^2 - a^2)}$$

Hence the required chord of curvature

$$= 2\rho \sin \phi = 2\sqrt{(r^2 - a^2)} \cdot \frac{\sqrt{(r^2 - a^2)}}{r}$$

$$= \frac{2(r^2 - a^2)}{r}.$$

2. If C_x and C_y be the chords of curvature parallel to the axes to any point of the curve $y = ae^{x/a}$, prove that

$$\frac{1}{C_x^2} + \frac{1}{C_y^2} = \frac{1}{2aC_x}$$

Solution. We have

$$y = ae^{x/a}$$

$$\therefore \quad \frac{dy}{dx} = ae^{x/a}\left(\frac{1}{a}\right) = e^{x/a} = y/a$$

and

$$\frac{d^2y}{dx^2} = \frac{1}{a} \cdot \frac{dy}{dx} = \frac{y}{a^2}$$

$$\therefore \quad \rho = \frac{\left[1 + \left(\frac{dy}{dx}\right)^2\right]^{3/2}}{d^2y/dx^2} = \frac{\left(1 + \frac{y^2}{a^2}\right)^{3/2}}{y/a^2}$$

$$= \frac{(a^2 + y^2)^{3/2}}{ay}$$

Again, $\tan \psi = \frac{dy}{dx} = \frac{y}{a}$. Hence, $\sin \psi = \frac{y}{\sqrt{(a^2 + y^2)}}$ and $\cos \psi = \frac{a}{\sqrt{(a^2 + y^2)}}$

Now, $\quad C_x = 2\rho \sin \psi = \frac{2(a^2 + y^2)^{3/2}}{ay} \cdot \frac{y}{\sqrt{(a^2 + y^2)}} = \frac{2}{a}(a^2 + y^2)$

and $\quad C_y = 2\rho \cos \psi = \frac{2(a^2 + y^2)^{3/2}}{ay} \cdot \frac{y}{\sqrt{(a^2 + y^2)}} = \frac{2}{y}(a^2 + y^2)$

$$\therefore \quad \frac{1}{C_x^2} + \frac{1}{C_y^2} = \frac{a^2}{4(a^2 + y^2)^2} + \frac{y^2}{4(a^2 + y^2)^2} = \frac{1}{4(a^2 + y^2)}$$

$$= \frac{1}{2a} \cdot \frac{a}{2(a^2 + y^2)} = \frac{1}{2a} \cdot \frac{1}{C_x}.$$

EXERCISES

1. Find the chord of curvature through the pole of the curve $r^2 \cos 2\theta = a^2$. [Ans. $2r$]

2. If in the cardioid $r = a(1 + \cos\theta)$, C_r and C_θ be the chords of curvature respectively along and perpendicular to the radius vector, show that
$$C_r + C_\theta = \frac{8}{3} a\, C_r.$$

3. Find the chord of curvature through the pole of the cardioid $r = a(1 - \cos\theta)$. [Ans. $\frac{4}{3}r$]

4. Show that the chord of curvature, through the focus, of a parabola is four times the focal distance of the point, and the chord of curvature parallel to the axis has the same length.

5. Prove that the common chord of the parabola and the circle of curvature at any point is of length $8\sqrt{\{r(r-a)\}}$, where r is the distance of the point from the focus of the parabola.

6. Show that the chord of curvature through the pole of the curve $r^n = a^n \cos n\theta$ is $2r/n+1$.

6.6 CENTRE OF CURVATURE

Let (α, β) be the coordinates of the centre of curvature C at $P(x, y)$ and the radius of curvature be ρ. Let the tangent at P makes angle ψ with x-axis so that position direction of the normal makes angle

$$\psi + \frac{\pi}{2} \text{ with } x\text{-axis}.$$

Then the equation of normal PC is

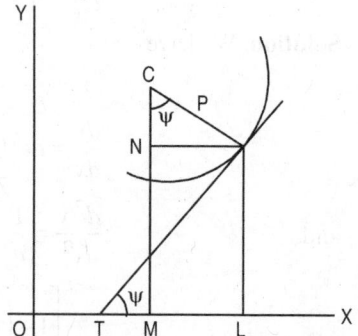

$$Y - y = \tan\left(\psi + \frac{\pi}{2}\right) \cdot (X - x)$$

or
$$\frac{Y - y}{\sin\left(\frac{\pi}{2} + \psi\right)} = \frac{X - x}{\cos\left(\frac{\pi}{2} + \psi\right)} = r$$

or
$$\frac{X - x}{-\sin\psi} = \frac{Y - y}{\cos\psi} = r,$$

Where x, y are the current-coordinates of any point on the normal and r is the variable distance of the variable point (X, Y) from (x, y)

Thus the coordinates (X, Y) of the point on the normal are
$$X = -r\sin\psi + x \text{ and } Y = r\cos\psi + y$$

For the centre of curvature (α, β), $r = \rho$,

$\therefore$ $\quad \alpha = -\rho\sin\psi + x$ and $\beta = \rho\cos\psi + y$.

Also $\quad \tan\psi = \dfrac{dy}{dx} = y_1$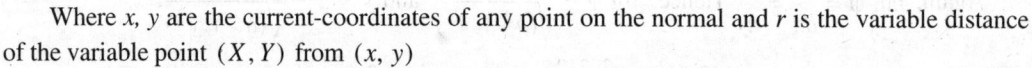

$\therefore \quad \sin\psi = \dfrac{y_1}{\sqrt{1+y_1^2}}$ and $\cos\psi = \dfrac{1}{\sqrt{(1+y_1^2)}}$

Then $\quad \alpha = x - \dfrac{\rho y_1}{\sqrt{(1+y_1^2)}}$ and $\beta = y + \dfrac{\rho}{\sqrt{(1+y_1^2)}}$

Curvature

But $\rho = \dfrac{(1+y_1^2)^{3/2}}{y_2}$

Hence $\alpha = x - \dfrac{y_1(1+y_1^2)}{y_2}$

and $\beta = y + \dfrac{(1+y_1^2)}{y_2}$

Aliter. $PC = \rho$ and $\angle PCN = \psi$

Then $\alpha = OL - PN = x - \rho \sin \psi$

$= x - \dfrac{(1+y_1^2)^{3/2}}{y_2} \cdot \dfrac{y_1}{\sqrt{(1+y_1^2)}}$

$= x - \dfrac{(1+y_1^2) y_1}{y_2}$

and $\beta = CM = CN + NM = \rho \cos \psi + PL$

$= \rho \cos x + y$

$= y + \dfrac{(1+y_1^2)^{3/2}}{y_2} \cdot \dfrac{1}{\sqrt{(1+y_1^2)}}$

$= y + \dfrac{1+y_1^2}{y_2}$

6.7 CIRCLE OF CURVATURE

The circle with radius equal to radius of curvature ρ and its centre the centre of curvature (α, β) is called the *circle of curvature*.

Then the equation of the circle of curvature is

$$(x-\alpha)^2 + (y-\beta)^2 = \rho^2$$

6.8 EVOLUTES AND INVOLUTES

If the centre of curvature for each point on a curve be taken, we get a new curve called the *evolute* of the original one.

Also the original curve, when considered with respect to its evolute, is called an *involute*.

The connection between these curves is represented in the adjoining figure. Let P_1, P_2, P_3 .... etc. represents a series of infinitely near points on a curve; $C_1, C_2, C_3, ...$ etc. the corresponding centres of curvature; then lines P_1C_1, P_2C_2, P_3C_3 etc. are normals to the curve and the lines C_1C_2, C_2C_3, C_3C_4 etc. may be regarded in the limit and consecutive elements of the evolute; also since each of the normals P_1C_1, P_2C_2, P_3C_3 etc. passes through two

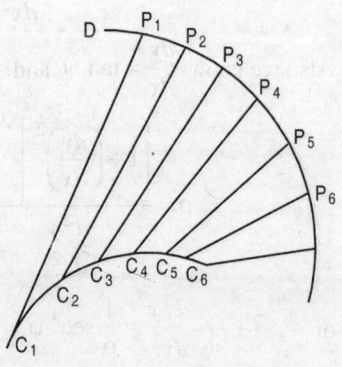

consecutive points on the evolute, they are tangents to that curve on the limit. Again if P_1, P_2, P_3, P_4

etc. denote the length of the radii of curvature at the points P_1, P_2, P_3, etc. we have

$$\rho_1 = P_1C_1, \rho_2 = P_2C_2, \rho_3 = P_3C_3, \ldots$$

Therefore $\rho_1 - \rho_2 = P_1C_1 - P_2C_2$ (since the points P_1, P_2 are nearer)

Also $\rho_2 - \rho_3 = C_2C_3, \rho_3 - \rho_4 = C_3C_4 \ldots$

$$\rho_{n-1} - \rho_n = C_{n-1}C_n$$

Hence by addition, we have

$$\rho_1 - \rho_n = C_1C_2 + C_2C_3 + C_3C_4 + \ldots + C_{n-1}C_n$$

This result still holds when the number n is increased indefinitely and we infer the length of any arc of the evolute is equal, in general, to the difference between the radii of curvature at its extremities. In other words, *the evolute is the locus of the, centre of curvature.*

6.8.1 To Determine the Equation of an Evolute

We know that the corrdinates of the centre of curvature are given by (α, β) where as

$$\alpha = x - \frac{y_1(1 + y_1^2)}{y_2}$$

and $$\beta = y + \frac{(1 + y_1^2)}{y_2}$$

Now we eliminate x and y (parameters in this case), and then a relation between α and β is the required equation of the evolute.

EXAMPLES

1. *Prove that the coordinates of the centre of curvature at any point (x, y) can be expressed in the form $x - \frac{dy}{d\psi}$ and $y + \frac{dx}{d\psi}$.*

Solution. We know if (α, β) be the centre of curvature of any point (x, y) of a curve then

$$\alpha = x - \frac{\left[1 + \left(\frac{dy}{dx}\right)^2\right] \cdot \frac{dy}{dx}}{\frac{d^2y}{dx^2}} \text{ and } \beta = y + \frac{\left\{1 + \left(\frac{dy}{dx}\right)^2\right\}}{\frac{d^2y}{dx^2}}$$

Also we know $\frac{dy}{dx} = \tan\psi$ and

$$\rho = \frac{\left[1 + \left(\frac{dy}{dx}\right)^2\right]^{3/2}}{\frac{d^2y}{dx^2}} = \frac{\sec^3\psi}{d^2y/dx^2}$$

or $$\frac{d^2y}{dx^2} = \frac{1}{\rho}\sec^3\psi$$

$\therefore$ $$\alpha = x - \frac{(1 + \tan^2\psi)\tan\psi}{\left(\frac{1}{\rho}\right)\sec^3\psi}$$

Curvature

$$= x - \rho \sin \psi = x - \frac{ds}{d\psi} \cdot \frac{dy}{ds} \qquad \left[\because \rho = \frac{ds}{d\psi} \text{ and } \sin \psi = \frac{dy}{ds}\right]$$

$$= x - \frac{dy}{d\psi}$$

Similarly, $\quad \beta = y + \dfrac{1+\tan^2 \psi}{\left(\dfrac{1}{\rho}\right)\sec^3 \psi} = y + \rho \cos \psi$

$$= y + \frac{ds}{d\psi} \cdot \frac{dx}{ds} = y + \frac{dx}{d\psi}$$

2. *Prove that the points on the curve $r = f(\theta)$, the circle of curvature at which passes through the origin are given by the equation.*

$$f(\theta) + f''(\theta) = 0$$

Solution. Let $P(r, \theta)$ be a point on the given curve, the circle of curvature, with centre C, at which passes through the origin O.

Let C be the centre of curvature. Join OP and let PC produced meet the circle of curvature at P in D. Let PT be the tangent to the curve at P, then

$$\angle OPT = \phi = \angle PDO$$

Now, OP = radius vector r and $PD = 2\rho$.

Hence in $\triangle POD$, we get $PO = PD \sin \phi$

or $\qquad r = 2\rho \sin \phi \qquad$ ...(i)

Also $\qquad \tan \phi = r \dfrac{d\theta}{dr} = \dfrac{r}{r_1}, \qquad$ where $r_1 = \dfrac{dr}{d\theta}$

$\therefore \qquad \sin \phi = \dfrac{r}{(r^2 + r_1^2)^{1/2}} \qquad$ ...(ii)

Also $\qquad \rho = \dfrac{(r^2 + r_1^2)^{3/2}}{r^2 + 2r_1^2 - rr_2} \qquad$ ...(iii)

Hence from (i), (ii) and (iii), we get

$$r = \frac{2(r^2 + r_1^2)^{3/2}}{r^2 + 2r_1^2 - rr_2} \cdot \frac{r}{\sqrt{(r^2 + r_1^2)}}$$

$$= \frac{2r(r^2 + r_1^2)}{r^2 + 2r_1^2 - rr_2}$$

or $\qquad r^2 + 2r_1^2 - rr_2 = 2r^2 + 2r_1^2 \qquad$ or $\qquad r^2 + rr_2 = 0$

or $\qquad r + r_2 = 0,\qquad$ where $r = f(\theta)$

or $\qquad f(\theta) + f''(\theta) = 0$

3. *Obtain the evolute of the parabola $y^2 = 4ax$.*

Solution. Here the point (x, y) may be taken as $x = at^2$, $y = 2at$, which satisfy the equation of the parabola. Hence

$$\frac{dx}{dt} = 2at \quad \text{and} \quad \frac{dy}{dt} = 2a$$

∴ $$\frac{dy}{dx} = \frac{1}{t} \quad \text{and} \quad \frac{d^2y}{dx^2} = -\frac{1}{t^2}\frac{dt}{dx}$$

$$= -\frac{1}{2at^3}.$$

Then if (α, β) be the cordiantes of the centre of curvature, we have

$$\alpha = x - \frac{y_1(1 + y_1^2)}{y_2} = at^2 - \frac{\frac{1}{t}\left(1 + \frac{1}{t^2}\right)}{-\frac{1}{2at^3}}$$

$$= at^2 + 2at^2\left(1 - \frac{1}{t^2}\right) = 3at^2 + 2a$$

and $$\beta = y + \frac{(1 + y_1^2)}{y_2} = 2at + \frac{\left(1 + \frac{1}{t^2}\right)}{-\frac{1}{2at^3}}$$

$$= 2at + (t^2 + 1)(-2at)$$

$$= -2at^3$$

Now, $\alpha = 3at^2 + 2a,$

$\beta = -2at^3$

Eliminating t between these relations, we obtain

$$\left(\frac{\alpha - 2a}{3a}\right)^{1/2} = \left(-\frac{\beta}{2a}\right)^{1/3}$$

or $$4(\alpha - 2a)^3 = 27 a\beta^2$$

Hence the locus is

$$27 ay^2 = 4(x - 2a)^3$$

EXERCISES

1. In the parabola $x^2 = 4ay$ prove that the coordinates of the centre of curvature are

$$\left(-\frac{x^3}{4a^2}, 2a + \frac{3x^2}{4a}\right).$$

2. In the curve $y = c \cosh\left(\frac{x}{c}\right)$, show that coordinates of the centre of curvature are

$$\alpha = x - y\left(\frac{y^2}{c^2} - 1\right)^{1/2}, \beta = 2y$$

Prove also that the radius of curvature is equal to the part of the normal intercepted between the curve and the axis of x.

Curvature

3. Show that the centre of curvature at the point determined by t on the ellipse $x = a \cos t$, $y = b \sin t$ is given by
$$x = \frac{a^2 - b^2}{a} \cos^3 t, \quad y = -\frac{a^2 - b^2}{b} \sin^3 t.$$

4. Find the evolute of the curve
$$x^{2/3} + y^{2/3} = a^{2/3}$$
[**Ans.** $(x+y)^{2/3} + (x-y)^{2/3} = 2a^{2/3}$]

5. If (α, β) be the coordinates of the centre of curvature of the parabola $\sqrt{x} + \sqrt{y} = \sqrt{a}$ at (x, y), then prove that
$$\alpha + \beta = 3(x + y)$$

7

Concavity, Convexity and Singular Points

7.1 INTRODUCTION

To discuss the concavity and convexity we require the concept of Multiple points. *Multiple points* are those points through which more than one branches of a curve pass. As a particular case, if two branches pass through a single point, then there is a case of *double points*.

The *point of inflexion* is that point where the curve crosses the tangent. In this case the curve is concave on one side and convex on the other side. As in the adjoining figure P is a point of inflexion, because the curve is concave on one side and convex on the other side with respect to the line MN.

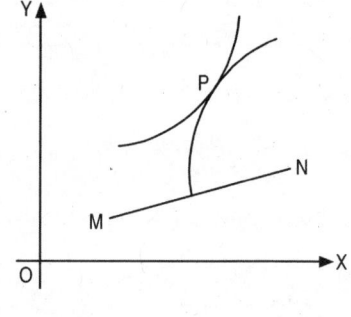

The point of inflexion is a kind of a singular point. The second kind is multiple points.

7.2 CONCAVITY

P is any point on a curve as shown in figure and AB is any other straight line which does not pass through P. The curve is said to be concave with respect to line AB, if the arc of a sufficiently small measure which contains P lies entirely within the acute angle formed by the line AB and tangent at P to the curve.

7.3 CONVEXITY

P is any point on the curve and AB is a given straight line. The curve is said to be convex at P with respect to AB if the arc of a sufficiently small measure lies without the acute angle subtended by AB and the tangent at P, as shown in figure.

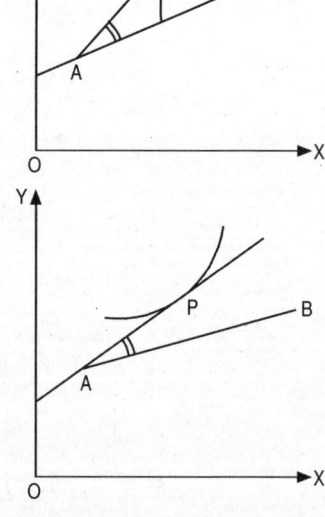

7.4 A TEST OF CONCAVITY AND CONVEXITY

In order to test the concavity and convexity, we shall take the help of the system of rectangular axes.

Le the equation of the curve be $y = f(x)$ and P be any point (x, y) and Q any other point $(x+h, y+k)$ in the neighbourhood of the point P (on both the sides) and a perpendicular from Q be dropped on the axes of x intersecting the tangent at P at the point R.

The curve is convex at P [fig. (a)], if $QS > RS$, i.e., if $QS - RS$ is positive, and is concave at P [fig. (b)], if $RS > QS$, i.e., $QS - RS$ is negative.

Now the equation of the tangent at P is $Y - y = f'(x)(X - x)$.

Concavity, Convexity and Singular Points

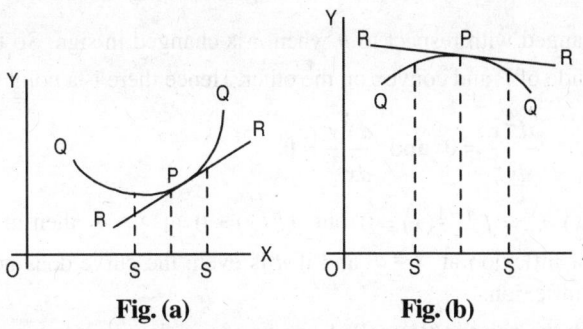

Fig. (a)   Fig. (b)

Now for point Q, put $X = x + h$, then
$$QS = Y = y + hf'(x) = f(x) + hf'(x)$$
Also by Taylor's Theorem,
$$f(x+h) = f(x) + hf'(x) + \frac{h^2}{2!}f''(x) + \ldots + \frac{1}{(n-1)!}f^{n-1}(x) + \frac{1}{n!}h^n f^n(x+\theta h),$$
where $0 < \theta < 1$.

Then $\quad f(x+h) = f(x) - hf'(x) = \frac{h^2}{2!}f''(x) + \ldots + \frac{1}{n!}h^n f^n(x+\theta h)$

$\therefore \quad QS - RS = \dfrac{h^2}{2!}f''(x) + \ldots$

Now if $f''(x)$ is not zero and h is taken sufficiently small, the sign of the right hand side will be the same as that of $f''(x)$ whether h is negative or positive as h^2 is always positive.

Then the curve is convex or concave according as $\dfrac{d^2 y}{dx^2}$ or $y\dfrac{d^2 y}{dx^2}$ (y being positive for the point P above the axis of x) is positive or negative.

Now if the point P is below the axis of x, the case is reversed, i.e., the curve is convex or concave according as $\dfrac{d^2 y}{dx^2}$ is negative or positive, but for the point below the x-axis y is negative.

Thus the curve is convex or concave at P according as $y\dfrac{d^2 y}{dx^2}$ is positive or negative. Hence

(i) A curve is convex at P to the axis of x if $y\dfrac{d^2 y}{dx^2}$ is positive at P.

(ii) A curve is concave at P to the axis of x if $y\dfrac{d^2 y}{dx^2}$ is negative at P.

7.4.1 Test for the Point of Inflexion

Now again $\quad QS - RS = \dfrac{1}{2!}h^2 f''(x) + \dfrac{h^3}{3!}f'''(x) + \ldots$

If at P, $f''(x)$ is zero but not $f'''(x)$ then the sign of the right hand side will be governed by $h^3 f'''(x)$ which is changed with respect to h when h is changed in sign. So the curve is concave to the axis of x on one side of P and convex on the other. Hence there is a point of inflexion at P if

$$\frac{d^2y}{dx^2} = 0 \text{ and } \frac{d^3y}{dx^3} \neq 0.$$

Let $f''(x) = f'''(x) = ... = f^{n-1}(x) = 0$ but $f^n(x) \neq 0$ at $x = a$; then in general, if n is odd, the curve has a point of inflexion at $x = a$ and if n is even, the curve does not cross the tangent, i.e., there is no point of inflexion.

Note. If $n > 2$, then the point is also called a point of undulation.

Point of inflexion in case of polar curves.

Let P be any point on the polar curve. If p, the length of the perpendicular from the pole on the tangent increases with the radius vector r, then the curve will be concave at P with respect to the

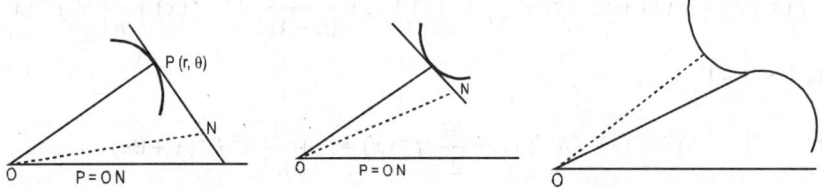

pole, i.e., the curve is concave at P with regard to the pole. If $\frac{dp}{dr}$ is positive there. Similarly the curve will be convex at P with regard to the pole if $\frac{dp}{dr}$ is negative at P.

If $\frac{dp}{dr}$ if zero at P, positive for the points on one side of P and negative for the points on the other side of P, then there must be a point of inflexion at P.

Now, $r \frac{dr}{dp} = \rho$ (radius of curvature)

$$= \frac{\left[r^2 + \left(\frac{dr}{d\theta}\right)^2\right]^{3/2}}{\left[r^2 + 2\left(\frac{dr}{d\theta}\right)^2 - r\frac{d^2r}{d\theta^2}\right]}$$

It follows that if $r^2 + 2\left(\frac{dr}{d\theta}\right)^2 - r\frac{d^2r}{d\theta^2} = 0$ at P, there is in general a point of inflexion at P.

Note. If $u = \frac{1}{r}$, then $u + \frac{d^2u}{d\theta^2} = 0$ holds good for the point of inflexion.

Concavity, Convexity and Singular Points

EXAMPLES

1. *Find the points of inflexion of the curves*
$$y = 3x^4 - 4x^3 + 1.$$

Solution. The given curve is
$$y = 3x^4 - 4x^3 + 1$$

Differentaiting with respect to x, we get
$$\frac{dy}{dx} = 12x^3 - 12x^2$$

and
$$\frac{d^2y}{dx^2} = 36x^2 - 24x$$

Now, for the point of inflexion,
$$\frac{d^2y}{dx^2} = 0, \quad i.e., \quad 36x^2 - 24x = 0$$

or
$$12x(3x-2) = 0,$$

whence $x = 0$ and $(3x - 2) = 0$,

Again
$$\frac{d^3y}{dx^3} = 72x - 24$$

Now (i) is not zero for $x = 0$ or $x = \frac{2}{3}$, hence points of inflexion exist at these points.

Now, when $\quad x = 0, y = 1$

and when $\quad x = \frac{2}{3}, y = \frac{11}{27}$.

Hence the points of inflexion are $(0, 1)$ and $\left(\frac{2}{3}, \frac{11}{27}\right)$.

2. *Show that the points of inflexion on the curve $r = b\theta^n$ are given by $r = b[-n(n+1)]^{n/2}$.*

Solution. We have $\quad r = b\theta^n$

Now,
$$\frac{dr}{d\theta} = bn\theta^{n-1}$$

and
$$\frac{d^2r}{d\theta^2} = bn(n-1)\theta^{n-2}$$

for the point of inflexion
$$\frac{d^2r}{d\theta^2} = bn(n-1)\theta^{n-2}$$

i.e., $\quad b^2\theta^{2n} + b^2n^2\theta^{2n-2} - b^2n(n-1)\theta^{2n-2} = 0$

i.e., $\quad \theta^2 + 2n^2 - n(n-1) = 0$

or $\quad \theta^2 + n(n+1) = 0$

or $\theta = [-n(n+1)]^{1/2}$

Therefore $r = b[-n(n+1)]^{n/2}$

3. *For a curve given by its polar equation show that the points of inflexion are given by* $u + \dfrac{d^2u}{d\theta^2} = 0$, *where* $u = \dfrac{1}{r}$.

Solution. Here $u = \dfrac{1}{r}$, then $\dfrac{dr}{d\theta} = -u^{-2}\dfrac{du}{d\theta}$

$$\dfrac{d^2r}{d\theta^2} = 2u^{-3}\left(\dfrac{du}{d\theta}\right) - u^{-2}\dfrac{d^2u}{d\theta^2}$$

Then
$$\rho = \dfrac{\left[r^2 + \left(\dfrac{dr}{d\theta}\right)^2\right]^{3/2}}{r^2 + 2\left(\dfrac{dr}{d\theta}\right)^2 - r\dfrac{d^2r}{d\theta^2}}$$

$$= \dfrac{\left[u^2 + \left(\dfrac{du}{d\theta}\right)^2\right]^{3/2}}{u^3\left(u + \dfrac{d^2u}{d\theta^2}\right)}$$

At the point of inflexion, ρ is infinite

i.e., $\quad u^3\left(u + \dfrac{d^2u}{d\theta^2}\right) = 0$

or $\quad u + \dfrac{d^2u}{d\theta^2} = 0$.

EXERCISES

1. Find the point of inflexion of the curve $y = 4x^3 - 18x^2 + 27x - 7$. [**Ans.** $x = \dfrac{3}{2}$]

2. Investigate the points of inflexion of the curve
$$y = (x-2)^6(x-3)^5$$
[**Ans.** Points of inflexion at $x = 3, 28 \pm \sqrt{3}/11$, points of undulation at $x = 2$.]

3. Show that the curve $y = \dfrac{1-x}{1+x^2}$ has three points of inflexion which lie in a straight line.

4. Show that the curve $y = e^x$ is at every point convex to the foot of the corresponding ordinate.

5. In the curve $a^{m-1}y = x^m$, prove that the origin is a point of inflexion if m be odd and greater than 2.

Concavity, Convexity and Singular Points

7.5 MULTIPLE POINTS

Although we have discussed in § 7.1 what are the multiple points, but we shall again study them in details.

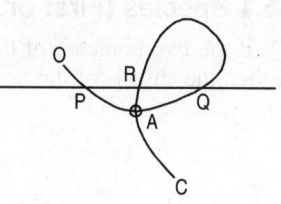

In the adjoining figure, the curve intersects again at the point A. A is called *Double point* becuase two branches C and D of the curve pass through it. If PQ is a secant of the curve, it intersects the curve in one more point as in R. As P and Q tend to A, so does R. Therefore in the limit when the secant PQ becomes a tangent to the branch D of the curve, the three points P, Q, R coincide at A. The tangent and the branch D have, therefore, three coincident points in common. If a curve is of the nth degree, the tangent at its double point will intersect the curve in $(n-3)$ points.

Classification of Double Points

(a) **Node.** The tangents at a double point of a curve are real and distinct, two real branches of the curves intersect at it. The double point is then called a *node*.

(b) **Cusp.** If the tangents are coincident, the two branches of the curve must themselves be tangential to each other at the point then the double point is called *cusp*.

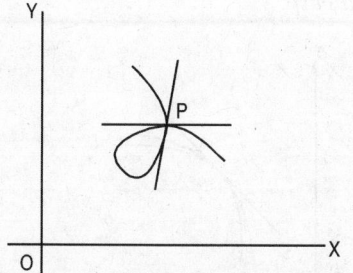
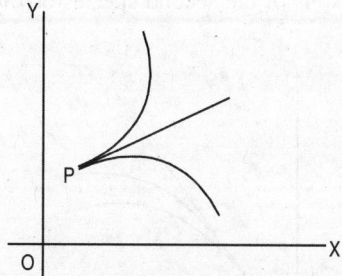

(c) **Conjugate point or isolated point.** If the tangents at double points are imaginary, though the coordinates of the point satisfy the equation of the curve, but the curve has no real points consecutive to this point, the double point is called *conjugate point*.

Note. The point lies altogether outside the curve itself.

Types of the Cusps

A cusp may be single or double according as the curve lies entirely on one side of the normal or on both sides.

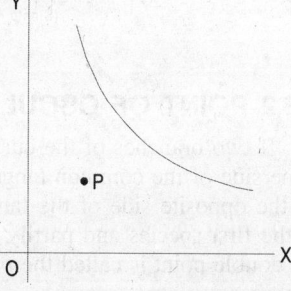

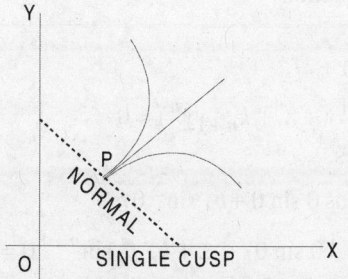
SINGLE CUSP

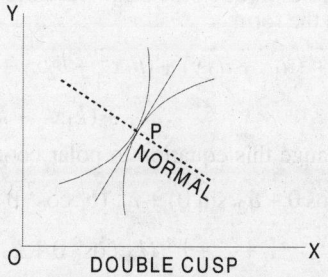
DOUBLE CUSP

7.5.1 Species (First or Second)

If the two branches of the curve at the cusp lie on opposite sides of the common tangent there at, then the cusp is of the first species or *keratoid cusp*.

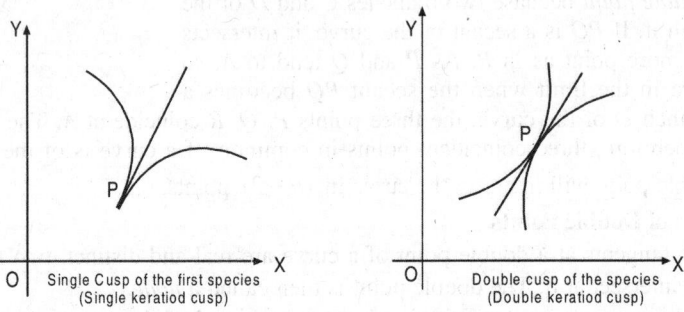

If the two branches of the curve at the cusp lie on the same side of the common tangent there at, then the cusp is of the second species or *rhampoid cusp*.

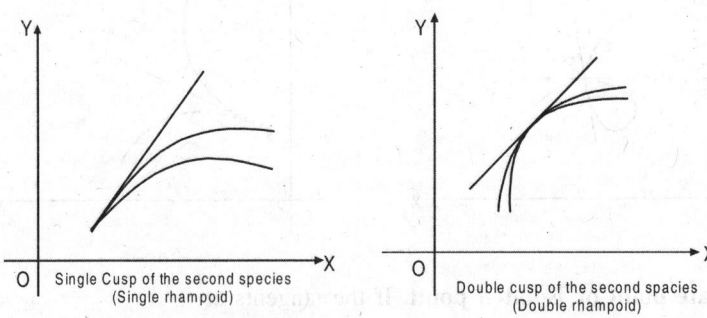

7.5.2 POINT OF OSCUL-INFLEXION

If two branches of the curve above the cusp are on the same side of the common tangent while below the cusp are on the opposite side of the tangent, then the cusp is partly of the first species and partly of the second sxpecies; then the double point is called the *point of oscul-inflextion*.

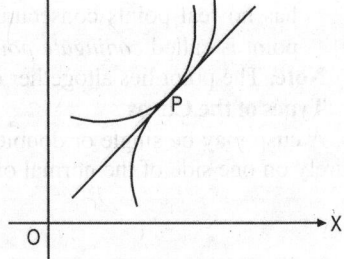

7.6 TANGENTS AT THE ORIGIN

If a curve passes through the origin its cartesian equation can be written in the form

$$(a_1 x + a_2 y) + (b_1 x^2 + b_2 xy + b_2 y^2) + ...$$
$$+ (k_1 x^n + k_2 x^{n-1} y + ... + k_{n+1} y^n) = 0 \qquad ...(i)$$

We now change this equation to polar coordinates.

$$r(a_1 \cos\theta + a_2 \sin\theta) + r^2 (b_1 \cos^2\theta + b_2 \cos\theta \sin\theta + b_3 \sin^2\theta)$$
$$+ ... + r^n (k_1 \cos^n\theta + k_2 \cos^{n-1}\theta \sin\theta) + ... + k_{n+1} \sin^{n+1}\theta = 0 \qquad ...(ii)$$

Since r is a common factor in all terms on the left hand side of the equation one root of the equation is zero.

If further $a_1 \cos\theta + a_2 \sin\theta = 0$ two roots of the equation become zero.

Concavity, Convexity and Singular Points

Therefore a line in the direction $\theta = \tan^{-1}(-a_1/a_2)$ meets the curve in two coincident points at the origin; the line is therefore a tangent there. Its equation is

$$a_1 x + a_2 y = 0$$

If $a_1 = 0$, the axis of x is a tangent. If $a_2 = 0$, the axis of y is a tangent.

Such is the case if $a_1 = a_2 = 0$.

The straight line in the direction of θ is given by

$$b_1 \cos^2 \theta + b_2 \cos \theta \sin \theta + b_3 \sin^2 \theta = 0 \qquad \text{...(iii)}$$

will intersect the curve in three coincident points. The equation (iii) will have two distinct roots, provided $b_2^2 > 4b_1 b_3$ in which case the origin is a node. If $b_2^2 = 4b_1 b_2$ the tangents are coincident and the origin is a cusp.

If $b_2^2 < 4b_1 b_3$, real tangents do not exist and the origin is then a conjugate point.

In each case the equation to the tangent is

$$b_1 x^2 + b_2 xy + b_3 y^2 = 0.$$

Thus we conclude that if a curve passes through the origin, the equation to the tangents at the origin is obtained by equating to zero the terms of the lowest degree in x and y.

7.7 SEARCH FOR DOUBLE POINTS AND THE NECESSARY CONDITION FOR THEIR EXISTENCE

Let P be any point (x, y) on the curve

$$f(x, y) = 0 \qquad \text{...(i)}$$

Differentiating (i), we obtain

$$\frac{\partial f}{\partial x} + \frac{\partial f}{\partial y} \cdot \frac{dy}{dx} = 0 \qquad \text{...(ii)}$$

At a multiple point of a curve, the curve has atleast two tangents, and accordingly $\dfrac{dy}{dx}$ must have at least two values at multiple point.

The equation (ii) is of the first degree in $\dfrac{dy}{dx}$ which can be satisfied by more than one value of $\dfrac{dy}{dx}$ if and only if

$$\frac{\partial f}{\partial x} = 0, \frac{\partial f}{\partial y} = 0$$

Thus we see that the necessary and sufficient conditions for any point (x, y) on $f(x, y) = 0$ to be a multiple point are that $\partial f/\partial x = 0, \partial f/\partial y = 0$ and $f(x, y) = 0$.

Differentiating (ii) with respect to x, we obtain

$$\frac{d}{dx}\left(\frac{\partial f}{\partial x}\right) + \frac{d}{dx}\left[\left(\frac{\partial f}{\partial y}\right) \cdot \frac{dy}{dx}\right] = 0$$

or

$$\left[\frac{\partial^2 f}{\partial x^2} + \frac{\partial^2 f}{\partial x \partial y} \cdot \frac{dy}{dx}\right] + \left[\frac{\partial^2 f}{\partial x \partial y} \cdot \frac{dy}{dx} + \frac{\partial f}{\partial y} \cdot \frac{d^2 y}{dx^2} + \frac{\partial^2 f}{\partial y^2}\left(\frac{dy}{dx}\right)^2\right] = 0$$

or

$$\frac{\partial^2 y}{\partial y^2}\left(\frac{dy}{dx}\right)^2 + 2\frac{\partial^2 y}{\partial x \partial y}\left(\frac{dy}{dx}\right) + \frac{\partial^2 f}{\partial x^2} = 0$$

which is a quadratic equation in $\dfrac{dy}{dx}$ and $\dfrac{\partial f}{\partial y} = 0, \dfrac{\partial f}{\partial x} = 0$ at the multiple point.

Therefore the two tangents are distinct, coincident or imaginary according as the values of $\dfrac{dy}{dx}$ are real and distinct, equal or imaginary i.e., according as

$$\left(\dfrac{\partial^2 f}{\partial x\,\partial y}\right)^2 - \dfrac{\partial^2 f}{\partial x^2} \cdot \dfrac{\partial^2 f}{\partial y^2} > = < 0$$

If $\dfrac{\partial^2 f}{\partial x^2} = \dfrac{\partial^2 f}{\partial y^2} = \dfrac{\partial^2 f}{\partial x\,\partial y} = 0$, the point (x, y) will be multiple point of order higher than second.

Nature of cusp. Once the cusp point is located we transfer the origin to the cusp point. The transferred equation shall be of the form

$$(ax + by)^2 + \phi(x, y) = 0 \qquad \ldots(i)$$

The order of terms in $\phi(x, y)$ will be higher than two. If p is the perpendicular from the point (x, y) in the neighbourhood of the origin on the tangent line $ax + by = 0$, then

$$p = \dfrac{ax + by}{\sqrt{a^2 + b^2}} \qquad \ldots(ii)$$

If we eliminate y between (i) and (ii) we obtain an equation of the form

$$f(x, p) = 0 \qquad \ldots(iii)$$

Now, we solve (iii) for p; then the sign of p will be determined for very small values of x or y on both sides of the origin.

If for small positive values of x, the sign of p is different, then the cusp at the origin is of the first species on the right of y-axis and if the sign of p is the same then it is of the second species on the right of y-axis.

(Tangents at the origin)

Working rule. Equate to zero the lowest degree terms in the curve.

Example. Find the tangents at the origin of the curve

$$x^2(x^2 + y^2) = a(x - y)$$

Solution.

Equating to zero the lowest degree terms we obtain the equation of the tangents at origin as $x - y = 0$.

EXERCISES

Find tangents at the origin :

1. $y^2(2a - x) = x^3$ [Ans. $y = 0$]
2. $y^3 = x^3 + ax^2$ [Ans. $x = 0$]
3. $y^2(a + x) = x^2(3a - x)$ [Ans. $y = \pm\sqrt{3}x$]
4. $(x^2 + y^2)(2a - x) = b^2 x$.   [Ans. $x = 0$]

7.8 TANGENTS AT THE MULTIPLE POINTS

Working rule

(i) Differentiate the given function $f(x, y)$ partially with respect to x, i.e., $\dfrac{\partial f}{\partial x}$ and with respect to y, i.e., $\dfrac{\partial f}{\partial y}$.

Concavity, Convexity and Singular Points

(ii) Put $\frac{\partial f}{\partial x} = 0$ and $\frac{\partial f}{\partial y} = 0$. Solve them to determine x and y.

(iii) Then (x, y) are multiple points (only those which satisfy the curve)

(iv) Transfer the origin to multiple points.

(v) Then equate to zero the lowest degree terms in the transformed equation.

Example. *Find the multiple points on the curve*
$$x^4 - 2ay^3 - 3a^2y^2 - 2a^2x^2 + a^4 = 0.$$
Also find the tangents at the multiple points.

Solution. We have $f(x, y) = x^4 - 2ay^3 - 3a^2y^2 - 2a^2x^2 + a^4$

$\therefore \quad \frac{\partial f}{\partial x} = 4x^3 - 4a^2x$ and $\frac{\partial f}{\partial y} = -6ay^2 - 6a^2y$

By putting $\frac{\partial f}{\partial x} = 0$, *i.e.*, $4x^3 - 4a^2x = 0$ we get $x = 0, a, -a$.

and $\frac{\partial f}{\partial y} = 0$, *i.e.*, $-6ay^2 - 6a^2y = 0$ gives $y = 0, -a$

Hence the points for which the partial derivative are zero.
$$(0, 0), (0 - a), (a, 0), (a, -a), (-a, 0)(-a, -a)$$
Only the points which satisfy the curve are
$$(a, 0), (-a, 0) \text{ and } (0, -a)$$

Now to determine the tangents, we shall shift the origin to these points. First we shift the origin at $(a, 0)$. The transformed equation shall be given by replacing x and y by $x + a$ and $y + 0$.

$\therefore \quad (X+a)^4 - 2a(Y+0)^3 - 3a^2(Y+0)^2 - 2a^2(X+a)^2 + a^4 = 0,$

i.e., $\quad X^4 + 4aX^3 - 2aY^3 + 4a^2X^2 - 3a^2Y^2 = 0$

The tangent at the new origin are
$$4a^2X^2 - 3a^2Y^2 = 0$$

i.e., $\quad Y = \pm(2/\sqrt{3})X.$

Hence the tangents at $(a, 0)$ are
$$y = \pm\left(\frac{2}{\sqrt{3}}\right)(x - a).$$

Similarly at other points, the tangents can be determined.

EXERCISES

Find the multiple points and also the tangents there at for the following curves:

1. $(x-2)^2 = y(y-1)^2$ [**Ans.** $(2, 1)$; $y = x - 1$ and $y = -x + 3$]
2. $(x+y)^3 = \sqrt{2}(y - x + 2)^2$ [**Ans.** $(1, -1)$; $y - x + 2 = 0, y - x + 2 = 0$]
3. $y(y-6) = x^2(x-2)^3 - 9.$ [**Ans.** $(0, 3)$ and $(2, 3)$; imaginary and $y = 3$ coincident]

7.9 PROBLEMS ON THE POSITIONS AND CHARACTER OF THE MULTIPLE POINTS

Working Rule

1. Differentaite $f(x, y)$ with respect to x and y partially, *i.e.*, $\frac{\partial f}{\partial x}$ and $\frac{\partial f}{\partial y}$.

2. Put $\frac{\partial f}{\partial x} = 0$ and $\frac{\partial f}{\partial y} = 0$.
3. Solve the equations obtained in (2) for x and y.
4. Then determine $\frac{\partial^2 f}{\partial x^2}, \frac{\partial^2 f}{\partial x \partial y}$ and $\frac{\partial^2 f}{\partial y^2}$.
5. Apply the test $\left(\frac{\partial^2 f}{\partial x \partial y}\right)^2 - \frac{\partial^2 f}{\partial x^2} \cdot \frac{\partial^2 f}{\partial y^2} > = < 0$ according as there is a node, cusp or a conjugate point.

EXAMPLE

Determine the position and character of double points on the curve,

$$x^3 + x^2 + y^2 - x - 4y + 3 = 0.$$

Solution. We have $f(x, y) = x^3 + x^2 + y^2 - x - 4y + 3$

$\therefore \quad \frac{\partial f}{\partial x} = 3x^2 + 2x - 1$ and $\frac{\partial f}{\partial y} = 2y - 4$

For multiple points $\frac{\partial f}{\partial x} = 0, \frac{\partial f}{\partial y} = 0$ and $f(x, y) = 0$, we get

$$3x^2 + 2x - 1 = 0, \text{ i.e., } x = -1, \frac{1}{3}.$$

and $\qquad 2y - 4 = 0, i.e., y = 2.$

Thus the points are $(-1, 2)$ and $\left(\frac{1}{3}, 2\right)$. But only point which satisfies the curve is $(-1, 2)$.

Therefore $(-1, 2)$ is a double point.

Again $\qquad \frac{\partial^2 f}{\partial x^2} = 6x + 2, \frac{\partial^2 f}{\partial y^2} = 2$ and $\frac{\partial^2 f}{\partial x \partial y} = 0$

At $\qquad (-1, 2), \frac{\partial^2 f}{\partial x^2} = -4; \frac{\partial^2 f}{\partial y^2} = 2$ and $\frac{\partial^2 f}{\partial x \partial y} = 0.$

Here $\left(\frac{\partial^2 f}{\partial x \partial y}\right)^2 - \frac{\partial^2 f}{\partial x^2} \cdot \frac{\partial^2 f}{\partial y^2} = 0 - (-4) \cdot 2 = 8.$

which is positive and greater than zero. Therefore the point $(-1, 2)$ is a node.

7.9.1 PROBLEMS HAVING SPECIES AND CONJUGATE POINTS ETC.

Nature of the cusp at the origin when the x-axis is a tangent.

When x-axis, *i.e.*, $y = 0$ is common tangent to the two branches of the curve at the origin, the origin is cusp and the following procedure shall help :
 (i) Solve for y.
 (ii) If the signs of the roots are the same, the second species; and if the signs are different, the first species.
 (iii) If the roots of y are real for positive and negative values of x; then double; if imaginary for one value (either positive or negative) then single cusp.

Concavity, Convexity and Singular Points

EXAMPLES

1. *Show that the curve* $ay^2 = x^2 y + x^3$ *has a cusp of first species at the origin.*

Solution. The given curve is $\quad ay^2 = x^2 y + x^3$

The curve passes throguh the origin. Tangents at the origin can be determined by equating the lowest degree term to zero *i.e.*, $y^2 = 0$ (two coincident tangents) is the equation of common tangent (x-axis is the tangent). Hence the origin may be a cusp or a conjugate point.

Now solve the equation for y :

$$ay^2 - x^2 y - x^2 = 0$$

$$\therefore \quad y = \frac{x^2 \pm \sqrt{(x^4 + 4ax^3)}}{2a}$$

$$= \frac{x^2 \pm x^2 \sqrt{(1 + 4a/x)}}{2a}$$

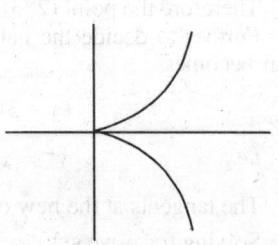

Now x is positive and small and y has two real values, one positive and the other negative. This show that there is a cusp of 1st species.

Now x is negative and small, y is imaginary, therefore no part of the curve is on left hand side of the normal at origin.

This shows that there is a single cusp of first species. The figure shows single cusp of first species.

2. *Locate the double points of the curve* $y(y-6) = x^2(x-2)^3 - 9$, *and ascertain their nature.*

Solution. Here $f(x, y) = x^2 (x-2)^3 - y(y-6) - 9 = 0$ ...(i)

$$\frac{\partial f}{\partial x} = 2x(x-2)^3 + x^3 \cdot 3(x-2)^2$$

$$= x(x-2)^2 (5x-4) \qquad \text{...(ii)}$$

$$\frac{\partial f}{\partial y} = 6 - 2y \qquad \text{...(iii)}$$

$$\frac{\partial^2 f}{\partial x^2} = (x-2)^2 (5x-4) + 2x(x-2)(5x-4) + 5x(x-2)^2 \qquad \text{...(iv)}$$

$$\frac{\partial^2 f}{\partial y^2} = -2 \qquad \text{...(v)}$$

and $\quad \dfrac{\partial^2 f}{\partial x \, \partial y} = 0 \qquad$ ...(vi)

Now $\dfrac{\partial f}{\partial x} = 0 \quad$ gives $\quad x = 0, 2, \dfrac{4}{5}$

and $\dfrac{\partial f}{\partial y} = 0$ gives $y = 3$.

Hence possible double points are $(0, 3)$, $(2, 3)$ and $\left(\dfrac{4}{5}, 3\right)$.

The point $\left(\dfrac{4}{5}, 3\right)$ does not satisfy $f(x, y) = 0$, and therefore there are only two double points.

Now at (0, 3), $\dfrac{\partial^2 f}{\partial x^2} = -16, \dfrac{\partial^2 f}{\partial y^2} = -2, \dfrac{\partial^2 f}{\partial x\, \partial y} = 0.$

∴ $\left(\dfrac{\partial^2 f}{\partial x\, \partial y}\right)^2 < \dfrac{\partial^2 f}{\partial x^2} \cdot \dfrac{\partial^2 f}{\partial y^2}.$

Hence (0, 3) is a conjugate point.

At (2, 3), $\dfrac{\partial^2 f}{\partial x^2} = 0, \dfrac{\partial^2 f}{\partial y^2} = -2$ and $\dfrac{\partial^2 f}{\partial x\, \partial y} = 0$

∴ $\left(\dfrac{\partial^2 f}{\partial x\, \partial y}\right)^2 = \dfrac{\partial^2 f}{\partial x^2} \cdot \dfrac{\partial^2 f}{\partial y^2}$

Therefore the point (2, 3) is a cusp.

Further to decide the nature of the cusp, transfer the origin to the point (2, 3). The equation then becomes

$$(y+3)(y-3) = x^3(x+2)^2 - 9$$

i.e., $$y^2 = x^3(x+2)^2.$$

The tangents at the new origin are $y^2 = 0$, i.e., x-axis is the common tangent.

Solving for y, we get

$$y = \pm (x+2)\sqrt{x^3},$$

which shows that when x is negative, y is imaginary and when x is positive, y has two values one positive other negative. Thus near the new origin the curve lies on both sides of x-axis (tangent) and only on one side of y-axis (normal). Hence the new origin (2, 3) is a single cusp of first species.

EXERCISES

Find the nature of the origin for the following curves :

1. $y^2 + 3ax^2 + x^3 = 0.$ [Ans. Conjugate]
2. $y^3 - ax^2 - x^3 = 0.$ [Ans. Single cusp of first species]
3. $x^4 - 2ax^2 y - axy^2 + a^2 y^2 = 0.$ [Ans. Single cusp of secodn species]
4. $a^3 y^2 - 2a^2 x^2 y - x^5 = 0.$ [Ans. Point of Oscul-inflexian]
5. Determine the position and character of the double points of the curve
$x^3 - y^2 - 7x^2 + 4y + 15x - 13 = 0$. [Ans. (3, 2), node]

8
Curve Tracing

8.1 INTRODUCTION

The object of Curve Tracing is to find the approximatelyu shape of the curve. There are three important forms of the curve.

(i) Cartesian equations, *i.e.,* those equations which are functions of cartesian coordinates x and y *e.g.,* $y^2 = 4ax$, $x^3 + y^3 = 3ax^2y$ etc. are cartesian equations of curves.

(ii) Polar equations, *i.e.,* those equations which are functions of polar coordinates r and θ, *e.g.,* $r = a(1 + \cos\theta)$, $r = a \sin 3\theta$ are polar equations of curves.

(iii) Parametric equations, *i.e.,* those equations in which x and y are functions of third parameter, say t, *i.e.,* $x = a \cos t$, $y = b \sin t$; $x = a(t + \sin t)$, $y = (1 - \cos t)$, are the parametric equations.

8.2 TRACING OF CARTESIAN CURVES

(1) Symmetry : We determine the symmetry of the curve, by applying the following rules :

(a) The curve is symmetrical about the x-axis, if there are even and only even powers of y in the equation, *i.e.,* if the equation is $y^2 = 4ax$, the curve lies uniformly on both the sides of x-axis.

(b) The curve is symmetrical about the y-axis, if there are even and only even powers of x in the equation, *i.e.,* $x^2 = 4ay$ contains evey powers of x. A curve may of course by symmetrical about both axes, *i.e.,* $x^2 + y^2 = a^2$ contains symmetry about x-axis and y-axis.

(c) The curve is symmetrical about the line $y = x$, if its equation remains unaltered on interchanging x and y, in $x^2 + y^2 = a^2$, if we interchange x and y, then $y^2 + x^2 = a^2$, the equation is the same. Hence the curve is symmetrical about $y = x$.

(d) The curve is symmetrical in opposite quadrants, if its equation remains unaltered on changing the signs of x and y, *i.e.,* in $x^3 + y^3 = 3ax^2y$, if we interchange the signs of x and y, we observe $(-x)^3 + (-y)^3 = 3a(-x)^2(-y)$; *i.e.,* $x^3 + y^3 = 3ax^2y$, that the equation remains unaffected. Hence the curve is symmetrical in opposite quadrants.'

(2) Points : Find the points where the curve cuts the axes.

(a) Put $y = 0$ in the equation of the curve and solve for x, and thus we will find the points where the curve cuts the x-axis, *e.g.,*

$$a^2x^2 = y^3(2a - y)$$

Putting $y = 0$, we get $x^2 = 0$ or $x = 0$.

Thus origin is a point where the curve cuts the x-axis.

(b) Similarly, put $x = 0$ in the equation of the curve and solve for y and thus we will find the points where the curve cuts the y-axis *e.g.,* $a^2x^2 = y^3(2a - y)$.

Putting $x = 0$, we get

$$y^3(2a - y) = 0 \therefore y = 0 \quad \text{or} \quad y = 2a$$

Therefore, the curve cuts the y-axis at the origin and $y = 2a$.

(3) Shape at the origin : If the curve passes through the origin, its shape at the origins should be determined in the following terms :
(i) whether there is a node,
(ii) whether there is a cusp,
(iii) or there is a conjugate point.

The following procedure should be adopted :

Write down the equations of *tangents at the origin*. This can be done by equating the lowest degree terms to zero.
(i) For node, there will be two real and non-coincident tangents.
(ii) for cusp there will be two real and coincident tangents,
(iii) for conjugate points, there will be two imaginary tangents.

If we have only one tangent, there will be either a point of inflexion $\left(\dfrac{d^2y}{dx^2} = 0, \text{ but } \dfrac{d^3y}{dx^3} \neq 0 \right)$ or the parabolic shape or only the upper half or the lower half of the parabola depending upon the regions in which the curve lies and does not lie.

(4) Shape at the points where the curve intersects the axes.

If the curve intersects any axis (x-axis or y-axis), we determine its coordinates, and then we shift the origin to this point and write down the equation of the curve with respect of this new origin. And then we determine the shape of the curve (cusp, node etc.) at this new origin.

Note : If the origin is shifted to a point whose coordinates are (a, b), the given curve is transformed from $f(x, y) = 0$ to $f(x + a, y + b) = 0$. If the curve intersects the axes at the origin and a number of other points, its shape at each and all of those points is determined one by one.

(5) Region : We find the region in which the curve does not exist, *i.e.*, if y is imaginary between two values of x, then the curve does not exist in the region bounded by the lines shown by the values of x.

(6) Asymptotes : We determine all the real asymptotes of the given curves and mark them on the co-ordinate axes, if any.

Asymptotes which are parallel to the axes : For the asymptotes of the curve which are parallel to the axes, we first of all determine the degree of the equation of curve, which is suppose n. If the equation of the curve contains x^n and y^n both, then there will be no asymptotes which will be parallel to the axes. It may, however, have oblique asymptotes. Suppose the curve does not contain x^n; then equate to zero the total coefficient of next lower degree of x, *i.e.*, x^{n-1} and we shall get asymptotes parallel to x-axis, it may be just possible that x^{n-1} is also absent; then equate the coefficient of next lower degree term, *i.e.*, x^{n-2} to zero. Similar explanation can be given for the asymptotes parallel to y-axis.

The following examples shall help us in understanding the tracing of certesian curves.

EXAMPLES

1. *Trace the curve* $y^2 (a + x) = x^2 (b - x)$.

Sol. (a) The curve has only one asymptotes *i.e.*, $a + x = 0$, which can be determined as the coefficient of the highest degree term of either y or $x = 0$, *i.e.*, $x = -a$ parallel to y-axis.

(b) The curve is symmetrical only about x-axis (there are even powers of y only).

(c) The tangents at the origin are $y^2 a = x^2 b$, i.e.,

$$y = \pm \sqrt{\left(\frac{b}{a}\right)} x$$

(which are obtained by equating the lowest degree terms of the equation).

This shows that there are two real and non-coincident tangents, hence there is a node at the origin.

(d) The curve intersects the x-axis, i.e., $y = 0$; we have $x = b$; thus at $(b, 0)$ curve intersects the x-axis.

Also, if we shift the origin to this new point, we have

$$y^2 (a + b + x) = (x + b)^2 (-x)$$

and the tangent at this point is $x = 0$ with respect to new origin and $x = b$ with respect to old origin which means that at $(b, 0)$ the tangent is parallel to y-axis.

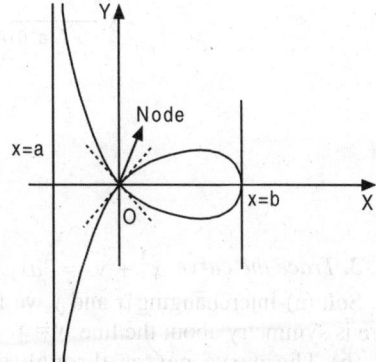

(e) The curve can also be written as

$$y^2 = x^2 \frac{b - x}{a + x}.$$

If x is negative and $> a$ (numerically); y is imaginary, hence the part of the curve does not lie in the region for which x is negative and $x > a$.

Also for $x > b$, y is imaginary, the curve does not lie beyond $x = b$.

For these considerations we have the curve as shown in the figure.

2. Trace the curve

$$y^2 (a^2 + x^2) = x^2 (a^2 - x^2)$$

Sol. (a) The curve is symmetrical about both the axes.

(b) The curve passes through the origin where the tangents are

$$y^2 = x^2, \quad i.e., \quad y = \pm x$$

(obtained by equating the lowest degree terms to zero).

These two tangents are real and different. Hence origin is a node.

(c) The curve does not meet the y-axis, except at origin and cuts x-axis at $(\pm a, 0)$. If we transfer the origin to $(a, 0)$, the equation to the curve will become

$$y^2 \{a^2 + (x + a)^2\} = (x + a)^2 (-x^2 - 2ax)$$

and the tangent at the new origin will be $x = 0$, obtained by equating to zero the terms of teh lowest degree.

(d) The curve has no real asymptote.

(e) Solving for y, and considering only the positive value,

$$y = x \left(\frac{a^2 - x^2}{a^2 + x^2}\right)^{1/2}$$

(i) When $x = 0$, $y = 0$.
(ii) When x lies between 0 and a, y is real;
(iii) When $x = a$, $y = 0$.
(iv) When $x > a$, y is imaginary.

The approximate shape of the curve will be, as shown below.

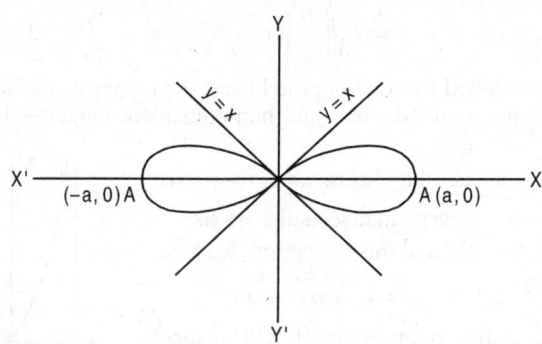

3. *Trace the curve* $x^3 + y^3 = 3axy$.

Sol. (a) Interchanging x and y we find that the equation of the curve remains unaltered, hence there is symmetry about the line $y = x$.

(b) The curve passes through the origin and the tangents at the origin are given by $xy = 0$, i.e., $x = 0$ and $y = 0$, i.e., the coordinate axes.

(c) Solving the equation of the curve and $y = x$ we find that the line $y = x$ meets the curve in $(0, 0)$ and $\left(\dfrac{3a}{2}, \dfrac{3a}{2}\right)$.

(d) The equation of the curve is $x^3 + y^3 = 3axy$.
Differentiating with respect to x, we get

$$3x^2 + 3y^2 \frac{dy}{dx} = 3a\left[x\frac{dy}{dx} + y.1\right]$$

or $$\frac{dy}{dx} = \frac{ay - x^2}{y^2 - ax}$$

$$\therefore \quad \left(\frac{dy}{dx}\right) \text{ at } \left(\frac{3a}{2}, \frac{3a}{2}\right) = \frac{a(3a/2) - (3a/2)^2}{(3a/2)^2 - (3a/2)a} = -1.$$

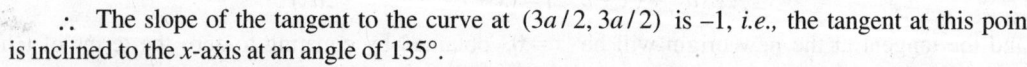

∴ The slope of the tangent to the curve at $(3a/2, 3a/2)$ is -1, i.e., the tangent at this point is inclined to the x-axis at an angle of $135°$.

(e) No asymptote parallel to the axes. But $x + y + a = 0$ is an oblique asymptote.
The shape of the curve is as shown in figure.

4. *Trace the curve* $y^2 x = a^2 (x - a)$.

Sol. (a) $x = 0$ and $y = \pm a$ are asymptotes.
(b) Symmetry about x-axis.
(c) Does not pass through the origin.
(d) It intersects x-axis at $(a, 0)$ and does not cut y-axis.

(e) Shifting the origin to $(a, 0)$ the curve is $y^2 (x + a) = a^2 x$.

The tangent at new origin is $x = 0$, i.e., with reference to old origin $x = a$ is the tangent.

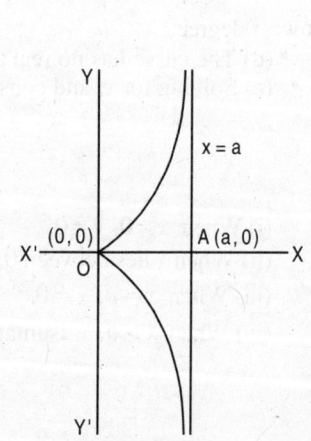

Curve Tracing

(f) $y = \pm a \sqrt{\{(x-a)/x\}}$, when $x < a$, y is imaginary which shows the curve does not exist between $x = a$ and $y = a$ but for negative values of x, y has two real values and also when $x \to 0$ from L.H.S., y tends to $+\infty$ and $-\infty$.

Hence the shape of the curve is as shown in the figure.

EXERCISE

Trace the following curves :

1. $y^2 (a + x) = x^2 (a - x)$
2. $y^2 (a + x) = x^2 (3a - x)$
3. $ay^2 = x^2 (a - x)$
4. $y^2 = x^3$
5. $x^2 = y^2 (x + 1)^3$
6. $27 ay^2 = 4(x - 2a)^3$
7. $y^2 (1 - x^2) = x^2 (1 + x^2)$
8. $y^2 (a + x) = (a - x) x^2$

ANSWERS

1.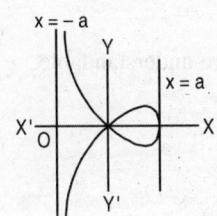
2.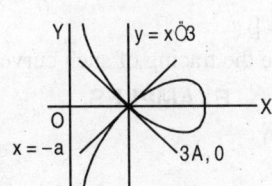
3.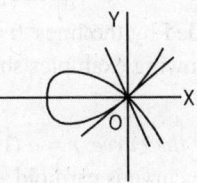
4.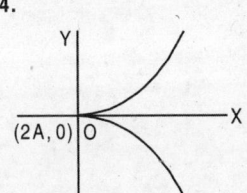
5.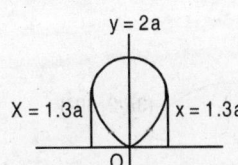
6.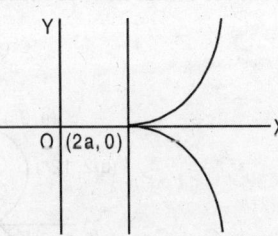
7.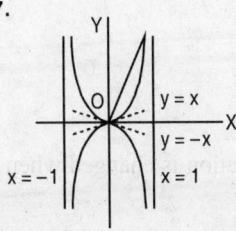
8.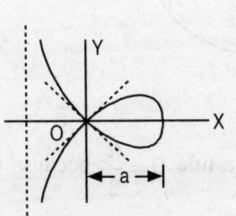

382
8.3 TRACING OF POLAR CURVES

The following procedure should be adopted to trace the curve whose polar equation is given :

(1) Symmetry

(i) The curve is symmetrical about the initial line if the equation remains unaltered on changing the sign of θ.

(ii) The curve is symmetrical about the pole if and only if the even powers of r occur in the equation, and then the pole is the centre.

(iii) The curve is symmetrical about the line $\theta = \dfrac{\pi}{4}$, if the equation remains unaltered when θ is changed to $-\theta$, and r into $-r$.

(2) Values of r and θ.

(i) If for some real value of θ, r is zero, then pole lies on the curve. This value of θ gives the tangent at the pole.

(ii) We prepare a table for certain known values of r corresponding to θ, then plot these points on the curve. The values of θ between 0 and 2π should only be considered.

(3) Asymptotes

The curve will have an infinite branch if r has an infinite value for some value of θ. In such a case we find the asymptote.

(4) If the value of r is imaginary when θ lines between α and β, the curve does not lie in the region bounded by the lines $\theta = \alpha, \theta = \beta$.

The following examples shall make the tracing of such curves dmore understandable.

EXAMPLES

1. *Trace the curve* $r = a(1 + \cos\theta)$.

Sol. The curve is cardioid.

(a) The curve is symmetrical.

(i) about the initial line, because θ is changed to $-\theta$, but the equation remains unaltered.

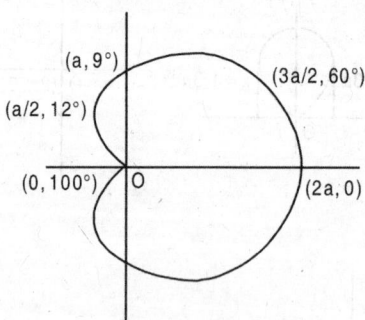

(ii) The curve is not symmetrical about a line $\theta = \dfrac{\pi}{2}$ because the equation is changed when θ is changed to $\pi - \theta$.

Curve Tracing

(b) The pole lies on the curve because for $\theta = \pi$, $r = 0$. The tangent at the pole is $\theta = \pi$.

$\theta = 0$	$\dfrac{\pi}{4}$	$\dfrac{\pi}{2}$	$\dfrac{3}{2}\pi$	π	$\dfrac{5}{4}\pi$	$\dfrac{3}{2}\pi$	2π
$r = 2a$	$\dfrac{a(\sqrt{2}+1)}{\sqrt{2}}$	a	$\dfrac{1}{2}a$	0	$\dfrac{a(\sqrt{2}-1)}{\sqrt{2}}$	a	$2a$.

Hence the shape of the curve is as given in the figure.

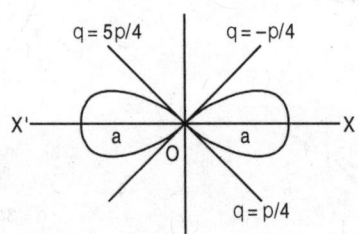

2. *Trace the curve* $r^2 = a^2 \cos 2\theta$.

Sol. (a) The curve is symmetrical about the initial about.

(b) The curve is symmetrical about the pole because the equation remains unaltered when $-r$ is put for r.

(c) The curve is symmetrical about $\theta = \dfrac{\pi}{2}$.

(d) $r = 0$ when $\theta = \dfrac{\pi}{4}$ and $\dfrac{3}{4}\pi$.

Hence the curve passes through pole and tangents at pole are given by $\theta = \dfrac{\pi}{4}$ and $\theta = \dfrac{3}{4}\pi$.

(e) When $\theta = 0$, $r = \pm a$, so the points $(a, 0)$ and $(-a, 0)$ lie on the curve.

(f) r^2 decreases from a^2 to 0 as θ increases from 0 to $\dfrac{\pi}{4}$.

$\theta = 0°$	$30°$	$45°$	$90°$	$135°$	$150°$	$180°$
$r = \pm a$	$\pm a/\sqrt{2}$	0	imaginary	0	$\pm a/\sqrt{2}$	$\pm a$

The above points give the shape as in figure.

EXERCISE

Trace the following curves :

1. $r\theta = a$
2. $r = a(1 - \cos\theta)$
3. $r^2 \cos 2\theta = a^2$
4. $r = a \cos 2\theta$
5. $r = a \sin 4\theta$
6. $r = \dfrac{a \sin^2\theta}{\cos\theta}$

ANSWERS

1.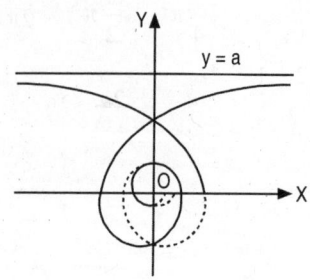

2.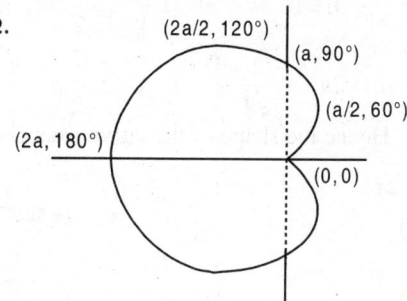

3.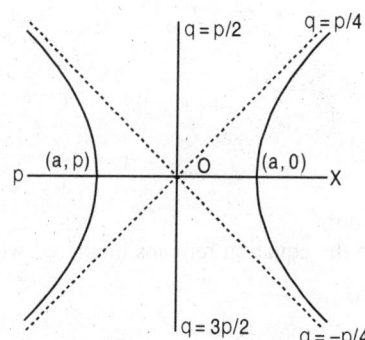

4.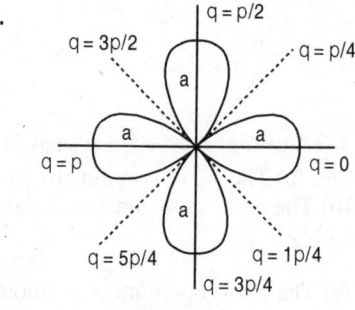

5.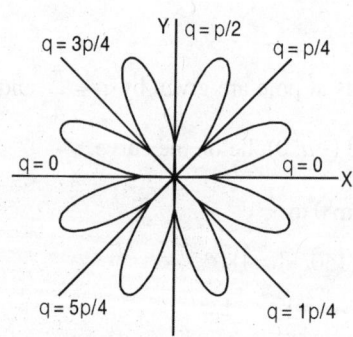

6.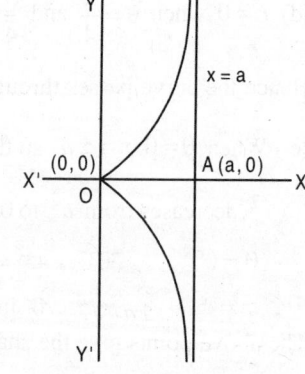

8.4 TRACING OF PARAMETRIC EQUATIONS

There are two methods to trace the curve which have parametric relations :

(1) We eliminate the parameter between the given relations and obtain the cartesian equation of the curve.

(2) We assign different values to the parameter and find the corresponding values of x and y which are marked.

We find $\dfrac{dy}{dx}$ for the values.

The followng curves can be traced by the first method :

(i) $x = a \cos \theta$, $y = a \sin \theta$ (Circle)

(ii) $x = a \cos \phi$, $y = b \sin \phi$ (Ellipse)

Curve Tracing

(iii) $x = at^2$, $y = 2at$ (Parabola)

(iv) $x = a \cos^3 t$, $y = b \sin^3 t$ (Hypo cycloid)

(v) $x = a \sin^2 t$, $y = a \dfrac{\sin^3 t}{\cos t}$ (Cissoid)

$y^2 (a - x) = a^3$

(vi) $x = \dfrac{3at}{1+t^2}$, $y = \dfrac{2t}{1+t^2}$ (Folium of Descrates)

$x^3 + y^3 = 3axy$.

The following curves are those curves in which the parameter can not be eliminated.

EXAMPLES

1. *Trace the curve*

$$x = a\left(\cos t + \log \tan \dfrac{t}{2}\right)$$

$$y = a \sin t.$$

Sol. To trace the given curve, we have

(a) $\dfrac{dy}{dt} = a \cos t$; $\dfrac{dx}{dt} = a\left(-\sin t + \dfrac{1}{\tan \dfrac{t}{2}} \cdot \sec^2 \dfrac{t}{2} \cdot \dfrac{1}{2}\right)$

$$= \dfrac{a}{\sin t}(1 - \sin^2 t) = \dfrac{a \cos^2 t}{\sin t}$$

$\therefore$ $\dfrac{dy}{dx} = \dfrac{\dfrac{dy}{dt}}{\dfrac{dx}{dt}} = \dfrac{a \cos t \cdot \sin t}{a \cos^2 t} = \tan t$

(b) The following table gives the corresponding values of t, x, y, $\dfrac{dy}{dx}$.

$t = 0$	$\pi/2$	π	$-\pi/2$	$-\pi$
$x = -\infty$	0	∞	0	∞
$y = 0$	a	0	$-a$	0
$dy/dx = 0$	∞	0	$-\infty$	0

From the table, we observe that t varies from 0 to π, x increases from $-\infty$ to 0, and then increases from 0 to ∞, x increases from 0 to a and then decreases from a to 0.

This given the shape of the curve in the first and second quadrants. Similarly, the shape of the curve is obtained in third and fourth quadrants.

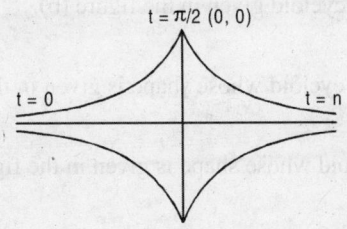

2. *Trace the curve* $x = a(t - \sin t)$.
$$y = a(1 - \cos t), \text{ when } 0 \le t \le 2\pi.$$

Sol. (a) $\dfrac{dy}{dt} = a \sin t,\quad \dfrac{dx}{dt} = a(1 - \cos t)$

$\therefore\quad \dfrac{dy}{dx} = \dfrac{a \sin t}{a(1 - \cos t)} = \cos \dfrac{t}{2}.$

(b)

	$t = 0$	$\dfrac{\pi}{2}$	π	$\dfrac{3}{2}\pi$	2π
$x = 0$		$a\left(\dfrac{\pi}{2} - 1\right)$	$a\pi$	$a\left(\dfrac{3}{2}\pi + 1\right)$	$2a\pi$
$y = 0$		a	$2a$	a	0
$\dfrac{dy}{dx} = \infty$		1	0	-1	∞

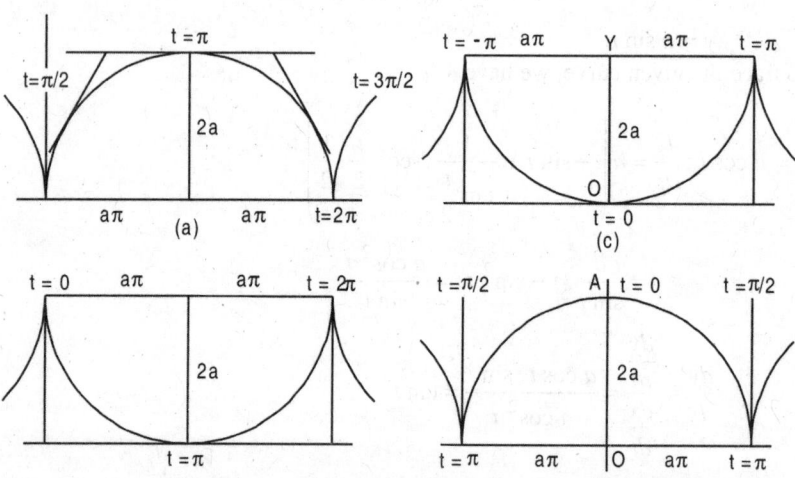

From the above table, when $t = \pi,\ \dfrac{dy}{dx} = 0$, i.e., the tangent is parallel to x-axis at the point $(a\pi, 2a)$ and $\dfrac{dy}{dx} = \infty$ when $t = 0$ or 2π. Hence the tangents are perpendicular to the x-axis at the points $(a, 0)$ and $(2a\pi, 0)$.

Also y increases when t increases from 0 to π and decreases when t increases from π to 2π. Hence, the shape of the curve is as given in the figure (a).

(b) Similarly,
$x = a(t - \sin t),$
$y = a(1 + \cos t)$ is also a cycloid given in the figure (b).

(c) $x = a(t + \sin t)$
$y = a(1 - \cos t)$ is also a cycloid whose shape is given in the figure (c).

(d) $x = a(t + \sin t),$
$y = a(1 + \cos t)$ is a cycloid whose shape is given in the figure (d).

INTEGRAL CALCULUS

Integration of irrational algebraic functions and transcendental functions. Reduction formulae. Definite Integrals. Quadrature. Rectification. Volumes and Surfaces of solids of revolution.

1

Integration of Irrational Algerabic Functions

1.1 INTEGRATION OF RATIONAL FUNCTIONS OF X

$(ax+b)^{1/n}$ and $(ax+b)^{1/m}$, etc.

(1) Rational Functions of x, $(ax+b)^{1/n}$ can be integrated by the substitution $ax+b=t^n$.

(2) Rational functions of x, $(ax+b)^{1/n}$ and $(ax+b)^{1/m}$ can be integrated by putting $ax+b=t^p$, where p is the L.C.M. of m and n.

(3) If the given function contains fractional powers of x, then the integration is effected by putting $x=t^n$, where n is the L.C.M., of the denominators of the various powers of x.

Remark : From (1) it is clear that expressions of the form

$$\frac{1}{(ax+b)\sqrt{(px+q)}} \quad \text{and} \quad \frac{1}{(ax^2+bx+c)\sqrt{(px+q)}}$$

can be integrated by the substitution $px+q=t^2$.

Furthermore, if $f(x)$ be any rational function of x, then the expressions of the form

$$\frac{f(x)}{(ax+b)\sqrt{(px+q)}} \quad \text{and} \quad \frac{f(x)}{(ax^2+bx+c)\sqrt{(px+q)}}$$

can also be integrated by the substitution $px+q=t^2$.

EXAMPLES

1. Evaluate : $\int \dfrac{x}{(x-3)\sqrt{(x+1)}}\,dx$.

Sol. Here, we put $x+1=t^2$; $\therefore dx=2t\,dt$. Hence

$$\int \frac{x\,dx}{(x-3)\sqrt{(x+1)}} = \int \frac{(t^2-1)\,2t\,dt}{(t^2-4)\cdot t}$$

$$= 2\int \frac{t^2-1}{t^2-4}\,dt = 2\int \frac{(t^2-4)+3}{t^2-4}\,dt$$

$$= 2\int \left(1+\frac{3}{t^2-4}\right) dt = 2t + \frac{3}{2}\log\frac{t-2}{t+2}$$

$$= 2\sqrt{(x+1)} + \frac{3}{2}\log\frac{\sqrt{(x+1)}-2}{\sqrt{(x+1)}+2}$$

2. Evaluate : $\int \dfrac{dx}{(1+x)^{1/2} - (1+x)^{1/3}}$.

Sol. Here the L.C.M. of 2 and 3 is 6. Therefore, putting $1 + x = t^6$ and so $dx = 6t^5\, dt$, we have

$$\text{given integral} = \int \dfrac{6t^5\, dt}{t^3 - t^2} = 6\int \dfrac{t^3}{t-1}\, dt$$

$$= \int\left(t^2 + t + 1 + \dfrac{1}{t-1}\right) dt, \text{ by division}$$

$$= \dfrac{1}{2}t^3 + \dfrac{1}{2}t^2 + t + \log(t-1), \text{ where } t = (1+x)^{1/6}.$$

EXERCISE

Integrate with respect to x :

1. $\dfrac{1}{(x+1)\sqrt{(x+2)}}$ $\left[\text{Ans. } \log \dfrac{\sqrt{(x+2)} - 1}{\sqrt{(x+2)} + 1}\right]$

2. $\dfrac{x^3}{(x-1)\sqrt{(x+2)}}$ $\left[\text{Ans. } \dfrac{2}{5}t^5 - 2t^3 + 6t + \dfrac{1}{\sqrt{3}}\log\left\{\dfrac{(t-\sqrt{3})}{(t+\sqrt{3})}\right\}; t = \sqrt{(x+2)}\right]$

3. $\dfrac{x}{(1+x)^{1/3} - (1+x)^{1/2}}$ $\left[\text{Ans. } -\left(\dfrac{2}{3}t^9 + \dfrac{3}{4}t^8 + \dfrac{6}{7}t^7 + \dfrac{6}{5}t^5 + \dfrac{3}{2}t^4\right); t = (1+x)^{1/6}\right]$

4. Evaluate $\int_8^{15} \dfrac{dx}{(x-3)\sqrt{(x+1)}}$. $\left[\text{Ans. } \dfrac{1}{2}\log \dfrac{5}{3}\right]$

1.2 INTEGRATION OF $\dfrac{1}{\sqrt{(ax^2 + bx + c)}}$

We can integrate $\dfrac{1}{\sqrt{(ax^2 + bx + c)}}$ by transforming $ax^2 + bx + c$ into the form $a\{(x+\alpha)^2 \pm \beta^2\}$> The final result with depend upon the sign of a and the sign *which precedes* β^2.

For instance, let a be *negative* and so $(-a)$ is positive. Therefore, we take out $(-a)$ from the radical sign and write.

$$\int \dfrac{dx}{\sqrt{(ax^2 + bx + c)}} = \dfrac{1}{\sqrt{(-a)}} \int \dfrac{dx}{\sqrt{[-c/a - (x^2 + bx/a)]}}$$

$$= \dfrac{1}{\sqrt{(-a)}} \int \dfrac{dx}{\sqrt{[(b^2/4a^2 - c/a) - (x + b/2a)^2]}}$$

$$= \dfrac{1}{\sqrt{(-a)}} \sin^{-1} \dfrac{(x + b/2a)}{\sqrt{(b^2/4a^2 - c/a)}}$$

$$= \dfrac{1}{\sqrt{(-a)}} \sin^{-1} \dfrac{-(2ax + b)}{\sqrt{(b - 4ac)}}$$

Integration of Irrational Algebraic Functions

since the positive square root of $\dfrac{b^2}{4a^2} - \dfrac{c}{a}$ is $\dfrac{\sqrt{(b^2-4ac)}}{-2a}$ when a is negative. Again if a is positive, we write the integral as

$$\frac{1}{\sqrt{a}} \int \frac{dx}{\sqrt{[x^2 + bx/a + c/a]}}$$

which may be written as

$$\frac{1}{\sqrt{a}} \int \frac{dx}{\sqrt{(x+b/2a)^2 + (4ac-b^2)/4a^2}}$$

or

$$\frac{1}{\sqrt{a}} \int \frac{dx}{\sqrt{(x+b/2a)^2 - (b^2-4ac)/4a^2}}$$

according as $b^2 < 4ac$ or $b^2 > 4ac$; therefore, the real form of the integral is

$$\frac{1}{\sqrt{a}} \sinh^{-1} \frac{2ax+b}{\sqrt{(4ac-b^2)}} \quad \text{or} \quad \frac{1}{\sqrt{a}} \cosh^{-1} \frac{2ax+b}{\sqrt{(b^2-4ac)}}$$

according as $b^2 < 4ac$ or $b^2 > 4ac$.

1.2.1 INTEGRATION OF $\sqrt{(ax^2 + bx + c)}$

This function can be integrated by reducing $ax^2 + bx + c$ into the form $a\{(x+\alpha)^2 \pm \beta^2\}$. The final result will naturally depend upon the sign of a and the sign which precedes β^2.

1.2.2 INTEGRATION OF $\dfrac{px+q}{\sqrt{(ax^2+bx+c)}}$

To integrate this function, we express $px+q$ as

$$px+q = A \frac{d}{dx}(ax^2+bx+c) + B$$

i.e., $\qquad px+q = A(2ax+b) + B$

Equating the coefficients of x and constant terms from the two sides gives

$$A = \frac{p}{2a} \quad \text{and} \quad B = q - \frac{bp}{2a}.$$

$$\therefore \quad \int \frac{(px+q)\,dx}{\sqrt{(ax^2+bx+c)}} = \frac{p}{2a} \int \frac{(2ax+b)}{\sqrt{(ax^2+bx+c)}} + \int \frac{(q-bp/2a)\,dx}{\sqrt{(ax^2+bx+c)}}$$

The first integral on the right is evaluated by the substitution $ax^2+bx+c = t^2$ and for the second integral we proceed as in § 1.2.

Remark : The constants A and B can also be found by mere inspection, for we readily see that

$$px+q = \frac{p}{2a}(2ax+b) + \left(q - \frac{bp}{2a}\right).$$

1.2.3 INTEGRATION OF $(px+q)\sqrt{(ax^2+bx+c)}$

To integrate this function we express $(px+q)$ as in § 1.2.2.
Therefore,

$$\int (px+q)\sqrt{ax^2+bx+c}\ dx = \frac{p}{2a}\int (2ax+b)\sqrt{ax^2+bx+c}\ dx$$

$$+\left(q-\frac{bq}{2a}\right)\int \sqrt{ax^2+bx+c}\ dx$$

The first integral on the right can be evaluated by the substitution $ax^2+bx+c = t^2$ and for the second we proceed as in § 1.2.1.

EXAMPLES

1. Integrate $\dfrac{1}{\sqrt{(1-4x-2x^2)}}$.

Sol. Here $1-4x-2x^2 = 2\left[\dfrac{1}{2}-(x^2+2x)\right] = 2\left[\dfrac{3}{2}-(x+1)^2\right]$

$$\therefore \int \frac{dx}{\sqrt{(1-4x-2x^2)}} = \frac{1}{\sqrt{2}}\int \frac{dx}{\sqrt{\left[\dfrac{3}{2}-(x+1)^2\right]}}$$

$$= \frac{1}{\sqrt{2}}\cdot \sin^{-1}\frac{x+1}{\sqrt{\left(\dfrac{3}{2}\right)}} = \frac{1}{\sqrt{2}}\sin^{-1}\frac{\sqrt{2}\,(x+1)}{\sqrt{3}}.$$

2. Evaluate $\int \sqrt{(15-2x-x^2)}\ dx$.

Sol. Here, we have

$$\text{integral} = \int \sqrt{[15-(x^2+2x)]}\ dx$$

$$= \int \sqrt{\{16-(x^2+2x+1)\}}\ dx = \int \sqrt{[16-(x+1)^2]}\ dx$$

$$= \frac{1}{2}(x+1)\sqrt{\{16-(x+1)^2\}} + \frac{16}{2}\sin^{-1}\{(x+1)/4\}$$

$$= \frac{1}{2}(x+1)\sqrt{(15-2x-x^2)} + 8\sin^{-1}\frac{1}{4}(x+1).$$

3. Evaluate : $\int \dfrac{(x+1)}{\sqrt{(x^2-x+1)}}\ dx$.

Sol. Notice that $\dfrac{d}{dx}(x^2-x+1) = (2x-1)$. Therefore,

$$\int \frac{(x+1)\ dx}{\sqrt{(x^2-x+1)}} = \int \frac{\dfrac{1}{2}(2x-1)+\dfrac{3}{2}}{\sqrt{(x^2-x+1)}}\ dx$$

$$= \frac{1}{2}\int \frac{(2x-1)\ dx}{\sqrt{(x^2-x+1)}} + \frac{3}{2}\int \frac{dx}{\sqrt{(x^2-x+1)}} \qquad ...(i)$$

Integration of Irrational Algebraic Functions

To evaluate the first integral on the right hand side, let
$$x^2 - x + 1 = t^2 \text{ ; so that } (2x-1)\,dx = 2t\,dt.$$

Therefore, $\dfrac{1}{2}\int \dfrac{(2x-1)\,dx}{\sqrt{(x^2-x+1)}} = \dfrac{1}{2}\int \dfrac{2t\,dt}{t} = \int dt$

$$= t = \sqrt{(x^2 - x + 1)} \qquad \ldots(ii)$$

Also $\displaystyle\int \dfrac{dx}{\sqrt{(x^2-x+1)}} = \int \dfrac{dx}{\sqrt{\left[\left(x-\dfrac{1}{2}\right)^2 + \dfrac{3}{4}\right]}} = \sinh^{-1}\dfrac{x-\dfrac{1}{2}}{\sqrt{\dfrac{3}{2}}}$ ...(iii)

Substituting these values in (i), we obtain

$$\int \dfrac{(x+1)\,dx}{\sqrt{(x^2-x+1)}} = \sqrt{(x^2-x+1)} + \dfrac{3}{2}\sinh^{-1}\dfrac{2x-1}{\sqrt{3}}$$

4. Evaluate : $\int (3x-2)\sqrt{(x^2+x+1)}\,dx.$

Sol. Here $d(x^2+x+1) = (2x+1)\,dx$. Therefore, let
$$3x - 2 = A(2x+1) + B.$$
Comparing the two sides gives

$$2A = 3,\ A+B = -2;\qquad \therefore\ A = \dfrac{3}{2},\ B = -\dfrac{7}{2}$$

Therefore, $\int (3x-2)\sqrt{(x^2+x+1)}\,dx$

$$= \dfrac{3}{2}\int (2x+1)\sqrt{(x^2+x+1)}\,dx - \dfrac{7}{2}\int \sqrt{(x^2+x+1)}\,dx, \qquad \ldots(i)$$

Putting $x^2+x+1 = t^2$, and $(2x+1)\,dt = 2t\,dt$ gives

$$\int (2x+1)\sqrt{(x^2+x+1)}\,dx = \int 2t^2\,dt = \dfrac{2}{3}t^3 = \dfrac{2}{3}(x^2+x+1)^{3/2}$$

Also $\displaystyle\int \sqrt{(x^2+x+1)}\,dx = \int \sqrt{\left[\left(x+\dfrac{1}{2}\right)^2 + \dfrac{3}{4}\right]}\,dx$

$$= \dfrac{1}{2}\left\{\left(x+\dfrac{1}{2}\right)\sqrt{\left(x+\dfrac{1}{2}\right)^2 + \dfrac{3}{4}} + \dfrac{3}{4}\sinh^{-1}\dfrac{x+\dfrac{1}{2}}{\sqrt{\dfrac{3}{2}}}\right\}$$

Substituting these values in (i), we obtain

$$\int (2x-2)\sqrt{(x^2+x+1)}\,dx = (x^2+x+1)^{3/2} - \dfrac{7}{4}\left(x+\dfrac{1}{2}\right)\sqrt{(x^2+x+1)}$$
$$- \dfrac{21}{16}\sinh^{-1}\dfrac{(2x+1)}{\sqrt{3}}.$$

EXERCISE

1. $\dfrac{1}{\sqrt{(x^2+2x+3)}}$

2. $\dfrac{1}{\sqrt{(1-x-x^2)}}$

3. $\sqrt{(2x^2+3x+4)}$

4. $\dfrac{x}{\sqrt{(x^2+x+1)}}$

5. $\dfrac{2x+5}{\sqrt{(x^2+3x+1)}}$

6. $(x+1)\sqrt{(x^2-x+1)}$

ANSWERS

1. $\sinh^{-1}\left\{\dfrac{(x+1)}{\sqrt{2}}\right\}$

2. $\sinh^{-1}\left\{\dfrac{(2x+1)}{\sqrt{5}}\right\}$

3. $\dfrac{1}{8}(4x+3)\sqrt{(2x^2+3x+4)} + \dfrac{23}{32}\sqrt{2}\log\left\{x+\dfrac{3}{4}+\sqrt{\left(x^2+\dfrac{3}{2}x+2\right)}\right\}$

4. $\sqrt{(x^2+x+1)} - \dfrac{1}{2}\sinh^{-1}\left\{\dfrac{(2x+1)}{\sqrt{3}}\right\}$

5. $2\sqrt{(x^2+3x+1)} + 2\cosh^{-1}\left\{\dfrac{(2x+3)}{\sqrt{5}}\right\}$

6. $\dfrac{1}{24}(8x^2+10x-1)\sqrt{(x^2-x+1)} + \dfrac{9}{16}\sinh^{-1}\left\{\dfrac{(2x-1)}{\sqrt{3}}\right\}$

1.2.4 INTEGRATION OF $\dfrac{px^2+qx+r}{\sqrt{ax^2+bx+c}}$

Here we express px^2+qx+r as follows:

$$px^2+qx+r = A(ax^2+bx+c) + B\dfrac{d}{dx}(ax^2+bx+c) + C$$

$$= A(ax^2+bx+c) + B(2ax+b) + C$$

where A, B, C are arbitrary constants and their values can be obtained by comparing the coefficients of like powers of x from both sides.

Hence the given integral

$$= A\int\dfrac{ax^2+bx+c}{\sqrt{ax^2+bx+c}}dx + B\int\dfrac{(2ax+b)\,dx}{\sqrt{ax^2+bx+c}} + C\int\dfrac{dx}{\sqrt{ax^2+bx+c}}$$

$$= A\int\sqrt{ax^2+bx+c}\,dx + B\int\dfrac{(2ax+b)\,dx}{\sqrt{(ax^2+bx+c)}} + C\int\dfrac{dx}{\sqrt{(ax^2+bx+c)}}$$

These integrals are of the standard forms of previous sections and their values can be obtained by the methods explained there.

Integration of Irrational Algebraic Functions

1.2.5 INTEGRATION OF $\dfrac{c_0 x^n + c_1 x^{n-1} + \ldots + c_{n-1} x + c_n}{\sqrt{ax^2 + bx + c}}$

To integrate this function, we assume

$$\int \dfrac{c_0 x^n + c_1 x^{n-1} + \ldots + c_{n-1} x + c_n}{\sqrt{ax^2 + bx + c}}\, dx$$

$$= (C_0 x^{n-1} + C_1 x^{n-2} + \ldots + C_{n-2} x + C_{n-1}) \sqrt{ax^2 + bx + c} + C_n \int \dfrac{dx}{\sqrt{ax^2 + bx + c}}$$

where $C_0, C_1, C_2, \ldots, C_{n-1}, C_n$ are arbitrary constants.

Differentiating both sides and multiplying by $\sqrt{ax^2 + bx + c}$ gives

$$c_0 x^n + c_1 x^{n-1} + c_2 x^{n-2} + \ldots + c_{n-1} x + c_n$$

$$= [(n-1) C_0 x^{n-2} + (n-2) C_1 x^{n-3} + \ldots + C_{n-1}] (ax^2 + bx + c)$$

$$+ \dfrac{1}{2} (C_0 x^{n-1} + C_1 x^{n-2} + \ldots + C_{n-1}) (2ax + b) + C_n$$

The constants $C_0, C_1, C_2, \ldots, C_n$ are evaluated by comparing the coefficients of like powers of x from both sides.

Also, $\int \dfrac{dx}{\sqrt{(ax^2 + bx + c)}}$ can be evaluated by the method explained in § 1.2. Hence, the integral is completely determined.

EXAMPLE

Ex. Evaluate : $\int \dfrac{6x^3 + 15x^2 - 7x + 6}{\sqrt{(2x^2 - 2x + 1)}}\, dx$.

Sol. Let $\int \dfrac{6x^3 + 15x^2 - 7x + 6}{\sqrt{(2x^2 - 2x + 1)}}\, dx$.

$$= (C_0 x^2 + C_1 x + C_2) \sqrt{(2x^2 - 2x + 1)} + C_3 \int \dfrac{dx}{\sqrt{(2x^2 - 2x + 1)}}$$

Differentiating both sides with respect to x and multiplying by $\sqrt{(2x^2 - 2x + 1)}$, we have

$$6x^3 + 15x^2 - 7x + 6 = (2C_0 x + C_1)(2x^2 - 2x + 1)$$

$$+ (C_0 x^2 + C_1 x + C_2) \cdot \dfrac{1}{2}(4x - 2) + C_3$$

Comparing the coefficients of x^3, x^2, x and constant terms respectively from the two sides, we get

$$6C_0 = 6$$
$$-4C_0 + 2C_1 - C_0 + 2C_1 = 15$$
$$2C_0 - 2C_1 - C_1 + 2C_2 = -7$$

and
$$C_1 - C_2 + C_3 = 6$$
Solving, we get $C_0 = 1, C_1 = 5, C_2 = 3$ and $C_3 = 4$.

Also,
$$\int \frac{dx}{\sqrt{(2x^2 - 2x + 1)}} = \frac{1}{\sqrt{2}} \int \frac{dx}{\sqrt{\left(x^2 - x + \frac{1}{2}\right)}}$$

$$= \frac{1}{\sqrt{2}} \int \frac{dx}{\sqrt{\left\{\left(x - \frac{1}{2}\right)^2 + \frac{1}{4}\right\}}}$$

$$= \frac{1}{\sqrt{2}} \sinh^{-1} \frac{x - \frac{1}{2}}{\frac{1}{2}} = \frac{1}{\sqrt{2}} \sinh^{-1} (2x - 1).$$

Hence, $\int \frac{6x^3 + 15x^2 - 7x + 6}{\sqrt{(2x^2 - 2x + 1)}} dx = (x^2 + 5x + 3) \sqrt{(2x^2 - 2x + 1)} + 2\sqrt{2} \sinh^{-1} (2x - 1).$

EXERCISE

Integrate with respect to x:

1. $\dfrac{x^2 + 1}{\sqrt{(x^2 + 4)}}$

2. $\dfrac{x^2 - 2}{\sqrt{(3 - x^2)}}$

3. $\dfrac{x^2 - x + 1}{\sqrt{(2x^2 - x + 2)}}$

4. $\dfrac{x^3 + x^2 + x + 1}{\sqrt{(x^2 + 2x + 3)}}$

ANSWERS

1. $\dfrac{1}{2} x \sqrt{(x^2 + 4)} - \sinh^{-1} \left(\dfrac{1}{2} x\right)$

2. $-\dfrac{1}{2} x \sqrt{(3 - x^2)} - \dfrac{1}{2} \sin^{-1} \left(\dfrac{x}{\sqrt{3}}\right)$

3. $\dfrac{1}{16} (4x - 5) \sqrt{(2x^2 - x + 2)} + \dfrac{11}{64} \sqrt{2} \log \left\{x - \dfrac{1}{4} \sqrt{\left(x^2 - \dfrac{1}{2} x + 1\right)}\right\}$

4. $\dfrac{1}{3} (x^2 + 2x + 3)^{3/2} - (x + 1) \sqrt{(x^2 + 2x + 3)} + 2 \sinh^{-1} \left(\dfrac{x + 1}{\sqrt{2}}\right)$

1.2.6 INTEGRATION OF $\dfrac{1}{(px + q) \sqrt{(ax^2 + bx + c)}}$

To integrate such functions, we put
$$px + q = \frac{1}{t}.$$

Integration of Irrelational Algebraic Functions

Taking logarithms of both sides and differentiating, we find

$$\frac{p\,dx}{px+q} = -\frac{dt}{t}.$$

Also, $ax^2 + bx + c = \dfrac{a}{p^2}\left(\dfrac{1}{t} - q\right)^2 + \dfrac{b}{p}\left(\dfrac{1}{t} - q\right) + c$

$$\equiv \frac{At^2 + Bt + C}{t^2}, \text{ say.}$$

Hence, the integral reduces to the known form

$$\int \frac{dx}{(px+q)\sqrt{(ax^2+bx+c)}} = -\frac{1}{p}\int \frac{dt}{\sqrt{(At^2+Bt+C)}}$$

which we have already discussed earlier.

Remark: If $f(x)$ be a rational integral algebraic function of x, then

$$\int \frac{f(x)}{(px+q)\sqrt{(ax^2+bx+c)}}\,dx$$

can be evauated by expressing $\dfrac{f(x)}{(px+q)}$ in the form (by common division)

$$Ax^n + Bx^{n-1} + \ldots + Kx + L\frac{M}{px+q}$$

where $Ax^n + \ldots + Kx + L$ is the quotient and M the remainder.

Thus, the integral reduces to the sum of the integrals

$$\int \frac{Ax^n + Bx^{n-1} + \ldots + L}{\sqrt{(ax^2+bx+c)}}\,dx + M\int \frac{dx}{(px+q)\sqrt{(ax^2+bx+c)}}$$

The first integral can be evaluated by the method given in § 1.2.5, whereas the second integral can be comuted by the method discussed above.

1.2.7 INTEGRATION OF $\dfrac{1}{(x-k)^r \sqrt{ax^2+bx+c}}$

Let $x > k$, then putting $x - k = \dfrac{1}{t}$, we have (however, if $x < k$, then we put $k - x = 1/t$)

given integral $= \displaystyle\int \frac{(-1/t^2)\,dt}{(1/t)^r \sqrt{(ax^2+bx+c)}}$

$$= -\int \frac{t^{r-1}\,dt}{\sqrt{(ax^2 t^2 + bxt^2 + ct^2)}}$$

$$= -\int \frac{t^{r-1}\,dt}{\sqrt{[a(1+kt)^2 + bt(1+kt) + ct^2]}}$$

$$= -\int \frac{r^{r-1}\,dt}{\sqrt{[(ak^2+bk+c)t^2 + (2ak+b)t + a]}}$$

$$= -\int \frac{t^{r-1} dt}{\sqrt{(At^2 + Bt + C)}},$$

where $A = ak^2 + bk + c$; $B = 2ak + b$; $C = a$.

This can be integrated by the method of § 1.2, if $r = 1$ and by method of § 1.2.5, if $r > 1$.

Remark : If $f(x)$ is a rational integral algebraic function and its denominator can be resolved into real linear factors, then clearly $\dfrac{f(x)}{\sqrt{(ax^2 + bx + c)}}$ can be integrated by first resolving $f(x)$ into partial functions.

EXAMPLES

1. Evaluate : $\displaystyle\int \frac{dx}{(x-1)\sqrt{(x^2 + x + 1)}}, \, x > 1.$

Sol. Put $x - 1 = \dfrac{1}{t}$, so that $dx = -\dfrac{dt}{t^2}$. Therefore,

$$\int \frac{dx}{(x-1)\sqrt{(x^2 + x + 1)}} = \int \frac{-dt/t^2}{(1/t)\sqrt{(x^2 + x + 1)}}$$

$$= -\int \frac{dt}{\sqrt{(t^2 x^2 + t^2 x + 1)}}$$

$$= -\int \frac{dt}{\sqrt{[(t+1)^2 + t(t+1) + t^2]}}$$

$$= -\int \frac{dt}{\sqrt{(3t^2 + 3t + 1)}}$$

$$= -\frac{1}{\sqrt{3}} \int \frac{dt}{\sqrt{\left[\left(t + \frac{1}{2}\right)^2 + \frac{1}{3} - \frac{1}{4}\right]}}$$

$$= -\frac{1}{\sqrt{3}} \sinh^{-1} \frac{t + \frac{1}{2}}{\sqrt{(1/12)}}$$

$$= -\frac{1}{\sqrt{3}} \sinh^{-1} \left\{\sqrt{3}\left(\frac{x+1}{x-1}\right)\right\}$$

2. Evaluate : $\displaystyle\int \frac{dx}{(x-1)^2 \sqrt{(1-x^2)}}.$

Sol. Clearly, $\sqrt{(1-x^2)}$ will be real if $x < 1$. Therefore, putting

$1 - x = \dfrac{1}{t}$ and so $dx = \dfrac{dt}{t^2}$, we have

Integration of Irrelational Algebraic Functions

$$\int \frac{dx}{(x-1)^2 \sqrt{(1-x^2)}} = \int \frac{dx}{(1-x)^2 \sqrt{(1-x^2)}}$$

$$= \int \frac{dt/t^2}{(1/t^2)\sqrt{(1-x^2)}} = \int \frac{t \, dt}{\sqrt{(t^2 - t^2 x^2)}}$$

$$= \int \frac{t \, dt}{\sqrt{[t^2 - (t-1)^2]}} = \int \frac{t \, dt}{\sqrt{(2t-1)}}.$$

Again, put $2t - 1 = u^2, \therefore dt = u \, du$. Hence,

$$\int \frac{dx}{(x-1)^2 \sqrt{(1-x^2)}} = \int \frac{\frac{1}{2}(u^2+1) u \, du}{u} = \frac{1}{2} \int (u^2 + 1) \, du$$

$$= \frac{1}{2}\left(\frac{1}{3}u^3 + u\right) = \frac{1}{6} u (u^2 + 3)$$

$$= \frac{1}{6}\sqrt{2t-1}.(2t+2), \text{ since } u^2 = 2t-1$$

$$= \frac{1}{3}\sqrt{\frac{2}{1-x} - 1}.\left(\frac{1}{1-x} + 1\right), \text{ as } t = \frac{1}{1-x}$$

$$= \frac{1}{3}\sqrt{1+x}\, \frac{(2-x)}{(1-x)} \sqrt{1-x}.$$

EXERCISE

Integrate with respet to x :

1. $\dfrac{1}{(x-a)\sqrt{(x^2-a^2)}}$ $\qquad \left[\text{Ans. } -\dfrac{1}{a}\sqrt{\left(\dfrac{x+a}{x-1}\right)}\right]$

2. $\dfrac{1}{(x+1)\sqrt{(1+2x-x^2)}}$ $\qquad \left[\text{Ans. } \dfrac{1}{\sqrt{2}} \sin^{-1} \dfrac{x\sqrt{2}}{x+1}\right]$

3. $\dfrac{x}{(x+1)\sqrt{(x^2+1)}}$ [Hint : Write $\dfrac{x}{x+1} = 1 - \dfrac{1}{x+1}$.] $\left[\text{Ans. } \sinh^{-1} x + \dfrac{1}{\sqrt{2}} \sinh^{-1} \dfrac{1-x}{x+1}.\right]$

4. $\dfrac{1}{(x^2-1)\sqrt{(1+x^2)}}$ [Hint : Break $1/(x^2 - 1)$ into its partial fractions.]

$$\left[\text{Ans. } \frac{1}{2\sqrt{2}} \sinh^{-1} \frac{1+x}{1-x} + \sinh^{-1}\frac{1-x}{x+1}.\right]$$

1.3 INTEGRATION OF $\dfrac{1}{(Ax^2 + B)\sqrt{(Cx^2 + D)}}$

Here we first put $x = \dfrac{1}{t}$, and so Thus,

$$\int \dfrac{dx}{(Ax^2 + B)\sqrt{(Cx^2 + D)}} = \int \dfrac{(-1/t^2)\, dt}{\left(\dfrac{A}{t^2} + B\right)\sqrt{\dfrac{C}{t^2} + D}}$$

$$= -\int \dfrac{t\, dt}{(A + Bt^2)\sqrt{(C + Dt^2)}}$$

We now susbtitute $C + Dt^2 = u^2$. This reduces the integral to the form $\int \dfrac{du}{u^2 \pm \alpha^2}$.

Accordingly, the final result will depend upon the sign which percedes α^2.

EXAMPLES

Ex. *Integrate* $\dfrac{1}{(1 + x^2)\sqrt{(1 - x^2)}}$.

Sol. Putting $x = \dfrac{1}{t}$ and so $dx = -\dfrac{1}{t^2}\, dt$, we have

$$\int \dfrac{dx}{(1 + x^2)\sqrt{(1 - x^2)}} = \int \dfrac{(-1/t^2)\, dt}{(1 + 1/t^2)\sqrt{(1 - 1/t^2)}} = -\int \dfrac{t\, dt}{(t^2 + 1)\sqrt{(t^2 - 1)}}$$

Again, let $t^2 - 1 = u^2$, so that $t\, dt = u\, du$. Then

$$\int \dfrac{dx}{(1 + x^2)\sqrt{(1 - x^2)}} = -\int \dfrac{u\, du}{(u^2 + 2)\cdot u} = -\int \dfrac{du}{u^2 + 2}$$

$$= -\dfrac{1}{\sqrt{2}} \tan^{-1} \dfrac{u}{\sqrt{2}} = -\dfrac{1}{\sqrt{2}} \tan^{-1} \dfrac{\sqrt{(t^2 - 1)}}{\sqrt{2}}$$

$$= -\dfrac{1}{\sqrt{2}} \tan^{-1} \dfrac{\sqrt{(1 - x^2)}}{x\sqrt{2}}, \text{ since } t = \dfrac{1}{x}.$$

EXERCISES

Integrate with respect to x :

1. $\dfrac{1}{x^2\sqrt{(x^2 - 1)}}$ $\qquad\qquad\left[\text{Ans. } \dfrac{\sqrt{(x^2 - 1)}}{x}\right]$

2. $\dfrac{1}{(2x^2 + 3)\sqrt{(x^2 - 4)}}$ $\qquad\left[\text{Ans. } \left(\dfrac{1}{2\sqrt{33}}\right) \log \left[\dfrac{\{x\sqrt{11} + \sqrt{(3x^2 - 12)}\}}{\{x\sqrt{11} - \sqrt{(3x^2 - 12)}\}}\right]\right]$

3. Show that $\int_0^{1/\sqrt{3}} \dfrac{dx}{(1 + x^2)\sqrt{(1 - x^2)}} = \dfrac{\pi}{4\sqrt{2}}$.

Integration of Irrational Algebraic Functions

1.4 GENERAL CASE

Certain irrational functions which are not already in one or other of the standard forms considered previously can be changed into an integral form by rationalising the numerator or the denominator. Some expressions can be broken up into two or more, each of which is integrable or can easily be reduced to an integrable form. A few illustrations are given below.

EXAMPLES

1. Integrate $\dfrac{1}{\{\sqrt{(x+a)} + \sqrt{(x+b)}\}}$.

Sol. $\displaystyle\int \dfrac{1}{\{\sqrt{(x+a)} + \sqrt{(x+b)}\}} = \int \dfrac{\sqrt{(x+a)} - \sqrt{(x+b)}}{(x+a) - (x+b)}\, dx,$

$= \dfrac{1}{(a-b)} \int [\sqrt{(x+a)} - \sqrt{(x+b)}]\, dx$

$= \dfrac{1}{(a-b)} \left[\dfrac{2}{3}(x+a)^{3/2} - \dfrac{2}{3}(x+b)^{3/2} \right]$

$= \dfrac{2}{3(a-b)} [(x+a)^{3/2} - (x+b)^{3/2}]$

2. Integrate $x\sqrt{\dfrac{1+x}{1-x}}$.

Sol. $\displaystyle\int x\sqrt{\dfrac{1+x}{1-x}}\, dx = \int \dfrac{x(1+x)\, dx}{\sqrt{(1-x^2)}},$

Multiplying numerator and denominator by $\sqrt{(1+x)}$.

$= \displaystyle\int \dfrac{x\, dx}{\sqrt{(1-x^2)}} + \int \dfrac{x^2\, dx}{\sqrt{(1-x^2)}} = \dfrac{1}{2}\int \dfrac{2x\, dx}{\sqrt{(1-x^2)}} - \int \dfrac{(1-x^2) - 1}{\sqrt{(1-x^2)}}\, dx$

$= \dfrac{1}{2}\int \dfrac{2x\, dx}{\sqrt{(1-x^2)}} - \int \sqrt{(1-x^2)}\, dx + \int \dfrac{dx}{\sqrt{(1-x^2)}}$

$= -\dfrac{1}{2}\sqrt{(1-x^2)} - \dfrac{1}{2}x\sqrt{(1-x^2)} - \dfrac{1}{2}\sin^{-1} x + \sin^{-1} x$

$= -\dfrac{1}{2}\sqrt{(1-x^2)} - \dfrac{1}{2}x\sqrt{(1+x^2)} + \dfrac{1}{2}\sin^{-1} x.$

EXERCISE

Integrate with respect to x :

1. $x\sqrt{\dfrac{1-x}{1+x}}$ $\left[\text{Ans.} \left(\dfrac{1}{2}x - 1\right)\sqrt{(1-x^2)} - \dfrac{1}{2}\sin^{-1} x \right]$

2. $\dfrac{1}{x + \sqrt{(x^2+1)}}$ $\left[\text{Ans.} \dfrac{1}{2}[x^2 - x + \sqrt{(x^2-1)} + \cosh^{-1} x] \right]$

3. $\dfrac{\sqrt{(1+x^2)}}{1-x^2}$ $\quad$ [Ans. $-\log\{x+\sqrt{(1+x^2)}\} - \dfrac{1}{\sqrt{2}}\log\left[\dfrac{\sqrt{(1+x^2)} - x\sqrt{2}}{\sqrt{(1+x^2)} + x\sqrt{2}}\right]$]

4. $\dfrac{x+1}{(x^2+4)\sqrt{(x^2+9)}}$ $\quad$ [**Hint :** Ratonalize the numerator.]

$$\left[\text{Ans.} \quad \dfrac{1}{2\sqrt{5}}\log\dfrac{\sqrt{(x^2+9)} - \sqrt{5}}{\sqrt{(x^2+9)} + \sqrt{5}} - \dfrac{1}{2\sqrt{5}}\tan^{-1}\dfrac{2\sqrt{(x^2+9)}}{2\sqrt{5}}\right]$$

1.5 SUBSTITUTIONS

Certain irrational functions can often be integrated more conveniently by a suitable substitution. If the function to be integrated is not of one or other of the standard forms dealt so far, the substitution is to be found by inspection.

For instance, functions involving

$$\sqrt{(a^2 - x^2)} \quad \text{or} \quad \sqrt{(a^2 + x^2)} \quad \text{or} \quad \sqrt{(x^2 - a^2)}$$

and no other radical can be integrated by a trigonometrical substitution. We can substitute

$$x = a\sin\theta \quad \text{or} \quad x = a\sin\theta \quad \text{or} \quad x = a\sec\theta$$

respectively in the above cases to get rid of the square root.

Moreover, functions involving

$$\sqrt{(a^2 - b^2 x^2)} \quad \text{or} \quad \sqrt{(a^2 + b^2 x^2)} \quad \text{or} \quad \sqrt{(b^2 x^2 - a^2)}$$

and no other irrational factor may be integrated by the trigonometric substitution

$$bx = a\sin\theta \quad \text{or} \quad bx = a\tan\theta \quad \text{or} \quad bx = a\tan\theta \quad \text{respectively.}$$

Since $\sqrt{(ax^2 + bx + c)}$ can easily be reduced to one of the above forms, trigonometric substitutions are also applicable in the case of functions involving $\sqrt{(ax^2 + bx + c)}$.

If some power of x, say x^{n-1} is a factor of the integrated and the remaining part is a function of x^n alone, the substitution $x^n = t$ often simplifies the integration.

EXAMPLES

1. *Integrate* $\dfrac{1}{x^2\sqrt{(1-x^2)}}$.

Sol. Putting $x = \sin\theta$, and so $dx = \cos\theta\, d\theta$, we have

$$\int \dfrac{dx}{x^2\sqrt{(1-x^2)}} = \int \dfrac{\cos\theta\, d\theta}{\sin^2\theta \cos\theta}$$

$$= \int \operatorname{cosec}^2\theta\, d\theta = -\cot\theta$$

$$= -\sqrt{(1-x^2)}/x, \text{ since } \sin\theta = x.$$

2. *Evaluate :* $\int_0^a x\sqrt{\dfrac{(a-x)}{(a+x)}}\, dx$

Sol. Here, we put $x = a\sin\theta, \therefore dx = a\cos\theta\, d\theta$. For the limits we see that when $x = a, \sin\theta = 1$ or $\theta = \dfrac{1}{2}\pi$; and when $x = 0, \sin\theta = 0$ or $\theta = 0$. Therefore,

Integration of Irrational Algebraic Functions

$$\int_0^a x \sqrt{\frac{a-x}{a+x}} \, dx = \int_0^{1/2\pi} a \sin\theta \sqrt{\frac{a(1-\sin\theta)}{a(1+\sin\theta)}} \cdot a\cos\theta \, d\theta$$

$$= a^2 \int_0^{1/2\pi} \frac{\sin\theta \cos\theta \cdot (1-\sin\theta)}{\sqrt{(1-\sin^2\theta)}} \, d\theta$$

$$= a^2 \int_0^{1/2\pi} (\sin\theta - \sin^2\theta) \, d\theta$$

$$= a^2 [-\cos\theta]_0^{1/2\pi} - \frac{a^2}{2} \int_0^{1/2\pi} (1-\cos 2\theta) \, d\theta$$

$$= a^2 \left(-\cos\frac{1}{2}\pi + \cos 0\right) - \frac{a^2}{2} \left[\theta - \frac{\sin 2\theta}{2}\right]_0^{\frac{1}{2}\pi}$$

$$= a^2 \left(-\cos\frac{1}{2}\pi + \cos 0\right) - \frac{a^2}{2} \left[\theta - \frac{\sin 2\theta}{2}\right]_0^{\frac{1}{2}\pi}$$

Remark : Another suitable substitution is $x = a \cos 2\theta$.

EXERCISE

Integrate with respect to x :

1. $\dfrac{1}{x^2 \sqrt{(1+x^2)}}$ $\qquad\qquad$ [Ans. $-\{\sqrt{(1+x^2)}\}/x$]

2. $\dfrac{1}{(a^2 - b^2 x^2)^{3/2}}$ $\qquad\qquad$ [Ans. $x/a^2 \sqrt{(a^2 - b^2 x^2)}$]

3. $\dfrac{1}{x\sqrt{(a^n + x^n)}}$ $\qquad\qquad$ [Ans. $\dfrac{1}{na^{n/2}} \log \dfrac{\sqrt{(a^n + x^n)} - a^{n/2}}{\sqrt{(a^n + x^n)} + a^{n/2}}$]

4. Evaluate $\int_0^a x \sqrt{\dfrac{a^2 - x^2}{a^2 + x^2}} \, dx.$ $\qquad\qquad$ [Ans. $\dfrac{1}{2} a^2 \left(\dfrac{1}{2}\pi - 1\right)$]

1.6 INTEGRATION OF $x^m (a+bx^n)^p$

Case 1. When p is a positive integer.

In this case, expand $(a + bx^n)^p$ by Binomial theorem into a finite series. This transforms the integrand into a sum of a finite number of terms each of which is easily integrable.

Example : *Evaluate :* $\int x^{1/4} (2 + 3x^{1/3})^2 \, dx.$

Sol. Here $x^{1/4} (2 + 3x^{1/3})^2 = x^{1/4} (4 + 12 x^{1/3} + 9 x^{2/3})$

$$= 4x^{1/4} + 12 x^{7/12} + 9 x^{11/12}$$

Therefore, $\int x^{1/4} (2 + 3x^{1/3})^2 \, dx = \int (4x^{1/4} + 12 x^{7/12} + 9 x^{11/12}) \, dx$

$$= 4 \cdot \frac{4}{5} x^{5/4} + 12 \cdot \frac{12}{19} x^{19/12} + 9 \cdot \frac{12}{23} x^{23/12}$$

$$= \frac{16}{5} x^{5/4} + \frac{144}{19} x^{19/12} + \frac{108}{23} x^{23/12}$$

Case 2. When $\dfrac{m+1}{n}$ is an integer.

In this case if p is not an integer, then it must be of the form $p = r/s$ and the integration can be effected by the substitution

$$a + bx^n = t^s.$$

Case 3. When $p + \dfrac{m+1}{n}$ is an integer, p is not an integer.

In this case, we put $x = \dfrac{1}{t}$. Thus,

$$\int x^m (a+bx^n)^p \, dx = -\int \dfrac{1}{t^{m+2}} \left(a + \dfrac{b}{t^n}\right)^p dt$$

$$= -\int t^{-(m+np+2)} (b + at^n)^p \, dt.$$

This integral comes under case 2, since

$$\dfrac{-(m+np+2)+1}{n} \quad i.e., \quad -\left(p + \dfrac{m+1}{n}\right) \text{ is an integer.}$$

Hence to evaluate the transformed integral, put

$$b + at^n = u^s, \quad \text{where } p = \dfrac{r}{s}.$$

Example : Evaluate : $\int x^{-2/3} (1+x^{1/2})^{-5/3} \, dx$.

Sol. Here $p + \dfrac{m+1}{n} = -\dfrac{5}{3} + \dfrac{-\dfrac{2}{3}+1}{\dfrac{1}{2}} = -1$ which is an integer. Hence, putting $x = 1/t$, we obtain

$$\int x^{-2/3} (1+x^{1/2})^{-5/3} \, dx = -\int t^{2/3 + 5/6 - 2} (t^{1/2}+1)^{-5/3} dt = -\int t^{-1/2} (1+t^{1/2})^{-5/3} dt.$$

Again, putting $1 + t^{1/2} = u^3$ and $\dfrac{1}{2} t^{-1/2} dt = 3u^2 du$, we have

$$\int x^{-2/3} (1+x^{1/2})^{-5/3} \, dx = -6 \int u^{-5} \cdot u^2 du = 3u^{-2}$$

$$= 3(1+t^{1/2})^{-2/3} = 3(1+x^{-1/2})^{-2/3}$$

EXERCISE

Integrate with respect to x :

1. $x^{2/3} (1+x^{6/5})^2$ $\qquad \left[\text{Ans. } \dfrac{3}{5} x^{5/3} + \dfrac{30}{43} x^{43/15} + \dfrac{15}{61} x^{61/15}\right]$

2. $x^{1/3} (1+x^{3/4})^3$ $\qquad \left[\text{Ans. } \dfrac{3}{4} x^{4/3} + \dfrac{36}{25} x^{25/12} + \dfrac{18}{17} x^{17/6} + \dfrac{12}{43} x^{43/12}\right]$

3. $x(1+x^3)^{1/3}$

$$\left[\text{Ans. } \dfrac{u}{3(u^3-1)} + \dfrac{1}{9} \log \dfrac{\sqrt{(u^2+u+1)}}{u-1} + \dfrac{1}{3\sqrt{3}} \tan^{-1} \dfrac{2u+1}{\sqrt{3}}, \text{ where } u = x^{-1}(1+x^3)^{1/3}\right]$$

4. $x^3 (1+x^2)^{1/3}$ $\qquad \left[\text{Ans. } \dfrac{3}{56} (1+x^2)^{4/3} (4x^2-3)\right]$

Integration of Irrelational Algebraic Functions

MISCELLANEOUS EXAMPLES

1. Evaluate : $\int \dfrac{dx}{\sqrt{(x-\alpha)(x-\beta)}}$.

Sol. One method is to proceed as in Sec. 1.2 However, an easier method is to put $x - \alpha = t^2$, so that $dx = 2t\, dt$. Then

$$\int \frac{dx}{\sqrt{(x-\alpha)(x-\beta)}} = \int \frac{2t\, dt}{\sqrt{t^2(t^2+\alpha-\beta)}} = \int \frac{2d}{\sqrt{t^2+\alpha-\beta}}$$

$$= 2\log[t + \sqrt{t^2+\alpha-\beta}]$$

$$= 2\log[\sqrt{x-\alpha} + \sqrt{x-\beta}]$$

2. Evaluate : $\int_\alpha^\beta \sqrt{(x-\alpha)(\beta-x)}\, dx,\ \beta > \alpha$.

Sol. Such definite integrals are of frequent occurrence in the applications of Integral Calculus. Whenever $x-\alpha$ and $\beta-x$ occur as a *product* or *quotient* under a radical sign in a definite (or indefinite) integral, the best substitution is

$$x = \alpha\cos^2\theta + \beta\sin^2\theta$$

Therefore, $x = \alpha\cos^2\theta + \beta\sin^2\theta$

$$x - \alpha = (\alpha\cos^2\theta + \beta\sin^2\theta) - \alpha = (\beta-\alpha)\sin^2\theta$$

and $\qquad \beta - x = \beta - (\alpha\cos^2\theta + \beta\sin^2\theta)$

$$= (\beta-\alpha)\cos^2\theta.$$

Also, for the limits, we observe :

when $x = \alpha$, i.e., $x - \alpha = 0$, we have $\sin^2\theta = 0$ or $\theta = 0$

and when $x = \beta$, i.e., $\beta - x = 0$, we have $\cos^2\theta = 0$ or $\theta = \dfrac{1}{2}\pi$

$\therefore\qquad$ the integral $= \int_\alpha^\beta \sqrt{(x-\alpha)(\beta-x)}\, dx$

$$= \int_0^{\pi/2} \sqrt{(\beta-\alpha)^2 \sin^2\theta \cos^2\theta}\,.\, 2(\beta-\alpha)\sin\theta\cos\theta\, d\theta$$

$$= (\beta-\alpha)^2 \int_0^{\pi/2} 2\sin^2\theta\cos^2\theta\, d\theta$$

$$= \frac{1}{2}(\beta-\alpha)^2 \int_0^{\pi/2} \sin^2 2\theta\, d\theta$$

$$= \frac{1}{4}(\beta-\alpha)^2 \int_0^{\pi/2} (1-\cos 4\theta)\, d\theta$$

$$= \frac{(\beta-\alpha)^2}{4}\left[\theta - \frac{\sin 4\theta}{4}\right]_0^{\pi/2} = \frac{\pi(\beta-\alpha)^2}{8}.$$

3. Evaluate $\int_0^a x \sqrt{\left(\dfrac{a^2-x^2}{a^2+x^2}\right)}\, dx$.

Sol. Put $x^2 = a^2\cos 2\theta$

Then $\qquad\qquad 2x\, dx = -2a^2 \sin 2\theta\, d\theta$

or $\qquad\qquad x\, dx = -a^2 \sin 2\theta\, d\theta$

Also $x = 0$ given $\cos 2\theta = 0$ or $\theta = \pi/4$
and $x = a$ gives $\cos 2\theta = 1$ or $\theta = 0$

$$\therefore \int_0^a x \sqrt{\frac{a^2 - x^2}{a^2 + x^2}}\, dx = \int_{\pi/4}^0 \sqrt{\frac{a^2 - a^2 \cos 2\theta}{a^2 + a^2 \cos 2\theta}}\, (-a^2 \sin 2\theta)\, d\theta$$

$$= a^2 \int_0^{\pi/4} \sqrt{\frac{1 - \cos 2\theta}{1 + \cos 2\theta}}\, \sin 2\theta\, d\theta$$

$$= a^2 \int_0^{\pi/4} \sqrt{\frac{2 \sin^2 \theta}{2 \cos^2 \theta}}\, 2 \sin \theta \cos \theta\, d\theta$$

$$= a^2 \int_0^{\pi/4} 2 \sin^2 \theta\, d\theta = a^2 \int_0^{\pi/4} (1 - \cos 2\theta)\, d\theta$$

$$= a^2 \left[\theta - \frac{1}{2} \sin 2\theta \right]_0^{\pi/4} = a^2 \left[\frac{1}{4}\pi - \frac{1}{2} \sin \frac{\pi}{2} \right]$$

$$= \frac{1}{4} a^2 (\pi - 2).$$

EXERCISE

Evaluate:

1. $\int \sqrt{\frac{a-x}{x}}\, dx.$ (put $x = \sin^2 \theta$). $\quad$ [Ans. $a \sin^{-1} \sqrt{\left(\frac{x}{a}\right)} + \sqrt{(ax - x^2)}$]

2. $\int \frac{dx}{\sqrt{(x - \alpha)(\beta - x)}}$ $\quad$ [Ans. $a \sin^{-1} \sqrt{\left(\frac{x}{a}\right)} + \sqrt{(ax - x^2)}$]

3. $\int \frac{dx}{(x - a)\sqrt{(x - a)(b - x)}}$ $\quad$ [Ans. $\left(\frac{2}{a - b}\right) \sqrt{\frac{(b - x)}{(x - a)}}$]

4. $\int \sqrt{\frac{x + a}{x + b}} \cdot \frac{dx}{x + c}$

[**Hint**: Rationalize the numerator and apply § 1.2.7.]

[Ans. $\cosh^{-1} \left\{ \frac{(2x + a + b)}{(a - b)} \right\} - \sqrt{\left(\frac{c - a}{c - b}\right)}; \cosh^{-1} \frac{2(c - b)(c - a) + (a + b - 2c)(x + c)}{(a - b)(x + c)}$]

5. $\int_\alpha^\beta \frac{dx}{\sqrt{\{(x - \alpha)(\beta - x)\}}}, \beta > \alpha.$ $\quad$ [Ans. π]

☐☐☐

2
Integration of Transcendental Functions

2.1 TRANSCENDENTAL FUNCTIONS

A function $y = f(x)$ which satisfies an equation of the form

$$p_0(x) y^n + p_1(x) y^{n-1} + \ldots + p_{n-1}(x) + p_n(x) = 0 \qquad \ldots(i)$$

where $p_0(x), \ldots, p_n(x)$ are polynomials in x and n is a positive integer, is called an *algebraic function*.

Functions which are not algebraic, *i.e.*, do not satisfy equations of the form (i) are called *transcendental functions*.

Some special transcendentral functions are :

1. Trigonometric functions : $f(x) = \sin x, \cos x, \tan x$, etc.

2. Inverse trigonometric functions : $f(x) = \sin^{-1} x, \cos^{-1} x$, etc.

3. Exponential functions : $f(x) = a^x$ where $a \neq 0, 1$.

4. Logarithmic functions : $f(x) = \log_a x$ where $a \neq 0, 1$. This and the exponential function are inverse functions. If $a = e = 2.71828, \ldots$, we write $f(x) = \log_e x = \ln x$, called the natural logarithm of x.

5. Hyperbolic functions : $f(x) = \sinh x, \cosh x, \tanh x$, etc.

6. Inverse hyperbolic functions : $f(x) = \sinh^{-1} x, \cosh^{-1} x$, etc.

INTEGRATION OF TRIGONOMETRIC FUNCTIONS

2.2 INTEGRATION OF $\sin^m x$ or $\cos^n x$, WHERE n IS AN ODD POSITIVE INTEGER

Let $n = 2m + 1$, where m is a positive integer, then

$$\int \sin^n x \, dx = \int \sin^{2m+1} x \, dx$$
$$= \int \sin^{2m} x \cdot \sin x \, dx$$
$$= \int (1 - \cos^2 x)^m \sin x \, dx$$

Now put $\cos x = t$ and $-\sin x \, dx = dt$. Then

$$\int \sin^n x \, dx = -\int (1 - t^2)^m \, dt.$$

The integrand $(1 - t^2)^m$ can be expanded by Binomial Theorem into a finite number of terms and each term of the expansion can be integrated separately.

Similarly, when n is an *odd positive integer*, then $\cos^n x$ can be integrated by the substitution $\sin x = t$.

Note : If n be an even positive integer, then $\sin^n x$ or $\cos^n x$ may be integrated by the method of successive reduction which will be explained in the next chapter. However, if the answer is required in terms of a series of *sines of cosines of multiples of the angle*, then we use trigonometrical transformations which will be explained in § 2.4.

Example : *Integrate* $\cos^6 x$.

Sol. Let
$$\cos x + i \sin x = t \qquad \ldots(1)$$
$$\Rightarrow \qquad \cos x - i \sin x = \frac{1}{t} \qquad \ldots(2)$$

Now adding (1) and (2)
$$2\cos x = t + \frac{1}{t}$$
$$\Rightarrow \qquad 2^6 \cos^6 x = \left(t + \frac{1}{t}\right)^6$$
$$= t^6 + {}^6c_1 t^4 + {}^6c_2 t^2 + {}^6c_3 + {}^6c_4 \frac{1}{t^2} + {}^6c_4 \frac{1}{t^4} + \frac{1}{t^6}$$
$$= \left(t^6 + \frac{1}{t^6}\right) + 6\left(t^4 + \frac{1}{t^4}\right) + \frac{6.5}{1.2}\left(t^2 + \frac{1}{t^2}\right) + 20$$
$$= 2\cos 6x + 6.2 \cos 4x + \cos 2x + 20$$

$$\therefore \quad \int \cos^6 x \, dx = \frac{1}{2^5} \int (\cos 6x + 6 \cos 4x + 15 \cos 2x + 10) \, dx$$
$$= \frac{1}{2^5}\left[\frac{\sin 6x}{4} + \frac{6 \sin 4x}{4} + \frac{15 \sin 2x}{2} + 10x\right]$$
$$= \frac{1}{192} \sin 6x + \frac{3}{64} \sin 4x + \frac{15}{64} \sin 2x + 5\frac{x}{16}.$$

EXERCISE

Integrate with respect to x :

1. $\sin^3 x$ $\qquad\qquad$ [Ans. $-\cos x + 1/3 \cos^3 x$]
2. $\cos^3 x$ $\qquad\qquad$ [Ans. $\sin x - 1/3 \sin^3 x$]
3. $\sin^5 x$ $\qquad\qquad$ $\left[\text{Ans. } -\cos x + \frac{2}{3}\cos^3 x - \frac{1}{5}\cos^5 x\right]$
4. $\cos^7 x$ $\qquad\qquad$ $\left[\text{Ans. } \sin x - \sin^3 x + \frac{3}{5}\sin^5 x - \frac{1}{7}\sin^7 x\right]$

2.2.1 INTEGRATION OF $\sin^m x \cos^n x$.

Case 1. When m or n is an odd positive integer.

Let $m = 2r + 1$, where r is a positive integer, then for all values of n, integration can be effected by the substitution
$$\cos x = t \qquad \text{and} \qquad -\sin x \, dx = dt$$

Hence $\qquad \int \sin^m x \cos^n x \, dx = \int \sin^{2r+1} x \cos^n x \, dx$
$$= \int \sin^{2r} x \cos^n x . \sin x \, dx$$

Integration of Transcendental Functions

$$= \int (1 - \cos^2 x)^r \cos^n x \cdot \sin x \, dx$$
$$= \int (1 - t^2)^r t^n \, (-dt)$$

The integral on the right hand side can be eavluated by expanding $(1-t^2)^r$ by Binomial Theorem.

Case 2. When $m + n$ is an even negative integer.
In this case the integration is effected by the substitution

$$\tan x = t$$

Hence, if $m + n = -2r$, when r is any positive integer, we have

$$\int \sin^m x \cos^n x \, dx = \int \frac{\sin^m x}{\cos^m x} \cdot \cos^{m+n} x \, dx$$

$$= \int \tan^m x \cos^{-2r} x \, dx$$

$$= \int \tan^m x \sec^{2r} x \, dx$$

$$= \int \tan^m x \sec^{2(r-1)} x \sec^2 x \, dx$$

$$= \int \tan^m x \, (1 + \tan^2 x)^{r-1} \sec^2 x \, dx$$

$$= \int t^m \, (1 + t^2)^{r-1} \, dt, \text{ where } \tan x = t.$$

The integral on the right hand side can be evaluated by expanding $(1+t^2)^{r-1}$ by Binomial Theorem.

Note : In every case in which m and n are positive integers, $\int \sin^m x \cos^n x \, dx$ may be evaluated by the method of successive reduction (see next chapter), or by expressing $\sin^m x \cos^n x$ as the sum of sines or cosines of multiples of x.

EXAMPLES

1. Evaluate $\int \sin^2 x \cos^3 x \, dx$.

Sol. Here the power of $\cos x$ is 3 which is an odd positive integer.
Therefore, putting $\sin x = t$ and $\cos x \, dx = dt$, we have

$$\int \sin^2 x \cos^3 x \, dx = \int \sin^2 x \cos^2 x \cdot \cos x \, dx$$

$$= \int \sin^2 x \, (1 - \sin^2 x) \cos x \, dx$$

$$= \int t^2 \, (1 - t^2) \, dt = \int (t^2 - t^4) \, dt$$

$$= \frac{1}{3} t^3 - \frac{1}{5} t^5, = \frac{1}{3} \sin^3 x - \frac{1}{5} \sin^5 x \text{ where } t = \sin x.$$

2. Evaluate : $\int \sec^{2/3} x \, \text{cosec}^{4/3} x \, dx$.

Sol. We have $\sec^{2/3} x \, \text{cosec}^{4/3} x = \sin^{-4/3} x \cos^{-2/3} x$, so that $m + n = -\frac{4}{3} - \frac{2}{3} = -2$
which is an even negative integer. Thus, we write

$$\int \sec^{2/3} x \, \text{cosec}^{4/3} x \, dx = \int \frac{\sec^{2/3} x}{\sin^{4/3} x} \, dx = \int \frac{\sec^2 x}{\tan^{4/3} x} \, dx.$$

Putting $\tan x = t$, and $\sec^2 x \, dx = dt$, we obtain

$$\int \sec^{2/3} x \cosec^{4/3} x \, dx = \int \frac{dt}{t^{4/3}} = \int t^{-4/3} \, dt$$

$$= \frac{t^{-1/3}}{-1/3} = -3 \tan^{-1/3} x.$$

3. *Integrate* $\sqrt{(\tan x)} \sec x \cosec x$.

Sol. $\int \sqrt{(\tan x)} \sec x \cosec x \, dx$

$$= \int \sqrt{\left(\frac{\sin x}{\cos x}\right)} \cdot \frac{1}{\cos x \sin x} \, dx$$

$$= \int \frac{dx}{\cos^{3/2} x \sin^{1/2} x} = \int \frac{\cos^{1/2} x \, dx}{\cos^2 x \sin^{1/2} x}$$

multiplying num. and denom. by $\cos^{1/2} x$.

$$= \int \frac{\sec^2 x}{\tan^{1/2} x} \, dx = \int \frac{dt}{t^{1/2}}, \text{ where } t = \tan x$$

$$= 2 t^{1/2} = 2 \sqrt{(\tan x)}$$

EXERCISE

Integrate with respect to x :

1. $\sin^3 x \cos^2 x$. $\qquad \left[\text{Ans. } -\frac{1}{3} \cos^3 x + \frac{1}{5} \cos^5 x \right]$

2. $\sin^5 x \cos^2 x$ $\qquad \left[\text{Ans. } -\frac{1}{3} \cos^3 x + \frac{2}{5} \cos^5 x - \frac{1}{7} \cos^7 x \right]$

3. $\sin^5 x \cos^{3/4} x$ $\qquad \left[\text{Ans. } -4 \cos^{7/4} x \left(\frac{1}{7} - \frac{2}{15} \cos^2 x + \frac{1}{23} \cos^4 x \right) \right]$

4. $\cosec^{2/3} x \cos^3 x$ $\qquad \left[\text{Ans. } 3 \sin^{1/3} x \left(1 - \frac{1}{7} \sin^2 x \right) \right]$

5. $\sec x \tan^3 x$ $\qquad \left[\text{Ans. } \frac{1}{3} \sec^3 x - \sec x \right]$

6. $\frac{1}{\sin x} \cos^3 x$. $\qquad \left[\text{Ans. } \log (\tan x) + \frac{1}{2} \tan^2 x \right]$

2.2.2 USE OF MULTIPLE ANGLES

Any positive integral power of a *sine* or cosine or their product can be expressed with the help of Trigonometry in a series of sine or cosines of multiples of the angle, and then each term can be integrated at once by applying the formulae

$$\int \cos nx \, dx = \frac{\sin nx}{n} \quad \text{and} \quad \int \sin nx \, dx = -\frac{\cos nx}{n}.$$

EXAMPLES

1. *If m and n are integers, prove that*

$$\int_0^\pi \cos mx \sin nx \, dx = \frac{2n}{n^2 - m^2} \text{ or } 0$$

according as $n - m$ *is odd or even.*

Integration of Transcendental Functions

Sol. We know that
$$2 \sin nx \cos mx = \sin(m+n)x + \sin(n-m)x$$

Therefore,
$$I = \int_0^\pi \sin x \cos mx \, dx$$

$$= \frac{1}{2} \int_0^\pi \{\sin(m+n)x + \sin(n-m)x\} \, dx$$

$$= -\frac{1}{2} \left[\frac{\cos(m+n)x}{m+n} + \frac{\cos(n-m)x}{n-m} \right]_0^\pi$$

$$= -\frac{1}{2} \left[\frac{(-1)^{m+n}}{m+n} + \frac{(-1)^{n-m}}{n-m} \right] = \frac{1}{2} \left[\frac{1}{m+n} + \frac{1}{n-m} \right] = \frac{2n}{n^2 - m^2}.$$

Again, if $n - m$ is even, then $n + m$ is also even. Therefore,
$$I = -\frac{1}{2} \left[\frac{1}{n+m} + \frac{1}{n-m} \right] + \frac{1}{2} \left[\frac{1}{n+m} + \frac{1}{n-m} \right] = 0$$

2. Evaluate $\int_0^a \dfrac{x^4 \, dx}{(a^2 + x^2)^4}$.

Sol. $\int_0^a \dfrac{x^4 \, dx}{(a^2 + x^2)^4} = \int_0^{\pi/4} \dfrac{a^4 \tan^4 \theta \cdot a \sec^2 \theta \, d\theta}{(a^2 \sec^2 \theta)^4}$ putting $x = a \tan \theta$

$$= \frac{1}{a^2} \int_0^{\pi/4} \frac{\tan^4 \theta \, d\theta}{\sec^6 \theta} = \frac{1}{a^3} \int_0^{\pi/4} \sin^4 \theta \cos^2 \theta \, d\theta$$

$$= \frac{1}{a^3} \left[\frac{1}{32} \left\{ \frac{1}{6} \sin 6\theta - \frac{1}{2} \sin 4\theta - \frac{1}{2} \sin 2\theta + 2\theta \right\} \right]_0^{\pi/4}$$

$$= \frac{1}{32a^3} \left[\frac{1}{6} \sin\left(\frac{3}{2}\pi\right) - \frac{1}{2} \sin(\pi) - \frac{1}{2} \sin\left(\frac{\pi}{2}\right) + \frac{\pi}{2} \right]$$

$$= \frac{1}{32a^3} \left[\frac{1}{6}(-1) - \frac{1}{2}(0) - \frac{1}{2}(1) + \frac{\pi}{2} \right]$$

$$= \frac{1}{32a^3} \left(\frac{\pi}{2} - \frac{2}{3} \right) = \frac{1}{16a^3} \left(\frac{\pi}{4} - \frac{1}{3} \right)$$

EXERCISE

Integrate by expressing the following functions into a series of sines or cosines of multiples of the angle:

1. Evaluate $\int_0^{\pi/4} \sin^4 \theta \, d\theta$. $\left[\text{Ans. } \dfrac{(3\pi - 8)}{32} \right]$

2. Evaluate $\int_0^\pi \cos^6 x \, dx$. $\left[\text{Ans. } \dfrac{5\pi}{16} \right]$

3. $\cos x \cos 2x \cos 3x$ $\left[\text{Ans. } \dfrac{1}{48} [2 \sin 6x + 3 \sin 4x + 6 \sin 2x + 12x] \right]$

4. $\dfrac{1}{\sin^{1/2} x} \cos^{7/2} x$ $\left[\text{Ans. } (\tan\theta)^{1/2} + \dfrac{2}{5}(\tan\theta)^{5/2}\right]$

5. $\dfrac{1}{(\sin^2 x \cos^2 x)}$ [Ans. $\tan x - \cot x$]

INTEGRATION OF RATIONAL FUNCTIONS OF SINES AND COSINES

2.3 INTEGRATION OF

(a) $\dfrac{1}{a + b \cos^2 x}$

(b) $\dfrac{1}{a + b \sin^2 x}$

(c) $\dfrac{1}{a + b \sin^2 x}$

In each of these cases we divide the numerator and denominator by $\cos^2 x$ (or multiply by $\sec^2 x$) and then put $\tan x = t$. For instance,

$$\int \dfrac{dx}{a + b \cos^2 x} = \int \dfrac{\sec^2 x\, dx}{a \sec^2 x + b}$$

$$= \int \dfrac{\sec^2 x\, dx}{a(1 + \tan^2 x) + b} = \int \dfrac{\sec^2 x\, dx}{a + b + a \tan^2 x}$$

Now put $\tan x = t$ and $\sec^2 x\, dx = dt$. Therefore,

$$\int \dfrac{dx}{a + b \cos^2 x} = \int \dfrac{dt}{(a+b) + at^2}$$

This integral is of the form $\dfrac{1}{a}\int \dfrac{dt}{t^2 \pm \alpha^2}$ and hence can be evaluated readily.

EXAMPLES

1. Evaluate : $\int \dfrac{1}{1 + 3\sin^2 x}\, dx.$

Sol. Dividing the Numerator and Denominator by $\cos^2 x$ gives

$$\int \dfrac{dx}{1 + 3\sin^2 x} = \int \dfrac{\sec^2 x\, dx}{\sec^2 x + 3\tan^2 x} = \int \dfrac{\sec^2 x\, dx}{1 + 4\tan^2 x}$$

$$= \int \dfrac{dt}{1 + 4t^2}, \text{ where } \tan x = t, \sec^2 x\, dx = dt.$$

$$= \dfrac{1}{2}\tan^{-1}(2t) = \dfrac{1}{2}\tan^{-1}(2\tan x).$$

Integration of Transcendental Functions

2. Integrate $\dfrac{1}{\sin x + \sin 2x}$.

Sol. $I = \displaystyle\int \dfrac{dx}{\sin x + 2 \sin x \cos x} = \int \dfrac{dx}{\sin x (1 + 2 \cos x)}$

$$= \int \dfrac{\sin x \, dx}{\sin^2 x (1 + \cos x)} = \int \dfrac{\sin x \, dx}{(1 - \cos^2 x)(1 + 2 \cos x)}$$

Putting $\cos x = t \Rightarrow -\sin x \, dx = dt$

$$= \int \dfrac{-dt}{(1 - t^2)(1 + 2t)}$$

$$= \int \dfrac{-dt}{(1 - t)(1 + t)(1 + 2t)}$$

Let $\dfrac{1}{(1-t)(1+t)(1+2t)} = \dfrac{A}{1-t} + \dfrac{B}{1+t} + \dfrac{C}{1+2t}$

$I \equiv A(1+t)(1+2t) + B(1-t)(1+2t) + C(1-t^2)$

$t = 1 \quad \Rightarrow \quad A = \dfrac{1}{6}$

$t = -1 \quad \Rightarrow \quad B = -\dfrac{1}{2}$

$t = -\dfrac{1}{2} \quad \Rightarrow \quad C = -\dfrac{4}{3}$

$I = \displaystyle\int \dfrac{dt}{6(1-t)} - \int \dfrac{dt}{2(1+t)} + \dfrac{4}{3} \int \dfrac{dt}{(1+2t)}$

$= \dfrac{1}{6} \log(1-t) - \dfrac{1}{2} \log(1+t) + \dfrac{2}{3} \log(1+2t)$

$\dfrac{1}{6} \log(1-\cos x) - \dfrac{1}{2} \log(1+\cos x) + \dfrac{2}{3} \log(1+2\cos x)$.

EXERCISE

Integrate with respect to x:

1. $\dfrac{1}{1 + \cos^2 x}$ $\qquad\left[\text{Ans. } \dfrac{1}{\sqrt{2}} \tan^{-1}\left(\dfrac{\tan x}{\sqrt{2}}\right)\right]$

2. $\dfrac{1}{2 + \sin 2x}$ $\qquad\left[\text{Ans. } \dfrac{1}{\sqrt{3}} \tan^{-1}\left(\dfrac{2 \tan x + 1}{\sqrt{3}}\right)\right]$

3. $\dfrac{1}{a^2 - b^2 \cos^2 x}, a > b$ $\qquad\left[\text{Ans. } \dfrac{1}{a\sqrt{(a^2 - b^2)}} \tan^{-1} \dfrac{a \tan x}{\sqrt{(a^2 - b^2)}}\right]$

4. $\dfrac{1}{a^2 \cos^2 x + b^2 \sin^2 x}$ $\qquad\left[\text{Ans. } \dfrac{1}{ab} \tan^{-1}\left(\dfrac{b \tan x}{a}\right)\right]$

5. $\dfrac{1}{a \cos^2 x + b}$ $\qquad\left[\text{Ans. } \dfrac{1}{\sqrt{b}\sqrt{a+b}} \tan^{-1}\left(\dfrac{\sqrt{b} \tan x}{\sqrt{a+b}}\right)\right]$

2.4 INTEGRATION OF $\dfrac{1}{a + b \cos x}$

To integrate $\dfrac{1}{a + b \cos x}$, we write

$$a + b \cos x = a\left(\cos^2 \frac{x}{2} + \sin^2 \frac{x}{2}\right) + b\left(\cos^2 \frac{x}{2} - \sin^2 \frac{x}{2}\right)$$

$$= (a + b) \cos^2 \frac{x}{2} + (a - b) \sin^2 \frac{x}{2}$$

$$= (a - b) \cos^2 \frac{x}{2} \left(\frac{a + b}{a - b} + \tan^2 \frac{x}{2}\right)$$

$$\therefore \int \frac{dx}{a + b \cos x} = \int \frac{dx}{(a - b) \cos^2 \frac{x}{2} \left(\frac{a + b}{a - b} + \tan^2 \frac{x}{2}\right)}$$

$$= \frac{2}{a - b} \int \frac{\frac{1}{2} \sec^2 \frac{x}{2} \, dx}{\frac{a + b}{a - b} + \tan^2 \frac{x}{2}}$$

Now, put $\tan \dfrac{x}{2} = t$ and $\dfrac{1}{2} \sec^2 \dfrac{x}{2} \, dx = dt$. Then

$$\int \frac{dx}{a + b \cos x} = \frac{2}{a - b} \int \frac{dt}{t^2 + (a + b)/(a - b)}. \qquad \ldots (i)$$

Case 1. If $a > b$, then

$$\int \frac{dx}{a + b \cos x} = \frac{2}{a - b} \cdot \sqrt{\frac{a - b}{a + b}} \tan^{-1}\left\{\sqrt{\frac{a - b}{a + x}} \, t\right\}$$

$$= \frac{2}{\sqrt{(a^2 - b^2)}} \tan^{-1}\left\{\sqrt{\frac{a - b}{a + b}} \tan \frac{x}{2}\right\}$$

Case 2. If $a < b$, we have from (i)

$$\int \frac{dx}{a + b \cos x} = \frac{2}{b - a} \int \frac{dt}{(b + a)/(b - a) - t^2}$$

$$= \frac{2}{b - a} \cdot \frac{1}{2\sqrt{\dfrac{b + a}{b - a}}} \log \frac{\sqrt{\dfrac{b + a}{b - a}} + t}{\sqrt{\dfrac{b + a}{b - a}} - t}$$

$$= \frac{1}{\sqrt{(b^2 - a^2)}} \log \frac{\sqrt{(b + a)} + \sqrt{b - a} \tan \dfrac{1}{2} x}{\sqrt{b + a} - \sqrt{b - a} \tan \dfrac{1}{2} x}$$

Note : Here both a and b have been assumed to be *positive*. If one of a and b negative, the results must be modified accordingly.

Integration of Transcendental Functions

2.4.1 Integration of $\dfrac{1}{a+b\sin x}$

To integrate this function, we write

$$a + b \sin x = a\left(\cos^2\frac{x}{2} + \sin^2\frac{x}{2}\right) + 2b \sin\frac{x}{2}\cos\frac{x}{2}$$

$$= a\cos^2\frac{x}{2}\left(1 + \frac{2b}{a}\tan\frac{x}{2} + \tan^2\frac{x}{2}\right)$$

$$\therefore \int\frac{dx}{a+b\sin x} = \int\frac{dx}{a\cos^2\frac{1}{2}x\left[\tan^2\frac{1}{2}x + \left(\frac{2b}{a}\right)\tan\frac{1}{2}x + 1\right]}$$

$$= \frac{2}{a}\int\frac{\frac{1}{2}\sec^2\frac{1}{2}x\, dx}{\tan^2\frac{1}{2}x + \left(\frac{2b}{a}\right)\tan\frac{1}{2}x + 1}$$

Putting $\tan\frac{1}{2}x = t$ and $\frac{1}{2}\sec^2\frac{1}{2}x\, dx = dt$ gives

$$\int\frac{dx}{a+b\sin x} = \frac{2}{a}\int\frac{dt}{t^2 + \left(\frac{2b}{a}\right)t + 1}$$

$$= \frac{2}{a}\int\frac{dt}{\left(t+\frac{b}{a}\right)^2 + \frac{(a^2-b^2)}{a^2}}$$

The integral on the right takes the standard forms $\int\dfrac{du}{u^2\pm\alpha^2}$ according as $a>b$ or $a<b$ and hence can be evaluated readily.

2.4.2 Integration of $\dfrac{1}{a+b\sin x+c\cos x}$

To integrate this function we may write

$$a = a\left(\cos^2\frac{x}{2} + \sin^2\frac{x}{2}\right)$$

$$\sin x = 2\sin\frac{x}{2}\cos\frac{x}{2}$$

and

$$\cos x = \cos^2\frac{x}{2} - \sin^2\frac{x}{2}$$

Then divide the numerator and denominator by $\cos^2\frac{1}{2}x$ and put $\tan\frac{1}{2}x = t$ to obtain

$$\int\frac{dx}{a+b\sin x+c\cos x} = \int\frac{2dt}{(a-c)t^2 + 2bt + (a+c)}$$

The integral on the right hand side may now be evaluated at once.

EXAMPLES

1. *Prove that* $\int_0^a \dfrac{d\theta}{\cos\alpha + \cos\theta} = \operatorname{cosec}\alpha \log(\sec\alpha)$.

Sol. $\displaystyle\int \dfrac{d\theta}{\cos\alpha + \cos\theta}$

$$= \int \dfrac{d\theta}{\cos\alpha\left(\cos^2\dfrac{\theta}{2} + \sin^2\dfrac{\theta}{2}\right) + \left(\cos^2\dfrac{\theta}{2} - \sin^2\dfrac{\theta}{2}\right)}$$

$$= \int \dfrac{d\theta}{(1+\cos\alpha)\cos^2\dfrac{\theta}{2} - (1-\cos\alpha)\sin^2\dfrac{\theta}{2}}$$

$$= \int \dfrac{\sec^2\dfrac{\theta}{2}\, d\theta}{(1+\cos\alpha) - (1-\cos\alpha)\tan^2\dfrac{\theta}{2}}$$

$$= \dfrac{2}{(1-\cos\alpha)} \int \dfrac{dt}{\dfrac{(1+\cos\alpha)}{(1-\cos\alpha)} - t^2} \qquad \text{when } t = \tan\dfrac{\theta}{2}$$

$$= \dfrac{2}{2\sin^2\dfrac{\alpha}{2}} \int \dfrac{dt}{\left(\cot^2\dfrac{\alpha}{2}\right) - t^2}$$

$$= \dfrac{1}{\sin^2\dfrac{\alpha}{2}} \cdot \dfrac{1}{2\cot\dfrac{\alpha}{2}} \log\left(\dfrac{\cot\dfrac{\alpha}{2} + t}{\cot\dfrac{\alpha}{2} - t}\right)$$

$$= \dfrac{1}{\sin\alpha} \log\left(\dfrac{\cot\dfrac{\alpha}{2} + \tan\dfrac{\theta}{2}}{\cot\dfrac{\alpha}{2} - \tan\dfrac{\theta}{2}}\right)$$

$$= \dfrac{1}{\sin\alpha} \log\left(\dfrac{\cos\dfrac{\alpha}{2}\cos\dfrac{\theta}{2} + \sin\dfrac{\theta}{2}\sin\dfrac{\alpha}{2}}{\cos\dfrac{\alpha}{2}\cos\dfrac{\theta}{2} - \sin\dfrac{\theta}{2}\sin\dfrac{\alpha}{2}}\right)$$

$$= \operatorname{cosec}\alpha \log\left\{\dfrac{\cos\dfrac{1}{2}(\alpha - \theta)}{\cos\dfrac{1}{2}(\theta + \alpha)}\right\}$$

$\therefore \displaystyle\int_0^a \dfrac{d\theta}{\cos\alpha + \cos\theta} = \operatorname{cosec}\alpha \left[\log \dfrac{\cos\dfrac{1}{2}(\alpha - \theta)}{\cos\dfrac{1}{2}(\theta + \alpha)}\right]_0^\alpha$

Integration of Transcendental Functions

$$= \operatorname{cosec} \alpha \left[\log\left(\frac{1}{\cos \alpha}\right) - \log (1) \right]$$

$$= \operatorname{cosec} \alpha \, \log \sec \alpha.$$

2. Evaluate $\int_0^\pi \dfrac{dx}{1 - 2a \cos x + a^2}$.

$$= \frac{\pi}{1-a^2} \text{ or } \frac{\pi}{a^2-1}, \text{ according as } a < \text{ or } > 1.$$

Sol. $\int_0^\pi \dfrac{dx}{1 + a^2 - 2a \cos x}$

$$= \int_0^\pi \frac{dx}{(1+a^2)\left(\cos^2 \frac{x}{2} + \sin^2 \frac{x}{2}\right) - 2a\left(\cos^2 \frac{x}{2} - \sin^2 \frac{x}{2}\right)}$$

$$= \int_0^\pi \frac{dx}{(1 + a^2 - 2a)\cos^2 \frac{x}{2} + (1 + a^2 + 2a)\sin^2 \frac{x}{2}}$$

$$= \int_0^\pi \frac{dx}{(1-a)^2 \cos^2 \frac{x}{2} + (1-a)^2 \sin^2 \frac{x}{2}}$$

$$= \int_0^\pi \frac{\sec^2 \frac{x}{2} \, dx}{(1-a)^2 + (1+a)^2 \tan^2 \frac{x}{2}}$$

$$= \frac{2}{(1+a)^2} \int_0^\infty \frac{dt}{\left(\frac{1-a}{1+a}\right)^2 + t^2}, \text{ where } t = \tan \frac{x}{2}$$

$$= \frac{2}{(1+a)^2} \int_0^\infty \frac{dt}{t^2 + \left(\frac{1-a}{1+a}\right)^2}$$

If $a < 1$, $\int_0^\pi \dfrac{dx}{(1+a^2 - 2a \cos x)} = \dfrac{2}{(1+a^2)} \cdot \dfrac{1+a}{(1-a)} \left[\tan^{-1}\left\{\dfrac{1+a}{1-a}\right\} \right]_0^\infty$

$$= \frac{2}{1-a^2}[\tan^{-1} \infty - \tan^{-1} 0] = \frac{\pi}{1-a^2}$$

If $a > 1$, $\int_0^\pi \dfrac{dx}{1 + a^2 - 2a \cos x} = \dfrac{2}{(1+a)^2} \cdot \dfrac{1+a}{(1-a)} \left[\tan^{-1}\left\{\left(\dfrac{1+a}{1-a}\right)\right\} t \right]_0^\infty$

$$= -\frac{2}{(a+1)(a-1)} \left[\tan^{-1}\left\{-\left(\frac{a+1}{a-1}\right) t\right\}\right]_0^\infty \qquad \because a > 1$$

$$= -\frac{2}{(a^2-1)}[\tan^{-1}(-\infty) - \tan^{-1} 0]$$

$$= -\frac{2}{(a^2-1)}\left(-\frac{\pi}{2} - 0\right) = \frac{\pi}{(a^2-1)}.$$

EXERCISES

Prove that:

1. $\int_0^{\pi/2} \dfrac{d\theta}{1+2\cos\theta} = \dfrac{\log(2+\sqrt{2})}{\sqrt{3}}$

2. $\int_0^{\pi} \dfrac{dx}{a+b\cos x} = \dfrac{\pi}{\sqrt{(a^2-b^2)}}$, if $a>b$.

3. $\int_0^{\pi} \dfrac{dx}{3+2\sin x + \cos x} = \dfrac{\pi}{4}$.

4. $\int_5^{\pi/2} \dfrac{dx}{4+5\cos x} = \dfrac{1}{3}\log\dfrac{2\left(3-\tan\dfrac{1}{2}\right)}{\left(3+\tan\dfrac{1}{2}\right)}$

2.5 INTEGRATION OF $\dfrac{1}{a\sin x + b\cos x}$

This function may be integrated by reducing it to a rational algebraic function of t by the substitution $\tan\dfrac{1}{2}x = t$ as in the last three sections. However, we may integrate the function more simply by combining the two terms in the denominator into one by changing the constants. Thus by writing $a = r\cos\alpha$ and $b = r\sin\alpha$, we find that

$$a\sin x + b\cos x = r\sin(x+\alpha) = \sqrt{a^2+b^2}\,\sin\left(x + \tan^{-1}\dfrac{b}{a}\right)$$

and therefore, the integral of the given function can be written down at once.

2.5.1 Integration of $\dfrac{p\sin x + q\cos x}{a\sin x + b\cos x}$

Such rational functions of $\sin x$ and $\cos x$ may be integrated by expressing the numerator of the integrated as follows:

$$p\sin x + q\cos x = A\,(\text{Den.}^r) + B\,(\text{Diff.coeff. of Den.}^r)$$
$$= A\,(a\sin x + b\cos x) + B\,(a\cos x - b\sin x)$$

The arbitrary constants A and B are found by comparing the coefficients of $\sin x$ and $\cos x$ from the two sides of the above identity. Then

$$\text{Given integral} = \int \dfrac{A\,(a\sin x + b\cos x) + B\,(a\cos x - b\sin x)}{a\sin x + b\cos x}\,dx$$

$$= A\int dx + B\int \dfrac{d\,(a\sin x + b\cos x)}{a\sin x + b\cos x}$$

$$= Ax + B\log(a\sin x + b\cos x),\text{ etc.}$$

Integration of Transcendental Functions

2.5.2 Integration of $\dfrac{l + m \sin x + n \cos x}{a + b \sin x + c \cos x}$

For such functions we express the numerator as follows :

$l + m \sin x + n \cos x = A\,(\text{Den.}^r) + B\,(\text{diff .coeff .of Den.}^r) + C$

$\qquad\qquad\qquad\qquad = A\,(a + b \sin x + c \cos x) + B\,(b \cos x - c \sin x) + C$

The arbitratary constants A, B, C are found by comparing the coefficients of $\sin x, \cos x$ and constant terms from the two sides of the above identity. Then
Given integral

$$= A \int dx + B \int \dfrac{d\,(a + b \sin x + c \cos x)}{a + b \sin x + c \cos x} + C \int \dfrac{dx}{a + b \sin x + c \cos x}$$

$$= Ax + b \log (a + b \sin x + c \cos x) + C \int \dfrac{dx}{a + b \sin x + c \cos x}$$

The integral on the right hand side is evaluated by the method of Sec. 2.4.2.

2.6 INTEGRATION OF ANY RATIONAL FUNCTIONS OF $\sin x$ AND $\cos x$

Let $\tan \dfrac{1}{2} x = t$, then

$$\sin x = \dfrac{2 \sin \dfrac{1}{2} x \cos \dfrac{1}{2} x}{\cos^2 \dfrac{1}{2} x + \sin^2 \dfrac{1}{2} x} = \dfrac{2 \tan \dfrac{1}{2} x}{1 + \tan^2 \dfrac{1}{2} x} = \dfrac{2t}{1 + t^2}$$

$$\cos x = \dfrac{\cos^2 \dfrac{1}{2} x - \sin^2 \dfrac{1}{2} x}{\cos^2 \dfrac{1}{2} x + \sin^2 \dfrac{1}{2} x} = \dfrac{1 - \tan^2 \dfrac{1}{2} x}{1 + \tan^2 \dfrac{1}{2} x} = \dfrac{1 - t^2}{1 + t^2}.$$

Also, $\qquad \dfrac{dt}{dx} = \dfrac{1}{2} \sec^2 \dfrac{1}{2} x = \dfrac{1}{2}\left(1 + \tan^2 \dfrac{1}{2} x\right); \qquad \therefore\ dx = \dfrac{2dt}{1 + t^2}.$

Thus, the substitution $\tan \dfrac{1}{2} x = t$ will transform a rational function of $\sin x$ and $\cos x$ into a rational algebraic function of t.

This method, however, is not very convenient in practice, because the resulting function of t is a fraction whose denominator is generally of a high degree in t. Sometimes other substitutions are found more convenient.

Very often it is possible to device some alternative method. For instance, dividing the numerator and denominator of a function by a suitable power of $\sin x$ or $\cos x$ converts the given function into the product of $\sec^2 x$ and a rational function of $\tan x$, or $\operatorname{cosec}^2 x$ and a rational function of $\cot x$. The resulting function is then integrated by putting $\tan x$ or $\cot x$ equal to new variable t.

In some cases, the integrated is broken up into two or more parts each of which can be integrated readily.

EXAMPLES

1. Evaluate $\int \dfrac{d\theta}{(a \sin^2 \theta + b \cos^2 \theta)^2}.$

Sol. Denote the given integral by I. Then dividing the numerator and denominator by $\cos^4 \theta$ gives

$$I = \int \frac{d\theta}{(a \sin^2 \theta + b \cos^2 \theta)^2} = \int \frac{\sec^4 \theta \, d\theta}{(b + a \tan^2 \theta)^2}$$

$$= \int \frac{(1+t^2) \, dt}{(b + at^2)^2} \qquad \text{whose } t = \tan \theta$$

$$= \frac{1}{a} \int \frac{(b + at^2) + a - b}{(b + at^2)}$$

$$= \frac{1}{a^2} \int \frac{(b + at^2) + a - b}{(b + at^2)} \, dt.$$

$$= \frac{1}{a^2} \int \frac{dt}{c + t^2} + \frac{a-b}{a^3} \int \frac{dt}{(c + t^2)^2}, \qquad \text{where } c = \frac{b}{a}$$

$$= \frac{1}{a^2} \int \frac{dt}{c + t^2} + \frac{a-b}{a^3} \left\{ \frac{1}{2c \, (c + t^2)} + \frac{1}{2c} \int \frac{dt}{c + t^2} \right\}$$

$$= \text{etc.}$$

2. Integrate $\dfrac{1}{(a + b \tan \theta)}$.

Sol. $\displaystyle \int \frac{1}{(a + b \tan \theta)} = \int \frac{\cos \theta \, d\theta}{a \cos \theta + b \sin \theta}$

Let $\cos \theta = A \, (a \cos \theta + b \sin \theta) + B \dfrac{d}{d\theta} (a \cos \theta + b \sin \theta)$

or $\cos \theta = A \, (a \cos \theta + b \sin \theta) + B \, (-a \sin \theta + b \cos \theta)$ ...(i)

Equating the coeff. of $\sin \theta$ and $\cos \theta$ on both sides, we get

$$0 = Ab - Ba \quad \text{and} \quad 1 = Aa + Bb$$

$$\therefore \quad A = \frac{a}{(a^2 + b^2)} \quad \text{and} \quad B = \frac{b}{(a^2 + b^2)}$$

$$\therefore \quad \cos \theta = \frac{a}{(a^2 + b^2)} (a \cos \theta + b \sin \theta) + \frac{b}{(a^2 + b^2)} (-a \sin \theta + b \cos \theta)$$

$$\therefore \quad \int \frac{d\theta}{a + b \tan \theta}$$

$$= \int \left\{ \frac{\dfrac{a}{(a^2 + b^2)} (a \cos \theta + b \sin \theta) + \dfrac{b}{(a^2 + b^2)} (-a \sin \theta + b \cos \theta)}{(a \cos \theta + b \sin \theta)} \right\} d\theta$$

$$= \frac{a}{(a^2 + b^2)} \int d\theta + \frac{b}{(a^2 + b^2)} \int \frac{(-a \sin \theta + b \cos \theta) \, d\theta}{a \cos \theta + b \sin \theta}$$

$$= \frac{a}{(a^2 + b^2)} \theta + \frac{b}{(a^2 + b^2)} \log (a \cos \theta + b \sin \theta)$$

Integration of Transcendental Functions 421

EXERCISE

Evaluate the following integrals :

1. $\int \dfrac{\cos x}{2 \sin x + 3 \cos x} dx.$ $\left[\text{Ans. } \dfrac{3}{13} x + \dfrac{2}{13} \log (2 \sin x + 3 \cos x)\right]$

2. $\int \dfrac{\cos x}{2 + \cos x} dx.$ $\left[\text{Ans. } x - \dfrac{4}{\sqrt{3}} \tan^{-1}\left(\dfrac{\tan \dfrac{x}{2}}{\sqrt{3}}\right)\right]$

[**Hint :** By division, $\cos x/(2 + \cos x) = 1 - 2/(2 + \cos x)$.]

3. $\int \dfrac{\cos x}{a + b \cos x} dx.$ $\left[\text{Ans. } \dfrac{x}{b} - \dfrac{a}{b} \int \dfrac{dx}{(a + b \cos x)}\right]$

4. $\int \dfrac{1}{a \sin x + b \cos x} dx.$ $\left[\text{Ans. } (a^2 + b^2)^{-1/2} \log \tan \dfrac{1}{2}\left\{x + \tan^{-1}\left(\dfrac{b}{a}\right)\right\}\right]$

5. $\int \dfrac{3 + 4 \sin x + 2 \cos x}{3 + 2 \sin x + \cos x} dx.$ $\left[\text{Ans. } 2x - 3 \tan^{-1}\left(1 + \tan^{-1} \dfrac{1}{2} x\right)\right]$

2.7 INTEGRATION OF OTHER TRANSCENDENTRAL FUNCTIONS

There are no general propositions which will enable us to integrate every function of $\sin x$, $\cos x$, etc. or of e^x, $\log x$, etc. or of hyperbolic functions. In such cases the method of integration by parts or the method of substitution should be tried. In the case of a rational function of e^x, the substitution $e^x = t$ should be tried. Sometimes the transformation of the integrand also facilitate the integration.

EXAMPLE

Ex : *Evaluate* $\int \dfrac{e^x (1 + x)}{(2 + x)^2} dx.$

Sol. Breaking the integrand into two parts, we have

$$\int \dfrac{e^x (1 + x)}{(2 + x)^2} dx = \int \dfrac{e^x}{2 + x} dx - \int \dfrac{e^x}{(2 + x)^2} dx.$$

Integrating the first integral on the right by parts, this gives

$$\int \dfrac{e^x (1 + x)}{(2 + x)^2} dx = \left\{e^x \cdot \dfrac{1}{2 + x} - \int e^x \dfrac{-1}{(2 + x)^2} dx\right\} - \int \dfrac{e^x dx}{(2 + x)^2}$$

$$= \dfrac{e^x}{2 + x}.$$

EXERCISES

Integrated with respet to x :

1. $\dfrac{1}{e^x + 1}$ $\left[\text{Ans. } x - \log (e^x + 1)\right]$

2. $\dfrac{1}{e^x - 1}$ $\left[\text{Ans. } \log (e^x - 1) - x\right]$

3. $\dfrac{1}{(1+e^x)(1+e^{-x})}$ $\qquad\left[\text{Ans. } -\dfrac{1}{(1+e^x)}\right]$

4. $\dfrac{1}{(e^x-1)(e^x+3)}$ $\qquad\left[\text{Ans. } -\dfrac{1}{3}x+\dfrac{1}{4}\log(e^x-1)+\dfrac{1}{12}\log(e^x+3)\right]$

5. $\dfrac{1}{(e^x-1)^2}$ $\qquad\left[\text{Ans. } x-\log(e^x-1)-\dfrac{1}{(e^x-1)}\right]$

6. $\dfrac{\{\log(x+1)\}}{x^2}$ $\qquad\left[\text{Ans. } \log x-\left(1+\dfrac{1}{x}\right)\log(1+x)\right]$

7. $\cos x \cosh x$ $\qquad\left[\text{Ans. } \dfrac{1}{2}\sin x \cosh x + \dfrac{1}{2}\cos x \sinh x\right]$

2.8 INTEGRATION OF EXPANSION

Sometimes we encounter functions which are not integrable by the methods discussed so far. In some cases the given function of x may be expanded in powers of x and the result integrated *term by term*. The student, however, must remember that this process of integration is not always justified. The consideration of the conditions under which this can be done is beyond the scope of the present book. We shall consider here only those functions for which the necessary conditions hold.

The student would do well to remember the sums of some standard series given below :

(1) $1 - \dfrac{1}{2} + \dfrac{1}{3} - \dfrac{1}{4} + \dfrac{1}{5} - \ldots$ an inf. $= \log_e 2$.

(2) $\dfrac{1}{1^2} + \dfrac{1}{2^2} + \dfrac{1}{3^2} + \dfrac{1}{4^2} + \ldots$ ad. inf. $= \dfrac{\pi^2}{6}$

(3) $\dfrac{1}{1^2} - \dfrac{1}{2^2} + \dfrac{1}{3^2} - \dfrac{1}{4^2} + \ldots$ ad. inf. $= \dfrac{\pi^2}{12}$

(4) $\dfrac{1}{1^2} + \dfrac{1}{3^2} + \dfrac{1}{5^2} + \dfrac{1}{7^2} + \ldots$ ad. inf. $= \dfrac{\pi^2}{8}$.

(5) $\dfrac{1}{2^2} + \dfrac{1}{4^2} + \dfrac{1}{6^2} + \dfrac{1}{8^2} + \ldots$ ad. inf. $= \dfrac{\pi^2}{24}$.

EXAMPLES

1. Evaluate : $\int_0^1 \dfrac{\log(1-x)}{x} dx$.

Sol. Expanding the logarithm, we have

$$\int_0^1 \dfrac{\log(1-x)}{x} dx = \int_0^1 \dfrac{1}{x}\left(-x - \dfrac{x^2}{2} - \dfrac{x^3}{3} + \ldots\right) dx$$

$$= -\int_0^1 \left(1 + \dfrac{x}{2} + \dfrac{x^2}{3} + \ldots \text{ad. inf.}\right) dx$$

$$= -\left[x + \dfrac{x^2}{2^2} + \dfrac{x^3}{3^2} + \dfrac{x^4}{4^2} + \ldots\right]_0^1$$

Integration of Transcendental Functions

$$= -\left(1 + \frac{1}{2^2} + \frac{1}{3^2} + \frac{1}{4^2} + \ldots\right) = -\frac{\pi^2}{6}. \qquad \ldots(i)$$

If we put $x = 1 - y$ and therefore, $dx = -dy$, we have

$$\int_0^1 \frac{\log(1-x)}{x} dx = \int_1^0 \frac{\log y}{1-y}(-dy) = \int_0^1 \frac{\log y}{1-y} dy.$$

Changing the dummy variable y to x and using (i), we have also

$$\int_0^1 \frac{\log x}{1-x} dx = -\frac{\pi^2}{6}.$$

EXERCISE

Evaluate the following integrals:

1. $\int_0^1 \frac{\log(1+x)}{x} dx.$ $\left[\text{Ans. } \dfrac{\pi^2}{12}\right]$

2. $\int_0^1 \frac{\log x}{1+x} dx$ $\left[\text{Ans. } -\dfrac{\pi^2}{12}\right]$

3. $\int_0^1 \frac{1}{x} \log \frac{1+x}{1-x} dx$ $\left[\text{Ans. } \dfrac{\pi^2}{4}\right]$

4. $\int_0^\infty \frac{x}{1+e^x} dx$ $\left[\text{Ans. } \dfrac{\pi^2}{12}\right]$

5. $\int_0^1 \log\left(\dfrac{1}{x} - 1\right) dx$ [Ans. 0]

6. $\int_0^1 x^2 \sin^{-1} x \, dx$ $\left[\text{Ans. } \dfrac{\pi}{6} - \dfrac{2}{9}\right]$

MISCELLANEOUS EXAMPLES

1. *If m and n are integers, show that*

$$\int_0^\pi \sin mx \sin nx \, dx = 0 \text{ if } m \neq n \text{ and } \frac{\pi}{2} \text{ if } m = n$$

Sol. Since $2 \sin A \sin B = \cos(A - B) - \cos(A + B)$, we have

$$\text{integral} = \frac{1}{2} \int_0^\pi [\cos(m-n)x - \cos(m+n)x] dx$$

$$= \frac{1}{2}\left[\frac{\sin(m-n)x}{m-n} - \frac{\sin(m+n)x}{m+n}\right]_0^\pi$$

$$= \frac{1}{2}\left[\frac{\sin(m-n)\pi}{m-n} - \frac{\sin(m+n)\pi}{m+n} - \frac{\sin 0}{m-n} + \frac{\sin 0}{m+n}\right] \qquad \ldots(i)$$

Case 1. Since m and n are integers, $m - n$ and $m + n$ are also integers. Moreover, sine of an integral multiple of π is zero. Hence, if $m \neq n$, each term in (i) is zero, and so

$$\int_0^\pi \sin mx \sin nx \, dx = 0$$

Case 2. If $m=n$, then $m-n=0$ and so the 1st and 3rd terms in (i) become indeterminate. Accordingly in this case, the value of the integral cannot be found from (i). Thus, putting $m=n$ in the given integral we have

$$\text{integral} = \int_0^\pi \sin^2 nx\, dx = \frac{1}{2}\int_0^\pi (1-\cos 2nx)\, dx$$

$$= \frac{1}{2}\left[x - \frac{\sin 2nx}{2n}\right]_0^\pi = \frac{\pi}{2}$$

2. Show that $\int_0^\infty \left(\dfrac{\log x}{1-x}\right)^2 dx = \dfrac{2\pi^2}{3}$.

Sol. Breaking the given integral into sum of two integral, we have

$$\int_0^\infty \left(\frac{\log x}{1-x}\right)^2 dx = \int_0^1 \left(\frac{\log x}{1-x}\right)^2 dx + \int_1^\infty \left(\frac{\log x}{1-x}\right)^2 dx. \qquad \ldots(i)$$

Putting $x = \dfrac{1}{t}$ and $dx = -\dfrac{dt}{t^2}$, we see that

$$\int_1^\infty \left(\frac{\log x}{1-x}\right)^2 dx = \int_1^0 \frac{\left(\log \frac{1}{t}\right)^2}{\left(1-\frac{1}{t}\right)^2}\left(-\frac{dt}{t^2}\right) = -\int_1^0 \frac{(\log t)^2}{(1-t)^2}\, dt$$

$$= \int_0^1 \left(\frac{\log x}{1-x}\right)^2 dx,$$

Substituting this in (i), we have

$$\int_0^\infty \left(\frac{\log x}{1-x}\right)^2 dx = 2\int_0^1 \left(\frac{\log x}{1-x}\right)^2 dx \qquad \ldots(ii)$$

Now, integrating by parts gives

$$\int \frac{(\log x)^2}{(1-x)^2}\, dx = \frac{(\log x)^2}{1-x} - 2\int \frac{\log x}{x(1-x)}\, dx$$

$$= \frac{(\log x)^2}{1-x} - 2\int \frac{\log x}{x}\, dx - 2\int \frac{\log x}{1-x}\, dx$$

$$= \frac{(\log x)^2}{1-x} - (\log x)^2 - 2\int \frac{\log x}{1-x}\, dx$$

Therefore, $\int_0^1 \left(\dfrac{\log x}{1-x}\right)^2 dx = \left[\dfrac{x(\log x)^2}{1-x}\right]_0^1 - 2\int_0^1 \dfrac{\log x}{1-x}\, dx \qquad \ldots(iii)$

By L' Hospital's rule one can easily see that

$$\lim_{x\to 1} \frac{(\log x)^2}{1-x} = 0, \text{ and } \lim_{x\to 0} x(\log x)^2 = 0$$

$\therefore$ (iii) gives $\int_0^1 \left(\dfrac{\log x}{1-x}\right)^2 dx = -\int_0^1 \dfrac{\log x}{1-x}\, dx$

Integration of Transcendental Functions 425

$$= -2\left(-\frac{\pi^2}{6}\right),$$

Hence, from (ii) $\int_0^\infty \left(\frac{\log x}{1-x}\right)^2 dx = 2 \cdot \frac{\pi^2}{3} = \frac{2\pi^2}{3}.$

EXERCISE

1. $\int \frac{dx}{(a \sin x + b \cos x)^2}$ $\left[\text{Ans.} -\frac{1}{a} \cdot \frac{1}{a \tan x + b}\right]$

   [Hint : Integrand $= \sec^2 x / (a \tan x + b)^2.$]

2. $\int \frac{\sin 2x}{(a + b \cos x)^2} dx.$ $\left[\text{Ans.} -\frac{2}{b^2}\left\{\log(a + b \cos x) + \frac{a}{a + b \cos x}\right\}\right]$

3. $\int \sin^{-1}\left(\frac{2x}{1+x^2}\right) dx$ [Ans. $2x \tan^{-1} x - \log(1 + x^2)$]

4. $\int \cos^{-1}\left(\frac{1-x}{1+x}\right)^2 dx$ [Ans. $2x \tan^{-1} x - \log(1 + x^2)$]

5. $\int \sin^{-1}\sqrt{\frac{x}{a+x}} dx$ $\left[\text{Ans. } (a+x)\tan^{-1}\sqrt{\left(\frac{x}{a}\right)} - \sqrt{ax}\right]$

   [Hint : Put $x = \tan^2 \theta.$]

6. $\int \frac{dx}{\sqrt{\sin^3 x \sin(x + \alpha)}}$ $\left[\text{Ans.} -\frac{2}{\sin \alpha}\sqrt{\frac{\sin(x+\alpha)}{\sin x}}\right]$

   [Hint : Integrand $= \frac{\cosec^2 x}{\sqrt{[\cos \alpha + \sin \alpha \cot \alpha]}}, \therefore$ Put $\cot x = t.$]

7. $\int \sqrt{\frac{\sin(x-\alpha)}{\sin(x+\alpha)}} dx.$

   [Hint : Integrand $= \sqrt{\frac{\sin(x-\alpha)\sin(x-\alpha)}{\sin(x+\alpha)\sin(x-\alpha)}}$

   $= \frac{\sin(x-\alpha)}{\sqrt{\sin^2 x - \sin^2 \alpha}} = \frac{\sin x \cos \alpha - \cos x \sin \alpha}{\sqrt{\sin^2 x - \cos^2 \alpha}}$

   Now break the integrand into sum of two terms. Put $\cos x = t$ in the first integral and $\sin x = u$ in the second.]

   [Ans. $\cos \alpha \cos^{-1}(\cos x \sec \alpha) - \sin \alpha \cosh^{-1}(\sin x \cosec \alpha)$]

8. $\int \dfrac{x^2\,dx}{(x\sin x + \cos x)^2}.$ $\left[\text{Ans. } \dfrac{x^5}{5}\left\{(\log x)^2 - \dfrac{2}{5}\log x + \dfrac{2}{25}\right\}\right]$

[**Hint :** Notice that $d(x\sin x + \cos x) = x\cos x\,dx.$ Therefore, write $x^2 = x\cos x \cdot x\sec x$ and integrate by parts, taking $x\sec x$ as the first function.]

9. $\int e^x \{\log(\sec x + \tan x) + \sec x\}\,dx.$ $\left[\text{Ans. } e^x \log(\sec x + \tan x)\right]$

10. $\int \dfrac{dx}{\cosh^3 x}$ (Put $\cosh x = \sec t.$) $\left[\text{Ans. } \dfrac{1}{2}(\sec^{-1}\cosh x + \operatorname{sech} x \tanh x)\right]$

□□□

3
Reduction Formulae

3.1 REDUCTION FORMULAE

In integral calculus sometimes functions occur whose integrals are not immediately reducible to one or other of the standard forms already known, and whose integrals are not directly obtainable. In such cases, howevery, integrals may be linearly connected by some algebraic formula with the integrals of another expression whcih itself may be either immediately integrable or comparatively easier to integrate. Such connecting algebraical relations are called **Reduction Formulae.**

A reduction formulae is generally derived by the method of *integration by parts*. The method of integration of a function by repeated application of a reduction formula is called **Integration by Successive Reduction.**

3.2 REDUCTION FORMULA FOR $\int \sin^n x \, dx$

Let
$$I_n = \int \sin^n x \, dx = \int \sin x \sin^{n-1} x \, dx$$
$$= (-\cos x) \sin^{n-1} x - \int (n-1) \sin^{n-2} x \cos x \, (-\cos x) \, dx$$
$$= -\cos x \sin^{n-1} x + (n-1) \int \sin^{n-2} x \cos^2 x \, dx$$
$$= -\cos x \sin^{n-1} x + (n-1) \int \sin^{n-2} x \, (1 - \sin^2 x) \, dx$$
$$= -\cos x \sin^{n-1} x + (n-1) \int \sin^{n-2} x \, dx - (n-1) \int \sin^n x \, dx$$
$$= -\cos x \sin^{n-1} x + (n-1) \int \sin^{n-1} x \, dx - (n-1) I_n$$

Therefore $\quad I_n (1 + n - 1) = -\cos x \sin^{n-1} x + (n-1) I_{n-2}$

or $\quad n I_n = -\cos x \sin^{n-1} x + (n-1) I_{n-2}$

or $\quad I_n = -\dfrac{\cos x \sin^{n-1} x}{n} + \dfrac{(n-1)}{n} I_{n-2}$

where $\quad I_{n-2} = \int \sin^{n-2} x \, dx$

3.3 REDUCTION FORMULA FOR $\int \cos^n x \, dx$

Let
$$I_n = \int \cos^n x \, dx$$
$$= \int \cos x \cdot \cos^{n-1} x \, dx$$
$$= \sin x \cos^{n-1} x - \int (n-1) \sin x \cos^{n-2} x \, (-\sin x) \, dx$$
$$= \sin x \cos^{n-1} x + (n-1) \int \cos^{n-1} x \sin^2 x \, dx$$
$$= \sin x \cos^{n-1} x + (n-1) \int \cos^{n-2} x \, dx - (n-1) \int \cos^n x \, dx$$

or $\quad I_n (1 + n - 1) = \sin x \cos^{n-1} x + \cos^{n-1} x + (n-1) \int \cos^{n-2} x \, dx$

or
$$I_n = \frac{\sin x \cos^{n-1} x}{n} + \frac{n-1}{n} I_{n-2}$$

3.3.1 EVALUATION OF $\int_0^{\pi/2} \sin^n x \, dx$ and $\int_0^{\pi/2} \cos^n x \, dx$

Firstly, we obtain the reduction formula of

$$\int \sin^n x \, dx = -\frac{1}{n} \sin^{n-1} x \cos x + \frac{n-1}{n} \int \sin^{n-2} x \, dx$$

This gives

$$\int_0^{\pi/2} \sin^n x \, dx = \left[-\frac{1}{n} \sin^{n-1} x \cos x \right]_0^{\pi/2} + \frac{n-1}{n} \int_0^{\pi/2} \sin^{n-2} x \, dx$$

$$= 0 + \frac{n-1}{n} \int_0^{\pi/2} \sin^{n-2} x \, dx$$

Writing I_n for $\int_0^{\pi/2} \sin^n x \, dx$, we get the connecting formula as

$$I_n = \frac{n-1}{n} I_{n-2}$$

With the help of this formula, we may successively connect I_{n-2} with I_{n-4}, I_{n-4} with I_{n-6} etc. and finally, I_3 with I_1 or I_2 with I_0 according as n is odd or even. Thus, we have

$$I_n = \frac{n-1}{n} I_{n-2}$$

$$I_{n-2} = \frac{n-3}{n-2} I_{n-4}$$

$$I_{n-4} = \frac{n-5}{n-4} I_{n-6}$$

....
....

$$I_3 = \frac{2}{3} I_1, \text{ if } n \text{ is odd}$$

or
$$I_2 = \frac{1}{2} I_0, \text{ if } n \text{ is even}$$

From these we get

$$I_n = \begin{cases} \dfrac{n-1}{n} \cdot \dfrac{n-3}{n-2} \cdot \dfrac{n-5}{n-4} \cdots \dfrac{2}{3} I_1, & \text{when } n \text{ is odd} \\ \dfrac{n-1}{n} \cdot \dfrac{n-3}{n-2} \cdot \dfrac{n-5}{n-4} \cdots \dfrac{1}{2} I_0, & \text{when } n \text{ is even.} \end{cases}$$

But $I_1 = \int_0^{\pi/2} \sin x \, dx = \left[-\cos x \right]_0^{\pi/2} = 1$

and $I_0 = \int_0^{\pi/2} \sin^0 x \, dx = \int_0^{\pi/2} 1 \, dx$

$$= [x]_0^{\pi/2} = \frac{\pi}{2}$$

Hence, $\int_0^{\pi/2} \sin^n x \, dx = \dfrac{n-1}{n} \cdot \dfrac{n-3}{n-2} \cdot \dfrac{n-5}{n-4} \cdots \dfrac{2}{3}$, when n is odd

Reduction Formulae

$$= \frac{n-1}{n} \cdot \frac{n-3}{n-2} \cdot \frac{n-5}{n-4} \cdots \frac{1}{2} \cdot \frac{\pi}{2} \quad \text{when } n \text{ is even}$$

Proceeding similarly we can find that

$$\int_0^{\pi/2} \cos^n x \, dx = \frac{n-1}{n} \cdot \frac{n-3}{n-2} \cdot \frac{n-5}{n-4} \cdots \frac{2}{3}, \text{ when } n \text{ is odd}$$

$$= \frac{n-1}{n} \cdot \frac{n-3}{n-2} \cdot \frac{n-5}{n-4} \cdots \frac{1}{2} \cdot \frac{\pi}{2} \quad \text{when } n \text{ is even}$$

The above formulae are know as **Willi's Formulae**.

EXAMPLES

1. Evaluate $\int \sin^6 x \, dx$

Sol. We have $\quad I_6 = \int \sin^6 x \, dx$

We know that $\quad I_n = -\dfrac{\cos x \sin^{n-1} x}{n} + \dfrac{n-1}{n} I_{n-2}$

Therefore, $\quad I_6 = -\dfrac{\cos x \sin^5 x}{6} + \dfrac{5}{6} I_4$

$$= -\frac{\cos x \sin^5 x}{6} + \frac{5}{6}\left(-\frac{\cos x \sin^3 x}{4} + \frac{3}{4} I_2\right)$$

$$= -\frac{\cos x \sin^5 x}{6} - \frac{5}{24} \cos x \sin^3 x + \frac{5}{8} \int \sin^2 x \, dx$$

$$= -\frac{\cos x \sin^5 x}{6} - \frac{5}{24} \cos x \sin^3 x + \frac{5}{16} \int (1 - \cos 2x) \, dx$$

$$= -\frac{1}{6} \cos x \sin^5 x - \frac{5}{25} \cos x \sin^3 x + \frac{5}{16} x - \frac{5}{32} \sin 2x$$

2. Find the values of

(i) $\int_0^3 \sqrt{\left(\dfrac{x^3}{3-x}\right)} \, dx$

(ii) $\int_0^\infty \dfrac{dx}{(1+x^2)^4}$

Sol. (i) Put $x = 3 \sin^2 \theta$, so that $dx = 6 \sin \theta \cos \theta \, d\theta$

$$\therefore \int_0^3 \sqrt{\left(\frac{x^3}{3-x}\right)} dx = \int_0^{\pi/2} \left\{\frac{27 \sin^6 \theta}{3(1-\sin^2 \theta)}\right\} 6 \sin \theta \cos \theta \, d\theta$$

$$= 18 \int_0^{\pi/2} \frac{\sin^3 \theta}{\cos \theta} \cdot \sin \theta \cos \theta \, d\theta$$

$$= 18 \int_0^{\pi/2} \sin^4 \theta \, d\theta$$

$$= 18 \cdot \frac{3}{4} \cdot \frac{1}{2} \cdot \frac{\pi}{2} = \frac{27}{8} \pi$$

(ii) Put $x = \tan \theta$, so that $dx = \sec^2 \theta \, d\theta$

$$\therefore \int_0^\infty \frac{dx}{(1+x^2)^4} = \int_0^{\pi/2} \frac{\sec^2\theta\, d\theta}{(1+\tan^2\theta)^4}$$

$$= \int_0^{\pi/2} \frac{\sec^2\theta\, d\theta}{\sec^8\theta} = \int_0^{\pi/2} \cos^6\theta\, d\theta$$

$$= \frac{5}{6}\cdot\frac{3}{4}\cdot\frac{1}{2}\cdot\frac{\pi}{2} = \frac{5\pi}{32}$$

3. Evaluate $\int_0^{\pi/4} (\cos 2\theta)^{3/2} \cos\theta\, d\theta$

Sol. Here $\int_0^{\pi/4} (1-2\sin^2\theta)^{3/2} \cos\theta\, d\theta$

Now, put $\sqrt{2}\sin\theta = \sin\phi$, so that

$$\cos\theta\, d\theta = \frac{1}{\sqrt{2}} \cos\phi\, d\phi$$

and when $\theta = 0$, then $\phi = 0$, when $\theta = \pi/4$, then $\sin\phi = \sqrt{2}\sin\frac{\pi}{4} = 1$

$$\therefore \qquad \phi = \pi/2$$

Hence $\int_0^{\pi/4}(\cos 2\theta)^{3/2} \cos\theta\, d\theta = \int_0^{\pi/2}(1-\sin^2\phi)^{3/2} \frac{1}{\sqrt{2}} \cos\phi\, d\phi$

$$= \frac{1}{\sqrt{2}} \int_0^{\pi/2} \cos^4\phi\, d\phi$$

$$= \frac{1}{\sqrt{2}} \cdot \frac{3}{4}\cdot\frac{1}{2}\cdot\frac{\pi}{2} = \frac{3\pi}{16\sqrt{2}}$$

EXERCISES

1. Evaluate $\int \cos^7 x\, dx$ $\qquad\left[\text{Ans. } \frac{1}{7}\sin x\left[\cos^6 x + \frac{6}{5}\cos^4 x + \frac{8}{5}\cos^2 x + \frac{16}{5}\right]\right]$

2. Evaluate $\int \sin^7 x\, dx$ $\qquad\left[\text{Ans. } -\frac{1}{3}\cos x\left[\sin^6 x + \frac{6}{5}\sin^4 x + \frac{8}{5}\sin^2 x + \frac{16}{5}\right]\right]$

3. Find the reduction formula $\int \sin^n dx$ for and hence evaluate $\int \sin^4 x\, dx$.

$$\left[\text{Ans. } -\frac{1}{4}\left[\sin^3 x + \cos x + \frac{3}{2}\sin x \cos x - \frac{3}{2}x\right]\right]$$

Evaluate the following integrals :

4. $\int_0^{\pi/2} \cos^4 x\, dx$ $\qquad\left[\text{Ans. } \frac{3\pi}{16}\right]$

5. $\int_0^{\pi/2} \sin^6 x\, dx$ $\qquad\left[\text{Ans. } \frac{5\pi}{32}\right]$

6. $\int_0^{\pi/2} \cos^{10} x\, dx$ $\qquad\left[\text{Ans. } \frac{63\pi}{512}\right]$

7. $\int_0^{2a} \frac{x^{3/2}}{\sqrt{(2a-x)}}\, dx$ $\qquad\left[\text{Ans. } \frac{63}{8}\pi a^2\right]$

Reduction Formulae

8. $\int_0^1 x^5 \sqrt{\dfrac{1+x^2}{1-x^2}}\, dx$ $\qquad\qquad\qquad$ $\left[\text{Ans. } \dfrac{3\pi + 8}{24}\right]$

9. $\int_0^a x^2 \sqrt{\dfrac{a-x}{a+x}}\, dx$ $\qquad\qquad\qquad$ $\left[\text{Ans. } \dfrac{3\pi - 8}{12} a^3\right]$

10. (a) $\int_0^\infty \dfrac{dx}{(a^2 + x^2)^{n+1/2}}$, n being a positive integer

    (b) $\int_0^\infty \dfrac{1}{(1+x^2)^5}\, dx$ $\qquad$ $\left[\text{Ans. (a) } \dfrac{4^{n-1}[(n-1)!]}{a^{2n}(2n-1)!},\ \text{(b) } \dfrac{35\pi}{256}\right]$

3.4 REDUCTION FORMULA FOR $\int \tan^n x\, dx$, n BEING A POSITIVE INTEGER

Let $\qquad I_n = \int \tan^n x\, dx$

$\qquad\qquad = \int \tan^{n-2} x \cdot \tan^2 x\, dx$

$\qquad\qquad = \int \tan^{n-2} x\, (\sec^2 x - 1)\, dx$

$\qquad\qquad = \int \tan^{n-2} x \sec^2 x\, dx - \int \tan^{n-2} x\, dx$

$\qquad\qquad = \dfrac{\tan^{n-1} x}{n-1} - \int \tan^{n-2} x\, dx$

$\therefore \qquad I_n = \dfrac{\tan^{n-1}}{n-1} - I_{n-2}$

where $\quad I_{n-2} = \int \tan^{n-2} x\, dx$

3.5 REDUCTION FORMULA FOR $\int \cot^n x\, dx$ n BEING A POSITIVE INTEGER

Let $\qquad I_n = \int \cot^n x\, dx$

$\qquad\qquad = \int \cot^{n-2} x \cdot \cot^2 x\, dx$

$\qquad\qquad - \int \cot^{n-2} x\, (\text{cosec}^2 x - 1)\, dx$

$\qquad\qquad = \int \cot^{n-2} \text{cosec}^2 x\, dx - \int \cot^{n-2} x\, dx$

$\qquad\qquad = -\dfrac{\cot^{n-1} x}{n-1} - I_{n-2}$

$\therefore \qquad I_n = -\dfrac{\cot^{n-1} x}{n-1} - I_{n-2}$

3.6 REDUCTION FORMULA FOR $\int \sec^n x\, dx$

Let $\qquad I_n = \int \sec^n x\, dx = \int \sec^{n-2} x \sec^2 x\, dx$

$\qquad\qquad = \sec^{n-2} \tan x - (n-2) \int \sec^{n-3} x \sec x \tan^2 x\, dx$

$\qquad\qquad = \sec^{n-2} x \tan x - (n-2) \int \sec^{n-2} x\, (\sec^2 x - 1)\, dx$

$$= \sec^{n-2} x \tan x - (n-2) \int \sec^n x\, dx + (n-2) \int \sec^{n-2} x\, dx$$
$$= \sec^{n-2} x \tan x - (n-2) I_n + (n-2) I_{n-2}$$

or $\quad I_n(n-2+1) = \sec^{n-2} x \tan x + (n-2) I_{n-2}$

$\therefore \quad I_n = \dfrac{1}{(n-1)} \sec^{n-2} x \tan x + \dfrac{n-2}{n-1} I_{n-2}$

3.7 REDUCTION FORMULA FOR $\int \operatorname{cosec}^n x\, dx$

Let $\quad I_n = \int \operatorname{cosec}^n dx = \int \operatorname{cosec}^{n-2} x \operatorname{cosec}^2 x\, dx$

$$= (-\cot x) \operatorname{cosec}^{n-2} x - (n-2) \int \operatorname{cosec}^{n-2} x \cot^2 x\, dx$$
$$= -\cot x \operatorname{cosec}^{n-2} x - (n-2) \int \operatorname{cosec}^{n-2} x (\operatorname{cosec}^2 x - 1)\, dx$$
$$= -\cot x \operatorname{cosec}^{n-2} x - (n-2) I_n + (n-2) I_{n-2}$$

Therefore
$$I_n + (n-2) I_n = -\cot x \operatorname{cosec}^{n-2} x + (n-2) I_{n-2}$$
$$I_n = -\dfrac{1}{(n-1)} \cot x \operatorname{cosec}^{n-2} x + \dfrac{n-2}{n-1} I_{n-2}$$

EXAMPLES

1. If $I_n = \int_0^{\pi/4} \tan^n x\, dx$, show that
$$I_n + I_{n-2} = \dfrac{1}{(n-1)}$$

Sol. From 3.4, we have
$$\int \tan^n x\, dx = \dfrac{\tan^{n-1} x}{(n-1)} - \int \tan^{n-2} x\, dx$$

$\therefore \quad \int_0^{\pi/4} \tan^n x\, dx = \dfrac{1}{(n-1)} \left[\tan_n^{x-1}\right]_0^{\pi/4} - \int_0^{\pi/4} \tan^{n-2} dx$

or $\quad I_n = \dfrac{1}{(n-1)}[1-0] - I_{n-2}.$

or $\quad I_n + I_{n-2} = \dfrac{1}{n-1}$

2. Evaluate $\int_0^a (a^2 + x^2)^{5/2} dx$

Sol. Let $x = a \tan \theta$, so that $dx = a \sec^2 \theta\, d\theta$.

$\therefore \quad \int_0^a (a^2 + x^2)^{5/2} dx = \int_0^{\pi/4} (a^2 + a^2 \tan^2 \theta)^{5/2} a \sec^2 \theta\, d\theta$

$$= a^6 \int_0^{\pi/4} \sec^7 \theta\, d\theta$$

Now, we have
$$\int \sec^n x\, dx = \dfrac{\sec^{n-2} x \tan x}{(n-1)} + \dfrac{n-2}{n-1} \int \sec^{n-2} x\, dx$$

Reduction Formulae

Then $\int_0^a (a^2 + x^2)^{5/2} dx = a^6 \left[\left\{ \frac{\sec^5 x \tan x}{6} \right\}_0^{\pi/4} + \frac{5}{6} \int_0^{\pi/4} \sec^5 x \, dx \right]$

$= a^6 \left[\frac{(\sqrt{2})^5}{6} + \frac{5}{6} \left(\frac{\sec^3 x \tan x}{4} \right)_0^{\pi/4} + \frac{3}{4} \int_0^{\pi/4} \sec^3 x \, dx \right]$

$= a^6 \left[\frac{4\sqrt{2}}{6} + \frac{5}{6} \frac{(\sqrt{2})^3}{4} + \frac{5}{8} \left\{ \left(\frac{\sec x \tan x}{2} \right)_0^{\pi/4} + \frac{1}{2} \int_0^{\pi/4} \sec x \, dx \right\} \right]$

$= a^6 \left[\frac{4\sqrt{2}}{6} + \frac{5\sqrt{2}}{12} + \frac{5}{8} \cdot \frac{\sqrt{2}}{2} + \frac{5}{16} \{ \log(\sec x + \tan x) \}_0^{\pi/4} \right]$

$= a^6 \left[\frac{67\sqrt{2}}{48} + 15 \log(\sqrt{2} + 1) \right]$

$= \frac{a^6}{48} [67\sqrt{2} + 15 \log(\sqrt{2} + 1)]$

EXERCISES

Integrate the following functiosn with respect to x:

1. $\tan^3 x$ $\qquad \left[\text{Ans. } \frac{1}{2} \tan^2 x + \log \cos x \right]$

2. $\cot^6 x$ $\qquad \left[\text{Ans. } -\frac{1}{5} \cot^5 x + \frac{1}{2} \cot^3 x - \cot x - x \right]$

3. $\sec^3 x$ $\qquad \left[\text{Ans. } \frac{1}{2} [\sec x + \tan x] \right]$

4. $\csc^3 x$ $\qquad \left[\text{Ans. } \frac{1}{2} [-\csc x \cot x + \log(\csc x - \cot x)] \right]$

5. If $\phi(n) = \int_0^{\pi/4} \tan^4 \theta \, d\theta$ show that $n[\phi(n-1) + \phi(n+1)] = 1$

Hence prove that $\int_0^a \frac{x^5 \, dx}{(2a^2 - x^2)^3} = \frac{1}{4} (2 \log 2 - 1)$

3.8 REDUCTION FORMULAE FOR $\int \sin^m x \cos^n dx$ WE HAVE

$I_{m,n} = \int \sin^m \cos^n x \, dx$

$= \int \sin^m x \cos^{n-1} x \cos x \, dx$

$= \frac{\sin^{m+1} x}{m+1} \cos^{n-1} x + \frac{(n-1)}{(m+1)} \int \sin^{m+1} x \cos^{n-1} x \sin x \, dx$

$= \frac{\sin^{m+1} x}{m+1} \cos^{n-1} x + \frac{n-1}{m+1} \int \sin^m x \cos^{n-2} x (1 - \cos^2 x) \, dx$

$$= \frac{1}{m+1} \sin^{m+1} x \cos^{n-1} x + \frac{n-1}{m+1} \int \sin^m x \cos^{n-2} x \, dx$$
$$- \frac{n-1}{m+1} \int \sin^m x \cos^n x \, dx$$

or $\quad I_{m,n}\left(1 + \frac{n-1}{m+1}\right) = \frac{\sin^{m+1} x \cos^{n-1} x}{m+1} + \frac{n-1}{m+1} I_{m,n-2}$

or $\quad I_{m,n}\left(\frac{m+n}{m+1}\right) = \frac{\sin^{m+1} x \cos^{n-1} x}{m+1} + \frac{n-1}{m+1} I_{m,n-2}$

or $\quad I_{m,n} = \frac{\sin^{m+1} x \cos^{n-1} x}{m+n} + \frac{(n-1)}{m+n} I_{m,n-2}$

Therefore
$$\int \sin^m x \cos^n x \, dx$$
$$= \frac{\sin^{m+1} x \cos^{n-1} x}{m+n} + \frac{n-1}{m+n} \int \sin^m x \cos^{n-2} x \, dx$$

Note 1. When n is even, by repeatedly using this formula $I_{m,n}$ can ber reduced to I_m only, i.e., $\int \sin^m x \, dx$ which further can be done by 3.2. As

$$I_m = \int \sin^m x \, dx = -\frac{\sin^{m-1} x \cos x}{m} + \frac{m-1}{m} I_{m-2}$$

Note 2. By writting $\int \sin^m x \cos^n x \, dx$ as $\int \sin^{m-1} x \cos^n x \sin x \, dx$ and integrating by parts the reduction formula can be obtained as

$$\int \sin^m x \cos^n x \, dx = -\frac{\sin^{m-1} x \cos^{n+1} x}{m+n} + \frac{m-1}{m+2} \int \sin^{m-2} x \cos^n x \, dx$$

or $\quad I_{m,n} = -\frac{\sin^{m-1} x \cos^{n+1} x}{m+n} + \frac{m-1}{m+n} I_{m-2},$

Similarly other four formulae can also be obtained as

$$I_{m,n} = -\frac{\sin^{m+1} x \cos^{n+1} x}{m+1} + \frac{m+n+2}{n+1} I_{m,n+2}$$

$$I_{m,n} = \frac{\sin^{m+1} x \cos^{n+1} x}{m+1} + \frac{m+n+2}{m+1} I_{m+2,n}$$

$$I_{m,n} = -\frac{\sin^{m+1} x \cos^{n+1} x}{n+1} + \frac{m-1}{n+1} I_{m-2,n+2}$$

and $\quad I_{m,n} = \frac{\sin^{m+1} x \cos^{n-1} x}{m+1} + \frac{n-1}{m+1} I_{m+2,n-2}$

3.8.1 TO EVALUATE $\int_0^{\pi/2} \sin^m x \cos^n x \, dx$

From the § 3.8 we have

Reduction Formulae

$$I_{m,n} = -\left[\frac{\sin^{m-1}x \cos^{n+1}}{m+n}\right]_0^{\pi/2} + \frac{m-1}{m+1}\int_0^{\pi/2}\sin^{m-2}x \cos^n x\, dx$$

Now, $\quad I_{m,n} = \dfrac{m-1}{m+n} I_{m-2,n}$

where $\quad I_{m-2,n} = \int_0^{\pi/2}\sin^{m-2}x \cos^n x\, dx$

$$= \frac{m-1}{m+n}\cdot\frac{(m-2)-1}{(m-2)+n} I_{m-4,n}$$

Case I. Let m and n be even positive integers, then

$$I_{m,n} = \frac{(m-1)(m-3)\ldots 3.1}{(m+n)(m+n-3)\ldots(n+2)} I_{0,n}$$

and $\quad I_{0,n} = \int_0^{\pi/2}\cos^n x\, dx$

$$= \left[\frac{\cos^{n-1}x \sin x}{n}\right]_0^{\pi/2} + \frac{n-1}{n}\int_0^{\pi/2}\cos^{n-2}x\, dx$$

$$= \frac{n-1}{n} I_{n-2}$$

$$= \frac{n-1}{n}\cdot\frac{n-3}{n-2}\ldots\frac{1}{2} I_0$$

and $\quad I_0 = \int_0^{\pi/2}(\cos x)^0 dx = \dfrac{\pi}{2}$

Therefore, $\quad I_n = \dfrac{n-1}{n}\cdot\dfrac{n-3}{n-2}\ldots\dfrac{1}{2}\cdot\dfrac{\pi}{2}$

Then $\quad I_{m,n} = \dfrac{[(m-n)(m-3)\ldots 1][(n-1)(n-3)\ldots 1]}{[(m+n)(m+n-2)\ldots(n+2)][n(n-2)\ldots 2]}\dfrac{\pi}{2}$

$$= \frac{\Gamma\left(\dfrac{m+1}{2}\right)\cdot\Gamma\left(\dfrac{n+1}{2}\right)}{2\Gamma\left(\dfrac{m+n+2}{2}\right)}$$

Case II. Let m be even and n be odd $m = 2k$ and $n = 2r-1$. Then $I_{m,n} = I_{2k, 2r-1}$

Now $\quad I_{2k,2r-1} = \dfrac{(2k-1)(2k-3)}{(2k+2r-1)(2k+2r-3)} I_{2k-4,2r-1}$

But $\quad I_{0,2r-1} = \int_0^{\pi/2}\cos^{2r-1}x\, dx$

$$= \frac{(2r-2)(2r-4)\ldots 2}{(2r-1)(2r-3)\ldots 3} \qquad \text{By § 3.3.1}$$

Then $\quad I_{2k,2r-1} = \dfrac{[(2k-1)(2k-3)\ldots 1][(2r-2)(2r-4)\ldots 2]}{[(2k+2r-1)(2k+2r-3)\ldots(2r+1)][(2r-1)(2r-3)\ldots 3]}$

$$= \frac{1.3.5\ldots(2k-3)(2k-1)\, 2.4.6\ldots(2r-4)(2r-2)}{1.3.5\ldots(2k+2r-1)}$$

$$= \frac{\dfrac{2^k}{\sqrt{\pi}} \Gamma\left(\dfrac{2k+1}{2}\right) 2^{r-1} \Gamma\left(\dfrac{2r}{2}\right)}{\dfrac{2^{k+r}}{\sqrt{\pi}} \Gamma\left(\dfrac{2k+2r+1}{2}\right)}$$

Hence, $\quad I_{m,n} = \dfrac{\Gamma\left(\dfrac{m+1}{2}\right) \pm \Gamma\left(\dfrac{n+1}{2}\right)}{2\Gamma\left(\dfrac{m+n+2}{2}\right)}$

Case III. Let m he odd and n be even *i.e.*, $m = 2k-1$, $n = 2r$.

Then $\quad I_{2k-1,2r} = \dfrac{(2k-2)(2k-4)\ldots 2}{(2k+2r-1)(2k+2r-3)\ldots(2r+3)} I_{1,2r}$

But $\quad I_{1,2r} = \int_0^{\pi/2} \sin x \cos^{2r} x\, dx$

$$= \left[\dfrac{\cos^{2r+1}}{2r+1}\right]_0^{\pi/2} = \dfrac{1}{2r+1}$$

Thereby $I_{2k-1,2r} = \dfrac{(2k-2)(2k-4)\ldots 2}{(2k+2r-1)(2k+2r-3)\ldots(2r+3)(2r+1)}$

$$= \dfrac{[(2k-2)(2k-4)\ldots(2r-1)(2r-3)\ldots 3.1]}{[(2k+2r-1)(2k+2r-3)\ldots(2r+3)(2r+1)][(2r-1)(2r-3)\ldots 3.1]}$$

$$= \dfrac{\dfrac{2^{k-1}}{\sqrt{\pi}} \Gamma\left(\dfrac{2k}{2}\right) \dfrac{2^r}{\sqrt{\pi}} \Gamma\left(\dfrac{2r+1}{2}\right)}{\dfrac{2^{k+r}}{\sqrt{\pi}} \Gamma\left(\dfrac{2k+2r+1}{2}\right)}$$

$$I_{m,n} = \dfrac{\Gamma\left(\dfrac{m+1}{2}\right) \Gamma\left(\dfrac{n+1}{2}\right)}{2\Gamma\left(\dfrac{m+n+2}{2}\right)}$$

Case IV. Let m and n be odd.
Then $m = 2k-1$, $n = 2r-1$.

Therefore $I_{m,n}$ becomes $I_{2k-1, 2r-1}$

Then $\quad I_{2k-1,2r-1} = \dfrac{(2k-2)(2k-4)\ldots 2}{(2k+2r-2)(2k+2r-4)-(2r+1)} I_{1,2r-1}$

But $\quad I_{1,2r-1} = \int_0^{\pi/2} \sin x \cos^{2r-1} x\, dx$

$$= -\left[\dfrac{\cos^{2r} x}{2r}\right]_0^{\pi/2} = \dfrac{1}{2r}$$

Thereby $I_{2k-1,2r-1} = \dfrac{(2k-2)(2k-4)\ldots 2}{(2k+2r-2)(2k+2r-4)\ldots(2r+2)(2r)}$

Reduction Formulae

$$= \frac{[(2k-2)(2k-4)\ldots 2][(2r-2)(2r-4)\ldots 4.2]}{[(2k+2r-2)(2k+2r-4)]\ldots[(2r+2)(2r)][(2r-2)\ldots 4.2]}$$

$$= \frac{2^{k-1}\Gamma\left(\frac{2k}{2}\right) 2^{r-1}\Gamma\left(\frac{2r}{2}\right)}{2^{k+r-2}\Gamma\left(\frac{2k+2r}{2}\right)}$$

$$\therefore \quad I_{m,n} = \frac{-[(m-1)(m-3)\ldots 2][(n-1)(n-3)\ldots 2]}{[(m+n)(m+n-2)\ldots(n+1)][(n-1)(-3)\ldots 2]}$$

$$= \frac{\Gamma\left(\frac{m+1}{2}\right)\Gamma\left(\frac{n+1}{2}\right)}{2\Gamma\left(\frac{m+n+2}{2}\right)}$$

Therefore the integration of $\int_0^{\pi/2} \sin^m x \cos^n x \, dx$ is given by

$$\int_0^{\pi/2} \sin^m x \cos^n x \, dx = \frac{\Gamma\left(\frac{m+1}{2}\right)\Gamma\left(\frac{n+1}{2}\right)}{2\Gamma\left(\frac{m+n+2}{2}\right)}$$

EXAMPLE

If I_n denotes $\int_0^1 (a^2 - x^2)^n \, dx, n > 0$ then prove that

$$I_n = \frac{2na^2}{2n+1} I_{n-1}$$

Sol. Put $x = a \sin \theta$, so that $dx = a \cos \theta \, d\theta$.

$$\therefore \quad I_n = \int_0^{\pi/2} a^{2n} \cos^{2n} \theta \cdot a \cos \theta \, d\theta$$

$$= a^{2n+1} \int_0^{\pi/2} \cos^{2n+1} \theta \, d\theta$$

$$= a^{2n+1} \int_0^{\pi/2} \sin^0 \theta \cos^{2n+1} \theta \, d\theta$$

$$= a^{2n+1} \frac{\Gamma\left(\frac{2n+2}{n}\right)\Gamma\left(\frac{1}{2}\right)}{2\Gamma\left(\frac{2n+3}{2}\right)} \quad \ldots(i)$$

Putting $(n-1)$ in place of n in (i), we get

$$I_{n-1} = a^{2n+1} \frac{\Gamma(n)\Gamma\frac{1}{2}}{2\Gamma\left(\frac{2n+1}{2}\right)} \quad \ldots(ii)$$

$$\frac{I_n}{I_{n-1}} = a^2 \cdot \frac{\Gamma{n+1}}{\Gamma\left(\frac{2n+3}{2}\right)} \cdot \frac{\Gamma\left(\frac{2n+1}{2}\right)}{\Gamma(n)}$$

$$= a^2 \cdot \frac{n\overline{\lceil(n)}}{\overline{\lceil\left(\frac{2n+1}{2}\right)}\overline{\lceil\left(\frac{2n+1}{2}\right)}} \cdot \frac{\overline{\lceil\left(\frac{2n+1}{2}\right)}}{\overline{\lceil(n)}} \qquad [\because \overline{\lceil n+1} = n\overline{\lceil n}]$$

$$= \frac{2n\, a^2}{2n+1}$$

$$\therefore \quad I_n = \frac{2n\, a^2}{2n+1} \cdot I_{n-1}$$

EXERCISES

Integrate the following functions with respect to x:

1. $\sin^2 x \cos^6 x$
   $\left[\text{Ans. } \frac{1}{8}\sin x \left[-\cos^7 x + \frac{1}{6}\cos^5 x + \frac{5}{24}\cos^3 x + \frac{5}{16}\cos x\right] + \frac{5x}{128}\right]$

2. $\sin^4 x \cos^2 x$
   $\left[\text{Ans. } \frac{1}{6}\cos x \left[\sin^5 x - \frac{1}{4}\sin^3 x - \frac{3}{8}\sin x\right] + \frac{1}{16}x\right]$

3. $\int_0^{\pi/2} \sin^4 x\,(1+\cos x)^3\, dx$
   $\left[\text{Ans. } \dfrac{9\pi}{16}\right]$

4. $\int_0^a x^4 \sqrt{(a^2 - x^2)}\, dx$
   $\left[\text{Ans. } \dfrac{\pi a^6}{32}\right]$

5. $\int_0^{2a} x^3 \sqrt{(2ax - x^2)}\, dx$
   $\left[\text{Ans. } \dfrac{7\pi a^5}{8}\right]$

6. $\int_0^{2a} \dfrac{x^4\, dx}{(a^2 + x^2)^4}$
   $\left[\text{Ans. } \dfrac{3\pi - 4}{192} \cdot \dfrac{1}{a^3}\right]$

7. Prove that $\int_a^b (x-a)^m (b-x)^n\, dx = \dfrac{m!\, n!}{(m+n+1)!} (b-a)^{m+n+1}$

   where m and n are positiver integers.

3.9 REDUCTION FORMULA FOR $\int x^n \sin mx\, dx$

We have $\quad I_{n,m} = \int x^n \sin mx\, dx$

$$= x^n \left(-\frac{\cos mx}{m}\right) + \frac{n}{m} \int x^{n-1} \cos mx\, dx$$

$$= -\frac{1}{m} x^n \cos mx + \frac{n}{m}\left[x^n \cdot \frac{1}{m}\sin mx - \frac{n-1}{m}\int x^{n-2} \sin mx\, dx\right]$$

$$= -\frac{1}{m} x^n \cos mx + \frac{n}{m^2} x^{n-1} \sin mx - \frac{(n-1)n}{m^2} I_{n-2m}$$

Similarly the reduction formula for $\int x^n \cos mx\, dx$ can be evaluated, then

$$I_{n,m} = \frac{x^n \sin mx}{m} + \frac{n}{m^2} x^{n-1} \cos mx - \frac{n(n-1)}{m^2} \int x^{n-2} \cos mx\, dx$$

Reduction Formulae

$$= \frac{x^n \sin mx}{m} + \frac{n}{m^2} x^{n-1} \cos mx - \frac{n(n-1)}{m^2} I_{n-2m}$$

EXAMPLES

1. If $u = \int_0^{\pi/2} x^n \sin x \, dx$ and n is greater than 1, prove that $u_n + n(n-1) u_{n-2} = n \left(\frac{1}{2}\pi\right)^{n-1}$

Sol. We have

$$u_n = \int_0^{\pi/2} x^n \sin x \, dx = [-\cos x \cdot x^n]_0^{\pi/2} + n \int_0^{\pi/2} \cos x \cdot x^{n-1} dx$$

$$= n \int_0^{\pi/2} \cos x \cdot x^{n-1} dx$$

$$= n [\sin x \cdot x^{n-1}]_0^{\pi/2} - n(n-1) \int_0^{\pi/2} \sin x \cdot x^{n-2} dx$$

$$= n \cdot \left(\frac{1}{2}\pi\right)^{n-1} - n(n-1) u_{n-2}$$

or $\quad = u_n + n(n-1) u_{n-2} = n \left(\frac{1}{2}\pi\right)^{n-1}$

2. If $I_n = \int_0^{\pi/2} x^n \sin(2p+1) x \, dx$, prove that

$$I_n + \frac{n(n-1)}{(2p+1)^2} I_{n-2} = (-1)^p \frac{n}{(2p+2)^2} \left(\frac{\pi}{2}\right)^{n-2}$$

n and p being positive integers.

Sol. We have

$$I_n = \int_0^{\pi/2} x^n \sin(2p+1) x \, dx$$

$$= -\left[\frac{x^n \cos(2p+1)x}{(2p+1)}\right]_0^{\pi/2} + \frac{n}{(2p+1)} \int_0^{\pi/2} x^{n-1} \cos(2p+1) x \, dx$$

$$= \frac{n}{(2p+1)^2} [x^{n-1} \sin(2p+1) x]_0^{\pi/2} - \frac{n(n-1)}{(2p+1)^2} \int_0^{\pi/2} x^{n-2} \sin(2p+1) x \, dx$$

or $\quad I_n = \frac{n}{(2p+1)^n} (-1)^n \left(\frac{\pi}{2}\right)^{n-1} - \frac{n(n-1)}{(2p+1)^2} I_{n-2}$

as $\sin(2p+1) x$ when $x = \frac{\pi}{2}$ is $\sin(2p+1)\frac{\pi}{2}$

or $\sin\left(p\pi + \frac{\pi}{2}\right)$ or $\sin p\pi \cos\frac{\pi}{2} + \cos p\pi \sin\frac{\pi}{2}$, i.e., $\cos p\pi$ or $(-1)^p$.

EXERCISES

1. Evaluate $\int_0^{\pi/2} x^3 \sin x \, dx$ $\qquad \left[\text{Ans. } \frac{2}{27} - \frac{1}{12}\pi^2\right]$

2. Evaluate $\int x^2 \sin 2x \, dx$ $\qquad \left[\text{Ans. } \frac{1}{4}[\cos 2x + 2x \sin 2x - 2x^2 \cos 2x]\right]$

3. If $u_n = \int_0^{\pi/2} x^n \sin mx \, dx$, prove that $u_n = \frac{n\pi^{n-1}}{m^2 \cdot 2^{n-1}} - \frac{n(n-1)}{m^2} u_{n-2}$

3.10 REDUCTION FORMULA FOR $\int x \sin^n x \, dx$

We have $I_n = \int x \sin^n x \, dx = \int x \sin^{n-1} x \cdot \sin x \, dx$

$$= (x \sin^{n-1} x)(-\cos x) - \int (-\cos x)\{x(n-1)\sin^{n-2} x \cos x + \sin^{n-1} x\} \, dx$$

$$= -x \sin^{n-1} x \cos x + (n-1) \int x \sin^{n-2} x \cos^2 x \, dx + \int \sin^{n-1} x \cos x \, dx$$

$$= -x \sin^{n-1} x \cos x + (n-1) \int x \sin^{n-2} x (1 - \sin^2 x) \, dx + \frac{\sin^n x}{n}$$

$$= -x \sin^{n-1} x \cos x (n-1) \int x \sin^{n-2} x \, dx - (n-1) \int x \sin^n x \, dx + \frac{\sin^n x}{n}$$

$$= -x \sin^{n-1} x \cos x + (n-1) I_{n-2} - (n-1) I_n + \frac{\sin^n x}{n}$$

Therefore $I_n (1+n-1) = -x \sin^{n-1} x \cos x + \frac{\sin^n x}{n} + (n-1) I_{n-2}$

$$\therefore \quad I_n = -\frac{x \sin^{n-1} x \cos x}{n} + \frac{\sin^n x}{n^2} + \frac{n-1}{n} I_{n-2}$$

Similarly, we can derive reduction formula for $\int x \cos^n x \, dx$

$$I_n = \frac{x \cos^{n-1} x \sin x}{n} + \frac{\cos^n x}{n^2} + \frac{n-1}{n} I_{n-2}$$

where $I_{n-2} = \int x \cos^{n-2} x \, dx$

EXAMPLE

If $U_n = \int_0^{\pi/2} \theta \sin^n \theta \, d\theta$ and $n > 1$ prove that

$$U_n = \frac{1}{n^2} + \frac{n-1}{n} U_{n-2}. \text{ Deduce that } U_5 = \frac{149}{225}$$

Sol. We have

$$U_n = \int_0^{\pi/2} \theta \sin^n \theta \, d\theta = \int_0^{\pi/2} (\theta \sin^{n-1} \theta) \sin \theta \, d\theta$$

$$= [\theta \sin^{n-1} \theta (-\cos \theta)]_0^{\pi/2} - \int_0^{\pi/2} \{\sin^{n-1} \theta + \theta (n-1) \sin^{n-2} \theta \cos \theta\}(-\cos \theta) \, d\theta$$

$$= 0 + \int_0^{\pi/2} \sin^{n-1} \theta \cos \theta \, d\theta + (n-1) \int_0^{\pi/2} \theta \sin^{n-2} \theta \cos^2 \theta \, d\theta$$

$$= \left[\frac{\sin^2 \theta}{n}\right]_0^{\pi/2} + (n-1) \int_0^{\pi/2} \theta \sin^{n-2} \theta (1 - \sin^2 \theta) \, d\theta$$

$$= \frac{1}{n} + (n-1) U_{n-2} - (n-1) U_n$$

or $[1+(n-1)] U_n = \frac{1}{n} + (n-1) U_{n-2}$

$$\therefore \quad U_n = \frac{1}{n^2} + \frac{n-1}{n} U_{n-2} \qquad \ldots(i)$$

Reduction Formulae

Deduction. Putting $n = 5, 3$ successively in (i), we have

$$U_5 = \frac{1}{25} + \frac{4}{5} U_3 \quad \text{and} \quad U_3 = \frac{1}{9} + \frac{2}{3} U_1$$

But
$$U_1 = \int_0^{\pi/2} \theta \sin \theta \, d\theta$$
$$= [\theta(-\cos\theta)]_0^{\pi/2} - \int_0^{\pi/2} 1 \cdot (-\cos\theta) \, d\theta$$
$$= 0 + [\sin\theta]_0^{\pi/2} = 1$$

$$\therefore \quad U_5 = \frac{1}{25} + \frac{4}{5}\left(\frac{1}{9} + \frac{2}{3}\right)$$
$$= \frac{1}{25} + \frac{28}{45} = \frac{149}{225}$$

EXERCISES

1. Evaluate $\int x \cos^5 x \, dx$

   $\left[\text{Ans. } \frac{1}{5}\left[x \sin x \cos^4 x + \frac{\cos^5 x}{3} + \frac{4}{3} x \sin x \cos^3 x + \frac{4}{9}\cos^3 x + \frac{8}{3}(x \sin x + \cos x)\right]\right]$

2. Obtain a reduction formula for $\int x \cos^m x \, dx$ and deduce that

   $$\int_0^{\pi/2} x \cos^3 x \, dx = \frac{1}{9}(3\pi - 7)$$

3.11 REDUCTION FORMULA FOR $\int e^{ax} \sin^n bx \, dx$

Let $I_n = \int e^{ax} \sin^n bx \, dx$

$$= \frac{e^{ax} \sin^n bx}{a} - \frac{nb}{a}\left[\frac{e^{ax}}{a} \sin^{n-1} bx \cos bx - \frac{1}{a}\int e^{ax}\{b(n-1)\sin^{n-2} bx \cos^2 bx\right.$$
$$\left. - \sin^{n-1}(\sin bx) b\} \, dx\right]$$

$$= \frac{e^{ax} \sin^n bx}{a} - \frac{nb}{a^2} e^{ax} \sin^{n-1} bx \cos x$$
$$+ \frac{nb}{b^2} \int e^{ax} [b(n-1)\sin^{n-2} bx (1 - \sin^2 bx) - b \sin^{n-1} bx \sin bx] \, dx$$

$$= \frac{e^{ax} \sin^n bx}{a} - \frac{nb}{a^2} e^{ax} \sin^{n-1} bx \cos bx + n(n-1)\frac{b^2}{a^2} \int e^{ax} \sin^{n-2} bx \, dx$$

$$= \frac{e^{ax} \sin^n bx}{a} - \frac{nb}{a^2} e^{ax} \sin^{n-1} bx \cos bx$$
$$+ n(n-1)\frac{b^2}{a^2} I_{n-2} - n(n-1)\frac{b^2}{a^2} I_n - \frac{nb^2}{a^2} I_n$$

$$= \frac{e^{ax} \sin^n bx}{a} - \frac{nb}{a^2} e^{ax} \sin^{n-1} bx \cos bx + n(n-1)\frac{b^2}{a^2} I_{n-2} - n^2 \frac{b^2}{a^2} I_n$$

Therefore, $I_n \left(1 + \dfrac{n^2 b^2}{a^2}\right)$

$$= \dfrac{e^{ax} \sin^n bx}{a} - \dfrac{nb}{a^2} e^{ax} \sin^{n-1} bx \cos bx + n(n-1) \dfrac{b^2}{a^2} I_{n-2}$$

$\therefore \quad I_n = \dfrac{e^{ax}}{a^2 + n^2 b^2} (a \sin^n bx - nb \sin^{n-1} bx \cos bx) + \dfrac{n(n-1) b^2}{a^2 + n^2 b^2} I_{n-2}$

Similarly a reduction formula for $\int e^{ax} \cos^n bx \, dx$ can be determined which is as follow:

$$\int e^{ax} \cos^n bx \, dx = \dfrac{e^{ax}}{a^2 + n^2 b^2} (a \cos^n bx + nb \sin bx \cos^{n-1} bx) + \dfrac{n(n-1)}{a^2 + n^2 b^2} I_{n-2}$$

EXAMPLE

Integrating by parts twice, or otherwise, obtain a reduction formula for $I_m = \int_0^\infty e^{-x} \sin^m x \, dx$ where $m \geq 2$ in the form $(1 + m^2) I_m = m(m-1) I_{m-2}$ and hence evaluate I_4.

Sol. On integrating twice by parts taking e^{-x} as second function we have

$$I_m = \left[-\sin^m x e^{-x}\right]_0^\infty - \int_0^\infty m \sin^{m-1} x \cos x (-e^{-x}) \, dx$$

$$= 0 + m \int_0^\infty (\sin^{m-1} x \cos x) e^{-x} dx$$

$$= m [\{\sin^{m-1} x \cos x\} (-e^{-x})]_0^\infty$$

$$\qquad + m \int_0^\infty [(m-1) \sin^{m-2} x \cos^2 x - \sin^m x] e^{-x} \, dx$$

$$= 0 + m(m-1) \int_0^\infty \sin^{m-2} x (1 - \sin^2 x) e^{-x} \, dx - m \int_0^\infty e^{-x} \sin^m x \, dx$$

$$= m(m-1) \int_0^\infty e^{-x} \sin^{m-2} x \, dx - m(m-1) \int_0^\infty e^{-x} \sin^m x \, dx$$

$$\qquad\qquad\qquad\qquad\qquad - m \int_0^\infty e^{-x} \sin^m x \, dx$$

$\qquad I_m = m(m-1) I_{m-2} - m^2 I_m$

or $\qquad (1 + m^2) I_m = m(m-1) I_{m-2}$

To evaluate I_4, put $m = 4, 2$ successively in (i), so that

$$17 I_4 = 12 I_2 \quad \text{or} \quad I_4 = \dfrac{12}{17} I_2$$

and $\qquad 5 I_2 = 2 I_0, \quad \text{i.e.,} \quad I_2 = \dfrac{2}{5} I_0$

But $\qquad I_0 = [-e^{-x}]_0^\infty = 1$

$\therefore \qquad I_4 = \dfrac{12}{17} \times \dfrac{2}{5} \times 1 = \dfrac{24}{85}$

EXERCISES

1. Evaluate $\int e^{2x} \cos^3 x \, dx \qquad \left[\text{Ans.} \ \dfrac{e^{2x}}{4} \left[\dfrac{1}{\sqrt{13}} \cos\left(3x - \tan^{-1} \dfrac{3}{2}\right) + \dfrac{3}{\sqrt{5}} \cos\left(x - \tan^{-1} \dfrac{1}{2}\right)\right]\right]$

Reduction Formulae

2. If $I_n = \int e^{ax} \sin n \, bx \, dx$, find a reduction formula for I_n in terms of I_{n-2}. Hence evaluate $\int_0^\infty e^{-x} e^{-x} \sin^4 x \, dx$. $\left[\text{Ans. } \dfrac{24}{85}\right]$

3. Find a reduction formula for $\int e^{-ax} \sin^n x \, dx$; hence deduce the reduction formula for $\int \cos ax \sin^n x \, dx$.

$$\left[\text{Ans. } -\frac{e^{-ax} \sin^{n-1} x}{a^2 + x^2} [a \sin x + n \cos x] + \frac{n(n-1)}{(a^2 + n^2)} \int e^{-ax} \sin^{n-2} x \, dx \right.$$

$$\left. -\frac{\sin^{n-1} x}{n^2 - a^2}[a \sin x \sin ax + n \cos x \cos ax] + \frac{n(n-1)}{n^2 - a^2} \int \cos ax \sin^{n-2} x \, dx \right]$$

3.12 REDUCTION FORMULA FOR $\int \cos^m x \cos nx \, dx$

Let $I_{m,n} = \int \cos^m x \cos nx \, dx$

$$= \frac{\cos^m x \sin nx}{n} - \frac{m}{n} \int \cos^{m-1} x (-\sin x) \sin nx \, dx$$

But $\cos(n-1)x = \cos nx \cos x + \sin nx \sin x$

Therefore, $\sin nx \sin x = \cos(n-1)x - \cos x \cos nx$

Thereby,

$$I_{m,n} = \frac{\cos^m x \sin nx}{n} + \frac{m}{n} \int \cos^{m-1} x [\cos(n-1)x - \cos nx \cos x] \, dx$$

$$= \frac{\cos^m x \sin nx}{n} + \frac{m}{n} \int \cos^{m-1} x \cos(n-1)x \, dx$$

$$ \frac{\cos^m x \sin nx}{n} + \frac{m}{n} I_{m-1, n-1} - \frac{m}{n} I_{m,n}$$

or $\quad I_{m,n}\left(1 + \dfrac{m}{n}\right) = \dfrac{\cos^m x \sin nx}{n} + \dfrac{m}{n} I_{m-1, n-1}$

$\therefore \quad I_{m,n} = \dfrac{\cos^m x \sin nx}{m+n} + \dfrac{m}{m+n} I_{m-1, n-1}$

Similarly a reduction formula for

$$\int \cos^m x \sin nx \, dx$$

can be determined which is as follows :

$$I_{m,n} = \int \cos^m x \sin nx \, dx$$

$$= -\frac{\cos^m x \cos nx}{m+n} + \frac{m}{m+n} I_{m-1, n-1}$$

3.12.1 REDUCTION FORMULA FOR $\int \dfrac{\sin nx}{\sin x} dx$

We know that
$$\sin nx - \sin(n-2)x = 2\cos(n-1)x \sin x$$
or $$\sin nx = \sin(n-2)x + 2\cos(n-1)x \sin x$$

$\therefore \quad \int \dfrac{\sin nx}{\sin x} dx = \int \dfrac{2\cos(n-1)x \sin x + \sin(n-2)x}{\sin x} dx$

$$= 2\int \cos(n-1)x \, dx + \int \dfrac{\sin(n-2)x}{\sin x} dx$$

or $\quad \int \dfrac{\sin nx}{\sin x} dx = \dfrac{2\sin(n-1)x}{n-1} + \int \dfrac{\sin(n-2)x}{\sin x} dx$

EXAMPLES

1. *Prove that* $\int_0^{\pi/2} \cos^n x \cos nx \, dx = \dfrac{\pi}{2^{n+1}}$, *n being a positive integer.*

Sol. Let $\quad I_{n,n} = \int_0^{\pi/2} \cos^{nx} \cos nx \, dx$

By the reduction formula $\quad I_{m,n} = \dfrac{\cos^m x \sin nx}{m+n} + \dfrac{m}{m+n} I_{m-1, n-1}$

Here $\quad I_{n,n} = \left[\dfrac{\cos^n x \sin nx}{2n}\right]_0^{\pi/2} + \dfrac{1}{2}\int_0^{\pi/2} \cos^{n-1} x \cos(n-1)x \, dx$

$$= \dfrac{1}{2} I_{n-1, n-1} = \dfrac{1}{2}\left[\dfrac{1}{2} I_{n-2, n-2}\right]$$

$$= \left[\dfrac{1}{2} \cdot \dfrac{1}{2} \ldots \text{to } n \text{ factors}\right] I_{0,0}$$

$$= \dfrac{1}{2^n} \int_0^{\pi/2} \cos^0 x \cos(0 \cdot x) \, dx$$

$$= \dfrac{1}{2^n} \int_0^{\pi/2} dx = \dfrac{1}{2^n}[x]_0^{\pi/2}$$

$$= \dfrac{\pi}{2^{n+1}}$$

2. *Prove that* $\int_0^\pi \dfrac{\sin nx}{\sin x} dx = \pi$ *or 0 as n is an odd or even positive integer.*

Sol. Let $\quad I_n = \int_0^\pi \dfrac{\sin nx}{\sin x} dx$

$$= \left[\dfrac{2\sin(n-1)x}{n-1}\right]_0^\pi + \int_0^\pi \dfrac{\sin(n-2)x}{\sin x} dx$$

[using reduction formula of § 3.12.1]

$$= 0 + I_{n-2} = I_{n-2}$$

Reduction Formulae

similarly, $I_n = I_{n-2} = I_{n-4} = \ldots (I_2 \text{ or } I_1)$

Case I. If n is odd positive integer, we have

$$I_n = I_1 = \int_0^\pi \frac{\sin x}{\sin x} dx = \int_0^\pi dx = \pi$$

Case II. If n is even positive integer, we have

$$I_n = I_2 = \int_0^\pi \frac{2 \sin 2x}{\sin x} d_2 = 2 \int_0^\pi \cos x \, dx$$

$$= 2 [\sin x]_0^\pi = 0$$

EXERCISES

1. Evaluate $\int_0^{\pi/2} \sin^2 x \cos 3x \, dx$ $\qquad \left[\text{Ans. } -\frac{7}{15}\right]$

2. Prove that if $I_{m,n} = \int \cos^m x \sin nx \, dx$,

$$(m+n) I_{m,n} = -\cos^m x \cos nx + I_{m-1, n-1}$$

   Hence or otherwise evaluate

   $$\int_0^{\pi/2} \cos^5 x \sin 3x \, dx$$

3. If $I_{m,n} = \int_0^{\pi/2} \cos^m x \sin nx \, dx$, show that $I_{1,2} = \frac{12}{35}$

4. Find the reduction formula for $I_{m,n}$ where $I_{m,n} = \int_0^{\pi/2} (\cos x)^m \sin nx \, dx$.

   Deduce that $\qquad I_{m,n} = \frac{1}{2^{m+1}} \left[2 + \frac{2^2}{2} + \frac{2^3}{3} + \ldots + \frac{2^m}{m} \right]$

5. Prove that $I_n = \int \frac{\sin nx}{\sin x} dx$, prove that reduction formula

   $$I_n = \frac{2 \sin (n-1) x}{n-1} + I_{n-2}$$

   Hence, evaluate $\int_0^{\pi/2} \frac{\sin 7x}{\sin x} dx$. $\qquad$ [Ans. π]

3.13 REDUCTION FORMULA FOR $\int x^n e^{ax} \sin bx \, dx$

We know that $\qquad \int e^{ax} \sin bx \, dx = \frac{1}{r} e^{ax} \sin (bx - \phi)$

where, $\qquad r = \sqrt{(a^2 + b^2)}$ and $\phi = \tan^{-1}\left(\frac{b}{a}\right)$

Then $\qquad I_n = x^n \frac{e^{ax}}{r} \sin (bx - \phi) - \int n x^{n-1} \frac{e^{ax}}{r} \sin (bx - \phi) \, dx$

$= x^n \frac{e^{ax}}{r} \sin (bx - \phi) - \frac{n}{r} \left[x^{n-1} \frac{e^{ax}}{r} \sin (bx - 2\phi) - (n-1) \int x^{n-2} \frac{e^{ax}}{r} \sin (bx - 2\phi) \, dx \right]$

$$= x^n \frac{e^{ax}}{r} \sin(bx - \phi) - \frac{n}{r^2} x^{n-1} e^{ax} \sin(bax - 2\phi)$$

$$+ \frac{n(n-1)}{r^2} \int e^{ax} \sin(bx - 2\phi) x^{n-2} dx$$

Continuing this process, we have

$$\int x^n (e^{ax} \sin bx) dx = x^n \left\{ \frac{e^{ax}}{r} \sin(bx - \phi) \right\}$$

$$- nx^{n-1} \left\{ \frac{e^{ax}}{r^2} \sin(bx - 2\phi) \right\} + n(n-1) x^{n-2}$$

$$\left\{ \frac{e^{ax}}{r^3} \sin(bx - 3\phi) \right\} \ldots + (-1)^n n!$$

$$\left[\frac{e^{ax}}{r^{n+1}} \sin\{bx - (n+1)\phi\} + \ldots \right]$$

Similarly the reduction formula for $\int x^n e^{ax} \cos bx \, dx$ can be determined.

3.14 REDUCTION FORMULA FOR $\int x^n (\log x)^m \, dx$

We have $\quad I_{n,m} = \int x^n (\log x)^m \, dx$

$$= \frac{x^{n+1}}{n+1} (\log x)^m - \frac{m}{n+1} \int x^n (\log x)^{m-1} dx$$

$$= \frac{x^{n+1}}{n+1} (\log x)^m - \frac{m}{n+1} I_{n, m-1}$$

(ii) Reduction Formula for $\int \dfrac{x^n}{(\log x)^m} dx$

We have $\quad \int \dfrac{x^n}{(\log x)^m} dx = \int x^{n+1} \left[\dfrac{1}{(\log x)^m} \cdot \dfrac{1}{x} \right] dx$

Now integrating by parts, taking x^{n+1} as the first function we have

$$\int \frac{x^n}{(\log x)^m} dx = \int x^{n+1} \frac{1}{x (\log x)^m} dx$$

$$= x^{n+1} \frac{(\log x)^{-m+1}}{-m+1} - \int (n+1) x^n \frac{(\log x)^{-m+1}}{-m+1} dx$$

$$= -\frac{x^{n+1}}{(m-1)(\log x)^{m-1}} + \frac{n+1}{m-1} \int \frac{x^n}{(\log x)^{m-1}} dx$$

Reduction Formulae

EXAMPLES

1. Show that $\int_0^\infty e^{-x} x^n \, dx = n!$

Sol. $\int_0^\infty e^{-x} x^n \, dx = \left[x^n \left(\dfrac{e^{-x}}{-1} \right) \right]_0^\infty - \int_0^\infty n \, x^{n-1} \left(\dfrac{e^{-x}}{-1} \right) dx$ ...(1)

Integrating by parts taking x^n as first function.

Now $\lim\limits_{x \to \infty} x^n e^{-x} = \lim\limits_{x \to \infty} \dfrac{x^n}{e^x}$, which is of the form $\dfrac{\infty}{\infty}$.

$$= \lim_{x \to \infty} \left\{ \dfrac{x^n}{1 + x + x^2 2! + \ldots + \dfrac{x^{n-1}}{(n-1)!} + \dfrac{x^n}{n!} + \dfrac{x^{n+1}}{(n+1)!} + \ldots} \right\}$$

$$= \lim_{x \to \infty} \left[\dfrac{1}{\dfrac{1}{x^n} + \dfrac{1}{x^{n-1}} + \dfrac{1}{21 x^{n-2}} + \ldots + \dfrac{1}{x(n-1)!} + \dfrac{1}{n!} + \dfrac{x}{(n+1)!} + \ldots} \right]$$

$$= \dfrac{1}{\infty} = 0.$$

∴ From (1), $\int_0^\infty e^{-x} x^n \, dx = [0] + n \int_0^\infty x^{n-1} e^{-x} \, dx$

or $\int_0^\infty e^{-x} x^n \, dx = n \int_0^\infty x^{n-1} e^{-x} \, dx$ ...(2)

$$= n \left[(n-1) \int_0^\infty x^{n-2} e^{-x} \, dx \right]$$

∴ By repeated application of (2), we have

$$\int_0^\infty e^{-x} x^n \, dx = n(n-1)(n-2) \ldots 2 \cdot \int_0^\infty x^0 e^{-x} \, dx$$

$$= n(n-1)(n-2) \ldots 2 \cdot 1 \left[\dfrac{e^{-x}}{-1} \right]_0^\infty$$

$$= n(n-1)(n-2) \ldots 2 \cdot 1 (0+1) = n!$$

2. If m and n are positive integers and

$$f(m, n) = \int_0^1 x^{n-1} (\log x)^m \, dx, \text{ Prove } f(m, n) = -\dfrac{m}{n} f(m-1, n)$$

Deduce that $f(m, n) = (-1)^m \dfrac{m!}{n^{m+1}}$

Sol. Integrating by parts, taking x^{n-1} as the second function, we have

$$f(m, n) = \int_0^1 x^{n-1} (\log x)^m \, dx$$

$$= \left[(\log x)^m \cdot \dfrac{x^n}{n} \right]_0^1 - \int_0^1 n (\log x)^{m-1} \dfrac{1}{x} \cdot \dfrac{x^n}{n} \, dx$$

$$= 0 - \dfrac{m}{n} \int_0^1 x^{n-1} (\log x)^{m-1} \, dx$$

since $\lim_{x \to 0} x^n (\log x)^m = \lim_{x \to 0} \frac{(\log x)^m}{x^{-n}} = 0$ (by Hospital's Rule)

$\therefore \qquad f(m, n) = -\frac{m}{n} f(m-1, n)$ ...(i)

Now to deduce $f(m, n) = \frac{(-1)^m m!}{n^{m+1}}$, we

Use the reduction formula (i) successively, so that

$$f(m, n) = \left(-\frac{m}{n}\right)\left(-\frac{m-1}{n}\right)\left(-\frac{m-2}{n}\right)\cdots\left(-\frac{1}{n}\right)\int_0^1 x^{n-1} dx$$

$$= (-1)^m \frac{m(m-1)(m-2)\cdots 1}{n^m} \left[\frac{x^n}{n}\right]_0^1$$

$$= -(1)^m \frac{m!}{n^m} \cdot \frac{1}{n}$$

EXERCISES

Evaluate:

1. $\int x^2 e^{2x \cos \alpha} \sin(2x \sin \alpha)\, dx$ $\qquad \left[\text{Ans. } \frac{1}{2} e^x [x(\cos x + \sin x) - \cos x]\right]$

2. $\int_0^\infty x e^{-2x} \cos x\, dx$ $\qquad \left[\text{Ans. } \frac{3}{25}\right]$

3. $\int x^m (\log x)^3\, dx$ $\qquad \left[\text{Ans. } \frac{1}{3} x^3 \left[(\log x)^2 - \frac{2}{3} \log x + \frac{2}{9}\right]\right]$

4. $\int_0^1 x^m (\log x)^4\, dx$ $\qquad \left[\text{Ans. } \frac{24}{(m+1)^4}\right]$

5. (i) $\int e^{2x} x^3\, dx$; (ii) $\int e^{mx} x^{-2}\, dx$

$\left[\text{Ans. (i) } \frac{1}{4} e^{2x}(2x^3 - 32x^2 + 3x - 1); \text{(ii) } m\left[\log x + \frac{mx}{1!} + \frac{m^2 x^2}{2 \cdot 2!} + \frac{m^3 x^3}{3 \cdot 3!} + \cdots\right]\right]$

6. If $I_n = \int_0^\infty e^{-x} x^{n-1} \log x\, dx$, show that

$$I_{n+2} - (2n+1) I_{n+1} + n^2 I_n = 0$$

[**Hint.** Integrate by parts taking e^{-x} as the second function]

3.15 REDUCTION FORMULA FOR $\int \frac{dx}{(x^2 + a^2)^n}$, n BEING A POSITIVE INTEGER

Let $I_n = \int \frac{dx}{(x^2 + a^2)^n}$, then

$$I_{n-1} = \int \frac{dx}{(x^2 + a^2)^{n-1}}$$

Reduction Formulae

Taking $I_{n-1} = \int 1 \cdot \dfrac{1}{(x^2 + a^2)^{n-1}} dx$

and integrating by parts taking 1 as second function, we get

$$I_{n-1} = \dfrac{x}{(x^2+a^2)^{n-1}} - \int -(n-1)\dfrac{2x}{(x^2+a^2)} \, x \, dx$$

$$= \dfrac{x}{(x^2+a^2)^{n-1}} + 2(n-1)\int \dfrac{x^2}{(x^2+a^2)^n} dx$$

$$= \dfrac{x}{(x^2+a^2)^{n-1}} + 2(n-1)\int \dfrac{x^2 + a^2 - a^2}{(x^2+a^2)^n} dx$$

$$= \dfrac{x}{(x^2+a^2)^{n-1}} + 2(n-1)\int \dfrac{dx}{(x^2+a^2)^{n-1}} - 2(n-1)a^2 \int \dfrac{dx}{(x^2+a^2)^n}$$

or $\quad I_{n-1}(1 - 2n + 2) = \dfrac{2x}{(x^2+a^2)^{n-1}} - 2(n-1)a^2 I_n$

or $\quad 2(n-1)a^2 I_n = \dfrac{x}{(x^2+a^2)^{n-1}} + (2n-3) I_{n-1}$

or $\quad I_n = \dfrac{x}{a^2(2n-2)(x^2+a^2)^{n-1}} + \dfrac{(2n-3)}{2(n-1)a^2} I_{n-1}$

This is the required reduction formula.

3.16 REDUCTION FORMULA FOR $\int x^m(a+bx^n)^p dx$ WHERE m, n, p ARE POSITIVE OR NEGATIVE INTGERS OR FRACTIONS

Let $\quad I_m = \int x^m (a + bx^n)^p dx = \int x^{m-n+1} x^{n-1} (a+bx^n)^p dx$

as $\quad d(a+bx^n) = nbx^{n-1} dx$

So $\quad I_n = \dfrac{1}{nb} \int x^{m-n+1}(nbx^{n-1})(a+bx^n)^p dx$

$$= \dfrac{1}{nb}\left[x^{m-n+1} \dfrac{(a+bx)^{p+1}}{p+1} - \dfrac{(m-n+1)}{p+1} \int x^{m-n} (a+bx)^{p+1} dx \right]$$

(by using integration by parts method taking $(nbx^{n-1})(a+bx^n)^p$ as second function and x^{m-n+1} as first function).

$$= \dfrac{1}{nb} \cdot \dfrac{x^{m-n+1}(a+bx^n)^{p+1}}{(p+1)} \int x^{m-n}(a+bx^n)^p (a+bx^n) dx$$

$$= \dfrac{1}{nb} \cdot \dfrac{x^{m-n+1}}{(p+1)} (a+bx^n)^{p+1} - \dfrac{(m-n+1)}{(p+1)nb} \int x^{m-n}(a+bx^n)^p dx$$

$$- \dfrac{(m-n+1)}{(p+1)nb} \int x^m (a+bx^n)^p dx$$

Therefore, $\quad I_m = \dfrac{1}{nb} \cdot \dfrac{x^{m-n+1}}{(p+1)}(a+bx^n)^{p+1} - \dfrac{(m-n+1)}{(p+1)nb} \int x^{m-n}(a+bx^n)^p dx$

$$- \dfrac{(m-n+1)}{(p+1)nb} I_m$$

or $I_m\left\{1+\dfrac{m-n+1}{(p+1)n}\right\} = \dfrac{x^{m-n+1}(a+bx^n)^{p+1}}{nb(p+1)} - \dfrac{(m-n+1)}{(p+1)nb}I_{m-n}$

Hence $\int x^m(a+bx^n)^p\, dx = \dfrac{x^{m-n+1}(a+bx^n)^{p+1}}{b(nb+m+2)}$

$\qquad\qquad\qquad\qquad - \dfrac{a(m-n+1)}{b(np+m+1)}\int x^{m-n}(a+bx^n)^p\, dx$

Note. $\int x^m(a+bx^n)^p\, dx$ can be connected with six possible relations as given below.

1. $\int x^n(a+bx^n)^p\, dx$

    $= \dfrac{x^{m-n+1}(a+bx^n)^{p+1}}{b(np+m-1)} - \dfrac{a(m-n+1)}{b(np+m+1)}\int x^{m-n}(a+bx^n)^p\, dx$

2. $\int x^m(a+bx^n)^p\, dx$

    $= \dfrac{x^{m+1}(a+bx^n)^p}{np+m+1} - \dfrac{(n\,p\,a)}{np+m+1}\int x^m(a+bx^n)^{p-1}\, dx$

3. $\int x^m(a+bx^n)^p\, dx$

    $= -\dfrac{x^{m+1}(a+bx^n)^{p+1}}{(p+1)na} + \dfrac{(m+np+n+1)}{(p+1)na}\int x^m(a+bx^n)^{p+1}\, dx$

4. $\int x^m(a+bx^n)^p\, dx$

    $= -\dfrac{x^{m+1}(a+bx^n)^{p+1}}{(m+1)a} - \dfrac{b(m+np+n+a)}{(m+1)a}\int x^{m+n}(a+bx^n)^p\, dx$

5. $\int x^m(a+bx^n)^p\, dx$

    $= \dfrac{x^{m-n+1}(a+bx^n)^{p+1}}{np(p+1)} - \dfrac{(m-n+1)}{nb(p+1)}\int x^{m-n}(a+bx^n)^{p+1}\, dx$

6. $\int x^m(a+bx^n)^p\, dx$

    $= -\dfrac{x^{m+1}(a+bx^n)^p}{(m+1)} - \dfrac{bnp}{m+1}\int x^{m+n}(a+bx^n)^{p+1}\, dx$

3.17 REDUCTION FORMULA FOR $\int (a^2+x^2)^{n/2}dx$

Let $I = \int(a^2+x^2)^{n/2}dx$

$= \int(a^2+x^2)^{n/2}\cdot 1\, dx$

$= x(a^2+x^2)^{n/2} - \dfrac{n}{2}\cdot 2\int(a^2+x^2)^{n/2-1}\cdot 1\, dx$

$= x(a^2+x^2)^{n/2} - n\cdot\int(a^2+x^2)^{n/2-1}x^2\, dx$

$= x(a^2+x^2)^{n/2} - n\int(a^2+x^2)^{n/2-1}(a^2+x^2-a^2)\, dx$

$= x(a^2+x^2)^{n/2} - n\int(a^2+x^2)^{n-2}dx + na^2\int(a^2+x^2)^{n/2-1}dx$

Reduction Formulae

Therefore

$$I(n+1) = x(a^2+x^2)^{n/2} + na^2 \int (a^2+x^2)^{n/2-1} dx$$

$$\therefore \quad I = \frac{x(a^2+x^2)^{n/2}}{1+n} + \frac{na^2}{1+n} \int (a^2+x^2)^{n/2-1} dx$$

EXAMPLES

1. If $I_n = \int x^n (a-x)^{1/2} dx$, prove that $(2n+3) I_n = 2an\, I_{n-1} - 2x^n (a-x)^{3/2}$.

Sol. We have

$$I_n = \int x^n (a-x)^{1/2} dx$$

$$= x^n \{-(2/3)(a-x)^{3/2}\} - \int nx^{n-1} \left\{ -\frac{2}{3}(a+x^{3/2}) \right\} dx$$

$$= -\frac{2}{3} x^n (a-x)^{3/2} + \frac{2}{3} n \int x^{n-1}(a-x)(a-x)^{1/2} dx$$

or $\quad I_n = -\frac{2}{3} x^n (a-x)^{3/2} + \frac{2}{3} na \int x^{n-2}(a-x)^{1/2} dx$

or $\quad \qquad -\frac{2}{3} n \int x^n (a-x)^{1/2} dx$

or $\quad I_n = -\frac{2}{3} x^n (a-x)^{3/2} + \frac{2}{3} na\, I_{n-1} - \frac{2}{3} n\, I_n$

or $\quad \left(1 + \frac{2}{3} n\right) I_n = \frac{2}{3} na\, I_{n-1} - \frac{2}{3} x^n (a-x)^{3/2}$

or $\quad (2n+3) I_n = 2na\, I_{n-1} - 2x^n (a-x)^{3/2}$

2. Evaluate $\int (a^2+x^2)^{5/2} dx$.

Sol. Let

$$\int (a^2+x^2)^{5/2} dx = \int (a^2+x^2)^{5/2} \cdot 1 \cdot dx$$

$$= (a^2+x^2)^{5/2} - \int \frac{5}{2}(a^2+x^2)^{3/2} \cdot 2x \cdot x\, dx$$

$$= x(a^2+x^2)^{5/2} - 5\int (a^2+x^2)^{3/2} \cdot x^2\, dx$$

$$= x(a^2+x^2)^{5/2} - 5\int (a^2+x^2)^{3/2} \cdot (x^2+a^2-a^2)\, dx$$

$$= x(a^2+x^2)^{5/2} + 5a^2 \int (a^2+x^2)^{3/2} dx - 5\int (a^2+x^2)^{5/2} dx$$

or $\quad 6\int (a^2+x^2)^{5/2} dx = x(a^2+x^2)^{5/2} + 5a^2 \int (a^2+x^2)^{3/2} dx$

or $\quad \int (a^2+x^2)^{5/2} dx = \frac{x(a^2+x^2)^{5/2}}{6} + \frac{5a^2}{6} \int (a^2+x^2)^{3/2} dx$ ...(i)

Similarly, we can find

$$\int (a^2+x^2)^{3/2} dx = \frac{x(a^2+x^2)^{3/2}}{4} + \frac{3a^2}{4} \int (a^2+x^2)^{1/2} dx$$

$$= \frac{x(a^2+x^2)^{3/2}}{4} + \frac{3a^2}{4} \left[\frac{x}{2}\sqrt{(a^2+x^2)} + \frac{a^2}{2} + \sinh^{-1}\frac{x}{a} \right]$$

Putting in (i), we get

$$\int (a^2 + x^2)^{5/2} \, dx = \frac{x(a^2+x^2)^{5/2}}{6} + \frac{5a^2}{6} \left[\frac{x(a^2+x^2)^{3/2}}{4} \right.$$

$$\left. + \frac{3a^2}{8} \left\{ x\sqrt{(a^2+x^2)} + a^2 \sinh^{-1}\frac{x}{a} \right\} \right]$$

$$= \frac{x(a^2+x^2)^{5/2}}{6} + \frac{5a^2}{24} x(a^2+x^2)^{3/2} + \frac{5a^4}{16} x\sqrt{(a^2+x^2)} + \frac{5a^4}{16} \sinh^{-1}\frac{x}{a}$$

3. *Prove that* $\int_0^\infty \frac{dx}{(1+x^2)^4} = \frac{5\pi}{32}$

Sol. From the reduction formula, we have

$$\int \frac{dx}{(x^2+a^2)^n} = \frac{x}{2(n-1)a^2 (x^2+a^2)^{n-1}} + \frac{2n-3}{2(n-1)a^2} \int \frac{dx}{(x^2+a^2)^{n-1}}$$

Now putting $a^2 = 1$ and taking limits from 0 to ∞, we get

$$\int_0^\infty \frac{dx}{(x^2+1)^n} = \left[\frac{x}{2(n-1)(x^2+1)^{n-1}} \right]_0^\infty + \frac{2n-3}{2(n-1)} \int_0^\infty \frac{dx}{(x^2+1)^{n-1}}$$

$$= \frac{2n-3}{2(n-1)} \int_0^\infty \frac{dx}{(x^2+1)^{n-1}}$$

i.e., $I_n = \frac{2n-3}{2(n-1)} I_{n-1}$

Putting successively $n = 4, 3, 2$ we get

$$I_4 = \frac{5}{6} I_3 = \frac{5}{6} \cdot \frac{3}{4} I_2 = \frac{5}{8} I_2$$

$$= \frac{5}{8} \cdot \frac{1}{2} I_1 = \frac{5}{16} I_1 \Rightarrow I_4 = \frac{5}{16} I_1$$

But $\quad I_1 = \int_0^\infty \frac{dx}{x^2+1} = [\tan^{-1} x]_0^\infty = \frac{\pi}{2}$

$\therefore \quad I_4 = \frac{5}{16} \cdot \frac{\pi}{2} = \frac{5\pi}{32}$

Hence $\quad \int_0^\infty \frac{dx}{(x^2+1)^2} = \frac{5\pi}{32}$

EXERCISES

1. Show that $\int_0^a (x^2+a^2)^{5/2} = \frac{67\sqrt{2}}{48} a^6 + \frac{5a^6}{16} \log(1+\sqrt{2})$

2. If $I_n = \int_0^a (a^2-x^2)^n \, dx$ and $n > 0$ prove that $I_n = \frac{2na^2}{2n+1} I_{n-1}$

Hence show that $\int_0^a (a^2-x^2)^3 \, dx = \frac{16}{35} a^7$

Reduction Formulae

3. If $U_n = \int x^n \sqrt{(a^2 - x^2)}\, dx$, prove that

$$U_n = -\frac{x^{n-1}(a^2 - x^2)^{3/2}}{n+2} + \frac{(n-1)\, a^2}{n+2} U_{n-2}$$

Hence or otherwise evaluate $\int_0^a x^4 \sqrt{(a^2 - x^2)}\, dx$ $\qquad$ $\left[\text{Ans. } \dfrac{\pi a^6}{32}\right]$

4. Prove that

(i) $\int_0^\infty \dfrac{dx}{(a^2 + x^2)^4} = \dfrac{5\pi}{32 a^7}$ $\qquad$ (ii) $\int_0^\infty \dfrac{dx}{(1 + x^2)^5} = \dfrac{35\pi}{256}$

5. Find a reduction formula for $\int \dfrac{x^m\, dx}{(a^3 + x^3)^{n/3}}$ and obtain the value of $\int x^8 (x^3 - 1)^{-1/3}\, dx$.

$\left[\text{Ans. } \dfrac{x^{m-2}}{(m+n+1)(a^3 x^3)^{n/3 - 1}} - \dfrac{(m-2)\, a^3}{m-n+1} \int \dfrac{x^{m-3}}{(a^3 + x^3)^{n/3}}\, dx;\ \dfrac{(x^3 - 1)^{2/3}}{40}(5x^6 + 6x^2 + 9)\right]$

6. If I_n denotes $\int_0^1 x^p (1 - x^q)^n\, dx$ where p, q and n are positive, prove that
$(nq + p + 1)\, I_n = nq\, I_{n-1}$

   Evaluate I_n when n is a positive integer.

7. If $U_{m,n} = \int_0^1 x^m (1 - x)^n\, dx$ prove that $U_{m,n} = U_{n,m}$ and $(m + n + 1)\, U_{m,n} = n\, U_{m, n-1}$

   Hence deduce that $U_{m,n} = \dfrac{m!\, n!}{(m + n + 1)!}$

8. If $\phi(n) = a^{2n} \int_0^\infty \dfrac{dx}{(x^2 + a^2)^n}$ prove that $\phi(n) = \dfrac{2n - 3}{2n - 2} \phi(n - 1)$.

❏❏❏

4
Definite Integrals

4.1 DEFINITE INTEGRALS

$\int_a^b \phi(x)\,dx$ is defined as a definite integral which is equal to $F(b) - F(a)$, $F(x)$ is an integral of $\phi(x)$.

It can be shown as follows :

$$\int_a^b \phi(x)\,dx = [F(x) + c]_a^b$$
$$= [F(b) + c] - [F(a) + c]$$
$$= F(b) - F(a)$$

Thus the value of $\int_a^b \phi(x)\,dx$ does not depend upon the value of the arbitrary constant.

It is generally read as the *"Integral from a to b of $\phi(x)$"* and b is called the *upper limit* and a, *lower limit*. The interval (a, b) *is called the range of integration.*

4.2 PROPERTIES OF DEFINITE INTEGRALS

1. Prove that $\int_a^b \phi(x)\,dx = \int_a^b \phi(t)\,dt$

Let $\int \phi(x)\,dx = F(x) + c_1$, there by

$\int \phi(t)\,dt = F(t) + c_2$

Therefore $\int_a^b \phi(x)\,dx = [F(x) + c_1]_a^b$

$= F(b) - F(a)$

Similarly, $\int_a^b \phi(t)\,dt = [F(t) + c_2]_a^b$

$= F(b) - F(a)$

$\therefore \quad \int_a^b \phi(x)\,dx = \int_a^b \phi(t)\,dt$

2. Prove that $\int_a^b \phi(x)\,dx = \int_a^c \phi(x)\,dx + \int_c^b \phi(x)\,dx$

R.H.S. We have $\int_a^c \phi(x)\,dx + \int_c^b \phi(x)\,dx$

$= [F(x) + c]_a^c + [F(x) + c]_c^b$
$= F(c) - F(a) + F(b) - F(c)$
$= F(b) - F(a) = \int_a^b \phi(x)\,dx$

$\Rightarrow$ R.H.S. = L.H.S.

3. Prove that $\int_a^b \phi(x)\,dx = -\int_b^a \phi(x)\,dx$

R.H.S. $= -\int_b^a \phi(x)\,dx = -[F(x) + c]_b^a = -[\{F(a) + c\} - \{F(b) + c\}]$

$= -F(a) + F(b)$

$= \int_a^b \phi(x)\,dx.$ $\Rightarrow$ R.H.S. = L.H.S.

Definite Integrals

4. Prove that $\int_0^a \phi(x)\, dx = \int_0^a \phi(a-x)\, dx$

R.H.S. $= \int_0^a \phi(a-x)\, dx$

Put $a - x = t, -dx = dt$

Upper limit for $x = a, t = 0$

Lower limit for $x = 0, t = a$

∴ R.H.S. $= -\int_a^0 \phi(t)\, dt$

$\quad\quad\quad\quad = \int_0^a \phi(t)\, dt$ \quad\quad [by Prop. 3]

$\quad\quad\quad\quad = \int_0^a \phi(x)\, dx$ \quad\quad [by Prop. 1]

$\quad\quad\quad\quad = $ L.H.S.

5. Prove that $\int_0^{2a} \phi(x)\, dx = 2 \int_0^a \phi(x)\, dx$ if $\phi(2a - x) = \phi(x)$ and $= 0$ if $\phi(2a - x) = -\phi(x)$.

We have

L.H.S. $= \int_0^{2a} \phi(x)\, dx = \int_0^a \phi(x)\, dx + \int_a^{2a} \phi(x)\, dx$ \quad\quad ...(i)

Now consider $\int_a^{2a} \phi(x)\, dx$.

Put $2a - x = t$ or $2a - t = x, -dx = dt$, Upper limit for $x = 2a, t = 0$ and lower for $x = a, t = a$

∴ $\int_a^{2a} \phi(x)\, dx = -\int_a^0 \phi(2a - t)\, dt$

$\quad\quad\quad\quad = \int_0^a \phi(2a - t)\, dt$

$\quad\quad\quad\quad = \int_0^a \phi(2a - x)\, dx$

Then \quad L.H.S. $= \int_0^a \phi(2a - x)\, dx + \int_0^a \phi(x)\, dx$ \quad\quad ...(ii)

Now, if $\phi(2a - x) = \phi(x)$,

Then \quad $\int_0^{2a} \phi(x)\, dx = \int_0^a \phi(x)\, dx + \int_0^a \phi(x)\, dx$

$\quad\quad\quad\quad = 2 \int_0^a \phi(x)\, dx.$

Again, if $\quad \phi(2a - x) = -\phi(x)$

Then \quad $\int_0^{2a} \phi(x)\, dx = \int_0^a \phi(x)\, dx - \int_0^a \phi(x)\, dx$

$\quad\quad\quad\quad = 0.$

6. Prove that $\int_{-a}^a \phi(x)\, dx = 0$, if $\phi(-x) = -\phi(x)$

$\quad\quad\quad\quad\quad\quad = 2 \int_0^a \phi(x)\, dx$, if $\phi(-x) = \phi(x)$

We have $\quad \int_{-a}^a \phi(x)\, dx = \int_{-a}^0 \phi(x)\, dx + \int_0^a \phi(x)\, dx.$

New consider $\int_{-a}^0 \phi(x)\, dx$ Put $x = -t, dx = -dt$, upper limit for $x = 0, t = 0$, lower limit for $x = -a, t = a$.

Now $\quad \int_{-a}^0 \phi(x)\, dx = -\int_a^0 \phi(-t)\, dt = \int_0^a \phi(-t)\, dt$

$\quad\quad\quad\quad = \int_0^a \phi(-x)\, dx$ \quad\quad\quad\quad\quad\quad (by prop. 1)

∴ $\quad \int_{-a}^a \phi(x)\, dx = \int_0^a \phi(-x)\, dx + \int_0^a \phi(x)\, dx$

If $\phi(-x) = -\phi(x)$, then

$$\int_{-a}^{a} \phi(x)\,dx = -\int_0^a \phi(x)\,dx + \int_0^a \phi(x)\,dx$$

$$= 0$$

and if $\phi(-x) = \phi(x)$, than

$$\int_{-a}^{a} \phi(x)\,dx = \int_0^a \phi(x)\,dx + \int_0^a \phi(x)\,dx$$

$$= 2\int_0^a \phi(x)\,dx.$$

Note. The following results should be committed to memory.

(i) $\int_0^{\pi/2} f(\sin x)\,dx = \int_0^{\pi/2} f\left\{\sin\left(\dfrac{\pi}{2} - x\right)\right\}dx$

$$= \int_0^{\pi/2} f(\cos x)\,dx$$

(ii) $\int_0^{\pi/2} f(\sin 2x) \cos x\,dx$

$$= \int_0^{\pi/2} f\sin\left\{2\left(\dfrac{\pi}{2} - x\right)\right\}\cos\left(\dfrac{\pi}{2} - x\right)dx$$

$$= \int_0^{\pi/2} f(\sin 2x) \sin x\,dx$$

(iii) $\int_0^{\pi/2} \sin^n x\,dx = \int_0^{\pi/2} \sin^n\left(\dfrac{\pi}{2} - x\right)dx$

$$= \int_0^{\pi/2} \cos^n x\,dx.$$

EXAMPLES

1. Evaluate $\int_0^\pi \dfrac{x\sin x}{1 + \cos^2 x}\,dx$

Sol. Let $I = \int_0^\pi \dfrac{x\sin x}{1 + \cos^2 x}\,dx$

$$= \int_0^\pi \dfrac{(\pi - x)\sin(\pi - x)}{1 + \cos^2(\pi - x)}\,dx$$

$$= \int_0^\pi \dfrac{(\pi - x)\sin x}{1 + \cos^2 x}\,dx$$

$$= \pi\int_0^\pi \dfrac{\sin x}{1 + \cos^2 x}\,dx - I$$

$$\therefore\quad 2I = 2\pi\int_0^\pi \dfrac{\sin x}{1 + \cos^2 x}\,dx$$

$$\therefore\quad I = -\pi[\tan^{-1}(\cos x)]_0^{\pi/2}\ \dfrac{\pi^2}{4}.$$

2. Evaluate $\int_0^\pi \dfrac{\sin^2 x}{\sin x + \cos x}\,dx$

Sol. $I = \int_0^\pi \dfrac{\sin^2 x}{\sin x + \cos x}\,dx$

Definite Integrals

$$I = \int_0^\pi \frac{\sin^2\left(\frac{\pi}{2} - x\right)}{\sin\left(\frac{\pi}{2} - x\right) + \cos\left(\frac{\pi}{2} - x\right)} dx$$

$$= \int_0^\pi \frac{\cos^2 x}{\sin x + \cos x} dx$$

$$\therefore \quad 2I = \int_0^\pi \frac{\sin^2 x + \cos^2 x}{\sin x + \cos x} dx$$

$$= \int_0^\pi \frac{dx}{\sin x + \cos x}$$

$$= \frac{1}{\sqrt{2}} \int_0^\pi \frac{dx}{\frac{1}{\sqrt{2}} \sin x + \frac{1}{\sqrt{2}} \cos x}$$

$$= \frac{1}{\sqrt{2}} \int_0^\pi \frac{dx}{\sin\left(x + \frac{\pi}{4}\right)}$$

$$= \frac{1}{\sqrt{2}} \int_0^\pi \mathrm{cosec}\left(x + \frac{\pi}{4}\right) dx$$

$$= \frac{1}{\sqrt{2}} \left[\log \tan\left(\frac{1}{2} x + \frac{1}{8}\pi\right)\right]_0^{\pi/2}$$

$$= \frac{1}{\sqrt{2}} \left[\log \tan \frac{3}{8}\pi - \log \tan \frac{1}{8}\pi\right]$$

$$= \frac{1}{\sqrt{2}} \left[\log \frac{\tan \frac{3}{8}\pi}{\tan \frac{1}{8}\pi}\right]$$

$$= \frac{1}{\sqrt{2}} \left[\log \frac{\tan\left(\frac{\pi}{2} - \frac{1}{8}\pi\right)}{\tan \frac{1}{8}\pi}\right]$$

$$= \frac{1}{\sqrt{2}} \log \frac{\cot \frac{\pi}{8}}{\tan \frac{\pi}{8}} = \frac{1}{2} \log \frac{\cos^2 \frac{1}{8}\pi}{\sin^2 \frac{1}{8}\pi}$$

$$= \frac{1}{\sqrt{2}} \log \frac{1 + \cos \frac{\pi}{4}}{1 - \cos \frac{\pi}{4}} = \frac{1}{\sqrt{2}} \log \left(\frac{\sqrt{2}+1}{\sqrt{2}-1}\right)$$

$$= \sqrt{2} \log (\sqrt{2} + 1).$$

$$\therefore \quad I = \frac{1}{\sqrt{2}} \log(\sqrt{2} + 1).$$

EXERCISES

Evaluate

1. $\int_0^\pi \cos^6 x \, dx$ $\qquad\left[\text{Ans. } \dfrac{5\pi}{16}\right]$

2. $\int_0^\pi \sin^4 x \, dx$ $\qquad\left[\text{Ans. } \dfrac{3}{8}\pi\right]$

3. $\int_0^\pi \theta \sin^3 \theta \, d\theta$ $\qquad\left[\text{Ans. } \dfrac{2}{3}\pi\right]$

Evaluate the following :

4. $\int_0^{\pi/2} \dfrac{\sqrt{(\sin x)}}{\sqrt{(\sin x)} + \sqrt{(\cos x)}} \, dx$ $\qquad$ [Ans. $\pi/4$]

5. $\int_0^\pi \dfrac{x \, dx}{1 + \sin x}$ $\qquad$ [Ans. π]

6. $\int_0^\pi \dfrac{x}{1 + \cos^2 x} \, dx$ $\qquad\left[\text{Ans. } \dfrac{1}{4}\pi^2\sqrt{2}\right]$

7. $\int_0^\pi \dfrac{x \, dx}{a^2 - \cos^2 x}, \, a > 1$ $\qquad\left[\text{Ans. } \pi^2 / \sqrt{(a^2 - 1)\, 2a}\right]$

8. $\int_0^\pi \dfrac{x \, dx}{(a^2 \cos^2 x + b^2 \sin^2 x)}$ $\qquad\left[\text{Ans. } \dfrac{\pi^2}{2ab}\right]$

9. $\int_0^\pi \dfrac{\sin x}{\sin x + \cos x} \, dx$ $\qquad\left[\text{Ans. } \dfrac{1}{4}\pi\right]$

10. $\int_0^\pi \dfrac{dx}{1 + \cot x}$ $\qquad\left[\text{Ans. } \dfrac{1}{4}\pi\right]$

MORE EXAMPLES

1. *Evaluate* $I = \int_0^{\pi/2} \log \sin x \, dx$

Sol. We have $\qquad I = \int_0^{\pi/2} \log \sin x \, dx \qquad\qquad\qquad …(i)$

$\qquad\qquad\qquad = \int_0^{\pi/2} \log \sin\left(\dfrac{\pi}{2} - x\right) dx$

$\therefore \qquad\qquad\qquad I = \int_0^{\pi/2} \log \cos x \, dx \qquad\qquad\qquad …(ii)$

Adding (i) and (ii), we get

$\qquad\qquad\qquad I + I = \int_0^{\pi/2} \log \sin x \, dx + \int_0^{\pi/2} \log \cos x \, dx$

or $\qquad\qquad\qquad 2I = \int_0^{\pi/2} \log \sin x \cos x \, dx$

$\qquad\qquad\qquad = \int_0^{\pi/2} \log\left(\dfrac{\sin 2x}{2}\right) dx$

$\qquad\qquad\qquad = \int_0^{\pi/2} \log \sin 2x \, dx - \int_0^{\pi/2} \log 2 \, dx$

$\qquad\qquad\qquad = \int_0^{\pi/2} \log \sin 2x \, dx - \log 2 \, [x]_0^{\pi/2}$

$\qquad\qquad\qquad = \int_0^{\pi/2} \log \sin 2x \, dx - \dfrac{\pi}{2} \log 2$

Definite Integrals

Put $\quad 2x = u \quad \therefore \quad 2\,dx = du$

for $\quad x = \dfrac{\pi}{2},\, u = \pi$ (upper limit)

for $\quad x = 0,\, u = 0$ (lower limit)

Then $\quad 2I = \dfrac{1}{2}\int_0^{\pi} \log \sin u\, du - \dfrac{\pi}{2}\log 2$

$\qquad\qquad = \dfrac{2}{2}\int_0^{\pi/2} \log \sin u\, du - \dfrac{\pi}{2}\log 2$

$\qquad\qquad = \int_0^{\pi/2} \log \sin x\, dx - \dfrac{\pi}{2}\log 2$

$\qquad\qquad = I - \dfrac{\pi}{2}\log 2$

$\therefore \qquad I = -\dfrac{\pi}{2}\log 2$

$\therefore \qquad \int_0^{\pi/2} \log \sin x\, dx = -\dfrac{\pi}{2}\log 2 = \dfrac{\pi}{2}\log \dfrac{1}{2}$

Similarly, $\quad \int_0^{\pi/2} \log \cos x\, dx = -\dfrac{\pi}{2}\log 2$

$\qquad\qquad\qquad\qquad = \dfrac{\pi}{2}\log \dfrac{1}{2}$

Hence $\int_0^{\pi/2} \log \sin x\, dx = \int_0^{\pi/2} \log \cos x\, dx = -\dfrac{1}{2}\pi \log 2$

2. Evalaute $\int_0^{\infty} \log\left(x + \dfrac{1}{x}\right) \dfrac{dx}{1+x^2}$

Sol. We have $I = \int_0^{\infty} \log\left(x + \dfrac{1}{x}\right) \dfrac{dx}{1+x^2}$.

Put $x = \tan \theta,\, dx = \sec^2 \theta\, d\theta$,

Upper limit $\infty = \tan \theta,\, \theta - \dfrac{\pi}{2}$

Lower limit $0 = \tan \theta,\, \theta = 0$.

$\therefore \quad I = \int_0^{\pi/2} \log\left(\tan \theta + \dfrac{1}{\tan \theta}\right) \dfrac{\sec^2 \theta\, d\theta}{1 + \tan^2 \theta}$

$\qquad = \int_0^{\pi/2} \log\left(\dfrac{\tan^2 \theta + 1}{\tan \theta}\right) d\theta$

$\qquad = \int_0^{\pi/2} \log \dfrac{1}{\sin \theta \cos \theta}\, d\theta$

$\qquad = -\int_0^{\pi/2} \log \sin \theta\, d\theta - \int_0^{\pi/2} \log \cos \theta\, d\theta$

$\qquad = \dfrac{\pi}{2}\log 2 + \dfrac{\pi}{2}\log 2$

$\qquad = \pi \log 2.$ $\hfill$ (from example 1)

EXERCISES

Evalaute the following :

1. $\int_0^\pi x \log(\sin x)\, dx$ $\left[\text{Ans. } -\dfrac{1}{2}\pi^2 \log 2\right]$

2. $\int_0^{\pi/2} \log \tan x\, dx$ [Ans. 0]

3. $\int_0^1 \dfrac{\sin^{-1} x}{x}\, dx$ $\left[\text{Ans. } \dfrac{1}{2}\pi \log 2\right]$

4. $\int_0^{\pi/2} x^2 \csc^2 x\, dx$ $[\text{Ans. } \pi \log 2]$

5. $\int_0^{\pi/2} \left(\dfrac{\theta}{\sin \theta}\right)^2 d\theta$ $[\text{Ans. } \pi \log 2]$

6. $\int_0^{\pi/2} \log \cot x\, dx$ [Ans. 0]

7. Prove that (i) $\int_0^{\pi/4} \log(1+\tan\theta)\, d\theta = \dfrac{\pi}{8} \log_e 2$

   (ii) $\int_1^0 \dfrac{\log(1+x)}{1+x^2}\, dx = -\dfrac{\pi}{8} \log_e 2$

8. Show that $\int_0^\pi \dfrac{x \tan x\, dx}{\sec x + \tan x} = \pi\left(\dfrac{\pi}{2} - 1\right)$

9. Prove that $\int_0^\pi \dfrac{x \tan x}{\sec x + \tan x}\, dx = \dfrac{\pi^2}{4}$

10. Prove that $\int_0^\pi \dfrac{x\, dx}{1 + \cos\alpha \sin x} = \dfrac{\pi\alpha}{\sin\alpha}$

11. Prove that $\int_0^\pi \dfrac{x \sin x}{1 + \sin x}\, dx = \pi\left(\dfrac{\pi}{2} - 1\right)$

4.3 SUMMATION OF SERIES

So far integration has simply been regarded as the inverse of differentaition. Now, we shall define the definite integral as

$$\int_a^b f(x)\, dx = \lim_{h \to 0} h\{f(a) + f(a+h) + f(a+2h) + \ldots + f[a+(n-1)h]\}$$

$$= \lim_{h \to 0} h \sum_{r=0}^{n-1} f(a+rh)$$

i.e., if $f(x)$ is a continuous and single valued function in the interval (a, b), where a and b are finite and $a < b$, and if the interval (a, b) be divided into n equal parts, each of width h, we have

$nh = b - a$, then

$$\int_a^b f(x)\, dx = \lim_{h \to 0} h \sum_{r=0}^{n-1} f(a + rh)$$

Putting $a = 0$ and $b = 1$, the above reduces to

$$\int_0^1 f(x)\, dx = \lim_{h \to 0} h \sum_{r=0}^{n-1} f(rh),$$

Definite Integrals

where $nh = 1$ or $h = \dfrac{1}{n}$

$$= \lim_{h \to 0} \frac{1}{n} \sum_{r=0}^{n-1} f\left(\frac{r}{n}\right) \qquad \because h = \frac{1}{n}$$

$$= \lim_{n \to \infty} \sum_{r=0}^{n-1} \frac{1}{n} f\left(\frac{r}{n}\right)$$

Hence $\quad \lim_{n \to \infty} \sum_{r=0}^{n-1} \dfrac{1}{n} f\left(\dfrac{r}{n}\right) = \int_0^1 f(x)\, dx$

Note 1. In order that a series may be summed up by this formula, it must possess the following properties :

(i) Each term of the series must be multiplied by $\dfrac{1}{n}$.

(ii) All the terms should be some function of $\dfrac{r}{n}$ which vary from term to term in A.P. with common difference $\dfrac{1}{n}$.

(iii) The number of terms in the given series should be n, but if, however, the number of terms is less by one or two, then required sum will not be changed because each term tends to zero.

Note 2. Working Rule. To write down the corresponding definite integral, replace $\dfrac{r}{n}$ by x, $\dfrac{1}{n}$ by dx and $\underset{n \to \infty}{\mathrm{Lt}} \Sigma$ by the sign of integration, *i.e.*, by $\int$. To choose the limits of integration insert the values of $\dfrac{r}{n}$ for the first and the last terms, as $n \to \infty$. These will be the lower and upper limits respectively.

EXAMPLES

1. *Evaluate the limit when* $n \to \infty$ *of the series*

$$\frac{1}{1+n^3} + \frac{4}{8+n^3} + \frac{9}{27+n^3} + \ldots + \frac{r^2}{r^3+n^3} + \ldots + \frac{1}{2n}.$$

Sol. rth term $= \dfrac{r^2}{r^3 + n^3} = \dfrac{1}{n^3}\left[\dfrac{r^2}{(r/n)^3 + 1}\right]$

$$= \frac{1}{n}\left[\frac{(r/n)^2}{(r/n)^3 + 1}\right]$$

$\therefore$ The given limit $= \lim_{x \to \infty} \sum_{r=1}^{n} \dfrac{1}{n}\left\{\dfrac{(r/n)^2}{(r/n)^3 + 1}\right\}$

$$= \int_0^1 \frac{x^2\, dx}{x^3 + 1} = \left[\frac{1}{3} \log(x^3 + 1)\right]_0^1 = \frac{1}{3} \log 2$$

2. *Evaluate :* $\lim_{x \to \infty} \sum_{r=1}^{n-1} \dfrac{1}{n} \sqrt{\left(\dfrac{n+r}{n-r}\right)}$

Sol. $\lim_{x \to \infty} \sum_{r=1}^{n-1} \frac{1}{n} \sqrt{\left(\frac{n+r}{n-r}\right)} = \lim_{x \to \infty} \sum_{r=1}^{n-1} \frac{1}{n} \sqrt{\left(\frac{1+r/n}{1-r/n}\right)}$

$= \int_0^1 \sqrt{\left(\frac{1+x}{1-x}\right)} dx = \int_0^1 \frac{(1+x)\, dx}{\sqrt{(1-x^2)}}$

$= \int_0^1 \frac{dx}{\sqrt{(1-x^2)}} + \int_0^1 \frac{x\, dx}{\sqrt{(1-x^2)}}$

$= [\sin^{-1} x]_0^1 - [-(1-x^2)^{1/2}]_0^1$

$= \frac{\pi}{2} + (0+1) = \frac{\pi+2}{2}$

3. *Prove that*

$= \underset{n \to \infty}{\text{Lim}} \left\{ \left(1 + \frac{1}{n^2}\right)\left(1 + \frac{2^2}{n^2}\right)\left(1 + \frac{3^2}{n^2}\right) \cdots \left(1 + \frac{n^2}{n^2}\right) \right\}^{1/n}$

is equal to $2e^{(\pi-r)/2}$.

Sol. Let the given product be A.

Then $\log A = \underset{n \to \infty}{\text{Lim}} \sum_{r=1}^{n} \frac{1}{n} \log\left(1 + \frac{r^2}{n^2}\right)$

$= \int_0^1 \log(1+x^2)\, dx$

$= [x \log(1+x)^2]_0^1 - \int \frac{2x^2}{1+x^2} dx$

$= \log 2 - 2 \int_0^1 \frac{x^2+1-1}{1+x^2} dx$

$= \log 2 - 2[x - \tan^{-1} x]_0^1$

$= \log 2 - 2 + \frac{\pi}{2} = \log 2 + \log e^{(\pi-4)/2}$

∴ $\log A = \log 2e^{(\pi-4)/2}$

Hence $A = 2e^{(\pi-r)/2}$

4. *Apply the definition of a definite integral as the limit of a sum to evaluate.*

$\underset{n \to \infty}{\text{Lim}} \left(\frac{n!}{n^n}\right)^{1/n}$

Sol. Let $A = \underset{n \to \infty}{\text{Lim}} \left(\frac{n!}{n^n}\right)^{1/n}$

$= \underset{n \to \infty}{\text{Lim}} \left\{\frac{1.2.3.4.5\ldots n}{n^n}\right\}^{1/n}$

$= \underset{n \to \infty}{\text{Lim}} \left\{\left(\frac{1}{n}\right)\left(\frac{2}{n}\right)\left(\frac{3}{n}\right)\cdots\left(\frac{n}{n}\right)\right\}^{1/n}$

Definite Integrals

$$\therefore \quad \log A = \underset{n\to\infty}{\text{Lim}} \frac{1}{n}\left[\log\left(\frac{1}{n}\right) + \log\left(\frac{2}{n}\right) + \log\left(\frac{3}{n}\right) + \ldots + \log\left(\frac{n}{n}\right)\right]$$

$$= \underset{n\to\infty}{\text{Lim}} \sum_{r=1}^{n} \frac{1}{n} \log\left(\frac{r}{n}\right) = \int_0^1 \log x \, dx$$

$$= [(\log x) \cdot x]_0^1 - \int_0^1 \frac{1}{x} \cdot x \, dx$$

integrating by parts

$$= 0 - \int_0^1 dx = -[x]_0^1 = -1$$

$$\therefore \quad A = e^{-1} = \frac{1}{e}.$$

5. *Show that the limit of the sum* $\dfrac{1}{n} + \dfrac{1}{n+1} + \dfrac{1}{n+2} + \ldots + \dfrac{1}{3n}$, *when n is indefinitely increased is log 3.*

Sol. In the given series, we have $(3n - n) + 1 = (2n + 1)$ terms.

Now, $\quad r$th term $= \dfrac{1}{n+r} = \dfrac{1}{n}\left\{\dfrac{1}{1+(r/n)}\right\}$

$\therefore$ The required limit $= \underset{n\to\infty}{\text{Lim}} \sum_{r=0}^{2n} \dfrac{1}{n}\left\{\dfrac{1}{1+(r/n)}\right\}$

Here for the corresponding definite integral the lower limit $= \underset{n\to\infty}{\text{Lim}}\left(\dfrac{r}{n}\right)$ for the first term

$$= \underset{n\to\infty}{\text{Lim}}\left(\frac{0}{n}\right) = 0 \quad (\because r = 0 \text{ for the last term})$$

Upper limit $= \underset{n\to\infty}{\text{Lim}}\left(\dfrac{r}{n}\right)$ for the last term

$$= \underset{n\to\infty}{\text{Lim}}\left(\frac{2n}{n}\right) \quad (\because r = 2n \text{ for the last term})$$

$$= 2$$

$\therefore$ The required limit $= \int_0^2 \dfrac{dx}{(1+x)}$

$$= [\log(1+x)]_0^2 = \log 3.$$

EXERCISES

Find the limit when $n \to \infty$ of the series

1. $\dfrac{1}{n+1} + \dfrac{1}{n+2} + \dfrac{1}{n+3} + \ldots + \dfrac{1}{2n}$ $\quad$ [Ans. log 2]

2. $\dfrac{1}{n} + \dfrac{1}{n+1} + \dfrac{1}{n+2} + \ldots + \dfrac{1}{n+(n+1)}$. $\quad$ [Ans. log 2]

3. $\dfrac{\sqrt{n}}{n^{3/2}} + \dfrac{\sqrt{n}}{(n+3)^{3/2}} + \dfrac{\sqrt{n}}{(n+6)^{3/2}} + \ldots + \dfrac{\sqrt{n}}{\{n+3(n-1)\}^{3/2}}$ $\quad \left[\text{Ans. } \dfrac{1}{3}\right]$

4. $\dfrac{1}{n^3}+\dfrac{4}{n^3}+\dfrac{9}{n^3}+\dfrac{16}{n^3}+\ldots+\dfrac{n^2}{n^3}$ $\left[\text{Ans. } \dfrac{1}{3}\right]$

5. $\dfrac{n}{n^2}+\dfrac{n}{n^2+1^2}+\dfrac{n}{n^2+2^2}+\ldots+\dfrac{n}{n^2+(n+1)^2}$. $\left[\text{Ans. } \dfrac{1}{4}\pi\right]$

6. Evaluate $\displaystyle\lim_{n\to\infty}\sum_{r=1}^{n-1}\dfrac{1}{\sqrt{(n^2-r^2)}}$ $\left[\text{Ans. } \dfrac{1}{2}\pi\right]$

7. $\dfrac{1}{n}\left\{\sin^{2K}\dfrac{\pi}{2n}+\sin^{2K}\dfrac{2\pi}{2n}+\sin^{2K}\dfrac{3\pi}{2n}+\ldots+\sin^{2K}\dfrac{\pi}{2}\right\}$. $\left[\text{Ans. } \dfrac{(2k)!}{(2^k \cdot k!)^2}\right]$

8. $\displaystyle\lim_{n\to\infty}\left[\left(1+\dfrac{1}{n}\right)\left(1+\dfrac{2}{n}\right)\left(1+\dfrac{3}{n}\right)\ldots\left(1+\dfrac{n}{n}\right)\right]^{1/n}$ $\left[\text{Ans. } \dfrac{4}{e}\right]$

9. $\displaystyle\lim_{n\to\infty}\left[\left(1+\dfrac{1}{n}\right)\left(1+\dfrac{2}{n}\right)^{1/2}\left(1+\dfrac{3}{n}\right)^{1/3}\ldots\left(1+\dfrac{n}{n}\right)^{1/n}\right]$ $\left[\text{Ans. } e^{\pi^2/12}\right]$

10. $\displaystyle\lim_{n\to\infty}\left[\tan\dfrac{\pi}{2n}\tan\dfrac{2\pi}{2n}\tan\dfrac{3\pi}{2n}\ldots\tan\dfrac{n\pi}{2n}\right]^{1/n}$ [Ans. 1]

☐☐☐

5
Area of Curves (Quadrature)

5.1 QUADRATURE
The process of finding the area bounded by any portion of a plane curve is called **quadrature**.

5.2 AREA IN CARTESIAN CO-ORDINATES

Let $y = f(x)$ is the equation of a curve and it is required to find the area enclosed by the curve, the axis of x and the ordinates $x = a$ and $x = b$. This area may be called the area 'under the curve' between two ordinates.

Let A be the area from $x = a$ upto any ordinate $x = x$. When x increases by δx, A increases by δA, the area of the strip of width δx. Even when a small interval δx in x, y varies as the strip is not of uniform height. Let y_1 and y_2 be the least and greatest values

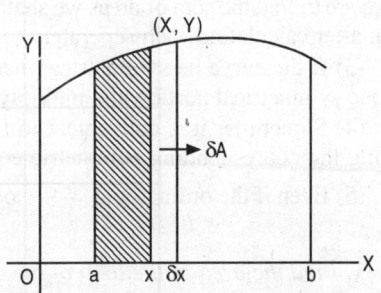

of y in the interval δx. Then δA is greater than $y_1 \delta x$ and less than $y_2 \delta x$. Hence $\dfrac{\delta A}{\delta x}$ is between y_1 and y_2.

Now as $\delta x \to 0$ all values of y in the interval tend to the value at $x = x$, i.e., y_1 and y_2 have a common limit which must therefore be the limit of $\dfrac{\delta A}{\delta x}$.

Hence $$\dfrac{\delta A}{\delta x} = y = f(x).$$

Let $\phi(x)$ be any function whose derivative is $f(x)$.

Then $\qquad A = \phi(x) + c$

But $A = 0$, when

Therefore $\qquad 0 = \phi(a) + c, c = -\phi(a).$

There by $\qquad A = \phi(x) - \phi(a)$

Tus we have obtained an expression for the area under the curve from $x = a$ upto any stage $x = x$.

Now this area from $x = a$ upto $x = b$ is $\phi(b) - \phi(a)$

This is also the result of definite integral $\int_a^b f(x)\,dx$.

Therefore the area of the curve bounded by the curve $y = f(x)$, the x-axis is and the ordinates $x = a$ and $x = b$ is given by

$$A = \int_a^b f(x)\,dx = \int_a^b y\,dx.$$

466 Integral Calculus

Similarly, it can be shown that the area bounded by the curve $x = f(y)$, the y-axis and the lines $y = a$ and $y = b$ is

$$\int_a^b f(y)\, dy = \int_a^b x\, dy.$$

5.2.1 Some Important Notes

(1) In choosing the limits of integration the smaller value of x at which the ordinate is drawn will be taken as the lower limit and the greater as the as the upper limit. In other words, while choosing the values of x for limits of integration we shall move from left to right on x-axis and from depth to height while choosing the values of y for limits of integration.

(2) We take into consideration only the numerical value of the areas and not their algebraical values and as such we shall discard the negative sign if some area comes out to be negative. Similarly if we are to find the sum of areas we shall find their arithmetical, *i.e.*, numerical sum and not algebraical sum after calculating each seperately.

(3) If the curve be symmetrical then to avoid all confusion we shall endeavour to find the area of one symmetrical portion and multiply it by n if there be n such symmetrical portions.

(4) Sometimes it is convenient to find the area of curves whose cartesian equation is given by taking the cooresponding parametric equations.

(5) Even if the ordinates at $x = a$ or $x = b$ be infinite the above formulae shall be true.

EXAMPLES

1. *Find the area of the loop of the curve*

$$xy^2 + (x + a)^2 (x + 2a) = 0.$$

Sol. The equation of the curve is $y^2 = \dfrac{-(x+a)^2 (x+2a)}{x}$.

The following conclusions may be drawn from the equation of the curve

(i) The curve contains even degree of y, *i.e.*, y^2, thereby the curve is symmetrical about x-axis.

(ii) The curve passes through $(-a, 0)$ and $(-2a, 0)$ when $y = 0$ is substituted in the curve.

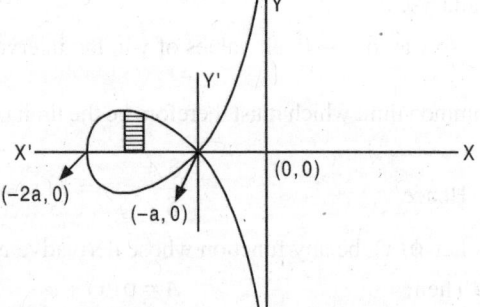

Hence the required area of the loop
$= 2 \int_{-2a}^{-a} y\, dx$, as the limits are from $-2a$ to $-a$, the loop lies between these points.

$\therefore \quad A = 2 \int_{-2a}^{-a} \sqrt{\left\{ -\dfrac{(x+a)^2 (x+2a)}{x} \right\}}\, dx$

Put $x + 2a = t$, then $dx = dt$, then

$A = 2 \int_0^a \sqrt{\dfrac{-(t-a)^2}{(t-2a)}}\, dt$

$= 2 \int_0^a (t-a) \sqrt{\dfrac{t}{2a-t}}\, dt$

$= -2 \int_0^a (a-t) \sqrt{\left\{ \dfrac{a-a+t}{a+(a-t)} \right\}}\, dt$

Area of Curves (Quadrature)

Now, put $a - t = a \sin \theta$, then $-dt = a \cos \theta \, d\theta$

$$\therefore \quad A = -2a^2 \int_0^{\pi/2} \sqrt{\left(\frac{1-\sin\theta}{1+\sin\theta}\right)} \sin\theta \cos\theta \, d\theta$$

$$= -2a^2 \int_0^{\pi/2} \frac{(1-\sin\theta)}{\sqrt{1-\sin^2\theta}} \cdot \sin\theta \cos\theta \, d\theta$$

$$= -2a^2 \int_0^{\pi/2} \sin\theta (1-\sin\theta) \, d\theta$$

$$= -2a^2 \int_0^{\pi/2} \sin\theta \, d\theta + 2a^2 \int_0^{\pi/2} \sin^2\theta \, d\theta$$

$$= -2a^2 \left[\frac{\lceil 1 \cdot \frac{1}{2}}{2 \lceil \frac{3}{2}} - \frac{\lceil \frac{3}{2} \cdot \lceil \frac{1}{2}}{2 \lceil 2} \right]$$

$$= -2a^2 \left(1 - \frac{\pi}{4}\right) = 2\left(\frac{\pi}{4} - 1\right) a^2.$$

2. *Find the area included between the cycloid* $x = a(\theta - \sin\theta)$ *and* $y = a(1 - \cos\theta)$ *and its base.*

Sol. The limits are from 0 to π.

$$\therefore \quad \text{Area} = 2 \int y \, dx = 2 \int_0^y y \frac{dx}{d\theta} \, d\theta$$

Now, we have $\dfrac{dx}{d\theta} = a(1 - \cos\theta)$

$$\therefore \quad \text{Area} = 2 \int_0^\pi a^2 (1 - \cos^2\theta)^2 \, d\theta$$

$$= 2a^2 \int_0^\pi d\theta + 2a^2 \int_0^\pi \cos^2\theta \, d\theta - 4a^2 \int_0^\pi \cos\theta \, d\theta$$

$$= 2a^2 \pi + 4a^2 \int_0^{\pi/2} \cos^2\theta \, d\theta - 4a^2 \times 0$$

$$= 2a^2 \pi + a^2 \pi = 3\pi a^2.$$

3. *Find the area of the portion bounded by the curve* $x(x^2 + y^2) = a(x^2 - y^2)$ *and its asymptote.*

or

find the area of the loop of the curve $y^2(a - x) = x^2(a + x)$

Sol. We have

$$y^2(a - x) = x^2(a + x)$$

or $\quad x(x^2 + y^2) = a(x^2 - y^2)$

i.e., the both curves are same.

We have
(i) The curve is symmetrical about x-axis.
(ii) Curve crosses the x-axis at $(0, 0)$ and $(a, 0)$, hence loop is formed between $(0, 0)$ and $(a, 0)$.

(iii) $x + a = 0$ or $x = -a$ is the asymptote.
The curve is as shown in figure.
Now the area between the curve and its asymptote.

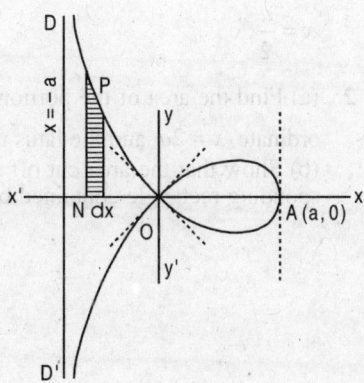

$$= 2 \int_{-a}^{0} y \, dx$$

$$= 2 \int_{-a}^{0} x \sqrt{\left(\frac{a-x}{a+x}\right)} \, dx$$

putting $x = a \sin \theta$, $dx = a \cos \theta \, d\theta$,

$$= 2 \int_{-\pi/2}^{0} a \sin \theta \sqrt{\left(\frac{a - a \sin \theta}{a + a \sin \theta}\right)} \cdot a \cos \theta \, d\theta$$

$$= 2a^2 \int_{-\pi/2}^{0} \sqrt{\left(\frac{1 - \sin \theta}{1 + \sin \theta}\right)} \cdot \sin \theta \cos \theta \, d\theta$$

$$= 2a^2 \int_{-\pi/2}^{0} \frac{(1 - \sin \theta) \sin \theta}{\sqrt{(1 - \sin^2 \theta)}} \cdot \cos \theta \, d\theta$$

$$= 2a^2 \int_{-\pi/2}^{0} (\sin \theta - \sin^2 \theta) \, d\theta$$

$$= 2a^2 \left[-\cos \theta - \frac{\theta}{2} + \frac{\sin 2\theta}{4} \right]_{-\pi/2}^{0}$$

$$= 2a^2 \left[-1 + \frac{\pi}{4} \right] = 2a^2 \left(\frac{1}{4} \pi - 1 \right)$$

and the area of its loop

$$= 2 \int_{0}^{a} y \, dx = 2 \int_{0}^{a} x \sqrt{\left(\frac{a-x}{a+x}\right)} \, dx$$

$$= 2a^2 \left[-\cos \theta - \frac{\theta}{2} + \frac{\sin 2\theta}{4} \right]_{0}^{\pi/2}$$

$$= 2a^2 \left[-\frac{1}{4} \pi + 1 \right] = \frac{1}{2} a^2 (4 - \pi).$$

EXERCISES

1. (a) Find the whole area of the circle $x^2 + y^2 = a^2$.

   (b) Find the areas of the portions into which the circle $x^2 + y^2 = a^2$ is divided by the line $x = \frac{a}{2}$.

   $\left[\text{Ans. (a) } \pi^2 a; \text{(b) } \frac{a^2}{12}(4\pi - 3\sqrt{2}), \frac{a^2}{12}(8\pi + 3\sqrt{3}) \right]$

2. (a) Find the area of the portion of the parabola $y^2 = 4ax$ included between the x, axis, the ordinate $x = 2a$ and the latus rectum.

   (b) Show that the area cut off a parabola by any double ordinate is two-thrids of the corresponding rectangle contained by the double ordinate and its distance from the vertex.

   $\left[\text{Ans. (a) } \frac{4}{3} a^2 (2\sqrt{2} - 1) \right]$

Area of Curves (Quadrature)

3. Find the area bounded by the ellipse $\dfrac{x^2}{a^2} + \dfrac{y^2}{b^2} = 1$, the ordinates $x = c$ and $x = d$.

$$\left[\text{Ans. } \dfrac{b}{a}\left\{ d\sqrt{(a^2-d^2)} - c\sqrt{(a^2-c^2)} + a^2\left(\sin^{-1}\dfrac{d}{a} - \sin^{-1}\dfrac{c}{a}\right)\right\}\right]$$

4. Show that the curve $a^2 y = x^2(x+a)$ includes with the x-axis an area $\dfrac{a^2}{12}$.

5. Find the area of the loop of the following curves :
   (a) $a^4 y^2 = x^4(a^2 - x^2)$.
   (b) $ay^2 = x^2(a-x)$.
   (c) $a^3 y^2 = x^4(b+x)$.
   (d) $a^2 y^2 = x^3(2a - x)$.

$$\left[\text{Ans. (a) } \dfrac{\pi a^2}{8}; \text{ (b) } \dfrac{8a^2}{5}; \text{ (c) } \dfrac{32}{105} b^{-1/2} a^{-3/2}, \text{ (d) } \pi a^2 \right]$$

6. Find the whole area of the curves :
   (a) $a^2 x^2 = y^3(2a - y)$
   (b) $a^2 y^2 = x^2(a^2 - x^2)$.
   (c) $a^2 y^2 = x^2(a^2 - x^2)$.

$$\left[\text{Ans. (a) } \pi a^2; \text{ (b) } \dfrac{4}{3} a^2; \text{ (c) } 3\sqrt{3}\, a^2 \right]$$

7. (i) Find the area of the astroid $x^{2/3} + y^{2/3} = a^{2/3}$ or
   $x = a\cos^3 t,\ y = a\sin^3 t$.

   (ii) Find the area of the hypocycloid $\left(\dfrac{x}{a}\right)^{2/3} + \left(\dfrac{y}{b}\right)^{2/3} = 1$ or the curve
   $x = a\cos^3 t,\ y = b\sin^3 t$.

$$\left[\text{Ans. (i) } \dfrac{3}{8}\pi a^2; \text{ (ii) } \dfrac{3}{8}\pi a b \right]$$

8. Find the areas between the following curves and their asymptotes :
   (a) $y^2(a - x) = x^3$
   (b) $y^2(a - x) = (a - x)^3$
   (c) $y^2(2a - x) = x^3$

$$\left[\text{Ans. (a) } \dfrac{3}{4}\pi a^2; \text{ (b) } 3\pi a^2 \text{ (c) } 3\pi a^2 \right]$$

9. Find the area bounded by the rectangular hyperbola $xy = c^2$, the x-axis, and the ordinates $x = a,\ x = b$.

$$\left[\text{Ans. } c^2 \log\left(\dfrac{b}{a}\right) \right]$$

10. Show that the ordinate $x = a$ divides the area between $y^2(2a - x) = x^3$ and its asymptote into two parts in the ratio $3\pi - 8 : 3\pi + 8$.

5.3 AREA ENCLOSED BY TWO CURVES

Let the two curves $y = \phi_1(x)$ and $y = \phi_2(x)$ intersect in the two points $(a, c), (b, d)$ and let between these points, the first curve lies above the second curve as in fig. (a).

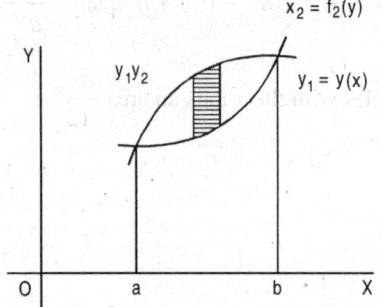
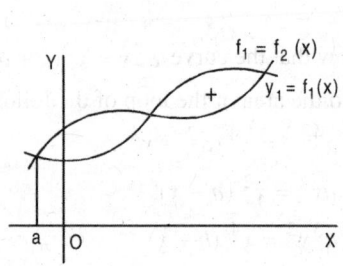

The area between the curves
= area under first curve – area under second curve
$= \int_a^b \phi_1(x)\,dx - \int_a^b \phi_2(x)\,dx$
$= \int_a^b [\phi_1(x) - \phi_2(x)]\,dx$
$= \int_a^b (y_1 - y_2)\,dx.$

Note 1. The curves intersect between (a, c) and (b, d) as in fig. (b) the definite integral $\int_a^b \{\phi_1(x) - \phi_2(x)\}\,dx$ gives the algebraic sum of areas in which each part is assigned positive or negative sign according as the curve $y = \phi_1(x)$ is above or below the curve $y = \phi_2(x)$.

EXAMPLES

1. *Find the area common to two curves* $y^2 = ax$ *and* $x^2 + y^2 = 4ax$.

Sol. $y^2 = ax$ is the equation of the parabola and $x^2 + y^2 = 4ax$ is the equation of the circle. These curves can be shown as in the figure. Now, we have $y^2 = ax, x^2 + y^2 = 4ax$, solving these, we have $x = 0$ or $x = 3a$.

Therefore these two curves intersect at P, where $x = 3a$.

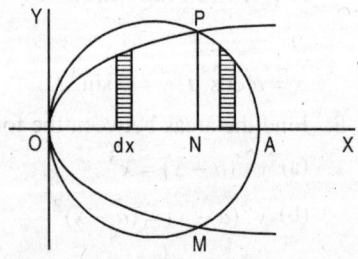

Now the required area
$= 2 \times$ area OPA
$= 2$ (area OPN + area PNA)
[are OPN is completely enclosed by parabola
while area PAN is enclosed by circle]

$= 2 \int_0^{3a} \sqrt{(ax)}\,dx = + 2 \int_0^{4a} \sqrt{(4ax - x^2)}\,dx$

$= \sqrt{a} \left[\dfrac{2}{3} x^{3/2}\right]_0^{3a} + 2 \int_0^{4a} \sqrt{\{4a^2 - (x - 2a^2)\}}\,dx$

$= 4\sqrt{3}\, a^2 + 2 \left[\dfrac{1}{2}(x - 2a)\sqrt{\{4a^2 - (x - 2a)^2\}} + \dfrac{1}{2} 4a^2 \sin^{-1} \dfrac{x - 2a}{2a}\right]_{3a}^{4a}$

Area of Curves (Quadrature)

$$= 4\sqrt{3}\, a^2 + a\left[2a^2 \sin^{-1} 1 - \frac{a^2}{2}\sqrt{3} - 2a^2 \sin^{-1}\frac{1}{2}\right]$$

$$= 4\sqrt{3}a^2 + 4a^2\frac{\pi}{2} - a^2\sqrt{3} - \frac{2\pi a^2}{3}$$

$$= 3\sqrt{3}\,a^2 + \frac{4}{3}\pi a^2 = a^2\left[3\sqrt{3} + \frac{4}{3}\pi\right]$$

2. *Find the area bounded by the parabola $y^2 = 4ax$ and $x^2 = 4ay$.*

Sol. Solving the equations $y^2 = 4ax$ and $x^2 = 4ay$, the points of intersection are $(0, 0)$ and $(4a, 4a)$. The limits of integration are from $x = 0$ to $x = 4a$.

The area bounded between the two parabolas = Area OAPM

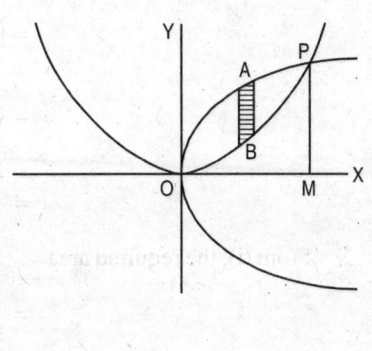

$$= \text{Area } OAPM - \text{area } OBPM$$

$$= \int_0^{4a} \sqrt{(4ax)}\, dx$$

$$= \int_0^{4a} \sqrt{(4ax)}\, dx - \int_0^{4a}\frac{x^2}{4a}\, dx$$

$$= \frac{2}{3}\sqrt{(4a)}\left[x^{3/2}\right]_0^{4a} - \frac{1}{4a}\left[\frac{x^3}{3}\right]_0^{4a}$$

$$= \frac{2}{3}\sqrt{4a}\,[4a\sqrt{4a}] - \frac{1}{4a}\frac{64a^3}{3}$$

$$= \frac{32\,a^2}{3} - \frac{16a^2}{3} = \frac{16a^2}{3}.$$

3. *Find the area of the segment cut off from the parabola $y^2 = 2x$ by the straight line $y = 4x - 1$.*

Sol. In the figure AB is the straight line $y = 4x - 1$, which cuts the parabola at A and B, and the x-axis at C.

Solving the equation of the parabola and equation of line, we get

$$(4x - 1)^2 = 2x$$

or $\qquad 16x^2 - 10x + 1 = 0$

or $\qquad x = \dfrac{1}{2}, \dfrac{1}{8}$, whence $y = 1, -\dfrac{1}{2}$

Therefore the coordinates of A and B are $\left(\dfrac{1}{2}, 1\right)$ and $\left(\dfrac{1}{8}, -\dfrac{1}{2}\right)$

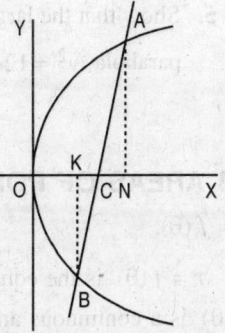

respectively. Also, the coordinates of C are $\left(\dfrac{1}{4}, 0\right)$. AN and BK are perpendiculars from A and B on the x-axis.

$\therefore \qquad AN = 1,\ BK = \dfrac{1}{2}$

Required Area = area OCA + area OCB

Now $\qquad$ area OCA = area OAN − area $\triangle ACN$

$$= \int_0^{1/2} \sqrt{(2x)} - \frac{1}{2} \times CN \times AN$$

$$= \sqrt{2} \left[\frac{2}{3} x^{3/2} \right]_0^{1/2} - \frac{1}{2} \times (ON - OC) \times AN$$

$$= \frac{2\sqrt{2}}{3} \left(\frac{1}{2\sqrt{2}} \right) - \frac{1}{2} \times \left(\frac{1}{2} - \frac{1}{4} \right) \times 1$$

$$= \frac{1}{3} - \frac{1}{8} = \frac{5}{24}.$$

Again, area OCB = area OBK + area of $\triangle BKC$

$$= \int_0^{1/8} \sqrt{(2x)}\, dx + \frac{1}{2} \times KC \times BK$$

$$= \sqrt{2} \left[\frac{2}{3} x^{3/2} \right]_0^{1/8} + \frac{1}{2} \times (OC - OK) \times BK$$

$$= \frac{2\sqrt{2}}{3} \left(\frac{1}{8\sqrt{2}} \right) + \frac{1}{2} \times \left(\frac{1}{4} - \frac{1}{8} \right) \cdot \frac{1}{2} = \frac{1}{24} + \frac{1}{32} = \frac{7}{96}.$$

∴ From (i), the required area

$$= \frac{5}{24} + \frac{7}{96} = \frac{27}{96} = \frac{9}{32}.$$

EXERCISES

1. Find the area enclosed by the parabolas $y^2 = 4ax$ and $x^2 = 4by$. [**Ans.** $\frac{16}{3} ab$]

2. Prove that the area enclosed by the parabolas $y^2 = 4a(x+a)$ and $y^2 = 4b(b-x)$ is $\frac{8}{3}(a+b)\sqrt{ab}$.

3. Prove that the area of the region bounded by the parabolas $y = x^2$ and $y = 4 - x^2$ is $\frac{16}{3}\sqrt{2}$.

4. Find the area lying above the x-axis and included between the circle $x^2 + y^2 = 2ax$ and the parabola $y^2 = ax$. [**Ans.** $a^2 \left(\frac{1}{4}\pi - \frac{2}{3} \right)$]

5. Show that the larger of the two areas into which the circle $x^2 + y^2 = 64a^2$ is divided by the parabola $y^2 = 12ax$ is

$$\frac{16}{3} a^2 (8\pi - \sqrt{3}).$$

5.4 AREAS OF POLAR CURVES
$r = f(\theta)$.

$r = f(\theta)$ is the equation of the curve where $f(\theta)$ is a continuous and single valued function of θ in the interval (α, β); OA and OB are the radii vectors $\theta = \alpha$ and $\theta = \beta$.

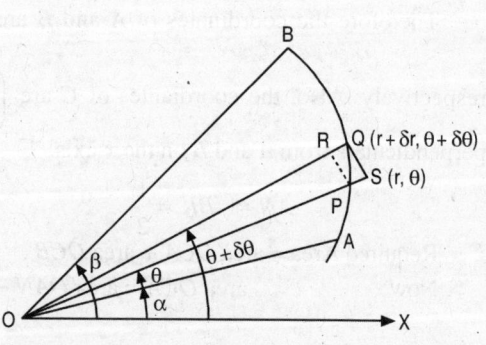

Area of Curves (Quadrature)

Let P be any point whose polar coordinates are (r, θ) antoher point having coordinates $(r + \delta r, \theta + \delta\theta)$ very near to P. Now with O as centre and OP, OQ as radii we draw the arcs of circles to meet OQ in R and OP produced in S respectively.

Area OPQ is equal to δA when θ changes from θ to $\theta + \delta\theta$

This area OPQ lies between the area of the circular sections OPR and OQS, arc area $OPR <$ area $OPQ <$ area OQS.

or $$\frac{1}{2} r^2 \delta\theta < \delta A < \frac{1}{2}(r + \delta\theta)^2 \delta\theta$$

or $$\frac{1}{2} r^2 < \frac{\delta A}{\delta\theta} < \frac{1}{2}(r + \delta r)^2$$

as Q approaches P as a limit, $\delta\theta \to 0$.

Therefore $$\frac{\delta A}{\delta\theta} = \frac{1}{2} r^2$$

Therefore $$A = \frac{1}{2} \int r^2 \, d\theta + C$$

$$= \frac{1}{2} \int_\alpha^\beta r^2 \, d\theta$$

When limit are α to β.

Hence $$A = \frac{1}{2} \int_\alpha^\beta r^2 \, d\theta = \frac{1}{2} \int_\alpha^\beta \{f(\theta)\}^2 \, d\theta.$$

Note. Consider the values of θ which makes $r = 0$. These values are the limits of integration in the case of a loop.

EXAMPLES

1. *Find the area of the cardioid $r = a(1 + \cos\theta)$.*

Sol. The curve is $r = a(1 + \cos\theta)$.

It is symmetrical about the initial line as by putting $-\theta$ for θ, the equation of curve does not change.

Again $\theta = \pi$ makes $r = 0$, hence the limits are from 0 to π.

$\therefore$ The required area $= 2 \int_0^\pi \frac{1}{2} r^2 \, d\theta$

$$= 2 \cdot \frac{1}{2} \int_0^\pi a^2 (1 + \cos\theta)^2 \, d\theta$$

$$= 4a^2 \int_0^\pi \cos^4 \frac{\theta}{2} \, d\theta$$

$$= 8a^2 \int_0^{\pi/2} \cos^4 \phi \, d\phi, \text{ putting } \frac{\theta}{2} = \phi \text{ and } d\theta = 2d\phi.$$

$$= 8a^2 \cdot \frac{\overline{\frac{5}{2}} \cdot \overline{\frac{1}{2}}}{2 \overline{3}} = \frac{4a^2}{2} \cdot \frac{3}{2} \cdot \frac{1}{2} \pi = \frac{3\pi a^2}{2}.$$

2. *Find the area of one loop of the curve $r = \sqrt{3} \cos 3\theta + \sin 3\theta$.*

Sol. The equation of the curve is $r = \sqrt{3} \cos 3\theta + \sin 3\theta$

$$= 2\left[\left(\frac{\sqrt{3}}{2}\right)\cos 3\theta + \frac{1}{2}\sin 3\theta\right]$$

$$= 2\left[\left(\frac{\sqrt{3}}{2}\right)\cos 3\theta + \frac{1}{2}\sin 3\theta\right]$$

$$= 2\cos\left(3\theta - \frac{\pi}{6}\right) = 2\cos 3\left(\theta - \frac{\pi}{18}\right).$$

Turn the initial line through an angle $\frac{1}{18}\pi$ and putting θ as $\theta + \frac{1}{18}\pi$, the above equation reduces to $r = 2\cos 3\theta$ and the tracing of this curve is as in figure.

Also for $r = 2\cos 3\theta$, when $r = 0$ we get $\cos 3\theta = 0$ or $3\theta = \pm\frac{\pi}{6}$ and there is also symmetry about the initial line.

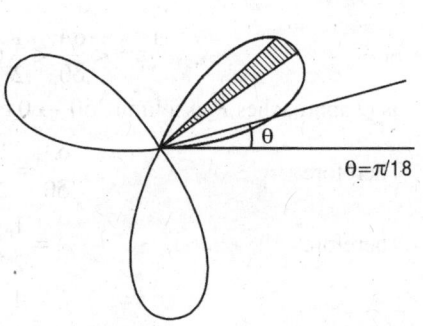

$\theta = \pi/18$

$\therefore$ Required area

$$= 2 \times \frac{1}{2}\int_0^{\pi/6} r^2\, d\theta$$

$$= 4\int_0^{\pi/6}\cos^2 3\theta\, d\theta, \text{ as } r = 2\cos 3\theta$$

$$= \frac{4}{3}\int_0^{\pi/6}\cos^2\phi\, d\phi, \qquad\qquad \text{putting } 3\theta = \phi$$

$$= \frac{4}{3}\cdot\frac{\left\lceil\frac{3}{2}\cdot\right\lceil\frac{1}{2}}{2\lceil 2} = \frac{1}{3}\pi$$

EXERCISES

1. Find the area enclosed by the curve $r = a(1 - \cos\theta)$. $\left[\text{Ans. } \frac{3}{2}\pi a^2\right]$

2. Find the area of a loop of the curve $r^2 = a^2 \cos 2\theta$. $\left[\text{Ans. } \frac{a^2}{2}\right]$

3. Find the area of the limacon $r = a + b\cos\theta\ (a > b)$.
   Hence or otherwise find the area of the cardioids
   $r = a(1 + \cos\theta)$ and $r = a(1 - \cos\theta)$ $\left[\text{Ans. } \pi\left(a^2 + \frac{1}{2}b^2\right), \frac{3}{2}\pi a^2, \frac{3}{2}\pi a^2\right]$

4. Find the area of the curve $r = 3 + 2\cos\theta$. $[\text{Ans. } 11\pi]$

5. Find the ratio of the area of the larger to the area of the smaller loop of the curve
   $$r = \frac{1}{2} + \cos 2\theta.$$ $\left[\text{Ans. } \dfrac{4\pi + 3\sqrt{2}}{2\pi - 3\sqrt{3}}\right]$

6. Show that the whole area of the curve $r^2 = a^2\cos^2\theta + b^2\sin^2\theta$ is $\frac{1}{2}\pi(a^2 + b^2)$.

Area of Curves (Quadrature)

7. Find the area of the loop of the curve $r = a \sin 3\theta$.
$$\left[\text{Ans. } \frac{\pi a^2}{12}\right]$$

7. (i) Find the whole area of the curve $r = a \cos 2\theta$
   (ii) Find the whole area of the curve $r = a \sin 2\theta$.

5.5 AREA BOUNDED BY TWO POLAR CURVES

Area bounded by two polar curve $r = f_1(\theta)$ and $r = f_2(\theta)$ is

$$= \frac{1}{2} \int_\alpha^\beta [\{f_1(\theta)\}^2 - \{f_2(\theta)\}^2] d\theta$$

EXAMPLES

1. *Find the area common to the circles $r = a\sqrt{2}$ and $r = 2a \cos\theta$.*

Sol. The given curves are

$$r = a\sqrt{2} \quad ...(i)$$
$$r = 2a \cos\theta \quad ...(ii)$$

From (i) and (ii)

$$a\sqrt{2} = 2a \cos\theta$$

$$\cos\theta = \frac{1}{\sqrt{2}}, \text{ or } \theta = \frac{\pi}{4}$$

The point of intersection gives $\theta = \frac{\pi}{4}$

∴ The required area

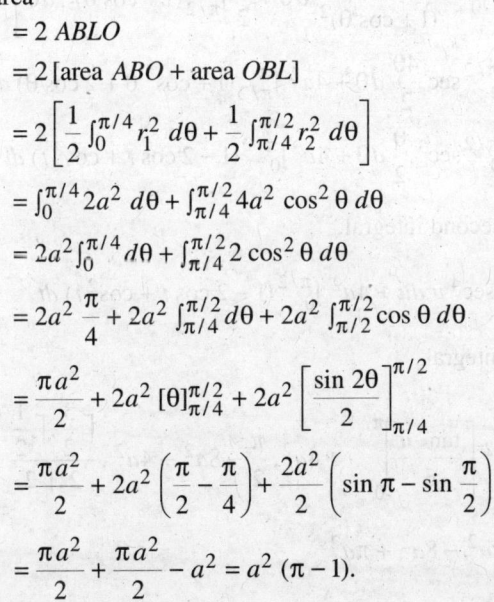

$r = a\sqrt{2},\ r = 2a\cos\theta$

$$= 2\ ABLO$$
$$= 2\ [\text{area } ABO + \text{area } OBL]$$
$$= 2\left[\frac{1}{2}\int_0^{\pi/4} r_1^2\, d\theta + \frac{1}{2}\int_{\pi/4}^{\pi/2} r_2^2\, d\theta\right]$$
$$= \int_0^{\pi/4} 2a^2\, d\theta + \int_{\pi/4}^{\pi/2} 4a^2 \cos^2\theta\, d\theta$$
$$= 2a^2 \int_0^{\pi/4} d\theta + \int_{\pi/4}^{\pi/2} 2\cos^2\theta\, d\theta$$
$$= 2a^2 \frac{\pi}{4} + 2a^2 \int_{\pi/4}^{\pi/2} d\theta + 2a^2 \int_{\pi/2}^{\pi/2} \cos\theta\, d\theta$$
$$= \frac{\pi a^2}{2} + 2a^2\, [\theta]_{\pi/4}^{\pi/2} + 2a^2 \left[\frac{\sin 2\theta}{2}\right]_{\pi/4}^{\pi/2}$$
$$= \frac{\pi a^2}{2} + 2a^2 \left(\frac{\pi}{2} - \frac{\pi}{4}\right) + \frac{2a^2}{2}\left(\sin\pi - \sin\frac{\pi}{2}\right)$$
$$= \frac{\pi a^2}{2} + \frac{\pi a^2}{2} - a^2 = a^2\,(\pi - 1).$$

2. *Find the ratio of the two parts into which the parabola $2a = r(1 + \cos\theta)$ divides the area of the cardiod $r = 2a(1 + \cos\theta)$.*

Sol. Solving the given equations $2a = r(1 + \cos\theta)$ and $r = 2a(1 + \cos\theta)$.
We have

$$(1 + \cos\theta)^2 = 1$$

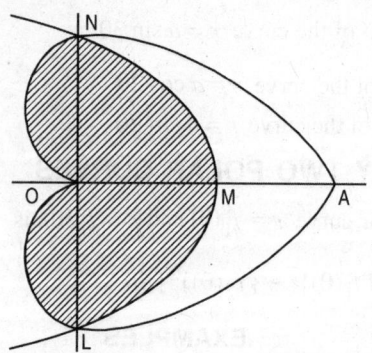

$\therefore \qquad \cos\theta = 0$

or $\qquad \theta = \dfrac{\pi}{2}$

Limits are 0 to $\dfrac{\pi}{2}$ and $\dfrac{\pi}{2}$ to π.
Therefore the required area

$= \text{area } OLMN$

$= 2 \text{ area } OMNO$

$= 2 [\text{area } OMN + \text{area } ONO]$

$= 2\left[\dfrac{1}{2}\int_0^{\pi/2} \dfrac{2a}{(1+\cos\theta)^2} d\theta + \dfrac{1}{2}\int_{\pi/2}^{\pi}(1+\cos\theta)^2 \, d\theta\right]$

$= a^2 \int_0^{\pi/2} \sec\dfrac{4\theta}{2} \, d\theta + 4a^2 \int_{\pi/2}^{\pi}(1+\cos^2\theta + 2\cos\theta) \, d\theta$

$= a^2 \int_0^{\pi/2} \sec^3\dfrac{\theta}{2} \, d\theta + 4a^2 \int_0^{\pi/2}(1 - 2\cos t + \cos^2 t) \, dt$

By putting $\theta = \dfrac{\pi}{2} + t$ in second integral.

$= 2a^2 \int_0^{\pi/4} \sec^4 u \, du + 4a^2 \int_0^{\pi/2}(1 - 2\cos t + \cos^2 t) \, dt$

By putting $\dfrac{\theta}{2} = u$ in 1st integral.

$= 2a^2 + 2a^2 \left[\dfrac{\tan^3 u}{3}\right]_0^{\pi/4} + 4a^2 \cdot \dfrac{\pi}{2} - 8a^2 + 4a^2 \cdot \dfrac{\lceil \frac{3}{2} \cdot \lceil \frac{1}{2}}{2 \cdot \lceil 2}$

$= \dfrac{8a^2}{3} + 2\pi a^2 - 8a^2 + \pi a^2$

$= 3\pi a^2 - \dfrac{16a^2}{3}.$

and unshaded area = whole cardioid − shaded area

$= 6\pi a^2 - \left(3\pi a^2 - \dfrac{16a^2}{3}\right)$

Area of Curves (Quadrature)

$$= \frac{a^2}{3}(9\pi + 16)$$

∴ The required ratio

$$= \frac{\frac{a^2}{3}(9\pi - 16)}{\frac{a^2}{3}(9\pi + 16)} = \frac{9\pi - 16}{9\pi + 16}$$

EXERCISES

1. Find the area outside the circle $r = 2a \cos\theta$ and inside the cardioid $r = a(1 + \cos\theta)$.

$$\left[\text{Ans. } \frac{\pi a^2}{2}\right]$$

2. Show that the area contained between the circle $r = a$ and the curve $r = a \cos 5\theta$ is equal to three fifth of the area of the circle.

3. Show that the area included between the curves $r = a(1 + \cos\theta)$, $r = a(1 - \cos\theta)$ is $\frac{1}{2}a^2(3\pi - 8)$.

4. Find the area common to the circle $r = a$ and the cardioid $r = a(1 + \cos\theta)$.

$$\left[\text{Ans. } \left(\frac{5}{4}\pi - 2\right)a^2\right]$$

5. Find the total area inside the circle $r = \sin\theta$ and outside the cardiod $r = 1 - \cos\theta$.

$$\left[\text{Ans. } 1 - \frac{\pi}{4}\right]$$

5.6 CARTESIAN EQUATIONS CHANGED TO POLAR EQUATIONS

Example. *Find the area of the loop of the curve* $x^3 + y^3 = 3axy$.

Sol. Putting $x = r\cos\theta$ and $y = r\sin\theta$, the given equation transforms to $r^2(\cos^3\theta + \sin^3\theta) = 3ar^2 \cos\theta \sin\theta$

or $$r = \frac{3a\cos\theta \sin\theta}{\cos^3\theta + \sin^3\theta}$$

From the above equation it is clear that $r = 0$, when $\theta = 0$ and $\theta = \frac{\pi}{2}$. Also when $0 < \theta < \frac{\pi}{2}$, r is positive. Hence the loop of the curve lies between $\theta = 0$ and $\theta = \frac{\pi}{2}$.

∴ Required area of the loop

$$= \frac{1}{2}\int_0^{\pi/2} r^2 \, d\theta$$

$$= \frac{9a^2}{2} \int_0^{\pi/2} \frac{\cos^2\theta \sin^2\theta \, d\theta}{(\cos^3\theta + \sin^3\theta)^2}$$

$$= 9\frac{a^2}{2} \int_0^{\pi/2} \frac{\tan^2\theta \sec^2\theta \, d\theta}{(1 + \tan^3\theta)^2},$$

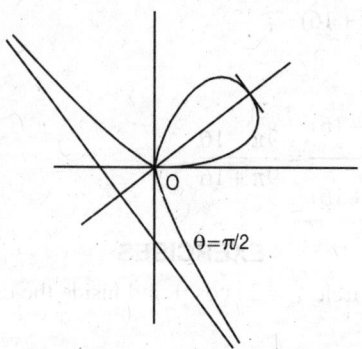

$$= \frac{9a^2}{2} \int_1^\infty \frac{1}{t^2} \frac{dt}{3},$$

where $1 + \tan^3 \theta = t$, $3 \tan^2 \theta \sec^2 \theta \, d\theta \, d\theta = dt$.

$$= \frac{3a^2}{2} \left[-\frac{1}{t} \right]_1^\infty = \frac{3a^2}{2}$$

EXERCISES

1. Show that the area of the loop of the curve $x^4 + y^4 = 4a^2 xy$ is $\frac{1}{2} \pi a^2$.

2. Show that the area of the loop of the curve $x^4 + y^4 = 2a^2 xy$ is $\frac{1}{4} \pi a^2$.

3. (a) Show that the area of the loop of the curve $x^5 + y^5 = 5a x^2 y^2$ is $\frac{5}{2} a^2$.

   (b) Show that the arc of the loop of the curve $x^5 + y^5 = 5ax^2 y^2$ is five times the area of the loop of the curve $r^2 = a^2 \cos 2\theta$.

❑❑❑

6
Lengths of Curves (Rectification)

6.1 RECTIFICATION

The process of determining the lengths of arcs of plane curves whose equations are given in cartesian, parametric or polar form is called *rectification*. If s denotes the length of the arc of the curve, then the formulae, known in differential calculus for arcs, shall enable us to determine it.

6.2 LENGTHS OF THE CARTESIAN CURVES [$y = f(x)$]

We know from differential calculus, that

$$\frac{ds}{dx} = \sqrt{1 + \left(\frac{dy}{dx}\right)^2}$$

It follows that

$$s = \int_a^b \sqrt{1 + \left(\frac{dy}{dx}\right)^2}\, dx,$$

Where a is the abscissa of the point A from which s is measured and b is the abscissa of the point B to which it is measured,

Similarly,

$$\frac{ds}{dy} = \sqrt{\left[1 + \left(\frac{dx}{dy}\right)^2\right]}$$

gives

$$s = \int_a^b \sqrt{\left[1 + \left(\frac{dx}{dy}\right)^2\right]}\, dy,$$

where a and b are the ordinates of A and B.

EXAMPLES

1. *Find the whole length of the astroid* $x^{2/3} + y^{2/3} = a^{2/3}$.

Sol. The curve is symmetrical about both the axes and crosses the x-axis at $(\pm a, 0)$ and y-axis at $(0, \pm a)$

Differentiating, $x^{2/3} + y^{2/3} = a^{2/3}$, we get

$$\frac{2}{3} x^{-1/3} + \frac{2}{3} y^{-1/3} \frac{dy}{dx} = 0$$

or

$$\frac{dy}{dx} = -\left(\frac{y}{x}\right)^{1/3}$$

$$\therefore \quad 1 + \left(\frac{dy}{dx}\right)^2 = 1 + \left(\frac{y}{x}\right)^{2/3} = \frac{x^{2/3} + y^{2/3}}{x^{2/3}} = \frac{a^{2/3}}{x^{2/3}}$$

∴ Required length $= 4 \int_0^a \left[1 + \left(\frac{dy}{dx}\right)^2\right]^{1/2} dx$

$= 4 \int_0^a \left(\frac{a^{2/3}}{x^{2/3}}\right)^{1/2} dx$

$= 4a^{1/3} \int_0^a x^{-1/3} dx = 4a^{1/3} \times \frac{3}{2} [x^{2/3}]_0^a$

$= 6a^{1/3} \cdot a^{2/3} = 6a$

2. Show that the length of an arc parabola $y^2 = 4ax$ cut-off by the line $3y = 8x$ is a $\log\left(2 + \frac{15}{16}\right)$.

Sol. Solving $y^2 = 4ax$ and $3y = 8x$ for y, we get

$$y^2 = 4a \cdot \frac{3y}{8} \quad \text{or} \quad 2y^2 - 3ay = 0$$

or $\qquad y = 0, \frac{3a}{2}.$

Thus the parabola and the straight line intersect at the points where $y = 0$ and $3a/2$.

Now, differentiating $y^2 = 4ax$, we get

$$2y \cdot \frac{dy}{dx} = 4a \quad \text{or} \quad \frac{dy}{dx} = \frac{2a}{y}$$

or $\qquad \frac{dx}{dy} = \frac{y}{2a}$

∴ If s denotes the arc length of the parabola measured in the direction of y increasing, then

$$\frac{ds}{dy} = \sqrt{1 + \left(\frac{dx}{dy}\right)^2} = \frac{1}{2a}\sqrt{(4a^2 + y^2)}$$

Let s denotes the required arc length, then

$$s = \int_0^{3a/2} \frac{1}{2a} \sqrt{(4a^2 + y^2)} \, dy$$

$$= \frac{1}{2a}\left[\frac{y}{2}\sqrt{4a^2 + y^2} + \frac{4a^2}{2} \log\{y + \sqrt{(4a^2 + y^2)}\}\right]_0^{3a/2}$$

$$= \frac{1}{4a}\left[\frac{3a}{2}\sqrt{\left(\frac{25a^2}{4}\right)} + 4a^2 \log\left\{\frac{3a}{2} + \sqrt{\left(\frac{25a^2}{4}\right)}\right\} - 4a^2 \log 2a\right]$$

$$= a\left(\frac{15}{16} + \log 2\right)$$

Lengths of Curves (Rectification)

EXERCISES

1. Find the length of the arc of the curve $ay^2 = x^3$ from the vertex to the point (a, a).

$$\left[\text{Ans. } \frac{a}{27}[13\sqrt{(13)} - 8]\right]$$

2. Find the arc of the curve $y = ae^x$ from the point $(0, a)$ to the point (x_1, y_1).

$$\left[\text{Ans. } \log\left[\frac{y}{1 + \sqrt{(1 + y^2)}}\right] + \sqrt{(1 + a^2)}\right]$$

3. Show that the length of the curve $y = \log \sec x$ betwen the points, where $x = 0$ and $x = \pi/3$, is $\log(2 + \sqrt{3})$.

4. Find the length of the arc of $y = x^2$ from the vertex to the point (x, y).

$$\left[\text{Ans. } \frac{1}{2}x\sqrt{(4x^2 + 1)} + \frac{1}{4}\log\left[x + \frac{1}{2}\sqrt{(4x^2 + 1)}\right] + \frac{1}{4}\log 2\right]$$

5. Find the length of the curve $y = \log\dfrac{e^x - 1}{e^x + 1}$ from $x = 1$ to $x = 2$. $\quad\left[\text{Ans. } \log(e + e^{-1})\right]$

6. Show that the whole length of $(ax)^{2/3} + (by)^{2/3} = (a^2 - b^2)^{2/3}$ is $\left(\dfrac{a^2}{b} + \dfrac{b^2}{a}\right)$.

7. Find the length of the parabola $y^2 = 4ax$ from the vertex to an extremity of the latus-rectum.

$$\left[\text{Ans. } a[\sqrt{2} + \log(1 + \sqrt{2})]\right]$$

8. (i) Find the perimeter of the loop of the curve $3ay^2 = x^2(a - x)$.

   (ii) Find the length of the loop of the curve $3ay^2 = x(x - a)^2$. $\quad\left[\text{Ans. (i) } \dfrac{4a}{\sqrt{3}}, \text{ (ii) } \dfrac{4a}{\sqrt{3}}\right]$

9. Find the length of one quadratnt of the curve
$$\left(\frac{x}{y}\right)^{2/3} + \left(\frac{y}{b}\right)^{2/3} = 1.$$

$$\left[\text{Ans. } \frac{a^2 + ab + b^2}{a + b}\right]$$

6.3 LENGTH OF PARAMETRIC CURVES

If the equation of the curve is given in the form
$$x = f(t), \quad y = \phi(t),$$

then
$$\frac{ds}{dt} = \sqrt{\left\{\left(\frac{dx}{dt}\right)^2 + \left(\frac{dy}{dt}\right)^2\right\}}$$

and hence the length of arc of the curve will be

$$s = \int_a^b \left[\left(\frac{dx}{dt}\right)^2 + \left(\frac{dy}{dt}\right)^2\right]^{1/2} dt.$$

where the arc of the curve extends from $t = a$ to $t = b$

EXAMPLES

1. *Find the whole length of the cycloid* $x = a(\theta + \sin\theta)$, $y = a(1 - \cos\theta)$, $-\pi \leq \theta \leq \pi$.

Sol. We have

$$x = a(\theta + \sin\theta), \qquad \therefore \quad \frac{dx}{d\theta} = a(1 + \cos\theta)$$

$$y = a(1 - \cos\theta), \qquad \therefore \quad \frac{dy}{d\theta} = a\sin\theta$$

The limits are from 0 to π.

$\therefore$ The required length

$$= 2\int_0^\pi \sqrt{\{a^2(1+\cos\theta)^2 + a^2\sin^2\theta\}}\, d\theta$$

$$= 4a\int_0^\pi \cos\frac{\theta}{2}\, d\theta = 8a\left[\sin\frac{\theta}{2}\right]_0^\pi$$

$$= 8a.$$

2. *Show that the length of an arc of the curve* $x\sin\theta + y\cos\theta = f'(\theta)$, $x\sin\theta - y\cos\theta = f''(\theta)$ *is given by* $s = f(\theta) + f''(\theta) + c$.

Sol. The given relations are

$$x\sin\theta + y\cos\theta = f'(\theta) \qquad \ldots(i)$$

$$x\sin\theta + y\cos\theta = f''(\theta) \qquad \ldots(ii)$$

Multiplying (i) by $\sin\theta$ and (ii) by $\cos\theta$ and adding, we have

$$x = f'(\theta)\sin\theta + f''(\theta)\cos\theta \qquad \ldots(iii)$$

Similarly, multiplying (i) by $\cos\theta$ and (ii) by $\sin\theta$ and subtracting, we get

$$y = f'(\theta)\cos\theta - f''(\theta)\sin\theta \qquad \ldots(iv)$$

From (ii) and (iv), we get

$$\frac{dx}{d\theta} = f''(\theta)\sin\theta + f'(\theta)\cos\theta - f'''(\theta)\sin\theta + f'''(\theta)\cos\theta$$

$$= -\sin\theta\{f'(\theta) + f'''(\theta)\}$$

Therefore $\dfrac{ds}{d\theta} \sqrt{\left\{\left(\dfrac{dx}{d\theta}\right)^2 + \left(\dfrac{dy}{d\theta}\right)^2\right\}}$

$$= f'(\theta) + f'''(\theta).$$

Thus the required length is

$$s = \int\{f'(\theta) + f'''(\theta)\}\, d\theta + c$$

$$= f(\theta) + f''(\theta) + c.$$

EXERCISES

1. Find the length of the arc of the curve $x = e^\theta \sin\theta$, $y = e^\theta \cos\theta$ from $\theta = 0$ to $\theta = \dfrac{\pi}{2}$.

$$\left[\text{Ans. } \sqrt{2}\,(e^{\pi/2} - 1)\right]$$

2. Find the length of the curve given by the equations $x = a(\cos t + t\sin t)$ and $y = a(\sin t - t\cos t)$, between $t = 0, t = 2$.

$$[\text{Ans. } 2a]$$

Lengths of Curves (Rectification)

3. Prove that the length of the arc of the curve $x = a \sin 2\theta (1 + \cos 2\theta)$, $y = a \cos 2\theta (1 - \cos 2\theta)$ from $(0, 0)$ to (x, y) is $\dfrac{4a}{3} \sin 3\theta$.

4. Find the length of an arc of the cycloid $x = a(\theta - \sin \theta)$, $y = a(1 - \cos \theta)$. **[Ans. $8a$]**

5. Find the length of an arc of the cycloid $x = a(\theta - \sin \theta)$, $y = a(1 - \cos \theta)$. **[Ans. $8a$]**

6. Find the length of the loop of the curve $x = t^2$, $y = t - \dfrac{t^3}{3}$. **[Ans. $4\sqrt{3}$]**

7. Show that the length of an arc of the epi-cycloid

$$x = (a+b)\cos\theta - b\cos\dfrac{a+b}{b}\theta$$

$$y = (a+b)\sin\theta - b\sin\dfrac{a+b}{b}\theta$$

is given by $s = \dfrac{4b(a+b)}{2} \cos \dfrac{a\theta}{2b}$, s being measured from the point at which $\theta = \pi b/a$.

6.4 LENGTH OF THE POLAR CURVES

The length of the arc of the curve $r = f(\theta)$ between $\theta = \alpha$ and $\theta = \beta$ is

$$\int_\alpha^\beta \sqrt{\left\{r^2 + \left(\dfrac{dr}{d\theta}\right)^2\right\}}\, d\theta$$

Similarly the length of the arc of the curve $\theta = f(r)$ between $r = r_1$ and $r = r_2$ is

$$\int_{r_1}^{r_2} \sqrt{\left\{r^2 \left(\dfrac{d\theta}{dr^2}\right)^2 + 1\right\}}\, dr$$

EXAMPLES

1. *Show that the perimeter of the cardiod $r = a(1 + \cos \theta)$ is $8a$.*

Sol. The curve is symmetrical about the initial line; so the perimeter $= 2 \times$ length of the arc, $\theta = 0$ to $\theta = \pi$.

But $\qquad r = a(1 + \cos \theta)$, then $\dfrac{dr}{d\theta} = -a \sin \theta$.

Now $\qquad \dfrac{ds}{d\theta} = \sqrt{\{a^2(1+\cos\theta)^2 + a^2 \sin^2 \theta\}}$

$$= 2a \cos \dfrac{\theta}{2}.$$

Hence the required length

$$= 2 \int_0^\pi 2a \cos \dfrac{\theta}{2}\, d\theta = 8a \left[\sin \dfrac{\theta}{2}\right]_0^\pi$$

$$= 8a.$$

2. *Prove that the cardiod $r = a(1 - \cos \theta)$ is divided by the line $4r \cos \theta = 3a$ into two parts, such that the lengths of the arcs on either side of this line are equal.*

Sol. The given curve and straight line are $r = a(1 + \cos \theta)$, and $4r \cos \theta = 3a$, respectively. These curves cut one another at B and C.

Thereby $\dfrac{3a}{4\cos\theta} = a(1+\cos\theta)$

or $\qquad 4\cos^2\theta + 4\cos\theta - 3 = 0$

$\therefore \quad \cos\theta = \dfrac{1}{2}$ or $\theta = \dfrac{\pi}{3}$, the value $\cos\theta = -\dfrac{3}{2}$ is inadmissible.

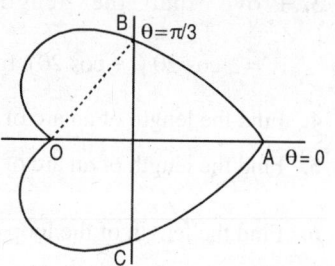

The arc BAC is to be shown equal to the arc BOC.

Now $r = a(1+\cos\theta)$, $\dfrac{dr}{d\theta} = -a\sin\theta$

$\therefore \quad \dfrac{ds}{d\theta} = \sqrt{[a^2(1+\cos\theta)^2 + a^2\sin^2\theta]} = 2a\cos\dfrac{\theta}{2}$

Therefore the length of the arc AB

$$= 2a\int_0^{\pi/3}\cos\dfrac{\theta}{2}\,d\theta = 4a\left[\sin\dfrac{\theta}{2}\right]_0^{\pi/3}$$

$$= 4a \cdot \dfrac{1}{2} = 2a.$$

Hence arc $BAC = 2$ arc $AB = 4a$

also whole length $= 2\int_0^\pi 2a\cos\dfrac{\theta}{2}\,d\theta$

$$= 8a\left[\sin\dfrac{\theta}{2}\right]_0^\pi = 8a.$$

Therefore the arc BOC = whole length − arc BAC

$$= 8a - 4a = 4a.$$

EXERCISES

Find the length of–

1. The arc of the cardiod $r = a(1-\cos\theta)$ between the points whose vertical angles arc α and β.

   $\left[\text{Ans. } 4a\left(\cos\dfrac{1}{2}\alpha - \cos\dfrac{1}{2}\beta\right)\right]$

2. The arc of the spiral $r = a\theta$ between the points $r = r_1$ and $r = r_2$.

   $\left[\text{Ans. } f(r_2/a) - f(r_1/a) \text{ where } f(\theta) = \dfrac{1}{2}a[\theta\sqrt{(1+\theta^2)}] + \log\{\theta + \sqrt{(1+\theta^2)}\}\right]$

3. The perimeter of the curve $r = a\cos\theta$. $\qquad$ [Ans. πa]

4. Show that the length of the arc of the upper half of the cardioid $r = a(1+\cos\theta)$ is bisected by $\theta = \pi/3$.

5. Prove that the perimeter of the curve $r = a + b\cos\theta$ is approximately $= 2\pi a(2 + b^2/4a^2)$ if b/a is small.

6.5 INTRINSIC EQUATION

A relation between s and ψ is called the *intrinsic equation* of the curve if s be the length of the arc of the curve measured from a fixed point to any other point at which the tangent to the curve subtends an angle ψ with the fixed line.

Lengths of Curves (Rectification)

6.5.1 To find the instrinsic equation from the cartesian equation

Suppose the curve is $y = f(x)$

$$\tan \psi = \frac{dy}{dx} \qquad \text{...(i)}$$

and

$$s = \int_a^x \sqrt{\left\{1+\left(\frac{dy}{dx}\right)^2\right\}} \, dx,$$

where s is measured from a fixed point whose abscissa is a.
This gives $\quad s = f(x)$
elimination of x, between (i) and (ii) gives the intrinsic equation

$$s = \phi(\psi).$$

6.5.2 To find the intrinsic equation from the polar equation

Suppose the curve is

$$r = f(\theta), \frac{dr}{d\theta} = f'(\theta)$$

Now, $\qquad \psi = \theta + \phi \qquad \text{...(i)}$

$$\tan \phi = r \frac{d\theta}{dr} = \frac{f(\theta)}{f'(\theta)}$$

$$s = \int_\alpha^\theta \sqrt{\left\{r^2 + \left(\frac{dr}{d\theta}\right)^2\right\}} \, d\theta$$

$$= \int_\alpha^\theta \sqrt{\{f(\theta)\}^2 + \left\{\frac{f(\theta)}{f'(\theta)}\right\}^2} \, d\theta$$

$$= f(\theta) \qquad \text{...(iii)}$$

where α is the vectroial angle of the point from which s is measured. Eliminating θ and ϕ between (i), (ii) and (iii) the relation between s and ψ shall be known.

EXAMPLES

1. *Find the intrinsic equation to a parabola in its simplest form. Deduce that length of the arc intercepted between the vertex and an extremity of the latus rectum is $a [\sqrt{2} + \log (\sqrt{2}+1)]$, $4a$ being the latus rectum.*

Solution. We have the equation of parabola as

$$y^2 = 4ax; \qquad \therefore \qquad \frac{dy}{dx} = 4a$$

Thereby $\qquad \dfrac{dy}{dx} = \dfrac{2a}{y} = \tan \psi,$

$\therefore \qquad y = 2a \cot \psi$

Now $\qquad s = \int_0^y \sqrt{\left\{1+\left(\frac{dy}{dx}\right)^2\right\}} \, dy$

$$= \frac{1}{2a} \int_0^y \sqrt{(4a^2 + y^2)} \, dy$$

$$= \frac{1}{2a}\left[\frac{1}{2}y\sqrt{(y^2+4a^2)} + \frac{1}{2}\cdot 4a^2 \log\{y+\sqrt{(4y^2+4a^2)}\}\right]_0^y dy$$

or
$$S = \frac{1}{2a}\left[\frac{1}{2}y\sqrt{(y^2+4a^2)} + 2a^2 \log\{y+\sqrt{(y^2+4a^2)}\} - 2a^2 \log 2a\right]$$

By putting $y = 2a \cot \psi$, we get

$$S = \frac{1}{4a}\cdot 2a\cot\psi \cdot 2a\,\mathrm{cosec}\,\psi + \frac{2a^2}{2a}\log 2a(\cot\psi + \mathrm{cosec}\,\psi) - \frac{2a^2}{2a}\log 2a$$
$$= [a\cot\psi\,\mathrm{cosec}\,\psi + a\log(\cot\psi + \mathrm{cosec}\,\psi)]$$

At the extremity of the latus rectum $x = a$ and $y = 2a$.

∴ $y = 2a\cot\psi$ gives $2a = 2a\cot\psi$,

$\cot\psi = 1$ or $\psi = \pi/4$

Therefore the required length

$$s = a[\sqrt{2} + \log(\sqrt{2}+1)]$$

2. *Show that the intrinsic equation of the cycloid* $x = a(t + \sin t)$, $y = a(1 - \cos t)$, *is* $s = 4a \sin \psi$.

Sol. We have

$$x = a(t+\sin t); \qquad \therefore \quad \frac{dx}{dt} = a(1+\cos t)$$

$$\frac{dx}{dt} = a(1+\cos t) \qquad \therefore \quad \frac{dy}{dt} = a\sin t$$

∴ $$\frac{dy}{dx} = \frac{a\sin t}{a(1+\cos t)} = \tan\frac{t}{2} = \tan\psi$$

Therefore $\psi = \frac{1}{2}t$ ...(i)

Length of the arc

$$s = \int_0^t \sqrt{\left\{\left(\frac{dx}{dt}\right)^2 + \left(\frac{dy}{dt}\right)^2\right\}}\,dt$$

$$= a\int_0^t [(1+\cos t)^2 + \sin^2 t]^{1/2}\,dt$$

$$= 2a\int_0^t \cos\frac{t}{2}\,dt = 4a\left[\sin\frac{t}{2}\right]_0^t$$

$$= 4a\sin\frac{t}{2}$$

∴ $s = 4a\sin\psi$, from (i).

3. *Find the intrinsic equation of the cardiod* $r = a(1-\cos\theta)$.

Sol. We have

$$r = a(1-\cos\theta),\quad \frac{dr}{d\theta} = a\sin\theta,$$

Now, $$\tan\phi = r\frac{d\theta}{dr} = \frac{a(1-\cos\theta)}{a\sin\theta}$$

Lengths of Curves (Rectification)

$$= \tan\frac{\theta}{2}; \qquad \therefore \phi = \frac{\theta}{2}.$$

$$\psi = \theta + \phi = \theta + \frac{\theta}{2}$$

$$= \frac{3\theta}{2}$$

Now,
$$s = \int_0^\theta \sqrt{\left\{r^2 + \left(\frac{dr}{d\theta}\right)^2\right\}}\, d\theta$$

$$= \int_0^\theta \sqrt{\{a^2(1-\cos\theta)^2 + a^2 \sin^2\theta\}}\, d\theta$$

$$= \int_0^\theta 2a \sin\frac{\theta}{2}\, d\theta$$

$$= 2a\left[-2\cos\frac{\theta}{2}\right]_0^\theta$$

$$= 4a\left(1 - \cos\frac{\theta}{2}\right)$$

Therefore
$$s = 4a\left(1 - \cos\frac{\psi}{3}\right)$$

$$= 8a \sin^2\frac{\psi}{6}, \text{ as } \psi = \frac{3\theta}{2}.$$

EXERCISES

1. Find the intrinsic equation of catenary $y = a \cosh x/a$. **[Ans.** $s = a \tan \psi$**]**
2. Find the intrinsic equation of the curve $y^3 = ax^2$. **[Ans.** $27s = 8a(\text{cosec}^3 \psi - 1)$**]**
3. Find the intrinsic equation of the parabola $y^2 = 4ax$.
   [Ans. $s = a \cot \psi \, \text{cosec}\, \psi + a \log(\text{cosec}\,\psi + \cot\psi)$**]**
4. Find the intrinsic equation of $y = a \log \sec(x/a)$. **[Ans.** $s = a \log(\sec\psi + \tan\psi)$**]**
4. Show that the intrinsic equation of the curve $ay^2 = x^3$, taking its cusp as the fixed point, is $27s = 8a(\sec^3 \psi - 1)$.
5. Show that the intrinsic equation of $x = a(\theta + \sin\theta)$, $y = a(1 - \cos\theta)$, $-\pi \le \theta \le \pi$ is $s = 4a \sin\psi$.
6. Show that the intrinsic equation of the curve
   $x = a(2\cos t - \cos 2t)$, $y = a(2\sin t - \sin 2t)$ is $s = 16 \sin^2 \psi/6$
7. Find the intrinsic equation of the curve $r = a(1 + \cos\theta)$. **[Ans.** $s = 4a \sin\left(\frac{\psi}{3} - \frac{\pi}{6}\right)$**]**
8. Show that the intrinsic equation of the curve $r = ae^{m\theta}$, when the arc is measured from $(a, 0)$ is $s = \dfrac{a\sqrt{1+m^2}}{m}[e^{m(\psi - \beta)} - 1]$, where $\beta = \tan^{-1}\left(\dfrac{1}{m}\right)$.

7

Volume and Surface of Solid of Revolution

7.1 VOLUME OF A SOLID OF REVOLUTION

If the area, from $x = a$ to $x = b$ under a curve $y = f(x)$ is rotated through four right angles or one revolution about the x-axis, it generates a solid with plane circular ends as shown in figure. Divide the volume by planes perpendicular to the x-axis. The area of the cross-section by any such planes cutting the curve at (x, y) is πy^2.

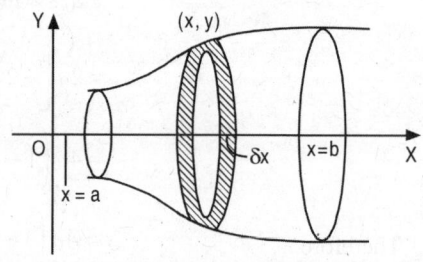

The infinitesimal element of volume is the volume between this plane and a parallel plane at a further distance δx. The element of volume is $A\, \delta x$ correct to the first order.

Thus the required volume $= \int_a^b A\, dx$ between suitable limits a and b. But $A = \pi y^2$ hence

$$V = \int_a^b \pi y^2\, dx$$

Note 1. The element of volume is $A\, \delta x$ correct to the first order. This would be absolutely correct if the cross-sectional areas remained constant in the interval δx. A is liable to small variations in the small interval but the consequent correction to the element of volume is of the second order of infinitesimals.

Note 2. The volume of the solid generated by revolution about the y-axis of the area bounded by the curve $x = f(y)$, the y-axis and the abscissae $y = a$, $y = b$, is

$$\int_a^b \pi x^2 dy$$

Note 3. If the curve is $r = f(\theta)$ and it revolves about the initial line, the volume generated is

$\pi \int_a^b y^2 dx = \pi \int_{\theta_1}^{\theta_2} y^2 \dfrac{dx}{d\theta} \cdot d\theta$, where θ_1 and θ_1 are values of θ correspnding to the points where $x = a$, $x = b$ respectively.

$$x = r \cos \theta \quad \text{and} \quad y = r \sin \theta.$$

Hence $\quad$ volume $= \int_{\theta_1}^{\theta_2} r^2 \sin^2 \theta \dfrac{d}{d\theta} (r \cos \theta)\, d\theta$

Note 4. The curve is given by $x = \phi(t)$ and $y = \psi(t)$ and $x = a$ and $x = b$ given by t_1 and t_2. The volume of the solid generated by the revolution of the curve about x-axis

$$= \pi \int_{t_1}^{t_2} y^2 \dfrac{dx}{dt}\, dt$$

$$= \pi \int_{t_1}^{t_2} \{\psi(t)\}^2 \dfrac{d}{dt} \{\phi(t)\}\, dt$$

Volume and Surface of Solid of Revolution

7.2 VOLUME OF CARTESIAN CURVES

1. *If the hyperbola $\dfrac{x^2}{a^2} - \dfrac{y^2}{b^2} = 1$ revolves about the x-axis, show that the volume included between the surface thus generated, the cone generated by the asymptote and two planes perpendicular to the axis of x, at a distance h apart, is equal to that of a circular cylinder of height h and radius b.*

Sol. The equation of the hyperbola is $\dfrac{x^2}{a^2} - \dfrac{y^2}{b^2} = 1$

The equation of asymptotes is

$$\dfrac{x^2}{a^2} - \dfrac{y^2}{b^2} = 0$$

or $\quad y = \pm \dfrac{b}{a} x.$

OK is an asymptote which is $y = \dfrac{b}{a} x.$

Let us take two perpendicular planes $x = l$ and $x = l + h$ from the origin which are at a distance h.

The volume of the part of the cone generated by asymptote between the planes $x = l$ and $x = l + h$

$$= \int_l^{l+h} \pi y^2 \, dx = \pi \int_l^{l+h} \dfrac{b^2}{a^2} x^2 \, dx$$

$$= \dfrac{\pi b^2}{3a^2} \left[\dfrac{x^2}{3} \right]_l^{l+h}$$

$$= \dfrac{\pi b^2}{3a^2} [(l+h)^3 - l^3]$$

$$= \dfrac{\pi b^2}{3a^2} [3l^2 h + 3lh^2 + h^3] \qquad \ldots(i)$$

Now the volume of the portion of the solid generated by the hyperboloid between two planes

$$= \int_l^{l+h} \pi y^2 \, dx \text{ for the curve } \dfrac{x^2}{a^2} - \dfrac{y^2}{b^2} = 1$$

$$= \int_l^{l+h} b^2 \pi \left(\dfrac{x^2}{a^2} - 1 \right) dx$$

$$= \dfrac{\pi b^2}{a^2} \int_l^{l+h} (x^2 - a^2) \, dx = \dfrac{\pi b^2}{a^2} \left[\dfrac{x^3}{3} - a^2 x \right]_l^{l+h}$$

$$= \dfrac{\pi b^2}{3a^2} [(l+h)^3 - 3a^2(l+h) - l^2 + 3a^2 l]$$

$$= \dfrac{\pi b^2}{3a^2} [3l^2 h + 3lh^2 + h^3 + 3a^2 h] \qquad \ldots(ii)$$

The required volume

$$= (i) - (ii) = \frac{\pi b^2}{3a^2}(3a^2 h)$$

$$= \pi b^2 h$$

= volume of the cylinder of radius b and height h.

2. *Find the volume of the solid generated by the revolution of the curve* $y = \dfrac{a^3}{a^2 + x^2}$ *about its asymptote.*

Sol. The equation of the curve is

$$y = \frac{a^3}{a^2 + x^2}$$

or $\qquad x^2 y = a^2(a - y)$

∴ The equation of the asymptote is $y = 0$, i.e., x-axis (equating the coefficient of highest degree term of x to zero).

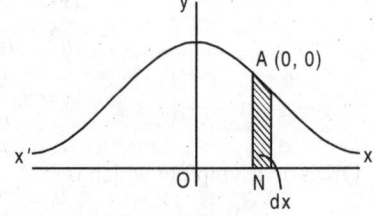

∴ Required volume $= \int_{-\infty}^{\infty} \pi y^2 dx$ (As x varies from $-\infty$ to ∞)

$$= \pi a^3 \int_{-\infty}^{\infty} \frac{dx}{(a^2 + x^2)^2}$$

$$= 2\pi a^6 \int_0^{\pi/2} \frac{a \sec^2 \theta \, d\theta}{a^4 (1 + \tan^2 \theta)^2}$$

$$= 2\pi a^3 \int_0^{\pi/2} \cos^2 \theta \, d\theta = 2\pi a^3 \frac{\overline{\left|\frac{3}{2}\right.} \overline{\left|\frac{1}{2}\right.}}{2 \overline{\left|2\right.}}$$

$$= 2\pi a^3 \cdot \frac{1}{2} \cdot \frac{1}{2} \sqrt{\pi} \cdot \sqrt{\pi} = \frac{1}{2}\pi^2 a^3.$$

EXERCISES

1. Determine the volume of a hemisphere and then show that the volume of a sphere is $\dfrac{4}{3}\pi a^3$.

$$\left[\text{Ans. } \frac{2}{3}\pi a^3\right]$$

2. Find the volume of the cone of semivertical angle α and height h. $\left[\text{Ans. } \dfrac{1}{3}\pi h^3 \tan^2 \alpha\right]$

3. Find the volume of a segment of height h of a sphere of radius r. $\left[\text{Ans. } \pi h^2 \left(r - \dfrac{1}{3}h\right)\right]$

4. A segment is cut off from a sphere of radius a by its plane of a distance $a/2$ from the centre. Show that the volume of the segment is $\dfrac{5}{32}$ of the volume of the sphere.

5. Find the volume generated by the revolution of the ellipse $\dfrac{x^2}{a^2} + \dfrac{y^2}{b^2} = 1$ about the major axis. $\left[\text{Ans. } \dfrac{4}{3}\pi ab^2\right]$

Volume and Surface of Solid of Revolution

6. The part of the ellipse $\dfrac{x^2}{a^2} + \dfrac{y^2}{b^2} = 1$ cut off by a latus rectum revolves about the tangent at the nearest vertex. Find the volume of the reel thus generated. $\left[\text{Ans. } \dfrac{2\pi b}{3a}[6ba^2 - 3a^2 b]\right]$

7. The ellipse $b^2 x^2 + a^2 y^2 = a^2 b^2$ is divided into two parts by the line $x = a/2$, and the smaller part is rotated through four right angles about this line. Prove that the volume generated is $\pi a^2 b \left(\dfrac{3}{4}\sqrt{3} - \dfrac{\pi}{3}\right)$.

8. Find the volume of the prolate and oblate spheroids generated by the ellipse whose major and minor axes are $(24)^{1/3}$ and $(3\pi)^{1/3}$ respectively. $\left[\text{Ans. } \pi^2, 2\pi^2\right]$

9. Find the volume of the paraboloid generated by the revolution of the parabola $y^2 = 4ax$ about the x-axis from $x = 0$ to $x = b$. $[\text{Ans. } 2\pi ah]$

10. The area of the parabola $y^2 = 4ax$ lying between the vertex and the latus rectum is revolved about x-axis. Find the volume generated. $\left[\text{Ans. } 2\pi a^3\right]$

11. The part of the parabola $y^2 = 4ax$ cut off by the latus rectum revolves about the tangent at the vertex. Find the volume of the reel thus generated. $\left[\text{Ans. } \dfrac{4}{5}\pi a^2\right]$

12. Find the volume of the solid generated by the revolution of an arc of the catenary $y = c \cosh \dfrac{x}{c}$ about the x-axis. $\left[\text{Ans. } \dfrac{\pi c^2}{2}\left[x + \dfrac{c}{2}\sinh\dfrac{2x}{c}\right]\right]$

13. Find the volume of the spindle-shaped solid generated by revolving the astroid $x^{2/3} + y^{2/3} = a^{2/3}$ about the x-axis. $\left[\text{Ans. } \dfrac{32\pi a^3}{105}\right]$

14. Find the volume of the solid generated by the revolution of the area between the curve $xy^2 = 4a^2(2a - x)$ and its asymptote about the asymptote. $\left[\text{Ans. } 4\pi^2 a^3\right]$

15. The curve $y^2(a + x) = x^2(3a - x)$ revolves about the x-axis. Find the volume generated by the loop. $\left[\text{Ans. } \pi a^3 (8 \log 2 - 3)\right]$

7.3 VOLUME OF POLAR CURVES

To find the volume of the solid generated by the revolution about the initial line of the area bounded by the curve $r = f(\theta)$ and the radii vectors $\theta = \alpha, \theta = \beta$.

Let the sectorial area *OAB* revolves about the initial line *OX* and this area be divided into small elements *OPQ* whose area is clearly $\dfrac{1}{2}r^2 \delta\theta$ to the first order of infinitesimals and its centroid is at a distance $\dfrac{2}{3}r$ from *O*.

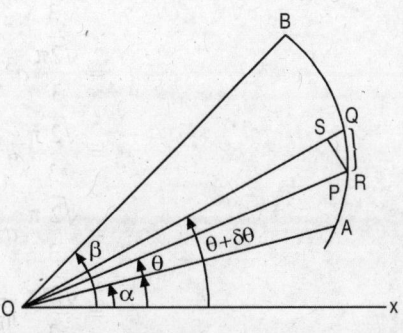

Therefore the volume of the solid generated by the revolution about the initial line is

$$\left(\frac{1}{2}r^2\delta\theta\right).2\pi\left(\frac{2}{3}r\sin\theta\right)$$

Hence the required volume

$$=\frac{2}{3}\int_\alpha^\beta r^3 \sin\theta\, d\theta$$

Note. The volume of the solid generated by the revolution of the area bounded by the curve $r = f(\theta)$ and the radii vectors $\theta = \alpha, \theta = \beta$ about the line $\theta = \dfrac{\pi}{2}$ is

$$=\frac{2}{3}\pi \int_\alpha^\beta r^3 \cos\theta\, d\theta$$

EXAMPLE

Find the volume of the solid formed by revolving one loop of the curve $r^2 = a^2 \cos 2\theta$,
(i) about the initial line

(ii) about the line $\theta = \dfrac{\pi}{2}$.

Sol. The equation of generating curve is

$$r^2 = a^2 \cos 2\theta$$

(i) The required volume

$$=\frac{2\pi}{3}\int_0^{\pi/4} r^3 \sin\theta\, d\theta$$

$$=\frac{2\pi}{3}\int_0^{\pi/4} a^3 \cos^{3/2} 2\theta \sin\theta\, d\theta$$

$$=\frac{2\pi}{3}\int_0^{\pi/4} (2\cos^2\theta - 1)^{3/2} \sin\theta\, d\theta$$

Putting $\sqrt{2}\cos\theta = t, -\sqrt{2}\sin\theta\, d\theta = dt$

$$=-\frac{2\pi}{3\sqrt{2}} a^3 \int_{\sqrt{2}}^1 (t^2 - 1)^{3/2} dt$$

$$=\frac{2\pi}{3\sqrt{2}} a^3 \int_1^{\sqrt{2}} (t^2 - 1)^{3/2} dt$$

put $t = \sec\phi, dt = \sec\phi\tan\phi\, d\phi$

$$=\frac{\sqrt{2}\,\pi}{3} a^3 \int_0^{\pi/4} (\sec^2\phi - 1)^{3/2} \sec\phi\tan\phi\, d\phi$$

$$=\frac{\sqrt{2}\pi}{3} a^3 \int_0^{\pi/4} \tan^4\phi \sec\phi\, d\phi$$

$$=\frac{\sqrt{2}\,\pi}{3} a^3 \int_0^{\pi/4} (\sec^2\phi - 1)^2 \sec\phi\, d\phi$$

$$=\frac{\sqrt{2}\,\pi}{3} a^3 \int_0^{\pi/4} (\sec^4\phi + 1 - 2\sec^2\phi) \sec\phi\, d\phi$$

$$=\frac{\sqrt{2}\,\pi}{3} \int_0^{\pi/4} (\sec^5\phi + \sec\phi - 2\sec^3\phi)\, d\phi \qquad \ldots(i)$$

Volume and Surface of Solid of Revolution

For $\int \sec^n \phi \, d\phi$, the reduction formula is

$$\int \sec^n \phi \, d\phi = \frac{\sec^{n-2} \phi \tan \phi}{n-1} + \frac{n-2}{n-1} \int \sec^{n-2} \phi \, d\phi$$

$$\int_0^{\pi/4} \sec^5 \phi \, d\phi = \left[\frac{\sec^3 \phi \tan \phi}{4} \right]_0^{\pi/4} + \frac{3}{4} \int_0^{\pi/4} \sec^3 \phi \, d\phi$$

$$= \frac{\sec^3 \frac{\pi}{4} \tan \frac{\pi}{4}}{4} + \frac{3}{4} \left[\frac{\sec \phi \tan \phi}{2} \right]_0^{\pi/4} + \frac{3}{4} \cdot \frac{1}{2} \int_0^{\pi/4} \sec \phi \, d\phi$$

$$= \frac{2\sqrt{2}}{4} + \frac{3\sqrt{2}}{8} + \frac{3}{8} [\log (\sec \phi + \tan \phi)]_0^{\pi/4}$$

$$= \frac{\sqrt{2}}{2} + \frac{3\sqrt{2}}{8} + \frac{3}{8} (\sqrt{2} + 1)$$

$$\int_0^{\pi/4} \sec^3 \phi \, d\phi = \left[\frac{\sec \phi \tan \phi}{2} \right]_0^{\pi/4} + \frac{1}{2} \int_0^{\pi/4} \sec \phi \, d\phi$$

$$= \frac{\sqrt{2}}{2} + \frac{1}{2} [\log (\sec \phi + \tan \phi)]_0^{\pi/4}$$

$$= \frac{\sqrt{2}}{2} + \frac{1}{2} \log (\sqrt{2} + 1)$$

and $\int_0^{\pi/4} \sec \phi \, d\phi = [\log (\sec \phi + \tan \phi)]_0^{\pi/4}$

$$= \log (\sqrt{2} + 1)$$

By putting these values in (i), we get the required volume

$$= \frac{\sqrt{2} \pi a^3}{3} \left[\frac{\sqrt{2}}{2} + \frac{3\sqrt{2}}{8} + \frac{3}{8} \log (\sqrt{2} + 1) + \log (\sqrt{2} + 1) - \frac{2\sqrt{2}}{2} - \log (\sqrt{2} + 1) \right]$$

$$= \frac{\sqrt{2} \pi a^3}{3} \left[\frac{3}{8} \log (\sqrt{2} + 1) - \frac{\sqrt{2}}{8} \right]$$

$$= \pi a^3 \left[\frac{\sqrt{2}}{8} \log (\sqrt{2} + 1) - \frac{1}{12} \right]$$

(ii) About $\theta = \frac{\pi}{2}$,

$$V = \frac{2}{3} \pi \int_{-\pi/4}^{\pi/4} r^3 \cos \theta \, d\theta$$

$$= \frac{4}{3} \pi \int_0^{\pi/4} a^3 \cos^{3/2} 2\theta \cos \theta \, d\theta$$

$$= \frac{4\pi a^3}{3} \int_0^{\pi/4} (1 - 2 \sin^2 \theta)^{3/2} \cos \theta \, d\theta$$

Put $\sqrt{2} \sin \theta + \sin \phi$, $\sqrt{2} \cos \theta \, d\theta = \cos \phi \, d\phi$

$$= \frac{4\pi a^3}{3\sqrt{2}} \int_0^{\pi/2} (1 - \sin^2 \phi)^{3/2} \cos \phi \, d\phi$$

$$= \frac{2\sqrt{2}}{3}\pi a^3 \int_0^{\pi/2} \cos^4 \phi\, d\phi$$

$$= \frac{2\sqrt{2}}{3}\pi a^3 \cdot \frac{\overline{\left|\frac{5}{2}\right.}\cdot\overline{\left|\frac{1}{2}\right.}}{2\overline{\left|3\right.}} = \frac{\pi^2 a^3}{4/\sqrt{2}}$$

EXERCISES

1. The arc of the cardioid $r = a(1+\cos\theta)$ included between $-\frac{1}{2}\pi \le \theta \le \frac{1}{2}\pi$ is rotated about the line $\theta = 0$. Prove that its volume in $\frac{5}{2}a^2\pi$.

2. Show that the volume of the solid formed by the revolution of the limacon $r = a + b\cos\theta\ (a < b)$ about the initial line is
$$\frac{4}{3}\pi a(a^2 + b^2)$$

3. Prove that the volume generated by the revolution of the curve $r = a + b\sec\theta,\ -\frac{\pi}{4} < \theta < \frac{\pi}{2}$ about its asymptote is $2\pi a^2\left(\frac{8}{3}a + \frac{1}{2}\pi b\right)$.

4. Show that the volume generated by the revolution of the curve $r\cos\theta = a\cos 2\theta$ about the initial line is $2\pi a^3\left(\log 2 - \frac{2}{3}\right)$.

5. Find the volume of the solid generated by the revolution of the cardioid $r = a(1-\cos\theta)$ about the initial line. $\left[\text{Ans. } \frac{8}{3}\pi a^3\right]$

6. Find the volume generated by revolution of the lemniscate $r^2 = a^2\cos 2\theta$ about the line $\theta = \pi/2$. $\left[\text{Ans. } \frac{\pi^2 a^2}{4\sqrt{2}}\right]$

7.4 VOLUME OF PARAMETRIC CURVES

Example. *Find the volume of the solid generated by the revolution of the tractrix* $x = a\cos t + \frac{1}{2}a\log\tan^2 t/2,\ y = a\sin t$ *about its asymptote.*

Sol. The asymptote is x-axis, i.e., $y = 0$ for the curve. The required volume

$$= 2\int_0^{\pi/2} \pi y^2 \frac{dx}{dt}\, dt,$$

We have
$$x = a\cos t + \frac{a}{2}\log\tan^2 t/2$$

$$\therefore \quad \frac{dx}{dt} = -a\sin t + \frac{2a\sec^2 \frac{t}{2}}{2\tan t/2}$$

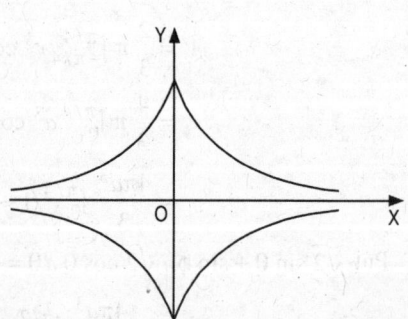

$$= -a\sin t + \frac{a}{\sin t} = \frac{a\cos^2 t}{\sin t}$$

$$\therefore \quad \text{Volume} = 2\pi \int_0^{\pi/2} a^2 \sin^2 t \cdot \frac{a\cos^2 t}{\sin t} dt$$

$$= 2\pi a^3 \int_0^{\pi/2} \sin t \cos^2 t \, dt$$

$$= 2\pi a^3 \cdot \frac{\lceil 1 \cdot \lceil \frac{3}{2}}{2\lceil \frac{5}{2}} = \frac{2}{3}\pi a^3$$

EXERCISES

1. Find the volume generated by the revolution of the loop of the curve $x = t^2$, $y = t - \frac{1}{3}t^3$ about the x-axis. $\left[\text{Ans. } \frac{3\pi}{4}\right]$

2. Show that the volume of the solid generted by the revolution of the cycloid $x = a(\theta + \sin\theta)$, $y = a(1 - \cos\theta)$, $0 \le \theta \le \pi$ about the y-axis is $\pi a^3 \left(\frac{3}{2}\pi^2 - \frac{8}{3}\right)$.

3. Find the volume of the solid generated by the curve $x = 2a\sin^2 t$, $y = 2a\sin^3 t/\cos t$ about its asysuptote. $\left[\text{Ans. } 2\pi^2 a^2\right]$

7.5 SURFACE OF REVOLUTION

Let AB be a portion of a curve $y = f(x)$ which revolves about OX, generating a solid of revolution. It is required to find an expression for the surface of this solid. Now let PQ be a small part of the curve which on revolving generates a portion of the surface.

Let $PQ = \delta s$ and the coordinates of P be (x, y) and $(x + \delta x, y + \delta y)$ be the coordinates of Q, then y and $y + \delta y$ are the lengths of the ordinates of P and Q. As PQ is small, the portion of the surface which it genrates may be taken as surface of the frustum of a cone and then its area is $2\pi \frac{y + (y + \delta y)}{2} \delta s$. If PQ becomes indefinitely small so that $\delta y \to 0$, then in the limit, area of the strip generated by $PQ = 2\pi y\, \delta s$.

Hence the whole surface

$$= 2\pi \int y \, ds$$

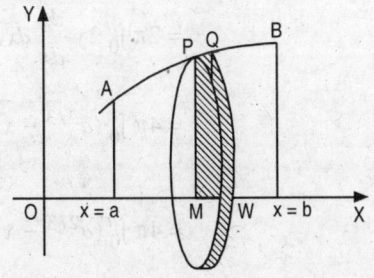

This may be written as

$$\int 2\pi y \frac{ds}{dx} dx, \int 2\pi y \frac{ds}{dy} dy,$$

$$\int 2\pi y \frac{ds}{dt} dt, \int 2\pi y \frac{ds}{2\theta} d\theta,$$

$$\int 2\pi y \frac{ds}{dr} dr, \text{ where}$$

$$\frac{ds}{dx} = \sqrt{\left\{1 + \left(\frac{dy}{dx}\right)^2\right\}}, \frac{ds}{dt} = \sqrt{\left\{\left(\frac{dx}{dt}\right)^2 + \left(\frac{dy}{dt}\right)^2\right\}}$$

$$\frac{ds}{dx} = \sqrt{\left\{1+\left(\frac{dx}{dy}\right)^2\right\}}, \quad \frac{ds}{d\theta} = \sqrt{\left\{r^2+\left(\frac{dr}{d\theta}\right)^2\right\}},$$

$$\frac{ds}{dr} = \sqrt{\left\{r^2\left(\frac{d\theta}{dr}\right)^2+1\right\}}$$

Note. If the curve is rotated about y-axis, then the whole-surface of revolution shall be given by

$$\int 2\pi x \, ds$$

EXAMPLES

1. *Find the area of the surface of a cone whose semi vertical angle is α and base a circle of radius a.*

Sol. The cone is formed by the revolution of the line $y = x \tan \alpha$ about x-axis, where

$$\frac{a}{h} = \tan \alpha, \, h \text{ is the height of the cone}$$

$$\therefore \quad \frac{dy}{dx} = \tan \alpha$$

$$\therefore \quad S = \int_0^h 2\pi y \frac{ds}{dx} dx$$

$$= 2\pi \int_0^h x \tan \alpha \sqrt{(1+\tan^2 \alpha)} \, dx$$

$$= 2\pi \tan \alpha \sec \alpha \left[\frac{x^2}{2}\right]_0^h$$

$$= 2\pi \frac{\sin \alpha}{\cos^2 \alpha} \cdot \frac{h^2}{2} = \frac{\pi a^2}{\sin \alpha}$$

$$= \pi a^2 \csc \alpha$$

2. *Find the surface of the solid generated by the revolution of the astroid $x^{2/3} + y^{2/3} = a^{2/3}$ about the x-axis.*

Sol. The limits for x and 0 to a.
The required surface

$$= 2\pi \int_0^a 2y \frac{ds}{dx} dx$$

$$= 4\pi \int_0^a (a^{2/3} - x^{2/3})^{3/2} \sqrt{\left\{1+\left(\frac{dy}{dx}\right)^2\right\}} \, dx$$

$$= 4\pi \int_0^a (a^{2/3} - x^{2/3})^{3/2} \sqrt{\left(1+\frac{y^{2/3}}{x^{2/3}}\right)} \, dx$$

$$= 4\pi \int_0^a (a^{2/3} - x^{2/3})^{3/2} \frac{\sqrt{(x^{2/3}+y^{2/3})}}{x^{1/3}} dx$$

$$= 4\pi \int_0^a (a^{2/3} - x^{2/3})^{3/2} a^{1/3} x^{-1/3} dx$$

Put $x = a \sin^3 \theta, \, dx = 3a \sin^2 \theta \cos \theta \, d\theta$

$$= 4\pi a^{1/3} \int_0^{\pi/2} (a^{2/3} - a^{2/3} \sin^2 \theta)^{3/2} (a \sin \theta)^{-1/3} \cdot 3a \sin^2 \theta \cos \theta \, d\theta$$

$$= 12 \pi a^{1/3} \cdot a^2 \cdot a^{-1/3} \int_0^{\pi/2} \cos^4 \theta \sin \theta \, d\theta$$

$$= 12 \pi a^2 \frac{\overline{\left|\frac{5}{2}\right.} \cdot \overline{\left|1\right.}}{\overline{\left|7/2\right.}} = \frac{12 \pi a^2}{5}$$

EXERCISES

1. Find the surface of a sphere of radius a. $\left[\text{Ans. } 4\pi a^2\right]$

2. Find the curved surface of the solid generated by the revolution about x-axis of the area bounded by the parabola $y^2 = 4ax$, the ordinate $x = 3a$ and the x-axis. $\left[\text{Ans. } \frac{56}{3}\pi a^2\right]$

3. Find the surface of a zone cut off from a sphere of radius a and parallel planes having distance h between them. $[\text{Ans. } \pi ah]$

4. Find the surface of the prolate spheroid by the revolution of the ellipse $\frac{x^2}{a^2} + \frac{y^2}{b^2} = 1$ about major axis. $\left[\text{Ans. } \pi a^2 [3\sqrt{2} - \log(1+\sqrt{2})]\right]$

5. Find the curved surface of the solid genrated by the revolution about the x-axis of the arc bounded by the parabola $y^2 = 4ax$ the x-axis and the ordinate $x = h$.
$$\left[\text{Ans. } \frac{8}{3}\pi\sqrt{a}\{(a+h)^{3/2} - a^{3/2}\}\right]$$

6. Find the surface area of the solid generated by the revolution of the arc of the parabola $y^2 = 4ax$ bounded by its latus rectum about the x-axis. $\left[\text{Ans. } \frac{8\pi a^2}{3}[2\sqrt{2} - 1]\right]$

7. The part of the parabola $y^2 = 4ax$ cut off by the latus-rectum revolves about the tangent at the vertex. Find the curved surface of the reel thus generated.
$$\left[\text{Ans. } \pi a^2 [3\sqrt{2} - \log(1+\sqrt{2})]\right]$$

8. Find the surface generated by the revolution of the curve $y = c \cosh \frac{x}{c}$ about the x-axis, between the planes $x = a$ and $x = b$.
$$\left[\text{Ans. } \pi c (b-a) + \frac{1}{2} c \left\{\sinh\left(\frac{2b}{c}\right) - \sinh\left(\frac{2a}{c}\right)\right\}\right]$$

9. Find the area of the solid formed by the revolution, about the axis of y, of the part of the curve $ay^2 = x^3$ from $x = 0$ to $x = 4a$ which is above the x-axis.
$$\left[\text{Ans. } \frac{128 \pi a^2}{1215}(125\sqrt{10} + 1)\right]$$

10. Prove that the surface of the prolate spheroid formed by the revolution of an ellipse of eccentricity e about its major axis is equal to 2 area of the ellipse $\left\{\sqrt{(1-e^2)} + \frac{\sin^{-1} e}{e}\right\}$.

7.6 SURFACE OF PARAMETRIC CURVES

EXAMPLES

1. *Prove that the surface generated by the revolution of the tractrix* $x = a \cos t + \frac{a}{2} \log \tan^2 \frac{t}{2}$, $y = a \sin t$ *about its asymptote is equal to the surface of sphere of radius* a.

Sol. Here $x = a \cos t + \frac{a}{2} \log \tan^2 \frac{t}{2}$

$$\frac{dx}{dt} = -a \sin t + \frac{a}{2} \cdot \frac{2 \tan \frac{t}{2} \sec^2 \frac{t}{2} \cdot \frac{1}{2}}{\tan^2 \frac{t}{2}}$$

$$= -a \sin t + \frac{a}{2 \sin \frac{t}{2} \cos \frac{t}{2}} = \frac{a \cos^2 t}{\sin t}$$

$$y = a \sin t, \quad \frac{dy}{dt} = a \cos t$$

Therefore

$$\frac{ds}{dt} = \sqrt{\left\{\left(\frac{dx}{dt}\right)^2 + \left(\frac{dy}{dt}\right)^2\right\}}$$

$$= a \sqrt{\left(\frac{\cos^4 t}{\sin^2 t} + \cos^2 t\right)}$$

$$= \frac{a \cos t}{\sin t} = a \cot t$$

The required surface

$$= 2 \times 2\pi \int_0^{\pi/2} y \frac{ds}{dt} dt$$

$$= 4\pi \int_0^{\pi/2} a \sin t \cdot a \cot t \, dt$$

$$= 4\pi a^2 \int_0^{\pi/2} \cos t \, dt$$

$$= 4\pi a^2 [\sin t]_0^{\pi/2} = 4\pi a^2.$$

2. *Evaluate the area of the surface generated by the revolution of the cycloid* $x = a(\theta - \sin \theta)$, $y = a(1 - \cos \theta)$ *about x-axis.*

Sol. The parametric equations of the curve are

$$x = a(\theta - \sin \theta), \quad y = a(1 - \cos \theta)$$

$$\therefore \quad \frac{dx}{dt} = a(1 - \cos \theta), \quad \frac{dy}{d\theta} = a \sin \theta$$

$$\therefore \quad \frac{ds}{d\theta} = \sqrt{\left\{\left(\frac{dx}{d\theta}\right)^2 + \left(\frac{dy}{d\theta}\right)^2\right\}}$$

$$= a \sqrt{\{(1 - \cos \theta)^2 + \sin^2 \theta\}}$$

Volume and Surface of Solid of Revolution

$$= a\sqrt{(1-2\cos\theta + \cos^2\theta + \sin^2\theta)}$$

$$= a\sqrt{2}\cdot\sqrt{(1-\cos\theta)} = 2a\sin\frac{\theta}{2}$$

For one arc of the curve θ varies from 0 to 2π.
the required area of the surface

$$= 2\pi \int_0^{2\pi} y \frac{ds}{d\theta} d\theta$$

$$= 2\pi \int_0^2 a(1-\cos\theta), 2a\sin\frac{\theta}{2} d\theta$$

$$= 8\pi a^3 \int_0^{2\pi} \sin^3\frac{\theta}{2} d\theta$$

$$= 16\pi a^2 \int_0^{\pi} \sin^3\frac{\theta}{2} d\theta$$

$$= 32\pi a^2 \int_0^{\pi/2} \sin^3 t \, dt \qquad \text{putting } \frac{1}{2}\theta = t \text{ or } d\theta = 2\, dt$$

$$= 32\pi a^2 \cdot \frac{\left\lceil 2\cdot\left\lceil \frac{1}{2}\right.\right.}{2\left\lceil \frac{5}{2}\right.} = \frac{64}{3}\pi a^2$$

EXERCISES

1. Prove that the surface area of the solid generated by the revolution about the x-axis of the loop of the curve $x = t^2$, $y = t - \frac{t^3}{3}$ is 3π.
2. The portion between the two consecutive cusps of the cycloid $x = a(\theta + \sin\theta)$, $y = a(1+\cos\theta)$ is revolved about the x-axis. Show that the area of the surface so formed to the area of the cycloid as $64:9$.
3. Prove that the surface and volume generated by the revolution of the tractrix

$$x = a\cos t + \frac{1}{2}a\log\tan^2\frac{t}{3}, \; y = a\sin t.$$

about its asymptote are respectively equal to the surface and half the volume of a sphere of radius a.

7.7 SURFACE OF POLAR CURVES

EXAMPLES

1. *Find the surface of the solid formed by the revolution of the cardioid $r = a(1+\cos\theta)$ about the initial line.*

Sol. The given curve is

$$r = a(1+\cos\theta)$$

Limits are 0 to π and $\frac{dr}{d\theta} = -a\sin\theta$

and $\qquad y = r\sin\theta = a(1+\cos\theta)\sin\theta$

$\therefore\quad$ The required surface

$$= 2\pi \int_0^{\pi} a(1+\cos\theta)\sin\theta \sqrt{\left\{r^2 + \left(\frac{dr}{d\theta}\right)^2\right\}} d\theta$$

$$= 2\pi a \int_0^\pi (1+\cos\theta)\sin\theta \cdot a \sqrt{\{(1+\cos\theta)^2 + \sin^2\theta\}} \, d\theta$$

$$= 2\pi a^2 \int_0^\pi \sin\theta (1+\cos\theta) \cdot 2\cos\frac{\theta}{2} \, d\theta$$

$$= 4a^2 \int_0^\pi 2\sin\frac{\theta}{2}\cos^2\frac{\theta}{2} \cdot 2\cos^2\frac{\theta}{2} \, d\theta$$

$$= 16\pi a^2 \int_0^\pi \cos^4\frac{\theta}{2}\sin\frac{\theta}{2} \, d\theta$$

$$= \frac{32\pi a^2}{5}\left[-\cos^5\frac{\theta}{2}\right]_0^\pi$$

$$= \frac{32}{5}\pi a^2.$$

2. *Find the surface of the solid generated by the revolution of the lemniscate $r^2 = a^2 \cos 2\theta$ about the initial line.*

Sol. There are two loops in the curve. For the upper half of the loop on the right, θ varies from 0 to $\frac{1}{4}\pi$.

The equation of the curve is

$$r^2 = a^2 \cos 2\theta$$

differentiating, we get

$$2r\frac{dr}{d\theta} = -2a^2 \sin 2\theta$$

or

$$\frac{dr}{d\theta} = -\frac{a^2 \sin 2\theta}{r}$$

$$\therefore \quad \frac{ds}{d\theta} = \sqrt{\left[r^2 + \left(\frac{dr}{d\theta}\right)^2\right]}$$

$$= \sqrt{\left[a^2\cos 2\theta + \frac{a^4 \sin^2 2\theta}{r^2}\right]}$$

$$= a\sqrt{\left(\cos 2\theta + \frac{\sin^2 2\theta}{\cos 2\theta}\right)}$$

$$= \frac{a}{\sqrt{(\cos 2\theta)}}$$

$\therefore$ The required surface $= 2 \times$ surface generated by one loop

$$= 2 \times 2\pi \int_0^{\pi/4} y \frac{ds}{d\theta} \cdot d\theta$$

$$= 4\pi \int_0^{\pi/4} r \sin\theta \frac{a}{\sqrt{(\cos 2\theta)}} \, d\theta$$

$$= 4\pi \int_0^{\pi/4} a\sqrt{(\cos 2\theta)} \sin\theta \frac{a}{\sqrt{(\cos 2\theta)}} \, d\theta$$

$$= 4\pi a^2 \int_0^{\pi/4} \sin\theta \, d\theta = 4a^2[-\cos\theta]_0^{\pi/4}$$
$$= 4\pi a^2 \left(1 - \frac{1}{\sqrt{2}}\right).$$

EXERCISES

1. Show that the area of the surface of revolution formed by the revolution of the curve $r = 2a \cos\theta$ about the initial line is $4\pi a^2$.

2. The lemnicate $r^2 = a^2 \cos 2\theta$ revolves about a tangent at the pole. Show that the surface of the solid generated is $4\pi a^2$.

3. Find the surface of the solid generated by the revolution of the cardioid $r = a(1 - \cos\theta)$ about the initial line. $\left[\text{Ans. } \dfrac{32}{5}\pi a^2\right]$

ORDINARY DIFFERENTIAL EQUATIONS

Degree and order of a differential equation. Equations of first order and first degree. Equations in which the variables are separable. Homogeneous equations. Linear equations and equations reducible to the linear form. Exact differential equations. First order higher degree equations solvable for x, y, p. Clairaut's form and singular solutions. Geometrical meaning of a differential equation. Orthogonal trajectories. Linear differential equations with constant coefficients. Homogeneous linear ordinary differential equations.

Linear differential equations of second order. Transformation of the equation by changing the dependent variable/the independent variable. Method of variation of parameters.

Ordinary simultaneous differential equations.

1
Differential Equations : An Introduction

1.1 DIFFERENTIAL EQUATION

An equation involving differentials or differential coefficients or derivatives is called a *differential equation*.

For example,
$$\frac{dy}{dx} = x + 3 \qquad \text{...(i)}$$

$$\frac{d^2y}{dx^2} + 3\frac{dy}{dx} + 2y = 0, \qquad \text{...(ii)}$$

$$(y+b)^2 \frac{dx}{dz} + z\frac{dy}{dz} - (y+c) = 0 \qquad \text{...(iii)}$$

$$y + x\frac{dy}{dx} = k\sqrt{\left\{1 + \left(\frac{dy}{dx}\right)^2\right\}} \qquad \text{...(iv)}$$

$$\rho = \frac{\left[1 + \left(\frac{dy}{dx}\right)^2\right]^{3/2}}{d^2y/dx^2} \qquad \text{...(v)}$$

$$y\frac{dy}{dx} = x\left(\frac{dy}{dx}\right)^2 + a \qquad \text{...(vi)}$$

all are differential equations.

These are called *ordinary differential equations*. They can be defined as those equations in which all the differential coefficients have reference to a single independent variabe. Secondly, there are *partial differential equations* which are beyond our scope.

1.2 ORDER AND DEGREE OF DIFFERENTIAL EQUATION

The order of a differential equation is the order of the highest derivative which appears in it.
For example, the order of the equation

$$\left[1 + \left(\frac{dy}{dx}\right)^2\right]^{3/2} = \rho \cdot \frac{d^2y}{dx^2}$$

is two, as the highest derivative is $\frac{d^2y}{dx^2}$ which appears in it.

Thus, this differential equation is called a *differential equation of second order*.

The degree of a differential equation is the degree of the highest derivative which appears in it, after the equation has been made clear of the radicals and the fractions.

From the examples (i) to (vi), we see that (i) is of the first order and first degree, (ii) is of the second order and first degree, (iv) is of 1st order and 2nd degree, (v) is of the second order and second degree and (vi) is of the first order and second degree.

1.2.1 Solutions of Differential Equations

A *solution* or *integral* of a differential equation is a relation that exists between the dependent and independent variables and at the same time the derivatives obtained there from satisfy the given equation. For example, $y = \sin x + c$ is a solution of differential equation $\dfrac{dy}{dx} = \cos x$.

1.2.2 Kinds of Solutions

(a) General Solution or the Complete Primitive : The solution of a differential equation which contains a number of arbitrary constants equal to the order of the differential equation is called the *general solution* or the *complete primitive*. For example, $y = c_1 \cos x + c_2 \sin x$ is a general solution of the differential equation.

$$\dfrac{d^2 y}{dx^2} + y = 0$$

We observe here that the differential equation under consideration is of the second order; therefore, its general solution shall contain two arbitrary constants. Such constants which appear in general solution, depending upon the order of the equation, are called *arbitrary constants*.

(b) Particular Solution : If particular values are given to the arbitrary constants in the general solution, then the solution so obtained is called the *particular solution*.

For example, the complete primitive of (i) is

$$y = A \cos(x + B) = a \cos x + b \sin x$$

If we assign the particular values 1, 0 to these constants a and b, then $y = \cos x$ is defined as a *Particular Integral*. Another possible Particular Integral is $y = \sin x$.

(c) Singular Solution : The solution which cannot be derived from the complete primitive or general solution by assigning particular values to the arbitrary constants are called the *Singular Solutions*. For example, $y^2 = 4ax$ is a singular solution of $y \dfrac{dy}{dx} = x \left(\dfrac{dy}{dx}\right)^2 + a$.

EXAMPLES

1. *Find the differential equation of the family of the curves $y = Ae^{2x} + Be^{-2x}$, for different values of A and B.*

Solution. We have

$$y = Ae^{2x} + Be^{-2x} \qquad \ldots(1)$$

Differentiating, we get

$$\dfrac{dy}{dx} = 2Ae^{2x} - 2Be^{-2x}$$

and

$$\dfrac{d^2 y}{dx^2} = 4Ae^{2x} + 4Be^{-2x} = 4(Ae^{2x} + Be^{-2x}) = 4y$$

∴ Required differential equation is

$$\dfrac{d^2 y}{dx^2} = 4y$$

Differential Equations : An Introduction

2. Find the differential equation of the system of curves $y = ax^2 + b \cos nx + c$, where a, b, c are arbitrary constants.

Solution. We have

$$y = ax^2 + b \cos nx + c \qquad \text{...(i)}$$

Differentiating,

$$\frac{dy}{dx} = 2ax - bn \sin nx \qquad \text{...(ii)}$$

$$\frac{d^2y}{dx^2} = 2a - bn^2 \cos nx \qquad \text{...(iii)}$$

and

$$\frac{d^3y}{dx^3} = bn^3 \sin nx \qquad \text{...(iv)}$$

Eliminating a between (ii) and (iii), we get

$$x \frac{d^2y}{dx^2} - \frac{dy}{dx} = -bn(x \cos nx - \sin nx) \qquad \text{...(v)}$$

Eliminating b between (iv) and (v) by dividing, we get

$$\frac{\dfrac{d^3y}{dx^3}}{x \dfrac{d^2y}{dx^2} - \dfrac{dy}{dx}} = \frac{-x^2 \sin nx}{x \cos nx - \sin nx}$$

This is the required differential equation. Note that here three constants have been eliminated and we have got a differential equation of third order.

EXERCISES

1. Find the differential equation from $x^2 - y^2 + 2\lambda xy = 1$, where λ is a parameter.

2. Determine a differential equation from the equation $ax^2 + by^2 = 1$, where a and b are parameters.

3. From $x^2 + y^2 + 2ax + 2by + c = 0$, derive a differential equation not containing a, b and c.

4. Form the differential equation of all circles of radius r.

5. Show that $v = \dfrac{A}{r} + B$ is a solution of $\dfrac{d^2v}{dr^2} + \dfrac{2}{r} \dfrac{dv}{dr} = 0$.

6. Find the differential equation corresponding to

$$y = ae^{2x} + be^{-3x} + ce^x$$

ANSWERS

1. $(x^2 - y^2 - 1)\left(x \dfrac{dy}{dx} + y\right) + 2xy\left(y \dfrac{dy}{dx} - x\right) = 0$;

2. $x\left\{y\dfrac{d^2y}{dx^2} + \left(\dfrac{dy}{dx}\right)^2\right\} = y\dfrac{dy}{dx}$;

3. $\left[1 + \left(\dfrac{dy}{dx}\right)^2\right]\dfrac{d^3y}{dx^3} - 3\dfrac{dy}{dx}\left(\dfrac{d^2y}{dx^2}\right) = 0$;

4. $\left[1 + \left(\dfrac{dy}{dx}\right)^2\right]^3 = y^2\left(\dfrac{d^2y}{dx^2}\right)^2$;

5. $\dfrac{d^3y}{dx^3} - 7\dfrac{dy}{dx} + 6y = 0.$

2
Differential Equations of First Order and First Degree

2.1 EQUATIONS OF THE FIRST ORDER AND FIRST DEGREE

The general form of an ordinary differential equation of the first order and first degree is $M + N\frac{dy}{dx} = 0,$ where M and N are functions of x, y i.e., $f(x, y)$ and $\phi(x, y)$ or constants. This equation can also be expressed as

$$\frac{dy}{dx} = \frac{f(x, y)}{\phi(x, y)} = -\frac{M}{N} \qquad [M \text{ and } N \text{ are } -f(x, y) \text{ and } \phi(x, y)]$$

It is some times found convenient to write the equation in this form.

$$\phi(x, y)\, dy = f(x, y)\, dx$$

or
$$-f(x, y)\, dx + \phi(x, y)\, dy = 0$$

or
$$M\, dx + N\, dy = 0$$

2.2 EQUATIONS IN WHICH THE VARIABLES ARE SEPARABLE

If any differential equation can be written as

$$f(x)\, dx + \phi(y)\, dy = 0$$

We say that variables are seperable. Such equations can be solved as given below :

$$f(x)\, dx + \phi(y)\, dy = 0 \qquad \ldots(i)$$

or
$$f(x) + \phi(y)\frac{dy}{dx} = 0$$

Integrating with respect to x, we have

$$\int f(x)\, dx + \int \phi(y)\frac{dy}{dx}\, dx = c,$$

Where c is an arbitrary cosntant.

or
$$\int f(x)\, dx + \int \phi(y)\, dy = c$$

Thus the solution of (i) is obtained by summing up the integrals of $f(x)$ and $\phi(y)$ with respect to x and y respectively and equating the sum to a constant.

WORKING RULE

1. *Arrange all the terms containing x together with dx on one side and all the terms of y with dy on the other side or put them inside the crocked brackets on one side only.*
2. *Integrate each term seperately with respect to either x or y as associated with dx or dy.*
3. *Add an arbitrary constant on one side only.*

EXAMPLES

1. Solve: $y - x\dfrac{dy}{dx} = a\left(y^2 + \dfrac{dy}{dx}\right)$

Solution. The equation may be written as

$$(y - ay^2) = (x + a)\frac{dy}{dx}$$

or $\quad \dfrac{dy}{y(1-ay)} = \dfrac{dx}{(x+a)}$

or $\quad \left(\dfrac{1}{y} + \dfrac{a}{1-ay}\right)dy = \dfrac{dx}{(x+a)}$

Integrating, we get

$$\log y - \log(1 - ay) = \log(x + a) + \log c$$

or $\quad \log \dfrac{y}{1 - ay} = \log c(x + a)$

$\therefore \quad y = c(1 - ay)(x + a).$

2. Solve: $3e^x \tan y + (1 - e^x)\sec^2 y \dfrac{dy}{dx} = 0.$

Solution. Put $\quad \tan y = v \Rightarrow \sec^2 y \dfrac{dy}{dx} = \dfrac{dv}{dx}$

The given equation becomes $3e^x v + (1 - e^x)\dfrac{dv}{dx} = 0 \quad$ or $\quad \dfrac{3e^x dx}{e^x - 1} = \dfrac{dv}{v}.$

On integration $\quad 3\log(e^x - 1) = \log v + \log c = \log vc$

or $\quad (e^x - 1)^3 = cv \quad$ or $\quad (e^x - 1)^3 = c\tan y$

EXERCISE

Solve the following differential equations :

1. $\dfrac{dy}{dx} = \dfrac{1+y^2}{1+x^2}$ [**Ans.** $y - x = c(1 + yx)$]

2. $\dfrac{dy}{dx} = e^{x-y} + x^2 e^{-y}$ [**Ans.** $e^y = e^x + \dfrac{x^3}{3} + c$]

3. $\dfrac{dy}{dx} = e^{x+y} + x^2 e^{x^3} + y$ [**Ans.** $-e^{-y} = e^x + \dfrac{1}{3}e^{x^3} + c$]

4. $a\left(x\dfrac{dy}{dx} + 2y\right) = xy\dfrac{dy}{dx}$ [**Ans.** $yx^2 = ce^{y/a}$]

5. (a) $\dfrac{dy}{dx} = \dfrac{x(2\log x + 1)}{\sin y + y\cos y}$

   (b) $\dfrac{dy}{dx} = \dfrac{\sin x + x\cos x}{y(2\log y + 1)}$ [**Ans.** (a) $y\sin y = x^2 \log x + c$ (b) $x\sin x = y^2 \log y + c$]

6. $(xy^2 + x)dx + (yx^2 + y)dy = 0.$ [**Ans.** $(x^2 + 1)(y^2 + 1) = c$]

Differential Equations of First Order and First Degree

2.3 CHANGE OF VARIABLES

Sometimes a substitution or the change of variables reduces a given differential equation to the form in which the variables are separable. The following examples will make the students familiar with the process.

EXAMPLES

1. *Solve :* $(x + y)^2 \dfrac{dy}{dx} = a^2.$

Sol. Put $x + y = v$, then $1 + \dfrac{dy}{dx} = \dfrac{dv}{dx}$

or $\dfrac{dy}{dx} = \dfrac{dv}{dx} - 1$

Putting the values in given equation, we have

$$v^2 \left(\dfrac{dv}{dx} - 1 \right) = a^2$$

or $$v^2 \dfrac{dv}{dx} = (v^2 + a^2)$$

or $$dx = \dfrac{v^2}{v^2 + a^2} dv = \dfrac{v^2 + a^2 - a^2}{v^2 + a^2} dv$$

$$= \left(1 - \dfrac{a^2}{v^2 + a^2} \right) dv.$$

Integrating, we get

$$x = v - a \tan^{-1} \dfrac{v}{a} + c.$$

or $$x = x + y - a \tan^{-1} \dfrac{x + y}{a} + c$$

or $$a \tan \left(\dfrac{y - c}{a} \right) = x + y$$

2. *Solve :* $\dfrac{x\,dx + y\,dy}{x\,dy - y\,dx} = \sqrt{\left(\dfrac{a^2 - x^2 - y^2}{x^2 + y^2} \right)}$

Sol. Let

$x = r \cos \theta;$ $\therefore\ \dfrac{dx}{d\theta} = \dfrac{dr}{d\theta} \cos \theta - r \sin \theta$

$y = r \sin \theta;$ $\therefore\ \dfrac{dy}{d\theta} = \dfrac{dr}{d\theta} \sin \theta + r \cos \theta$

$\dfrac{dy}{dx} = \dfrac{dy/d\theta}{dx/d\theta}$ Also $x^2 + y^2 = r^2$

By these substitutions the given equation reduces to

$$\dfrac{r \cos \theta \left(\dfrac{dr}{d\theta} \cos \theta - r \sin \theta \right) + r \sin \theta \left(\dfrac{dr}{d\theta} \sin \theta + r \cos \theta \right)}{r \cos \theta \left(\dfrac{dr}{d\theta} \sin \theta + r \cos \theta \right) - r \sin \theta \left(\dfrac{dr}{d\theta} \cos \theta - r \sin \theta \right)}$$

$$= \sqrt{\left(\frac{a^2 - r^2}{r^2}\right)}$$

or
$$\frac{r \frac{dr}{d\theta}(\cos^2\theta + \sin^2\theta)}{r^2(\cos^2\theta + \sin^2\theta)} = \frac{(a^2 - r^2)^{1/2}}{r}$$

or
$$\frac{dr}{d\theta} = (a^2 - r^2)^{1/2}$$

or
$$\frac{dr}{(a^2 - r^2)^{1/2}} = d\theta \text{ or } \sin^{-1}\frac{r}{a} = \theta + c,$$

$$\therefore \quad r = a \sin(\theta + c)$$

By putting the values of r and θ, we get

$$(x^2 + y^2)^{1/2} = a \sin\left\{\tan^{-1}\left(\frac{y}{x}\right) + c\right\}$$

EXERCISE

Solve the following differential equations :

1. $(x - y)^2 \dfrac{dy}{dx} = a^2$ $\quad\left[\text{Ans. } 2y + c = a \log \dfrac{x - y - a}{x - y + a}\right]$

2. $\dfrac{dy}{dx} = \sin(x + y) + \cos(x + y)$ $\quad\left[\text{Ans. } \log\left(1 + \tan \dfrac{x + y}{2}\right) = x + c.\right]$

3. $\sin^{-1}\left(\dfrac{dy}{dx}\right) = x + y$ $\quad\left[\text{Ans. } (x + c)\left\{1 + \tan\left(\dfrac{x + y}{2}\right)\right\} + 2 = 0\right]$

4. $x\,dy - y\,dx = (x^2 + y^2)\,dx$ $\quad[\text{Ans. } y = x \tan(c - ay)]$

5. $\dfrac{dy}{dx} - x \tan(y - x) = 1$ $\quad\left[\text{Ans. } \sin(y - x) = ce^{1/2 x^2}\right]$

6. $\left(\dfrac{x + y - b}{x + y - b}\right) \dfrac{dy}{dx} = \dfrac{x + y + a}{x + y + b}$ $\quad\left[\text{Ans. } (b - a) \log\{(x + y)^2 - ab\} = 2(x - y) + c\right]$

7. $\sin y \dfrac{dy}{dx} = \cos y (1 - x \cos y)$ $\quad[\text{Ans. } \sec y = (x + 1) + ce^x]$

8. $\sec^2 y \dfrac{dy}{dx} + 2x \tan y = x^3$ $\quad\left[\text{Ans. } \tan y = ce^{-x^2} + \dfrac{1}{2}(x^2 - 1)\right]$

9. $\dfrac{dy}{dx} = e^{x - y}(e^x - e^y)$ $\quad\left[\text{Ans. } e^y = ce^{-e^x} + e^x - 1\right]$

2.4 HOMOGENEOUS EQUATIONS

All the differential equations of the type

$$\frac{dy}{dx} = \frac{f_1(x, y)}{f_2(x, y)}$$

where $f_1(x, y)$ and $f_2(x, y)$ are homogeneous functions of x and y of the same degree, are called *homogeneous differential equations*.

Differential Equations of First Order and First Degree

Such equations can be solved by putting $y = vx$, where v is a function of x.
Then differentiating it, we get

$$\frac{dy}{dx} = v + x\frac{dv}{dx}$$

Now the homogeneous equation takes up the new form as

$$v + x\frac{dv}{dx} = \frac{f_1(x, vx)}{f_2(x, vx)} = f(v)$$

[as there functions $f_1(x, y), f_2(x, y)$ are of the same degree, on simplification of $\frac{f_1(x, vx)}{f_2(x, vx)}$ we would expect a function of v only i.e., $f(v)$.

Therefore, $$x\frac{dv}{dx} = f(v) - v$$

or $$\frac{dv}{f(v) - v} = \frac{dx}{x}$$

Integrating it, we have

$$F(v) = \log x + c, \text{ if } c = \log\frac{1}{a} = -\log c$$

or $$\log\frac{x}{a} = F\left(\frac{y}{x}\right)$$

WORKING RULE :

1. Put $y = vx$ (or $x = vy$).
2. Transform the given equation in terms of v and x (or v and y).
3. Apply the method of separation of variables.
4. In the end, after integration replace v by y/x (or x/y).

EXAMPLES

1. Solve : $x^2 y\, dx - (x^3 + y^3)\, dy = 0$.

Sol. The given equation is

$$\frac{dy}{dx} = \frac{x^2 y}{x^3 + y^3}$$

This is a homogenous differential equation. To solve it, put

$$y = vx \quad \therefore \quad \frac{dy}{dx} = v + x\frac{dv}{dx}$$

Then given equation becomes

$$v + x\frac{dv}{dx} = \frac{x^2 \cdot vx}{x^3 + v^3 \cdot x^3} = \frac{v}{1 + v^3}$$

$$\therefore \quad x\frac{dv}{dx} = \frac{v}{1 + v^3} - v = \frac{-v^4}{1 + v^3}$$

or $$\left(-\frac{1}{v^4} - \frac{1}{v}\right) dv = \frac{dx}{x}$$

Integrating it, we get

$$\frac{1}{3v^3} - \log v = \log x + \log c$$

∴ $$\log cxv = \frac{1}{3v^3} \quad \text{or} \quad \log cx \cdot \frac{y}{x} = \frac{x^3}{3y^3}$$

or $$cy = e^{x^3/3y^3}$$

2. Solve : $x \dfrac{dy}{dx} - y = x\sqrt{(x^2 + y^2)}$.

Sol. The given equation is

$$\frac{dy}{dx} = \frac{y + x\sqrt{(x^2 + y^2)}}{x}$$

Put $y = vx$, $\dfrac{dy}{dx} = v + x \dfrac{dv}{dx}$, we get

$$v + x \frac{dv}{dx} = \frac{vx + x\sqrt{(x^2 + v^2 x^2)}}{x}$$

$$= v + x\sqrt{1 + v^2}$$

or $$\frac{dv}{\sqrt{(1 + v^2)}} = dx$$

Integrating, we get

$$\sinh^{-1} v = x + c$$

or $$v = \sinh(x + c)$$

or $$y = x \sinh(x + c)$$

EXERCISE

Solve the following differential equations :

1. $(x^2 - y^2) dx + 2xy\, dy = 0$ [Ans. $x^2 + y^2 = cx$]
2. $2xy\, dx + (y^2 - x^2) dy = 0$ [Ans. $y = c(x^2 + y^2)$]
3. $(x^2 + y^2) dx + 2xy\, dy = 0$ [Ans. $x(x^2 + 3y^2) = c$]
4. $(x^2 + xy) dy = (x^2 + y^2) dx$ [Ans. $c(y - x)^2 = xe^{-y/x}$]
5. $x(x - y) dy + y^2 dx = 0$ [Ans. $xy = e^{y/x}$]
6. $cxy = e^{-x/y}$ [Ans. $cxy = e^{-x/y}$]
7. $\left\{x\sqrt{(x^2 + y^2)} - y^2\right\} dx + xy\, dy = 0$ $\left[\text{Ans. } \sqrt{(x^2 + y^2)} = x \log\left(\dfrac{c}{x}\right)\right]$
8. $y^2 + x^2 \dfrac{dy}{dx} = xy \dfrac{dy}{dx}$ $\left[\text{Ans. } cy = e^{y/x}\right]$
9. $x^2 dy + y(x + y) dx = 0$ [Ans. $y + 2x = cx^2 y$]
10. $x^2 \dfrac{dy}{dx} = \dfrac{y(x + y)}{2}$ [Ans. $(y - x)^2 = cy^2 x$]

Differential Equations of First Order and First Degree 515

11. $(2x - y) dx + (x - 2y) dy = 0$ [Ans. $(x+ y)^3 (x - y) = c$]

12. $(1+ e^{x/y}) dx + e^{x/y} \left(1 - \dfrac{x}{y}\right) dy = 0$ [Ans. $x + y e^{x/y} = c$]

2.5 EQUATIONS RADUCIBLE TO A HOMOGENEOUS FORM

All the differential equations of the type

$$\dfrac{dy}{dx} = \dfrac{ax + by + c}{Ax + By + C}$$

Case (A) when $\dfrac{a}{A} \neq \dfrac{b}{B}$

can be reduced to the homogeneous form as below :

Put $X = x - h$ and $Y = y - k \Rightarrow dX = dx$ and $dY = dy$

Therefore, $\dfrac{dY}{dX} = \dfrac{dy}{dx}$; then the equation $\dfrac{dy}{dx} = \dfrac{ax + by + c}{Ax + By + C}$ takes up the form as

$$\dfrac{dY}{dX} = \dfrac{aX + bY + (ah + bk + c)}{AX + BY + (Ah + Bk + C)}$$

Now choose h and k, so that

$$ah + bk + c = 0$$
$$Ah + Bk + C = 0$$

Then $\dfrac{dY}{dX} = \dfrac{aX + bY}{AX + BY}$

This is a homogeneous differential equation.
It can be solved as in the previous article.

Working Rule

1. *Put $x = X + h$, $y = Y + k$ in the given differential equation.*
2. *Equate the constant terms in the numerator and denominator to zero.*
3. *Solve these resulting equations of h and k obtained in (2) for h and k.*
4. *Solve the resultnig homogeneous equations in X and Y by method of 2.4.*
5. *Replace X by x-h and Y by y-k.*
6. *Substitute the values of h and k.*

Case (B) : When $\dfrac{a}{A} = \dfrac{b}{B}$.

1. *Put $ax + by = v$.*
2. *Solve the resulting equation by using 2.2 (variables are separable method.)*

EXAMPLES

1. Solve $(2x + y + 3) dx = (2y + x + 1) dy$.

Sol. The given equation is

$$\dfrac{dy}{dx} = \dfrac{2x+ y + 3}{2y + x + 1}. \text{ Here } \dfrac{a}{A} = \dfrac{2}{1} = 2, \dfrac{b}{B} = \dfrac{1}{2} \Rightarrow \dfrac{a}{A} \neq \dfrac{b}{B} \text{ (case A above)}$$

This is the equation reducible to homogeneous form. To solve it, put $x = X + h$ and $y = Y + k$,

we get $\dfrac{dy}{dY} = \dfrac{2X + Y + (2h + k + 3)}{2Y + X + (h + 2k + 1)}$

Putting $2h + k + 3 = 0$

$$h + 2k + 1 = 0 \quad i.e., \quad h = -\frac{5}{3}, k = \frac{1}{3}$$

Hence $x = X - \frac{5}{3}$ and $y = Y + \frac{1}{3} \Rightarrow X = x + \frac{5}{3}$ and $Y = y - \frac{1}{3}$

we get
$$\frac{dY}{dX} = \frac{2X + Y}{2Y + X}$$

Put $Y = vX, \frac{dY}{dX} = v + X \frac{dv}{dX}$,

$$\therefore \qquad v + X \frac{dv}{dX} = \frac{2 + v}{2v + 1}$$

or $$X \frac{dv}{dX} = \frac{2 + v}{2v + 1} - v = \frac{2(1 - v^2)}{2v + 1}$$

or $$\frac{2v + 1}{(1 - v)(1 + v)} dv = 2 \frac{dX}{X}$$

or $$\left[\frac{3}{2} \cdot \frac{1}{1 - v} - \frac{1}{2(1 + v)} \right] dv = 2 \frac{dX}{X}$$

Integrating, we get

$$-3 \log(1 - v) - \log(1 + v) = 4 \log X - \log c$$

or $$\log X^4 (1 + v)(1 - v)^3 = \log c$$

or $$X^4 \left(1 + \frac{Y}{X}\right) \left(1 - \frac{Y}{X}\right)^3 = c$$

or $$(X + Y)(X - Y)^3 = c$$

Putting the values of X, Y and h, k we get

$$\left(x + y + \frac{4}{3}\right)(x - y + 2)^3 = c$$

2. Solve : $\frac{dy}{dx} = \frac{x - y + 3}{2x - 2y + 5}$.

Sol. The given equation can be put as

$$\frac{dy}{dx} = \frac{x - y + 3}{2(x - y) + 5}$$

Put $x - y = v, \therefore 1 - \frac{dy}{dx} = \frac{dv}{dx}$

$$\therefore \qquad 1 - \frac{dv}{dx} = \frac{v + 3}{2v + 5}$$

or $$\frac{dv}{dx} = 1 - \frac{v + 3}{2v + 5} = \frac{v + 2}{2v + 5}$$

$$\therefore \qquad \frac{2v + 5}{v + 2} dv = dx$$

Differential Equations of First Order and First Degree 517

or
$$\left(2 + \frac{1}{v+2}\right) dv = dx$$

On integrating, we get

$$2v + \log(v+2) = x + c$$

or
$$2(x-y) + \log(x-y+2) = x + c$$

or
$$x - 2y + \log(x-y+2) = c.$$

EXERCISE

Solve following differential equations :

1. $\dfrac{dy}{dx} = \dfrac{2y - x - 4}{y - 3x + 3}$

$$\left[\text{Ans. } \left\{ \log(v^2 - 5v + 1) - \frac{1}{\sqrt{21}} \log \frac{v - \frac{5}{2} - \frac{\sqrt{21}}{2}}{v - \frac{5}{2} + \frac{\sqrt{21}}{2}} \right\} = -\log(x-2) + c, \text{ where } v = \frac{y-3}{x-2} \right]$$

2. $\dfrac{dy}{dx} = \dfrac{3y - x + 7}{x - 7y - 3}$ $\quad\left[\text{Ans. } \dfrac{3}{14} \log \dfrac{y - x + 1}{y + x - 1} = \log c \sqrt{\{y^2 - (x-1)^2\}} \right]$

3. $(6x - 5y + 4) dy + (y - 2x - 1) dx = 0$

$$\left[\text{Ans. } \log c \left(x + \frac{1}{4}\right)^{-2} = \left[\log(5v - 2)(v - 1) - \frac{5}{3} \log \frac{5(v-1)}{(5v-2)} \right] \text{ where } v = \frac{y - \frac{1}{2}}{x + \frac{1}{4}} \right]$$

4. $(2x + 3y - 5)\dfrac{dy}{dx} + (3x + 2y - 5) = 0$ $\quad$ [Ans. $3(x^2 + y^2) + 4xy - 10(x+y) = c$]

5. $\dfrac{dy}{dx} = \dfrac{2x + 9y - 20}{6x + 2y - 10}$ $\quad$ [Ans. $(y - 2x)^2 = c(2y + x - 5)$]

6. $(y + x + 5) dy = (y - x + 1) dx$ $\quad\left[\text{Ans. } \log(x^2 + y^2 + 4x + 6y + 13) + 2\tan^{-1} \dfrac{y+3}{x+2} = c \right]$

7. $\dfrac{dy}{dx} = \dfrac{2x - y + 1}{x + 2y - 3}$ $\quad$ [Ans. $xy + y^2 - 3y - (x^2 + x) = c$]

8. $(2x - y + 1) dx + (2y - x - 1) = 0$ $\quad$ [Ans. $x^2 + y^2 - xy + x - y = c$]

9. $\dfrac{dy}{dx} = \dfrac{2x - 6y + 7}{x - 3y + 4}$ $\quad\left[\text{Ans. } 2x - y + \dfrac{1}{5}\log(5x - 15y + 17) = c \right]$

10. $\dfrac{dy}{dx} = \dfrac{2y + x - 1}{2x + 4y + 3}$ $\quad\left[\text{Ans. } 2y - x + \dfrac{1}{4}\log(8y + 4x + 5) = c \right]$

11. $\dfrac{dy}{dx} = \dfrac{4x + 6y + 5}{3y + 2x + 4}$ $\quad\left[\text{Ans. } y - 2x + \dfrac{3}{8}\log(24y + 16x + 23) = c \right]$

12. $\dfrac{dy}{dx} = \dfrac{x + y + 1}{2x + 2y + 3}$ $\quad$ [Ans. $6y - 3x + \log(3x + 3y + 4) = c$]

2.6 LINEAR DIFFERENTIAL EQUATIONS

The equation $\dfrac{dy}{dx} + Py = Q$, whose left side is linear in both the dependent variable and its derivative, is called a *linear differential equation of the first order*, when P and Q are the functions of x only or constants. The differential equation $\dfrac{dx}{dy} + P_1 x = Q_1$ where P_1, Q_1 are functions of y only or constants is also a linear differential equation (where x is dependent variable).

2.6.1 SOLUTION OF LINEAR EQUATIONS

The given equation is

$$\dfrac{dy}{dx} + Py = Q \qquad \ldots(i)$$

Since
$$\dfrac{d}{dx}\{y e^{(\int P dx)}\} = \dfrac{dy}{dx} e^{(\int P dx)} + y P e^{(\int P dx)}$$

or
$$\dfrac{d}{dx}\{y e^{(\int P dx)}\} = e^{(\int P dx)} \left(\dfrac{dy}{dx} + Py\right)$$

$e^{\int P dx}$ is an integrating factor of (i) and its primitive or general solution is

$$y e^{(\int P dx)} = \int \{Q \cdot e^{\int P dx}\}\, dx + c,$$

where c is an arbitrary constant.

Working Rule :

(1) *Determine the integrating factor $e^{\int P dx}$ which is represented by I.F.*

(2) *Then multiply the differential equation by I.F. ($e^{\int P dx}$). Now integrate by using integration by parts method taking $e^{\int P dx}$ as first function and y as second function.*

Then we get

Dependent variable $\times$ I.F. $= \int \{Q \cdot e^{\int P dx}\}\, dx + c$

or
$$y \times \text{I.F.} = \int \{Q \cdot e^{\int P dx}\}\, dx + c$$

(3) *If x is independent variable then I.F. $= e^{\int P_1 dy}$ and solution is*

$$x \times \text{I.F.} = \int \{Q_1 e^{P_1 dy}\}\, dy + c$$

EXAMPLES

1. Solve : $(x + 2y)^3 \dfrac{dy}{dx} = y.$

Sol. The given equation can be written as

$$y \dfrac{dx}{dy} = x + 2y^3$$

or
$$\dfrac{dx}{dy} - \dfrac{1}{y} x = 2y^2$$

It is of the form $\dfrac{dx}{dy} + Px = Q.$

Hence, I.F. $= e^{-\int \frac{1}{y} dy} = e^{-\log y} = \dfrac{1}{y}$. (since x is dependent variable)

Differential Equations of First Order and First Degree

Therefore, the solution of the given equation will be

$$\frac{1}{y} \cdot x = \int 2y^2 \cdot \frac{1}{y} dy + c$$

or
$$\frac{x}{y} = y^2 + c$$

or
$$x = y^3 + cy.$$

2. Solve : $(1 + y^2) + (x - e^{\tan^{-1} y}) \frac{dy}{dx} = 0.$

Sol. The given equation is

$$(1 + y^2) \frac{dx}{dy} + x = e^{\tan^{-1} y}$$

or
$$\frac{dx}{dy} + \frac{1}{1 + y^2} \cdot x = \frac{1}{1 + y^2} e^{\tan^{-1} y}$$

$$\text{I.F.} = e^{\int \frac{1}{1 + y^2} dy} = e^{\tan^{-1} y} \qquad \text{(since } x \text{ is dependent variable)}$$

∴ Solution of the equation will be

$$x \cdot e^{\tan^{-1} y} = \int \frac{1}{1 + y^2} e^{\tan^{-1} y} \cdot e^{\tan^{-1} y} dy + c$$

$$= \int \frac{1}{1 + y^2} e^{2 \tan^{-1} y} dy + c$$

Let
$$\tan^{-1} y = t, \quad \frac{1}{1 + y^2} dy = dt$$

Then
$$x \cdot e^{\tan^{-1} y} = \int e^{2t} dt + c$$

$$= \frac{1}{2} e^{2t} + c$$

or
$$x e^{\tan^{-1} y} = \frac{1}{2} e^{2 \tan^{-1} y} + c$$

3. Solve : $\sqrt{(a^2 + x^2)} \frac{dy}{dx} + y = \sqrt{(a^2 + x^2)} - x.$

Sol. The given equation can be put in the form

$$\frac{dy}{dx} + \frac{y}{\sqrt{a^2 + x^2}} = \frac{\sqrt{(a^2 + x^2)} - x}{\sqrt{(a^2 + x^2)}}$$

∴
$$\text{I.F.} = e^{\int \frac{1}{(a^2 + x^2)^{1/2}} dx} = e^{\log \frac{x + \sqrt{(a^2 + x^2)}}{a}}$$

$$= \frac{\sqrt{(a^2 + x^2)} + x}{a}$$

Hence the solution of the given equation is

$$\frac{\sqrt{(a^2 + x^2)} + x}{a} \cdot y = \int \frac{\sqrt{(a^2 + x^2)} - x}{\sqrt{a^2 + x^2}} \cdot \frac{\sqrt{(a^2 + x^2)} + x}{a} \cdot dx$$

$$= \int \frac{(a^2 + x^2) - x^2}{a\sqrt{(a^2 + x^2)}} dx$$

$$= \int \frac{a}{\sqrt{(a^2 + x^2)}} dx = a \sinh^{-1} \frac{x}{a} + c$$

EXERCISE

Solve the following differential equation :

1. $\dfrac{dy}{dx} + \dfrac{2x}{1+x^2} y = \dfrac{1}{(1+x^2)^2}$ [**Ans.** $y(1+x^2) = \tan^{-1} x + c$]

2. $(x+y+1)\dfrac{dy}{dx} = 1$ [**Ans.** $x = ce^y - (y+2)$]

3. $\dfrac{dy}{dx} + y \sec x = \tan x$ [**Ans.** $y(\sec x + \tan x) = \sec x + \tan x - x + c$]

4. $\dfrac{dy}{dx} + y \tan x - \sec x = 0$ [**Ans.** $y \sec x = \tan x + c$]

5. $\dfrac{dy}{dx} - y \tan x = -2 \sin x$ $\left[\textbf{Ans.}\ y \cos x = \dfrac{1}{2} \cos 2x + c\right]$

6. $\dfrac{dy}{dx} + y \cot x = x\, 2 \cos x$ $\left[\textbf{Ans.}\ y \sin x = c - \dfrac{1}{2} \cos 2x\right]$

7. $\dfrac{dy}{dx} + \dfrac{y}{x} = x^2$, if $y = 1$ when $x = 1$ $\left[\textbf{Ans.}\ xy = \dfrac{x^4}{4} + \dfrac{3}{4}\right]$

8. $(1+x^2)\dfrac{dy}{dx} + y = e^{\tan^{-1} x}$ $\left[\textbf{Ans.}\ 2y\, e^{\tan^{-1} x} = e^{2 \tan^{-1} x} + c\right]$

9. $(1+y^2)(x - e^{-\tan^{-1} y})\dfrac{dy}{dx} = 0$ [**Ans.** $xe^{\tan^{-1} y} = \tan^{-1} y + c$]

10. $(1+y^2) dx = (\tan^{-1} y - x)\, dy$ [**Ans.** $x = \tan^{-1} y - 1 + ce^{-\tan^{-1} y}$]

11. $x \log x \dfrac{dy}{dx} + y = 2 \log x$ [**Ans.** $y \log x = (\log x)^2 + c$]

12. $(2x - 10y^3)\dfrac{dy}{dx} + y = 0$ [**Ans.** $x = 2y^3 + cy^{-2}$]

13. $x\dfrac{dy}{dx} - y = 2x^2 \operatorname{cosec} 2x$ [**Ans.** $y = x \log \tan x + cx$]

14. $\dfrac{dy}{dx} + \dfrac{y}{(1-x^2)^{3/2}} = \dfrac{x + \sqrt{(1-x^2)}}{(1-x^2)^2}$ $\left[\textbf{Ans.}\ y = t + ce^{-t}\ \text{where}\ t = \dfrac{-x}{\sqrt{(1-x^2)}}\right]$

15. $\cos^3 x \dfrac{dy}{dx} + y \cos x = \sin x$ [**Ans.** $y = \tan x - 1 + ce^{-\tan x}$]

Differential Equations of First Order and First Degree

2.7 DIFFERENTIAL EQUATIONS REDUCIBLE TO THE LINEAR FORM (BERNOULLI'S EQUATION)

A differential equation, having the form

$$\frac{dy}{dx} + Py = Qy^n$$

is called an *equation reducible to the linear form (or Bernoulli's equation)* where P and Q are functions of x only.

Solution

The given equation is

$$y^{-n}\frac{dy}{dx} + Py^{-n+1} = Q \qquad \ldots(i)$$

Putting $y^{-n+1} = v$,

$$(-n+1)y^{-n}\frac{dy}{dx} = \frac{dv}{dx}$$

then (i) becomes

$$-\frac{1}{(n-1)}\frac{dv}{dx} + Pv = Q$$

or

$$\frac{dv}{dx} - (n-1)Pv = Q(1-n) \qquad \ldots(ii)$$

(ii) is a linear equation of the first order

Integrating factor is $e^{\{\int -(n-1)Pdx\}}$

Therefore, required solution is

$$ve^{\{\int -(n-1)Pdx\}} = \int [(1-n)Qe^{\{\int -(n-1)Pdx\}}]dx + c$$

where c is an arbitrary constant.

Similarly, $\frac{dx}{dy} + P_1 x = Q_1 x^n$, where P_1 and Q_1 are the functions of y alone can be reduced to the form as shown below :

$$ue^{\{\int -(n-1)P_1 dy\}} = \int [Q_1(1-n)e^{\{\int -(n-1)P_1 dy\}}]dy + c,$$

where $u = x^{-n+1}$.

Working Rule

A. 1. *Write the equation in the form*

$$\frac{dy}{dx} + Py = Qy^n$$

2. *Divide by y^n and bring it to the form as in § 2.7 equation (i)*

3. *Put $y^{n+1} = v$ and change the equation to the form*

$$\frac{dv}{dx} - (n-1)Pv = Q(1-n)$$

4. Then apply the § 2.7.

B. *In case of the equation*

$$\frac{dx}{dy} + P_1 x = Q_1 x^n$$

apply the similar line, i.e., first divide by x^n, then put substitution $x^{1-n} = V$ and this brings a linear differential equatino (§ 2.6.1).

EXAMPLES

1. Solve : $\dfrac{dy}{dx} - \dfrac{\tan y}{1+x} = (1+x) e^x \sec y.$

Sol. The given equation is

$$\cos y \dfrac{dy}{dx} - \dfrac{\sin y}{1+x} = (1+x) e^x$$

Let $\sin y = v$, $\cos y \dfrac{dy}{dx} = \dfrac{dv}{dx}$

we get $\dfrac{dv}{dx} - \dfrac{v}{1+x} = (1+x) e^x$

$$\text{I.F.} = e^{\int -\frac{1}{1+x} dx} = e^{-\log(1+x)} = \dfrac{1}{1+x}$$

Hence the solution is

$$\dfrac{1}{1+x} \cdot v = \int e^x dx + c$$

$$= e^x + c$$

or $\sin y = (1+x)(e^x + c)$

2. Solve $y - x \dfrac{dy}{dx} = a\left[y^2 + \dfrac{dy}{dx}\right]$

Sol. The equation can be written as

$$(x+a)\dfrac{dy}{dx} - y = -ay^2 \quad \text{or} \quad -\dfrac{1}{y^2}\dfrac{dy}{dx} + \dfrac{1}{y}\cdot\dfrac{1}{x+a} = \dfrac{a}{x+a}$$

(by dividing with $-y^2(x+a)$)

Put $\dfrac{1}{y} = v, \; -\dfrac{1}{y^2}\dfrac{dy}{dx} = \dfrac{dv}{dx}$

$\therefore \quad \dfrac{dv}{dx} + \dfrac{1}{x+a} v = \dfrac{a}{x+a}$

$$\text{I.F.} = e^{\int \frac{1}{x+a} dx} = e^{\log(x+a)} = (x+a)$$

Hence the solution of the equation is

$$v(x+a) = \int \dfrac{a}{x+a} \cdot (x+a)\, dx + c$$

$$= ax + c$$

or $\dfrac{1}{y}(x+a) = ax + c$

or $(x+a) = y(ax+c)$

Differential Equations of First Order and First Degree

3. Solve : $\dfrac{dz}{dx}+\dfrac{z}{x}\log z=\dfrac{z}{x^2}(\log z)^2$.

Sol. The given equation is

$$\dfrac{1}{z(\log z)^2}\dfrac{dz}{dx}+\dfrac{1}{x}\cdot\dfrac{1}{\log z}=\dfrac{1}{x^2}$$

Put $-\dfrac{1}{\log z}=v,\quad \therefore\ \dfrac{1}{\log z^2}\cdot\dfrac{1}{z}\cdot\dfrac{dz}{dx}=\dfrac{dv}{dx}$

Hence equation becomes

$$\dfrac{dv}{dx}-\dfrac{1}{x}v=\dfrac{1}{x^2}$$

This is linear equation where

$$\text{I.F.}=e^{-\int\frac{1}{x}dx}=\dfrac{1}{x}$$

Hence solution is

$$\dfrac{1}{x}\cdot v=\int\dfrac{1}{x^3}dx+c$$

or $\quad \dfrac{1}{x}\left(-\dfrac{1}{\log z}\right)=-\dfrac{1}{2x^2}+c$

or $\quad \dfrac{1}{x\log z}=\dfrac{1}{2x^2}-c$

EXERCISE

Solve the following differential equations :

1. $2\dfrac{dy}{dx}-\dfrac{y}{x}=\dfrac{y^2}{x^2}$ $\qquad$ [**Ans.** $x=y-cy\sqrt{x}$]

2. $\dfrac{dy}{dx}+\dfrac{y}{x}=\dfrac{y^2}{x^2}$ $\qquad$ $\left[\textbf{Ans. } \dfrac{1}{xy}=\dfrac{1}{2x^2}+c\right]$

3. $\dfrac{dy}{dx}+2y\tan x=y^2$ $\qquad$ $\left[\textbf{Ans. } \cos^2 x+\left(\dfrac{1}{2}x+\dfrac{1}{4}\sin 2x\right)y=cy\right]$

4. $\dfrac{dy}{dx}+y\cot x=y^2\sin^2 x$ $\qquad$ $\left[\textbf{Ans. } \dfrac{1}{y}\operatorname{cosec} x=\cos x-c\right]$

5. $\dfrac{dy}{dx}(x^2y^3+xy)=1$ $\qquad$ $\left[\textbf{Ans. } -\dfrac{1}{x}=y^2-2+ce^{-y^2/2}\right]$

6. $(x^3y^2+xy)\,dx=dy$ $\qquad$ $\left[\textbf{Ans. } -\dfrac{1}{y}=x^2-2+ce^{-x^2/2}\right]$

7. $\dfrac{dy}{dx}+yx=y^2 e^{x^2/2}\sin x$ $\qquad$ [**Ans.** $ye^{x^2/2}(x\log x-x+c)+1=0$]

8. $x\dfrac{dy}{dx}+y\log y=\dfrac{1}{x^2}\tan y\sin y$ $\qquad$ [**Ans.** $x.\log y=e^x(x-1)+c$]

9. $y(2xy + e^x)\,dx - e^x\,dy = 0$ [Ans. $y(x^2 + c) + e^x = 0$]

10. $\dfrac{dy}{dx} + \dfrac{1}{x}\tan y = \dfrac{1}{x^2}\tan y \sin y$ [Ans. $2x = \sin y\,(1 - 2cx^2)$]

2.8 EXACT DIFFERENTIAL EQUATION

Any differential equation which can be derived from its primitive by direct differention without any further transformation, such as elimination or reduction, is called an *exact differential equation*. For example, the differential equation

$$(ax + hy + g)\,dx + (hx + by + f)\,dy = 0$$

is exact as it is derived from its primitive.

$$\frac{ax^2}{2} + hxy + gx + \frac{by^2}{2} + fy + c = 0$$

by direct differentiation.

2.8.1 Theorem

The necessary and sufficient condition for the ordinary differential equation $Mdx + Ndy = 0$ to be exact is

$$\frac{\partial M}{\partial N} = \frac{\partial N}{\partial x}$$

The condition is necessary.

Let $f(x, y) = c$ be primitive of the equation $Mdx + Ndy = 0$, where c is the arbitrary constant. Now,

$$f(x, y) = c \qquad \ldots(i)$$

differentiating (i) with respect to x, we have

$$\frac{\partial f}{\partial x} + \frac{\partial f}{\partial y} \cdot \frac{dy}{dx} = 0 \qquad \ldots(ii)$$

But it is identical with an exact differential equation

$$Mdx + Ndy = 0 \qquad \ldots(iii)$$

Therefore, $M = \dfrac{\partial f}{\partial x} \qquad \ldots(iv)$

$$N = \frac{\partial f}{\partial y} \qquad \ldots(v)$$

Differentiating (iv) and (v) partially with respect to y and x respectively we have

$$\frac{\partial M}{\partial y} = \frac{\partial^2 f}{\partial x\,\partial y}$$

and

$$\frac{\partial N}{\partial x} = \frac{\partial^2 f}{\partial y\,\partial x}$$

since

$$\frac{\partial^2 f}{\partial x\,\partial y} = \frac{\partial^2 f}{\partial y\,\partial x}$$

∴

$$\frac{\partial M}{\partial y} = \frac{\partial N}{\partial x}$$

Differential Equations of First Order and First Degree

If the equation $Mdx + Ndy = 0$ be exact. Then it is a necessary condition that.
$$\frac{\partial M}{\partial y} = \frac{\partial N}{\partial x}$$

The condition is sufficient.

We have to show that if $\frac{\partial M}{\partial y} = \frac{\partial N}{\partial x}$, then the differential equation $Mdx + Ndy = df$, where f is $f(x, y)$ which is equal to c, i.e.,
$$f = f(x, y) - c = 0$$

Let $\quad\quad\quad\quad \int Mdx = P$

Therefore, $\quad\quad\quad M = \frac{\partial P}{\partial x}$

or $\quad\quad \frac{\partial N}{\partial x} = \frac{\partial M}{\partial y} = \frac{\partial}{\partial y}\left(\frac{\partial P}{\partial x}\right) = \frac{\partial^2 P}{\partial y \, \partial x}$

Therefore, $\quad\quad \frac{\partial N}{\partial x} = \frac{\partial}{\partial x} \cdot \left(\frac{\partial P}{\partial y}\right)$

or $\quad\quad\quad N = \frac{\partial P}{\partial y} + f(y)$

using the values of M and N, we have
$$Mdx + Ndy = \frac{\partial P}{\partial x} dx + \frac{\partial P}{\partial y} \cdot dy + f(y) \, dy$$
$$= d(p) + d\{F_1(y)\}$$

where $F_1(y) = \int f(y) \, dy$

Therefore, $\quad Mdx + Ndy = d[P + \{F_1(y)\}]$

writing $P + F_1(y) = f(x, y)$

we have $Mdx + Ndy = d(f)$

where $d(f) = 0$ [as $f(x, y) = c$ and $d(f)$ is the differential of $f(x, y)$ which is zero.]

Therefore, the equation is $Mdx + Ndy = 0$.

Working Rule :

1. *First show* $\frac{\partial M}{\partial y} = \frac{\partial N}{\partial x}$, *M and N are the respective coefficients of dx and dy.*
2. *Integrate the coefficient of dx (i.e., M) with respect to x, regarding y to be constant.*
3. *Omit the terms containing x in N and find the integral of the coefficient of dy with respect to y.*
4. *Add the above two results and equate this sum to an arbitrary constant, i.e.,*
$$\int Mdx + \int (\text{terms in } y \text{ not containing } x) \, dy = c.$$
(y constant)

EXAMPLES

1. Solve : $x \, dx + y \, dy = a^2 \dfrac{x \, dy - y \, dx}{x^2 + y^2}$.

Sol. The given differential equation can be written as

$$\left(x + \frac{a^2 y}{x^2 + y^2}\right) dx + \left(y - \frac{a^2 x}{x^2 + y^2}\right) dy = 0$$

Here $M = x + \dfrac{a^2 y}{x^2 + y^2}$ and $N = y - \dfrac{a^2 x}{x^2 + y^2}$

$$\frac{\partial M}{\partial y} = 0 + \frac{a^2(x^2 + y^2) - a^2 y \cdot 2y}{(x^2 + y^2)^2} = \frac{a^2(x^2 - y^2)}{(x^2 + y^2)^2}$$

$$\frac{\partial N}{\partial x} = 0 - \frac{a^2[(x^2 + y^2) - x \cdot 2x]}{(x^2 + y^2)^2} = \frac{a^2(x^2 - y^2)}{(x^2 + y^2)^2}$$

Thus, $\dfrac{\partial M}{\partial N} = \dfrac{\partial N}{\partial x}$ and hence the given equation is exact. Therefore, its solution is

$$\int \left(x + \frac{a^2 y}{x^2 + y^2}\right) dx + \int \left(y - \frac{a^2 x}{(x^2 + y^2)}\right) dy = c$$

(y constant)   (leaving the terms contains x)

or $\dfrac{x^2}{2} + a^2 y \cdot \dfrac{1}{y} \tan^{-1}\left(\dfrac{x}{y}\right) + \dfrac{y^2}{2} = c$

or $x^2 + y^2 + 2a^2 \tan^{-1}\left(\dfrac{x}{y}\right) = 2c = k$

2. Solve : $(1 + e^{x/y}) dx + e^{x/y}\left(1 - \dfrac{x}{y}\right) dy = 0.$

Sol. Comparing the given equation with $Mdx + Ndy = 0$, we get

$$M = 1 + e^{x/y}, \quad N = e^{x/y}\left(1 - \frac{x}{y}\right)$$

$\therefore \quad \dfrac{\partial M}{\partial N} = e^{x/y}\left(-\dfrac{x}{y^2}\right)$

$\dfrac{\partial N}{\partial x} = e^{x/y} \dfrac{1}{y}\left(1 - \dfrac{x}{y^2}\right) + e^{x/y}\left(-\dfrac{1}{y}\right) = e^{x/y}\left(-\dfrac{x}{y^2}\right)$

Hence $\dfrac{\partial M}{\partial y} = \dfrac{\partial N}{\partial x}$ and the equation is exact. Its solution is

$$\int (1 + e^{x/y}) dx + \int 0 \cdot dy = 0$$

(y constant)

or $x + y e^{x/y} = c$

EXERCISE

Solve following differential equations :

1. $(x + 2y - 3) dy - (2x - y + 1) dx = 0.$ [Ans. $y^2 - x^2 + xy - x - 3y = c$]

2. $(x^3 + 3xy^2) dx + (3x^2 y + y^3) dy = 0$ [Ans. $x^4 + y^4 + 6x^2 y^2 = c$]

3. $(e^y + 1) \cos x \, dx + e^y \sin x \, dy = 0$ [Ans. $(e^y + 1) \sin x = c$]

Differential Equations of First Order and First Degree 527

4. $y \sin 2x \, dx - (1 + y^2 + \cos^2 x) \, dy = 0$

$$\left[\text{Ans. } y \cos 2x + 2y + \frac{2}{3} y^3 = c\right]$$

5. $(\sin x \cos y + e^{2x}) \, dx + (\cos x \sin y + \tan y) \, dy = 0$

$$\left[\text{Ans. } -\cos x \cos y + \frac{1}{2} e^{2x} + \log \sec y = c\right]$$

6. $\left\{y\left(1 + \frac{1}{x}\right) + \cos y\right\} dx + (x + \log x - x \sin y) \, dy = 0$

$$[\text{Ans. } y(x + \log x) + x \cos y = c]$$

2.8 INTEGRATING FACTORS

By integrating factors we mean those functions of x and y which when multiplied with the differential equation will make it exact.

Hence some equations which are not exact can be made exact by multiplication with the integrating factor and then we can solve it either by re-arranging the terms and making them exact differential or by the method of exact equations.

2.9.1 INTEGRATING FACTORS FOUND BY INSPECTION

The following exact differentials may be committed to memory :

1. $d\left(\dfrac{x}{y}\right) = \dfrac{y\,dx - x\,dy}{y^2}$

2. $d\left(\dfrac{y}{x}\right) = \dfrac{x\,dy - y\,dx}{x^2}$

3. $d\left(\tan^{-1} \dfrac{x}{y}\right) = \dfrac{y\,dx - x\,dy}{x^2 + y^2}$

4. $d\left(\tan^{-1} \dfrac{y}{x}\right) = \dfrac{x\,dy - y\,dx}{x^2 + y^2}$

5. $d\left(\log \dfrac{x}{y}\right) = \dfrac{y\,dx - x\,dy}{xy}$

6. $d\left(\log \dfrac{y}{x}\right) = \dfrac{x\,dy - y\,dx}{xy}$

7. $d(xy) = x\,dy + y\,dx$

8. $d\left(\dfrac{1}{xy}\right) = -\dfrac{x\,dy + y\,dx}{x^2 y^2}$

9. $d\left(\dfrac{x^2}{y}\right) = \dfrac{2xy\,dx - x^2 dy}{y^2}$

10. $d\left(\dfrac{y^2}{x}\right) = \dfrac{2xy\,dy - y^2 dx}{x^2}$

11. $d\left(\dfrac{x^2}{y}\right) = \dfrac{2x^2 y\,dy - 2y^2 x\,dx}{x^4}$

12. $d\left(\dfrac{x^2}{y^2}\right) = \dfrac{2xy^2 dx - 2yx^2 dy}{y^4}$

13. $d\left(\dfrac{e^x}{y}\right) = \dfrac{y e^x dx - e^x dy}{y^2}$

14. $d[\log(x^2 + y^2)] = \dfrac{2x\,dx + 2y\,dy}{x^2 + y^2}$

EXAMPLES

1. Solve : $(y^2 e^x + 2xy) \, dx - x^2 dy = 0.$

Sol. In this equation

$$M = y^2 e^x + 2xy, \quad N = -x^2$$

$$\frac{\partial M}{\partial y} = 2y e^x + 2x, \quad \frac{\partial N}{\partial x} = -2x$$

Clearly, $\dfrac{\partial M}{\partial y} \neq \dfrac{\partial N}{\partial x}$.

Hence the equation is not exact.

Now, if in the solution e^x is multiplied by some other function, then it must occur twice in the differential equation. But since it is occurring only once, therefore, we should divide by y^2.

$$\therefore \quad e^x dx + \frac{2xy\, dx - x^2 dy}{y^2} = 0$$

or

$$\left(e^x + \frac{2x}{y}\right) dx + \left(-\frac{x^2}{y^2}\right) dy = 0$$

Now,

$$M = e^x + \frac{2x}{y}, \quad N = -\frac{x^2}{y^2}$$

$$\frac{\partial M}{\partial y} = -\frac{2x}{y^2}, \quad \frac{\partial N}{\partial x} = -\frac{2x}{y^2}$$

Now, the equation is exact and its solution is

$$\int \left(e^x + \frac{2x}{y}\right) dx = c$$

(y constant)

or

$$e^x + \frac{x^2}{y} = c.$$

2. Solve : $(x^2 + y^2 + a^2) y\, dy + (x^2 + y^2 - a^2) x\, dx = 0$.

Sol. The given equation can be written as

$$(x^2 + y^2)(2x\, dx + 2y\, dy) + 2a^2 y\, dy - 2a^2 x\, dx = 0$$

or

$$(x^2 + y^2)\, d(x^2 + y^2) + 2a^2 y\, dy - 2ax\, dx = 0$$

In the first term only putting $(x^2 + y^2) = v$, we get

$$v\, dv + 2a^2 y\, dy - 2a^2 x\, dx = 0$$

Integrating each term, we get

$$\frac{1}{2}v^2 + 2a^2 \cdot \frac{1}{2} y^2 - 2a^2 \cdot \frac{1}{2} x^2 = c$$

or

$$\frac{1}{2}(x^2 + y^2)^2 + a^2 (y^2 - x^2) = c$$

or

$$(x^2 + y^2)^2 + 2a^2 (y^2 - x^2) = 2c$$

EXERCISE

Solve following differential equations :

1. $x\, dy - y\, dx + 2x^3 dx = 0$ $\qquad$ [**Ans.** $y + x^3 = cx$]

2. $x\, dy - (y - x)\, dx = 0$ $\qquad$ $\left[\textbf{Ans.}\ \dfrac{y}{x} + \log x = c\right]$

3. $y(axy + e^x)\, dx - e^x dy = 0$ $\qquad$ $\left[\textbf{Ans.}\ \dfrac{ax^2}{2} + \dfrac{e^x}{y} = c\right]$

4. $e^y dx + (xe^y + 2y)\, dy = 0$ $\qquad$ [**Ans.** $xe^y + y^2 = c$]

5. $a(x\, dy + 2y\, dx) = xy\, dy$ $\qquad$ [**Ans.** $a \log y + 2a \log x - y = c$]

Differential Equations of First Order and First Degree

2.9.2 Some Rules for finding Integrating Factor of Equation $Mdx + Ndy = 0$ to make it exact.

We have already done some questions in which I.F. was found by inspection. Below we shall give some rules for finding I.F. of differential equation $Mdx + Ndy = 0$, which is not exact and where M, N are functions of x, y.

Rule 1. If $Mx + Ny \neq 0$ and the equation is homogeneous, then $\dfrac{1}{Mx + Ny}$ is an I.F.

Rule 2. If the equation $Mdx + Ndy = 0$ is not exact but is of the form $f_1(xy)\, ydx + f_2(xy)\, x\, dy = 0$, then $\dfrac{1}{Mx - Ny}$ is an integrating factor provided $Mx - Ny \neq 0$.

EXAMPLES

1. Solve : $(x^2y - 2xy^2)\, dx - (x^3 - 3x^2y)\, dy = 0$.

Sol. In this equation
$$M = x^2y - 2xy^2, \quad N = 3x^2y - x^3$$
$$\frac{\partial M}{\partial y} = x^2 - 4y, \quad \frac{\partial N}{\partial x} = 6xy - 3x^2$$

Clearly, $\dfrac{\partial M}{\partial y} \neq \dfrac{\partial N}{\partial x}$, hence the given equation is not exact.

In this equation M and N are homogeneous functions of x and y and $\dfrac{1}{Mx + Ny} = \dfrac{1}{x^2 y^2} \neq 0$.

Hence, $\quad$ I.F. $= \dfrac{1}{x^2 y^2}$.

Multiplying equation by I.F., we get
$$\left(\frac{1}{y} - \frac{2}{x}\right) dx + \left(\frac{3}{y} - \frac{x}{y^2}\right) dy = 0$$

in this equation
$$M = \frac{1}{y} - \frac{2}{x}, \quad N = \frac{3}{y} - \frac{x}{y^2}$$
$$\frac{\partial M}{\partial y} = -\frac{1}{y^2}, \quad \frac{\partial N}{\partial x} = -\frac{1}{y^2}$$

Hence this equation is exact and its solution is $\dfrac{x}{y} + \log \dfrac{y^3}{x^2} = c$.

2. Solve : $(x^3y^3 + x^2y^2 + xy + 1)\, y\, dx + (x^3y^3 - x^2y^2 - xy + 1)\, x\, dy = 0$.

Sol. The given equation is not exact. Also, we have
$$Mx = xy\,(x^3y^3 + x^2y^2 + xy + 1)$$
$$Ny = xy\,(x^3y^3 - x^2y^2 - xy + 1)$$

∴ $\quad Mx - Ny = xy\,(2x^2y^2 + 2xy) = 2x^2y^2\,(xy + 1) \neq 0$

∴ $\quad$ I.F. $= \dfrac{1}{2x^2y^2\,(xy + 1)}$

Multiplying the equation by I.F., we get

$$\frac{(xy+1)(x^2y^2+1)}{2x^2y^2(xy+1)} y\, dx + \frac{(xy+1)(x^2y^2-2xy+1)}{2x^2y^2(xy+1)} x\, dy = 0$$

or

$$\frac{1}{2}\left(\frac{x^2y^2+1}{x^2y^2}\right) y\, dx + \frac{1}{2}\left(\frac{x^2y^2-2xy+1}{x^2y^2}\right) x\, dy = 0$$

or

$$\frac{1}{2}\left(y+\frac{1}{x^2y}\right) dx + \frac{1}{2}\left(x-\frac{2}{y}+\frac{1}{xy^2}\right) dy = 0$$

Now, $M = \dfrac{1}{2}\left(y+\dfrac{1}{x^2y}\right)$

$N = \dfrac{1}{2}\left(x-\dfrac{2}{y}+\dfrac{1}{xy^2}\right)$

$\dfrac{\partial M}{\partial y} = \dfrac{1}{2}\left(1-\dfrac{1}{x^2y^2}\right)$, $\dfrac{\partial N}{\partial x} = \dfrac{1}{2}\left(1-\dfrac{1}{x^2y^2}\right)$

Now, $\dfrac{\partial M}{\partial y} = \dfrac{\partial N}{\partial x}$, hence the equation is exact and its solution is

$$\int \frac{1}{2}\left(y+\frac{1}{x^2y}\right) dx + \frac{1}{2}\int \frac{-2}{y}\, dy = c$$

(y constant)

or $\dfrac{1}{2}\left(xy-\dfrac{1}{xy}-2\log y\right) = c$ or $xy - \dfrac{1}{xy} - \log y^2 = 2c = k$

EXERCISE

Solve following differential equations :

1. $x^2y\,dx - (x^3+y^3)\,dy = 0$ [**Ans.** $y = ce^{x^3/3y^3}$]

2. $y(xy+2x^2y^2)\,dx + x(xy-x^2y^2)\,dy = 0$ $\left[\textbf{Ans. } \log\dfrac{x^2}{y}-\dfrac{1}{xy} = c\right]$

3. $(xy \sin xy + \cos xy)\,y\,dx + (xy \sin xy - \cos xy)\,x\,dy = 0$ [**Ans.** $x = cy \cos xy$]

4. $(x^4y^4+x^2y^2+xy)\,y\,dx + (x^4y^4-x^2y^2+xy)\,xdy = 0$ $\left[\textbf{Ans. } \dfrac{1}{2}x^2y^2-\dfrac{1}{xy}+\log\dfrac{x}{y} = c\right]$

2.9.3

Rule 3. When $\dfrac{\dfrac{\partial M}{\partial y}-\dfrac{\partial N}{\partial x}}{N}$ is a function of x alone say $f(x)$, then I.F. $= e^{\int f(x)\,dx}$.

Rule 4. When $\dfrac{\dfrac{\partial M}{\partial x}-\dfrac{\partial M}{\partial y}}{M}$ is a function of y alone, say $f(y)$, then I.F. $= e^{\int f(y)\,dy}$.

Differential Equations of First Order and First Degree

EXAMPLES

1. *Solve* : $(x^2 + y^2 + 2x) dx + 2y \, dy = 0$.

Sol. We have
$$M = x^2 + y^2 + 2x ; \quad N = 2y$$
$$\frac{\partial M}{\partial y} = 2y ; \quad \frac{\partial N}{\partial x} = 0$$

But
$$\frac{\frac{\partial M}{\partial y} - \frac{\partial N}{\partial x}}{N} = \frac{2y - 0}{2y} = 1,$$

Which may be regarded as function of x.

$\therefore \qquad \text{I.F.} = e^{\int 1 \, dx} = e^x$

Multiplying by e^x, the equation becomes
$$e^x (x^2 + y^2 + 2x) dx + 2y e^x dy = 0$$

Now, $\qquad M = e^x (x^2 + y^2 + 2x), \qquad N = 2y e^x$

$\therefore \qquad \frac{\partial M}{\partial y} = 2 y e^x, \quad \frac{\partial N}{\partial x} = 2 y e^x$

Hence, this equation is exact and its solution as usual shall be
$$e^x (x^2 + y^2) = c$$

2. *Solve* : $(3x^2 y^4 + 2xy) \cdot dx + (2x^3 y^3 - x^2) dy = 0$.

Sol. In the given equation, we have
$$M = 3x^2 y^4 + 2xy, \quad N = 2x^3 y^3 - x^2$$
$$\frac{\partial M}{\partial y} = 12x^2 y^3 + 2x, \quad \frac{\partial N}{\partial x} = 6x^2 y^3 - 2x$$

Hence the equation is not exact, but
$$\frac{\frac{\partial N}{\partial x} - \frac{\partial M}{\partial y}}{M} = \frac{6x^2 y^3 - 2x - 12x^2 y^3 - 2x}{xy(3xy^3 + 2)}$$
$$= -\frac{2x(3xy^3 + 2)}{xy(3xy^2 + 2)} = -\frac{2}{y}$$

Which is a function of y alone.

$\therefore \qquad \text{I.F.} = e^{\int -\frac{2}{y} dy} = \frac{1}{y^2}$

Multiplying by $\frac{1}{y^2}$, we get
$$\left(3x^2 y^2 + \frac{2x}{y}\right) dx + \left(2x^3 y - \frac{x^2}{y^2}\right) dy = 0$$

which can be shown to be exact.

∴ Solution is

$$\int \left(3x^2 y^2 + \frac{2x}{y}\right) dx = c$$

(y constant)

as there being no term in N free from x.

$$x^2 y^2 + \frac{x^2}{y} = c$$

or $\qquad x^2 \cdot y^3 + x^2 = cy$

EXERCISES

Solve following differential equations :

1. $(x^3 + xy^4) dx + 2y^3 dy = 0$ [Ans. $e^{x^2}(x^2 - 1 + y^4) = c$]
2. $(x^2 + y^2) dx - 2xy\, dy = 0$ [Ans. $x^2 - y^2 = cx$]
3. $(x^2 + y^2 + 1) dx - 2xy\, dy = 0$ [Ans. $x^2 - y^2 - 1 = cx$]
4. $(xy^3 + y) dx + 2(xy^2 + x + y^4) dy = 0$ [Ans. $3x^2 y^4 + 6xy^3 + 2y^6 = c$]

2.9.4 RULES 5. IF AN EQUATION IS OF THE FORM

$$x^a y^b (my\, dx + nx\, dy) + x^r y^s (py\, dx + qx\, dy) = 0$$

Where a, b, m, n, r, s, p and q are all constants having any value, then the integrating factor will be $x^h y^k$ where h and k are such that after multiplying by the integrating factor the condition of exactness is satisfied.

EXAMPLE

1. Solve : $(y^2 + 2x^2 y) dx + (2x^3 - xy) dy = 0.$

Sol. The given equation is not exact. It can be put in the form

$$y(y\, dx - x\, dy) + x^2 (2y\, dx + 2x\, dy) = 0$$

Let $x^h y^k$ is an I.F. Multiplying the equation by it, we get

$$(x^h y^{k+2} + 2x^{h+2} y^{k+1}) dx + (2x^{h+3} y^k - x^{h+1} y^{k+1}) dy = 0$$

Here $\quad M = x^h y^{k+2} + 2x^{h+2} y^{k+1}$

$\quad N = 2x^{h+3} y^k - x^{h+1} y^{k+1}$

∴ $\quad \dfrac{\partial M}{\partial y} = (k+2) x^h y^{k+1} + 2(k+1) x^{h+2} y^k$

and $\quad \dfrac{\partial N}{\partial x} = -(h+1) x^h y^{k+1} + 2(h+3) x^{h+2} y^k$

These two must be equal for exactness and hence comparing we get

$\qquad k + 2 = -(h+1)$

and $\qquad 2(k+1) = 2(h+3)$

Solving these equations, we get

$$h = -\frac{5}{2}$$

and $\qquad k = -\frac{1}{2}$

Differential Equations of First Order and First Degree

$\therefore \quad x^{-5/2} y^{-1/2}$ is the integrating factor.

Multiplying, we get

$$(x^{-5/2} y^{3/2} + 2x^{-1/2} y^{1/2}) \, dx + (2x^{1/2} y^{-1/2} - x^{3/2} y^{1/2}) \, dy = 0$$

Which is now exact and its solution as usual is

$$\int (x^{-5/2} y^{3/2} + 2x^{-1/2} y^{1/2}) \, dx = c$$

(y constant)

there being no term in N free from x.

$$4x^{1/2} y^{1/3} + \frac{2}{3} x^{-3/2} y^{3/2} = c$$

EXERCISE

Solving following differential equations :

1. $(2y\,dx + 3x\,dy) + 2xy\,(3y\,dx + 4x\,dy) = 0$ [Ans. $x^2 y^3 + 2x^2 y^4 = c$]

2. $(3x + 2y^2)\, y\,dx + 2x\,(2x + 3y^2)\, dy = 0$ [Ans. $x^2 y^4 (x + y^2) = c$]

3

Differential Equations of First Order but Not of First Degree

3.1 DEFINITION

Any differential equation of the form

$$p^n + A_1 p^{n-1} + A_2 p^{n-2} + A_3 p^{n-3} + \ldots + A_n = 0$$

is called a *differential equation of the first order and nth degree* where p is a symbol for $\dfrac{dy}{dx}$ and $A_1, A_2, A_3, \ldots, A_n$ all are functions of x and y.

3.2 EQUATIONS SOLVABLE FOR p.

The given differential equation is

$$p^n + A_1 p^{n-1} + A_2 p^{n-2} + A_3 p^{n-3} + \ldots + A_{n-1} p + A_n = 0$$

where $p = \dfrac{dy}{dx}$ and $A_1, A_2, \ldots, A_n$ are functions of x and y. Suppose it can be solved for p and the solution is of the form

$$(p - f_1)(p - f_2) \ldots (p - f_n) = 0$$

where $f_1, f_2, \ldots, f_n$ are functions of x and y.

Now each factor can be equated to zero and the resulting equations of the first degree and first order can be solved. Let the general solutions of the equation

$$p - f_1 = 0, \; p - f_2 = 0, \ldots, p - f_n = 0 \text{ be}$$

$$\phi_1(x, y, c_1) = 0, \phi_2(x, y, c_2) = 0, \ldots, \phi_n(x, y, c_n) = 0 \qquad \ldots(i)$$

Therefore, the most general solution is

$$[\phi_1(x, y, c_1)][\phi_2(x, y, c_2)] \ldots [\phi_n(x, y, c_n)] = 0 \qquad \ldots(ii)$$

There will be no loss of generally if n arbitrary constants $c_1, c_2, \ldots, c_n$ in (ii) are replaced by one arbitrary constant c, because every particular solution obtained from the equations (i) can also be obtained from (ii) by assigning a suitable value of c.

Therefore, the required solution is

$$[\phi_1(x, y, c)][\phi_2(x, y, c)] \ldots [\phi_n(x, y, c)] = 0$$

Working Rule

1. *Put p for $\dfrac{dy}{dx}$.*
2. *Factorize or solve for p.*
3. *Solve the equations by using the mehods explained in chapter 2.*

Differential Equations of First Order but Not of First Degree

4. Write down the result in closed brackets ni the form of product and put this resulting equation equal to zero by replacing $c_1, c_2, c_3, ..., c_n$ by c.

EXAMPLES

1. Solve : $p^2 + 2py \cot x - y^2 = 0.$

Sol. From given equation, we have

$$p = \frac{-2y \cot x \pm \sqrt{4y^2 \cot^2 x + 4y^2}}{2}$$

$$= -y \cot x \pm y \cosec x$$

∴ $\qquad p = y (\cosec x - \cot x)$ ...(i)

and $\qquad p = -y (\cosec x + \cot x)$ ...(ii)

From (i) $\qquad \dfrac{dy}{y} = (\cosec x - \cot x) \, dx$

On integration, we get

$$\log y = \log \tan \frac{x}{2} - \log \sin x + \log c$$

$$= -\log (1 + \cos x) + \log c$$

Hence, $\qquad y (1 + \cos x) = c$ ...(iii)

Similarly, the solution of (ii) is

$$y (1 - \cos x) = c \qquad \text{...(iv)}$$

Hence the required solution of the given equation is

$$y (1 \pm \cos x) = c$$

2. Solve : $\left(1 - y^2 + \dfrac{y^4}{x^2}\right) p^2 - 2 \dfrac{y}{x} p + \dfrac{y^2}{x^2} = 0.$

Sol. The given equation can be written as

$$p^2 - \frac{2y}{x} p + \frac{y^2}{x^2} = p^2 y^2 \left(1 - \frac{y^2}{x^2}\right)$$

or $\qquad (px - y)^2 = p^2 y^2 (x^2 - y^2)$

or $\qquad px - y = \pm py \sqrt{x^2 - y^2}$

∴ $\qquad p [x \mp y \sqrt{(x^2 - y^2)}] = y$

or $\qquad \dfrac{dx}{dy} = \dfrac{x \mp y \sqrt{x^2 - y^2}}{y}$

Put $x = vy$, $\dfrac{dx}{dy} = v + y \cdot \dfrac{dv}{dy}$

$$\therefore \quad v + \frac{dv}{dy} = \frac{vy \mp y^2 \sqrt{(v^2-1)}}{y} = v \mp y\sqrt{(v^2-1)}$$

or $$\frac{dv}{dy} = \mp \sqrt{(v^2-1)}$$

or $$\frac{dv}{\sqrt{(v^2-1)}} = \mp dy$$

$$\therefore \quad \cosh^{-1} v = (c \mp y)$$

or $$\log[v + \sqrt{(v^2-1)}] = c \mp y$$

$$\therefore \quad \log \frac{x + \sqrt{(x^2-y^2)}}{y} = c \mp y$$

EXERCISE

Solve the following differential equations:

1. $p^2 x^2 - xyp - y^2 = 0$ [Ans. $y^2 = cx^{(1\pm\sqrt{5})}$]

2. $x^2 p^2 + xyp - 6y^2 = 0$ [Ans. $(y - cx^2)(yx^3 - c) = 0$]

3. $xp^2 + (y-x)p - y = 0$ [Ans. $(y - x - c)(xy - c) = 0$]

4. $xy(p^2 + 1) = (x^2 + y^2) p$ [Ans. $(y^2 - x^2 - c)(y - cx) = 0$]

5. $4y^2 p^2 + 2pxy(3x+1) + 3x^3 = 0$ [Ans. $x^2 + 2y^2 = c, x^3 + y^2 = c$]

6. $xyp^2 + (3x^2 - 2y^2)p - 6xy = 0$ [Ans. $(y - cx^2)(3x^2 + y^2 - c) = 0$]

3.3 EQUATIONS SOLVABLE FOR y.

The given differential equation is

$$p^n + A_1 p^{n-1} + A_2 p^{n-2} + A_3 p^{n-3} + \ldots + A_{n-1} p + A_n = 0$$

where $p, A_1, A_2, \ldots$, have their usual meanings. Let us supose that it can be expressed in the form

$$y = \phi(x, p) \qquad \ldots(i)$$

Differentiating (i) with respect to x, we get

$$p = \frac{dy}{dx} = f\left(x, p, \frac{dy}{dx}\right) \qquad \ldots(ii)$$

(ii) is a new differential equation with variable x and p.

Let the general solution of (ii) be

$$\psi(x, p, c) = 0 \qquad \ldots(iii)$$

Eliminating p between (i) and (iii), we obtain $F(x, y, c)$ as the required solution of in the parametric form,

$x = F_1(p, c)$ and $y = F_2(p, c)$, where p is the parameter, in case the elimination is tedious.

Differential Equations of First Order but Not of First Degree

Working Rule

1. *Solve for y and put it as $\phi(x, p)$ only.*
2. *Differentiate it with respect to x.*
3. *Put p for $\dfrac{dy}{dx}$ and use the methods as employed previously.*
4. *Eliminate p between (1) and (3) and get the general solution.*

EXAMPLES

1. Solve $y + px = x^4 p^2$.

Sol. We have
$$y + px = x^4 p^2$$
or
$$y = -px + x^4 p^2 \qquad \qquad \text{...(i)}$$

Differentiating with respect to x and denoting $\dfrac{dy}{dx}$ by p, we get

$$p = -p - x\frac{dp}{dx} + x^4 \cdot 2p \cdot \frac{dp}{dx} + p^2 \cdot 4x^3$$

or
$$2p + x\frac{dp}{dx} = 2px^3\left(2p + x\frac{dp}{dx}\right)$$

or
$$\left(2p + x\frac{dp}{dx}\right)(1 - 2px^3) = 0$$

$\therefore$
$$2p + x\frac{dp}{dx} = 0$$

or
$$\frac{2}{x}dx + \frac{dp}{p} = 0$$

or
$$\log x^2 + \log p = \log c$$

or
$$px^2 = c, \quad \therefore \quad p = \frac{c}{x^2} \qquad \qquad \text{...(ii)}$$

Eliminating between (i) and (ii), we get
$$y = -x \cdot \frac{c}{x^2} + x^4 \cdot \frac{c^4}{x^4}$$

or
$$xy = -c + c^2 x$$

2. Solve : $y = \sin p - p \cos p$.

Sol. The equation is already solved for y, hence, differentiating it w.r.t., x, we get,

$$p = (\cos p - \cos p + p \sin p)\frac{dp}{dx}$$

or
$$1 = \sin p \frac{dp}{dx} \qquad \qquad \text{...(i)}$$

which is in variables separable for hence, integrating it, we get

$$x + c_1 = \int \sin p \, dp$$

or $$x + c_1 = -\cos p$$

or $$\cos p = c - x \qquad ..(ii)$$

Eliminating p from (ii) and the given differential equation, we get

$$y = \sqrt{1 - (c-x)^2} - (c-x).\cos^{-1}(c-x)$$

or $$\cos[\sqrt{(1-c^2+2cx-x^2)} - y]/(c-x) = c - x$$

which is the general solution of the given difdferential equation.

3. Solve : $9(y + xp \log p) = (2 + 3 \log p) p^3$.

Sol. Differentiating given equation w.r.t. x, we get

$$9\left\{p + p \log p + x\left(p\cdot\frac{1}{p} + \log p\right)\frac{dp}{dx}\right\} = 2.3p^2 \frac{dp}{dx} + 3\left(p^3 \cdot \frac{1}{p} + 3p^2 \log p\right)\frac{dp}{dx}$$

$$\Rightarrow \quad p(1 + \log p) + x(1 + \log p)\frac{dp}{dx} = p^2 (1 + \log p)\frac{dp}{dx}$$

$$\Rightarrow \quad \left\{p + (x - p)^2 \frac{dp}{dx}\right\}(1 + \log p) = 0$$

$$\Rightarrow \quad p + (x - p^2)\frac{dp}{dx} = 0 \text{ or } (1 + \log p) = 0 \qquad ...(1)$$

The second equation of eqn. (1) is free from differentials hence it will yield singular solution. From first equation

$$p\frac{dx}{dp} + x - p^2 = 0 \quad \text{or} \quad \frac{dx}{dp} + \frac{1}{p}.x = p \qquad ...(2)$$

This is linear equation in x, for which

$$\text{I.F.} = e^{\int 1/p \, dp} = e^{\log p} = p$$

Hence solution of eqn. (2) is

$$xp = \int p.p \, dp + c$$

$$\Rightarrow \quad xp = \frac{p^3}{3} + c \qquad ...(3)$$

Equation (3) and the give equation give the combined solution.

EXERCISE

Solve following differential equations :

1. $x^3 p^3 + x^2 py + a^3 = 0$ [**Ans.** $c^2 + cxy + a^3 x = 0$]

2. $y = 2px + p^4 x^3$ [**Ans.** $(y - c^2)^2 = 4cx$]

3. $\dfrac{x}{p} - ap$ $\left[\textbf{Ans. } x\dfrac{p}{\sqrt{1-p^2}}(c + a\sin^{-1}), y = -ap + \dfrac{1}{\sqrt{1-p^2}}(c + a\sin^{-1} p)\right]$

4. $y = 2px - p^2$ $\left[\textbf{Ans. } x = cp^{-2} + \dfrac{2}{3}p, y = 2cp^{-1} + \dfrac{1}{3}p^2\right]$

Differential Equations of First Order but Not of First Degree 539

5. $y = y = 2px + p^2$ [Ans. $4(y^2 - 3cx)(x^2 + y^2) = (xy + 3c)^2$]

6. $x - yp = ap^2$ $\left[\text{Ans. } y = \frac{1}{\sqrt{(1-p^2)}}(c + a \sin^{-1} p) - ap, \ x = \frac{p}{\sqrt{(1-p^2)}}(c + a \sin^{-1} p) \right]$

7. $y - x = x\dfrac{dy}{dx} + \left(\dfrac{dy}{dx}\right)^2$ [Ans. $y - x = xp + p^2, \ x = ce^{-p} - 2(p-1)$]

8. $x + yp = ap^2$ $\left[\text{Ans. } x = \dfrac{\sqrt{(p^2+1)}}{p} = (a \sinh^{-1} p + c), \ y = -\dfrac{x}{p} + \dfrac{a}{p} \right]$

9. $y = 3x + a \log p$ $\left[\text{Ans. } y = 3x + a \log \dfrac{3}{1 - ce^{3x/a}} \right]$

3.4 EQUATIONS SOLVABLE FOR x.

The given differential equation is

$$p^n + A_1 p^{n-1} + A_2 p^{n-2} + \dots + A_{n-1} p + A_n = 0$$

where p and $A_1, A_2, \dots, A_n$ have their usual meanings. Let us suppose that it can be expressed in the form

$$x = f(y, p) \qquad \dots(i)$$

Differentiating (i) with respect to y, we have

$$\frac{1}{p} = \frac{dx}{dy} = \phi\left(y, p, \frac{dp}{dy}\right) \qquad \dots(ii)$$

Now, (ii) is a new differential equation with variables y and p.
Let the general solution of (ii) be

$$\psi(x, p, c) = 0 \qquad \dots(iii)$$

Eliminating p between (i) and (iii), we get required solution either in $F(x, y, c) = 0$ or in parametric form.

WORKING RULE

1. *Solve for x and put it as $f(y, p)$.*
2. *Differentiate it with respect to y.*
3. *Put $1/p$ for $\dfrac{dx}{dy}$.*
4. *Eliminate p between (1) and general solution.*

EXAMPLES

1. Solve : $ap^2 + py - x = 0$.

Sol. Solving the given differential equation for x, we get

$$x = py + ap^2 \qquad \dots(1)$$

Differentiating (1) w.r.t. y, and writing $1/p$ for dx/dy, we get

$$\frac{1}{p} = p + y\frac{dp}{dy} + 2ap\frac{dp}{dy}$$

or $$\frac{1-p^2}{p} = y\frac{dp}{dy} + 2ap\frac{dp}{dy}$$

or $$\frac{1-p^2}{p}\frac{dy}{dp} - y = 2ap, \text{ multiplying both sides by } \frac{dy}{dp}$$

or $$\frac{dy}{dp} + \frac{p}{p^2-1}y = -\frac{2ap^2}{p^2-1} \qquad ...(2)$$

which is linear differential equation.

Hence, $\quad\text{I.F.} = e^{\int (p/p^2-1)\,dp} = e^{[1/2 \log(p^2-1)]}$

$$= \sqrt{(p^2-1)}$$

∴ Solution of equation (2) will be

$$y\sqrt{(p^2-1)} = \int \frac{-2ap^2}{(p^2-1)}\sqrt{(p^2-1)}\,dp + c$$

$$= -2a\int \frac{(p^2-1)+1}{\sqrt{(p^2-1)}}\,dp + c$$

$$= -2a\left[\int \sqrt{(p^2-1)} + \frac{1}{\sqrt{(p^2-1)}}\right]dp + c$$

$$= -2a\left[\frac{1}{2}p\sqrt{(p^2-1)} - \frac{1}{2}\cosh^{-1}p + \cosh^{-1}p\right] + c$$

$$= -ap\sqrt{(p^2-1)} - a\cosh^{-1}p + c$$

or $$y = \frac{c - a\cosh^{-1}p}{\sqrt{p^2-1}} - ap \qquad ...(3)$$

Substituting this value of y in (1), we get

$$x = p\left(\frac{c - a\cosh^{-1}p}{\sqrt{(p^2-1)}} - ap\right) + ap^2$$

or $$x = \frac{p(c - a\cosh^{-1}p)}{\sqrt{(p^2-1)}} \qquad ...(4)$$

The equations (3) and (4) constitute the parametric equations of the required solution.

Differential Equations of First Order but Not of First Degree

2. Solve : $ayp^2 + (2x - b)p - y = 0$.

Sol. The given equation can be written as

$$2x = \frac{y}{p} + b - ayp$$

Differentiating with respect to y, we get

$$2 \cdot \frac{1}{p} = \frac{1}{p} - \frac{1}{p^2} y \frac{dp}{dy} - a\left(p + y\frac{dp}{dy}\right)$$

or

$$\frac{1}{p^2}\left(p + y\frac{dp}{dy}\right) + a\left(p + y\frac{dp}{dy}\right) = 0$$

or

$$\left(\frac{1}{p^2} + a\right)\left(p + y\frac{dp}{dy}\right) = 0$$

$\therefore$

$$\frac{y}{p} \cdot \frac{dp}{dy} + 1 = 0 \quad \text{or} \quad \frac{dp}{p} + \frac{dy}{y} = 0$$

or

$$\log p + \log y = \log c \quad \text{or} \quad py = c \quad \text{or} \quad p = \frac{c}{y}.$$

on putting value of p in given equation, we get $2x = \frac{y^2}{c} + b - ac$

or

$$2xc = y^2 + bc - ac^2$$

$$ac^2 + (2x - b)c - y^2 = 0$$

EXERCISE

Solve following differential equations :

1. $y = 2px + y^2 p^3$ [**Ans.** $y^2 = 2cx + c^3$]

2. $yp^2 - 2xp + y = 0$ [**Ans.** $y^2 = 2cx - c^2$]

3. $y = 2px + p^2 y$ [**Ans.** $y^2 = 2cy + c^2$]

4. $p^3 - 4xyp + 8y^2 = 0$ [**Ans.** $y = c(c - x)^2$]

5. $y = yp^2 + 2px$ [**Ans.** $y^2 = 2cx + c^2$]

6. $p = \tan\left(x - \frac{p}{1+p^2}\right)$ $\left[\textbf{Ans. } y = c - \frac{1}{1+p^2}, x = \tan^{-1} p + \frac{p}{1+p^2}\right]$

3.5 CLAIRAUT'S EQUATIONS

The equation $y = px + f(p)$ is known as Clairaut's equation, where p has its usual meaning. It can be solved in the following manner.

The given equation is

$$y = px + f(p) \qquad \qquad ...(i)$$

Differentiating with respect to x, we have

$$\frac{dy}{dx} = p + x\frac{dp}{dx} + f'(p)\frac{dp}{dx}$$

or
$$p = p + x\frac{dp}{dx} + f'(p)\frac{dp}{dx}$$

or
$$\frac{dp}{dx}\{x + f'(p)\} = 0$$

Now, either
$$\frac{dp}{dx} = 0 \text{ or } x + f'(p) = 0$$

But $\frac{dp}{dx} = 0$ gives $p = c$ ...(ii)

and elimination of p between (i) and (ii), gives
$$y = cx + f(c) \qquad ...(iii)$$
where c is arbitrary constant.

(iii) is the general solution or complete solution of (i).

If we eliminate p between the following equations :
$$y = px + f(p) \qquad ...(iv)$$
and
$$x + f'(p) = 0 \qquad ...(v)$$

We shall get another solution which the general solution (iii) does not contain. This solution does not contain any arbitrary constant, nor can it be derived from the general solution by giving particular values to the arbitrary constant. It is known as **Singular Solution**.

EXAMPLES

1. *Solve* $y = px + \dfrac{a}{p}$ *and obtain the singular solution.*

Sol. Differentiating with respect to x, we get
$$p = p + x\frac{dp}{dx} - \frac{a}{p^2}\cdot\frac{dp}{dx}$$

∴
$$\frac{dp}{dx}\left(x - \frac{a}{p^2}\right) = 0$$

Taking $\dfrac{dp}{dx} = 0$, we get $p = c$ and putting in the given equation, we get
$$y = cx + \frac{a}{c}$$
as the solution.

Taking
$$x - \left(\frac{a}{p^2}\right) = 0 \quad \text{or} \quad p = \sqrt{\left(\frac{a}{x}\right)}$$

Eliminating p between this and the given equation we shall get the singular solution as
$$y = x\sqrt{\left(\frac{a}{x}\right)} + a\sqrt{\left(\frac{x}{a}\right)}$$
or
$$y = 2\sqrt{(ax)} \quad \text{or} \quad y^2 = 4ax.$$

2. *Solve :* $p^2 x(x-2) + p(2y - 2xy - x + 2) + y^2 + y = 0$.

Sol. The given equation can be written as

$$(y^2 - 2pxy + p^2 y^2) + 2p(y - px) + (y - px + 2p) = 0$$

or $\qquad (y - px)^2 + (y - px)(2p + 1) + 2p = 0$

or $\qquad (y - px + 1)(y - px + 2p) = 0$

or $\qquad y = px - 1 \quad \text{and} \quad y = px - 2p$.

Each of them is of Clairaut's form and hence the solution is obtained by putting $p = c$ and is

$$(y - cx + 2c)(y - cx + 1) = 0$$

EXERCISE

1. Solve $y = px + p - p^2$ and obtain the singular solution as well.

[**Ans.** $y = cx + c - c^2, \, 4y = (x+1)^2$]

2. Solve $y = px + ap(1 - p)$. [**Ans.** $y = cx + ac(1 - c)$]

3. Solve $y = x \dfrac{dy}{dx} + \left(\dfrac{dy}{dx}\right)^2$. [**Ans.** $y = cx + c^2$]

4. Solve $y = px - \dfrac{1}{p^2}$. $\left[\text{**Ans.** } y = cx - \dfrac{1}{c^2}\right]$

5. Solve $xp^2 - yp + a = 0$. [**Ans.** $y = cx + a/c$]

6. Solve : $xp^2 - yp + 2 = 0$. [**Ans.** $y = cx + 2/c$]

7. Solve : $p^2 x = py - 1$. $\left[\text{**Ans.** } y = cx + \dfrac{1}{c}\right]$

8. Solve : $(y - px)(p - 1) = p^2 \log p$. $\left[\text{**Ans.** } y = cx + \dfrac{c^2 \log c}{c - 1}\right]$

3.5.1 EQUATIONS REDUCIBLE TO CLAIRAUT'S FORM

EXAMPLES

1. *Solve :* $x^2(y - px) = p^2 y$.

Sol. We shall reduce the above equation to Clairaut's form by change of variables.

Put $\qquad x^2 = u \qquad$ and $\qquad y^2 = v$

$\therefore \qquad 2x \, dx = du \qquad$ and $\qquad 2y \, dy = dv$

or $\qquad \dfrac{y}{x} \dfrac{dy}{dx} = \dfrac{dv}{du} \qquad$ and $\qquad \dfrac{y}{x} p = \dfrac{dv}{du}$

Putting for p in the given equations, we get

$$x^2 \left(y - x \cdot \dfrac{x}{y} \dfrac{dv}{du}\right) = \dfrac{x^2}{y^2} \left(\dfrac{dv}{du}\right)^2 y$$

or
$$\left(y^2 - x^2 \cdot \frac{dv}{du}\right) = \left(\frac{dv}{du}\right)^2 \qquad [\text{Put } x^2 = u \text{ and } y^2 = v]$$

or
$$v = u\frac{dv}{du} + \left(\frac{dv}{du}\right)^2$$

If we take $\dfrac{dv}{du} = P$, then

or $v = Pu + P^2$, which is Clairaut's form, and its solution is

$$v = cu + c^2 \quad \text{or} \quad y^2 = cx^2 + c^2$$

2. *Use the transformation $u = x^2$ and $v = y^2$ to solve*

$$(px - y)(py + x) = h^2 p$$

Sol. Just as in Example 1, putting, $x^2 = u$ and $y^2 = v$, we have

$$\frac{y}{x} p = \frac{dv}{du} \quad \text{or} \quad p = \frac{x}{y}\frac{dv}{du}$$

Putting for p in the given equation, we get

$$p^2 xy + p(x^2 - y^2 - h^2) - xy = 0$$

or
$$xy \cdot \frac{x^2}{y^2}\left(\frac{dv}{du}\right)^2 + \frac{x}{y}\left(\frac{dv}{du}\right)(x^2 - y^2 - h^2) - xy = 0$$

or
$$x^2 \left(\frac{dv}{du}\right)^2 + \left(\frac{dv}{du}\right)(x^2 - y^2 - h^2) - y^2 = 0$$

Put $\dfrac{dv}{du} = P$, then

$$uP^2 + P(u - v - h^2) - v = 0$$

or
$$uP(P + 1) - v(P + 1) - Ph^2 = 0$$

or
$$v = uP - \frac{P}{(P+1)} h^2$$

which is of the form o $y = Px + f(P)$,
i.e., Clairaut's form.

Hence the solution is $v = uc - \dfrac{c}{c+1} h^2$

or $y^2 = cx^2 - \dfrac{c}{c+1} h^2$

Differential Equations of First Order but Not of First Degree

EXERCISE

1. Using substitution $x^2 = u$ and $y^2 = v$, solve the equation
$$axyp^2 + (x^2 - ay^2 - b)p - xy = 0$$
$$\left[\text{Ans. } y^2 = cx^2 - \frac{bc}{ac+1}\right]$$

Change the following equations to Clairaut's from and.

2. Solve: $xyp^2 - (x^2 - y^2 - 1)p + xy = 0.$ $\left[\text{Ans. } y^2 = cx^2 - \frac{c}{c-1}\right]$

3. Solve: $y = px + \dfrac{p}{x}.$ [Ans. $y^2 = cx^3 + c$]

4. Solve the equation $(px - y)(x - py) = 2p.$ $\left[\text{Ans. } y^2 = cx^2 - \dfrac{2c}{1-c}\right]$

5. Solve $y = 2px + y^2 p^3.$ $\left[\text{Ans. } y^2 = cx + \dfrac{c^3}{8}\right]$

3.6 DEFINITIONS

We have already discussed Clairaut's Equation in § 3.5. The equation $y = px + f(p)$ is known as Clairaut's Equation where p has its usual meaning.

Discriminants:

For a differential equation $f(x, y, p) = 0$ the p-discriminant is defined as the triplet (x, y, p) satisfying the two equations

$$f(x, y, p) = 0$$

$$\frac{\partial f(x, y, p)}{\partial p} = 0 \qquad \text{...(i)}$$

Let $\phi(x, y, c) = 0$ be the solution of the given differential equation $f(x, y, p) = 0$, where c is an arbitrary constant. The c-discriminant is determined by eliminating c between two equations

$$\phi(x, y, c) = 0 \quad \text{and} \quad \frac{\partial \phi(x, y, c)}{\partial c} = 0 \qquad \text{...(ii)}$$

obviously, the c-discriminant is the locus for each point of which $\phi(x, y, c) = 0$ has equal values of c and p-discriminant is the locus for each point of which $f(x, y, p)$ has equal values of p.

Envelope

The envelope is the locus of points of intersection of the consecutive curves of the system which are obtained by assigning different values of c in $\phi(x, y, c) = 0$. We can obtain this locus by eliminating c between the equation (ii). Thus this locus is contained in c-discriminant. Also the envelope is touched at any point on it by same curve of the system.

Therefore, its x, y, p at each point are identical with x, y, p of same point on one of the curves of the system.

Therefore, the equation to the envelope is also a solution of the differential equation $f(x, y, p) = 0.$

Singular Solution

The solution which does not contain any arbitrary constant, nor can it be derived from the general solution by giving particular values to the arbitrary constant is known as *singular solution*.

Both the c-discriminant and the p-discriminant contain the equation to the envelope which is said to be singular solution.

Tac-locus

The name tac-locus is given to a curve consisting of points at which two curves of a family have the same tangent. The p-discriminant has equal values of p, but these values may belong to two curves of the system that are not consecutive, *i.e.*, these curves have different c's. Locus of such point is called the *tac-locus*. For illustration, we consider a family of circles, all haveing the same radii and their centres lying on a straight line.

The two circles, which are not consecutive touch (they have the same p) but the direction of the tangent at point of contact is not the direction of the line of centres which is the locus of the points of contact of the circles. The line of centres is the tac-locus.

Nodal locus.

The c-discriminant gives equal values of c but these values may belong to the nodes which are also ultimate points of intersection of the consecutive curves. Locus of such points is called the *nodal-locus*. The p for the nodal locus may be different from p of the curves passing through the node, hence the x, y, p belonging to the nodal locus will not satisfy the differential equation.

However, in few cases they may satisfy the differential equation in which case the nodal locus would also be an envelope.

Cuspoidal locus.

Similar to the case of nodes, the cusps of also appear in c-discriminant, as they are also the ultimate points of intersection of the consecutive curves. Their locus is called the *cuspoidal-locus*.

EXAMPLES

1. *Find the singular solution of*

$$y = x\frac{dy}{dx} + a\sqrt{1 + \left(\frac{dy}{dx}\right)^2}$$

or $y = xp + a\sqrt{1 + p^2}.$

Sol. The equation is of the Clairaut's form, so that its solution is

$$y = cx + a\sqrt{1 + c^2} \qquad \text{...(i)}$$

or $c^2(a^2 - x) + 2xyc + a^2 - y^2 = 0 \qquad \text{...(ii)}$

The required singular solution will be included in the envelope of this equation. Since equation (ii) is a quadratic in c, its envelope is

$$(B^2 - 4AC = 0)$$

$$4x^2y^2 - 4(a^2 - x^2)(a^2 - y^2) = 0$$

i.e., $x^2 + y^2 = a^2$

This relation satisfies the given equation, and hence it is the required singular solution.

Geometrical Interpretetion.

Here the general solution represents the system of straight lines

$$y = cx + a\sqrt{(1 + c^2)}$$

all of which touch the circle $x^2 + y^2 = a^2$.

Differential Equations of First Order but Not of First Degree

2. *Find general and singular solution of the differential equation* $p^3 - 4pxy + 8y^2 = 0$.

Sol. Solving the given equation for x, we get

$$x = \left(\frac{2y}{p}\right) + \left(\frac{p^2}{4y}\right) \qquad \ldots(1)$$

Differentiating both sides of (1) w.r.t. y, we get

$$\frac{1}{p} = \frac{2}{p} - \frac{2y}{p^2}\cdot\frac{dp}{dy} - \frac{p^2}{4y^2} + \frac{p}{2y}\cdot\frac{dp}{dy}$$

$$\Rightarrow \frac{2y}{p}\left(\frac{1}{p} - \frac{p^2}{4y^2}\right)\frac{dp}{dy} = \left(\frac{1}{p} - \frac{p^2}{4y^2}\right)$$

$$\Rightarrow \frac{2y}{p}\frac{dp}{dy} = 1$$

$$\Rightarrow 2\left(\frac{dp}{p}\right) = \frac{dy}{y} \qquad \ldots(2)$$

On integration

$$2 \log p = \log c + \log y$$

$$\therefore \qquad p^2 = cy \qquad \ldots(3)$$

Again from eqn. (1)

$$p^2 (p^2 - 4xy)^2 = (-8y^2)^2 \qquad \ldots(4)$$

Eliminating p from eqn. (3) and (4), we get the required general solution as :

$$cy(cy - 4xy)^2 = 64y^4$$

i.e., $\qquad c(c - 4x)^2 = 64y$

i.e., $\qquad c(c - x)^2 = y,\qquad$ where $c = c/4 \qquad \ldots(5)$

Singular Solution : As singular solution is included in the envelope of the general solution of the given equation, hence differentiating eqn.(5) partially w.r.t. c, we get

$$1.(c - x)^2 + c.2(c - x) = 0$$

or $\qquad (c - x)(3c - x) = 0$

or $\qquad c = x, c = \frac{1}{3}x \qquad \ldots(6)$

By putting the values of c in eqn. (5) we get the equation of required envelope as

$$y = 0 \quad \text{or} \quad 27y = 4x^3$$

It can easily be verified that these both relations satisfy the given differential equation. Hence, required singular solutions are

$$y = 0,\ 27y = 4x^3.$$

3. *Solve and examine for singular solution*

$$(8p^3 - 27)x = 12p^2 y$$

Sol. The given differential equation is

$$(8p^3 - 27) x = 12p^2 y \qquad \text{...(i)}$$

Solving the above differential equation for y, we get

$$y = \frac{8p^3 x - 27x}{12p^2}$$

or
$$y = \frac{2}{3} px - \frac{9}{4} \frac{x}{p^2} \qquad \text{...(ii)}$$

Differentiating (ii) with respect to x, we get

$$p = \frac{2}{3} p + \frac{2}{3} x \frac{dp}{dx} - \frac{9}{4} \left(\frac{p^2 - x \cdot 2p \frac{dp}{dx}}{p^4} \right)$$

or
$$\left(\frac{4p^3 + 27}{12p^2} \right) = \left(\frac{4p^3 + 27}{6p^3} \right) x \frac{dp}{dx}$$

which gives
$$4p^3 + 27 = 0 \qquad \text{...(iii)}$$

and
$$\frac{1}{2} = \frac{x}{p} \frac{dp}{dx} \qquad \text{...(iv)}$$

Solving (iv), we get, $p^2 = c_1 x$ ...(v)

Eliminating p from (i) and (v), we get, general solution of (i) as

$$[8(c_1 x)^{3/2} - 27] x = 12 c_1 xy$$

or
$$[8(c_1 x)^{3/2} - 27 - 12 c_1 y] x = 0$$

giving, $x = 0$

and
$$8(c_1 x)^{3/2} - 27 - 12 c_1 y = 0$$

or
$$8(c_1 x)^{3/2} = 27 + 12 c_1 y$$

or
$$64 c_1^3 x^3 = (12 c_1 y + 27)^2$$

or
$$x^3 = \frac{144 c_1^2}{64 c_1^3} \left(y + \frac{27}{12 c_1} \right)^2 = \frac{9}{4 c_1} \left(y + \frac{9}{4 c_1} \right)^2$$

replacing $\frac{9}{4c_1}$ by c, we get

$$x^3 = c(y + c)^2$$

Hence, the general solution of (i) is given by

$$x [x^2 - c(y + c)^2] = 0$$

Hence, the general solution of (i) is given by

$$x [x^2 - c(y + c)^2] = 0 \qquad \text{...(vi)}$$

Differential Equations of First Order but Not of First Degree

Differentiating (vi) partially w.r.t. c, we get

$$x[-(y+c)^2 - c.2(y+c)] = 0$$

giving, $\qquad c = -\dfrac{y}{3}$ ...(vii)

Eliminating c from (vi) and (vii), we get the c-discriminant as

$$x\left[x^3 - \left(-\dfrac{y}{3}\right)\left(y - \dfrac{y}{3}\right)^2\right] = 0$$

or $\qquad 4y^3 = (27x^3) = 0$ ...(viii)

Also, differentiating (i) partially with respect to p, we get

$$24p^2 x = 24py$$

or $\qquad p = \dfrac{y}{x}$ ...(ix)

Hence, eliminating p from (i) and (ix), we get, the p-discriminant given by

$$\left(\dfrac{8y^3}{x^3} - 27\right)x = 12\left(\dfrac{y}{x}\right)^2 \cdot y$$

or $\qquad 8y^3 - 27x^3 = 12y^3$

or $\qquad 4y^3 + 27x^3 = 0$ ...(x)

Since, non-repeated common factor in p-discriminant gives

$$4y^3 + 27x^3 = 0 \qquad \text{...(xi)}$$

which satisfies, the given differential equation (i) (it can be verified). Hence, it gives the singular solution of the differential equation (i).

EXERCISE

Find the general and singular solutions of :

1. $9p^2(1-y)^2 = 4(2-y)$ [**Ans.** $(x+c)^2 = (y+1)^2(2-y); y = 2.$]
2. $y^2 - 2pxy + p^2(x^2 - 1) = m^2$ [**Ans.** $(y - cx)^2 = m^2 + c^2; y^2 + m^2 x^2 = m^2$]
3. $3xy = 2px^2 - 2p^2$ [**Ans.** $(3y + c)^2 = 2cx^3; 6y = x^3$]
4. $(xp - y)^2 = p^2 - 1$ [**Ans.** $(cx - y)^2 = c^2 - 1; x^2 - y^2 = 1$]
5. Find the complete primitive and singular solution of

$$y = px + \sqrt{b^2 + a^2 p^2}$$

Interpret your result geometrically.

[**Ans.** $y = cx + \sqrt{b^2 + a^2 c^2}$; $\dfrac{x^2}{a^2} + \dfrac{y^2}{b^2} = 1$. The complete primitive is a system of straight lines, each member touching the ellipse $\dfrac{x^2}{a^2} + \dfrac{y^2}{b^2} = 1$.]

6. Find the singular solution of $p^2 + y^2 = 1$. [**Ans.** No singular solution.]

7. Find the general and singular solution of $y^2(1+p^2) = a^2$.

[**Ans.** $(x+c)^2 + y^2 = a^2$, $y^2 - a^2 = 0$.]

8. Obtain the singular solution of the equation

$$p^2 y^2 \cos^2 \alpha - 2pxy \sin^2 \alpha + y^2 - x^2 \sin^2 \alpha = 0$$

directly from the equation and also from its complete primitive explaining the geometrical significance of the irrelevant factors that present themselves.

[**Ans.** $y = \pm c \tan \alpha$; $y = 0$ represents the tac–locus.]

9. Reduce the equation

$$xp^2 - 2yp + x + 2y = 0$$

to Clairaut's form by putting $y - x = v$ and $x^2 = u$. Hence, obtain the interpret the primitive and singular solution of the equation. Show that the given equation represents a family of parabolas touching a pair of straight lines.

[**Ans.** $2c^2 x^2 - 2c(y-x) + 1 = 0$; $y - x = \pm\sqrt{2x}$. The complete primitive represents a family of parabolas, each member touching the line $y - x = \pm\sqrt{2x}$.]

10. Change the equation

$$x^2 p^2 + yp(2x+y) + y = 0$$

in Clairaut's form by putting $y = x$, $xy = v$. Then find its general and singular solution.

[**Ans.** $xy = cy + c^2$, $y = 0$ and $y + 4x = 0$]

4

Geometrical Interpretation and Orthogonal Trajectories

4.1 GEOMETRICAL MEANING OF A DIFFERENTIAL EQUATION

(i) Differential Equation of First Order and First Degree:

Let
$$f\left(x, y, \frac{dy}{dx}\right) = 0 \qquad \ldots(1)$$

be the given differential equation. It is to be noted that $\frac{dy}{dx}$ represents the gradient of the tangent to a curve which is represented by an equation in Cartesian form.

Now, let $A_1(x_1, y_1)$ be any point on the Cartesian plane and $A_2(x_2, y_2)$ be another point on the same plane, which is very close to A_1 and such that gradient of $A_1 A_2$ is given by m_1, i.e., by the value of $\frac{dy}{dx}$ given by (1) at the point $A_1(x_1, y_1)$.

Now, consider another point $A_3(x_3, y_3)$ very close to A_2 and the gradient of $A_2 A_3$ be $m_2 = \frac{dy}{dx}$ at (x_2, y_2) given by (1). Similarly, take other points $P, A_4, A_5, \ldots$, In the limiting case, the different directed segments $A_1 A_2, A_2 A_3, A_3 A_4, \ldots$ tend to a curve C, say, and the co-ordinates (x, y) of every point on this curve and the value of $\frac{dy}{dx}$ i.e., gradient of the tangent to the curve at the point satisfy the differential equation (1).

Similarly, if we start with any other point, in plane, which does not lie on the above obtained curve, we shall get another curve possessing the above property.

Hence, through every point in the plane, there will pass a particular curve possessing the above property. Therefore, the differential equation (1) representsd a family of curves, which is confirmed by the fact that the solution of (1) is of the form

$$F(x, y, c) = 0 \qquad \ldots(2)$$

where c is an arbitrary constant. The equation of each curve is a particular solution of the differential equation and the general solution (2) represents the system, itself, of the curves.

All the curves represented by the general solution (2) taken together make the **locus of the differential equation.**

(ii) Differential Equation of First Order and Second Degree: It will be quadratic in $\frac{dy}{dx}$ giving its two values at every point on the plane. Therefore, two members of the family of curves represented by it will pass through every point, consequently, the general solution of it, i.e., $F(x, y, c) = 0$, will have two different values of c at each point and hence, the general solution will involve c in the second degree.

4.2 FAMILY OF CURVES

We have already discussed in § 4.1 that if

$$f\left(x, y, \frac{dy}{dx}\right) = 0 \qquad \text{...(1)}$$

be a differential equation and

$$F(x, y, c) = 0 \qquad \text{...(2)}$$

be its general solution, then for different values of c, we get different curves forming a family of curves. Equation (2) represents that family and c is called the parameter of the family.

Equation (1) is the differential equation of the family, which is obtained by eliminating c from (2) and its differential coefficient with respect to x.

Thus, $y^2 = 4ax$ represents a family of parabolas having common vertex and axis, a is parameter of the family. To find out the differential equation of this family of parabolas, differentiate it w.r.t. x, to get $2y \dfrac{dy}{dx} = 4a$. Eliminate the parameter from the two equations, we get

$$y^2 = 2y \frac{dy}{dx} \cdot x$$

or
$$2x \frac{dy}{dx} = y$$

which is the differential equation of the family.

Sometimes the equation of the family is not given directly but the rule of formation of the curves (members of the family) are given. According to those rules we construct the equation of the family.

EXAMPLES

1. *Find the family of curves in which the length of polar subnormal is constant.*

Sol. Let the length of polar subnormal is a. We know that the point $P(r, \theta)$ is any member of family of curves, then polar subnormal

$$= \frac{dr}{d\theta}$$

Hence by the given condition

$$\frac{dr}{d\theta} = a \qquad \text{...(i)}$$

which is the differential equation of family of curves.

$\therefore \qquad dr = a\, d\theta$

On integrating of equation (i)

$$r = a\theta + c$$

where c is arbitrary constant.

Hence, it is the required polar subnormal of the family of cure.

2. *Find the equation of the family of curves in which the length of the tangent between the point of contact and the axis of x is of constant length equal to a.*

Sol. Let $P(x, y)$ be any point on one member of the family and PT be the tangent at P meeting at X-axis in T. PM be the ordinate and ψ be the angle which the tangent at P makes with the axis of x.

Then,
$$\tan \psi = \frac{dy}{dx} \qquad ...(i)$$

Now, from right angled triangle PTM,
$$PT = PM \, \text{cosec} \, \psi$$

Hence,
$$a = y\sqrt{1 + \cot^2 \psi} \qquad [\because PT = a, \text{given}]$$

$$= y\sqrt{1 + \left(\frac{dx}{dy}\right)^2}$$

or
$$a^2 = y^2 \left[1 + \left(\frac{dx}{dy}\right)^2\right] \qquad ...(ii)$$

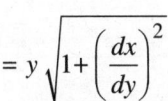

which is the differential equation of family of curves, required.
Re-arranging (ii), we get
$$\frac{dx}{dy} = \pm \frac{\sqrt{a^2 - y^2}}{y}$$

which is in variables separable form.

Hence,
$$\pm dx = \frac{\sqrt{a^2 - y^2}}{y} dy$$

which on integration gives,
$$c \pm x = \int \frac{\sqrt{a^2 - y^2}}{y} dy \quad [\text{Putting } y = a \sin \theta \text{ and } dy = a \cos \theta \, d\theta]$$

$$= \int \frac{a \cos \theta}{a \sin \theta} \, a \cos \theta \, d\theta$$

$$= a \int \text{cosec} \, \theta \, d\theta - a \int \sin \theta \, d\theta$$

$$= a \log \tan \frac{\theta}{2} + a \cos \theta$$

Putting back the value of θ, we get the equation of the family required curves as,
$$c \pm x = a \log \tan \left(\frac{1}{2} \sin^{-1} \frac{y}{a}\right) + \sqrt{a^2 - x^2}$$

c being the parameter of the family.

4.3 TRAJECTORY

Any curve which cuts every member of a given family of curves according to some given law is called a *Trajectory*.

In case, a curve acts at a constant angle α, it is called an α-*trajectory*.

If this angle α is $90°$, then the curve is called an *orthogonal trajectory* of the family of curves.

4.4 DIFFERENTIAL EQUATION OF ORTHOGONAL TRAJECTORY

(a) Let the differential equation of the given family of curves be

$$f\left(x, y, \frac{dy}{dx}\right) = 0 \qquad \ldots(i)$$

and suppose that X, Y are the current coordinates of any point on an orthogonal trajectory of (i).

At a point of intersection of (i) and orthogonal trajectory,

$$x = X, \ y = Y, \ \frac{dY}{dX} = \frac{1}{dy/dx}$$

Therefore, (i) gives

$$f\left(X, Y, -\frac{dX}{dY}\right) = 0$$

which is the required equation of the orthogonal trajectory.

(b) Let the differential equation of the given family of curves be

$$f\left(r, \theta, \frac{dr}{d\theta}\right) = 0 \qquad \ldots(i)$$

Let ϕ be the angle between radius vector and tangent at point (r, θ) on a given family of curves.

Therefore, $\qquad \tan \phi = r \dfrac{d\theta}{dr}$

Let ϕ_1 be the angle between the radius vector and tangent at any point (r_1, θ_1) on the orthogonal trajectory $r = r_1; \theta = \theta_1$ and $\phi_1 = \dfrac{\pi}{2} + \phi$ or $\tan \phi_1 = -\cot \phi$.

or $\qquad \tan \phi_1 \tan \phi = -1$

or $\qquad \left(r_1 \dfrac{d\theta_1}{dr_1}\right)\left(r \dfrac{d\theta}{dr}\right) = -1$

Therefore, differential equation of the given curves by writing $-\dfrac{1}{r}\dfrac{dr}{d\theta}$ for $r\dfrac{d\theta}{dr}$.

Therefore, $f\left(r, \theta, -r^2 \dfrac{d\theta}{dr}\right) = 0$ is the required equation of the orthognoal trajectory.

Working Rule

(1) *Form the differential equation of the family of curves.*

(2) *Write* $-\dfrac{1}{dy/dx}$ *for* $\dfrac{dy}{dx}$ *or* $-r^2 \dfrac{d\theta}{dr}$ *for* $\dfrac{dr}{d\theta}$.

(3) *Solve this new differential equation to get the equation of orthogonal trajectory.*

4.4.1 Self Orthogonal Families

If every member of a family of curves cuts all other members of the family at right angle, then the family is the system of orthogonal trajectories of itself and hence it is called self-orthogonal family.

Thus, if the differential equation of the family is identical with the differential equation to its orthogonal trajectory then the family is self-orthogonal.

Geometrical Interpretation and Orthogonal Trajectories

EXAMPLES

1. *Find the orthogonal trajectory of the family of semi-cubical parabolas $ay^2 = x^3$, where a is a variable parameter.*

Sol. The given curve is

$$ay^2 = x^3, \qquad \therefore \quad 2ay\frac{dy}{dx} = 3x^2$$

To eliminate a, dividing these two equations, we get

$$\frac{2}{y}\frac{dy}{dx} = \frac{3}{x}$$

This is the differential equation of given family of curves and in order to find the differential equation of orthogonal trajectory, replace

$$\frac{dy}{dx} \quad \text{by} \quad -\frac{1}{dy/dx}$$

Hence, the differential equation of orthogonal trajectory will be

$$\frac{2}{y} = -\frac{3}{x}\frac{dy}{dx}$$

or
$$2x\,dx + 3y\,dy = 0$$

On integration, this gives

$$x^2 + 3 \cdot \frac{y^2}{2} = k$$

or
$$2x^2 + 3y^2 = c^2$$

2. *Find the orthogonal trajectory of $r^n = a^n \cos n\theta$.*

Sol. Taking log and differentiating, we get

$$n \cdot \frac{1}{r}\frac{dr}{d\theta} = 0 + n \cdot \frac{1}{\cos n\theta}(-\sin n\theta)$$

or
$$r\frac{d\theta}{dr} = -\cot n\theta$$

Replacing $r\dfrac{d\theta}{dr}$ by $-\dfrac{1}{r}\dfrac{dr}{d\theta}$, we get the differential equation of orthogonal trajectory as

$$-\frac{1}{r}\frac{dr}{d\theta} = -\cot n\theta$$

or
$$\frac{dr}{r} = \cot n\theta \, d\theta$$

Integrating, we get

$$\log r = \frac{1}{n}\log \sin n\theta + \log k$$

or
$$n \log \frac{r}{k} = \log \sin n\theta$$

or $$\frac{r^n}{k^n} = \sin n\theta$$

or $$r^n = k^n \sin n\theta$$

3. *Find the orthogonal trajectories of the system of circles touching a given straight line at a given point.*

Sol. Let the circle touches the x-axis at the point (0, 0) so that its centre is (0, a) on y-axis, where a is the radius. Hence its equation is $x^2 + (y-a)^2 = a^2$ or $x^2 + y^2 - 2ay = 0$, where a is the parameter.

Differentiating, we get

$$2x + (2y - 2a)\frac{dy}{dx} = 0$$

or $$x + \left(y - \frac{x^2+y^2}{2y}\right)\frac{dy}{dx} = 0$$

or $$2xy + (y^2 - x^2)\frac{dy}{dx} = 0$$

This is the differentiating equation of the family of circles. Its orthogonal trajectory is obtained by replacing $\frac{dy}{dx}$ by $-\frac{1}{dy/dx}$ whose equation becomes,

$$2xy - (y^2 - x^2)\frac{dx}{dy} = 0$$

or $$2xy\frac{dy}{dx} - (y^2 - x^2) = 0$$

or $$2y\frac{dy}{dx} - \frac{y^2}{x^2} = -x$$

Put $y^2 = v$

$\therefore$ $$2y\frac{dy}{dx} = \frac{dv}{dx}$$

Hence the equation becomes

$$\frac{dv}{dx} - \frac{1}{x}v = -x$$

This is linear equation whose I.F. $= e^{-\int \frac{1}{x}dx} = e^{-\log x} = \frac{1}{x}$.

Thus the solution of the equation will be

$$v \cdot \frac{1}{x} = \int \frac{1}{x}(-x)\,dx + c$$

or $$y^2 \cdot \frac{1}{x} = -x + c$$

Geometrical Interpretation and Orthogonal Trajectories

or $\qquad y^2 + x^2 = cx$

or $\qquad x^2 + y^2 - 2bx = 0$

where $c = 2b$

or $\qquad (x - b)^2 + y^2 = b^2$

This represents a family of circles touching y-axis at (0, 0) and whose radius is b so that its centre is $(b, 0)$.

4. *Prove that the system of confocal conics* $\dfrac{x^2}{a^2 + \lambda} + \dfrac{y^2}{b^2 + \lambda} = 1$ *is self-orthogonal.*

Solution. Differentiating the given relation,
We get

$$\frac{2x}{a^2 + \lambda} + \frac{2y}{b^2 + \lambda} \cdot \frac{dy}{dx} = 0$$

or $\qquad x(b^2 + \lambda) + y(a^2 + \lambda)\dfrac{dy}{dx} = 0$

or $\qquad \lambda = -\dfrac{b^2 x + a^2 y \dfrac{dy}{dx}}{x + y \dfrac{dy}{dx}}$

$\therefore \qquad a^2 + \lambda = a^2 - \dfrac{b^2 x + a^2 y \, dy/dx}{x + y \dfrac{dy}{dx}} = -\dfrac{(a^2 - b^2) y \dfrac{dy}{dx}}{x + y \dfrac{dy}{dx}}$

and $\qquad b^2 + \lambda = b^2 - \dfrac{b^2 x + a^2 y \left(\dfrac{dy}{dx}\right)}{x + y \dfrac{dy}{dx}}$

$\qquad = -\dfrac{(a^2 - b^2) y \dfrac{dy}{dx}}{x + y \dfrac{dy}{dx}}$

Putting the values of $a^2 + \lambda$ and $b^2 + \lambda$ in the given equation, we get

$$\left(x - \frac{y}{dy/dx}\right)\left(x + y\frac{dy}{dx}\right) = a^2 - b^2 \qquad \ldots (i)$$

In order to find its trajectory, we have to replace $\dfrac{dy}{dx}$ by $-\left(\dfrac{1}{dy/dx}\right)$ and we get

$$\left(x + y\frac{dy}{dx}\right)\left(x - \frac{y}{dy/dx}\right) = a^2 - b^2$$

which is same as (i), and hence the system of conics is self-orthogonal.

EXERCISE

1. Find the orthogonal trajectories of the family of parabolas $y = ax^2$.

   [Ans. $x^2 + 2y^2 = c^2$]

2. Find the equation of the family of curves which is orthogonal to the family $y = ax^3$.

   [Ans. $x^2 + 3y^2 = c^2$]

3. Find the orthogonal trajectory of the family of circles $x^2 + y^2 + 2fy + 1 = 0$, where f is parameter.

   [Ans. $x^2 + y^2 + 2gx - 1 = 0$]

4. Find the orthogonal trajectories of the cardiodes $r = a(1 + \cos \theta)$.

   [Ans. $r = c(1 - \cos \theta)$]

5. Find the equation of the system of orthogonal trajectories of the parabolas $r = \dfrac{2a}{(1 + \cos \theta)}$, where a is the parameter.

   $\left[\text{Ans. } r = \dfrac{2b}{1 - \cos \theta}\right]$

6. Find the orthogonal trajectories of the curves $y = \dfrac{x^3 - a^3}{3x}$, a being the parameter.

   $\left[\text{Ans. } x^2 = y - \dfrac{1}{2} + ke^{-2y}\right]$

7. Find the orthogonal trajecotory of the system of curves given by $\left(\dfrac{dy}{dx}\right)^2 = \dfrac{a}{x}$.

   $\left[\text{Ans. } (x^{3/2} + c^{3/2})^2 = \dfrac{9}{4} ay^2\right]$

8. Find the orthogonal trajectories of the family of astroids $x^{2/3} + y^{2/3} = a^{2/3}$.

   [Ans. $x^{4/3} - y^{4/3} = c^{4/3}$]

9. Find the orthogonal trajectories of $\dfrac{x^2}{a^2} + \dfrac{y^2}{a^2 + \lambda} = 1$, where λ is parameter.

10. Find the orthogonal trajectory of $r^n \sin n\theta = a^n$.

5
Linear Differential Equations with Constant Coefficients

5.1 LINEAR DIFFERENTIAL EQUATIONS

The equation

$$P_0 \frac{d^n y}{dx^n} + P_1 \frac{d^{n-1} y}{dx^{n-1}} + P_2 \frac{d^{n-2} y}{dx^{n-2}} + \ldots + P_n y = Q \qquad \ldots(i)$$

where $P_0, P_1, P_3, \ldots, P_n$ and Q are either constants or functions of x, is called a *linear differential equation of the nth order*.

If $P_0, P_1, P_2, \ldots, P_n$ are all constants (Q may not be constant), then the equation is said to be a *linear differential equation with constant coefficients*.

If $y = f_1(x)$ is solution of the equation

$$P_0 \frac{d^n y}{dx^n} + P_1 \frac{d^{n-1} y}{dx^{n-1}} + \ldots + P_n y = 0 \qquad \ldots(ii)$$

then $y = c f_1(x)$ where c is an arbitrary constant, is also a solution.

If $y = f_1(x)$ and $y = f_2(x)$ are any two solutions of (ii) then $y = c_1 f_1(x) + c_2 f_2(x)$ is also a solution of (ii), c_1 and c_2 being arbitrary constants.

Proof : Since $f_1(x)$ and $f_2(x)$ are solutions of (ii), we have

$$P_0 \frac{d^n [f_1(x)]}{dx^n} + P_1 \frac{d^{n-1} [f_1(x)]}{dx^{n-1}} + \ldots + P_n [f_1(x)] = 0$$

and

$$P_0 \frac{d^n [f_2(x)]}{dx^n} + P_1 \frac{d^{n-1} [f_2(x)]}{dx^{n-1}} + \ldots + P_n [f_2(x)] = 0$$

If c_1 and c_2 are any two arbitrary constants, we have

$$P_0 \frac{d^n . [c_1 f_1(x) + c_2 f_2(x)]}{dx^n} + P_1 \frac{d^{n-1} [c_1 f_1(x) + c_2 f_2(x)]}{dx^{n-1}} + \ldots + P_n [c_1 f_1(x) + c_2 f_2(x)]$$

$$= c_1 \left[\frac{P_0 d^n \{f_1(x)\}}{dx^n} + P_1 \frac{d^{n-1} \{f_1(x)\}}{dx^{n-1}} + \ldots + P_n \{f_1(x)\} \right]$$

$$+ c_2 \left[P_0 \frac{d^n \{f_2(x)\}}{dx^n} + P_1 \frac{d^{n-1} \{f_2(x)\}}{dx^{n-1}} + \ldots + P_n \{f_2(x)\} \right]$$

$$= c_1 . 0 + c_2 . 0 = 0.$$

Now, generalising this statements the following theorem can be established :

If $y = f_1(x), y = f_2(x), y = f_3(x), ..., y = f_n(x)$ are n linearly independent solutions of the equation

$$P_0 \frac{d^n y}{dx^n} + P_1 \frac{d^{n-1} y}{dx^{n-1}} + ... + P_n y = 0,$$

then the complete or general solution of this equation is

$$y = c_1 f_1(x) + c_2 f_2(x) + ... + c_n f_n(x)$$

where $c_1, c_2, ..., c_n$ are arbitrary constants.

The functions $f_1(x), f_2(x), f_3(x), ..., f_n(x)$ are said to be linearly independent if we can not find constants $b_1, b_2, b_3, ..., b_n$ such that

$$b_1 f_1(x) + b_2 f_2(x) + b_3 f_3(x) + ... + b_n f_n(x) = 0 \text{ identically}.$$

5.2 SOLUTION OF THE LINEAR EQUATION WITH CONSTANT COEFFICIENTS

To solve the equation

$$P_0 \frac{d^n y}{dx^n} + P_1 \frac{d^{n-1} y}{dx^{n-1}} + ... + P_n y = 0$$

where $P_0, P_1, P_2, ..., P_n$ are constants.

The given equation is

$$P_0 \frac{d^n y}{dx^n} + P_1 \frac{d^{n-1} y}{dx^{n-1}} + P_2 \frac{d^{n-2} y}{dx^{n-2}} + ... + P_3 y = 0 \qquad ...(i)$$

Substitute $y = e^{mx}$ on a trial basis,

Then

$$e^{mx} (P_0 m^n + P_1 m^{n-1} + ... + P_n) = 0 \qquad ...(ii)$$

Now, e^{mx} is a solution of (i) if m is a root of the algebraic equation

$$P_0 m^n + P_1 m^{n-1} + ... + P_n = 0 \qquad ...(iii)$$

5.3 AUXILLIARY EQUATION

The equation (iii) is called the *auxilliary equation*. Therefore, if m have a value say m_1 that satisfies (iii), $y = e^{m_1 x}$ is an integral of (i), and if the n roots of (iii) be $m_1, m_2, m_3, ..., m_n$ the complete solution of (i) is

$$y = c_1 e^{m_1 x} + c_2 e^{m_2 x} + ... + c_n e^{m_n x}$$

This will be the case when all the roots $m_1, m_2, m_3, ..., m_n$ of the auxilliary equation are real, distinct and different.

5.3.1 Auxilliary Equation having Equal Roots

If the auxilliary equation has two roots equal, say m_1 and m_2 the solution of the given equation

$$P_0 \frac{d^n y}{dx^n} + P_1 \frac{d^{n-1} y}{dx^{n-1}} + ... + P_n y = 0$$

Linear Differential Equations with Constant Coefficients

will be
$$y = (c_1 + c_2) e^{m_1 x} + c_3 e^{m_2 x} + \ldots + c_n e^{m_n x}$$

or
$$y = c e^{m_1 x} + c_3 e^{m_2 x} + \ldots + c_n e^{m_n x}$$

where $c_1 + c_2 = c$

This is not the general solution of (i), because it contains $(n-1)$ arbitrary constants while the order of the equation is n. To obtain the general solution of (i) in this case, we proceed as follows:

Consider the repeated factor as $\left(\dfrac{dy}{dx} - m_1\right)^2 y = 0$. This can be written as $(D - m_1)^2 y = 0$, where $D \equiv \dfrac{d}{dx}$.

Put
$$(D - m_1) y = v;$$

then
$$(D - m_1) v = 0.$$

Therefore,
$$\frac{dv}{dx} = m_1 v$$

or
$$\frac{dv}{v} = m_1 dx$$

Integrating, we have $\log v - \log c_2 = m_1 x$ or $\log \dfrac{v}{c_2} = m_1 x$

Hence,
$$\frac{v}{c_2} = e^{m_1 x} \quad \text{or} \quad v = c_2 e^{m_1 x}$$

or
$$(D - m_1) y = c_2 e^{m_1 x}$$

or
$$\frac{dy}{dx} - m_1 y = c_2 e^{m_1 x}$$

This is a linear differential equation and we will have

$$y e^{-m_1 x} = c_1 + \int c_2 e^{m_1 x} \cdot e^{-m_1 x} dx$$
$$= c_1 + c_2 x$$

$$\therefore \quad y = (c_1 + c_2 x) e^{m_1 x}.$$

This consequently means that if two roots of the auxilliary equation are equal, the general solution of (i) will be

$$y = (c_1 + c_2 x) e^{m_1 x} + c_3 e^{m_3 x} + \ldots + c_n e^{m_n x}$$

Note : In general, if r roots of the auxilliary equation $P_0 m^n + P_1 m^{n-1} + \ldots + P_n = 0$ are equal to m_1 (say), the general solution of (i) will be

$$y = (c_1 + c_2 x + c_3 x^2 + \ldots + c_r x^{r-1}) e^{m_1 x} + c_{r+1} e^{m_{r+1} x} + \ldots + c_n e^{m_n x}$$

5.3.2 Auxilliary Equation having complex roots

If some of the roots of auxilliary equation are complex, then we shall follow the procedure as given below :

Let $\alpha \pm i\beta$ be the roots of the auxilliary equation; then the corresponding part of the solution shall become

$$= c_1 e^{(\alpha + i\beta)x} + c_2 e^{(\alpha - i\beta)x}$$

$$= c_1 e^{\alpha x} e^{i\beta x} + c_2 e^{\alpha x} e^{-i\beta x}$$

$$= e^{\alpha x} (c_1 \cos \beta x + i c_1 \sin \beta x) + e^{\alpha x} (c_2 \cos \beta x - i c_2 \sin \beta x)$$

$$= e^{\alpha x} [(c_1 + c_2) \cos \beta x + (i c_1 - i c_2) \sin \beta x]$$

$$= e^{\alpha x} [A \cos \beta x + B \sin \beta x]$$

where A and B are arbitrary constants.
Therefore, the solution is

$$y = e^{\alpha x} (c_1 \cos \beta x + c_2 \sin \beta x) + c_3 e^{m_3 x} + \dots + c_n e^{m_n x}$$

Note 1. The expression $e^{\alpha x} (A \cos \beta x + B \sin \beta x)$ can also be written as

$$c_1 e^{\alpha x} \cos (\beta x \pm c_2) \quad \text{or} \quad c_1 e^{\alpha x} \sin (\beta x \pm c_2)$$

Note 2. If the auxilliary equation has two equal pairs of complex roots, say $\alpha \pm i\beta$ occurring twice, then the portion of the solution corresponding to these roots, is

$$e^{\alpha x} [(c_1 + c_2 x) \cos \beta x + (c_3 + c_4 x) \sin \beta x]$$

Note 3. If the auxilliary equation has the roots as $\alpha \pm \sqrt{\beta}$, then the portion of the solution corresponding to these roots is

$$c_1 e^{\alpha x} \cosh (x\sqrt{\beta} + c_2) \quad \text{or} \quad c_1 e^{\alpha x} \sinh (x\sqrt{\beta} + c_2)$$

Aid to memory : Solution of equations of the form

$$P_0 \frac{d^n y}{dx^n} + P_1 \frac{d^{n-1} y}{dx^{n-1}} + \dots + P_n y = 0$$

Nature of the roots	Solution
1. Real and distinct, i.e., $m_1, m_2, \dots, m_n$.	$y = c_1 e^{m_1 x} + c_2 e^{m_2 x} + \dots + c_n e^{m_n x}$
2. Real and equal, each m_1 (say)	$y = (c_1 + c_2 x + c_3 x^2 + \dots + c_n x^{n-1}) e^{m_1 x}$
3. Non-repeated roots as $\alpha \pm i\beta$	$y = (c_1 \cos \beta x + c_2 \sin \beta x) e^{\alpha x}$
	or $\quad y = c_1 e^{\alpha x} \cos (\beta x + c_2)$
	or $\quad y = c_1 e^{\alpha x} \sin (\beta x + c_2)$
4. Repeated roots $\alpha \pm i\beta$, r times	$y = [(c_1 + c_2 x + \dots + c_r x^{r-1}) \cos \beta x$
	$+ (c'_1 + c'_2 + \dots c'_r x^{r-1}) \sin \beta x] e^{\alpha x}$
5. Irrational roots as $\alpha \pm \sqrt{\beta}$	$y = c_1 e^{\alpha x} \cosh (x\sqrt{\beta} + c_2)$
	or $\quad y = c_1 e^{\alpha x} \sinh (x\sqrt{\beta} + c_2)$

Linear Differential Equations with Constant Coefficients 563

The symbol D is used for $\dfrac{d}{dx}$ and D^n for $\dfrac{d^n}{dx^n}$. It should be kept in mind that D and D^{-1} are the inverse operations, i.e., as D means differentiation, D^{-1} means integration.

EXAMPLES

1. Solve $\dfrac{d^4 y}{dx^4} + m^4 y = 0$.

Sol. The given equation is
$$(D^4 + m^4) y = 0$$
The auxilliary equation is
$$\lambda^4 + m^4 = 0$$
Making it a perfect square,
$$(\lambda^4 + 2\lambda^2 m^2 + m^4) - 2\lambda^2 m^2 = 0$$
or
$$(\lambda^2 + m^2)^2 - (\sqrt{2}\lambda m)^2 = 0$$
or
$$(\lambda^2 + \sqrt{2}\lambda m + m^2)(\lambda^2 - \sqrt{2}\lambda m + m^2) = 0$$
$$\therefore \quad \lambda = -\dfrac{m}{\sqrt{2}} \pm \dfrac{im}{\sqrt{2}} \quad \text{and} \quad \dfrac{m}{\sqrt{2}} \pm i\dfrac{m}{\sqrt{2}}.$$

Hence general solution of the equation will be
$$y = e^{(-m/\sqrt{2})x}\left[c_1 \cos \dfrac{m}{\sqrt{2}} x + c_2 \sin \dfrac{m}{\sqrt{2}} x\right] + e^{(m/\sqrt{2})x}\left[c_3 \cos \dfrac{m}{\sqrt{2}} x + c_4 \sin \dfrac{m}{\sqrt{2}} x\right]$$

2. Solve : $\dfrac{d^4 y}{dx^4} + 2k \dfrac{d^2 y}{dx^2} + k^2 y = 0$.

Sol. The auxilliary equation is
$$m^4 + 2km^2 + k^2 = 0$$
or
$$(m^2 + k)^2 = 0$$
$$\therefore \quad m = 0 \pm \sqrt{ki}, 0 \pm \sqrt{ki}.$$

Thus, the imaginary roots are repeated twice. Hence the solution is
$$y = (c_1 + c_2 x) \cos \sqrt{k}x + (c_3 + c_4 x) \sin \sqrt{k}x$$

EXERCISE

Solve the following differential equations :

1. $\dfrac{d^2 y}{dx^2} + 3 \dfrac{dy}{dx} + 2y = 0$
[**Ans.** $y = c_1 e^{-x} + c_2 e^{-2x}$]

2. $9 \dfrac{d^2 x}{dz^2} + 18 \dfrac{dx}{dz} - 16x = 0$
[**Ans.** $x = c_1 e^{2/3z} + c_2 e^{-8/3z}$]

3. $\dfrac{d^3 y}{dx^3} - 9 \dfrac{d^2 y}{dx^2} + 23 \dfrac{dy}{dx} - 15y = 0$
[**Ans.** $y = c_1 e^x + c_2 e^{3x} + c_3 e^{5x}$]

4. $\dfrac{d^2y}{dx^2} - 2\dfrac{dy}{dx} + y = 0.$ [Ans. $y = (c_1 + c_2 x)\, e^x$]

5. $(D^4 + 2D^3 + D^2)\, y = 0$ [Ans. $y = (c_1 + c_2 x) + (c_3 + c_4 x)\, e^{-x}$]

6. $(D^4 - D^3 - 9D^2 - 11D - 4)\, y = 0$ [Ans. $y = e^{-x}(c_1 + c_2 x + c_3 x^2) + c_4 e^{4x}$]

7. $\dfrac{d^2y}{dx^2} + 8\dfrac{dy}{dx} + 25 y = 0$ [Ans. $y = e^{-4x}(c_1 \cos 3x + c_2 \sin 3x)$]

8. $\dfrac{d^4y}{dx^4} + y = 0$ [Ans. $y = c_1 e^{x/\sqrt{2}} \cos(x/\sqrt{2} + c_2) + c_3 e^{-x/\sqrt{2}} \cos(x/\sqrt{2} + c_4)$]

5.4 THE DIFFERENTIAL EQUATION OF THE TYPE

$$\dfrac{d^n y}{dx^n} + P_1 \dfrac{d^{n-1} y}{dx^{n-1}} + \ldots + P_n y = 0 \qquad \ldots(i)$$

Contains its solution as

$$c_1 f_1(x) + c_2 f_2(x) + \ldots + c_n f_n(x) + f(x)$$

Thus the general solution of (i) consists of two parts, one of which contains n arbitrary constants and is a solution of the equation obtained from (i), by putting second member equal to zero, and other contains no arbitrary constants. The former is called the *complementary function* (C.F.) and the latter the *particular integral* (P.I.). We have already determined the complementary functions in the previous examples.

5.5 PARTICULAR INTEGRAL

Let

$$\dfrac{1}{f(D)} Q \qquad \ldots(i)$$

denote some function of x which when operated upon by $f(D)$ gives Q. This function of x is a particular solution of the differential equation

$$f(D)\, y = Q \qquad \ldots(ii)$$

As $f(D)$ and $\{f(D)\}^{-1}$ are inverse operations, therefore,

$$D\{D^{-1}(Q)\} = Q \qquad \text{(particular case)}$$

or

$$\dfrac{d}{dx}\{D^{-1}(Q)\} = Q$$

or

$$D^{-1}(Q) = \int Q\, dx.$$

Note : Properties of $\dfrac{1}{f(D)}$.

1. If $Q = u_1 + u_2 + u_3 + \ldots + u_n$, then

$$\dfrac{1}{f(D)} Q = \dfrac{1}{f(D)} u_1 + \dfrac{1}{f(D)} u_2 + \ldots + \dfrac{1}{f(D)} u_n$$

Linear Differential Equations with Constant Coefficients

2. $\dfrac{1}{f(D)}(kQ) = k \cdot \dfrac{1}{f(D)} Q$, where k is a constant.

3. $\dfrac{1}{f(D)}$ can be resolved into factors.

4. $\dfrac{1}{f(D)}$ can be broken into partial fractions.

5. $\dfrac{1}{f(D)} Q$ is a particular integration.

5.5.1 To show that $\dfrac{1}{D-\alpha} Q = e^{\alpha x} \int e^{-\alpha x} Q \, dx$

Let $\qquad \dfrac{1}{(D-\alpha)} Q = V$

Therefore, $\qquad (D-\alpha) V = Q$

or $\qquad \dfrac{dV}{dx} - \alpha V = Q$

This is a linear differential equation. The solution is

$$Ve^{-\alpha x} = \int Q e^{-\alpha x} \, dx + c$$

or $\qquad V = e^{\alpha x} \int Q e^{-\alpha x} \, dx + c e^{\alpha x}$

Now c can be taken zero, for we want only a particular solution.

Hence, $\qquad V = e^{\alpha x} \int Q e^{-\alpha x} \, dx.$

or $\qquad \dfrac{1}{(D-x)} Q = e^{\alpha x} \int Q e^{-\alpha x} \, dx.$

We are now in a position to evaluate

$$\{f(D)\}^{-1} Q.$$

Let on factorisation

$$f(D) \equiv (D-\alpha_1)(D-\alpha_2)\ldots(D-\alpha_n)$$

Then $\qquad (D-\alpha_1)(D-\alpha_2)\ldots(D-\alpha_n) = Q$

It follows that

$$(D-\alpha_2)(D-\alpha_3)\ldots(D-\alpha_n) y = (D-\alpha_1)^{-1} Q$$
$$= e^{\alpha_1 x} \int e^{-\alpha_1 x} Q \, dx$$

Therefore,

$$(D-\alpha_3)\ldots(D-\alpha_n) y = (D-\alpha_2)^{-1} e^{\alpha_1 x} \int e^{-\alpha_1 x} Q \, dx$$

or $\qquad (D-\alpha_3)\ldots(D-\alpha_n) y = e^{\alpha_2 x} \int e^{(\alpha_1 - \alpha_2) x} \int e^{-\alpha_1 x} Q \, dx$

and so on hence, we get generally.

$$y = e^{\alpha_n x} \int e^{(\alpha_{n-1} - \alpha_n)x} \int \ldots \int e^{(\alpha_1 - \alpha_2)x} \int e^{-\alpha_1 x} Q \, dx \ldots dx.$$

This is the required particular integral.

Note : In case $f(D)$ fails to give real linear factors, we may use imaginary factors and use the above method and finally put the result in a real form.

Let $\dfrac{1}{f(D)}$ be capable of resolving into partial fractions. Thus

$$\frac{1}{f(D)} = \frac{A_1}{D - \alpha_1} + \frac{A_2}{D - \alpha_2} + \ldots + \frac{A_n}{D - \alpha_n}$$

Now, particular integral

$$= \frac{1}{f(D)} Q = \frac{A_1}{D - \alpha_1} Q + \frac{A_2}{D - \alpha_2} Q + \ldots + \frac{A_n}{D - \alpha_n} Q.$$

$$= A_1 e^{\alpha_1 x} \int e^{-\alpha_1 x} Q \, dx + A_2 e^{\alpha_2 x} \int e^{-\alpha_2 x} Q \, dx + \ldots + A_n e^{\alpha_n x} \int e^{-\alpha_n x} Q \, dx.$$

5.5.2 To evaluate $\dfrac{1}{f(D)} e^{ax}$, where

$$f(D) \equiv P_0 D^n + P_1 D^{n-1} + \ldots + P_n$$

and $f(\alpha) \neq 0.$

We know that
$$D(e^{ax}) = a e^{ax}$$
$$D^2(e^{ax}) = a^2 e^{ax}$$
$$\ldots\ldots\ldots\ldots\ldots$$
$$\ldots\ldots\ldots\ldots\ldots$$
$$D^n(e^{ax}) = a^n e^{ax}$$

Therefore,
$$f(D) e^{ax} = (P_0 D^n + P_1 D^{n-1} + \ldots + P_n) e^{ax}$$
$$= P_0 D^n e^{ax} + P_1 D^{n-1} e^{ax} + \ldots + P_n e^{ax}$$
$$= P_0 a^n e^{ax} + P_1 a^{n-1} e^{ax} + \ldots + P_n e^{ax}$$
$$= (P_0 a^n + P_1 a^{n-1} + \ldots + P_n) e^{ax}$$

Now, $\quad f(D) e^{ax} = f(a) e^{ax}.$

Operating upon both sides with $\dfrac{1}{f(D)}$ we have

$$\frac{1}{f(D)} f(D) e^{ax} = \frac{1}{f(D)} f(a) e^{ax}$$

$$e^{ax} = f(a) \frac{1}{f(D)} e^{ax}$$

$$\therefore \quad \frac{e^{ax}}{f(a)} = \frac{1}{f(D)} e^{ax}, \text{ provided } f(a) \neq 0. \quad \text{or} \quad \frac{1}{f(D)} e^{ax} = \frac{1}{f(a)} e^{ax}.$$

Linear Differential Equations with Constant Coefficients

EXAMPLE

1. Solve : $\dfrac{d^2y}{dx^2} - 5\dfrac{dy}{dx} + 6y = e^{4x}.$

Sol. The auxilliary equation is

$$m^2 - 5m + 6 = 0$$

or $\quad\quad (m-2)(m-3) = 0 \quad$ or $\quad m = 2, 3$

$\therefore \quad\quad\quad \text{C.F.} = c_1 e^{2x} + c_2 e^{3x}$

$$\text{P.I.} = \dfrac{1}{D^2 - 5D + 6} e^{4x}$$

$$= \dfrac{1}{(4)^2 - 5(4) + 6} e^{4x} = \dfrac{1}{16 - 20 + 6} e^{4x}$$

$$= \dfrac{1}{2} e^{4x}$$

$\therefore$ Complete Solution is

$$y = \text{C.F.} + \text{P.I.}$$

or $\quad\quad y = c_1 e^{2x} + c_2 e^{3x} + \dfrac{1}{2} e^{4x}$

EXERCISE

Solve the following differential equations :

1. $(D^2 + 5D + 6) y = e^{2x}$

$$\left[\text{Ans. } y = c_1 e^{-2x} + c_2 e^{-3x} + \dfrac{1}{20} e^{2x}\right]$$

2. $(D^2 - 4D + 1) y = e^{2x} - e^{-x}$

$$\left[\text{Ans. } y = e^{2x}(c_1 \cosh x\sqrt{3} + c_2 \sinh x\sqrt{3}) - \dfrac{1}{3} e^{2x} - \dfrac{1}{6} e^{-x}\right]$$

3. $\dfrac{d^2y}{dx^2} + 31 \dfrac{dy}{dx} + 240 y = 272 e^{-x}$

$$\left[\text{Ans. } y = c_1 e^{-15x} + c_2 e^{-16x} + \dfrac{136}{105} e^{-x}\right]$$

4. $(D^3 + 1) y = (e^x + 1)^2.$

$$\left[\text{Ans. } y = c_1 e^{-x} + e^{x/2} \left(c_2 \cos \dfrac{1}{2} \sqrt{3} x + c_2 \sin \dfrac{1}{2} \sqrt{3} x \right) + \dfrac{1}{4} e^{2x} + e^x + 1\right]$$

5.5.3 To evaluate $\dfrac{1}{f(D)} \sin ax,$ **where** $f(D) \equiv P_0 D^n + P_1 D^{n-1} + ... P_n.$

Case 1. *When f(D) contains even powers of D*

Let $\quad\quad f(D^2) \equiv P_0 (D^2)^n + P_1 (D^2)^{n-1} + ... + P_n$

We notice that $D^2 \sin ax = -a^2 \sin ax.$

$$D^4 \sin ax = (-a^2)^2 \sin ax$$

$$D^6 \sin ax = (-a^2)^3 \sin ax$$

....
....

$$(D^2)^n \sin ax = (-a^2)^n \sin ax$$

Therefore, $f(D^2) \sin ax = P_0 (D^{2n} + P_1 D^{2n-2} + \ldots + P_n) \sin ax$

or $f(D^2) \sin ax$

$$= P_0 D^{2n} \sin ax + P_1 D^{2n-2} \sin ax + \ldots + P_n \sin ax$$

$$= P_0(-a^2)^n \sin ax + P_1 (-a^2)^{n-1} \sin ax + \ldots + P_n \sin ax$$

$$= f(-a^2) \sin ax.$$

Operating on both sides with $\dfrac{1}{f(D^2)}$, we have

$$\dfrac{1}{f(D^2)} f(D^2) \sin ax = \dfrac{1}{f(D^2)} f(-a^2) \sin ax \text{ or } \sin ax = f(-a^2) \cdot \dfrac{1}{f(D^2)} \sin ax.$$

Dividing both sides by $f(-a^2)$, we have

$$\dfrac{1}{f(D^2)} \sin ax = \dfrac{1}{f(-a^2)} \sin ax.$$

Similarly, we can show that $\dfrac{1}{f(D^2)} \cos ax = \dfrac{1}{f(-a^2)} \cos ax.$

Case II. When $f(D)$ contains odd powers of D.

Let it be put in the from $f_1(D^2) + D f_2(D^2)$; then

$$\dfrac{1}{f(D)} \sin ax = \dfrac{1}{f_1(D^2) + D f_2(D^2)} \sin ax.$$

$$= \dfrac{1}{f_1(-a^2) + D f_2(-a^2)} \sin ax.$$

$$= \dfrac{1}{m + nD} \sin ax \text{ say} \qquad [\text{where } m = f_1(-a^2), n = f_2(-a^2)]$$

$$= (m - nD) \left\{ \dfrac{1}{(m - nD)} \cdot \dfrac{1}{m + nD} \sin ax \right\}$$

$$(\text{since } (m - nD), \dfrac{1}{(m - nD)} \text{ are inverse operations.})$$

$$= (m - nD) \dfrac{1}{m^2 + n^2 a^2} \sin ax = \dfrac{1}{(m^2 + n^2 a^2)} (m - nD) \sin ax$$

Linear Differential Equations with Constant Coefficients

$$= \frac{m \sin ax - na \cos ax}{m^2 + n^2 a^2}$$

$$= \frac{f_1(-a^2) \sin ax - f_2(-a^2) a \cos ax}{\{f_1(-a^2)\}^2 + a^2 \{f_2(-a^2)\}^2}$$

Note : Similar results are true for $\dfrac{1}{f(D)} \cos ax$.

EXAMPLE

1. Solve $(D^3 + D^2 - D - 1) y = \cos 2x$.

Sol. Here auxilliary equation is

$$m^3 + m^2 - m - 1 = 0 \quad \text{or} \quad (m-1)(m+1)^2 = 0$$

$\therefore \quad m = 1, -1, -1$.

Hence, $\text{C.F.} = c_1 e^x + (c_2 + c_3 x) e^{-x}$

$$\text{P.I.} = \frac{1}{D^3 + D^2 - D - 1} \cos 2x \qquad [\text{Put } D^2 = 4]$$

$$= \frac{1}{-4D - 4 - D - 1} \cos 2x = \frac{1}{-5(D+1)} \cos 2x$$

$$= -\frac{1}{5} \cdot \frac{D-1}{D^2 - 1} \cos 2x = -\frac{1}{5} \frac{D(\cos 2x) - \cos 2x}{-4 - 1}$$

$$= \frac{1}{25}(-2 \sin 2x - \cos 2x) = -\frac{1}{25}(2 \sin 2x + \cos 2x)$$

Hence general solution of the equation is

$$y = \text{C.F.} + \text{P.I.}$$

or $\qquad y = c_1 e^x + (c_2 + c_3 x) e^{-x} - \dfrac{1}{25}(2 \sin 2x + \cos 2x)$

EXERCISE

Solve the following differential equations :

1. $(D^2 - D - 2) y = \sin 2x$

$$\left[\text{Ans. } y = c_1 e^{2x} + c_2 e^{-x} + \frac{1}{20}(\cos 2x - 3 \sin 2x) \right]$$

2. $\dfrac{d^2 y}{dx^2} - 2 \dfrac{dy}{dx} + 5y = \sin 3x$

$$\left[\text{Ans. } y = e^x (c_1 \cos 2x + c_2 \sin 2x) + \frac{1}{26}(3 \cos 3x - 2 \sin 3x) \right]$$

3. $(D^4 + 2D^3 - 3D^2) y = 3e^{2x} + 4 \sin x$

$$\left[\text{Ans. } y = (c_1 + c_2 x) + c_3 e^x + c_4 e^{-3x} + \frac{2}{5}(2 \sin x + \cos x) \right]$$

4. $(D^3 + 1) y = \cos 2x$

$$\left[\text{Ans. } y = c_1 e^{-x} + e^{x/2} \left(c_2 \cos \frac{\sqrt{3}}{2} x + c_3 \sin \frac{\sqrt{3}}{2} x \right) + \frac{1}{65} (\cos 2x - 8 \sin 2x) \right]$$

5. $(D^3 - 2D^2 + 3) y = \cos x$

$$\left[\text{Ans. } y = c_1 e^{-x} + e^{3x/2} \left\{ c_2 \cos \frac{\sqrt{3}}{2} + c_3 \sin \frac{\sqrt{3}}{2} x \right\} + \frac{1}{26} (5 \cos x - \sin x) \right]$$

6. $(D^2 - 3D + 2) y = 6e^{3x} + \sin 2x$

$$\left[\text{Ans. } y = c_1 e^x + c_2 e^{2x} + 3e^{3x} + \frac{1}{40} (6 \cos 2x - 2 \sin 2x) \right]$$

7. $(D^2 + 1) y = \sin x \sin 2x$

$$\left[\text{Ans. } y = (c_1 \cos x + c_2 \sin x) + \frac{x}{4} \sin x + \frac{1}{16} \cos 3x \right]$$

5.5.4 To evaluate $\dfrac{1}{f(D)} x^m$, where m is a positive integer and

$$f(D) = P_0 D^n + P_1 D^{n-1} + P_2 D^{n-2} + \ldots + P_n$$

Let us consider first $(D - a)^{-1} x^m$, i.e.,

$$\frac{1}{(D-a)} x^m = e^{ax} \int e^{-ax} x^m \, dx$$

$$= e^{ax} \left\{ \frac{-e^{-ax} x^m}{a} - \frac{mx^{m-1} e^{-ax}}{a^2} - \frac{m(m-1) x^{m-2} e^{-ax}}{a^3} - \frac{m(m-1)\ldots 2.1}{a^{m+1}} e^{-ax} \right\} \quad \ldots(i)$$

by repeatedly integration by parts.

But if we expand $\dfrac{1}{D-a}$ in powers of D, we get

$$\frac{1}{(D-a)} x^m = \frac{1}{a\left(1 - \dfrac{D}{a}\right)} x^m$$

$$= -\frac{1}{a} \left(1 + \frac{D}{a} + \frac{D^2}{a^2} + \ldots \right) x^m$$

$$= -\frac{1}{a} \left\{ x^m + \frac{mx^{m-1}}{a} + \frac{m(m-1) x^{m-2}}{a^2} + \ldots + \frac{m(m-1)\ldots 2.1}{a^m} \right\} \quad \ldots(ii)$$

We see that (i) and (ii) are the same.

Linear Differential Equations with Constant Coefficients 571

Now, we have shown in the previous article that we can break up $\dfrac{1}{f(D)}$ in $\dfrac{1}{f(D)} Q$ into partial fractions and expand each by binonmial in ascending powers of D, taking the expansinon only as far as the term in D^m. Then we operate on x^m by every term in the expansion so obtained and add the results of the operation. The following examples shall make the process clear.

EXAMPLES

1. Solve the equation $\dfrac{d^2y}{dx^2} = a + bx + cx^2$, given that $\dfrac{dy}{dx} = 0$ when $x = 0$ and $y = d$ when $x = 0$.

Sol. Auxilliary equation is

$$m^2 = 0, \quad \therefore \quad m = 0, 0$$

$\therefore \quad$ C.F. $= (c_1 + c_2 x) e^{0 \cdot x} = (c_1 + c_2 x)$

$$\text{P.I.} = \dfrac{1}{D^2}(a + bx + cx^2)$$

$$= \dfrac{1}{D}\left(ax + x\dfrac{b}{2}x^2 + \dfrac{c}{3}x^3\right)$$

$$= \left(\dfrac{ax^2}{2} + \dfrac{b}{6}x^3 + \dfrac{c}{12}x^4\right)$$

$\therefore \quad y = c_1 + c_2 x + \dfrac{1}{2}ax^2 + \dfrac{1}{6}bx^3 + \dfrac{1}{12}cx^4$

$\dfrac{dy}{dx} = c_2 + ax + \dfrac{1}{2}bx^2 + \dfrac{1}{3}cx^3$ ...(i)

Putting $x = 0$ and $y = d$, $\dfrac{dy}{dx} = 0$ in (i) and (ii) we get

$$c_2 = 0 \quad \text{and} \quad c_1 = d$$

$\therefore \quad y = d + \dfrac{1}{2}ax^2 + \dfrac{1}{6}bx^3 + \dfrac{1}{12}cx^4$

2. Solve : $\dfrac{d^2y}{dx^2} - 4\dfrac{dy}{dx} + 4y = x^2 + e^x + \cos 2x$.

Sol. The auxilliary equation is

$$m^2 - 4m + 4 = 0$$

or $\quad (m - 2)^2 = 0 \Rightarrow m = 2, 2$

Hence, $\quad$ C.F. $= (c_1 + c_2 x) e^{2x}$

$$\text{P.I.} = \dfrac{1}{(D^2 - 4D + 4)}(x^2 + e^x + \cos 2x)$$

$$= \frac{1}{(D^2 - 4D + 4)} x^2 + \frac{1}{D^2 - 4D + 4} e^x + \frac{1}{D^2 - 4D + 4} \cos 2x$$

$$= \frac{1}{(D-2)^2} x^2 + \frac{1}{1^2 - 4.1 + 4} e^x + \frac{1}{-4 - 4D + 4} \cos 2x$$

$$= \frac{1}{4}\left(1 + \frac{D}{2}\right)^{-2} x^2 + e^x - \frac{1}{4}\cdot\frac{1}{D}\cos 2x$$

$$= \frac{1}{4}\left(1 + D + \frac{3}{4} D^2 + ...\right) x^2 + e^x - \frac{1}{8}\sin 2x$$

$$= \frac{1}{4}\left(x^2 + 2x + \frac{3}{2}\right) + e^x - \frac{1}{8}\sin 2x$$

∴ The complete solution is
$$y = C.F. + P.I.$$

or
$$y = (c_1 + c_2 x) e^{2x} + \frac{1}{4}\left(x^2 + 2x + \frac{3}{2}\right) + e^x - \frac{1}{8}\sin 2x$$

EXERCISE

Solve following differential equations :

1. $(D^2 + D - 6) y = x$
   $\left[\text{Ans. } y = c_1 e^{2x} + c_2 e^{-3x} - \frac{1}{36}(6x + 1)\right]$

2. $\dfrac{d^2 y}{dx^2} - 4y = x^3$
   $\left[\text{Ans. } y = c_1 e^{2x} + c_2 e^{-2x} - \frac{1}{8} x (2x^2 + 3)\right]$

3. $\dfrac{d^2 y}{dx^2} - 3\dfrac{dy}{dx} + 2y = \sin 3x + x^2 + x + e^{4x}$
   $\left[\text{Ans. } y = c_1 e^{2x} + c_2 e^x + \frac{1}{130}(9 \cos 3x - 7 \sin 3x) + \frac{1}{2}(x^2 + 4x + 5) + \frac{1}{6} e^{4x}\right]$

4. $\dfrac{d^2 y}{dx^2} - 4\dfrac{dy}{dx} + y = 73 \sin 2x + x + 13 e^{-x/2}$
   $[\text{Ans. } y = c_1 e^{(2+\sqrt{3})x} + c_2 e^{(2-\sqrt{3})x} + 8 \cos 2x - 3 \sin 2x + x + 4 + 4 e^{-x/2}]$

5. $\dfrac{d^2 y}{dx^2} + \dfrac{dy}{dx} - 2y = x + \sin x$
   $\left[\text{Ans. } y = c_1 e^{-2x} + c_2 e^x - \frac{1}{4}(2x + 1) - \frac{1}{10}(\cos x + 3 \sin x)\right]$

6. $\dfrac{d^2 y}{dx^2} - 4y = e^x + \sin 2x + \cos^2 x$
   $\left[\text{Ans. } y = c_1 e^{2x} + c_2 e^{-2x} - \frac{1}{3} e^x - \frac{1}{8}\sin 2x - \frac{1}{16}\cos 2x - \frac{1}{8}\right]$

Linear Differential Equations with Constant Coefficients

5.5.5 To evaluate $\dfrac{1}{f(D)} e^{ax} V$, where V where V is any function of x and

$$f(D) = P_0 D^n + P_1 D^{n-1} + \ldots + P_n$$

Let us consider any function V_1 of x.

We know that
$$D(e^{ax} V_1) = e^{ax} D(V_1) + V_1 a e^{ax}$$

or
$$D(e^{ax} V_1) = e^{ax}(D+a) V_1$$

$$D^2 (e^{ax} V_1) = D\{e^{ax}(D+a) V_1\}$$

$$= a e^{ax}(D+a) V_1 + e^{ax}(D^2 + aD) V_1$$

$$= e^{ax}(D^2 + 2aD + a^2) V_1$$

$$= e^{ax}(D+a)^2 V_1$$

....
....

In general
$$D^n (e^{ax} V_1) = e^{ax}(D+a)^n V_1$$

Now,
$$f(D)[e^{ax} V_1] = [P_0 D^n + P_1 D^{n-1} + \ldots + P_n][e^{ax} V_1]$$

$$= P_0 e^{ax}(D+a)^n V_1 + P_1 e^{ax}(D+a)^{n-1} V_1 + \ldots + P_n e^{ax} V_1$$

$$= e^{ax}[P_0 (D+a)^n + P_1 (D+a)^{n-1} + \ldots + P_n] V_1$$

Let us now assume that V_1 is given by

$$f(D+a) V_1 = V$$

$\therefore$
$$V_1 = \dfrac{1}{f(D+a)} V$$

or
$$f(D) \left\{ e^{ax} \dfrac{1}{f(D+a)} V \right\} = e^{ax} V$$

operating on both sides by $\dfrac{1}{f(D)}$, we get

$$\dfrac{1}{f(D)} f(D) \left\{ e^{ax} \dfrac{1}{f(D+a)} V \right\} = \dfrac{1}{f(D)} e^{ax} V$$

or
$$e^{ax} \cdot \dfrac{1}{f(D+a)} V = \dfrac{1}{f(D)} e^{ax} V$$

Hence,
$$\dfrac{1}{f(D)} e^{ax} \cdot V = e^{ax} \dfrac{1}{f(D+a)} V.$$

EXAMPLE

1. Solve : $(D^2 - 2D + 5) y = e^{2x} \sin x$.

Sol. The auxilliary equation is

$$m^2 - 2m + 5 = 0$$

$$\therefore \quad m = \frac{2 \pm \sqrt{(4-20)}}{2} = 1 \pm 2i$$

$$\therefore \quad \text{C.F.} = e^{ax}(c_1 \cos 2x + c_2 \sin 2x)$$

$$\text{P.I.} = \frac{1}{(D-2D+5)} e^{2x} \sin x$$

$$= e^{2x} \cdot \frac{1}{(D+2)^2 - 2(D+2) + 5} \sin x$$

$$= e^{2x} \cdot \frac{1}{D^2 + 2D + 5} \sin x$$

$$= e^{2x} \cdot \frac{1}{-(1)^2 + 2D + 5} \sin x$$

$$= \frac{1}{2} e^{2x} \frac{1}{D+2} \sin x = \frac{1}{2} e^{2x} \frac{D-2}{(D^2-4)} \sin x$$

$$= \frac{1}{2} e^{ax} \frac{(D-2) \sin x}{-1-4}$$

$$= -\frac{1}{10} e^{2x} [D(\sin x) - 2 \sin x]$$

$$= -\frac{1}{10} e^{2x} (\cos x - 2 \sin x)$$

Hence, the complete solution is
$$y = \text{C.F.} + \text{P.I.}$$

or
$$y = e^x (c_1 \cos 2x + c_2 \sin 2x) - \frac{1}{10} e^{2x} (\cos x - 2 \sin x)$$

EXERCISE

Solve following differential equations :

1. $\dfrac{d^2 y}{dx^2} - 2 \dfrac{dy}{dx} + y = x^2 e^x$ $\left[\text{Ans. } y = (c_1 + c_2 x) e^x + \dfrac{1}{12} x^4 e^x \right]$

2. $(D^2 + 2D + 1) y = x^2 e^{3x}$ $\left[\text{Ans. } y = (c_1 + c_2 x) e^{-x} + \dfrac{1}{16} e^{3x} \left(x^2 - x + \dfrac{3}{8} \right) \right]$

3. $\dfrac{d^2 y}{dx^2} - 5 \dfrac{dy}{dx} + 6y = x(x + e^x)$

$\left[\text{Ans. } y = c_1 e^{2x} + c_2 e^{3x} + \dfrac{1}{108}(18x^2 + 30x + 19) + \dfrac{1}{4} e^x (2x+3) \right]$

4. $\dfrac{d^2 y}{dx^2} + 3 \dfrac{dy}{dx} + 2y = \cosh 2x \sin x$

$\left[\text{Ans. } y = c_1 e^{-x} + c_2 e^{-2x} - \dfrac{1}{340} e^{2x} (7 \cos x - 11 \sin x) + \dfrac{1}{4} e^{-2x} (\cos x - \sin x) \right]$

Linear Differential Equations with Constant Coefficients

5. $(D^2+ 3D + 2) y = e^{2x} \sin x$ $\left[\text{Ans. } y = c_1e^{-x} + c_2e^{-2x} - \dfrac{1}{170} e^{2x} (\cos x - 11\sin x)\right]$

6. $(D^2 - 2D + 4) y = e^x \cos x$ $\left[\text{Ans. } y = c_1e^x \cos (\sqrt{3}x + c_2) + \dfrac{1}{2} e^x \cos x\right]$

7. $(D^3 + 1) y = e^{2x} \sin x + e^{x/2} \sin\left(\dfrac{\sqrt{3}}{2} x\right)$

$\left[\text{Ans. } y = c_1e^{-x} + e^{x/2}\left(c_2 \cos \dfrac{\sqrt{3}}{2} x + c_3 \sin \dfrac{\sqrt{3}}{2} x\right) - \dfrac{1}{16} xe^{x/2}\left(\sin \dfrac{\sqrt{3}}{2} x + \sqrt{3} \cos \dfrac{\sqrt{3}}{2} x\right)\right]$

8. $(D^3 - D^2 + 3D + 5) y = x^2 + e^x \cos 2x$

$\left[\text{Ans. } y = c_1e^{-x} + e^x (c_2 \cos 2x + c_3 \sin 2x) + \dfrac{1}{16} xe^x (\sin 2x - \cos 2x)\right]$

5.5.6 To evalaute $\dfrac{1}{f(D)} e^{ax}$, when $f(D) = 0$.

In this case, if we calculate P.I. by previous method, then

$$\text{P.I.} = \dfrac{1}{f(D)} e^{ax} = \dfrac{1}{f(a)} e^{ax} = \dfrac{e^{ax}}{0},$$

hence this method fails.

Now, since a is a root of $f(D)$, $(D - a)$ must be a factor of $f(D)$ which can be written in the form $(D - a) \phi (D)$.

$$\therefore \quad \text{P.I.} = \dfrac{1}{(D - a) \phi (D)} e^{ax}$$

$$= \dfrac{1}{(D- a) \phi(a)} e^{ax} \text{ where } \phi(a) \text{ is now a constant only.}$$

$$= \dfrac{1}{\phi(a)} \cdot \dfrac{e^{ax}.1}{(D - a)}$$

$$= \dfrac{1}{\phi(a)} e^{ax} \cdot \dfrac{1}{(D + a - a)} .1$$

$$= \dfrac{1}{\phi(a)} e^{ax} \left(\dfrac{1}{D}\right) 1.$$

$$= \dfrac{1}{\phi(a)} e^{ax} \cdot x.$$

Similarly, if $f(D) = (D - a)^p \phi(D)$ then P.I of e^{ax} is

$$\dfrac{e^{ax}}{f(D)} = \dfrac{e^{ax}}{(D- a)^p \phi(D)} = \dfrac{e^{ax} . 1}{\phi(a) (D - a)^p}$$

$$= \dfrac{e^{ax}}{\phi(a)} \cdot \left[\dfrac{1}{(D + a - a)^p}\right] 1$$

$$= \frac{e^{ax}}{\phi(a)}\left(\frac{1}{D^p}\right)1$$

$$= \frac{e^{ax}}{\phi(a)} \cdot \frac{x^p}{p!}$$

Rule : *Put $D = a$ in those factors of $f(D)$ which do not vanish for $D = a$ and then make the question as P.I. of a product of e^{ax} and 1 which is calculate buy § 5.5.6 and reduces to the calculation of $\frac{1}{D}.1$ or $\frac{1}{D^2}.1$ or $\frac{1}{D^3}.1$ and so on which are $x, \frac{x^2}{2!}, \frac{x^3}{2!}$ respectively.*

EXAMPLES

1. *Solve :* $(D^3 - 5D^2 + 7D - 3)y = e^{2x} \cosh x$.

Sol. The auxilliary equation is

$$m^3 - 5m^2 + 7m - 3 = 0$$

or $\qquad (m-1)(m^2 - 4m + 3) = 0$

or $\qquad (m-1)(m-1)(m-3) = 0$

or $\qquad m = 1, 1, 3$.

$\therefore \quad$ C.F. $= (c_1 + c_2 x)e^x + c_3 e^{3x}$

$$\text{P.I} = \frac{1}{(D^3 - 5D^2 + 7D - 3)} e^{2x}\left(\frac{e^x + e^{-x}}{2}\right)$$

$$= \frac{1}{(D-1)^2(D-3)} e^{2x}\left(\frac{e^x + e^{-x}}{2}\right)$$

$$= \frac{1}{2} \cdot \frac{1}{(D-1)^2(D-3)} e^{3x} + \frac{1}{2} \cdot \frac{1}{(D-1)^2(D-3)} e^x$$

Now, $\qquad \dfrac{1}{2} \cdot \dfrac{1}{(D-1)^2(D-3)} e^{3x} = \dfrac{1}{2} \cdot \dfrac{1}{(3-1)^2(D-3)} e^{3x}$,

putting 3 for D in the factor $(D-1)^2$

$$= \frac{1}{8} \frac{1}{D-3} e^{3x} \cdot 1 = \frac{1}{8} e^{3x} \cdot \frac{1}{D+3-3} \cdot 1$$

$$= \frac{1}{8} e^{3x} \frac{1}{D} \cdot 1 = \frac{1}{8} x e^{3x}$$

Again $\qquad \dfrac{1}{2} \cdot \dfrac{1}{(D-1)^2(D-3)} e^x = \dfrac{1}{2(D-1)^2(1-3)} e^x$,

putting 1 for D in the factor $D-3$

$$= -\frac{1}{4} \cdot \frac{1}{(D-1)^2} e^x \cdot 1 = -\frac{1}{4} e^x \cdot \frac{1}{(D+1-1)^2} \cdot 1$$

$$= -\frac{1}{4} e^x \frac{1}{D^2} \cdot 1 = -\frac{1}{4} e^x \frac{1}{D} x = -\frac{1}{4} e^x \frac{1}{2} x^2$$

Linear Differential Equations with Constant Coefficients

$$= -\frac{1}{8} x^2 e^x.$$

$$\therefore \quad \text{P.I.} = \frac{1}{8} x e^{3x} - \frac{1}{8} x^2 e^x$$

$\therefore$ The complete solution is $y = $ (C.F.) + (P.I.)

or $$y = (c_1 + c_2 x) e^x + c_3 e^{3x} + \frac{1}{8} x e^{3x} - \frac{1}{8} x^2 e^x$$

2. *Solve the differential equation*

$$(D^2 - 1) y = \cosh x + \sin x + x^2$$

Sol. Auxilliary equation is

$$m^2 - 1 = 0 \text{ or } m = \pm 1$$

$$\therefore \quad \text{C.F.} = c_1 e^x + c_2 e^{-x}$$

$$\text{P.I.} = \frac{1}{(D^2 - 1)} \cosh x + \frac{1}{(D^2 - 1)} \sin x + \frac{1}{(D^2 - 1)} x^2$$

$$= \frac{1}{(D^2 - 1)} \left(\frac{e^x + e^{-x}}{2} \right) + \frac{1}{-1^2 - 1} \sin x - \frac{1}{(1 - D^2)} x^2$$

$$= \frac{1}{2} \cdot \frac{1}{(D-1)(D+1)} e^x + \frac{1}{2} \cdot \frac{1}{(D-1)(D+1)} e^{-x} - \frac{1}{2} \sin x - (1 + D^2 + ...) x^2$$

$$= \frac{1}{4} e^x \frac{1}{D} \cdot 1 - \frac{1}{4} e^{-x} \frac{1}{D} 1 - \frac{1}{2} \sin x - (x^2 + 2x)$$

$$= \frac{1}{4} x e^x + \frac{1}{4} x e^{-x} - \frac{1}{2} \sin x - (x^2 + 2x)$$

$$= \frac{1}{2} x \left(\frac{e^x + e^{-x}}{2} \right) - \frac{1}{2} \sin x - (x^2 + 2x)$$

$$= \frac{1}{2} x \cosh x - \frac{1}{2} \sin x - x^2 - 2x$$

$\therefore$ Solution of the equation is

$$y = \text{C.F.} + \text{P.I.}$$

or $$y = c_1 e^x + c_2 e^{-x} + \frac{1}{2} x \cosh x - \frac{1}{2} \sin x - x^2 - 2x$$

EXERCISE

Solve following differential equations :

1. $\dfrac{d^2 y}{dx^2} - 9 \dfrac{dy}{dx} + 18 y = \cosh 3x$ $\left[\text{Ans. } y = c_1 e^{3x} + c_2 e^{6x} - \dfrac{1}{6} x e^{3x} + \dfrac{1}{54} e^{-3x} \right]$

2. $(D-1)^2 (D^2 + 1)^2 y = e^x$

$\left[\text{Ans. } y = (c_1 + c_2 x) e^x + (c_3 + c_4 x) \cos x + (c_5 + c_6 x) \sin x + \dfrac{1}{8} x^2 e^x \right]$

3. $(D^2 - 6D + 9) y = 4 e^{3x}$ [Ans. $y = (c_1 + c_2) e^{3x} + 2 x^2 e^{3x}$]

4. $(D-1)^3 (D+1) y = e^x + e^{-x}$

$$\left[\text{Ans. } y = (c_1 + c_2 x + c_3 x^2) e^x + c_4 e^{-x} + \frac{1}{12} x^3 e^x - \frac{1}{8} e^{-x}\right]$$

5. $(D^2 + 6D + 9) y = 2e^{-3x}$ \hspace{2cm} [Ans. $y = (c_1 + c_2 x + x^2) e^{-3x}$]

5.5.7 To evaluate $\dfrac{1}{f(D^2)}$ sin ax or cos ax, when $f(-a^2) = 0$

In such cases instead of calculating P.I. for sin ax or cos ax we shall calculate P.I. for e^{iax}.

$$e^{iax} = \cos ax + i \sin ax$$

Thus, \hspace{1cm} P.I. for e^{iax} = P.I. for (cos ax + i sin ax)

$\therefore$ real part of P.I. for e^{iax} = P.I. for cos ax

and imaginary part of P.I. for e^{iax} = P.I. for sin ax

$\therefore \dfrac{\cos ax}{D^2 + a^2}$ and $\dfrac{\sin ax}{D^2 + a^2}$ are respectively real and imaginary parts of

$$\dfrac{e^{iax}}{D^2 + a^2}$$

$$= \dfrac{e^{iax}}{(D + ai)(D - ai)} = \dfrac{e^{aix}}{(ai + ai)(D - ai)}$$

$$= \dfrac{e^{iax}}{2ai} \left(\dfrac{1}{(D + ai - ai)}\right).1$$

$$= \dfrac{e^{iax}}{2ai} \cdot \dfrac{1}{D} .1 = \dfrac{x}{2ai} (e^{aix})$$

$$= -\dfrac{ix(\cos ax + i \sin ax)}{2a}$$

$$= -\dfrac{ix}{2a} \cos ax + \dfrac{x}{2a} \sin ax$$

$\therefore$ Real part $\dfrac{x}{2a} \sin ax = \dfrac{\cos ax}{D^2 + a^2}$

Imaginary part $= \dfrac{x}{2a} \cos ax = \dfrac{\sin ax}{D^2 + a^2}$

EXAMPLE

1. Solve: $(D^3 + a^2 D) y = \sin ax.$
Sol. Auxilliary equation is

$$m^3 + a^2 m = 0 \quad \text{or} \quad m = 0, \pm ai$$

$\therefore$ \hspace{1cm} C.F. = $c_1 + c_2 \cos ax + c_3 \sin ax$

$$\text{P.I.} = \dfrac{1}{D(D^2 + a^2)} \sin ax$$

Linear Differential Equations with Constant Coefficients

$$= \frac{1}{(D^2+a^2)} \int \sin ax \, dx = -\frac{1}{a} \frac{1}{(D^2+a^2)} \cos ax \, dx$$

$$= -\frac{1}{a} \cdot \frac{x}{2} \int \cos ax \, dx = -\frac{x}{2a^2} \sin ax$$

Hence, general solution is

$$c_1 + c_2 \cos ax + c_3 \sin ax - \frac{x}{2a^2} \sin ax$$

EXERCISE

1. $(D^2+4)(D^2+1) y = \cos 2x + \sin x$

   $\left[\text{Ans. } y = c_1 \cos 2x + c_2 \sin 2x + c_3 \cos x + c_4 \sin x - \frac{1}{12} x \sin 2x - \frac{1}{6} x \cos x\right]$

2. $(D^2+4) y = e^x + \sin 2x$.

   $\left[\text{Ans. } y = (c_1 \cos 2x + c_2 \sin 2x) + \frac{1}{5} e^x - \frac{1}{4} x \cos 2x\right]$

3. $\dfrac{d^2 y}{dx^2} + a^2 y = \cos ax$.

   $\left[\text{Ans. } y = c_1 \sin ax + c_2 \cos ax + \dfrac{x}{2a} \sin ax\right]$

5.5.8 To evaluate $\dfrac{1}{f(D)} xV$, where V is any function of x except e^{ax}

Here,
$$f(D) = P_0 D^n + P_1 D^{n-1} + \ldots + P_n$$

Let us consider
$$D^n (xV) = xD^n V + {}^nC_1 D^{n-1} V \quad \text{(by applying Leibnitz's theroem)}$$

we have $\quad f(D)(xV) = xf(D) V + f'(D) V$

Taking the inverse operator, we have

$$\frac{1}{f(D)}(xV) = x \frac{1}{f(D)} V + \left[\frac{d}{dD} + \frac{1}{f(D)}\right] V$$

But
$$\frac{d}{dD}\left[\frac{1}{f(D)}\right] = -\frac{f'(D)}{\{f(D)\}^2}$$

∴
$$\frac{1}{f(D)}(xV) = x \frac{1}{f(d)} V - \frac{f'(D)}{\{f(D)\}^2}$$

EXAMPLES

1. Solve : $(D^2 - 2D + 1) y = xe^x \sin x.$

Sol. The auxilliary equation is

$$m^2 - 2m + 1 = 0 \text{ or } (m-1)^2 = 0$$

∴ $\quad m = 1, 1$

∴ $\quad$ C.F. $= (c_1 + c_2 x) e^x$

$$\text{P.I.} = \frac{1}{D^2 - 2D + 1} xe^x \sin x = \frac{1}{(D-1)^2} e^x (\sin x)$$

$$= e^x \frac{1}{(D+1-1)^2} x \sin x$$

$$= e^x \cdot \frac{1}{D^2} (x \sin x)$$

$$= e^x \cdot \frac{1}{D} \int x \sin x \, dx$$

$$= e^x \cdot \frac{1}{D} (-x \cos x + \sin x)$$

$$= e^x [\int -x \cos x \, dx + \int \sin x \, dx]$$

$$= e^x [-x \sin x + \int \sin x \, dx - \cos x]$$

$$= e^x [-x \sin x - 2 \cos x]$$

Hence, the required solution is

$$y = C.F. + P.I.$$

or

$$y = (c_1 + c_2 x) e^x - e^x (x \sin x + 2 \cos x)$$

2. Solve: $(D^2 - 4D + 4) y = 8x^2 e^{2x} \sin 2x$.

Sol. We have

$$C.F. = (c_1 + c_1 x) e^{2x}$$

$$P.I. = 8 \cdot \frac{1}{(D-2)^2} e^{2x} (x^2 \sin 2x)$$

$$= 8 e^{2x} \frac{1}{(D+2-2)^2} x^2 \sin 2x$$

$$= 8 e^{2x} \cdot \frac{1}{D^2} (x^2 \sin 2x)$$

$$= 8 e^{2x} \cdot \frac{1}{D^2} (x^2 \sin 2x)$$

$$= 8 e^{2x} \cdot I$$

where $I = \frac{1}{D^2} x^2 \sin 2x$.

$$= \text{imaginary part of } \frac{1}{D^2} x^2 e^{2ix}$$

$$= I.P. \text{ of } e^{2ix} \frac{1}{(D+2i)^2} x^2$$

$$= I.P. \text{ of } \frac{e^{2ix}}{4i^2} \left(1 + \frac{D}{2i}\right)^{-2} x^2$$

$$= I.P. \text{ of } \frac{e^{2ix}}{-4} \left(1 - \frac{iD}{2}\right)^{-2} x^2$$

Linear Differential Equations with Constant Coefficients

$$= \text{I.P. of } \frac{e^{2ix}}{-4}\left[1 + 2\cdot\left(\frac{iD}{2}\right) + 3\left(\frac{iD}{2}\right)^3 + \ldots\right]x^2$$

$$= \text{I.P. of } \frac{e^{2ix}}{-4}\left[1 + Di - \frac{3}{4}D^2 + \ldots\right]x^2$$

$$= \text{I.P. of } \frac{e^{2ix}}{-4}\left[x^2 + 2xi - \frac{3}{2}\right]$$

$$= \text{I.P. of } -\frac{1}{4}(\cos 2x + i \sin 2x)\left(x^2 + 2xi - \frac{3}{2}\right)$$

$$= -\frac{1}{4}\left[\left(x^2 - \frac{3}{2}\right)\sin 2x + 2x \cos 2x\right]$$

$$= -\frac{1}{8}[(2x^2 - 3)\sin 2x + 4x \cos 2x]$$

$$\therefore \quad \text{P.I.} = 8e^{2x}\cdot\text{I.} = 8e^{2x}\left[-\frac{1}{8}\{(2x^2 - 3)\sin 2x + 4x \cos 2x\}\right]$$

$$= -e^{2x}[(2x^2 - 3)\sin 2x + 4x \cos 2x]$$

Hence, general solution is

$$y = e^{2x}[c_1 + c_2 x + 3 \sin 2x - 2x^2 \sin 2x - 4x \cos 2x]$$

EXERCISE

Solve the following differential equations :

1. $\dfrac{d^2 y}{dx^2} + 4y = x \sin x$ $\quad\left[\text{Ans. } y = c_1 \cos 2x + c_2 \sin 2x + \left(\dfrac{1}{9}\right)(3x \sin x - 2 \cos x)\right]$

2. $(D^2 + a^2) y = x \cos ax$ $\quad\left[\text{Ans. } y = c_1 \cos ax + c_2 \sin ax + \dfrac{1}{4a^2}(ax^2 \sin ax + x \cos ax)\right]$

3. $(D - 1)^2 y = x \sin x$ $\quad\left[\text{Ans. } y = (c_1 + c_2 x) e^x + \dfrac{x}{2}\cos x + \dfrac{1}{3}(\cos x - \sin x)\right]$

4. $\dfrac{d^4 y}{dx^4} - y = x \sin x$ $\quad\left[\text{Ans. } y = c_1 e^x + c_2 e^{-x} + c_3 \cos x + c_4 \sin x + \dfrac{1}{8}(x^2 \cos x - 3x \sin x)\right]$

5. $\dfrac{d^2 y}{dx^2} - y = xe^x \sin x$ $\quad\left[\text{Ans. } y = c_1 e^x + c_2 e^{-x} + \dfrac{1}{25}e^x[(14 - 5x)\sin x - 2(5x + 1)\cos x]\right]$

6. $(D^2 - 1) y = x \sin x + (1 + x^2) e^x$ $\quad\left[\text{Ans. } y = c_1 e^x + c_2 e^{-x} - \dfrac{1}{2}(x \sin x + \cos x)\right]$

MISCELLANEOUS EXAMPLES

1. Solve : $\dfrac{d^2 y}{dx^2} + a^2 y = \sec ax.$

Sol. C.F. $= c_1 \cos ax + c_2 \sin ax$

$$\text{P.I.} = \frac{1}{D^2 + a^2}\sec ax = \frac{1}{(D + ai)(D - ai)}\sec ax.$$

$$= \frac{1}{2ai}\left[\frac{1}{D-ai} - \frac{1}{D+ai}\right]\sec ax \qquad ...(i)$$

Now, $\quad \dfrac{1}{D-ai}\sec ax = \dfrac{1}{(D-ai)}e^{iax}\cdot\dfrac{e^{-iax}}{\cos ax}$

using $\quad \dfrac{1}{f(D)}(e^{ax}V) = e^{ax}\left[\dfrac{1}{f(D+a)}\right]V,$

we get $\quad \dfrac{1}{f(D)}(e^{ax}V) = e^{iax}\cdot\dfrac{1}{D}\left(\dfrac{e^{-iax}}{\cos ax}\right)$

$$= e^{iax}\int\dfrac{\cos ax - i\sin ax}{\cos ax}dx$$

$$= e^{iax}\int(1 - i\tan ax)\,dx = e^{iax}\left[x + \dfrac{i}{a}\log\cos ax\right]$$

Similarly, $\quad \dfrac{1}{D+ai}\cdot\sec ax = e^{-iax}\left[x - \dfrac{i}{a}\log\cos ax\right]$

Putting in (i), we get the required P.I. as

$$\dfrac{1}{2ai}\left[e^{iax}\left\{x + \dfrac{i}{a}\log\cos ax\right\} - e^{-iax}\left\{x - \dfrac{i}{a}\log\cos ax\right\}\right]$$

$$= \dfrac{x}{a}\cdot\dfrac{e^{iax}-e^{-iax}}{2i} + \dfrac{1}{a^2}\log(\cos ax)\dfrac{e^{iax}+e^{-iax}}{2}$$

$$= \dfrac{x}{a}\sin ax + \dfrac{1}{a^2}\log(\cos ax)\cdot\cos ax$$

Hence general solution is

$$y = c_1\cos ax + c_2\sin ax + \dfrac{1}{a}\sin ax + \dfrac{1}{a^2}\cos ax\,\log(\cos ax)$$

2. Solve : $\dfrac{d^2y}{dx^2} + 4y = 4\tan 2x.$

Sol. C.F. $= c_1\cos 2x + c_2\sin 2x$

$$\text{P.I.} = \dfrac{4\tan 2x}{D^2 + 4} = \dfrac{1}{i}\left[\dfrac{1}{D-2i} - \dfrac{1}{D+2i}\right]\tan 2x \qquad ...(i)$$

Now, let $\quad \dfrac{1}{D-2i}\tan 2x = v.$

$\therefore \qquad (D - 2i)v = \tan 2x \quad \text{or} \quad \dfrac{dv}{dx} - 2iv = \tan 2x$

$$\text{I.F.} = e^{-2ix}$$

Hence, $\quad v.e^{-2ix} = \int e^{-2ix}\tan 2x\,dx$

$$= \int(\cos 2x - i\sin 2x)\dfrac{\sin 2x}{\cos 2x}dx$$

Linear Differential Equations with Constant Coefficients

$$= \int \sin 2x \, dx - i \int \frac{1 - \cos^2 2x}{\cos 2x} \, dx$$

$$= -\frac{\cos 2x}{2} - i \int (\sec 2x - \cos 2x) \, dx$$

$$= -\frac{\cos 2x}{2} - i \cdot \frac{1}{2} \cdot \log \tan\left(\frac{\pi}{4} + x\right) + \frac{i}{2} \sin 2x$$

$$= -\frac{1}{2}(\cos 2x - i \sin 2x) - \frac{i}{2} \log \tan\left(\frac{\pi}{4} + x\right)$$

or $\qquad ve^{-2ix} = -\frac{1}{2} e^{-2ix} - \frac{i}{2} \log \tan\left(\frac{\pi}{4} + x\right)$

$\therefore \qquad v = -\frac{1}{2} - \frac{i}{2} \log \tan\left(\frac{\pi}{4} + x\right) e^{2ix}$

Similarly, if we calculate $\dfrac{1}{D + 2i} \tan 2x = u$, we will get

$$u = -\frac{1}{2} + \frac{i}{2} \log \tan\left(\frac{\pi}{4} + x\right) e^{-2ix} \quad \text{(on replacing } i \text{ by } -i \text{ in above value of } v)$$

$\therefore \qquad \text{P.I.} = \frac{1}{i}(v - u)$

$$= \frac{1}{i}\left[-\frac{i}{2} \log \tan\left(\frac{\pi}{4} + x\right)(e^{-2ix} + e^{2ix})\right]$$

$$= -\frac{1}{2} \log \tan\left(\frac{\pi}{4} + x\right) \cdot 2\cos 2x$$

$$= -\log \tan\left(\frac{\pi}{4} + x\right) \cos 2x$$

$\therefore$ General solution is

$$y = c_1 \cos 2x + c_2 \sin 2x - \log \tan\left(\frac{\pi}{4} + x\right) \cos 2x$$

EXERCISE

Solve following differential equations:

1. $\dfrac{d^2 y}{dx^2} + y = \operatorname{cosec} x$ $\qquad$ [**Ans.** $y = c_1 \cos x + c_2 \sin x - x \cos x + \log (\sin x) \sin x$]

2. $(D^2 + 4) y = \sec 2x$ $\qquad$ $\left[\textbf{Ans. } y = c_1 \cos 2x + c_2 \sin 2x + \dfrac{x}{2} \sin 2x + \dfrac{1}{4} \cos 2x \log (\cos 2x)\right]$

3. $(D^2 + a^2) y = \tan ax$ $\qquad$ $\left[\textbf{Ans. } y = c_1 \cos ax + c_2 \sin ax - \dfrac{1}{a^2} \log \tan\left(\dfrac{\pi}{a} + \dfrac{ax}{2}\right) \cos ax\right]$

◻◻◻

6
Homogeneous Linear Differential Equations

6.1 DEFINITION

Any differential equation of the form

$$x^n \frac{d^n y}{dx^n} + p_1 x^{n-1} \frac{d^{n-1} y}{dx^{n-1}} + \ldots + p_n y = X$$

is called a *homogeneous differential equation* where $p_1, p_2, \ldots, p_n$ are constants and X is a funftion of x.

Method of Solution

To solve homogeneous linear differential equations, we change the independent variable x to z by the substitution

$$z = \log x \quad \text{or} \quad x = e^z \quad \text{and} \quad \frac{dz}{dx} = \frac{1}{x}$$

Then

$$\frac{dy}{dx} = \frac{dy}{dz} \cdot \frac{dz}{dx} = \frac{dy}{dz} \cdot \frac{1}{x}$$

or

$$x \frac{dy}{dx} = \frac{dy}{dz}$$

$$\frac{d^2 y}{dx^2} = \frac{d}{dx}\left(\frac{dy}{dx}\right) = \frac{d}{dx}\left(\frac{1}{x} \cdot \frac{dy}{dx}\right) = \frac{1}{x} \frac{d}{dx}\left(\frac{dy}{dz}\right) + \frac{dy}{dz}\left(-\frac{1}{x^2}\right)$$

$$= \frac{1}{x} \frac{d^2 y}{dz^2} \cdot \frac{dz}{dx} - \frac{1}{x^2} \frac{dy}{dz} = \frac{1}{x^2} \frac{d^2 y}{dz^2} - \frac{1}{x^2} \frac{dy}{dz}$$

or

$$x^2 \frac{d^2 y}{dx^2} = \frac{d^2 y}{dz^2} - \frac{dy}{dz}.$$

Continuing this process, we have

$$x^n \frac{d^n y}{dx^n} = \left[\frac{d^n y}{dz^n} - \frac{n(n-1)}{2!} \frac{d^{n-1} y}{dz^{n-1}} + \ldots + (-1)^{n-1} n! \frac{dy}{dz}\right]$$

Now, write down $D \equiv \dfrac{d}{dz}$.

Then

$$x \frac{dy}{dz} = Dy; \quad x^2 \frac{d^2 y}{dx^2} = (D^2 - D)y = D(D-1)y$$

Homogeneous Linear Differential Equations

$$x^3 \frac{d^3y}{dx^3} = (D^3 - 3D^2 + 2d)\, y = D(D-1)(D-2)\, y$$

Similarly, $\quad x^n \dfrac{d^n y}{dx^n} = D(D-1)(D-2)\ldots(D-n+1)\, y.$

Let us consider this transformed equation

$$f(D)\, y = X$$

The general solution of (i) is the sum of a particular solution of (i) and general solution of $f(D)\, y = 0$.

Now it can be solved by the methods given in chapter 5.

EXAMPLES

1. Solve : $x^2 \dfrac{d^2y}{dx^2} - x \dfrac{dy}{dx} + y = 2 \log x.$

Sol. Putting $x = e^z$ and denoting $\dfrac{d}{dz}$ by D, the given differential equation becomes

$$[D(D-1) - D + 1]\, y = 2z$$

or $\qquad (D^2 - 2D + 1)\, y = 2z$

Auxilliary equation is $\quad m^2 - 2m + 1 = 0$

or $\qquad (m-1)^2 = 0$

$\therefore \qquad m = 1, 1$

$\therefore \qquad$ C.F. $= (c_1 + c_2 z)\, e^z$

Also, $\quad$ P.I. $= \dfrac{1}{D^2 - 2D + 1}\, 2z = 2 \cdot \dfrac{1}{(1-D)^2}\, z$

$\qquad = 2(1-D)^{-2}\, z = 2[1 + 2D + \ldots]\, z$

$\qquad = 2(z + 2)$

Hence, the general solution is

$$y = \text{C.F.} + \text{P.I.}$$

$$y = e^z (c_1 + c_2 z) + 2z + 4$$

$$y = x(c_1 + c_2 \log x) + 2 \log x + 4.$$

2. Solve : $x^2 \dfrac{d^2 y}{dx^2} + 4x \dfrac{dy}{dx} + 2y = e^x.$

Sol. Putting $x = e^z$, $z = \log x$ and $\dfrac{d}{dz} \equiv D$, we get

$$\{D(D-1) + 4D + 2\}\, y = e^{e^z}$$

or $\qquad (D^2 + 3D + 2)\, y = e^{e^z}$

Auxilliary equation is
$$m^2 + 3m + 2 = 0 \therefore m = -1, -2$$

$\therefore$ C.F. $= c_1 e^{-z} + c_2 e^{-2z} = c_1 x^{-1} + c_2 x^{-2}$

$$\text{P.I.} = \frac{1}{(D+2)}\left(\frac{1}{D+1} e^{e^z}\right)$$

Let $\dfrac{1}{D+1} e^{e^z} = u$

$\therefore \dfrac{du}{dz} + u = e^{e^z}$

This is linear equation whose I.F. $= e^z$.

$\therefore \quad u.e^z = \int e^{e^z} . e^z\, dz = e^{e^z}$

$\therefore \quad u = e^{e^z} . e^{-z}$

$\therefore \quad \text{P.I.} = \dfrac{1}{D+2} . (e^{e^z} . e^{-z}) = v$ (say)

$\therefore \dfrac{dv}{dz} + 2v = e^{e^z} . e^{-z}.$

This is again a linear equation where I.F. $= e^{2z}$.

$\therefore \quad v.e^{2z} = \int e^{e^z} . e^{-z} . e^{2z}\, dz = \int e^{e^z} . e^z\, dz . e^{e^z}$

or $\quad v = e^{e^z} . e^{-2z} = \dfrac{e^x}{x^2}$

$\therefore$ General solution of the equation is

$$y = c_1 x^{-1} + c_2 x^{-2} + \frac{e^x}{x^2}.$$

3. Solve $x^2 \dfrac{d^2 y}{dx^2} + x \dfrac{dy}{dx} - y = x^2 e^x.$

Sol. The given differential equation

$$x^2 \frac{d^2 y}{dx^2} + x \frac{dy}{dx} - y = x^2 e^x$$

is a homogeneous equation of second degree.

Hence substitute $x = e^z$ or $z = \log_e x$, reduces in the form

$\therefore \quad [D(D-1) + D - 1] y = e^{2z} . e^{e^z}$

or $\quad (D^2 - 1) y = e^{2z} . e^{e^z}$ ...(ii)

which is linear differential equation, with constant coefficient where

Homogeneous Linear Differential Equations

$$D = \frac{d}{dz}.$$

Hence, its auxilliary equation is

$$m^2 - 1 = 0 \quad \text{or} \quad m = \pm 1$$

$\therefore$ Its C.F. $= c_1 e^z + c_2 e^{-z} = c_1 x + c_2 x^{-1}$

and $\quad$ P.I. $= \dfrac{1}{D^2 - 1} e^{z^2} . e^{e^z} = \dfrac{1}{(D-1)(D+1)} e^{2z} . e^{e^z}$

$= \dfrac{1}{2} \left\{ \dfrac{1}{D-1} - \dfrac{1}{D+1} \right\} e^{2z} e^{e^z}$

$= \dfrac{1}{2} \left[\dfrac{1}{D-1} e^{2z} e^{e^z} - \dfrac{1}{D+1} e^{2z} e^{e^z} \right]$

$= \dfrac{1}{2} [e^z \int e^{-z} (e^{2z} e^{e^z}) \, dz - e^{-z} \int e^z (e^{2z} e^{e^z}) \, dz]$

$= \dfrac{1}{2} [e^z \int e^z e^{e^z} \, dz - e^{-z} \int (e^z)^2 e^{e^z} . e^z \, dz] \quad \left[\because z = \log x \therefore dz = \dfrac{1}{x} dx \right]$

$= \dfrac{1}{2} \left[x \int xe^x . \dfrac{1}{x} dx - x^{-1} \int x^2 e^x . x . \dfrac{1}{x} dx \right]$

$= \dfrac{1}{2} [xe^x - x^{-1} \int x^2 e^x \, dx]$

$= \dfrac{1}{2} [xe^x - x^{-1} (x^2 e^x - \int 2x \, e^x \, dx)]$

$= \dfrac{1}{2} [xe^x - x^{-1} x^2 e^x - 2 \int xe^x \, dx]$

$= \dfrac{1}{2} [xe^x - xe^x + 2x^{-1} (xe^x - \int 1 . e^x \, dx)]$

$= \dfrac{1}{2} [xe^x - xe^x + 2x^{-1} (xe^x - e^x)]$

$= e^x (x^{-1} x - x^{-1})$

$= e^x (1 - x^{-1})$

Hence, the general solution of the equation (i).

$$y = \text{C.F.} + \text{P.I.} = c_1 x + c_2 x^{-1} + e^x (1-x)^{-1}$$

EXERCISE

Solve the following differential equations :

1. $x^3 \dfrac{d^3 y}{dx^3} - x^2 \dfrac{d^2 y}{dx^2} + 2x \dfrac{dy}{dx} - 2y = x^3 + 3x.$

$\left[\text{Ans. } y = (c_1 + c_2 \log x) x + c_3 x^2 + \dfrac{x^3}{4} - \dfrac{3}{2} x (\log x)^2 \right]$

2. $x^3 \dfrac{d^3y}{dx^3} + 2x^2 \dfrac{d^2y}{dx^2} + 3x \dfrac{dy}{dx} - 3y = x + x^2$

$$\left[\text{Ans. } y = c_1 x + c_2 \cos(\sqrt{3} \log x) + c_3 \sin(\sqrt{3} \log x) + \dfrac{1}{7} x^2 + \dfrac{1}{4} x \log x\right]$$

3. $(x^3 D^3 + 3x^2 D^2 - 2xD + 2) y = 0$ $[\text{Ans. } y = x(c_1 + c_2 \log x) + c_3 x^{-2}]$

4. $x^3 \dfrac{d^3y}{dx^3} + 2x^2 \dfrac{d^2y}{dx^2} + 2y = 10\left(x + \dfrac{1}{x}\right)$

$$\left[\text{Ans. } y = \dfrac{c_1}{x} + x\{c_2 \cos(\log x) + c_3 \sin(\log x)\} + 5x + \dfrac{2 \log x}{x}\right]$$

5. $x^2 \dfrac{d^2y}{dx^2} + 2x \dfrac{dy}{dx} = \log x$ $\left[\text{Ans. } y = c_1 + \dfrac{c_2}{x} + \dfrac{(\log x)^2}{2} - (\log x)\right]$

6. $x^3 \dfrac{d^3y}{dx^3} + 2x^2 \dfrac{d^2y}{dx^2} + x \dfrac{dy}{dx} - y = \cos(\log x)$

$$\left[\text{Ans. } y = c_1 x + c_2 \cos(\log x) + c_3 \sin(\log x) + \dfrac{1}{4} \log x \{\cos(\log x) - \sin(\log x)\}\right]$$

7. $x^2 \dfrac{d^2y}{dx^2} + 6x \dfrac{dy}{dx} + 6y = (\log x)^2$

$$\left[\text{Ans. } y = c_1 e^{-2x} + c_2 e^{-3x} + \dfrac{1}{108}\{18(\log x)^2 - 30(\log x) + 19\}\right]$$

8. $x^2 \dfrac{d^2y}{dx^2} + 5x \dfrac{dy}{dx} + 4y = x \log x$ $\left[\text{Ans. } y = (c_1 + c_2 \log x) x^{-2} + \dfrac{x}{x}\left(\log x - \dfrac{2}{3}\right)\right]$

9. $(x^4 D^4 + 6x^3 D^3 + 9x^2 D^2 + 3xD + 1) y = (1 + \log x)^2$

[Ans. $y = (c_1 + c_2 \log x) \cos(\log x) + (c_3 + c_4 \log x) \sin(\log x) + (\log x)^3 + 2(\log x) - 3$]

10. $x^2 D^2 y - 3xDy + 5y = x^2 \sin \log x$.

$$\left[\text{Ans. } y = x^2 \{c_1 \cos(\log x) + c_2 \sin(\log x)\} - \dfrac{x^2}{2} \log x \cos(\log x)\right]$$

6.2 EQUATIONS REDUCIBLE TO HOMOGENEOUS FORM

Any equation of the form

$$(a+bx)^n \dfrac{d^n y}{dx^n} + P_1(a+bx)^{n-1} \dfrac{d^{n-1} y}{dx^{n-1}} + \ldots + P_{n-1}(a+bx) \dfrac{dy}{dx} + P_n y = F(x) \quad \ldots(i)$$

where the coefficients $P_1, P_2, \ldots, P_n$ are constants, can be transformed into homogeneous linear equation with constant coefficients by changing independent variable from x to z, by the substitution $a + bx = z$. For, on letting $a + bx = z$, we have

Homogeneous Linear Differential Equations

$$\frac{dy}{dx} = \frac{dy}{dz} \cdot \frac{dz}{dx} = b \frac{dy}{dz}$$

$$\frac{d^2 y}{dx^2} = \frac{d}{dx}\left(\frac{dy}{dx}\right) = \frac{d}{dz}\left(b \frac{dy}{dz}\right) \frac{dz}{dx}$$

$$= b^2 \frac{d^2 y}{dz^2}$$

$$\cdots\cdots\cdots\cdots = \cdots\cdots\cdots\cdots$$

$$\frac{d^n y}{dx^n} = b^n \frac{d^n y}{dz^n}$$

Substituting these in (i) and dividing throughout by b^n, we obtain

$$z^n \frac{d^n y}{dz^n} + \frac{P_1}{b} z^{n-1} \frac{d^{n-1} y}{dz^{n-1}} + \frac{P_2}{b^2} z^{n-2} \frac{d^{n-2} y}{dz^{n-2}} + \ldots + \frac{P_{n-1}}{b^{n-1}} z \frac{dy}{dz} + \frac{P_n}{b^n} y$$

$$= \frac{1}{b^n} F.\left(\frac{z-a}{b}\right) \qquad \ldots(ii)$$

This equation can be solved by the method of § 6.1.

EXAMPLES

1. *Solve the differential equation :*

$$(1+2x)^2 \frac{d^2 y}{dx^2} - 6(1+2x) \frac{dy}{dx} + 16 y = 8(1+2x)^2$$

given that $y(0) = 0$, $y'(0) = 2$.

Sol. Putting $\quad e^z = 1 + 2x \quad$ or $\quad z = \log(1 + 2x)$

and denoting $\dfrac{d}{dz}$ by D, we get

$$(1+2x) \frac{dy}{dx} = 2Dy \quad \text{and} \quad (1+2x)^2 \frac{d^2 y}{dx^2} = 4D(D-1) y$$

$\therefore$ The given equation transforms to

$$[4D(D-1) - 12D + 16] y = 8e^{2z}$$

or $\qquad (D^2 - 4D + 4) y = 2e^{2z}$

The auxilliary equation is

$$m^2 - 4m + 4 = 0$$

or $\qquad (m-2)^2 = 0, \quad \therefore \quad m = 2, 2$

$\therefore \qquad$ C.F. $= (c_1 + c_2 z) e^{2z} = [c_1 + c_2 \log(1+2x)](1+2x)^2$

Also, $\qquad$ P.I. $= \dfrac{1}{D^2 - 4D + 4} 2e^{2z}$

$$= 2 \cdot \frac{1}{(D-2)^2} \cdot e^{2z} \cdot 1$$

$$= 2e^{2z} \cdot \frac{1}{(D+2-2)^2} \cdot 1 = 2e^{2z} \cdot \frac{1}{D^2} \cdot 1$$

$$= 2e^{2z} \cdot \frac{z^2}{2} = z^2 e^{2z} = (1+2x)^2 \, [\log(1+2x)]^2$$

∴ The general solution is

$$y = \text{C.F.} + \text{P.I.}$$

or $\quad y = [c_1 + c_2 \log(1+2x)](1+2x)^2 + (1+2x)^2 \,[\log(1+2x)^2]$ ...(1)

Now, it is given that $y(0) = 0$ and $y'(0) = 2$

we have $\quad \dfrac{dy}{dx} = 2c_2 (1+2x) + 4[c_1 + c_2 \log(1+2x)](1+2x)$

$$+ 4(1+2x)[\log(1+2x)]^2 + 4(1+2x)\log(1+2x)$$

Hence, by given condition, we have

$$0 = c_1 \text{ and } 2 = 2c_2 + 4c_1$$

From these, we get $c_1 = 0$ and $c_2 = 1$

By putting these values in (1), we get the required solution as

$$y = (1+2x)^2 \log(1+2x) + (1+2x)^2 \,[\log(1+2x)]^2$$

$$= (1+2x)^2 \log(1+2x) \,[1 + \log(1+2x)]$$

2. Solve : $(1+x)^2 \dfrac{dy}{dx} + (1+x)\dfrac{dy}{dx} + y = 4 \cos \log(1+4)$.

Sol. Putting $(1+x) = e^z$ and denoting $\dfrac{d}{dz}$ by D, the given differential equation reduces to

$$[D(D-1) + D + 1] y = 4 \cos z$$

or $\quad (D^2 + 1) y = 4 \cos z$

Auxilliary equation is

$$m^2 + 1 = 0, \therefore m = \pm i$$

∴ $\quad \text{C.F.} = e^{0.z}(c_1 \cos z + c_2 \sin z) = c_1 \cos \log(1+x) + c_2 \sin \log(1+x)$

$$\text{P.I.} = \frac{1}{D^2 + 1} 4 \cos z$$

$$= \text{Real part in } 4 \cdot \frac{1}{D^2 + 1}(\cos z + i \sin z) = \text{Real part in } 4 \cdot \frac{1}{D^2 + 1} e^{iz}$$

Now, $\quad \dfrac{1}{(D^2 + 1)} e^{iz} = \dfrac{1}{(D+i)(D-1)} e^{iz}$

Homogeneous Linear Differential Equations

$$= \frac{1}{(i+i)(D-i)} e^{iz} = \frac{1}{2i} e^{iz} \frac{1}{D+i-i} \cdot 1 = \frac{e^{iz}}{2i} \frac{1}{D} \cdot 1$$

$$= \frac{ze^{iz}}{2i} = \frac{z}{2i}(\cos z + i \sin z) = -\frac{iz}{2}(\cos z + i \sin z) = -\frac{iz}{2}\cos z + \frac{z}{2}\sin z$$

$$\therefore \qquad \text{P.I.} = \text{Real part of } 4 \left[\frac{-iz}{2}\cos z + \frac{z}{2}\sin z \right]$$

$$= 2z \sin z = 2 \log(1+x) \sin \log(1+x)$$

Hence the complete solution of the given equation is

$$y = c_1 \cos \log(1+x) + c_2 \sin \log(1+x) + 2 \log(1+x) \sin \log(1+x)$$

EXERCISE

Solve following differential equations:

1. $(5+2x)^2 \dfrac{d^2y}{dx^2} - 6(5+2x)\dfrac{dy}{dx} + 8y = 0$

   $\left[\text{Ans. } y = (5+2x)^2 \left[c_1 (5+2x)^{\sqrt{2}} + c_2 (5+2x)^{-\sqrt{2}} \right] \right]$

2. $16(x+1)^4 \dfrac{d^4y}{dx^4} + 96(x+1)^3 \dfrac{d^3y}{dx^3} + 104(x+1)^2 \dfrac{d^2y}{dx^2} + 8(x+1)\dfrac{dy}{dx} + y = x^2 + 4x + 3$

   $\left[\text{Ans. } y = (c_1 + c_2 z) e^{z/2} + (c_3 + c_4 z) e^{-z/2} + \dfrac{e^{2z}}{225} + \dfrac{2}{9} e^z, \text{ where } z = \log(1+x) \right]$

3. $(x-a)^2 \dfrac{d^2y}{dx^2} - 4(x+a)\dfrac{dy}{dx} + 6y = x$

   $\left[\text{Ans. } y = c_1 (x+a)^2 + c_2 (x+a)^3 + \dfrac{1}{2}(x+a) - \dfrac{1}{6}a \right]$

4. $(3x+2)^2 \dfrac{d^2y}{dx^2} + 5(3x+2)\dfrac{dy}{dx} - 3y = x^2 + x + 1$

   $\left[\text{Ans. } y = c_1 (3x+2)^{1/3} + c_2 (3x+2)^{-1} + \dfrac{(3x+2)^2}{405} - \dfrac{(3x+2)}{108} - \dfrac{7}{27} \right]$

7

Linear Differential Equations of Second Order with Variable Coefficients

7.1 LINEAR EQUATIONS OF SECOND ORDER

Differential equations of the form

$$\frac{d^2y}{dx^2} + P\frac{dy}{dx} + Qy = X,$$

where P, Q and X are functions of x alone, are called *linear equation of the second order*. There is no general method to solve such type of equations. Here we have discussed certain methods by which solutions of such equations can be found.

7.2 THE COMPLETE SOLUTION IN TERMS OF A KNOWN INTEGRAL

Let $y = u$ be a known integral in the complementary function of

$$\frac{d^2y}{dx^2} + P\frac{dy}{dx} + Qy = X \qquad \ldots(1)$$

i.e, it is a solution of

$$\frac{d^2y}{dx^2} + P\frac{dy}{dx} + Qy = 0$$

$$\therefore \quad \frac{d^2u}{dx^2} + P\frac{du}{dx} + Qu = 0 \qquad \ldots(2)$$

On substituting $y = uv$, we get

$$\frac{dy}{dx} = v\frac{du}{dx} + u\frac{dv}{dx}$$

and

$$\frac{d^2y}{dx^2} = v\frac{d^2u}{dx^2} + 2\frac{du}{dx}\cdot\frac{dv}{dx} + u\frac{d^2v}{dx^2}$$

Substituting these values in equation (i), we get

$$\left(v\frac{d^2u}{dx^2} + 2\frac{du}{dx}\frac{dv}{dx} + u\frac{d^2v}{dx^2}\right) + P\left(v\frac{du}{dx} + u\frac{dv}{dx}\right) + Quv = X$$

or

$$u\frac{d^2v}{dx^2} + \frac{dv}{dx}\left(2\frac{du}{dx} + Pu\right) + v\left(\frac{d^2u}{dx^2} + P\frac{du}{dx} + Qu\right) = X$$

or

$$u\frac{d^2v}{dx^2} + \frac{dv}{dx}\left(2\frac{du}{dx} + Pu\right) = X \qquad \text{(from (2))}$$

Linear Differential Equations of Second Order with Variable Coefficients

or
$$\frac{d^2v}{dx^2} + \left(P + \frac{2}{u}\frac{du}{dx}\right)\frac{dv}{dx} = \frac{X}{u} \qquad ...(3)$$

Putting $\quad \dfrac{dv}{dx} = p, \dfrac{d^2v}{dx^2} = \dfrac{dp}{dx},\quad$ (3) becomes

$$\frac{dp}{dx} + \left(P + \frac{2}{u}\frac{du}{dx}\right) p = \frac{X}{u} \qquad ...(4)$$

This is a linear differential equation with p as dependent variable.

$$\text{I.F.} = e^{\int\left(P + \frac{2}{u}\frac{du}{dx}\right)dx}$$

$$= e^{\{2\log u + \int P\,dx\}} = u^2 e^{\int P\,dx}$$

Hence solution of (4) will be

$$p u^2 e^{\int P\,dx} = \int \left[\frac{X}{u} \cdot u^2 e^{\int P\,dx}\right] dx + C_1 \qquad ...(5)$$

$\therefore \qquad p = \dfrac{dv}{dx} = \dfrac{C_1 e^{-\int P\,dx}}{u^2} + \dfrac{e^{-\int P\,dx}}{u^2}\int uX e^{\int P\,dx}\,dx$

Integrating this, we get

$$v = C_2 + C_1 \int \frac{e^{-\int P\,dx}}{u^2}\,dx + \int\left[\frac{e^{-\int P\,dx}}{u^2}\int uX e^{\int P\,dx}\,dx\right]dx$$

Hence the solution of (1) is

$$y = uv = C_2 u + C_1 u \int \frac{e^{-\int P\,dx}}{u^2}\,dx + u\int\left[\frac{e^{-\int P\,dx}}{u^2}\int uX e^{\int P\,dx}\,dx\right]dx \qquad ..(6)$$

Since it conatins two arbitrary constants, hence it is the complete primitive of the given equation in terms of the known integral.

Note : It is worthwhile to find one integral belonging to the complementary function by inspection. In this connectino it may be remembered that :

$y = e^x$ is a part of complementary function if $P + Q + 1 = 0$

$y = e^{-x}$ is a part of complementary function if $1 - P + Q = 0$

$y = e^{ax}$ is a part of complementary function if $1 + \dfrac{P}{a} + \dfrac{Q}{a^2} = 0$

$y = x$ is a part of complementary function if $P + Qx = 0$

$y = x^2$ is a part of complementary function if $2 + 2Px + Qx^2 = 0$

EXAMPLES

1. Solve : $(1 - x^2)\dfrac{d^2y}{dx^2} + x\dfrac{dy}{dx} - y = x(1 - x^2)^{3/2}.$

Sol. The given equation can be written as

$$\frac{d^2y}{dx^2} + \frac{x}{1-x^2}\frac{dy}{dx} - \frac{1}{1-x^2} y = x(1-x^2)^{1/2} \qquad ...(1)$$

Here $P + Qx = 0$, $\therefore y = x$ is a part of the C.F. of the solution of (1).

Putting $y = vx$, $\dfrac{dy}{dx} = v + x\dfrac{dv}{dx}$ and $\dfrac{d^2y}{dx^2} = x\dfrac{d^2v}{dx^2} + 2\dfrac{dv}{dx}$ in (1), we get

$$\frac{d^2v}{dx^2} + \left(\frac{2}{x} + \frac{x}{1-x^2}\right)\frac{dv}{dx} = (1-x^2)^{1/2}$$

or $\qquad \dfrac{dp}{dx} + \left(\dfrac{2}{x} + \dfrac{x}{1-x^2}\right) p = (1-x^2)^{1/2} \qquad \left(\text{where } p = \dfrac{dv}{dx}\right)$

This is a linear equation in p.

$$\text{I.F.} = e^{\int \left(\frac{2}{x} + \frac{x}{1-x^2}\right)dx} = e^{\left\{2\log x - \frac{1}{2}\log(1-x^2)\right\}}$$

$$= \frac{x^2}{\sqrt{(1-x^2)}}$$

$\therefore \qquad p \cdot \dfrac{x^2}{\sqrt{(1-x^2)}} = \int \sqrt{(1-x^2)} \cdot \dfrac{x^2}{\sqrt{(1-x^2)}} dx + C_1$

$$= \frac{1}{3}x^3 + C_1$$

$\therefore \qquad p = \dfrac{dv}{dx} = \dfrac{1}{3} x\sqrt{(1-x^2)} + \dfrac{C_1}{x^2}\sqrt{(1-x^2)}$

Integrating, we get

$$v = -\frac{1}{9}(1-x^2)^{3/2} + C_1 (1-x^2)^{1/2}\left(-\frac{1}{x}\right) - C_1 \int \frac{dx}{\sqrt{(1-x^2)}} + C_2$$

or $\qquad v = -\dfrac{1}{9}(1-x^2)^{3/2} - \dfrac{C_1}{x}(1-x^2)^{1/2} - C_1 \sin^{-1} x + C_2$

Hence the complete solution of (1) is

$$y = vx = -\frac{1}{9} x(1-x^2)^{3/2} - C_1 \{(1-x^2)^{1/2} + x \sin^{-1} x\} + C_2 x$$

2. Solve : $\dfrac{d^2y}{dx^2} - \cot x \dfrac{dy}{dx} - (1 - \cot x) y = e^x \sin x.$

Sol. The given differential equation is

$$\frac{d^2y}{dx^2} - \cot x \frac{dy}{dx} - (1 - \cot x) y = e^x \sin x. \qquad ...(1)$$

Here, we have $1 + P + Q = 0$

Linear Differential Equations of Second Order with Variable Coefficients

Hence $y = e^x$ is a part of C.F. of the solution of (1).

Putting $y = ve^x$, $\dfrac{dy}{dx} = \dfrac{dv}{dx} e^x + ve^x$ and $\dfrac{d^2y}{dx^2} = \dfrac{d^2v}{dx^2} e^x + 2 \dfrac{dv}{dx} e^x + ve^x$ in (1), we get

$$\dfrac{d^2v}{dx^2} + (2 - \cot x) \dfrac{dv}{dx} = \sin x$$

or $\quad \dfrac{dp}{dx} + (2 - \cot x) p = \sin x$, where $p = \dfrac{dv}{dx}$

This is a linear differential equation in p.

$$\text{I.F.} = e^{\int (2 - \cot x) dx} = e^{(2x - \log \sin x)}$$

$$= \dfrac{e^{2x}}{\sin x}.$$

Hence solution will be

$$p \cdot \dfrac{e^{2x}}{\sin x} = \int \dfrac{e^{2x}}{\sin x} \sin x \, dx + C_1$$

$$= \dfrac{1}{2} e^{2x} + C_1$$

$\therefore \quad p = \dfrac{dv}{dx} = \dfrac{1}{2} \sin x + C_1 e^{-2x} \cdot \sin x$

Integrating this, we get

$$v = -\dfrac{1}{2} \cos x + \dfrac{C_1}{5} e^{-2x}(-2 \sin x - \cos x) + C_2$$

Hence the complete solution of (1) is

$$y = ve^x = -\dfrac{1}{2} e^x \cos x - \dfrac{C_1}{5} e^{-x} (2 \sin x + \cos x) + C_2 e^x$$

3. Solve: $\dfrac{d^2y}{dx^2} + \left(1 + \dfrac{2}{x} \cot x - \dfrac{2}{x^2}\right) y = x \cos x$ given that $\dfrac{\sin x}{x}$ is a C.F.

Sol. Let $y = v \cdot \dfrac{\sin x}{x}$, then

$$\dfrac{dy}{dx} = \dfrac{dv}{dx} \cdot \dfrac{\sin x}{x} + v \left(\dfrac{x \cos x - \sin x}{x^2}\right) = \dfrac{dv}{dx} \cdot \dfrac{\sin x}{x} + v \left(\dfrac{\cos x}{x} - \dfrac{\sin x}{x^2}\right),$$

$$\dfrac{d^2y}{dx^2} = \dfrac{d^2v}{dx^2} \cdot \dfrac{\sin x}{x} + 2 \dfrac{dv}{dx} \cdot \left(\dfrac{\cos x}{x} - \dfrac{\sin x}{x^2}\right) + v \left(-\dfrac{\sin x}{x} - \dfrac{2 \cos x}{x^2} + \dfrac{2 \sin x}{x^3}\right)$$

Putting these values in the given equation, we get

$$\dfrac{d^2v}{dx^2} \cdot \dfrac{\sin x}{x} + 2 \left(\dfrac{\cos x}{x} - \dfrac{\sin x}{x^2}\right) \dfrac{dv}{dx} + \left(-\dfrac{\sin x}{x} - \dfrac{2 \cos x}{x^2} + \dfrac{2 \sin x}{x^3}\right) v$$

$$+ \left(1 + \dfrac{2 \cot x}{x} - \dfrac{2}{x^2}\right) v \dfrac{\sin x}{x} = x \cos x$$

or
$$\frac{d^2v}{dx^2} + 2\left(\cot x - \frac{1}{x}\right)\frac{dv}{dx} = x^2 \cot x.$$

Now, putting $\frac{dv}{dx} = p$, we get

$$\frac{dp}{dx} + 2\left(\cot x - \frac{1}{x}\right)p = x^2 \cot x$$

which is a linear equation in p.

$$\text{I.F.} = e^{\int 2\left(\cot x - \frac{1}{x}\right)dx} = \frac{\sin^2 x}{x^2}$$

$$\therefore \quad p \cdot \frac{\sin^2 x}{x^2} = \int x^2 \cot x \cdot \frac{\sin^2 x}{x^2} dx + C_1$$

$$= \frac{1}{2}\int \sin 2x \, dx + C_1 = -\frac{1}{4}\cos 2x + C_1$$

$$= -\frac{1}{4} + \frac{1}{2}\sin^2 x + C_1$$

$$\therefore \quad p = \frac{dv}{dx} = -\frac{1}{4}x^2 \csc^2 x + \frac{1}{2}x^2 \csc^2 x$$

Integrating, we get

$$v = \left(C_1 - \frac{1}{4}\right)\left[-x^2 \cot x + 2x \log x - 2\int \log \sin x \, dx\right] + \frac{x^3}{6} + C_2$$

Hence the complete solution is

$$y = v \cdot \frac{\sin x}{x}$$

$$= \left(C_1 - \frac{1}{4}\right)\left[-x \cos x + 2\sin x \log \sin x - \frac{2 \sin x}{x} \cdot \int \log \sin x \, dx\right] + \frac{x^2 \sin x}{6} + C_2 \frac{\sin x}{x}$$

EXERCISE

Solve the following different equations :

1. $x\frac{d^2y}{dx^2} - 2(x+1)\frac{dy}{dx} + (x+2)y = (x-2)e^x$ $\quad\left[\text{Ans. } y = \left(-\frac{x^2}{2} + x + \frac{C_1}{3}x^3 + C_2\right)e^x\right]$

2. $(3-x)\frac{d^2y}{dx^2} - (9-4x)\frac{dy}{dx} + (6-3x)y = 0.$

$$\left[\text{Ans. } y = \frac{C_1}{8}e^{3x}(183 - 150x + 42x^2 - 4x^3) + C_2 e^x\right]$$

3. $x\frac{dy}{dx} - y = (x-1)\left(\frac{d^2y}{dx^2} - x + 1\right)$ $\quad[\text{Ans. } y = C_1 e^x + C_2 x - (1 + x^2)]$

Linear Differential Equations of Second Order with Variable Coefficients 597

4. $x(x \cos x - 2 \sin x) \dfrac{d^2y}{dx^2} + (x^2 + 2) \sin x \dfrac{dy}{dx} - 2(x \sin x + \cos x) y = 0.$

[**Ans.** $y = C_1 \sin x + C_2 x^2$]

5. Solve $x^2 \dfrac{d^2y}{dx^2} + x \dfrac{dy}{dx} - y = 0,$ given that $x + \dfrac{1}{x}$ is one integral.

$$\left[\text{Ans. } y = -\dfrac{C_1}{2x} + C_2\left(x + \dfrac{1}{x}\right)\right]$$

6. Solve $x^2 \dfrac{d^2y}{dx^2} + x \dfrac{dy}{dx} - 9y = 0,$ given that $y = x^3$ is a solution.

$$\left[\text{Ans. } y = -\dfrac{C_1}{6} x^{-3} + C_2 x^3\right]$$

7. Solve $(\sin x - x \cos x) \dfrac{d^2y}{dx^2} - x \sin x \dfrac{dy}{dx} + y \sin x = 0,$ given that $y = \sin x$ is a solution.

[**Ans.** $y = Ax + B \sin x$]

8. $x \dfrac{d^2y}{dx^2} - \dfrac{dy}{dx} - 4x^3 y = -4x^5,$ given that $y = e^{x^2}$ is a solution of the L.H.S. equated to zero.

$$\left[\text{Ans. } y = x^2 - \dfrac{C_1}{4} e^{-x^2} + C_2 e^{x^2}\right]$$

7.3 METHOD OF REMOVAL OF THE FIRST DERIVATIVE (REDUCTION TO NORMAL FORM)

It is not always possible to get the part of C.F. of the solution of the differential equation

$$\dfrac{d^2y}{dx^2} + P \dfrac{dy}{dx} + Qy = X$$

Hence the method given in § 7.2 cannot be used. In some cases the equation can be solved by reducing in into the normal form in which the term containing the first derivative is absent.

For this, let the general solution of the equation be $y = uv$, where u is some function of x and is not a part of its C.F. Now,

$$\dfrac{dy}{dx} = v \dfrac{du}{dx} + u \dfrac{dv}{dx} \quad \text{and} \quad \dfrac{d^2y}{dx^2} = v \dfrac{d^2u}{dx^2} + 2 \dfrac{dv}{dx} \cdot \dfrac{du}{dx} + u \dfrac{d^2v}{dx^2}$$

Putting these values in equation (1), we get

$$\left\{ v \dfrac{d^2u}{dx^2} + 2 \dfrac{dv}{dx} \cdot \dfrac{du}{dx} + u \dfrac{d^2v}{dx^2} \right\} + P \left\{ v \dfrac{du}{dx} + u \dfrac{dv}{dx} \right\} + Quv = X$$

or
$$u \dfrac{d^2v}{dx^2} + u \dfrac{dv}{dx} \left[P + \dfrac{2}{u} \dfrac{du}{dx} \right] + v \left[\dfrac{d^2u}{dx^2} + P \dfrac{du}{dx} + Qu \right] = X \qquad \text{...(2)}$$

To remove the term of first derivative, we will choose u such that

$$P + \frac{2}{u}\frac{du}{dx} = 0 \quad \text{or} \quad \frac{du}{u} = -\frac{P}{2}dx$$

Integrating this, we get

$$\log u = -\int \frac{P}{2}dx \quad \text{or} \quad u = e^{\{-\int \frac{P}{2}dx\}} \qquad ...(3)$$

Now, equation (2) becomes

$$u\frac{d^2u}{dx^2} + v\left[\frac{d^2u}{dx^2} + P\frac{du}{dx} + Qu\right] = X$$

or

$$\frac{d^2v}{dx^2} + \frac{v}{u}\left[\frac{d^2u}{dx^2} + P\frac{du}{dx} + Qu\right] = \frac{X}{u} \qquad ...(4)$$

But from above, we get

$$\frac{du}{dx} = -\frac{P}{2}u, \quad \frac{d^2u}{dx^2} = -\frac{1}{2}\left[P\frac{du}{dx} + u\frac{dP}{dx}\right]$$

$$= -\frac{1}{2}\left[P\left(-\frac{P}{2}u\right) + u\frac{dP}{dx}\right]$$

$$= \frac{1}{4}P^2 u - \frac{u}{2}\frac{dP}{dx}$$

Again putting these values in (3), we get

$$\frac{d^2u}{dx^2} + v\left[\frac{1}{4}P^2 - \frac{P^2}{2} + Q - \frac{1}{2}\frac{dP}{dx}\right] = X \cdot e^{\{\int \frac{1}{2} Pdx\}}$$

or

$$\frac{d^2v}{dx^2} + v\left[Q - \frac{1}{4}P^2 - \frac{1}{2}\frac{dP}{dx}\right] = Xe^{\{\int \frac{1}{2} Pdx\}} \qquad ...(5)$$

This reduced equation can easily be integrated.
The equation (5) is called the **normal form** of the equation (1).
Note : *Students are advised to remember equation (5), so that it may be written directly.*

EXAMPLES

1. *Solve :*

$$x\frac{d}{dx}\left(x\frac{dy}{dx} - y\right) - 2x\frac{dy}{dx} + 2y + x^2 y = 0$$

Sol. The given equation can be written as

$$\frac{d^2y}{dx^2} - \frac{2}{x}\frac{dy}{dx} + \left(1 + \frac{2}{x^2}\right)y = 0$$

Here,

$$P = -\frac{2}{x}, Q = \left(1 + \frac{2}{x^2}\right) \quad \text{and} \quad X = 0.$$

Now, if we take

Linear Differential Equations of Second Order with Variable Coefficients

$$u = e^{\left(-\frac{1}{2}\int P dx\right)} = e^{\left\{-\frac{1}{2}\int\left(-\frac{2}{x}\right)dx\right\}} = x.$$

On substituting uv for y in the original equation, it reduces to

$$\frac{d^2v}{dx^2} + v\left(Q - \frac{1}{4}P^2 - \frac{1}{2}\frac{dP}{dx}\right) = 0$$

or

$$\frac{d^2v}{dx^2} + v\left(1 + \frac{2}{x^2} - \frac{1}{x^2} - \frac{1}{x^2}\right) = 0$$

or

$$\frac{d^2v}{dx^2} + v = 0,$$

whose solution is

$$v = C_1 \cos x + C_2 \sin x.$$

Hence the complete solution is

$$y = vu = (C_1 \cos x + C_2 \sin x)\cdot x$$

2. Solve : $\dfrac{d^2y}{dx^2} - \dfrac{1}{x^{1/2}}\dfrac{dy}{dx} + \dfrac{1}{4x^2}(x + x^{1/2} - 8)\, y = 0.$

Sol. Here $P = -\dfrac{1}{x^{1/2}}, Q = \dfrac{1}{4x^2}(x + x^{1/2} - 8), X = 0.$

Now, we take

$$u = e^{\left(-\frac{1}{2}\int P dx\right)} = e^{\left(\frac{1}{2}\int x^{-1/2} dx\right)} = e^{x^{1/2}}$$

On substituting vu for y, we get

$$\frac{d^2v}{dx^2} + v\left[Q - \frac{1}{4}P^2 - \frac{1}{2}\frac{dP}{dx}\right] = 0$$

or

$$\frac{d^2v}{dx^2} + v\left[\frac{1}{4x^2}(x + x^{1/2} - 8) - \frac{1}{4}\cdot\frac{1}{x} - \frac{1}{2}\cdot\frac{1}{2}\cdot\frac{1}{x^{3/2}}\right] = 0$$

or

$$\frac{d^2v}{dx^2} - \frac{2}{x^2}v = 0 \quad \text{or} \quad x^2\frac{d^2v}{dx^2} - 2v = 0.$$

This is a homogeneous equation. Hence in order to solve, we put $x = e^z$ and putting $\dfrac{d}{dz} \equiv D$, we have

$$D(D-1)v - 2v = 0 \quad \text{or} \quad (D^2 - D - 2)v = 0$$

A.E. is $\quad m^2 - m - 2 = 0, \quad$ giving $m = 2, -1.$

$\therefore \quad v = C_1 e^{2z} + C_2 e^{-z} = C_1 x^2 + \dfrac{C_2}{x}$

Hence the complete solution will be

$$y = uv = e^{\sqrt{x}}\left(C_1 x^2 + \frac{C_2}{x}\right)$$

3. Solve $\dfrac{d^2y}{dx^2} - 2\tan x \dfrac{dy}{dx} + 5y = \sec x \cdot e^x$.

Sol. Let $y = vz$

be the solution of the given equation, when v and z are functions of x to be determined.

Then $\quad\quad\quad\quad y_1 = v_1 z + v z_1, \quad y_2 = v_2 z + 2 v_1 z_1 + v z_2$

Substituting these values of y_2, y_1 and y in the given equation, we get

$$(v_2 z + 2v_1 z_1 + v z_2) - 2\tan x\,(v_1 z + v z_1) + 5vz = e^x \sec x$$

or $\quad\quad v_2 + \left(\dfrac{2}{z} z_1 - 2\tan x\right) v_1 + \dfrac{1}{z}[z_2 - 2\tan x\, z_1 + 5z]v = \dfrac{e^x \sec x}{z}\quad\quad$ ...(2)

In order to reduce it to normal form, its second term should be removed. So, we take

$$\dfrac{2}{z} z_1 - 2\tan x = 0$$

or $\quad\quad \dfrac{dz}{z} = \tan x\, dx$

Integrating $\log z = \sec x$

or $\quad\quad\quad\quad z = \sec x \quad\quad\quad$ ...(3)

$\therefore\quad\quad z_1 = \sec x \tan x,\ z_2 = \sec x (\sec^2 x) + \tan x (\sec x \tan x)$

$$= \sec x (\sec^2 x + \tan^2 x)$$

$\therefore$ From (2), coefficient of v

$$= \dfrac{1}{z}[z_2 - 2\tan x\, z_1 + 5z]$$

$$= \dfrac{1}{\sec x}[\sec x (\sec^2 x + \tan^2 x) - 2\tan x \sec x \tan x + 5 \sec x]$$

$$= \sec^2 x + \tan^2 x - 2\tan^2 x + 5 = 6$$

$\therefore$ From (2), we get

$$v_2 + 6v = e^x$$

or $\quad\quad\quad\quad (D^2 + 6) v = e^x \quad\quad\quad$ ...(4)

Its auxilliary equation is $m^2 + 6 = 0$

or $\quad\quad\quad\quad m = \pm i\sqrt{6}$

$\therefore\quad\quad$ C.F. $= c_1 \cos(x\sqrt{6} + c_2)$

$$\text{P.I.} = \dfrac{1}{D^2 + 6} e^x = \dfrac{1}{1+6} e^x$$

$$= \dfrac{1}{7} e^x$$

Linear Differential Equations of Second Order with Variable Coefficients 601

∴ Solution of (4) is

$$v = c_1 \cos(x\sqrt{6} + c_1) + \frac{1}{7} e^x$$

Hence from (1), with the help of (3), the required complete solution of the given equation is

$$y = vz = \left[c_1 \cos(x\sqrt{6} + c_2) + \frac{1}{7} e^x\right] \sec x$$

EXERCISE

Solve the following differential equations :

1. $\dfrac{d^2y}{dx^2} + 2x \dfrac{dy}{dx} + (x^2 + 5) y = xe^{-x^2/2}$ $\quad\left[\text{Ans. } y = \left(C_1 \cos 2x + C_2 \sin 2x + \dfrac{x}{4}\right) . e^{-x^2/2}\right]$

2. $\dfrac{d^2y}{dx^2} - 4x \dfrac{dy}{dx} + (4x^2 - 3) y = e^{x^2}$ $\quad [\text{Ans. } y = (C_1 e^x + C_2 e^{-x} - 1) . e^{x^2}]$

3. $\dfrac{d}{dx}\left(\cos^2 x \dfrac{dy}{dx}\right) + \cos^2 x . y = 0$ $\quad [\text{Ans. } y = \sec x (C_1 \cos \sqrt{2}x + C_2 \sin \sqrt{2}x)]$

4. $\dfrac{d^2y}{dx^2} - \dfrac{2}{x}\dfrac{dy}{dx} + \left(a^2 + \dfrac{2}{x^2}\right) y = 0$ $\quad [\text{Ans. } y = (C_1 \cos ax + C_2 \sin ax) . x]$

5. $\left(\dfrac{d^2y}{dx^2} + y\right) \cot x + 2\left(\dfrac{dy}{dx} + y \tan x\right) = \sec x$ $\quad \left[\text{Ans. } y = \dfrac{1}{2} \sin x + (C_1 x + C_2) \cos x\right]$

6. $\dfrac{d^2y}{dx^2} + 2x \dfrac{dy}{dx} + (x^2 - 8) y = x^2 e^{-x^2/2}$

$$\left[\text{Ans. } y = \left[C_1 e^{3x} + C_2 e^{-3x} - \frac{1}{9}\left(x^2 + \frac{2}{9}\right)\right] e^{-x^2/2}\right]$$

7.4 TRANSFORMATION OF THE EQUATION BY CHANGING THE INDEPENDENT VARIABLE

Sometimes an equation may be solved by changing the independent variable by some suitable substitution.

Let the equation of second order be

$$\frac{d^2y}{dx^2} + P\frac{dy}{dx} + Qy = X \qquad \qquad ...(1)$$

Now, if we change the independent variable from x to z where z is some function of x, then

$$\frac{dy}{dx} = \frac{dy}{dz} \cdot \frac{dz}{dx}$$

and

$$\frac{d^2y}{dx^2} = \frac{d}{dx}\left(\frac{dy}{dz} \cdot \frac{dz}{dx}\right) = \frac{d}{dz}\left(\frac{dy}{dz} \cdot \frac{dz}{dx}\right) \frac{dz}{dx}$$

$$= \frac{d^2y}{dz^2}\left(\frac{dz}{dx}\right)^2 + \frac{dy}{dz} \cdot \frac{d^2z}{dx^2}$$

Putting these values in the equation (1), we get

$$\left\{\frac{d^2y}{dz^2}\left(\frac{dz}{dx}\right)^2+\frac{dy}{dz}\cdot\frac{d^2z}{dx^2}\right\}+P\left\{\frac{dy}{dz}\cdot\frac{dz}{dx}\right\}+Q.y=X$$

or

$$\frac{d^2y}{dz^2}+\left\{\frac{\frac{d^2z}{dx^2}+P\frac{dz}{dx}}{\left(\frac{dz}{dx}\right)^2}\right\}\frac{dy}{dz}+\frac{Q}{\left(\frac{dz}{dx}\right)^2}y=\frac{X}{\left(\frac{dz}{dx}\right)^2}$$

or

$$\frac{d^2z}{dz^2}+P_1\frac{dy}{dz}+Q_1y=X_1 \qquad \ldots(2)$$

where,

$$P_1=\frac{\frac{d^2z}{dx^2}+P\frac{dz}{dx}}{\left(\frac{dz}{dx}\right)^2},\ Q_1=\frac{Q}{\left(\frac{dz}{dx}\right)^2}\ \text{and}\ X_1=\frac{X}{\left(\frac{dz}{dx}\right)^2} \qquad \ldots(3)$$

P_1, Q_1, X_1 are functions of x but can be expressed in terms of z by the given relation between z and x.

Now, we can solve equation (2) in following two ways :

(i) Let we choose z in such a way that P_1 vanishes, i.e.,

$$\frac{d^2z}{dx^2}+P\frac{dz}{dx}=0$$

or

$$z=\int e^{\{-\int Pdx\}}.dx$$

then the equation (1) is changed into

$$\frac{d^2y}{dx^2}+Q_1y=X_1$$

This is integrable provided Q_1 comes out to be a constant or of the form $\left(\frac{\text{constant}}{z^2}\right)$.

(ii) We may choose z in such a way that $Q_1=\left\{\frac{Q}{(dz/dx)^2}\right\}$ is a constant (say C^2), then

$$\frac{Q}{\left(\frac{dz}{dx}\right)^2}=C^2 \quad\text{or}\quad C\frac{dz}{dx}=\sqrt{Q}$$

∴ $$C_z=\int\sqrt{(Q)}\,dx.$$

with this substitution of z in (1), we have

$$\frac{d^2y}{dx^2}+P_1\frac{dy}{dz}+C^2y=X_1$$

This differential equation can be integrated provided P_1 also comes out to be a constant.

Linear Differential Equations of Second Order with Variable Coefficients

EXAMPLES

1. *Solve* : $\cos x \dfrac{d^2y}{dx^2} + \dfrac{dy}{dx} \sin x - 2y \cos^3 x = 2 \cos^5 x.$

Sol. The given equation can be written as

$$\dfrac{d^2y}{dx^2} + \tan x \dfrac{dy}{dx} - 2\cos^2 x \cdot y = 2\cos^4 x$$

Comparing this with the standard form, we get

$$P = \tan x, \quad Q = -2\cos^2 x, \quad X = 2\cos^4 x.$$

Changing the independent variable x to z, the given equation is transformed into

$$\dfrac{d^2y}{dz^2} + P_1 \dfrac{dy}{dz} + Q_1 y + X_1 \qquad \qquad \ldots(1)$$

where

$$P_1 = \dfrac{\dfrac{d^2z}{dx^2} + P\dfrac{dz}{dx}}{\left(\dfrac{dz}{dx}\right)^2} = \dfrac{\dfrac{d^2z}{dx^2} + \tan x \cdot \dfrac{dz}{dx}}{\left(\dfrac{dz}{dx}\right)^2}$$

$$Q_1 = \dfrac{Q}{\left(\dfrac{dz}{dx}\right)^2} = -\dfrac{2\cos^2 x}{\left(\dfrac{dz}{dx}\right)^2} \quad \text{and} \quad X_1 = \dfrac{X}{\left(\dfrac{dz}{dx}\right)^2} = \dfrac{2\cos^4 x}{\left(\dfrac{dz}{dx}\right)^2}$$

Now choose z in such a way that Q_1 is equal to -2 (constant)

$$\therefore \qquad -\dfrac{2\cos^2 x}{\left(\dfrac{dz}{dx}\right)^2} = -2$$

or

$$\dfrac{dz}{dx} = \cos x,$$

which gives $z = \sin x$.

With this substitution the equation (1) is changed into

$$\dfrac{d^2y}{dz^2} + \dfrac{-\sin x + \cos x \cdot \tan x}{\cos^2 x} \dfrac{dy}{dz} + (-2)y = \dfrac{2\cos^4 x}{\cos^2 x}$$

or

$$\dfrac{d^2y}{dz^2} - 2y = 2\cos^2 x = 2(1-z^2) = 2 - 2z^2 \qquad \qquad \ldots(2)$$

A.E. is $m^2 - 2 = 0$,

giving $m = \pm\sqrt{2}$

$\therefore \qquad \text{C.F.} = C_1 e^{(\sqrt{2}z)} + C_2 e^{(-\sqrt{2}x)}$

Also, $\quad$ P.I. $= \dfrac{1}{D^2 - 2}(2 - 2z^2) = \left(1 - \dfrac{D^2}{2}\right)^{-1} \cdot (z^2 - 1)$

$$= \left(1 + \dfrac{D^2}{2} + \dfrac{D^2}{4} + ...\right)(z^2 - 1)$$

$$= z^2 - 1 + 1 = z^2.$$

Hence the completely solution is given by

$$y = C_1 e^{(\sqrt{2}z)} + C_2 e^{(-\sqrt{2}z)} + z^2$$

$$= C_1 e^{(\sqrt{2}\sin x)} + C_2 e^{(-\sqrt{2}\sin x)} + \sin^2 x$$

2. Solve :

$$\dfrac{d^2 y}{dx^2} + \left(1 - \dfrac{1}{x}\right)\dfrac{dy}{dx} + 4x^2 y e^{-2x} = 4(x^2 + x^3) e^{-3x}$$

Sol. Here,

$$P = \left(1 - \dfrac{1}{x}\right), Q = 4x^2 e^{-2x}, X = 4x^2 (1 + x) e^{-3x}$$

Changing the independent variable x to z, the given equation is transformed into

$$\dfrac{d^2 y}{dz^2} + P_1 \dfrac{dy}{dz} + Q_1 y = X_1 \qquad ...(1)$$

where $\quad P_1 = \dfrac{\dfrac{d^2 z}{dx^2} + P \dfrac{dz}{dx}}{\left(\dfrac{dz}{dx}\right)^2}, \ Q_1 = \dfrac{Q}{\left(\dfrac{dz}{dx}\right)^2} \ $ and $\ X_1 = \dfrac{X}{\left(\dfrac{dz}{dx}\right)^2}$

Now choose z in such a way that $Q_1 = 4$.

$\therefore \quad 4 = \dfrac{4x^2 e^{-2x}}{\left(\dfrac{dz}{dx}\right)^2} \quad$ or $\quad \dfrac{dz}{dx} = xe^{-x}$

so that $z = -e^{-x}(1 + x)$.

With this substitution, the equation (1) is changed into

$$\dfrac{d^2 y}{dz^2} + \dfrac{e^{-x}(1 - x) + \dfrac{x - 1}{x} \cdot xe^{-x}}{x^2 e^{-2x}} \cdot \dfrac{dy}{dz} + 4y$$

$$= 4e^{-x}(1 + x) = -4z$$

or $\qquad \dfrac{d^2 y}{dz^2} + 4y = -4z$

A.E. is $m^2 + 4 = 0$,

giving $m = \pm 2i$

$\therefore$ $\quad$ C.F. $= C_1 \cos 2z + C_2 \sin 2z$

$= C_1 \cos \{-2e^{-x}(1+x)\} + C_2 \sin \{-2e^{-x}(1+x)\}$

$= C_1 \cos \{2e^{-x}(1+x)\} - C_2 \sin \{2e^{-x}(1+x)\}$

$$\text{P.I.} = \frac{1}{D^2 + 4}(-4z) = -\left(1 + \frac{D^2}{4}\right)^{-1} z$$

$$= -\left(1 - \frac{D^2}{4} + \ldots\right) z = -z = e^{-x}(1+x).$$

Hence the completely solution is given by

$$y = \text{C.F.} + \text{P.I.} = C_1 \cos \{2e^{-x}(1+x)\} - C_2 \sin \{2e^{-x}(1+x)\} + e^{-x}(1+x).$$

EXERCISE

Solve the following differential equations :

1. $\dfrac{d^2 y}{dx^2} + (3 \sin x - \cot x) \dfrac{dy}{dx} + 2 \sin^2 x . y = e^{-\cos x} . \sin^2 x$

$\left[\textbf{Ans. } y = C_1 e^{\cos x} + C_2 e^{2\cos x} + \dfrac{1}{6} e^{-\cos x}\right]$

2. $(1+x^2)^2 \dfrac{d^2 y}{dx^2} + 2x(1+x^2) \dfrac{dy}{dx} + 4y = 0$ $\quad$ [**Ans.** $(1+x^2) y = C_1 (1 - x^2) + 2C_2 x$]

3. $\dfrac{d^2 y}{dx^2} - \cot x \dfrac{dy}{dx} - \sin^2 x . y = \cos x - \cos^3 x$ $\quad$ [**Ans.** $y = C_1 e^{-\cos x} + C_2 e^{\cos x} - \cos x$]

4. $x^6 \dfrac{d^2 y}{dx^2} + \dfrac{dy}{dx} \tan x + y \cos^2 x = 0$ $\quad \left[\textbf{Ans. } y = C_1 \cos \dfrac{a}{2x^2} - C_2 \sin \dfrac{a}{2x^2} + \dfrac{1}{a^2 x^2}\right]$

5. $\dfrac{d^2 y}{dx^2} + \dfrac{dy}{dx} \tan x + y \cos^2 x = 0$ $\quad$ [**Ans.** $y = C_1 \cos (\sin x) + C_2 \sin (\sin x)$]

7.5 OPERATIONAL METHOD

In some equations it is possible to factorise the operator.
Let the equation

$$\frac{d^2 y}{dx^2} + P \frac{dy}{dx} + Qy = X,$$

be written as

$$\{f(D)\} y = X$$

or $\quad \phi_1(D) . \phi_2(D) y = X$

which obviously means that $\phi_2(D)$ operates and y and $\phi_1(D)$ on the obtained result and we get the same result as if $f(D)$ operates upon y.

Note : *A great care is to be taken in writing the factors in the right order, as they are* **not commutative.**

The process will be clear by the following example.

EXAMPLE

1. Solve : $3x^2 \dfrac{d^2y}{dx^2} + (2 + 6x - 6x^2) \dfrac{dy}{dx} - 4y = 0.$

Sol. The given equation can be written as

$$\left(3x^2 \dfrac{d^2y}{dx^2} + 6x \dfrac{dy}{dx} + 2 \dfrac{dy}{dx}\right) - \left(6x^2 \dfrac{dy}{dx} + 4y\right) = 0$$

or $\qquad \dfrac{d}{dx}\left(3x^2 \dfrac{dy}{dx} + 2y\right) - 2\left(3x^2 \dfrac{dy}{dx} + 2y\right) = 0$

or $\qquad (D-2)(3x^2 D + 2)y = 0 \qquad \qquad \qquad ...(1)$

Now, let $(3x^2 D + 2)y = v$

Hence (1) becomes

$$(D - 2)v = 0$$

or $\qquad v = C_1 e^{2x}$

$\therefore \qquad (3x^2 D + 2)y = C_1 e^{2x}$

or $\qquad 3x^2 \dfrac{dy}{dx} + 2y = C_1 e^{2x}$

or $\qquad \dfrac{dy}{dx} + \dfrac{2}{3x^2} y = C_1 \cdot \dfrac{e^{2x}}{3x^2}$

Integrating factor of this equation is

$$e^{\left(\frac{2}{3}\int \frac{1}{x^2} dx\right)} = e^{-\frac{2}{3x}}$$

$\therefore \qquad y.e^{\left(-\frac{2}{3x}\right)} = a_2 + a_1 \int x^{-2} e^{\left(2x - \frac{2}{3x}\right)} . dx$

or $\qquad y = a_2 e^{\left(\frac{2}{3}x\right)} + a_1 e^{\left(\frac{2}{3}x\right)} \int x^{-2} e^{\left(-\frac{2}{3}x\right)} dx$

EXERCISE

Solve following differential equations by operational method :

1. $3x^2 \dfrac{d^2y}{dx^2} + (2 - 6x^2) \dfrac{dy}{dx} - 4y = 0 \qquad \left[\text{Ans. } y = C_1 e^{2x} + C_2 e^{2x} \int e^{\left(\frac{2}{3x} - 2x\right)} dx\right]$

2. $[xD^2 - (x+2)D + 2]y = x^2$ [**Ans.** $y = C_2 e^x - x^3 + C_1(x^2 + 2x - 2)$]

3. $x\dfrac{d^2y}{dx^2} + (1-x)\dfrac{dy}{dx} - y = e^x$ [**Ans.** $y = C_1 x + C_2 e^x \int e^{-x} \cdot x^{-1} dx + e^x \log x$]

7.6 METHOD OF VARIATION OF PARAMETERS

Let $y = A\phi(x) + B\psi(x)$ be the general solution of the equation

$$\dfrac{d^2y}{dx^2} + P\dfrac{dy}{dx} + Qy = 0 \qquad ...(1)$$

Now let A and B be the functions of x and let

$$y = A\phi(x) + B\psi(x) \qquad ...(2)$$

satisfy the equation

$$\dfrac{d^2y}{dx^2} + P\dfrac{dy}{dx} + Qy = X \qquad ...(3)$$

and A and B satisfy some conditions.

Differentiating (2), we get

$$\dfrac{dy}{dx} = A\phi'(x) + \phi(x)\dfrac{dA}{dx} + B\psi'(x)\dfrac{dB}{dx}$$

Let

$$\phi(x)\dfrac{dA}{dx} + \psi(x)\dfrac{dB}{dx} = 0 \qquad ...(4)$$

be the first condition which A and B satisfy. Now,

$$\dfrac{dy}{dx} = A\phi'(x) + B\psi'(x)$$

and

$$\dfrac{d^2y}{dx^2} = A\phi''(x) + B\psi''(x) + \phi'(x)\dfrac{dA}{dx} + \psi'(x)\dfrac{dB}{dx}$$

Substituting these values in (3), we get

$$A\phi''(x) + B\psi''(x) + \phi'(x)\dfrac{dA}{dx} + \psi'(x)\dfrac{dB}{dx} + P[A\phi'(x) + B\psi'(x)]$$
$$+ Q[A\phi(x) + B\psi(x)] = X \qquad ...(5)$$

Now since $y = A\phi(x) + B\psi(x)$ satisfies (1), hence

$$A\phi''(x) + A\psi''(x) + P[A\phi'(x) + B\psi'(x)] + Q[A\phi(x) + B\psi(x)] = 0$$

By putting these equation (5) reduces to

$$\phi'(x)\dfrac{dA}{dx} + \psi'(x)\dfrac{dB}{dx} = X \qquad ...(6)$$

Multiplying (4) by $\psi'(x)$ and (6) by $\psi(x)$ and subtracting, we get

$$\dfrac{dA}{dx}[\phi(x)\psi'(x) - \psi(x)\phi'(x)] = -X\psi(x)$$

$$\therefore \quad \frac{dA}{dx} = \frac{X\psi(x)}{\psi(x)\phi'(x) - \phi(x)\psi'(x)}$$

$$\therefore \quad A = \int \frac{X \cdot \psi(x)}{\psi(x)\phi'(x) - \phi(x)\psi'(x)} dx + C_1$$

Similarly B can be determined.

Hence putting the values of A and B, the solution will be obtained.

Since in this method the general solution is obtained by varying the arbitrary constants of the complementary function, hence this method is called as *method of variation of parameters*.

Note : *The method of variation of parameters must be used only if instructed to do so.*

EXAMPLES

1. *Solve the method of variation of parameters, the equation*

$$\frac{d^2 y}{dx^2} + a^2 y = \sec ax$$

Sol. The solution of the equation

$$\frac{d^2 y}{dx^2} + a^2 y = 0$$

is, $\qquad y = A \cos ax + B \sin ax$ ...(1)

Now, let A and B be the functions of x and let (1) satisfy the given equation.

$$\therefore \quad \frac{dy}{dx} = -A \cdot a \sin ax + B \cdot a \cos ax + \cos ax \frac{dA}{dx} + \sin ax \frac{dB}{dx}$$

Now, choose A and B such that

$$\cos ax \frac{dA}{dx} + \sin ax \frac{dB}{dx} = 0 \qquad \ldots(2)$$

then $\qquad \dfrac{dy}{dx} = -A \cdot a \sin ax + B \cdot a \cos ax$

Now, $\qquad \dfrac{d^2 y}{dx^2} = -Aa^2 \cos ax - Ba^2 \sin ax - \dfrac{dA}{dx} \cdot a \sin ax + \dfrac{dB}{dx} \cdot a \cos ax$

If (1) satisfies the given equation, when A and B are functions of x, we have

$$-\frac{dA}{dx} \cdot a \sin ax + a \cos ax \frac{dB}{dx} = \sec ax \qquad \ldots(3)$$

Solving (2) and (3), we get

$$a \frac{dB}{dx} = 1, \quad \text{and} \quad a \frac{dA}{dx} = -\tan ax$$

$$\therefore \quad A = \frac{1}{a^2} \log \cos ax + C_1 \quad \text{and} \quad B = \frac{x}{a} + C_2$$

Hence the solution of the given equation is

$$y = C_1 \cos ax + C_2 \sin ax + \frac{1}{a^2} (\log \cos ax) \cos ax + \frac{x}{a} \sin ax.$$

Linear Differential Equations of Second Order with Variable Coefficients

2. *Solve by the method of variation of parameters :*

$$(1-x^2)\frac{d^2y}{dx^2} - 4x\frac{dy}{dx} - (1+x^2)y = x.$$

Sol. Here we have,

$$P = \frac{4x}{(1-x^2)}, \quad Q = -\frac{(1+x^2)}{(1-x^2)} \quad \text{and} \quad X = \frac{x}{(1-x^2)}$$

Put
$$y = v \cdot e^{\left(-\frac{1}{2}\int P\,dx\right)} = v e^{\left\{\int 2x/(1-x^2)\,dx\right\}}$$

$$= \frac{v}{(1-x^2)}$$

Then the given relation reduces to

$$\frac{d^2v}{dx^2} + v = x \qquad \qquad \ldots(1)$$

The solution of the equation

$$\frac{d^2v}{dx^2} + v = 0$$

is $\qquad v = A\cos x + B\sin x \qquad \ldots(2)$

If A and B are functions of x,

$$\frac{dv}{dx} = -A\sin x + B\cos x + \cos x \frac{dA}{dx} + \sin x \frac{dB}{dx}$$

Let us choose A and B such that

$$\cos x \frac{dA}{dx} + \sin x \frac{dB}{dx} = 0 \qquad \ldots(3)$$

Then
$$\frac{dv}{dx} = -A\sin x + B\cos x$$

and
$$\frac{d^2v}{dx^2} = -A\cos x - B\sin x - \sin x \frac{dA}{dx} + \cos x \frac{dB}{dx}$$

If (2) satisfies the given equation, then substituting these values in (1), we get

$$-\sin x \frac{dA}{dx} + \cos x \frac{dB}{dx} = x \qquad \ldots(4)$$

Solving (3) and (4), we get

$$\frac{dB}{dx} = x\cos x$$

i.e., $\qquad B = x\sin x + \cos x + C_1$

$$\frac{dA}{dx} = -x\sin x$$

i.e., $\qquad A = x\cos x - \sin x + C_2$

$\therefore \qquad y = \dfrac{v}{1-x^2}[(x\cos x - \sin x + C_2)\cos x + (x\sin x + \cos x + C_1)\sin x]$

3. Apply the method of variation of parameters to solve $\dfrac{d^2 y}{dx^2} - y = \dfrac{2}{1+e^x}$.

Sol. The C.F., i.e., solution of $\dfrac{d^2 y}{dx^2} - y = 0$ is

$$y = c_1 e^x + c_2 e^{-x}$$

Now let $y = A e^x + B e^{-x}$,

where A and B are functions of x, be a solution of the given equation.

Then $\dfrac{dy}{dx} = A e^x - B e^{-x} + A_1 e^x + B_1 e^{-x}$.

Choose A, B, such that $\qquad A_1 e^x + B_1 e^{-x} = 0 \qquad$ ...(1)

Then $\dfrac{dy}{dx} = A e^x - B e^{-x}$.

and $\dfrac{d^2 y}{dx^2} = A e^x + B e^{-x} + A_1 e^x - B_1 e^{-x} = y + A_1 e^x - B_1 e^{-x}$

Putting these values in given equation, we get

$$(y + A_1 e^x - B_1 e^{-x}) - y = \dfrac{2}{(1+e^x)}$$

or $\qquad A_1 e^x - B_1 e^{-x} = \dfrac{1}{(1+e^x)}$

From (1), $B e^{-x} = - A e^x$, $\therefore$ (2) gives

$$2 A_1 e^x = \dfrac{2}{1+e^x} \quad \text{or} \quad A_1 = \dfrac{e^{-x}}{1+e^x}$$

Also, $B_1 = -\dfrac{e^x}{1+e^x}$, so that $B = -\int \dfrac{e^x}{1+e^x} = -\log(1+e^x) + c_2$

and $\quad A = \int A_1 \, dx = \int \dfrac{e^{-x}}{1+e^x} = \int \dfrac{1}{z^2 (1+z)} \, dz$, where $z = e^x$

$= \int \left(\dfrac{1}{z^2} - \dfrac{1}{z} + \dfrac{1}{1+z} \right) dz = -\dfrac{1}{z} - \log z + \log(1+z) + c_1$

$= \log \dfrac{1+z}{z} - \dfrac{1}{z} + c_1 = \log \dfrac{1+e^x}{e^x} - e^{-x} + c_1$.

Hence the complete solution is $y = A e^x + B e^{-x}$.

$$= e^x \log \dfrac{1+e^x}{e^x} - 1 + c e^x - e^{-x} \log(1+e^x) + c_2 e^{-x}$$

EXERCISE

Solve following differential equations by the method of variation of parameters :

1. $\dfrac{d^2y}{dx^2} + y = \operatorname{cosec} x.$ [Ans. $y = (C_1 - x) \cos x + (C_2 + \log \sin x) \sin x$]

2. $\dfrac{d^3y}{dx^3} - 6\dfrac{d^2y}{dx^2} + 11\dfrac{dy}{dx} - 6y = e^{2x}$ [Ans. $y = C_1 e^x + (C_2 - x) e^{2x} + C_3 e^{3x}$]

3. $x^2 \dfrac{d^2y}{dx^2} - 2x(1+x)\dfrac{dy}{dx} + 2(x+1) y = x^3$ $\left[\text{Ans. } y = -\dfrac{1}{4} x - \dfrac{1}{2} x^2 + C_1 x + C_2 x e^{2x}\right]$

4. $\dfrac{d^2y}{dx^2} + y = x$ [Ans. $y = c_1 \cos x + c_2 \sin x + x$]

5. $(1-x)\dfrac{d^2y}{dx^2} + x\dfrac{dy}{dx} - y = (1-x)^2$ [Ans. $y = C_1 e^x + C_2 x + (x^2 + x + 1)$]

7.7 SELECTION OF AN APPROPRIATE METHOD TO SOLVE LINEAR DIFFERENTIAL EQUATIONS OF SECOND ORDER

The solution of a linear differential equation of second order becomes easier if we hav a proper guideline to select the appropriate method to solve that particular equation. The following guidelines will really help the students to solve the equation :

1. First, put the differential equation in the standard form

$$\dfrac{d^2y}{dx^2} + P\dfrac{dy}{dx} + Qy = X$$

2. Then try to find by inspection an integral belonging to the complementary function of the given differential equation. If it is found then solve it as in § 7.2.

3. If a part of the C.F. cannot be found by inspection, then find the value of $Q - \dfrac{1}{2}\dfrac{dP}{dx} - \dfrac{1}{4} P^2$.

   If it is a constant or of the form $\dfrac{\text{constant}}{x^2}$, then the normal form of the equation is easily integrable and proceed as in § 7.3

4. If the methods (2) and (3) both fail, the method of change of variable should be tried. In this method choose z such that $\dfrac{Q}{(dz/dx)} = $ a suitable constant.

5. In some cases the method of operational factors may be useful in solving the equation easily.

6. The method of variation of parameters must be used only when it is instructed to use the same.

8
Simultaneous Ordinary Differential Equations

8.1 SIMULTANEOUS DIFFERENTIAL EQUATIONS

So far we hae been considering equations containing two variables. In this chapter we shall discuss the case where there are as many simultaneous equations as there are dependent variables (all the equations being linear).

Let
$$f_1(D) x + f_2(D) y = X_1 \qquad \ldots(1)$$
$$\phi_1(D) x + \phi_2(D) y = X_2 \qquad \ldots(2)$$

be the equations in which $f_1(D), \phi_1(D), f_2(D)$ and $\phi_2(D)$ are all rational integral functions of D with constant coefficients. Also D denotes the operator $\dfrac{d}{dt}$, and X_1, X_2 the functions of the independnet variable t.

Now operating both sides of (1) by $\phi_2(D)$, both sides of (2) by $f_2(D)$ and subtracting, we get

$$[f_1(D) \phi_2(D) - \phi_1(D) f_2(D)] x = \phi_2(D) \cdot f_2(D) X_2 \qquad \ldots(3)$$

This equation (3) can be solved by the methods considered in previous chapters. The solution will contain as many mutually independent arbitrary constants as the highest index of D in (3), *i.e.*, equal to the order of the equations.

Now let the solution be
$$x = \phi(t, c_1, c_2) \qquad \ldots(4)$$

Substitute this value of x in either (1) or (2), and solve it for y.

Again this value of y can also be found by eliminating x between (1) and (2) and integrating the resulting equation; suppose the solution is $y = \psi(t, c_1', c_2' \ldots)$.

The relation between the two sets of arbitrary constants can be found by substituting the values of x and y so obtained in either of the equations (1) and (2) and then equating the coefficients of eveyr different function of t on the two sides of the resulting identity.

EXAMPLES

1. Solve
$$\dfrac{dx}{dt} - 7x + y = 0, \qquad \ldots(1)$$
$$\dfrac{dy}{dt} - 2x - 5y = 0. \qquad \ldots(2)$$

Sol. Here the given equations can be written as
$$(D - 7) x + y = 0. \qquad \ldots(3)$$
or
$$(D - 5) y - 2x = 0; \qquad \ldots(4)$$

Eliminating y, we get
$$(D - 7).(D - 5) x + 2x = 0,$$

i.e.,
$$\dfrac{d^2 x}{dt^2} - 12 \dfrac{dx}{dt} + 37x = 0.$$

Simultaneous Ordinary Differential Equations

∴ A.E. is $\quad m^2 - 12m + 37 = 0$

or $\quad m = [12 \pm \sqrt{(144-148)}]/2$

$\quad = 6 \pm i;$

∴ $\quad x = e^{6t}(A\cos t + B\sin t).$

Putting this value of x in (1).

$$e^{6t}(-A\cos t + B\cos t) + 6e^{6t}(A\cos t + B\sin t)$$
$$- 7e^{6t}(A\cos t + B\sin t) + y = 0$$

or $\quad y = e^{6t}[(A-B)\cos t + (A+B)\sin t].$

So the solution is $\quad x = e^{6t}(A\cos t + B\sin t),$

$\quad y = e^{6t}[(A-B)\cos t + (A+B)\sin t].$

2. Solve $\quad \dfrac{dx}{dt} + 4x + 3y = t,$ ...(1)

$\quad \dfrac{dy}{dt} + 2x + 5y = e^t.$ ...(2)

Sol. Writing the given equation as

$\quad (D+4)x + 2y = t$ ...(3)

$\quad (D+5)y + 2x = e^t$ ...(4)

Eliminating y from (3) and (4),

$\quad (D+4)(D+5)x - 6x = (D+5)t - 3e^t.$

$\quad (D^2 + 9D + 20)x - 6x = 1 + 5t - 3e^t.$

or $\quad (D^2 + 9D + 14)x = 1 + 5t - 3e^t.$

A.E. is $\quad m^2 + 9m + 14 = 0$

or $\quad (m+7)(m+2) = 0.$

∴ C.F. is $c_1 e^{-7t} + c_2 e^{-2t}.$

Also $\quad$ P.I. $= \dfrac{1}{14 + 9D + D^2}(1+5t) - \dfrac{1}{14 + 9D + D^2} 3e^t$

$\quad = \dfrac{1}{14\left(1 + \dfrac{9D}{14} + \dfrac{D^2}{14}\right)}(1+5t) - \dfrac{3e^t}{24}$

$\quad = \dfrac{1}{14}\left(1 - \dfrac{9D}{14} - \dfrac{D^2}{14} \ldots\right)(1+5t) - \dfrac{3e^t}{24}.$

$\quad = \dfrac{1}{14}\left(1 + 5t - \dfrac{45}{14}\right) - \dfrac{3e^t}{24}.$

∴ $x = c_1 e^{-7t} + c_2 e^{-2t} + \dfrac{5t}{14} - \dfrac{31}{196} - \dfrac{e^t}{8}$

$$\therefore \quad \frac{dx}{dt} = -7c_1 e^{-7t} - 2c_2 e^{-2t} + \frac{5}{14} - \frac{e^t}{8}.$$

Putting these values in (1),

$$-7c_1 e^{-7t} - 2c_2 e^{-2t} + \frac{5}{14} - \frac{e^t}{8} + 4c_1 e^{-7t} + 4c_2 e^{-2t} + \frac{10}{7} t - \frac{31}{49} - \frac{e^t}{2} + 3y = t.$$

$$\therefore \quad y = \frac{1}{3}\left[3c_1 e^{-7t} - 2c_2 e^{-2t} - \frac{3t}{7} + \frac{5}{8} e^t + \frac{27}{98} \right]$$

and $\quad x = c_1 e^{-7t} + c_2 e^{-2t} + \frac{15}{14} t - \frac{31}{196} - \frac{1}{8} e^t.$

3. Solve

$$\frac{d^2 x}{dt^2} + m^2 y = 0, \qquad \ldots(1)$$

$$\frac{d^2 y}{dt^2} - m^2 x = 0. \qquad \ldots(2)$$

Sol. Eliminating y, we get

$$D^4 x + m^4 x = 0$$

or $\quad (D^4 + m^4) x = 0.$

A.E. is $\quad M^4 + m^4 = 0$

$$(M^2 + m^2) - 2 M^2 m^2 = 0$$

or $\quad (M^2 + m^2 - \sqrt{2} Mm)(M^2 + m^2 + \sqrt{2} Mm) = 0$

$\therefore$ when $\quad M^2 - \sqrt{2} Mm + m^2 = 0,$

$$M = \frac{\sqrt{2} m \pm \sqrt{(2m^2 - 4m^2)}}{2}$$

$$= \frac{m}{\sqrt{2}} \pm \frac{m}{\sqrt{2}} i \quad \text{and so on.}$$

$$\therefore \quad x = \exp\left(\frac{mt}{\sqrt{2}}\right)\left[A \cos\frac{mt}{\sqrt{2}} + B \sin\frac{mt}{\sqrt{2}} \right] + \exp\left(-\frac{mt}{\sqrt{2}}\right)\left[C \cos\frac{mt}{\sqrt{2}} + D \sin\frac{mt}{\sqrt{2}} \right].$$

$$\therefore \quad \frac{dx}{dt} = \exp\left(\frac{m}{\sqrt{2}}\right)\left[-A\frac{m}{\sqrt{2}} \sin\frac{mt}{\sqrt{2}} + B\frac{m}{\sqrt{2}} \cos\frac{mt}{\sqrt{2}} \right] + \frac{m}{\sqrt{2}} \exp\left(\frac{mt}{\sqrt{2}}\right)\left[A \cos\frac{mt}{\sqrt{2}} + B \sin\frac{mt}{\sqrt{2}} \right]$$

$$+ \exp\left(-\frac{mt}{\sqrt{2}}\right)\left[-C\frac{mt}{\sqrt{2}} \sin\frac{mt}{\sqrt{2}} + \frac{Dm}{\sqrt{2}} \cos\frac{mt}{\sqrt{2}} \right] - \frac{m}{\sqrt{2}} \exp\left(-\frac{mt}{\sqrt{2}}\right)\left[C \cos\frac{mt}{\sqrt{2}} + D \sin\frac{mt}{\sqrt{2}} \right]$$

$$= \frac{m}{\sqrt{2}} \exp\left(\frac{mt}{\sqrt{2}}\right)\left[(A+B) \cos\frac{mt}{\sqrt{2}} + (B-A) \sin\frac{mt}{\sqrt{2}} \right]$$

Simultaneous Ordinary Differential Equations

$$-\frac{m}{\sqrt{2}}\exp\left(-\frac{mt}{\sqrt{2}}\right)\left[(C-D)\cos\frac{mt}{\sqrt{2}}+(C+D)\sin\frac{mt}{\sqrt{2}}\right]$$

$$\therefore \quad y = -\frac{1}{m^2}\frac{d^2x}{dt^2} = \frac{1}{m^2}\left[\frac{m^2}{2}\exp\left(\frac{mt}{\sqrt{2}}\right)\left\{2B\cos\frac{mt}{\sqrt{2}}-2A\sin\frac{mt}{\sqrt{2}}\right\}\right]$$

$$-\frac{1}{m^2}\left[-\frac{m^2}{2}\exp\left(-\frac{mt}{\sqrt{2}}\right)\left\{-2D\cos\frac{mt}{\sqrt{2}}+2C\sin\frac{mt}{\sqrt{2}}\right\}\right]$$

$$=\exp\left(\frac{mt}{\sqrt{2}}\right)\left[-B\cos\frac{mt}{\sqrt{2}}+A\sin\frac{mt}{\sqrt{2}}\right]+\exp\left(-\frac{mt}{\sqrt{2}}\right)\left[-C\sin\frac{mt}{\sqrt{2}}+D\cos\frac{mt}{\sqrt{2}}\right]$$

and $\quad x = \exp\left(\frac{mt}{\sqrt{2}}\right)\left[A\cos\frac{mt}{\sqrt{2}}+B\sin\frac{mt}{\sqrt{2}}\right]+\exp\left(-\frac{mt}{\sqrt{2}}\right)\left[C\cos\frac{mt}{\sqrt{2}}+D\sin\frac{mt}{\sqrt{2}}\right]$

4. Solve
$$\frac{dx}{dt}+\frac{dy}{dt}-2y = 2\cos t - 7\sin t.$$
$$\frac{dx}{dt}-\frac{dy}{dt}+2x = 4\cos t - 3\sin t.$$

Sol. Denoting d/dt by D, the given equation can be written as
$$Dx + (D-2)y = 2\cos t - 7\sin t.$$
$$(D+2)x - Dy = 4\cos t - 3\sin t.$$

Eliminating y, we have
$$D^2 x + (D-2)(D+2)x.$$
$$= D[2\cos t - 7\sin t] + (D-2)(4\cos t - 3\sin t)$$
$$(2D^2 - 4)x = -2\sin t - 7\cos t - 4\sin t - 3\cos t - 8\cos t + 6\sin t$$

or $\quad (D^2 - 2)x = -9\cos t.$

$\therefore \quad$ C.F. $= c_1 e^{\sqrt{2}t} + c_2 e^{-\sqrt{2}t}$

P.I. $= \dfrac{-9\cos t}{D^2 - 2} = \dfrac{-9\cos t}{-3} = 3\cos t$

$\therefore \quad x = c_1 e^{\sqrt{2}t} + c_2 e^{\sqrt{2}t} + 3\cos t$

and $\quad \dfrac{dx}{dt} = c_1\sqrt{2}e^{-\sqrt{2}t} - c_2\sqrt{2}e^{-\sqrt{2}t} - 3\sin t$

Also adding the given equations, we have
$$2\frac{dx}{dt} - 2y + 2x = 6\cos t - 10\sin t$$

or $\quad y = \dfrac{dx}{dt} + x - 3\cos t + 5\sin t$

$\quad = c e^{\sqrt{2}t}(\sqrt{2}+1) + c_2 e^{-\sqrt{2}t}(1-\sqrt{2}) + 2\sin t$

EXERCISES

Solve:

1. $\dfrac{d^2x}{dt^2} - \dfrac{dy}{dt} = 2x + 2t, \quad \dfrac{dx}{dt} + 4\dfrac{dy}{dt} = 3y.$

[Ans. $x = Ae^{-3/2t} + (B + ct)e^t - t, \quad y = -\dfrac{A}{6}e^{-3t/2} + e^t(3c - B - ct) - \dfrac{1}{3}$]

2. $\dfrac{dx}{dt} = 3x + 2y, \quad \dfrac{dy}{dt} = 5x + 3y,$

[Ans. $x = c_1 e^{(3+\sqrt{10})t} + c_2 e^{(3-\sqrt{10})t}; \quad y = \dfrac{1}{2}\sqrt{10}\,[c_1 e^{(3+\sqrt{10})t} - c_2 e^{(3-\sqrt{10})t}]$]

3. $\dfrac{dx}{dt} = x - 2y, \quad \dfrac{dy}{dt} = 5x + 3y$

[Ans. $x = e^{2t}(c_1 \cos 3t + c_2 \sin 3t); \quad y = \dfrac{1}{2}[e^{2t}\cos 3t\,(c_1 + 3c_2) - e^{2t}\sin 3t\,(3c_1 - c_2)]$]

4. $\dfrac{d^2x}{dt^2} - 2\dfrac{dy}{dt} - x = e^t \cos t, \quad \dfrac{d^2y}{dt^2} + 2\dfrac{dx}{dt} - y = e^t \sin t.$

[Ans. $x = (a + bt)\cos t + (c + dt)\sin t + \dfrac{e^t}{25}(4\sin t - 3\cos t)$

$y = -(a + bt)\sin t + (c + dt)\cos t - \dfrac{e^t}{25}(3\sin t + 4\cos t).$]

5. $\dfrac{d^2x}{dt^2} + 4x + y = te^{3t}, \quad \dfrac{d^2y}{dt^2} + y - 2x = \cos^2 t.$

[Ans. $x = c_1 \cos\sqrt{3}t + c_2 \sin\sqrt{3}t + c_3 \cos\sqrt{2}t + c_4 \sin\sqrt{2}t$

$-\dfrac{49}{1452}e^{3t} + \dfrac{5}{66}te^{3t} - \dfrac{1}{12} - \dfrac{1}{4}\cos 2t;$

$y = -3c_1 \cos\sqrt{3}t - c_2 \sin\sqrt{3}t - 2c_3 \cos\sqrt{2}t - 2c_4 \sin\sqrt{2}t + \dfrac{1}{66}te^{3t} + \dfrac{1}{3} - \dfrac{23}{2452}e^{3t}.$]

6. $2\dfrac{d^2y}{dx^2} - \dfrac{dz}{dx} - 4y = 2x, \quad \dfrac{dy}{dx} + 4\dfrac{dz}{dx} - 3z = 0.$

[Ans. $y = (c_1 + c_2 x)e^x + 3c_2 e^{-3x/2} - \dfrac{x}{2}; \quad z = 2(3c_2 - c_1 - c_2 x)e^x = c_3 e^{-\frac{3}{2}x} - \dfrac{1}{3}$]

8.2 THE EQUATION OF THE TYPE

$$P_1\, dx + Q_1\, dy + R_1\, dz = 0 \qquad \ldots(1)$$
$$P_2\, dx + Q_2\, dy + R_2\, dz = 0$$

when P_1, P_2 etc. are functions of x, y, z. We can write these equations as

$$P_1 \dfrac{dx}{dz} + Q_1 \dfrac{dy}{dz} + R_1 = 0,$$

$$P_2 \dfrac{dx}{dz} + Q_2 \dfrac{dy}{dz} + R_2 = 0.$$

Simultaneous Ordinary Differential Equations

Solving them for $\dfrac{dx}{dz}, \dfrac{dy}{dz}$

$$\dfrac{dx}{dz} = \dfrac{Q_1 R_2 - Q_2 R_1}{P_1 Q_2 - Q_1 P_2} ; \dfrac{dy}{dz} = \dfrac{P_1 R_2 - P_1 R_2}{P_1 Q_2 - Q_1 P_2}$$

whence
$$\dfrac{dx}{Q_1 R_2 - Q_2 R_1} = \dfrac{dy}{R_1 P_2 - R_2 P_1} = \dfrac{dz}{P_1 Q_2 - P_2 Q_1} \qquad \text{...(2)}$$

Hence forth the equations (2) will be taken as the standard form of a pair of ordinary simultaneous equations of the first order and of the first degree.

8.3 SOLUTION OF $\dfrac{dx}{P} = \dfrac{dy}{Q} = \dfrac{dz}{R}$.

We have
$$\dfrac{dy}{P} = \dfrac{dy}{Q} = \dfrac{dz}{R}$$

$$= \dfrac{l\,dx + m\,dy + n\,dz}{lP + mQ + nR}$$

and if $\quad lP + mQ + nR = 0$

then $\quad l\,dx + m\,dy + n\,dz = 0$,

and if it is an exact differential, say du, then $u = a$ is one equation of the complete solution.

Similarly choosing l', m', n' such that

$$l'P + m'Q + n'R = 0$$

then $\quad l'\,dx + m'\,dy + n'\,dz = dv = 0$

whence $v = b$ is another equation of the complete solution.

Note. Also to find a solution we shoose l, m, n such that $l\,dx + m\,dy + n\,dz$ is differential of $lP + mQ + nR$.

8.4 THE GENERAL EXPRESSION OF INTEGRALS

If $l\,dx + m\,dy + n\,dz$ is an exact differential du, and since

$$du = \dfrac{\partial u}{\partial x} dx + \dfrac{\partial u}{\partial y} dy + \dfrac{\partial u}{\partial z} dz,$$

we have
$$\dfrac{l}{\partial u / \partial x} = \dfrac{m}{\partial u / \partial y} = \dfrac{n}{\partial u / \partial z}.$$

$\therefore \quad lP + mQ + nR = 0$ can be written as

$$P \dfrac{\partial u}{\partial x} + Q \dfrac{\partial u}{\partial y} + R \dfrac{\partial u}{\partial z} = 0 \qquad \text{...(1)}$$

Hence if $u = a$ is integral of $\dfrac{dx}{P} = \dfrac{dy}{Q} = \dfrac{dz}{R}$, then $u = a$ also satisfies (1).

Conversely. Suppose that $u = a$ satisfies

$$P \dfrac{\partial u}{\partial x} + Q \dfrac{\partial u}{\partial y} + R \dfrac{\partial u}{\partial z} = 0.$$

Also
$$\dfrac{dx}{P} = \dfrac{dy}{Q} = \dfrac{dz}{R}$$

$$= \frac{\frac{\partial u}{\partial x} dx + \frac{\partial u}{\partial y} dy + \frac{\partial u}{\partial z} dz}{P \frac{\partial u}{\partial x} + Q \frac{\partial u}{\partial y} + R \frac{\partial u}{\partial z}}$$

and since the denominator is zero, the numerator is

$$\frac{\partial u}{\partial x} dx + \frac{\partial u}{\partial y} dy + \frac{\partial u}{\partial z} dz = 0$$

This is total differential of u; so $u = a$ is an integral of the given equation.

Hence if $u = a, v = b$ are the two independent integrals of the given equations, then $\phi(u, v) = \phi(a, b) = c$ or $\phi(u, v) = 0$, where ϕ is an arbitrary function, is a solution of (1). This can be verified directly. So

$$\phi(u, v) = 0 \quad \text{or} \quad u = \phi(v)$$

is the general expression for the integrals of these equations.

Note. $\phi(u, v) = 0$ is equally generally since the constants can be involved in the arbitrary function.

8.5 GEOMETRICAL MEANING OF $\frac{dx}{P} = \frac{dy}{Q} = \frac{dz}{R}$

We know the direction ratios of the tangent to a curve at any point (x, y, z) on it are proportional to dx, dy, dz at that point. Hence geometrically the given equations represent a systems of curves in space, such that the directions ratios of the tangent to any one of these curves in space, at a point (x, y, z) on it are proportioanl to P, Q and R at that point. If $u = a, v = b$ are the general solutions of $\frac{dx}{P} = \frac{dy}{Q} = \frac{dz}{R}$, then system of curves must be the curve of intersection of the surfaces $u = a, v = b$. It is also clear that since a, b are arbitrary constants, the system of curves represented by the equations is doubly infinite.

EXAMPLES

1. Solve $\frac{dx}{x^2 + y^2} = \frac{dy}{2xy} = \frac{dz}{(x+y)z}$.

Sol. Clearly $\frac{dx + dy}{(x+y)^2} = \frac{dz}{(x+y)z}$ or $dx - dy = dz$

i.e., $x - y - z = c_1$.

Also $\frac{x\,dx - y\,dy}{x^3 - xy^2 - x^2y + y^3} = \frac{dz}{z(x-y)}$

or $\frac{x\,dx - y\,dy}{x^2(x-y) - y^2(x-y)} = \frac{dz}{z(x-y)}$ or $\frac{x\,dx - y\,dy}{x^2 - y^2} = \frac{dz}{z}$

or $\frac{2x\,dx - 2y\,dy}{x^2 - y^2} = \frac{2dz}{z}$

$\therefore \quad \log(x^2 - y^2) = 2 \log z + \log c_2$

$\therefore \quad (x^2 - y^2) = c_2 z^2$.

Hence the general solution is

$$\phi\left(x - y - z, \frac{x^2 - y^2}{z^2}\right) = 0.$$

2. Solve $\quad \dfrac{dx}{y+z} = \dfrac{dy}{z+x} = \dfrac{dz}{x+y}$

Sol. The given equation is

$$\dfrac{dx}{y+z} = \dfrac{dy}{z+x} = \dfrac{dz}{x+y} \qquad \ldots(i)$$

Choosing $1, -1$ and $0, 1, -1$ as multipliers each fraction

$$\dfrac{dx - dy}{(y+z) - (z+x)} \text{ and } \dfrac{dy - dz}{(z+x) - (x+y)} \qquad \ldots(ii)$$

$\therefore \qquad \dfrac{dx - dy}{-(x-y)} = \dfrac{dy - dz}{-(y-z)}$

or $\qquad \dfrac{dx - dy}{x - y} = \dfrac{dy - dz}{-y - z}$

Integrating $\log(x-y) - \log(y-z) = \log c_1$

or $\qquad \dfrac{x-y}{y-z} = c_1 \qquad \ldots(iii)$

Choosing $1, 1.1$ as multipliers, each fraction in (i)

$$\dfrac{dx + dy + dz}{2(x+y+z)} \qquad \ldots(iv)$$

Combining the first fraction in (ii) with fraction (iv)

$$\dfrac{dx - dy}{-(x-y)} = \dfrac{dx + dy + dz}{2(x+y+z)}$$

or $\qquad \dfrac{dx - dy}{x - y} + \dfrac{dx + dy + dz}{2(x+y+z)} = 0$

Integrating, $\quad \log(x-y) + \dfrac{1}{2}\log(x+y+z) = \log c_2$

or $\qquad (x-y)(x+y+z)^{1/2} = c_2$

$\therefore$ The required solution is given by (iii) and (v).

3. Solve $\quad \dfrac{dx}{1} = \dfrac{dy}{-2} = \dfrac{dz}{3x^2 \sin(y+2x)}.$

Sol. From first two members,

$$2x + y = c_1 \qquad \ldots(1)$$

Taking the first and the last members,

$$\dfrac{dx}{1} = \dfrac{dz}{3x^2 \sin(y+2x)} = \dfrac{dz}{3x^2 \sin c_1}, \quad \text{using (1)}$$

$\therefore \qquad dz = 3x^2 \sin c_1 \, dx$

Integrating, $\quad dz = 3x^2 \sin c_1 \, dx$

or $\qquad z - x^3 \sin(y+2x) = c_2 \qquad \ldots(2)$

The required solution consists of the equations (1) and (2).

4. Solve $\dfrac{dx}{\cos(x+y)} = \dfrac{dy}{\sin(x+y)} = \dfrac{dz}{z}$

Sol. From the given equations, we have

$$\dfrac{dx+dy}{\cos(x+y)+\sin(x+y)} = \dfrac{dx-dy}{\cos(x+y)-\sin(x+y)} = \dfrac{dz}{z}$$

Taking the first two members, we get

$$\dfrac{\cos(x+y)-\sin(x+y)}{\cos(x+y)+\sin(x+y)}(dx+dy) = dx-dy$$

Integrating,

$$\log[\cos(x+y)+\sin(x+y)] = x - y + \log c_1$$

or $\quad [\cos(x+y)+\sin(x+y)]e^{y-x} = c_1$ ...(1)

Again taking the first and third members, we get

$$\dfrac{dx+dy}{\cos(x+y)+\sin(x+y)} = \dfrac{dz}{z}$$

or $\quad \dfrac{dz}{z} = \dfrac{(1/\sqrt{2})(dx+dy)}{\sin(x+y+\pi/4)}$

or $\quad \sqrt{2}\,\dfrac{dz}{z} = \operatorname{cosec}\left(x+y+\dfrac{\pi}{4}\right)(dx+dy)$

Integrating,

$$\sqrt{2}\log z = \log \tan \dfrac{1}{2}\left(x+y+\dfrac{1}{4}\pi\right) + \log c_2$$

or $\quad z^{\sqrt{2}} \cot\left(\dfrac{x}{2}+\dfrac{y}{2}+\dfrac{\pi}{8}\right) = c_2$ ...(2)

The required solution consists of the equation (1) and (2).

EXERCISES

Solve following differential equations :

1. $\dfrac{dx}{y} = \dfrac{dy}{x} = \dfrac{dz}{xyz^2(x^2-y^2)}$ $\quad\left[\text{Ans. } \phi\left[\dfrac{2+x^2z(x^2-y^2)}{z}, x^2-y^2\right]=0\right]$

2. $\dfrac{dx}{y} = \dfrac{dy}{zx} = \dfrac{dz}{xy}$ $\quad\left[\text{Ans. } \phi[x^2-y^2, x^2-z^2]=0\right]$

3. $\dfrac{dx}{x^2-y^2-z^2} = \dfrac{dy}{2xy} = \dfrac{dz}{2zx}$ $\quad\left[\text{Ans. } \phi\left[\dfrac{x^2+y^2+z^2}{z}, \dfrac{y}{z}\right]=0\right]$

4. $\dfrac{dx}{mz-my} = \dfrac{dy}{nx-lz} = \dfrac{dz}{ly-mx}$ $\quad\left[\text{Ans. } \phi(x^2+y^2+z^2, lx+my+nz)=0\right]$

5. $\dfrac{dx}{y-zx} = \dfrac{dy}{yz+x} = \dfrac{dz}{x^2+y^2}$ $\quad\left[\text{Ans. } \phi(x^2-y^2+z^2, xy-z)=0\right]$

6. $\dfrac{dx}{x} = \dfrac{dy}{z} = \dfrac{dz}{z-a\sqrt{(x^2+y^2+z^2)}}$ $\quad\left[\text{Ans. } \phi\left[x/y, y^{1-a}/\{\sqrt{(x^2+y^2+z^2)}\,z\}\right]=0\right]$

Simultaneous Ordinary Differential Equations

7. $\dfrac{dx}{1} = \dfrac{dy}{3} = \dfrac{dz}{5z + \tan(y - 3x)}$ $\left[\text{Ans. } \phi\left[y - 3x, \dfrac{5z + \tan(y - 3x)}{e^{5x}}\right] = 0\right]$

8. $\dfrac{dx}{xz(z^2 + xy)} = \dfrac{dy}{-yz(z^2 + xy)} = \dfrac{dz}{x^4}$ $\left[\text{Ans. } \phi[xy, (z^2 + xy)^2 - x^4] = 0\right]$

9. $\dfrac{a\,dx}{(b-c)\,yz} = \dfrac{b\,dy}{(c-a)\,zx} = \dfrac{c\,dz}{(a-b)\,xy}$ $\left[\text{Ans. } \phi(ax^2 + by^2 + cz^2, a^2x^2 + b^2y^2 + c^2z^2) = 0\right]$

10. $\dfrac{dx}{y^3 x - 2x^4} = \dfrac{dy}{2y^4 - x^3 y} = \dfrac{dz}{9z(x^3 - y^3)}$ $\left[\text{Ans. } \phi\left[\dfrac{y}{x^2} + \dfrac{x}{y^2}, xyz^{1/3}\right] = 0\right]$

BMH 103 (a & b) Vector Analysis and Geometry

- **Vector Analysis**
- **Geometry**
  (a) **Two-Dimensional**
  (b) **Three-Dimensional**

VECTOR ANALYSIS

Scalar and vector product of three vectors. Product of four vectors. Reciprocal Vectors. Vector differentiation. Gradient, divergence and curl. Vector Integration. Theorems of Gauss, Green, Stokes and problems based on these.

1
Multiple Products

INTRODUCTION

The scalar product of two vectors **a** and **b** is a scalar and the vector product (or cross product) of two vectors **a** and **b** is a vector, *i.e.*,

$$\mathbf{a} \cdot \mathbf{b} = |\mathbf{a}||\mathbf{b}| \cos \theta,$$

where θ is the angle between the directions of vectors **a** and **b**, and

$$\mathbf{a} \times \mathbf{b} = |\mathbf{a}||\mathbf{b}| \sin \theta \, \hat{\mathbf{n}}$$

where $\hat{\mathbf{n}}$ is a unit vector perpendicular to the plane of the given vectors. Now, if we want to multiply third vector **c** to **a** and **b**, then we should multiply two vectors **a** and **b** first. For this **a . b** will be of no use as it is a scalar. On the other hand $(\mathbf{a} \times \mathbf{b}) \cdot \mathbf{c}$ and $(\mathbf{a} \times \mathbf{b}) \times \mathbf{c}$ and are of immense use. These products are known as triple products of vectors.

1.1 SCALAR TRIPLE PRODUCT

Let **a, b, c** be any three non-coplanar vectors. The vector product or cross-product of two vectors **a** and **b** is a vector, which can be used in combination with a third vector **c** to form the scalar triple product of three vectors, which is denoted as $(\mathbf{a} \times \mathbf{b}) \cdot \mathbf{c}$. This product is a number, *i.e.*, scalar product of two vectors $(\mathbf{a} \times \mathbf{b})$ and **c** must be a scalar. Hence it is called a scalar product.

Scalar triple product is also denoted as [**a b c**]. It can also be denoted in the following manner:

$$(\mathbf{a} \times \mathbf{b}) \cdot \mathbf{c} = [\mathbf{a}, \mathbf{b}, \mathbf{c}] = [\mathbf{a} \ \mathbf{b} \ \mathbf{c}]$$
$$= \mathbf{a} \cdot (\mathbf{b} \times \mathbf{c}) = \mathbf{b} \cdot (\mathbf{c} \times \mathbf{a})$$
$$= \mathbf{c} \cdot (\mathbf{a} \times \mathbf{b})$$
$$= (\mathbf{b} \times \mathbf{c}) \cdot \mathbf{a}$$
$$= (\mathbf{c} \times \mathbf{a}) \cdot \mathbf{b}$$

Note. *The value of the triple product depends only on the cyclic order of the factors and is independent of the position of the dot and cross between them.*

1.1.1 Geometrical Significance of Scalar Triple Product

Let *OA, OB, OC* be the coterminous edges of a parallelopiped having magnitudes and directions of **a, b** and **c** respectively. Let ϕ be the angle between the vectors $\mathbf{a} \times \mathbf{b}$ and **c**. But $\mathbf{a} \times \mathbf{b} = |\mathbf{a}||\mathbf{b}| \sin \theta \, \hat{\mathbf{n}}$ in a direction perpendicular to the plane of the vectors **a** and **b**.

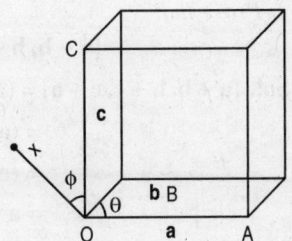

Then
$$(\mathbf{a} \times \mathbf{b}) \cdot \mathbf{c} = |\mathbf{a}||\mathbf{b}| \sin \theta \, \hat{\mathbf{n}} \cdot \mathbf{c}$$
$$= |\mathbf{a}||\mathbf{b}| \sin \theta \cdot |\mathbf{c}| \cos \phi$$

where ϕ is the angle between directions of $\mathbf{a} \times \mathbf{b}$ and **c**

$$= |\mathbf{a}||\mathbf{b}||\mathbf{c}| \sin \theta \cos \phi$$

But $|\mathbf{a}||\mathbf{b}||\mathbf{c}| \sin \theta \cos \phi$ is the volume of parallelopiped having its sides *OA, OB, OC*.

Hence, $\qquad V = (\mathbf{a} \times \mathbf{b}) \cdot \mathbf{c}$

The scalar triple product is also called the *box product*.

Note. (i) $(\mathbf{a} \times \mathbf{b}) \cdot \mathbf{c} = -(\mathbf{b} \times \mathbf{a}) \cdot \mathbf{c}$, because $\mathbf{a} \times \mathbf{b} = -(\mathbf{b} \times \mathbf{a})$.

(ii) *The cyclic arrangement of three vectors does not change the value of the scalar triple product, but the arrangement which is not in accordance with the cyclic order gives rise to the change of sign but not magnitude.*

Corollary 1. If the vector $\mathbf{a}$, $\mathbf{b}$ and $\mathbf{c}$ are coplanar then the scalar product
$$[\mathbf{a} \ \mathbf{b} \ \mathbf{c}] = 0$$

Corollary 2. If either two vectors of the three are equal, then the scalar triple product $[\mathbf{a} \ \mathbf{a} \ \mathbf{b}]$ or $[\mathbf{a} \ \mathbf{b} \ \mathbf{b}]$ or $[\mathbf{a} \ \mathbf{c} \ \mathbf{c}]$ etc. vanish.

1.1.2 To express $\mathbf{a} \cdot (\mathbf{b} \times \mathbf{c})$ in the determinant form

Let $\mathbf{a}, \mathbf{b}, \mathbf{c}$ be three vectors such that
$$\mathbf{a} = a_1 \mathbf{i} + a_2 \mathbf{j} + a_3 \mathbf{k},$$
$$\mathbf{b} = b_1 \mathbf{i} + b_2 \mathbf{j} + b_3 \mathbf{k},$$
$$\mathbf{c} = c_1 \mathbf{i} + c_2 \mathbf{j} + c_3 \mathbf{k},$$

then
$$\mathbf{b} \times \mathbf{c} = \begin{vmatrix} \mathbf{i} & \mathbf{j} & \mathbf{k} \\ b_1 & b_2 & b_3 \\ c_1 & c_2 & c_3 \end{vmatrix}$$
$$= \mathbf{i}(b_2 c_3 - b_3 c_2) - \mathbf{j}(b_1 c_3 - b_3 c_1) + \mathbf{k}(b_1 c_2 - b_2 c_1)$$

and
$$\mathbf{a} \cdot (\mathbf{b} \times \mathbf{c}) = a_1(b_2 c_3 - b_3 c_2) - a_2(b_1 c_3 - b_3 c_1) + a_3(b_1 c_2 - b_2 c_1)$$

$$\therefore \quad \mathbf{a} \cdot (\mathbf{b} \times \mathbf{c}) = \begin{vmatrix} a_1 & a_2 & a_3 \\ b_1 & b_2 & b_3 \\ c_1 & c_2 & c_3 \end{vmatrix}$$

EXAMPLES

1. *Find the volume of parallelopiped whose edges are represented by $\mathbf{a}, \mathbf{b}, \mathbf{c}$ where $\mathbf{a}, \mathbf{b}$ and $\mathbf{c}$ are given below*
$$\mathbf{a} = \mathbf{i} + \mathbf{j} + \mathbf{k}, \mathbf{b} = 2\mathbf{i} - \mathbf{j} - 3\mathbf{k} \text{ and } \mathbf{c} = 3\mathbf{i} - \mathbf{k}$$

Sol. We have, volume of parallelopiped
$$= [\mathbf{a} \ \mathbf{b} \ \mathbf{c}] = \begin{vmatrix} 1 & 1 & 1 \\ 2 & -1 & -3 \\ 3 & 0 & -1 \end{vmatrix}$$
$$= 1(1 - 0) - 1 \cdot (-2 + 9) + 1(0 + 3)$$
$$= 1 - 7 + 3 = -3 = 3 \text{ units.}$$

2. *Prove that*
$$[\mathbf{a} + \mathbf{b}, \mathbf{b} + \mathbf{c}, \mathbf{c} + \mathbf{a}] = 2[\mathbf{a} \ \mathbf{b} \ \mathbf{c}]$$

Sol. $[\mathbf{a} + \mathbf{b}, \mathbf{b} + \mathbf{c}, \mathbf{c} + \mathbf{a}] = (\mathbf{a} + \mathbf{b}) \cdot \{(\mathbf{b} + \mathbf{c}) \times (\mathbf{c} + \mathbf{a})\}$

$\qquad = (\mathbf{a} + \mathbf{b}) \cdot \{\mathbf{b} \times \mathbf{c} + \mathbf{b} \times \mathbf{a} + \mathbf{c} \times \mathbf{c} + \mathbf{c} \times \mathbf{a}\}$

$\qquad = (\mathbf{a} + \mathbf{b}) \cdot \{\mathbf{b} \times \mathbf{c} + \mathbf{b} \times \mathbf{a} + \mathbf{c} \times \mathbf{a}\}$ as $\mathbf{c} \times \mathbf{c} = 0$

$\qquad = \mathbf{a} \cdot (\mathbf{b} \times \mathbf{c}) + \mathbf{a} \cdot (\mathbf{b} \times \mathbf{a}) + \mathbf{a} \cdot (\mathbf{c} \times \mathbf{a}) + \mathbf{b} \cdot (\mathbf{b} \times \mathbf{c})$

$\qquad \qquad + \mathbf{b} \cdot (\mathbf{b} \times \mathbf{a}) + \mathbf{b} \cdot (\mathbf{c} \times \mathbf{a})$

$\qquad = [\mathbf{a} \ \mathbf{b} \ \mathbf{c}] + [\mathbf{b} \ \mathbf{c} \ \mathbf{a}],$

all other scalar triple products are zero.

Multiple Products

$$= [\mathbf{a}\ \mathbf{b}\ \mathbf{c}] + [\mathbf{a}\ \mathbf{b}\ \mathbf{c}] \text{ as } [\mathbf{b}\ \mathbf{c}\ \mathbf{a}] = [\mathbf{a}\ \mathbf{b}\ \mathbf{c}]$$
$$= 2[\mathbf{a}\ \mathbf{b}\ \mathbf{c}]$$

3. *Prove that* $[\mathbf{l}\ \mathbf{m}\ \mathbf{n}][\mathbf{a}\ \mathbf{b}\ \mathbf{c}]$.

$$= \begin{vmatrix} \mathbf{l}.\mathbf{a} & \mathbf{l}.\mathbf{b} & \mathbf{l}.\mathbf{c} \\ \mathbf{m}.\mathbf{a} & \mathbf{m}.\mathbf{b} & \mathbf{m}.\mathbf{c} \\ \mathbf{n}.\mathbf{a} & \mathbf{n}.\mathbf{b} & \mathbf{n}.\mathbf{c} \end{vmatrix}$$

Sol. Let $\mathbf{a} = a_1\mathbf{i} + a_2\mathbf{j} + a_3\mathbf{k}, \mathbf{b} = b_1\mathbf{i} + b_2\mathbf{j} + b_3\mathbf{k}$,
$\mathbf{c} = c_1\mathbf{i} + c_2\mathbf{j} + c_3\mathbf{k}, \mathbf{l} = l_1\mathbf{i} + l_2\mathbf{j} + l_3\mathbf{k}$,
$\mathbf{m} = m_1\mathbf{i} + m_2\mathbf{j} + m_3\mathbf{k}$ and $\mathbf{n} = n_1\mathbf{i} + n_2\mathbf{j} + n_3\mathbf{k}$

Now, $[\mathbf{l}\ \mathbf{m}\ \mathbf{n}][\mathbf{a}\ \mathbf{b}\ \mathbf{c}] = \begin{vmatrix} l_1 & l_2 & l_3 \\ m_1 & m_2 & m_3 \\ n_1 & n_2 & n_3 \end{vmatrix} \begin{vmatrix} a_1 & a_2 & a_3 \\ b_1 & b_2 & b_3 \\ c_1 & c_2 & c_3 \end{vmatrix}$

$$= \begin{vmatrix} l_1a_1 + l_2a_2 + l_3a_3 & l_1b_1 + l_2b_2 + l_3b_3 & l_1c_1 + l_2c_2 + l_3c_3 \\ m_1a_1 + m_2a_2 + m_3a_3 & m_1b_1 + m_2b_2 + m_3b_3 & m_1c_1 + m_2c_2 + m_3c_3 \\ n_1a_1 + n_2a_2 + n_3a_3 & n_1b_1 + n_2b_2 + n_3b_3 & n_1c_1 + n_2c_2 + n_3c_3 \end{vmatrix}$$

$$= \begin{vmatrix} \mathbf{l}.\mathbf{a} & \mathbf{l}.\mathbf{b} & \mathbf{l}.\mathbf{c} \\ \mathbf{m}.\mathbf{a} & \mathbf{m}.\mathbf{b} & \mathbf{m}.\mathbf{c} \\ \mathbf{n}.\mathbf{a} & \mathbf{n}.\mathbf{b} & \mathbf{n}.\mathbf{c} \end{vmatrix}$$

[Since $\mathbf{l}.\mathbf{a} = (l_1\mathbf{i} + l_2\mathbf{j} + l_3\mathbf{k}).(a_1\mathbf{i} + a_2\mathbf{j} + a_3\mathbf{k}) = l_1a_1 + l_2a_2 + l_3a_3$ etc.]

EXERCISES

1. Find the volume of the parallelopiped whsoe edges are given below :
   (a) $2\mathbf{i} - 3\mathbf{j} + 4\mathbf{k}, \mathbf{i} + 2\mathbf{j} - \mathbf{k}, 3\mathbf{i} - \mathbf{j} + 2\mathbf{k}$ [Ans. 7]
   (b) $\mathbf{i} + 2\mathbf{j} + 3\mathbf{k}, 3\mathbf{i} + 7\mathbf{j} - 4\mathbf{k}, \mathbf{i} - 5\mathbf{j} + 3\mathbf{k}$ [Ans. 91]
2. Prove that if $\mathbf{l}, \mathbf{m}, \mathbf{n}$ be the three non-coplanar vectors, then

$$[\mathbf{l}\ \mathbf{m}\ \mathbf{n}][\mathbf{a}\times\mathbf{b}] = \begin{vmatrix} \mathbf{l}.\mathbf{a} & \mathbf{l}.\mathbf{b} & \mathbf{l} \\ \mathbf{m}.\mathbf{a} & \mathbf{m}.\mathbf{b} & \mathbf{m} \\ \mathbf{n}.\mathbf{a} & \mathbf{n}.\mathbf{b} & \mathbf{n} \end{vmatrix}$$

3. Prove that
   (i) the four points $4\mathbf{i} + 5\mathbf{j} + \mathbf{k}, -\mathbf{j} - \mathbf{k}, 3\mathbf{i} + 9\mathbf{j} + 4\mathbf{k}$ and $-4\mathbf{i} + 4\mathbf{j} + 4\mathbf{k}$ are coplanar.
   (ii) $4\mathbf{i} + 8\mathbf{j} + 12\mathbf{k}, 2\mathbf{i} + 4\mathbf{j} + 6\mathbf{k}, 3\mathbf{i} + 5\mathbf{j} + 4\mathbf{k}$ and $5\mathbf{i} + 8\mathbf{j} + 5\mathbf{k}$ are coplanar.

1.2 VOLUME OF TETRAHEDRON

Let A, OBC be the tetrahedron of three conterminous edges OA, OB, OC, representing the vectors $\mathbf{a}, \mathbf{b}, \mathbf{c}$ respectively.

We have, volume of tetrahedron

$$= \frac{1}{3} \text{ area of the base} \times \text{height}$$

$$= \frac{1}{3} \times \text{Area of } \triangle OBC \times \text{perp. height}$$

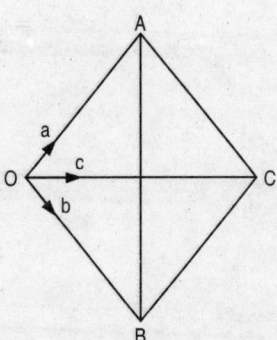

$$= \frac{1}{3} \times (\frac{1}{2} \mathbf{b} \times \mathbf{c}) \cdot \mathbf{a}$$

$$= \frac{1}{6} [\mathbf{a} \ \mathbf{b} \ \mathbf{c}]$$

1.2.1 Volume of the tetrahedron in terms of position vectors of the Vertices

Let A, B, C, D be the vertices of the tetrahedron and let $\mathbf{a}, \mathbf{b}, \mathbf{c}, \mathbf{d}$ be the position vectors of these vertices respectively. Then

$\overrightarrow{DA} = \mathbf{a} - \mathbf{d},$

$$\overrightarrow{DB} = \mathbf{b} - \mathbf{d}, \overrightarrow{DC} = \mathbf{c} - \mathbf{d}.$$

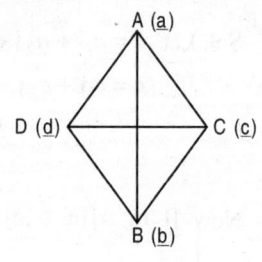

∴ The volume of the tetrahedron

$$= \frac{1}{6} [\mathbf{a} - \mathbf{d}, \mathbf{b} - \mathbf{d}, \mathbf{c} - \mathbf{d}]$$

$$= \frac{1}{6} [(\mathbf{a} - \mathbf{d}) \cdot (\mathbf{b} - \mathbf{d}) \times (\mathbf{c} - \mathbf{d})].$$

EXAMPLE

Three concurrent edges of tetrahedron OA, OB, OC have lengths a, b, c and angles BOC, COA, AOB are respectively θ, ψ, ϕ. *Show that the volume V of the tetrahedron is given by the formula*

$$V^2 = \frac{a^2 b^2 c^2}{36} \begin{vmatrix} 1 & \cos\phi & \cos\psi \\ \cos\phi & 1 & \cos\theta \\ \cos\psi & \cos\theta & 1 \end{vmatrix}$$

Sol. Let O be the origin and $\mathbf{a}, \mathbf{b}, \mathbf{c}$ the position vectors of A, B, C respectively. Also, let

$$\mathbf{a} = a_1 \mathbf{i} + a_2 \mathbf{j} + a_3 \mathbf{k}, \ \mathbf{b} = b_1 \mathbf{i} + b_2 \mathbf{j} + b_3 \mathbf{k}, \ \mathbf{c} = c_1 \mathbf{i} + c_2 \mathbf{j} + c_3 \mathbf{k}.$$

Hence,

$$V = \frac{1}{6} [\mathbf{a} \ \mathbf{b} \ \mathbf{c}] = \frac{1}{6} \begin{vmatrix} a_1 & a_2 & a_3 \\ b_1 & b_2 & b_3 \\ c_1 & c_2 & c_3 \end{vmatrix}$$

This gives,

$$V^2 = \frac{1}{36} \begin{vmatrix} a_1 & a_2 & a_3 \\ b_1 & b_2 & b_3 \\ c_1 & c_2 & c_3 \end{vmatrix} \times \begin{vmatrix} a_1 & a_2 & a_3 \\ b_1 & b_2 & b_3 \\ c_1 & c_2 & c_3 \end{vmatrix}$$

$$= \frac{1}{36} \begin{vmatrix} a_1^2 + a_2^2 + a_3^2 & a_1 b_1 + a_2 b_2 + a_3 b_3 & a_1 c_1 + a_2 c_2 + a_3 c_3 \\ a_1 b_1 + a_2 b_2 + a_3 b_3 & b_1^2 + b_2^2 + b_3^2 & b_1 c_1 + b_2 c_2 + b_3 c_3 \\ a_1 c_1 + a_2 c_2 + a_3 c_3 & b_1 c_1 + b_2 c_2 + b_3 c_3 & c_1^2 + c_2^2 + c_3^2 \end{vmatrix}$$

$$= \frac{1}{36} \begin{vmatrix} a^2 & \mathbf{a} \cdot \mathbf{b} & \mathbf{a} \cdot \mathbf{c} \\ \mathbf{a} \cdot \mathbf{b} & b^2 & \mathbf{b} \cdot \mathbf{c} \\ \mathbf{a} \cdot \mathbf{c} & \mathbf{b} \cdot \mathbf{c} & c^2 \end{vmatrix}$$

Multiple Products

$$= \frac{1}{36} \begin{vmatrix} a^2 & ab\cos\phi & ac\cos\psi \\ ab\cos\phi & b^2 & bc\cos\theta \\ ac\cos\psi & bc\cos\theta & c^2 \end{vmatrix}$$

$$= \frac{a^2 b^2 c^2}{36} \begin{vmatrix} 1 & \cos\phi & \cos\psi \\ \cos\phi & 1 & \cos\theta \\ \cos\psi & \cos\theta & 1 \end{vmatrix}$$

EXERCISES

1. Show that the points whose position vectors are **a, b, c, d** will be coplanar if
$$[\mathbf{a}\ \mathbf{b}\ \mathbf{c}] - [\mathbf{a}\ \mathbf{b}\ \mathbf{d}] + [\mathbf{a}\ \mathbf{c}\ \mathbf{d}] - [\mathbf{b}\ \mathbf{c}\ \mathbf{d}] = 0$$
   [**Hint.** Show that the volume of the tetrahedron formed by these vertices is zero]

2. Show that the volume of the tetrahedron is
$$= \frac{1}{6} \begin{vmatrix} a_1 & a_2 & a_3 & 1 \\ b_1 & b_2 & b_3 & 1 \\ c_1 & c_2 & c_3 & 1 \\ d_1 & d_2 & d_3 & 1 \end{vmatrix}$$
where $(a_1, a_2, a_3), (b_1, b_2, b_3), (c_1, c_2, c_3), (d_1, d_2, d_3)$ are the co-ordinates of the vertices.

1.3 VECTOR TRIPLE PRODUCT

The cross product of **a** and **b** × **c** is written as $\mathbf{a} \times (\mathbf{b} \times \mathbf{c})$ which is it-self a vector. The vector $\mathbf{a} \times (\mathbf{b} \times \mathbf{c})$ is perpendicular to **b** × **c** and therefore coplanar with **b** and **c**. Hence it can be expressed in terms of **b** and **c**.

Let
$$\mathbf{P} = \mathbf{a} \times (\mathbf{b} \times \mathbf{c})$$

Then we have
$$\mathbf{P} = \mathbf{a} \times (\mathbf{b} \times \mathbf{c}) = l\mathbf{b} + m\mathbf{c} \qquad \ldots(i)$$

Multiplying both sides of (i) scalarly by **a**, we get
$$\mathbf{P} \cdot \mathbf{a} = l\mathbf{b} \cdot \mathbf{a} + m\mathbf{c} \cdot \mathbf{a}$$

$$\therefore \quad \frac{l}{\mathbf{c} \cdot \mathbf{a}} = -\frac{m}{\mathbf{b} \cdot \mathbf{a}} = \lambda \text{ (say)}$$

Putting these values of l, m in (i) we get
$$\mathbf{P} = \mathbf{a} \times (\mathbf{b} \cdot \mathbf{c}) = \lambda \{(\mathbf{c} \cdot \mathbf{a})\mathbf{b} - (\mathbf{b} \cdot \mathbf{a})\} \qquad \ldots(ii)$$

To find the value of λ, let us suppose
$$\mathbf{a} = a_1\mathbf{i} + a_2\mathbf{j} + a_3\mathbf{k}$$
$$\mathbf{b} = \qquad b_2\mathbf{j} + b_3\mathbf{k}$$
$$\mathbf{c} = \qquad\qquad c_3\mathbf{k}$$

From this we obtain
$$\mathbf{a} \times (\mathbf{b} \times \mathbf{c}) = (a_1\mathbf{i} + a_2\mathbf{j} + a_3\mathbf{k}) \times (b_2 c_3)\mathbf{i}$$
$$= -a_1 b_2 c_3 \mathbf{k} + a_3 b_2 c_3 \mathbf{j}$$

Also
$$(\mathbf{c} \cdot \mathbf{a})\mathbf{b} - (\mathbf{b} \cdot \mathbf{a})\mathbf{c}$$
$$= (a_3 c_3)(b_2\mathbf{j} + b_3\mathbf{k}) - (a_2 b_2 + a_3 b_3) c_3 \mathbf{k}$$
$$= a_3 b_2 c_3 \mathbf{j} - a_2 b_2 c_3 \mathbf{k}$$

By putting these value in (ii), we get
$$\lambda = 1.$$

Hence, $\quad \mathbf{a} \times (\mathbf{b} \times \mathbf{c}) = (\mathbf{a} \cdot \mathbf{c})\mathbf{b} - (\mathbf{a} \cdot \mathbf{b})\mathbf{c}$

Similarly, we can prove that
$$\mathbf{b} \times (\mathbf{c} \times \mathbf{a}) = (\mathbf{b} \cdot \mathbf{a})\mathbf{c} - (\mathbf{b} \cdot \mathbf{c}) \cdot \mathbf{a}$$
and $\quad \mathbf{c} \times (\mathbf{a} \times \mathbf{b}) = (\mathbf{c} \cdot \mathbf{b})\mathbf{a} - (\mathbf{c} \cdot \mathbf{a})\mathbf{b}$

Note. *In a vector triple product, the position of the brackets can not be altered as*
$$\mathbf{a} \times (\mathbf{b} \times \mathbf{c}) = (\mathbf{a} \cdot \mathbf{c})\mathbf{b} - (\mathbf{a} \cdot \mathbf{b})\mathbf{c}$$
but $\quad (\mathbf{a} \times \mathbf{b}) \times \mathbf{c} = (\mathbf{c} \cdot \mathbf{a})\mathbf{b} - (\mathbf{c} \cdot \mathbf{b})\mathbf{a}$

EXAMPLES

1. *Prove that*
$$\mathbf{i} \times (\mathbf{a} \times \mathbf{i}) + \mathbf{j} \times (\mathbf{a} \times \mathbf{j}) + \mathbf{k} \times (\mathbf{a} \times \mathbf{k}) = 2\mathbf{a}$$

Sol. $\mathbf{i} \times (\mathbf{a} \times \mathbf{i}) + \mathbf{j} \times (\mathbf{a} \times \mathbf{j}) + \mathbf{k} \times (\mathbf{a} \times \mathbf{k})$

$= (\mathbf{i} \cdot \mathbf{i})\mathbf{a} - (\mathbf{i} \cdot \mathbf{a})\mathbf{i} + (\mathbf{j} \cdot \mathbf{j})\mathbf{a} - (\mathbf{j} \cdot \mathbf{a})\mathbf{j} + (\mathbf{k} \cdot \mathbf{k})\mathbf{a} - (\mathbf{k} \cdot \mathbf{a})\mathbf{k}$

$= 3\mathbf{a} - \{(\mathbf{i} \cdot \mathbf{a})\mathbf{i} + (\mathbf{j} \cdot \mathbf{a})\mathbf{j} + (\mathbf{k} \cdot \mathbf{a})\mathbf{k}\}$

$= 3\mathbf{a} - [\{\mathbf{i} \cdot (a_1\mathbf{i} + a_2\mathbf{j} + a_3\mathbf{k})\}\mathbf{i} + \{\mathbf{j} \cdot (a_1\mathbf{i} + a_2\mathbf{j} + a_3\mathbf{k})\}\mathbf{j}$
$\qquad\qquad\qquad + \{\mathbf{k} \cdot (a_1\mathbf{i} + a_2\mathbf{j} + a_3\mathbf{k})\}\mathbf{k}]$

$= 3\mathbf{a} - (a_1\mathbf{i} + a_2\mathbf{j} + a_3\mathbf{k})$

$= 3\mathbf{a} - \mathbf{a} = 2\mathbf{a}$.

2. *If $\mathbf{a}, \mathbf{b}, \mathbf{c}$ be three unit vectors such that $\mathbf{a} \times (\mathbf{b} \times \mathbf{c}) = \frac{1}{2}\mathbf{b}$ find the angles which $\mathbf{a}$ makes with $\mathbf{b}$ and $\mathbf{c}$, $\mathbf{b}$ and $\mathbf{c}$ being non-parallel.*

Sol. We have
$$\mathbf{a} \times (\mathbf{b} \times \mathbf{c}) = \frac{1}{2}\mathbf{b}$$
i.e., $\quad (\mathbf{a} \cdot \mathbf{c})\mathbf{b} - (\mathbf{a} \cdot \mathbf{b})\mathbf{c} = \frac{1}{2}\mathbf{b}$

or $\quad \{(\mathbf{a} \cdot \mathbf{c}) - \frac{1}{2}\}\mathbf{b} = (\mathbf{a} \cdot \mathbf{b})\mathbf{c}$

Which shows that $\mathbf{b}$ and $\mathbf{c}$ are paralell, but given that they are non-parallel. Hence their coefficients must be zero.

i.e., $\quad \mathbf{a} \cdot \mathbf{c} - \frac{1}{2} = 0$ or $(\mathbf{a} \cdot \mathbf{c}) = \frac{1}{2}$

and $\quad \mathbf{a} \cdot \mathbf{b} = 0$

If the angle between $\mathbf{a}$ and $\mathbf{c}$ is α and the angle between $\mathbf{a}$ and $\mathbf{b}$ is β then

$$|\mathbf{a}||\mathbf{c}|\cos\alpha = \frac{1}{2} \quad \text{and} \quad |\mathbf{a}||\mathbf{b}|\cos\beta = 0$$

i.e., $\quad \cos\alpha = \frac{1}{2}$ and $\cos\beta = 0$ $\qquad$ [$\because \mathbf{a}, \mathbf{b}$ and $\mathbf{c}$ are unit vectors]

i.e., $\quad \alpha = 60°$ and $\beta = 90°$

3. *Show that*
$$\mathbf{a} \times (\mathbf{b} \times \mathbf{c}) + \mathbf{b} \times (\mathbf{c} \times \mathbf{a}) + \mathbf{c} \times (\mathbf{a} \times \mathbf{b}) = 0$$

Sol. We have
$$\mathbf{a} \times (\mathbf{b} \times \mathbf{c}) = (\mathbf{a} \cdot \mathbf{c})\mathbf{b} - (\mathbf{a} \cdot \mathbf{b})\mathbf{c}$$
$$\mathbf{b} \times (\mathbf{c} \times \mathbf{a}) = (\mathbf{b} \cdot \mathbf{a})\mathbf{c} - (\mathbf{b} \cdot \mathbf{c})\mathbf{a}$$
$$\mathbf{c} \times (\mathbf{a} \times \mathbf{b}) = (\mathbf{c} \cdot \mathbf{b})\mathbf{a} - (\mathbf{c} \cdot \mathbf{a})\mathbf{b}$$

Multiple Products

Adding these we get
$$\mathbf{a} \times (\mathbf{b} \times \mathbf{c}) + \mathbf{b}(\mathbf{c} \times \mathbf{b})\mathbf{a} + \mathbf{c} \times (\mathbf{a} \times \mathbf{b}) = 0$$

4. *If $\mathbf{a}, \mathbf{b}, \mathbf{c}$ are three non-coplanar vector, show that $[\mathbf{a} \times \mathbf{b}, \mathbf{b} \times \mathbf{c}, \mathbf{c} \times \mathbf{a}] = [\mathbf{a}\,\mathbf{b}\,\mathbf{c}]^2$*

Sol. Let $\mathbf{a} = a_1\mathbf{i} + a_2\mathbf{j} + a_3\mathbf{k}$, $\mathbf{b} = b_1\mathbf{i} + b_2\mathbf{j} + b_3\mathbf{k}$
$\mathbf{c} = c_1\mathbf{i} + c_2\mathbf{j} + c_3\mathbf{k}$

Then $\mathbf{a} \times \mathbf{b} = (a_1\mathbf{i} + a_2\mathbf{j} + a_3\mathbf{k}) \times (b_1\mathbf{i} + b_2\mathbf{j} + b_3\mathbf{k})$
$= (a_2 b_3 - a_3 b_2)\mathbf{i} + (a_3 b_1 - a_1 b_3)\mathbf{j} + (a_1 b_2 - a_2 b_1)\mathbf{k}$

Similarly, $\mathbf{b} \times \mathbf{c} = (b_2 c_3 - b_3 c_2)\mathbf{i} + (b_3 c_1 - b_1 c_3)\mathbf{j} + (b_1 c_2 - b_2 c_1)\mathbf{k}$

and $\mathbf{c} \times \mathbf{a} = (c_2 a_3 - c_3 a_2)\mathbf{c} + (c_3 a_1 - c_1 a_3)\mathbf{j} + (c_1 a_2 - c_2 a_1)\mathbf{k}$

$\therefore \quad [\mathbf{a} \times \mathbf{b}, \mathbf{b} \times \mathbf{c}, \mathbf{c} \times \mathbf{a}]$

$$= \begin{vmatrix} a_2 b_3 - a_3 b_2 & a_3 b_1 - a_1 b_3 & a_1 b_2 - a_2 b_1 \\ b_2 c_3 - b_3 c_2 & b_3 c_1 - b_1 c_3 & b_1 c_2 - b_2 c_1 \\ c_2 a_3 - c_3 a_2 & c_3 a_1 - c_1 c_3 & c_1 a_2 - c_2 a_1 \end{vmatrix}$$

$$= \begin{vmatrix} C_1 & C_2 & C_3 \\ A_1 & A_2 & A_3 \\ B_1 & B_2 & B_3 \end{vmatrix}$$

where capital letters denote the cofactors of the corresponding small letters in the determinant
$\begin{vmatrix} a_1 & a_2 & a_3 \\ b_1 & b_2 & b_3 \\ c_1 & c_2 & c_3 \end{vmatrix}$.

$$\therefore \quad [\mathbf{a} \times \mathbf{b}, \mathbf{b} \times \mathbf{c}, \mathbf{c} \times \mathbf{a}] = \begin{vmatrix} A_1 & A_2 & A_3 \\ B_1 & B_2 & B_3 \\ C_1 & C_2 & C_3 \end{vmatrix}, \text{ by interchanging rows.}$$

$$= \begin{vmatrix} a_1 & b_1 & c_1 \\ a_2 & b_2 & c_2 \\ a_3 & b_3 & c_3 \end{vmatrix}^2 \quad \text{By property of determinant.}$$

$= [\mathbf{a}\,\mathbf{b}\,\mathbf{c}]^2$

EXERCISES

1. If $\mathbf{a} = 2\mathbf{i} - 10\mathbf{j} + 2\mathbf{k}$, $\mathbf{b} = 3\mathbf{i} + \mathbf{j} + 2\mathbf{k}$, $\mathbf{c} = 2\mathbf{i} + \mathbf{j} + 3\mathbf{k}$. Find the value of $\mathbf{a} \times (\mathbf{b} \times \mathbf{c})$. **[Ans. 0]**

2. If $\mathbf{a} = \mathbf{i} - \mathbf{j} - \mathbf{k}$, $\mathbf{b} = \mathbf{i} - 3\mathbf{j} + 4\mathbf{k}$. Find the value of $\mathbf{a} \times (\mathbf{b} \times \mathbf{c})$. **[Ans. $3(\mathbf{i} - 3\mathbf{j} + 4\mathbf{k})$]**

3. Prove that
   (i) $\mathbf{a} \times (\mathbf{b} \times \mathbf{a}) = (\mathbf{a} \times \mathbf{b}) \times \mathbf{a}$
   (ii) $(\mathbf{a} \times \mathbf{b}) \times (\mathbf{a} \times \mathbf{c}) \cdot \mathbf{d} = (\mathbf{a} \cdot \mathbf{d})[\mathbf{a}\,\mathbf{b}\,\mathbf{c}]$
   (iii) $(\mathbf{b} \times \mathbf{c}) \times (\mathbf{c} \times \mathbf{a}) = [\mathbf{a}\,\mathbf{b}\,\mathbf{c}]\mathbf{c}$

4. Show that $\mathbf{i} \times (\mathbf{j} \times \mathbf{k}) = 0$.

5. Show that $\mathbf{a} \times (\mathbf{b} \times \mathbf{c}), \mathbf{b} \times (\mathbf{c} \times \mathbf{a})$ and $\mathbf{c} \times (\mathbf{a} \times \mathbf{b})$ are coplanar.
   [**Hint:** (i) The sum of given vectors is a zero vector. Hence, any one of the given vectors can be expressed as linear combination of the other two vectors. or (ii) Show that the scalar triple product of three given vectors is zero.]

1.4.1 Vector Product of Four Vectors

The vector product or cross-product of $(\mathbf{a} \times \mathbf{b})$ and $(\mathbf{c} \times \mathbf{d})$ is called a vector product of four vectors, which is denoted as $(\mathbf{a} \times \mathbf{b}) \times (\mathbf{c} \times \mathbf{d})$. This is a vector which is perpendicular to $(\mathbf{a} \times \mathbf{b})$ and therefore coplanar with $\mathbf{a}$ and $\mathbf{b}$. Similarly, this is perpendicular to $(\mathbf{c} \times \mathbf{d})$ and therefore coplanar with $\mathbf{c}$ and $\mathbf{d}$. Therefore it can be expressed in terms of $\mathbf{a}$ and $\mathbf{b}$ as well as in terms of $\mathbf{c}$ and $\mathbf{d}$.

Now, suppose $\mathbf{c} \times \mathbf{d} = \mathbf{m}$

then
$$(\mathbf{a} \times \mathbf{b}) \times (\mathbf{c} \times \mathbf{d}) = (\mathbf{a} \times \mathbf{b}) \times \mathbf{m}$$
$$= (\mathbf{a} \cdot \mathbf{m})\mathbf{b} - (\mathbf{b} \cdot \mathbf{m})\mathbf{a}$$
$$= \{\mathbf{a} \cdot (\mathbf{c} \times \mathbf{d})\}\mathbf{b} - \{\mathbf{b} \cdot (\mathbf{c} \times \mathbf{d})\}\mathbf{a}$$
$$= [\mathbf{a}\,\mathbf{c}\,\mathbf{d}]\mathbf{b} - [\mathbf{b}\,\mathbf{c}\,\mathbf{d}]\mathbf{a} \qquad \ldots(i)$$

Similarly, taking $\mathbf{a} \times \mathbf{b} = \mathbf{n}$, we get
$$(\mathbf{a} \times \mathbf{b}) \times (\mathbf{c} \times \mathbf{d}) = \mathbf{n} \times (\mathbf{c} \times \mathbf{d})$$
$$= (\mathbf{n} \cdot \mathbf{d})\mathbf{c} - (\mathbf{n} \cdot \mathbf{c})\mathbf{d}$$
$$= \{(\mathbf{a} \times \mathbf{b}) \cdot \mathbf{d}\}\mathbf{c} - \{(\mathbf{a} \times \mathbf{b}) \cdot \mathbf{c}\}\mathbf{d}$$
$$= [\mathbf{a}\,\mathbf{b}\,\mathbf{d}]\mathbf{c} - [\mathbf{a}\,\mathbf{b}\,\mathbf{c}]\mathbf{d} \qquad \ldots(ii)$$

Thus, form (i) and (ii)
$$[\mathbf{a}\,\mathbf{c}\,\mathbf{d}]\mathbf{b} - [\mathbf{b}\,\mathbf{c}\,\mathbf{d}]\mathbf{a} = [\mathbf{a}\,\mathbf{b}\,\mathbf{d}]\mathbf{c} - [\mathbf{a}\,\mathbf{b}\,\mathbf{c}]\mathbf{d}.$$

If we write $\mathbf{r}$ for $\mathbf{d}$, then
$$[\mathbf{b}\,\mathbf{c}\,\mathbf{r}]\mathbf{a} - [\mathbf{a}\,\mathbf{c}\,\mathbf{r}]\mathbf{b} + [\mathbf{a}\,\mathbf{b}\,\mathbf{r}]\mathbf{c} - [\mathbf{a}\,\mathbf{b}\,\mathbf{c}]\mathbf{r} = 0$$

$$\therefore \quad \mathbf{r} = \frac{[\mathbf{r}\,\mathbf{b}\,\mathbf{c}]\mathbf{a} + [\mathbf{r}\,\mathbf{c}\,\mathbf{a}]\mathbf{b} + [\mathbf{r}\,\mathbf{a}\,\mathbf{b}]\mathbf{c}}{[\mathbf{a}\,\mathbf{b}\,\mathbf{c}]}$$

which is valid except when the denominator $[\mathbf{a}\,\mathbf{b}\,\mathbf{c}]$ vanishes that is except when $\mathbf{a}, \mathbf{b}, \mathbf{c}$ are coplanar.

EXAMPLES

1. *Prove that*
$$\mathbf{d} \cdot [\mathbf{a} \times \{\mathbf{b} \times (\mathbf{c} \times \mathbf{d})\}] = (\mathbf{b} \cdot \mathbf{d})[\mathbf{a}\,\mathbf{c}\,\mathbf{d}]$$

Sol. We have
$$\mathbf{a} \times \{\mathbf{b} \times (\mathbf{c} \times \mathbf{d})\} = \mathbf{a} \times \{(\mathbf{b} \cdot \mathbf{d})\mathbf{c} - (\mathbf{b} \cdot \mathbf{c})\mathbf{d}\}$$
$$= (\mathbf{b} \cdot \mathbf{d})(\mathbf{a} \times \mathbf{c}) - (\mathbf{b} \cdot \mathbf{c})(\mathbf{a} \times \mathbf{d})$$
$$\therefore \quad \mathbf{d} \cdot [\mathbf{a} \times \{\mathbf{b} \times (\mathbf{c} \times \mathbf{d})\}] = \mathbf{d} \cdot [(\mathbf{b} \cdot \mathbf{d})(\mathbf{a} \times \mathbf{c}) - (\mathbf{b} \cdot \mathbf{c})(\mathbf{a} \times \mathbf{d})]$$
$$= (\mathbf{b} \cdot \mathbf{d})[\mathbf{d}\,\mathbf{a}\,\mathbf{c}] - (\mathbf{b} \cdot \mathbf{c})[\mathbf{d}\,\mathbf{a}\,\mathbf{d}]$$
$$= (\mathbf{b} \cdot \mathbf{d})[\mathbf{a}\,\mathbf{c}\,\mathbf{d}], \text{ Since } [\mathbf{d}\,\mathbf{a}\,\mathbf{d}] = 0$$

2. *Prove that*
$$[\mathbf{a} \times \mathbf{b}, \mathbf{c} \times \mathbf{d}, \mathbf{e} \times \mathbf{f}] = [\mathbf{a}\,\mathbf{b}\,\mathbf{d}][\mathbf{c}\,\mathbf{e}\,\mathbf{f}] - [\mathbf{a}\,\mathbf{b}\,\mathbf{c}][\mathbf{d}\,\mathbf{e}\,\mathbf{f}]$$
$$= [\mathbf{a}\,\mathbf{b}\,\mathbf{e}][\mathbf{f}\,\mathbf{c}\,\mathbf{d}] - [\mathbf{a}\,\mathbf{b}\,\mathbf{f}][\mathbf{e}\,\mathbf{c}\,\mathbf{d}]$$
$$= [\mathbf{c}\,\mathbf{d}\,\mathbf{a}][\mathbf{b}\,\mathbf{e}\,\mathbf{f}] - [\mathbf{c}\,\mathbf{d}\,\mathbf{b}][\mathbf{a}\,\mathbf{e}\,\mathbf{f}]$$

Sol. We have
$$[\mathbf{a} \times \mathbf{b}, \mathbf{c} \times \mathbf{d}, \mathbf{e} \times \mathbf{f}] = (\mathbf{a} \times \mathbf{b}) \cdot \{(\mathbf{c} \times \mathbf{d}) \times (\mathbf{e} \times \mathbf{f})\}$$
$$= (\mathbf{a} \times \mathbf{b}) \cdot \{[\mathbf{c}\,\mathbf{d}\,\mathbf{f}]\mathbf{e} - [\mathbf{c}\,\mathbf{d}\,\mathbf{e}]\mathbf{f}\}$$
$$= [\mathbf{a}\,\mathbf{b}\,\mathbf{e}][\mathbf{c}\,\mathbf{d}\,\mathbf{f}] - [\mathbf{c}\,\mathbf{d}\,\mathbf{e}][\mathbf{a}\,\mathbf{b}\,\mathbf{f}]$$

Multiple Products

Again,
$$[a \times b, c \times d, e \times f] = (c \times d) \cdot \{(e \times f) \times (a \times b)\}$$
$$= (c \times d) \cdot \{[e\,f\,d]a - [e\,f\,a]b\}$$
$$= [c\,d\,a][b\,e\,f] - [c\,d\,b][a\,e\,f]$$

and
$$[a \times b, c \times d, e \times f] = (e \times f) \cdot \{(a \times b) \times (c \times d)\}$$
$$= (e \times f) \cdot \{[a\,b\,d]c - [a\,b\,c]d\}$$
$$= [a\,b\,d][c\,e\,f] - [a\,b\,c][d\,e\,f]$$

EXERCISES

1. Prove that $a \times \{b \times (c \times d)\}$
$$= (b \cdot d)(a \times c) - (b \cdot c)(c \times d)$$
2. $(a \times b) \cdot (c \times d) + (a \times c) \cdot (d \times b) + (a \times d) \cdot (b \times c) = 0$
3. If a, b, c are coplanar then show that $(a \times b) \times (c \times d) = 0$
4. Prove that $[a \times p, b \times q, c \times r] + [a \times q, b \times r, r \times p] + [a \times r, b \times p, c \times q] = 0$

1.5 RECIPROCAL SYSTEM OF VECTORS

If a, b, c are three non-coplanar vectors so that $[a\,b\,c] \neq 0$ then the three vectors a', b', c' are called reciprocals of a, b, c, where

$$a' = \frac{b \times c}{[a\,b\,c]}, \quad b' = \frac{c \times a}{[a\,b\,c]}$$

$$c' = \frac{a \times b}{[a\,b\,c]}.$$

1.5.1 Properties of Reciprocal System of Vectors

(i) $a \cdot a' = b \cdot b' = c \cdot c' = 1$

$$a \cdot a' = a \cdot \frac{b \times c}{[a\,b\,c]} = \frac{[a\,b\,c]}{[a\,b\,c]} = 1$$

Similarly, $b \cdot b' = c \cdot c' = 1$

(ii) $a \cdot b' = b \cdot c' = c \cdot a' = 0$

since $a \cdot b' = a \cdot \frac{c \times a}{[a\,b\,c]} = \frac{[a\,c\,a]}{[a\,b\,c]} = 0$

Similarly, $b \cdot c' = c \cdot a' = 0$

EXAMPLES

1. *Find a set of vectors reciprocal to the set* $2i + 3j - k, i - j - 2k, -i + 2j + 2k$.

Sol. Let
$$a = 2i + 3j - k, b = i - j - 2k,$$
$$c = -i + 2j + 2k$$

$\therefore \quad [a\,b\,c] = \begin{vmatrix} 2 & 3 & -1 \\ 1 & -1 & -2 \\ -1 & 2 & 2 \end{vmatrix} = 3$

$$(\mathbf{b} \times \mathbf{c}) = \begin{vmatrix} \mathbf{i} & \mathbf{j} & \mathbf{k} \\ 1 & -1 & -2 \\ -1 & 2 & 2 \end{vmatrix} = 2\mathbf{i} + \mathbf{k}$$

$$\therefore \quad \mathbf{a}' = \frac{\mathbf{b} \times \mathbf{c}}{[\mathbf{a}\,\mathbf{b}\,\mathbf{c}]} = \frac{2\mathbf{i} + \mathbf{k}}{3}.$$

Similarly, $\mathbf{b}' = \dfrac{\mathbf{c} \times \mathbf{a}}{[\mathbf{a}\,\mathbf{b}\,\mathbf{c}]} = \dfrac{-8\mathbf{i} + 3\mathbf{j} - 7\mathbf{k}}{3}$

and $\mathbf{c}' = \dfrac{\mathbf{a} \times \mathbf{b}}{[\mathbf{a}\,\mathbf{b}\,\mathbf{c}]} = \dfrac{-7\mathbf{i} + 3\mathbf{j} - 5\mathbf{k}}{3}$

2. *Prove that*

$$\mathbf{a}' \times \mathbf{b}' + \mathbf{b}' \times \mathbf{c}' + \mathbf{c}' \times \mathbf{a}' = \frac{\mathbf{a} + \mathbf{b} + \mathbf{c}}{[\mathbf{a}\,\mathbf{b}\,\mathbf{c}]}$$

Sol. We have $\mathbf{a}' \times \mathbf{b}' = \dfrac{\mathbf{b} \times \mathbf{c}}{[\mathbf{a}\,\mathbf{b}\,\mathbf{c}]} \times \dfrac{\mathbf{c} \times \mathbf{a}}{[\mathbf{a}\,\mathbf{b}\,\mathbf{c}]}$

$$= \frac{(\mathbf{b} \times \mathbf{c}) \times (\mathbf{c} \times \mathbf{a})}{[\mathbf{a}\,\mathbf{b}\,\mathbf{c}]^2} = \frac{[\mathbf{b}\,\mathbf{c}\,\mathbf{a}]\mathbf{c} - [\mathbf{b}\,\mathbf{c}\,\mathbf{c}]\mathbf{a}}{[\mathbf{a}\,\mathbf{b}\,\mathbf{c}]^2}$$

$$= \frac{[\mathbf{a}\,\mathbf{b}\,\mathbf{c}]\mathbf{c}}{[\mathbf{a}\,\mathbf{b}\,\mathbf{c}]^2} = \frac{\mathbf{c}}{[\mathbf{a}\,\mathbf{b}\,\mathbf{c}]} \qquad \ldots(i)$$

Similarly, $\mathbf{b}' \times \mathbf{c}' = \dfrac{\mathbf{a}}{[\mathbf{a}\,\mathbf{b}\,\mathbf{c}]}$ $\qquad \ldots(ii)$

and $\mathbf{c}' \times \mathbf{a}' = \dfrac{\mathbf{b}}{[\mathbf{a}\,\mathbf{b}\,\mathbf{c}]}$ $\qquad \ldots(iii)$

Adding (i), (ii) and (iii), we have

$$\mathbf{a}' \times \mathbf{b}' + \mathbf{b}' \times \mathbf{c}' + \mathbf{c}' \times \mathbf{a}' = \frac{\mathbf{a} + \mathbf{b} + \mathbf{c}}{[\mathbf{a}\,\mathbf{b}\,\mathbf{c}]}$$

EXERCISES

1. Prove that $[\mathbf{a}'\,\mathbf{b}'\,\mathbf{c}'] = \dfrac{1}{[\mathbf{a}\,\mathbf{b}\,\mathbf{c}]}$.

2. Prove that $\mathbf{a}\cdot\mathbf{a}' + \mathbf{b}\cdot\mathbf{b}' + \mathbf{c}\cdot\mathbf{c}' = 3$

3. Find the set of vectors reciprocal to $\mathbf{i}, \mathbf{i} + \mathbf{j}, \mathbf{i} + \mathbf{j} + \mathbf{k}$. [**Ans.** $\mathbf{i} - \mathbf{j}, \mathbf{j} - \mathbf{k}, \mathbf{k}$]

4. Prove that the set reciprocal to $\mathbf{a}, \mathbf{b}, \mathbf{a} \times \mathbf{b}$ is $\dfrac{\mathbf{b} \times (\mathbf{a} \times \mathbf{b})}{c^2}, \dfrac{(\mathbf{a} \times \mathbf{b}) \times \mathbf{a}}{c^2}$ and $\dfrac{\mathbf{a} \times \mathbf{b}}{c^2}$, where $c = |\mathbf{a} \times \mathbf{b}|$.

2
Differentiation of Vectors

2.1 INTRODUCTION

Let the vector **r** be a function of a scalar variable t, then
$$\mathbf{r} = \mathbf{f}(t)$$
If only one value of **r** corresponds to each value of t, then **r** is defined as a *single valued function* of the scalar variable t. If t varies continuously, so does **r** then the end point of **r** describes a continuous curve.

The folloiwng illustrations make the point clear :

$\mathbf{r} = a \cos t\, \mathbf{i} + a \sin t\, \mathbf{j}$ (Circle)

$\mathbf{r} = a \cos t\, \mathbf{i} + b \sin t\, \mathbf{j}$ (Ellipse)

$\mathbf{r} = at^2\, \mathbf{i} + 2at\, \mathbf{j}$ (Parabola)

$\mathbf{r} = a \sec t\, \mathbf{i} + b \tan t\, \mathbf{j}$ (Hyperbola)

These all are the vector equations of the curves. For different values of t, the end point of the vector describes the curve as mentioned above.

Note. *Any vector $f(t)$ can also be written in the form*
$$\mathbf{f}(t) = f_1(t)\, \mathbf{i} + f_2(t)\, \mathbf{j} + f_3(t)\, \mathbf{k}$$
where $f_1(t), f_2(t), f_3(t)$ are three scalar functions of t.

2.2 DIFFERENTIATON OF VECTOR

Let the vector **r** be a continuous and single velued function of a scalar variable t (*i.e.*, length of the vector can be determined if a value of t is given). With O as origin, let the vector **r** be represented by $\overrightarrow{OA}$ for a certain value of t and let $\mathbf{r} + \delta\mathbf{r}$ be represented by $\overrightarrow{OB}$ corresponding to the value $t + \delta t$ where δt is a small increment in t. Then δt produces an increment $(\mathbf{r} + \delta\mathbf{r} - \mathbf{r})$ *i.e.*, $\delta \mathbf{r}$ in **r**.

The increment $\delta \mathbf{r}$ is equal to $\overrightarrow{AB}$. Then also the quotient $\dfrac{\partial \mathbf{r}}{\partial t}$ is a vector. If $\delta t \to 0$ then $\delta \mathbf{r} \to 0$ and the point B moves towards A to coincide with it and then chord AB coincides with the tangent at P to the curve.

If the limiting value of the quotient $\dfrac{\delta \mathbf{r}}{\delta t}$ as $\delta t \to 0$ exists, then this value is defined as the differential coefficient of **r** with respect to t and the vector **r** is to be differentiable and is denoted by $\dfrac{d\mathbf{r}}{dt}$.

This process is known as *differentiation*, and the differential coefficient is known as the *derivative* or the derived function.

Thus,
$$\frac{d\mathbf{r}}{dt} = \lim_{\delta t \to 0} \frac{\delta \mathbf{r}}{\delta t} = \lim_{\delta t \to 0} \frac{(\mathbf{r} + \delta \mathbf{r}) - \mathbf{r}}{\delta t}$$

2.2.1 Successive Derivative

$\frac{d\mathbf{r}}{dt}$ may also have a derivative which is denoted by $\frac{d^2\mathbf{r}}{dt^2}$ and is known as a second derivative of **r**. Similarly, $\frac{d^3\mathbf{r}}{dt^3}$ is known as third derivative, *i.e.*, derivative of $\frac{d^2\mathbf{r}}{dt^2}$. Also $\frac{d\mathbf{r}}{dt}, \frac{d^2\mathbf{r}}{dt}, \ldots$ are written symbolically as $\dot{\mathbf{r}}, \ddot{\mathbf{r}}, \ldots$ respectively.

2.2.2 Derivative of a Constant Vector

Let **a** be the constant vector. Now, if there is an increment of δt in t, then there occurs no change in **a**, *i.e.*, $\delta \mathbf{a} = 0$.

$$\therefore \quad \frac{d\mathbf{a}}{dt} = \lim_{\delta t \to 0} \frac{\delta \mathbf{a}}{\delta t} = 0.$$

Hence the *derivative of a constant vector* **a** *is zero*.

2.2.3 Derivative of a sum of vectors

Let **a** and **b** be two vectors which are functions of variable scalar t and $\mathbf{a} + \mathbf{b}$ is their vector sum.

Let $\delta \mathbf{a}$ and $\delta \mathbf{b}$ be the increments due to the increment δt.

Then
$$\frac{d}{dt}(\mathbf{a} + \mathbf{b}) = \lim_{\delta t \to 0} \frac{(\mathbf{a} + \delta \mathbf{a} + \mathbf{b} + \delta \mathbf{b}) - (\mathbf{a} + \mathbf{b})}{\delta t}$$
$$= \lim_{\delta t \to 0} \frac{\delta \mathbf{a} + \delta \mathbf{b}}{\delta t} = \lim_{\delta t \to 0} \frac{\delta \mathbf{a}}{\delta t} + \lim_{\delta t \to 0} \frac{\delta \mathbf{b}}{\delta t}$$
$$= \frac{d\mathbf{a}}{dt} + \frac{d\mathbf{b}}{dt}.$$

Hence the *derivative of the sum of two vectors is equal to the sum of their derivatives*.

2.3 FUNCTION OF A FUNCTION

Let **r** be a differentiable function of a scalar s and s be the continuous function of another scalar variable t.

Then an increment δt in t produces increment $\delta r, \delta s$ in r and s respectively which tend to zero as $\delta t \to 0$.

Also,
$$\frac{\partial \mathbf{r}}{\partial t} = \frac{\partial \mathbf{r}}{\partial s} \cdot \frac{\partial s}{\partial t}$$
$$\frac{\partial \mathbf{r}}{\partial t} = \lim_{\delta t \to 0} \frac{\partial \mathbf{r}}{\partial s} \cdot \frac{\partial s}{\partial t} = \frac{d\mathbf{r}}{ds} \cdot \frac{ds}{dt},$$

Note. *If the scalar variable t be the time and r be the position vector of a moving particle P relative to the origin O, then δr is the displacement of the point in time δt, $\frac{\partial \mathbf{r}}{\delta t}$ is the average velocity.*

The limiting value of this average velocity as $\delta t \to 0$ is the instantaneous velocity **v** of the particle.

Differentiation of Vectors

Thus $\quad v = \dfrac{dr}{dt} = \dot{r}$

Again, if δv be the increment in the velocity vector v during the interval of time δt, then $\dfrac{\delta v}{\delta t}$ is the instantaneous acceleration of the particle. Thus

$$\text{acceleration} = \dfrac{dv}{dt} = \dfrac{d^2r}{dt^2} = \ddot{r}$$

2.4 DERIVATIVE OF SCALAR PRODUCT OF VECTORS

Let a and b two differentiable vector functions of the scalar t. If δa and δb be the incremenets in a and b respectively due to the increment δt in t, then the increment in the product $a \cdot b$ is

$$\delta(a \cdot b) = (a + \delta a) \cdot (b + \delta b) - (a \cdot b)$$
$$= \delta a \cdot b + a \cdot \delta b + \delta a \cdot \delta b$$

Since δa and δb are very small, hence their product $\delta a \cdot \delta b$ may be neglected. Dividing by δt and taking the limits as $\delta t \to 0$, we obtain

$$\dfrac{d}{dt}(a \cdot b) = \dfrac{da}{dt} \cdot b + a \cdot \dfrac{db}{dt} \qquad \text{...(i)}$$

Corollay 1. If $a = b$, then the abvoe formula gives

$$\dfrac{d}{dt}(a \cdot b) = \dfrac{da}{dt} \cdot a + a \cdot \dfrac{db}{dt} \quad \text{or} \quad \dfrac{d}{dt}(a \cdot a) = \dfrac{d}{dt}(a^2) = 2a \cdot \dfrac{da}{dt} \qquad \text{...(ii)}$$

But $a^2 = a^2$, where a is the modulus of a. Therefore

$$\dfrac{d}{dt}(a^2) = \dfrac{d}{dt}(a^2)$$

so that $\quad 2a \cdot \dfrac{da}{dt} = 2a \dfrac{da}{dt}$

Hence $\quad a \cdot \dfrac{da}{dt} = a \dfrac{da}{dt} \qquad \text{...(iii)}$

Corollary 2. In the case of a vector a of constant length, a^2 is constant and therefore

$$\dfrac{d}{dt} a^2 = 0 \quad \text{or} \quad a \cdot \dfrac{da}{dt} = 0,$$

hence the derivative of a vector of constant length is perpendicular to the vector.

2.4.1 Derivative of Vector Product of Vectors

Here the increment in the vector product $a \times b$ is

$$\delta(a \times b) = (a + \delta a) \times (b + \delta b) - (a \times b)$$
$$= \delta a \times b + a \times \delta b + \delta a \times \delta b$$

Neglecting $\delta a \times \delta b$ and dividing by δt and taking limits as $\delta t \to 0$ we obtain

$$\dfrac{d}{dt}(a \times b) = \dfrac{da}{dt} \times b + a \times \dfrac{db}{dt} \qquad \text{...(i)}$$

Note. *In the formula, the order of the factors in any term must not be changed unless the sign is changed at the same time, as both the terms are vector products of two vectors.*

Corollary 1. Replacing b by $\dfrac{da}{dt}$, we have from (i)

$$\dfrac{d}{dt}\left(a \times \dfrac{da}{dt}\right) = \dfrac{da}{dt} \times \dfrac{da}{dt} + a \times \dfrac{d^2a}{dt}$$

$$\therefore \quad \dfrac{d}{dt}\left(a \times \dfrac{da}{dt}\right) = a \times \dfrac{d^2a}{dt^2} \qquad \text{...(ii)}$$

(Since the cross product of two equal vectors $\dfrac{d\mathbf{a}}{dt}$ is zero).

Corollary 2. If $\mathbf{a} = a\hat{\mathbf{a}}$, where $\hat{\mathbf{a}}$ is the unit-directional unit vector, then

$$\mathbf{a} \times \dfrac{d\mathbf{a}}{dt} = a\hat{\mathbf{a}} \times \dfrac{d}{dt}(a\hat{\mathbf{a}})$$

$$= a\hat{\mathbf{a}} \times \left(\dfrac{da}{dt}\hat{\mathbf{a}} + a\dfrac{d\hat{\mathbf{a}}}{dt} \right) = \left(a\dfrac{da}{dt} \right)(\hat{\mathbf{a}} \times \hat{\mathbf{a}}) + a^2\hat{\mathbf{a}} \dfrac{d\hat{\mathbf{a}}}{dt}$$

$$= a^2\hat{\mathbf{a}} \times \dfrac{d\hat{\mathbf{a}}}{dt}, \text{ as } \hat{\mathbf{a}} \times \hat{\mathbf{a}} = \mathbf{0}$$

$$= \mathbf{0}, \text{ since } \dfrac{d\hat{\mathbf{a}}}{dt} = 0 \qquad \ldots\text{(iii)}$$

Hence $\mathbf{a} \times \dfrac{d\mathbf{a}}{dt} = \mathbf{0}$

2.4.2 Derivative of Scalar triple Product [abc]

$$\dfrac{d}{dt}[\mathbf{a}\ \mathbf{b}\ \mathbf{c}] = \dfrac{d}{dt}\{\mathbf{a} \cdot (\mathbf{b} \times \mathbf{c})\}$$

$$= \dfrac{d\mathbf{a}}{dt} \cdot (\mathbf{b} \times \mathbf{c}) + \mathbf{a} \cdot \dfrac{d}{dt}(\mathbf{b} \times \mathbf{c})$$

$$= \dfrac{d\mathbf{a}}{dt} \cdot (\mathbf{b} \times \mathbf{c}) + \mathbf{a} \cdot \left\{ \dfrac{d\mathbf{b}}{dt} \times \mathbf{c} + \mathbf{b} \times \dfrac{d\mathbf{c}}{dt} \right\}$$

$$= \dfrac{d\mathbf{a}}{dt} \cdot (\mathbf{b} \times \mathbf{c}) + \mathbf{c} \cdot \left\{ \dfrac{d\mathbf{b}}{dt} \times \mathbf{c} + \mathbf{b} \times \dfrac{d\mathbf{c}}{dt} \right\}$$

$$= \dfrac{d\mathbf{a}}{dt} \cdot (\mathbf{b} \times \mathbf{c}) + \mathbf{a} \cdot \left\{ \dfrac{d\mathbf{b}}{dt} \times \mathbf{c} \right\} + \mathbf{a} \cdot \left(\mathbf{b} \times \dfrac{d\mathbf{c}}{dt} \right)$$

$$= \left[\dfrac{d\mathbf{a}}{dt}\ \mathbf{b}\ \mathbf{c} \right] + \left[\mathbf{a}\ \dfrac{d\mathbf{b}}{dt}\ \mathbf{c} \right] + \left[\mathbf{a}\ \mathbf{b}\ \dfrac{d\mathbf{c}}{dt} \right]$$

Where the cyclic order in each term is maintained throughout.

2.4.3 Derivative of $\mathbf{a} \times (\mathbf{b} \times \mathbf{c})$

$$\dfrac{d}{dt}\{\mathbf{a} \times (\mathbf{b} \times \mathbf{c})\} = \dfrac{d\mathbf{a}}{dt} \times (\mathbf{b} \times \mathbf{c}) + \mathbf{a} \times \dfrac{d}{dt}(\mathbf{b} \times \mathbf{c})$$

$$= \dfrac{d\mathbf{a}}{dt} \times (\mathbf{b} \times \mathbf{c}) + \mathbf{a} \times \left[\dfrac{d\mathbf{b}}{dt} \times \mathbf{c} + \mathbf{b} \times \dfrac{d\mathbf{c}}{dt} \right]$$

$$= \dfrac{d\mathbf{a}}{dt} \times (\mathbf{b} \times \mathbf{c}) + \mathbf{a} \times \left(\dfrac{d\mathbf{b}}{dt} \times \mathbf{c} \right) + \mathbf{a} \times \left(\mathbf{b} \times \dfrac{d\mathbf{c}}{dt} \right)$$

Where the order of the factors is maintained in each term.

EXAMPLES

1. *If* $\mathbf{r} = a\cos t\ \mathbf{i} + a\sin t\ \mathbf{j} + t\ \mathbf{k}$, *find* $\dfrac{d\mathbf{r}}{dt}, \dfrac{d^2\mathbf{r}}{dt^2}, \left| \dfrac{d^2\mathbf{r}}{dt^2} \right|$.

Sol. We have

$$\dfrac{d\mathbf{r}}{dt} = \dfrac{d}{dt}(a\cos t)\mathbf{i} + \dfrac{d}{dt}(a\sin t)\mathbf{j} + \dfrac{d}{dt}(t)\mathbf{k}$$

$$= -a\sin t\ \mathbf{i} + a\cos t\ \mathbf{j} + \mathbf{k}$$

Differentiation of Vectors

$$\frac{d^2\mathbf{r}}{dt^2} = \frac{d}{dt}(-a\sin t)\mathbf{i} + \frac{d}{dt}(a\cos t)\mathbf{j} + \frac{d}{dt}(\mathbf{k})$$

$$= -a\cos t\,\mathbf{i} - a\sin t\,\mathbf{j}$$

Hence $\left|\dfrac{d^2\mathbf{r}}{dt^2}\right| = \sqrt{(-a\cos t)^2 + (-a\sin t)^2} = a$

2. *A particle moves along the curve*

$$x = t^3 + 1,\ y = t^2,\ z = 2t + 5,$$

where t is the time. Find the components of its velocity and acceleration at time $t = 1$ *in the direction* $\mathbf{i} + \mathbf{j} + 3\mathbf{k}$.

Sol. Unit vector in the direction of $\mathbf{i} + \mathbf{j} + 3\mathbf{k}$ is

$$= \frac{\mathbf{i} + \mathbf{j} + 3\mathbf{k}}{\sqrt{1+1+9}} = \frac{\mathbf{i} + \mathbf{j} + 3\mathbf{k}}{\sqrt{11}} \qquad\qquad ...(i)$$

Now, velocity $= \dfrac{d\mathbf{r}}{dt} = \dfrac{d}{dt}(x\mathbf{i} + y\mathbf{j} + 3\mathbf{k})$

$$= \frac{d}{dt}\{(t^3+1)\mathbf{i} + t^2\mathbf{j} + (2t+5)\mathbf{k}\}$$

$$= 3t^2\mathbf{i} + 2t\,\mathbf{j} + 2\mathbf{k}$$

$$= 3\mathbf{i} + 2\mathbf{j} + 2\mathbf{k} \text{ at } t = 1 \qquad\qquad ...(ii)$$

Hence the component of velocity at $t = 1$ in the direction of the vector $\mathbf{i} + \mathbf{j} + 3\mathbf{k}$ is

$$= (3\mathbf{i} + 3\mathbf{j} + 2\mathbf{k}) \cdot (\text{unit vector in the direction of } \mathbf{i} + \mathbf{j} + 3\mathbf{k})$$

$$= (3\mathbf{i} + 2\mathbf{j} + 2\mathbf{k}) \cdot \frac{\mathbf{i} + \mathbf{j} + 3\mathbf{k}}{\sqrt{11}}$$

$$= \frac{3 + 2 + 6}{\sqrt{11}} = \sqrt{11}$$

Again, acceleration $= \dfrac{d^2\mathbf{r}}{dt^2}$

$$= \frac{d}{dt}\left(\frac{d\mathbf{r}}{dt}\right) = \frac{d}{dt}(3t^2\mathbf{i} + 2t\mathbf{j} + 2\mathbf{k})$$

$$= 6t\,\mathbf{i} + 2\mathbf{j}$$

$$= 6\mathbf{i} + 2\mathbf{j} \text{ at } t = 1 \qquad\qquad ...(iii)$$

Hence, the component of acceleration at $t = 1$ in the direction of the vector $\mathbf{i} + \mathbf{j} + 3\mathbf{k}$ is

$$= (6\mathbf{i} + 2\mathbf{j}) \cdot (\text{unit vector along } \mathbf{i} + \mathbf{j} + 3\mathbf{k})$$

$$= (6\mathbf{i} + 2\mathbf{j}) \cdot \left(\frac{\mathbf{i} + \mathbf{j} + 3\mathbf{k}}{\sqrt{11}}\right)$$

$$= \frac{6+2}{\sqrt{11}} = \frac{8}{\sqrt{11}}$$

EXERCISES

1. If $\mathbf{r} = 3\mathbf{i} - 6t^2\mathbf{j} + 4t\,\mathbf{k}$, find $\dfrac{d\mathbf{r}}{dt}$ and $\dfrac{d^2\mathbf{r}}{dt^2}$. [Ans. $-12t\,\mathbf{j} + 4\mathbf{k}, -12\mathbf{j}$]

2. Show that if $\mathbf{a}, \mathbf{b}, \mathbf{c}$ are constant vectors, then

$$\mathbf{r} = \mathbf{a}t^2 + \mathbf{b}t + \mathbf{c}$$

is the path of a point moving with constant acceleration.

3. If $\mathbf{r} = (1 - \cos t)\mathbf{i} + (t - \sin t)\mathbf{j}$, find $\dfrac{d\mathbf{r}}{dt}$ and $\dfrac{d^2\mathbf{r}}{dt^2}$.

[Ans. $(\sin t)\mathbf{i} + (1 - \cos t)\mathbf{j}, (\cos t)\mathbf{i} + (\sin t)\mathbf{j}$]

4. If $\mathbf{r} = t^2\mathbf{i} + (\sin t)\mathbf{j} + (1 - t)\mathbf{k}$, find $\dfrac{d\mathbf{r}}{dt}$ and $\dfrac{d^2\mathbf{r}}{dt^2}$.

5. A particle moves along the curve $x = 2t^2$, $y = t^2 - 4t$, $z = 3t - 5$ where t is the time. Find the components of its velocity and acceleration at time $t = 1$ in the direction $\mathbf{i} - 3\mathbf{j} + 2\mathbf{k}$.

$$\left[\text{Ans. } v = \frac{16}{\sqrt{14}}, a = \frac{-2}{\sqrt{14}} \right]$$

MORE EXAMPLES

1. *If $\hat{\mathbf{r}}$ be the unit vector in the direction of $\mathbf{r}$, show that*

$$\hat{\mathbf{r}} \times d\hat{\mathbf{r}} = \frac{\mathbf{r} \times d\mathbf{r}}{r^2}$$

Sol. We have

$$d\hat{\mathbf{r}} = d\left(\frac{\mathbf{r}}{r}\right) = \frac{1}{r} d\mathbf{r} - \frac{\mathbf{r}}{r^2} dr$$

$$\therefore \quad \hat{\mathbf{r}} \times d\hat{\mathbf{r}} = \hat{\mathbf{r}} \times \left(\frac{1}{r} d\mathbf{r} - \frac{\mathbf{r}}{r^2} dr\right)$$

$$= \frac{\mathbf{r}}{r} \times \left(\frac{1}{r} d\mathbf{r} - \frac{\mathbf{r}}{r^2} dr\right)$$

$$= \frac{\mathbf{r} \times d\mathbf{r}}{r^2} - \frac{\mathbf{r} \times \mathbf{r}}{r^3} dr$$

$$= \frac{\mathbf{r} \times d\mathbf{r}}{r^2} \quad \text{as } \mathbf{r} \times \mathbf{r} = 0$$

2. *If $\mathbf{r} = (a \cos t)\mathbf{i} + (a \sin t)\mathbf{j} + (at \tan \alpha)\mathbf{k}$ find $\left|\dfrac{d\mathbf{r}}{dt} \times \dfrac{d^2\mathbf{r}}{dt^2}\right|$ and $\left[\dfrac{d\mathbf{r}}{dt} \dfrac{d^2\mathbf{r}}{dt^2} \dfrac{d^3\mathbf{r}}{dt^3}\right]$.*

Sol. We have,

$$\mathbf{r} = (a \cos t)\mathbf{i} + (a \sin t)\mathbf{j} + (at \tan \alpha)\mathbf{k}$$

Then $\dfrac{d\mathbf{r}}{dt} = (-a \sin t)\mathbf{i} + (a \cos t)\mathbf{j} + (a \tan \alpha)\mathbf{k}$

$$\dfrac{d^2\mathbf{r}}{dt^2} = (-a \cos t)\mathbf{i} + (-a \sin t)\mathbf{j}$$

$$\dfrac{d^3\mathbf{r}}{dt^3} = (a \sin t)\mathbf{i} - (a \cos t)\mathbf{j}$$

$$\therefore \quad \frac{d\mathbf{r}}{dt} \times \frac{d^2\mathbf{r}}{dt^2} = \begin{vmatrix} \mathbf{i} & \mathbf{j} & \mathbf{k} \\ -a \sin t & a \cos t & a \tan \alpha \\ -a \cos t & -a \sin t & 0 \end{vmatrix}$$

$$= \mathbf{i}(0 + a^2 \tan \alpha \sin t) + \mathbf{j}(-a^2 \cos t \tan \alpha - 0)$$
$$+ \mathbf{k}(a^2 \sin^2 t + a^2 \cos^2 t)$$

$$= a^2 \tan \alpha \sin t \, \mathbf{i} - a^2 \cos \alpha \, \mathbf{j} + a^2 \mathbf{k}$$

Differentiation of Vectors

$$\therefore \quad \left| \frac{d\mathbf{r}}{dt^2} \times \frac{d^2\mathbf{r}}{dt^2} \right| = \sqrt{a^4 \tan^2 \alpha \sin^2 t + a^4 \cos^2 t \tan^2 \alpha + a^4}$$

$$= a^2 \sqrt{\tan^2 \alpha (\sin^2 t + \cos^2 t) + 1}$$

$$= a^2 \sqrt{\tan^2 \alpha + 1} = a^2 \sec \alpha.$$

$$\left[\frac{d\mathbf{r}}{dt} \; \frac{d^2\mathbf{r}}{dt^2} \; \frac{d^3\mathbf{r}}{dt^3} \right] = \begin{vmatrix} -a \sin t & a \cos t & a \tan \alpha \\ -a \cos t & -a \sin t & 0 \\ a \sin t & -a \cos t & 0 \end{vmatrix}$$

$$= a \tan \alpha [a^2 \cos^2 t + a^2 \sin^2 t], \text{ expanding w.r.t. } c_3.$$

$$= a^3 \tan \alpha.$$

3. *If* $\dfrac{d\mathbf{a}}{dt} = \mathbf{c} \times \mathbf{a}, \; \dfrac{d\mathbf{b}}{dt} = \mathbf{c} \times \mathbf{b},$ *show that* $\dfrac{d}{dt}(\mathbf{a} \times \mathbf{b}) = \mathbf{c} \times (\mathbf{a} \times \mathbf{b}).$

Sol. We have

$$\frac{d}{dt}(\mathbf{a} \times \mathbf{b}) = \frac{d\mathbf{a}}{dt} \times \mathbf{b} + \mathbf{a} \times \frac{d\mathbf{b}}{dt}$$

$$= (\mathbf{c} \times \mathbf{a}) \times \mathbf{b} + \mathbf{a} \times (\mathbf{c} \times \mathbf{b})$$

$$= -\mathbf{b} \times (\mathbf{c} \times \mathbf{a}) + \mathbf{a} \times (\mathbf{c} \times \mathbf{b})$$

$$= -\{(\mathbf{b} \cdot \mathbf{a}) \mathbf{c} - (\mathbf{b} \cdot \mathbf{c}) \mathbf{a}\} + (\mathbf{a} \cdot \mathbf{b}) \mathbf{c} - (\mathbf{a} \cdot \mathbf{c}) \mathbf{b}$$

$$= (\mathbf{b} \cdot \mathbf{c}) \mathbf{a} - (\mathbf{a} \cdot \mathbf{c}) \mathbf{b}$$

$$= (\mathbf{c} \cdot \mathbf{b}) \mathbf{a} - (\mathbf{c} \cdot \mathbf{a}) \mathbf{b} = \mathbf{c} \times (\mathbf{a} \times \mathbf{b}).$$

4. *Prove that the necessary and sufficient condition for the vector* $\mathbf{a}(t)$ *to have constant magnitude is*

$$\mathbf{a} \cdot \frac{d\mathbf{a}}{dt} = 0.$$

Sol. *The condition is necessary.* If $\mathbf{a}$ is a vector of constant magnitude, then

$$\mathbf{a} \cdot \mathbf{a} = \mathbf{a}^2 = a^2 = \text{constant}$$

Differentiating with respect to t, we get

$$\mathbf{a} \cdot \frac{d\mathbf{a}}{dt} + \frac{d\mathbf{a}}{dt} \cdot \mathbf{a} = 0$$

or

$$\mathbf{a} \cdot \frac{d\mathbf{a}}{dt} = 0$$

Condition is sufficient. If $\mathbf{a} \cdot \dfrac{d\mathbf{a}}{dt} = 0$, then

$$\frac{1}{2} \frac{d}{dt}(\mathbf{a} \cdot \mathbf{a}) = 0$$

or

$$(\mathbf{a} \cdot \mathbf{a}) = \text{constant}$$

or

$$a^2 = \text{constant}.$$

EXERCISES

1. If $\mathbf{r}$ is a vector function of a scalar t and $\mathbf{a}$ is a constant vector, differentiate the following expressions with respect to t :

(i) $\mathbf{r} \cdot \mathbf{a}$, (ii) $\mathbf{r} \times \mathbf{a}$, (iii) $\mathbf{r} \cdot \mathbf{r}$, (iv) $\mathbf{r} \times \mathbf{r}$.

[**Ans.** (i) $(\dot{\mathbf{r}} \cdot \mathbf{a})$, (ii) $(\dot{\mathbf{r}} \times \mathbf{a})$, (iii) $(2\mathbf{r} \cdot \dot{\mathbf{r}})$, (iv) 0.]

2. If $\mathbf{r} = (\cos nt)\mathbf{i} + (\sin nt)\mathbf{j}$, show that (a) $\mathbf{r} \times \dfrac{d\mathbf{r}}{dt} = n\mathbf{k}$, (b) $\mathbf{r} \cdot \dfrac{d\mathbf{r}}{dt} = 0$, (c) $\dfrac{d^2\mathbf{r}}{dt^2} = -n^2\mathbf{r}$.

3. Show that
$$\dfrac{d^2}{dt^2}\left(\mathbf{r} \times \dfrac{d\mathbf{r}}{dt}\right) = \dfrac{d\mathbf{r}}{dt} \times \dfrac{d^2\mathbf{r}}{dt^2} + \mathbf{r} \times \dfrac{d^3\mathbf{r}}{dt^3}$$

4. If $\mathbf{r} = \mathbf{a}\cos\omega t + \mathbf{b}\sin\omega t$, where $\mathbf{a}, \mathbf{b}, \omega$ are constants, show that $\mathbf{r} \times \dfrac{d\mathbf{r}}{dt} = \omega\mathbf{a} \times \mathbf{b}$ and $\dfrac{d^2\mathbf{r}}{dt^2} + \omega^2 \mathbf{r} = 0$.

[**Ans.** $a^2 \sec\alpha, a^3 \tan\alpha$]

5. If $\mathbf{r} \times d\mathbf{r} = 0$, show that $\hat{\mathbf{r}} = $ constant.

6. Show that the necessary and sufficient condition for the vector $\mathbf{a}(t)$ to have a fixed direction $\mathbf{a} \times \dfrac{d\mathbf{a}}{dt} = 0$.

7. If $\mathbf{r}$ is the position vecor of a moving point and r is the modulus of $\mathbf{r}$, show that
$$\mathbf{r} \cdot \dfrac{d\mathbf{r}}{dt} = r\dfrac{dr}{dt}$$
and interpret the relations
$$\mathbf{r} \cdot \dfrac{d\mathbf{r}}{dt} = 0 \text{ and } \mathbf{r} \times \dfrac{d\mathbf{r}}{dt} = 0$$

8. If $\mathbf{r}$ is a unit vector, then prove that
$$\left|\mathbf{r} \times \dfrac{d\mathbf{r}}{dt}\right| = \left|\dfrac{d\mathbf{r}}{dt}\right|$$

3
Differential Operators

3.1 PARTIAL DERIVATIVES

In chapter 2 we have considered a vector function of a single scalar variable t i.e., $\mathbf{f}(t)$. Now we shall consider a vector function of several scalar variables. A vector function of two scalar variables say u, v is $\mathbf{f}(u, v)$ and $\dfrac{\partial \mathbf{f}}{\partial u}$ is the partial derivative of $\mathbf{f}(u, v)$ with respect to u,

i.e.,
$$\frac{\partial \mathbf{f}}{\partial u} = \lim_{\delta u \to 0} \frac{\mathbf{f}(u + \delta u, v) - \mathbf{f}(u, v)}{\delta u}$$

Similarly,
$$\frac{\partial \mathbf{f}}{\partial v} = \lim_{\delta v \to 0} \frac{\mathbf{f}(u, v + \delta v) - \mathbf{f}(u, v)}{\delta v}$$

In case $\mathbf{f}(u, v) = f_1(u, v)\mathbf{i} + f_2(u, v)\mathbf{j} + f_3(u, v)\mathbf{k}$

then
$$\frac{\partial \mathbf{f}}{\partial u} = \frac{\partial f_1}{\partial u}\mathbf{i} + \frac{\partial f_2}{\partial u}\mathbf{j} + \frac{\partial f_3}{\partial u}\mathbf{k}$$

and
$$\frac{\partial \mathbf{f}}{\partial v} = \frac{\partial f_1}{\partial v}\mathbf{i} + \frac{\partial f_2}{\partial v}\mathbf{j} + \frac{\partial f_3}{\partial v}\mathbf{k}$$

Partial derivatives of second and higher order.

Again $\dfrac{\partial f}{\partial u}$ and $\dfrac{\partial f}{\partial v}$ both are vector functions of two scalar variables u and v and these possesss partial derivatives with respect to u and v.

$$\frac{\partial^2 \mathbf{f}}{\partial u^2} = \frac{\partial}{\partial u}\left(\frac{\partial \mathbf{f}}{\partial u}\right), \quad \frac{\partial^2 \mathbf{f}}{\partial v^2} = \frac{\partial}{\partial v}\left(\frac{\partial \mathbf{f}}{\partial v}\right)$$

$$\frac{\partial^2 \mathbf{f}}{\partial u \, \partial v} = \frac{\partial}{\partial u}\left(\frac{\partial \mathbf{f}}{\partial v}\right), \quad \frac{\partial^2 \mathbf{f}}{\partial v \, \partial u} = \frac{\partial}{\partial v}\left(\frac{\partial \mathbf{f}}{\partial u}\right)$$

Also,
$$\frac{\partial^2 \mathbf{f}}{\partial u \, \partial v} = \frac{\partial^2 \mathbf{f}}{\partial v \, \partial u}.$$

Again, if $\mathbf{r} = \mathbf{f}(u, v)$, where $u = \phi(p, q)$, $v = \psi(p, q)$, i.e., u and v are scalar functions of two scalar variables p and q then

$$\frac{\partial \mathbf{r}}{\partial p} = \frac{\partial \mathbf{f}}{\partial u} \cdot \frac{\partial u}{\partial p} + \frac{\partial \mathbf{f}}{\partial v} \cdot \frac{\partial v}{\partial q}$$

The total change in $\mathbf{f}$ due to simultaneous change in variables u and v is given by

$$d\mathbf{f} = \frac{\partial \mathbf{f}}{\partial u} du + \frac{\partial \mathbf{f}}{\partial v} dv.$$

In partial differentiation of vectors the same laws are followed as in ordinary calculus for scalar functions.

If $\mathbf{r}$ and $\mathbf{s}$ be two vectors functions of x, y, z then we have

(i) $\quad \dfrac{\partial}{\partial x}(\mathbf{r} + \mathbf{s}) = \dfrac{\partial \mathbf{r}}{\partial x} + \dfrac{\partial \mathbf{s}}{\partial x}$

(ii) $\quad \dfrac{\partial}{\partial x}(\mathbf{r}\cdot\mathbf{s}) = \mathbf{r}\cdot\dfrac{\partial \mathbf{s}}{\partial x} + \dfrac{\partial \mathbf{r}}{\partial x}\cdot\mathbf{s}$

(iii) $\quad \dfrac{\partial}{\partial x}(\mathbf{r}\times\mathbf{s}) = \mathbf{r}\times\dfrac{\partial \mathbf{s}}{\partial x} + \dfrac{\partial \mathbf{r}}{\partial x}\times\mathbf{s}$

(iv) $\quad \dfrac{\partial^2}{\partial y\,\partial x}(\mathbf{r}\cdot\mathbf{s}) = \dfrac{\partial}{\partial y}\left\{\dfrac{\partial}{\partial x}(\mathbf{r}\cdot\mathbf{s})\right\}$

$\qquad = \dfrac{\partial}{\partial y}\left\{\mathbf{r}\cdot\dfrac{\partial \mathbf{s}}{\partial x} + \dfrac{\partial \mathbf{r}}{\partial x}\cdot\mathbf{s}\right\}$

$\qquad = \dfrac{\partial}{\partial y}\left(\mathbf{r}\cdot\dfrac{\partial \mathbf{s}}{\partial x}\right) + \dfrac{\partial}{\partial y}\left(\dfrac{\partial \mathbf{r}}{\partial x}\cdot\mathbf{s}\right)$

$\qquad = \mathbf{r}\cdot\dfrac{\partial^2 \mathbf{s}}{\partial y\,\partial x} + \dfrac{\partial \mathbf{r}}{\partial y}\cdot\dfrac{\partial \mathbf{s}}{\partial x} + \dfrac{\partial^2 \mathbf{r}}{\partial y\,\partial x}\cdot\mathbf{s} + \dfrac{\partial \mathbf{r}}{\partial x}\cdot\dfrac{\partial \mathbf{s}}{\partial y}$

(v) $\quad \dfrac{\partial^2}{\partial y\,\partial x}(\mathbf{r}\times\mathbf{s}) = \mathbf{r}\times\dfrac{\partial^2 \mathbf{s}}{\partial y\,\partial x} + \dfrac{\partial \mathbf{r}}{\partial y}\times\dfrac{\partial \mathbf{s}}{\partial x} + \dfrac{\partial \mathbf{r}}{\partial x}\times\dfrac{\partial \mathbf{s}}{\partial y} + \dfrac{\partial^2 \mathbf{r}}{\partial y\,\partial x}\times\mathbf{s}$

EXAMPLE

If $\mathbf{a} = xyz\,\mathbf{i} + x^2 z\,\mathbf{j} - y^3\mathbf{k}$ and $\mathbf{b} = x^3\,\mathbf{i} - xyz\,\mathbf{j} + x^3 z\,\mathbf{k}$, calculate $\dfrac{\partial^2 \mathbf{a}}{\partial y^2}\times\dfrac{\partial^2 \mathbf{b}}{\partial x^2}$ at the point $(1, 1, 0)$.

Sol. We have

$$\dfrac{\partial \mathbf{a}}{\partial y} = xz\,\mathbf{i} - 3y^2\,\mathbf{k}$$

$\therefore\quad \dfrac{\partial^2 \mathbf{a}}{\partial y^2} = -6y\,\mathbf{k}$

$\dfrac{\partial \mathbf{b}}{\partial x} = 3x^2\mathbf{i} - yz\,\mathbf{j} + 2xz\,\mathbf{k}$

$\therefore\quad \dfrac{\partial^2 \mathbf{b}}{dx^2} = 6x\,\mathbf{i} + 2x\,\mathbf{k}$

$\therefore\quad \dfrac{\partial^2 \mathbf{a}}{\partial y^2}\times\dfrac{\partial^2 \mathbf{b}}{\partial x^2} = (-6y\,\mathbf{k})\times(6x\,\mathbf{i} + 2z\,\mathbf{k})$

$\qquad = -36xy\,\mathbf{j}$

$\qquad = -36\,\mathbf{j}$ at the point $(1, 1, 0)$

EXERCISES

1. If $\mathbf{a} = x^2 yz\,\mathbf{i} - 2xz^3\mathbf{j} + xz^2\mathbf{k}$ and $\mathbf{b} = 2z\,\mathbf{i} + y\,\mathbf{j} - x^2\mathbf{k}$, find $\dfrac{\partial^2}{\partial x\,\partial y}(\mathbf{a}\times\mathbf{b})$ at $(1, 0, -2)$.

[Ans. $-4\,(\mathbf{i} + 2\mathbf{j})$]

2. If $\mathbf{P} = e^{xy}\mathbf{i} + (x - 2y)\,\mathbf{j} + (x\sin y)\,\mathbf{k}$, find the values of $\dfrac{\partial \mathbf{P}}{\partial x}, \dfrac{\partial \mathbf{P}}{\partial y}, \dfrac{\partial^2 \mathbf{P}}{\partial x^2}, \dfrac{\partial^2 \mathbf{P}}{\partial y^2}$ and $\dfrac{\partial^2 \mathbf{P}}{\partial x\,\partial y}$.

[Ans. $\dfrac{\partial \mathbf{P}}{\partial x} = (ye^{xy})\,\mathbf{i} + \mathbf{y} + (\sin y)\,\mathbf{k}$, $\quad \dfrac{\partial \mathbf{P}}{\partial y} = (xe^{xy})\,\mathbf{i} - 2\mathbf{j} + (x\cos y)\,\mathbf{k}$

$\dfrac{\partial^2 \mathbf{P}}{\partial x^2} = (y^2 e^{xy})\,\mathbf{i},\qquad \dfrac{\partial^2 \mathbf{P}}{\partial y^2} = x^2 e^{xy}\,\mathbf{i} - (x\sin y)\,\mathbf{k}$]

Differential Operators

3. If $\phi(x, y, z) = xy^2z$ and $\mathbf{A} = xz\mathbf{i} - xy\mathbf{j} + yz^2\mathbf{k}$ find $\dfrac{\partial^3}{\partial x^2 \partial z}(\phi \mathbf{A})$ at $(2, -1, -1)$.

[Ans. $4\mathbf{i} + 2\mathbf{j}$]

3.2 THE OPERATOR DEL (∇)

$$\nabla \equiv \mathbf{i}\frac{\partial}{\partial x} + \mathbf{j}\frac{\partial}{\partial y} + \mathbf{k}\frac{\partial}{\partial z},$$

and it operates distributively.

Hence
$$\nabla f = \mathbf{i}\frac{\partial f}{\partial x} + \mathbf{j}\frac{\partial f}{\partial y} + \mathbf{k}\frac{\partial f}{\partial z}$$

may be thought of as ∇ operating on f, i.e.,

$$\nabla f = \left(\mathbf{i}\frac{\partial}{\partial x} + \mathbf{j}\frac{\partial}{\partial y} + \mathbf{k}\frac{\partial}{\partial z}\right) f$$

∇ is a vector operator and is called differential operator. As ∇ is made up of three symbolic components along the three axes $\mathbf{i}, \mathbf{j}, \mathbf{k}$ and the symbolic magnitude of these are $\dfrac{\partial}{\partial x}, \dfrac{\partial}{\partial y}, \dfrac{\partial}{\partial z}$ respectively. Hence it may be looked upon as symbolic vector itself.

3.3 SCALAR POINT FUNCTION

If corresponding to each point P of a region R of space there corresponds a scalar denoted by $\phi(P)$ then ϕ is said to be a scalar point function for the region R. If the co-ordinates of P be (x, y, z) then

$$\phi(P) = \phi(x, y, z).$$

As an example, the density $\phi(P)$ at any point P of a certain body occupying given region R is a scalar piont function. Similarly the temperature $\phi(P)$ at any point P of a fluid occupying a certain region is a scalar point function. As another example we may say that the distance of any point P in space from a fixed point P_0 is scalar function.

$$\phi(P) = [(x - x_0)^2 + (y - y_0)^2 + (z - z_0)^2]^{1/2}$$

3.3.1 Vector Point Function

If corresponding to each point P of a region P of space there corresponds a vector defined by $\mathbf{f}(P)$ then $\mathbf{f}$ is called a vector point function for the region R.

If the coordinates of P be (x, y, z) then

$$\mathbf{f}(P) = \mathbf{f}(x, y, z) = f_1(x, y, z)\mathbf{i} + f_2(x, y, z)\mathbf{j} + f_3(x, y, z)\mathbf{k}$$

For example, if the velocity of a particle at any time t occupying the postiion P in a certain region is $\mathbf{f}(P)$ then $\mathbf{f}(P)$ is a vector point function for that region.

3.4 GRADIENT OR SLOPE OF A SCALAR POINT-FUNCTION

If $f(x, y, z)$ be a scalar point function and continuously differentiable then the vector

$$\nabla f = \mathbf{i}\frac{\partial f}{\partial x} + \mathbf{j}\frac{\partial f}{\partial y} + \mathbf{k}\frac{\partial f}{\partial z}$$

is called the gradient of f and is written as grad f.

It should be noted that ∇f is a vector whose three components are $\dfrac{\partial f}{\partial x}, \dfrac{\partial f}{\partial y}, \dfrac{\partial f}{\partial z}$. Thus if f is a scalar point function, then ∇f is a vector point function.

3.4 OPERATOR $\mathbf{a}.\nabla$, $\mathbf{a}$ BEING ANY VECTOR

The operator $\mathbf{a}.\nabla$ is defined by the quantity
$$\mathbf{a}.\nabla = \mathbf{a}.\mathbf{i}\frac{\partial}{\partial x} + \mathbf{a}.\mathbf{j}\frac{\partial}{\partial y} + \mathbf{a}.\mathbf{k}\frac{\partial}{\partial z}$$
where $\mathbf{a} = a_1\mathbf{i} + a_2\mathbf{j} + a_3\mathbf{k}$.
so that we have
$$(\mathbf{a}.\nabla)f = \mathbf{a}.\mathbf{i}\frac{\partial f}{\partial x} + \mathbf{a}.\mathbf{j}\frac{\partial f}{\partial y} + \mathbf{a}.\mathbf{k}\frac{\partial f}{\partial z}$$
For a unit vector $\mathbf{a}$,
$$\mathbf{a}.\mathbf{i}, \mathbf{a}.\mathbf{j}, \mathbf{a}.\mathbf{k}$$
are the direction cosines of $\mathbf{a}$ and hence
$$(\mathbf{a}.\nabla)f, (\mathbf{a}.\nabla)\mathbf{f}$$
stand for the directional derivatives of the respective functions along the directions of the unit vector $\mathbf{a}$.

3.4 TOTAL DIFFERENCE df, WHERE f IS A SCALAR POINT FUNCTION

We have
$$df = \frac{\partial f}{\partial x}dx + \frac{\partial f}{\partial y}dy + \frac{\partial f}{\partial z}dz \qquad \ldots(i)$$
Also, $d\mathbf{r}.(\nabla f) = (\mathbf{i}\,dx + \mathbf{j}\,dy + \mathbf{k}\,dz).\left(\mathbf{i}\frac{\partial f}{\partial x} + \mathbf{j}\frac{\partial f}{\partial y} + \mathbf{k}\frac{\partial f}{\partial z}\right)$
$$= \frac{\partial f}{\partial x}dx + \frac{\partial f}{\partial y}dy + \frac{\partial f}{\partial z}dz \qquad \ldots(ii)$$
From (i) and (ii), we have
$$df = d\mathbf{r}.\nabla f = d\mathbf{r}.\text{grad } f.$$
where total differential is df, and $\mathbf{f}$ is a vector point function.
We have
$$d\mathbf{f} = \frac{\partial \mathbf{f}}{\partial x}dx + \frac{\partial \mathbf{f}}{\partial y}dy + \frac{\partial \mathbf{f}}{\partial z}dz$$
$$= d\mathbf{r}.\mathbf{i}\frac{\partial \mathbf{f}}{\partial x} + d\mathbf{r}.\mathbf{j}\frac{\partial \mathbf{f}}{\partial y} + d\mathbf{r}.\mathbf{k}\frac{\partial \mathbf{f}}{\partial z}$$
$$= (d\mathbf{r}.\nabla)\mathbf{f}$$

3.4.3 Theorem

If f and g are two functions then
$$\text{grad}(f \pm g) = \left(\mathbf{i}\frac{\partial}{\partial x} + \mathbf{j}\frac{\partial}{\partial y} + \mathbf{k}\frac{\partial}{\partial z}\right)(f \pm g)$$
$$= \mathbf{i}\frac{\partial}{\partial x}(f \pm g) + \mathbf{j}\frac{\partial}{\partial y}(f \pm g) + \mathbf{k}\frac{\partial}{\partial z}(f \pm g)$$
$$= \left(\mathbf{i}\frac{\partial f}{\partial x} + \mathbf{j}\frac{\partial f}{\partial y} + \mathbf{k}\frac{\partial f}{\partial z}\right) \pm \left(\mathbf{i}\frac{\partial g}{\partial x} + \mathbf{j}\frac{\partial g}{\partial y} + \mathbf{k}\frac{\partial g}{\partial z}\right)$$
$$= \text{grad } f \pm \text{grad } g$$

Differential Operators

3.4.4 Gradient of a Scalar Product

We have
$$\text{grad}(fg) = \nabla(fg) = \left(\mathbf{i}\frac{\partial}{\partial x} + \mathbf{j}\frac{\partial}{\partial y} + \mathbf{k}\frac{\partial}{\partial z}\right) fg$$

$$= \mathbf{i}\frac{\partial}{\partial x}(fg) + \mathbf{j}\frac{\partial}{\partial y}(fg) + \mathbf{k}\frac{\partial}{\partial z}(fg)$$

$$= \mathbf{i}\left(f\frac{\partial g}{\partial x} + g\frac{\partial f}{\partial x}\right) + \mathbf{j}\left(f\frac{\partial g}{\partial y} + g\frac{\partial f}{\partial y}\right) + \mathbf{k}\left(f\frac{\partial g}{\partial z} + g\frac{\partial f}{\partial z}\right)$$

$$= f\left(\mathbf{i}\frac{\partial g}{\partial x} + \mathbf{j}\frac{\partial g}{\partial y} + \mathbf{k}\frac{\partial g}{\partial z}\right) + g\left(\mathbf{i}\frac{\partial f}{\partial x} + \mathbf{j}\frac{\partial f}{\partial x} + \mathbf{k}\frac{\partial f}{\partial z}\right)$$

$$= f(\text{grad }g) + g(\text{grad }f)$$

or $\quad \nabla(fg) = f\nabla(g) + g\nabla(f)$

3.4.5 Gradient of a Quotient

We have
$$\nabla\left(\frac{f}{g}\right) = \mathbf{i}\frac{\partial}{\partial x}\left(\frac{f}{g}\right) + \mathbf{j}\frac{\partial}{\partial y}\left(\frac{f}{g}\right) + \mathbf{k}\frac{\partial}{\partial z}\left(\frac{f}{g}\right)$$

$$= \mathbf{i}\left(\frac{g\frac{\partial f}{\partial x} - f\frac{\partial g}{\partial x}}{g^2}\right) + \mathbf{j}\left(\frac{g\frac{\partial f}{\partial y} - f\frac{\partial g}{\partial y}}{g^2}\right) + \mathbf{k}\left(\frac{g\frac{\partial f}{\partial z} - f\frac{\partial g}{\partial z}}{g^2}\right)$$

$$= \frac{1}{g^2}\left[g\left(\mathbf{i}\frac{\partial f}{\partial x} + \mathbf{j}\frac{\partial f}{\partial y} + \mathbf{k}\frac{\partial f}{\partial z}\right) - f\left(\mathbf{i}\frac{\partial g}{\partial x} + \mathbf{j}\frac{\partial g}{\partial y} + \mathbf{k}\frac{\partial g}{\partial z}\right)\right]$$

$$= \frac{1}{g^2}[g\nabla f - f\nabla g]$$

$\therefore \quad \text{grad}\left(\frac{f}{g}\right) = \dfrac{g \cdot \text{grad } f - f \cdot \text{grad } g}{g^2}$

3.5 GRADIENT IN POLAR CO-ORDINATES

Let $\mathbf{r}$ be the position vector of a point $P(r, \theta)$. Let $\hat{\mathbf{e}}_r$ be the unit vector along $\mathbf{r}$ (in the sense of r increasing). Let $\hat{\mathbf{e}}_\theta$ be the unit vector along perpendicular to $\mathbf{r}$ (in the sense of θ increasing). Then the distance ds in the direction of $\mathbf{r}$ is dr and directional derviative along $\hat{\mathbf{e}}_r$ is $\hat{\mathbf{e}}_r \cdot \nabla \phi$ where $\phi(x, y, z) = 0$, is the level surface.

Hence, $\quad \dfrac{\partial \phi}{\partial r} = \hat{\mathbf{e}}_r \cdot \nabla \phi \quad$ ...(1)

Again, the distance ds in the direction perpendicular to $\mathbf{r}$ is $r\,d\theta$. Also, directional derivative along $\hat{\mathbf{e}}_\theta \cdot \nabla \phi$.

Hence, $\quad \dfrac{\partial \phi}{r\,\partial \theta} = \hat{\mathbf{e}}_\theta \cdot \nabla \phi. \quad$ ...(2)

Clearly, the components of $\nabla\phi$ along $\hat{\mathbf{e}}_r$ and along $\hat{\mathbf{e}}_\theta$ are respectively $\hat{\mathbf{e}}_r \cdot \nabla\phi$ and $\hat{\mathbf{e}}_\theta \cdot \nabla\phi$.

Hence, $\quad \nabla\phi = (\hat{\mathbf{e}}_r \cdot \nabla\phi)\,\hat{\mathbf{e}}_r + (\hat{\mathbf{e}}_\theta \cdot \nabla\theta)\,\hat{\mathbf{e}}_\theta$

$$\nabla\phi = \frac{\partial\phi}{\partial r}\hat{\mathbf{e}}_r + \frac{1}{r}\frac{\partial\phi}{\partial\theta}\hat{\mathbf{e}}_\theta \qquad \ldots(3)$$

[using (1) and (2)]

Equation (3) expresses gradient in polar coordinates.

EXAMPLES

1. *Find grad ϕ, if $\phi = r^n = (x^2 + y^2 + z^2)^{n/2}$.*

Sol. We have
$$\frac{\partial\phi}{\partial x} = \frac{n}{2}(x^2 + y^2 + z^2)^{\frac{n}{2}-1}\cdot 2x$$
$$= n\cdot(r^2)^{(n-2)/2}\cdot x$$
$$= nr^{n-2}\cdot x$$

Similarly, $\dfrac{\partial\phi}{\partial y} = nr^{n-2}y, \dfrac{\partial\phi}{\partial z} = nr^{n-2}z$

$\therefore$ $\operatorname{grad}\phi = \mathbf{i}\dfrac{\partial\phi}{\partial x} + \mathbf{j}\dfrac{\partial\phi}{\partial y} + \mathbf{k}\dfrac{\partial\phi}{\partial z}$

$\qquad = nr^{n-2}x\,\mathbf{i} + nr^{n-2}y\,\mathbf{j} + nr^{n-2}z\,\mathbf{k}$

$\qquad = nr^{n-2}(\mathbf{i}\,x + \mathbf{j}\,y + \mathbf{k}\,z)$

$\qquad = nr^{n-2}\mathbf{r}.$

2. *Prove that $\operatorname{grad} f(r) \times \mathbf{r} = 0$.*

Sol. We have
$$\operatorname{grad}\{f(r)\} = \mathbf{i}\frac{\partial}{\partial x}f(r) + \mathbf{j}\frac{\partial}{\partial y}f(r) + \mathbf{k}\frac{\partial}{\partial z}f(r)$$
$$= \mathbf{i}\cdot f'(r)\frac{\partial r}{\partial x} + \mathbf{j}\,f'(r)\frac{\partial r}{\partial y} + \mathbf{k}\,f'(r)\frac{\partial r}{\partial z}$$
$$= f'(r)\left[\mathbf{i}\frac{\partial r}{\partial x} + \mathbf{j}\frac{\partial r}{\partial y} + \mathbf{k}\frac{\partial r}{\partial z}\right] \qquad \ldots(i)$$

We have $r^2 = x^2 + y^2 + z^2$

$\therefore$ $2r\dfrac{\partial r}{\partial x} = 2x \qquad$ or $\qquad \dfrac{\partial r}{\partial x} = \dfrac{x}{r}.$

Similarly $\dfrac{\partial r}{\partial y} = \dfrac{y}{r}, \dfrac{\partial r}{\partial z} = \dfrac{z}{r}.$

Hence from (i)
$$\operatorname{grad}\{f(r)\} = f'(r)\left[\mathbf{i}\frac{x}{r} + \mathbf{j}\frac{y}{r} + \mathbf{k}\frac{z}{r}\right]$$
$$= \frac{f'(r)}{r}\mathbf{r}$$

Hence $\operatorname{grad} f(r) \times \mathbf{r} = \dfrac{f'(r)}{r}\mathbf{r} \times \mathbf{r}$

$\qquad\qquad = 0 \qquad [\because \mathbf{r} \times \mathbf{r} = 0]$

3. *If $\phi(x, y) = \log\sqrt{(x^2 + y^2)}$, show that*
$$\operatorname{grad}\phi = \frac{\mathbf{r} - (\mathbf{k}\cdot\mathbf{r})\mathbf{k}}{\{\mathbf{r} - (\mathbf{k}\cdot\mathbf{r})\mathbf{k}\}\cdot\{\mathbf{r} - (\mathbf{k}\cdot\mathbf{r})\mathbf{k}\}}$$

Sol. We have

$\mathbf{r} = x\mathbf{i} + y\mathbf{j} + z\mathbf{k} \qquad \therefore \quad \mathbf{r}\cdot\mathbf{k} = z \qquad \ldots(i)$

Differential Operators

Now, $\phi = \frac{1}{2} \log (x^2 + y^2)$

$\therefore \quad \frac{\partial \phi}{\partial x} = \frac{1}{2(x^2 + y^2)} \cdot 2x = \frac{x}{x^2 + y^2}$

Similarly, $\frac{\partial \phi}{\partial y} = \frac{y}{x^2 + y^2}, \frac{\partial \phi}{\partial z} = 0$

$\therefore \quad \text{grad } \phi = \mathbf{i} \frac{\partial \phi}{\partial x} + \mathbf{j} \frac{\partial \phi}{\partial y} + \mathbf{k} \frac{\partial \phi}{\partial z}$

$= \frac{x}{x^2 + y^2} \mathbf{i} + \frac{y}{(x^2 + y^2)} \mathbf{j} + 0 \cdot \mathbf{k}$

$= \frac{x\mathbf{i} + y\mathbf{j}}{x^2 + y^2} = \frac{\mathbf{r} - z\mathbf{k}}{(x\mathbf{i} + y\mathbf{j}) \cdot (x\mathbf{i} + y\mathbf{j})}$

$= \frac{\mathbf{r} - z\mathbf{k}}{(\mathbf{r} - z\mathbf{k}) \cdot (\mathbf{r} - z\mathbf{k})}$, By (i).

Now, by replacing z by $\mathbf{r.k}$, we get

$$\text{grad } \phi = \frac{\mathbf{r} - (\mathbf{k} \cdot \mathbf{r}) \mathbf{k}}{\{\mathbf{r} - (\mathbf{k} \cdot \mathbf{r}) \mathbf{k}\} \cdot \{\mathbf{r} - (\mathbf{k} \cdot \mathbf{r}) \mathbf{k}\}}$$

4. *If $\phi = \log |\mathbf{r}|$, then show that* $\text{grad } \phi = \frac{\mathbf{r}}{r^2}$.

Sol. Let $|\mathbf{r}| = r$, then $r^2 = x^2 + y^2 + z^2$

$\Rightarrow \quad 2r \frac{\partial r}{\partial x} = 2x \quad \text{or} \quad \frac{\partial r}{\partial x} = \frac{x}{r}$

Similarly, $\frac{\partial r}{\partial y} = \frac{y}{r}, \frac{\partial r}{\partial z} = \frac{z}{r}$

Now grad $\{\log |\mathbf{r}|\} = \text{grad} (\log r)$

$= \mathbf{i} \frac{\partial}{\partial x} (\log r) + \mathbf{j} \frac{\partial}{\partial y} (\log r) + \mathbf{k} \frac{\partial}{\partial z} (\log r)$

$= \mathbf{i} \left(\frac{1}{r} \frac{\partial r}{\partial x} \right) + \mathbf{j} \left(\frac{1}{r} \frac{\partial r}{\partial y} \right) + \mathbf{k} \left(\frac{1}{r} \frac{\partial r}{\partial z} \right)$

$= \mathbf{i} \left(\frac{1}{r} \cdot \frac{x}{r} \right) + \mathbf{j} \left(\frac{1}{r} \cdot \frac{y}{r} \right) + \mathbf{k} \left(\frac{1}{r} \cdot \frac{z}{r} \right)$

$= \frac{(\mathbf{i} x + \mathbf{j} y + \mathbf{k} z)}{r^2} = \frac{\mathbf{r}}{r^2}$.

5. *If $\mathbf{a}$ and $\mathbf{b}$ be constant vectors then show that* $\text{grad } [\mathbf{r} \, \mathbf{a} \, \mathbf{b}] = \mathbf{a} \times \mathbf{b}$.

Sol. Let $\mathbf{r} = x\mathbf{i} + y\mathbf{j} + z\mathbf{k}, \mathbf{a} = a_1\mathbf{i} + a_2\mathbf{j} + a_3\mathbf{k}$

and $\mathbf{b} = b_1 \mathbf{i} + b_2 \mathbf{j} + b_3 \mathbf{k}$

Then

$$\phi = [\mathbf{r} \, \mathbf{a} \, \mathbf{b}] = \begin{vmatrix} x & y & z \\ a_1 & a_2 & a_3 \\ b_1 & b_2 & b_3 \end{vmatrix}$$

$= x (a_2 b_3 - a_3 b_2) + y (a_3 b_1 - a_1 b_3) + z (a_1 b_2 - a_2 b_1)$

Now,
$$\frac{\partial \phi}{\partial z} = (a_2b_3 - a_3b_2), \quad \frac{\partial \phi}{\partial y} = a_3b_1 - a_1b_3$$

and
$$\frac{\partial \phi}{\partial z} = (a_1b_2 - a_2b_1)$$

∴
$$\text{grad } \phi = \text{grad } [\mathbf{r\ a\ b}] = \mathbf{i}\frac{\partial \phi}{\partial x} + \mathbf{j}\frac{\partial \phi}{\partial y} + \mathbf{k}\frac{\partial \phi}{\partial z}$$
$$= \mathbf{i}(a_2b_3 - a_3b_2) + \mathbf{j}(a_3b_1 - a_1b_3) + \mathbf{k}(a_1b_2 - a_2b_1)$$
$$= \mathbf{a} \times \mathbf{b}.$$

EXERCISES

1. Prove the grad $\left(\dfrac{1}{r}\right) = -\dfrac{\mathbf{r}}{r^3}$.

2. Show that grad $(\mathbf{r} \cdot \mathbf{r}) = 2\mathbf{r}$.

3. If $\phi = f(r)$, then show that grad $\phi = \dfrac{f'(r)}{r}\mathbf{r}$.

4. If $f = 3x^2y - y^3z^2$ find grad f at the point $(1, -2, -1)$. [Ans. $-12\mathbf{i} - 9\mathbf{j} - 16\mathbf{k}$]

5. If $\phi = (3r^2 - 4\sqrt{r} + 6r^{-1/3})$ find $\nabla \phi$. [Ans. $2(3 - r^{-3/2} - r^{-7/3})\mathbf{r}$]

3.6 SCALAR AND VECTORS FIELDS

If to every point in a region, finite or infinite there corresponds a definite value of some physical property, the region is called a **field**.

If this property is a scalar, the field is called a *scalar field* for example density at all points, or potential at all points, or temperature at any given instant are scalar fields.

If this property is a vector, the field is known as *vector field*. For example the velocity at all pionts of a fluid or intensity of electric field at all points are the vector fields.

Equipotential or Level Surfaces

Let $\phi(x, y, z)$ be a scalar point function over a certain region. All those points which satisfy an equation of the type.

$\phi(x, y, z) = \text{constant} = c$ will constitute a family of surfaces which are called *level surfaces*. For all points on a member of the above family of surfaces the function $\phi(x, y, z)$ will be the same.

3.7 DIRECTIONAL DERIVATIVE OF A FUNCTION

If s represents a distance from any point $P(x, y, z)$ on the level surface $f(x, y, z) = 0$, in the direction of a unit vector $\hat{\mathbf{a}}$ then $\dfrac{df}{ds}$ is defined as the directional derivative of f in the direction of $\hat{\mathbf{a}}$.

The directional derivatives of $f(x, y, z)$ along the positive directions of x, y and z axes are $\dfrac{\partial f}{\partial x}, \dfrac{\partial f}{\partial y}$ and $\dfrac{\partial f}{\partial z}$ respectively.

Also the directional derivatives of a vector function $\mathbf{f}$ along the coordinate axes are $\dfrac{\partial \mathbf{f}}{\partial x}, \dfrac{\partial \mathbf{f}}{\partial y}, \dfrac{\partial \mathbf{f}}{\partial z}$.

3.8 SOME THEOREMS

Theorem I. *grad $f (= \nabla f)$ is vector normal to the surface $f(x, y, z) = c$, where c is a constant.*

Let $A(x, y, z)$ be a point on the surface $f(x, y, z) = c$, and $B(x + \delta x, y + \delta y, z + \delta z)$ be another point in the neighbourhood of point A.

Differential Operators

Let
$$r = x\mathbf{i} + y\mathbf{j} + z\mathbf{k}$$
$$\therefore \quad r + \delta r = (x + \delta x)\mathbf{i} + (y + \delta y)\mathbf{j} + (z + \delta z)\mathbf{k}$$

on subtracting, we get
$$\overrightarrow{AB} = \delta r = \delta x\,\mathbf{i} + \delta y\,\mathbf{j} + \delta z\,\mathbf{k}$$

When $B \to A$, AB tends to the tangent at A to the given surface. Therefore in the limit, (i) becomes
$$d\mathbf{r} = dx\,\mathbf{i} + dy\,\mathbf{j} + dz\,\mathbf{k},$$
and this lies in the tangent plane to the surface at A. But we know that
$$df = \frac{\partial f}{\partial x} dx + \frac{\partial f}{\partial y} dy + \frac{\partial f}{\partial z} dz$$
$$= \left(\mathbf{i}\frac{\partial f}{\partial x} + \mathbf{j}\frac{\partial f}{\partial y} + \mathbf{k}\frac{\partial f}{\partial z} \right) \cdot (dx\,\mathbf{i} + dy\,\mathbf{j} + dz\,\mathbf{k})$$
$$= \nabla f \cdot d\mathbf{r}.$$

Since $f(x, y, z) = $ constant, $df = 0$, hence
$$\nabla f \cdot d\mathbf{r} = 0.$$

Hence ∇f is perpendicular to $d\mathbf{r}$, i.e., perpendicular to the tangent plane at A. i.e., normal to the surface $f(x, y, z) = c$.

Theorem II. *The directional derivatives of a scalar field f at a point $A(x, y, z)$ in the direction of a unit vector $\hat{a}$ is given by*
$$\frac{df}{ds} = \hat{a} \cdot \text{grad } f = (\hat{a} \cdot \nabla) f$$

i.e., directional derivative $\dfrac{df}{ds}$ is the resolved part of ∇f (or grad f) in the direction of $\hat{a}$.

As $\hat{a}$ is unit vector at $A(x, y, z)$.

Hence
$$\hat{a} = \mathbf{i}\frac{dx}{ds} + \mathbf{j}\frac{dy}{dz} + \mathbf{k}\frac{dz}{ds}.$$

Where s is a length in the direction of $\hat{a}$

Also,
$$\hat{a} \cdot \text{grad } f = \left(\mathbf{i}\frac{dx}{dx} + \mathbf{j}\frac{dy}{ds} + \mathbf{k}\frac{dz}{ds} \right) \cdot \left(\mathbf{i}\frac{\partial f}{\partial x} + \mathbf{j}\frac{\partial f}{\partial y} + \mathbf{k}\frac{\partial f}{\partial z} \right)$$
$$= \frac{\partial f}{\partial x} \cdot \frac{dx}{ds} + \frac{\partial f}{\partial y} \cdot \frac{dy}{dz} + \frac{\partial f}{\partial z} \cdot \frac{dz}{ds}$$
$$= \frac{df}{ds}.$$

But $\mathbf{a} \cdot \text{grad } f = \text{grad } f \cos \theta$, where θ is the angel grad f makes with $\mathbf{a}$
Hence the second result.
Similarly the directional derivative of a vector field $\mathbf{f}$ at a point (x, y, z) in the direction of unit vector $\hat{a}$ is $(\hat{a} \nabla) \mathbf{f}$.

Theorem III. *If $\hat{n}$ be a unit vector normal to the level surface $f(x, y, z) = c$ at a point A in the direction f increasing and n be a distance along the normal, then*
$$\text{grad } f = \frac{df}{dn} \hat{n}$$

Since grad f is normal to $f(x, y, z) = c$, hence grad f is of the form
$$\text{grad } f = A\hat{n}$$
where A is some constant and $\hat{n}$ is the unit vector along the tangent.

Now from theorem III,
$$\hat{n} \cdot \operatorname{grad} f = \frac{df}{dn}$$
By using (i), (ii) becomes
$$\hat{n} \cdot A\hat{n} = \frac{df}{dn} \quad \text{or} \quad A = \frac{df}{dn}.$$
Hence from (i)
$$\operatorname{grad} f = \frac{df}{dn} \mathbf{n}$$
Hence the magnitude of ∇f is equal to $\frac{df}{dn}$.

Thus, *the gradient of scalar field f is a vector nromal to the surface f = constant and having a magnitude equal to the rate of change of f along the normal.*

Theorem IV. *grad f is a vector in the direction in which the maximum value of $\frac{df}{ds}$ occurs.*

The directional derivative in the direction of $\hat{a}$ is given by
$$\frac{df}{ds} = \hat{a} \cdot \operatorname{grad} f = \hat{a} \cdot \frac{df}{ds} \hat{n}, \text{ by theorem III}$$
$$= \frac{df}{ds} \hat{a} \cdot \hat{n} = \frac{df}{ds} \cos \theta,$$
where θ is the angle between $\hat{a}$ and $\hat{n}$.

This value will be maximum when $\cos \theta = 1$, *i.e.*, the angle between $\hat{a}$ and $\hat{n}$ is zero, *i.e.*, $\hat{a}$ is along the normal.

Thus the directional derivative is maximum along the normal to the surface. Its maximum value is $\frac{df}{dn}$ *i.e.*, $|\operatorname{grad} f|$.

3.9 TANGENT PLANE AND NORMAL LINE

To find the vector equations of the tangent plane and normal line to the surface $f(x, y, z) = k$ where k is a constant.

(i) Tangent Plane : Let the point of contact of the tangent plane with the given surface be A. Let $\mathbf{r}_0$ be the position vector of A. Let P be any point on the tangent plane. Let $\mathbf{r}$ be the position vector of P. Then the vector $\mathbf{r} - \mathbf{r}_0$ will lie in the tangent plane. Again, we know that grad f is in a direction normal to the tangent plane. Hence, $\mathbf{r} - \mathbf{r}_0$ will be perpendicular to grad f [If a line is perpendicular to a plane, then, it is perpendicular to every line lying in the plane.]

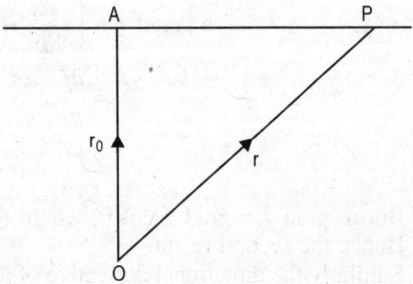

$$\therefore \quad (\mathbf{r} - \mathbf{r}_0) \cdot \operatorname{grad} f = 0. \quad \ldots(1)$$

This equation is satisfied by any piont $\mathbf{r}$ lying in the tangent plane and is not satisfied by any other point. Hence, (1) is the required equation of the tangent plane to the given surface at the point $\mathbf{r}_0$.

(ii) Normal Line : Let P be any point on the normal line at $\mathbf{r}_0$. Let $\mathbf{r}$ be the position vector of P. Then $\mathbf{r} - \mathbf{r}_0$ lies along the normal line.

Since, grad f is normal to the tangent plane, hence, $\mathbf{r} - \mathbf{r}_0$, is parallel to grad f.

$$\therefore \quad (\mathbf{r} - \mathbf{r}_0) \times \operatorname{grad} f = 0.$$

Differential Operators

A discussed in (1), we can show that equation (2) represents the equation of the normal line to the given surface at r_0.

3.9.1 Tangent Line and Normal Plane

To find the vector equations of the tangent line and normal plane at a given point of the curve represented by the intersection of the two surfaces $f_1(x, y, z) = 0$ and $f_2(x, y, z) = 0$.

(i) Tangent Line : Let r_0 be the position vector of the given point of the curve and r be the position vector of any point on the tangent line. Then $r - r_0$ is a vector along the tangent line. Hence, $r - r_0$ is perpendicular to both grad f_1 and grad f_2. Hence, $r - r_0$ will be parallel to grad $f_1 \times$ grad f_2.

$\therefore \qquad (r - r_0) \times (\text{grad } f_1 \times \text{grad } f_2) = 0.$ ...(1)

This equation is satisfied by any piont r lying on the tangent line and is not satisfied by any other point. Hence, equation (1) represents the equation of the tangent line at r_0.

(iii) Normal Plane : Let r be the position vector of any point on the normal plane through r_0. Then the vector $r - r_0$ will lie in the normal plane. Hence, the vector $r - r_0$ will be parallel to the plane through grad f_1 and grad f_2, i.e., perpendicular to grad $f_1 \times$ grad f_2.

$(r - r_0) \cdot (\text{grad } f_1 \times \text{grad } f_2) = 0.$ ...(2)

Obviously, (2) represents the equation of the normal plane at r_0.

EXAMPLES

1. *Find the directional derivative of the function* $\phi = x^2 - y^2 + 2z^2$ *at the point* $P(1, 2, 3)$ *in the direction of the line PQ where Q is the point* $(5, 0, 4)$.

Sol. Here,

$$\text{grad } \phi = \mathbf{i} \cdot 2x - \mathbf{j} \cdot 2y + \mathbf{k} \cdot 4z$$
$$= 2\mathbf{i} - 4\mathbf{j} + 12\mathbf{k} \text{ at } P(1, 2, 3).$$

and $\qquad \mathbf{a} = \overrightarrow{PQ} = 4\mathbf{i} - 2\mathbf{j} + \mathbf{k}$

$\therefore \qquad \hat{\mathbf{a}} = \dfrac{4\mathbf{i} - 2\mathbf{j} + \mathbf{k}}{\sqrt{(21)}}$

$\therefore$ directional derivative along the given direction $= \mathbf{a} \cdot \text{grad } \phi$

$$= \frac{1}{\sqrt{(21)}} (4\mathbf{i} - 2\mathbf{j} + \mathbf{k}) \cdot (2\mathbf{i} - 4\mathbf{j} + 12\mathbf{k})$$

$$= \frac{1}{\sqrt{(21)}} (8 + 8 + 12) = \frac{28}{\sqrt{(21)}}$$

$$= 4\sqrt{\left(\frac{7}{3}\right)}.$$

2. *Find the directional derivative of* $\phi = x^2 yz + 4xz^2$ *at* $(1, -2, -1)$ *in the direction* $2\mathbf{i} - \mathbf{j} - 2\mathbf{k}$. *In what direction the directional derivative will be maximum and what is its magnitude ? Also find a unit normal to the surface* $x^2 yz + 4xz^2 = 6$ *at the point* $(1, -2, -1)$.

Sol. $\qquad \phi = x^2 yz + 4xz^2$

$\therefore \qquad \dfrac{\partial \phi}{\partial x} = 2xyz + 4z^2$

$\qquad \dfrac{\partial \phi}{\partial y} = x^2 z,$

$$\frac{\partial \phi}{\partial z} = x^2 y + 8xz$$

$$\therefore \quad \text{grad } \phi = \mathbf{i}\frac{\partial \phi}{\partial x} + \mathbf{j}\frac{\partial \phi}{\partial y} + \mathbf{k}\frac{\partial \phi}{\partial z}$$

$$= (2xyz + 4z^2)\,\mathbf{i} + (x^2 z)\,\mathbf{j} + (x^2 y + 8xy)\,\mathbf{k}$$

$$= 8\,\mathbf{i} - \mathbf{j} - 10\,\mathbf{k} \text{ at the point } (1, -2, -1)$$

Let $\hat{\mathbf{a}}$ be the unit vector in the given direction.

Then
$$\hat{\mathbf{a}} = \frac{2\,\mathbf{i} - \mathbf{j} - 2\,\mathbf{k}}{\sqrt{4+1+4}} = \frac{2\,\mathbf{i} - \mathbf{j} - 2\,\mathbf{k}}{3}.$$

$\therefore$ Directional derivative $= \dfrac{\partial \phi}{\partial s} = \hat{\mathbf{a}} \cdot \text{grad } \phi$

$$= \left(\frac{2\,\mathbf{i} - \mathbf{j} - \mathbf{k}}{3}\right) \cdot (8\,\mathbf{i} - \mathbf{j} - 10\,\mathbf{k})$$

$$= \frac{16 + 1 + 20}{3} = \frac{37}{3}.$$

Again, we know that the directional derivative is maximum in the direction of normal which is the direction of grad ϕ. Hence, the directional derivative is maximum along grad $\phi = 8\,\mathbf{i} - \mathbf{j} - 10\,\mathbf{k}$.

Further, maximum value of the directional derivative

$$= |\text{grad } \phi| = |8\,\mathbf{i} - \mathbf{j} - 10\,\mathbf{k}|$$

$$= \sqrt{64 + 1 + 100} = \sqrt{165}.$$

Again, a unit vector normal to the surface $= \dfrac{\text{grad } \phi}{|\text{grad } \phi|} = \dfrac{8\,\mathbf{i} - \mathbf{j} - 10\,\mathbf{k}}{\sqrt{165}}$

3. *Find the equation of the tangent plane and normal line to the surface $x^2 + y^2 + z^2 = 25$ at the point $(4, 0, 3)$.*

Sol. Let $f = x^2 + y^2 + z^2 - 25$

Then
$$\text{grad } f = \frac{\partial f}{\partial x}\mathbf{i} + \frac{\partial f}{\partial y}\mathbf{j} + \frac{\partial f}{\partial z}\mathbf{k}$$

$$= 2x\,\mathbf{i} + 2y\,\mathbf{j} + 2z\,\mathbf{k}$$

$$= 8\,\mathbf{i} + 2\,\mathbf{k}, \text{ at the point } (4, 0, 3)$$

Also $\qquad \mathbf{r} = x\,\mathbf{i} + y\,\mathbf{j} + z\,\mathbf{k}, \ \mathbf{r}_0 = 4\,\mathbf{i} + 0\,\mathbf{j} + 3\,\mathbf{k}$

$\therefore \qquad \mathbf{r} - \mathbf{r}_0 = (x-4)\,\mathbf{i} + 4\,\mathbf{j} + (z-3)\,\mathbf{k}$

Equation of tangent plane is

$$(\mathbf{r} - \mathbf{r}_0) \cdot \text{grad } f = 0$$

$\Rightarrow \qquad [(x-4)\,\mathbf{i} + y\,\mathbf{j} + (z-3)\,\mathbf{k}] \cdot [8\,\mathbf{i} + 6\,\mathbf{k}] = 0$

$\Rightarrow \qquad 8(x-4) + 6(z-3) = 0 \ \Rightarrow \ 4x + 3z = 25$

The equation of normal line is $(\mathbf{r} - \mathbf{r}_0) \times \text{grad } f = 0$

$\Rightarrow \qquad [(x-4)\,\mathbf{i} + y\,\mathbf{j} + (z-3)\,\mathbf{k}] \times [8\,\mathbf{i} + 6\,\mathbf{k}] = 0 \Rightarrow \begin{vmatrix} \mathbf{i} & \mathbf{j} & \mathbf{k} \\ x-4 & y & z-3 \\ 8 & 0 & 6 \end{vmatrix} = 0$

$\Rightarrow \qquad 3y\,\mathbf{i} + [4(z-3) - 3(x-4)]\,\mathbf{j} + (-4y)\,\mathbf{k} = 0 = 0$

Equating the coefficients of $\mathbf{i}, \mathbf{j}, \mathbf{k}$ from both sides, we get

$$3y = 0, \ 4(z-3) - 3(x-4) = 0,$$

$$-4y = 0$$

Differential Operators

$$\Rightarrow \quad y = 0, \frac{x-4}{4} = \frac{z-3}{3}$$

∴ Required equation of normal is $\dfrac{x-4}{4} = \dfrac{y}{0} = \dfrac{z-3}{3}$

4. Find the directional derivative of $f = x^2 + y^2 + z^2$ at $(1, 2, 3)$ in the direction of line $x/3 = y/4 = z/5$.

Sol. We have directional derivative $= \hat{a} \cdot \text{grad } f$

Now, vector in direction of line $x/3 = y/4 = z/5$,

$$\mathbf{a} = 3\mathbf{i} + 4\mathbf{j} + 5\mathbf{k}$$

∴ $\hat{a} = \dfrac{3\mathbf{i} + 4\mathbf{j} + 5\mathbf{k}}{\sqrt{9+16+25}} = \dfrac{3\mathbf{i} + 4\mathbf{j} + 5\mathbf{k}}{5\sqrt{2}}$

and $\text{grad } f = \mathbf{i}\dfrac{\partial f}{\partial x} + \mathbf{j}\dfrac{\partial f}{\partial y} + \mathbf{k}\dfrac{\partial f}{\partial z}$

$= \mathbf{i}(2x) + \mathbf{j}(2y) + \mathbf{k}(2z)$
$= 2\mathbf{i} + 4\mathbf{j} + 6\mathbf{k}$ (1, 2, 3)

∴ directional deerivative $= \hat{a} \cdot \text{grad } f$

$$= \frac{1}{5\sqrt{2}}(3\mathbf{i} + 4\mathbf{j} + 5\mathbf{k}) = \frac{52}{5\sqrt{2}}$$

$$= \frac{52\sqrt{2}}{10}.$$

EXERCISES

1. (a) Find the directional derivative of $\phi = x^3 + y^3 + z^3$ at the point $(1, 1, -2)$ in the direction of the vector $\mathbf{i} + 2\mathbf{j} + \mathbf{k}$. [**Ans.** $21/\sqrt{6}$]

(b) Find the directional derivative of $\phi = 3x^2 - 2y - 3z$ at the point $(1, 1, 1)$ in the direction specified by $2\mathbf{i} + 2\mathbf{j} - \mathbf{k}$. [**Ans.** 19/3]

2. In what direction from the point $(2, 1, -1)$ is the directional derivative of $\phi = x^2 y z^2$ is a maximum and what is the magnitude? [**Ans.** $-4\mathbf{i} - 4\mathbf{j} + 12\mathbf{k}; 4\sqrt{(11)}$]

3. (a) Find the maxtimum value of the directional derivative of $\phi = 2x^2 + 3y^2 + 5z^2$ at the point $(1, 1, -4)$.

(b) Find the maximum value of the directional derivative of $\phi = xy + yz + zx$ at the point $(1, 0, 2)$.

4. (a) Find the unit vector normal to $\phi = x^2 + y^2 + z$ at the point $(1, -1, 2)$.

$$\left[\textbf{Ans. } \frac{1}{3}(2\mathbf{i} + 2\mathbf{j} + \mathbf{k})\right]$$

(b) Find a unit ector normal to the surface $\phi = x^2 + y^2 - z^2$ at the point $(1, 1, 1)$.

$$\left[\textbf{Ans. } \frac{2(\mathbf{i} + \mathbf{j} - \mathbf{k})}{\sqrt{3}}\right]$$

5. Find the equation of tangent plane and normal line to the surface $xyz = 4$ at the point $(2, -1, 5)$.

$$\left[\textbf{Ans. } 2x + y + z = 6, \frac{x-1}{2} = \frac{y-2}{1} = \frac{z-2}{1}\right]$$

3.10 DIVERGENCE OF A VECTOR

If $\mathbf{f}(x, y, z)$ is any given continuously differentiable vector point function, then the scalar function defined by

$$\nabla \cdot \mathbf{f} = \left(\mathbf{i}\frac{\partial}{\partial x} + \mathbf{j}\frac{\partial}{\partial y} + \mathbf{k}\frac{\partial}{\partial z}\right) \cdot (\mathbf{i}\, f_1 + \mathbf{j}\, f_2 + \mathbf{k}\, f_3)$$

is called the *divergence of* $\mathbf{f}$, and is written as div. $\mathbf{f}$. We read it as del dot $\mathbf{f}$ or divergence of $\mathbf{f}$. It is clear that *div* $\mathbf{f}$ *is scalar*.

Solenoidal vector. A vector $\mathbf{f}$ is called a solenoidal vector if div $\mathbf{f}$ vanishes. *i.e.*, where div. $\mathbf{f} = 0$.

3.10.1 Divergence of a Sum

Let $\mathbf{f}$ and $\mathbf{F}$ be two vector functions, then

$$\text{div } (\mathbf{f} + \mathbf{F}) = \mathbf{i} \cdot \frac{\partial}{\partial x}(\mathbf{f} + \mathbf{F}) + \mathbf{j} \cdot \frac{\partial}{\partial y}(\mathbf{f} + \mathbf{F}) + \mathbf{k} \cdot \frac{\partial}{\partial z}(\mathbf{f} + \mathbf{F})$$

$$= \mathbf{i} \cdot \left(\frac{\partial \mathbf{f}}{\partial x} + \frac{\partial \mathbf{F}}{\partial x}\right) + \mathbf{j} \cdot \left(\frac{\partial \mathbf{f}}{\partial y} + \frac{\partial \mathbf{F}}{\partial y}\right) + \mathbf{k} \cdot \left(\frac{\partial \mathbf{f}}{\partial z} + \frac{\partial \mathbf{F}}{\partial z}\right)$$

$$= \left(\frac{\partial f_1}{\partial x} + \frac{\partial F_1}{\partial x}\right) + \left(\frac{\partial f_2}{\partial y} + \frac{\partial F_2}{\partial y}\right) + \left(\frac{\partial f_3}{\partial z} + \frac{\partial F_3}{\partial z}\right).$$

where $\qquad \mathbf{f} = \mathbf{i}\, f_1 + \mathbf{j}\, f_2 + \mathbf{k}\, f_3$

and $\qquad \mathbf{F} = \mathbf{i}\, F_1 + \mathbf{j}\, F_2 + \mathbf{k}\, F_3$

$$= \left(\frac{\partial f_1}{\partial x} + \frac{\partial f_2}{\partial y} + \frac{\partial f_3}{\partial z}\right) + \left(\frac{\partial F_1}{\partial x} + \frac{\partial F_2}{\partial y} + \frac{\partial F_3}{\partial z}\right)$$

Note. *If* $\mathbf{a} = \mathbf{i}\, a_1 + \mathbf{j}\, a_2 + \mathbf{k}\, a_3$ *is a constant vector, then* $\dfrac{\partial a_1}{\partial x}, \dfrac{\partial a_2}{\partial y}, \dfrac{\partial a_3}{\partial z}$ *are all zero. Hence for a constant vector* $\mathbf{a}$*,*

$$\text{div } \mathbf{a} = 0.$$

3.11 CURL OF A VECTOR

If $\mathbf{f}$ is any given continuously differentiable vector point function, then the vector function defined by

$$\nabla \times \mathbf{f} = \mathbf{i} \times \frac{\partial \mathbf{f}}{\partial x} + \mathbf{j} \times \frac{\partial \mathbf{f}}{\partial y} + \mathbf{k} \times \frac{\partial \mathbf{f}}{\partial z}$$

is called the curl of $\mathbf{f}$ and is written as curl $\mathbf{f}$. We read it as *del cross* $\mathbf{f}$ or *curl of* $\mathbf{f}$.

3.11.1 Expression of Curl f in terms of the components of f

Let $\qquad \mathbf{f} = \mathbf{i}\, f_1 + \mathbf{j}\, f_2 + \mathbf{k}\, f_3$

Then curl $\qquad \mathbf{f} = \nabla \times \mathbf{f}$

$$= \left(\mathbf{i}\frac{\partial}{\partial x} + \mathbf{j}\frac{\partial}{\partial y} + \mathbf{k}\frac{\partial}{\partial z}\right) \times (\mathbf{i}\, f_1 + \mathbf{j}\, f_2 + \mathbf{k}\, f_3)$$

$$= \mathbf{i}\left(\frac{\partial f_3}{\partial y} - \frac{\partial f_2}{\partial z}\right) + \mathbf{j}\left(\frac{\partial f_1}{\partial z} - \frac{\partial f_3}{\partial x}\right) + \mathbf{k}\left(\frac{\partial f_2}{\partial x} - \frac{\partial f_1}{\partial y}\right)$$

$$= \begin{vmatrix} \mathbf{i} & \mathbf{j} & \mathbf{k} \\ \partial/\partial x & \partial/\partial y & \partial/\partial y \\ f_1 & f_2 & f_3 \end{vmatrix}$$

Obviously, the components of curl $\mathbf{f}$ along the co-ordinate axes are

$$\left(\frac{\partial f_3}{\partial y} - \frac{\partial f_2}{\partial z}\right), \left(\frac{\partial f_1}{\partial z} - \frac{\partial f_3}{\partial x}\right), \text{ and } \left(\frac{\partial f_2}{\partial x} - \frac{\partial f_1}{\partial y}\right).$$

Differential Operators

3.11.2 Curl of Sums

$$\text{Curl}(\mathbf{f} + \mathbf{F}) = \nabla \times \{\mathbf{i}(f_1 + F_1) + \mathbf{j}(f_1 + F_2) + \mathbf{k}(f_3 + F_3)\}$$

Here $\mathbf{f} = \mathbf{i} f_1 + \mathbf{j} f_2 + \mathbf{k} f_3$

and $\mathbf{F} = \mathbf{i} F_1 + \mathbf{j} F_2 + \mathbf{k} F_3$

so curl $(\mathbf{f} + \mathbf{F}) = \begin{vmatrix} \mathbf{i} & \mathbf{j} & \mathbf{k} \\ \partial/\partial x & \partial/\partial y & \partial/\partial z \\ f_1 + F_1 & f_2 + F_2 & f_3 + F_3 \end{vmatrix}$

$$= \begin{vmatrix} \mathbf{i} & \mathbf{j} & \mathbf{k} \\ \partial/\partial x & \partial/\partial y & \partial/\partial z \\ f_1 & f_2 & f_3 \end{vmatrix} + \begin{vmatrix} \mathbf{i} & \mathbf{j} & \mathbf{k} \\ \partial/\partial x & \partial/\partial y & \partial/\partial z \\ F_1 & F_2 & F_3 \end{vmatrix}$$

By the property of determinants

$= \text{curl} \, \mathbf{f} + \text{curl} \, \mathbf{F}.$

Note. If $\mathbf{a}$ is a constant vector then all differentials of its components are zero. Hence for a constant vector

$$\text{Curl} \, \mathbf{a} = 0.$$

Irrotational Vector

A vector $\mathbf{f}$ is said to be irrotational if curl $\mathbf{f} = 0$. (or if curl $\mathbf{f}$ vanishes).

3.12 THE LAPLACIAN OPERATOR ∇^2

The operator ∇^2 is defined by the equation

$$(\nabla \cdot \nabla) f = \nabla \cdot (\nabla f)$$

The operator ∇^2 is called the **Laplacian**

Now, $\nabla^2 f = \nabla(\nabla f)$

$$= \left(\mathbf{i}\frac{\partial}{\partial x} + \mathbf{j}\frac{\partial}{\partial y} + \mathbf{k}\frac{\partial}{\partial z}\right) \cdot \left(\mathbf{i}\frac{\partial f}{\partial x} + \mathbf{j}\frac{\partial f}{\partial y} + \mathbf{k}\frac{\partial f}{\partial z}\right)$$

$$= \frac{\partial^2 f}{\partial x^2} + \frac{\partial^2 f}{\partial y^2} + \frac{\partial^2 f}{\partial z^2} \qquad \text{...(i)}$$

since $\mathbf{i} \cdot \mathbf{i} = \mathbf{j} \cdot \mathbf{j} = \mathbf{k} \cdot \mathbf{k} = 1$ and $\mathbf{i} \cdot \mathbf{j} = \mathbf{j} \cdot \mathbf{k} = 0$

Again, $\nabla \cdot \nabla f = \left(\mathbf{i}\dfrac{\partial}{\partial x} + \mathbf{j}\dfrac{\partial}{\partial y} + \mathbf{k}\dfrac{\partial}{\partial z}\right)\left(\mathbf{i}\dfrac{\partial}{\partial x} + \mathbf{j}\dfrac{\partial}{\partial y} + \mathbf{k}\dfrac{\partial}{\partial z}\right) f$

$$= \left(\frac{\partial^2}{\partial x^2} + \frac{\partial^2}{\partial y^2} + \frac{\partial^2}{\partial z^2}\right) f$$

$$= \frac{\partial^2 f}{\partial x^2} + \frac{\partial^2 f}{\partial y^2} + \frac{\partial^2 f}{\partial z^2} \qquad \text{...(ii)}$$

Hence from (i) and (ii) $\nabla \cdot \nabla f = (\nabla f)$

$\nabla^2 f$ is read as *del square of f*. Again if $\nabla^2 f = 0$, then $\nabla^2 f_1 = \nabla^2 f_2 = \nabla^2 f_3 = 0$.

EXAMPLES

1. *If* $\mathbf{f} = (x + y + 1)\mathbf{i} + \mathbf{j} + (-x - y)\mathbf{k}$, *find curl* $\mathbf{f}$ *and* $\mathbf{f}$. *curl* $\mathbf{f}$.

Sol. We have

$$\text{curl} \, \mathbf{f} = \begin{vmatrix} \mathbf{i} & \mathbf{j} & \mathbf{k} \\ \partial/\partial x & \partial/\partial y & \partial/\partial z \\ x+y+1 & 1 & -x-y \end{vmatrix}$$

$$= \mathbf{i}(-1-0) - \mathbf{j}(-1-0) + \mathbf{k}(0-1)$$

$$= -\mathbf{i} + \mathbf{j} + \mathbf{k}$$

$$\therefore \quad \mathbf{f} \cdot \text{curl } \mathbf{f} = [(x+y+1)\mathbf{i} + \mathbf{j} + (-x-y)\mathbf{k}] \cdot (-\mathbf{i} + \mathbf{j} - \mathbf{k})$$
$$= -(x+y+1) + 1 - (-x-y)$$
$$= -x-y-1+1+x+y = 0.$$

2. *If* $\mathbf{F} = xy^2\mathbf{i} + 2x^2yz\,\mathbf{j} - 3yz^2\mathbf{k}$, *find div $\mathbf{f}$ and curl $\mathbf{f}$ at* $(1, -1, 1)$.

Sol. $\quad \text{div}\,\mathbf{f} = \dfrac{\partial}{\partial x}(xy^2) + \dfrac{\partial}{\partial y}(2x^2yz) + \dfrac{\partial}{\partial z}(-3yz^2)$

$\qquad\qquad = y^2 + 2x^2z - 6yz.$

$(\text{div}\,\mathbf{f})_{(1,-1,1)} = 1 + 2 + 6 = 9.$

$$\text{curl } \mathbf{f} = \begin{vmatrix} \mathbf{i} & \mathbf{j} & \mathbf{k} \\ \partial/\partial x & \partial/\partial y & \partial/\partial z \\ xy^2 & 2x^2yz & -3yz^2 \end{vmatrix}$$

$$= \mathbf{i}\left[\dfrac{\partial}{\partial y}(-3yz^2) - \dfrac{\partial}{\partial z}(2x^2yz)\right] + \mathbf{j}\left[\dfrac{\partial}{\partial z}(xy^2) - \dfrac{\partial}{\partial x}(-3yz^2)\right]$$
$$+ \mathbf{k}\left[\dfrac{\partial}{\partial x}(2x^2yz) - \dfrac{\partial}{\partial y}(xy^2)\right]$$

$$= \mathbf{i}[-3z^2 - 2x^2y] + \mathbf{j}[0+0] + \mathbf{k}[4xyz - 2xy]$$
$$= -\mathbf{i}(3z^2 + 2x^2y) + \mathbf{k}(4xyz - 2xy)$$
$$= -\mathbf{i}(3-2) + \mathbf{k}(-4+2)$$

$(\text{curl } \mathbf{f})_{(1,-1,1)} = -\mathbf{i} - 2\mathbf{k}$

3. *If* $\mathbf{u} = \dfrac{x\mathbf{i} + y\mathbf{j} + z\mathbf{k}}{\sqrt{(x^2+y^2+z^2)}} = \hat{\mathbf{r}}$, *show that* $\nabla \cdot \mathbf{u} = \dfrac{2}{\sqrt{(x^2+y^2+z^2)}} = \dfrac{2}{r}$ *and* $\nabla \times \mathbf{u} = 0$.

Sol. We have $r^2 = x^2 + y^2 + z^2$, then $\dfrac{\partial r}{\partial x} = \dfrac{x}{r}, \dfrac{\partial r}{\partial y} = \dfrac{y}{r}$ and $\dfrac{\partial r}{\partial z} = \dfrac{z}{r}$.

Now $\qquad \mathbf{u} = \dfrac{x}{r}\mathbf{i} + \dfrac{y}{r}\mathbf{j} + \dfrac{z}{r}\mathbf{k}$

$$\nabla \cdot \mathbf{u} = \text{div}\,\mathbf{u} = \dfrac{\partial}{\partial x}\left(\dfrac{x}{r}\right) + \dfrac{\partial}{\partial y}\left(\dfrac{y}{r}\right) + \dfrac{\partial}{\partial z}\left(\dfrac{z}{r}\right)$$

$$= \dfrac{r - x\dfrac{\partial r}{\partial x}}{r^2} + \dfrac{r - y\dfrac{\partial r}{\partial y}}{r^2} + \dfrac{r - z\dfrac{\partial r}{\partial z}}{r^2}$$

$$= \dfrac{1}{r^2}\left[3r - \left(x \cdot \dfrac{x}{r} + y \cdot \dfrac{y}{r} + z \cdot \dfrac{z}{r}\right)\right]$$

$$= \dfrac{1}{r^2}\left[3r - \left(\dfrac{x^2+y^2+z^2}{r}\right)\right] = \dfrac{1}{r^2}\left[3r - \dfrac{r^2}{r}\right]$$

$$= \dfrac{1}{r^2}(3r - r) = \dfrac{1}{r^2} \cdot 2r = \dfrac{2}{r}.$$

$$\nabla \cdot \mathbf{u} = \begin{vmatrix} \mathbf{i} & \mathbf{j} & \mathbf{k} \\ \partial/\partial x & \partial/\partial y & \partial/\partial z \\ x/r & y/r & z/r \end{vmatrix}$$

Differential Operators

$$= \mathbf{i}\left\{\frac{\partial}{\partial y}(z/r) - \frac{\partial}{\partial z}(y/r)\right\} + \mathbf{j}\left\{\frac{\partial}{\partial z}(x/r) - \frac{\partial}{\partial x}(z/r)\right\}$$
$$+ \mathbf{k}\left\{\frac{\partial}{\partial x}(y/r) - \frac{\partial}{\partial y}(x/r)\right\}$$

$$= \mathbf{i}\left\{-\frac{1}{r^2}\frac{\partial r}{\partial y}\cdot z + \frac{1}{r^2}\frac{\partial r}{\partial z}\cdot y\right\} + \mathbf{j}\left\{-\frac{1}{r^2}\frac{\partial r}{\partial z}\cdot x + \frac{1}{r^2}\frac{\partial r}{\partial x}\cdot z\right\}$$
$$+ \mathbf{k}\left\{-\frac{1}{r^2}\frac{\partial r}{\partial x}\cdot y + \frac{1}{r^2}\frac{\partial r}{\partial y}\cdot x\right\}$$

$$= \mathbf{i}\left\{-\frac{yz}{r^3} + \frac{yz}{r^3}\right\} + \mathbf{j}\left\{-\frac{xz}{r^3} + \frac{xz}{r^3}\right\} + \mathbf{k}\left\{-\frac{xy}{r^3} + \frac{xy}{r^3}\right\}$$

$$= \mathbf{i}(0) + \mathbf{j}(0) + \mathbf{k}(0) = 0$$

4. Find the constants a, b, c, so that

$$\mathbf{f} = (x + 2y + az)\mathbf{i} + (bx - 3y - z)\mathbf{j} + (4x + cy + 2z)\mathbf{k} \text{ is irrotational.}$$

Sol. A vector $\mathbf{f}$ is said to be irrotational if curl $\mathbf{f} = 0$.

Now,
$$\text{curl } \mathbf{f} = \begin{vmatrix} \mathbf{i} & \mathbf{j} & \mathbf{k} \\ \partial/\partial x & \partial/\partial y & \partial/\partial z \\ f_1 & f_2 & f_3 \end{vmatrix}$$

$$= \begin{vmatrix} \mathbf{i} & \mathbf{j} & \mathbf{k} \\ \partial/\partial x & \partial/\partial y & \partial/\partial z \\ x+2y+ax & bx-3y-z & 4x+cy+2z \end{vmatrix}$$

$$= (c+1)\mathbf{i} + (a-4)\mathbf{j} + (b-2)\mathbf{k}$$

Now curl $\mathbf{f} = 0$ if $c + 1 = 0, a - 4 = 0$ and $b - 2 = 0$.

∴ $a = 4, b = 2, c = -1$.

5. If the vector $\mathbf{f} = 3x\mathbf{i} + (x + y)\mathbf{j} - ax\mathbf{k}$ is solenoidal, find **a**.

Sol. A vector $\mathbf{f}$ is said to be solenoidal if div $\mathbf{f} = 0$

∴ $\text{div } \mathbf{f} = \frac{\partial}{\partial x}(3x) + \frac{\partial}{\partial y}(x+y) + \frac{\partial}{\partial z}(-az) = 3 + 1 - a = 0$

∴ $a = 4$.

EXERCISES

1. Find the curl of the vector function $\mathbf{f} = y(x + z)\mathbf{i} + z(x + y)\mathbf{j} + x(y + z)\mathbf{k}$ and hence find the value of curl (curl **f**). [Ans. curl $\mathbf{f} = -y\mathbf{i} + z\mathbf{j} - x\mathbf{k}$, curl (curl **f**) $= \mathbf{i} + \mathbf{j} + \mathbf{k}$]

2. (a) If, $\mathbf{f} = z\mathbf{i} + x\mathbf{j} + y\mathbf{k}$, then show that curl (cur **f**) = 0.
   (b) If $\mathbf{f} = x^2 y\mathbf{i} + y^2 z\mathbf{j} - z^2 x\mathbf{k}$, find $\nabla \times \mathbf{f}$ at (1, 2, 3). [Ans. $-4\mathbf{i} + 9\mathbf{j} - \mathbf{k}$]
   (c) If $\mathbf{f} = xy^2\mathbf{i} - 2y^2 z^3\mathbf{j} + xyz^2\mathbf{k}$, find div. **f** at (1, -1, 1). [Ans. 3]

3. (a) If $\mathbf{f} = (x^2 - y^2)\mathbf{i} + 2xy\mathbf{j} + (y^2 - 2xy)\mathbf{k}$, find div. **f** and curl **f**.
   [Ans. div. $\mathbf{f} = 2x + 2y$, curl $\mathbf{f} = (2y - 2x)\mathbf{i} + 2y\mathbf{j} + 4y\mathbf{k}$]
   (b) Find the divergence and curl of the vector function $\mathbf{f} = (2z - 3y)\mathbf{i} + (3x - z)\mathbf{j} + (y - 2x)\mathbf{k}$. [Ans. div $\mathbf{f} = 0$, curl $\mathbf{f} = 2\mathbf{i} + 4\mathbf{j} + 6\mathbf{k}$]
   (c) Compute divergence and curl of the vector $\mathbf{f} = x^2 y\mathbf{i} + xz\mathbf{j} + 2yz\mathbf{k}$ at $(-1, 1, 1)$.
   [Ans. div. $\mathbf{f} = 0$, curl $\mathbf{f} = 3\mathbf{i}$]

4. (a) If $\mathbf{f} = 3xy\mathbf{i} + 20yz^2\mathbf{j} - 15xz\mathbf{k}$ and $\phi = y^2 - xz$, than find div. ($\phi \mathbf{f}$).
   [Ans. $3y^3 - 20xz^3 - 15xy^2 + 30xz^2 + 60y^2 z^2 - 6xyz$]

(b) If $\mathbf{f} = x^2\mathbf{i} + y\mathbf{j} + z\mathbf{k}$ and $\phi = xy^2$, find div $(\phi \mathbf{f})$. [Ans. $xy^2(3x+4y)$]

5. Determine the constant a so that the vector
$\mathbf{V} = (x+3y)\mathbf{i} + (y-2z)\mathbf{j} + (x+az)\mathbf{k}$ is solenoidal. [Ans. -2]

6. Prove that $\phi = 2x^2 - 5y^2 + 3z^2$ satisfies Laplace's equation $\nabla^2 \phi = 0$.

7. Show that the vectors
   (a) $\quad\mathbf{f} = (4xy - z^3)\mathbf{i} + 2x^2\mathbf{j} - 3xz^2\mathbf{k}$
   and (b) $\quad\mathbf{f} = (y^2\cos x + z^3)\mathbf{i} + (2y\sin x - 4)\mathbf{j} + (3xz^2 + 2)\mathbf{k}$
   are irrotational.

8. Determine a, b, c so that the vector $\mathbf{f}$ given by
$\mathbf{f} = (2x + 3y + az)\mathbf{i} + (bx + 2y + 3z)\mathbf{j} + (2x + cy + 3z)\mathbf{k}$ is irrotational.
[Ans. $a=2, b=3, c=3$]

EXAMPLES

1. *Prove that* $\operatorname{div} \hat{\mathbf{r}} = \dfrac{2}{r}$

Sol. We have

$$\hat{\mathbf{r}} = \frac{x\mathbf{i} + y\mathbf{j} + z\mathbf{k}}{r} = \frac{x}{r}\mathbf{i} + \frac{y}{r}\mathbf{j} + \frac{z}{r}\mathbf{k}$$

$$\therefore\quad \operatorname{div} \hat{\mathbf{r}} = \frac{\partial}{\partial x}\left(\frac{x}{r}\right) + \frac{\partial}{\partial y}\left(\frac{y}{r}\right) + \frac{\partial}{\partial z}\left(\frac{z}{r}\right)$$

$$= \frac{r \cdot 1 - x \cdot \dfrac{\partial r}{\partial x}}{r^2} + \frac{r \cdot 1 - y \cdot \dfrac{\partial r}{\partial y}}{r^2} + \frac{r \cdot 1 - z \cdot \dfrac{\partial r}{\partial z}}{r^2}$$

$$= \frac{1}{r^2}\left[3r - \left(x\frac{\partial r}{\partial x} + y\frac{\partial r}{\partial y} + z\frac{\partial r}{\partial z}\right)\right] \quad\ldots(i)$$

we have $r^2 = x^2 + y^2 + z^2$

$$\therefore\quad \frac{\partial r}{\partial x} = \frac{x}{r},\ \frac{\partial r}{\partial y} = \frac{y}{r}\ \text{and}\ \frac{\partial r}{\partial z} = \frac{z}{r}$$

hence from (i)

$$\operatorname{div} \hat{\mathbf{r}} = \frac{1}{r^2}\left[3r - \left(x \cdot \frac{x}{r} + y \cdot \frac{y}{r} + z \cdot \frac{z}{r}\right)\right]$$

$$= \frac{1}{r^2}\left[3r - \frac{x^2 + y^2 + z^2}{r}\right]$$

$$= \frac{1}{r^2}\left[3r - \frac{r^2}{r}\right]$$

$$= \frac{1}{r^2} \cdot 2r = \frac{2}{r}$$

2. *Prove that* $\operatorname{div} r^n \mathbf{r} = (n+3)r^n$.

Sol. We have

$$r^n \mathbf{r} = r^n x\mathbf{i} + r^n y\mathbf{j} + r^n z\mathbf{k}$$

$$\therefore\quad \operatorname{div} r^n\mathbf{r} = \frac{\partial}{\partial x}(r^n x) + \frac{\partial}{\partial y}(r^n y) + \frac{\partial}{\partial z}(r^n z)$$

$$= r^n \cdot 1 + nr^{n-1}x \cdot \frac{\partial r}{\partial x} + r^n \cdot 1 + nr^{n-1}y \cdot \frac{\partial r}{\partial y} + r^n \cdot 1 + nr^{n-1}z \cdot \frac{\partial r}{\partial z}$$

Differential Operators

$$= 3r^n + nr^{n-1}\left(x\frac{\partial r}{\partial x} + y\frac{\partial r}{\partial y} + z\frac{\partial r}{\partial z}\right)$$

$$= 3r^n + nr^{n-1}\left(x\frac{x}{r} + y\frac{y}{r} + z\frac{z}{r}\right)$$

$$= 3r^n + nr^n = (n+3)r^n.$$

3. Prove that curl $(r^n \mathbf{r}) = \mathbf{0}$, i.e., $r^n \mathbf{r}$ is irrotational.

Sol. We have, $\mathbf{r} = \mathbf{i}\,x + \mathbf{j}\,y + \mathbf{k}\,z, r^2 + y^2 + z^2.$

Then $\dfrac{\partial r}{\partial x} = \dfrac{x}{r}, \dfrac{\partial r}{\partial y} = \dfrac{y}{r}, \dfrac{\partial r}{\partial z} = \dfrac{z}{r}$

$$\text{curl }(r^n \mathbf{r}) = \nabla \times (\mathbf{i}\,xr^n + \mathbf{j}\,yr^n + \mathbf{k}\,zr^n)$$

$$= \begin{vmatrix} \mathbf{i} & \mathbf{j} & \mathbf{k} \\ \partial/\partial x & \partial/\partial y & \partial/\partial z \\ xr^n & yr^n & zr^n \end{vmatrix}$$

$$= \mathbf{i}\left[\frac{\partial}{\partial y}(zr^n) - \frac{\partial}{\partial z}(yr^n)\right] + \mathbf{j}\left[\frac{\partial}{\partial z}(xr^n) - \frac{\partial}{\partial x}(zr^n)\right]$$

$$+ \mathbf{k}\left[\frac{\partial}{\partial x}(yr^n) - \frac{\partial}{\partial y}(xr^n)\right]$$

$$= \mathbf{i}\left[nr^{n-1}\frac{\partial r}{\partial y}\cdot z - nr^{n-1}\frac{\partial r}{\partial z}\cdot y\right] + \mathbf{j}\left[nr^{n-1}\frac{\partial r}{\partial z}x - nr^{n-1}\frac{\partial r}{\partial x}\cdot z\right]$$

$$+ \mathbf{k}\left[nr^{n-1}\frac{\partial r}{\partial x}y - nr^{n-1}\frac{\partial x}{\partial y}\cdot x\right]$$

$$= nr^{n-1}\left[\mathbf{i}\left(\frac{yz}{r} - \frac{zy}{r}\right) + \mathbf{j}\left(\frac{zx}{r} - \frac{xz}{r}\right) + \mathbf{k}\left(\frac{xy}{r} - \frac{yx}{r}\right)\right]$$

$$= nr^{n-1}[\mathbf{i}\cdot 0 + \mathbf{j}\cdot 0 + \mathbf{k}\cdot 0] = \mathbf{0}.$$

4. Prove that $\text{div }(\mathbf{r}\times\mathbf{a}) = 0$ or $\text{div }(\mathbf{a}\times\mathbf{r}) = 0.$

Sol. We know, $\mathbf{r} = x\mathbf{i} + y\mathbf{j} + z\mathbf{k}$

$$\mathbf{a} = a_1\mathbf{i} + a_2\mathbf{j} + a_3\mathbf{k}$$

$$\mathbf{r}\times\mathbf{a} = \begin{vmatrix} \mathbf{i} & \mathbf{j} & \mathbf{k} \\ x & y & z \\ a_1 & a_2 & a_3 \end{vmatrix}$$

$$= \mathbf{i}(a_3 y - a_2 z) - \mathbf{j}(a_3 x - a_1 z) + \mathbf{k}(a_2 x - a_1 y)$$

Now, $\text{div}(\mathbf{r}\times\mathbf{a}) = \nabla\cdot(\mathbf{r}\times\mathbf{a})$

$$= \left\{\mathbf{i}\frac{\partial}{\partial x} + \mathbf{j}\frac{\partial}{\partial y} + \mathbf{k}\frac{\partial}{\partial z}\right\}$$

$$\{\mathbf{i}(a_3 y - a_2 z) - \mathbf{j}(a_3 x - a_1 z) + \mathbf{k}(a_2 x - a_1 c)\}$$

$$= \frac{\partial}{\partial x}(a_3 y - a_2 z) - \frac{\partial}{\partial y}(a_3 x - a_1 z) + \frac{\partial}{\partial z}(a_2 x - a_1 y) = 0.$$

5. If $\mathbf{a}$ and $\mathbf{b}$ are constant vectors, then show that $\text{curl}[(\mathbf{r}\times\mathbf{a})\times\mathbf{b}] = \mathbf{b}\times\mathbf{a}$.

Sol. We have

$$\text{curl}[(\mathbf{r}\times\mathbf{a})\times\mathbf{b}] = \text{curl}[(\mathbf{b}\cdot\mathbf{r})\mathbf{a} - (\mathbf{b}\cdot\mathbf{a})\mathbf{r}]$$

$$= \text{curl}\,[(b_1 x + b_2 y + b_3 z)(a_1\,\mathbf{i} + a_2\,\mathbf{j} + a_3\,\mathbf{k}) - (a_1 b_1 + a_2 b_2 + a_3 b_3)\cdot(x\,\mathbf{i} + y\,\mathbf{j} + z\,\mathbf{k})]$$

$$= \begin{vmatrix} \mathbf{i} & \mathbf{j} & \mathbf{k} \\ \partial/\partial x & \partial/\partial y & \partial/\partial z \\ a_1(b_1 x + b_2 y + b_3 z) & a_2(b_1 x + b_2 y + b_3 z) & a_3(b_1 x + b_2 y + b_3 z) \end{vmatrix}$$

$$-\begin{vmatrix} \mathbf{i} & \mathbf{j} & \mathbf{k} \\ \partial/\partial x & \partial/\partial y & \partial/\partial z \\ (a_1 b_1 + a_2 b_2 + a_3 b_3)x & (a_1 b_1 + a_2 b_2 + a_3 b_3)y & (a_1 b_1 + a_2 b_2 + a_3 b_3)z \end{vmatrix}$$

$$= \mathbf{i}(a_3 b_2 - a_2 b_3) - \mathbf{j}(a_3 b_1 - a_1 b_3) + \mathbf{k}(a_2 b_1 - a_1 b_2) = \mathbf{b} \times \mathbf{a}$$

EXERCISES

1. If $\mathbf{f} = \hat{\mathbf{r}}$, then prove that curl $\mathbf{f} = 0$.
2. Show that div $\dfrac{\mathbf{r}}{r^3} = 0$.
3. If $\mathbf{v} = \boldsymbol{\omega} \times \mathbf{r}$ prove that $\boldsymbol{\omega} = \dfrac{1}{2}\,\text{curl}\,\mathbf{v}$ where $\boldsymbol{\omega}$ is a constant vector.
4. Prove that div $[(\mathbf{r} \times \mathbf{a}) \times \mathbf{b}] = -2\mathbf{b}\cdot\mathbf{a}$, where $\mathbf{a}$ and $\mathbf{b}$ are constant vectors.
5. Prove that div $\dfrac{\mathbf{a} \times \mathbf{r}}{r^3} = 0$ and curl $\dfrac{\mathbf{a} \times \mathbf{r}}{r^3} = -\dfrac{\mathbf{a}}{r^3} + \dfrac{3\mathbf{r}}{r^5}(\mathbf{a}\cdot\mathbf{r})$
6. Prove that $\mathbf{a}\cdot\nabla\left(\dfrac{1}{r}\right) = -\dfrac{\mathbf{a}\cdot\mathbf{r}}{r^3}$.

3.13 SECOND ORDER DIFFERENTIAL FUNCTIONS

The three quantities which have been specially studied in this chapter are grad ϕ, div $\mathbf{f}$ and curl $\mathbf{f}$. For these quantities certain improtant points are to be noted :

(i) grad is associated with a *scalar point function* ϕ and grad ϕ is a *vector*.
(ii) Divergence is associated with vector point function $\mathbf{f}$ and divergence $\mathbf{f}$ is a *scalar*.
(iii) Curl is associated with vector function $\mathbf{f}$ and curl $\mathbf{f}$ is a *vector*.

Since grad ϕ and curl $\mathbf{f}$ are vector points functions and as such we can find their divergence as well as curl. Also div $\mathbf{f}$ is a scalar point function we can find its grad we may thus form the following functions :

$$\text{curl grad }\phi \equiv \nabla \times (\nabla\phi)$$
$$\text{div curl }\mathbf{f} \equiv \nabla\cdot(\nabla \times \mathbf{f})$$
$$\text{div grad }\phi \equiv \nabla\cdot(\nabla\phi)$$
$$\text{curl curl }\mathbf{f} = \nabla \times (\nabla \times \mathbf{f})$$
$$\text{grad div }\mathbf{f} = \nabla(\nabla\cdot\mathbf{f})$$

These are called *second order differential functions*.

Properties of Second Order Differential Operators

Property 1. Curl (grad ϕ) $\equiv \nabla \times (\nabla\phi) = 0$.

$$\nabla \times (\nabla\phi) = \nabla \times \left(\mathbf{i}\frac{\partial\phi}{\partial x} + \mathbf{j}\frac{\partial\phi}{\partial y} + \mathbf{k}\frac{\partial\phi}{\partial z}\right)$$

$$= \begin{vmatrix} \mathbf{i} & \mathbf{j} & \mathbf{k} \\ \dfrac{\partial}{\partial x} & \dfrac{\partial}{\partial y} & \dfrac{\partial}{\partial z} \\ \dfrac{\partial\phi}{\partial x} & \dfrac{\partial\phi}{\partial y} & \dfrac{\partial\phi}{\partial z} \end{vmatrix}$$

Differential Operators

$$= \mathbf{i}\left(\frac{\partial^2 \phi}{\partial y \, \partial z} - \frac{\partial^2 \phi}{\partial z \, \partial y}\right) + \mathbf{j}\left(\frac{\partial^2 \phi}{\partial z \, \partial x} - \frac{\partial^2 \phi}{\partial x \, \partial z}\right) + \mathbf{k}\left(\frac{\partial^2 \phi}{\partial x \, \partial y} - \frac{\partial^2 \phi}{\partial y \, \partial x}\right)$$

$$= 0, \text{ as } \frac{\partial^2 \phi}{\partial y \, \partial z} = \frac{\partial^2 \phi}{\partial z \, \partial y} \text{ etc.}$$

Hence $\nabla \times (\nabla \phi) = 0$.

Property 2. div (curl f) $\equiv \nabla \cdot (\nabla \times \mathbf{f}) = 0$

Let $\mathbf{f} = \mathbf{i} f_1 + \mathbf{j} f_2 + \mathbf{k} f_3$

$$\therefore \quad \nabla \times \mathbf{f} = \begin{vmatrix} \mathbf{i} & \mathbf{j} & \mathbf{k} \\ \frac{\partial}{\partial x} & \frac{\partial}{\partial y} & \frac{\partial}{\partial z} \\ f_1 & f_2 & f_3 \end{vmatrix}$$

$$= \mathbf{i}\left(\frac{\partial f_3}{\partial y} - \frac{\partial f_2}{\partial z}\right) + \mathbf{j}\left(\frac{\partial f_1}{\partial z} - \frac{\partial f_3}{\partial x}\right) + \mathbf{k}\left(\frac{\partial f_2}{\partial x} - \frac{\partial f_1}{\partial y}\right)$$

Therefore

$$\nabla \cdot (\nabla \times \mathbf{f}) = \frac{\partial}{\partial x}\left(\frac{\partial f_3}{\partial y} - \frac{\partial f_2}{\partial z}\right) + \frac{\partial}{\partial y}\left(\frac{\partial f_1}{\partial z} - \frac{\partial f_3}{\partial x}\right) + \frac{\partial}{\partial z}\left(\frac{\partial f_2}{\partial x} - \frac{\partial f_1}{\partial y}\right)$$

$$= \frac{\partial^2 f_3}{\partial x \, \partial y} - \frac{\partial^2 f_2}{\partial x \, \partial z} + \frac{\partial^2 f_1}{\partial y \, \partial z} - \frac{\partial^2 f_3}{\partial y \, \partial x} + \frac{\partial^2 f_2}{\partial z \, \partial x} - \frac{\partial^2 f}{\partial z \, \partial y}.$$

Hence $\nabla \cdot (\nabla \times \mathbf{f}) = 0$

Property 3. div (grad ϕ) $\equiv \nabla \cdot (\nabla \phi) = \dfrac{\partial^2 \phi}{\partial x^2} + \dfrac{\partial^2 \phi}{\partial y^2} + \dfrac{\partial^2 \phi}{\partial z^2}$

We have

$$\nabla \cdot (\nabla \phi) = \left(\mathbf{i}\frac{\partial}{\partial x} + \mathbf{j}\frac{\partial}{\partial y} + \mathbf{k}\frac{\partial}{\partial y}\right) \cdot \left(\mathbf{i}\frac{\partial \phi}{\partial x} + \mathbf{j}\frac{\partial \phi}{\partial y} + \mathbf{k}\frac{\partial \phi}{\partial z}\right)$$

$$= \frac{\partial^2 \phi}{\partial x^2} + \frac{\partial^2 \phi}{\partial y^2} + \frac{\partial^2 \phi}{\partial z^2} = \nabla^2 \phi$$

$$\therefore \quad \nabla \cdot (\nabla \phi) = \nabla^2 \phi$$

Property 4. Curl (curl f) = grad (div) f $- \nabla^2 \mathbf{f}$

i.e., $\nabla \times (\nabla \times \mathbf{f}) = (\nabla \cdot \mathbf{f}) - \nabla^2 \mathbf{f}$

Let $\mathbf{f} = \mathbf{i} f_1 + \mathbf{j} f_2 + \mathbf{k} f_3$

$$\therefore \quad \nabla \times \mathbf{f} = \mathbf{i}\left(\frac{\partial f_3}{\partial y} - \frac{\partial f_2}{\partial z}\right) + \mathbf{j}\left(\frac{\partial f_1}{\partial z} - \frac{\partial f_3}{\partial x}\right) + \mathbf{k}\left(\frac{\partial f_2}{\partial x} - \frac{\partial f_1}{\partial y}\right)$$

$$\therefore \quad \nabla \times (\nabla \times \mathbf{f}) = \begin{vmatrix} \mathbf{i} & \mathbf{j} & \mathbf{k} \\ \partial/\partial x & \partial/\partial y & \partial/\partial z \\ \frac{\partial f_3}{\partial y} - \frac{\partial f_2}{\partial z} & \frac{\partial f_1}{\partial z} - \frac{\partial f_3}{\partial x} & \frac{\partial f_2}{\partial x} - \frac{\partial f_1}{\partial y} \end{vmatrix}$$

$$= \mathbf{i}\left\{\frac{\partial}{\partial y}\left(\frac{\partial f_2}{\partial x} - \frac{\partial f_1}{\partial y}\right) - \frac{\partial}{\partial z}\left(\frac{\partial f_1}{\partial z} - \frac{\partial f_3}{\partial x}\right)\right\}$$

$$+ \mathbf{j}\left\{\frac{\partial}{\partial z}\left(\frac{\partial f_3}{\partial y} - \frac{\partial f_2}{\partial z}\right) - \frac{\partial}{\partial x}\left(\frac{\partial f_2}{\partial x} - \frac{\partial f_1}{\partial y}\right)\right\}$$

$$+ \mathbf{k}\left\{\frac{\partial}{\partial x}\left(\frac{\partial f_1}{\partial z} - \frac{\partial f_3}{\partial x}\right) - \frac{\partial}{\partial y}\left(\frac{\partial f_3}{\partial y} - \frac{\partial f_2}{\partial z}\right)\right\}$$

$$= \mathbf{i}\left\{\frac{\partial^2 f_2}{\partial y\,\partial x} - \frac{\partial^2 f_1}{\partial y^2} - \frac{\partial^2 f_1}{\partial z^2} + \frac{\partial^2 f_3}{\partial z\,\partial x}\right\} + \mathbf{j}\left\{\frac{\partial^2 f_3}{\partial z\,\partial y} - \frac{\partial^2 f_2}{\partial z^2} - \frac{\partial^2 f_2}{\partial x^2} + \frac{\partial^2 f_1}{\partial x\,\partial y}\right\}$$

$$+ \mathbf{k}\left\{\frac{\partial^2 f_1}{\partial x\,\partial z} - \frac{\partial^2 f_3}{\partial x^2} - \frac{\partial^2 f_3}{\partial y^2} + \frac{\partial^2 f_2}{\partial y\,\partial z}\right\}$$

$$= \mathbf{i}\,\frac{\partial}{\partial x}\left(\frac{\partial f_1}{\partial x} + \frac{\partial f_2}{\partial y} + \frac{\partial f_3}{\partial z}\right) + \mathbf{j}\,\frac{\partial}{\partial y}\left(\frac{\partial f_1}{\partial x} + \frac{\partial f_2}{\partial y} + \frac{\partial f_3}{\partial z}\right)$$

$$+ \mathbf{k}\,\frac{\partial}{\partial z}\left(\frac{\partial f_1}{\partial x} + \frac{\partial f_2}{\partial y} + \frac{\partial f_3}{\partial z}\right) - (\mathbf{i}\,\nabla^2 f_1 + \mathbf{j}\,\nabla^2 f_2 + \mathbf{k}\,\nabla^2 f_3)$$

$$= \nabla(\nabla\cdot\mathbf{f}) - \nabla^2(\mathbf{i}\,f_1 + \mathbf{j}\,f_2 + \mathbf{k}\,f_3)$$

$$= \nabla(\nabla\cdot\mathbf{f}) - \nabla^2\mathbf{f}$$

so $\quad \nabla\times(\nabla\times\mathbf{f}) = \nabla(\nabla\cdot\mathbf{f}) - \nabla^2\mathbf{f}.$

EXAMPLES

1. *Prove that* div grad $r^n = \nabla^2 r^n = n(n+1)r^{n-2}$.

Sol. Let $\mathbf{r} = x\mathbf{i} + y\mathbf{j} + z\mathbf{k}$

$\therefore \qquad r = \sqrt{(x^2 + y^2 + z^2)}$

$\therefore \qquad \dfrac{\partial r}{\partial x} = \dfrac{2x}{2\sqrt{(x^2+y^2+z^2)}} = \dfrac{x}{r},\; \dfrac{\partial r}{\partial y} = \dfrac{y}{r} \;\text{and}\; \dfrac{\partial r}{\partial z} = \dfrac{z}{r}$ ...(i)

Let $\phi = r^n$, then

$$\frac{\partial\phi}{\partial x} = nr^{n-1}\,\frac{\partial r}{\partial x} = nr^{n-1}\cdot\frac{x}{r} = nxr^{n-2} \qquad ...(ii)$$

$$\frac{\partial^2\phi}{\partial x^2} = n\left[1\cdot r^{n-2} + x(n-2)r^{n-3}\cdot\frac{x}{r}\right]$$

$$= nr^{n-2}\left[1 + \frac{(n-2)}{r^2}x^2\right]$$

Similarly,

$$\frac{\partial^2\phi}{\partial y^2} = nr^{n-2}\left[1 + \frac{(n-2)}{r^2}y^2\right]$$

and $\quad \dfrac{\partial^2\phi}{\partial z^2} = nr^{n-2} = nr^{n-2}\left[1 + \dfrac{(n-2)}{r^2}z^2\right]$

$\therefore \qquad$ div grad $r^n = \nabla^2\phi = \dfrac{\partial^2\phi}{\partial x^2} + \dfrac{\partial^2\phi}{\partial y^2} + \dfrac{\partial^2\phi}{\partial z^2}$

$$= nr^{n-2}\left[1 + \frac{(n-2)}{r^2} + x^2\right] + nr^{n-2}\left[1 + \frac{(n-2)}{r^2}y^2\right]$$

$$+ nr^{n-2}\left[1 + \frac{(n-2)}{r^2}z^2\right]$$

$$= nr^{n-2}\left[3 + \frac{(n-2)}{r^2}(x^2 + y^2 + z^2)\right]$$

$$= nr^{n-2}\left[3 + \frac{(n-2)}{r^2}\cdot r^2\right]$$

Differential Operators

$$= nr^{n-2}(3+n-2)$$
$$= n(n+1)r^{n-2}$$

2. *Prove that curl grad* $r^n = 0$.

Sol. We have : $\mathbf{r} = \mathbf{i}\,x + \mathbf{j}\,y + \mathbf{k}\,z$ and $r^2 = x^2 + y^2 + z^2$.

Also, $\dfrac{\partial r}{\partial x} = \dfrac{x}{r}, \dfrac{\partial r}{\partial y} = \dfrac{y}{r}, \dfrac{\partial r}{\partial z} = \dfrac{z}{r}$.

$$\text{grad } r^n = \mathbf{i}\,\frac{\partial}{\partial x}(r^n) + \mathbf{j}\,\frac{\partial}{\partial y}(r^n) + \mathbf{k}\,\frac{\partial}{\partial z}(r^n)$$

$$= \mathbf{i}\cdot nr^{n-1}\frac{\partial r}{\partial x} + \mathbf{j}\cdot nr^{n-1}\frac{\partial r}{\partial y} + \mathbf{k}\cdot nr^{n-1}\frac{\partial r}{\partial z}$$

$$= nr^{n-1}\frac{x}{r}\mathbf{i} + nr^{n-1}\frac{y}{r}\mathbf{j} + nr^{n-1}\frac{z}{r}\mathbf{k}$$

$$= nr^{n-2}(\mathbf{i}\,x + \mathbf{j}\,y + \mathbf{k}\,z) = nr^{n-2}\mathbf{r}.$$

$$\therefore \quad \text{curl grad } r^n = \begin{vmatrix} \mathbf{i} & \mathbf{j} & \mathbf{k} \\ \partial/\partial x & \partial/\partial y & \partial/\partial z \\ nr^{n-2}x & nr^{n-2}y & nr^{n-2}z \end{vmatrix}$$

$$= n\left[\mathbf{i}\left\{\frac{\partial}{\partial y}(r^{n-2}z) - \frac{\partial}{\partial z}(r^{n-2}y)\right\} + \mathbf{j}\left\{\frac{\partial}{\partial z}(r^{n-2}x) - \frac{\partial}{\partial x}(r^{n-2}z)\right\}\right.$$
$$\left. + \mathbf{k}\left\{\frac{\partial}{\partial x}(r^{n-2}y) - \frac{\partial}{\partial y}(r^{n-2}x)\right\}\right]$$

$$= n\left[\mathbf{i}\left\{(n-2)r^{n-3}\frac{\partial r}{\partial y}z - (n-2)r^{n-3}\frac{\partial r}{\partial z}y\right\}\right.$$
$$+ \mathbf{j}\left\{(n-2)r^{n-3}\frac{\partial r}{\partial z}x - (n-2)r^{n-3}\frac{\partial r}{\partial x}z\right\}$$
$$\left. + \mathbf{k}\left\{(n-2)r^{n-2}\frac{\partial r}{\partial x}y - (n-2)r^{n-3}\frac{\partial r}{\partial y}x\right\}\right]$$

$$= n(n-2)r^{n-3}\left[\mathbf{i}\left(\frac{y}{r}z - \frac{z}{r}y\right) + \mathbf{j}\left(\frac{z}{r}x - \frac{x}{r}z\right) + \mathbf{k}\left(\frac{x}{r}y - \frac{y}{r}x\right)\right]$$

$$= n(n-2)r^{n-3}[\mathbf{i}\cdot 0 + \mathbf{j}\cdot 0 + \mathbf{k}\cdot 0]$$
$$= 0.$$

3. *Prove that* $\nabla^2\left(\dfrac{x}{r^2}\right) = -\dfrac{2x}{r^4}$.

Sol. We have

$$\text{L.H.S.} = \left(\frac{\partial^2}{\partial x^2} + \frac{\partial^2}{\partial y^2} + \frac{\partial^2}{\partial z^2}\right)\left(\frac{x}{r^2}\right) \quad ...(i)$$

We have

$$\frac{\partial^2}{\partial x^2}\left(\frac{x}{r^2}\right) = \frac{\partial}{\partial x}\left[\frac{\partial}{\partial x}\left(\frac{x}{r^2}\right)\right]$$

$$= \frac{\partial}{\partial x}\left[1\cdot\frac{1}{r^2} - \frac{2x}{r^3}\cdot\frac{\partial r}{\partial x}\right]$$

$$= \frac{\partial}{\partial x}\left[\frac{1}{r^2} - \frac{2x}{r^3}\cdot\frac{x}{r}\right] = \frac{\partial}{\partial x}\left[\frac{1}{r^2} - \frac{2x^2}{r^4}\right]$$

$$= \left[-\frac{2}{r^3}\frac{x}{r} - \frac{4x}{r^4} + \frac{8x^2}{r^5}\cdot\frac{x}{r} \right]$$

$$= \left[-\frac{2}{r^4}x - \frac{4x}{r^4} + \frac{8x^3}{r^6} \right]$$

$$= -\frac{6x}{r^4} + \frac{8x^3}{r^6} = -2x\left[\frac{3}{r^4} - \frac{4x^2}{r^6}\right] \qquad ...(i)$$

Now
$$\frac{\partial}{\partial y^2}\left(\frac{x}{r^2}\right) = \frac{\partial}{\partial y}\left[\frac{\partial}{\partial y}\left(\frac{x}{r^2}\right)\right] = \frac{\partial}{\partial y}\left[\frac{-2x}{r^3}\cdot\frac{\partial r}{\partial y}\right]$$

$$= \frac{\partial}{\partial y}\left[\frac{-2x}{r^3}\cdot\frac{y}{r}\right] = \frac{\partial}{\partial y}\left(\frac{-2xy}{r^4}\right)$$

$$= -2x\left[\frac{1}{r^4} - \frac{4y}{r^5}\cdot\frac{y}{r}\right]$$

$$= -2x\left(\frac{1}{r^4} - \frac{4y^2}{r^6}\right) \qquad ...(ii)$$

Similarly,
$$\frac{\partial^2}{\partial z^2}\left(\frac{x}{r^2}\right) = -2x\left(\frac{1}{r^4} - \frac{4z^2}{r^6}\right) \qquad ...(iii)$$

$$\therefore \quad \nabla^2\left(\frac{x}{r^2}\right) = -2x\left[\frac{3}{r^4} + \frac{1}{r^4} + \frac{1}{r^4} - \frac{4}{r^6}(x^2+y^2+z^2)\right]$$

$$= -2x\left[\frac{5}{r^4} - \frac{4}{r^4}\right] = \frac{-2x}{r^4}.$$

EXERCISES

1. Verify that curl grad $\phi = 0$ and div $\phi \mathbf{A} = \text{grad}\,\phi\cdot\mathbf{A} + \phi\,\text{div}\,\mathbf{A}$ given that $\phi = x^3 + y^3 + z^3 + 3xyz$ and $= x^2\mathbf{i} + y^2\mathbf{j} + z^2\mathbf{k}$.
2. If $\mathbf{f} = x^2y\,\mathbf{i} + xz\,\mathbf{j} + 2yz\,\mathbf{k}$ prove that div (curl $\mathbf{f}$) = 0.
3. Prove that div grad $\left(\dfrac{1}{r}\right) = 0$ or $\nabla^2\left(\dfrac{1}{r}\right) = 0$.
4. Show that $\nabla^2 f(r) = f''(r) + \dfrac{2}{r}f'(r)$
5. Prove that curl $\{f(r)\mathbf{r}\} = 0$.
6. Prove that curl curl $\mathbf{F} = 0$ where $\mathbf{F} = z\mathbf{i} + x\mathbf{j} + y\mathbf{k}$.

3.14 VECTOR IDENTITIES

If u and v be two scalar functions and $\mathbf{a}$ and $\mathbf{b}$ be two vector functions, then we can have the products uv and $\mathbf{a}\cdot\mathbf{b}$ both scalar. So we shall find

grad (uv) and grad $(\mathbf{a}\cdot\mathbf{b})$. Similarly products $u\,\mathbf{a}$ and $\mathbf{a}\times\mathbf{b}$ are vetors, so we can find btoh their divergence as well as curl, *i.e.*,

div. $(u\,\mathbf{a})$, div $(\mathbf{a}\times\mathbf{b})$ and curl $(u\,\mathbf{a})$, curl $(\mathbf{a}\times\mathbf{b})$

These results are known as *vector identities* and we shall find these one by one.

(I) **grad (u v) = u grad v + u grad v**

Differential Operators

We have, $\quad \operatorname{grad}(uv) = \Sigma \mathbf{i} \dfrac{\partial}{\partial x}(uv)$

$$= \Sigma \mathbf{i}\left(u\dfrac{\partial v}{\partial x} + v\dfrac{\partial u}{\partial x}\right)$$

$$= u\left(\mathbf{i}\dfrac{\partial v}{\partial x} + \mathbf{j}\dfrac{\partial v}{\partial y} + \mathbf{k}\dfrac{\partial v}{\partial z}\right) + v\left(\mathbf{i}\dfrac{\partial u}{\partial x} + \mathbf{j}\dfrac{\partial u}{\partial y} + \mathbf{k}\dfrac{\partial u}{\partial z}\right)$$

$$= u \operatorname{grad} v + v \operatorname{grad} u$$

i.e., $\quad \nabla(uv) = u\nabla v + v\nabla u$

(II) grad (a . b) = a × curl b + b × curl a + (a . ∇) b + (b . ∇) a

We have $\quad \operatorname{grad}(\mathbf{a.b}) = \Sigma \mathbf{i} \dfrac{\partial}{\partial x}(\mathbf{a.b})$

$$= \Sigma \mathbf{i}\left(\mathbf{a}.\dfrac{\partial \mathbf{b}}{\partial x} + \mathbf{b}.\dfrac{\partial \mathbf{a}}{\partial x}\right)$$

$$= \Sigma \mathbf{i}\left(\mathbf{a}.\dfrac{\partial \mathbf{b}}{\partial x}\right) + \Sigma \mathbf{i}\left(\mathbf{b}.\dfrac{\partial \mathbf{a}}{\partial x}\right) \qquad \ldots(i)$$

Now, $\quad \mathbf{a}\times\left(\mathbf{i}\times\dfrac{\partial \mathbf{b}}{\partial x}\right) = \left(\mathbf{a}.\dfrac{\partial \mathbf{b}}{\partial x}\right)\mathbf{i} - (\mathbf{a.i})\dfrac{\partial \mathbf{b}}{\partial x}$

or $\quad \left(\mathbf{a}.\dfrac{\partial \mathbf{b}}{\partial x}\right)\mathbf{i} = \mathbf{a}\times\left(\mathbf{i}\times\dfrac{\partial \mathbf{b}}{\partial x}\right) + (\mathbf{a.i})\dfrac{\partial \mathbf{b}}{\partial x}$

∴ $\quad \Sigma\left(\mathbf{a}.\dfrac{\partial \mathbf{b}}{\partial x}\right)\mathbf{i} = \mathbf{a}\times\Sigma\left(\mathbf{i}\times\dfrac{\partial \mathbf{b}}{\partial x}\right) + \Sigma(\mathbf{a.i})\dfrac{\partial \mathbf{b}}{\partial x}$

$$= \mathbf{a}\times\operatorname{curl}\mathbf{b} + (\mathbf{a}.\nabla)\mathbf{b} \qquad \ldots(ii)$$

Similarly,

$$\Sigma\left(\mathbf{b}.\dfrac{\partial \mathbf{a}}{\partial x}\right)\mathbf{i} = \mathbf{b}\times\operatorname{curl}\mathbf{a} + (\mathbf{b}.\nabla)\mathbf{b} \qquad \ldots(iii)$$

Hence from (i), (ii), and (iii), we obtain

$$\nabla(\mathbf{a.b}) = \mathbf{a}\times\operatorname{curl}\mathbf{b} + \mathbf{b}\times\operatorname{curl}\mathbf{a} + (\mathbf{a}.\nabla)\mathbf{b} + (\mathbf{b}.\nabla)\mathbf{a}$$

(III) div (u a) = u div a + a . grad u

We have

$$\operatorname{div}(u\mathbf{a}) = \mathbf{i}.\dfrac{\partial}{\partial x}(u\mathbf{a}) + \mathbf{j}.\dfrac{\partial}{\partial y}(u\mathbf{a}) + \mathbf{k}.\dfrac{\partial}{\partial z}(u\mathbf{a})$$

$$= \mathbf{i}.\left(\mathbf{a}\dfrac{\partial u}{\partial x} + u\dfrac{\partial \mathbf{a}}{\partial x}\right) + \mathbf{j}.\left(\mathbf{a}\dfrac{\partial u}{\partial y} + u\dfrac{\partial \mathbf{a}}{\partial y}\right) + \mathbf{k}\left(\mathbf{a}\dfrac{\partial u}{\partial x} + u\dfrac{\partial \mathbf{a}}{\partial z}\right)$$

$$= \mathbf{i}.\left(\mathbf{a}\dfrac{\partial u}{\partial x}\right) + \mathbf{j}.\left(\mathbf{a}\dfrac{\partial u}{\partial y}\right) + \mathbf{k}\left(\mathbf{a}\dfrac{\partial u}{\partial z}\right) + u\left(\mathbf{i}\dfrac{\partial \mathbf{a}}{\partial x} + \mathbf{j}\dfrac{\partial \mathbf{a}}{\partial y} + \mathbf{k}\dfrac{\partial \mathbf{a}}{\partial z}\right)$$

$$= \mathbf{a}.\left(\mathbf{i}\dfrac{\partial u}{\partial x} + \mathbf{j}\dfrac{\partial u}{\partial y} + \mathbf{k}\dfrac{\partial u}{\partial z}\right) + u\left(\mathbf{i}\dfrac{\partial \mathbf{a}}{\partial x} + \mathbf{j}\dfrac{\partial \mathbf{a}}{\partial y} + \dfrac{\partial \mathbf{a}}{\partial z}\right)$$

$$= \mathbf{a}.\operatorname{grad} u + u\operatorname{div}\mathbf{a}.$$

(IV) div (a × b) = b . curl a − a . curl b.

We have $\operatorname{div}(\mathbf{a}\times\mathbf{b}) = \mathbf{i}.\dfrac{\partial}{\partial x}(\mathbf{a}\times\mathbf{b}) + \mathbf{j}.\dfrac{\partial}{\partial y}(\mathbf{a}+\mathbf{b}) + \mathbf{k}.\dfrac{\partial}{\partial z}(\mathbf{a}\times\mathbf{b})$

$$= \Sigma \mathbf{i}\left(\dfrac{\partial \mathbf{a}}{\partial x}\times\mathbf{b} + \mathbf{a}\times\dfrac{\partial \mathbf{b}}{\partial x}\right)$$

$$= \Sigma \mathbf{i} \frac{\partial \mathbf{a}}{\partial x} + \mathbf{b} + \Sigma \mathbf{i} \cdot \mathbf{a} \times \frac{\partial \mathbf{b}}{\partial x}$$

$$= \left(\Sigma \mathbf{i} \times \frac{\partial \mathbf{a}}{\partial x}\right) \cdot \mathbf{b} - \left(\Sigma \mathbf{i} \times \frac{\partial \mathbf{b}}{\partial x}\right) \cdot \mathbf{a}$$

$$= \mathbf{b} \cdot \text{curl } \mathbf{a} - \mathbf{a} \cdot \text{curl } \mathbf{b}$$

(V) curl $(u\,\mathbf{a}) = (\text{grad } u) \times \mathbf{a} + u\,\text{curl }\mathbf{a}$

We have

$$\text{curl }(u\,\mathbf{a}) = \mathbf{i} \times \frac{\partial}{\partial x}(u\,\mathbf{a}) + \mathbf{j} \times \frac{\partial}{\partial y}(u\,\mathbf{a}) + \mathbf{k} \times \frac{\partial}{\partial z}(u\,\mathbf{a})$$

$$= \Sigma \mathbf{i} \times \left(\frac{\partial u}{\partial x}\mathbf{a} + u \frac{\partial \mathbf{a}}{\partial x}\right)$$

$$= \Sigma \mathbf{i} \times \left(\frac{\partial u}{\partial x}\mathbf{a}\right) + \Sigma \mathbf{i} \times u \frac{\partial \mathbf{a}}{\partial x}$$

$$= \Sigma \left(\mathbf{i}\frac{\partial u}{\partial x}\right) \times \mathbf{a} + \left(\Sigma \mathbf{i} \times \frac{\partial \mathbf{a}}{\partial x}\right) u$$

$$= (\text{grad } u) \times \mathbf{a} + u\,\text{curl }\mathbf{a}$$

(VI) curl $(\mathbf{a} \times \mathbf{b}) = \mathbf{a}\,\text{div }\mathbf{b} - \mathbf{b}\,\text{div }\mathbf{a} + (\mathbf{b}.\nabla)\mathbf{a} - (\mathbf{a}.\nabla)\mathbf{b}$

We have

$$\text{curl }(\mathbf{a} \times \mathbf{b}) = \mathbf{i} \times \frac{\partial}{\partial x}(\mathbf{a} \times \mathbf{b}) + \mathbf{j} \times \frac{\partial}{\partial y}(\mathbf{a} \times \mathbf{b}) + \mathbf{k} \times \frac{\partial}{\partial z}(\mathbf{a} \times \mathbf{b})$$

$$= \Sigma \mathbf{i} \times \left(\mathbf{a} \times \frac{\partial \mathbf{b}}{\partial x} + \frac{\partial \mathbf{a}}{\partial x} \times \mathbf{b}\right)$$

$$= \Sigma \mathbf{i} \times \left(\mathbf{a} \times \frac{\partial \mathbf{b}}{\partial x}\right) + \Sigma \mathbf{i} \times \left(\frac{\partial \mathbf{a}}{\partial x} \times \mathbf{b}\right)$$

$$= \Sigma \left(\mathbf{i}.\frac{\partial \mathbf{b}}{\partial x}\right)\mathbf{a} - \Sigma(\mathbf{i}.\mathbf{a})\frac{\partial \mathbf{b}}{\partial x} + \Sigma(\mathbf{i}.\mathbf{b})\frac{\partial \mathbf{a}}{\partial x} - \Sigma\left(\mathbf{i}.\frac{\partial \mathbf{a}}{\partial x}\right)\mathbf{b}$$

as $\mathbf{a} \times (\mathbf{b} \times \mathbf{c}) = (\mathbf{a}.\mathbf{c})\mathbf{b} - (\mathbf{a}.\mathbf{b})\mathbf{c}$

$$= \left(\Sigma \mathbf{i}.\frac{\partial \mathbf{b}}{\partial x}\right)\mathbf{a} - \left(\Sigma \mathbf{i}.\frac{\partial \mathbf{a}}{\partial x}\right)\mathbf{b} + \Sigma(\mathbf{i}.\mathbf{b})\frac{\partial \mathbf{a}}{\partial x} - \Sigma(\mathbf{i}.\mathbf{a})\frac{\partial \mathbf{b}}{\partial x}$$

$$= \mathbf{a}\,\text{div }\mathbf{b} - \mathbf{b}\,\text{div }\mathbf{a} + (\mathbf{b}.\nabla)\mathbf{a} - (\mathbf{a}.\nabla)\mathbf{b}$$

EXAMPLES

1. *If $\mathbf{a}$ be a constant vector find* $\text{grad }(\mathbf{a}.\mathbf{f})$, $\text{div }(\mathbf{a} \times \mathbf{f})$ *and* $\text{curl }(\mathbf{a} \times \mathbf{f})$.

Sol. We have

$\text{grad }(\mathbf{a}.\mathbf{f}) = \mathbf{a} \times \text{curl }\mathbf{f} + \mathbf{f} \times \text{curl }\mathbf{a} + (\mathbf{a}.\nabla)\mathbf{f} + (\mathbf{f}.\nabla)\mathbf{a}$, by Identity II

$\qquad = \mathbf{a} \times \text{curl }\mathbf{a} + (\mathbf{a}.\nabla)\mathbf{f}$

as $\mathbf{a}$ is constant vector, hence

$\qquad \text{curl }\mathbf{a} = 0$ and $(\mathbf{f}.\nabla)\mathbf{a} = 0$

$\text{div }(\mathbf{a} \times \mathbf{f}) = (\text{curl }\mathbf{a}).\mathbf{f} - (\text{curl }\mathbf{f}).\mathbf{a}$ by identity IV

$\qquad = -(\text{curl }\mathbf{f}).\mathbf{a}$ as $\text{curl }\mathbf{a} = 0$

Hence $\text{curl }(\mathbf{a} \times \mathbf{f}) = \mathbf{a}\,\text{div }\mathbf{a} + (\mathbf{f}.\nabla)\mathbf{a} - (\mathbf{a}.\nabla)\mathbf{f}$

by Identity VI as $\text{div }\mathbf{a} = 0$ and $(\mathbf{f}.\nabla)\mathbf{a} = 0$

2. *If* $\mathbf{f} = \psi\,\text{grad }\phi$, *show that* $\mathbf{f}.\text{curl }\mathbf{f} = 0$.

Sol. We have

$\qquad \text{curl }\mathbf{f} = \text{curl }(\psi\,\text{grad }\phi)$

Differential Operators

$$= (\text{grad } \psi) \times (\text{grad } \phi) + \psi \text{ curl } (\text{grad } \phi)$$
$$[\text{as curl } u\mathbf{a} = (\text{grad } u) \times \mathbf{a} + u \text{ curl } \mathbf{a}]$$
$$= (\text{grad } \psi) \times (\text{grad } \phi) \quad \text{as curl } (\text{grad } \phi) = 0$$
$$\therefore \quad \mathbf{f} \cdot \text{curl } \mathbf{f} = \psi (\text{grad } \phi) \cdot \{(\text{grad } \psi) \times (\text{grad } \phi)\} = 0$$
as scalar triple product in which two vectors are equal is zero.

3. If
$$\mathbf{F} = \left(y \frac{\partial f}{\partial z} - z \frac{\partial f}{\partial y} \right) \mathbf{i} + \left(z \frac{\partial f}{\partial x} - x \frac{\partial f}{\partial z} \right) \mathbf{j} + \left(x \frac{\partial f}{\partial y} - y \frac{\partial f}{\partial x} \right) \mathbf{k}$$
then prove that
(i) $\mathbf{F} = \mathbf{r} \times \nabla f$, (ii) $\mathbf{F} \cdot \mathbf{r} = 0$ and (iii) $\mathbf{F} \cdot \text{grad} = 0$.

Sol. We have
$$\mathbf{r} = \mathbf{i} x + \mathbf{j} y + \mathbf{k} z$$
and
$$\nabla f = \mathbf{i} \frac{\partial f}{\partial x} + \mathbf{j} \frac{\partial f}{\partial y} + \mathbf{k} \frac{\partial f}{\partial z}$$

(i) $\therefore$
$$\mathbf{r} \times \nabla f = \begin{vmatrix} \mathbf{i} & \mathbf{j} & \mathbf{k} \\ x & y & z \\ \frac{\partial f}{\partial x} & \frac{\partial f}{\partial y} & \frac{\partial f}{\partial z} \end{vmatrix}$$
$$= \mathbf{i}\left(y \frac{\partial f}{\partial z} - z \frac{\partial f}{\partial y} \right) + \mathbf{j}\left(z \frac{\partial f}{\partial x} - x \frac{\partial f}{\partial z} \right) + \mathbf{k}\left(x \frac{\partial f}{\partial y} - y \frac{\partial f}{\partial x} \right)$$
$$= \mathbf{F}$$

(ii) $\mathbf{F} \cdot \mathbf{r} = \left[\left(t \frac{\partial f}{\partial z} - z \frac{\partial f}{\partial x} \right) \mathbf{i} + \left(z \frac{\partial f}{\partial y} - x \frac{\partial f}{\partial z} \right) \mathbf{j} + \left(x \frac{\partial f}{\partial y} - y \frac{\partial f}{\partial x} \right) \mathbf{k} \right] \cdot (\mathbf{i} x + \mathbf{j} y + \mathbf{k} z)$
$$= \left(xy \frac{\partial f}{\partial z} - xz \frac{\partial f}{\partial y} + yz \frac{\partial f}{\partial x} - xy \frac{\partial f}{\partial z} + zx \frac{\partial f}{\partial y} - yz \frac{\partial f}{\partial x} \right)$$
$$= 0$$

(ii) $\mathbf{F} \cdot \text{grad } f = \left[\left(y \frac{\partial f}{\partial z} - z \frac{\partial f}{\partial y} \right) \mathbf{i} + \left(z \frac{\partial f}{\partial x} - x \frac{\partial f}{\partial z} \right) \mathbf{j} \right.$
$$\left. + \left(x \frac{\partial f}{\partial y} - y \frac{\partial f}{\partial x} \right) \mathbf{k} \right] \cdot \left(\mathbf{i} \frac{\partial f}{\partial x} + \mathbf{j} \frac{\partial f}{\partial y} + \mathbf{k} \frac{\partial f}{\partial z} \right)$$
$$= y \frac{\partial f}{\partial x} \cdot \frac{\partial f}{\partial z} - z \frac{\partial f}{\partial x} \cdot \frac{\partial f}{\partial y} + z \frac{\partial f}{\partial x} \cdot \frac{\partial f}{\partial y} - x \frac{\partial f}{\partial y} \cdot \frac{\partial f}{\partial z} + x \frac{\partial f}{\partial y} \cdot \frac{\partial f}{\partial z} - y \frac{\partial f}{\partial x} \cdot \frac{\partial f}{\partial z}$$
$$= 0.$$

EXERCISES

1. If $\mathbf{f}$ and $\mathbf{g}$ are irrotational, show that $\mathbf{f} \times \mathbf{g}$ is solenoidal.
2. Prove that $\nabla \cdot \left[r \nabla \left(\frac{1}{r^3} \right) \right] = \frac{3}{r^4}$
3. Prove that $\nabla^2 \left[\left(\frac{\mathbf{r}}{r^2} \right) \right] = 2r^{-4}$
4. Prove that $\nabla \left(\mathbf{a} \cdot \frac{\mathbf{r}}{r^n} \right) = \frac{\mathbf{a}}{r^n} - n \frac{(\mathbf{a} \cdot \mathbf{r}) \mathbf{r}}{n+2}$
5. Prove that $\text{div } \frac{\mathbf{a} \times \mathbf{r}}{r^3} = 0$ and $\text{curl } \frac{\mathbf{a} \times \mathbf{r}}{r^3} = -\frac{\mathbf{a}}{r^3} + \frac{3\mathbf{r}}{r^5} (\mathbf{a} \cdot \mathbf{r})$.

6. Prove that $\nabla \cdot \left(\mathbf{a} \times \dfrac{\mathbf{r}}{r^n} \right) = 0$
7. Show that curl $(\mathbf{a} \cdot \mathbf{r})\mathbf{a} = 0$
8. Prove that curl $(\psi \nabla \phi) = \nabla \psi \times \nabla \phi = -\text{curl}\,(\phi \nabla \psi)$.
9. Show that $\mathbf{F} = x\hat{\mathbf{i}} + y\hat{\mathbf{j}} + z\hat{\mathbf{k}}$ is conservative and find ϕ such that $F = \nabla \phi$.
10. (i) Prove that $\mathbf{v} \cdot \nabla \mathbf{v} = \dfrac{1}{2} \nabla v^2 - \mathbf{v} \times \text{curl}\,\mathbf{v}$.
    (ii) If $\mathbf{a}$ is a constant unit vector, prove that
    $$\hat{\mathbf{a}} \cdot [\nabla (\mathbf{v} \cdot \mathbf{a}) - \text{curl}\,(\mathbf{v} \times \hat{\mathbf{a}})] = \text{div}\,\mathbf{v}.$$

❑❑❑

4
Integration of Vectors

4.1 INTEGRATION OF VECTORS

Integration is the inverse process of differentiation. If two functions **B** (*t*) and **f** (*t*) are connected together such that

$$\frac{d}{dt}\{\mathbf{F}(t)\} = f(t),$$ then **F**(*t*) is called integral of **f**(*t*) and in symbol

$$\mathbf{F}(t) = \int \mathbf{f}(t)\, dt$$

f (*t*), the function to be integrated is called the integrand and *t* is the variable of integration.

If **c** is an arbitrary constant vector, then we have

$$\frac{d}{dt}[\mathbf{F}(t) \pm \mathbf{c}] = \mathbf{f}(t)$$

i.e., $\int \mathbf{f}(t)\, dt = \mathbf{F}(t) \pm \mathbf{c}$

The arbitrary constant **c** is called the constant of integration.
The following standard results have been derived :

$$\frac{d}{dt}(\mathbf{r}.\mathbf{s}) = \frac{d\mathbf{r}}{dt}.\mathbf{s} + \mathbf{r}.\frac{d\mathbf{s}}{dt}$$

$$\therefore \int \left(\frac{d\mathbf{r}}{dt}.\mathbf{s} + \mathbf{r}.\frac{d\mathbf{s}}{dt}\right) dt = \mathbf{r}.\mathbf{s} + \mathbf{c} \qquad \ldots(i)$$

Particularly, if **s** = **r**,

$$\int \left(2\mathbf{r}.\frac{d\mathbf{r}}{dt}\right) dt = r^2 + \mathbf{c} \qquad \ldots(ii)$$

Since the derivative of $\left(\frac{d\mathbf{r}}{dt}\right)^2$ is $2\left(\frac{d\mathbf{r}}{dt}.\frac{d^2\mathbf{r}}{dt^2}\right)$

hence $\int 2\left(\frac{d\mathbf{r}}{dt}.\frac{d^2\mathbf{r}}{dt^2}\right) dt = \left(\frac{d\mathbf{r}}{dt}\right)^2 + \mathbf{c} \qquad \ldots(iii)$

Again, the derivative of the unit vector $\hat{\mathbf{r}}$ may be written as

$$\frac{d}{dt}(\hat{\mathbf{r}}) = \frac{d}{dt}\left(\frac{\mathbf{r}}{r}\right) = \frac{1}{r}\frac{d\mathbf{r}}{dt} - \frac{1}{r^2}\frac{dr}{dt}\mathbf{r}$$

$$\therefore \int \left(\frac{1}{r}\frac{d\mathbf{r}}{dt} - \frac{1}{r^2}\frac{dr}{dt}\mathbf{r}\right) = \frac{\mathbf{r}}{r} + \mathbf{c} = \hat{\mathbf{r}} + \mathbf{c} \qquad \ldots(iv)$$

Note : *It should be borne in mind that the constant of integration is of the same nature as the integrand, i.e., if integrand is a vector **c** is a vector and if integrand is a scalar **c** is a scalar.*

EXAMPLES

1. *Find the value of* **r** *satisfying the equation* $\frac{d^2\mathbf{r}}{dt^2} = \mathbf{a}t + \mathbf{b}$, *where* **a** *and* **b** *are constant vectors.*

Sol. Integrating the $\frac{d^2\mathbf{r}}{dt^2} = \mathbf{a}t + \mathbf{b},$

we get
$$\frac{d\mathbf{r}}{dt} = \frac{1}{2}\mathbf{a}t^2 + \mathbf{b}t + \mathbf{c},$$
Where **c** is a constant.

Again integration, we get $\mathbf{r} = \frac{1}{6}\mathbf{a}t^3 + \frac{1}{2}\mathbf{b}t^2 + \mathbf{c}t + \mathbf{d}$

Where **d** is a constant.

2. If $\mathbf{r}(t) = 5t^2\mathbf{i} + t\mathbf{j} - t^3\mathbf{k}$,
$$\int_1^2 \left(\mathbf{r} \times \frac{d^2\mathbf{r}}{dt^2}\right) dt = -14\mathbf{i} + 75\mathbf{j} - 15\mathbf{k}$$

Sol. We have $\int_1^2 \left(\mathbf{r} \times \frac{d^2\mathbf{r}}{dt^2}\right) dt = \mathbf{r} \times \frac{d\mathbf{r}}{dt} + \mathbf{c}$

∴ $\int_1^2 \left(\mathbf{r} \times \frac{d^2\mathbf{r}}{dt^2}\right) dt = \left[\mathbf{r} \times \frac{d\mathbf{r}}{dt}\right]_1^2$

Now, $\frac{d\mathbf{r}}{dt} = 10t\,\mathbf{i} + \mathbf{j} - 3t^2\,\mathbf{k}$

∴ $\mathbf{r} \times \frac{d\mathbf{r}}{dt} = \begin{vmatrix} \mathbf{i} & \mathbf{j} & \mathbf{k} \\ 5t^2 & t & -t^3 \\ 10t & 1 & -3t^2 \end{vmatrix}$

$$= -2t^3\,\mathbf{i} + 5t^4\,\mathbf{j} - 5t^2\,\mathbf{k}$$

$\int_1^2 \left(\mathbf{r} \times \frac{d^2\mathbf{r}}{dt^2}\right) dt = \left[-2t^3\,\mathbf{i} + 5t^4\,\mathbf{j} - 5t^2\,\mathbf{k}\right]_1^2$

$$= -14\mathbf{i} + 75\mathbf{j} - 15\mathbf{k}$$

3. Given that
$\mathbf{r}(t) = 2\mathbf{i} - \mathbf{j} + 2\mathbf{k}$, when $t = 2$
$= 4\mathbf{i} - 2\mathbf{j} + 3\mathbf{k}$, when $t = 3$

Show that $\int_2^3 \left(\mathbf{r} \cdot \frac{d\mathbf{r}}{dt}\right) dt = 10$.

Sol. We have $\int \left(\mathbf{r} \cdot \frac{d\mathbf{r}}{dt}\right) dt = \frac{1}{2}r^2 + c$

∴ $\int_2^3 \left(\mathbf{r} \cdot \frac{d\mathbf{r}}{dt}\right) dt = \left[\frac{1}{2}r^2\right]_{t=2}^{t=3}$

When $t = 3$, $\mathbf{r} = 4\mathbf{i} - 2\mathbf{j} + 3\mathbf{k}$
$r^2 = (4\mathbf{i} - 2\mathbf{j} + 3\mathbf{k}) \cdot (4\mathbf{i} - 2\mathbf{j} + 3\mathbf{k})$
$= 16 + 4 + 9 = 29$

When $t = 2$, $\mathbf{r} = 2\mathbf{i} - \mathbf{j} + 2\mathbf{k}$
$r^2 = (2\mathbf{i} - \mathbf{j} + 2\mathbf{k}) \cdot (2\mathbf{i} - \mathbf{j} + 2\mathbf{k})$
$= 4 + 1 + 4 = 9$

∴ $\int_2^3 \left(\mathbf{r} \times \frac{d\mathbf{r}}{dt}\right) dt = \frac{1}{2}[29 - 9] = 10.$

4. If $\mathbf{r} \times d\mathbf{r} = 0$, show that $\hat{\mathbf{r}} = $ constant.

Sol. Let $\mathbf{r} = x\mathbf{i} + y\mathbf{j} + z\mathbf{k}$

Integration of Vectors

then $d\mathbf{r} = dx\,\mathbf{i} + dy\,\mathbf{j} + dz\,\mathbf{k}$

$\therefore \qquad \mathbf{r} \times d\mathbf{r} = \mathbf{0}$

$\Rightarrow \qquad (x\mathbf{i} + y\mathbf{j} + z\mathbf{k}) \times (dx\,\mathbf{i} + dy\,\mathbf{j} + dz\,\mathbf{k}) = \mathbf{0}$

$\Rightarrow \qquad \begin{vmatrix} \mathbf{i} & \mathbf{j} & \mathbf{k} \\ x & y & z \\ dx & dy & dz \end{vmatrix} = \mathbf{0}$

$\Rightarrow \qquad (y\,dz - z\,dy)\mathbf{i} + (z\,dx - x\,dz)\mathbf{j} + (x\,dy - y\,dx)\mathbf{k} = \mathbf{0}$
$\qquad = 0\mathbf{i} + 0\mathbf{j} + 0\mathbf{k}.$

Equating the coefficients of **i, j, k** on both sides, we get

$$y\,dz - z\,dy = 0 \Rightarrow \frac{dy}{y} = \frac{dz}{z}$$

$$z\,dx - x\,dz = 0 \Rightarrow \frac{dz}{z} = \frac{dx}{x}$$

$$x\,dy - y\,dx = 0 \Rightarrow \frac{dx}{x} = \frac{dy}{y}.$$

The three results on combination admit

$$\frac{dx}{x} = \frac{dy}{y} = \frac{dz}{z}.$$

Taking first two,
Integrating both sides, we get
$$\log x = \log y + \log c_1,$$

where $\log c_1$ is an arbitrary scalar constant of integration.

$\Rightarrow \qquad \log x = \log(c_1 y)$

$\Rightarrow \qquad x = c_1 y.$

Taking last two,
Integrating yields $\qquad \log z = \log y + \log c_2,$

where $\log c_2$ is an arbitrary scalar constant of integration.

$\Rightarrow \qquad z = c_2 y.$

Hence, $\qquad \hat{\mathbf{r}} = \dfrac{\mathbf{r}}{|\mathbf{r}|} = \dfrac{x\mathbf{i} + y\mathbf{j} + z\mathbf{k}}{\sqrt{x^2 + y^2 + z^2}}$

$\qquad\qquad = \dfrac{c_1 y\,\mathbf{i} + y\,\mathbf{j} + c_2 y\,\mathbf{k}}{\sqrt{c_1^2 y^2 + y^2 + c_2^2 y^2}}$

$\qquad\qquad = \dfrac{c_1 \mathbf{i} + \mathbf{j} + c_2 \mathbf{k}}{\sqrt{c_1^2 + 1 + c_2^2}}$

which is clearly a constant vector being independent of x, y and z.

5. *Show that necessary and sufficient condition that direction of given vector* **r** *is constant is that*

$$\mathbf{r} \times \frac{d\mathbf{r}}{dt} = 0.$$

Sol. Let $|\mathbf{r}| = r_1$

$\qquad\qquad \hat{\mathbf{r}} = \mathbf{R}$

Hence, $\qquad \mathbf{r} = r_1 \mathbf{R}$ ...(1)

Necessary condition : Given that direction of **r** is constant, we have to prove that

$$\mathbf{r} \times \frac{d\mathbf{r}}{dt} = 0$$

$$\mathbf{r} \times \frac{d\mathbf{r}}{dt} = r_1 \mathbf{R} + \frac{d}{dt}\{r_1 \mathbf{R}\}$$

$$= r_1 \mathbf{R} \times \left\{\frac{dr_1}{dt} \mathbf{R} + r_1 \frac{d\mathbf{R}}{dt}\right\}$$

$$= r_1 \frac{dr_1}{dt}(\mathbf{R} \times \mathbf{R}) + r_1^2 \left(\mathbf{R} \times \frac{d\mathbf{R}}{dt}\right)$$

$$= r_1^2 \left(\mathbf{R} \times \frac{d\mathbf{R}}{dt}\right) \quad [\because \mathbf{R} \times \mathbf{R} = 0]$$

$$= 0. \text{ as } \mathbf{R} \text{ is constant, so } \frac{d\mathbf{R}}{dt} = 0.$$

Sufficient Condition : Given that $\mathbf{r} \times \frac{d\mathbf{r}}{dt} = 0$.

We have to prove direction of $\mathbf{r}$ is constant

$$\mathbf{r} \times \frac{d\mathbf{r}}{dt} = \begin{vmatrix} \mathbf{i} & \mathbf{j} & \mathbf{k} \\ x & y & z \\ \frac{dx}{dt} & \frac{dy}{dt} & \frac{dz}{dt} \end{vmatrix}$$

$$\Rightarrow \qquad = \mathbf{i}\left\{y\frac{dz}{dt} - z\frac{dy}{dt}\right\} - \mathbf{j}\left\{x\frac{dz}{dt} - z\frac{dx}{dt}\right\} + \mathbf{k}\left\{x\frac{dy}{dt} - y\frac{dx}{dt}\right\}$$

$$\Rightarrow \qquad = 0\mathbf{i} + 0\mathbf{j} + 0\mathbf{k}$$

Comparing the coefficient of $\mathbf{i}, \mathbf{j}, \mathbf{k}$

$$y\frac{dz}{dt} - z\frac{dy}{dt} = 0$$

$$\Rightarrow \qquad \frac{dz}{z} = \frac{dy}{y}$$

$$\Rightarrow \qquad \log z = \log y + \log c_1$$

$$\Rightarrow \qquad z = c_1 y$$

$$x\frac{dz}{dt} - z\frac{dx}{dt} = 0$$

$$\frac{dx}{x} = \frac{dz}{z}$$

$$\Rightarrow \qquad \log z = \log x + \log c_2$$

$$\Rightarrow \qquad z = c_2 x$$

$$x\frac{dy}{dt} - y\frac{dx}{dt} = 0$$

$$\frac{dy}{y} = \frac{dx}{x}$$

$$\Rightarrow \qquad \log y = \log x + \log c_3$$

$$\Rightarrow \qquad y = c_3 x$$

Hence, $\qquad \mathbf{r} = \dfrac{\mathbf{r}}{|\mathbf{r}|} = \dfrac{x\mathbf{i} + y\mathbf{j} + z\mathbf{k}}{\sqrt{x^2 + y^2 + z^2}}$

$$= \dfrac{x\mathbf{i} + c_3 x\mathbf{j} + c_2 x\mathbf{k}}{\sqrt{x^2 + c_3^2 x^2 + c_2^2 x^2}}$$

Integration of Vectors

$$= \frac{\mathbf{i} + c_2\,\mathbf{j} + c_3\,\mathbf{k}}{\sqrt{1 + c_3^2 + c_2^2}}$$

which is independent of x, y, z so it is constnat vector.

EXERCISES

1. If $\mathbf{f}(t) = t\,\mathbf{i} + (t^2 - 2t)\,\mathbf{j} + (2t^2 + 3t^3)\,\mathbf{k}$,

   find $\int_0^1 \mathbf{f}(t)\,dt$. $\left[\text{Ans.}\ \dfrac{1}{2}\mathbf{i} - \dfrac{2}{3}\mathbf{j} + \dfrac{7}{4}\mathbf{k}\right]$

2. If $\mathbf{r} = t\,\mathbf{i} - t^2\,\mathbf{j} + (t-1)\,\mathbf{k}$ and $\mathbf{s} = 2t^2\,\mathbf{i} + 6t\,\mathbf{k}$, evaluate

   (i) $\int_0^2 (\mathbf{r}\cdot\mathbf{s})\,dt$,   (ii) $\int_0^2 (\mathbf{r}\times\mathbf{s})\,dt$. $\left[\text{Ans. (i) } 12,\ \text{(ii)} -24\,\mathbf{i} - \dfrac{40}{3}\mathbf{j} + \dfrac{64}{5}\mathbf{k}\right]$

3. Evaluate $\int_0^1 (e^t\,\mathbf{i} + e^{-2t}\,\mathbf{j} + t\,\mathbf{k})\,dt$. $\left[\text{Ans. } (e-1)\,\mathbf{i} - \dfrac{1}{2}(e^{-2}-1)\,\mathbf{j} + \dfrac{1}{2}\mathbf{k}\right]$

4. The acceleration of a particle at any time t is $e^t\,\mathbf{i} + e^{2t}\,\mathbf{j} + \mathbf{k}$ find $\mathbf{v}$ given that $\mathbf{v} = \mathbf{i} + \mathbf{j}$, at $t = 0$. $\left[\text{Ans. } e^t\,\mathbf{i} + \dfrac{1}{2}(2^{2t}+1)\,\mathbf{j} + t\,\mathbf{k}\right]$

5. If $\mathbf{a} = t\,\mathbf{i} - 3\,\mathbf{j} + 2t\,\mathbf{k}$, $\mathbf{b} = \mathbf{i} - 2\,\mathbf{j} + 2\,\mathbf{k}$, $\mathbf{c} = 3\,\mathbf{i} + t\,\mathbf{j} - \mathbf{k}$
   show that
   (i) $\int_1^2 \mathbf{a}\cdot(\mathbf{b}\times\mathbf{c})\,dt = 0$
   (ii) $\int_1^2 \mathbf{a}\times(\mathbf{b}\times\mathbf{c})\,dt = -\dfrac{87}{2}\mathbf{i} - \dfrac{44}{3}\mathbf{j} + \dfrac{15}{2}\mathbf{k}$

4.2 LINE INTEGRALS

Let $\mathbf{r} = \mathbf{f}(t)$ represents, a continuously differentiable curve denoted by C and $\mathbf{f}(\mathbf{r})$ be a continuous vector point function. Then $\dfrac{d\mathbf{r}}{ds}$ is a unit vector function along the tangent at and point P on the curve. The component of the vector function $\mathbf{F}$ along this tangent is $\mathbf{F}\cdot\dfrac{d\mathbf{r}}{ds}$ which is a function of s for points on the curve. Then

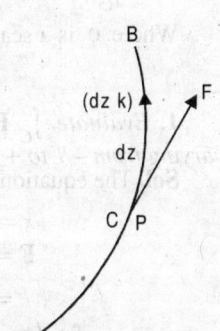

$$\int_C \mathbf{F}\cdot\dfrac{d\mathbf{r}}{ds}\,ds = \int_C \mathbf{F}\cdot d\mathbf{r},$$

is called the line integral or tangent line integral of $\mathbf{F}(\mathbf{r})$ along C.

Let $\quad \mathbf{F} = \mathbf{i}\,F_1 + \mathbf{j}\,F_2 + \mathbf{k}\,F_3$
and $\quad \mathbf{r} = \mathbf{i}\,x + \mathbf{j}\,y + \mathbf{k}\,z$
$\therefore\quad d\mathbf{r} = \mathbf{i}\,dx + \mathbf{j}\,dy + \mathbf{k}\,dz$

$\therefore\quad \int_C \mathbf{F}\cdot d\mathbf{r} = \int (F_1\,\mathbf{i} + F_2\,\mathbf{j} + F_3\,\mathbf{k})\cdot(dx\,\mathbf{i} + dy\,\mathbf{j} + dz\,\mathbf{k})$

$\qquad = \int (F_1\,dx + F_2\,dy + F_3\,dz)$

$\qquad = \int \left(F_1\dfrac{dx}{dt} + F_2\dfrac{dy}{dt} + F_3\dfrac{dz}{dt}\right)dt$

$\therefore\quad \int_C \mathbf{F}\cdot d\mathbf{r} = \int_{t_1}^{t_2} \left(F_1\dfrac{dx}{dt} + F_2\dfrac{dy}{dt} + F_3\dfrac{dz}{dt}\right)dt,$

Where t_1 and t_2 are the values of the parameter t for extremities A and B of the arc of the curve C.

Again, if $\mathbf{r} = x\,\mathbf{i} + y\,\mathbf{j} + z\,\mathbf{k}$

$$\therefore \qquad \frac{d\mathbf{r}}{ds} = \frac{dx}{ds}\mathbf{i} + \frac{dy}{ds}\mathbf{j} + \frac{dz}{ds}\mathbf{k}$$

$$\therefore \qquad \int_C \mathbf{F} \cdot d\mathbf{r} = \int_C \mathbf{F} \cdot \frac{d\mathbf{r}}{ds} ds$$

$$= \int_{s_1}^{s_2} \left(F_1 \frac{dx}{dt} + F_2 \frac{dy}{dt} + F_3 \frac{dz}{dt} \right) ds$$

Where s_1 and s_2 are the values of s for the extremities of A and B of the arc C.

Physical Interpretation of $\int_C \mathbf{F} \cdot d\mathbf{r}$

If $\mathbf{F}$ represents a force acting on a particle moving along the curve C then the line integral $\int_C \mathbf{F} \cdot d\mathbf{r}$ represents the *work done by the force*. If $\mathbf{F}$ represents the velocity of fluid, it is called the *circulation of* $\mathbf{F}$ about C.

Other types of line integrals

(i) $\quad \int_C \mathbf{F} \cdot d\mathbf{r} = \int_C \mathbf{F} \times \frac{d\mathbf{r}}{ds} \cdot ds = \int_{s_1}^{s_2} \mathbf{F} \times \mathbf{t}\, ds,$

Where $\mathbf{t}$ is a unit tangent vector.

Now, $\quad \mathbf{F} \cdot d\mathbf{r} = \begin{vmatrix} \mathbf{i} & \mathbf{j} & \mathbf{k} \\ F_1 & F_2 & F_3 \\ dx & dy & dz \end{vmatrix}$

$$= \mathbf{i}\,(F_2\,dz - F_3\,dy) + \mathbf{j}\,(F_3\,dx - F_1\,dz) + \mathbf{k}\,(F_1\,dy - F_2\,dx)$$

$\therefore \quad \int_C \mathbf{F} \times d\mathbf{r} = \mathbf{i} \int_C (F_2\,dz - F_3\,dy) + \mathbf{j} \int_C (F_3\,dx - F_1\,dz) + \mathbf{k} \int_C (F_1\,dy - F_2\,dx)$

(ii) $\quad \int_C \phi\, d\mathbf{r} = \mathbf{i} \int_C \phi\, dx + \mathbf{j} \int_C \phi\, dy + \mathbf{k} \int_C \phi\, dz,$

Where ϕ is a scalar point function.

EXAMPLES

1. Evaluate $\int_C \mathbf{F} \cdot d\mathbf{r}$, where $\mathbf{F} = xy\,\mathbf{i} + yz\,\mathbf{j} + zx\,\mathbf{k}$ and where C is $\mathbf{r} = \mathbf{i}\,t + \mathbf{j}\,t^2 + \mathbf{k}\,t^3$, t varying from -1 to $+1$.

Sol. The equation of the curve in parameteric from is

$$x = t,\ y = t^2,\ z = t^3$$

$\therefore \qquad \mathbf{F} = xy\,\mathbf{i} + yz\,\mathbf{j} + zx\,\mathbf{k}$

$$= t^3\,\mathbf{i} + t^5\,\mathbf{j} + t^4\,\mathbf{k}$$

Also $\quad \dfrac{d\mathbf{r}}{dt} = \dfrac{dx}{dt}\mathbf{i} + \dfrac{dy}{dt}\mathbf{j} + \dfrac{dz}{dt}\mathbf{k}$

$$= \mathbf{i} + 2t\,\mathbf{j} + 3t^2\,\mathbf{k}$$

$\therefore \qquad \mathbf{F} \cdot \dfrac{d\mathbf{r}}{dt} = t^3 + 2t^6 + 3t^6 = t^3 + 5t^6$

$\therefore \qquad \int_C \mathbf{F} \cdot d\mathbf{r} = \int_C \cdot \dfrac{d\mathbf{r}}{dt} = \int_{-1}^{1} (t^3 + 5t^6)\, dt$

$$= \left[\frac{t^4}{4} + \frac{5t^7}{7} \right]_{-1}^{1} = \frac{10}{7}.$$

2. Evaluate $\int_C \mathbf{F} \cdot d\mathbf{r}$ where $\mathbf{F} = yz\,\mathbf{i} + zx\,\mathbf{j} + xy\,\mathbf{k}$ and C is the portion of the curve $\mathbf{r} = (a\cos t)\,\mathbf{i} + (b\sin t)\,\mathbf{j} + (ct)\,\mathbf{k}$ from $t = 0$ to $\pi/2$.

Sol. We have $\quad \mathbf{r} = (a\cos t)\,\mathbf{i} + (b\sin t)\,\mathbf{j} + (ct)\,\mathbf{k}.$

Hence, the parametric equations of the given curve are

Integration of Vectors

$$x = a\cos t$$
$$y = b\sin t$$
$$z = ct$$

Also, $\quad \dfrac{d\mathbf{r}}{dt} = (-a\sin t)\mathbf{i} + (b\cos t)\mathbf{j} + c\mathbf{k}$

Now, $\quad \int_C \mathbf{F} \cdot d\mathbf{r} = \int_C \mathbf{F} \cdot \dfrac{d\mathbf{r}}{dt} dt$

$= \int_C (yz\,\mathbf{i} + zx\,\mathbf{j} + xy\,\mathbf{k}) \cdot (-a\sin t\,\mathbf{i} + b\cos t\,\mathbf{j} + c\,\mathbf{k})\, dt$

$= \int_C (bct\sin t\,\mathbf{i} + act\cos t\,\mathbf{j} + ab\sin t\cos t\,\mathbf{k})$
$\qquad\qquad \cdot (-a\sin t\,\mathbf{i} + b\cos t\,\mathbf{j} + c\,\mathbf{k})\, dt$

$= \int_C (-abc\,t\sin^2 t + abc\,t\cos^2 t + abc\sin t\cos t)\, dt$

$= abc \int_C [t(\cos^2 t - \sin^2 t) + \sin t\cos t]\, dt$

$= abc \int_C \left(t\cos 2t + \dfrac{\sin 2t}{2}\right) dt$

$= abc \int_0^{\pi/2} \left(t\cos 2t + \dfrac{\sin 2t}{2}\right) dt$

$= abc \left[t\,\dfrac{\sin 2t}{2} + \dfrac{\cos 2t}{4} - \dfrac{\cos 2t}{4}\right]_0^{\pi/2}$

$= \dfrac{abc}{2} (t\sin 2t)_0^{\pi/2}$

$= 0.$

3. *Evaluate* $\int_C \mathbf{F} \cdot d\mathbf{r}$, *where* $\mathbf{F} = (x^2 + y^2)\mathbf{i} - 2xy\,\mathbf{j}$ *and the curve C is the rectangle in the xy plane bounded by* $y = 0, x = a, y = b, x = 0$.

Sol.
In the xy-plane,
$\quad z = 0$
$\therefore \quad \mathbf{r} = x\mathbf{i} + y\mathbf{j}$
or $\quad d\mathbf{r} = dx\,\mathbf{i} + dy\,\mathbf{j}$
$\therefore \quad \int_C \mathbf{F} \cdot d\mathbf{r} = \int_C [(x^2 + y^2)\,dx - 2xy\,dy] \quad \ldots(i)$

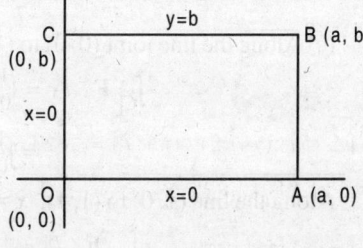

Now,
$\int_C \mathbf{F} \cdot d\mathbf{r} = \int_{OA} \mathbf{F} \cdot d\mathbf{r} + \int_{AB} \mathbf{F} \cdot d\mathbf{r} + \int_{BC} \mathbf{F} \cdot d\mathbf{r} + \int_{CO} \mathbf{F} \cdot d\mathbf{r}$

Along OA, $y = 0$
$\therefore \quad dy = 0$ and x varies from 0 to a.
Along AB, $x = a$
$\therefore \quad dx = 0$, and y varies from 0 to b.
Along BC, $y = b$
$\therefore \quad dy = 0$ and x varies from a to 0
Along CO, $x = 0$
$\therefore \quad dx = 0$ and y varies from b to 0.
Hence from (i) and (ii), we get

$\int_C \mathbf{F} \cdot d\mathbf{r} = \int_0^a x^2\,dx - \int_0^b 2ay\,dy + \int_a^0 (x^2 + b^2)\,dx + \int_b^0 0 \cdot dy$

$$= \frac{a^3}{3} - 2a \cdot \frac{b^2}{2} + \left[\frac{x^3}{3} + b^2 x\right]_a^0 + 0$$

$$= \frac{a^3}{3} - ab^2 - \frac{a^3}{3} - b^2 a = -2ab^2.$$

4. *If* $\mathbf{F} = (x^2 + y^3)\mathbf{i} + (x^3 - y^2)\mathbf{j}$, *Evaluate* $\int_C \mathbf{F} \cdot d\mathbf{r}$ *the following paths :*

(a) $y^2 = x$, *joining* $(0, 0)$ *to* $(1, 1)$

(b) $x^2 = y$, *joining* $(0, 0)$ *to* $(1, 1)$

(c) *Along the straight line joining* $(0, 0)$ *to* $(1, 0)$ *and then to* $(1, 1)$.

(d) *Along the straight line joining* $(0, 0)$ *to* $(2, -2)$ *then to* $(0, -1)$ *and then to* $(1, 1)$.

Sol. Here we have

$$\mathbf{r} = x\mathbf{i} + y\mathbf{j} \text{ so that } d\mathbf{r} = dx\,\mathbf{i} + dy\,\mathbf{j}$$

Hence $\quad \int_C \mathbf{F} \cdot d\mathbf{r} = \int_C (x^2 + y^3)\,dx + (x^3 - y^2)\,dy \qquad \ldots(i)$

(a) We have $y^2 = x$, $\therefore dx = 2y\,dy$ and y varies from 0 to 1

$\therefore \quad \int_C \mathbf{F} \cdot d\mathbf{r} = \int_0^1 (y^4 + y^3)(2y\,dy) + (y^6 - y^2)\,dy$

$$= \int_0^1 (y^6 + 2y^5 + 2y^4 - y^2)\,dy$$

$$= \frac{1}{7} + \frac{1}{3} + \frac{2}{5} - \frac{1}{3} = \frac{19}{35}$$

(b) We have $x^2 = y$ $\therefore 2x\,dy = dx$ and x varies from 0 to 1

$\therefore \quad \int_C \mathbf{F} \cdot d\mathbf{r} = \int_0^1 (x^2 + x^6)\,dx + (x^3 - x^4)\,2x\,dx$

$$= \int_0^1 (x^6 - 2x^5 + 2x^4 + x^2)\,dx$$

$$= \frac{1}{7} - \frac{1}{3} + \frac{2}{5} + \frac{1}{3} = \frac{19}{35}$$

(c) Along the line joint $(0, 0)$ to $(1, 0)$ $y = 0$, $\therefore dy = 0$ and x varies from 0 to 1

$\therefore \quad \int_{C_1} \mathbf{F} \cdot d\mathbf{r} = \int_0^1 (x^2 + y^3)\,dx$

$$= \int_0^1 x^2\,dx = \frac{1}{3}, \text{ because } y = 0$$

Along the line $(1, 0)$ to $(1, 1)$, $x = 1$ $\therefore dx = 0$ and y varies form 0 to 1.

$\therefore \quad \int_{C_2} \mathbf{F} \cdot d\mathbf{r} = \int_0^1 (x^3 - y^2)\,dy$

$$= \int_0^1 (1 - y^2)\,dy, \quad \text{as } x = 1$$

$$= \left[y - \frac{y^3}{3}\right]_0^1 = \frac{2}{3}$$

$\therefore \quad \int_C \mathbf{F} \cdot d\mathbf{r} = \int_{C_1} \mathbf{F} \cdot d\mathbf{r} + \int_{C_2} \mathbf{F} \cdot d\mathbf{r}$

$$= \frac{1}{3} + \frac{2}{3} = 1.$$

(d) The equation of the line joining $(0, 0)$ and $(2, -2)$ is

$$y = -x$$

$\therefore \quad dy = -dx$ and x varies from 0 to 2.

Now put $y = -x$ and $dy = -dx$ in (i) and integrate with in limits 0 to 2.

Integration of Vectors

$$\therefore \quad \int_{C_1} \mathbf{F} \cdot d\mathbf{r} = \int_0^2 (x^2 - x^3) \, dx - (x^3 - x^2) \, dx$$

$$= 2\int_0^2 (x^2 - x^3) \, dx = -\frac{8}{3}.$$

Along C_2 the line joining $(2, -2)$ to $(0, -1)$ has the equation

$$y + 1 = \frac{-2+1}{2+0}(x - 0)$$

or $\qquad y = -\frac{(x+2)}{2}$

$\therefore \quad dy = -\dfrac{1}{2} dx$ and x varies from 2 to 0.

Putting the above data in (i) and integrating with respect to x within the above limits, we get

$$\int_{C_2} \mathbf{F} \cdot d\mathbf{r} = -\frac{9}{2},$$

Similarly, $\quad \int_{C_3} \mathbf{F} \cdot d\mathbf{r} = -\dfrac{1}{6}$

$\therefore \quad \int_C \mathbf{F} \cdot d\mathbf{r} = -\dfrac{8}{3} - \dfrac{9}{2} - \dfrac{1}{6} = -\dfrac{22}{3}.$

EXERCISES

1. Evaluate $\int_C \mathbf{F} \cdot d\mathbf{r}$, where $\mathbf{F} = x^2 \mathbf{i} - xy \mathbf{j}$ from the point $(0, 0)$ to $(1, 1)$ along the parabola $y^2 = x$. **[Ans. 1/12]**

2. Evaluate $\int_C \mathbf{F} \cdot d\mathbf{r}$, where $\mathbf{F} = xy \mathbf{i} + (x^2 + y^2) \mathbf{j}$ and curve C is the arc of $y = x^2 - 4$ from $(2, 0)$ to $(4, 12)$ in the xy-plane. **[Ans. 732]**

3. Evaluate $\int_C \mathbf{F} \cdot d\mathbf{r}$, where $\mathbf{F} = x^2 \mathbf{i} + y^2 \mathbf{j} + z^2 \mathbf{k}$ and C is the arc of the curve $\mathbf{r} = t\mathbf{i} + t^2 \mathbf{j} + t^3 \mathbf{k}$ from $t = 0$ to $t = 1$. **[Ans. 1]**

4. Evalute $\int_C \mathbf{F} \cdot d\mathbf{r}$, where $\mathbf{F} = xy \mathbf{i} + yz \mathbf{j} + zx \mathbf{k}$ and C is the arc of the curve $\mathbf{r} = (a \cos \theta) \mathbf{i} + (a \sin \theta) \mathbf{j} + (a \theta) \mathbf{k}$ to $\theta = \dfrac{\pi}{2}$. $\left[\text{Ans. } a^3 \left(\dfrac{5\pi}{8} - \dfrac{4}{3}\right)\right]$

5. If $\mathbf{F} = (2x + y) \mathbf{i} + (3y - x) \mathbf{j}$, evaluate $\int_C \mathbf{F} \cdot d\mathbf{r}$, where C is the curve in the xy plane consisting of straight line from $O(0, 0)$ to $A(2, 0)$ and then to $B(2, 2)$. **[Ans. 13]**

6. Evaluate $\int_C \mathbf{F} \cdot d\mathbf{r}$, where $\mathbf{F} = xy \mathbf{i} + (x^2 + y^2) \mathbf{j}$ and C is the curve in xy-plane consisting of $x = 2$ to $x = 4$ and $y = 12$. **[Ans. 768]**

7. If $\mathbf{F} = (2y + x) \mathbf{i} + xy \mathbf{j} + (yz - x) \mathbf{k}$, evaluate $\int_C \mathbf{F} \cdot d\mathbf{r}$ along the following C,
   (a) $x = 2t^2, y = t, z = t^3$ from $t = 0$ to $t = 1$.
   (b) The straight line form $(0, 0, 0)$ to $(0, 0, 1)$ then to $(0, 1, 1)$ and then to $(2, 1, 1)$.
   (c) The straight line joining $(0, 0, 0)$ to $(2, 1, 1)$. $\left[\text{Ans. (a) } \dfrac{106}{35}, \text{ (b) } 10, \text{ (c) } 8\right]$

4.3 NORMAL SURFACE INTEGRAL

Let $\mathbf{F}(\mathbf{r})$ be a continuous vector point function and $\mathbf{r} = \mathbf{f}(u, v)$ be a smooth surface such that $\mathbf{f}(u, v)$ possesses continuous first order partial derivatives.

Consider any portion of the surface which may be closed or not.

Divide this surface into a number of sub-surface $\delta S_1, \delta S_2, \delta S_3$ and so on. Let δS_p be one of the sub-surfaces.

Take any point P in this sub-surface and let n_p denote the positive unit normal vector to this sub-surface at the point P.

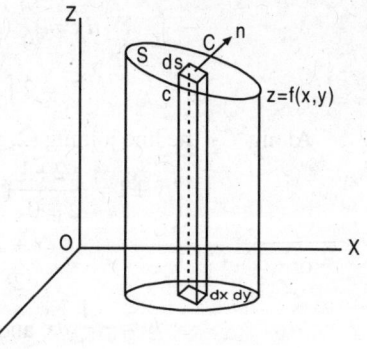

δS_p is the magnitude of the sub-surface and the corresponding vector area be denoted by δa_p.

$$\delta a_p = n_p \, \delta S_p$$

Consider the sum

$$\Sigma F_p \, \delta a_p = \Sigma F_p . n_p \, \delta S_p \qquad ...(i)$$

The summation extending to various sub-surfaces into which S has been divided. Also $\mathbf{F}_p . \mathbf{n}_p$ denotes the normal component of $\mathbf{F}_p$ at **P**.

The limit of the above sum when the number of sub-surface tends to infinity and the area of each sub-surface tends to zero is defined as the *normal surface integral* of $F(\mathbf{r})$ over S and is denoted as

$$\int_S \mathbf{F} . d\mathbf{a} = \int_S \mathbf{F} . \mathbf{n} \, dS.$$

the sign of the above integral will change if we choose the normal on the other side.

Cartesian Form

If F_1, F_2, F_3 be the components of **F** along the coordinate axes, then

$$\int_S \mathbf{F} . \mathbf{n} \, dS = \iint (F_1 \, dy \, dz + F_2 \, dz \, dx + F_3 \, dx \, dy).$$

The above formula can also be put into the form

$$\iint \left[F_1 \frac{\partial (y, z)}{\partial (u, v)} + F_2 \frac{\partial (z, x)}{\partial (u, v)} + F_3 \frac{\partial (x, y)}{\partial (u, v)} \right] du \, dv$$

When the integration is to be performed over the region in the $u - v$ plane. Corresponding to the surface S given by

$$\mathbf{r} = \mathbf{f}(u, v)$$

and $\dfrac{\partial (y, z)}{\partial (u, v)}$ is the Jacobian $= \begin{vmatrix} \dfrac{\partial y}{\partial u} & \dfrac{\partial y}{\partial v} \\ \dfrac{\partial z}{\partial u} & \dfrac{\partial z}{\partial v} \end{vmatrix}$ etc.

Other forms of surface integral are

$$\int_S \mathbf{F} \times d\mathbf{a} \quad \text{and} \quad \int_S \phi \, d\mathbf{a}.$$

Where **F** is a continuous vector point function and ϕ is a continuous scalar point function.

Various Other Forms of Surface Integral

$$\int_S \mathbf{F} . d\mathbf{a} = \int_S \mathbf{F} . \mathbf{n} \, dS$$

Now $\mathbf{n}_p \, \delta S_p$ is the vector area of δS_p and hence its projection on xy-plane whose unit normal is **k** is

$$n_p \, \delta S_p . \mathbf{k} = (n_p . \mathbf{k}) \, \delta S_p$$

But projection δS_p on xy plane is $\delta x \, \delta y$

$$\therefore \quad (n_p . \mathbf{k}) \, \delta S_p = \delta x \, \delta y,$$

$$\therefore \quad \delta S_p = \frac{\delta x \, \delta y}{n_p . \mathbf{k}}$$

$\therefore$ Surface Integral $\int_S \mathbf{F} . \mathbf{n} \, dS = \Sigma \mathbf{F}_p . \mathbf{n}_p \, \delta S_p$

Integration of Vectors

$$= \Sigma F_p . n_p \frac{\partial x \partial y}{n_p . k}$$

$$= \iint_{S_3} F . n \frac{dx\, dy}{n . k}$$

where S_3 is the projection of S on xy-plane.

Similarly, $\quad \int_S F . n\, dS = \iint_{S_1} F . n \frac{dy\, dz}{n . i}$

where S_1 is the projection of S on yz-plane or whose normal is i.

or $\quad \int_S F . n\, dS = \iint_{S_2} F . n \frac{dz\, dx}{n . j}$

where S_2 is the projection of S on zx-plane whose normal is j.

4.4 VOLUME INTEGRAL

Let $F(r)$ be a continuous vector point function and a volume V be enclosed by a surface given by $r = r(u, v)$. Divide the given volume into various $\partial v_1, \partial v_2, \ldots$ elements. Let δv_p be one such element and P be any point on it.

Consider the sum $\Sigma F_p \partial v_p$ ...(i)

Where the summation is to be extended to all the elements into which V has been divided. The limit of the above sum when the number of volume elements tends to infinity and each element tends to zero is defined as the *volume integral* and is written as

$$\int_V F\, dv.$$

In cartesian form

$$\int_V F\, dv = i \iiint_V F_1\, dx\, dy\, dz + j \iiint_V F_2\, dx\, dy\, dz + k \iiint_V F_3\, dx\, dy\, dz$$

EXAMPLES

1. Evaluate $\int_S \frac{r}{r^3} . d\mathbf{a}$, where S denotes the sphere of radius a with centre at the origin.

Sol. Let the equation to the sphere be

$$x^2 + y^2 + z^2 = a^2.$$

A normal to the above surface is given by

$$\text{grad } (x^2 + y^2 + z^2) = i \frac{\partial}{\partial x}(x^2 + y^2 + z^2) + j \frac{\partial}{\partial y}(x^2 + y^2 + z^2) + k \frac{\partial}{\partial z}(x^2 + y^2 + z^2)$$

$$= 2x\, i + 2y\, j + 2z\, k.$$

$\therefore \quad$ Unit normal $= \dfrac{2x\, i + 2y\, j + 2z\, k}{\sqrt{(4x^2 + 4y^2 + 4z^2)}}$

$$= \frac{x\, i + y\, j + z\, k}{a} = n$$

Again, $\quad F = \dfrac{r}{r^3} = \dfrac{x\, i + y\, j + z\, k}{(x^2 + y^2 + z^2)^{3/2}} = \dfrac{x\, i + y\, j + z\, k}{a^3}$

$\therefore \quad \int_S F . d\mathbf{a} = \int_S F . n\, dS$

$$= \int_S \frac{x\, i + y\, j + z\, k}{a^3} . \frac{x\, i + y\, j + z\, k}{a}\, dS$$

$$= \int \frac{x^2 + y^2 + z^2}{a^4}\, dS = \int_S \frac{a^2}{a^4}\, dS$$

$$= \frac{1}{a^2} \int_S dS = \frac{1}{a^2} \cdot 4\pi a^2 = 4\pi.$$

2. If $\mathbf{f} = y\mathbf{i} + (x - 2xz)\mathbf{j} - xy\mathbf{k}$, evaluate $\int_S (\nabla \times \mathbf{f}) \cdot \mathbf{n}\, dS$, where S is the surface of the sphere $x^2 + y^2 + z^2 = a^2$ above the xy-plane.

Sol. Let

$$\mathbf{F} = \nabla \times \mathbf{f} = \operatorname{curl} \mathbf{f} = \begin{vmatrix} \mathbf{i} & \mathbf{j} & \mathbf{k} \\ \dfrac{\partial}{\partial x} & \dfrac{\partial}{\partial y} & \dfrac{\partial}{\partial z} \\ y & x - 2xz & -xy \end{vmatrix} = x\mathbf{i} + y\mathbf{j} - 2z\mathbf{k}$$

Also, we know that the normal to the surface $x^2 + y^2 + z^2 = a^2$ will be

$$\operatorname{grad}(x^2 + y^2 + z^2) = 2x\mathbf{i} + 2y\mathbf{j} + 2z\mathbf{k}$$

$$\therefore \quad \mathbf{n} = \text{unit normal} = \frac{2x\mathbf{i} + 2y\mathbf{j} + 2z\mathbf{k}}{\sqrt{(4x^2 + 4y^2 + 4z^2)}}$$

$$= \frac{x\mathbf{i} + y\mathbf{j} + z\mathbf{k}}{a}$$

$$\therefore \quad \mathbf{F} \cdot \mathbf{n} = (x\mathbf{i} + y\mathbf{j} - 2z\mathbf{k}) \cdot \left(\frac{x\mathbf{i} + y\mathbf{j} + z\mathbf{k}}{a}\right)$$

$$= \frac{x^2 + y^2 - 2z^2}{a}$$

Also, we know that

$$\int_S \mathbf{F} \cdot \mathbf{n}\, dS = \int\int_{S_3} \mathbf{F} \cdot \mathbf{n}\, \frac{dx\, dy}{\mathbf{n} \cdot \mathbf{k}}$$

Where S_3 is the projection of S on xy-plane.

$$\mathbf{n} \cdot \mathbf{k} = \frac{x\mathbf{i} + y\mathbf{j} + z\mathbf{k}}{a} \cdot \mathbf{k} = \frac{z}{a}$$

$$= \frac{\sqrt{(a^2 - x^2 - y^2)}}{a}$$

Also, $\quad \mathbf{F} \cdot \mathbf{n} = \dfrac{x^2 + y^2 - 2z^2}{a}$

$$= \frac{x^2 + y^2 - 2(a^2 - x^2 - y^2)}{a}$$

$$= \frac{3(x^2 + y^2) - 2a^2}{a}$$

$$\therefore \quad \text{Surface Integral} = \int\int_{S_3} \frac{3(x^2 + y^2) - 2a^2}{a} \cdot \frac{dx\, dy}{\sqrt{(a^2 - x^2 - y^2)}} \, a \quad \ldots(i)$$

Now, S_3 is the projection of $x^2 + y^2 + z^2 = a^2$ in the xy-plane and is given by $x^2 + y^2 = a^2$.
In order to integrate (i), put $x = r\cos\theta$, $y = r\sin\theta$

$$\therefore \quad \int_S \mathbf{F} \cdot \mathbf{n}\, dS = \int_0^{2\pi} \int_0^a \frac{3r^2 - 2a^2}{\sqrt{(a^2 - r^2)}} r\, d\theta\, dr$$

$$= 2\pi \int_0^a \frac{3r^2 - 2a^2}{\sqrt{(a^2 - r^2)}} r\, dr$$

But $a^2 - r^2 = t^2$, ... $-2r\, dr = 2t\, dt$

Integration of Vectors

$$\therefore \quad \int_S \mathbf{F}\cdot\mathbf{n}\,dS = 2\pi \int_0^a \frac{3(a^2-t^2)-2a^2}{t}(-t)\,dt$$

$$= 2\pi \int_0^a (a^2 - 3t^2)\,dt = 2\pi [a^3 - a^3]_0^a$$

$$= 2\pi(a^3 - a^3) = 0.$$

3. If $\mathbf{F} = (2x^2 - 3z)\mathbf{i} - 2xy\mathbf{j} - 4x\mathbf{k}$, then evaluate $\int_V \nabla\cdot\mathbf{F}\,dV$ and $\iint\int_V \nabla\times\mathbf{F}\,dV$, where V is the closed region bounded by the plane $x = 0$, $y = 0$, $z = 0$ and $2x + 2y + z = 4$.

Sol. We have

$$\nabla\cdot\mathbf{f} = \frac{\partial}{\partial x}(2x^2 - 3z) + \frac{\partial}{\partial y}(-2xy) + \frac{\partial}{\partial z}(-4x)$$

$$= 4x - 2x - 0 = 2x$$

and $\quad dV = dx\,dy\,dz$.

Limits of z are from 0 to $4 - (2x + 2y)$, limits of y are from 0 to $2 - x$ and limits of x are from 0 to 2.

$$\therefore \quad \int_V = \nabla\cdot\mathbf{f}\,dV$$

$$= \int_0^{4-(2x+2y)} \int_0^{(2-x)} \int_0^2 2x\,dx\,dy\,dz$$

$$= \int_0^{(2-x)} \int_0^2 2x(4 - 2x - 2y)\,dx\,dy$$

$$= \int_0^2 \left[8xy - 4x^2y - 2xys\right]_0^{2-x}\,dx$$

$$= \int_0^2 [8x(2-x) - 4x^2(2-x) - 2x(2-x)^2]\,dx$$

$$= \int_0^2 (2x^3 - 8x^2 + 8x)\,dx = \left[2\cdot\frac{2^4}{4} - 8\cdot\frac{2^3}{3} + 8\cdot\frac{2^2}{2}\right] = \frac{8}{3}.$$

Again, $\quad \nabla\times\mathbf{F} = \begin{vmatrix} \mathbf{i} & \mathbf{j} & \mathbf{k} \\ \dfrac{\partial}{\partial x} & \dfrac{\partial}{\partial y} & \dfrac{\partial}{\partial z} \\ 2x^2 - 3x & -2xy & -4x \end{vmatrix}$

$$= \mathbf{j} - 2\mathbf{k}\,y$$

$$\therefore \quad \int_V \nabla\times\mathbf{f}\,dV = \iiint (\mathbf{j} - 2\mathbf{k}\,y)\,dx\,dy\,dz$$

$$= \int_0^{(4-2x-2y)} \int_0^{2-x} \int_0^2 (\mathbf{j} - 2\mathbf{k}\,y)\,dx\,dy\,dz$$

$$= \int_0^{2-x} \int_0^2 (\mathbf{j} - 2\mathbf{k}\,y)(4 - 2x - 2y)\,dx\,dy$$

$$= \int_0^2 \left[\mathbf{j}(4y - 2xy - y^2) - 2\mathbf{k}\left(2y^2 - xy^2 - \frac{2y^3}{3}\right)\right]_0^{2-x}\,dx$$

$$= \int_0^2 \left[\mathbf{j}(2-x)(4 - 2x - 2 + x) - 2\mathbf{k}(2-x^2)\left\{2 - x - \frac{2}{3}(2-x)\right\}\right]dx$$

$$= \int_0^2 \left[\mathbf{j}(2-x)^2 - \frac{2\mathbf{k}}{3}(2-x)^3\right]dx$$

$$= \left[\mathbf{j}\frac{(x-2)^3}{3} + \frac{2\mathbf{k}}{3}\frac{(x-2)^4}{4}\right]_0^2$$

$$= \mathbf{j}\cdot\frac{8}{3} + \frac{2\mathbf{k}}{3}\left(-\frac{16}{4}\right) = \frac{8}{3}(\mathbf{j} - \mathbf{k}).$$

EXERCISES

1. Evaluate $\int_S \mathbf{f} \cdot \mathbf{n}\, dS$ where $\mathbf{f} = y\mathbf{i} + 2x\mathbf{j} + z\mathbf{k}$ and S is the surface of the plane $2x + y = 6$ in the first octant cut off by the plane $z = 4$. [Ans. 108]

2. Evaluate $\int_S \mathbf{f} \cdot \mathbf{n}\, dS$ over the surface of the cylinder $x^2 + y^2 = 9$ included in the first octant between $z = 0$ and $z = 4$ where $\mathbf{f} = z\mathbf{i} + x\mathbf{j} - yz\mathbf{k}$. [Ans. 42]

3. Evaluate $\int_S \mathbf{f} \cdot \mathbf{n}\, dS$ where $\mathbf{f} = 4x\mathbf{i} - 2y^2\mathbf{j} + z^2\mathbf{k}$ taken over the region bounded by $x^2 + y^2 = 4, z = 0$ and $z = 3$. [Ans. 84p]

4. Evaluate $\int_S \mathbf{f} \cdot \mathbf{n}\, dS$ where $\mathbf{f} = 2xy\mathbf{i} - 2zy\mathbf{j} + x^2\mathbf{k}$ over the surface of cube bounded by the coordinate planes and the planes $x = a$, $y = a$, $z = a$. $\left[\text{Ans. } \dfrac{1}{2}a^4\right]$

5. Evaluate $\int_V \mathbf{f}\, dV$ for $\mathbf{f} = x\mathbf{i} + y\mathbf{j} + z\mathbf{k}$ where V is the region bounded by the surface $x = 0, y = 0, y = 6, z = 4$ and $z = x^2$. $\left[\text{Ans. } 24\mathbf{i} + 96\mathbf{j} + \dfrac{384}{5}\mathbf{k}\right]$

5

Gauss's, Green's and Stoke's Theorem

5.1 GAUSS'S DIVERGENCE THEOREM

Reduction of Surface Integral to Volume Integral

Statement : *The normal surface integral of a vector function **F** over the boundary of a closed region is equal to the volume integral of div **F** taken throughout the region.*

In symbols it may be stated as follows :

If **F** *be a continuously differentiable vector point function in a region V and S is a closed surface enclosing the region V, then*

$$\int_S \mathbf{F} \cdot \mathbf{n} \, dS = \int_V \operatorname{div} \mathbf{F} \, dV,$$

where **n** *is the unit outward drawn normal vector of the surface.*

In cartesian co-ordinates the Divergence theorem may be written as

$$\int_S \mathbf{F} \cdot \mathbf{n} \, dS = \int_V \operatorname{div} \mathbf{F} \, dV$$

or

$$\int\int_S (F_1 \, dy \, dz + F_2 \, dz \, dx + F_3 \, dx \, dy)$$

$$= \int\int\int_V \left(\frac{\partial F_1}{\partial x} + \frac{\partial F_2}{\partial y} + \frac{\partial F_3}{\partial z} \right) dx \, dy \, dz.$$

Proof : Let $\mathbf{F} = U\mathbf{i} + V\mathbf{j} + W\mathbf{k}$, where U, V, W and their derivatives in any direction are assumed to be uniform, finite and continuous.

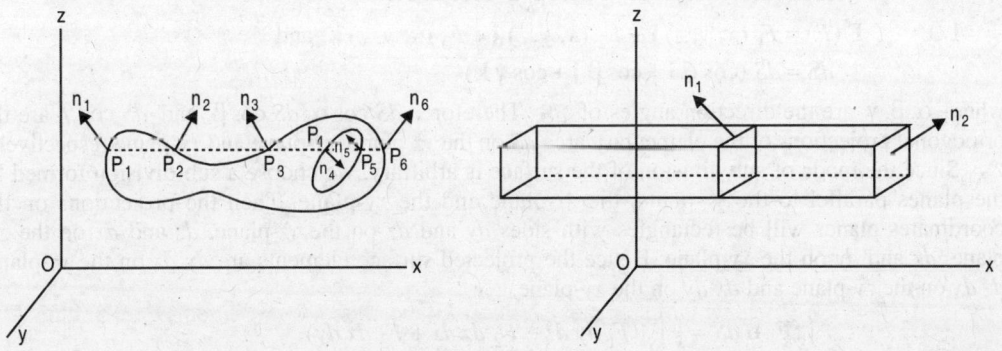

Let us consider the volume integral

$$I = \int\int\int \frac{\partial U}{\partial x} \, dx \, dy \, dz,$$

where $dx \, dy \, dz$ has been written for the volume element dV. For fixed values of y and z, take a rectangular prism parallel to x-axis, bounded by the planes $y, y + dy, z, z + dz$, the area of the normal section being $dy \, dz$.

The prism so formed cuts the boundary an even number of times at the points $P_1, P_2, \ldots, P_{2n}$.

If a point moves along the prism in the direction of x increasing, it enters the region at $P_1, P_3, ..., P_{2n-1}$ and leaves $P_2, P_4, ..., P_{2n}$.

Then taking the integral and integrating with respect to x, we obtain
$$I = \iint (-U_1 + U_2 - U_3 + ... - U_{2n-1} + U_{2n}) \, dy \, dz$$
where U_r is the value of U at that point P_r.

Let dS_r is the value of U at that point P_r.

Let dS_r be the area of the element of the boundary intercepted by the prism at the point P_r. Then
$$dy \, dz = \text{area of projection of } dS_r \text{ on the } yz\text{-plane}$$
$$= -\mathbf{i} \cdot \mathbf{n}_r \, dS_r \quad \text{for } r \text{ odd}$$
$$= \mathbf{i} \cdot \mathbf{n}_r \, dS_r \quad \text{for } r \text{ even},$$

as the angle $\mathbf{n}_r$ makes with $\mathbf{i}$ is acute or obtuse according as r is even or odd (when the line parallel to x-axis enters the surface, the outward normal makes an obtuse angle with it and acute angle when the line leaves the surface).

$\therefore \qquad I = \int \mathbf{i} \cdot (U_1 \, \mathbf{n}_1 \, dS_1 + U_1 \, \mathbf{n}_2 \, dS_2 + ... + U_{2n} \, \mathbf{n}_{2n} \, dS_{2n})$

On summing for all the rectangular prism, we obtain

$$\int \frac{\partial U}{\partial x} dv = \int U \mathbf{i} \cdot \mathbf{n} \, dS = \int U \mathbf{i} \cdot d\mathbf{S} \qquad ...(1)$$

Similarly $\qquad \int \frac{\partial V}{\partial y} dv = \int V \mathbf{j} \cdot \mathbf{n} \, dS \qquad ...(2)$

and $\qquad \int \frac{\partial W}{\partial z} dv = \int W \mathbf{k} \cdot \mathbf{n} \, dS \qquad ...(3)$

Adding (1), (2) and (3), we get
$$\int \left(\frac{\partial U}{\partial x} + \frac{\partial V}{\partial y} + \frac{\partial W}{\partial z} \right) dv = \int (U \mathbf{i} + V \mathbf{j} + W \mathbf{k}) \cdot \mathbf{n} \, dS$$

i.e., $\qquad \int \text{div } \mathbf{F} \, dv = \int \mathbf{F} \cdot \mathbf{n} \, dS$

Cartesian Representation of Gauss's Theorem

Let $\qquad \mathbf{F}(P) = F_1(x, y, z) \mathbf{i} + F_2(x, y, z) \mathbf{j} + F_3(x, y, z) \mathbf{k}$ and
$$d\mathbf{S} = dS (\cos \alpha \, \mathbf{i} + \cos \beta \, \mathbf{j} + \cos \gamma \, \mathbf{k})$$

where α, β, γ are the direction angles of $d\mathbf{S}$. Therefore, $dS \cos \alpha$, $dS \cos \beta$ and $dS \cos \gamma$ are the orthogonal projections of the elementary area dS on the yz-plane, zx-plane and xy-plane respectively.

Since the mode of sub division of the surface is arbitrary, we choose a sub-division formed by the planes parallel to the yz-plane, the zx-plane and the xy-plane. Then the projections on the coordinates planes will be rectangles with sides dy and dz on the yz-plane, dz and dx on the zx-plane, dx and dy on the xy-plane. Hence the projected surface elements are $dy \, dz$ on the yz-plane, $dz \, dx$ on the zy-plane and $dx \, dy$ on the xy-plane.

$\therefore \qquad \int_S \mathbf{F} \cdot \mathbf{n} \, dS = \iint_S (F_1 \, dy \, dz + F_2 \, dz \, dx + F_3 \, dx \, dy)$

Also by Gauss's divergence theorem, we have
$$\int_S \mathbf{F} \cdot \mathbf{n} \, dS = \int_V \text{div } \mathbf{F} \, dV$$

In cartesian coordinate $dV = dx \, dy \, dz$. Also
$$\text{div } \mathbf{F} = \nabla \cdot \mathbf{F} = \frac{\partial F_1}{\partial x} + \frac{\partial F_2}{\partial y} + \frac{\partial F_3}{\partial z}$$

$\therefore \qquad \int_V \text{div } \mathbf{F} \, dV = \iiint_V \left(\frac{\partial F_1}{\partial x} + \frac{\partial F_2}{\partial y} + \frac{\partial F_3}{\partial z} \right) dx \, dy \, dz$

Gauss's, Green's and Stoke's Theorem

Hence the Cartesian form of Gauss's theorem is

$$\iint_S (F_1\, dy\, dz + F_2\, dz\, dx + F_3\, dx\, dy) = \iiint_V \left(\frac{\partial F_1}{\partial x} + \frac{\partial F_2}{\partial y} + \frac{\partial F_3}{\partial z}\right) dx\, dy\, dz$$

EXAMPLES

1. *If* $\mathbf{F} = 2xy\,\mathbf{i} - yz\,\mathbf{j} + x^2\,\mathbf{k}$, *evaluate* $\int_S \mathbf{F}\cdot\mathbf{n}\, dS$, *where S denotes the entire surface of the cube bounded by the coordinate planes and the planes* $x = a, y = a, z = a$ *by the application of Gauss's theorem.*

Sol. We have
$$\mathbf{F} = 2xy\,\mathbf{i} - yz\,\mathbf{j} + x^2\,\mathbf{k}$$

$\therefore \quad \operatorname{div} \mathbf{F} = \dfrac{\partial}{\partial x}(2xy) + \dfrac{\partial}{\partial y}(-yz) + \dfrac{\partial}{\partial x}(x^2)$

$\qquad\qquad = 2y - z$

$\therefore \quad \int_S \mathbf{F}\cdot\mathbf{n}\, dS = \int_V \operatorname{div} \mathbf{F}\, dV$, by Gauss's divergence theorem

$\qquad = \int_0^a \int_0^a \int_0^a (2y - z)\, dx\, dy\, dz$

$\qquad = \int_0^a \int_0^a \left[2yz - \dfrac{z^2}{2}\right]_0^a dx\, dy$

$\qquad = \int_0^a \int_0^a \left(2ay - \dfrac{a^2}{y}\right) dx\, dy$

$\qquad = \int_0^a \left[ay^2 - \dfrac{1}{2}a^2 y\right]_0^a dx$

$\qquad = \int_0^a \left(a^3 - \dfrac{1}{2}a^3\right) dx = \dfrac{1}{2}a^3 [x]_0^a$

$\qquad = \dfrac{1}{2}a^4.$

2. *Verify Gauss divergence theorem for*
$$\iint_S \{(x^3 - yz)\, dy\, dz - 2x^2 y\, dz\, dx + z\, dx\, dy\}$$
over the surface of cube bounded by coordinate planes and the planes $x = y = z = a$.

Sol. Let $\mathbf{F} = F_1\,\mathbf{i} + F_2\,\mathbf{j} + F_3\,\mathbf{k}$.
From Gauss divergence theorem, we know
$$\int_S \mathbf{F}\cdot\mathbf{n}\, dS = \int_S [F_1\, dy\, dz + F_2\, dz\, dx + F_3\, dx\, dy] = \int_V \operatorname{div}\mathbf{F}\, dV. \qquad\qquad ...(i)$$

Here, $\quad F_1 = x^3 - yz,\ F_2 = -2x^2 y,\ F_3 = z$

Hence, $\quad \mathbf{F} = (x^3 - yz)\,\mathbf{i} - 2x^2 y\,\mathbf{j} + z\,\mathbf{k}$

$\operatorname{div} \mathbf{F} = \left\{\mathbf{i}\dfrac{\partial}{\partial x} + \mathbf{j}\dfrac{\partial}{\partial y} + \mathbf{k}\dfrac{\partial}{\partial z}\right\}\cdot\{(x^3 - yz)\mathbf{i} - 2x^2 y\,\mathbf{j} + z\,\mathbf{k}\}$

$\qquad = \dfrac{\partial}{\partial x}(x^3 - yz) + \dfrac{\partial}{\partial y}(-2x^2 y) + \dfrac{\partial}{\partial z}(z)$

$\qquad = 3x^2 - 2x^2 + 1 = x^2 + 1$

Hence, $\int_S \mathbf{F}\cdot\mathbf{n}\, dS = \int_V (x^2 + 1)\, dV$

$\qquad = \int_0^a \int_0^a \int_0^a (x^2 + 1)\, dx\, dy\, dz$

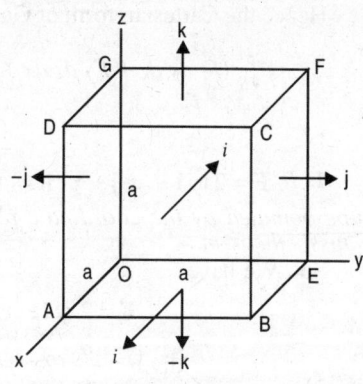

$$= \int_0^a \int_0^a (x^2+1)\{z\}_0^a \, dx \, dy$$
$$= a \int_0^a \int_0^a (x^2+1) \, dx \, dy$$
$$= a \int_0^a (x^2+1)\{y\}_0^a \, dx$$
$$= a^2 \int_0^a (x^2+1) \, dx$$
$$= a^2 \left\{ \frac{x^3}{3} + x \right\}_0^a$$
$$= a^2 \left\{ \frac{a^3}{3} + a \right\} = \frac{a^5}{3} + a^3 \qquad ...(ii)$$

Verification by Direct Integral : Outward drawn unit vector normal to face $OEFG$ is $-\mathbf{i}$ and dS is $dy \, dz$.

If I_1 is integral along this face,

$$I_1 = \int_S \mathbf{F} \cdot \mathbf{n} \, dS = \int \int_S \mathbf{F} \cdot (-\mathbf{i}) \, dy \, dz$$
$$= \int \int_S (x^3 - yz) \, dy \, dz \quad [\text{as } x=0 \text{ for this face}]$$
$$= \int_0^a \int_0^a yz \, dy \, dz = \int_0^a y \left\{ \frac{z^2}{2} \right\}_0^a dy$$
$$= \frac{a^2}{2} \int_0^a y \, dy = \frac{a^2}{2}\left[\frac{y^2}{2}\right]_0^a = \frac{a^4}{4}$$

For face $ABCD$, its equation is $x=a$ and $\mathbf{n} \, dS = \mathbf{i} \, dy \, dz$,

If I_2 is integral along this face

$$I_2 = \int \int_S \mathbf{F} \cdot \mathbf{i} \, dy \, dz$$
$$= \int \int_S (x^3 - yz) \, dy \, dz$$
$$= \int_0^a \int_0^a (a^3 - yz) \, dy \, dz$$
$$= \int_0^a \left\{ a^3 z - y\frac{a^2}{2} \right\}_0^a dy$$
$$= \left[a^4 y - \frac{a^2}{2} \frac{y^2}{2} \right]_0^a = a^5 - \frac{a^4}{4}$$

If I_3 is integral along face $OGDA$ whose equation is
$$y = 0$$
$$\mathbf{n} \, dS = -\mathbf{j} \, dx \, dz$$

Hence, $$I_3 = \int \int_S \mathbf{F} \cdot (-\mathbf{j}) \, dx \, dz$$
$$= -\int \int_S -2x^2 y \, dx \, dz$$
$$= 0, \text{ as } y=0.$$

If I_4 is integral along face $BEFC$ whose equation is
$$y = a$$
$$\mathbf{n} \, dS = \mathbf{j} \, dx \, dz$$

Gauss's, Green's and Stoke's Theorem

Then
$$I_4 = -\iint_S 2x^2 y \, dx \, dz$$
$$= -2a \int_0^a \int_0^a x^2 \, dx \, dz$$
$$= -2a \int_0^a x^2 \{z\}_0^a \, dx$$
$$= -2a \int_0^a x^2 \, dx$$
$$= -2a^2 \left[\frac{x^3}{3}\right]_0^a = -\frac{2}{3} a^5.$$

If I_5 is integral along face $OABF$ whose equation is
$$z = 0$$
$$\mathbf{n} \, dS = -\mathbf{k} \, dx \, dy$$
$$I_5 = \iint_S \mathbf{F} \cdot (-\mathbf{k} \, dx \, dy)$$
$$= -\iint_S z \, dx \, dy = 0 \text{ as } z = 0,$$

If I_6 is integral along face $OFGD$ whose equation is
$$z = a \quad \mathbf{n} \, dS = \mathbf{k} \, dx \, dy$$
$$I_6 = \iint_S z \, dx \, dy = \int_0^a \int_0^0 a \, dx \, dy$$
$$= a \int_0^a [y]_0^a \, dx = a^2 \int_0^a dx = a^3$$

Total surface
$$I = I_1 + I_2 + I_3 + I_4 + I_5 + I_6$$
$$= \frac{a^4}{4} + a^5 - \frac{a^4}{4} + 0 - \frac{2}{3} a^5 + 0 + a^3$$
$$= \frac{a^5}{3} + a^3 \qquad \qquad ...(iii)$$

which is equal to volume integral. Hence Gauss theorem is verified.

3. *Show that*
$$\int_S (ax\mathbf{i} + by\mathbf{j} + cz\mathbf{k}) \cdot \mathbf{n} \, dS = \frac{4}{3} \pi (a + b + c)$$

where S is the surface of the sphere $x^2 + y^2 + z^2 = 1$.

Sol. We have by Gauss's divergence theorem
$$\int_S \mathbf{F} \cdot \mathbf{n} \, dS = \int_V \text{div } \mathbf{F} \, dV$$

Now,
$$\text{div } \mathbf{F} = \frac{\partial}{\partial x}(ax) + \frac{\partial}{\partial y}(by) + \frac{\partial}{\partial z}(cz) = a + b + c$$

$\therefore$
$$\int_S \mathbf{F} \cdot \mathbf{n} \, dS = \int_V (a + b + c) \, dV$$
$$= (a + b + c) V.$$

Now $\quad V = $ Volume of sphere of unit radius
$$= \frac{4}{3} \cdot \pi \cdot 1^3 = \frac{4}{3} \pi$$

$\therefore \quad \int_S (ax\mathbf{i} + by\mathbf{j} + cz\mathbf{k}) \cdot dS = (a+b+c) \cdot \frac{4}{3}\pi = \frac{4}{3}(a+b+c)\pi.$

EXERCISES

1. Show that $\frac{1}{3}\int_S \mathbf{r}\cdot\mathbf{n}\, dS = V$.

2. Evaluate $\int_S \mathbf{F}\cdot\mathbf{n}\, ds$ when $\mathbf{F} = 4xy\,\mathbf{i} + yz\,\mathbf{j} - xz\,\mathbf{k}$ and S is the surface of the cube bounded by the plane $x=0, x=2, y=2, y=0$ and $z=0, z=2$. [Ans. 32]

3. Evaluate $\int_S (x\mathbf{i} + y\mathbf{j} + z\mathbf{k})\cdot\mathbf{n}\, dS$ where S denotes the surface of the cube bounded by the planes $x=0, x=a, y=0, y=a, z=0, z=a$ by the application of Gauss's theorem. [Ans. $3a^3$]

4. Verify Divergence theorem for $\mathbf{f} = 4x\mathbf{i} - 2y^2\mathbf{j} + z^2\mathbf{k}$ taken over the region bounded by $x^2 + y^2 = 4, z=0$ and $z=3$.

5. Verify the divergence theorem for the function $\mathbf{F} = y\mathbf{i} + x\mathbf{j} + z^2\mathbf{k}$ over the cylindrical region S bounded by $x^2 + y^2 = a^2, z=0$ and $z=h$.

6. Evaluate $\int_S (y^2z^2\mathbf{i} + z^2x^2\mathbf{j} + z^2y^2\mathbf{k})\cdot\mathbf{n}\, dS$, where S is the part of the sphere $x^2 + y^2 + z^2 = 1$ above the xy-plane. [Ans. 3/12]

5.2 GREEN'S THEOREM IN THE PLANE

Relation between Plane Surface Integral and Line Integral

If S is a closed region in the xy-plane bounded by a simple closed curve C and if $\phi(x, y)$ and $\psi(x, y)$ are continuous functions having continuous partial derivatives in R, then

$$\oint_C (\psi\, dx + \phi\, dy) = \iint_S \left(\frac{\partial \phi}{\partial x} - \frac{\partial \psi}{\partial y}\right) dx\, dy,$$

where C is traversed in the positive (anti-clockwise) direction.

In vector notation, Green's theorem is

$$\int_C \mathbf{F}\cdot d\mathbf{r} = \int_S \text{curl}\,\mathbf{F}\cdot\mathbf{n}\, dS$$

where $\mathbf{n} = \mathbf{k}$ for xy-plane and $dS = dx\, dy$ and $\text{curl}\,\mathbf{F}\cdot\mathbf{n} = \left\{\dfrac{\partial \phi}{\partial x} - \dfrac{\partial \psi}{\partial y}\right\}$ and $\mathbf{F} = i\psi + j\phi$

Proof: By Stoke's theorem, we have

$$\iint_S \text{curl}\,\mathbf{F}\cdot\mathbf{n}\, dS = \int_C \mathbf{F}\cdot d\mathbf{r} \qquad ...(1)$$

Let $\mathbf{F} = \mathbf{i}\psi + \mathbf{j}\phi$, then we have

$$\text{curl}\,\mathbf{F} = \begin{vmatrix} \mathbf{i} & \mathbf{j} & \mathbf{k} \\ \partial/\partial x & \partial/\partial y & \partial/\partial z \\ \psi & \phi & 0 \end{vmatrix}$$

$$= -\mathbf{i}\frac{\partial \phi}{\partial z} + \mathbf{j}\frac{\partial \psi}{\partial z} + \mathbf{k}\left(\frac{\partial \phi}{\partial x} - \frac{\partial \psi}{\partial y}\right)$$

Also, since $\mathbf{n} = \mathbf{k}$, we have

$$\text{curl}\,\mathbf{F}\cdot\mathbf{n} = \text{curl}\,\mathbf{F}\cdot\mathbf{k} = \frac{\partial \phi}{\partial x} - \frac{\partial \psi}{\partial y}.$$

$\therefore \qquad \iint_S \text{curl}\,\mathbf{F}\cdot\mathbf{n}\, dS = \iint_S \left(\frac{\partial \phi}{\partial x} - \frac{\partial \psi}{\partial y}\right) dx\, dy \qquad ...(2)$

Also, $\qquad \int_C \mathbf{F}\cdot d\mathbf{r} = \int_C (\psi\mathbf{i} + \phi\mathbf{j})\cdot(\mathbf{i}\, dx + \mathbf{j}\, dy + \mathbf{k}\, dz)$

Gauss's, Green's and Stoke's Theorem

$$= \int_C (\psi\, dx + \phi\, dy)$$

Hence from (1), (2) and (3), we get

$$\int\int_S \left(\frac{\partial \phi}{\partial x} - \frac{\partial \psi}{\partial y}\right) dx\, dy = \int_C (\psi\, dx + \phi\, dy)$$

EXAMPLES

1. *Verify Green's theorem in the plane for*

$$\oint_C [(x^2 + y^2)\, dx - 2xy\, dy],$$

where C is the rectangle bounded by $y = 0, x = 0, y = b, x = a$.

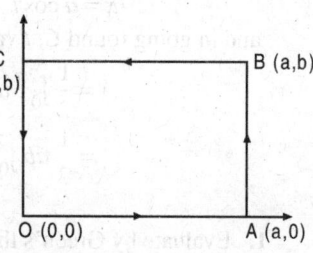

Sol. By Green's theorem we have

$$\oint_C [(x^2 + y^2)\, dx - 2xy\, dy]$$

$$= \int\int_S \left[\frac{\partial}{\partial x}(-2xy) - \frac{\partial}{\partial y}(x^2 + y^2)\right] dx\, dy$$

$$= \int\int_S (-2y - 2y)\, dx\, dy - \int\int -4y\, dx\, dy - \int\int -4y\, dx\, dy \qquad ...(i)$$

Now to verify Green's theorem, first we shall evaluate Line integral of L.H.S. along the rectangle OABC.

For this, if $\quad \mathbf{F} = (x^2 + y^2)\mathbf{i} - 2xy\, \mathbf{j}$

then $\quad \int_C \mathbf{F} \cdot d\mathbf{r} = \int_C \{(x^2 + y^2)\, dx - 2xy\, dy\}$

$$= \int_{OA} \mathbf{F} \cdot d\mathbf{r} + \int_{AB} \mathbf{F} \cdot d\mathbf{r} + \int_{BC} \mathbf{F} \cdot d\mathbf{r} + \int_{CO} \mathbf{F} \cdot d\mathbf{r} \qquad ...(ii)$$

Along *OA*, $y = 0, \therefore dy = 0$ and y varies from 0 to a
Along *AB*, $x = a, \therefore dx = 0$ and y varies from 0 to b
Along *BC*, $y = b, \therefore dy = 0$ and x varies from a to 0
Along *CO*, $x = 0, \therefore dx = 0$ and y varies from b to 0

$\therefore$ From (ii), we get

$$\int_C \mathbf{F} \cdot d\mathbf{r} = \int_0^a x^2\, dx - \int_0^b 2ay\, dy + \int_a^0 (x^2 + b^2)\, dx \int_b^0 0 \cdot dy$$

$$= \frac{a^3}{3} - 2a \cdot \frac{b^2}{2} + \left[\frac{x^3}{3} + b^2 x\right]_a^0 = -2ab^2. \qquad ...(iii)$$

R.H.S. $= \int_{x=0}^a \int_{y=0}^b (-4y)\, dx\, dy$

$$= -2b \int_0^a dx = -2ab^2 \qquad ...(iv)$$

Hence the Green's theorem is verified.

2. *Show that the area bounded by a simple closed curve C is given by* $\dfrac{1}{2}\int_C (x\, dy - y\, dx)$ *and hence find the area of an ellipse.*

Sol. We know that

$$\int_C (\psi\, dx + \phi\, dy) = \int\int_S \left(\frac{\partial \phi}{\partial x} - \frac{\partial \psi}{\partial y}\right) dx\, dy$$

where S is the plane area A enclosed by a curve C. Choosing

$$\psi = -y \text{ and } \phi = x$$

$\therefore \quad \dfrac{\partial \psi}{\partial y} = -1 \text{ and } \dfrac{\partial \phi}{\partial x} = 1$

$$\therefore \quad \int_C (-y\, dx + x\, dy) = \iint_S 2\, dx\, dy$$
$$= 2 \iint_S dx\, dy = 2A$$
$$\therefore \quad A = \frac{1}{2} \int (x\, dy - y\, dx) \qquad \ldots(i)$$

Let the parametric equation of the ellipse be
$$x = a \cos t,\ y = b \sin t$$
and in going round C, t varies form 0 to 2π.

$$\therefore \quad A = \frac{1}{2} \int_0^{2\pi} a \cos t\, (b \cos t\, dt) - (b \sin t)(-a \sin t)\, dt$$
$$= \frac{1}{2} ab \int_0^{2\pi} (\cos^2 t + \sin^2 t)\, dt = \frac{1}{2} ab \cdot 2\pi = \pi\, ab$$

EXERCISES

1. Evaluate by Green's theorem in the plane
$$\oint_C [(x^2 - \cosh y)\, dx + (y + \sin x)\, dy]$$
where C is the rectangle with vertices $(0, 0)$, $(\pi, 0)$, $(\pi, 1)$, $(0, 1)$. [Ans. $\pi(\cosh 1 - 1)$]

2. Verify Green's theorem in the plane for $\int_C (xy + y^2)\, dx + x^2 dy$, where C is the closed curve of the region bounded by $y = x$ and $y = x^2$.

3. Verify Green's theorem in the plane for $\int_C (3x^2 - 8y^2)\, dx + (4y - 6xy)\, dy$, where C is the boundary of the region defined by $x = 0,\ y = 0,\ x + y = 1$.

4. Evaluate by Green's theorem $\int_C (e^{-x} \sin y\, dx + e^{-x} \cos y\, dy)$, where C is the rectangle with vertices $(0, 0)$, $(\pi, 0)$, $\left(\pi, \dfrac{\pi}{2}\right)$, $(0, \pi/2)$ and hence verigfy Green's Theorem.

[Ans. $2(e^{-\pi} - 1)$]

5.3 STOKES THEOREM

Relation between Line integral and Surface Integral.

Statement : *The line integral of the tangential component of a vector function $\mathbf{F}$ taken around a simple closed curve C is equal to the normal surface integral of curl $\mathbf{F}$ taken over any surface S having C as its boundary.*

In symbolic form we can state the above theorem as follows :

If $\mathbf{F}$ is any continuously differentiable vector function and S is a surface bounded by a curve C then,
$$\int_C \mathbf{F} \cdot d\mathbf{r} = \int_S \text{curl } \mathbf{F} \cdot \mathbf{n}\, dS$$
where $\mathbf{n}$ is the unit normal vector at any point of S, which is drawn in the sense in which a right handed screw would move when rotated in the sense of description of C.

Proof : Consider a space curve C bounding an open surface S. Divide S into m sub-regions so small that they may be assumed to be planar with areas $\Delta S_1, \Delta S_2, \ldots, \Delta S_m$. Choose any point (ξ_r, η_r, ζ_r) inside or on the boundary C_r of ΔS_r.

Assume that C is described in the positive sense. Then an orientation for each C_r is determined as follows :

(i) If C_r and C have an edge in common, this edge is described in the same direction along both boundaries, and

Gauss's, Green's and Stoke's Theorem

(ii) If C_r and C_s have an edge in common, this edge is described in opposite directions.

Let the unit normal vector at (ξ_r, η_r, ζ_r) be $\mathbf{n}_r$ with positive direction such that this and the direction of C_r are related by the right hand screw rule.

From the definition of curl $\mathbf{F}$ as a limit. We have

$$\mathbf{n}_r \cdot \text{curl } \mathbf{F}(\xi_r, \eta_r, \zeta_r) \Delta S_r z = \int_{C_r} \mathbf{F} \cdot d\mathbf{r} + \epsilon_r \Delta S_r,$$

where ϵ_r tends to zero as ΔS_r tends to zero. Addition of these equations for $r = 1, 2, 3, ..., m$ gives

$$\sum_{r=1}^{m} \mathbf{n}_r \cdot \text{curl } \mathbf{F}(\xi_r, \eta_r, \zeta_r) \Delta S_r = \sum_{r=1}^{m} \int_{C_r} \mathbf{F} \cdot d\mathbf{r} + \sum_{r=1}^{m} \epsilon_r \Delta S_r$$

Now $\sum_{r=1}^{m} \epsilon_r \Delta S_r \leq S (\max \epsilon_r)$, where S is the total area of the surface and hence this term tends to zero as m tends to infinity in such a way that each ΔS_r shrinks to a point.

Further, the contribution of the circulation from the two adjacent boundary curves is zero as they are traversed in opposite directions. Hence in the limit, we have

$$\int_S \mathbf{n} \cdot \text{curl } \mathbf{F} \, dS = \int_A \mathbf{F} \cdot d\mathbf{r} \quad ...(1)$$

where $\mathbf{n}$ is the vector field of positive unit normals to the surface S. We have thus Stoke's theorem :

$$\int_C \mathbf{F} \cdot d\mathbf{r} = \int_S (\nabla \times F) \cdot dS \quad ...(2)$$

5.4 STOKES THEOREM IN CARTESIAN FORM

Let F_1, F_2, F_3 be the components of vector point function $\mathbf{F}$ so that $\mathbf{F} = F_1 \mathbf{i} + F_2 \mathbf{j} + F_3 \mathbf{k}$ and an unit outward drawn normal be $\mathbf{r} = \mathbf{i} l + \mathbf{j} m + \mathbf{k} n$ where l, m, n are direction cosines.

Again $\mathbf{r} = x\mathbf{i} + y\mathbf{j} + z\mathbf{k}$, or $d\mathbf{r} = dx\mathbf{i} + dy\mathbf{j} + dz\mathbf{k}$

$\therefore \quad \int_C \mathbf{F} \cdot d\mathbf{r} = \int_C F_1 \, dx + F_2 \, dy + F_3 \, dz$ ...(i)

Now,
$$\text{curl } \mathbf{F} = \begin{vmatrix} \mathbf{i} & \mathbf{j} & \mathbf{k} \\ \frac{\partial}{\partial x} & \frac{\partial}{\partial y} & \frac{\partial}{\partial z} \\ F_1 & F_2 & F_3 \end{vmatrix}$$

$$= \left(\frac{\partial F_3}{\partial y} - \frac{\partial F_2}{\partial z}\right)\mathbf{i} + \left(\frac{\partial F_1}{\partial z} - \frac{\partial F_3}{\partial x}\right)\mathbf{j} + \left(\frac{\partial F_2}{\partial x} - \frac{\partial F_1}{\partial y}\right)\mathbf{k}$$

$\therefore \quad \mathbf{n} \cdot \text{curl } \mathbf{F} = \left(\frac{\partial F_3}{\partial y} - \frac{\partial F_2}{\partial z}\right) l + \left(\frac{\partial F_1}{\partial z} - \frac{\partial F_3}{\partial x}\right) m + \left(\frac{\partial F_2}{\partial x} - \frac{\partial F_1}{\partial y}\right) n$

$\therefore \quad \mathbf{n} \cdot \text{curl } \mathbf{F} \, dS = \Sigma \left(\frac{\partial F_3}{\partial y} - \frac{\partial F_2}{\partial z}\right) l \, dS$

$$= \Sigma \left(\frac{\partial F_3}{\partial y} - \frac{\partial F_2}{\partial z}\right) dy \, dz$$

$\because \quad l \, dS = \cos \alpha \cdot dS = dy \, dz$

Now by Stoke's Theorem

$$\int_C \mathbf{F} \cdot d\mathbf{r} = \int_S \mathbf{n} \cdot \text{curl } \mathbf{F} \, dS \quad \text{or} \quad \int_C (F_1 \, dx + F_2 \, dy + F_3 \, dz)$$

$$= \int_S \left[\left(\frac{\partial F_3}{\partial y} - \frac{\partial F_2}{\partial z}\right) dy \, dz + \left(\frac{\partial F_1}{\partial y} - \frac{\partial F_3}{\partial x}\right) dz \, dz + \left(\frac{\partial F_2}{\partial x} - \frac{\partial F_1}{\partial y}\right) dx \, dy\right]$$

This is cartesian equivalent of Stokes Theorem.

EXAMPLES

1. *Verify Stokes Theorem for* $\mathbf{F} = (2x - y)\mathbf{i} - yz^2 \mathbf{j} - y^2 z\mathbf{k}$ *where S is the upper half surface of the sphere* $x^2 + y^2 + z^2 = 1$ *and C is its boundary.*

Sol. We have the Stoke's Theorem as

$$\int_C \text{curl } \mathbf{F} \cdot \mathbf{n} \, dS = \int_C \mathbf{F} \cdot d\mathbf{r}$$

Clearly, C the boundary of the upper half of the sphere is a circle $x^2 + y^2 = 1$ in the xy plane whose parametric equations be taken as

$$x = \cos t, \quad y = \sin t$$

$$\therefore \quad \int_C \mathbf{F} \cdot d\mathbf{r} = \int F_1 \, dx + F_2 \, dy + F_3 \, dz$$

$$= \int (2x - y) \, dx - yz^2 \, dy - y^2 z \, dz$$

Put $z = 0$

$$= \int_0^{2\pi} (2x - y) \frac{dx}{dt} \, dt$$

$$= \int_0^{2\pi} (2\cos t - \sin t)(-\sin t) \, dt$$

$$= \int_0^{2\pi} (-2\cos t \sin t + \sin^2 t) \, dt$$

$$= \left[-\cos^2 t\right]_0^{2\pi} + 4\int_0^{\pi/2} \sin^2 t \, dt$$

$$= 0 + 4 \cdot \frac{1}{2} \cdot \frac{\pi}{2} = \pi. \qquad \ldots(i)$$

Again

$$\text{curl } \mathbf{F} = \begin{vmatrix} \mathbf{i} & \mathbf{j} & \mathbf{k} \\ \frac{\partial}{\partial x} & \frac{\partial}{\partial y} & \frac{\partial}{\partial z} \\ 2x - y & -yz^2 & -y^2 z \end{vmatrix}$$

$$= \mathbf{i}(-2yz + 2yz) + \mathbf{j}(0 - 0) + \mathbf{k}(0 + 1) = \mathbf{k}$$

$$\therefore \quad \text{curl } \mathbf{F} \cdot \mathbf{n} = \mathbf{k} \cdot \mathbf{n} = \mathbf{n} \cdot \mathbf{k}$$

$$\therefore \quad \int_S \text{curl } \mathbf{F} \cdot \mathbf{n} \, dS = \int_S \mathbf{n} \cdot \mathbf{k} \, dS$$

$$= \iint_R \mathbf{n} \cdot \mathbf{k} \, \frac{dx \, dy}{\mathbf{n} \cdot \mathbf{k}}$$

where R is the projection of S and xy-plane.

$$\therefore \quad \int_S \text{curl } \mathbf{F} \cdot \mathbf{n} \, dS = \int_{x=-1}^{1} \int_{y=-\sqrt{(1-x^2)}}^{\sqrt{(1-x^2)}} dx \, dy$$

$$= 4 \int_0^1 \int_0^{\sqrt{(1-x^2)}} dx \, dy = 4 \int_0^1 \sqrt{(1-x^2)} \, dx$$

$$= 4 \left[\frac{4}{x}\sqrt{(1-x)^2} + \frac{1}{2}\sin^{-1} x\right]_0^1$$

$$= 4 \cdot \frac{1}{2} \cdot \frac{\pi}{2} = \pi \qquad \ldots(iii)$$

From (i) and (ii), we get $\quad \int_S \text{curl } \mathbf{F} \cdot \mathbf{n} \, dS = \int_C \mathbf{F} \cdot d\mathbf{r}$

Hence the Stokes Theorem.

2. *Evaluate by Stokes theorem* $\int_C (e^x \, dx + 2y \, dy - dz)$ *where C is the curve* $x^2 + y^2 = 4, z = 2.$

Gauss's, Green's and Stoke's Theorem

Sol. We have $\int_C \mathbf{F} \cdot d\mathbf{r} = \int_C (F_1\, dx + F_2\, dy + F_3\, dz)$

$$= \int_C (e^x\, dx + 2y\, dy - dz)$$

where $\mathbf{F} = e^x \mathbf{i} + 2y\mathbf{j} - \mathbf{k}$, $d\mathbf{r} = \mathbf{i}\, dx + \mathbf{j}\, dy + \mathbf{k}\, dz$

By Stokes Theorem $\quad \int_C \mathbf{F} \cdot d\mathbf{r} = \int_{S_1} \mathbf{n} \cdot \text{curl } \mathbf{F}\, dS$ ...(i)

Where S_1 the surface whose boundary C is given by the circle $x^2 + y^2 = 4, z = 2$, *i.e.*, a circle with centre $(0, 0, 2)$ and radius 2. Clearly $\mathbf{n} = \mathbf{k}$. Now,

$$\text{curl } \mathbf{F} = \begin{vmatrix} \mathbf{i} & \mathbf{j} & \mathbf{k} \\ \dfrac{\partial}{\partial x} & \dfrac{\partial}{\partial y} & \dfrac{\partial}{\partial z} \\ e^x & 2y & -1 \end{vmatrix} = 0$$

$\therefore \qquad \mathbf{n} \cdot \text{curl } \mathbf{F} = 0$

Hence, from (i) $\quad \int_C \mathbf{F} \cdot d\mathbf{r} = 0$.

3. *Verify Stoke's Theorem for* $\mathbf{F} = (x^2 + y^2)\mathbf{i} - 2xy\mathbf{j}$ *taken round the rectangle bounded by* $x = \pm a, y = 0, y = b$.

Sol. Clearly

$$\mathbf{F} \cdot d\mathbf{r} = (x^2 + y^2)\, dx - 2xy\, dy \qquad \qquad \text{...(i)}$$

$\therefore \quad \int_C \mathbf{F} \cdot d\mathbf{r} = \int_{AD} \mathbf{F} \cdot d\mathbf{r} + \int_{DC} \mathbf{F} \cdot d\mathbf{r} + \int_{CB} \mathbf{F} \cdot d\mathbf{r} + \int_{BA} \mathbf{F} \cdot d\mathbf{r}$

$\qquad = I_1 + I_2 + I_3 + I_4$

For $I_1, y = b, dy = 0$ and x varies from a to $-a$.

$\therefore \quad I_1 = \int [(x^2 + y^2)\, dx - 2xy\, dy]$

$\qquad = \int_a^{-a} (x^2 + b^2)\, dx + 0 \qquad \because y = b$

$\qquad = \left[\dfrac{1}{3}x^3 + b^2 x\right]_a^{-a} = -\left(\dfrac{2}{3}a^3 + 2b^2 a\right)$

Similarly, $\quad I_2 = -ab^2, I_3 = \dfrac{2}{3}a^3$ and $I_4 = -ab^2$

$\therefore \quad \int_C \mathbf{F} \cdot d\mathbf{r} = -\dfrac{2}{3}a^3 - 2b^2 a - ab^2 + \dfrac{2}{3}a^3 - ab^2 = -4ab^2$ ...(ii)

Again, we have $\text{curl } \mathbf{F} = -4y\mathbf{k}, \mathbf{n} = \mathbf{k}$

$\therefore \quad \mathbf{n} \cdot \text{curl } \mathbf{F} = \mathbf{k} \cdot (-4y\mathbf{k}) = -4y$.

$dS = \dfrac{dx\, dy}{\mathbf{n} \cdot \mathbf{k}} = \dfrac{dx\, dy}{\mathbf{k} \cdot \mathbf{k}} = dx\, dy$

$\therefore \quad \int_S \mathbf{n} \cdot \text{curl } \mathbf{F}\, dS = \int_{-a}^{a} \int_0^b -4x\, dx\, dy$

$\qquad \qquad = -4\int_{-a}^{0} \left[\dfrac{1}{2}y^2\, dx\right]_0^b$

$\qquad \qquad = -2b^2 [x]_{-a}^{a} = -4ab^2$ ...(iii)

Hence from (ii) and (iii), we get

$\int_C \mathbf{F} \cdot d\mathbf{r} = \int_S \mathbf{n} \cdot \text{curl } \mathbf{F}\, dS = -4ab^2$

EXERCISES

1. Prove that $\int_C \mathbf{r} \cdot d\mathbf{r} = 0$
2. Verify Stokes Theorem for the function $\mathbf{F} = z\mathbf{i} + x\mathbf{j} + y\mathbf{k}$ where curve is the unit circle in the xy-plane bounding the hemisphere $z = \sqrt{(1 - x^2 - y^2)}$.
3. Prove that $\int_C \mathbf{r} \times d\mathbf{r} = 2\int_S \mathbf{n}\, d\mathbf{r}$ where symbols have their usual meanings.
4. Verify Stokes Theorem for the function $\mathbf{F} = x^2\mathbf{i} + xy\mathbf{j}$ integrated round the square in the plane $z = 0$ whose sides are along the lines $x = 0$, $x = a$, $y = 0$, $y = a$.
5. Verify Stokes Theorem for the vector field defined by $\mathbf{F} = (x^2 - y^2)\mathbf{i} + 2xy\mathbf{j}$ in the rectangular region in the xy-plane bounded by lines $x = 0$, $x = a$, $y = 0$, $y = b$.
6. Verify Stokes Theorem where $\mathbf{F} = (y - z)\mathbf{i} + yz\mathbf{j} - xz\mathbf{k}$ and S is given by $x = 0$, $y = 0$, $z = 0$, $x = 1$, $y = 1$, $z = 1$.
7. Verify Stokes Theorem given that $\mathbf{F} = y\mathbf{i} + 2x\mathbf{j} + z\mathbf{k}$ and C is the circle $x^2 + y^2 = 1$ in the xy-plane and S the plane area bounded by C.
8. Verify the Stoke's Theorem for the function $\mathbf{F} = y\mathbf{i} + z\mathbf{j}$ over the plane surface $2x + 2y + z = 2$ lying in the first octant.
9. Verify the Stoke's theorem for the function $\mathbf{F} = x^2y^2\mathbf{i} + 2xy\mathbf{j}$ when the integration is taken around the volume enclosed by the rectangle $x = \pm a$, $y = 0$, $y = b$.

GEOMETRY

(a) Two Dimensional

General equation of second degree. Tracing of conics. System of conics. Confocal conics. Polar equation of a conic.

1

General Equation of Second Degree and Tracing of Conics

1.1 CONIC SECTIONS

The curves which come under the category of conic sections are : a pair of straight lines, a circle, a parabola, an ellipse and a hyperbola. The name conic section is derived from the fact that these curves were first obtained by cutting a cone in various ways.

Analytically a conic section is defined as follows :

A conic section or conic is the locus of a point which moves so that its distance from a fixed point is in a constant ratio to its perpendicular distance from a fixed straight line.

The fixed point is called the *focus*. The fixed straight line is called the *directrix*. The constant ratio is called the *eccentricity*, and is denoted by e.

The straight line passing through the focus and perpendicular to the directrix is called the *axis*.

A point of intersection of a conic with its axis is called a *vertex*.

When the eccentricity e is equal to unity the conic section is called a *parabola*. when e is less than unity, it is called an *ellipse*.

When e is greater than unity, it is called a *hyperbola*.

1.2 THE GENERAL EQUATION OF SECOND DEGREE

To show that the general equation of the second degree always represents a conic section.

The general equation of the second degree is

$$ax^2 + 2hxy + by^2 + 2gx + 2fy + c = 0 \qquad \ldots(1)$$

To remove the term of xy from (1), let us turn the coordinate axes through an angle θ, the origin remaining the same. So we have to substitute $x \cos\theta - y \sin\theta$ for x and $x \sin\theta + y \cos\theta$ for y in (1). Then equation (1) will give

$$a(x\cos\theta - y\sin\theta)^2 + 2h(x\cos\theta - y\sin\theta)(x\sin\theta + y\cos\theta) + b(x\sin\theta + y\cos\theta)^2$$
$$+ 2g(x\cos\theta - y\sin\theta) + 2f(x\sin\theta + y\cos\theta) + c = 0$$

or $\quad x^2 (a\cos^2\theta + 2h\cos\theta\sin\theta + b\sin^2\theta)$
$$+ 2xy \{h(\cos^2\theta - \sin^2\theta) + (b-a)\sin\theta\cos\theta\}$$
$$+ y^2 (a\sin^2\theta - 2h\cos\theta\sin\theta + b\cos^2\theta)$$
$$+ 2x(g\cos\theta + f\sin\theta) + 2y(f\cos\theta - g\sin\theta) + c = 0 \ldots(2)$$

Now we choose θ so that the coefficient of xy in (2) becomes zero. For this we have

$$h(\cos^2\theta - \sin^2\theta) + (b - a)\sin\theta\cos\theta = 0$$

or
$$h \cos 2\theta - \frac{1}{2}(b-a) \sin 2\theta = 0$$

or
$$\tan 2\theta = \frac{2h}{a-b} \qquad \ldots(3)$$

The relation (3) always gives real values of θ for all vlaues of a, b and h. If we substitute the values of $\cos \theta$ and $\sin \theta$ found from (3) in (2), the term of xy is removed and the equation (2) takes the form

$$Ax^2 + By^2 + 2Gx + 2Fy + C = 0 \qquad \ldots(4)$$

Now the following cases arise :

Case I. Let $A \neq 0$ and $B \neq 0$. The equation (4) may be written as

$$A\left(x^2 + \frac{2Gx}{A} + \frac{G^2}{A^2}\right) + B\left(y^2 + \frac{2Fy}{B} + \frac{F^2}{B^2}\right) - \frac{G^2}{A} - \frac{F^2}{B} + C = 0$$

or
$$A\left(x + \frac{G}{A}\right)^2 + B\left(y + \frac{F}{B}\right)^2 = \frac{G^2}{A} + \frac{F^2}{B} - C = K \ (say).$$

Shifting the origin to $(-G/A, -F/B)$, this equation becomes

$$Ax^2 + By^2 = K \qquad \ldots(5)$$

Now the following sub cases arise :

(i) If $K = 0$, the equation (5) becomes $Ax^2 + By^2 = 0$ and this represents a **pair of straight lines.** These straight lines are real if A and B are of opposite signs and are imaginary if A and B are both of the same sign.

(ii) If $K \neq 0$, the equation (5) may be written as

$$\frac{x^2}{K/A} + \frac{y^2}{K/B} = 1 \qquad \ldots(6)$$

If K/A and K/B are both positive, the equation (6) represents an **ellipse** which becomes a **circle** if in addition to being positive K/A and K/B are also equal. Again the equation (6) represents a **hyperbola** if K/A and K/B are of opposite signs. If K/A and K/B are both negative, the equation (6) is said to represent an **imaginary ellipse.**

Case II. One of A and B is zero while the other is not zero. Without loss of generality we can take $A = 0$ and $B \neq 0$ because if $B = 0$ and $A \neq 0$, the procedure and the result are similar.

Now if $A = 0$ and $B \neq 0$, the equation (4) becomes

$$By^2 + 2Gx + 2Fy + C = 0$$

or
$$y^2 + \frac{2F}{B} y = -\frac{2G}{B} x - \frac{C}{B}$$

or
$$\left(y + \frac{F}{B}\right)^2 = -\frac{2G}{B} x - \frac{C}{B} + \frac{F^2}{B^2} \qquad \ldots(7)$$

If $G = 0$, the equation (7) represents two **parallel straight lines** which are coincident if $F^2 - BC$ is zero.

If $G \neq 0$, the equation (7) can be written as

$$\left(y + \frac{F}{B}\right)^2 = -\frac{2G}{B}\left(x - \frac{F^2}{2BG} + \frac{C}{2G}\right)$$

General Equation of Second Degree and Tracing of Conics

Shifting the origin to $\left(\dfrac{F^2}{2BG} - \dfrac{C}{2G}, -\dfrac{F}{B}\right)$, this equation becomes

$$y^2 = -(2G/B)x,$$

which represents a parabola.

Hence in every case the general equation of second degree represents a conic section.

1.3 CENTRE OF A CONIC

To find the coordinates of the centre of the conic represented by the general eqaution.

Definition. *The centre of a conic section is a point which bisects all those chords of the conic that pass through it.*

Let the equation of the conic be

$$ax^2 + 2hxy + by^2 + 2gx + 2fy + c = 0 \qquad \ldots(1)$$

Let (x_1, y_1) be the centre of this conic.

Transfer the origin to the point (x_1, y_1) without altering the direction of the axes, then the new equation is

$$a(x+x_1)^2 + 2h(x+x_1)(y+y_1) + b(y+y_1)^2 + 2g(x+x_1) + 2f(y+y_1) + c = 0$$

i.e., $ax^2 + 2hxy + by^2 + 2x(ax_1 + hy_1 + g) + 2y(hx_1 + by_1 + f)$

$$+ ax_1^2 + 2hx_1y_1 + by_1^2 + 2fy_1 + c = 0 \qquad \ldots(2)$$

Since the centre of the conic is now the origin of coordinates, corresponding to every point (x', y') on the conic there must be the diametrically opposite point $(-x', -y')$ also on it. In other words, on substituting in turn x', y' and $-x', -y'$ for x and y in (2), there should be no change of sign, but the first degree terms in x and y change sign. Therefore the first degree terms must be absent, *i.e.*, the coefficients of x and y must vanish, so that we have

$$ax_1 + hy_1 + g = 0 \qquad \ldots(3)$$

and $$hx_1 + by_1 + f = 0 \qquad \ldots(4)$$

Solving (3) and (4) we get x_1 and y_1, in general as

$$x_1 = \frac{hf - bg}{ab - h^2} \text{ and } y_1 = \frac{gh - af}{ab - h^2}$$

Hence the coordinates of the centre are

$$\left(\frac{hf - bg}{ab - h^2}, \frac{gh - af}{ab - h^2}\right) \qquad \ldots(5)$$

Note 1. Let $f(x, y) \equiv ax^2 + 2hxy + by^2 + 2gx + 2gy + c$

Then $$\frac{\partial f}{\partial x} = 2(ax + hy + g)$$

and $$\frac{\partial f}{\partial y} = 2(hx + by + f)$$

The equation (3) and (4) are equivalent to

$$ax + hy + g = 0$$

and $$hx + by + f = 0$$

i.e., $$\frac{\partial f}{\partial x} = 0 \text{ and } \frac{\partial f}{\partial y} = 0.$$

Hence to find the coordinates of the centre of conic

$$f(x, y) \equiv ax^2 + 2hxy + by^2 + 2gx + 2fy + c = 0,$$

solve the equations

$$\frac{\partial f}{\partial x} = 0 \text{ and } \frac{\partial f}{\partial y} = 0.$$

Note 2. It is obvious from eqaution (5) that the coordinates of the centre become infinite if $ab - h^2 = 0$. But if $ab - h^2 = 0$, the second degree terms in the given equation (1) from a perfect square and therefore the equation represents a parabola. Thus the centre of a parabola is at infinity.

If, however, $hf - bg = 0$ when $ab - h^2 = 0$, then

$$\frac{a}{h} = \frac{h}{b} = \frac{g}{f} \qquad \ldots(6)$$

So that the equations (3) and (4) represents the same equation, viz.,

$$ax_1 + hy_1 + g = 0,$$

i.e., we have only one equation to determine the centre, and therefore we get an infinite number of centres which all lie on the line

$$ax + hy + g = 0 \qquad \ldots(7)$$

In this case the conic section consists a pair of parallel straight lines, both parallel to the line of centres (7).

These parallel lines can be obtained as follows :

From (6)

$$\left(\frac{a}{h}\right)\left(\frac{h}{b}\right) = \left(\frac{g}{f}\right)^2$$

i.e.,

$$\frac{\sqrt{a}}{\sqrt{b}} = \frac{g}{f};$$

so that equation (1) can be written as

$$(x\sqrt{a} + y\sqrt{b})^2 + \frac{2g}{\sqrt{a}}(x\sqrt{a} + y\sqrt{b}) + c = 0,$$

which can be solved for $x\sqrt{a} + y\sqrt{b}$, and hence this represents two parallel straight lines.

1.4 EQUATION OF THE CONIC REFERRED TO CENTRE AS ORIGIN

To find the equation of the conic referred to centre as origin and to axes through the centre parallel to the origin axes.

Let the equation of the conic be

$$ax^2 + 2hxy + by^2 + 2gx + 2fy + c = 0 \qquad \ldots(1)$$

Let the centre of the conic (1) be (x_1, y_1)
so that the equation for finding these coordinates are

$$ax_1 + hy_1 + g = 0 \qquad \ldots(2)$$

and

$$hx_1 + by_1 + f = 0 \qquad \ldots(3)$$

Solving these equations, the coordinates of the centre are

$$\left(\frac{hf - bg}{ab - h^2}, \frac{gh - af}{ab - h^2}\right)$$

When the origin is transferred to (x_1, y_1) then the transformed equation becomes

$$ax^2 + 2hxy + by^2 + c_1 = 0 \qquad \ldots(4)$$

General Equation of Second Degree and Tracing of Conics

where
$$c_1 = ax_1^2 + 2hx_1y_1 + by_1^2 + 2gx_1 + 2fy_1 + c$$
$$= gx_1 + fy_1 + c \qquad ...(5)$$
[From (2) and (3)]

Substituting the coordinates of centre, we have
$$c_1 = g\left(\frac{hf - bg}{ab - h^2}\right) + f\left(\frac{gh - af}{ab - h^2}\right) + c$$
$$= \frac{abc + 2fgh - af^2 - bg^2 - ch^2}{ab - h^2}$$
$$= \frac{\Delta}{ab - h^2}$$

where $\Delta \equiv abc + 2fgh - af^2 - bg^2 - ch^2$.

The equation (4) can therefore be written in the form
$$ax^2 + 2gxy + by^2 + \frac{\Delta}{ab - h^2} = 0, \qquad ...(6)$$

which is the required equation of the conic referred to the centre as origin.

Note. The value of c_1 can also be obtained as follows :

On eliminating x_1, y_1 from the equations (2), (3) and (5), we have
$$\begin{vmatrix} a & h & g \\ h & b & f \\ g & f & c - c_1 \end{vmatrix} = 0$$

or
$$\begin{vmatrix} a & h & g \\ h & b & f \\ g & f & c \end{vmatrix} + \begin{vmatrix} a & h & 0 \\ h & b & 0 \\ g & f & -c_1 \end{vmatrix} = 0.$$

or
$$\Delta - c_1(ab - h^2) = 0$$

i.e.,
$$c_1 = \frac{\Delta}{ab - h^2},$$

where
$$\Delta \equiv \begin{vmatrix} a & h & g \\ h & b & f \\ g & f & c \end{vmatrix} = abc + 2fgh - af^2 - bg^2 - ch^2$$

1.5 DISCRIMINANT

The expression
$$abc + 2fgh - af^2 - bg^2 - ch^2$$
usually denoted by Δ is called the **discriminant** of
$$ax^2 + 2hxy + by^2 + 2gx + 2fy + c = 0 \qquad ...(1)$$

If $\Delta = 0$, the general equation of second degree represents a pair of straight lines.

EXAMPLE

Find the coordinates of the centre of the conic
$$x^2 - 2xy + y^2 + 10x - 10y + 21 = 0.$$

and find the equation of the conic referred to the centre as origin.

Sol. Let
$$f(x, y) \equiv x^2 - 3xy + y^2 + 10x - 10y + 21$$

The equations giving the centre are
$$\frac{\partial f}{\partial x} = 2x - 3y + 10 = 0$$

and
$$\frac{\partial f}{\partial y} = -3x + 2y - 10 = 0$$

Solving these equations, we have
$$x = -2, \ y = 2.$$

Hence the coordinates of the centre are $(-2, 2)$.

Now,
$$g = \frac{1}{2}(\text{coeff. of } x) = 5$$

$$f = \frac{1}{2}(\text{coeff. of } y) = -5 \text{ and } c = 21$$

Hence now constant
$$c_1 = gx_1 + fy_1 + c$$
$$= 5(-2) + (-5)2 + 21$$
$$= -20 + 21 = 1.$$

Hence the equation of the conic referred to the centre as origin is
$$x^2 - 3xy + y^2 + 1 = 0.$$

EXERCISES

Find the coordinates of the centre, and also the equations, referred to the origin, of the following conics :

1. $2x^2 + y^2 - 3xy - 5x + 4y + 6 = 0.$ [**Ans.** $(2, 1); 2x^2 + y^2 - 3xy + 3 = 0$]

2. $3x^2 - 8xy - 3y^2 + 10x - 13y + 8 = 0.$ $\left[\textbf{Ans. } \left(-\frac{41}{25}, \frac{1}{50}\right); 3x^2 - 8xy - 3y^2 = \frac{33}{100}\right]$

3. $6x^2 - 5xy - 6y^2 + 14x + 5y + 4 = 0$ $\left[\textbf{Ans. } \left(-\frac{11}{13}, \frac{10}{13}\right); 6x^2 - 5xy - 63y^2 = 0.\right]$

4. $13x^2 - 18xy + 37y^2 + 2x + 14y - 2 = 0.$ $\left[\textbf{Ans. } \left(-\frac{1}{4}, \frac{1}{4}\right); 13x^2 - 18xy - 37y^2 - 4 = 0.\right]$

5. $2x^2 - 72xy + 23y^2 - 4x - 28y - 48 = 0.$

$$\left[\textbf{Ans. } \left(-\frac{11}{25}, -\frac{2}{25}\right); 2x^2 - 72xy - 23y^2 - 46 = 0.\right]$$

1.6 ASYMPTOTE OF A CONIC

To find the equation of the asymptotes of the conic section represented by the general equation of the second degree.

Let the equation to the conic be
$$ax^2 + 2hxy + by^2 + 2gx + 2fy + c = 0 \qquad \ldots(1)$$

Since the equation to the asymptotes differ from the equation of the conic only in its constant term, so let the equation to the asymptotes be
$$ax^2 + 2hxy + by^2 + 2gx + 2fy + c + \lambda = 0, \qquad \ldots(2)$$

General Equation of Second Degree and Tracing of Conics

where λ is a constant and is to be so chosen that (2) may represent a pair of straight lines.
Now the condition that (2) may represents a pair of straight lines is

$$\begin{vmatrix} a & h & g \\ h & b & f \\ g & f & c+\lambda \end{vmatrix} = 0$$

i.e.,
$$\begin{vmatrix} a & h & g \\ h & b & f \\ g & f & c \end{vmatrix} + \begin{vmatrix} a & h & 0 \\ h & b & 0 \\ g & f & \lambda \end{vmatrix} = 0$$

i.e., $\Delta + \lambda(ab - h^2) = 0$

or $\lambda = -\dfrac{\Delta}{(ab-h)^2}$.

Hence from (2), the equation of the asymptotes to (1) is

$$ax^2 + 2hxy + by^2 + 2gx + 2fy + c - \frac{\Delta}{(ab-h^2)} = 0 \qquad ...(3)$$

Also we know $\Delta \equiv abc + 2fgh - af^2 - bg^2 - ch^2$

$\therefore \quad c - \dfrac{\Delta}{(ab-h^2)} = \dfrac{[c(ab-h^2) - (abc + 2fgh - af^2 - bg^2 - ch^2)]}{(ab-h^2)}$

$= (af^2 + bg^2 - 2fgh)/(ab-h^2)$

Note. Conjugate hyperbola. The equation of conjugate hyperbola differs from that of the asymptotes by the same constant by which the equation of the asymptotes differ from that of the conic.

Hence the equation of the conjugate hyperbola to (1) is

$$ax^2 + 2hxy + by^2 + 2gx + 2fy + c - 2\left\{\frac{\Delta}{(ab-h^2)}\right\} = 0$$

i.e., Eqn. of asymptotes – Eqn. of conic
= Eqn. of conjugate hyperboal – Eqn. of asymptotes

EXAMPLE

Find the asymptoates of the hyperbola

$$6x^2 - 7xy - 3y^2 - 2x - 8y - 6 = 0.$$

Also find its conjugate hyperbola.

Sol. Let the asymptotes of the given hyperbola be

$$6x^2 - 7xy - 3y^2 - 2x - 8y - 6 + \lambda = 0 \qquad ...(1)$$

If (1) gives the asymptotes, then the must represent a pair of straight lines, the condition for which is

$$abc + 2fgh - af^2 - bg^2 - ch^2 = 0$$

$\therefore \quad 6(-3)(\lambda - 6) + 2(-4)(-1)(-7/2) - 6(-4)^2 + 3(-1)^2 - (\lambda - 6)(-7/2)^2 = 0$

or $\quad -18(\lambda - 6) - 21 - 96 + 3 - (\lambda - 6)(49/4) = 0$

or $\quad \lambda = 2.$

Hence the required equation of the asymptotes is

$$6x^2 - 7xy - 3y^2 - 2x - 8y - 4 = 0 \qquad ...(2)$$

or $\qquad 6x^2 - x(7y+2) = 3y^2 + 8y + 4$

or $\qquad 36x^2 - 2.6x\left(\dfrac{7y+2}{2}\right) + \left(\dfrac{7y+2}{2}\right)^2 = \left(\dfrac{7y+2}{2}\right)^2 + 18y^2 + 48y + 24$

or $\qquad \left(6x - \dfrac{7y+2}{2}\right)^2 = \dfrac{1}{4}[(7y+2)^2 - 72y^2 + 192y + 96]$

or $\qquad (12x - 7y - 2)^2 = 121y^2 + 220y + 100$

$\qquad\qquad = (11y+10)^2$

or $\qquad 12x - 7y - 2 = \pm(11y+10)$

or $\qquad 2x - 3y - 2 = 0 \ \text{ and } \ 3x + y + 2 = 0.$

Also, we know that the equation of the conjugate hyperbola differs from that of the asymptotes by the same quantity as the equation of the asymptotes differs from that of the hyperbola.

Hence from (2) the equation of the conjugate hyperbola is

$$6x^2 - 7xy - 3y^2 - 2x - 8y - 4 + 2 = 0$$

or $\qquad 6x^2 - 7xy - 3y^2 - 2x - 8y - 2 = 0.$

EXERCISES

Find the asymptotes of the following conics, and also the equations of their conjugate hyperbolas.

1. $y^2 - xy - 2x^2 - 5y + x - 6 = 0$

    [Ans. $(y + x - 2)(y - 2x - 3) = 0$; $y^2 - xy - 2x^2 - 5y + x + 18 = 0$]

2. $x^2 - 4xy - 5y^2 + 6x + 42y - 63 = 0$

    [Ans. $x^2 - 4xy - 5y^2 + 6x + 42y - 72 = 0$; $x^2 - 4xy - 5y^2 + 6x + 42y - 81 = 0$]

3. Find the equation of the asymptotes of the conic

    $$3x^2 - 2xy - 5y^2 + 7x - 9y = 0$$

    and find the equation of the conic which has the same asymptotes and which passes through the point (2, 2).

    [Ans. $3x^2 - 2xzy - 5y^2 + 7x - 9y + 2 = 0$; $3x^2 - 2xy - 5y^2 + 7x - 9y + 20 = 0$]

4. Find the equation of the hyperbola whose asymptotes are parallel to $2x + 3y = 0$ and $3x + 2y = 0$, whose centre is at (1, 2) and which passes through (5, 3).

    [Ans. $(2x + 3y - 8)(3x + 2y - 7) = 154$]

5. If (x_1, y_1) be the centre of the hyperbola

    $$f(x, y) \equiv ax^2 + 2hxy + by^2 + 2gx + 2fy + c = 0,$$

    show that the equation of the asymptotes is $f(x, y) = f(x_1, y_1)$

1.7 NATURE OF A CONIC

To determine the tests for detecting the nature of the conic by an examination of the general equation of the second degree.

Let the equation of the conic be

$$ax^2 + 2hxy + by^2 + 2gx + 2fy + c = 0 \qquad \text{...(1)}$$

If asymptotes of an ellipse or parabola be found, it wil be observed that the asymptotes of ellipse are imaginary whereas the parabola has one asymptote, which is at infinite distance and parallel to its axis.

General Equation of Second Degree and Tracing of Conics

Thus, the criterion of finding the asymptotes enables us to specify the nature of the conic represented by the equation (1).

The equation of the asymptotes of the conic (1) is given by

$$ax^2 + 2hxy + by^2 + 2gx + 2fy + c - \frac{\Delta}{ab - h^2} = 0 \qquad ...(2)$$

Clearly, the equation (2) represents a pair of straight lines (two asymptotes) parallel to the lines given by the equation

$$ax^2 + 2hxy + by^2 = 0, \qquad ...(3)$$

and therefore the angle between the asymptotes will be equal to the angle between the lines (3).

Let θ be the angle between the lines (3)

$$\therefore \qquad \tan\theta = \frac{2\sqrt{h^2 - ab}}{a + b} \qquad ...(4)$$

The following cases arise :

Case I. If $h^2 - ab > 0$, $\tan\theta$ is real. Therefore θ is also real, and then lines (3) are real.

Therefore, the asymptotes of (1) are real and different. Hence the conic (1) represents a *hyperbola*.

If in addition to the above condition

$$a + b = 0.$$

$\tan\theta =$ infinite and then $\theta = \pi/2$, *i.e.*, the asymptotes are at right angles.

Hence the hyperbola is a **rectangular hyperbola**.

Case II. If $h^2 - ab = 0$, $\tan\theta = 0$ and then the lines (3) are real and coincident.

Therefore the expression $ax^2 + 2hxy + by^2$ becomes a perfect square, so that the two asymptotes of the conic coincide and hence the conic is a **parabola**.

Case III. If $h^2 - ab < 0$, $\tan\theta$ is imaginary and then the lines (3) are imaginary.

Therefore the two asymptotes are imaginary.

Hence the conic is an **ellipse**.

Case IV. If $h = 0$, and $a = b$, the general equation represents a **circle**.

Case V. If $\Delta = 0$, then general equation represents a **pair of straight lines**. If in addition to this condition $h^2 = ab$, then straight lines are pair of **parallel straight lines**.

For reference the results for the general equation of second degree, viz. $ax^2 + 2hxy + by^2 + 2gx + 2fy + c = 0$ are given below :

	Condition	Nature of the Conic
1.	$\Delta = 0, h^2 \neq ab$	A pair of intersecting straight lines.
2.	$\Delta = 0, h^2 = ab$	A pair of parallel straight lines.
3.	$\Delta = 0, a + b = 0$	A pair of perpendicular lines
4.	$\Delta \neq 0, h^2 - ab < 0$	An ellipse
5.	$\Delta \neq 0, h^2 - ab = 0$	A parabola
6.	$\Delta \neq 0, h^2 - ab > 0$	A hyperbola
7.	$\Delta \neq 0, h^2 - ab > 0, a + b = 0$	A rectangular hyperbola
8.	$\Delta \neq 0, h = 0, a = b$	A circle

1.8 AXIS, LATUS RECTUM, VERTEX AND FOCUS OF PARABOLA

If the general equation of the second degree

$$ax^2 + 2hxy + by^2 + 2gx + 2fy + c = 0$$

represents a parabola, to find the equation of the axis, the length of the latus rectum and the coordinates of vertex and focus and also to trace it.

The given general equation of the second degree is

$$ax^2 + 2hxy + by^2 + 2gx + 2fy + c = 0 \qquad \ldots(1)$$

We know that if the equation (1) represents a parabola, the second degree terms form a perfect square. So we can put $ax^2 + 2hxy + by^2 = (\alpha x + \beta y)^2$ where $\alpha^2 = a, \beta^2 = b$ and $\alpha\beta = h$.

The equation (1) then takes the form

$$(\alpha x + \beta y)^2 + 2gx + 2fy + c = 0$$

or $\qquad (\alpha x + \beta y)^2 = -2gx - 2fy - c$

or $\qquad (\alpha x + \beta y + \lambda)^2 = 2x(\lambda\alpha - g) + 2y(\lambda\beta - f) + (\lambda^2 - c) \qquad \ldots(2)$

where λ is an arbitrary constant.

We choose λ such that the lines

$$\alpha x + \beta y + \lambda = 0 \text{ and } 2x(\lambda\alpha - g) + 2y(\lambda\beta - f) + (\lambda^2 - c) = 0$$

are at right angles. For this we must have the product of the slopes of these lines $= -1 \ (m_1 m_2 = -1)$.

i.e., $\qquad \left(-\dfrac{\alpha}{\beta}\right)\left\{-\dfrac{(\lambda\alpha - g)}{(\lambda\beta - f)}\right\} = -1$

or $\qquad \lambda\alpha^2 + \lambda\beta^2 = \alpha g + \beta f, \text{ or } \lambda = \dfrac{\alpha g + \beta f}{\alpha^2 + \beta^2} \qquad \ldots(3)$

For this value of λ, the coefficient in the right hand side of (2)

$$= \dfrac{\alpha^2 g + \alpha\beta f}{\alpha^2 + \beta^2} - g = \dfrac{\beta(\alpha f - \beta g)}{\alpha^2 + \beta^2}$$

and the coefficients of $2y$

$$= \dfrac{\alpha\beta g + \beta^2 f}{\alpha^2 + \beta^2} - f = \dfrac{-\alpha(\alpha f - \beta g)}{\alpha^2 + \beta^2}$$

Hence the equation (2) becomes

$$(\alpha x + \beta y + \lambda)^2 = \dfrac{2(\alpha f - \beta g)}{\alpha^2 + \beta^2}(\beta x - \alpha y) + \lambda^2 - c$$

or $\qquad (\alpha x + \beta y + \lambda)^2 = \dfrac{2(\alpha f - \beta g)}{\alpha^2 + \beta^2}(\beta x - \alpha y + c'),$

where $\qquad c' = \dfrac{(\lambda^2 - c)(\alpha^2 + \beta^2)}{2(\alpha f - \beta g)} \qquad \ldots(4)$

or $\qquad \left\{\dfrac{\alpha x + \beta y + \lambda}{\sqrt{(\alpha^2 + \beta^2)}}\right\}^2 = \dfrac{2(\alpha f - \beta g)}{(\alpha^2 + \beta^2)^{3/2}}\left\{\dfrac{\beta x - \alpha y + c'}{\sqrt{\alpha^2 + \beta^2}}\right\} \qquad \ldots(5)$

General Equation of Second Degree and Tracing of Conics

In the equation (5)

$$\frac{\alpha x + \beta y + \lambda}{\sqrt{(\alpha^2 + \beta^2)}} \quad \text{and} \quad \frac{\beta x - \alpha y + c'}{\sqrt{(\alpha^2 + \beta^2)}}$$

are the perpendicular distances of the point (x, y) on the curve from the mutually perpendicular straight lines $\alpha x + \beta y + \lambda = 0$ and $\beta x - \alpha y + c' = 0$ respectively.

Let us transform the coordinate axes so that the straight lines

$$\alpha x + \beta y + \lambda = 0 \qquad \ldots(6)$$

and

$$\beta x - \alpha y + c' = 0 \qquad \ldots(7)$$

become the new axes of x and y respectively.

If (X, Y) are the coordinates of the point (x, y) with respect to these new coordinates axes, then

$X =$ the perpendicular distance of the point (x, y) from the new y-axis, $i.e.$, the line (7).

$$= (\beta x - \alpha y + c') / \sqrt{(\alpha^2 + \beta^2)}$$

and $Y =$ the perpendicular distance of the point (x, y) from the new x-axis, $i.e.$, the line (6)

$$= (\alpha x + \beta y + \lambda) / \sqrt{(\alpha^2 + \beta^2)}.$$

Therefore, the equation (5) transform into

$$Y^2 = 4pX, \qquad \ldots(8)$$

where

$$4p = \frac{2(\alpha f - \beta g)}{(\alpha^2 + \beta^2)^{3/2}} \qquad \ldots(9)$$

The equation (8) is the equation of the parabola in the standard form. The equation of the **axis of the parabola** $Y = 0$, $i.e.$, $\alpha x + \beta y + \lambda = 0$, $i.e.$, the equation (6) and the equation of the **tangent at the vertex** is $X = 0$, $i.e.$, $\beta x - \alpha y + c' = 0$, $i.e.$, the equation (7). The **length of the latus rectum** of the parabola $= 4p$, given by (9).

The **vertex** of the parabola is the point of intersection of the lines (6) and (7).

The **equation of the latus rectum** of the parabola is $X = p$,

i.e.,

$$\frac{(\beta x - \alpha y + c')}{\sqrt{(\alpha^2 + y^2)}} = \frac{(\alpha f - \beta g)}{2(\alpha^2 + \beta^2)^{3/2}}$$

i.e.,

$$\beta x - \alpha y + c' = \frac{(\alpha f - \beta g)}{2(\alpha^2 + \beta^2)} \qquad \ldots(10)$$

The **focus** of the parabola is the point of intersection of the axis of the parabola, $i.e.$, the line (6) and the latus rectum of the parabola, $i.e.$, the line (10).

The **equation of the directrix** of the parabola is $X = -p$.

i.e.,

$$\beta x - \alpha y + c' = -\frac{(\alpha f - \beta g)}{2(\alpha^2 + \beta^2)} \qquad \ldots(11)$$

The **foot of the directrix** is the point of intersection of the lines (6) and (11).

Tracing of parabola

(1) First draw the rectangular axes OX and OY and plot the vertex A.

(2) Draw the axis $\alpha x + \beta y + \lambda = 0$ and then draw a line perpendicular to it through the vertex A. This straight line is the tangent at the vertex given by $\beta x - \alpha y + c' = 0$.

(3) Find the points where the given parabola meets the coordinate axes. The points will enable us to know that on what side of the tangent at the vertex the curve lies.

(4) Now on the axis of the parabola, towards the side the curve lies, take a point S (i.e., the focus) such that

$$AS = p = \frac{1}{4} \cdot \text{latus rectum}.$$

Draw LSL' perpendicular to the axis of the parabola and mark off $SL = SL' = 2p$.

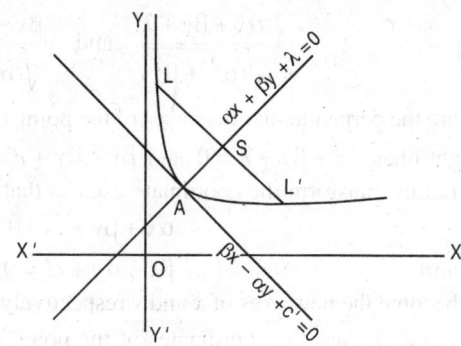

(5) Now draw the figure of the curve touching the line $\beta x - \alpha y + c' = 0$ at the point A, symmetrical about the line $\alpha x + \beta y + \lambda = 0$ and passing through the points L, L' and also the points where the curve meets the coordinate axes. If necessary, find some more points satisfying the equation of the curve.

EXAMPLES

1. *Trace the parabola*

$$16x^2 - 24xy + 9y^2 + 77x - 64y + 95 = 0.$$

Also find the coordinates of its vertex and focus.

Sol. The equation of the given conic is

$$16x^2 - 24xy + 9y^2 + 77x - 64y + 95 = 0. \qquad \ldots(1)$$

Comparing with the general equation of second degree, we have

$$a = 16, h = -12, b = 9, g = \frac{77}{2}, f = -32, c = 95.$$

Nature. $h^2 - ab = 144 - 144 = 0$

and

$$\Delta = abc + 2fgh - af^2 - bg^2 - ch^2$$

$$= 13680 + 29568 - 16384 - 9\left(\frac{77}{2}\right)^2 - 95 \times 144 \neq 0.$$

Therefore, the equation (1) represents a parabola.
Write the given equation as

$$(4x - 3y)^2 = -77x + 64y - 95 \qquad \ldots(2)$$

Introducing an arbitrary constant λ, the equation (2) is written in the following form

$$(4x - 3y + \lambda)^2 = (8\lambda - 77)x + (64 - 6\lambda)y + \lambda^2 - 95 \qquad \ldots(3)$$

Choose λ such that the lines

$$4x - 3y + \lambda = 0,$$

and

$$(8\lambda - 77)x + (64 - 6\lambda)y + \lambda^2 - 95 = 0$$

are at right angles.
The condition for this is

$$\left(\frac{4}{3}\right)\left[-\frac{8\lambda - 77}{64 - 6\lambda}\right] = -1$$

or $\qquad 4(8\lambda - 77) - 3(64 - 6\lambda) = 0$

or $\qquad 50\lambda = 308 + 192 = 500$

i.e., $\lambda = 10$.

Substituting this value of λ in (3), we have
$$(4x - 3y + 10)^2 = 3x + 4y + 5.$$
This equation can be written as
$$25 \cdot \left(\frac{4x - 3y + 10}{\sqrt{(4^2 + 3^2)}} \right)^2 = 5 \cdot \left(\frac{3x + 4y + 5}{\sqrt{(3^2 + 4^2)}} \right)$$
or
$$\left(\frac{4x - 3y + 10}{5} \right)^2 = \frac{1}{5} \left(\frac{3x + 4y + 5}{5} \right)$$

Writing this equation in the form of standard equation of a parabola, $Y^2 = 4aX$, we see that
$$Y \equiv \frac{4x - 3y + 10}{5} = 0, \text{ gives the axis}$$
and
$$X \equiv \frac{3x + 4y + 5}{5} = 0, \text{ gives the tangent at the vertex.}$$
i.e., the equation of the axis of the parabola is
$$4x - 3y + 10 = 0, \qquad \ldots(4)$$
and the equation of the **tangent at the vertex** is
$$3x + 4y + 5 = 0 \qquad \ldots(5)$$

The latus rectum of the parabola, $4a = \dfrac{1}{5}$.

Vertex. Solving (4) and (5), the coordinates of vertex are $\left(-\dfrac{11}{5}, \dfrac{2}{5} \right)$.

Focus. The focus is obtained by solving the equations
$$Y = 0 \quad \text{and} \quad X = a$$
i.e.,
$$4x - 3y + 10 = 0,$$
and
$$\frac{3x + 4y + 5}{5} = \frac{1}{20}$$
i.e., $4x - 3y + 10 = 0$, and $12x + 16y + 19 = 0$.

Hence, coordinates of focus are
$$\left(-\frac{217}{100}, \frac{11}{25} \right).$$

Directrix. The equation of directrix is
$$x + a = 0$$
i.e.,
$$\frac{3x + 4y + 5}{5} + \frac{1}{20} = 0$$
or
$$12x + 16y + 21 = 0.$$

Construction. To trace the parabola, find the points of intersection of the equation (1) and the coordinate axes.

Now, putting $y = 0$ in (1), we get
$$16x^2 + 77x + 95 = 0.$$
It gives two imaginary roots, therefore, the parabola intersects the x-axis in imaginary points.

Again, putting $x = 0$, we get

$$9y^2 - 64y + 95 = 0$$

i.e., $(9y - 19)(y - 5) = 0$

i.e., $y = \dfrac{19}{9}$, and $y = 5$,

i.e., the curve cuts the y-axis in two points $\left(0, \dfrac{19}{9}\right)$ and $(0, 5)$.

Now plot the vertex $A\left(-\dfrac{11}{5}, \dfrac{2}{5}\right)$. Trace the axis (4) and tangent at vertex (5). Plot the points $\left(0, \dfrac{19}{9}\right)$ and $(0, 5)$ and then trace the parabola as shown in the figure.

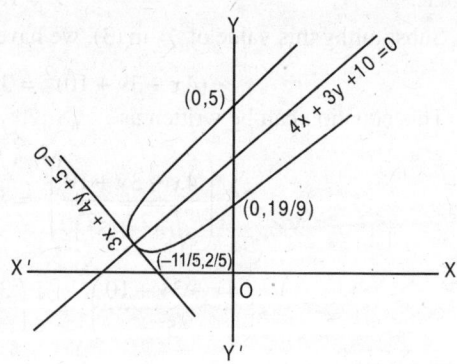

2. *Trace the conic* $(x + 2y)^2 = 6(y - 2x)$. *Also find its focus and the equation of its latus rectum.*

Sol. The given equation is

$$(x + 2y)^2 = 6(y - 2x) \qquad \text{...(1)}$$

Since the second degree terms in the given equation form a perfect square and $\Delta \neq 0$, so the given conic is a parabola.

The given equation can be written as

$$\left[\dfrac{x + 2y}{\sqrt{(1^2 + 2^2)}}\right]^2 = \dfrac{6}{\sqrt{5}} \left[\dfrac{y - 2x}{\sqrt{(1^1 + 2^2)}}\right]$$

or $\left(\dfrac{x + 2y}{\sqrt{5}}\right)^3 = \dfrac{6}{\sqrt{5}}(y - 2x\sqrt{5})$

or $Y^2 = \dfrac{6}{\sqrt{5}} X$,

where $X = \dfrac{y - 2x}{\sqrt{5}}$ and $Y = \dfrac{x + 2y}{\sqrt{5}}$.

∴ Latus reactum of the parabola "$4a$" $= \dfrac{6}{\sqrt{5}}$ or $a = \dfrac{3}{2\sqrt{5}}$.

Axis of the parabola is $Y = 0$ i.e., $x + 2y = 0$ ...(2)

Tangent at the vertex is $X = 0$, i.e., $y - 2x = 0$. ...(3)

Solving (2) and (3), we get the coordinates of the vertex as $(0, 0)$.
Again the focus of the parabola is given by

$$X = a, Y = 0$$

or $\dfrac{y - 2x}{\sqrt{5}} = \dfrac{3}{2\sqrt{5}}$

and $x + 2y = 0$

or $2y - 4x = 3$

and $x + 2y = 0$.

Solving these wer get focus as $\left(-\dfrac{3}{5}, \dfrac{3}{10}\right)$.

Also, the equation of the latus rectum is $X = a$,

i.e., $\dfrac{y-2x}{\sqrt{5}} = \dfrac{3}{2\sqrt{5}}$ or $2y - 4x = 3$,

Tracing. Solving (1) and $y = 0$, we get $x^2 + 12x = 0$, or $x = 0, -12$, i.e., the parabola cuts the x-axis at $(0, 0)$ and $(-12, 0)$.

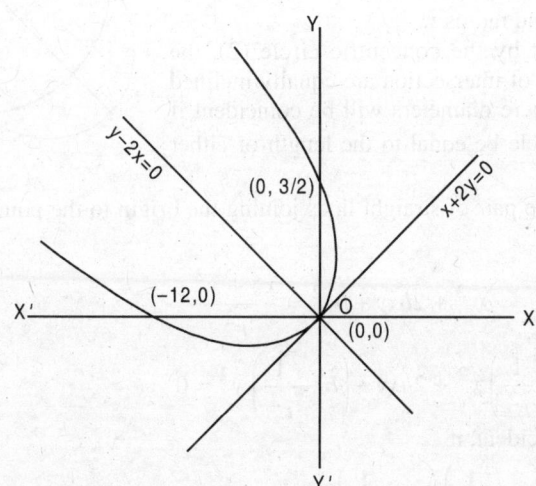

Again solving (1) and $x = 0$, we get $4y^2 = 6y$ or $y = 0, \dfrac{3}{2}$, i.e., the parabola passes through the points $(0, 0)$ and $\left(0, \dfrac{3}{2}\right)$.

For tracing of the parabola, take vertex at $(0, 0)$. Through $(0, 0)$ draw two lines mutually at right angles, one of them inclined at an angle $\tan^{-1} \sqrt{2}$ to the x-axis. This is tangent at the vertex, other being the axis of the parabola.

The shape of the conic is as shown in the figure.

EXERCISES

Trace the Conics :

1. $9x^2 - 24xy + 16y^2 - 18x - 101y + 19 = 0$ and find the coordinates of its focus.
2. $x^2 + 2xy + y^2 - 2x - 1 = 0.$
3. $x^2 - 4xy + 4y^2 - 12x - 6y - 39 = 0.$
4. $9x^2 + 24xy + 16y^2 - 2x + 14y + 1 = 0.$
5. $16x^2 - 24xy + 9y^2 - 104x - 172y + 44 = 0$
6. $x^2 + 4y^2 - 4xy - 32x + 4y + 16 = 0$
7. $9x^2 + 6xy + y^2 + 20x - 4y + 8 = 0.$
8. $4x^2 + 12xy + 9y^2 + 6x + 2 = 0.$

1.9 LENGTHS AND POSITIONS OF THE AXIS OF A CONIC

To find the lengths and positions of the axes of ther central conic represented by the equation
$$ax^2 + 2hxy + by^2 = 1.$$

Let the given central conic be
$$ax^2 + 2hxy + by^2 = 1 \qquad ...(1)$$

Its centre is evidently the origin. Let this conic be cut by a concentric circle

$$x^2 + y^2 = r^2, \qquad ...(2)$$

whose centre is $(0, 0)$ and radius r.

If the conic (1) be cut by the concentric circle (2), the diameter through the points of intersection are equally inclined to the axes of the conic. There diameters will be coincident if $2r$, the diameter of the circle be equal to the length of either axies of the conic.

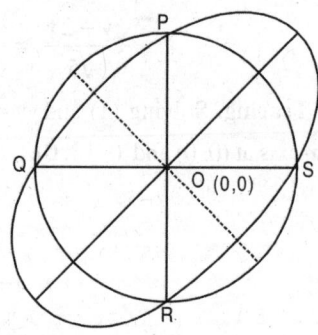

Now the equation of the pair of straight lines joining the origin to the points of intersection of (1) and (2) is

$$ax^2 + 2hxy + by^2 = \frac{x^2 + y^2}{r^2}$$

or
$$\left(a - \frac{1}{r^2}\right)x^2 + 2hxy + \left(b - \frac{1}{r^2}\right)y^2 = 0 \qquad ...(3)$$

These lines will be coincident, if

$$\left(a - \frac{1}{r^2}\right)\left(b - \frac{1}{r^2}\right) = h^2 \qquad ...(4)$$

Hence the lengths of the semi-axes of the conic are the roots of the equation (4). Let the roots of this quadratic equation in r^2 be r_1^2 and r_2^2. Now two cases arise :

Case I. If r_1^2 and r_2^2 are both positive, the conic is an ellipse. If $r_1 > r_2$ then $2r_1$ and $2r_2$ are the lengths of the major and minor axes of the ellipse.

Case II. If r_1^2 is positive whereas r_2^2 is negative, then the conic is hyperbola. In this case $2r_1$ is the length of the transverse axis and $2\{|r_1^2|\}^{1/2}$ is the length of the conjugate axis of the hyperbola.

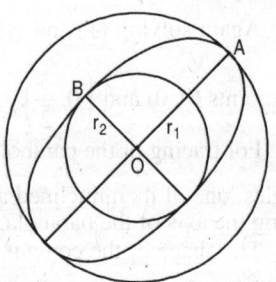

Equation of the axes

Multiplicying (3) by $(a - 1/r^2)$, we have

$$\left(a - \frac{1}{r^2}\right)^2 x^2 + 2h\left(a - \frac{1}{r^2}\right)xy + \left(a - \frac{1}{r^2}\right)\left(b - \frac{1}{r^2}\right) = 0$$

or
$$\left(a - \frac{1}{r^2}\right)^2 x^2 + 2h\left(a - \frac{1}{r^2}\right)xy + h^2 y^2 = 0,$$

from (4).

or
$$\left[\left(a - \frac{1}{r^2}\right)x + hy\right]^2 = 0$$

General Equation of Second Degree and Tracing of Conics

or $$\left(a - \frac{1}{r^2}\right)x + hy = 0 \qquad \ldots(3)$$

Substituting r_1^2 and r_2^2 for r^2 by turns in (5), we can obtain the equations of the axes of the conic.

1.10 ECCENTRICITY, FOCI AND DIRECTRICES OF A CENTRAL CONIC

To find the eccentricity, foci and directrices of a central conic represented by
$$ax^2 + 2hxy + by^2 = 1 \qquad \ldots(1)$$

Eccentricity of the conic. The lengths of the semi-axes of the conic (1) are given by the equation
$$\left(a - \frac{1}{r^2}\right)\left(b - \frac{1}{r^2}\right) = h^2$$

or $$\frac{1}{r^4} - (a+b)\frac{1}{r^2} + ab - h^2 = 0 \qquad \ldots(2)$$

This equation is quadratic in r^2. Let the roots of this quadratic equation be r_1^2 and r_2^2. Then lengths of the semi-axes of the conic are r_1 and r_2.

Case I. When r_1^2 and r_2^2 are both positive.

In this case the conic is an ellipse. If $r_1^2 > r_2^2$, then r_1^2 and r_2^2 are the squares of the semi-major and semi-minor axes.

Therefore the eccentricity of the conic is given by
$$r_2^2 = r_1^2(1 - e^2)$$

i.e., $$e^2 = 1 - \frac{r_2^2}{r_1^2}$$

or $$e = \sqrt{\left(1 - \frac{r_2^2}{r_1^2}\right)} \qquad \ldots(3)$$

Case II. When r_1^2 is positive and r_2^2 is negative. In this case the conic is hyperbola r_1^2 and r_2^2 are the square of semi-transverse and semi-conjugate axes.

Therefore the eccentricity e of the conic is given by
$$-r_2^2 = r_1^2(e^2 - 1) \qquad \ldots(4)$$

Foci of the conic. The foci are the points on the major axis (or transverse axis) at a distance er_1 from the centre of the conic, where e is the eccentricity and r_1 the length of semi-major axis or (semi transverse axis).

Let θ be the inclination of the major axis or transverse axis with the x-axis then we have
$$\tan 2\theta = \frac{2h}{a-b}$$

Hence the coordinates of focii are
$$(x_1 \pm er_1 \cos\theta,\ y_1 \pm er_1 \sin\theta)$$

where $er_1 = \sqrt{(r_1^2 - r_2^2)}$, where (x_1, y_1) are the coordinates of centre.

Directrices of the conic. The directrices are the lines perpendicular to the major axis at a distance r_1/e, i.e., $r_1^2/\sqrt{(r_1^2 - r_2^2)}$, from the centre C. Their equations are therefore

$$(x - x') \cos \theta + (y - y') \sin \theta = \pm r_1^2 / \sqrt{(r_1^2 - r_2^2)}$$

In case the conic is hyperbola the same formulae hold but in that case the value of r_2^2 is negative.

1.11 TRACING OF CENTRAL CONIC

Let the conic be

$$F(x, y) \equiv ax^2 + 2hxy + by^2 + 2gx + 2fy + c = 0 \qquad \ldots(1)$$

Here we are giving the summary of the method to trace an ellipse or a hyperbola;

(i) Centre : Find the centre (x_1, y_1) of the conic (1) by solving the equations $\frac{\partial F}{\partial x} = 0$ and $\frac{\partial F}{\partial y} = 0$.

(ii) New constant term and the equation of the conic referred to the centre as origin.

Transferring the origin to the centre (x_1, y_1), the equation (1) becomes

$$ax^2 + 2hxy + by^2 + c_1 = 0 \qquad \ldots(2)$$

where c_1 = the new constant term = $gx_1 + fy_1 + c$

In case $c_1 = 0$, then equation (1) will represent a pair of straight lines.

(iii) Standard form of the equation of the conic.

If $c_1 \neq 0$, the equation (2) may be rewritten as

$$-\frac{a}{c_1} x^2 - \frac{2h}{c_1} xy - \frac{b}{c_1} y^2 = 1,$$

or $\qquad Ax^2 + 2Hxy + By^2 = 1$, where $A = -a/c_1$ etc. $\qquad \ldots(3)$

(iv) Length of the axes of the conic. The squares of the lengths of the semi-axes of the conic are the roots of the equation

$$\left(A - \frac{1}{r^2}\right)\left(B - \frac{1}{r^2}\right) = H^2.$$

which is a quadratic in r^2.

If r_1^2 and r_2^2 are its roots and both are positive, the conic will be an ellipse and if they are of opposite signs, the conic will be a hyperbola.

We shall always take $r_1^2 > r_2^2$ in the case of an ellipse so that $2r_1$ and $2r_2$ are the lengths of the major and minor axes respectively. In the case of a hyperbola the negative value of r^2 will be represented by r_2^2 and the lengths of the transverse and conjugate axes are therefore $2r_1$ and $2\sqrt{|r_2^2|}$.

(v) Equation of the axes. The equation of the major axis (in the case of an ellipse) or that of the transverse axis (in the case of a hyperbola) referred to the centre as origin is

$$\left(A - \frac{1}{r_1^2}\right) x + Hy = 0 \qquad \ldots(4)$$

The equation of the other axis is

$$\left(A - \frac{1}{r_2^2}\right) x + Hy = 0 \qquad \ldots(5)$$

General Equation of Second Degree and Tracing of Conics

(vi) The eccentricity e *(if required)* is given by

$$e = \sqrt{\left(1 - \frac{r_2^2}{r_1^2}\right)}$$

(vii) The coordinates of the foci *(if required)*

If θ is the inclination of the major (or transverse) axis to the x-axis, then the coordinates of the foci with respect to the original coordiante axes are

$$(x_1 + er_1 \cos\theta, y_1 + er_1 \sin\theta) \text{ and } (x_1 - er_1 \cos\theta, y_1 - er_1 \sin\theta),$$

where $\quad er_1 = \sqrt{(r_1^2 - r_2^2)}.$

(viii) The length of the latus rectum *(if required)* is

$$= \frac{2r_2^2}{r_1^2}$$

(ix) Special Points. Find the points where the given conic (1) meets the original x and y axes. This will enable as to draw the figure of conic more accurately

(x) Actual tracing

(a) Draw rectangular axes.

(b) Plot the centre (x_1, y_1) as calculated in step (i).

(c) Draw the axes through the centre making angles θ_1 and θ_2 with x-axis as calculated in steps (iv) and (v).

(d) If the conic is an ellipse mark off lengths r_1 and r_2 on both sides of the centre along major and minor axes, or if the conic is a hyperbola mark points at distance r_1 on both sides of the centre along transverse axis.

Through one of these points draw a line at right angles to the transverse axis and mark points at distance $\sqrt{|r_2^2|}$ on both sides of the transverse axis along this line. Join these points with the centre and produce. These are the asymptotes of the hyperbola.

Now with the help of step (ix) roughly draw the curve.

EXAMPLES

1. *Trace the conic*

$$17x^2 - 12xy + 8y^2 + 46x + 28y + 17 = 0.$$

Also find the eccentricity, foci and the equations of its directrices.

Sol. Let the equation of the conic be

$$F(x, y) \equiv 17x^2 - 12xy + 8y^2 + 46x - 28y + 17 = 0 \qquad ...(1)$$

On comparison with the general equation of second degree, we have

$$a = 17, h = -6, b = 8, g = 23, f = -14 \text{ and } c = 17.$$

1. Nature : $h^2 - ab = 36 - 136 = -100$, which is negative and

$$\Delta = abc + 2fgh - af^2 - bg^2 - ch^2$$
$$= 2312 + 3864 - 3332 - 4232 - 612$$
$$= -2000 \neq 0.$$

Hence the given conic is an ellipse.

2. Centre : The coordinates of the centre are obtained by solving the equations.

$$\frac{\partial F}{\partial x} = 0 \quad \text{and} \quad \frac{\partial F}{\partial y} = 0$$

i.e., $\qquad 17x - 6y + 23 = 0$

and
$$3x - 4y + 7 = 0$$
On solving these, we get the coordiantes of centre as $(-1, 1)$.

3. New cosntant term : Equation of the conic (1) referred to the centre $(-1, 1)$ as origin is
$$17x^2 - 12xy + 8y^2 + c_1 = 0 \qquad ...(2)$$
where
$$c_1 = gx_1 + fy_1 + c$$
$$= 23(-1) - 14(1) + 17 = -20.$$

4. Standard Form. Substituting the value of c_1 in eqn. (2), we get
$$17x^2 - 12xy + 8y^2 - 20 = 0$$
or
$$\frac{17}{20}x^2 - \frac{12}{20}xy + \frac{8}{20}y^2 = 1$$
or
$$\frac{17}{20}x^2 - \frac{3}{5}xy + \frac{2}{5}y^2 = 1$$

Here
$$A = \frac{17}{20}, H = \frac{-3}{10}, B = \frac{2}{5}.$$

5. Lengths and equations of axes : The length of semi axes are the roots of the equation
$$\left(A - \frac{1}{r^2}\right)\left(B - \frac{1}{r^2}\right) = H^2$$

or
$$\left(\frac{17}{20} - \frac{1}{r^2}\right)\left(\frac{2}{5} - \frac{1}{r^2}\right) = \left(\frac{-3}{10}\right)^2$$

or
$$\frac{1}{r^4} - \left(\frac{17}{20} + \frac{2}{5}\right)\frac{1}{r^2} + \frac{17}{50} - \frac{9}{100} = 0$$

or
$$\frac{1}{r^4} - \frac{5}{4}\frac{1}{r^2} + \frac{1}{4} = 0$$

or
$$r^4 - 5r^2 + 4 = 0$$

or
$$(r^2 - 4)(r^2 - 1) = 0$$

or
$$r^2 = 4, 1.$$

If r_1^2 and r_2^2 are the squares of the semi-major and semi-major axes of the ellipse

∴ $\qquad r_1^2 = 4 \quad$ and $\quad r_2^2 = 1,$

i.e., major axis $= 2r_1 = 4$, minor axis $= 2r_2 = 2$
The equation of the major axis is given by
$$\left(A - \frac{1}{r_1^2}\right)x + Hy = 0$$

or
$$\left(\frac{17}{20} - \frac{1}{4}\right)x + \left(-\frac{3}{10}\right)y = 0$$

or
$$\frac{3}{5}x - \frac{3}{10}y = 0, \text{ or } 2x - y = 0,$$

Similarly equation of the minor axis is $x + 2y = 0$.

General Equation of Second Degree and Tracing of Conics

6. Eccentricity : The eccentricity e of the ellipse is given by

$$e = \sqrt{\left(1 - \frac{r_2^2}{r_1^2}\right)} = \sqrt{\left(1 - \frac{1}{4}\right)} = \frac{\sqrt{3}}{2}.$$

7. Foci : Let the major axis make an angle θ with the x-axis, so that its slope is
$$\tan \theta = 2.$$

The coordinates of the foci referred to the original axes are

$$(x_1 \pm er_1 \cos \theta, \; y_1 \pm er_1 \sin \theta)$$

or
$$\left(-1 \pm \sqrt{3} \cdot \frac{1}{\sqrt{5}}, \; 1 \pm \sqrt{3} \cdot \frac{2}{\sqrt{5}}\right)$$

or
$$\left(-1 \pm \frac{\sqrt{15}}{5}, \; 1 \pm \frac{2\sqrt{15}}{5}\right).$$

8. Equations of directrices : The equations of the directrices are given by

$$x \cos \theta + y \sin \theta = \pm \frac{r_1}{e},$$

where θ is the inclination of the major axis with the x-axis.

Here $\quad \tan \theta = 2,\;$ and $\;\dfrac{r_1}{e} = \dfrac{2}{(\sqrt{3}/2)} = \dfrac{4}{\sqrt{3}} = \dfrac{4\sqrt{3}}{3}$

Therefore equations of the directrices are

$$x \frac{1}{\sqrt{5}} + y \frac{2}{\sqrt{53}} = \pm \frac{4\sqrt{3}}{3}$$

or
$$x + 2y = \pm \frac{4\sqrt{15}}{3}$$

9. Tracing of the conic

The curve (1) meets the x-axis when $y = 0$.

Therefore putting $y = 0$ in (1), we have

$$17x^2 + 46x + 17 = 0$$

i.e., $\quad x = \dfrac{-46 \pm \sqrt{(46)^2 - 4 \times 17 \times 17}}{2 \times 17}$

$$= \frac{-46 \pm \sqrt{960}}{34} = \frac{-46 \pm 8\sqrt{15}}{34}$$

$$= \frac{-23 \pm 4\sqrt{15}}{17}$$

These points are

$$\left[\frac{-23 + 4\sqrt{15}}{17}, 0\right] \;\text{and}\; \left[\frac{-23 - 4\sqrt{15}}{17}, 0\right]$$

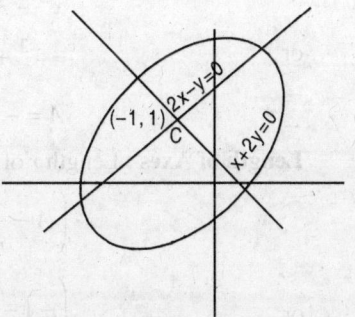

Again the curve meets the y-axis, when $x = 0$.

$\therefore \quad\quad\quad 8y^2 - 28y + 17 = 0$

i.e., $\quad\quad y = 28 \pm \dfrac{\sqrt{(28)^2 - 4 \times 8 \times 17}}{16}$

$$= \frac{28 \pm \sqrt{240}}{16} = \frac{7 \pm \sqrt{15}}{4}$$

These points are

$$\left[0, \frac{7+\sqrt{15}}{4}\right] \text{ and } \left[0, \frac{7-\sqrt{15}}{4}\right]$$

Now plot the centre $(-1, 1)$ and the points on the coordinates axes where the ellipse meets them. Draw the major and minor axes. The shape of the curve is as shown in the figure.

2. *Trace the hyperbola*

$$x^2 - 3xy + y^2 + 10x - 10y + 21 = 0$$

and find the equations of its axes and asymptotes.

Sol. The given conic is

$$f(x, y) = x^2 - 3xy + y^2 + 10x - 10y + 21 = 0. \qquad ...(1)$$

Nature. Here

$$h^2 - ab = \left(-\frac{3}{2}\right)^2 - 1.1 = \frac{5}{4} > 0$$

Also $\Delta \neq 0$.

$\therefore$ eqn. (1) represents a hyperbola.

Centre : The coordiantes of centre may be found by

$$\frac{\partial F}{\partial x} = 0, \text{ and } \frac{\partial F}{\partial y} = 0$$

i.e., $\qquad 2x - 3y + 10 = 0 \text{ and } -3x + 2y - 10 = 0$

giving $x = -2$ and $y = 2$

Hence coordinate of the centre are $(-2, 2)$.

New Constant and Standard form of the equation.

The equation of the conic referred to the parallel axes through the centre is

$$x^2 - 3xy + y^2 + c' = 0 \qquad ...(2)$$

where $\qquad c' = gx_1 + fy_1 + c$

$$= 5(-2) + (-5)2 + 21 = 1.$$

$\therefore$ The equation (2) becomes

$$x^2 - 3xy + y^2 = -1$$

or $\qquad -x^2 + 3xy - y^2 = 1$

$\therefore \qquad A = -1, H = \frac{3}{2}, B = -1.$

Length of Axes : Length r of the equation is given by

$$\left(A - \frac{1}{r^2}\right)\left(B - \frac{1}{r^2}\right) = H^2$$

or $\qquad \left(-1 - \frac{1}{r^2}\right)\left(-1 - \frac{1}{r^2}\right) = \left(\frac{3}{2}\right)^2$

or $\qquad \left(1 + \frac{1}{r^2}\right)^2 = \left(\frac{3}{2}\right)^2$

General Equation of Second Degree and Tracing of Conics

or
$$1 + \frac{1}{r^2} = \pm \frac{3}{2}$$

giving $\frac{1}{r^2} = \frac{1}{2}, -\frac{5}{2}, i.e.,$

$$r_1^2 = 2 \text{ and } r_2^2 = -\frac{2}{5}.$$

Hence transverse axis is of length $2\sqrt{2}$ and length of the conjugate axis $= 2\sqrt{|r_1^2|} = 2\sqrt{(2/5)}$.

Equation of Axes : The equation of the transverse axis referred to centre as origin is

$$\left(A - \frac{1}{r_1^2}\right)x + Hy = 0, \text{ or } \left(-1 - \frac{1}{2}\right)x + \frac{3}{2}y = 0$$

or $\qquad x - y = 0.$

i.e., it makes an angles of 45° with both the coodiante axes.

Tracing : Putting $x = 0$ in (1), we get $y^2 - 10y + 21 = 0$, which gives $y = 3, 7, i.e.,$ the curve meets y-axis at points (0, 3) and (0, 7).

Putting $y = 0$ in (1), we get

$x^2 + 10x + 21 = 0$ which gives $x = -3, -7$, *i.e.*, the curve meets x-axis at points $(-3, 0)$ and $(-7, 0)$.

The shape of the hyperbola is as shown in figure.

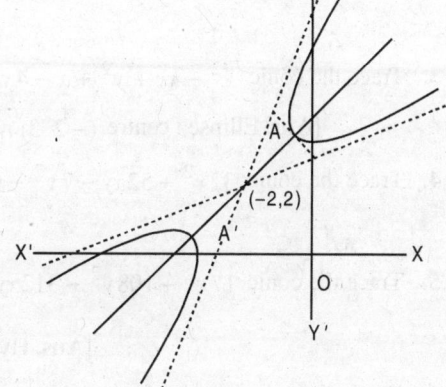

Asymptotes : We know that the equation of the asymptotes differ from the equation of the hyperbola only by a constant term. So let

$$x^2 - 3xy + y^2 + 10x - 10y + \lambda = 0$$

be the equation of the asymptotes of the given hyperbola.

If will represents a pair of straight lines if $\Delta = 0$

i.e., if $\qquad abc + 2fgh - af^2 - bg^2 - ch^2 = 0$

i.e., if $\qquad 1.1\lambda + 2(-5)(5)\left(-\frac{3}{2}\right) - 1.(-5)^2$

$$- 1(5)^2 - \lambda \left(-\frac{3}{2}\right)^2 = 0$$

i.e., $\qquad \lambda = 20.$

Hence the combined equation of the asymptotes is

$$x^2 - 3xy + y^2 + 10x - 10y + 20 = 0.$$

Solving it as a quadratic in x, we have

$$x^2 - x(3y - 10) + (y^2 - 10y + 20) = 0$$

or $\qquad x = \dfrac{(3y - 10) \pm \sqrt{[(3y - 10)^2 - 4.1(y^2 - 10y + 20)]}}{2}$

or $\quad 2x = (3y - 10) \pm \sqrt{5y^2 - 20y + 20}$

or $\quad 2x = 3y - 10 \pm \sqrt{5}\sqrt{(y-2)^2}$

Hence $2x = 3y - 10 \pm \sqrt{5}(y-2)$ are the required asymptotes.

EXERCISES

1. Trace the conic $36x^2 + 24xy + 28y^2 - 72x + 126y + 81 = 0$ and find the equation of the axes.

   [**Ans.** Ellipse : centre $(2, -3)$, $r_1^2 = 4$, $r_2^2 = 9$, major axis $4x + 3y + 1 = 0$, minor axis $4y - 3x + 18 = 0$]

2. Trace the curve $8x^2 - 4xy + 5y^2 - 16x - 14y + 17 = 0$ and find the equation of its axes.

   [**Ans.** Ellipse : centre $\left(\dfrac{3}{2}, 2\right)$, $r_1^2 = \dfrac{9}{4}$, $r_2^2 = 1$ major axis $2x - y - 1 = 0$, minor axis $2x + 4y - 11 = 0$]

3. Trace the conic $x^2 + xy + y^2 + x - 4y + 1 = 0$. Also find the equation of its axes.

   [**Ans.** Ellipse : centre $(-2, 3)$ $r_1^2 = 12$, $r_2^2 = 4$, the axes $x + y = 0$ and $x - y + 5 = 0$]

4. Trace the conic $32x^2 + 52xy - 7y^2 - 64x - 52y - 148 = 0$.

   [**Ans.** Hyperbola : centre $(1, 0)$, $r_1^2 = 4$, $r_1^2 = -9$]

5. Trace the conic $17x^2 + 108y^2 - 312xy - 1146x - 72y + 39 = 0$.

   [**Ans.** Hyperbola : centre $(-3, -4)$, $r_1^2 = \dfrac{1902}{100}$, $r_2^2 = -\dfrac{1902}{225}$]

6. Find the position of the centre and the lengths of the axes of the conic $97x^2 - 60xy + 72y^2 - 314x + 348y + 37 = 0$. Also trace the conic.

   [**Ans.** Ellipse : centre $(1, -2)$, $r_1^2 = 9$, $r_2^2 = 4$]

7. Trace the conic $41x^2 + 24xy + 9y^2 - 130ax - 60ay + 116a^2 = 0$.

   [**Ans.** Ellipse : centre $(a, 2a)$, $r_1^2 = \dfrac{9a^2}{5}$, $r_2^2 = \dfrac{a^2}{5}$]

8. Trace the conic $17x^2 + 12xy + 8y^2 - 46x - 28y + 33 = 0$ find the its eccentricity.

   [**Ans.** Ellipse : centre $(1, 1)$; $r_1^2 = \dfrac{4}{5}$, $r_2^2 = \dfrac{1}{5}$]

9. Trace the curve $x^2 - 4xy - 2y^2 + 10x + 4y = 0$. Also find the equation of its axes.

   [**Ans.** Hyperbola : centre $(-1, 2)$, $r_1^2 = \dfrac{1}{2}$, $r_2^2 = -\dfrac{1}{3}$, transverse axis $x + 2y = 3$, conjugate axis $2x - y + 4 = 0$]

General Equation of Second Degree and Tracing of Conics

10. Trace the conics:

   (a) $x^2 + y^2 - 4xy - 2x - 20y - 11 = 0$.

   [Ans. Hyperbola : centre $(-7, -4)$; $r_1^2 = 36, r_2^2 = -12$]

   (b) $x^2 - 5xy + y^2 + 8x - 20y + 15 = 0$.

   [Ans. Hyperbola : centre $(-4, 0)$; $r_1^2 = \dfrac{1}{7}, r_2^2 = -\dfrac{2}{3}$]

11. Show that the product of the semi-axes of the ellipse whose equation is $x^2 - xy - 2y^2 - 2x - 6y + 7 = 0$ is $2/\sqrt{7}$ and that the equation of the axes is $x^2 - y^2 - 2xy + 8y - 8 = 0$.

2
System of Conics : Confocal Conics

2.1 CONIC THROUGH FIVE POINTS

Let the most general equation of the conic be

$$ax^2 + 2hxy + by^2 + 2gx + 2fy + c = 0 \qquad \ldots(1)$$

This equation seems to contain six constants a, h, b, g, f and c but the actual number of constants is five as can easily be seen on dividing by one of the constants. Thus on dividing by c, the equation (1) contains five independent constants

$$\frac{a}{c}, \frac{h}{c}, \frac{b}{c}, \frac{g}{c} \text{ and } \frac{f}{c}, \qquad \ldots(2)$$

which are fixed for any particular conic.

The five constants thus obtained can be determined completely if there are five independent equations between them. A conic therefore can be made to satisfy five conditions and no more. For example, a conic can be made to pass through five given points, or to pass through four given points, and to touch a given straight line.

The equation may however give more than one set of values of the ratios (2), and therefore more than one conic may satisfy the given conditions; but the number of such conics will be finite if the conditions are really independent.

If there are only four (or less than four) conditions given, an infinite number of conics will satisfy them.

2.2 CONIC THROUGH FOUR POINTS AND TOUCHING A STRAIGHT LINE

The co-ordinates of the four points lead to four independent relations between the constants. The fifth relation is obtained by the condition of tangency of the straight line. But the fifth will give a quadratic relation. Evaluation of the arbitrary constants from these five relations will therefore lead to more than one set of values for them and consequently more than one conic will be found satisfying these conditions.

2.3 TO FIND THE GENERAL EQUATION OF THE CONIC PASSING THROUGH THE POINT OF INTERSECTION OF A CURVE AND A STRAIGHT LINE

The general equation of the conic is

$$S \equiv ax^2 + 2hxy^2 + by^2 + 2gx + 2fy + c = 0 \qquad \ldots(1)$$

and the equation of the straight line is

$$u \equiv lx + my + n = 0 \qquad \ldots(2)$$

Consider the equation

$$ax^2 + 2hxy + by^2 + 2gx + 2fg + c + \lambda \, (lx + my + n) = 0 \qquad \ldots(3)$$

This equation being of second degree represents a conic. The co-ordinates of the points which satisfy (1) and (2) also satisfy (3). Therefore the points of intersection of (1) and (2) lie on (3).

Therefore, (3) is the equation of a conic passing through the points of intersection of (1) and (2)

i.e., $$S + \lambda u = 0$$

System of Conics : Confocal Conics

2.4

To find the equations to the conic sections passing through the intersection of a conic and two given straight lines.

Let the equation of the conic section be

$$S \equiv ax^2 + by^2 + 2hxy + 2gx + 2fy + c = 0 \qquad ...(1)$$

and the equation of the straight line be

$$u_1 \equiv l_1 x + m_1 y + n_1 = 0 \qquad ...(2)$$

and

$$u_2 \equiv l_2 x + m_2 y + n_2 = 0 \qquad ...(3)$$

Conic (1) is cut by the line (2) in the points P and Q and by (3) in points R and T respectively. The equation of a conic passing through their four points of intersection can be written as

$$S + \lambda u_1 u_2 = 0 \qquad ...(4)$$

For different values of λ, we shall obtain different conics passing through these above points.

2.5

To show that a central conic has four and only four foci, two of which are real and two imaginary.

Let the equation of the central conic be

$$ax^2 + by^2 - 1 = 0 \qquad ...(1)$$

Let (x_1, y_1) be a focus, and let

$$x \cos \alpha + y \sin \alpha - p = 0$$

be the equation of the corresponding directrix; so that if e be the eccentricity of the conic, its equation will be

$$(x - x_1)^2 + (y - y_1)^2 - e^2 (x \cos \alpha + y \sin \alpha - p)^2 = 0. \qquad ...(2)$$

Since equations (1) and (2) represent the same conic, and the coefficient of xy must be zero in (2), so that

$$\sin \alpha \cos \alpha = 0$$

$\therefore \qquad \alpha = 0 \text{ or } \dfrac{\pi}{2}.$

Hence a directrix is parallel to one or the other of the coordinate axes.
Also the coefficients of x and y are zero in (1), so must be in (2)

$\therefore \qquad x_1 = e^2 p \cos \alpha \quad \text{and} \quad y_1 = e^2 p \sin \alpha.$

If $\alpha = 0$, then

$$x_1 = e^2 p \quad \text{and} \quad y_1 = 0. \qquad ...(3)$$

Also, comparing the other coefficients in (1) and (2), we have

$$\frac{a}{1 - e^2} = \frac{b}{1} = \frac{-1}{x_1^2 - e^2 p^2}$$

$\therefore \qquad 1 - e^2 = \dfrac{a}{b} \quad \text{or} \quad e = \sqrt{\left(1 - \dfrac{a}{b}\right)} \qquad ...(4)$

and

$$a(x_1^2 - e^2 p^2) = -(1 - e^2)$$

or

$$a(x_1^2 - x_1 p) = -\left(1 - \frac{x_1}{p}\right), \text{ using (3)}$$

or

$$apx_1 = 1 \qquad ...(5)$$

Also from (3) and (4)

or
$$1 - \frac{x_1}{p} = \frac{a}{b}$$

or
$$\frac{x_1}{p} = 1 - \frac{a}{b}$$

or
$$ax_1^2 = 1 - \frac{a}{b} \qquad \text{using (5)}$$

or
$$x_1^2 = \frac{1}{a} - \frac{1}{b}. \qquad ...(6)$$

From (6), we see that there are two foci on the axis of x (since $y_1 = 0$) whose distances from the centre are

$$\pm \sqrt{\frac{1}{a} - \frac{1}{b}}. \qquad ...(7)$$

Again, from (5) we have

$$x_1 = \frac{1}{ap}.$$

Therefore the coordinates of a focus are $(x_1, 0)$ i.e., $\left(\frac{1}{ap}, 0\right)$.

The polar of this point with respect to the given conic (1) is

$$ax \cdot \frac{1}{ap} + by \cdot 0 - 1 = 0$$

or
$$x = p.$$

Clearly, it is the equation of the corresponding directrix.

If $\alpha = \frac{\pi}{2}$, we can show in a similar manner that there are two foci on the axis of y whose distances from the centre are

$$\pm \sqrt{\frac{1}{b} - \frac{1}{a}} \qquad ...(8)$$

From (7) and (8) we see that out of the two points of foci one is clearly real and the other imaginary, whatever the value of a and b (assumed real) may be.

Further, the eccentricity of a conic referred to a focus on the axis of x is k, from (4), equal to $\sqrt{\left(1 - \frac{a}{b}\right)}$; and the eccentricity referred to a focus on the axis of y will similarly be $\sqrt{\left(1 - \frac{b}{a}\right)}$. If the curves is an ellipse, a and b have the same sign, then clearly one of these eccentricities is real and the other imaginary. If however, the curve is a hyperbola, a and b have different signs then both eccentricities are real.

In any conic, if e_1 and e_2 be the two eccentricities, then from above we have

$$\frac{1}{e_1^2} + \frac{1}{e_2^2} = \frac{a}{a-b} + \frac{b}{b-a} = 1.$$

2.6 INTERSECTION OF TWO CONICS

To show that two conics, in general, intersect in four points, real or imaginary.

Let the equations of the two given conics be

$$a_1 x^2 + 2h_1 xy + b_1 y^2 + 2g_1 x + 2f_1 y + c_1 = 0, \qquad ...(1)$$

and
$$a_2 x^2 + 2h_2 xy + b_2 y^2 + 2g_2 x + 2f_2 y + c_2 = 0. \qquad ...(2)$$

System of Conics : Confocal Conics

Rewriting these equations, we have

$$a_1 x^2 + 2x(h_1 y + g_1) + (b_1 y^2 + 2f_1 y + c_1) = 0$$

and

$$a_2 x^2 + 2x(h_2 y + g_2) + (b_2 y^2 + 2f_2 y + c_2) = 0.$$

Eliminating x between these equations, we get an equation of the fourth degree in y, giving four values, real or imaginary, for y. Also if we eliminate x^2 between the equations, we find that there is only one value of x for each value of y.

Therefore the two conics, in general, intersect in four points, real or imaginary.

2.7

To find the equation of any conic passing through the points of intersection of two given conics.

Let the two conics be

$$S_1(x, y) \equiv a_1 x^2 + 2h_1 xy + b_1 y^2 + 2g_1 x + 2f_1 y + c_1 = 0, \quad \ldots(1)$$

and

$$S_2(x, y) \equiv a_2 x^2 + 2h_2 xy + b_2 y^2 + 2g_2 x + 2f_2 y + c_2 = 0. \quad \ldots(2)$$

Now consider the equation

$$S_1 + \lambda S_2 = 0, \quad \ldots(3)$$

where λ is an arbitrary constant.

Equation (3) represents the equation of any conic passing through the intersections of (1) and (2).

Since S_1 and S_2 are both of the second degree in x and y, the equation (3) is of second degree, and hence represents a conic.

Also, since (3) is satisfied when both S_1 and S_2 are zero, it is satisfied by the points (real or imaginary) which are common to (1) and (2).

Hence (3) is the equation of a conic which passes through the intersection of (1) and (2).

Corollary 1. *Equations of the straight lines passing through the intersections of two conics given by the general conics.*

From equation (3) above, we have the equation of the any conic passing through the intersections of the two conics $S_1 = 0$ and $S_2 = 0$ is

$$(a_1 + \lambda a_2) x^2 + 2(h_1 + \lambda h_2) xy + (b_1 + \lambda b_2) y^2 + 2(g_1 + \lambda g_2) x$$
$$+ 2(f_1 + \lambda f_2) y + (c_1 + \lambda c_2) = 0. \quad \ldots(1)$$

It will represents a pair of straight lines, if

$$(a_1 + \lambda a_2)(b_1 + \lambda b_2)(c_1 + \lambda c_2) + 2(f_1 + \lambda f_2)(g_1 + \lambda g_2)(h_1 + \lambda h_2)$$
$$- (a_1 + \lambda a_2)(f_1 + \lambda f_2)^2 - (b_1 + \lambda b_2)(g_1 + \lambda g_2)^2$$
$$- (c_1 + \lambda c_2)(h_1 + \lambda h_2^2) = 0. \quad \ldots(2)$$

This is a cubic equation in λ and will give in general three pairs of straight lines, when substituted successively in (1), passing through the intersections (real or imaginary) of the two conics.

Also, since a cubic equation always has atleast one real root, one value of λ always real, and it will be shown that there can always be drawn atleast one pair of real straight lines through the intersections of two conics.

Corollary 2. *All conics which pass through the intersections of two rectangular hyperbolas are themselves rectangular hyperbolas.*

Let the conics $S_1 = 0$ and $S_2 = 0$ be rectangular hyperbolas, so that

$$a_1 + b_1 = 0 \quad \text{and} \quad a_2 + b_2 = 0. \quad \ldots(1)$$

In the conic $S_1 + \lambda S_2 = 0$, the sum of the coefficients of x^2 and y^2 is

$$a_1 + \lambda a_2 + b_1 + \lambda b_2 = 0$$

or $\qquad (a_1 + b_1) + \lambda (a_2 + b_2) = 0,$ ...(2)

which is identically equal to zero in virtue of the above equations (1).

Hence the conic $S_1 + \lambda S_2 = 0$ which passes through the intersection of $S_1 = 0$ and $S_2 = 0$, the rectangular hyperbolas, is a rectangular hyperbola.

EXAMPLES

1. *Prove that in general two parabolas can be drawn to pass through the intersections of the conics*

$$ax^2 + 2hxy + by^2 + 2gx + 2fy + c = 0,$$

and $\qquad a'x^2 + 2h'xy + b'y^2 + 2g'x + 2f'y + c' = 0,$

and that their axes are at right angles if

$$h(a' - b') = h'(a - b).$$

Sol. Let the equations of the two given conics be

$$ax^2 + 2hxy + by^2 + 2gx + 2fy + c = 0, \qquad ...(1)$$

and $\qquad a'x^2 + 2h'xy + b'y^2 + 2g'x + 2f'y + c' = 0.$ ...(2)

Equation of any conic passing through the intersections of the given conics (1) and (2) is

$$ax^2 + 2hxy + by^2 + 2gx + 2fy + c$$
$$+ \lambda(a'x^2 + 2h'xy + b'y^2 + 2g'x + 2f'y + c') = 0$$

or $\qquad (a + \lambda a')x^2 + 2(h + \lambda h')xy + (b + \lambda b')y^2 + 2(g + \lambda g')x$
$$+ 2(f + \lambda f')y + (c + \lambda c') = 0 \qquad ...(3)$$

This equation will represent a parabola if the second degree terms, *viz.*

$$(a + \lambda a')x^2 + 2(h + \lambda h')xy + (b + \lambda b')y^2$$

form a perfect square. The condition for this is

$$[2(h + \lambda h')]^2 = 4[(a + \lambda')(b + \lambda b')]$$

or $\qquad (h + \lambda h')^2 = (a + \lambda a')(b + \lambda b')$

or $\qquad \lambda^2 (h'^2 - a'b') + \lambda (2hh' - ab' - a'b) + (h^2 - ab) = 0.$ ...(4)

which is a quadratic equation in λ, it will have two roots and therefore equation (3) will represent two parabolas.

Thus in general two parabolas can be drawn through the intersection of the two given conics (1) and (2).

If λ_1, λ_2 be the roots of the equation (4), then

$$\lambda_1 + \lambda_2 = -\frac{2hh' - ab' - a'b}{h'^2 - a'b'}$$

and $\qquad \lambda_1 \lambda_2 = \dfrac{h^2 - ab}{h'^2 - a'b'}.$

Now the quadratic expression of (3) can be written as

$$(\sqrt{a + \lambda a'}x + \sqrt{b + \lambda b'}y)^2.$$

Therefore slope of the axes of the parabolas represented by (3) is given by

$$-\frac{\sqrt{a + \lambda a'}}{\sqrt{b + \lambda b'}}.$$

System of Conics : Confocal Conics

If the axes of the two parabolas for $\lambda = \lambda_1, \lambda_2$ be at right angles, then

$$-\frac{\sqrt{a+\lambda_1 a'}}{\sqrt{b+\lambda_1 b'}} \times -\frac{\sqrt{a+\lambda_2 a'}}{\sqrt{b+\lambda_2 b'}} = -1.$$

Squaring, we get

$$(a+\lambda_1 a')(a+\lambda_2 a') = (b+\lambda_1 b')(b+\lambda_2 b')$$

or

$$(a^2 + b^2) + \lambda_1 \lambda_2 (a'^2 - b'^2) + (\lambda_1 + \lambda_2)(aa' - bb') = 0.$$

Substituting the value of $(\lambda_1 + \lambda_2)$ and $(\lambda_1 \lambda_2)$, and simplifying, we get

$$(ah' - ba')^2 = (bh' - b'h)^2$$

or
$$ah' - ha' = bh' - b'h$$

or
$$h'(a-b) = h(a'-b').$$

2. *Prove that the locus of centres of conics which pass through four given points is a conic whose asymptotes are parallel to the axes of the two parabolas through the four points.*

Sol. Let us take the straight line joining two of the given points as the axis of x and the line joining he other two points are the axes of y.

Let the straight lines
$$ax + by + 1 = 0, \text{ and } a'x + b'y + 1 = 0$$

cut the coordinate axes (which are oblique) in the four given points.
Then the equations
$$xy = 0, \text{ and } (ax + by + 1)(a'x + b'y + 1) = 0$$

represent two conics (actually pairs of straight lines) through the four given points.
Therefore any conic passing throguh the four given points is

$$(ax + by + 1)(a'x + b'y + 1) + \lambda xy = 0. \qquad ...(1)$$

Equation (1) can be written as

$$aa'x^2 + (ab' + ba' + \lambda) xy + bb'y^2 + (a + a')x + (b + b')y + 1 = 0. \qquad ...(2)$$

This will represent a parabola if the second degree terms form a perfect square. The condition for this is

$$(ab' + a'b + \lambda)^2 = 4(aa')(bb'). \qquad ...(3)$$

Equation is quadratic in λ and gives two values of λ and consequently two parabolas will pass through the four given points.

Now writing the second degree terms of (2) as a perfect square, we have

$$(x\sqrt{aa'} \pm y\sqrt{bb'})^2 + (a + a')x + (b + b')y + 1 = 0$$

The axes of the two parabolas will thus be parallel to the lines

$$x\sqrt{aa'} \pm y\sqrt{bb'} = 0,$$

or the pair of lines

$$aa'x^2 - bb'y^2 = 0. \qquad ...(4)$$

Now the centre of the conic (1) is obtained from the equations

$$a(a'x + b'y + 1) + a'(ax + by + 1) + \lambda y = 0,$$

and
$$b(a'x + b'y + 1) + b'(ax + by + 1) + \lambda x = 0.$$

Eliminating λ, between these equations, we get the locus of the locus of the centre of the conic (1) as

$$2aa'x^2 - 2bb'y^2 + (a + a')x - (b + b')y = 0.$$

The asymptotes of this locus are parallel to the lines

$$aa'x^2 - bb'y^2 = 0,$$

Geometry

i.e., from (4), parallel to the axes of the parabolas through the four given points.

2.8 CONTACT OF CONCIS

In general two conics intersect in four points. The four points of intersection may be either all real, or two real and two imaginary or all the four imaginary.

Let P, Q, R, S be the four real points in which two conics intersect. Sometimes it may happen that two or more of these points coincide, then the two conics are said to have contact of different orders.

Now we discuss the contacts of different orders.

(i) Contact of the zeroth order.

The contact is said to be of the *zeroth order* if all the four points of intersection namely P, Q, R and S, are distinct. [Fig. (i)]

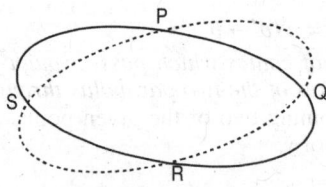

(ii) Contact of the first order : If two of the four points, say P and Q coincide at P and the other two *i.e.,* R and S are distinct, the contact at P is said to be of the *first order* [Fig. (ii)]. In this case, we say that the two conics touch at the point P.

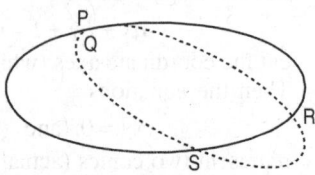

(iii) Double Contact : Suppose now that the points P and Q coincide, as also R and S, but P and R do not coincide. The conics then touch at two points P and R and are said to have a *double contact.* [Fig. (iii)]

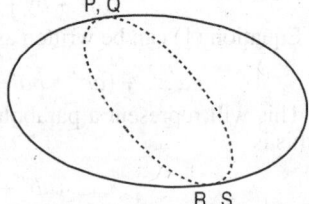

(iv) Contact of the second order : If three points say P, Q, R coincide at P and the fourth points S is distinct from them, the contact is said to be of the *second order.* [Fig. (iv)]

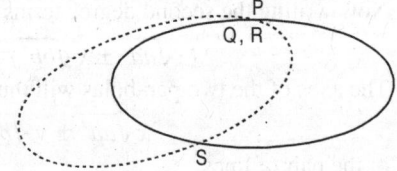

(v) Contact of the third order : If all the four points P, Q, R, S coincide at P, the contact is said to be of the *third order.* [Fig. (v)].

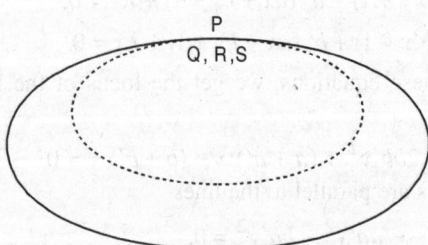

System of Conics : Confocal Conics

Osculating Curve : If two curves have a contact of the second order at a point, they are said to have the same curvature there. By the curvature of a curve we mean the rate at which the tangent is deflecting. When the two curves have a contact of the second order they may be regarded as the limiting case of the two curves having three very near points in common. But the two curves with three consecutive points in common will have two consecutvie tangents in common and thus two curves will have the same curvatrue because the tangent is deflecting at the same rate for both of them.

A curve which has with a given curve a contact of the highest possible order is said to be an *osculating curve*.

Hence in the case of contact of conics a conic which has a contact of the third order with a given conic at a point, is the osculating curve at that point.

Circle of curvature or osculating curve : A circle which has a contact of the second order (three point contact) with a given conic at a given point, is called the *circle of curvature* at that point and its radius is called the *radius of curvature*.

2.9 DOUBLE CONTACT OF CONICS

Let there be two conics $S = 0$ and $S' = 0$. Consider the equation

$$S + \lambda S' = 0 \qquad \ldots(i)$$

where λ is an arbitrary constant. This equation being of second degree, represents a conic. Further, this equation is satisfied by all such points which satisfy $S = 0$ and $S' = 0$. Hence (i) represents a conic passing through the four points of intersection of the conics $S = 0$ and $S' = 0$.

If the conic $S' = 0$ represents a pair of straight lines given by

$$u \equiv a_1 x + b_1 y + c_1 = 0$$

and
$$v \equiv a_2 x + b_2 y + c_2 = 0,$$

then $S + \lambda uv = 0$ is the equation of a conic passing through the points of intersection of the conic $S = 0$ and the lines $u = 0, v = 0$.

We know, that two conics are said to have double contact if they touch each other at two different points. Let the points P and Q coincide at P and R and S at R, i.e., the straight lines $u = 0$ and $v = 0$ coincide, then

$$S + \lambda u^2 = 0 \qquad \ldots(ii)$$

represents a conic passing through P and Q, i.e., the conic (ii) has double contact with the conic $S = 0$ at the points where the line $u = 0$ intersects the conic $S = 0$.

2.9.1 Tangents from an External Point to a Conic Found by the Method of Double Contact

To find by the method of double contact the equation of the pair of tangents that can be drawn from a given point to a conic.

Let the equation to the conic be

$$S \equiv ax^2 + 2hxy + by^2 + 2gx + 2fy + c = 0 \qquad \ldots(i)$$

Let (x_1, y_1) be the given point A from which the tangents AP and AQ are drawn to the conic (i). Then the equation of the chord of contact PQ of the tangents drawn from A to the conic (i) is

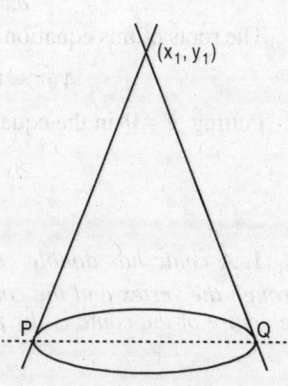

$$T \equiv axx_1 + h(xy_1 + yx_1) + byy_1 + g(x + x_1) + f(y + y_1) + c = 0 \qquad \ldots(ii)$$

Now the equation of any conic having double contact with the given conic (i) at P and Q is

$$ax^2 + 2hxy + by^2 + 2gx + 2fy + c - \lambda [axx_1 + h(xy_1 + yx_1) + byy_1 + g(x + x_1)$$
$$+ f(y + y_1) + c]^2 = 0 \qquad \ldots(iii)$$

Since the pair of tangents AP, AQ is a conic having double contact with the given conic at P and Q, therefore the equation of the pair of tangents must be a particular case of (iii) for some value of λ.

Obviously, we can draw only one conic which passes through A and touches the given conic at P and Q. Therefore if we find λ so that (iii) passes through $A(x_1, y_1)$, we shall get the required equation of the pair of tangents AP, AQ.

Now the point (x_1, y_1) satisfied the equation (iii) if

$$S_1 - \lambda S_1^2 = 0, \qquad \ldots(iv)$$

where S_1 stands for what the left hand side of (i) becomes when x and y in it are replaced by x_1 and y_1 respectively.

The equation (iv) gives $\lambda = \dfrac{1}{S_1}$. Putting $\lambda = \dfrac{1}{S_1}$ in (iii), we get the equation of the pair of tangents AP and AQ as

$$S - \frac{1}{S_1} T^2 = 0$$

or $$SS_1 = T^2.$$

2.10 EQUATION OF A CONIC REFERRED TO TANGENT AND NORMAL AS COORDINATE AXES

Since the tangent and normal at a point on the conic are taken as the coordinate axes, therefore the point of contact is the origin. Thus the origin must lie on the conic and, therefore, the equation of the conic will not contain the constant term. So its equation can be written as

$$ax^2 + 2hxy + by^2 + 2gx + 2fy = 0 \qquad \ldots(i)$$

Again let us choose x-axis along the tangent, so that $y = 0$ meets (i) in two coincident points at origin given by

$$ax^2 + 2gx = 0.$$

The roots of this equation will be coincident if its discriminant is zero, *i.e.*,

$$4g^2 = 0, \qquad i.e., \qquad g = 0.$$

Putting $g = 0$ in the equation (i), the required equation of the conic is given by

$$ax^2 + 2hxy + by^2 + 2fy = 0.$$

EXAMPLES

1. *A conic has double contact with the parabola $y^2 = 4ax$. If the chord of contact passes through the vertex and the conic passes through the focus of the parabola, prove that the locus of the centre of the conic is the parabola*

$$y^2 = a(2x - a).$$

System of Conics : Confocal Conics

Sol. Let the chord of contact passing through the vertex be $y = mx$. Then any conic having double contact with the given parabola at the points of intersection of the parabola and the chord $y = mx$, is

$$y^2 - 4ax + \lambda (y - mx)^2 = 0 \qquad \ldots(i)$$

Since this passes through the focus $(a, 0)$ of the given parabola, we must have

$$0 - 4a^2 + \lambda (0 - ma)^2 = 0,$$

i.e.,
$$\lambda = \frac{4}{m^2}.$$

Hence the conic (i) becomes

$$m^2 (y^2 - 4ax) + 4 (y - mx)^2 = 0 \qquad \ldots(ii)$$

The coordiantes of the centre of (ii) can be obtained by differentiating (ii) partially w.r.t. x and y. Thus we get

$$-4am^2 - 8m (y - mx) = 0,$$

i.e.,
$$2 (y - mx) + am = 0 \qquad \ldots(iii)$$

and
$$2m^2 y + 8 (y - mx) = 0,$$

i.e.,
$$4 (y - mx) + m^2 y = 0 \qquad \ldots(iv)$$

Multiplying (iii) by 2 and subtracting from (iv), we get

$$m^2 y - 2am = 0$$

i.e.,
$$m = 2a / y.$$

Putting the value of m in (iii), we find the required locus as

$$2\left(y - \frac{2ax}{y}\right) + \frac{2a^2}{y} = 0$$

i.e.,
$$y^2 = a (2x - a).$$

2. *Prove that the conic $x^2 + 2y^2 - 1 = 0$ and $3x^2 + 8xy + 10y^2 - 4x - 8y + 1 = 0$ have double contact with each other. Find the coordinates of the point of intersection of the tangent at the two points of contact.*

Sol. The given conics are

$$x^2 + 2y^2 - 1 = 0, \qquad \ldots(i)$$

and
$$3x^2 + 8xy + 10y^2 - 4x - 8y + 1 = 0, \qquad \ldots(ii)$$

Multiplying (i) by 3 and subtracting from (ii), we have

$$4y^2 + 8xy - 4x - 8y + 4 = 0$$

i.e.,
$$x (2y - 1) = 2y - y^2 - 1,$$

i.e.,
$$x = \frac{2y - y^2 - 1}{2y - 1}$$

Substituting this value of x in (i), we get

$$2y^2 = 1 - \left(\frac{2y - y^2 - 1}{2y - 1}\right)^2,$$

Which after simplification gives

$$y^2 (3y-2)^2 = 0,$$

i.e., $\qquad y = 0 \text{ or } \dfrac{2}{3}.$

Putting $y = 0$ in (i), we get $x = \pm 1$. But the point $(-1, 0)$ does not satisfy the equation (ii). Therefore one of the point of intersection of (i) and (ii) is $(1, 0)$.

Now, putting $y = \dfrac{2}{3}$ in (i), we get $x = \pm \dfrac{1}{3}$. But $\left(\dfrac{1}{3}, \dfrac{2}{3}\right)$ does not satisfy (ii), whereas $\left(-\dfrac{1}{3}, \dfrac{2}{3}\right)$ lies on (ii). Hence the other point of intersection of (i) and (ii) is $\left(-\dfrac{1}{3}, \dfrac{2}{3}\right)$.

The equation of the tangents at $(1, 0)$ on (i) is
$$x.1 + 2y.0 - 1 = 0.$$
or $\qquad x - 1 = 0 \qquad$ ...(iii)

The equation of the tangent at $(1, 0)$ on (ii)
$$3x.1 + 4(x.0 + y.1) + 10y.0 - 2(x+1) - 4(y+0) + 1 = 0,$$
i.e., $\qquad x - 1 = 0 \qquad$ ...(iv)

The equations (iii) and (iv) show that the conics (i) and (ii) touch each other at $(1, 0)$.

Again, the tangent at $\left(-\dfrac{1}{3}, \dfrac{2}{3}\right)$ on (i) is
$$x\left(-\dfrac{1}{3}\right) + 2y\left(\dfrac{2}{3}\right) - 1 = 0,$$
i.e., $\qquad x - 4y + 3 = 0$

the tangent at $\left(-\dfrac{1}{3}, \dfrac{2}{3}\right)$ on (ii) is
$$3x\left(-\dfrac{1}{3}\right) + 4\left\{x.\dfrac{2}{3} + y\left(-\dfrac{1}{3}\right)\right\} + 10y.\dfrac{2}{3} - 2\left(x - \dfrac{1}{3}\right) - 4\left(y + \dfrac{2}{3}\right) + 1 = 0,$$
i.e., $\qquad x - 4y + 3 = 0 \qquad$ ...(vi)

The equations (v) and (vi) show that the conic (i) touches (ii) at $\left(-\dfrac{1}{3}, \dfrac{2}{3}\right)$.

Solving (iii) and (iv) for x and y, we get the point of intersection of the two tangents, as the point $(1, 1)$.

3. *If a circle has double contact with a conic, show that the chord of contact is parallel to one or other of the axes.*

Sol. Referred to the axes of the conic as the coordinate axes, let the equation of the conic be
$$ax^2 + by^2 = 1. \qquad ...(i)$$

Let the equation of the chord joining the points of contact of the conic and the circle be given by
$$lx + my + n = 0 \qquad ...(ii)$$

The equation of a conic having double contact with the conic (i) at the points of intersection of (i) and (ii) is given by
$$(ax^2 + by^2 - 1) + \lambda (lx + my + n)^2 = 0 \qquad ...(iii)$$

The equation (iii) will represent a circle if the coefficient of $xy = 0$, i.e., $2\lambda lm = 0$, i.e., $l = 0$ or $m = 0$, since $\lambda \neq 0$, and

System of Conics : Confocal Conics

The coeff. of x^2 = the coeff. of y^2

or $\qquad a + \lambda l^2 = b + \lambda m^2 \qquad$ ...(iv)

If $l = 0$, the equation (ii) of the chord of contact is given by $my + n = 0$, which is a straight line parallel to x-axis.

And if $m = 0$ the equation (ii) of the chord of contact becomes $lx + n = 0$ and hence it is parallel to y-axis.

The value of λ corresponding to $l = 0$ or $m = 0$ is determined from the relation (iv).

Hence the equation (iii) will represent a circle if the chord of contact (ii) of the conic (i) and the circle is parallel to one or the other axis of the conic (ii).

EXERCISES

1. Find the equation of a conic which has double contact with a given conic.
2. A family of conics have double contact with a given conic at the ends of given chord. Prove that the locus of the centres of conic of the family is the diameter of the given conic conjugate to the given chord.
3. A rectangular hyperbola has double contact with the parabola $y^2 = 4ax$. Prove that the centre of the hyperbola and the pole of the chord of contact are equidistant from the directrix of parabola.
4. Show that the conic $c(x^2 + y^2) + 2xy\sqrt{(a-c)(b-c)} = 1 \, (a > b > c > 0)$ has a double contact with both conics
$$ax^2 + by^2 = 1 \quad \text{and} \quad bx^2 + ay^2 = 1.$$
5. Prove that the locus of the centres of conics which touch the coordinate axes at distance h and k from the origin is the straight line $hy = kx$.
6. Prove that there will be two circles passing through the origin which have double contact with the conic
$$(x^2 + y^2) \cos 2\alpha + 2xy \sin 2\alpha + 2 = 0.$$
7. Two conics $S_1 = 0, S_2 = 0$ have a pair of common chords $u = 0, v = 0$, such that $S_1 - S_2 \equiv uv$. Prove that the conic
$$\lambda^2 u^2 - 2\lambda (S_1 + S_2) + v^2 = 0$$
has double contact with both $S_1 = 0$ and $S_2 = 0$.
8. Two circles each have double contact with a parabola and touch each other. Prove that the difference between their radii is equal to the latus rectum of the parabola.
9. Prove that the common chords of a conic and a circle taken in pair are equally inclined to the axes of the conic.
10. Prove that the general equation to the ellipse, having double contact with the circle $x^2 + y^2 = a^2$ and touching the axis of x at the origin is
$$c^2 x^2 + (a^2 + c^2) y^2 - 2a^2 cy = 0.$$
11. If a circle and ellipse have double contact with one another, prove that the length of the tangent drawn from any point of the ellipse to the circle varies as the distance of the point from the chord of contact.

12. Prove that the conics
$$x^2 + 3y^2 - 1 = 0$$
and $$2x^2 + 12xy + 39y^2 - 2x - 12y = 0$$ [Ans. (1, 2)]

have double contact with each other. Find the coordinates of the point of intersection of the tangents at the two points of contact.

13. If the chords of contact of two circles with a conic with which they have double contact be parallel, prove that the radical axis of the circle bisects the distance between these chords.

14. A circle has double contact with a hyperbola. If the chord of contact be paralell to the transverse axis, prove that the ratio of the length of tangent to the circle from any point of the hyperbola to the distance of the point from the chord of contact is equal to the eccentricity of the conjugate hyperbola.

2.11 CONFOCAL CONICS

Two conics are said to be confocal when they have both foci common.

Since one axis of a conic passes through the foci and the other is perpendicular bisector of the line joining them, therefore, confocal conics must have the same centre and same principal axes.

2.11.1 Conics Confocal with Ellipse

To find equation to conics which are confocal with ellipse.
Let the equation of ellipse be
$$\frac{x^2}{a^2} + \frac{y^2}{b^2} = 1 \qquad ...(i)$$

We know that conics having the same foci, have the same centre and axis.
Now equation of any conic having the same centre and axes as the given conic is
$$\frac{x^2}{P} + \frac{y^2}{Q} = 1 \qquad ...(ii)$$

The focii of (i) at the points
$$(\pm ae, 0) \qquad \text{or} \qquad [\pm \sqrt{(a^2 - b^2)}, 0].$$

The focii of (ii) are at the points $[\pm \sqrt{(P-Q)}, 0]$. Now they will be same, *i.e.*,

if $$P - Q = a^2 - b^2$$

i.e., $$P - a^2 = Q - b^2 = \lambda$$

$\therefore \qquad P = a^2 + \lambda \qquad \text{and} \qquad Q = b^2 + \lambda.$

Hence the (ii) equation beomes
$$\frac{x^2}{a^2 + \lambda} + \frac{y^2}{b^2 + \lambda} = 1 \qquad ...(iii)$$

This is the equation of the system of conics confocal with (i). For different values of λ, we have different confocal conics. Here λ is the parameter of the system of confocal conics.

2.11.2 Confocal Parabolas

Parabolas having comon focus and common axis are said to be confocal. If the common focus is at the origin and the axis is x-axis, then the equation
$$y^2 = 4\lambda (x + \lambda).$$

represents a system of confocal parabolas. For different values of parameter λ, we get different confocal parabolas.

System of Conics : Confocal Conics

2.12 CONFOCALS THROUGH A POINT

To show that through any point in the plane of a given conic there can be drawn two conics confocal with it, one of which is an ellipse and the other a hyperbola.

Or

There pass two confocals through any point in the plane of an ellipse, one of which is ellipse and the other one is a hyperbola.

Let the given ellipse be

$$\frac{x^2}{a^2} + \frac{y^2}{b^2} = 1 \qquad \ldots(i)$$

and the given point be (h, k).

Any conic confocal with (i) can be written as

$$\frac{x^2}{a^2 + \lambda} + \frac{y^2}{b^2 + \lambda} = 1 \qquad \ldots(ii)$$

If this passes through (h, k), then we have

$$\frac{h^2}{a^2 + \lambda} + \frac{k^2}{b^2 + \lambda} = 1,$$

This equation can be written as

$$f(\lambda) = (a^2 + \lambda)(b^2 + \lambda) - h^2(b^2 + \lambda) - k^2(a^2 + \lambda) = 0 \qquad \ldots(iii)$$

The equation (iii) gives, in general, two values of λ, corresponding to which, we get two confocals through (h, k).

Now assuming $a > b$, we have

$$f(-a^2) = -h^2(b^2 - a^2), \text{ which is positive}$$

$$f(-b^2) = -k^2(a^2 - b^2), \text{ which is negative}$$

and $\qquad f(+\infty) = +\infty$, which is also positive.

Thus, the equation (iii) has two real roots for λ. These roots λ_1 and λ_2 are such that

$$-a^2 < \lambda_1 < -b^2 < \lambda_2 < +\infty,$$

which imply that

(i) $\qquad a^2 + \lambda_1 > 0, b^2 + \lambda_1 < 0$

and (ii) $\qquad a^2 + \lambda_2 > 0, b^2 + \lambda_2 > 0$

Hence for $\lambda = \lambda_1$, the conic (ii) represents a hyperbola and for $\lambda = \lambda_2$, the conic (ii) is an ellipse.

2.12.1 Confocals Cut at Right Angles

To show that the two confocals through a given point cut at right angles.

Let the two conics confocal with the ellipse

$$\frac{x^2}{a^2} + \frac{y^2}{b^2} = 1$$

be

$$\frac{x^2}{a^2 + \lambda_1} + \frac{y^2}{b^2 + \lambda_1} = 1$$

and

$$\frac{x^2}{a^2 + \lambda_2} + \frac{y^2}{b^2 + \lambda_2} = 1.$$

If these pass through (h, k), we have

$$\frac{h^2}{a^2+\lambda_1} + \frac{k^2}{b^2+\lambda_1} = 1 \qquad \text{...(i)}$$

and
$$\frac{h^2}{a^2+\lambda_2} + \frac{k^2}{b^2+\lambda_2} = 1 \qquad \text{...(ii)}$$

The tangents to the two confocals cut at (h, k) are

$$\frac{hx}{a^2+\lambda_1} + \frac{ky}{b^2+\lambda_1} = 1$$

and
$$\frac{hx}{a^2+\lambda_2} + \frac{ky}{b^2+\lambda_2} = 1.$$

These cut at right angles, if

$$\frac{h^2}{(a^2+\lambda_1)(a^2+\lambda_2)} + \frac{k^2}{(b^2+\lambda_1)(b^2+\lambda_2)} = 0 \qquad \text{...(iii)}$$

Now subtracting equation (ii) from (i), we get

$$(\lambda_2 - \lambda_1)\left\{\frac{h^2}{(a^2+\lambda_1)(a^2+\lambda_2)} + \frac{k^2}{(b^2+\lambda_1)(b^2+\lambda_1)}\right\} = 0.$$

But, since $\lambda_1 \ne \lambda_2$, we have

$$\frac{h^2}{(a^2+\lambda_1)(a^2+\lambda_2)} + \frac{k^2}{(b^2+\lambda_1)(b^2+\lambda_2)} = 0,$$

which is the same as (iii), therefore the two confocals cut at right angles.

2.12.2 Propositions on Confocal Conics

(I) *One and only one conic of a confocal system will touch a given straight line*

Let the given line be

$$x \cos \alpha + y \sin \alpha = p, \qquad \text{...(i)}$$

and the confocal system be

$$\frac{x^2}{a^2+\lambda} + \frac{y^2}{b^2+\lambda} = 1, \qquad \text{...(ii)}$$

The condition that (i) touches (ii), is

$$(a^2+\lambda)\cos^2\alpha + (b^2+\lambda)\sin^2\alpha = p^2,$$

which gives

$$\lambda = p^2 - a^2 \cos^2\alpha - b^2 \sin^2\alpha.$$

Since, we get only one value of λ, therefore one and only one conic of the given system will touch the given line.

(II) *The difference of the squares of the perpendiculars drawn from the centre on any two parallel tangents to two given confocals is constant.*

Let the confocals be,

$$\frac{x^2}{a^2+\lambda_1} + \frac{y^2}{b^2+\lambda_1} = 1$$

and
$$\frac{x^2}{a^2+\lambda_2} + \frac{y^2}{b^2+\lambda_2} = 1.$$

System of Conics : Confocal Conics

Let the paralell tangents to these be
$$x \cos \alpha + y \sin \alpha = p_1$$
and
$$x \cos \alpha + y \sin \alpha = p_2$$
respectively. Then we have
$$p_1^2 = (a^2 + \lambda_1) \cos^2 \alpha + (b^2 + \lambda_1) \sin^2 \alpha$$
and
$$p_2^2 = (a^2 + \lambda_2) \cos^2 \alpha + (b^2 + \lambda_2) \sin^2 \alpha.$$
Hence,
$$p_1^2 - p_2^2 = (\lambda_1 - \lambda_2)(\cos^2 \alpha + \sin^2 \alpha) = \lambda_1 - \lambda_2.$$
which being independent of α, is constant.

(III) *The locus of the point from whcih two perpendicular tangents can be drawn one to each of two given confocals is a circle.*

Let the two confocals be
$$\frac{x^2}{a^2 + \lambda_1} + \frac{y^2}{b^2 + \lambda_1} = 1 \qquad \ldots(i)$$
and
$$\frac{x^2}{a^2 + \lambda_2} + \frac{y^2}{b^2 + \lambda_2} = 1 \qquad \ldots(ii)$$

Let a tangent of (i) be
$$x \cos \alpha + y \sin \alpha = p_1, \qquad \ldots(iii)$$
then, we have
$$p_1^2 = (a^2 + \lambda_1) \cos^2 \alpha + (b^2 + \lambda_1) \sin^2 \alpha$$
Let the tangent on (ii), perpendicular to (iii) be
$$x \cos\left(\alpha + \frac{\pi}{2}\right) + y \sin\left(\alpha + \frac{\pi}{2}\right) = p_2,$$
i.e.,
$$-x \sin \alpha + y \cos \alpha = p_2, \qquad \ldots(iv)$$
$$p_2^2 = (a^2 + \lambda_2) \sin^2 \alpha + (b^2 + \lambda_2) \cos^2 \alpha.$$

The locus of point of intersection of (iii) and (iv) is obtained by eleminating α from these equations. Hence squaring and adding (iii) and (iv), we get
$$x^2 + y^2 = p_1^2 + p_2^2$$
or
$$x^2 + y^2 = a^2 + b^2 + \lambda_1 + \lambda_2,$$
which is the equation of a circle.

(IV) *The locus of the pole of a given straight line with respect to a system of confocal conics is a straight line.*

Let the system of confocals be
$$\frac{x^2}{a^2 + \lambda} + \frac{y^2}{b^2 + \lambda} = 1 \qquad \ldots(i)$$
and the given straight line be
$$lx + my + 1 = 0 \qquad \ldots(ii)$$

Let the pole of (ii) with respect to (i) be (x', y'). Then the polar of (x', y') is
$$\frac{xx'}{a^2 + \lambda} + \frac{yy'}{b^2 + \lambda} = 1 \qquad \ldots(iii)$$

The equation (ii) and (iii) are identical, so, comparing these, we get

$$\frac{x'/(a^2+\lambda)}{l} = \frac{y'/(b^2+\lambda)}{m} = -1$$

This gives $\quad \dfrac{x'}{l} = -(a^2+\lambda)$

and $\quad \dfrac{y'}{m} = -(b^2+\lambda)$

Hence $\quad \dfrac{x'}{l} - \dfrac{y'}{m} = b^2 - a^2$

Therefore, the locus of the pole (x', y') is

$$\frac{x}{l} - \frac{y}{m} = b^2 - a^2$$

which is a straight line.

EXAMPLES

1. *Prove that the confocal hyperbola through the point on the ellipse* $\dfrac{x^2}{a^2} + \dfrac{y^2}{b^2} = 1$ *whose eccentric angle is* α, *is*

$$\frac{x^2}{\cos^2\alpha} - \frac{y^2}{\sin^2\alpha} = a^2 - b^2.$$

Sol. Let the confocal

$$\frac{x^2}{a^2+\lambda} + \frac{y^2}{b^2+\lambda} = 1 \qquad \ldots(i)$$

pass through $(a\cos\alpha, b\sin\alpha)$, then we have

$$\frac{a^2\cos^2\alpha}{a^2+\lambda} + \frac{b^2\sin^2\alpha}{b^2+\lambda} = 1,$$

i.e., $\quad (a^2+\lambda)(b^2+\lambda) - a^2\cos^2\alpha\,(b^2+\lambda) - b^2\sin^2\alpha\,(a^2+\lambda) = 0,$

i.e., $\quad \lambda^2 + \lambda(a^2\sin^2\alpha + b^2\cos^2\alpha) = 0,$

$\therefore \qquad \lambda = 0 \qquad$ or $\qquad \lambda = -(a^2\sin^2\alpha + b^2\cos^2\alpha).$

$\lambda = 0$ gives the given ellipse and so putting

$\lambda = -(a^2\sin^2\alpha + b^2\cos^2\alpha)$ in (i), we get the required confocal hyperbola as

$$\frac{x^2}{\cos^2\alpha} - \frac{y^2}{\sin^2\alpha} = a^2 - b^2.$$

2. *Show that the ends of equal conjugate diameters of a series of confocal ellipses are on a confocal rectangular hyperbola.*

Sol. Let one member of the confocal ellipse be

$$\frac{x^2}{a^2+\lambda} + \frac{y^2}{b^2+\lambda} = 1 \qquad \ldots(i)$$

and let (α, β) be the one extremity of equal conjugate diameters of (i), then

$$\alpha = \sqrt{(a^2+\lambda)}\cos\frac{\pi}{4} \qquad \ldots(ii)$$

System of Conics : Confocal Conics

and
$$\beta = \sqrt{(b^2 + \lambda)} \sin \frac{\pi}{4}. \qquad ...(iii)$$

Squaring (ii) and (iii) and subtracting, we find
$$\alpha^2 - \beta^2 = \frac{a^2 - b^2}{2}.$$

The required locus of (α, β) is, therefore
$$x^2 - y^2 = \frac{a^2 - b^2}{2},$$
which is a rectangular hyperbola confocal with the system (i).

3. *Prove that the two conics* $ax^2 + 2hxy + by^2 = 1$ *and* $a'x^2 + 2h'xy + b'y^2 = 1$ *can be placed so as to be confocal if*
$$\frac{(a-b)^2 + 4h^2}{(ab - h^2)^2} = \frac{(a' - b')^2 + 4h'^2}{(a'b' - h'^2)^2}.$$

Sol. The equation of one conic is
$$ax^2 + 2hxy + by^2 = 1. \qquad ...(i)$$

The squares of the semi-axes of the conic (i) are given by
$$\left(a - \frac{1}{r^2}\right)\left(b - \frac{1}{r^2}\right) = h^2$$

or
$$(ab - h^2) r^4 - (a + b) r^2 + 1 = 0.$$

Let r_1^2 and r_2^2 be the roots of this equation, then we have
$$r_1^2 + r_2^2 = \frac{a+b}{ab - h^2} \quad \text{and} \quad r_1^2 r_2^2 = \frac{1}{(ab - h^2)}$$

We know that
$$(r_1^2 - r_2^2)^2 = (r_1^2 + r_2^2)^2 - 4 r_1^2 r_2^2$$

$$\therefore \quad r_1^2 - r_2^2 = \sqrt{\frac{(a+b)^2}{(ab - h^2)^2} - \frac{4}{(ab - h^2)}}$$

$$= \sqrt{\frac{\{(a+b)^2 - 4ab + 4h^2\}}{(ab - h^2)^2}}$$

$$= \frac{\sqrt{\{(a-b)^2 + 4h^2\}}}{(ab - h^2)} \qquad ...(ii)$$

The equation of the other conic is
$$a'x^2 + 2h'xy + b'y^2 = 1 \qquad ...(iii)$$

If R_1^2 and R_2^2 are the squares of the semi-axes of the conic (iii), then proceeding similarly as above, we get
$$R_1^2 - R_2^2 = \frac{\sqrt{\{(a'-b')^2 + 4h'^2\}}}{(a'b' - h'^2)} \qquad ...(iv)$$

Now if the conics (i) and (iii) are confocal, we have

i.e., $$r_1^2 - r_2^2 = R_1^2 - R_2^2,$$

i.e., $$(r_1^2 - r_2^2)^2 = (R_1^2 - R_2^2)^2$$

i.e., $$\frac{\{(a-b)^2 + 4h^2\}}{(ab-h^2)^2} = \frac{\{(a'-b')^2 + 4h'^2\}}{(a'b'-h'^2)^2}.$$

4. *If the confocals throguh* (x_1, y_1) *to the ellipse*

$$\frac{x^2}{a^2} + \frac{y^2}{b^2} = 1$$

are $\dfrac{x^2}{a^2 + \lambda_1} + \dfrac{y^2}{b^2 + \lambda_1} = 1$ *and* $\dfrac{x^2}{a^2 + \lambda_2} + \dfrac{y^2}{b^2 + \lambda_2} = 1,$

show that

(I) $\qquad \dfrac{x_1^2}{a^2} + \dfrac{y_1^2}{b^2} - 1 = \dfrac{-\lambda_1 \lambda_2}{a^2 b^2}.$

(II) $\qquad x_1^2 + y_1^2 - a^2 - b^2 = \lambda_1 + \lambda_2.$

Sol. The equation of any confocal to the given ellipse can be taken as

$$\frac{x^2}{a^2 + \lambda} + \frac{y^2}{b^2 + \lambda} = 1.$$

If this passes through (x_1, y_1), we have

i.e., $$\lambda^2 + \lambda(a^2 + b^2 - x_1^2 - y_1^2) + a^2 b^2 - x_1^2 b^2 - y_1^2 a^2 = 0.$$

Since the above equation has two roots λ_1 and λ_2, therefore, we have

$$\lambda_1 + \lambda_2 = x_1^2 + y_1^2 - a^2 - b^2,$$

which gives (II), and

$$\lambda_1 \lambda_2 = a^2 b^2 - x_1^2 b^2 - y_1^2 a^2$$

i.e., $\qquad \dfrac{x_1^2}{a^2} + \dfrac{y_1^2}{b^2} - 1 = -\dfrac{\lambda_1 \lambda_2}{a^2 b^2},$ which is (I).

EXERCISES

1. Show that one parameter family of curves $\dfrac{x^2}{a^2 - \lambda} + \dfrac{y^2}{b^2 - \lambda} = 1$ represents a system of confocal conics.

2. Show that the equation of the conic confocal with $x^2 + 2y^2 = 2$ and passing through the point $(1, 1)$ is

$$3x^2 - y^2 \pm \sqrt{(5)}(x^2 - y^2) = 2.$$

3. Show that only one of a given system of confocals can have a given straight line as a normal.

4. Show that points of contact of the tangents of the confocals $\dfrac{x^2}{a^2 + \lambda} + \dfrac{y^2}{b^2 + \lambda} = 1$, which are also tangents to the parabola $y^2 = 4x\sqrt{(a^2 - b^2)}$, lie on a straight line.

System of Conics : Confocal Conics

5. Prove that the locus of the points lying on a system of confocal ellipses, which have the same eccentric angle α is a confocal hyperbola whose asymptotes are inclined at an angle 2α.

6. Show that the locus of the point of contact of parallel tangents to a system of confocals is a rectangular hyperbola.

7. Normals are drawn from a point P to the ellipse $\dfrac{x^2}{a^2} + \dfrac{y^2}{b^2} = 1$. If λ, μ be the parameters of the confocals to the ellipse through P, show that the product of the nromals is
$$\dfrac{\lambda\mu\,(\lambda - \mu)}{a^2 - b^2}.$$

8. If λ, μ be the parameters of the confocals through two points P, Q on a given ellipse, show that if P, Q be the extremities of conjugate diameters, then $\lambda + \mu$ is constant and if the tangents at P and Q are at right angles, then $\dfrac{1}{\lambda} + \dfrac{1}{\mu}$ is constant.

9. If θ be the angle between the tangents from a given point P to the ellipse $\dfrac{x^2}{a^2} + \dfrac{y^2}{b^2} = 1$, show that $\tan\theta = \dfrac{2\sqrt{(-\lambda_1\lambda_2)}}{\lambda_1 + \lambda_2}$,

where λ_1 and λ_2 are the parameters of the confocals through P.

10. Show that the locus of points of contact of tangents drawn throguh a fixed point $(1, 1)$ to the system of conics confocal with $\dfrac{x^2}{a^2} + \dfrac{y^2}{b^2} = 1$, is the curve
$$\dfrac{x}{y - 1} + \dfrac{y}{x - 1} = \dfrac{a^2 - b^2}{y - x}.$$

□□□

3
Polar Equations

3.1 POLAR COORDINATES OF A POINT

Besides the cartesian system of corodinates, the position of a point in a plane can be indicated by its distance from a fixed point, and direction measured from a fixed line through the fixed point in the plane. In this system, the corodinates of the point are called the **polar coordinates**.

Let O be a fixed point and OX a fixed straight line in the plane of paper. Also let P be any point in this plane. Join P to O.

Let $OP = r$ and $\angle XOP = \theta$. Then the position of P is denoted by the symbol (r, θ).

The fixed point O is called the **pole** and the fixed straight line OX is called the **initial line**.

The length $OP = r$ is called the **radius vector** of P and the $\angle XOP = \theta$ is called the **vectorial angle** of P.

The symbol (r, θ) denotes the **polar coordinates** of the point P.

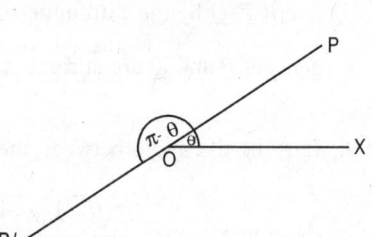

Sign of the radius vector

The radius vector is positive when measured from the origin along the line bounding the vectorial angle and negative when measured in the opposite direction.

If P be the point (r, θ), then the point $(r, \theta + 2\pi)$ is also P. In general $(r, \theta + 2n\pi)$ denotes the coordinates of the point $P(r, \theta)$, where n is a positive integer.

Produce PO to P' such that $OP = OP' = r$ and if the point P be (r, θ) then P' is the point $(-r, \theta)$.

Also if $\angle XOP' = \pi + \theta$, then the coordinates of P' are $(r, \pi + \theta)$ further (r, θ) and $(-r, \theta + \pi)$ denote the same point P. Thus we see that in polar coordinate system a point can be represented by infinitely large number of coordinates.

3.1.1 Cartesian-Polar Transformation

Take the initial line OX as the x-axis, and a line through the pole O perpendicular to OX as the y-axis.

Let (x, y) and (r, θ) be the cartesian and polar coordinates of any point P. Then

$$OM = x; \quad PM = y; \quad OP = r; \quad \angle POX = \theta.$$

$\therefore \qquad OM = OP \cos\theta \quad \text{and} \quad PM = OP \sin\theta,$

i.e., $\qquad x = r\cos\theta \quad \text{and} \quad y = r\sin\theta. \qquad …(i)$

From these relations, we have

$$x^2 + y^2 = r^2 \quad \text{and} \quad y/x = \tan\theta,$$

i.e., $\qquad r = \sqrt{(x^2 + y^2)} \quad \text{and} \quad \theta = \tan^{-1}(y/x). \qquad …(ii)$

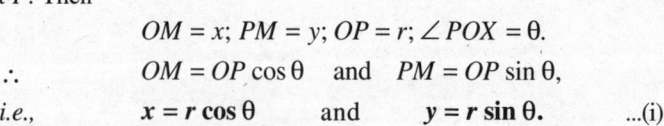

Polar Equations

3.2 DISTANCE BETWEEN TWO POINTS

Let $P(r_1, \theta_1)$ and $Q(r_2, \theta_2)$ be the given points. Join OP, OQ and PQ. Then $OP = r_1$, $OQ = r_2$ and $\angle POQ = \theta_2 - \theta_1$.

Hence, from $\triangle OPQ$,

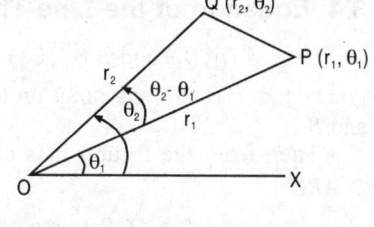

$$\cos(\theta_2 - \theta_1) = \cos POQ$$

$$= \frac{OP^2 + OQ^2 - PQ^2}{2OP \cdot OQ}$$

$$= \frac{r_1^2 + r_2^2 - PQ^2}{2r_1 r_2}$$

$$\Rightarrow \quad 2r_1 r_2 \cos(\theta_2 - \theta_1) = r_1^2 + r_2^2 - PQ^2$$

$$\Rightarrow \quad PQ^2 = r_1^2 + r_2^2 - 2r_1 r_2 \cos(\theta_2 - \theta_1).$$

3.3 POLAR EQUATION OF A STRAIGHT LINE

Let OX be the intial line and APM the straight line whose equation is required. Draw OM perpendicular upon the given line. Let $OM = p$ and $\angle XOM = \alpha$.

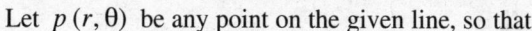

Let $p(r, \theta)$ be any point on the given line, so that

$$OP = r, \angle XOP = \theta, \quad \text{and} \quad \angle POM = \alpha - \theta.$$

Hence from the right angled triangle OMP, we have

$$OM = OP \cos POM \quad \text{or} \quad p = r \cos(\alpha - \theta)$$

i.e., $\quad r \cos(\theta - \alpha) = p$ ...(i)

which is the required equation of the straight line.

Remark 1. The polar equation of a straight line through the pole and inclined at an angle α to the initial line is

$$\theta = \alpha. \qquad ...(ii)$$

$$\Rightarrow \quad \tan \theta = \tan \alpha \Rightarrow y/x = \tan \alpha \Rightarrow y = x \tan \alpha$$

which is the cartesian equation of a line.

Remark 2. Equation of a line in cartesian form is

$$ax + by + c = 0$$

or $\quad ar \cos \theta + br \sin \theta + c = 0$

or $\quad a \cos \theta + b \sin \theta = -c/r$

which can be written as

$$l/r = A \cos \theta + B \sin \theta \quad \text{or} \quad 1/r = a \cos \theta + b \sin \theta.$$

which are equations of a straight line in polar coordinates.

Remark 3. Let a line be $l/r = A \cos \theta + B \sin \theta$ its equation is
In cartesian system, its equation is

$$ax + by = c$$

and a line perpendicular to this line is $bx - ay = \lambda$ which on changing back to polar system gives

$$r(b \cos \theta - a \sin \theta) = \lambda$$

i.e., $\quad \dfrac{\lambda}{r} = a \cos\left(\theta - \dfrac{\pi}{2}\right) + b \sin\left(\theta + \dfrac{\pi}{2}\right)$

Thus we find that the equation of a line perpendicular to a given line in polar coordinate system is obtained by replacing θ in the given equation to the line by $\theta + \frac{1}{2}\pi$ and changing the coefficient of $1/r$ to a new constant.

3.4 Equation of the Line Through two Points

Let $A(r_1, \theta_1)$ and $B(r_2, \theta_2)$ be the given points and $P(r, \theta)$ any point on the ling joining A and B.

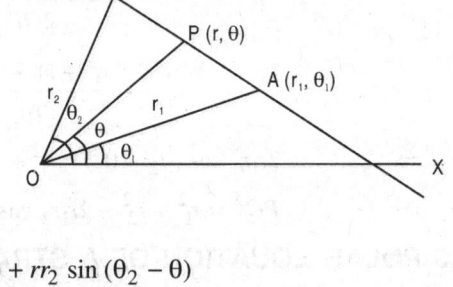

Then from the figure, it is clear that area of $\triangle ABC$

= area of $\triangle AOP$ + area of $\triangle POB$

i.e.,
$$\frac{1}{2} r_1 r_2 \sin AOB = \frac{1}{2} rr_1 \sin AOP + \frac{1}{2} rr_2 \sin POB$$

i.e., $r_1 r_2 \sin(\theta_2 - \theta_1) = rr_1 \sin(\theta - \theta_1) + rr_2 \sin(\theta_2 - \theta)$

i.e., $\dfrac{\sin(\theta_2 - \theta_1)}{r} = \dfrac{\sin(\theta - \theta_1)}{r_2} + \dfrac{\sin(\theta_2 - \theta)}{r_1}$.

3.5 POLAR EQUATION OF A CIRCLE

Let $C(R, \alpha)$ be the centre and a and the radius of the circle.

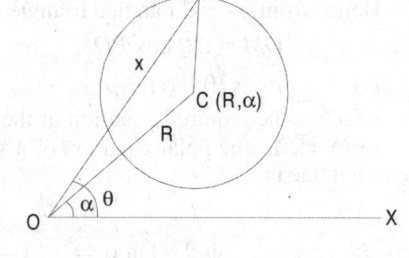

Take any point $P(r, \theta)$ on the circle, then

$OC = r, \angle XOC = \alpha$

$OP = r, \angle COP = \theta - \alpha$.

Hence from $\triangle COP$, we have

$$\cos(\theta - \alpha) = \frac{r^2 + R^2 - a^2}{2rR}$$

i.e. $r^2 - 2rR \cos(\theta - \alpha) + R^2 = a^2$

which is the required equation of the circle.

3.5.1 Particular Cases of the General Equation

(i) Center C on the initial line.

In this case $\alpha = 0$, and the equation of the circle becomes

$$r^2 - 2rR \cos\theta + R^2 = a^2. \qquad \ldots(i)$$

(ii) Pole O on the circle.

Then $OC = R = a$, and the equation of the cicle becomes

$$r^2 - 2ar \cos(\theta - \alpha) + a^2 = a^2$$

i.e., $r = 2a \cos(\theta - \alpha)$. ...(iii)

(iii) Pole on the circumference and centre on the initial line

Let $P(r, \theta)$ be any point on the circle, then

$\angle AOP = \theta;\quad \angle APO = 90°$

and $\quad OA = 2a$ (diameter).

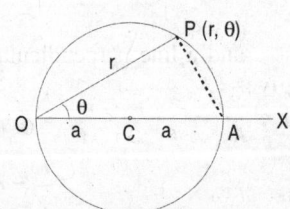

Polar Equations

Hence from $\triangle AOP$, we have
$$r = 2a \cos\theta$$
which is the required equation of the circle.

(iv) Initial line is tangent at the pole

Here $\angle XOP = \theta = \angle OBP$

and $\angle OPB = 90°$

Also $OB = 2a$ (diameter). Hence from $\triangle OPB$, we have
$$r = 2a \sin\theta \qquad \ldots(iv)$$
which is the required equation of the circle.

(v) Centre is the pole. In this case, for all positions of P, we have
$$r = a$$
which is therefore the required equation of the circle.

EXAMPLE

Find the condtiion that the straight line
$$1/r = a\cos\theta + b\sin\theta$$
may touch the circle $r = 2c \cos\theta$.

Sol. Eliminating θ between the given equations, the radii vectors of the points of intersection are given by

$$\frac{1}{r} = a \cdot \frac{r}{2c} + b\sqrt{1 - \left(\frac{r}{2c}\right)^2}$$

i.e., $\left(\dfrac{2c}{r} - ar\right)^2 = b^2(4c^2 - r^2)$

i.e., $(a^2 + b^2)r^4 - 4c(a + b^2 c)r^2 + 4c^2 = 0.$

This is a quadratic in r^2. Therefore the line will touch the circle, if the two values of r^2 in the above eqaution are equal,

i.e., if $\quad 16c^2(a + b^2 c)^2 - 16(a^2 + b^2)c^2 = 0$

i.e., if $\quad a^2 + b^4 c^2 + 2ab^2 c - a^2 - b^2 = 0$

i.e., if $\quad b^2 c^2 + 2ac = 1.$

which is the required condition.

EXERCISES

1. Prove that the straight line passing through the point of intersection of the lines
   $$1/r = a\cos\theta + b\sin\theta \quad \text{and} \quad 1/r = a'\cos\theta + b'\sin\theta$$
   is
   $$\frac{1+\lambda}{r} = (a + \lambda a')\cos\theta + (b + \lambda b')\sin\theta.$$
   Hence obtain the equation of the line passing through the pole and the point of intersection of the given lines.
   [**Hint.** Equation of line passing through the pole does not contain r]

2. Show that the condition that the chord cut off by the curve $l/r = 1 + e\cos\theta$ from the line $1/r = a\cos\theta + b\sin\theta$ may subtend a right angle at the pole is
   $$(la - e)^2 + l^2 b^2 = 2.$$

3. Find the polar equation to the circle described on the line joining the points (a, α) and (b, β) as diameter.

3.6 POLAR EQUATION OF A CONIC

To find the polar equation of a conic, the focus being the pole.

Let S be the focus, ZM the directrix and e the eccentricity of the conic.

Draw SZ perpendicular to ZM and take ZS (produced) as the positive direction of the initial line and S as the pole.

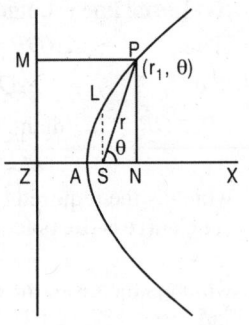

Let P be any point on the conic and let its coordinates be (r, θ), so that $SP = r$ and $\angle XSP = \theta$.

Draw PN perpendicular to the initial line, and PM perpendicular to the directrix.

If the length of the semi latus-rectum SL be l, then by definition of a conic, we have

$$l = SL = e \cdot SZ; \text{ giving } SZ = \frac{l}{e}.$$

Also $\quad SP = e \cdot PM = e \cdot NZ = e \cdot (SZ + SN)$

i.e., $\quad r = e\left(\dfrac{l}{e} + SP \cos\theta\right), \quad \begin{cases} \text{since } SZ = l/e \\ \text{and } SN = SP \cos\theta \end{cases}$

$$= e\left(\frac{l}{e} + r\cos\theta\right) = l + er\cos\theta$$

i.e., $\quad l = r(1 - e\cos\theta).$

Hence the required equation of the conic is

$$\frac{l}{r} = 1 - e\cos\theta. \qquad \text{...(i)}$$

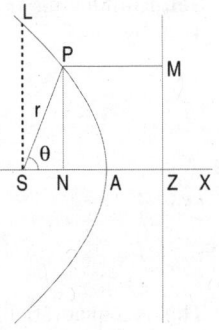

Note 1. In some books of coordinate geometry and in Astronomy the direction SZ of the initial line is taken to be positive. Then

$$l = SL = e \cdot SZ; \text{ giving } SZ = \frac{l}{e}.$$

Also $\quad r = SP = e \cdot PM$

i.e., $\quad r = e\left(\dfrac{l}{e} - r\cos\theta\right) = l - er\cos\theta.$

Hence, in this case, the equation to the conic is

$$\frac{l}{r} = 1 + e\cos\theta. \qquad \text{...(ii)}$$

Note 2. If the conic be a parabola, we have $e = 1$, so that equations (i) and (ii) become

$$\frac{l}{r} = 1 - \cos\theta = 2\sin^2\frac{\theta}{2} \qquad \text{...(iii)}$$

and $\quad \dfrac{l}{r} = 1 + \cos\theta = 2\cos^2\dfrac{\theta}{2}. \qquad \text{...(vi)}$

3.6.1 TO FIND THE POLAR EQUATION OF A CONIC WITH ITS FOCUS AS THE POLE AND ITS AXIS INCLINED AT AN ANGLE λ TO THE INITIAL LINE

Let the axis ZSR of the conic be inclined at an angle λ to the initial line SX (see Fig. (i)).

Polar Equations

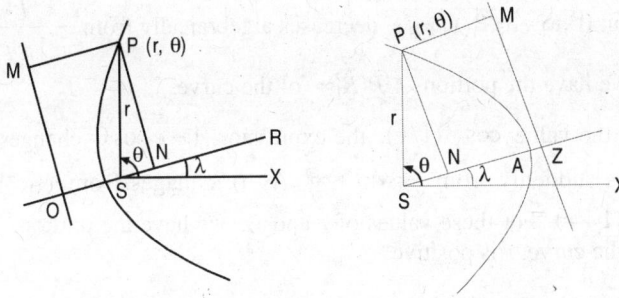

Let (r, θ) be the coordinates of any point P on the conic, then
$$\angle PSX = \theta \quad \text{and} \quad \angle RSX = \lambda; \quad \therefore \quad \angle PSR = \theta - \lambda.$$
Now
$$r = SP = e \cdot PM = e \cdot NZ = e \cdot (SZ + SN)$$
$$= e \left\{ \frac{l}{e} + r \cos(\theta - \lambda) \right\} = l + er \cos(\theta - \lambda)$$

i.e.
$$\frac{l}{r} = 1 - e \cos(\theta - \lambda). \qquad \ldots(ii)$$

3.6.2 To Trace the Conic $\frac{l}{r} = 1 - e \cos \theta$.

Case 1. *When the conic is a parabola*, we have $e = 1$, so that the equation to the conic becomes $l/r = 1 - \cos \theta$. When $\theta = 0$, we have $l/r = 0$, so that r is infinite. When θ change from $0°$ to $90°$, $\cos \theta$ decreases from 1 to 0, so that l/r increases from 0 to 1, *i.e.*, r decreases from infinity to l.

When θ changes from $90°$ to $180°$, $\cos \theta$ decreases from 0 to -1, so that l/r increases from 1 to 2, *i.e.*, r decreases from l to $\frac{1}{2}l$.

Also, since $\cos(-\theta) = \cos \theta$, the curve is symmetrical about the initial line; therefore the curve from $\theta = 180°$ to $\theta = 360°$ can be traced by symmetry.

Case 2. *When the conic is an ellipse*, $e < 1$. When $\theta = 0$, we have $l/r = 1 - e$, *i.e.*, $r = l/(1-e)$. These values of r and θ correspond to the point A'.

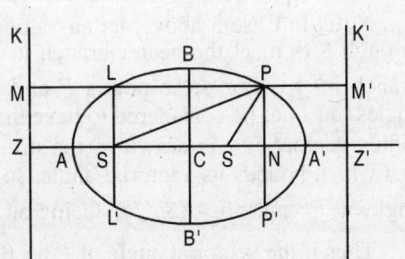

When θ increases from $0°$ to $90°$, $\cos \theta$ decreases from 1 to 0, so that $1 - e \cos \theta$ increases from $1 - e$ to 1, *i.e.*, r decreases from $l/(1-e)$ to l. These values of r and θ trace out the portion $A'PBL$.

When θ increases from $90°$ to $180°$, $\cos \theta$ decreases from 0 to -1 so that $1 - e \cos \theta$ increases from 1 to $1 + e$, *i.e.*, r decreases from l to $l/(1+e)$. For these values of r and θ we obtain the portion LA of the curve. The rest of the curve can be traced by symmetry.

Case 3. *When the conic is a hyperbola*, $e > 1$.

When $\theta = 0$, we have $r = \frac{l}{1-e} = -\frac{l}{e-1}$ (since $e > 1$). So that r is negative. These values of r and θ given the point A'. (See figure).

When θ increase from 0° to $\cos^{-1}(1/e)$, r decreases algebraically from $-\dfrac{l}{1-e}$ to $-\infty$. For these values of r and θ, we have the portion $A'P'_2R'_1\infty$ of the curve.

When θ just crosses the value $\cos^{-1}(1/e)$, the expression $1-e\cos\theta$ changes from -0 to $+0$ and hence r changes suddenly from $-\infty$ to $+\infty$. As θ chnages from $\cos^{-1}(1/e)$ to π, r decreases from $+\infty$ to $l/(1+e)$. For these values of r and θ, we have the portion ∞RPA of the curve. For this portion of the curve, r is positive.

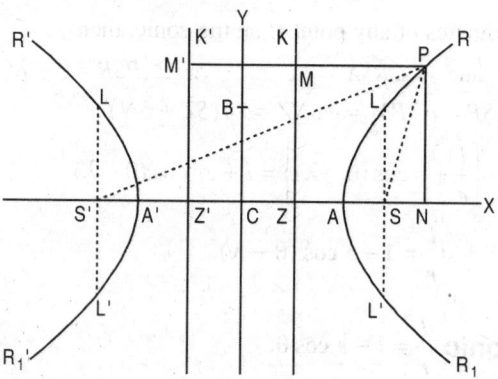

When θ changes from π to $2\pi - \cos^{-1}\dfrac{1}{e}$, r increases from $\dfrac{1}{1+e}$ to $+\infty$. For these values of r and θ, we have the portion $AL'R_1\infty$ of the curve. For this portion also, r is positive.

Again when θ crosses the value $2\pi - \cos^{-1}(1/e)$, the expression $1-e\cos\theta$ changes from $+0$ to -0 and hence r changes abruptly from $+\infty$ to $-\infty$. Also when θ changes from $2\pi - \cos^{-1}(1/e)$ to π, r increases algebraically from $-\infty$ to $-\dfrac{l}{e-1}$. For these values of r and θ, we have the portion $\infty R'A'$. For this portion of the curve, r is negative.

Note. In Case 3 above, let any straight line be drawn through S to meet the nearer branch in P and the further branch in P'. For these points P and P' the vectorial angles must not be considered to have the same value. The radius vector SP' is drawn in the direction *opposite* to that which bounds its vectorial angle, so that the vectorial angle of P' must be XSp', p' being on $P'S$ produced.

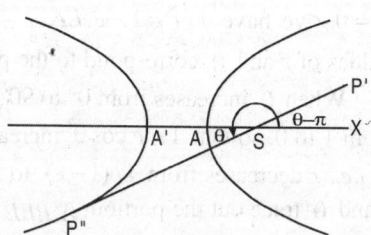

Thus if the vectorial angle of P be θ, then that of P' will be $\theta - \pi$. Therefore

$$SP = \frac{l}{1-e\cos\theta}$$

and

$$SP' = -\frac{l}{1-e\cos(\pi-\theta)} = -\frac{l}{1+e\cos\theta}$$

This is a relation connecting the distance SP' of any point P' on the further branch of the hyperbola with the angle XSP' $(=\theta)$ which it makes with the initial line.

3.7 EQUATION TO THE DIRECTRICES

Case 1. *Parabola.* In this case there is only one directrix and it is at a distance l on the *left* of the focus S, since
$$SZ = ML = SL = l.$$
Hence the equation of the directix is
$$x = -l$$
i.e.,
$$r \cos \theta = -l$$

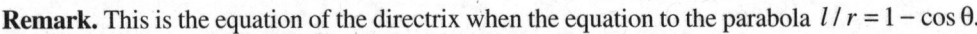

This can also be written as $\dfrac{l}{r} = -\cos \theta.$

Remark. This is the equation of the directrix when the equation to the parabola $l/r = 1 - \cos \theta$.

If the equation to the parabola be $l/r = 1 + \cos \theta$, then it is apparent that the equation of the directrix will be
$$\frac{l}{r} = \cos \theta.$$

Case 2. *Ellipse.* In this case there are two directrices. Let the equation to the ellipse be

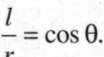

$$\frac{l}{r} = 1 - e \cos \theta.$$

Here S is the pole and ZM is the directrix on the left of S. Now
$$SL = e \cdot SZ; \quad \therefore \quad SZ = \frac{l}{e}.$$
Hence the equation to the directrix ZM (near the pole) is
$$x = -SZ \quad \text{or} \quad r \cos \theta = -\frac{l}{e}.$$
This can also be written as
$$\frac{l}{r} = -e \cos \theta. \qquad \ldots(i)$$

Again, $Z'M'$ is the other directrix corresponding to the focus S' (other than the pole) and it lies on the right of S (the pole). Now
$$SZ' = ZZ' - SZ$$
$$= 2CZ - SZ = 2 \cdot \frac{a}{e} - \frac{l}{e}$$
But
$$l = \frac{b^2}{a} = \frac{a^2(1-e^2)}{a} = a(1-e^2)$$
$$\therefore \quad SZ' = \frac{2}{e} \cdot \frac{l}{1-e^2} - \frac{l}{e} = \frac{l}{e} \cdot \frac{1+e^2}{1-e^2}$$

Hence the equation to the directrix $Z'M'$ is
$$x = -SZ \quad \text{or} \quad r \cos \theta = -\frac{l}{e}.$$
This can also be written as
$$\frac{l}{r} = -e \cos \theta. \qquad \ldots(i)$$

Again, $Z'M'$ is the other directrix corresponding to the focus S' (other than the pole) and it lies on the right of S (the pole). Now

$$SZ' = ZZ' - SZ$$
$$= 2CZ - SZ = 2 \cdot \frac{a}{e} \cdot \frac{l}{e}$$

But
$$l = \frac{b^2}{a} = \frac{a^2(1-e^2)}{a} = a(1-e^2)$$

∴
$$SZ' = \frac{2}{e} \cdot \frac{l}{1-e^2} - \frac{l}{e} = \frac{l}{e} \cdot \frac{1+e^2}{1-e^2}$$

Hence the equation to the directrix $Z'M'$ is

$$x = SZ' \quad \text{or} \quad r\cos\theta = \frac{l}{e} \cdot \frac{1+e^2}{1-e^2}.$$

This can also be written as

$$\frac{l}{r} = \frac{1-e^2}{1+e^2} \cdot e\cos\theta \qquad \text{...(ii)}$$

which is the directrix corresponding to the focus other than the pole.

Remark. If the equation to the conic be $\frac{l}{r} = 1 + e\cos\theta$, the corresponding equation of the directrices are [on changing e by $-e$ in equations (i) and (ii) above]

$$\frac{l}{r} = e\cos\theta \quad \text{and} \quad \frac{l}{r} = -\frac{1-e^2}{1+e^2} e\cos\theta$$

Case 3. *Hyperbola.* It can easily be seen that equations (i) and (ii) given above also hold for a hyperbola.

3.8 EQUATION TO THE ASYMPTOTES

Asymptotes exist only in the case of a *hyperbola*. If α be the angle which the asymptote CK makes with the x-axis, then

$$\tan\alpha = \frac{b}{a} = \frac{a\sqrt{(e^2-1)}}{a} = \sqrt{(e^2-1)}$$

so that $\cos\alpha = \frac{1}{e}$; giving $\alpha = \cos^{-1}\frac{1}{e}$.

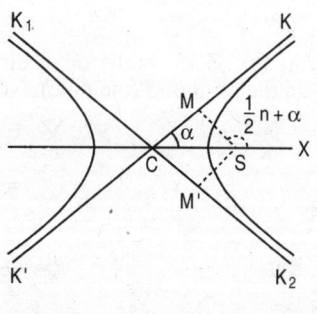

If SM be perpendicular upon the asymptote CK, then

$$SM = CS\sin\alpha = ae \cdot \frac{\sqrt{(e^2-1)}}{e} = a\sqrt{(e^2-1)} = b.$$

Also from the above figure it is clear that the perpendicular SM, SM' upon the asymptotes CK, CK' makes angles

$$\frac{\pi}{2} + \cos^{-1}\frac{1}{e} \quad \text{and} \quad \frac{3\pi}{2} - \cos^{-1}\frac{1}{e},$$

with the line CSX.

Hence, equation to the two asymptotes are

$$r\cos\left\{\theta - \left(\frac{\pi}{2} + \cos^{-1}\frac{1}{e}\right)\right\} = b$$

$$r \cos\left\{\theta - \left(\frac{3\pi}{2} - \cos^{-1}\frac{1}{e}\right)\right\} = b;$$

i.e.,
$$r \sin\left(\theta - \cos^{-1}\frac{1}{e}\right) = b \qquad \ldots(i)$$

and
$$r \sin\left(\theta + \cos^{-1}\frac{1}{e}\right) = -b. \qquad \ldots(ii)$$

EXAMPLES

1. *Show that the equations*

$$\frac{l}{r} = 1 - e\cos\theta \quad \text{and} \quad \frac{l}{r} = -1 - e\cos\theta$$

represent the same conic.

Sol. If the coordinates of a point, in polar coordinates, be (r, θ) then its coordinates may be taken as
$$(-r, \theta + \pi).$$

Thus, if in any polar equation to a curve we write $-r$ for r and $\theta + \pi$ for θ, its equation is altered but the curve remains unchanged. Hence changing r into $-r$ and θ into $\theta + \pi$ for first equation becomes

$$\frac{l}{-r} = 1 - e\cos(\theta + \pi), \ i.e., \ \frac{l}{r} = -1 - e\cos\theta$$

which is same as the second eqaution. Consequently the given equations represents the same conic.

Remark. This example show that in polar coordiantes system a curve may be represented by different equations.

2. *In any conic, prove that*
(a) The sum of the reciprocals of the segments of any focal chord is constant.
(b) The sum of the reciprocals of two perpendicular focal chords is constant.

Sol. Let the equation of the conic be

$$\frac{l}{r} = 1 + e\cos\theta.$$

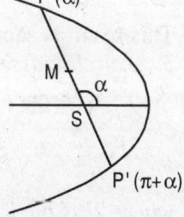

Also let PSP' be any focal chord of this conic. let the vectorial angle of P be α, so that the vectorial angle of P' is $\pi + \alpha$.

(a) Since $P(\alpha)$ and $P'(\pi + \alpha)$ lie on the given conic, we have (noting that when $\theta = \alpha, r = SP$ and when $\theta = \pi + \alpha, r = SP'$)

$$\frac{l}{SP} = 1 + e\cos\alpha \qquad \ldots(i)$$

and
$$\frac{l}{SP'} = 1 + e\cos(\pi + \alpha) \qquad \ldots(ii)$$

From (i) and (ii), we find that

$$\frac{1}{SP} + \frac{1}{SP'} = \frac{1 + e\cos\alpha}{l} + \frac{1 - e\cos\alpha}{l}$$

$$= \frac{2}{l} = \text{constant}$$

i.e., sum of the reciprocals of segments of a focal chord is constant.

Note. From the above result it also follows that the semi-latus-rectum l is the harmonic mean between the segments of any focal chord.

(b) Let QSQ' be the focal chord perpendicular to PSP', so that the vectorial angles of Q and Q' are

$$\frac{\pi}{2} + \alpha \quad \text{and} \quad \frac{3\pi}{2} + \alpha.$$

Since $Q\left(\frac{1}{2}\pi + \alpha\right)$ and $Q'\left(\frac{3}{2}\pi + \alpha\right)$ also lie on the given conic, we have

$$\frac{l}{SQ} = 1 + e \cos\left(\frac{\pi}{2} + \alpha\right) = 1 - e \sin\alpha \qquad \ldots(iii)$$

and

$$\frac{l}{SQ'} = 1 + e \cos\left(\frac{3\pi}{2} + \alpha\right) = 1 + e \sin\alpha \qquad \ldots(iv)$$

From (i), (ii), (iii) and (iv), we have

$$PP' = SP + SP'$$

$$= \frac{l}{1 + e \cos\alpha} + \frac{l}{1 - e \cos\alpha} = \frac{2l}{1 - e^2 \cos^2\alpha}$$

and

$$QQ' = SQ + SQ'$$

$$= \frac{l}{1 - e \sin\alpha} + \frac{l}{1 + e \sin\alpha} = \frac{2l}{1 - e^2 \sin^2\alpha}$$

$$\therefore \frac{1}{PP'} + \frac{1}{QQ'} = \frac{1 - e^2 \cos^2\alpha}{2l} + \frac{1 - e^2 \sin^2\alpha}{2l}$$

$$= \frac{1 - e^2(\cos^2\alpha + \sin^2\alpha)}{2l} = \frac{2 - e^2}{2l} = \text{const}.$$

This result is same for all such pairs of chords.

3. *A circle passes through the focus S of a conic and meets it in four points whose distances from S are r_1, r_2, r_3 and r_4. Prove that*

$$\frac{1}{r_1} + \frac{1}{r_2} + \frac{1}{r_3} + \frac{1}{r_4} = \frac{2}{l}$$

where $2l$ is the latus rectum of the conic.

Sol. Take the focus of the conic as the pole, and the axis of the conic as the initial line, so that its equation is

$$\frac{l}{r} = 1 + e \cos\theta. \qquad \ldots(i)$$

If the diameter of the circle, which passes throguh S, be inclined at an angle α to the axis, then its equation is,

$$r = d \cos(\theta - \alpha)$$

i.e.,

$$r = d (\cos\theta \cos\alpha + \sin\theta \sin\alpha), \qquad \ldots(ii)$$

where d is the diameter of the circle.

If we eliminate θ between (i) and (ii), we shall have an equation of fourth degree in r whose roots are r_1, r_2, r_3 and r_4 (the radii vectors of the points of intersection).

Polar Equations

From (i), $\cos\theta = \dfrac{l-r}{er}$ and $\sin\theta = \sqrt{1 - \left(\dfrac{l-r}{er}\right)^2}$.

Substituting these in (ii), we obtain

$$r = d\left\{\dfrac{l-r}{er}\cos\alpha + \sqrt{1 - \left(\dfrac{l-r}{er}\right)^2}\sin\alpha\right\}$$

i.e., $\{er^2 - d(l-r)\cos\alpha\}^2 = d^2\sin^2\alpha\,[e^2r^2 - (l-r)^2]$

i.e., $e^2r^4 + 2de\cos\alpha\, r^3 + (\ldots)r^2 - 2d^2lr + d^2l^2 = 0$.

Hence, from Theory of Equations, we have

$$\Sigma r_1 r_2 r_3 = -\dfrac{\text{coeff. of } r}{\text{coeff. of } r^4} = \dfrac{2d^2l}{e^2} \qquad \ldots\text{(iii)}$$

and $\quad r_1 r_2 r_3 r_4 = \dfrac{\text{const. term}}{\text{coeff. of } r^4} = \dfrac{d^2 l^2}{e^2} \qquad \ldots\text{(iv)}$

Dividing (iii) by (iv), we obtain

$$\dfrac{r_1 r_2 r_3 + r_1 r_2 r_4 + r_1 r_3 r_4 + r_2 r_3 r_4}{r_1 r_2 r_3 r_4} = \dfrac{2d^2 l / e^2}{d^2 l^2 / e^2}$$

i.e., $\quad \dfrac{1}{r_1} + \dfrac{1}{r_2} + \dfrac{1}{r_3} + \dfrac{1}{r_4} = \dfrac{2}{l}$.

4. *If a straight line drawn through the focus S of a hyperbola, parallel to an asymptote, meet the curve in P, prove that SP is one quarter of the latus rectum.*

Sol. Let the equation to the hyperbola be

$$\dfrac{l}{r} = 1 - e\cos\theta. \qquad \ldots\text{(i)}$$

Let SP be drawn parallel to the asymptote CK_1.

Clearly, SP makes an angle $\pi - \alpha$ with the initial line, where $\alpha = \cos^{-1}\dfrac{1}{e}$. Hence

$$\dfrac{l}{SP} = 1 - e\cos(\pi - \alpha), \text{ from (1)}$$

i.e., $\quad SP = \dfrac{l}{1 + e\cos\alpha} = \dfrac{l}{1 + e\cdot\dfrac{1}{e}} = \dfrac{l}{2} = \dfrac{1}{4}$ of the latus rectum.

EXERCISES

1. Show that the equations $\dfrac{l}{r} = 1 + e\cos\theta$ and $\dfrac{l}{r} = -1 + e\cos\theta$ represents the same conic.

2. Show that in a conic the semi-latus rectum is the harmonic mean between the segments of a focal chord.
   [**Hint.** See Note of illustrative Ex. 2 (a)]

3. Prove that perpendicular focal chords of a rectangular hyperbola are equal.
   [**Hint.** For a rectangular hyperbola, $e = \sqrt{2}$]

4. PSQ and $PS'R$ are two chords of an ellipse through the foci S and S'; prove that
$$\frac{SP}{QS} + \frac{S'P}{S'R} \text{ is independent of the position of } P.$$

5. If PSP' and QSQ' are perpendicular focal chords of a conic, prove that $\dfrac{1}{PS.SP'} + \dfrac{1}{QS.QS'}$ is constant.

6. Prove that the locus of the middle points of focal chords of a conic is a conic of the same kind as the original conic.

7. Show that the directrix of the conic $\dfrac{l}{r} = 1 + e\cos\theta$ corresponding to the focus other than the pole is
$$\frac{l}{r} = -\left(\frac{1-e^2}{1+e^2}\right) e\cos\theta.$$

8. A circle passes through the focus S of a conic and meets it in four points whose distances from S are r_1, r_2, r_3 and r_4. Prove that
$$r_1 r_2 r_3 r_4 = \frac{d^2 l^2}{e^2}$$
where $2l$ and e are the latus-rectum and eccentricity of the conic, and d is the diameter of the circle.

9. A circle of given radius passing through the focus S of a given conic intersects the conic in the points A, B, C, D; show that
$$SA \cdot SB \cdot SC \cdot SD = \text{constant}.$$

3.9 EQUATION OF THE CHORD JOINING TWO POINTS ON A CONIC

Let the equation to the conic be
$$\frac{l}{r} = 1 + e\cos\theta. \qquad \ldots(i)$$

Take two points P and Q on the conic and let their coordinates be (r_1, α) and (r_2, β) respectively.

Since these points lie on the conic, we have
$$\frac{l}{r_1} = 1 + e\cos\alpha \quad \text{and} \quad \frac{l}{r_2} = 1 + e\cos\beta. \qquad \ldots(ii)$$

Now, equation of any straight line is
$$\frac{l}{r} = A\cos\theta + B\sin\theta. \qquad \ldots(iii)$$

If it passes through $P(r_1, \alpha)$ and $Q(r_2, B)$ then
$$A\cos\alpha + B\sin\alpha = \frac{l}{r_1} = 1 + e\cos\alpha, \text{ by (ii)}$$

and
$$A\cos\beta + B\sin\beta = \frac{l}{r_2} = 1 + e\cos\beta, \text{ by (ii)}$$

Polar Equations

Therefore $\quad (A - e)\cos\alpha + B\sin\alpha = 1 \quad$ ...(iv)

and $\quad (A - e)\cos\beta + B\sin\beta = 1. \quad$ ...(v)

Multiplying (iv) by $\sin\beta$ and (v) by $\sin\alpha$ and subtracting, we get

$$(A - e)\sin(\beta - \alpha) = \sin\beta - \sin\alpha.$$

$\therefore \quad A - e = \dfrac{\sin\beta - \sin\alpha}{\sin(\beta - \alpha)}$

so that $\quad A = \dfrac{2\cos\frac{1}{2}(\beta + \alpha)\sin\frac{1}{2}(\beta - \alpha)}{2\sin\frac{1}{2}(\beta - \alpha)\cos\frac{1}{2}(\beta - \alpha)} + e$

$= \cos\frac{1}{2}(\alpha + \beta)\sec\frac{1}{2}(\beta - \alpha) + e.\quad$ ...(vi)

Again, multiplying (v) by $\cos\alpha$ and (iv) by $\cos\beta$ and subtracting, we get

$$B\sin(\beta - \alpha) = \cos\alpha - \cos\beta.$$

$\therefore \quad B = \dfrac{\cos\alpha - \cos\beta}{\sin(\beta - \alpha)}$

$= \dfrac{2\sin\frac{1}{2}(\alpha + \beta)\sin\frac{1}{2}(\beta - \alpha)}{2\sin\frac{1}{2}(\beta - \alpha)\cos\frac{1}{2}(\beta - \alpha)}$

$= \sin\frac{1}{2}(\alpha + \beta)\sec\frac{1}{2}(\beta - \alpha).\quad$ ...(vii)

Substituting these values in (iii), we find that the required equation of the chord joining the points P and Q is

$$l/r = \left\{\cos\frac{1}{2}(\alpha + \beta)\sec\frac{1}{2}(\beta - \alpha) + e\right\}\cos\theta$$

i.e., $\quad l/r = \sec\frac{1}{2}(\beta - \alpha)[\cos\theta\cos\frac{1}{2}(\alpha + \beta) + \sin\theta\sin\frac{1}{2}(\alpha + \beta)] + e\cos\theta$

or $\quad \dfrac{l}{r} = \sec\dfrac{\beta - \alpha}{2}\cos\left(\theta - \dfrac{\alpha + \beta}{2}\right) + e\cos\theta.\quad$ ...(viii)

Corollary 1. If the vectorial angles P and Q be $\alpha - \beta$ and $\alpha + \beta$, then

sum of the angles $= 2\alpha$; diff. of the angles $= 2\beta$.

Hence from (viii), the equation of the chord PQ becomes

$$\dfrac{l}{r} = \sec\beta\cos(\theta - \alpha) + e\cos\theta.$$

Corollary 2. If the equation to the conic be

$$\dfrac{l}{r} = 1 + e\cos(\theta - \lambda),$$

then the chord joining the points α and β is

$$\dfrac{l}{r} = \sec\dfrac{\beta - \alpha}{2}\cos\left(\theta - \dfrac{\alpha + \beta}{2}\right) + e\cos(\theta - \lambda).$$

3.9.1 Tangents to a Conic

To find the polar equation of the tangent at the point α on the conic $\dfrac{l}{r} = 1 + e\cos\theta$.

Let P be the point α. Take another point Q on the conic and let its vectorial angle be β.

Then proceeding as in the last section the equation to the chord PQ is (to be actually found for a complete derivation)

$$\frac{l}{r} = \sec\frac{\beta-\alpha}{2}\cos\left(\theta - \frac{\alpha+\beta}{2}\right) + e\cos\theta. \qquad \ldots\text{(i)}$$

When Q tends to P, so that $\beta \to \alpha$ and the chord PQ tends to the tangent at $P(\alpha)$.

Hence putting $\beta = \alpha$ in (i), we find that the equation of the tangent at the point 'α' is

$$\frac{l}{r} = \cos(\theta - \alpha) + e\cos\theta. \qquad \ldots\text{(ii)}$$

3.9.2 Asymptotes

To find the equation of the asymptotes of the conic

$$\frac{l}{r} = 1 + e\cos\theta$$

The conic is

$$\frac{l}{r} = 1 + e\cos\theta. \qquad \ldots\text{(i)}$$

Let (r', α) be a point on (i), then

$$\frac{l}{r'} = 1 + e\cos\alpha. \qquad \ldots\text{(ii)}$$

Now equation of the tangent at the point 'α' is

$$\frac{l}{r} = \cos(\theta - \alpha) + e\cos\theta. \qquad \ldots\text{(iii)}$$

This tangent will tend to an asymptote, when the point of contact moves to infinity, *i.e.*, $r' \to \infty$.

Therefore putting $r' = \infty$ in (ii) we have

$$0 = 1 + e\cos\alpha; \qquad \therefore \qquad \cos\alpha = -\frac{1}{e} \text{ and } \sin\alpha = \pm\sqrt{1 - \frac{1}{e^2}}$$

Now (iii) can be written as

$$\frac{l}{r} - \cos\theta\,(e + \cos\alpha) = \sin\theta\sin\alpha.$$

Substituting for $\sin\alpha$ and $\cos\alpha$ in this, the required asymptotes are

$$\frac{l}{r} - \cos\theta\left(e - \frac{1}{e}\right) = \pm\sqrt{1 - \frac{1}{e^2}}\sin\theta$$

i.e.,
$$\frac{el}{r} = (e^2 - 1)\cos\theta \pm \sqrt{(e^2 - 1)}\sin\theta.$$

These asymptotes will be real, if $e^2 - 1 > 0$ or $e > 1$.

Polar Equations

EXAMPLES

1. *Show that the condition that the line*

$$\frac{l}{r} = A \cos \theta + B \sin \theta$$

may touch the conic $\dfrac{l}{r} = 1 + e \cos \theta$ *is* $(A - e)^2 + B^2 = 1$.

Sol. Let the line

$$l/r = A \cos \theta + B \sin \theta \qquad \ldots(i)$$

touch the conic $l/r = 1 + e \cos \theta$ at the point whose vectorial angle is α.
Then line (i) must be identical with the tangent to the conic at the point α.
But the tangent to the given conic at the point α is

$$\frac{l}{r} = \cos (\theta - \alpha) + e \cos \theta$$

$$= \cos \theta (\cos \alpha + e) + \sin \theta \sin \alpha. \qquad \ldots(ii)$$

Comparing (i) and (ii) we obtain

$$A = \cos \alpha + e \quad \text{and} \quad B = \sin \alpha.$$

$\Rightarrow \qquad \cos \alpha = A - e \quad \text{and} \quad \sin \alpha = B$

Eliminating α between these equations (noting that $\cos^2 \alpha + \sin^2 \alpha = 1$), we see that the required condition of tangency is

$$(A - e)^2 + B^2 = 1.$$

2. *The tangents at two points P and Q of a conic meet in T; prove that the vectorial angle of T is the semi-sum of the vectorial angles of P and Q.*

Sol. Let the equation of the conic be

$$\frac{l}{r} = 1 + e \cos \theta \qquad \ldots(i)$$

Also let the vectorial angles of P and Q be α and β respectively. Then, equations of the tangents at P (the point α) and Q (the point β) are

$$\frac{l}{r} = \cos (\theta - \alpha) + e \cos \theta \qquad \ldots(ii)$$

and

$$\frac{l}{r} = \cos (\theta - \beta) + e \cos \theta \qquad \ldots(iii)$$

These tangents meet at T (given). To find the vectorial angle of T, we must eliminate r between (ii) and (iii). Therefore, subtracting (iii) from (ii) we obtain

$$\cos (\theta - \alpha) = \cos (\theta - \beta)$$

$\therefore \qquad \theta - \alpha = - (\theta - \beta), \text{ since } \alpha \neq \beta$

i.e., $\qquad 2\theta = \alpha + \beta$

so that $\qquad \theta = \dfrac{1}{2} (\alpha + \beta).$

This is the required *vectorial angle* of T and is clearly equal to the semi-sum of the vectorial angles of P and Q which are α and β respectively.

3. PSP' *is a focal chord of a conic. Prove that the angle between the tangents at P and P' is*

$$\tan^{-1} \left(\frac{2e \sin \alpha}{1 - e^2} \right)$$

where α *is the angle between the chord and the axis.*

Sol. Le the conic be $l/r = 1 + e \cos \theta$ ...(i)

Since the focal chord SP' makes an angle α with the axis, vectorial angles of P and P' must be α and $\pi + \alpha$ respectively.

Now equation of the tangent at $P(\alpha)$ is

$$\frac{l}{r} = \cos(\theta - \alpha) + e \cos \theta$$

i.e., $l = r \cos \theta \cdot (e + \cos \alpha) + r \sin \theta \cdot \sin \alpha$

i.e., $x(e + \cos \alpha) + y \sin \alpha = l$, on changing to cartesians.

$$\therefore \quad m_1 = \text{gradient of tangent at } P(\alpha) = -\frac{e + \cos \alpha}{\sin \alpha}$$

For P', writing $\pi + \alpha$ for α in the above result, we have

$$m_2 = \text{gradient of tangent at } P'(\pi + \alpha)$$

$$= -\frac{e + \cos(\pi + \alpha)}{\sin(\pi + \alpha)} = \frac{e - \cos \alpha}{\sin \alpha}$$

If ϕ = angle between the tangents at P and P', then

$$\tan \phi = \frac{m_1 \sim m_2}{1 + m_1 m_2} = \frac{\dfrac{e - \cos \alpha}{\sin \alpha} + \dfrac{e + \cos \alpha}{\sin \alpha}}{1 - \dfrac{e - \cos \alpha}{\sin \alpha} \cdot \dfrac{e + \cos \alpha}{\sin \alpha}} = \frac{2e \sin \alpha}{\sin^2 \alpha - (e^2 - \cos^2 \alpha)} = \frac{2e \sin \alpha}{1 - e^2}$$

Hence, $\phi = \tan^{-1}\left(\dfrac{2e \sin \alpha}{1 - e^2}\right)$.

4. *Show that the director circle of the conic*

$$\frac{l}{r} = 1 + e \cos \theta \text{ is } r^2 (1 - e^2) - 2elr \cos \theta - 2l^2 = 0.$$

Sol. The director circle of the conic is the locus of the point of intersection of perpendicular tangents. Now, tangents to the given conic at the points α and β are

$$\frac{l}{r} = \cos(\theta - \alpha) + e \cos \theta \qquad \text{...(i)}$$

and $$\frac{l}{r} = \cos(\theta - \beta) + e \cos \theta. \qquad \text{...(ii)}$$

These, when transformed to cartesian coordinates, become

$$l = x(e + \cos \alpha) + y \sin \alpha \qquad \text{...(iii)}$$

and $$l = x(e + \cos \beta) + y \sin \beta. \qquad \text{...(iv)}$$

The m's (slopes) of these straight lines are given by

$$m_1 = -\frac{e + \cos \alpha}{\sin \alpha} \quad \text{and} \quad m_2 = -\frac{e + \cos \beta}{\sin \beta}.$$

The tangents (i) and (ii) will be perpendicular if $m_1 m_2 = -1$,

i.e., $$\left(-\frac{e + \cos \alpha}{\sin \alpha}\right)\left(-\frac{e + \cos \beta}{\sin \beta}\right) = -1$$

i.e., $(e + \cos \alpha)(e + \cos \beta) + \sin \alpha \sin \beta = 0$

Polar Equations

i.e., $\quad e^2 + e(\cos\alpha + \cos\beta) + \cos\alpha\cos\beta + \sin\alpha\sin\beta = 0$

i.e., $\quad e^2 + 2e\cos\dfrac{\alpha+\beta}{2}\cos\dfrac{\alpha-\beta}{2} + \cos(\alpha-\beta) = 0$

i.e., $\quad e^2 + 2e\cos\dfrac{\alpha+\beta}{2}\cos\dfrac{\alpha-\beta}{2} + 2\cos^2\dfrac{\alpha-\beta}{2} - 1 = 0.$ ...(v)

Now in order to find the locus of the intersection of (i) and (ii), we must eliminate α and β between (i), (ii) and (v).

Subtracting (ii) from (i), we have

$$\cos(\theta - \alpha) = \cos(\theta - \beta) \Rightarrow \cos(\theta - \alpha) = \cos(\theta - \beta) \text{ or } \cos(\beta - \theta)$$

$\therefore \quad \theta - \alpha = (\theta - \beta) \text{ or } (\beta - \theta) \quad\quad$ since $\cos A = \cos(-A)$

$\Rightarrow$ either $\quad\quad \theta - \alpha = \theta - \beta \text{ or } \alpha = \beta$

or $\quad\quad\quad \theta - \alpha = \beta - \theta \text{ or } \theta = \dfrac{\alpha+\beta}{2}$

since $\alpha \ne \beta$, we have $\theta = \dfrac{\alpha+\beta}{2}$

Substituting this in (i), we obtain

$$\dfrac{l}{r} = \cos\left(\dfrac{\alpha+\beta}{2} - \alpha\right) + e\cos\theta$$

i.e., $\quad \cos\dfrac{\alpha-\beta}{2} = \dfrac{l}{r} - e\cos\theta.$ ...(vii)

Eliminating α, β between (v), (vi) and (vii), we obtain

$$e^2 + 2e\cos\theta\left(\dfrac{l}{r} - e\cos\theta\right) + 2\left(\dfrac{l}{r} - e\cos\theta\right)^2 - 1 = 0$$

i.e., $\quad r^2(e^2 - 1) + 2er\cos\theta(l - er\cos\theta) + 2(l - er\cos\theta)^2 = 0$

i.e., $\quad r^2(1 - e^2) + 2elr\cos\theta - 2l^2 = 0.$

5. Show that the two conics

$$l_1/r = 1 + e_1\cos\theta \quad\quad \text{and} \quad\quad l_2/r = 1 + e_2\cos(\theta - \alpha)$$

will touch one another if

$$l_1^2(1 - e_2)^2 + l_2^2(1 - e_1)^2 = 2l_1l_2(1 - e_1e_2\cos\alpha).$$

Sol. Let the given conic touch one another at the point whose vectorial angle is β. The equations of the tangents at the common point β to the two conics are

$$l_1/r = \cos(\theta - \beta) + e_1\cos\theta \quad\quad\quad\quad \text{...(i)}$$

and $\quad l_2/r = \cos(\theta - \beta) + e_2\cos(\theta - \alpha) \quad\quad\quad\quad$ ...(ii)

The equations (i) and (ii) may be written as

$$l_1/r = (\cos\beta + e_1)\cos\theta + \sin\beta\sin\theta \quad\quad\quad\quad \text{...(iii)}$$

and $\quad l_2/r = (\cos\beta + e_2\cos\alpha)\cos\theta + (\sin\beta + e_2\sin\alpha)\sin\theta \quad$ ...(iv)

The equations (iii) and (iv) should be identical. Hence comparing them, we get

$$\dfrac{l_1}{l_2} = \dfrac{\cos\beta + e_1}{\cos\beta + e_2\cos\alpha} = \dfrac{\sin\beta}{(\sin\beta + e_2\sin\alpha)}$$

or $\quad (l_1 - l_2)\cos\beta = -l_1e_2\cos\alpha + l_2e_1$

and
$$(l_1 - l_2) \sin \beta = l_1 e_2 \sin \alpha$$
Squaring and adding, we get
$$(l_1 - l_2)^2 = l_1^2 e_2^2 + l_2^2 e_1^2 - 2l_1 l_2 e_1 e_2 \cos \alpha$$
or
$$l_1^2 (1 - e_2^2) + l_2^2 (1 - e_1^2) = 2l_1 l_2 (1 - e_1 e_2 \cos \alpha)$$

EXERCISES

1. If the tangents at the extremities of a chord through the focus S of a hyperbola meet the focal axis at P and P', prove that $\dfrac{1}{SP} + \dfrac{1}{SP'} = $ constant.

2. Find the condition so that the straight line $\dfrac{l}{r} = A \cos \theta + B \sin \theta$ may be a tangent to the conic $l/r = 1 + e \cos (\theta - \theta')$.

3. If the tangent at any point P of the conic $\dfrac{l}{r} = 1 - e \cos \theta$ meets the directrix in K, then PSK is a right angle where S is the focus.

4. Prove that portion of the tangent intercepted between the conic and the directrix subtends a right angle at the corresponding focus.

5. PSP' is a focal chord of a conic. Prove that the tangent at P and P' intersects on the directrix.

6. Prove that the tangents drawn at the extremities of a focal chord of a parabola intersect at right angles.

7. If the tangent from P to the conic $\dfrac{l}{r} = 1 + e \cos \theta$ subtends a fixed angle β at the focus S, prove that the locus of the middle point of SP is a conic of eccentricity $e \sec \beta$.

8. The tangents at three points P, Q, R of a parabola form a triangle ABC. Show that $SP \cdot SQ \cdot SR = SA \cdot SB \cdot SC$.

9. A focal chord $P'SP$ of an ellipse is inclined at an angle α to the major axis. Show that the perpendicular from the focus of the tangent at P makes an angle $\tan^{-1} \dfrac{\sin \alpha}{e + \cos \alpha}$ with the axes.

10. A conic is described having the same focus and eccentricity as the conic $l/r = 1 + e \cos \theta$ and the two conics touch at the point $\theta = \alpha$. Prove that the length of its latus rectum is
$$\dfrac{2l (1 - e^2)}{1 + 2e \cos \alpha + e^2}$$
and that the angle between their axes is $2 \tan^{-1} \left(\dfrac{e + \cos \alpha}{\sin \alpha} \right)$.

11. Prove that the equation of the auxilliary circle of the conic $l/r = 1 + e \cos \theta$ is
$$r^2 (e^2 - 1) - 2elr \cos \theta + l^2 = 0.$$

Polar Equations

Hint. The auxillary circle is the locus of the foot of the pependicular from the focus upon any tangent at to the conic.

12. If PQ is the chord of contact of tangents drawn from a point T to a parabola whose focus is S, prove that $ST^2 = SP \cdot SQ$.

13. A chord subtends a constant angle 2λ at the focus of the conic
$$\frac{l}{r} = 1 + e\cos\theta,$$
Prove that the locus of the foot of the perpendicular on it from the focus is
$$r^2(e^2 - \sec^2\lambda) - 2elr\cos\theta + l^2 = 0.$$

14. A chord of the conic $\frac{l}{r} = 1 - e\cos\theta$ subtends a constant angle 2α at the focus. Prove that the locus of the point where it meets the internal bisector of the angle 2α is the conic section
$$\frac{l\cos\alpha}{r} = 1 - e\cos\alpha\cos\theta.$$

15. Find the locus of the poles of the chords of the conic
$$l/r = 1 + e\cos\theta$$
which subtend a constant angle 2α at the focus.

Hint. The pole of any chord PQ is same as the point of intersection of the tangents to the conic at P and Q.

16. S is a focus of a conic and PQ any chord subtending a right angle at S. Prove that the locus of intersection of tangents at P and Q is a conic whose focus is S. Find the latus rectum and eccentricity of the latter conic.

17. (a) Show that a chord of a conic which subtends a constant angle at a focus of the conic touches another conic.
    (b) Prove that the chords of a conic which subtend a constant angle at a focus envelope another fixed conic.

3.10 POLAR

To find the equation of the polar of any point (r_1, θ_1) with respect to the conic section
$$\frac{l}{r} = 1 + e\cos\theta. \qquad ...(i)$$

We know that the polar of the point $T(r_1, \theta_1)$ is same as the chord of contact of tangents drawn from the point T to the conic.

Let the vectorial angles of the points of contact of tangents from T be α and β.

Now the equations of the tangents at α and β to the conic (i) are
$$\frac{l}{r} = \cos(\theta - \alpha) + e\cos\theta$$
and
$$\frac{l}{r} = \cos(\theta - \beta) + e\cos\theta.$$

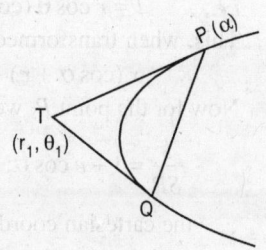

These tangents must pass through $T(r_1, \theta_1)$, so that

$$\frac{l}{r_1} = \cos(\theta_1 - \alpha) + e\cos\theta_1 \qquad \ldots(ii)$$

and
$$\frac{l}{r_1} = \cos(\theta_1 - \beta) + e\cos\theta_1. \qquad \ldots(iii)$$

Subtracting (iii) from (ii), we have
$$\cos(\theta_1 - \alpha) = \cos(\theta_1 - \beta)$$

Therefore, $\theta_1 - \alpha = -(\theta_1 - \beta)$, since $\alpha \neq \beta$

i.e.,
$$\frac{1}{2}(\alpha + \beta) = \theta_1. \qquad \ldots(iv)$$

Substituting this value of θ_1 in (ii) [or (iii)], we have
$$\frac{l}{r_1} = \cos\left(\frac{\alpha+\beta}{2} - \alpha\right) + e\cos\theta_1$$

so that
$$\cos\frac{\beta-\alpha}{2} = \frac{l}{r_1} - e\cos\theta_1. \qquad \ldots(v)$$

Also the equation to the line joining the points α and β is
$$\frac{l}{r} = \sec\frac{\beta-\alpha}{2}\cos\left(\theta - \frac{\alpha+\beta}{2}\right) + e\cos\theta$$

i.e.,
$$\left(\frac{l}{r} - e\cos\theta\right)\cos\frac{\beta-\alpha}{2} = \cos\left(\theta - \frac{\alpha+\beta}{2}\right)$$

Hence, from (ii) and (v), the equation to the polar of (r_1, θ_1) is
$$\left(\frac{l}{r} - e\cos\theta\right)\left(\frac{l}{r_1} - e\cos\theta_1\right) = \cos(\theta - \theta_1).$$

Note. The point $T(r_1, \theta_1)$ is called the *pole* of the chord PQ (See Fig.)

3.11 NORMAL TO A CONIC

To find the equation of the normal at the point α to the conic $\frac{l}{r} = 1 + e\cos\theta$.

The equation of the tangnet at the point α is

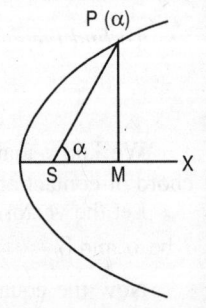

$$\frac{l}{r} = \cos(\theta - \alpha) + e\cos\theta$$

i.e., $l = r(\cos\theta\cos\alpha + \sin\theta\sin\alpha) + er\cos\theta$

i.e., $l = r\cos\theta(\cos\alpha + e) + r\sin\theta\sin\alpha.$

This, when transformed to cartesian coordinates becomes
$$x(\cos\alpha + e) + y\sin\alpha = l. \qquad \ldots(i)$$

Now for the point P, we have
$$\frac{l}{SP} = 1 + e\cos\alpha; \quad \text{giving } SP = \frac{l}{1+e\cos\alpha}$$

∴ the cartesian coordinates of P are $(SP\cos\alpha, SP\sin\alpha)$,

i.e.,
$$\left(\frac{l\cos\alpha}{1+e\cos\alpha}, \frac{l\sin\alpha}{1+e\cos\alpha}\right), \text{ as } SP = \frac{l}{1+e\cos\alpha}$$

Polar Equations

Now, *normal* at the point $P(\alpha)$ to the given conic is a straight line passing through $P(\alpha)$ and perpendicular to the tangent (i).
But, any line perpendicular to (i) is

$$y(\cos\alpha + e) - x\sin\alpha = \lambda. \qquad \ldots(ii)$$

If this passes through P, then

$$\frac{l\sin\alpha}{1+e\cos\alpha}(\cos\alpha + e) - \frac{l\cos\alpha}{1+e\cos\alpha}\sin\alpha = \lambda \quad \text{giving} \quad \lambda = \frac{el\sin\alpha}{1+e\cos\alpha}.$$

Substituting this in (ii), the required equation of the *normal* in cartesian coordinates is

$$y(\cos\alpha + e) - x\sin\alpha = \frac{el\sin\alpha}{1+e\cos\alpha}$$

and when transformed to polar coordinates, this becomes

$$r\sin\theta(\cos\alpha + e) - r\cos\theta\sin ga = \frac{el\sin\alpha}{1+e\cos\alpha}$$

i.e., $$\sin(\theta - \alpha) + e\sin\theta = \frac{e\sin\alpha}{1+e\cos\alpha}\cdot\frac{l}{r}.$$

EXAMPLES

1. *Find the equation of the chord of contact of tangents drawn from a point* (r_1, θ_1) *to the conic* $\dfrac{l}{r} = 1 + e\cos\theta.$

Sol. The equation of the chord of contact of tangents drawn from a point $T(r_1, \theta_1)$ to the given conic is same as the polar of T with respect to the conic. Hence the equation to the required chord of contact is

$$\left(\frac{l}{r} - e\cos\theta\right)\left(\frac{l}{r_1} - e\cos\theta_1\right) = \cos(\theta - \theta_1).$$

2. *If the normals at* α, β, γ *on* $\dfrac{l}{r} = 1 + \cos\theta$ *meet in the point* (ρ, ϕ) *show that* $2\phi = \alpha + \beta + \gamma.$

Sol. The equation of the parabola is

$$\frac{l}{r} = 1 + \cos\theta \qquad \ldots(i)$$

The normal at any point θ_1 is

$$\frac{l\sin\theta_1}{r(1+\cos\theta_1)} = \sin\theta + \sin(\theta - \theta_1) \qquad [\because e = 1]$$

If this normal passes through the point (ρ, ϕ) then

$$\frac{l\sin\theta_1}{\rho(1+\cos\theta_1)} = \sin\phi + \sin(\phi - \theta_1)$$

or $$\frac{2l\sin\frac{\theta_1}{2}\cos\frac{\theta_1}{2}}{\rho\cdot 2\cos^2\frac{\theta_1}{2}} = \sin\phi(1+\cos\theta_1) - \cos\phi\sin\theta_1$$

or $$\left(\frac{l}{\rho}\right)\tan\frac{1}{2}\theta_1 = \sin\phi(2\cos^2\theta_1) - \cos\phi\left(2\sin\frac{\theta_1}{2}\cos\frac{\theta_1}{2}\right)$$

or
$$\left(\frac{l}{\rho}\right) \tan \frac{\theta_1}{2} \left\{\sec^2 \frac{\theta_1}{2}\right\} = 2 \sin \phi - 2 \cos \phi \tan \frac{\theta_1}{2}$$

or
$$\left(\frac{l}{\rho}\right) \tan \frac{\theta_1}{2} \left\{1 + \tan^2 \frac{\theta_1}{2}\right\} = 2 \sin \phi - 2 \cos \phi \tan \frac{\theta_1}{2}$$

or
$$\left(\frac{l}{\rho}\right) \tan^3 \frac{\theta_1}{2} + \left\{\left(\frac{l}{\rho}\right) + 2 \cos \phi\right\} \tan \frac{\theta_1}{2} - 2 \sin \phi = 0 \qquad \text{...(iii)}$$

This is cubic equation is $\tan \frac{1}{2} \theta_1$ and hence there are three values of θ_1 on the parabola (i), the normals at which pass through (ρ, ϕ). These are given as α, β and γ.

∴ The roots of (ii) are $\tan\left(\frac{1}{2}\alpha\right)$, $\tan\left(\frac{1}{2}\beta\right)$ and $\tan\left(\frac{1}{2}\gamma\right)$

∴ From (ii) $s_1 \equiv \tan \frac{1}{2}\alpha + \tan \frac{1}{2}\beta + \tan \frac{1}{2}\gamma = 0,$

$$s_2 \equiv \tan \frac{1}{2}\alpha \tan \frac{1}{2}\beta + \tan \frac{1}{2}\beta \tan \frac{1}{2}\gamma + \tan \frac{1}{2}\gamma \tan \frac{1}{2}\alpha$$

$$= \frac{(l/\rho) + 2 \cos \phi}{(l/\rho)} = (l + 2\rho \cos \phi)/l$$

$$s_3 = \tan \frac{\alpha}{2} \tan \frac{\beta}{2} \tan \frac{\gamma}{2} = \frac{(2 \sin \phi)}{(l/\rho)}$$

$$= (2\rho \sin \phi)/l$$

Now,
$$\tan\left(\frac{\alpha}{2} + \frac{\beta}{2} + \frac{\gamma}{2}\right) = \frac{s_1 - s_3}{1 - s_2} = \frac{0 - \left(\frac{2\rho \sin \phi}{l}\right)}{1 - \frac{(l + 2\rho \cos \phi)}{l}} = \tan \phi$$

or $\quad \frac{1}{2}(\alpha + \beta + \gamma) = \phi \quad$ or $\quad \alpha + \beta + \gamma = 2\phi.$

3. *Find the locus of the pole of a chord of the conic $l/r = 1 + e \cos \theta$ which subtends a constant angle 2α at the focus.*

Sol. Let the chord of the conic,
$$l/r = 1 + e \cos \theta \qquad \text{...(i)}$$
be PQ.

Let the vectorial angles of P and Q be $\lambda - \alpha$ and $\lambda + \alpha$ respectively, so that
$$\angle QSP = 2\alpha.$$

The equation of the tangents to (i) at the points P and Q are respectively,
$$l/r = e \cos \theta + \cos(\theta - \lambda + \alpha) \qquad \text{...(ii)}$$
and
$$l/r = e \cos \theta + \cos(\theta - \lambda - \alpha). \qquad \text{...(iii)}$$

Since, the pole of the chord PQ is the point of intersection of the tangents at P and Q, hence the locus of the pole is obtained by eliminating λ between (ii) and (iii).

Subtracting (iii) and (ii) we get,
$$\cos(\theta - \lambda + \alpha) = \cos(\theta - \lambda - \alpha).$$

But since, $\alpha \neq 0.$

Polar Equations

Therefore, $\cos(\theta - \lambda + \alpha) = \cos(-\theta + \lambda + \alpha)$ [$\because \cos(-\theta) = \cos\theta$]

$\Rightarrow \quad \theta - \lambda + \alpha = -\theta + \lambda + \alpha$

$\Rightarrow \quad 2\alpha = 2\theta$

$\Rightarrow \quad \lambda = \theta.$

Substituting it in (ii), we get $l/r = e\cos\theta + \cos\alpha$

$\Rightarrow \quad \dfrac{l\sec\alpha}{r} = 1 + (e\sec\alpha)\cos\theta$

This is the required locus which is a conic of latus rectum $2l\sec\alpha$, focus being S.

EXERCISES

1. Given the focus and directrix of a conic, show that the polar of a given point with respect to it passes through a fixed point.

   [**Hint.** Refer to the second fig. of Sec. 3.6. Let S be the given focus and ZM the directrix. Also let $SZ = k$. Then by definition $l = SL = e \cdot SZ = ek$.

   Therefore, the polar of any point $P(r_1, \theta_1)$ with respect to the conic $l/r = 1 + e\cos\theta$ is (noting that $l = ek$)

   $$e^2\left(\dfrac{k}{r} - \cos\theta\right)\left(\dfrac{k}{r_1} - \cos\theta_1\right) = \cos(\theta - \theta_1).$$

   This line passes through the point of intersection of the lines

   $$\cos(\theta - \theta_1) = 0 \quad \text{and} \quad \dfrac{k}{r} - \cos\theta = 0.$$

   The first equation gives $\theta = \theta_1 - \dfrac{\pi}{2}$

   Thus the polar of P passes through the fixed point $r = k\csc\theta_1$, $\theta = \theta_1 - \dfrac{\pi}{2}$

   Geometrically this is the point where the line through the focus S and perpendicular to SP meets the directrix.

2. If a normal is drawn at one extremity of the latus rectum of the conic $l/r = 1 + e\cos\theta$, prove that the distance from the focus of the other point in which it meets the conic is

   $$\dfrac{1 + 3e^2 + e^4}{1 + e^2 - e^4}\, l.$$

3. If the normals at three points α, β, γ of the parabola

   $$\dfrac{2a}{r} = 1 - \cos\theta \quad \text{or} \quad r = a\csc^2\dfrac{\theta}{2}$$

   meet in a point whose vectorial angle is δ, prove that $2\delta = \alpha + \beta + \gamma - \pi$.

MISCELLENEOUS EXAMPLES

1. *Two conics have a common focus, prove that the two of their common chords pass through the intersection of their directrices.*

Sol. Taking the common focus as the pole, the equation of the two conics can be taken as

$$l/r = 1 + e \cos \theta \qquad \text{...(i)}$$

and
$$L/r = 1 + E \cos(\theta - \lambda). \qquad \text{...(ii)}$$

The equations of their directrices are respectively

$$l/r = e \cos \theta \qquad \text{...(iii)}$$

and
$$L/r = E \cos(\theta - \lambda). \qquad \text{...(iv)}$$

Subtracting (i) from (ii), the equation of the common chord is

$$\frac{L-l}{r} = E \cos(\theta - \lambda) - e \cos \theta. \qquad \text{...(v)}$$

Also subtracting (iii) from (iv), we get

$$\frac{L-l}{r} = E \cos(\theta - \lambda) - e \cos \theta$$

which is same as (v). Hence the common chord of (i) and (ii) passes throguh the intersection of the directrices.

Again, the equation of the conic (i) can be written as

$$\frac{l}{r} = -1 + e \cos \theta \qquad \text{...(vi)}$$

Adding (ii) and (vi), the equation of the other common chord is

$$\frac{L+l}{r} = E \cos(\theta - \lambda) + e \cos \theta. \qquad \text{...(vii)}$$

Also adding (iii) and (iv), we get

$$\frac{L+l}{r} = E \cos(\theta - \lambda) + e \cos \theta$$

which is same as (vii). Hence the second common chord of the two conics also passes through the intersection of the directrices.

2. *Prove the equation of a pair of tangents from a chord r_1, θ_1 to the conic $l/r = 1 + e \cos \theta$ is*

$$\left\{ \left(\frac{l}{r} - e \cos \theta \right)^2 - 1 \right\} \left\{ \left(\frac{l}{r_1} - e \cos \theta_1 \right)^2 - 1 \right\}$$

$$= \left\{ \left(\frac{l}{r} - e \cos \theta \right) \left(\frac{l}{r_2} - e \cos \theta_1 \right) - \cos(\theta - \theta_1) \right\}^2.$$

Sol. For convenience, let

$$S \equiv \frac{l}{r} - e \cos \theta \quad \text{and} \quad S_1 \equiv \frac{1}{r_1} - e \cos \theta_1.$$

Now tangent at any point 'α' is

$$\frac{l}{r} = \cos(\theta - \alpha) + e \cos \theta \quad \text{or} \quad S = \cos(\theta - \alpha). \qquad \text{...(i)}$$

If this passes through (r_1, θ_1), then

$$\frac{l}{r_1} = \cos(\theta_1 - \alpha) + e \cos \theta_1 \quad \text{or} \quad S_1 = \cos(\theta_1 - \alpha). \qquad \text{...(iii)}$$

Polar Equations

The required equation to the pair of tangents from (r_1, θ_1) will be obtained by eliminating α between (i) and (ii),

From (i) and (ii), we have

$$(S^2 - 1)(S_1^2 - 1) = \{\cos^2(\theta - \alpha) - 1\}\{\cos^2(\theta_1 - \alpha) - 1\}$$
$$= \{-\sin^2(\theta - \alpha)\}\{-\sin^2(\theta_1 - \alpha)\}$$
$$= \sin^2(\theta - \alpha)\sin^2(\theta_1 - \alpha), \qquad \ldots(iii)$$

and
$$SS_1 - \cos(\theta - \theta_1) = \cos(\theta - \alpha)\cos(\theta_1 - \alpha) - \cos(\theta - \theta_1)$$
$$= \frac{1}{2}[\cos(\theta + \theta_1 - 2\alpha) + \cos(\theta - \theta_1)] - \cos(\theta - \theta_1)$$
$$= \frac{1}{2}[\cos(\theta + \theta_1 - 2\alpha) - \cos(\theta - \theta_1)]$$
$$= \sin(\theta - \alpha)\sin(\alpha - \theta_1). \qquad \ldots(iv)$$

Hence from (iii) and (iv), the required equation to the pair of tangents from (r_1, θ_1) is

$$(S^2 - 1)(S_1^2 - 1) = \{SS_1 - \cos(\theta - \theta_1)\}^2.$$

EXERCISES

1. The tangents at two points P and Q of a conic meet in T and S is the focus, prove that if the conic be central, then

$$\frac{1}{SP \cdot SQ} - \frac{1}{ST^2} = \frac{1}{b^2}\sin^2\frac{PSQ}{2} \text{ where } b \text{ is the semi-minor axis.}$$

2. Show that two points having vectorial angles α and β on the conic

$$\frac{l}{r} = 1 + e\cos\theta$$

will be ends of a diameter, if $\tan\dfrac{\alpha}{2}\tan\dfrac{\beta}{2} = \dfrac{e+1}{e-1}$.

[Hint. The tangents at α and β must be paralell.]

3. PQ is a conic one of whose foci is S, and PQ passes through a fixed point O. Show that the product $\tan\dfrac{PSO}{2}\tan\dfrac{QSO}{2}$ is constant.

[Hint. Let the vectorial angles of P and Q be α and β respectively. If the conic be

$$l/r = 1 + e\cos\theta,$$

then equation to PQ is $\dfrac{l}{r} = \sec\dfrac{\alpha - \beta}{2}\cos\left(\theta - \dfrac{\alpha + \beta}{2}\right) + e\cos\theta.$

If this passes through the fixed point $O(\rho, \phi)$, then

$$\frac{l}{\rho} = \sec\frac{\alpha - \beta}{2}\cos\left(\phi - \frac{\alpha + \beta}{2}\right) + e\cos\phi \Rightarrow \frac{l}{\rho} = \frac{\cos\left(\phi - \dfrac{\alpha + \beta}{2}\right)}{\cos\left(\dfrac{\alpha - \beta}{2}\right)} + e\cos\phi \text{ so that}$$

$$\frac{\cos\left\{\phi - \frac{1}{2}(\alpha+\beta)\right\}}{\cos\frac{1}{2}(\alpha-\beta)} = \frac{l}{\rho} - e\cos\phi = \text{const} \quad \text{or} \quad \frac{\cos\left\{\left(\frac{\phi-\alpha}{2}\right)+\left(\frac{\phi-\beta}{2}\right)\right\}}{\cos\left\{\left(\frac{\phi-\alpha}{2}\right)-\left(\frac{\phi-\beta}{2}\right)\right\}} = \text{constant.}$$

or $\quad \dfrac{\cos(A-B)-\cos(A+B)}{\cos(A-B)+\cos(A+B)} = \text{constant.} \quad$ where $\dfrac{\phi-\alpha}{2} = A$ and $\dfrac{\phi-\beta}{2}$

or $\quad \dfrac{\cos(A-B)-\cos(A+B)}{\cos(A-B)+\cos(A+B)} = \text{constant}$

or $\quad \dfrac{2\sin A \sin B}{2\cos A \cos B} = \text{constant} \quad \text{or} \quad \tan A \tan B = \text{constant}$

or $\quad \tan\dfrac{\phi-\beta}{2} \tan\dfrac{\phi-B}{2} = \text{constant}$

4. One of two conics having a common focus is turned about the focus. Prove that the common chord touches another conic which has the focus and the eccentricity equal to the ratio of the eccentricities of the previous two.

[**Hint.** With the common focus as the pole, let the equations to the two conics be

$$l/r = 1 - e\cos\theta \qquad \ldots(\text{i})$$

and $\qquad L/r = 1 - E\cos(\theta - \lambda). \qquad \ldots(\text{ii})$

Subtracting (ii) from (i), equation to the common chord is

$$(l-L)/r = E\cos(\theta-\lambda) - e\cos\theta$$

i.e., $\qquad \dfrac{(l-L)/E}{r} = \cos(\theta-\lambda) - \dfrac{e}{E}\cos\theta.$

This line is always tangent to the conic whose eccentricity is e/E. This prove the assertion.]

□□□

GEOMETRY

(b) Three-Dimensional

Plane. The straight line and the plane. Sphere, Cone, Cylinder. Central Conicoids. Paraboloids. Plane Sections of Conicoids. Generating lines. Confocal Conicoids. Reduction of Second degree equations.

1
System of Co-ordinates

INTRODUCTION

In a plane the position of a point is determined by an ordered pair (x, y) of real numbers, obtained with reference to two straight lines in the plane generally at right angles. The position of a point in *space* is, however, determined by an ordered triad (x, y, z) of real numbers. We now proceed to explain as to how this is done.

1.1 CO-ORDINATES OF A POINT IN SPACE

Let $X'OX, Z'OZ$ be two perpendicular straight lines determining the XOZ-plane. Through O, their point of intersection, called the *origin*, draw the line $Y'OY$ perpendicular to the XOZ-plane so that we have three mutually perpendicular straight lines

$$X'OX, Y'OY, Z'OZ$$

known as *Rectangular Co-ordinate Axes**. The positive directions of the axes are indicated by arrow heads. these three axes, taken in pairs, determine the three planes,

$$XOY, YOZ \text{ and } ZOX$$

or briefly the XY, YZ, ZX planes mutually at right angles, known as *Rectangular Co-ordinate Planes*.

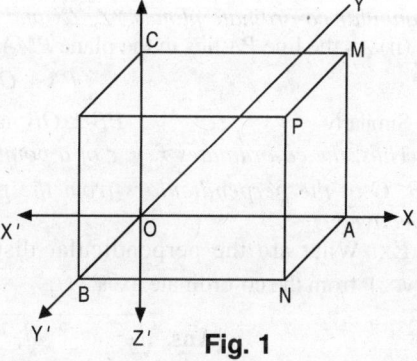

Fig. 1

Through *any* point, P, in space, draw three planes parallel to the three co-ordinate planes (being also perpendicular to the corresponding axes) to meet the axes in A, B, C.

Let $\qquad OA = x, \quad OB = y \text{ and } OC = z.$

These three numbers x, y, z taken in this order determined by the point P, are called the co-ordinates of the point P.

We refer to the ordered triad (x, y, z) formed of the co-ordinates of the point P as the point P itself.

Any one of these x, y, z will be positive or negative according as it is measured from O, along the corresponding axis, in the positive or the negative direction.

Conversely, given an ordered triad (x, y, z) of numbers, we can find the point whose co-ordinates are x, y, z. To do this, we proceed as follows:

(i) Measure OA, OB, OC along OX, OY, OZ equal to x, y, z respectively.
(ii) Through the points A, B, C draw planes paralell to the co-ordinate planes YZ, ZX, XY respectively.

The point of intersection of these three planes is the required point P.

---

* The plane XOZ containing the lines $X'OX$ and $Z'OZ$ may be imagined as the plane of the paper; the line OY as pointing towards the reader and OY' behind the paper.

Note. The three co-ordinate planes divide the whole space in eight compartments which are known as *eight octants* and since each of the co-ordinates of a point may be positive or negative, there are $2^3 (= 8)$ points whose co-ordinates have the same numerical values and which lie in the eight octants, one in each.

1.1.1 Further Explanation about Co-ordinates

In § 1.1 above, we have learnt that in order to obtain the co-ordinates of a point P, we have to draw three planes through P respectively parallel to the three co-ordinate planes. The three planes through P and the three co-ordinate planes determined a parallelopiped whose consideration leads to three other useful constructions for determining the co-ordinates of P.

The parallelopiped, in questions, has six rectangular faces consisting of three pairs of parallel planes, viz.,

$$PMAN, LCOB; PNBL, MAOC, PLCM, NBOA \qquad \text{(See fig. 1)}$$

(i) We have

$x = OA = CM = LP = $ perpendicular from P on the YZ-plane;

$y = OB = AN = MP = $ perpendicular from P on the ZX-plane;

$z = OC = AM = NP = $ perpendicular from P on the XY-plane.

Thus, the co-ordinates x, y, z of a point P, are the perpendicular distances of P from the three rectangular co-ordinate planes YZ, ZX and XY respectively.

(ii) As the line PA lies in the plane $PMAN$ which is perpendicular to the line OA^*, we have

$$PA \perp OA.$$

Similarly $\qquad PB \perp OB \quad \text{and} \quad PC \perp OC$

Thus, the co-ordinates x, y, z of a point P are also the distances from the origin O of the feet A, B, C of the perpendiculars from the point P to the co-ordinate axes $X'X, Y'Y$ and $Z'Z$ respectively.

Ex. What are the perpendicular distances of a point (x, y, z) from the co-ordinate axes?

[**Ans.** $\sqrt{y^2 + z^2}, \sqrt{z^2 + x^2}, \sqrt{x^2 + y^2}$]

(iii) We have (Fig. 1)

$NP = AM = OC = z;$

$AN = OB = y;$

$OA = x.$

Thus, (Fig. 2) if we draw the line PN perpendicular to the XY-plane meeting it at N and the line NA parallel to the line, OY meeting OX at A, we have

$$OA = x, AN = y, NP = z.$$

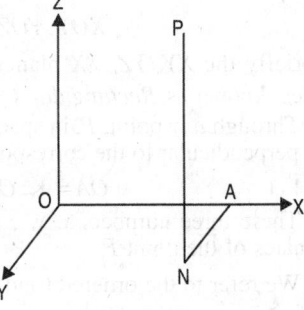

Fig. 2

EXAMPLE

In which octant the following points lie

(i) $(-1, -2, -3)$, (ii) $(a, b, -c)$, (iii) (a, b, c), (iv) $(-a, -b, c)$?

Sol. (i) Since all the three co-ordinates are negative hence, $(-1, -2, -3)$ lies in octant $OX' Y'Z'$ (Fig. 1).

(ii) Similarly $(a, b, -c)$ is a point in the octant $OXYZ'$.

(iii) It is a point in the octant $OXYZ$.

(iv) It is a point in the octant $OX'Y'Z$.

---

* A line perpendicular to a plane is perpendicular to every line is the plane.

System of Co-ordinates

EXERCISES

1. In Fig. 1 write down the co-ordinates of the point $A, B, C; L, M, N$ when the co-ordinates of P are (x, y, z).
2. Show that for every point (x, y, z) on the ZX-plane, $y = 0$.
3. Show that for every point (x, y, z) on the Y-axis, $x = 0, z = 0$.
4. What is the locus of a point (x, y, z) for which
   (i) $x = 0$, (ii) $y = 0$, (iii) $z = 0$,
   (iv) $x = a$, (v) $y = b$, (vi) $z = c$.
5. What is the locus of a point (x, y, z) for which
   (i) $y = 0, z = 0$, (ii) $z = 0, x = 0$, (iii) $x = 0, y = 0$
   (iv) $y = b, z = c$, (v) $z = c, x = a$, (vi) $x = a, y = b$.
6. P is any point (x, y, z), and α, β, γ are the angles which OP makes with X-axis, Y-axis and Z-axis respectively, show that
$$\cos\alpha = x/r, \cos\beta = y/r, \cos\gamma = z/r,$$
   where $r = OP$.
7. Find the length of the edges of the rectangular parallopiped formed by planes drawn through the points $(1, 2, 3)$ and $(4, 7, 6)$ parallel to the co-ordinate planes. [**Ans.** 3, 5, 3]

1.2 DISTANCE BETWEEN TWO POINTS

To find the distance between two given points $P(x_1, y_1, z_1)$ and $Q(x_2, y_2, z_2)$.

Through the points P, Q draw planes parallel to the co-ordinate planes to form a rectangular parallelopiped whose one diagonal is PQ.

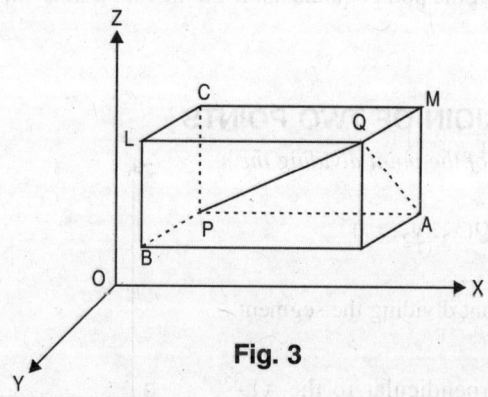

Fig. 3

Then
$$APCM, NBLQ; LCPB, QMAN; BPAN, LCMQ$$
are the three pairs of parallel faces of this parallelopiped.

Now, PA is the distance between the planes drawn through the points P and Q parallel to the YZ-plane and is, therefore, equal to the difference between their x-co-ordinates.

∴ $PA = x_2 - x_1,$
Similarly $AN = y_2 - y_1,$ ...(i)
and $NQ = z_2 - z_1.$

The line AQ lies in the plane $QMAN \perp PA$

$\Rightarrow$ $\quad AQ \perp PA$

$\Rightarrow$ $\quad PQ^2 = PA^2 + AQ^2$ ...(ii)

$\angle ANQ$ is a rt. angle

$\Rightarrow$ $\quad AQ^2 = AN^2 + NQ^2$ ...(iii)

From (i), (ii) and (iii), we obtain

$$PQ^2 = PA^2 + AQ^2 = PA^2 + AN^2 + NQ^2$$
$$= (x_2 - x_1)^2 + (y_2 - y_1)^2 + (z_2 - z_1)^2$$

Thus, the distance between the points (x_1, y_1, z_1) and (x_2, y_2, z_2) is

$$\sqrt{(x_2 - x_1)^2 + (y_2 - y_1)^2 + (z_2 - z_1)^2}$$

Cor. Distance from the origin. When P coincides with the origin O, we have $x_1 = y_1 = z_1 = 0$ so that we obtain,

$$OQ^2 = x_2^2 + y_2^2 + z_2^2$$

Note. The reader should notice the similarity of the formula obtained above for the distance between two points with the corresponding formula in plane co-ordinate geometry. Also refer 1.3.

EXERCISES

1. Find the distance between the points $(4, 3, -6)$ and $(-2, 1, -3)$. [Ans. 7]
2. Show that the point $(0, 7, 10)$, $(-1, 6, 6)$, $(-4, 9, 6)$ form an isosceles right-angled triangle.
3. Show that the three points $(-2, 3, 5)$, $(1, 2, 3)$, $(7, 0, -1)$ are collinear.
4. Show that the pints $(3, 2, 2)$, $(-1, 1, 3)$, $(0, 5, 6)$, $(2, 1, 6)$ lie on a sphere whose centre is $(1, 3, 4)$. Find also the radius of the sphere. [Ans. 3]
5. Find the co-ordinates of the point equidistant from the four points $(a, 0, 0)$, $(0, b, 0)$, $(0, 0, c)$ and $(0, 0, 0)$. $\left[\text{Ans. } \frac{1}{2}a, \frac{1}{2}b, \frac{1}{2}c\right]$

1.3 DIVISION OF THE JOIN OF TWO POINTS

To find the co-ordinates of the point dividing the segment joining the points

$P(x_1, y_1, z_1)$ and $Q(x_2, y_2, z_2)$

in the ratio $m : n$.

Let $R(x, y, z)$ be the piont dividing the segment PQ in the ratio $m : n$.

Draw PL, QM, RN perpendicular to the XY-plane.

The line PL, QM, RN clearly lie in one plane so that the points L, M, N lie in the straight line which is the intersection of this plane with the XY-plane.

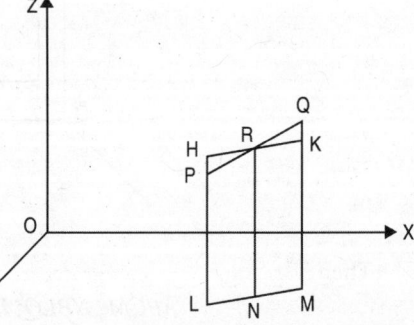

Fig. 4

The line throguh R parallel to the line LM shall lie in the same plane. Let it intersect PL and QM at H and K respectively.

The triangles HPR and QRK are similar.

$\Rightarrow \quad \dfrac{m}{n} = \dfrac{PR}{RQ} = \dfrac{PH}{KQ} = \dfrac{NR - LP}{MQ - NR} = \dfrac{z_2 - z_1}{z_2 - z}$

System of Co-ordinates

$$\Rightarrow \qquad z = \frac{mz_2 + nz_1}{m+n}.$$

Similarly, by drawing perpendiculars to the XY and YZ-planes, we obtain

$$y = \frac{my_2 + ny_1}{m+n} \quad \text{and} \quad x = \frac{mx_2 + nx_1}{m+n}$$

The point R divides PQ internally or externally according as the ratio $m:n$ is positive or negative.

Thus, the co-ordinates of the point which divides the join of the points (x_1, y_1, z_1) and (x_2, y_2, z_2) in the ratio $m:n$ are

$$\left(\frac{mx_2 + nx_1}{m+n}, \frac{my_2 + ny_1}{m+n}, \frac{mz_2 + nz_1}{m+n} \right)$$

Cor. 1. Co-ordinates of the middle point. In case R is the middle point of PQ, we have

$$m:n :: 1:1$$

$$\Rightarrow \qquad x = \frac{1}{2}(x_1 + x_2), \ y = \frac{1}{2}(y_1 + y_2), \ z = \frac{1}{2}(z_1 + z_2).$$

Cor. 2. Co-ordinates of a point on the join of two points : Putting k for m/n, we see that the co-ordinates of the point R which divides PQ in the ratio $k:1$ are

$$\left(\frac{kx_2 + x_1}{1+k}, \frac{ky_2 + y_1}{1+k}, \frac{kz_2 + z_1}{1+k} \right)$$

To every value of $k \neq -1$, ther corresponds a point R on the line PQ and to every point R on the line PQ corresponds some value of k, viz., PR/RQ.

Thus, we see that the point

$$\left(\frac{kx_2 + x_1}{1+k}, \frac{ky_2 + y_1}{1+k}, \frac{kz_2 + z_1}{1+k} \right) \qquad \ldots(i)$$

lies on the line PQ whatever value $k \neq -1$ may have and *conversely any* given point on the line PQ is obtained by giving *some* suitable value to k other than -1. This idea is sometimes expressed by saying that (i) is the *general co-ordiantes* of a point on the line joining $P(x_1, y_1, z_1)$ and $Q(x_2, y_2, z_2)$.

The set of points of the line joining the points (x_1, y_1, z_1) and (x_2, y_2, z_2) is

$$\left[\left(\frac{kx_2 + x_1}{1+k}, \frac{ky_2 + y_1}{1+k}, \frac{kz_2 + z_1}{1+k} \right); k \neq -1 \right]$$

In other words, the line joining the points (x_1, y_1, z_1) and (x_2, y_2, z_2) is the set

$$\left[\left(\frac{kx_2 + x_1}{1+k}, \frac{ky_2 + y_1}{1+k}, \frac{kz_2 + z_1}{1+k} \right); k \neq -1 \right]$$

EXAMPLES

1. *Find the ratio in which the lne joining the points* $(2, 4, 5), (3, 5, -4)$ *is divided by the xy-plane.*

Sol. Co-ordinates of a point that divides the line joining the given point in the ratio $k:1$ is

$$\left(\frac{3k+2}{k+1}, \frac{5k+4}{k+1}, \frac{-4k+5}{k+1} \right)$$

For a point on xy-plane, $z = 0$, i.e.,
$$\frac{-4k+5}{k+1} = 0, \text{ or } k = \frac{5}{4}.$$
Hence, the xy-plane divides the line in the ratio 5 : 4.

Putting $k = \frac{5}{4}$, the co-ordinates of the point are $\left(\frac{23}{9}, \frac{41}{9}, 0\right)$.

2. *Given that $P(3, 2, -4), Q(5, 4, -6), R(9, 8, -10)$ are collinear, find the ratio in which Q divides PR.*

Sol. Let Q divides PR in the ratio $k : 1$. Now x co-ordinate of Q should be
$$\frac{9k+3}{k+1} = 5, \text{ or } k = \frac{1}{2}.$$
Hence, the required ratio is 1 : 2.

3. *From any point $(1, -2, 3)$, lines are drawn to meet the sphere $x^2 + y^2 + z^2 = 4$ and they are divided in the ratio $2:3$. Prove that the points of section lie on a sphere.*

Sol. Let the line through $(1, -2, 3)$ meets the sphere in point (x_1, y_1, z_1). Hence,
$$x_1^2 + y_1^2 + z_1^2 = 4 \qquad \ldots(1)$$

Let the point (α, β, γ) divide the line joining the points $(1, -2, 3)$ and (x_1, y_1, z_1) in the ratio $2:3$.

Then
$$\alpha = \frac{2 \cdot x_1 + 3 \cdot 1}{2+3} \Rightarrow x_1 = \frac{5\alpha - 3}{2}$$

$$\beta = \frac{2 \cdot y_1 + 3(-2)}{2+3} \Rightarrow y_1 = \frac{5\beta + 6}{2}$$

$$\gamma = \frac{2 \cdot z_1 + 3 \cdot 3}{2+3} \Rightarrow z_1 = \frac{5\gamma - 9}{2}$$

From (1), we have
$$\frac{(5\alpha - 3)^2}{4} + \frac{(5\beta + 6)^2}{4} + \frac{(5\gamma - 9)^2}{4} = 4$$
$$\Rightarrow 25\alpha^2 + 25\beta^2 + 25\gamma^2 - 30\alpha + 60\beta - 90\gamma + 210 = 0$$
$$\Rightarrow 5\alpha^2 + 5\beta^2 + 5\gamma^2 - 6\alpha + 12\beta - 18\gamma + 22 = 0$$

Hence, locus of (α, β, γ) will be
$$x^2 + y^2 + z^2 - \frac{6}{5}x + \frac{12}{5}y - \frac{18}{5}z + \frac{22}{5} = 0,$$
which is a sphere.

EXERCISES

1. Find the co-ordinates of the points which divides the line joining the points $(2, -4, 3)$, $(-4, 5, -6)$ in the ratios.

(i) $(1:-4)$ and (ii) $(2:1)$. [**Ans.** (i) $(4, -7, 6)$; (ii) $(-2, 2, -3)$]

2. $A(3, 2, 0), B(5, 3, 2), C(-9, 6, -3)$ are three points forming a traingle, AD, the bisector of the angle BAC, meets BC at D. Find the co-ordinates of the point D.

$$\left[\textbf{Ans. } \frac{38}{16}, \frac{57}{16}, \frac{97}{16}\right]$$

3. Find the ratio in which the line joining the point
$$(2, 4, 5), (3, 5, -4)$$
is divided by the YZ-plane. [**Ans.** $-2:3$]

System of Co-ordinates

4. Find the ratio in which the *XY*-plane divides the join of
   $(-3, 4, -8)$ and $(5, -6, 4)$.
   Also obtain the point of intersection of the line with the plane.   [**Ans.** 2; $(7/3, -8/3, 0)$]

5. The three points $A(0, 0, 0)$, $B(2, -3, 3)$, $C(-2, 3, -3)$ are collinear. Find the ratio in which each point divides the segment joining the other two.
   [**Ans.** $AB/BC = -1/2$, $BC/CA = -2$, $CA/AB = 1$]

6. Show that the following triads of points are collinear :
   (i) $\{(2, 5, -4), (1, 4, -3), (4, 7, -6)\}$ (ii) $\{(5, 4, 2), (6, 2, -1), (8, -2, -7)\}$

7. Find the ratios in which the join of the points $(3, 2, 1)$, $(1, 3, 2)$ is divided by the locus of the equation $3x^2 - 72y^2 + 128z^2 = 3$.   [**Ans.** $-2 : 1; 1 : -2$]

8. $A(4, 8, 12)$, $B(2, 4, 6)$, $C(3, 5, 4)$, and $D(5, 8, 5)$ are four points; show that the lines AB and CD intersect.

9. Show that the point $(1, -1, 2)$, is common to the lines which join $(6, -7, 0)$ to $(16, -19, -4)$ and $(0, 3, -6)$ to $(2, -5, 10)$.

10. Show that the set of points on the plane determined by the three points $(x_1, y_1, z_1), (x_2, y_2, z_2)$ and (x_3, y_3, z_3) is
    $$\left[\left(\frac{lx_1 + mx_2 + nx_3}{l + m + n}, \frac{ly_1 + my_2 + ny_3}{l + m + n}, \frac{lz_1 + mz_2 + nz_3}{l + m + n}\right); l + m + n \neq 0\right].$$

11. Show that the centroid of the triangle with vertices $(x_r, y_r, z_r), r = 1, 2, 3$ is
    $$\left(\frac{x_1 + x_2 + x_3}{3}, \frac{y_1 + y_2 + y_3}{3}, \frac{z_1 + z_2 + z_3}{3}\right).$$

1.4 ANGLE BETWEEN TWO LINES

The meaning of the angle between two intersecting, *i.e.*, coplanar lines, is already known to the student. We now give the definition of the angle between two non-coplanar lines, also sometimes called *skew* lines.

Def. *The angle between two **non-coplanar**, i.e., non-intersecting lines is the angle between two intersecting lines drawn from any point parallel to each of the given lines.*

Note 1. To justify the definition of angle between two non-coplanar lines, as given above, it is necessary to show that this angle is independent of the position of the point through which the parallel lines are drawn, but here we simply assume this result.

Note 2. The angle between a given line and the co-ordinate axes are the angles which the line drawn through the origin parallel to the given lines makes with the axes.

1.5 DIRECTION COSINES OF A LINE

Let α, β, γ be the angles which any line makes with the positive directions of the co-ordinates axes. Then $\cos \alpha, \cos \beta, \cos \gamma$ are called the *direction cosines* of the given line and are generally denoted by l, m, n respectively.

Ex. What are the direction cosines of the axes of co-ordinates ?
[**Ans.** 1, 0, 0; 0, 1, 0; 0, 0, 1]

1.5.1 A Useful Relation

If O the origin of co-ordinates and (x, y, z) the co-ordinates of a point P, then
$$x = lr, y = mr, z = nr,$$
l, m, n being the direction cosines of the line OP and r, the length of the segment OP.

Through the point P draw the line PL perpendicular to the X-axis so that $OL = x$.

From the right-angled triangle OPL, we have

$$\frac{OL}{OP} = \cos \angle LOP \Rightarrow \frac{x}{r} = l \Rightarrow x = lr.$$

Similarly, we have

$$y = mr, \ z = nr.$$

1.6 RELATION BETWEEN DIRECTION COSINES

If l, m, n are the direction cosines of a line, then

$$l^2 + m^2 + n^2 = 1,$$

i.e., *the sum of the squres of the direction cosines of every line is one.*

Let OP be drawn through the origin parallel to the given lines so that l, m, n are the cosines of the angles which the line OP makes with the co-ordinate axes OX, OY, OZ respectively (Refer fig. 5).

Let (x, y, z) be the co-ordinates of any point P on this line.

Let $\qquad OP = r.$

We have $\qquad x = lr, \ y = mr, \ z = nr.$

Squaring and adding, we obtain

$$x^2 + y^2 + z^2 = (l^2 + m^2 + n^2) r^2$$

$\Rightarrow \qquad r^2 = OP^2 = x^2 + y^2 + z^2 = (l^2 + m^2 + n^2) r^2$

$\Rightarrow \qquad l^2 + m^2 + n^2 = 1.$

Cor. If a, b, c be three numbers *proportional* to the direction cosines l, m, n of a line, we have

$$\frac{l}{a} = \frac{m}{b} = \frac{n}{c} = \pm \frac{\sqrt{l^2 + m^2 + n^2}}{\sqrt{a^2 + b^2 + c^2}} = \pm \frac{1}{\sqrt{a^2 + b^2 + c^2}}$$

$\Rightarrow \qquad l = \pm \dfrac{a}{\sqrt{a^2 + b^2 + c^2}}, \ m = \pm \dfrac{b}{\sqrt{a^2 + b^2 + c^2}}, \ n = \pm \dfrac{c}{\sqrt{a^2 + b^2 + c^2}}$

where the same sign, positive or negative, is to be chosen throughout.

Direction Ratios : From above, we see that a set of three numbers which are proportional to the direction cosines of a line are sufficient to specify the direction of a line. Such numbers are called the **direction ratios** or **direction numbers** of the line. Thus, if a, b, c be the direction ratios of a line, its direction cosines are

$$\pm \frac{a}{\sqrt{\Sigma a^2}}, \ \pm \frac{b}{\sqrt{\Sigma a^2}}, \ \pm \frac{c}{\sqrt{\Sigma a^2}}.$$

Note. It is easy to see that if a line OP through the origin O makes angles α, β, γ with OX, OY, OZ then the line OP' obtained by producing OP backwards through O will make angles $\pi - \alpha, \ \pi - \beta, \ \pi - \gamma$ with the axes OX, OY, OZ. Thus, if

$$\cos \alpha = l, \ \cos \beta = m, \ \cos \gamma = n$$

are the direction cosines of OP, then

$$\cos (\pi - \alpha) = -l, \ \cos (\pi - \beta) = -m, \ \cos (\pi - \gamma) = -n$$

are the direction cosines of OP' i.e., of the line OP produced backwards.

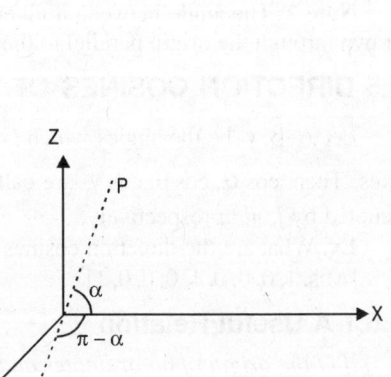

Fig. 6

System of Co-ordinates

Thus, if we ignore the two senses of a line, we can think of the direction cosines l, m, n or $-l, -m, -n$, determining the direction of one and the same line. This explains the ambiguity in the sign obtained above.

Note. The student should always make a distinction between direction cosines and direction ratios. It is only when l, m, n are direction cosines, that we have the relation

$$l^2 + m^2 + n^2 = 1.$$

EXAMPLES

1. *If α, β, γ be the angles which a line makes with the positive direction of the axes, prove that*

$$\sin^2 \alpha + \sin^2 \beta + \sin^2 \gamma = 2.$$

Sol. We have

$$l = \cos \alpha, \; m = \cos \beta, \; n = \cos \gamma$$

$$\therefore \quad \cos^2 \alpha + \cos^2 \beta + \cos^2 \gamma = 1$$

$$\Rightarrow \quad 1 - \sin^2 \alpha + 1 - \sin^2 \beta + 1 - \sin^2 \gamma = 1$$

$$\Rightarrow \quad \sin^2 \alpha + \sin^2 \beta + \sin^2 \gamma = 2.$$

2. *The direction cosines l, m, n, of two lines are connected by the relations*

$$l + m + n = 0 \quad \ldots(1)$$

$$2lm + 2ln - mn = 0 \quad \ldots(ii)$$

Find them.

Sol. We shall solve the two given equations one of which is of the first degree and the other of second degree in l, m, n.

Eliminating n between (i) and (ii), we get

$$2l^2 - lm - m^2 = 0$$

$$\Rightarrow \quad 2\left(\frac{l}{m}\right)^2 - \frac{l}{m} - 1 = 0 \quad \ldots(iii)$$

This equation gives two values of l/m implying that there are two lines. Let l_1, m_1, n_1; l_2, m_2, n_2 be the direction cosines of these lines.

The two roots of the quadratic equation (iii) in l/m are 1 and $-1/2$.

Also

$$l_1 + m_1 + n_1 = 0 \Rightarrow \frac{l_1}{m_1} + 1 + \frac{n_1}{m_1} = 0 \Rightarrow \frac{n_1}{m_1} = -2$$

$$l_2 + m_2 + n_2 = 0 \Rightarrow \frac{l_2}{m_2} + 1 + \frac{n_2}{m_2} = 0 \Rightarrow \frac{n_2}{m_2} = -\frac{1}{2}$$

Thus, we have

$$\frac{l_1}{1} = \frac{m_1}{1} = \frac{n_1}{-2} = \frac{1}{\sqrt{6}} \Rightarrow l_1 = \frac{1}{\sqrt{6}}, m_1 = \frac{1}{\sqrt{6}}, n_1 = -\frac{2}{\sqrt{6}}$$

$$\frac{l_2}{1} = \frac{m_2}{-2} = \frac{n_2}{1} = \frac{1}{\sqrt{6}} \Rightarrow l_2 = \frac{1}{\sqrt{6}}, m_2 = -\frac{2}{\sqrt{6}}, n_2 = -\frac{1}{\sqrt{6}}.$$

EXERCISES

1. 6, 2, 3 are direction ratios of a line. What are the direction cosines ?   [**Ans.** 6/7, 2/7, 3/7]

2. What are the direction cosines of lines equally inclined to the axes ? How many such lines are there ?   [**Ans.** $(1/\sqrt{3}, 1/\sqrt{3}, 1/\sqrt{3})$; 4]

3. The co-ordinates of a point P are (3, 12, 4). Find the direction cosines of the line OP.

[**Ans.** 3/13, 12/13, 4/13]

4. The direction cosines of two lines are determined by the relations
   (i) $l - 5m + 3n = 0$, $7l^2 + 5m^2 - 3n^2 = 0$;
   (ii) $l + m - n = 0$, $mn + 6ln - 12lm = 0$.

$$\text{Ans. } (i) \left[\frac{1}{\sqrt{14}}, \frac{2}{\sqrt{14}}, \frac{3}{\sqrt{14}}; -\frac{1}{\sqrt{6}}, \frac{1}{\sqrt{6}}, \frac{2}{\sqrt{6}}; \right.$$
$$(ii) \left. \frac{1}{\sqrt{26}}, \frac{3}{\sqrt{26}}, \frac{4}{\sqrt{26}}; \frac{1}{\sqrt{14}}, \frac{2}{\sqrt{14}}, \frac{3}{\sqrt{14}} \right]$$

1.7 PROJECTION ON A STRAIGHT LINE

1.7.1 Projection of a Point on a Line

The foot of the perpendicular from a given point on a given straight line is called the orthogonal projection (or simply projection) of the point on the line. This projection is the same point where the plane through the given point and perpendicular to the given line meets the line.

Thus, in Fig. 1, page 1, the point A is the projection of the point P on X-axis; also the points B and C are the projections of the point P on Y-axis and Z-axis respectively.

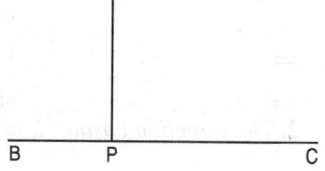

Fig. 7

1.7.2 Projection of a Segment on Another Line

The projection of a segment AB on a line CD is the segment $A'B'$ where A', B' are the projections of points A, B respectively on the line CD.

Clearly the projection $A'B'$ of the segment AB is the intercept made on CD by planes perpendicular to the line CD through the points A and B.

Ex. The co-ordinates of a point P are (x, y, z). What are the projections of the segment OP on the co-ordinate axes ?

Theorem. *The projection of a segment AB on a line CD is AB cos θ, where θ is the angle between the lines AB and CD.*

Let the planes through the points A and B perpendicular to the line CD meet the line CD in A', B' respectively so that $A'B'$ is the projection of AB. Through the point A draw a line $AP \parallel CD$ to meet the plane through the point B at P.

Now $\qquad AP \parallel CD$

$\Rightarrow \qquad \angle PAB = \theta.$

Also BP lies in the plane which is perpendicular to AP

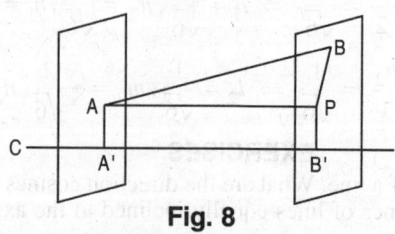

Fig. 8

$\Rightarrow \qquad \angle APB = 90°.$

Hence, $\qquad AP = AB \cos \theta$

Clearly $A'B'PA$ is a rectangle implying that
$$AP = A'B'$$

System of Co-ordinates

Hence, $A'B' = AB \cos \theta$.

Cor. Direction cosines of the join of two points.

To find the direction cosines of the line joining the two points
$$P(x_1, y_1, z_1) \text{ and } Q(x_2, y_2, z_2).$$

Let the points L, M be the feet of the perpendicular drawn from the points P, Q to the X-axis respectively so that we have
$$OL = x_1, OM = x_2$$

Projection of the segment PQ on X-axis $= LM$
$$= OM - OL = x_2 - x_1.$$

Also if l, m, n be the direction cosines of the line PQ, the projection of PQ on X-axis $= l \cdot PQ$.
$$l \cdot PQ = x_2 - x_1.$$

Similarly projecting PQ on Y-axis and Z-axis, we get
$$m \cdot PQ = y_2 - y_1,$$
$$n \cdot PQ = z_2 - z_1.$$

From these we obtain the relations*,
$$\frac{x_2 - x_1}{l} = \frac{y_2 - y_1}{m} = \frac{z_2 - z_1}{n} = PQ.$$

Thus, the direction cosines of the line joining the two points (x_1, y_1, z_1) and (x_2, y_2, z_2) are proportional to
$$x_2 - x_1, y_2 - y_1, z_2 - z_1.$$

EXAMPLE

The projection of a line on the line on the axes are 2, 3, 6. What is the length of the line?

Sol. Let PQ be the length of the line and $[l, m, n]$ be the direction cosines. Then,

projection on x-axes, $PQ \cdot l = 2$,
proejction on y-axes, $PQ \cdot m = 3$,
projection on z-axes, $PQ \cdot n = 6$.

Squaring and adding, we get
$$PQ^2(l^2 + m^2 + n^2) = 2^2 + 3^2 + 6^2$$
$$\Rightarrow PQ^2 = 49 \Rightarrow PQ = 7.$$

EXERCISES

1. Find the direction cosines of the lines joining the points
   (i) $(4, 3, -5)$ and $(-2, 1, -8)$. [Ans. 6/7, 2/7, 3/7]
   (ii) $(7, -5, 9)$ and $(5, -3, 8)$. [Ans. 2/3, -2/3, 1/3]
2. Show that the points $(1, -2, 3), (2, 3, -4), (0, -7, 10)$ are collinear.
3. The projections of a line on the co-ordinate axes are 12, 4, 3. Find the length and the direction cosines of the line. [Ans. 13; (12/13, 4/13, 3/13)]

1.7.3 Projection of a Broken Line

If $P_1, P_2, P_3, ..., P_n$ be a number of points in space, then the sum of the projections of the segments
$$P_1P_2, P_2P_3, ..., P_{n-1}P_n$$
on any line is equal to the projection of the segment P_1P_n on the same line.

Let $Q_1, Q_2, Q_3, ..., Q_n$

---

* The relations can be given in this form only if none of l, m, n is zero.

be the projections of the points
$$P_1, P_2, P_3, \ldots, P_n$$
on the given line. Then,
$$Q_1 Q_2 = \text{projection of } P_1 P_2$$
$$Q_2 Q_3 = \text{projection of } P_2 P_3$$
and so on.

Also $Q_1 Q_n = \text{projection of } P_1 P_n$.

As $Q_1, Q_2, Q_3, \ldots, Q_n$ lie on the same line, we have for all relative positions of these points on the line, the relation
$$Q_1 Q_2 + Q_2 Q_3 + \ldots + Q_{n-1} Q_n = Q_1 Q_n.$$
Hence, the result.

1.7.4 Projection of the Join of Two Points on a Line

To show that the projection of the segment joining the points
$$P(x_1, y_1, z_1) \text{ and } Q(x_2, y_2, z_2)$$
on a line with direction cosines, l, m, n is
$$(x_2 - x_1) l + (y_2 - y_1) m + (z_2 - z_1) n.$$

Through P, Q draw planes parallel to the co-ordinate planes to form a rectangular parallelopiped whose one diagonal is PQ (See Fig. 3, page 3).

Now $\quad PA = x_2 - x_1, \ AN = y_2 - y_1, \ NQ = z_2 - z_1.$

The lines PA, AN, NQ are respectively parallel to X-axis, Y-axis, Z-axis. Therefore, their respective projections on the line with direction cosines l, m, n are
$$(x_2 - x_1) l, (y_2 - y_1) m, (z_2 - z_1) n.$$
As the projection of the segment PQ on any line is equal to the sum of the projections of the segments PQ, AN, NQ on that line, the required projection is
$$(x_2 - x_1) l + (y_2 - y_1) m + (z_2 - z_1) n.$$

EXERCISES

1. $A(6, 3, 2), B(5, 1, 4), C(3, -4, 7), D(0, 2, 5)$ are four points. Find the projections of the segment AB on the line CD and the segment CD on the line AB. [**Ans.** $-13/7; -13/3$]
2. Show by projection that if P, Q, R, S are the points $(6, -6, 0), (-1, -7, 6), (3, -4, 4), (2, -9, 2)$ respectively, then $PQ \perp RS$.

1.8 ANGLE BETWEEN TWO LINES

To find the angle between lines whose direction cosines are (l_1, m_1, n_1) and (l_2, m_2, n_2).

Let OP_1, OP_2 be lines through the origin parallel to the given line so that the cosines of the angles which OP_1 and OP_2 make with the axes are l_1, m_1, n_1 and l_2, m_2, n_2 respectively and the angle between the given lines is the angle between OP_1 and OP_2. Let this angle be θ.

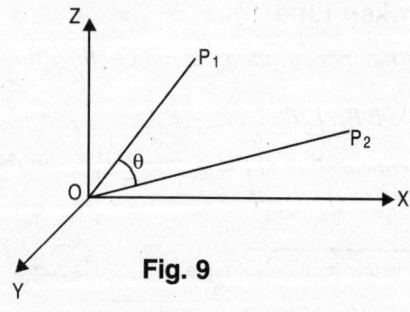

Fig. 9

System of Co-ordinates

Let the co-ordinates of P_2 be (x_2, y_2, z_2).

The projection of the segment OP_2 joining
$$O(0, 0, 0) \text{ and } P_2(x_2, y_2, z_2)$$
on the line OP_1 with direction cosines
$$l_1, m_1, n_1$$
is $\quad (x_2 - 0) l_1 + (y_2 - 0) m_1 + (z_2 - 0) n_1 = l_1 x_2 + m_1 y_2 + n_1 z_2.$

Also this projection is $OP_2 \cos \theta$.

$\therefore \quad OP_2 \cos \theta = l_1 x_2 + m_1 y_2 + n_1 z_2.$

But $\quad x_2 = l_2 \cdot OP_2, \; y_2 = m_2 \cdot OP_2, \; z_2 = n_2 \cdot OP_2 \qquad \qquad ...(1.5.1)$

$\therefore \quad OP_2 \cos \theta = (l_1 l_2 + m_1 m_2 + n_1 n_2) OP_2$

$\Rightarrow \quad \cos \theta = l_1 l_2 + m_1 m_2 + n_1 n_2.$

Second Method. Suppose $OP_1 = r_1, OP_2 = r_2$.

Let the co-ordinates of the points P_1, P_2 be (x_1, y_1, z_1) and (x_2, y_2, z_2) respectively, so that
$$x_1 = r_1 l_1, \; y_1 = r_1 m_1, \; z_1 = r_1 n_1 \qquad \qquad ...(1.5.2)$$
$$x_2 = r_2 l_2, \; y_2 = r_2 m_2, \; z_2 = r_2 n_2$$

We have
$$\begin{aligned} P_1 P_2^2 &= (x_2 - x_1)^2 + (y_2 - y_1)^2 + (z_2 - z_1)^2 \\ &= (x_2^2 + y_2^2 + z_2^2) + (x_1^2 + y_1^2 + z_1^2) - 2(x_1 x_2 + y_1 y_2 + z_1 z_2) \\ &= r_2^2 + r_1^2 - 2 r_2 r_1 (l_1 l_2 + m_1 m_2 + n_1 n_2) \end{aligned} \qquad ...(i)$$

Also from the cosine rule in Trigonometry, we have
$$P_1 P_2^2 = r_1^2 + r_2^2 - 2 r_1 r_2 \cos \theta \qquad \qquad ...(ii)$$

Therefore, from (i) and (ii), we obtain
$$r_1^2 + r_2^2 - 2 r_1 r_2 \cos \theta = P_1 P_2^2 = r_1^2 + r_2^2 - 2 r_1 r_2 (l_1 l_2 + m_1 m_2 + n_1 n_2)$$

$\Rightarrow \quad \cos \theta = l_1 l_2 + m_1 m_2 + n_1 n_2.$

Cor. 1. $\sin \theta$ and $\tan \theta$ The expressions for $\sin \theta$ and $\tan \theta$ in a convenient form are obtained as follows :

$$\begin{aligned} \sin^2 \theta &= 1 - \cos^2 \theta = 1 - (l_1 l_2 + m_1 m_2 + n_1 n_2)^2 \\ &= (l_1^2 + m_1^2 + n_1^2)(l_2^2 + m_2^2 + n_2^2) - (l_1 l_2 + m_1 m_2 + n_1 n_2)^2 \\ &= (l_1 m_2 - l_2 m_1)^2 + (m_1 n_2 - m_2 n_1)^2 + (n_1 l_2 - n_2 l_1)^2 \end{aligned}$$

$\Rightarrow \quad \sin \theta = \pm \sqrt{\Sigma (l_1 m_2 - l_2 m_1)^2}$

Also $\quad \tan \theta = \dfrac{\sin \theta}{\cos \theta} = \pm \dfrac{\sqrt{\Sigma (l_1 m_2 - l_2 m_1)^2}}{\Sigma l_1 l_2}$

Cor. 2. If the direction cosines of two lines be proportional to a_1, b_1, c_1 and a_2, b_2, c_2, so that their actual values are

$$\pm \dfrac{a_1}{\sqrt{a_1^2 + b_1^2 + c_1^2}}, \; \pm \dfrac{b_1}{\sqrt{a_1^2 + b_1^2 + c_1^2}}, \; \pm \dfrac{c_1}{\sqrt{a_1^2 + b_1^2 + c_1^2}};$$

$$\pm \dfrac{a_2}{\sqrt{a_2^2 + b_2^2 + c_2^2}}, \; \pm \dfrac{b_2}{\sqrt{a_2^2 + b_2^2 + c_2^2}}, \; \pm \dfrac{c_2}{\sqrt{a_2^2 + b_2^2 + c_2^2}},$$

and if θ be the angle between the given lines, we have

$$\cos \theta = \pm \frac{a_1 a_2 + b_1 b_2 + c_1 c_2}{\sqrt{a_1^2 + b_1^2 + c_1^2} \sqrt{a_2^2 + b_2^2 + c_2^2}},$$

$$\sin \theta = \pm \frac{\sqrt{(a_1 b_2 - a_2 b_1)^2 + (b_1 c_2 - b_2 c_1)^2 + (c_1 a_2 - c_2 a_1)^2}}{\sqrt{a_1^2 + b_1^2 + c_1^2} \sqrt{a_2^2 + b_2^2 + c_2^2}}$$

$$\tan \theta = \pm \frac{\sqrt{\Sigma (a_1 b_2 - a_2 b_1)^2}}{\Sigma a_1 a_2}.$$

The expression for $\tan \theta$ is the same whether we use direction cosines or direction ratios.

Cor. 3. Conditions for perpendicular and parallelism.
(i) The given lines are perpendicular

$\Rightarrow \qquad \theta = 90°$

$\Rightarrow \qquad \cos \theta = 0$

$\Rightarrow \qquad a_1 a_2 + b_1 b_2 + c_1 c_2 = 0.$

(ii) The given lines are parallel

$\Rightarrow$ the lines through the origin drawn parallel to the lines coincide.

$\Rightarrow$ the direction cosines of the lines are the same.

$\Rightarrow$ the direction ratios of the lines are proportional.

EXAMPLES

1. $l_1, m_1, n_1; l_2, m_2, n_2$ *are the direction cosines of two mutually perpendicular lines. Show that the direction cosines of the line perpendicular to them both are*

$$m_1 n_2 - m_2 n_1, n_1 l_2 - n_2 l_1, l_1 m_2 - l_2 m_1$$

Sol. If l, m, n be the direction cosines of the required line, we have

$$l l_1 + m m_1 + n n_1 = 0$$
$$l l_2 + m m_2 + n n_2 = 0$$

$$\Rightarrow \frac{l}{m_1 n_2 - m_2 n_1} = \frac{m}{n_1 l_2 - n_2 l_1} = \frac{n}{l_1 m_2 - l_2 m_1} = \frac{\sqrt{\Sigma l^2}}{\sqrt{\Sigma (m_1 n_2 - m_2 n_1)^2}} = \frac{1}{\sin \theta},$$

where θ is the angle between the given lines. As $\theta = 90°$, we have $\sin \theta = 1$. Hence, the result.

2. *A line makes angles* $\alpha, \beta, \gamma, \delta$ *with the four diagonals of a cube, prove that*

$$\cos^2 \alpha + \cos^2 \beta + \cos^2 \gamma + \cos^2 \delta = \frac{4}{3}.$$

Sol. Let a be the length of each side of the cube. Taking three coterminous edges OA, OB, OC as axes, the co-ordinates of various vertices will be $A(a, 0, 0), L(a, a, 0), B(0, a, 0), N(0, a, a), C(0, 0, a), M(a, 0, a), P(a, a, a)$ and $O(0, 0, 0)$.

d.c.s of diagonal OP are

$$\left[\frac{a}{\sqrt{a^2 + a^2 + a^2}}, \frac{a}{\sqrt{a^2 + a^2 + a^2}}, \frac{a}{\sqrt{a^2 + a^2 + a^2}} \right]$$

$$\Rightarrow \qquad \left[\frac{1}{\sqrt{3}}, \frac{1}{\sqrt{3}}, \frac{1}{\sqrt{3}} \right]$$

Similarly, d.c.'s of AN are $\left[-\frac{1}{\sqrt{3}}, \frac{1}{\sqrt{3}}, \frac{1}{\sqrt{3}} \right]$

System of Co-ordinates

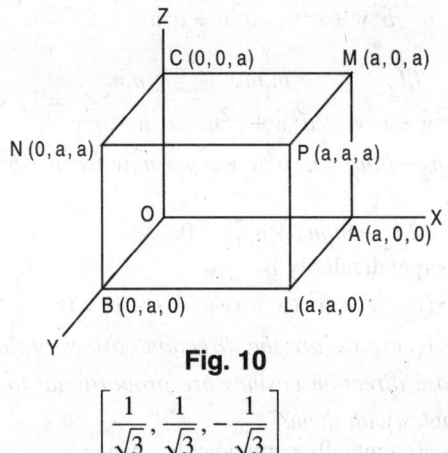

Fig. 10

of BM are $\left[\dfrac{1}{\sqrt{3}}, \dfrac{1}{\sqrt{3}}, -\dfrac{1}{\sqrt{3}}\right]$

and of CL are $\left[\dfrac{1}{\sqrt{3}}, \dfrac{1}{\sqrt{3}}, -\dfrac{1}{\sqrt{3}}\right].$

If $[l, m, n]$ be the d.c.'s of a line which makes angles $\alpha, \beta, \gamma, \delta$ with these four diagonals of the cube, then

$$\cos\alpha = \frac{l+m+n}{\sqrt{3}}, \cos\beta = \frac{-l+m+n}{\sqrt{3}}, \cos\gamma = \frac{l-m+n}{\sqrt{3}} \text{ and } \cos\delta = \frac{l+m-n}{\sqrt{3}}$$

Hence,
$$\cos^2\alpha + \cos^2\beta + \cos^2\gamma + \cos^2\delta$$
$$= \frac{1}{3}[(l+m+n)^2 + (-l+m+n)^2 + (l-m+n)^2 + (l+m-n)^2]$$
$$= \frac{4}{3}.$$

3. *Show that the straight lines whose direction cosines are given by the equations*
$$al + bm + cn = 0, \quad ul^2 + vm^2 + wn^2 = 0$$
are perpendicular or parallel according as
$$a^2(v+w) + b^2(w+u) + c^2(u+v) = 0 \text{ or } a^2/u + b^2/v + c^2/w = 0.$$

Sol. Eliminating l, between the given relations, we have
$$\frac{u(bm+cn)^2}{a^2} + vm^2 + wn^2 = 0$$
$$\Rightarrow \quad (b^2u + a^2v)m^2 + 2ubcmn + (c^2u + a^2w)n^2 = 0. \qquad ...(i)$$

If the lines be parallel, their direction cosines are equal so that the two values of m/n must be equal. The condition for this is
$$u^2b^2c^2 = (b^2u + a^2v)(c^2u + a^2w)$$
$$\Rightarrow \quad \frac{a^2}{u} + \frac{b^2}{v} + \frac{c^2}{w} = 0.$$

Again, if l_1, m_1, n_1 and l_2, m_2, n_2 be the direction cosines of the two lines then equation (i) gives
$$\frac{m_1}{n_1} \cdot \frac{m_2}{n_2} = \frac{m_1 m_2}{n_1 n_2} = \frac{c^2 u + a^2 w}{b^2 u + a^2 u}$$
$$\Rightarrow \quad \frac{m_1 m_2}{c^2 u + a^2 w} = \frac{n_1 n_2}{b^2 u + a^2 v}$$

Similarly the elimination of n, gives, (or by symmetry)

$$\frac{l_1 l_2}{b^2 w + c^2 v} = \frac{m_1 m_2}{a^2 w + c^2 u}$$

Thus, we have

$$\frac{l_1 l_2}{b^2 w + c^2 w} = \frac{m_1 m_2}{a^2 w + c^2 u} = \frac{n_1 n_2}{b^2 u + a^2 v} = k, \text{ say.}$$

$$\Rightarrow \quad l_1 l_2 + m_1 n_2 + n_1 n_2 = k (b^2 w + c^2 v + a^2 w + c^2 u + b^2 u + a^2 v)$$

For perpendicular lines

$$l_1 l_2 + m_1 m_2 + n_1 n_2 = 0.$$

Thus, the condition for perpendicularity is

$$a^2 (v + w) + b^2 (w + u) + c^2 (u + v) = 0.$$

5. *If $l_1, m_1, n_1; l_2, m_2, n_2; l_3, m_3, n_3$ are the direction cosines of three mutually perpendicular lines, show that the line whose direction cosines are proportional to $l_1 + l_2 + l_3, m_1 + m_2 + m_3, n_1 + n_2 + n_3$ makes equal angles with them.*

Sol. Since the given lines are mutaully perpendicular, hence

$$l_1 l_2 + m_1 m_2 + n_1 n_2 = 0 \qquad \text{...(1)}$$
$$l_2 l_3 + m_2 m_3 + n_2 n_3 = 0 \qquad \text{...(2)}$$
$$l_1 l_3 + m_1 m_3 + n_1 n_3 = 0 \qquad \text{..(3)}$$

Let θ be the angle between the line whsoe d.c.s are $[l_1, m_1, n_1]$ and the line whose d.c.'s are proportional to $l_1 + l_2 + l_3, m_1 + m_2 + m_3, n_1 + n_2 + n_3$; then

$$\cos \theta = \frac{l_1 (l_1 + l_2 + l_3) + m_1 (m_1 + m_2 + m_3) + n_1 (n_1 + n_2 + n_3)}{\sqrt{(l_1 + l_2 + l_3)^2 + (m_1 + m_2 + m_3)^2 + (n_1 + n_2 + n_3)^2}}$$

$$= \frac{1}{\sqrt{3}}, \text{ from (1), (2) and (3)}$$

which is independent of $l_1, m_1, n_1; l_2, m_2, n_2; l_3, m_3, n_3$.
Hence, the given line makes equal angles with the three lines.

6. *If a variable line in two adjacent positions had direction cosines l, m, n; $l + \delta l, m + \delta m, n + \partial n$, show that the small angle $\delta\theta$ between two positions is given by*

$$\delta\theta^2 = \delta l^2 + \delta m^2 + \delta n^2.$$

Sol. Since $[l, m, n]$ and $[l + \delta l, m + \delta m, n + \delta n]$ are d.c.'s, hence

$$l^2 + m^2 + n^2 = 1 \qquad \text{...(1)}$$

and $\quad (l + \delta l)^2 + (m + \delta m)^2 + (n + \delta n)^2 = 1$

$$\Rightarrow \quad \delta l^2 + \delta m^2 + \delta n^2 = -2 (l \delta l + m \delta m + n \delta n) \qquad \text{...(2)}$$

Now, $\quad \cos \delta\theta = l (l + \delta l) + m (m + \delta m) + n (n + \delta n)$

$$= l^2 + m^2 + n^2 + l \delta l + m \delta m + n \delta n$$

$$= 1 - \frac{1}{2} \{\delta l^2 + \delta m^2 + \delta n^2\}, \text{ from (1) and (2).}$$

$$\Rightarrow \quad \delta l^2 + \delta m^2 + \delta n^2 = 2 (1 - \cos \delta\theta) = 2 \cdot 2 \sin^2 \frac{\delta\theta}{2}$$

$$= 4 \left(\frac{1}{2} \delta\theta\right)^2 = \delta\theta^2$$

System of Co-ordinates

EXERCISES

1. Find the angles between the lines whose direction ratios are
   (i) $5, -12, 13; -3, 4, 5$. [**Ans.** $\cos^{-1}(1/65)$]
   (ii) $1, 1, 12; \sqrt{3}-1, -\sqrt{3}-1, 4$. [**Ans.** $\pi/3$]

2. If A, B, C, D are the points $(3, 4, 5), (4, 6, 3), (-1, 2, 4)$ and $(1, 0, 5)$, find the angle between CD and AB. [**Ans.** $\cos^{-1}(4/9)$]

3. Find the angle between any two diagonals of a cube. [**Ans.** $\cos^{-1}(1/3)$]

4. Show that a line can be found perpendicular to the three lines with direction cosines proportional to $(2, 1, 5), (4, -2, 2), (-6, 4, -1)$. Hence, show that if these three lines be concurrent, they are also coplanar.

5. Find the direction cosines of a line which is perpendicular to the lines whose direction ratios are $1, 2, 3; -1, 3, 5$. [**Ans.** $1/\sqrt{90}, -8/\sqrt{90}, 5/\sqrt{90}$]

6. l_1, m_1, n_1 and l_2, m_2, n_2 are the direction ratios of two intersecting lines. Show that lines through the intersection of these two with direction ratio
   $$l_1 + kl_2, m_1 + km_2, n_1 + kn_2$$
   are coplanar with them; k being a number whatsoever.
   (Show that they all have a common perpendicular direction)

7. Show that three concurrent lines with direction cosines
   $$(l_1, m_1, n_1), (l_2, m_2, n_2), (l_3, m_3, n_3)$$
   are coplanar if and only if
   $$\begin{vmatrix} l_1 & m_1 & n_1 \\ l_2 & m_2 & n_2 \\ l_3 & m_3 & n_3 \end{vmatrix} = 0$$

8. Show that the join of points $(1, 2, 3), (4, 5, 7)$ is parallel to the join of the points $(-4, 3, -6), (2, 9, 2)$.

9. Show that the points $(4, 7, 8), (2, 3, 4), (-1, -2, 1), (1, 2, 5)$ are the vertices of a parallelogram.

10. Show that the points $(5, -1, 1), (7, -4, 7), (1, -6, 10), (-1, -3, 4)$ are the vertices of a rhombus.

11. Show that the points $(0, 4, 1), (2, 3, -1), (4, 5, 0), (2, 6, 2)$ are the vertices of a square.

12. $A(1, 8, 4), B(0, -11, 4), C(2, -3, 1)$ are three points and D is the foot of the perpendicular from A on BC. Find the co-ordinates of D. [**Ans.** $4, 5, -2$]

13. Find the point in which the join of $(-9, 4, 5)$ and $(11, 0, -1)$ is met by the perpendicular from the origin. [**Ans.** $1, 2, 2$]

14. $A(-1, 2, -3), B(5, 0, -6), C(0, 4, -1)$ are three points. Show that the direction cosines of the bisectors of the angle BAC are proportional to $(25, 8, 5)$ and $(-11, 20, 23)$.
    [**Hint.** Find the co-ordinates of the points which divide BC in the ratio $AB : AC$]

15. Find the angle between the lines whose direction cosines are given by the equations $3l + m + 5n = 0$ and $6mn - 2nl + 5lm = 0$. [**Ans.** $\cos^{-1} 1/6$]

16. Show that the pair of lines whose direction cosines are given by $3lm - 4ln + mn = 0$, $l + 2m + 3n = 0$ are perpendicular.

17. Find the angle between the lines whose direction cosines satisfy the equations $l + m + n = 0$ and $2nl + 2lm - mn = 0$.

18. Show that the straight lines whose d.c.'s are given by $l + m + n = 0$, $2mn + 3nl - 5lm = 0$ are perpendicular to each other.

19. Find the angle between the lines $l + m + n = 0$, $\dfrac{mn}{q-r} + \dfrac{nl}{r-p} + \dfrac{lm}{p-q} = 0$. [Ans. $\pi/3$]

20. Show that the straight lines whose d.c.'s are given by $a^2 l + b^2 m + c^2 n = 0$, $mn + nl + lm = 0$ will be parallel if $a + b + c = 0$.

21. Show that the straight lines whose direction cosines are given by
$$al + bm + cn = 0, \; fmn + gnl + hlm = 0,$$
are perpendicular if
$$f/a + g/b + h/c = 0,$$
are parallel if
$$\sqrt{af} \pm \sqrt{bg} \pm \sqrt{ch} = 0.$$

22. If, in a tetrahedron $OABC$,
$$OA^2 + BC^2 = OB^2 + CA^2 = OC^2 + AB^2$$
then its pairs of opposite edges are at right angles.

23. l_1, m_1, n_1 and l_2, m_2, n_2 are two directions inclined at an angle φ, to each other. Show that
$$\dfrac{l_1 + l_2}{2\cos\dfrac{1}{2}\varphi}, \; \dfrac{m_1 + m_2}{2\cos\dfrac{1}{2}\varphi}, \; \dfrac{n_1 + n_2}{2\cos\dfrac{1}{2}\varphi}$$
are the direction cosines of the line which bisects the angle between these two directions.

24. Show that the direction equally inclined to the three mutually perpendicular directions
$$l_1, m_1, n_1; l_2, m_2, n_2; l_3, m_3, n_3$$
is given by the direction cosines
$$\dfrac{l_1 + l_2 + l_3}{\sqrt{3}}, \; \dfrac{m_1 + m_2 + m_3}{\sqrt{3}}, \; \dfrac{n_1 + n_2 + n_3}{\sqrt{3}}.$$

25. Show that the area of the triangle whose vertices are the origin and the points (x_1, y_1, z_1) and (x_2, y_2, z_2) is
$$\dfrac{1}{2}\sqrt{(y_1 z_2 - y_2 z_1)^2 + (z_1 x_2 - z_2 x_1)^2 + (x_1 y_2 - x_2 y_1)^2}.$$

❏❏❏

2
The Plane

GENERAL EQUATION OF FIRST DEGREE

An equation of the first degree in x, y, z is of the form
$$ax + by + cz + d = 0$$
where a, b, c are given real numbers and a, b, c are not all zero. The condition that a, b, c are not all zero is equivalent to the single condition $a^2 + b^2 + c^2 \neq 0$.

We are now interested in the locus of the points whose co-ordinates satsify an equation of first degree viz.,
$$ax + by + cz + d = 0, a^2 + b^2 + c^2 \neq 0.$$

It will be shown that this locus is a plane. To show this, we make use of the characteristic property of a plane which we given below :

A geometrical locus is a plane if it is such that if P and Q are any two points on the locus then every point of the line PQ is also a point on the locus.

2.1 THEOREM

Every equation of the first degree in x, y, z represents a plane.

Consider the equation
$$ax + by + cz + d = 0, a^2 + b^2 + c^2 \neq 0.$$

The locus of this equation will be a plane if *every* point of the line joining *any* two points on the locus also lies on the locus.

Let
$$P(x_1, y_1, z_1) \text{ and } Q(x_2, y_2, z_2)$$
be two points on the locus, so that we have
$$ax_1 + by_1 + cz_1 + d = 0 \qquad \ldots(i)$$
$$ax_2 + by_2 + cz_2 + d = 0 \qquad \ldots(ii)$$

Multiplying (ii) by k and adding to (i), we get
$$a(x_1 + kx_2) + b(y_1 + ky_2) + c(z_1 + kz_2) + d(1 + k) = 0$$
$$\Rightarrow \quad a\frac{x_1 + kx_2}{1+k} + b\frac{y_1 + ky_2}{1+k} + c\frac{z_1 + kz_2}{1+k} + d = 0$$

Assuming that $k \neq -1$, the relation (ii) shows that the point
$$\left(\frac{x_1 + kx_2}{1+k}, \frac{y_1 + ky_2}{1+k}, \frac{z_1 + kz_2}{1+k}\right)$$
is also a point on the locus for every value of $k \neq -1$.

Thus, every point on the straight line joining any two *arbitrary points* on the locus also lies on the locus. The given equation, therefore, represents a plane. Hence, every equation of the first degree in x, y, z represents a kplane.

Ex. Find the co-ordiantes of the point where the plane
$$ax + by + cz + d = 0, a^2 + b^2 + c^2 \neq 0$$
meets the three co-ordinate axes.

2.2 CONVERSE OF THE PRECEDING THEOREM

We shall now show that *the equation of every plane is of the first degree i.e., is of the form*
$$ax + by + cz + d = 0,$$

where
$$a^2 + b^2 + c^2 \neq 0.$$

Consider any plane. Let p be the length of the perpendicular from the origin to the plane and let l, m, n be the direction cosines of this perpendicular.

We shall show that for any point (x, y, z) on the plane, we have the relation
$$lx + my + nz = p$$
implying that the equation of the plane is of the first degree.

Let K be the foot of the perpendicular from the origin O to the plane. Let $OK = p$ and let l, m, n be its direction cosines. Take any point $P(x, y, z)$ on the plane.

Now PK lies in the plane
$$\Rightarrow \qquad PK \perp OK$$
$\Rightarrow$ the projection of OP on $OK = OK = p$.

Also the projection of the segment OP joining the points
$$O(0, 0, 0) \text{ and } P(x, y, z)$$
on the line OK with direction cosines
$$l, m, n$$
is $\qquad l(x - 0) + m(y - 0) + n(z - 0) = lx + my + nz.$

It follows that
$$lx + my + nz = p.$$

This equation, being satisfied by the co-ordinates of any point $P(x, y, z)$ on the given plane, is the equation of the plane.

Note 1. The equation
$$lx + my + nz = p,$$
is called the **normal** form of the equation of a plane.

Note 2. The plane whose equation is
$$ax + by + cz + d = 0$$
is referred to as
$$ax + by + cz + d = 0$$
itself *i.e.*, we often refer to an equation of the plane itself as the plane.

Ex. Find the equation of the plane containing the lines through the origin with direction cosines proportional to $(1, -2, 2)$ and $(2, 3, -1)$. [**Ans.** $4x - 5y - 7z = 0$]

2.3 TRANSFORMATION TO THE NORMAL FORM

To transform the equation
$$ax + by + cz + d = 0, a^2 + b^2 + c^2 \neq 0$$
to the normal form
$$lx + my + nz = p.$$

As these two equations represent the same plane, we have
$$-\frac{d}{p} = \frac{a}{l} = \frac{b}{m} = \frac{c}{n} = \pm \frac{\sqrt{a^2 + b^2 + c^2}}{\sqrt{l^2 + m^2 + n^2}} = \pm \sqrt{a^2 + b^2 + c^2}$$

Thus, $d/p = \pm \sqrt{a^2 + b^2 + c^2}$. As p, according to our convention, is always positive, we shall take positive or negative sign with the radical according as, d, is negative or positive.

Thus, if d be positive, we have
$$l = -\frac{a}{\sqrt{\Sigma a^2}}; m = -\frac{b}{\sqrt{\Sigma a^2}}; n = -\frac{c}{\sqrt{\Sigma a^2}}; p = +\frac{d}{\sqrt{\Sigma a^2}};$$
and if d be negative, we have

The Plane

$$l = \frac{a}{\sqrt{\Sigma a^2}}; m = \frac{b}{\sqrt{\Sigma a^2}}; n = \frac{c}{\sqrt{\Sigma a^2}}; p = -\frac{d}{\sqrt{\Sigma a^2}}$$

Thus, the normal form of the equation $ax + by + cz + d = 0$ is

$$-\frac{a}{\sqrt{\Sigma a^2}} x - \frac{b}{\sqrt{\Sigma a^2}} y - \frac{c}{\sqrt{\Sigma a^2}} z = \frac{d}{\sqrt{\Sigma a^2}},$$

if d be positive, and

$$+\frac{a}{\sqrt{\Sigma a^2}} x + \frac{b}{\sqrt{\Sigma a^2}} y + \frac{c}{\sqrt{\Sigma a^2}} z = -\frac{d}{\sqrt{\Sigma a^2}}$$

if d be negative.

2.3.1 Direction Cosines of the Normal to a Plane

From above we deduce that *the direction cosines of the normal to a plane are proportioanl to the coefficients of x, y, z in its equation or that the coefficients of x, y, z are direction ratios of the normal to the plane.*

Thus,
$$a, b, c$$
are direction ratios of the normal to the plane
$$ax + by + cz + d = 0.$$

Ex. 1. Find the direction cosines of the normals to the planes
(i) $2x - 3y + 6z = 7$,   (ii) $x + 2y + 2z - 1 = 0$.

[**Ans.** (i) $2/7, -3/7, 6/7$; (ii) $1/3, 2/3, 2/3$]

Ex. 2. Show that the normals to the planes
$$x - y + z = 1, \ 3x + 2y - z + 2 = 0$$
are perpendicular to each other.

2.3.2. Angle Between Two Planes

Angle between two planes is equal to the angle between the normals to them from any point.
It follows that the angle between the two planes
$$ax + by + cz + d = 0, \ a_1 x + b_1 y + c_1 z + d_1 = 0$$
being equal to the angle between the lines with direction ratios
$$a, b, c; a_1, b_1, c_1$$
is
$$\cos^{-1} \left\{ \frac{aa_1 + bb_1 + cc_1}{\sqrt{\Sigma a^2} \sqrt{\Sigma a_1^2}} \right\}.$$

Cor. Parallelism and perpendicularity of two planes. Two planes are parallel or perpendicular according as the normals to them are parallel or perpendicular.

Thus, the two planes
$$ax + by + cz + d = 0, \ a_1 x + b_1 y + c_1 z + d_1 = 0$$
will be parallel, if
$$a, b, c \ \text{ and } \ a_1, b_1, c_1$$
are direction ratios of the same line and will be perpendicular, if
$$aa_1 + bb_1 + cc_1 = 0.$$

EXERCISES

1. Find the angles between the following pairs of planes
    (i) $2x - y + 2z = 3$; $3x + 6y + 2z = 4$ [**Ans.** $\cos^{-1}(4/21)$]
    (ii) $2x - y + z = 6$; $x + y + 2z = 7$ [**Ans.** $\pi/3$]
    (iii) $3x - 4y + 5y = 0$; $2x - y - 2z = 5$ [**Ans.** $\pi/2$]

2. Show that the equations
$$ax + by + r = 0, by + cz + p = 0, cz + ax + q = 0$$
represents planes respectively perpendicular to the *XY, YZ, ZX* planes.
3. Show that $ax + by + cz + d = 0$ represents planes, perpendicular respectively to *YZ, ZX, XY* planes, if a, b, c separetely vanish (Similar to Ex. 2).
4. Show that the plane
$$x + 2y - 3z + 4 = 0$$
is perpendicular to each of the planes
$$2x + 5y + 4z + 1 = 0, 4x + 7y + 6z + 2 = 0.$$

2.4 DETERMINATION OF A PLANE UNDER GIVEN CONDITIONS

The *general* equation $ax + by + cz + d = 0$ of a plane contains *three* arbitrary constants (ratios of the coefficients a, b, c, d) and, therefore, a plane can be found to satisfy three conditions each giving rise to only one relation between the constants. The three constants can then be determined from the three resulting relations.

We give below a few sets of conditions which determine a plane :
(i) passing through *three* non-collinear points;
(ii) passing through *two* given points and perpendicular to a given plane;
(iii) passing through a given point and perpendicular to *two* given planes.

2.4.1. Intercept Form of the Equation of a Plane

To find the equation of a plane in terms of the intercepts a, b, c which it makes on the axes.

The intercept of a plane on any co-ordinate axis is the distance of the point where the plane meets the axis from the origin taken with the appropriate sign.

We of course, suppose here that the plane does not pass through the origin so that none of a, b, c is zero.

Let the equation of the plane be
$$Ax + By + Cz + D = 0. \qquad \text{...(i)}$$
The plane not passing through the origin, we have
$$D \neq 0.$$
The points $(a, 0, 0), (0, b, 0), (0, 0, c)$ lying on the plane (i), we have
$$aA + D = 0 \Rightarrow -\frac{A}{D} = \frac{1}{a}$$
$$bB + D = 0 \Rightarrow -\frac{B}{D} = \frac{1}{b}$$
$$cC + D = 0 \Rightarrow -\frac{C}{D} = \frac{1}{c}.$$
The equation (1) can be rewritten as
$$-\frac{A}{D}x - \frac{B}{D}y - \frac{C}{D}z = 1$$
so that after substitution, we obtain
$$\frac{x}{a} + \frac{y}{b} + \frac{z}{c} = 1,$$
as the required equation of the plane.

2.4.2. Plane Through Three Points

To find the equation of the plane through the three non-collinear points
$$(x_1, y_1, z_1), (x_2, y_2, z_2), (x_3, y_3, z_3)$$
Let the required equation of the palne be
$$ax + by + cz + d = 0. \qquad \text{...(i)}$$
As the given points lie on the plane (i), we have

The Plane

$$ax_1 + by_1 + cz_1 + d = 0, \quad \text{...(ii)}$$
$$ax_2 + by_2 + cz_2 + d = 0, \quad \text{...(iii)}$$
$$ax_3 + by_3 + cz_3 + d = 0. \quad \text{...(iv)}$$

Eliminating a, b, c, d from (i) – (iv), we have

$$\begin{vmatrix} x & y & z & 1 \\ x_1 & y_1 & z_1 & 1 \\ x_2 & y_2 & z_2 & 1 \\ x_3 & y_3 & z_3 & 1 \end{vmatrix} = 0$$

which is the required equation of the plane.

Cor. The equation of the plane which makes intercepts a, b, c respectively on the three co-ordiante axes is

$$\frac{x}{a} + \frac{y}{b} + \frac{z}{c} = 1$$

in that this is the plane through the 3 points $(a, 0, 0), (0, b, 0), (0, 0, c)$.

Note. In actual numerical exercises, the student would find it more convenient to follow the method of the first example below.

EXAMPLES

1. *Find the equation of the plane through the points*
$$P(2, 2, -1), Q(3, 4, 2), R(7, 0, 6)$$

Sol. The general equation of a plane through $P(2, 2, -1)$ is
$$a(x-2) + b(y-2) + c(z+1) = 0 \quad \text{...(i)}$$
It will pass through Q and R, if
$$a + 2b + 3c = 0$$
$$5a - 2b + 7c = 0.$$
These give
$$\frac{a}{20} = \frac{b}{8} = \frac{c}{-12} \quad \text{or} \quad \frac{a}{5} = \frac{b}{2} = \frac{c}{-3}.$$
Substituting these values in (i), we have
$$5(x-2) + 2(y-2) - 3(z+1) = 0$$
$$\Rightarrow \quad 5x + 2y - 3z - 17 = 0$$
as the required equation.

2. *Find the equation of the plane through the points*
$$(2, 2, 1) \text{ and } (9, 3, 6),$$
and perpendicular to the plane
$$2x + 6y + 6z = 9.$$

Sol. Any plane through $(2, 2, 1)$ is
$$a(x-2) + b(y-2) + c(z-1) = 0. \quad \text{...(i)}$$
It passes through $(9, 3, 6)$
$$\Rightarrow \quad a(9-2) + b(3-2) + c(6-1) = 0$$
$$\Rightarrow \quad 7a + b + 5c = 0 \quad \text{...(ii)}$$
The plane (i) is perpendicular to the given plane
$$\Rightarrow \quad 2a + 6b + 6c = 0 \quad \text{...(iii)}$$
From (ii) and (iii), we have
$$\frac{a}{-24} = \frac{b}{-32} = \frac{c}{40} \Rightarrow \frac{a}{3} = \frac{b}{4} = \frac{c}{-5}.$$
Substituting in (i), we see that the equation of the required plane is
$$3(x-2) + 4(y-2) - 5(z-1) = 0 \Leftrightarrow 3x + 4y - 5z = 9.$$

EXERCISES

1. Find the equation of the plane through the three points $(1, 1, 1), (1, -1, 1), (-7, -3, -5)$ and show that it is perpendicular to the XZ plane. **[Ans.** $3x - 4z + 1 = 0$**]**
2. Obtain the equation of the plane passing through the point $(-2, -2, 2)$ and containing.
3. If, from the point $P(a, b, c)$, perpendiculars PL, PM be drawn to YZ and ZX planes, find the equation of the plane OLM. **[Ans.** $bcx + cay - abz = 0$**]**
4. Show that the four points $(-6, 3, 2), (3, -2, 4), (5, 7, 3)$ and $(-13, 17, -1)$ are coplanar.
5. Show that the points $(6, -4, 4), (0, 0, -4)$ intersects the join of $(-1, -2, -3), (1, 2, -5)$.
6. Show that $(-1, 4, -3)$ is the circumcentre of the triangle formed by the points $(3, 2, -5), (-3, 8, -5), (-3, 2, 1)$.
7. Show that the equations of the three planes passing through the points, $(1, -2, 4), (3, -4, 5)$ and perpendicular to the XY, YZ, ZX planes are $x + y + 1 = 0; x - 2z + 7 = 0; y + 2z = 6$ respectively.
8. Obtain the equation of the plane which passes through the point $(-1, 3, 2)$ and is perpendicular to each of the two planes $x + 2y + 2z = 5; 3x + 2y + 2z = 8$.

    [Ans. $2x - 4y + 3z + 8 = 0$**]**
9. Find the equation of the plane which passes through $A(-1, 1, 1)$ and $B(1, -1, 1)$ and is perpendicular to the plane $x + 2y + 2z = 5$. **[Ans.** $2x + 2y - 3z + 3 = 0$**]**
10. Find the intercepts of the plane $2x - 3y + z = 12$ on the co-ordinate axes. **[Ans.** $6, -4, 12$**]**
11. A plane meets the co-ordiante axes A, B, C such that the centroid of the triangle ABC is the point (a, b, c), show that the equation of the plane is $x/a + y/b + z/c = 3$.
12. Find the equations of the two planes which pass through the points $(0, 4, -3), (6, -4, 3)$ other than the plane through the origin, which cut off from the axes intercepts wehose sum is zero. **[Ans.** $2x - 3y - 6z = 6; 6x + 3y - 2z = 18$**]**
13. A variable plane is at a constant distance p from the origin and meets the co-ordinate axes in A, B, C. Show that the locus of the centroid of the tetrahedron $OABC$ is $x^{-2} + y^{-2} + z^{-2} = 16p^{-2}$.

2.5 SYSTEMS OF PLANES

The equation of a plane satisfying two conditions will involve one arbitrary constant which can be chosen in an infinite number of ways, thus giving rise to an infinite number of planes, called a *System of planes*.

The arbitrary constant which is different for different members of the system is called a *Parameter*.

Similarly the equation of a plane satisfying one condition will involve two parameters.

The following are the equations of a few systems of planes involving one or two parameters :

1. The equation
$$ax + by + cz + k = 0$$
represents the system of planes parallel to a given plane
$$ax + by + cz + d = 0,$$
k being the parameter.

Thus, the set of planes parallel to a given plane
$$ax + by + cz + d = 0$$
is $\{ax + by + cz + k = 0; k$ is any number$\}$.

2. The equation
$$ax + by + cz + d = 0$$

The Plane

represents the system of planes perpendicular to given line with direction ratios a, b, c, d being the parameter.

3. The equation
$$(ax + by + cz + d) + k(a_1 x + b_1 y + c_1 z + d_1) = 0 \qquad \ldots(1)$$
represents the system of planes through the line of intersection of the plane
$$ax + by + cz + d = 0, \qquad \ldots(2)$$
$$a_1 x + b_1 y + c_1 z + d_1 = 0; \qquad \ldots(3)$$

k being the parameter, for the equation (1), being of the first degree in x, y, z represents a plane; and it is evidently satisfied by the co-ordinates of the points which satisfy (2) and (3), whatever value k may have.

4. The equation
$$A(x - x_1) + B(y - y_1) + C(z - z_1) = 0,$$
represents the system of planes passing through the point (x_1, y_1, z_1) where the required *two* parameters are the two ratios of the coefficients A, B, C; for the equation is of the first degree and is clearly satisfied by the point (x_1, y_1, z_1) whatever be the ratios of the coefficients.

EXAMPLES

1. *Find the equation of the plane passing through the lines of intersection of the planes*
$$2x - y = 0 \text{ and } 3z - y = 0$$
and perpendicular to the plane
$$4x + 5y - 3z = 8.$$

Sol. The plane
$$2x - y + k(3z - y) = 0 \Leftrightarrow 2x - (1 + k)y + 3kz = 0$$
passes through the line of intersection of the given planes whatever value k may have. This plane is perpendicular to
$$4x + 5y - 3z = 8$$
$$\Rightarrow \quad 2 \cdot 4 - (1 + k) \cdot 5 + 3k(-3) = 0 \Rightarrow 14k = 3 \Rightarrow k = 3/14.$$
Thus, the required equation is
$$2x - y + \left(\frac{3}{14}\right)(3z - y) = 0 \Leftrightarrow 28x - 17y + 9z = 0.$$

2. *A point P moves on a fixed plane $x/a + y/b + z/c = 1$. The plane through P perpendicular to OP meets the axes in A, B, C. The planes through A, B, C parallel to co-ordinate planes intersect in Q. Show that the locus of Q is*
$$\frac{1}{x^2} + \frac{1}{y^2} + \frac{1}{z^2} = \frac{1}{ax} + \frac{1}{by} + \frac{1}{cz}.$$

Sol. Let the point be $P = (\alpha, \beta, \gamma)$. Hence
$$\frac{\alpha}{a} + \frac{\beta}{b} + \frac{\gamma}{c} = 1 \qquad \ldots(1)$$
Equation of the plane perpendicular to OP is
$$\alpha x + \beta y + \gamma z = d.$$
But it passes through $P(\alpha, \beta, \gamma)$, we have
$$d = \alpha^2 + \beta^2 + \gamma^2$$
Hence, equation of plane through P and perpendicular to OP is
$$\alpha x + \beta y + \gamma z = \alpha^2 + \beta^2 + \gamma^2 \qquad \ldots(2)$$
$$\Rightarrow \quad OA = \frac{\alpha^2 + \beta^2 + \gamma^2}{\alpha}, \ OB = \frac{\alpha^2 + \beta^2 + \gamma^2}{\beta}, \ OC = \frac{\alpha^2 + \beta^2 + \gamma^2}{\gamma}.$$

So the planes through A, B, C parallel to the planes YOZ, ZOX, XOY intersect in the point Q whose co-ordinates are

$$x = \frac{\alpha^2 + \beta^2 + \gamma^2}{\alpha}, \quad y = \frac{\alpha^2 + \beta^2 + \gamma^2}{\beta}, \quad z = \frac{\alpha^2 + \beta^2 + \gamma^2}{\gamma}.$$

With the help of (1),
$$\frac{1}{x^2} + \frac{1}{y^2} + \frac{1}{z^2} = \frac{1}{\alpha^2 + \beta^2 + \gamma^2}$$

and
$$\frac{1}{ax} + \frac{1}{by} + \frac{1}{cz} = \frac{\frac{\alpha}{a} + \frac{\beta}{b} + \frac{\gamma}{c}}{\alpha^2 + \beta^2 + \gamma^2} = \frac{1}{\alpha^2 + \beta^2 + \gamma^2}, \text{ from (1)}.$$

Hence, required locus is
$$\frac{1}{x^2} + \frac{1}{y^2} + \frac{1}{z^2} = \frac{1}{ax} + \frac{1}{by} + \frac{1}{cz}.$$

3. *The plane $lx + my = 0$ is rotated about its line of intersection with the plane $z = 0$ through an angle α. Prove that the equation of the plane in its new position is*
$$lx + my \pm z \sqrt{l^2 + m^2} \tan \alpha = 0.$$

Sol. The equation of a plane through the line of intersection of the palnes $lx + my = 0$ and $z = 0$, is
$$lx + my + \lambda z = 0.$$

This plane makes an angle α with the plane $lx + my = 0$.

$$\therefore \quad \cos \alpha = \frac{l^2 + m^2}{\sqrt{(l^2 + m^2)(l^2 + m^2 + \lambda^2)}}$$

$$\therefore \quad \cos^2 \alpha = \frac{l^2 + m^2}{(l^2 + m^2 + \lambda^2)}$$

$$\Rightarrow \quad \lambda = \pm \sqrt{(l^2 + m^2)} \tan \alpha$$

Hence, required plane is
$$lx + my \pm z \sqrt{l^2 + m^2} \tan \alpha = 0.$$

4. *A triangle, the lengths of whose sides are a, b, and c is placed so that the middle points of the sides are on the axes. Show that the equation to the plane is*
$$x/\alpha + y/\beta + z/\gamma = 1.$$
where $\alpha^2 = \frac{(b^2 + c^2 - a^2)}{8}, \beta^2 = \frac{(c^2 - a^2 - b^2)}{8}, \gamma^2 = \frac{(a^2 + b^2 - c^2)}{8}.$

Sol. Let α, β, γ be the intercepts that the plane makes with the axes. E and F are the mid-points of AC and BC. Therefore, EF is parallel to and half of AB.

$$\therefore \quad EF^2 = OE^2 + OF^2 = \alpha^2 + \beta^2.$$

But $\quad EF = \frac{AB}{2} = \frac{c}{2} \Rightarrow \alpha^2 + \beta^2 = \frac{c^2}{4}$

Similarly, $\beta^2 + \gamma^2 = a^2/4$ and $\gamma^2 + \alpha^2 = b^2/4$.

Adding, $\alpha^2 + \beta^2 + \gamma^2 = \frac{a^2 + b^2 + c^2}{8}$

$$\Rightarrow \gamma^2 = \frac{a^2 + b^2 + c^2}{8} - \frac{c^2}{4} = \frac{a^2 + b^2 - c^2}{8}$$

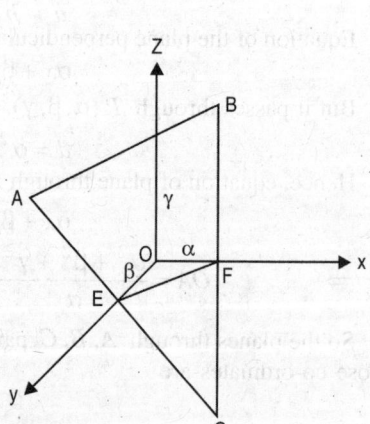

The Plane

Similarly,
$$\alpha^2 = \frac{b^2 + c^2 - a^2}{8}, \beta^2 = \frac{c^2 + a^2 - b^2}{8}.$$

Hence, the equation of plane is
$$x/\alpha + y/\beta + z/\gamma = 1, \text{ where } \alpha^2, \beta^2, \gamma^2 \text{ as given below.}$$

EXERCISES

1. Obtain the equation of the plane through the intersection of the planes
$$x + 2y + 3z + 4 = 0 \text{ and } 4x + 3y + 2z + 1 = 0$$
and the origin. [Ans. $3x + 2y + z = 0$]

2. Find the equation of the plane which is perpendicular to the plane
$$5x + 3y + 6z + 8 = 0$$
and which contains the line of intersection of the planes
$$x + 2y + 3z - 4 = 0, 2x + y - z + 5 = 0.$$
[Ans. $51x + 15y - 50z + 173 = 0$]

3. The plane $x - 2y + 3z = 0$ is rotated through a right angle about the line of intersection with the plane $2x + 3y - 4z - 5 = 0$, find the equation of the plane in its new position.
[Ans. $22x + 5y - 4z - 35 = 0$]

4. Find the equation of the plane through the line of intersection of the planes
$$ax + by + cz + d = 0, a_1 x + b_1 y + c_1 z = d_1 = 0$$
and perpendicular to the XY plane.
[Ans. $x(ac_1 - a_1 c) + y(bc_1 - b_1 c) + z(dc_1 - d_1 c) = 0$]

5. Find the equation of the plane through the point $(2, 3, 4)$ and parallel to the plane $5x - 6y + 7z = 3$. [Ans. $5x - 6y + 7z = 20$]

6. Find the equation of the plane through $(2, 3, -4)$ and $(1, -1, 3)$ parallel to the x-axis.
[Ans. $7y + 4z - 5 = 0$]

7. A variable plane is at a constant distance $3p$ from the origin and meets the axes in A, B and C. Show that the locus of the centroid of the triangle ABC is $x^{-2} + y^{-2} + z^{-2} = p^{-2}$.

8. Find the equation of the plane that passes through the point $(3, -3, 1)$ and is normal to the line joining the points $(3, 4, -1)$ and $(2, -1, 5)$. [Ans. $x + 5y - 6z + 18 = 0$]

9. Obtain the equation of the plane that bisects the segment joining the points $(1, 2, 3), (3, 4, 5)$, at right angles. [Ans. $x + y + z = 9$]

10. A variable plane passes through a fixed point (a, b, c) and meets the co-ordinate axes in A, B, C. Show that the locus of the point common to the planes through A, B, C parallel to the co-ordinate plane is
$$a/x + b/y + c/z = 1.$$

2.6 TWO SIDES OF A PLANE

Consider any plane. Two points P and Q which do not lie on the plane lie on the different or the same side of plane according as the segment PQ has or does not have a point in common with the plane.

We proceed to determine a criterion for two given points to lie on the same or different sides of a given plane and show that:

Two points $A(x_1, y_1, z_1), B(x_2, y_2, z_2)$ lie on the same or different sides of the plane
$$ax + by + cz + d = 0,$$

according as the expressions
$$ax_1 + by_1 + cz_1 + d, ax_2 + by_2 + cz_2 + d$$
are of the same or different signs.

Let the line AB meets the given plane in the point P and let P divides AB in the ratio $r:1$ so that r is positive or negative according as P divides AB internally or externally, *i.e.*, according as A and B lie on the opposite or the same side of the plane.

Since the point P whose co-ordinates are
$$\left(\frac{rx_2 + x_1}{r+1}, \frac{ry_2 + y_1}{r+1}, \frac{rz_2 + z_1}{r+1}\right)$$
lies on the same plane, we have
$$a\frac{rx_2 + x_1}{r+1} + b\frac{ry_2 + y_1}{r+1} + c\frac{rz_2 + z_1}{r+1} d = 0$$
$\Rightarrow \quad r(ax_2 + by_2 + cz_2 + d) + (ax_1 + by_1 + cz_1 + d) = 0,$
$\Rightarrow \quad r = -\dfrac{ax_1 + by_1 + cz_1 + d}{ax_2 + by_2 + cz_2 + d}$

This shows that r is negative or positive according as
$$ax_1 + by_1 + cz_1 + d, \ ax_2 + by_2 + cz_2 + d$$
are of the same or different signs.
Thus, the theorem is proved.

Ex. Show that the origin and the point $(2, -4, 3)$ lie on different sides of the plane $x + 3y - 5z + 7 = 0$.

2.7 LENGTH OF THE PERPENDICULAR FROM A POINT TO A PLANE

To find the perpendicular distance of the point
$$P(x_1, y_1, z_1)$$
from the plane
$$lx + my + nz = p.$$
The equation of the plane which passes through the point
$$P(x_1, y_1, z_1)$$
and is parallel to the given plane is
$$lx + my + nz = p_1$$
where
$$lx_1 + my_1 + nz_1 = p_1.$$
Let OKK' be the perpendicular from the origin O to the two parallel planes meeting them in K and K' so that
$$OK = p \quad \text{and} \quad OK' = p_1$$
Draw the line PL perpendicular to the given plane. We have
$$LP = OK' - OK$$
$$= p_1 - p = lx_1 + my_1 + nz_1 - p.$$

Cor. *To find the length of the perpendicular from the point*
$$(x_1, y_1, z_1)$$
to the plane as $ax + by + cz + d = 0.$

The normal form of the given equation being
$$\pm\frac{a}{\sqrt{\Sigma a^2}}x \pm \frac{b}{\sqrt{\Sigma a^2}}y \pm \frac{c}{\sqrt{\Sigma a^2}}z \pm \frac{d}{\sqrt{\Sigma a^2}} = 0.$$
the required length of the perpendicular is
$$\pm\frac{ax_1 + by_1 + cz_1 + d}{\sqrt{(a^2 + b^2 + c^2)}}.$$

The Plane

EXAMPLES

1. *Find the locus of a point, the sum of the squares of whose distances from the planes* $x+y+z=0$, $x-z=0$, $x-2y+z=0$ *is 9.*

Sol. Let the co-ordinates of the point be (α, β, γ). Its distances from the given planes are
$$\frac{\alpha+\beta+\gamma}{\sqrt{3}}, \frac{\alpha-\gamma}{\sqrt{2}}, \frac{\alpha-2\beta+\gamma}{\sqrt{6}}$$

We are given that
$$\left(\frac{\alpha+\beta+\gamma}{\sqrt{3}}\right)^2 + \left(\frac{\alpha-\gamma}{\sqrt{2}}\right)^2 + \left(\frac{\alpha-2\beta+\gamma}{\sqrt{6}}\right)^2 = 9$$

$\Rightarrow \quad 6\alpha^2 + 6\beta^2 + 6\gamma^2 = 54 \Rightarrow \alpha^2 + \beta^2 + \gamma^2 = 9$

Hence, the locus of (α, β, γ) is
$$x^2 + y^2 + z^2 = 9.$$

2. *Two systems of rectangular axes have the same origin. If a plane cuts them at distances a, b, c and a', b', c' respectively from the origin, prove that*
$$1/a^2 + 1/b^2 + 1/c^2 = 1/a'^2 + 1/b'^2 + 1/c'^2.$$

Sol. Equations of the plane w.r.t. two systems are
$$x/a + y/b + z/c = 1$$
and
$$x/a' + y/b' + z/c' = 1.$$

Since origin is common to both, hence the perpendicular distances of these planes from the origin must be equal. Hence,
$$\frac{1}{\sqrt{1/a^2 + 1/b^2 + 1/c^2}} = \frac{1}{\sqrt{1/a'^2 + 1/b'^2 + 1/c'^2}}$$

or
$$1/a^2 + 1/b^2 + 1/c^2 = 1/a'^2 + 1/b'^2 + 1/c'^2.$$

EXERCISES

1. Find the distances of the points $(2, 3, 4)$ and $(1, 1, 4)$ from the plane
$$3x - 6y + 2z + 11 = 0 \qquad \text{[Ans. 1; 16/7]}$$

2. Show that distances between the parallel planes
$$2x - 2y + z + 3 = 0 \text{ and } 4x - 4y + 2z + 5 = 0$$
is 1/6.
(The distance between two parallel planes is the distance of any point on one from the other).

3. Find the locus of the point whose distance from the origin is three times its distance from the plane $2x + 2z = 3$. [**Ans.** $3x^2 + 3z^2 - 4xy + 8xz - 4yz - 12x + 6y - 12z + 9 = 0$]

4. Show that $(1/8, 1/8, 1/8)$ is the incentre of the tetrahedron formed by the four planes $x = 0, y = 0, z = 0, x + 2y + 2z = 1.$

5. Sum of the distances of any number of fixed points from a variable plane is zero; show that the plane passes through a fixed point.

2.7.1 Bisectors of Angles Between Two Planes

Just as we have two bisectors between two given lines, we also have two bisectors between two given planes. Of course, the bisectors are now planes. We have proceed *to find the equations of the bisectors of the angle between the planes*
$$ax + by + cz + d = 0$$
$$a_1 x + b_1 y + c_1 z + d_1 = 0.$$

If (x, y, z) be a point on any one of the planes bisecting the angles between the planes, then the perpendiculars from this point to the two planes must be equal (in magnitude) so that

$$\frac{ax+by+cz+d}{\sqrt{(a^2+b^2+c^2)}} = \pm \frac{a_1x+b_1y+c_1z+d_1}{\sqrt{(a_1^2+b_1^2+c_1^2)}}$$

are the equations of the two bisecting planes.

Of these two bisecting planes, one bisects the acute and the other the obtuse angle between the given planes.

The bisector of the acute angle makes with either of the planes an angle which is less than $45°$ and the bisector of the obtuse angle makes with either of them an angle which is greater than $45°$. this gives a test for determining which angle, acute or obtuse, each bisecting plane bisects.

EXAMPLE

Find the equations of the planes bisecting the angles between the planes

$$x+2y+2z-3=0 \qquad ...(i)$$
$$3x+4y+12z+1=0 \qquad ...(ii)$$

and specify the one which bisects the acute angle.

Sol. The equations of the two bisecting planes are

$$\frac{x+2y+2z-3}{3} = \pm \frac{3x+4y+12z+1}{13}$$

$$\begin{cases} 2x+7y-5z-21=0, & ...(iii) \\ 11x+19y+31z-18=0 & ...(iv) \end{cases}$$

If θ be the angle between the planes (i) and (iii), we have

$$\cos\theta = \frac{2}{\sqrt{78}}$$

$\Rightarrow \qquad \tan\theta = \frac{\sqrt{74}}{2} > 1$

$\Rightarrow \theta$ is greater than $45°$.

$\Rightarrow$ the plane (iii) bisects the obtuse angle.

$\Rightarrow$ (iv) bisects the acture angle.

EXERCISES

1. Find the bisector of the acute angle between the planes
$$2x-y-2z+3=0, 3x-2y+6z+8=0.$$
[Ans. $23x-13y+32z+45=0$]

2. Show that the plane
$$14x-8y+13=0$$
bisects the obtuse angle between the planes
$$3x+4y-5z+1=0, 5x+12y-13z=0.$$

3. Find the bisector of that angle between the planes
$$3x-6y+2z+5=0, 4x-12y+3z-3=0$$
which contains the origin. [Ans. $67x-162y+47z+44=0$]

2.8 JOINT EQUATION OF TWO PLANES

Consider any two planes. We are interested in finding an equation which represents the two planes simultaneously. Thus, we propose to find an equation which will be satisfied if and only if a point lies on *either* of the two planes *i.e.*, either on one plane or the other or both.

Let
$$ax+by+cz+d=0 \qquad ...(i)$$
and
$$a_1x+b_1y+c_1z+d_1=0 \qquad ...(ii)$$
be the equations of two planes.

The Plane

Consider the equation
$$(ax + by + cz + d)(a_1x + b_1y + c_1z + d_1) = 0. \qquad \text{...(iii)}$$

A point (x_1, y_1, z_1) lies on (i)
$$\Rightarrow \quad ax_1 + by_1 + cz_1 + d = 0$$
$$\Rightarrow \quad (ax_1 + by_1 + cz_1 + d)(a_1x_1 + b_1y_1 + c_1z_1 + d_1) = 0.$$

A point (x_1, y_1, z_1) lies on (ii)
$$\Rightarrow \quad (ax_1 + by_1 + cz_1 + d)(a_1x_1 + b_1y_1 + c_1z_1 + d_1) = 0.$$

A point (x_1, y_1, z_1) lies on (iii)
$$\Rightarrow \quad ax_1 + by_1 + cz_1 + d = 0 \quad \text{or} \quad a_1x_1 + b_1y_1 + c_1z_1 + d_1 = 0.$$
$\Rightarrow \quad (x_1, y_1, z_1)$ lies on the plane (i) or on the plane (ii).

Thus, we have the following :

A point (x, y, z) either lies on the plane (i) or on the plane (ii)
$$\Leftrightarrow \quad (ax + by + cz + d)(a_1x + b_1y + c_1z + d_1) = 0.$$

We thus say that
$$(ax + by + cz + d)(a_1x + b_1y + c_1z + d_1) = 0$$
is the equation of the two planes.

2.8.1

Condition for the homogeneous second degree equation
$$ax^2 + by^2 + cz^2 + 2fyz + 2gzx + 2hxy = 0 \qquad \text{...(i)}$$
to represent two planes.

We suppose that the given equation represents two planes.
Let the equation of the two planes separately be
$$lx + my + nz = 0,$$
and
$$l'x + m'y + n'z = 0.$$

There cannot appear constant terms in the separate equations of the planes, for, otherwise, their joint equation will not be homogeneous.

We have
$$ax^2 + by^2 + cz^2 + 2fyz + 2gzx + 2hxy = (lx + my + nz)(l'x + m'y + n'z)$$
$$\Rightarrow \begin{cases} a = ll', \, b = mm', \, c = nn' \\ 2f = m'n + mn', \, 2g = ln' + l'n, \, 2h = lm' + l'm. \end{cases}$$

The required condition which is essentially the condition for the consistency of these equations is obtained on eliminating $l, m, n;\; l', m', n'$ from the above six relations and this can be easily effected as follows. We have

$$0 = \begin{vmatrix} l & l' & 0 \\ m & m' & 0 \\ n & n' & 0 \end{vmatrix} \times \begin{vmatrix} l' & l & 0 \\ m' & m & 0 \\ n' & n & 0 \end{vmatrix} = \begin{vmatrix} ll' + l'l & l'm + lm' & l'n + ln' \\ lm' + l'm & mm' + m'm & m'n + mn' \\ n'l + nl' & n'm + nm' & n'n + nn' \end{vmatrix}$$

$$= 8 \begin{vmatrix} a & h & g \\ h & b & f \\ g & f & c \end{vmatrix} = 8(abc + 2fgh - af^2 - bg^2 - ch^2).$$

Thus, if the equation
$$ax^2 + by^2 + cz^2 + 2fyz + 2gzx + 2hxy = 0,$$
represents two planes, we have the condition,
$$abc + 2fgh - af^2 - bg^2 - ch^2 = 0.$$

Cor. angle between planes. If θ be the angle between the planes represented by the equation (i), we have if $ll' + mm' + nn' \neq 0$

$$\tan \theta = \frac{\sqrt{(mn' - m'n)^2 + (nl' - n'l)^2 + (lm' - l'm)^2}}{ll' + mm' + nn'}$$

$$= \frac{2\sqrt{f^2 + g^2 + h^2 - ab - bc - ca}}{a + b + c}.$$

The planes will be at right angles if
$$ll' + mm' + nn' = 0 \Leftrightarrow a + b + c = 0.$$

Ex. *Prove that the equation* $2x^2 - 6y^2 - 12z^2 + 18yz + 2zx + xy = 0$ *represents a pair of planes. Also find the angle between them.*

Sol. $a = 2, b = -6, c = -12, f = 9, g = 1, h = 1/2$. These values satisfy the condition
$$abc + 2fgh - af^2 - bg^2 - ch^2 = 0.$$

$$\tan \theta = \frac{2\sqrt{f^2 + g^2 + h^2 - ab - bc - ca}}{a + b + c} = \frac{2\sqrt{185}}{2(-16)} = -\frac{\sqrt{185}}{16}$$

EXERCISES

Show that the following equations represent pairs of planes and also find the angles between each pair.

(i) $12x^2 - 2y^2 - 6z^2 - 2xy + 7yz + 6zx = 0$. [Ans. $\cos^{-1}(4/21)$]

(ii) $2x^2 - 2y^2 + 4z^2 + 6xz + 2yz + 3xy = 0$. [Ans. $\cos^{-1}(4/9)$]

(iii) Show that the equation
$$\frac{a}{y-z} + \frac{b}{z-x} + \frac{c}{x-y} = 0$$
represents a pair of planes.

2.9 ORTHOGONAL PROJECTION ON A PLANE

Corresponding to the notion of projection on a line, we also have that of projection on a plane whcih we now proceed to consider.

Def. Orthogonal projection on a plane. *The foot of the perpendicular from a point to a given plane is called the orthogonal projection of the point on the plane.*

This plane on which we project is called the plane of the projection.

Thus, (Fig. 1, page 1) *L, M, N* are respectively the orthogonal projections of the point *P* on the YZ, ZX and XY planes.

The proejction of a curve on a plane is the locus of the projections on the plane of any point on the curve.

The projection on a given plane of the area enclosed by a plane curve is the area enclosed by the projection of the curve on the plane.

In particular, the projection of a straight line on a given plane is the locus of the feet of the perpendiculars drawn from points on the line on the plane.

2.9.1

The following simple results in *Pure Solid Geometry* are assumed without proof :

(1) The projection of a straight line is a straight line.

(2) If a line *AB* in a plane be perpendicular to the line of intersection of this plane with the plane of projection, then the length of its projection is $AB \cos \theta$; θ being the angle between the two planes.

In case a segment *AB* is parallel to the plane of projection, then the length of the projection is the same as that of *AB*.

(3) The projection of the area, *A*, enclosed by a curve in a plane is $A \cos \theta$; θ being the angle between the plane of the curve containing the given area and the plane of projection.

The Plane

Theorem. *If A_x, A_y, A_z be the areas of the projections of an area, A, on the three co-ordinate planes, then*

$$A^2 = A_x^2 + A_y^2 + A_z^2$$

Let l, m, n be the direction cosines of the normal to the plane of the area A.
Since l is the cosine of the angle between the YZ plane and the plane of the area A, therefore,

$$A_x = lA.$$

Similarly, $A_y = mA,$

and $A_z = nA.$

Hence $A_x^2 + A_y^2 + A_z^2 = A^2 (l^2 + m^2 + n^2) = A^2.$

EXAMPLE

A plane makes intercepts $OA = a, OB = b$ and $OC = c$ respectively on the co-ordinate axes. Find the area of $\triangle ABC$.

Sol. Co-ordinates of the points A, B, C are $(a, 0, 0), (0, b, 0)$ and $(0, 0, c)$. Now, if A_x, A_y, A_z be the projections of the area of $\triangle ABC$ on the planes $x = 0, y = 0, z = 0$ respectively, then

A_x = area of $\triangle OBC$

$= \frac{1}{2} OB \cdot OC = \frac{1}{2} bc$

Similarly,

$A_y = \frac{1}{2} ac$ and $A_z = \frac{1}{2} ab.$

$\therefore$ Area of

$\triangle ABC = \sqrt{A_x^2 + A_y^2 + A_z^2}$

$= \frac{1}{2} \sqrt{a^2 b^2 + b^2 c^2 + c^2 a^2}.$

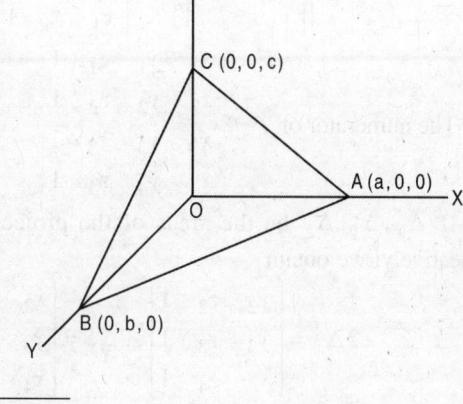

EXERCISES

1. Find the areas of the triangles whose vertices are the points :
   (i) $(a, 0, 0), (0, b, 0), (0, 0, c)$. (ii) $(x_1, y_1, z_1), (x_2, y_2, z_2), (x_3, y_3, z_3).$

2. From a point $P(x', y', z')$ a plane is drawn at right angles to OP to meet the co-ordinate axes at A, B, C; prove that the area of the triangle ABC is $r^5 / 2x'y'z'$, where r is the measure of OP.

2.10 VOLUME OF A TETRAHEDRON

To find the volume of a tetrahedron in terms of the co-ordinates $(x_1, y_1, z_1), (x_2, y_2, z_2), (x_3, y_3, z_3), (x_4, y_4, z_4)$ of its vertices A, B, C, D.

Let V be the volume of the tetrahedron $ABCD$
Then

$$V = \frac{1}{3} \Delta p, \qquad ...(i)$$

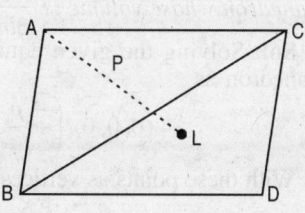

where p is the length of the perpendicular AL from a vertex A to the opposite face BCD; and Δ is the area of the triangle BCD.
The equation of the plane BCD is

$$\Leftrightarrow \quad x\begin{vmatrix} x & y & z & 1 \\ x_2 & y_2 & z_2 & 1 \\ x_3 & y_3 & z_3 & 1 \\ x_4 & y_4 & z_4 & 1 \end{vmatrix} = 0$$

$$\Leftrightarrow \quad x\begin{vmatrix} y_2 & z_2 & 1 \\ y_3 & z_3 & 1 \\ y_4 & z_4 & 1 \end{vmatrix} - y\begin{vmatrix} x_2 & z_2 & 1 \\ x_3 & z_3 & 1 \\ x_4 & z_4 & 1 \end{vmatrix} + z\begin{vmatrix} x_2 & y_2 & 1 \\ x_3 & y_3 & 1 \\ x_4 & y_4 & 1 \end{vmatrix} - \begin{vmatrix} x_2 & y_2 & z_2 \\ x_3 & y_3 & z_3 \\ x_4 & y_4 & z_4 \end{vmatrix} = 0 \qquad \ldots(i)$$

$$\therefore \quad \frac{x_1\begin{vmatrix} y_2 & z_2 & 1 \\ y_3 & z_3 & 1 \\ y_4 & z_4 & 1 \end{vmatrix} - y_1\begin{vmatrix} x_2 & z_2 & 1 \\ x_3 & z_3 & 1 \\ x_4 & z_4 & 1 \end{vmatrix} + z_1\begin{vmatrix} x_2 & y_2 & 1 \\ x_3 & y_3 & 1 \\ x_4 & y_4 & 1 \end{vmatrix} - \begin{vmatrix} x_2 & y_2 & z_2 \\ x_3 & y_3 & z_3 \\ x_4 & y_4 & z_4 \end{vmatrix}}{\left\{ \begin{vmatrix} y_2 & z_2 & 1 \\ y_3 & z_3 & 1 \\ y_4 & z_4 & 1 \end{vmatrix}^2 + \begin{vmatrix} x_2 & z_2 & 1 \\ x_3 & z_3 & 1 \\ x_4 & z_4 & 1 \end{vmatrix}^2 + \begin{vmatrix} x_2 & y_2 & 1 \\ x_3 & y_3 & 1 \\ x_4 & y_4 & 1 \end{vmatrix}^2 \right\}^{1/2}} \qquad \ldots(iii)$$

The numerator of $p = \begin{vmatrix} x_1 & y_1 & z_1 & 1 \\ x_2 & y_2 & z_2 & 1 \\ x_3 & y_3 & z_3 & 1 \\ x_4 & y_4 & z_4 & 1 \end{vmatrix}$.

If $\Delta_x, \Delta_y, \Delta_z$ be the areas of the projections of the triangle on the YZ, ZX, XY planes respectively, we obtain

$$2\Delta_x = \begin{vmatrix} y_2 & z_2 & 1 \\ y_3 & z_3 & 1 \\ y_4 & z_4 & 1 \end{vmatrix}, \; 2\Delta_y = \begin{vmatrix} x_2 & z_2 & 1 \\ x_3 & z_3 & 1 \\ x_4 & z_4 & 1 \end{vmatrix}, \; 2\Delta_3 = \begin{vmatrix} x_2 & y_2 & 1 \\ x_3 & y_3 & 1 \\ x_4 & y_4 & 1 \end{vmatrix}$$

Therefore the denominator of $p = [4(\Delta_x^2 + \Delta_y^2 + \Delta_z^2)]^{1/2} = 2\Delta$.

From (i) and (ii), we deduce that the required volume is

$$\frac{1}{3}\Delta p = \frac{1}{6}\begin{vmatrix} x_1 & y_1 & z_1 & 1 \\ x_2 & y_2 & z_2 & 1 \\ x_3 & y_3 & z_3 & 1 \\ x_4 & y_4 & z_4 & 1 \end{vmatrix}.$$

EXAMPLES

1. *Prove that the four planes $my + nz = 0$, $nz + lx = 0$, $lx + my = 0$, $lx + my + nz = p$ form a tetrahedron whose volume is $\dfrac{2p^3}{3lmn}$.*

Sol. Solving the given equations taking three planes at a time, we get the vertices of the tetrahedron as

$$(0, 0, 0), \left(\frac{-p}{l}, \frac{p}{m}, \frac{p}{n}\right), \left(\frac{p}{l}, \frac{-p}{m}, \frac{p}{n}\right) \text{ and } \left(\frac{p}{l}, \frac{p}{m}, \frac{-p}{n}\right).$$

With these points as vertices, the volume V of the terahedron is given by

The Plane

$$V = \frac{1}{6}\begin{vmatrix} 0 & 0 & 0 & 1 \\ -p/l & p/m & p/n & 1 \\ p/l & -p/m & p/n & 1 \\ p/l & p/m & -p/n & 1 \end{vmatrix} = \frac{-p^3}{6lmn}\begin{vmatrix} -1 & 1 & 1 \\ 1 & -1 & 1 \\ 1 & 1 & -1 \end{vmatrix}$$

$$= \frac{p^3}{6lmn}(4) = \frac{2}{3}\frac{p^3}{lmn}.$$

2. Find the volume of a tetrahedron in terms of the lengths of the three edges which meet in a point and of the angles which these edges make with each other in pairs.

Sol. Let $OABC$ be a tetrahedron.

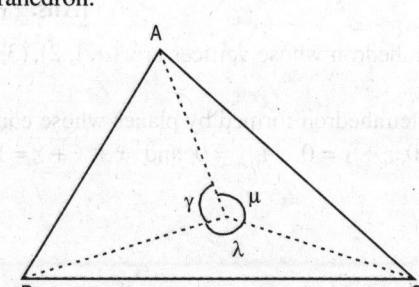

Let $OA = a, OB = b, OC = c$.
Let $\angle BOC = \lambda, \angle COA = \mu, \angle AOB = \nu$.

We take O as origin and any system of three mutually perpendicular lines through O as co-ordinate axes.

Let the direction cosines of the lines OA, OB, OC be

$$l_1, m_1, n_1; l_2, m_2, n_2; l_3, m_3, n_3.$$

Thus, the co-ordinates of A, B, C are

$$(l_1 a, m_1 a, n_1 a); (l_2 b, m_2 b, n_2 b); (l_3 c, m_3 c, n_3 c)$$

Therefore, the volume of the tetrahedron $OABC$

$$= \frac{1}{6}\begin{vmatrix} 0 & 0 & 0 & 1 \\ l_1 a & m_1 a & n_1 a & 1 \\ l_2 b & m_2 b & n_2 b & 1 \\ l_3 c & m_3 c & n_3 c & 1 \end{vmatrix} = \frac{1}{6}\begin{vmatrix} l_1 a & m_1 a & n_1 a \\ l_2 b & m_2 b & n_2 b \\ l_3 c & m_3 c & n_3 c \end{vmatrix} = \frac{abc}{6}\begin{vmatrix} l_1 & m_1 & n_1 \\ l_2 & m_2 & n_2 \\ l_3 & m_3 & n_3 \end{vmatrix}$$

Now

$$\begin{vmatrix} l_1 & m_1 & n_1 \\ l_2 & m_2 & n_2 \\ l_3 & m_3 & n_3 \end{vmatrix}^2 = \begin{vmatrix} l_1 & m_1 & n_1 \\ l_2 & m_2 & n_2 \\ l_3 & m_3 & n_3 \end{vmatrix} \times \begin{vmatrix} l_1 & m_1 & n_1 \\ l_2 & m_2 & n_2 \\ l_3 & m_3 & n_3 \end{vmatrix}$$

$$= \begin{vmatrix} \Sigma l_1^2 & \Sigma l_1 l_2 & \Sigma l_1 l_2 \\ \Sigma l_1 l_2 & \Sigma l_2^2 & \Sigma l_2 l_3 \\ \Sigma l_3 l_1 & \Sigma l_3 l_2 & \Sigma l_3^2 \end{vmatrix} = \begin{vmatrix} 1 & \cos \nu & \cos \mu \\ \cos \nu & 1 & \cos \lambda \\ \cos \mu & \cos \lambda & 1 \end{vmatrix}$$

Thus, the volume of the tetrahedron $OABC$

$$= \frac{abc}{6}\begin{vmatrix} 1 & \cos \nu & \cos \mu \\ \cos \nu & 1 & \cos \lambda \\ \cos \mu & \cos \lambda & 1 \end{vmatrix}^{1/2}$$

EXERCISES

1. The vertices of a tetrahedron are $(0, 1, 2)$, $(3, 0, 1)$, $(4, 3, 6)$, $(2, 3, 2)$; show that its volume is 6.
2. A, B, C are three fixed points and a variable point P moves so that the volume of the tetrahedron $PABC$ is constant; show that the locus of the point P is a plane parallel to the plane ABC.
3. A variable plane makes with the co-ordinate planes a tetrahedron of constant volume $64k^3$. Find

    (i) the locus of the centroid of the tetrahedron. **[Ans.** $xyz = 6k^3$**]**

    (ii) the locus of the foot of the perpendicular from the origin to the plane.

    [Ans. $(x^2 + y^2 + z^2)^3 = 384k^3 xyz$**]**

4. If the volume of the tetrahedron whose vertices are $(a, 1, 2), (3, 0, 1), (4, 3, 6), (2, 3, 2)$ is 6. Find the value of a. **[Ans. 0]**
5. Find the volume of the tetrahedron formed by planes whose equations are
$$y + z = 0, z + x = 0, x + y = 0 \text{ and } x + y + z = 1.$$ **[Ans. 2/3]**

3
The Straight Line

3.1 REPRESENTATION OF LINE

In this chapter, it is proposed to discuss the manner in which a straight line can be represented. We introduce the method analytically as follows :

Consider any two of the co-ordinate planes say YOZ and ZOX, whose equations are $x = 0$ and $y = 0$ respectively. These two planes intersect in Z-axis.

A point (x, y, z) lies on the Z-plane
$\Leftrightarrow$ {the point (x, y, z) lies on the YOZ plane **and** the point (x, y, z) lies on the ZOX plane
$\Leftrightarrow$ $x = 0$ **and** $y = 0$.

Thus, we see that a point (x, y, z) lies on the Z-axis if and only if, we simultaneously have $x = 0, y = 0$. We are thus, led to say that $x = 0, y = 0$ are the two equations of Z-axis.

Consider now any line whatsoever and any two planes through the line. Let
$$ax + by + cz + d = 0 \text{ and } a_1 x + b_1 y + c_1 z + d_1 = 0$$
be the equations of these two planes. Clearly, we have the following statement :

A point (x, y, z) lies on the given line if and only if we simultaneously have
$$ax + by + cz + d = 0 \text{ and } a_1 x + b_1 y + c_1 z + d_1 = 0$$
Thus, we say that
$$ax + by + cz + d = 0 \text{ and } a_1 x + b_1 y + c_1 z + d_1 = 0$$
are the two equations of the line.

It follows that a straight line is represented by **two** *equatons of the first degree in x, y, z.*

Of course any given line can be represented by *different pairs* of first degree equations, for we may take *any* pair of planes through the line and the equations of the same will constitute the equations of the line.

In particular, as the X-axis is the intersection of the XZ and XY planes, $y = 0, z = 0$ taken together are its equations. Similarly $x = 0, z = 0$ are the equations of the Y-axis and $x = 0, y = 0$ are the equations of the Z-axis.

EXERCISES

1. What is the locus of the point (x, y, z) which satisfies the following conditions :
   (i) $2x + 3y - 4z + 1 = 0$ and $3x - y + z + 2 = 0$
   (ii) $2x + 3y - 4z + 1 = 0$ or $3x - y + z + 2 = 0$
   (iii) $2x - 3y + 5z + 4 = 0$ and $2x + y + z - 8 = 0$
   (iv) $2x - 3y + 5z + 4 = 0$ or $2x + y + z - 8 = 0$.
2. Find the intersection of the line
   $$x - 2y + 4z + 4 = 0, \quad x + y + z - 8 = 0.$$
   with the plane
   $$x - y + 2z + 1 = 0.$$
   [Ans. 2, 5, 1]

3.1.1. Equaton of the Line Through a Given Point Drawn in a Given Direction

To find the equations of the line passing through a given point $A(x_1, y_1, z_1)$ *and having direction cosiens* l, m, n.

Let $P(x, y, z)$ be a point on the given line and let $AP = r$.

Projecting the segment AP on the co-ordinate axes, we obtain
$$x - x_1 = lr, \ y - y_1 = mr, \ z - z_1 = nr \qquad \ldots(i)$$
so that for all points (x, y, z) on the given line, we have
$$x = x_1 + lr, \ y = y_1 + mr, \ z = z_1 + nr.$$
Thus, the set of points on the given line is
$$\{(x_1 + lr, \ y_1 + mr, \ z_1 + nr)\};$$
r being any number.

In case none of l, m, n is zero, we have
$$\frac{x - x_1}{l} = \frac{y - y_1}{m} = \frac{z - z_1}{n} = r.$$
Thus, if $l \neq 0, m \neq 0, n \neq 0$, or equivalently $lmn \neq 0$,
$$\frac{x - x_1}{l} = \frac{y - y_1}{m} = \frac{z - z_1}{n} \qquad \ldots(ii)$$
are the *two* required equations of the line.

Clearly the equations (ii) of the line are not altered if we replace the direction cosines l, m, n by the three numbers proportional to them, so that it suffices to use direction ratios in place of direction cosines while writing down the equation of a line.

Cor. *From the relation (i)*, we have
$$x = x_1 + lr, \ y = y_1 + mr, \ z = z_1 + nr$$
so that the set of points on the line through the point (x_1, y_1, z_1) and having direction ratios l, m, n is
$$\{(x_1 + lr, \ y_1 + mr, \ z_1 + nr); \ r \text{ being any number}\}.$$
This statement does not depend upon the vanishing or otherwise of any of l, m, n

We may remark that r is what is known as the parameter here.

Note. The equation
$$\frac{x - x_1}{l} = \frac{y - y_1}{m}$$
of first degree, being free of z, represents the plane through the line drawn perpendicular to the XOY plane. Similar statements may be made about the equations
$$\frac{y - y_1}{m} = \frac{z - z_1}{n}, \ \frac{z - z_1}{n} = \frac{x - x_1}{l}.$$
The two equations
$$(x - x_1)/l = (y - y_1)/m, \ (y - y_1)/m = (z - z_1)/n$$
represents a pair of planes through the given line.

3.1.2. Equation of a Line Through Two Points

To find the equations of the line through two points
$$(x_1, y_1, z_1) \text{ and } (x_2, y_2, z_2)$$
Since
$$x_2 - x_1, \ y_2 - y_1, \ z_2 - z_1$$
are proportioanl to the direction cosines of the line, the required equations are
$$\frac{x - x_1}{x_2 - x_1} = \frac{y - y_1}{y_2 - y_1} = \frac{z - z_1}{z_2 - z_1}$$
Here we have assumed that none of
$$x_2 - x_1, \ y_2 - y_1, \ z_2 - z_1$$
is zero.

EXAMPLES

1. *If the axes are rectangular and if $l_1, m_1, n_1; l_2, m_2, n_2$ are direction cosines, show that the equations to the planes through the lines which bisect the angle between*
$$x/l_1 = y/m_1 = z/n_1; \ x/l_2 = y/m_2 = z/n_2$$
and at right angles to the plane containing them are

The Straight Line

$$(l_1 \pm l_2) x + (m_1 \pm m_2) y + (n_1 \pm n_2) z = 0.$$

Sol. The given lines pass through the origin. Co-ordinates of any two points, each of them at a distance r from the origin are (rl_1, rm_1, rn_1) and (rl_2, rm_2, rn_2). The co-ordinates of the middle point P of the line joining these two points are $\frac{1}{2} r (l_1 + l_2), \frac{1}{2} r (m_1 + m_2), \frac{1}{2} r (n_1 + n_2)$.

The point P clearly lies on one of the bisectors and since the two bisectors are at right angles to each other, hence, OP is normal to the plane passing throguh the other bisectors. The d.c.s of OP are proportional to

$$\frac{1}{2} (l_1 + l_2), \frac{1}{2} (m_1 + m_2), \frac{1}{2} (n_1 + n_2).$$

Hence, one of the required planes is

$$\frac{1}{2} (l_1 + l_2) x + \frac{1}{2} (m_1 + m_2) y + \frac{1}{2} (n_1 + n_2) z = 0.$$

i.e., $$(l_1 + l_2) x + (m_1 + m_2) y + (n_1 + n_2) z = 0.$$

Similarly, if P lies on the other bisector, its co-ordiantes will then be

$$\frac{1}{2} (l_1 - l_2) r, \frac{1}{2} (m_1 - m_2) r, \frac{1}{2} (n_1 - n_2) r.$$

The corresponding plane, therefore, will be

$$(l_1 - l_2) x + (m_1 - m_2) y + (n_1 - n_2) z = 0.$$

2. *Find the image of the point* $P (1, 3, 4)$ *in the plane*

$$2x - y + z + 3 = 0.$$

Sol. If two points P, Q be such that the line is bisected perpendicularly by a plane, then either of the points is the image of the other in the plane.

the line throguh P perpendicular to the given plane is

$$\frac{x-1}{2} = \frac{y-3}{-1} = \frac{z-4}{1},$$

so that the co-ordinates of Q are of the from

$$(2r + 1, -r + 3, r + 4).$$

Making use of the fact that the mid-point

$$\left(r + 1, -\frac{1}{2} r + 3, \frac{1}{2} r + 4 \right)$$

of PQ lies on the given plane, we see that

$$r = -2$$

so that the image of P is $(-3, 5, 2)$.

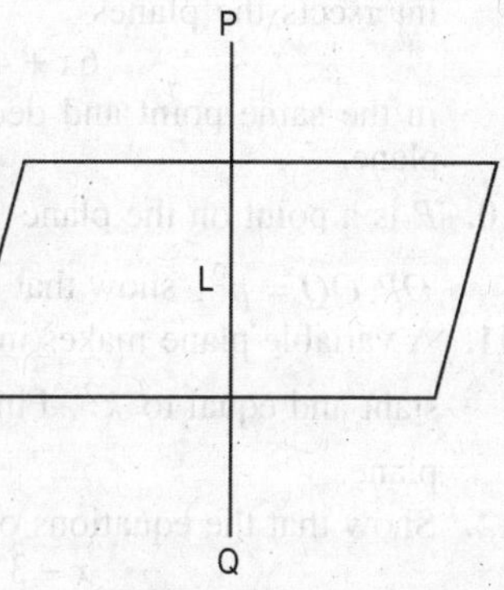

Fig. 1

EXERCISES

1. Find k so that the lines

$$\frac{x-1}{-3} = \frac{y-2}{2k} = \frac{z-3}{2}$$

$$\frac{x-1}{3k} = \frac{y-5}{1} = \frac{z-6}{-5}$$

may be perpendicular to each other. [Ans. $-10/7$]

2. Find two points on the line

$$\frac{x-2}{1} = \frac{y+3}{-2} = \frac{z-5}{2}$$

on either side of $(2, -3, -5)$ and at a distance 3 from it. [Ans. $(3, -5, -3); (1, -1, -7)$]

3. Find the point where the line joining $(2, -3, 1), (3, -4, -5)$ cuts the plane

$$2x + y + z = 7.$$

[Ans. $1, -2, 7$]

4. Find the distance of the point $(-1, -5, -10)$ from the point of intersection of the line $\frac{1}{2}(x-2) = \frac{1}{4}(y+1) = \frac{1}{12}(z-2)$ and the plane $x - y + z = 5$.

5. Find the distance of the point $(3, -4, 5)$ from the plane
$$2x + 5y - 6z = 16$$
measured along a line with direction cosines proportional to $(2, 1, -2)$. [Ans. 60/7]

6. Find the equations to the line through $(-1, 3, 2)$ and perpendicular to the plane $x + 2y + 2z = 3$, the length of the perpendicular and the co-ordinates of its foot.
[Ans. 2; $(-5/3, 5/3, 2/3)$]

7. Find the co-ordinates of the foot of the perpendicular drawn from the origin to the plane $2x + 3y - 4z + 1 = 0$; also find the co-ordinates of the point which is the image of the origin in the plane. [Ans. $(-2/29, -3/29, 4/29); (-4/29, -6/29, 8/29)$]

8. Find the equations to the line through (x_1, y_1, z_1) perpendicular to the plane $ax + by + cz + d = 0$ and the co-ordinates of its foot. Deduce the expression for the perpendicular distance of the given point from the given plane.

[Ans. $(ar + x_1, br + y_1, cr + z_1)$, where $r = -(ax_1 + by_1 + cz_1 + d)/(a^2 + b^2 + c^2)$]

9. Show that the line
$$\frac{1}{2}(x - 7) = -(y + 3) = (z - 4)$$
intersects the planes
$$6x + 4y - 5z = 4 \quad \text{and} \quad x - 5y + 2z = 12$$
in the same point and deduce that the line is coplanar with the line of intersection of the plane.

10. P is a point on the plane $lx + my + nz = p$ and a point Q is taken on the line OP such that $OP \cdot OQ = p^2$; show that the locus of the point Q is $p(lx + my + nz) = x^2 + y^2 + z^2$.

11. A variable plane makes intercepts on the co-ordinate axes the sum of whose squares is constant and equal to k^2. Find the locus of the foot of the perpendicular from the origin to the plane. [Ans. $(x^{-2} + y^{-2} + z^{-2})(x^2 + y^2 + z^2)^2 = k^2$]

12. Show that the equations of the lines bisecting the angles between the lines
$$\frac{x-3}{2} = \frac{y+4}{-1} = \frac{z-5}{-2}, \quad \frac{x-3}{4} = \frac{y+4}{-12} = \frac{z-5}{3}$$
are
$$\frac{x-3}{38} = \frac{y+4}{-49} = \frac{z-5}{-17}, \quad \frac{x-3}{14} = \frac{y+14}{23} = \frac{z-5}{-35}.$$

3.1.3. Two Forms of the Equation of a Line

It has been seen in 3.1.1, 3.1.2, that the equations of a straight line which we generally employ are of two forms.

One is the form deduced from the consideration that a straight line is completely determined when we know its direction ratios and the co-ordinates of any one point on it, or when any two points on the line are given. This is sometimes referred to as the **Symmetrical form** of the equations of a line.

The second form a is deduced from the consideration that a straight line is the locus of points common to any two planes through it. This is sometimes referred to as the **Unsymmetrical form** of the equations of a line.

In fact the symmetrical form takes note only of a special pair of planes through this line, viz., the pair of planes through the line perpendicular to two of the co-ordinate planes.

In the next section, it will be seen how one form of equations can be transferred into the other.

The Straight Line

3.1.4. Transformation from the Unsymmetrical to the Symmetrical Form

To transform the equations
$$ax + by + cz + d = 0, \; a_1 x + b_1 y + c_1 z + d_1 = 0,$$
of a line to the symmetrical form.
To transform these equations to the symmetrical form, we require :
(i) the direction ratios of the line, and
(ii) the co-ordinates of any one point on it.
Let l, m, n be the direction ratios of the line. Since the line lies in both the planes
$$ax + by + cz + d = 0 \text{ and } a_1 x + b_1 y + c_1 z + d_1 = 0,$$
it is perpendicular to the normals to both of them. The direction ratios of the normals to the planes being
$$a, b, c; a_1, b_1, c_1,$$
we have
$$\begin{cases} al + bm + cn = 0, \\ a_1 l + b_1 m + c_1 n = 0, \end{cases}$$
$$\Rightarrow \quad \frac{l}{bc_1 - b_1 c} = \frac{m}{ca_1 - c_1 a} = \frac{n}{ab_1 - a_1 b}.$$

Now, we require the co-ordinates of *any one* point on the line and there is an infinite number of points from which to choose. We, for the sake of convenience, find the point of intersection of the line with the plane $z = 0$. This point which is given by the equations
$$ax + by + d = 0 \text{ and } a_1 x + b_1 y + d_1 = 0,$$
is
$$\left(\frac{bd_1 - b_1 d}{ab_1 - a_1 b}, \frac{a_1 d - a d_1}{ab_1 - a_1 b}, 0 \right).$$

Thus, in the symmetrical form, the equations of the given line are
$$\frac{x - (bd_1 - b_1 d)/(ab_1 - a_1 b)}{bc_1 - b_1 c} = \frac{y - (a_1 d - a d_1)/(ab_1 - a_1 b)}{ca_1 - c_1 a} = \frac{z - 0}{ab_1 - a_1 b}.$$

EXAMPLES

1. Find the equation of the line through the point $(1, 2, 3)$ parallel to the line
$$x - y + 2z = 5, \; 3x + y + z = 6.$$

Sol. Let l, m, n be the direction ratios of the required line. Since it is parallel to the given line, the direction ratios of the given line are also l, m, n. But the given line is the intersection of the two planes $x - y + 2z = 5$ and $3x + y + z = 6$, and hence, lies in both the planes and is perpendicular to the normals of these planes.

$$l \cdot 1 - m \cdot 1 + n \cdot 2 = 0$$
and
$$l \cdot 3 + m \cdot 1 + n \cdot 1 = 0$$
$$\Rightarrow \quad \frac{l}{-3} = \frac{m}{5} = \frac{n}{4}$$

Thus, the equations of the line in symmetrical form are
$$\frac{x - 1}{-3} = \frac{y - 2}{5} = \frac{z - 3}{4}.$$

2. Prove that the equations to the line through (α, β, γ) at right angles to the lines
$$\frac{x}{l_1} = \frac{y}{m_1} = \frac{z}{n_1} ; \; \frac{x}{l_2} = \frac{y}{m_2} = \frac{z}{n_2}$$
are
$$\frac{x - \alpha}{m_1 n_2 - m_2 n_1} = \frac{y - \beta}{n_1 l_2 - n_2 l_1} = \frac{z - \gamma}{l_1 m_2 - l_2 m_1}.$$

Sol. Let the dc's of the required line be l, m, n. Since it is perpendicular to the given lines, hence
$$ll_1 + mm_1 + nn_1 = 0 \text{ and } ll_2 + mm_2 + nn_2 = 0.$$
Solving, we get
$$l/(m_1 n_2 - m_2 n_1) = m/(n_1 l_2 - n_2 l_1) = n/(l_1 m_2 - l_2 m_1).$$
Hence, the equations of the required line are
$$\frac{x - \alpha}{(m_1 n_2 - m_2 n_1)} = \frac{y - \beta}{(n_1 l_2 - n_2 l_1)} = \frac{z - \gamma}{(l_1 m_2 - l_2 m_1)}.$$

EXERCISES

1. Find, in a symmetrical form, the equations of the line
$$x + y + z = 1 = 0, \; 4x + y - 2z + 2 = 0$$
   and find its direction cosines. $\left[\text{Ans. } \frac{x + 1/3}{1} = \frac{y + 2/3}{-2} = \frac{z}{1}; \frac{1}{\sqrt{6}}, -\frac{2}{\sqrt{6}}, \frac{1}{\sqrt{6}}\right]$

2. Obtain the symmetrical form of the equations of the line
$$x - 2y + 3z = 4, \; 2x - 3y + 4z = 5.$$
   $\left[\text{Ans. } (x + 2) = \frac{1}{2}(y + 3) = z\right]$

3. Find the points of intersection of the line
$$x + y - z + 1 = 0 = 14x + 9y - 7z - 1$$
   with the XY and YZ planes, and hence put down the symmetrical form of its equations.
   $[\text{Ans. } -(x)/2 = (y - 4)/7 = (z - 5)/5]$

4. Find the equation of the plane through the point $(1, 1, 1)$ and perpendicular to the line
$$x - 2y + z = 2, \; 4x + 3y - z + 1 = 0.$$
   $[\text{Ans. } x - 5y - 11z + 15 = 0]$

5. Find the equation of the line through the point $(1, 2, 4)$ parallel to the line
$$3x + 2y - z = 4, \; x - 2y - 2z = 5.$$
   $[\text{Ans. } (x - 1)/6 = (2 - y)/5 = (z - 4)/8]$

6. Find the angle between the lines in which the planes
$$3x - 7y - 5z = 1, \; 5x - 13y + 3z + 2 = 0$$
   cut the plane $8x - 11y + z = 0$. $[\text{Ans. } 90°]$

7. Find the angle between the lines
$$3x + 2y + z - 5 = 0 = x + y - 2z - 3,$$
$$2x - y - z = 0 = 7x + 10y - 8z.$$
   $[\text{Ans. } 90°]$

8. Show that the condition for the lines
$$x = az + b, \; y = cz + d; \; x = a_1 z + b_1, \; y = c_1 z + d_1,$$
   to be perpendicular is
$$aa_1 + cc_1 + 1 = 0.$$

3.2 ANGLE BETWEEN A LINE AND A PLANE

To find the angle between the line
$$\frac{x - x_1}{l} = \frac{y - y_1}{m} = \frac{z - z_1}{n}$$
and the plane
$$ax + by + cz + d = 0.$$

The angle between a line and a plane is the complement of the angle between the line and the normal to the plane.

Since the direction cosines of the normal to the given plane and of the given line are proportional to a, b, c and l, m, n respectively, we have

The Straight Line

$$\sin \theta = \frac{al + bm + cn}{\sqrt{(a^2 + b^2 + c^2)} \sqrt{(l^2 + m^2 + n^2)}},$$

where θ is the required angle.
The straight line is *parallel to the plane*
$\Rightarrow \qquad \theta = 0$
$\Rightarrow \qquad al + bm + cn = 0.$

This codition is also evident from the fact that a *line will be parallel to a plane if and only if it is perpendicular to the normal to it.*

EXERCISES

1. Show that the line $\frac{1}{3}(x - 2) = \frac{1}{4}(y - 3) = \frac{1}{5}(z - 4)$ is parallel to the plane $2x + y - 2z = 3$.

2. Find the equations of the line through the point $(-2, 3, 4)$ and parallel to the planes $2x + 3y + 4z = 5$ and $3x + 4y + 5z = 6$.

$$\left[\text{Ans. } (x + 2) = -\frac{1}{2}(y - 3) = (z - 4) \right]$$

 [Hint. The direction ratios, l, m, n of the line are given by the relations $2l + 3m + 4n = 0 = 3l + 4m + 5n$.]

3. Find the equation of the plane through the points
$(1, 0, -1), (3, 2, 2)$
and parallel to the line.
$(x - 1) = (1 - y)/2 = (z - 2)/3.$ [Ans. $4x - y - 2z = 6$]

4. Show that the equations of the plane parallel to the join of
$(3, 2, -5)$ and $(0, -4, -11)$
and passing throguh the points
$(-2, 1, -3)$ and $(4, 3, 3)$
is
$4x + 3y - 5z = 10.$

5. Find the equation of the plane containing the line
$2x - 5y + 2z = 6, 2x + 3y - z = 5$
and parallel to the line $x = -y/6 = z/7$. [Ans. $6x + y - 16 = 0$]

6. Show that the equation of the plane through the line
$u_1 \equiv a_1 x + b_1 y + c_1 z + d_1 = 0, u_2 \equiv a_2 x + b_2 y + c_2 z + d_2 = 0$
and parallel to the line
$x/l = y/m = z/n$
is
$u_1 (a_2 l + b_2 m + c_2 n) = u_2 (a_1 l + b_1 m + c_1 n).$

7. Find the equation of the plane through the point (f, g, h) and parallel to the lines $x/l_r = y/m_r = z/n_r; r = 1, 2.$ [Ans. $\Sigma (x - f)(m_1 n_2 - m_2 n_1) = 0$]

8. Find the equations of the two planes through the origin which are parallel to the line
$(x - 1)/2 = -(y + 3) = -(z + 1)/2$
and distant 5/3 from it; show that the two planes are perpendicular.
[Ans. $2x + 2y + z = 0, x - 2y + 2z = 0$]

3.3 CONDITIONS FOR A LINE TO LIE IN A PLANE

To find the conditions for the line

$$\frac{x - x_1}{l} = \frac{y - y_1}{m} = \frac{z - z_1}{n}$$

to lie in the plane
$$ax + by + cz + d = 0.$$
The line would lie in the given plane if and only if every point of the line is a point of the plane, *i.e.*, the point
$$(lr + x_1, mr + y_1, nr + z_1)$$
lies on the plane for all values of r implying that the equation
$$r(al + bm + cn) + (ax_1 + by_1 + cz_1 + d) = 0$$
is true for every value of r.
This implies that
$$\begin{cases} al + bm + cn = 0 \\ ax_1 + by_1 + cz_1 + d = 0 \end{cases}$$
which are the required *two* conditions.

These coditions, when geometrically interpreted, state that a line lies in a given plane, if
(i) the normal to the plane is perpendicular to the line, and
(ii) any one point on the line lies in the plane.

Cor. *The general equation of a plane containing the line*
$$\frac{x - x_1}{l} = \frac{y - y_1}{m} = \frac{z - z_1}{n} \qquad \ldots(i)$$
is
$$A(x - x_1) + B(y - y_1) + C(z - z_1) = 0$$
where
$$Al + Bm + Cn = 0. \qquad \ldots(ii)$$
In other words, the set of planes containing the line (i) is
$$\{A(x - x_1) + B(y - y_1) + C(z - z_1) = 0, Al + Bm + Cn = 0\}$$

EXAMPLES

1. *Prove that the plane through (α, β, γ) and the line $x = py + q = rz + s$ is given by*
$$\begin{vmatrix} x & py + q & rz + s \\ \alpha & p\beta + q & r\gamma + s \\ 1 & 1 & 1 \end{vmatrix} = 0.$$

Sol. The given line can be written as
$$x/1 = y + q/p \, / \, 1/p = z + s/r \, / \, 1/r \qquad \ldots(1)$$
Let equation of any plane be
$$Ax + By + Cz + D = 0 \qquad \ldots(2)$$
It will pass through line (1), if
$$A.0 + B(-q/p) + C(-s/r) + D = 0 \qquad \ldots(3)$$
and
$$A \cdot 1 + B \cdot 1/p + C \cdot 1/r = 0 \qquad \ldots(4)$$
The plane will pass through (α, β, γ) if
$$A \cdot \alpha + B \cdot \beta + C \cdot \gamma + D = 0 \qquad \ldots(5)$$
Subtracting (3) from (2) and (5), we get
$$Ax + B(y + q/p) + C(z + s/r) = 0 \qquad \ldots(6)$$
$$A \cdot \alpha + B(\beta + q/p) + C(\gamma + s/r) = 0 \qquad \ldots(7)$$
Eliminating A, B, C from (6), (7) and (4)
$$\begin{vmatrix} x & y + q/p & z + s/r \\ \alpha & \beta + q/p & \gamma + s/r \\ 1 & 1/p & 1/r \end{vmatrix} = 0$$

$$\Rightarrow \begin{vmatrix} x & py + q & rz + s \\ \alpha & p\beta + q & r\gamma + s \\ 1 & 1 & 1 \end{vmatrix} = 0.$$

2. The axes are rectangular and the plane $\frac{x}{a} + \frac{y}{b} + \frac{z}{c} = 1$ meets them in A, B, C. Prove that the equations to BC are $x/0 = y/0 = z - c/-c$; that the equation to the plane through OX at right angles to BC is $by = cz$; that the three planes throguh OX, OY, OZ at right angles to BC, CA, AB respectively pass through the line $ax = by = cz$; and that the co-ordinates of the orthocentre of the triangle ABC are

$$\left[\frac{a^{-1}}{a^{-2} + b^{-2} + c^{-2}}, \frac{b^{-1}}{a^{-2} + b^{-2} + c^{-2}}, \frac{c^{-1}}{a^{-2} + b^{-2} + c^{-2}} \right].$$

Sol. The given plane meets the axes in points $A(a, 0, 0); B(0, b, 0)$ and $C(0, 0, c)$. Equations of the line through B and C are

$$\frac{x}{0} = \frac{y}{b} = \frac{z-c}{-c} \qquad \ldots(1)$$

Equation of any plane through OX is

$$y + \lambda z = 0.$$

If BC is perpendicular to above plane,

then, $\qquad b/1 = -c/\lambda \Rightarrow \lambda = -c/b$

Hence, the plane is $\qquad by = cz.$

Similarly the planes through OY and OZ and at right angles to CA and AB respectively are

$$cz = ax, \; ax = by.$$

Hence, the three planes pass through the line

$$ax = by = cz. \qquad \ldots(2)$$

The orthocentre of the triangle ABC lies where the line (2) meets the given plane.
Any point on (2) is $(r/a, r/b, r/c)$.
If it lies on the given plane, then

$$r = \frac{1}{a^{-2} + b^{-2} + c^{-2}}.$$

Hence, the co-ordinates of the orthocentre are

$$\left[\frac{a^{-1}}{a^{-2} + b^{-2} + c^{-2}}, \frac{b^{-1}}{a^{-2} + b^{-2} + c^{-2}}, \frac{c^{-1}}{a^{-2} + b^{-2} + c^{-2}} \right].$$

EXERCISES

1. Show that the line $x + 10 = (8 - y)/2 = z$ lies in the plane

$$x + 2y + 3z = 6$$

and the line

$$\frac{1}{3}(x - 2) = -(y + 2) = \frac{1}{4}(z - 3) \text{ in the plane}$$

$$2x + 2y - z + 3 = 0.$$

2. Find the equation of the plane containing the line

$$\frac{1}{2}(x + 2) + \frac{1}{3}(y + 3) = -\frac{1}{2}(z - 4)$$

and the point $(0, 6, 0)$. [**Ans.** $3x + 2y + 6z - 12 = 0$]

3. $\frac{x - x_1}{l_1} = \frac{y - y_1}{m_1} = \frac{z - z_1}{n_1}$ and $\frac{x - x_2}{l_2} = \frac{y - y_2}{m_2} = \frac{z - z_2}{n_2}$ and two straight lines. Find the equation of the plane containing the first line and parallel to the second.

[**Ans.** $\Sigma (x - x_1)(m_1 n_2 - m_2 n_1) = 0$]

4. Find the equation to the plane containing the line $y/b + z/c = 1, x = 0$ and parallel to the line $x/a + z/c = 1, y = 0$. [Ans. $x/a - y/b - z/c + 1 = 0$]

5. Find the equation to the plane which passes through the z-axis and is perpendicular to the line
$$\frac{x-1}{\cos \theta} = \frac{y+2}{\sin \theta} = \frac{z-3}{0}.$$
[Ans. $x \cos \theta + y \sin \theta = 0$]

6. Show that the equation of the plane which passes through the line
$$\frac{x-1}{3} = \frac{y+6}{4} = \frac{z+1}{2}$$
and is parallel to the line
$$\frac{x-2}{2} = \frac{y-1}{-3} = \frac{z+4}{5},$$
is $26x - 11y - 17z - 109 = 0$ and show that the point $(2, 1, -4)$ lies on it. What is the geometrical relation between the two lines and the plane?

7. Find the equation of the plane containing the line
$$-\frac{1}{3}(x+1) = \frac{1}{2}(y-3) = (z+2)$$
and the point $(0, 7, -7)$ and show that the line
$$x = \frac{1}{3}(7-y) = \frac{1}{2}(z+7)$$
lies in the same plane. [Ans. $x + y + z = 0$]

3.4 COPLANAR LINES, CONDITION FOR THE COPLANARITY OF LINES

To find the condition that two given straight lines
$$\frac{x-x_1}{l_1} = \frac{y-y_1}{m_1} = \frac{z-z_1}{n_1} \qquad \ldots(1)$$
$$\frac{x-x_2}{l_2} = \frac{y-y_2}{m_2} = \frac{z-z_2}{n_2} \qquad \ldots(2)$$
are coplanar.

Sol. First Method : Equation of *any* plane containing the line (1) is
$$A(x-x_1) + B(y-y_1) + C(z-z_1) = 0; \qquad \ldots(i)$$
A, B, C being numbers not all zero satisfying the condition
$$Al_1 + Bm_1 + Cn_1 = 0. \qquad \ldots(ii)$$
The plane (i) will contain the line (2) if

(a) the point (x_2, y_2, z_2) lies on it
$$\Rightarrow \qquad A(x_2 - x_1) + B(y_2 + y_1) + C(z_2 - z_1) = 0 \qquad \ldots(iii)$$
(b) the lines is perpendicular to the normal to the plane
$$\Rightarrow \qquad Al_2 + Bm_2 + Cn_2 = 0. \qquad \ldots(iv)$$

The two lines will be coplanar if the three linear homogeneous equations (ii), (iii), (iv) in A, B, C are consistent so that
$$\begin{vmatrix} x_2 - x_1 & y_2 - y_1 & z_2 - z_1 \\ l_1 & m_1 & n_1 \\ l_2 & m_2 & n_2 \end{vmatrix} = 0 \qquad \ldots(A)$$

which is thus, the required condition for the lines to intersect. Assuming this condition is satisfied, we see that the required equation of the plane is

The Straight Line

$$\begin{vmatrix} x-x_1 & y-y_1 & z-z_1 \\ l_1 & m_1 & n_1 \\ l_2 & m_2 & n_2 \end{vmatrix} = 0.$$

This is the equation of the plane contianing the two lines.

Second Method. Two lines are coplanar if and only if they intersect or are parallel. We first consider the case of intersection. The condition for intersection may also be obtained as follows :

$$(l_1 r_1 + x_1, m_1 r_1 + y_1, n_1 r_1 + z_1) \text{ and } (l_2 r_2 + x_2, m_2 r_2 + y_2, n_2 r_2 + z_2)$$

are the general co-ordinates of the points on the lines (1) and (2) respectively for all values of r_1 and r_2.

In case the lines intersect, these points should coincide for some values of r_1 and r_2. This requires that the following three equations

$$(x_1 - x_2) + l_1 r_1 - l_2 r_2 = 0,$$
$$(y_1 - y_2) + m_1 r_1 - m_2 r_2 = 0,$$
$$(z_1 - z_2) + n_1 r_1 - n_2 r_2 = 0.$$

in r_1, r_2 are consistent, so that we have the condition

$$\begin{vmatrix} x_1 - x_2 & l_1 & l_2 \\ y_1 - y_2 & m_1 & m_2 \\ z_1 - z_2 & n_1 & n_2 \end{vmatrix} = 0 \Leftrightarrow \begin{vmatrix} x_2 - x_1 & y_2 - y_1 & z_2 - z_1 \\ l_1 & m_1 & n_1 \\ l_2 & m_2 & n_2 \end{vmatrix} = 0$$

which is the same condition as (A).

This condition is clearly satisfied if the lines are parallel.

Note 1. In general, the equation

$$\begin{vmatrix} x-x_1 & y-y_1 & z-z_1 \\ l_1 & m_1 & n_1 \\ l_2 & m_2 & n_2 \end{vmatrix} = 0$$

represents the plane which passes through the line (1) and is parallel to the line (2), and the equation

$$\begin{vmatrix} x-x_2 & y-y_2 & z-z_2 \\ l_1 & m_1 & n_1 \\ l_2 & m_2 & n_2 \end{vmatrix} = 0$$

represents the plane which passes throgh the line (2) and is parallel to the line (1).

In case the lines are coplanar, the condition (A) shows that the point (x_2, y_2, z_2) lies on the first plane and the point (x_1, y_1, z_1) on the second. These two equations are then identical.

Thus, the plane containing two coplanar *lines* is the one which passes through one line and is parallel to the other *or*, through one line and any point on the other.

Note 2. Two lines will intersect if and only if, there exists a point whose co-ordinates satisfy the *four* equations, two of each line so that for intersection, we require that the four linear equations in three unknowns should be *consistent*.

It is sometimes comparatively more convenient to follow this method to obtain the condition of intersection or to prove the fact of intersection of two lines.

Note 3. The condition for the lines whose equations, given in the unsymmetrical form, are

$$a_1 x + b_1 y + c_1 z + d_1 = 0, a_2 x + b_2 y + c_2 z + d_2 = 0;$$
$$a_3 x + b_3 y + c_3 z + d_3 = 0, a_4 x + b_4 y + c_4 z + d_4 = 0;$$

to intersect, is the condition for the consistency of these four equations, *i.e.*,

$$\begin{vmatrix} a_1 & b_1 & c_1 & d_1 \\ a_2 & b_2 & c_2 & d_2 \\ a_3 & b_3 & c_3 & d_3 \\ a_4 & b_4 & c_4 & d_4 \end{vmatrix} = 0.$$

In case, this condition is satisfied, the co-ordinates of the point of intersection are obtained by solving any three of the four equations simultaneously.

EXAMPLES

1. *Show that the lines*

$$\frac{x+3}{2} = \frac{y+5}{3} = \frac{z-7}{-3}, \frac{x+1}{4} = \frac{y+1}{5} = \frac{z+1}{-1}$$

are coplanar and find the equation of the plane containing them.

Sol. The equation of the plane which contains the first line and is parallel to the second is

$$\begin{vmatrix} x+3 & y+5 & z-7 \\ 2 & 3 & -3 \\ 4 & 5 & -1 \end{vmatrix} = 0 \Leftrightarrow 6x - 5y - z = 0.$$

This plane, as may be easily seen, passes through the point $(-1, -1, -1)$ on the second line so that it also contains the second line.

Thus, the two lines are coplanar and the equation of the plane containing them is

$$6x - 5y - z = 0.$$

2. *If OA, OB, OC have direction ratios $l_r, m_r, n_r, r = 1, 2, 3$ and OA', OB', OC' bisect the angles BOC, COA, AOB, the planes AOA', BOB', COC' pass through the line*

$$\frac{x}{l_1 + l_2 + l_3} = \frac{y}{m_1 + m_2 + m_3} = \frac{z}{n_1 + n_2 + n_3}.$$

Sol. Let O be the origin, equations of OB and OC are

$$y/l_2 = y/m_2 = z/n_2$$

and $\quad x/l_3 = y/m_3 = z/n_3$

Points on these lines at unit distance are (l_2, m_2, n_2) and (l_3, m_3, n_3).

Corresponding point on bisector OA' is

$$\left[\frac{1}{2}(l_2 + l_3), \frac{1}{2}(m_2 + m_3), \frac{1}{2}(n_2 + n_3) \right]$$

∴ Equations of OA' are

$$\frac{x}{l_2 + l_3} = \frac{y}{m_2 + m_3} = \frac{z}{n_2 + n_3}$$

Now, equation of the plane containing OA and OA', i.e., AOA' is

$$\begin{vmatrix} x & y & z \\ l_1 & m_1 & n_1 \\ l_2 + l_3 & m_2 + m_3 & n_2 + n_3 \end{vmatrix} = 0$$

$$\Rightarrow \begin{vmatrix} x & y & z \\ l_1 & m_1 & n_1 \\ l_1 + l_2 + l_3 & m_1 + m_2 + m_3 & n_1 + n_2 + n_3 \end{vmatrix} \quad \text{(Operating } R_3 + R_2\text{)}$$

This plane clearly passes through the line

$$\frac{x}{l_1 + l_2 + l_3} = \frac{y}{m_1 + m_2 + m_3} = \frac{z}{n_1 + n_2 + n_3}.$$

Similarly plane BOB' and COC' pass through the same line.

The Straight Line

3. *A, A', B, B', C, C' are points on the axes. Show that the lines of intersection of the planes A'BC, AB'C, B'CA, BC'A', C'AB, CA'B' are coplanar.*

Sol. Let the points A, A', B, B', C, C' be $(a, 0, 0), (a', 0, 0), (0, b, 0), (0, b', 0), (0, 0, c), (0, 0, c')$ respectively.

Equations of the planes $A'BC$ and $AB'C'$ are
$$y/a' + y/b + z/c = 1 \text{ and } x/a + y/b' + z/c' = 1.$$
These equations taken together represent the line of intersection of the planes $A'BC$ and $AB'C'$.

Any plane through this line is
$$(x/a' + y/b + z/c - 1) + \lambda_1 (x/a + y/b' + z/c' - 1) = 0 \qquad ...(1)$$

Similarly, planes through the lines of intersection of $B'CA, BC'A'$; $C'AB, CA'B'$ are respectively
$$(x/a + y/b' + z/c - 1) + \lambda_2 (x/a' + y/b + z/c' - 1) = 0$$
and
$$(x/a + y/b + z/c' - 1) + \lambda_3 (x/a' + y/b' + z/c - 1) = 0$$

In case the three lines are coplanar then for some value of $\lambda_1, \lambda_2, \lambda_3$, the above equations must represent the same plane. This is obviously so, when $\lambda_1, \lambda_2, \lambda_3$, and then the plane becomes
$$x(1/a + 1/a') + y(1/b + 1/b') + z(1/c + 1/c') = 2.$$

EXERCISES

1. Show that the lines
$$\frac{1}{3}(x+4) = \frac{1}{5}(y+6) = -\frac{1}{2}(z-1)$$
$$3x - 2y + z + 5 = 0 = 2x + 3y + 4z - 4$$
are coplanar. Find also the co-ordinates of their point of intersection and the equation of the plane in which they lie.   [**Ans.** $(2, 4, -3); 45x - 17y + 25z + 53 = 0$]

2. Prove that the lines
$$\frac{x-1}{2} = \frac{y+1}{-3} = \frac{z+10}{8}; \frac{x-4}{1} = \frac{y+3}{-4} = \frac{z+1}{7}$$
intersect. Find also their point of intersection and the plane through them.
   [**Ans.** $(5, -7, 6); 11x = 6y + 5z + 67$]

3. Prove that the lines
$$\frac{x+1}{3} = \frac{y+3}{5} = \frac{z+5}{7}; \frac{x-2}{1} = \frac{y-4}{3} = \frac{z-6}{5}$$
intersect. Find their point of intersection and the plane in which they lie.
   [**Ans.** $(1/2, -1/2, 3/2); x - 2y + z = 0$]

4. Show that the lines
$$x + 2y - 5z + 9 = 0 = 3x - y + 2z - 5;$$
$$2x + 3y - z - 3 = 0 = 4x - 5y + z + 3$$
are coplanar.

5. Prove that the lines
$$x - 3y + 2z + 4 = 0 = 2x + y + 4z + 1;$$
$$3x + 2y + 5z - 1 = 0 = 2y + z$$
intersection and find the co-ordinates of their point of intersection.   [**Ans.** $(3, 1, -2)$]

6.
$$x - 2y - z - 3 = 0, 3x - y + 2z - 1 = 0,$$
$$2x - 2y + 3z - 2 = 0, x - y + z + 1 = 0$$
are two given pairs of planes. Show that the line of intersection of the first pair is coplanr with the line of intersection of the latter.

7. Show that the line of intersection of the planes
$$7x - 4y + 7z + 16 = 0,\ 4x + 3y - 2z + 3 = 0$$
is coplanar with the line of intersection of planes
$$x - 3y + 4z + 6 = 0,\ x - y + z + 1 = 0.$$
Obtain the equation of the plane through the two lines. [Ans. $3x - 7y + 9z + 13 = 0$]

8. Prove that the lines
$$\frac{x-a}{a'} = \frac{y-b}{b'} = \frac{z-c}{c'} \text{ and } \frac{x-a'}{a} = \frac{y-b'}{b} = \frac{z-c'}{c}$$
intersect and find the co-ordinates of the point of intersection and the equation of the plane in which by they lie. [Ans. $(a + a', b + b', c + c')$; $\Sigma x (bc' - b'c) = 0$]

3.5 NUMBER OF ARBITRARY CONSTANTS IN THE EQUATIONS OF A STRAIGHT LINE

We have already seen that the general equation of a plane contains **three** arbitrary constants and it will now be shown *that there are four arbitrary constants in the equations of a straight line.*

A given line PQ can be regarded as the intersection of *any* two planes through it. In particular, we may take the two planes perpendicular to two of the co-ordinate planes, say, YZ and ZX planes.

The equations of the planes through a line PQ perpendicular to the YZ and ZX planes are respectively of the forms
$$z = cy + d \text{ and } z = ax + b$$
which are, therefore, the equations of the line PQ and contain four arbitrary constants a, b, c, d.

Hence, the *equations of a straight line involve four arbitrary contants* as it is always possible to express them in the above form.

The fact that the general equations of a straight line contain four arbitrary constants may also be seen as follows :

We see that the equations
$$\frac{x - x_1}{l} = \frac{y - y_1}{m};\ \frac{y - y_1}{m} = \frac{z - z_1}{n}$$
are equivalent to
$$x = \frac{l}{m} y + \frac{(mx_1 - ly_1)}{m},\ y = \frac{m}{n} z + \frac{(ny_1 - mz_1)}{n}$$
respectively, so that
$$\frac{l}{m}, \frac{m}{n}, \frac{mx_1 - ly_1}{m}, \frac{ny_1 - mz_1}{n}$$
are the *four* arbitrary constants or parameters.

3.5.1 Determination of Lines Satisfying Given Conditions

We now consider the various *sets of conditions* which determine a line.

We know that the equations of a straight line involve four arbitrary constants and as such any four geometrical conditions, each of which gives rise to one relation between the constants, fix a straight line.

It may be noted that the conditions for a line to intersect a given line or be perpendicular to it separately involve one relation between the constants and hence, three more relations are required to fix the line.

A given conditions may sometimes give rise to two relations between the constants as, for instance, the conditions that the required line

(i) passes through a given point; (ii) has a given direction.

In such cases only two more relations will be required to fix the straight line.

We have already considered equations of a line which

(i) pass through a given point and have a given direction;
(ii) pass through two given points;

The Straight Line

(iii) pass through a point and are parallel to two given planes;
(iv) pass through a point and pependicular to two given lines.

Some further sets of conditions which determine a line are given below :

(v) passing throguh a given point and intersecting two given lines;
(vi) intersecting two given lines and having a given direction;
(vii) intersecting a given line at right angles and passing through a given point;
(viii) intersecting two given lines at right angles;
(ix) intersecting a given line parallel to a given line and passing through a given point;
(x) passing through a given point and perpendicular to two given lines; and so on.

An Important Note : *If*
$$u_1 = 0 = v_1 \quad \text{and} \quad u_2 = 0 = v_2$$
be two straight lines, then the general equations of a straight line intersecting them both are
$$u_1 + \lambda_1 v_1 = 0 = u_2 + \lambda_2 v_2,$$
where λ_1, λ_2 are any two numbers.

The line $u_1 + \lambda_1 v_1 = 0 = u_2 + \lambda_2 v_2$ lies in the plane $u_1 + \lambda_1 v_1 = 0$ which again contains the line $u_1 = 0 = v_1$.

The two lines
$$u_1 + \lambda_1 v_1 = 0 = u_2 + \lambda_2 v_2; u_1 = 0 = v_1$$
are, therefore, coplanar and hence they intersect.

Similarly, the same line intersects the line $u_2 = 0 = v_2$.

This conclusion will be found very helpful in what follows.

For the sake of illustration, we give below a few examples.

EXAMPLES

1. *Find the equations of the line which passes through the point $(2, -1, 1)$ and intersects the lines*
$$2x + y - 4 = 0 = y + 2z; \; x + 3z = 4, \; 2x + 5z = 8.$$

Sol. The line
$$2x + y - 4 + \lambda_1(y + 2z) = 0, \; x + 3z - 4 + \lambda_2(2x + 5z - 8) = 0$$
intersects the two given lines for all values of λ_1, λ_2.

The line will pass through the point $(2, -1, 1)$, if
$$-1 + \lambda_1 = 0 \; \text{and} \; 1 + \lambda_2 = 0,$$
$$\Rightarrow \quad \lambda_1 = 1, \lambda_2 = -1.$$

The required equations, therefore, are
$$x + y + z = 2 \; \text{and} \; x + 2z = 4.$$

2. *Find the equations of the line which passes through the point $(3, -1, 11)$ and is perpendicular to the line*
$$\frac{1}{2}x = \frac{1}{3}(y - 2) = \frac{1}{4}(z - 3).$$

Obtain also the foot of the perpendicular.

Sol. The co-ordinates of any point on the given line are
$$2r, 3r + 2, 4r + 3.$$

This will be the required foot of the perpendicular if the line joining it to the point $(3, -1, 11)$ be perpendicular to the given line. This requires
$$2(2r - 3) + 3(3r + 2 + 1) + 4(4r + 3 - 11) = 0 \Rightarrow r = 1.$$

Therefore, the required foot is $(2, 5, 7)$ and the required equations of the perpendiculars are
$$\frac{x - 3}{1} = \frac{y + 1}{-6} = \frac{z - 11}{4}.$$

EXERCISES

1. Find the equations of the perpendicular from
   (i) $(2, 4, -1)$ to $(x+5) = \frac{1}{4}(y+3) = \frac{1}{9}(z-6)$,
   (ii) $(-2, 2, -3)$ to $(x-3) = \frac{1}{2}(y+1) = -\frac{1}{4}(z-2)$,
   (iii) $(0, 0, 0)$ to $2x + y + z - 7 = 0 = 4x + z - 14$.
   Obtain also the feet of the perpendiculars.

   [Ans. (i) $\frac{1}{6}(x-2) = \frac{1}{3}(y-4) = \frac{1}{2}(z+1), (-4, 1, -3)$
   (ii) $\frac{1}{6}(x+2) = -(y-2) = (z+3), (4, 1, -2)$
   (iii) $-x/2 = y = z/4, (2/3, -1/3, -4/3)$
   (iv) $\frac{1}{6}(x+2) = -(y-2) = (z+3), (4, 1, -2)$]

2. A line with direction cosines proportional to $(7, 4, -1)$ is drawn to intersect the lines
   $$\frac{x-1}{3} = \frac{y-3}{-1} = \frac{z+2}{1}, \frac{x+3}{-3} = \frac{y-3}{2} = \frac{z-5}{4}.$$
   Find the points of intersection and the length intercepted on it.

   [Ans. $(7, 5, 0), (0, 1, 1), \sqrt{66}$]

3. Find the line which intersects the lines
   $$x + y + z = 1, 2x - y - z = 2; x - y - z = 3, 2x + 4y - z = 4$$
   and passes through the point $(1, 1, 1)$. Find also the points of intersection.

   $\left[\text{Ans. } x = 1, (y-1)/1 = (z-1)/3; \left(1, \frac{1}{2}, -\frac{1}{2}\right); (1, 0, -2)\right]$

4. Find the equations of the line which passes through the point $(-4, 3, 1)$, is parallel to the plane $x + 2y - z = 5$ and intersects the line
   $$-(x+1)/3 = (y-3)/2 = -(x-2)$$
   Find also the point of intersection. [Ans. $(x+4)/3 = -(y-3) = (z-1); (2, 1, 3)$]

5. Find the distance of the point $(-2, 3, -4)$ from the line
   $$(x+2)/3 = (2y+3)/4 = (3z+4)/5$$
   measured parallel to the plane
   $$4x + 12y - 3z + 1 = 0.$$ [Ans. $17/2$]

6. Find the equations of the straight line through the point $(2, 3, 4)$ perpendicular to the X-axis and intersecting the line $x = y = z$. [Ans. $x = 2, 2y - z = 2$]

3.6 THE SHORTEST DISTANCE BETWEEN TWO LINES

To show that the shortest distance between two lines lies along the line meeting them both at right angles.

Let AB, CD be two given lines.

A line is completely determined if it intersects two lines at right angles [See 3.5.1, Case (viii)].

Thus, there is one and only one line which intersects the two given lines at right angles, say, at G and H.

GH is, then, the shortest distance between the two lines for, if A, C be *any* two points, one on each of the two given lines,

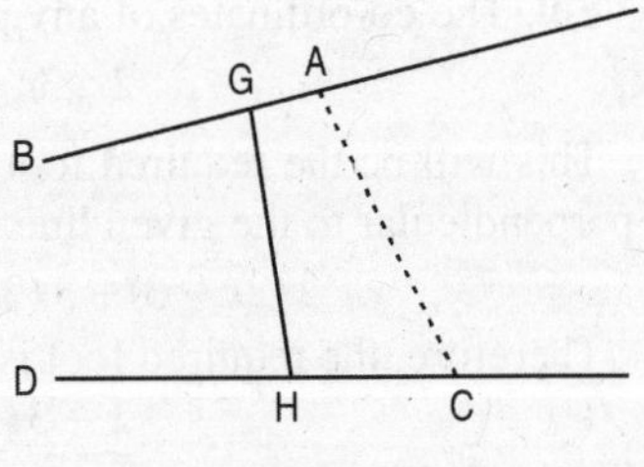

GH is the projection of AC on itself
$\Rightarrow \quad GH = AC \cos\theta$
$\Rightarrow \quad GH < AC$

Fig. 2

The Straight Line

θ, being the angle between GH and AC. Thus, GH is the shortest distance (S.D.) between the two lines AB and CD.

3.6.1

To find the magnitude and the equations of the line of shortest distance between two straight lines.

Let AB, CD be the two given lines, and GH, the line which meets them both at right angles at G and H. Then GH is the line of shortest distance between the given lines; the length of GH being the magnitude.

Let the equations of the given lines be

$$\frac{x-x_1}{l_1} = \frac{y-y_1}{m_1} = \frac{z-z_1}{n_1}, \qquad \text{...(i)}$$

$$\frac{x-x_2}{l_2} = \frac{y-y_2}{m_2} = \frac{z-z_2}{n_2} \qquad \text{...(ii)}$$

and let the shortest distance lie along the line

$$\frac{x-\alpha}{l} = \frac{y-\beta}{m} = \frac{z-\gamma}{n}, \qquad \text{...(iii)}$$

Line (iii) is perpendicular to both the lines (i) and (ii)

$$\Rightarrow \begin{cases} ll_1 + mm_1 + nn_1 = 0, \\ ll_2 + mm_2 + nn_2 = 0, \end{cases}$$

$$\Rightarrow \frac{l}{m_1 n_2 - m_2 n_1} = \frac{m}{n_1 l_2 - n_2 l_1} = \frac{n}{l_1 m_2 - l_2 m_1};$$

$$= \frac{1}{\sqrt{\Sigma (m_1 n_2 - m_2 n_1)^2}} \qquad \text{...(iv)}$$

The line of shortest distance is perpendicular to both the lines. Therefore, the magnitude of the shortest distance is the projection on the line of shortest distance of the line joining *any* two points, one on each of the given lines (i) and (ii).

Taking the projection of the join of $(x_1, y_1, z_1), (x_2, y_2, z_2)$ on the line with direction cosines l, m, n; we see that the shortest distance

$$= (x_2 - x_1) l + (y_2 - y_1) m + (z_2 - z_1) n,$$

where l, m, n have the values as given in (iv).

To find the equations of the line of shortest distance, we observe that it is coplanar with both the given lines.

The equations of the plane containing the coplanar lines (i) and (iii) is

$$\begin{vmatrix} x-x_1 & y-y_1 & z-z_1 \\ l_1 & m_1 & n_1 \\ l & m & n \end{vmatrix} = 0 \qquad \text{...(v)}$$

and that of the plane containing the coplanar lines (ii) and (iii) is

$$\begin{vmatrix} x-x_2 & y-y_2 & z-z_2 \\ l_2 & m_2 & n_2 \\ l & m & n \end{vmatrix} = 0. \qquad \text{...(vi)}$$

Thus, (v) and (vi) are the two equations of the line of shortest distance, where l, m, n are given in (iv).

Note. Other methods of determining the shortest distance are given below where an example has been solved by three different methods.

EXAMPLES

1. *Find the magnitude and the equations of the line of shortest distance between the lines :*

$$\frac{x-8}{3} = \frac{y+9}{-16} = \frac{z-10}{7}, \qquad \text{...(i)}$$

$$\frac{x-15}{3} = \frac{y-29}{8} = \frac{z-5}{-5}. \qquad \text{...(ii)}$$

Sol. First Method. Let l, m, n be the direction cosines of the line of shortest distance. As it is perpendicular to the two lines, we have

$$\begin{cases} 3l - 16m + 7n = 0, \\ 3l + 8m - 5n = 0. \end{cases}$$

$\Rightarrow \qquad \dfrac{l}{24} = \dfrac{m}{36} = \dfrac{n}{72},$

$\Rightarrow \qquad \dfrac{l}{2} = \dfrac{m}{3} = \dfrac{n}{6},$

$\Rightarrow \qquad l = \dfrac{2}{7}, m = \dfrac{3}{7}, n = \dfrac{6}{7}.$

The magnitude of the shortest distance is the projection of the join of the points $(8, -9, 10)$, $(15, 29, 5)$, on the line of the shortest distance and is, therefore,

$$= 7 \cdot \frac{2}{7} + 38 \cdot \frac{3}{7} - 5 \cdot \frac{6}{7} = 14$$

Again, the equation of the plane containing the first of the two given lines and the line of shortest distance is

$$\begin{vmatrix} x-8 & y+9 & z-10 \\ 3 & -16 & 7 \\ 2 & 3 & 6 \end{vmatrix} = 0 \Leftrightarrow 117x + 4y - 41z - 490 = 0.$$

Also the equation of the plane containing the second line and the shortest distance line is

$$\begin{vmatrix} x-15 & y-29 & z-5 \\ 3 & 8 & -5 \\ 2 & 3 & 6 \end{vmatrix} = 0 \Leftrightarrow 9x - 4y - z = 14.$$

Hence, the equations of the shortest distance line are

$$117x + 4y - 41z - 490 = 0 = 9x - 4y - z = 14.$$

Second Method

$$P(3r + 8, -16r - 9, 7r + 10), \quad P'(3r' + 15, 8r' + 29, 5r' + 5)$$

are the general co-ordinates of the points on the two lines respectively. The direction cosines of PP' are proportional to

$$3r - 3r' - 7, -16r - 8r' - 38, 7r + 5r' + 5.$$

Now PP' will be required line of shortest distance, if it is perpendicular to both are given lines, which requires

$$\begin{cases} 3(3r - 3r' - 7) - 16(-16r - 8r' - 38) + 7(7r + 5r' + 5) = 0, \\ 3(3r - 3r' - 7) + 8(-16r - 8r' - 38) - 5(7r + 5r' + 5) = 0. \end{cases}$$

$\Rightarrow \qquad 157r + 77r' + 311 = 0 \text{ and } 11r + 7r' + 25 = 0,$

$\Rightarrow \qquad r = -1, r' = -2.$

Therefore, the co-ordinates of the point P and P' are

$$(5, 7, 3) \text{ and } (9, 13, 15).$$

Hence, the shortest distance $PP' = 14$ and its equations are

$$\frac{x-5}{2} = \frac{y-7}{3} = \frac{z-3}{6}.$$

The Straight Line

This method is sometimes very convenient and is specially useful when we require also the points where the line of shortest distance meets the two lines.

2. *Find the shortest distance between the axis of z and the line*
$$ax + by + cz + d = 0, a'x + b'y + c'z + d' = 0$$

Sol. Now the general equation of the plane through the second given line is
$$ax + by + cz + d + k(a'x + b'y + c'z + d') = 0$$
$$\Leftrightarrow \quad (a + ka')x + (b + kb')y + (c + kc')z + (d + kd') = 0 \qquad \ldots(i)$$
k being the parameter.

It will be paralell to z-axis whose direction cosines are 0, 0, 1 if the normal to the plane is perpendicular to the z-axis, *i.e.*, if
$$0.(a + ka') + 0.(b + kb') + 1.(c + kc') = 0$$
$$\Rightarrow \qquad k = -c/c'.$$

Substituting this value of k in (i), we see that the equation of the plane through the second line parallel to the first is
$$(ac' - a'c)x + (bc' - b'c)y + (dc' - d'c) = 0 \qquad \ldots(ii)$$

The required S.D. is the distance of *any* point on the z-axis from the plane (ii) so that
S.D. = perpendicular from $(0, 0, 0)$, (a point on z-axis)
$$= \pm \frac{dc' - d'c}{\sqrt{(ac' - a'c)^2 + (bc' - b'c)^2}}.$$

3. *Prove that the S.D. between the diagonals of rectangular parallelopiped and the edges not meeting it are*
$$\frac{bc}{\sqrt{(b^2 + c^2)}}, \frac{ca}{\sqrt{(c^2 + a^2)}}, \frac{ab}{\sqrt{(a^2 + b^2)}}$$
where a, b, c are the lengths of the edges.

Sol. Let coterminous edges OA, OB, OC be taken as the axes of reference. We will find S.D. between the diagonal OP and edge BL (which does not meet OP). Equations of OP and BL are
$$\frac{x}{a} = \frac{y}{b} = \frac{z}{c} \quad \text{and} \quad \frac{x}{a} = \frac{y-b}{0} = \frac{z}{0}.$$

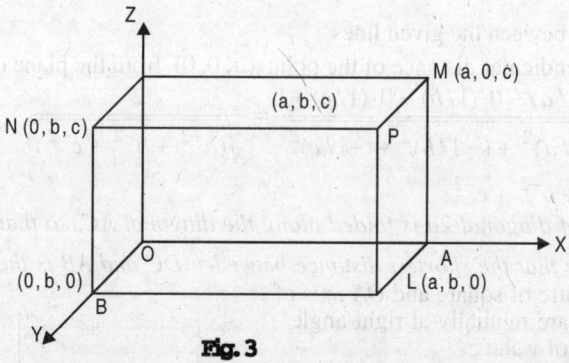

Fig. 3

Let l, m, n be the dc's of S.D., then
$$al + bm + cn = 0$$
and
$$al + 0 \cdot m + 0 \cdot n = 0$$
$$\Rightarrow \qquad l = 0, m = \frac{c}{\sqrt{b^2 + c^2}}, n = \frac{-b}{\sqrt{b^2 + c^2}}$$
$\therefore \qquad$ S.D. = Projection of OB on the line of S.D.

$$= (0-0)\, l + (b-0)\, m + (0-0)\, n = bm = \frac{bc}{\sqrt{b^2 + c^2}}.$$

Similarly, S.D.'s between OP and AL and OP and MC are $ca/\sqrt{c^2 + a^2}$ and $ab/\sqrt{a^2 + b^2}$.

4. Show that the equation of the plane containing the line
$$\frac{y}{b} + \frac{z}{c} = 1,\ x = 0$$
and parallel to the line
$$\frac{x}{a} - \frac{z}{c} = 1,\ y = 0$$
is
$$\frac{x}{a} - \frac{y}{b} - \frac{z}{c} + 1 = 0$$
and if $2d$ is the S.D. show that $d^{-2} = a^{-2} + b^{-2} + c^{-2}$.

Sol. The equation of the plane containing the line
$$\frac{y}{b} + \frac{z}{c} - 1 + \lambda x = 0$$
is
$$\left(\frac{y}{b} + \frac{z}{c} - 1\right) + \lambda x = 0$$
$\Rightarrow\qquad \lambda x + (1/b) y + (1/c) z - 1 = 0$ ...(i)

If it is parallel to the line
$$\frac{x}{a} - \frac{z}{c} = 1,\ y = 0,\ i.e.,\ \frac{x-a}{a} = \frac{y}{0} = \frac{z}{c},$$
then the normal to the plane (i) must be perpendicular to the line and so we have
$$\lambda \cdot (a) + (1/b) \cdot 0 + (1/c) \cdot c = 0 \Rightarrow \lambda = -1/a.$$

∴ From (1) the equation of the required palne is
$$\left(\frac{y}{b} + \frac{z}{c} - 1\right) - \frac{1}{a} x = 0 \Rightarrow \frac{x}{a} - \frac{y}{b} - \frac{z}{c} + 1 = 0 \qquad \text{...(ii)}$$

Now, any point on the line $\frac{x-a}{a} = \frac{y}{0} = \frac{z}{c}$ is $(a, 0, 0)$.

Therefore,
$2d = $ S.D. beween the given lines
$= $ perpendicular distance of the point $(a, 0, 0)$ from the plane (i)
$$= \frac{a \cdot (1/a) - 0 \cdot (1/b) - 0 \cdot (1/c) + 1}{\sqrt{(1/a)^2 + (-1/b)^2 + (-1/c)^2}} = \frac{2}{\sqrt{(a^{-2} + b^{-2} + c^{-2})}}$$
$\Rightarrow\qquad d^{-2} = a^{-2} + b^{-2} + c^{-2}.$

5. A square $ABCD$ of diagonal $2a$ is folded along the diagonal AC, so that planes DAC, BAC are at right angles. Show that the shortest distance betweden DC and AB is then $2a/\sqrt{3}$.

Sol. Let O be the centre of square and OA axis of x. Planes DAC and BAC are mutually at right angles. Take OB and OD as axes of y and z.

Then co-ordinates of A, B, C, D are $(a, 0, 0), (0, a, 0), (-a, 0, 0)$ and $(0, 0, a)$.

Equations of AB are $(x-a)/a = y/-a = z/0$ and of DC $x/a = y/0 = (z-a)/a$. Thus, a plane containing DC and parallel to AB is
$$\begin{vmatrix} x & y & z-a \\ a & 0 & a \\ a & -a & 0 \end{vmatrix} = 0$$

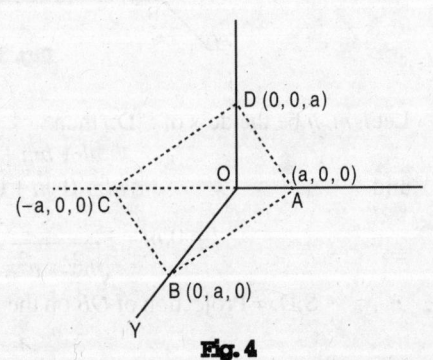

Fig. 4

The Straight Line

$$\Rightarrow \quad x+y+z+a=0$$

S.D. = Perpendicular distance of this plane from a point $(a, 0, 0)$ on AB

$$= \frac{a+a}{\sqrt{(1+1+1)}} = \frac{2a}{\sqrt{3}}.$$

EXERCISES

1. Find the magnitude and the equations of the line of shortest distance between the two lines :
   (i) $\dfrac{x-3}{2} = \dfrac{y+15}{-7} = \dfrac{z-9}{5}; \dfrac{x+1}{2} = \dfrac{y-1}{1} = \dfrac{z-9}{-3}.$
   (ii) $\dfrac{x-3}{-1} = \dfrac{y-4}{2} = \dfrac{z+2}{1}; \dfrac{x-1}{1} = \dfrac{y+7}{3} = \dfrac{z+2}{2}.$

   [**Ans.** (i) $x = y = z; 4\sqrt{3}$ (ii) $(x-4) = (y-2)/3 = -(z+3)/5; \sqrt{35}$]

2. Find the length and the equations of the shortest distance line between
   $$5x - y - z = 0 \qquad x - 2y + z + 3 = 0;$$
   $$7x - 4y - 2z = 0, \qquad x - y + z - 3 = 0.$$
   [**Hint.** Transform the equations to the symmetrical form]
   [**Ans.** $17x + 20y - 19z - 39 = 0 = 8x + 5y - 31z + 67; 13/\sqrt{75}$]

3. Find the magnitude and the position of the line of shortest distance between the lines
   (i) $2x + y - z = 0, x - y + 2z = 0; x + 2y - 3z = 4, 2x - 3y + 4z = 5$
   (ii) $\dfrac{x}{4} = \dfrac{y+1}{3} = \dfrac{z-2}{2}; 5x - 2y - 3z + 6 = 0; x - 3y + 2z - 3 = 0.$

   [**Ans.** (i) $3x + z = 0 = 22x - 5y + 4z - 67, 2\sqrt{14}/7$.
   (ii) $7x - 2y - 11z + 20 = 0 = 13x - 13z + 24; 17\sqrt{6}/39$]

4. Obtain the co-ordinates of the points where the shortest distance line between the lines
   $$\dfrac{x-23}{-6} = \dfrac{y-19}{-4} = \dfrac{z-25}{3}, \dfrac{x-12}{-9} = \dfrac{y-1}{4} = \dfrac{z-5}{2}$$
   meets them. [**Ans.** $(11, 11, 31)$ and $(3, 5, 7)$]

5. Find the co-ordinates of the points on the join of $(-3, 7, -13)$ and $(-6, 1, -10)$ which is nearest to the intersection of the planes
   $$3x - y - 3z + 32 = 0 \text{ and } 3x + 2y - 15z - 8 = 0. \qquad [\textbf{Ans. } (-7, -1, -9)]$$

6. Show that the shortest distance between the lines
   $$x + a = 2y = -12z \text{ and } x = y + 2a = 6z - 6a$$
   is $2a$.

3.7 LENGTH OF THE PERPENDICULAR FROM A POINT TO A LINE

To find the length of the perpendicular from a given point $P(x_1, y_1, z_1)$ *to a given line*
$$\dfrac{x-\alpha}{l} = \dfrac{y-\beta}{m} = \dfrac{z-\gamma}{n}.$$

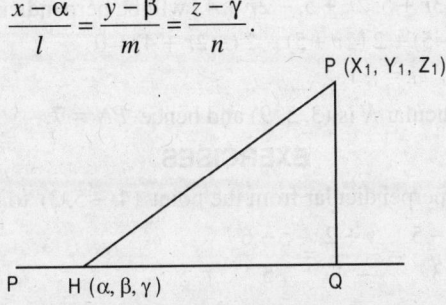

Fig. 5

Let H be the point (α, β, γ) on the given line and Q the foot of the perpendicular from the point P on it.

We have $PQ^2 = HP^2 - HQ^2$.

Also $HP^2 = (x_1 - \alpha)^2 + (y_1 - \beta)^2 + (z_1 - \gamma)^2$

and $HQ =$ projection of HP on the given line

$$= l(x_1 - \alpha) + m(y_1 - \beta) + n(z_1 - \gamma)$$

provided l, m, n are the actual direction cosines.

It follows that

$$PQ^2 = (x_1 - \alpha)^2 + (y_1 - \beta)^2 + (z_1 - \gamma)^2 - [l(x_1 - \alpha) + m(y_1 - \beta) + n(z_1 - \gamma)]^2.$$

EXAMPLE

Find the perpendicular distance of $P(1, 2, 3)$ from the line $\dfrac{x-6}{3} = \dfrac{y-7}{2} = \dfrac{z-7}{-2}$.

Sol. First Method. $A(6, 7, 7)$ will be a point on the line. Let the perpendicular from P meets the line in N. Then,

$$AP^2 = (6-1)^2 + (7-2)^2 + (7-3)^2 = 66$$

$AN =$ Projection of AP on the given line

$$= (6-1)\frac{3}{\sqrt{17}} + (7-2).\frac{2}{\sqrt{17}} + (7-3)\left(-\frac{2}{\sqrt{17}}\right)$$

$$= \sqrt{17}$$

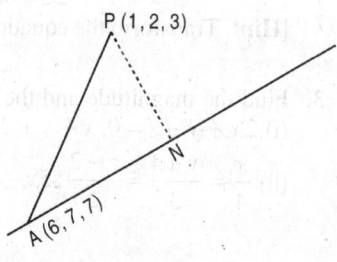

Fig. 6

∴ $PN^2 = AP^2 - AN^2 = 66 - 17 = 49$

⇒ $PN = 7$.

Second Method. The given line can be written as

$$\frac{x-6}{3/\sqrt{17}} = \frac{y-7}{2/\sqrt{17}} = \frac{z-7}{-2/\sqrt{17}}$$

∴ $PN^2 = \left\{\dfrac{2}{\sqrt{17}}(1-6) - \dfrac{3}{\sqrt{17}}(2-7)\right\}^2 + \left\{\dfrac{-2}{\sqrt{17}}(2-7) - \dfrac{2}{\sqrt{17}}(3-7)\right\}^2$

$$+ \left\{\dfrac{3}{\sqrt{17}}(3-7) + \dfrac{2}{\sqrt{17}}(1-6)\right\}^2 = 49$$

⇒ $PN = 7$.

Third Method. Any point N on the line is $(3r+6, 2r+7, -2r+7)$. Let this be the foot of the perpendicular.

Then PN whose dr's are $3r+5, 2r+5, -2r+4$, will be perpendicular to the given line.

∴ $3(3r+5) + 2(2r+5) - 2(-2r+4) = 0$

⇒ $r = -1$.

Thus, the foot of perpendicular N is $(3, 5, 9)$ and hence $PN = 7$.

EXERCISES

1. Find the length of the perpendicular from the point $(4, -5, 3)$ to the line

$$\frac{x-5}{3} = \frac{y+2}{-4} = \frac{z-6}{5}.$$

$\left[\text{Ans. } \dfrac{\sqrt{457}}{5}\right]$

The Straight Line

2. Find the locus of the point whcih moves so that its distance from the line $x = y = z$ is twice its distance from the plane $x + y + z = 1$.

[Ans. $x^2 + y^2 + z^2 + 5xy + 5yz + 5zx - 4x - 4y - 4z + 2 = 0$]

3. Find the length of the perpendicular from the point $P(5, 4, -1)$ upon the line
$$\frac{1}{2}(x-1) = \frac{1}{9}y = \frac{1}{5}z.$$
[Ans. $\sqrt{2109/110}$]

3.8 INTERSECTION OF THREE PLANES

Given three distinct planes such that no two of them are parallel.
We have the following three possibilities in respect of their intersection.
The three planes may :
 (i) have only one point in common (Fig. 7a);
 (ii) have a line in common so that the three planes are coaxial (Fig. 7b);
 (iii) form a triangular prism (Fig. 7c).
We shall in the following find conditions for each of these three possibilities.
Three planes are said to form a triangular prism if the three lines of intersection of the three planes, taken in pairs, are distinct and parallel.
Clearly, the three planes will form a triangular prism if the line of intersection of two of them be parallel to the third.

3.8.1

To find the condition that the three planes
$$a_r x + b_r y + c_r z + d_r = 0; \ (r = 1, 2, 3)$$
should form a prism or intersect in a line.

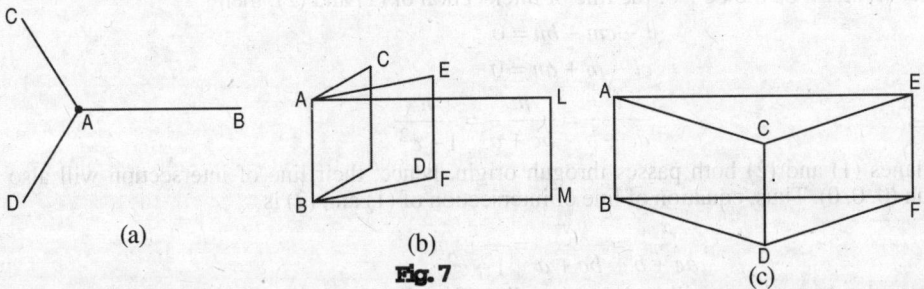

Fig. 7

We assume that the first two planes are *not* parallel.
The line of intersection of the first two planes is
$$\frac{x - (b_1 d_2 - b_2 d_1)/(a_1 b_2 - a_2 b_1)}{b_1 c_2 - b_2 c_1} = \frac{y - (a_2 d_1 - a_1 d_2)/(a_1 b_2 - a_2 b_1)}{a_2 c_1 - a_1 c_2} = \frac{z}{a_1 b_2 - a_2 b_1} \quad ...(i)$$

The three planes will form a triangular prism if this line is parallel to the third plane without lying in the same.
The line (i) will be parallel to the third plane, if
$$a_3(b_1 c_2 - b_2 c_1) + b_3(c_1 a_2 - c_2 a_1) + c_3(a_1 b_2 - a_2 b_1) = 0$$
$$\Rightarrow \begin{vmatrix} a_1 & b_1 & c_1 \\ a_2 & b_2 & c_2 \\ a_3 & b_3 & c_3 \end{vmatrix} = 0 \quad ...(ii)$$

Again, the planes will intersect in a line if and only if the line (i) lies in the plane
$$a_3 x + b_3 y + c_3 z + d_3 = 0.$$

This rquires :
(1) This line is parallel to the third plane *i.e.*, (ii) is satisfied, and

(2) the point $\left(\dfrac{b_1 d_2 - b_2 d_1}{a_1 b_2 - a_2 b_1}, \dfrac{a_2 d_1 - a_1 d_2}{a_1 b_2 - a_2 b_1}, 0\right)$ lies on it implying that

$$a_3(b_1 d_2 - b_2 d_1) + b_3(a_2 d_1 - a_1 d_2) + d_3(a_1 b_2 - a_2 b_1) = 0$$

$$\Rightarrow \quad \begin{vmatrix} a_1 & b_1 & d_1 \\ a_2 & b_2 & d_2 \\ a_3 & b_3 & d_3 \end{vmatrix} = 0 \qquad \ldots\text{(iii)}$$

Thus, the three planes will intersect in a line, if the condition
(ii) and (iii) hold
and will form a triangular prism, if
(ii) holds but (iii) does not hold.
The three planes will intersect in a unique finite point if the conditon (ii) does not hold.

EXAMPLES

1. *Prove that the planes* $x = cy + bz$, $y = az + cx$, $z = bx + ay$ *pass through one line if* $a^2 + b^2 + c^2 + 2abc = 1$. *Show that the equations of this line are*

$$\dfrac{x}{\sqrt{1-a^2}} = \dfrac{y}{\sqrt{1-b^2}} = \dfrac{z}{\sqrt{1-c^2}}.$$

Sol. The three planes can be written as

$$x - cy - bz = 0 \qquad \ldots(1)$$
$$cx - y + az = 0 \qquad \ldots(2)$$
$$bx + ay - z = 0 \qquad \ldots(3)$$

Let (l, m, n) be the dc's of the line of intersection of (1) and (2); then

$$l - cm - bn = 0$$
$$cl - m + an = 0$$

$$\Rightarrow \quad \dfrac{l}{ac+b} = \dfrac{m}{bc+a} = \dfrac{n}{1-c^2}$$

Planes (1) and (2) both passes throguh origin, hence, their line of intersection will also pass throguh (0, 0, 0). Thus, equation of line of intersection of (1) and (2) is

$$\dfrac{x}{ac+b} = \dfrac{y}{bc+a} = \dfrac{z}{1-c^2} \qquad \ldots(4)$$

Now the three planes will intersect in a line if (4) lies in (3). The point (0, 0, 0) of (4) already satisfies (3). Hence, the required condition is

$$b(ac+b) + a(bc+a) - (1-c^2) = 0$$

$$\Rightarrow \quad a^2 + b^2 + c^2 + 2abc = 1 \qquad \ldots(5)$$

We have

$$ac + b = \sqrt{(ac+b)^2} = \sqrt{a^2 c^2 + b^2 + 2abc} = \sqrt{a^2 c^2 + (1 - a^2 - c^2)} \qquad \text{[From (5)]}$$
$$= \sqrt{(1-a^2)(1-c^2)}.$$

Similarly, $bc + a = \sqrt{(1-b^2)(1-c^2)}$.

Putting these in (4), we get

$$\dfrac{x}{\sqrt{1-a^2}} = \dfrac{y}{\sqrt{1-b^2}} = \dfrac{z}{\sqrt{1-c^2}}.$$

2. *The plane* $x/a + y/b + z/c = 1$ *meets the axes OX, OY, OZ which are rectangular, in A, B, C. Prove that the plane through the axes and the internal bisector of the angles of the triangle ABC pass through the line*

The Straight Line

$$\frac{x}{a\sqrt{b^2+c^2}} = \frac{y}{b\sqrt{c^2+a^2}} = \frac{z}{c\sqrt{a^2+b^2}}.$$

Sol. We know that bisector of any angle of a triangle meets the opposite side at a point which divides it in the ratio of other two sides. Hence, if CF is the internal bisector of angle ABC, then

$$\frac{AF}{BC} = \frac{AC}{BC} = \frac{\sqrt{(a^2+c^2)}}{\sqrt{(b^2+c^2)}}$$

Thus, co-ordinates of F will become

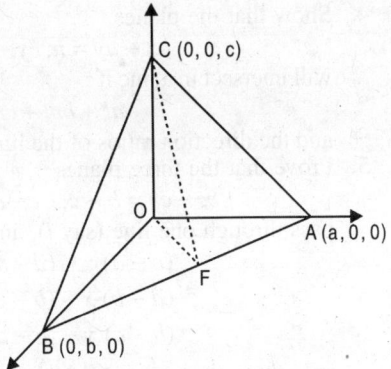

$$\left(\frac{a\sqrt{(b^2+c^2)}}{\sqrt{(a^2+c^2)}+\sqrt{(b^2+c^2)}}, \frac{b\sqrt{a^2+c^2}}{\sqrt{(a^2+c^2)}+\sqrt{(b^2+c^2)}}, 0 \right)$$

Plane through z-axis is
$$x + \lambda y = 0$$
If it passes through F, then
$$\lambda = -\frac{a\sqrt{(b^2+c^2)}}{b\sqrt{(a^2+c^2)}}$$

Hence, equation of plane OCF is
$$x - \frac{a\sqrt{(b^2+c^2)}}{b\sqrt{(a^2+c^2)}} y = 0$$

i.e., $$\frac{x}{a\sqrt{(b^2+c^2)}} = \frac{y}{b\sqrt{(a^2+c^2)}}.$$

Similarly, other planes through axes and lines bisecting the angles A and B are

$$\frac{y}{b\sqrt{a^2+c^2}} = \frac{z}{c\sqrt{(a^2+b^2)}} \quad \text{and} \quad \frac{x}{a\sqrt{(b^2+c^2)}} = \frac{z}{c\sqrt{(a^2+b^2)}}$$

The planes clearly pass through the line
$$\frac{x}{a\sqrt{(b^2+c^2)}} = \frac{y}{b\sqrt{(a^2+c^2)}} = \frac{z}{c\sqrt{(a^2+b^2)}}.$$

EXERCISES

1. Show that the following sets of planes intersect in lines :
   (i) $4x+3y+2z+7=0$, $2x+y-4z+1=0$, $x-7z-7=0$
   (ii) $2x+y+z+4=0$, $y-z+4=0$, $3x+2y+z+8=0$.
2. Show that the following sets of planes form triangular prisms :
   (i) $x+y+z+3=0$, $3x+y-2z+2=0$, $2x+4y+7z-7=0$.
   (ii) $x-z-1=0$, $x+y-2z-3=0$, $x-2y+z-3=0$.
3. Examine the nature of intersection of the following sets of planes :
   (i) $4x-5y-2z-2=0$, $5x-4y+2z+2=0$, $2x+2y+8z-1=0$.
   (ii) $2x+3y-z-2=0$, $3x+3y+z-4=0$, $x-y+2z-5=0$.
   (iii) $5x+3y+7z-4=0$, $3x+26y+2z-9=0$, $7x+2y+10z-5=0$.
   (iv) $2x+6y+11=0$, $6x+20y-6z+3=0$, $6y-18z+1=0$.

   [**Ans.** (i) prism, (ii) point, (iii) line, (iv) prism]

4. Show that the planes
$$bx - ay = n,\ cy - bz = l,\ az - cx = m,$$
will intersect in a line if
$$al + bm + cn = 0,$$
and the direction ratios of the line, then, are a, b, c.

5. Prove that the three planes
$$bz - cy = b - c,\ cx - az = c - a,\ ay - bx = a - b,$$
pass through one line (say l), and the three planes
$$(c - a) z - (a - b) y = b + c,$$
$$(a - b) x - (b - c) z = c + a,$$
$$(b - c) y - (c - a) z = a + b,$$
pass through another line (say l'). Show that the lines l and l' are at right angles to each other.

6. If the planes $x = y + z,\ y = az + x,\ z = x + ay$ pass through one line, find the value of a.

[**Ans.** $a = -1$]

7. Prove that the planes $ny - mz = \lambda,\ lz - nx = \mu$ and $mx - ly = \nu$ have a common line if $l\lambda + m\mu + n\nu = 0$. Show also that the distance of the line from the origin is
$$\left(\frac{\lambda^2 + \mu^2 + \nu^2}{l^2 + m^2 + n^2} \right)^{1/2}$$

4

The Sphere

4.1 DEFINITION

A **sphere** *is the locus of a point which remains at a constant distance from a fixed point.* The constant distance is called the *Radius* and the fixed point the *Centre* of the sphere.

4.1.1 Equation of a Sphere

Let (a, b, c) be the centre and r the radius of a given sphere.

Equating the radius r to the distance of any point (x, y, z) on the sphere from its centre (a, b, c) we have

$$(x-a)^2 + (y-b)^2 + (z-c)^2 = r^2$$

$$\Leftrightarrow \quad x^2 + y^2 + z^2 - 2ax - 2by - 2cz + (a^2 + b^2 + c^2 - r^2) = 0 \quad \text{...(A)}$$

which is the required equation of the given sphere.

Thus, the sphere whose centre is the point (a, b, c) and whose radius is r is the set

$$\{(x, y, z) = x^2 + y^2 + z^2 - 2ax - 2by - 2cz + (a^2 + b^2 + c^2 - r^2) = 0\}$$

We note the following *characteristics* of the equation (A) of the sphere :
1. It is of the second degree in x, y, z;
2. The coefficient of x^2, y^2, z^2 are all equal;
3. The product terms xy, yz, zx are absent.

Conversely, we consider the equation.

$$ax^2 + ay^2 + az^2 + 2ux + 2vy + 2wz + d = 0, a \neq 0 \quad \text{...(B)}$$

having the above three characteristics; a, u, v, w, d being given constants and $a \neq 0$.

The equation (B) can be written as

$$\left(x + \frac{u}{a}\right)^2 + \left(y + \frac{v}{a}\right)^2 + \left(z + \frac{w}{a}\right)^2 = \frac{u^2 + v^2 + w^2 - ad}{a^2}$$

This manner of rewriting shows that the distance between the variable point (x, y, z) and the fixed point

$$\left(-\frac{u}{a}, -\frac{v}{a}, -\frac{w}{a}\right)$$

is

$$\frac{\sqrt{u^2 + v^2 + w^2 - ad}}{|a|}, u^2 + v^2 + w^2 - ad \geq 0$$

and is, therefore, constant.

The locus of the equation (B) is thus a sphere, if

$$u^2 + v^2 + w^2 - ad \geq 0$$

4.1.2 General Equation of a Sphere

We write the equation (B) in the form

$$x^2 + y^2 + z^2 + \frac{2u}{a}x + \frac{2v}{a}y + \frac{2w}{a}z + \frac{d}{a} = 0, a \neq 0$$

$$\Leftrightarrow \quad x^2 + y^2 + z^2 + 2u'x + 2v'y + 2w'z + d' = 0$$

which is taken as the *general equation of a sphere.*

835

The family of spheres is thus given by the equation
$$x^2 + y^2 + z^2 + 2ux + 2vy + 2wz + d = 0$$
where u, v, w, d are parameters such that $u^2 + v^2 + w^2 - d \geq 0$.
The radius of the sphere is '0' if
$$u^2 + v^2 + w^2 - d = 0$$
In this case, the sphere is what we may call a *Point sphere*.

4.1.3 Equation to a sphere on line joining $(x_1, y_1, z_1), (x_2, y_2, z_2)$ as diameter.

Let $P(x, y, z)$ be a point on the sphere. Then APB is a right-angled triangle right-angled at P.

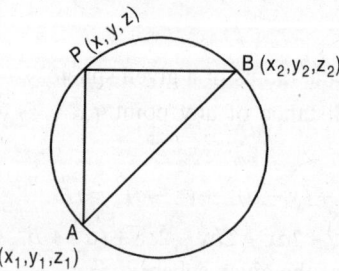

Now direction cosines of AP are proportional to $x - x_1, y - y_1, z - z_1$ and direction cosines of BP are proportional to $x - x_2, y - y_2, z - z_2$.
But AP and BP are at right angles to each other
$$\therefore \quad (x - x_1)(x - x_2) + (y - y_1)(y - y_2) + (z - z_1)(z - z_2) = 0$$
is the required equation of the sphere.

EXAMPLE

1. *A plane passes through a fixed point (a, b, c); show that the locus of the foot of the perpendicular to it from the origin is the sphere $x^2 + y^2 + z^2 - ax - by - cz = 0$.*

Sol. Any plane through (a, b, c) is
$$l(x - a) + m(y - b) + n(z - c) = 0 \quad \text{....(1)}$$
and the line perpendicular to it from the origin is

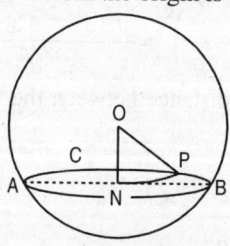

$$\frac{x}{l} = \frac{y}{m} = \frac{z}{n} \quad \text{....(2)}$$

The foot of the perpendicular is the point of intersection of (1) and (2). Thus, to find the locus of the foot of perpendicular, one should eliminate l, m, n between (1) and (2), *i.e.*,
$$x(x - a) + y(y - b) + z(z - c) = 0.$$
$$\Rightarrow \quad x^2 + y^2 + z^2 - ax - by - cz = 0$$

EXERCISES

1. Find the centres and the radii of the following spheres :
   (i) $x^2 + y^2 + z^2 - 6x + 8y - 10z + 1 = 0$,
   (ii) $x^2 + y^2 + z^2 + 2x - 4y - 6z + 5 = 0$,

The Sphere

(iii) $2x^2 + 2y^2 + 2z^2 - 2x + 4y + 2z + 3 = 0$.

$\left[\text{Ans. } (i)\ (3, -4, 5);\ 7\ (ii)\ (-1, 2, 3);\ 3\ (iii)\ \left(\frac{1}{2}, -1, -\frac{1}{2}\right);\ 0\right]$

2. Obtain the equation of the sphere described on the join of the points
$A(2, -3, 4)\ B(-5, 6, -7)$
as diameter.
[Ans. $x^2 + y^2 + z^2 + 3(x - y + z) - 56 = 0$]

3. Prove that the equation $ax^2 + ay^2 + az^2 + 2ux + 2vy + 2wz + d = 0$ represents a sphere. Find its radius and centre.

$\left[\text{Ans. } \frac{\sqrt{\Sigma u^2 - ad}}{a};\ \left(-\frac{u}{a}, -\frac{v}{a}, -\frac{w}{a}\right)\right]$

4. Through a point P three mutually perpendicular straight lines are drawn; one passes through a fixed point C on the z-axis, while the others intersect the x-axis and y-axis, respectively; show that the locus of P is a sphere of which C is the centre.

4.2 THE SPHERE THROUGH FOUR GIVEN POINTS

The general equation of a sphere contains *four* parameters and, as such a sphere can be uniquely determined so as to satisfy four conditions, each of which is such that it gives rise to one linear relation between the constants.

In particular, we can find a sphere through four non-coplanar points
$(x_1, y_1, z_1), (x_2, y_2, z_2), (x_3, y_3, z_3), (x_4, y_4, z_4)$

Let $\qquad x^2 + y^2 + z^2 + 2ux + 2vy + 2wz + d = 0 \qquad$ ...(i)

be the equation of the sphere through the four given points.
We have then the linear equation

$$x_1^2 + y_1^2 + z_1^2 + 2ux_1 + 2vy_1 + 2wz_1 + d = 0 \qquad \text{...(ii)}$$

and three more similar equations corresponding to the remaining three points so that we obtain a system of four linear equations in four unknowns u, v, w, d. We solve these equations and substituting the values thus obtained for u, v, w, d in (i); we get the required equation.

EXAMPLES

1. *Find the equation to the sphere through the points* $(0, 0, 0), (0, 1, -1), (-1, 2, 0), (1, 2, 3)$.

Sol. Let the equation of the sphere be

$$x^2 + y^2 + z^2 + 2ux + 2vy + 2wz + d = 0 \qquad \text{...(i)}$$

As it passes through given points, we have
$d = 0$;
$2 + 2v - 2w + d = 0$;
$2 + 2v - 2w + d = 0$;
$4 + 2u + 4v + 6w + d = 0$;

yielding $\quad u = -\frac{15}{14},\ v = -\frac{25}{14},\ w = -\frac{11}{14}\ $ and $d = 0$.

Hence, the equation of the sphere becomes

$$x^2 + y^2 + z^2 - \frac{15}{7}x - \frac{25}{7}y - \frac{11}{7}z = 0$$

or $\qquad 7(x^2 + y^2 + z^2) - 15x - 25y - 11z = 0$

2. *Prove that the centres of the spheres which touch the lines* $y = mx, z = c;\ y = -mx, z = -c;$ *lie upon the conicoid* $mxy + cz(1 + m^2) = 0$.

Sol. Let the equation of the sphere be
$$x^2 + y^2 + z^2 + 2ux + 2vy + 2wz + d = 0$$
Line $y = mx, z = c$ meets it where
$$x^2 + m^2x^2 + c^2 + 2ux + 2vmx + 2wc + d = 0$$
or
$$(1 + m^2)x^2 + 2(u + vm)x + (c^2 + 2wc + d) = 0$$

The line is given to be tangent to the sphere; hence the two values of x given by above equation must be coincident, which gives
$$(u + vm)^2 = (1 + m^2)(c^2 + 2wc + d) \qquad \ldots(1)$$
In the same way, line $y = -mx, z = -c$ will touch the sphere, if
$$(u - vm)^2 = (1 + m^2)(c^2 - 2wc + d) \qquad \ldots(2)$$
Subtracting (2) from (1), we have
$$4uvm = 4wc(1 + m^2)$$
or
$$uvm - wc(1 + m^2) = 0$$
Hence, the locus of centre $(-u, -v, -w)$ will be
$$xym + zc(1 + m^2) = 0$$

3. *A variable plane through a fixed point (a, b, c) cuts the co-ordinate axes in the point A, B, C. Show that the locus of the centres of the sphere $OABC$ is*
$$\frac{a}{x} + \frac{b}{y} + \frac{c}{z} = 2$$

Sol. Let the sphere $OABC$ be
$$x^2 + y^2 + z^2 + 2ux + 2vy + 2wz = 0$$
so that u, v, w are different for different spheres. The points A, B, C where it cuts the three axes are $(-2u, 0, 0), (0, -2v, 0), (0, 0, -2w)$. The equation of the plane ABC is
$$\frac{x}{-2u} + \frac{y}{-2v} + \frac{z}{-2w} = 1$$
Since this plane passes through (a, b, c) we have
$$\frac{a}{-2u} + \frac{b}{-2v} + \frac{c}{-2w} = 1 \qquad \ldots(2)$$
If x, y, z be the centre of the sphere (1), then
$$x = -u, y = -v, z = -w \qquad \ldots(3)$$
From (2) and (3), we obtain
$$\frac{a}{x} + \frac{b}{y} + \frac{c}{z} = 2$$

as the required locus.

4. *A sphere of constant radius k passes through the origin and cuts the axes in A, B and C. Find the locus of the centroid of the triangle ABC.*

Sol. Let the co-ordinates of A, B, C be $(a, 0, 0), (0, b, 0)$ and $(0, 0, c)$ respectively. The sphere also passes through the origin $(0, 0, 0)$.
Let the equation of the sphere be
$$x^2 + y^2 + z^2 + 2ux + 2vy + 2wz + d = 0$$
As it passes through $(0, 0, 0), (a, 0, 0), (0, b, 0)$ and $(0, 0, c)$, we have $d = 0$,
$$a^2 + 2ua + d = 0 \Rightarrow u = -\frac{1}{2}a, \ v = -\frac{1}{2}b, \ w = -\frac{1}{2}c$$

The Sphere

Let $P(x, y, z)$ be a point on the sphere described on the segment AB as diameter.

Since the section of the required sphere by the plane through the three points P, A, B is a great circle having AB as diameter, the point P lies on a semi-circle and, therefore,
$$PA \perp PB$$
The direction cosines of PA, PB being proportional to
$$x - x_1, y - y_1, z - z_1 \text{ and } x - x_2, y - y_2, z - z_2$$
they will be perpendicular, if
$$(x - x_1)(x - x_2) + (y - y_1) + (z - z_1)(z - z_2) = 0$$
which is the required equation of the sphere.

4.4 EQUATIONS OF A CIRCLE

A circle is the intersection of its plane with some sphere through it. As such, a circle can be represented by two equation, representing a sphere and the other a plane.

Thus, the two equations
$$x^2 + y^2 + z^2 + 2ux + 2vy + 2wz + d = 0, \quad lx + my + nz = p$$
taken together represent a circle.

A circle can also be represenetd by the equations of any two spheres through it.

Note : The reader may note that the equations
$$x^2 + y^2 + z^2 + 2fy + c = 0, \quad z = 0$$
also represented a circle which is the intersection of the cylinder.
$$x^2 + y^2 + z^2 + 2fy + c = 0$$
with the plane.

EXAMPLES

1. A variable plane is parallel to the given plane $\dfrac{x}{a} + \dfrac{y}{b} + \dfrac{z}{c} = 0$ and meets the axes in A, B, C. Prove that the circle ABC lies on the cone.
$$yz\left(\frac{b}{c} + \frac{c}{b}\right) + zx\left(\frac{c}{a} + \frac{a}{c}\right) + xy\left(\frac{a}{b} + \frac{b}{a}\right) = 0$$

Sol. Let the variable plane, which is parallel to the given plane, be
$$\frac{x}{a} + \frac{y}{b} + \frac{z}{c} = k$$
This meets the axes in $A(ak, 0, 0), B(0, bk, 0)$ and $C(0, 0, ck)$. Hence, equation of the sphere $OABC$ is
$$x^2 + y^2 + z^2 - akx - bky - ckz = 0$$
or
$$x^2 + y^2 + z^2 - k(ax + by + cz) = 0 \qquad ...(2)$$

The circle ABC lies on both, the plane (1) and the sphere (2). Hence, (1) and (2) together represented the circle ABC and the locus of the circle ABC will be obtained by eliminating k from (1) and (2). Thus, the locus of circle ABC is
$$(x^2 + y^2 + z^2) - \left(\frac{x}{a} + \frac{y}{b} + \frac{z}{c}\right)(ax + by + cz) = 0$$
or
$$yz\left(\frac{b}{c} + \frac{c}{b}\right) + zx\left(\frac{c}{a} + \frac{a}{c}\right) + xy\left(\frac{a}{b} + \frac{b}{a}\right) = 0$$

2. If r be the radius of the circle
$$x^2 + y^2 + z^2 + 2ux + 2vy + 2wz + d = 0, \quad lx + my + nz = 0$$
prove that
$$(r^2 + d)(l^2 + m^2 + d^2) = (mw - nv)^2 + (nu - hv)^2 + (lv - mu)^2$$

Sol. The equation of the given sphere is

$$x^2 + y^2 + z^2 + 2ux + 2vy + 2wz + d = 0 \qquad \ldots(1)$$

having centre at $(-u, -v, -w)$ and radius $CP = \sqrt{u^2 + v^2 + w^2 - d}$.

Now distance CN of centre of sphere from the plane is length of perpendicular from centre of sphere on the plane

$$lx + my + nz = 0$$

Hence, $\quad CN = \dfrac{lu + mv + nu}{\sqrt{l^2 + m^2 + n^2}}$

$$CP = \sqrt{u^2 + v^2 + w^2 - d}, \quad NP = r$$

$$r^2 = CP^2 - CN^2 = u^2 + v^2 + w^2 - d - \dfrac{(lu + mv + nw)^2}{l^2 + m^2 + n^2}$$

$\Rightarrow \quad (r^2 + d)(l^2 + m^2 + n^2) = (u^2 + v^2 + w^2)(l^2 + m^2 + n^2) - (lu + mv + nw)^2$

$\Rightarrow \quad (r^2 + d)(l^2 + m^2 + n^2) = (mw - nv)^2 + (nu - lw)^2 + (lv - mu)^2$

by Lagrange's identity.

3. *A is a point on OX and B on OY so that the angle OAB is constant* $(= \alpha)$. *On AB as diameter a circle is described whose plane is parallel to OZ. Prove that as AB varies, the circle generates the cone* $2xy - z^2 \sin 2\alpha = 0$.

Sol. Let A be the point $(a, 0, 0)$ and $B(0, b, 0)$.
Then since $\angle OAB = \alpha$, we have

$$\tan \alpha = \dfrac{b}{a}. \qquad \ldots(1)$$

Now a sphere on AB as diameter is

$$(x - a)x + (y - b)y + z^2 = 0$$

$\Rightarrow \qquad x^2 + y^2 + z^2 = ax + by \qquad \ldots(2)$

A plane through AB parallel to OZ is

$$\dfrac{x}{a} + \dfrac{y}{b} = 1 \qquad \ldots(3)$$

The required circle is given by intersection of (2) and (3). Now to determine the locus of the circle, we will eliminate a and b from (2), with the help of (1) and (3). So, we have

$$x^2 + y^2 + z^2 = (ax + by)\left(\dfrac{x}{a} + \dfrac{y}{b}\right)$$

$\Rightarrow \qquad x^2 + y^2 + z^2 = x^2 + y^2 + xy\left(\dfrac{a}{b} + \dfrac{b}{a}\right)$

$\Rightarrow \qquad z^2 = \dfrac{2xy}{\sin 2\alpha}$

$\Rightarrow \qquad 2xy - z^2 \sin 2\alpha = 0$

EXERCISES

1. Find the centre and the radius of the circle

$$x + 2y + 2z = 15, \quad x^2 + y^2 + z^2 - 2y - 4z = 11 \qquad [\text{Ans. } (1, 3, 4), \sqrt{7}\,]$$

2. Find the equations of that section of the sphere

$$x^2 + y^2 + z^2 = a^2$$

of which a given internal point (x_1, y_1, z_1) is the centre.

The Sphere

[**Hint** : The plane through (x_1, y_1, z_1) drawn perpendicular to the line joining this point to the centre $(0, 0, 0)$ of the sphere determines the required section.]

[**Ans.** $x^2 + y^2 + z^2 = a^2, xx_1 + yy_1 + zz_1 = x_1^2 + y_1^2 + z_1^2$]

3. Obtain the equations of the circle lying on the sphere
$$x^2 + y^2 + z^2 - 2x + 4y - 6z + 3 = 0$$
and having its centre at $(2, 3, -4)$.

[**Ans.** $x^2 + y^2 + z^2 - 2x + 4y - 6z + 3 = 0 - x + 5y - 7z - 45$]

4. O is the origin and A, B, C are the points.
$$(4a, 4b, 4c), (4b, 4c, 4a), (4c, 4a, 4b)$$
Show that the sphere
$$x^2 + y^2 + z^2 - 2(x + y + z)(a + b + c) + 8(bc + ca + ab) = 0$$
passes through the nine point circles of the faces of the tetrahedron $OABC$.

5. Find the equation of the diameter of the sphere $x^2 + y^2 + z^2 = 29$ such that a rotation about it will transfer the point $(4, -3, 2)$ to the point $(5, 0, -2)$ along a great circle of the sphere. Find also the angle through which the sphere must be so rotated.

$$\left[\textbf{Ans.} \frac{x}{2} = \frac{y}{6} = \frac{z}{6}, \cos^{-1}\left(\frac{16}{29}\right)\right]$$

6. Show that the following sets of points are concyclic :
   (i) $(5, 0, 2), (2, -6, 0), (7, -3, 8), (4, -9, 6)$
   (ii) $(-8, 5, 2), (-5, 2, 2), (-7, 6, 6), (-4, 3, 6)$.

4.4.1 Sphere Through a Given Circle

The equation
$$S + kU = 0$$
represents a sphere through the circle with equations
$$S = 0, \ U = 0$$
where $\quad S \equiv x^2 + y^2 + z^2 + 2ux + 2vy + 2wz + d$
$U \equiv lx + my + nz - p$
Thus, the set of spheres through the circle
$$S = 0, U = 0,$$
is $\qquad \{S + kU = 0; \ k \text{ is the parameter}\}$
Also the equation
$$S + kS' = 0$$
represents a sphere through the circle with equations
$$S = 0, S' = 0,$$
where
$$S' \equiv x^2 + y^2 + z^2 + 2u'x + 2v'y + 2w'z + d'$$
for all values of k,
The set of spheres through the circle
$$S = 0, \ S' = 0$$
is thus $\qquad \{S + kS' = 0; \ k \text{ is the parameter}\}$.

Here k is the a parameter which may be so chosen that these equations fulfil one more conditions.

Note 1. We notice that the equation of the plane of the circle through the two given spheres
$$S = 0, S' = 0$$
is $\qquad S - S' = 2(u - u')x + 2(v - v')y + 2(w - w')z + d - d' = 0$

From this we see that the equation of any sphere through the circle

is also of the form
$$S = 0, S' = 0$$
$$S + k(S - S') = 0$$
k, being the parameter.

This from sometimes proves comparatively more convenient.

Note 2. It is important to remember that the general equation of a sphere through the circle
$$x^2 + y^2 + 2gx + 2fy + c = 0, z = 0$$
is
$$x^2 + y^2 + z^2 + 2gx + 2fy - 2kz + c = 0,$$
where k is the parameter.

EXAMPLES

1. Find the equation of the sphere through the circle
$$x^2 + y^2 + z^2 = 9, \quad 2x + 3y + 4z = 5$$
and the point $(1, 2, 3)$.

Sol. The sphere
$$x^2 + y^2 + z^2 - 9 + k(2x + 3y + 4z - 5) = 0$$
passes through the given circle for all values of k.

It will pass through $(1, 2, 3)$ if
$$5 + 15k = 0 \implies k = -\frac{1}{3}$$

The required equation of the sphere, therefore, is
$$3(x^2 + y^2 + z^2) - 2x - 3y - 4z - 22 = 0$$

2. Show that the two circles
$$x^2 + y^2 + z^2 - y + 2z = 0, \quad x - y + z - 2 = 0;$$
$$x^2 + y^2 + z^2 + x - 3y + z - 5 = 0, \quad 2x - y + 4z - 1 = 0;$$
lie on the same sphere and find its equation.

Sol. The equation of any sphere through the first circle is
$$x^2 + y^2 + z^2 - y + 2z + k(x - y \; z - 2) = 0,$$
and that of *any* sphere through the second circle is
$$x^2 + y^2 + z^2 + x - 3y + z - 5 + k'(2x - y + 4z - 1) = 0$$

The equations (i) and (ii) will represent the same sphere, if k, k' can be chosen so as to satisfy the four lines equations.
$$k = 2k' + 1, -1 - k = -k' - 3$$
$$2 + k = 4k' + 1, \; -2k = -k' - 5$$

The first two of these equations give $k = 3, k' = 1$, and these values clearly satisfy the remaining two equations also. These four equations in k, k' being consistent, the two circles lie on the same sphere, viz.,
$$x^2 + y^2 + z^2 - y + 2z + 3(x - y + z - 2) = 0$$
$$\implies \quad x^2 + y^2 + z^2 + 3x - 4y + 5z - 6 = 0$$

3. Prove that the plane $x + 2y - z = 4$ cuts the sphere $x^2 + y^2 + z^2 - x + z - 2 = 0$ in a circle of radius unity and find the equation of sphere which has this circle for one of its great circle.

Sol. The centre of the given sphere is $(1/2, 0, -1/2)$ and its radius
$$= \sqrt{\left(\frac{1}{2}\right)^2 + 0^2 + \left(-\frac{1}{2}\right)^2 - (-1)} = \sqrt{\frac{5}{2}} = r.$$

Length of perpendicular from $(1/2, 0, -1/2)$ to the plane is

The Sphere

$$\frac{1}{2}\sqrt{6} = p \text{ (say)}$$

∴ Radius of circle $= \sqrt{r^2 - p^2} = \sqrt{\frac{5}{2} - \frac{6}{4}} = 1$.

Now, equation of a sphere through given circle is

$$x^2 + y^2 + z^2 - x + z - 2 + \lambda(x + 2y - z - 4) = 0$$

or $\qquad x^2 + y^2 + z^2 + (\lambda - 1)x + 2\lambda y + (1 - \lambda)z - (2 + 4\lambda) = 0 \qquad$ ...(1)

Its centre is $[-(\lambda - 1)/2, -\lambda, -(1 - \lambda)/2]$

If the circle is a great circle of the sphere (1), then its centre should lie on the plane

$$x + 2y - z - 4 = 0$$

of the circle.

$$-\frac{1}{2}(\lambda - 1) + 2(-\lambda) + \frac{1}{2}(1 - \lambda) - 4 = 0$$

or $\qquad -3\lambda - 3 = 0 \text{ or } \lambda = -1.$

From (1), the equation of required sphere is

$$x^2 + y^2 + z^2 - 2x - 2y + 2z + 2 = 0$$

EXERCISES

1. Find the equation of the sphere through the circle

   $$x^2 + y^2 + z^2 + 2x + 3y + 6 = 0, \quad x - 2y + 4z - 9 = 0,$$

   and the centre of the sphere

   $$x^2 + y^2 + z^2 - 2x + 4y - 6z + 5 = 0$$

   [Ans. $x^2 + y^2 + z^2 + 7y - 8z + 24 = 0$]

2. Find the equation to the sphere which passes through point (α, β, γ) and the circle $x^2 + y^2 = a^2, z = 0.$ [Ans. $(x^2 + y^2 + z^2 - a^2)\gamma + (a^2 - \alpha^2 - \beta^2 - \gamma^2)z = 0$]

3. Show that the equation of the sphere having its centre on the plane

   $$4x - 5y - z = 3$$

   and passing through the circle with equations

   $$x^2 + y^2 + z^2 - 2x - 3y + 4z + 8 = 0, \quad x^2 + y^2 + z^2 + 4x + 5y - 6z + 2 = 0$$

   is $\qquad x^2 + y^2 + z^2 + 7x + 9y - 11z - 1 = 0.$

4. Obtain the equation of the sphere having the circle

   $$x^2 + y^2 + z^2 + 10y - 4z - 8 = 0, \quad x + y + z = 3$$

   as the great circle.

   [Hint : The centre of the required sphere lies on the plane $x + y + z = 3$]

   [Ans. $x^2 + y^2 + z^2 - 4x + 6y - 8z + 4 = 0$]

5. A sphere S has point $(0, 1, 0), (3, -5, 2)$ at opposite ends of a diameter. Find the equation of the sphere having the intersection of the sphere S with the plane

   $$5x - 2y + 4z + 7 = 0$$

   as a great circle. [Ans. $x^2 + y^2 + z^2 + 2x + 2y + 2z + 2 = 0$]

6. Obtain the equation of the sphere which passes through the circle $x^2 + y^2 = 4, z = 0$ and is cut by the plane $x + 2y + 2z = 0$ in a circle of radius 3.

   [Ans. $x^2 + y^2 + z^2 = 6z - 4 = 0$]

7. Show that the two circles

$$2(x^2 + y^2 + z^2) + 8x - 13y + 17z - 17 = 0, \quad 2x + y - 3z + 1 = 0;$$

$$x^2 + y^2 + z^2 + 3x - 4y + 3z = 0, \quad x - y + 2z - 4 = 0;$$

lie on the same sphere and find its equation.

[**Ans.** $x^2 + y^2 + z^2 + 5x - 6y + 7z - 8 = 0$]

8. Prove that the circles

$$x^2 + y^2 + z^2 - 2x + 3y + 4z - 5 = 0; \quad 5y + 6z + 1 = 0;$$

$$x^2 + y^2 + z^2 - 3x - 4y + 5z - 6 = 0; \quad x + 2y - 7z = 0$$

lie on the same sphere and find its equation.

[**Ans.** $x^2 + y^2 + z^2 - 2x - 2y - 2z - 6 = 0$]

4.5 INTERSECTION OF A SPHERE AND A LINE

Let
$$x^2 + y^2 + z^2 + 2ux + 2vy + 2wz + d = 0 \qquad \ldots(1)$$

and
$$\frac{x - \alpha}{1} = \frac{y - \beta}{m} = \frac{z - \gamma}{n}, \qquad \ldots(2)$$

be the equations of a sphere and a line respectively.

The point $(lr + \alpha, mr + \beta, nr + \gamma)$ which lies on the given line (2) for all values of r, will also lie on the given sphere (1), for those of the values of r which satisfy the equation

$$r^2 (l^2 + m^2 + n^2) + 2r [l (\alpha + u) + m (\beta + v) + n (\lambda + w)]$$
$$+ (\alpha^2 + \beta^2 + \gamma^2 + 2u\alpha + 2v\beta + 2w\gamma + d) = 0 \qquad \ldots(A)$$

and this latter being a quadratic equation in r, gives two values say r_1, r_2 of r. We suppose that the equation has real roots so that r_1, r_2 are real. Then

$$(lr_1 + \alpha, mr_1 + \beta, nr_1 + \gamma), (lr_2 + \alpha, mr_2 + \beta, nr_2 + \gamma)$$

are the two points of intersection.

Example : Find the co-ordinates of the points where the line

$$\frac{1}{4}(x + 3) = \frac{1}{3}(y + 4) = -\frac{1}{5}(z - 8)$$

intersects the sphere

$$x^2 + y^2 + z^2 + 2x - 10y = 23 \qquad [\text{Ans. } (1, -1, 3); (5, 2, -2)]$$

4.5.1 Power of a Point

Let l, m, n be the direction cosines of the given line (2) in 6.5, so that $l^2 + m^2 + n^2 = 1$. Then, r_1, r_2 are the distances of the points $A(\alpha, \beta, \gamma)$ from the points of intersection P and Q and we have

$$AP.AQ = r_1 r_2 = \alpha^2 + \beta^2 + \gamma^2 + 2u\alpha + 2v\beta + 2w\gamma + d$$

which is independent of the direction cosines l, m, n.

Thus, *if from a fixed point A, chords be drawn in any direction to intersect a given sphere in P and Q, then AP.AQ is constant*. This constant is called the *Power* of the point A with respect to the sphere.

EXAMPLE

1. *Show that the sum of the squares of the intercepts made by a given sphere on any three mutually perpendicular straight lines through a fixed point is constant.*

Sol. Take the fixed point O as the origin and *any* three mutually perpendicular lines through it as the co-ordinate axes. With this choice of axes, let the equation of the given sphere be

The Sphere

$$x^2 + y^2 + z^2 + 2ux + 2vy + 2wz + d = 0$$

The X-axis, $(y = 0 = z)$ meets the sphere in points given by

$$x^2 + 2ux + d = 0$$

so that if x_1, x_2 be its roots, the two points of intersection are $(x_1, 0, 0), (x_2, 0, 0)$.
Also we have

$$x_1 - x_2 = -2u, \quad x_1 x_2 = d$$

(intercept on X-axis)$^2 = (x_1 - x_2)^2 = (x_1 + x_2)^2 - 4x_1 x_2 = 4(u^2 - d)$
Similarly,

(intercept on Y-axis)$^2 = 4(v^2 - d)$

(intercept on Z-axis)$^2 = 4(w^2 - d)$

The sum of the squares of the intecepts

$$= 4(u^2 + v^2 + w^2 - 3d)$$
$$= 4(u^2 + v^2 + w^2 - d) - 8d = 4r^2 - 8p,$$

where r is the radius of the given sphere and p is the power of the given point with respect to the sphere.
Since the sphere and the point are both given, r and p are both constants.
Hence, the result.

Note : The coefficeints u, v, w and d in the equation of the sphere will be different for different sets of mutually perpendicular lines through O as axes.
Since the sphere is fixed and the point O is also fixed the expression

$$r^2 = u^2 + v^2 + w^2 - d$$

for the square of the radius and

$$p = d,$$

for the power of the point, with respect to the sphere will be invariant.

EXERCISES

1. Find the locus of a point whose powers with respect, to two given spheres are in a constant ratio.
2. Show that the locus of the mid-points of a system of parallel chords of a sphere is a plane through its centre perpendicular to the given chords.

4.6 EQUATION OF A TANGENT PLANE

To find the equation of the tangent plane at any point (α, β, γ) of the sphere

$$x^2 + y^2 + z^2 + 2ux + 2vy + 2wz + d = 0$$

The point (α, β, γ) lies on the sphere.

$$\Rightarrow \quad \alpha^2 + \beta^2 + \gamma^2 + 2u\alpha + 2v\beta + 2w\gamma + d = 0 \quad \ldots(i)$$

The points of intersection of any line

$$\frac{x-\alpha}{1} = \frac{y-\beta}{m} = \frac{z-\gamma}{n} = r \quad \ldots(ii)$$

through (α, β, γ) with the given sphere are

$$(lr + \alpha, mr + \beta, nr + \gamma)$$

where the values of r are the roots of the quadratic equation

$$r^2 (l^2 + m^2 + n^2) + 2r[l(\alpha + u) + m(\beta + v) + n(\gamma + w)]$$
$$+ (\alpha^2 + \beta^2 + \gamma^2 + 2u\alpha + 2v\beta + 2w\gamma + d) = 0.$$

By virtue of the condition (i), one root of this quadratic equation is zero so that one of the points of intersection coincides with (α, β, γ).

In order that the second point of intersection may also coincide with (α, β, γ) the second value of r must also vanish and this requires,

$$l(\alpha + y) + m(\beta + v) + n(\gamma + w) = 0 \qquad \ldots\text{(iii)}$$

Thus, the line

$$\frac{x - \alpha}{l} = \frac{y - \beta}{m} = \frac{z - \gamma}{n}$$

meets the sphere in two coincident points at (α, β, γ) and so is a *tangent line* to it threat for any set of values l, m, n which satisfy the condition (iii).

The locus of the tangent lines at (α, β, γ) obtained by eliminating l, m, n between the condition (iii) and the equations (ii) of the line is

$$(x - \alpha)(\alpha + u) + (y - \beta)(\beta + v) + (z - \gamma)(\gamma + w) = 0$$

$$\Leftrightarrow \quad \alpha x + \beta y + \gamma z + u(x + \alpha) + v(y + \beta) + w(z + \gamma) + d$$

$$= \alpha^2 + \beta^2 + \gamma^2 + 2u\alpha + 2v\beta + 2w\gamma + d = 0 \qquad \text{[From (i)]}$$

which is a plane known as the *tangent plane* at (α, β, γ).

It follows that

$$(\alpha + u) x + (\beta + v) y + (\gamma + w) z + (u\alpha + v\beta + w\gamma + d) = 0$$

is the equation of the tangent plane to the given sphere at the given point (α, β, γ).

Cor. 1. *The line joining the centre of a sphere to any point on it is perpendicular to the tangent plane threat,* for the direction cosines of the line joining the centre $(-u, -v, -w)$ and the point (α, β, γ) on the sphere are proportional to

$$(\alpha + u, \beta + v, \gamma + w)$$

which are also the coefficients of x, y, z in the equation of the tangent plane at (α, β, γ). Hence, the result.

Cor. 2. If a plane or a line touches a sphere, then the length of the perpendicular from its centre to the plane or the line is equal to its radius.

Note : Any line in the tangent plane through its plane of contact touches the section of the sphere by any plane through the line.

EXAMPLES

1. Show that the plane $lx + my + nz = p$ will touch the sphere

$$x^2 + y^2 + z^2 + 2ux + 2vy + 2wz + d = 0$$

$$(ul + vm + wn + p)^2 = (l^2 + m^2 + n^2)(u^2 + v^2 + w^2 - d)$$

Sol. Equating the radius $\sqrt{u^2 + v^2 + w^2 - d}$ of the sphere to the length of the perpendicular from the centre $(-u, -v, -w)$ to the plane

$$lx + my + nz = p,$$

2. Find the equation of the sphere which touches the sphere

$$x^2 + y^2 + z^2 - x + 3y + 2z - 3 = 0$$

at the point $(1, 1, -1)$ *and passes through the origin.*

Sol. The tangent plane to the given sphere at $(1, 1, -1)$ is

$$x + 5y - 6 = 0$$

The equation of the required sphere is, therefore,

$$x^2 + y^2 + z^2 - x + 3y + 2z - 3 + k(x + 5y - 6) = 0$$

where k is a suitably chosen number.

This will pass through the origin if $k = -1/2$.

The Sphere

Thus, the required equation is
$$2(x^2 + y^2 + z^2) - 3x + y + 4z = 0$$

3. *Find the equation of the sphere through the circle,*
$$x^2 + y^2 + z^2 = 1, \quad 2x = 4y + 5z = 6$$
and touching the plane $z = 0$.

Sol. The sphere
$$x^2 + y^2 + z^2 - 1 + \lambda(2x + 4y - 5z - 6) = 0$$
passes through the given circle for all values of λ.

Its centre is $\left(-\lambda, -2\lambda, -\dfrac{5}{2}\lambda\right)$, and radius is $\left(\lambda^2 + 4\lambda^2 + \dfrac{25}{4}\lambda^2 + 1 + 6\lambda\right)^2$

Since it touches $z = 0$, we have by Cor. 2,
$$-\dfrac{5}{2}\lambda = \pm\left(5\lambda^2 - \dfrac{25}{4}\lambda^2 + 1 - 6\lambda\right)$$
$$\Rightarrow \quad 5\lambda^2 + 6\lambda + 1 = 0,$$

This gives $\lambda = -1$ or $-\dfrac{1}{5}$

The two required spheres, therefore, are
$$x^2 + y^2 + z^2 - 2x - 4y - 5z + 5 = 0$$
$$5(x^2 + y^2 + z^2) - 2x - 4y - 5z + 1 = 0$$

4. *If the tangent plane to the sphere* $x^2 + y^2 + z^2 = r^2$ *makes intercepts* a, b, c *on the co-ordinate axes, show that*
$$\dfrac{1}{a^2} + \dfrac{1}{b^2} + \dfrac{1}{c^2} = \dfrac{1}{r^2}.$$

Sol. The equation to the tangent plane at (α, β, γ) to the given sphere is
$$x\alpha + y\beta + z\gamma = r^2 \qquad \ldots(i)$$

Given that a is the intercept made by the palne (i) on x-axis. So,
$$a\alpha = r^2 \quad \Rightarrow \quad \alpha = \dfrac{r^2}{a}$$

Similarly, $\beta = \dfrac{r^2}{b}$ and $\gamma = \dfrac{r^2}{c}$.

Also, as (α, β, γ) is a point on the sphere, so we have
$$\alpha^2 + \beta^2 + \gamma^2 = r^2$$
$$\Rightarrow \quad \left(\dfrac{r^2}{a}\right)^2 + \left(\dfrac{r^2}{b}\right)^2 + \left(\dfrac{r^2}{c}\right)^2 = r^2$$
$$\Rightarrow \quad \dfrac{1}{a^2} + \dfrac{1}{b^2} + \dfrac{1}{c^2} = \dfrac{1}{r^2}.$$

EXERCISES

1. Find the equation of the tangent plane to the sphere
$$3(x^2 + y^2 + z^2) - 2x - 3y - 4z - 22 = 0$$
and the point $(1, 2, 3)$. **[Ans.** $4x + 9y + 14z - 64 = 0$**]**

2. Find the equation of the tangent line to the circle
$$x^2 + y^2 + z^2 + 5x - 7y + 2z - 8 = 0, \quad 3x - 2y + 4z + 3 = 0$$
and the point $(-3, 5, 4)$. **[Ans.** $(x + 3)/32 = (y - 5)/34 = -(z - 4)/7$**]**

3. Find the values of a for which the plane
$$x + y + z = a\sqrt{3}$$
touches the sphere $x^2 + y^2 + z^2 - 2x - 2y - 2z - 6 = 0$ [**Ans.** $\pm \sqrt{3}$]

4. Find the equation of the tangent planes to the sphere
$$x^2 + y^2 + z^2 + 2x - 4y + 6z - 7 = 0$$
which intersects the line
$$6x - 3y - 2z = 0 = 3z + 2$$ [**Ans.** $2x - y + 4z = 5, 4x - 2y - z = 16$]

5. Show that the plane $2x - 2y + z + 12 = 0$ touches the sphere
$$x^2 + y^2 + z^2 - 2x - 4y + 2z = 3$$
and find the point of contact. [**Ans.** $(-1, 4, -2)$]
[**Hint :** The point of contact of a tangent plane is the point where the line through the centre perpendicular to the plane meets the sphere.]

6. Find the co-ordinates of the points on the sphere
$$x^2 + y^2 + z^2 - 4x + 2y = 4$$
the tangent planes at which are parallel to the plane
$$2x - y + 2z = 1$$ [**Ans.** $(4, -2, 2), (0, 0, -2)$]

7. Show that the equation of the sphere which touches the sphere
$$4(x^2 + y^2 + z^2) + 10x - 25y - 2z = 0$$
at the point $(1, 2, -2)$ and passes through the point $(-1, 0, 0)$ is
$$x^2 + y^2 + z^2 + 2x - 6y + 1 = 0$$

8. Obtain the equations of the tangent planes to the sphere
$$x^2 + y^2 + z^2 + 6x - 2z + 1 = 0,$$
which pass through the line
$$3(16 - x) = 3z = 2y + 30$$
[**Ans.** $2x + 2y - z - 2 = 0, x + 2y - 2z + 14 = 0$]

9. Obtain the equations of the sphere which pass through the circle
$$x^2 + y^2 + z^2 - 2x + 2y + 4z - 3 = 0, \ 2x + y + z = 4$$
and touches the plane $3x + 4y = 14$.
[**Ans.** $x^2 + y^2 + z^2 + 2x + 4y + 6z - 11 = 0, x^2 + y^2 + z^2 - 2x + 2y + 4z - 3 = 0$]

10. Find the equation of the sphere which has its centre at the origin and which touches the line $2(x + 1) = 2 - y = z + 3$. [**Ans.** $9(x^2 + y^2 + z^2) = 5$]

11. Find the equation of the spheres of radius r which touch the three co-ordinates axes. How many such spheres are there ?
[**Ans.** $2(x^2 + y^2 + z^2) + 2\sqrt{2}(\pm x \pm y \pm z)r + r^2 = 0$; eight]

12. Prove that the equation of the sphere which lies in the octant $OXYZ$ and touches the co-ordinate planes is of the form
$$x^2 + y^2 + z^2 - 2\lambda(x + y + z) + 2\lambda^2 = 0.$$
Show that, in general, two spheres can be drawn throgh a given point to touch the co-ordinate planes and find for what positions of the point the spheres are (a) real, (b) coincident.
The distances of the centre from the co-ordinate planes are all equal to the radius so that we may suppose that λ is the radius and $(\lambda, \lambda, \lambda)$ is the centre, λ being the parameter.

The Sphere

4.6.1 Plane of Contact

To find the locus of the points of contact of the tangent planes which pass through a given point (α, β, γ) and touch the sphere.

$$x^2 + y^2 + z^2 + 2ux + 2vy + 2wz + d = 0$$

The tangent plane

$$x(x' + u) + y(y' + v) + z(z' + w) + (ux' + vy' + wz' + d) = 0$$

at (x', y', z') will pass through the point (α, β, γ), if

$$\alpha(x' + u) + \beta(y' + v) + \gamma(z' + w) + (ux' + vv' + wz' + d) = 0$$

$$\Leftrightarrow \quad x'(\alpha + u) + y'(\beta + v) + z'(\gamma + w) + (u\alpha + v\beta + w\gamma + d) = 0$$

which is the condition that the point (x', y', z') should lie on the plane

$$x(\alpha + u) + y(\beta + v) + z(\gamma + w) + (u\alpha + v\beta + w\gamma + d) = 0$$

It is called the *plane of contact* for the point (α, β, γ).

Thus, the locus of points of contact is the circle in which the plane cuts the sphere.

Ex. 1. Show that the line joining any point P to the centre of a given sphere is perpendicular to the plane of contact of P and if OP meets it in Q, then

$$OP \cdot OQ = (\text{radius})^2$$

Ex. 2. Show that the planes of contact of all points on the line

$$\frac{x}{2} = \frac{(y-a)}{3} = \frac{(z+3a)}{4}$$

with respect to the sphere $x^2 + y^2 + z^2 = a^2$ pass through the line

$$-(2x + 3a)/13 = (y - a)/3 = z/1.$$

4.6.2 The Polar Plane

*If a line drawn through a fixed point A meets a given sphere in points P, Q and a point R is taken on this line such that the segment AR is divided internally and externally by the points P, Q in the same ratio, then the locus of R is a plane called the **Polar Plane** of A w.r.t. the sphere,*

Consider the sphere

$$x^2 + y^2 + z^2 = a^2$$

and let A be the point (α, β, γ).

Let $R(x, y, z)$ be the co-ordinates of the point R on any line through A. The co-ordinates of the point dividing AR in the ratio $\lambda : 1$ are

$$\left[\left(\frac{\lambda x + \alpha}{\lambda + 1} \right), \left(\frac{\lambda y + \beta}{\lambda + 1} \right), \left(\frac{\lambda z + \gamma}{\lambda + 1} \right) \right]$$

This point will be on the sphere (1) for values of λ which are roots of the quadratic equation

$$\left(\frac{\lambda x + \alpha}{\lambda + 1} \right)^2 + \left(\frac{\lambda y + \beta}{\lambda + 1} \right)^2 + \left(\frac{\lambda z + \gamma}{\lambda + 1} \right)^2 = a^2,$$

$$\Leftrightarrow \quad \lambda^2(x^2 + y^2 + z^2 - a^2) + 2\lambda(\alpha x + \beta y + \gamma z - a^2) + (\alpha^2 + \beta^2 + \gamma^2 - a^2) = 0 \quad \ldots(2)$$

Its roots λ_1 and λ_2 are the ratios in which the points P, Q divide the segment AR.

Since P, Q divide th segment AR internally and externally in the same ratio, we have

$$\lambda_1 + \lambda_2 = 0$$

Thus, from (2), we have $\quad \alpha x + \beta y + \gamma z - a^2 = 0 \quad \ldots(3)$

which is the relation satisfied by the co-ordinates (x, y, z) of R.

Hence, (3) is the locus of R. Clearly, it is a plane.

Thus, we have seen here that the equation of the polar plane of the point (α, β, γ) with respect to the sphere
$$x^2 + y^2 + z^2 = a^2$$
is
$$\alpha x + \beta y + \gamma z = a^2$$
It may similarly be shown that the polar plane of (α, β, γ) with respect to the sphere
$$x^2 + y^2 + z^2 + 2ux + 2vy + 2wz + d = 0$$
is the plane $\quad (x + u) x + (\beta + v) y + (\gamma + w) z + (u\alpha + v\beta + w\gamma + d) = 0.$

On comparing the equation of the polar plane with that of the tangent plane (4.6) and the plane of contact (4.6.1), we see that the polar plane of a point lying on the sphere is the tangent plane at the point and that of a point, lying outside it, is its plane of contact.

Pole of a Plane : Def. *If π be the polar plane of a point P, then P is called the pole of the plane π.*

4.6.3 Pole of a Plane

To find the pole of the plane $\quad lx + my + nz = p \quad$ ...(i)

with respect to the sphere $\quad x^2 + y^2 + z^2 = a^2$

If (α, β, γ) be the tangent pole, then we see that the equation (i) is identical with
$$\alpha x + \beta y + \gamma z = a^2 \qquad \text{...(ii)}$$
so that, on comparing (i) and (ii), we obtain
$$\frac{\alpha}{l} = \frac{\beta}{m} = \frac{\gamma}{m} = \frac{a^2}{p},$$

$\Rightarrow \qquad \alpha = \dfrac{a^2 l}{p}, \beta = \dfrac{a^2 m}{p}, \gamma = \dfrac{a^2 n}{p}.$

Thus, the point
$$\left(\frac{a^2 l}{p}, \frac{a^2 m}{p}, \frac{a^2 n}{p} \right)$$
is the pole of the plane $lx + my + nz = p$, w.r.t. the sphere $x^2 + y^2 + z^2 = a^2$.

4.6.4 Some Results Concerning Poles and Polars

In the following discussion, we shall always take the equation of a sphere in the form
$$x^2 + y^2 + z^2 = a^2$$

1. *The line joining the centre O of a sphere to any point P is perpendicular to the polar plane of P.*

The direction ratios of the line joining the centre $O(0, 0, 0)$ to the point $P(\alpha, \beta, \gamma)$ are α, β, γ and these are also the direction ratios of the normal to the polar plane $\alpha x + \beta y + \gamma z = a^2$ of $P(\alpha, \beta, \gamma)$.

2. *If the line joining the centre O of a sphere to a point P meets the polar plane of P in Q, then*
$$OP \cdot OQ = a^2,$$
where a is the radius of the sphere.

We have $\qquad OP = \sqrt{\alpha^2 + \beta^2 + \gamma^2}$

Also, OQ, which is the length of perpendicular from the centre $O(0, 0, 0)$ to the polar plane $\alpha x + \beta y + \gamma z = a^2$ of P is given by

The Sphere

$$OQ = \frac{a^2}{\sqrt{\alpha^2 + \beta^2 + \gamma^2}}$$

Hence, the result.

3. *If the polar plane of a point P passes through a point Q, then the polar plane Q passes through P.*

The condition that the polar plane

$$\alpha_1 x + \beta_1 y + \gamma_1 z = a^2$$

of $P(\alpha_1, \beta_1, \gamma_1)$ passes through $Q(\alpha_2, \beta_2, \gamma_2)$ is

$$\alpha_1 \alpha_2 + \beta_1 \beta_2 + \gamma_1 \gamma_2 = a^2$$

which is also, by symmetry or directly the condition that the polar plane of Q passes through P.

Conjugate Points : *Two points such that the polar plane of either passes through the other are called conjugate points.*

4. *If the pole of a plane π_1 lies on another plane π_2, then the pole of π_2 also lies on π_1.*

The condition that the pole

$$\left(\frac{a^2 l_1}{p_1}, \frac{a^2 m_1}{p_1}, \frac{a^2 n_1}{p_1} \right)$$

of the plane π_1 $l_1 x + m_1 y + n_1 z = p_1$
lies on the plane π_2 $l_2 x + m_2 y + n_2 z = p_2$

is

$$a^2 (l_1 l_2 + m_1 m_2 + n_1 n_2) = p_1 p_2$$

which is also, clearly, the condition that the pole

$$\left(\frac{a^2 l_2}{p_2}, \frac{a^2 m_2}{p_2}, \frac{a^2 n_2}{p_2} \right)$$

of π_2 lies on π_1.

Conjugate planes : *Two planes such that the pole of either lies on the other are called conjugate planes.*

5. *The polar planes of all the points on a line l pass through another line l'.*

The polar plane of any point

$$(lr + \alpha, mr + \beta, nr + \gamma)$$

on the line l.

$$\frac{x - \alpha}{l} = \frac{y - \beta}{m} = \frac{z - \gamma}{n}$$

is $(lr + \alpha) x + (mr + \beta) y + (nr + \gamma) z = a^2$

$\Leftrightarrow$ $(\alpha x + \beta y + \gamma z - a^2) + r(lx + my + nz) = 0$

which clearly passes through the line

$$\alpha x + \beta y + \gamma z - a^2 = 0, \ lx + my + nz = 0,$$

whatever value, r, may have. Hence, the result.

Let this line be l'. We shall now prove that the polar plane of every point on l' also passes through the line l.

Now, as the polar plane of any arbitrary point P on l passes through every point of l', therefore, the polar plane of every point of l', passes through the point P on l and as, P is arbitrary, it passes through every point of l, *i.e.*, it passes through l.

Polar Lines : *Two lines such that the polar plane of every point on either passes through the other are called **Polar Lines**.*

EXAMPLE

1. Find the polar line of $(x-1)/2 = (y-2)/3 = (z-3)/4$ w.r.t. the sphere $x^2 + y^2 + z^2 = 16$.

Sol. Any point on given line is $(2r + 1, 3r + 2, 4r + 3)$. Polar plane of this point with respect to sphere is

$$x(2r + 1) + y(3r + 2) + z(4r + 3) = 16$$

i.e.,
$$(x + 2y + 3z - 16) + r(2x + 3y + 4z) = 0$$

This clearly, passes through the line

$$x + 2y + 3z - 16 = 0 = 2x + 3y + 4z$$

which is the required polar line.

EXERCISES

1. Show that the polar line of
$$\frac{(x+1)}{2} = \frac{(y-2)}{3} = (z+3)$$
with respect to the sphere
$$x^2 + y^2 + z^2 = 1;$$
is the line
$$\frac{7x+3}{11} = \frac{2-7y}{5} = \frac{z}{-1}.$$

2. Show that if a line l is coplanar with the polar line of a line l' then l' is coplanar with the polar line of l.
3. If PA, QB be drawn perpendicular to the planes of Q and P respectively, with respect to a sphere, with centre O, then
$$\frac{PA}{QB} = \frac{OP}{OQ}.$$
4. Show that, for a given sphere, there exist an unlimited number of tetraheda such that each vertex is the pole of the opposite face with respect to the sphere.
(Such a tetrahedron is known as a *self-conjugate* or *self-polar* tetrahedron).

4.7 ANGLE OF INTERSECTION OF TWO SPHERES

Def. *The angle of intersection of two spheres at a common point is the angle between the tangent planes to them at that point* and is, therefore, also equal to the angle between the radii of the spheres to the common point, the radii being perpendicular to the respective tangent planes at the point.

The angle of intersection at every common point of the spheres is the same, for if P, P' be any two common points of C, C' the centres of the spheres, the triangles $CC'P$ and $CC'P'$ are congruent and accordingly

$$\angle CPC' = \angle CP'C'$$

The spheres are said to be **orthogonal** if the angle of intersection of two spheres is a right angle. In this case

$$CC'^2 = CP^2 + C'P^2$$

4.7.1 Condition for the Orthogonality of Two Spheres

To find the condition for the two spheres
$$x^2 + y^2 + z^2 + 2u_1 x + 2v_1 y + 2w_1 z + d_1 = 0$$
$$x^2 + y^2 + z^2 + 2u_2 x + 2v_2 y + 2w_2 z + d_2 = 0$$
to be orthogonal.

The spheres will be orthogonal if the *square of the distance between their centres is equal to the sum of the squares of their radii* and this requires

$$(u_1 - u_2)^2 + (v_1 - v_2)^2 + (w_1 - w_2)^2 = (u_1^2 + v_1^2 + w_1^2 - d_1) + (u_2^2 + v_2^2 + w_2^2 - d_2)$$
$$\Leftrightarrow \quad 2u_1u_2 + 2v_1v_2 + 2w_1w_2 = d_1 + d_2$$

EXAMPLES

1. *If d is the distance between the centres of two spheres of radii r_1 and r_2, prove that the angle between them is $\cos^{-1}\{(r_1^2 + r_2^2 - d^2)/2r_1r_2\}$.*

Sol. The angle of intersection, *i.e.*, the angle between the tangents at P is the angle between the radii of the two spheres joining P. Thus, $\angle C_1PC_2 = \theta$.

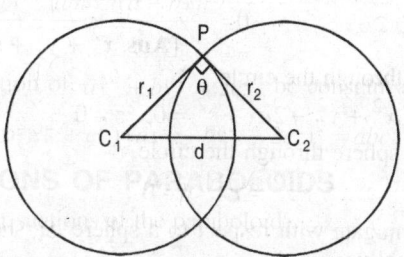

Applying cosine formulae to ΔC_1PC_2, we have
$$d^2 = r_1^2 + r_2^2 - 2r_1r_2 \cos\theta$$
$$\therefore \quad \cos\theta = (r_1^2 + r_2^2 - d^2)/2r_1r_2.$$

2. *Two spheres of radii r_1 and r_2 cut orthogonally. Prove that the radius of the common circle is $r_1r_2/\sqrt{r_1^2 + r_2^2}$.*

Sol. Let the common circle be
$$x^2 + y^2 = a^2, \quad z = 0$$
The sphere
$$x^2 + y^2 + z^2 + 2kz - a^2 = 0$$
passes through this circle for all values of k. Let the two given spheres through the circle be
$$x^2 + y^2 + z^2 + 2k_1z - a^2 = 0, \quad x^2 + y^2 + z^2 + 2k_2z - a^2 = 0$$
We have
$$r_1^2 = k_1^2 + a^2, \quad r_2^2 = k_2^2 + a^2 \qquad \ldots(i)$$
Since the spherses cut orthogonally, we have
$$2k_1k_2 = a^2 + a^2 = 2a^2 \qquad \ldots(ii)$$
From (i) and (ii), eliminating k_1, k_2 we have
$$(r_1^2 - a^2)(r_2^2 - a^2) = a^4 \Leftrightarrow a^2 = r_1^2 r_2^2/(r_1^2 + r_2^2).$$
Hence, the result.

EXERCISES

1. Find the equation of the sphere that passes through the circle
$$x^2 + y^2 + z^2 - 2x + 3y - 4z + 6 = 0, \quad 3x - 4y + 5z - 15 = 0$$
and cuts the sphere
$$x^2 + y^2 + z^2 + 2x + 4y - 6z + 11 = 0$$
orthogonally.

[**Ans.** $5(x^2 + y^2 + z^2) - 13x + 19y - 25z + 45 = 0$]

2. Find the equation of the sphere that passes through the two points
$$(0, 3, 0), (-2, -1, 4)$$
and cuts orthogonally the two spheres
$$x^2 + y^2 + z^2 + x - 3z - 2 = 0, \ 2(x^2 + y^2 + z^2) + x + 3y + 4 = 0$$
[**Ans.** $x^2 + y^2 + z^2 + 2x - 2y + 4z - 3 = 0$]

3. Find the equation of the sphere which touches the plane
$$3x + 2y - z + 2 = 0$$
at the point $(1, -2, 1)$ and cuts orthogonally the sphere
$$x^2 + y^2 + z^2 - 4x + 6y + 4 = 0$$
[**Ans.** $x^2 + y^2 + z^2 + 7x + 10y - 5z + 12 = 0$]

4. Show that every sphere through the circle
$$x^2 + y^2 - 2ax + r^2 = 0, \ z = 0$$
cuts orthogonally every sphere through the circle
$$x^2 + z^2 = r^2, \ y = 0$$

5. Two points P, Q are conjugate with respect to a sphere S; show that the sphere on PQ as diameter cuts S orthogonally.

6. If two spheres S_1 and S_2 are orthogonal, the polar plane of any point on S_1 with respect to S_2 passes through the other end of the diameter of S_1 through P.

4.8 RADICAL PLANE

The locus of a point whose powers with respect to two spheres are equal is a plane perpendicular to the line joining their centres :

The powers of the point $P(x, y, z)$ with respect to the spheres
$$S_1 \equiv x^2 + y^2 + z^2 + 2u_1 x + 2v_1 y + 2w_1 z + d_1 = 0;$$
$$S_2 \equiv x^2 + y^2 + z^2 + 2u_2 x + 2v_2 y + 2w_2 z + d_2 = 0,$$
are
$$x^2 + y^2 + z^2 + 2u_1 x + 2v_1 y + 2w_1 z + d_1,$$
and
$$x^2 + y^2 + z^2 + 2u_2 x + 2v_2 x + 2w_2 z + d_2,$$
respectively.

Equating these, we obtain
$$2x(u_1 - u_2) + 2y(v_1 - v_2) + 2z(w_1 - w_2) + (d_1 - d_2) = 0,$$
as the required locus. This locus being of the first degree in (x, y, z), represents a plane which is obviously perpendicular to the line joining the centres of the two spheres. This plane is called the *Radical plane* of the two spheres.

Thus, *the radical plane of the spheres*
$$S_1 = 0, \ S_2 = 0,$$
in both of which the coefficients of the second degree terms are equal to unity, is
$$S_1 - S_2 = 0$$

In case the two spheres intersect, the plane of their common circle is their radical plane (§4.3.2).

4.8.1 Radical Line

The three radical planes of three spheres intersect in a line,

If
$$S_1 = 0, \ S_2 = 0, \ S_3 = 0$$
be the three spheres, their radical planes
$$S_1 - S_2 = 0, \ S_2 - S_3 = 0, \ S_3 - S_1 = 0$$

The Sphere

clearly meet in the line
$$S_1 = S_2 = S_3$$
$$\Leftrightarrow S_1 - S_2 = 0, \ S_2 - S_3 = 0$$
This line is called the *Radical line* of the three spheres.

4.8.2 Radical Centre

The four radical lines of four spheres taken three by three intersect at a point.
The point common to the three planes
$$S_1 = S_2 = S_3 = S_4$$
is clearly common to the radical lines, taken three by three, of the four spheres
$$S_1 = 0, S_2 = 0, S_3 = 0, S_4 = 0$$
This point is the intersection of the two lines
$$S_1 - S_2 = 0, S_2 - S_3 = 0; \ S_1 - S_3 = 0, S_2 - S_4 = 0.$$
This point is called the *Radical centre* of the four spheres.

4.8.3 Theorem

If $S_1 = 0, S_2 = 0$, be two spheres, then the equation
$$S_1 + \lambda S_2 = 0$$
λ being the parameter, represents a system of spheres such that any two members, of the system have the same radical plane.

Let
$$S_1 + \lambda_1 S_2 = 0 \text{ and } S_1 + \lambda_2 S_2 = 0$$
be any two members of the system.

Making the coefficients of second degree terms unity, we write these equations in the form
$$\frac{S_1 + \lambda_1 S_2}{1 + \lambda_1} = 0, \ \frac{S_1 + \lambda_2 S_2}{1 + \lambda_2} = 0.$$

The radical plane of these two spheres is
$$\frac{S_1 + \lambda_1 S_2}{1 + \lambda_1} - \frac{S_1 + \lambda_2 S_2}{1 + \lambda_2} = 0,$$
$$\Leftrightarrow S_1 - S_2 = 0.$$

Since this equation is indepenent of λ_1 and λ_2, we see that every two members of the system have the same radical plane.

Co-axel Systems : Def. *A system of spheres any two members of which have the same radical plane is called a co-axal system of spheres.*

Thus, the system of spheres
$$S_1 + \lambda S_2 = 0$$
is co-axal and we say that it is determined, by the two spheres
$$S_1 = 0, S_2 = 0$$
The common radical plane is
$$S_1 - S_2 = 0$$
This co-axal system is also given by the equation
$$S_1 + k_2 (S_1 - S_2) = 0$$
Refer Note 1, § 4.4.1.

Note : It can similarly be proved that the system of spheres.
$$S + \lambda U = 0$$
is co-axal; $S = 0$ being a sphere and $U = 0$ a plane; the common radical plane is $U = 0$.

Cor. *The locus of the centres of spheres of a co-axal system is a line.*
For, if (x, y, z) be the centre of the sphere
$$S_1 + \lambda S_2 = 0,$$
we have
$$x = -\frac{u_1 + \lambda u_2}{1 + \lambda}, \quad y = -\frac{v_1 + \lambda v_2}{1 + \lambda}, \quad z = -\frac{w_1 + \lambda w_2}{1 + \lambda}$$
On eliminating λ, we find that it lies on the line
$$\frac{x + u_1}{u_1 - u_2} = \frac{y + v_1}{v_1 - v_2} = \frac{z + w_1}{w_1 - w_2}.$$
This result is also otherwise clear as the line joining the centres of any two spheres is perpendicular to their common radical plane.

4.9 A SIMPLIFIED FORM OF THE EQUATION OF TWO GIVEN SPHERES

By taking the line joining the centres of two given spheres as X-axis, their equations take the form
$$x^2 + y^2 + z^2 + 2u_1 x + d_1 = 0, \quad x^2 + y^2 + z^2 + 2u_2 x + d_2 = 0.$$
The radical plane of these spheres is
$$2x(u_1 - u_2) + (d_1 - d_2) = 0$$
Further, if we take this radical plane as the YZ plane, *i.e.*, $x = 0$, we have $d_1 = d_2 = d$ (say).

Thus, by taking the line joining the centres of two given spheres as X-axis and their radical plane as the YZ plane, their equations take the form
$$x^2 + y^2 + z^2 + 2u_1 x + d = 0, \quad x^2 + y^2 + z^2 + 2u_2 x + d = 0,$$
where u_1, u_2 are different.

Cor. 1. The equation
$$x^2 + y^2 + z^2 + 2kx + d = 0; \quad k, \text{ is a parameter.}$$
represents a co-axel system of spheres for different values of k, d being constant. The YZ plane is the common radical plane and X-axis the line of centres.

Cor. 2. Limiting Points : The equation
$$x^2 + y^2 + z^2 + 2kx + d = 0$$
can be written as $\quad (x + k)^2 + y^2 + z^2 = k^2 - d$

For $k = \pm \sqrt{d}$, we get spheres of the system with radius zero and thus the *system includes the two points spheres*
$$(-\sqrt{d}, 0, 0), (\sqrt{d}, 0, 0)$$
These two points, called the *limiting points*, exist only when d is positive, *i.e.*, when the spheres do not meet the radical plane is a real circle.

Def. Limiting points *of a co-axel system of spheres are the point spheres of the system.*

EXAMPLES

1. *Find the limiting points of the co-axal system defined by the sphere*
$$x^2 + y^2 + z^2 + 3x - 3y + 6 = 0, \quad x^2 + y^2 + z^2 - 6y - 6z + 6 = 0$$
Sol. The equation of the plane of the circle through the two given spheres is
$$3x + 3y + 6z = 0 \Rightarrow x + y + 2z = 0;$$
Then the equation of the co-axal system determined by the given sphere is
$$x^2 + y^2 + z^2 + 3x - 3y + 6 + \lambda(x + y + 2z) = 0,$$
$$\Leftrightarrow \quad x^2 + y^2 + z^2 + (3 + \lambda)x + (\lambda - 3)y + 2\lambda + 6 = 0. \quad \ldots(1)$$

The Sphere

λ being a parameter.

The centre of (1) is $\left(-\dfrac{3+\lambda}{2}, -\dfrac{\lambda-3}{2}, -\lambda\right)$

and its radius is $\left[\left(\dfrac{3+\lambda}{2}\right)^2 + \left(\dfrac{\lambda-3}{2}\right)^2 + \lambda^2 - 6\right]^{1/2}$

Equating this radius to zero, we obtain
$$6\lambda^2 - 6 = 0 \Leftrightarrow \lambda = \pm 1$$

The spheres corresponding to these values of λ become point spheres coinciding with their centres and are the limiting points of the given system of spheres.

The limiting points, therefore, are
$$(-1, 2, 1) \text{ and } (-2, 1, -1)$$

2. *Show that spheres which cut two given spheres along great circles all pass through two fixed points.*

Sol. With proper choice of axes, the equations of the given spheres take the form
$$x^2 + y^2 + z^2 + 2u_1 x + d = 0 \qquad \ldots(i)$$
$$x^2 + y^2 + z^2 + 2u_2 x + d = 0 \qquad \ldots(ii)$$

The equation of any other sphere is
$$x^2 + y^2 + z^2 + 2ux + 2vy + 2wz + c = 0 \qquad \ldots(iii)$$

where u, v, w are different for different spheres.

The plane
$$2x(u - u_1) + 2vy + 2wz + (c - d) = 0$$

If the circle common to the spheres (i) and (iii) will pass through the centre
$$(-u_1, 0, 0)$$

of (i) if, $\qquad -2u_1(u - u_1) + (c - d) = 0 \qquad \ldots(iv)$

which is thus the condition for the sphere (iii) to cut the sphere (i) along a great circle.

Similarly, $\qquad -2u_1(u - u_2) + (c - d) = 0 \qquad \ldots(v)$

is the condition for the sphere (iii) to cut the sphere (ii) along the great circle.

Solving the linear equations (iv) and (v) for u and c, we get
$$u = u_1 + u_2; \quad c_1 = 2u_1 u_2 + d,$$

so that u, c are constants, being dependent on u_1, u_2 d only.

The spheres (iii) cuts X-axis at points whose x-co-ordinates are the roots of the equation
$$x^2 + 2ux + c = 0$$

The roots of this equation are constant, depending as they do upon the constants u and c only.

Thus, every sphere (iii) meets the X-axis at the same two points and hence, the result.

EXERCISES

1. Show that the sphere
$$x^2 + y^2 + z^2 + 2vy + 2wz - d = 0$$
passes through the limiting points of the co-axal system
$$x^2 + y^2 + z^2 + 2kx + d = 0$$
and cuts every member of the system orthogonally, whatever be the values of v, w. Hence, deduce the every sphere that passes through the limiting points of a co-axel system cuts every sphere of the system orthogonally.

2. Show that the locus of the point spheres of the system
$$2x^2 + y^2 + z^2 + 2vy + 2wz - d = 0$$

is the common circle of the system
$$x^2 + y^2 + z^2 + 2ux + d = 0$$
u, v, w being the parameters and d a constant.

3. Show that the sphere which cuts two spheres orthogonally will cut every member of the co-axal system determined by them orthogonally.

4. Find the limiting points of the co-axal system of spheres
$$x^2 + y^2 + z^2 - 20x + 30y - 40z + 29 + \lambda(2x - 3y - 4z) = 0$$
[**Ans.** $(2, -3, -4), (-2, 3, -4)$]

5. Three spheres of radii r_1, r_2, r_3 have their centres A, B, C at the points $(a, 0, 0), (0, b, 0),$ $(0, 0, c)$ and $r_1^2 + r_2^2 + r_3^2 = a^2 + b^2 + c^2$. A fourth sphere passes through the origin and the points A, B, C. Show that the radical centre of the four spheres lies on the plane $ax + by + cz = 0$.

6. Show that the locus of a point from which equal tangents may be drawn to the three spheres
$$x^2 + y^2 + z^2 + 2x + 2y + 2z + 2 = 0, \quad x^2 + y^2 + z^2 + 4x + 4z = 0,$$
$$x^2 + y^2 + z^2 + x + 6y - 4z - 2 = 0$$
is the straight line
$$x/2 = (y-1)/5 = z/3$$

7. Show that there are, in general, two spheres of a co-axal system which touch a given plane. Find the equation to the two spheres of the co-axal system
$$x^2 + y^2 + z^2 - 5 + \lambda(2x + y + 3z - 3) = 0$$
which touch the plane $\quad 3x + 4y = 15$

[**Ans.** $x^2 + y^2 + z^2 + 4x + 2y - 6z - 11 = 0, 5(x^2 + y^2 + z^2) - 8x - 4y - 12z - 13 = 0$]

8. P is a variable point on a given line and A, B, C are its projections on the axes. Show that the $OABC$ passes through a fixed circle.

9. Show that the radical planes of the spheres of a co-axal system and of a given sphere pass through a line.

5

Cones, Cylinders

5.1 DEFINITIONS

A **cone** is a surface generated by a straight line which passes through a fixed point and satisfies one more condition; for instance, it may intersect a given curve or touch a given surface.

The fixed point is called the *Vertex* and the given curve the *Guiding curve* of the cone.

An individual straight line on the surface of a cone is called its *Generator*.

Thus, a cone is essentially a set of lines called *Generators* through a given point. Also, we may say that a cone is a set of points on its generators.

Whereas we can have cones with equations of any degree whatsoever depending upon the condition to be satisfied by its generators, we shall in this book be concerned only with *Quadratic cones*, i.e., cones with second degree equations.

It will be seen that the degree of the equation of a cone whose generators intersect a given conic or touch a given sphere is of the second degree.

5.1.1 Equation of a Cone with a Conic as Guiding Curve

To find the equation of the cone whose vertex is the point
$$(\alpha, \beta, \gamma)$$
are whose generators intersect the conic
$$ax^2 + 2hxy + by^2 + 2gx + 2fy + c = 0, \ z = 0 \qquad \ldots(i)$$

We have to find the locus of points on lines which pass through the given point (α, β, γ) and intersect the given curve.

The equations to *any* line through (α, β, γ) are
$$\frac{x-\alpha}{l} = \frac{y-\beta}{m} = \frac{z-\gamma}{n} \qquad \ldots(ii)$$

This line will be a generator of the cone if and only if it intersects the given curve.

This line meets the plane $z = 0$ in the point
$$\left(\alpha - \frac{l\gamma}{n}, \beta - \frac{m\gamma}{n}, 0\right)$$

which will lie on the given conic, if
$$a\left(\alpha - \frac{l\gamma}{n}\right)^2 + 2h\left(\alpha - \frac{l\gamma}{n}\right)\left(\beta - \frac{m\gamma}{n}\right) + b\left(\beta - \frac{m\gamma}{n}\right)^2$$
$$+ 2g\left(\alpha - \frac{l\gamma}{n}\right) + 2f\left(\beta - \frac{m\gamma}{n}\right) + c = 0 \qquad \ldots(iii)$$

This is the condition for the line (ii) to intersect the conic (i). Eliminating l, m, n between (ii) and (iii), we get
$$a\left(\alpha - \frac{x-\alpha}{z-\gamma}\gamma\right)^2 + 2h\left(\alpha - \frac{x-\alpha}{z-\gamma}\gamma\right)\left(\beta - \frac{y-\beta}{z-\gamma}\gamma\right) + b\left(\beta - \frac{y-\beta}{z-\gamma}\gamma\right)^2$$
$$+ 2g\left(\alpha - \frac{x-\alpha}{z-\gamma}\gamma\right) + 2f\left(\beta - \frac{y-\beta}{z-\gamma}\gamma\right) + c = 0$$

$$a(\alpha z - x\gamma)^2 + 2h(\alpha z - x\gamma)(\beta z - y\gamma) + (\beta z - y\gamma)^2$$
$$+ 2g(\alpha z - x\gamma)(z - \gamma) + 2f(\beta z - z\gamma)(z - \gamma) + c(z - \gamma)^2 = 0$$

which is the required equation of the cone.

EXAMPLES

1. *Find the equation of the cone whose vertex is (α, β, γ) and base $ax^2 + by^2 = 1, z = 0$.*

Solution. Any line through (α, β, γ) is

$$\frac{x - \alpha}{l} = \frac{y - \beta}{m} = \frac{z - \gamma}{n} \qquad \ldots(i)$$

This cuts $z = 0$, where

$$\frac{x - \alpha}{l} = \frac{y - \beta}{m} = \frac{-\gamma}{n}$$

$$\Rightarrow \qquad \left(\alpha - \frac{l\gamma}{n}, \beta - \frac{m\gamma}{n}, 0 \right)$$

which will lie on the given conic, if

$$a \left(\alpha - \frac{l\gamma}{n} \right)^2 + b \left(\beta - \frac{m\gamma}{n} \right)^2 = 1$$

Eliminating l, m, n with the help of (i), we get

$$a \left(\alpha - \frac{x - \alpha}{z - \gamma} \gamma \right)^2 + b \left(\beta - \frac{y - \beta}{z - \gamma} \gamma \right)^2 = 1$$

or $\qquad a(\alpha z - \gamma x)^2 + b(\beta z - z\gamma)^2 = (z - \gamma)^2$

This is the required cone.

2. *The vertex of cone is (a, b, c) and the yz-plane cuts it in the curve $F(y, z) = 0, x = 0$, show that xz-plane cuts it in the curve,*

$$y = 0, \quad F\left[\frac{bx}{x - a}, \frac{cx - az}{x - a} \right] = 0,$$

Solution. Any line through (a, b, c) is

$$\frac{x - a}{l} = \frac{y - b}{m} = \frac{z - c}{n} \qquad \ldots(i)$$

It meets $x = 0$ in the point $\left[0, b - \frac{am}{l}, c - \frac{an}{l} \right]$ and if it lies on the given curve $F(y, z) = 0$, then

$$F\left[b - \frac{am}{l}, c - \frac{an}{l} \right] = 0$$

Eliminating l, m, n between (i) and (ii), we get

$$F\left[b - a\left(\frac{y - b}{x - a} \right), c - a\left(\frac{z - c}{x - a} \right) \right] = 0$$

$$\Rightarrow \qquad F\left[\frac{bx - ay}{x - a}, \frac{cx - az}{x - a} \right] = 0$$

It meets zx-plane, *i.e.*, $y = 0$ in the curve.

$$F\left[\frac{bx}{x - a}, \frac{cx - az}{x - a} \right] = 0, \quad y = 0.$$

Cones, Cylinders

3. Find the equation of the cone with vertex at $(2a, b, c)$ and passing through the curve $x^2 + y^2 = 4a^2$ and $z = 0$. Find b and c if the cone also passes through the curve $y^2 = 4a(z + a)$, $x = 0$. Also, show that the cone is cut by the plane $y = 0$ in two straight lines and the angle θ between them is given by $\tan \theta = 2$.

Solution. Any line through $(2a, b, c)$ is

$$\frac{x - 2a}{l} = \frac{y - b}{m} = \frac{z - c}{n} \qquad \text{...(i)}$$

It meets $z = 0$ in the point $\left[2a - \dfrac{lc}{n}, b - \dfrac{mc}{n}, 0\right]$ and if it lies on the curve $x^2 + y^2 = 4a^2$, $z = 0$ then we get

$$\left(2a - \frac{lc}{n}\right)^2 + \left(b - \frac{mc}{n}\right)^2 = 4a^2 \qquad \text{...(ii)}$$

Eliminating l, m, n between (i) and (ii) we get the required cone as

$$\left[2a - \left(\frac{x - 2a}{z - c}\right)c\right]^2 + \left[b - \left(\frac{y - b}{z - c}\right)c\right]^2 = 4a^2$$

$$\Rightarrow \quad (2az - cx)^2 + (bz - yc)^2 = 4a^2(z - c)^2 \qquad \text{...(iii)}$$

If this cone passes through $y^2 = 4a(z + a)$, $x = 0$, then putting $x = 0$ in (iii), we get

$$(2az)^2 + (bz - yc)^2 = 4a^2(z - c)^2$$

$$\Rightarrow \quad b^2 z^2 + c^2 y^2 - 2bcyz - 4a^2 c^2 + 8a^2 cz = 0 \qquad \text{...(iv)}$$

If it is same as $y^2 = 4a(z + a)$, then comparing this with (iv), we have

$$b^2 = 0 \Rightarrow b = 0 \text{ which reduces (iv) to}$$

$$c^2 y^2 = 4a^2 c^2 - 8a^2 zc$$

$$\Rightarrow \quad y^2 = -\left(\frac{8a^2}{c}\right)\left(z - \frac{1}{2}c\right)$$

This gives $\quad -\dfrac{8a^2}{c} = 4a$ and $-\dfrac{1}{2}c = a$

$$\Rightarrow \quad c = -2a.$$

Hence, $b = 0$, $c = -2a$.

Substituting these values of b and c in (iii), the equation of cone intersecting the given cone reduces to

$$(2ax + 2az)^2 + 4a^2 y^2 = 4a^2(z + 2a)^2$$

$$\Rightarrow \quad x^2 + y^2 + 2zx - 4az - 4a^2 = 0$$

The plane $y = 0$ cuts cone (v) in

$$x^2 + 2zx - 4az - 4a^2 = 0, \; y = 0$$

$$\Rightarrow \quad (x^2 - 4a^2) + 2z(x - 2a) = 0, \; y = 0$$

$$\Rightarrow \quad (x - 2a)(x + 2a + 2z) = 0, \; y = 0$$

$$\Rightarrow \quad x - 2a = 0, \; y = 0$$

and $\quad x + 2a + 2z = 0, \; y = 0$

This gives the required lines. These lines lie in the plane $y = 0$ and their combined equation is
$$y = 0, \quad x^2 + 2zx - 4az - 4a^2 = 0$$
If θ be the angle between these lines, then
$$\tan\theta = \frac{2\sqrt{1^2 - 0}}{1 + 0} = 2.$$

EXERCISES

1. Find the equation of the cone whose generators pass through the point (α, β, γ) and have their direction cosines satisfying the relation $al^2 + bm^2 + cn^2 = 0$.

   [Ans. $a(x-\alpha)^2 + b(y-\beta)^2 + c(z-\gamma)^2 = 0$]

2. Find the equation of the cone whose vertex is the point $(1, 1, 0)$ and whose guiding curve is $y = 0, \ x^2 + z^2 = 4$. [Ans. $x^2 - 3y^2 + z^2 - 2xy + 8y - 4 = 0$]

3. Obtain the locus of the lines which pass through a point (α, β, γ) and through points of the conic.
$$\frac{x^2}{a^2} + \frac{y^2}{b^2} = 1, \ z = 0$$

   $$\left[\text{Ans.} \ \left(\frac{\alpha z - x\gamma}{a}\right)^2 + \left(\frac{\beta z - y\gamma}{b}\right)^2 = (z-\gamma)^2\right]$$

4. Show that he equation of the cone whose vertex is the origin and whose base is the circle through the three points $(a, 0, 0), (0, b, 0), (0, 0, c)$ is $\Sigma a(b^2 + c^2)yz = 0$.

5. Find the equation of the cone which vertex at $(1, 2, 3)$ and guiding curve
$$x^2 + y^2 + z^2 = 4, \ x + y + z = 1$$

   [Ans. $5x^2 + 3y^2 + z^2 - 2xy - 6yz - 4zx + 6x + 8y + 10z - 26 = 0$]

6. Find the equation of the cone whose vertex is at the point $(-1, 1, 2)$ and whose guiding curve is $3x^2 - y^2 = 1, z = 0$.

7. Find the equation of cone whose vertex is $(1, 2, 3)$ and base is $y^2 = 4ax, z = 0$.

5.1.2 Enveloping Cone of a Sphere

To find the equation of the cone whose vertex is at the point (α, β, γ) and whose generators touch the sphere
$$x^2 + y^2 + z^2 = a^2 \qquad \text{...(i)}$$
The equation to *any* line through (α, β, γ) are
$$\frac{x-\alpha}{l} = \frac{y-\beta}{m} = \frac{z-\gamma}{n} \qquad \text{...(ii)}$$
This line will be a generator of the given curve if and only if it touches the given sphere. The points of intersection of the line (ii) with the sphere (i) are given by
$$r^2(l^2 + m^2 + n^2) + 2r(l\alpha + m\beta + n\gamma) + (\alpha^2 + \beta^2 + \gamma^2 - a^2)$$
so that the line will touch the sphere, if the two roots of the quadratic equation in r are equal and this requires
$$(l\alpha + m\beta + n\gamma)^2 = (l^2 + m^2 + n^2)(\alpha^2 + \beta^2 + \gamma^2) \qquad \text{...(iii)}$$
This is the condition for the line (ii) to touch the sphere (i).

Cones, Cylinders

Eliminating l, m, n between (ii) and (iii), we get

$$[\alpha(x-\alpha) + \beta(y-\beta) + \gamma(z-\gamma)]^2$$
$$= [(x-\alpha) + (y-\beta) + (z-\gamma)]^2 (\alpha^2 + \beta^2 + \gamma^2 - a^2) \quad ...(iv)$$

which is the required equation of the cone.

If we write

$$S \equiv x^2 + y^2 + z^2 - a^2, \; S_1 \equiv \alpha^2 + \beta^2 + \gamma^2 - a^2, \; T \equiv \alpha x + \beta y + \gamma z - a^2$$

the equation (iv) can be rewritten as

$$(T - S_1)^2 = (S - 2T + S_1)S$$

$\Leftrightarrow \quad SS_1 = T^2$

$\Leftrightarrow \quad (x^2 + y^2 + z^2 - a^2)(\alpha^2 + \beta^2 + \gamma^2 - a^2) = (\alpha x + \beta y + \gamma z - a^2)^2$

Def. Enveloping Cone : *The cone formed by the tangent lines to a surface, drawn from a given point is called the Enveloping Cone of the surface with given point as its vertex.*

EXERCISES

1. Find the enveloping cone of the sphere

$$x^2 + y^2 + z^2 - 2x + 4z = 1$$

with its vertex at (1,1, 1). [**Ans.** $4x^2 + 3y^2 - 5z^2 - 6yz - 8x + 16z - 4 = 0$]

2. Show that the plane $z = 0$ cuts the enveloping cone of the sphere $x^2 + y^2 + z^2 = 11$ which has its vertex (2, 4, 1) in a rectangular hyperbola.

5.1.3 Quadratic Cones with Vertex at Origin

The equation of a cone whose vertex is the origin is homogeneous and conversely.

We take up the general equation

$$ax^2 + by^2 + cz^2 + 2fyz + 2gzx + 2hxy + 2ux + 2vy + 2wz + d = 0$$

of the second degree and show that it represents a cone with its vertex at the origin, if and only if

$$u = v = w = d = 0$$

Let the equation represents a cone with its vertex at the origin.

Let $P(x', y', z')$ be a point on the cone represented by equation (1). Then

$$(rx', ry', rz')$$

are the general co-ordinates of a point on the line OP joining the point P on the origin O.
Since the line OP is a generator of the cone (i), the point

$$(rx', ry', rz')$$

lies on it for every value of r implying that the equation

$$r^2(ax'^2 + by'^2 + cz'^2 + 2fy'z' + 2gz'x' + 2hx'y') + 2r(ux' + vy' + wz') + d = 0$$

is true for every value of r.

This implies that we have

$$ax'^2 + by'^2 + cz'^2 + 2fy'z' + 2gz'x' + 2hx'y' = 0 \quad ...(i)$$
$$ux' + vy' + wz' = 0 \quad ...(ii)$$
$$d = 0. \quad ...(iii)$$

From (ii), we see that if u, v, w be not all zero, then the co-ordinates x', y', z' of any point on the cone satisfy an equation of the first degree viz.,

$$ux + vy + wz = 0$$

so that the surface is a plane and we have a contradiction. Thus,

$$u = v = w = 0;\ d = 0$$

so that the equation of a cone with its veretx at the origin is necessarily homogeneous.

Conversely : *We show that every homogenous equation of the second degree represents cone with its vertex at the origin.*

It is clear from the nature of the equation that if the co-ordinates x', y', z' satisfy it, then so do also rx', ry', rz' for all values of r.

Hence, if any point P lies on the surface then every point on the line OP lies on it.

Thus, the surface is generated by lines through the origin O and hence, by definition is a cone with its vertex at O.

Note : A homogeneous equation of the second degree will represent a pair of planes, if the homogeneous expression can be factorized into linear factors. The condition for this has already been obtained in Chapter 2. A pair of intersecting planes can thus be thought of as a cone with any point on the line of intersection as a vertex thereof.

Cor. 1. If l, m, n be the direction ratios of any generator of the cone

$$ax^2 + by^2 + cz^2 + 2fyz + 2gzx + 2hxy = 0 \qquad \ldots(1)$$

so that the point (lr, mr, nr) lies on it for every value of r, we have

$$al^2 + bm^2 + cn^2 + 2fmn + 2gnl + 2hlm = 0 \qquad \ldots(2)$$

Conversely, it is obvious that if the result (2) be true then the line with direction ratios l, m, n is a generator of the cone whose equation is (1). The proof of this statement is straightforward.

Cor. 2. The general equation of a cone with its vertex at the point (α, β, γ) is

$$a(x-\alpha)^2 + b(y-\beta)^2 + c(z-\gamma)^2$$
$$+ 2f(z-\gamma)(y-\beta) + 2g(x-\alpha)(z-\gamma) + 2h(x-\alpha)(y-\beta) = 0,$$

an can easily be verified by transfering the origin to the point (α, β, γ).

EXAMPLES

1. *Find the equation of the cone whose vertex is at the origin and which passes through the curve given by the equations*

$$ax^2 + by^2 + cz^2 = 1,\ lx + my + nz = p$$

Solution. The required equation is the hoemogenous equation of the second degree satisfied by points satisfying the two given equations.

We have $\qquad lx + my + nz = p \Rightarrow \dfrac{lx + my + nz}{p} = 1$

Thus, the required equation is

$$ax^2 + by^2 + cz^2 = \left(\dfrac{lx + my + nz}{p}\right)^2$$

$\Leftrightarrow \qquad \Sigma(ap^2 - l^2)x^2 - 2\Sigma lmxy = 0$

2. *Show that the equation of the cone whose vertex is the origin and base curve $z = k$, $f(x, y) = 0$.*

$$f\left(\dfrac{xk}{z}, \dfrac{yk}{z}\right) = 0$$

Solution. Let $\qquad f(x, y) = ax^2 + by^2 + 2hxy + 2gx + 2fy + c = 0 \qquad \ldots(i)$

By making (i) homogeneous with the help of $z = k$, we get the equation of required cone as

Cones, Cylinders

$$ax^2 + by^2 + 2hxy + 2gx\left(\frac{z}{k}\right) + 2fy\left(\frac{z}{k}\right) + c\left(\frac{z}{k}\right)^2 = 0$$

Multiplying by $\dfrac{k^2}{z^2}$, we get

$$a\left(\frac{xk}{z}\right)^2 + b\left(\frac{yk}{z}\right)^2 + 2h\left(\frac{xk}{z}\right)\left(\frac{yk}{z}\right) + 2g\left(\frac{xk}{z}\right) + 2f\left(\frac{yk}{z}\right) + c = 0$$

$$\Rightarrow \quad f\left(\frac{xk}{z}, \frac{yk}{z}\right) = 0$$

EXERCISES

1. Find the equation of the cone whose vertex is at the origin and the direction cosines of whose generators satisfy the relation

    $$3l^2 - 4m^2 + 5n^2 = 0 \qquad \text{[Ans. } 3x^2 - 4y^2 + 5z^2 = 0]$$

2. Find the equation to the cones with vertex at the origin and which pass through the curves given by the equations
    (i) $z = 2,\ x^2 + y^2 = 4$
    (ii) $ax^2 + by^2 = 2z,\ lx + my + nz = p$
    (iii) $ax^2 + by^2 + cz^2 = 1,\ \alpha x^2 + \beta y^2 = 2z$
    (iv) $x^2 + y^2 + z^2 + x - 2y + 3z = 4;\ x^2 + y^2 + z^2 + 2x - 3y + 4z = 5$

    [Ans. (i) $x^2 + y^2 + z^2 = 0$, (ii) $p(ax^2 + by^2) = 2z(lx + my + nz)$,

    (iii) $(ax^2 + by^2 + cz^2) 4z^2 = (\alpha x^2 + \beta y^2)^2$, (iv) $2x^2 + y^2 - 5xy - 3yz + 4zx = 0]$

3. A sphere S and a plane α have, respectively, the equations

    $$\varphi + u + c = 0;\quad v = 1$$

    where $\varphi = x^2 + y^2 + z^2$, u and v are homogeneous linear functions of x, y, z and c is a constant. Find the equation of the cone whose generators join the origin O to the points of intersection of S and α.
    Show that this cone meets S again in points lying on a plane β and find the equation of β in terms of u, v and c.
    If the radius of S varies, while its centre, the plane α, at the point O remains fixed, prove that β passes through a fixed line.
    [The required conc, C, is given by

    $$C = \varphi + uv + cv^2.$$

    Now $\quad C - S \equiv (\varphi + uv + cv^2) - (\varphi + u + c) = (v - 1)(u + cv + c)$
    so that we see that the cone C meets the sphere S again in points lying on the palne $\beta \equiv u + cv + c = 0$.
    Since the radius of S varies and its centre remains fixed, we see that u is constant while c varies. Also v is constant. This shows that the plane $\beta \equiv u + c(v + 1)$ passes through the line of intersection of the fixed planes $u = 0, v + 1 = 0$.]

5.1.4 Determination of Quadratic Cones Under Given Conditions

As a general equation of quadratic cone with a given vertex contains *five arbitrary constants*, it follows that five conditions determine such a cone provided each condition gives rise to a single linear relation between the constants. For instance, *a cone can be determined so as to have any given five concurrent lines as generators*, no three of them being coplanar.

EXAMPLES

1. Show that the general equation to a cone which passes through the three axes is
$$fyz + gzx + hxy = 0$$
f, g, h being parameters.

Solution. The general equation of a cone with its vertex at the origin is
$$ax^2 + by^2 + cz^2 + 2fyz + 2gzx + 2hxy = 0$$
Now X-axis is a generator.
$\Rightarrow$ its direction cosines $(1, 0, 0)$ satisfy (i) $\Rightarrow$ $a = 0$.
Similarly, $b = c = 0$.

2. The plane $\dfrac{x}{a} + \dfrac{y}{b} + \dfrac{c}{z} = 1$ meets the co-ordinate axes in A, B, C. Prove that the equation to the cone generated by lines drawn from O to meet the circle ABC is
$$yz\left(\frac{b}{c} + \frac{c}{b}\right) + zx\left(\frac{c}{a} + \frac{a}{c}\right) + xy\left(\frac{a}{b} + \frac{b}{a}\right) = 0$$

Solution. Points A, B, C, are $(a, 0, 0)$, $(0, b, 0)$ and $(0, 0, c)$ respectively. Equation of the sphere OABC is
$$x^2 + y^2 + z^2 - ax - by - cz = 0 \qquad \text{...(i)}$$
The circle ABC is obtained by intersection of given plane with (i).
Making (i) homogeneous with the help of given plane, the required cone is
$$(x^2 + y^2 + z^2) - (ax + by + cz)\left(\frac{x}{a} + \frac{y}{b} + \frac{z}{c}\right) = 0$$

3. Planes through OX and OY include an angle α. Show that their line of intersection lies on the cone $z^2(x^2 + y^2 + z^2) = x^2 y^2 \tan^2 \alpha$.

Solution. Any plane through $OX (y = 0, z = 0)$ is
$$y + \lambda z = 0$$
Also a plane through OY is
$$x + \mu z = 0$$
The angle between two planes is α, i.e.,
$$\cos \alpha = \frac{0.1 + 1.0 + \lambda \mu}{\sqrt{1 + \lambda^2} \sqrt{1 + \mu^2}} = \frac{\mu \lambda}{\sqrt{1 + \lambda^2 + \mu^2 + \lambda^2 \mu^2}}$$

so that $\tan^2 \alpha = \dfrac{1 + \lambda^2 + \mu^2 + \lambda^2 \mu^2}{\lambda^2 \mu^2} - 1 = \dfrac{1 + \lambda^2 + \mu^2}{\lambda^2 \mu^2}$

Eliminating λ and μ from (i), (ii) and (iii), the required cone is
$$\tan^2 \alpha = \frac{1 + \dfrac{y^2}{z^2} + \dfrac{x^2}{z^2}}{\left(\dfrac{y^2}{z^2}\right)\left(\dfrac{x^2}{z^2}\right)} = \frac{z^2(x^2 + y^2 + z^2)}{x^2 y^2}$$

$\Rightarrow \qquad z^2(x^2 + y^2 + z^2) = x^2 y^2 \tan^2 \alpha.$

EXERCISES

1. Find the equation to the cone which passes through the three co-ordinate axes as well as the two lines
$$\frac{x}{1} = \frac{y}{-2} = \frac{z}{3}, \quad \frac{x}{3} = \frac{y}{-1} = \frac{z}{1} \qquad \text{[Ans. } 3yz + 16zx + 15xy = 0\text{]}$$

Cones, Cylinders

2. Find the equation of the cone which contains the three co-ordinate axes and the two lines through the origin with direction cosines (l_1, m_1, n_1) and (l_1, m_1, n_1).

[**Ans.** $\Sigma l_1 l_2 (m_1 n_2 - m_2 n_1) yz = 0$]

3. Find the equation of the quadratic cone which passes through the three co-ordinate axes and the three mutually perpendicular lines
$$\frac{1}{2} x = y - z, \quad x = \frac{1}{3} y = \frac{1}{5} z, \quad \frac{1}{8} z = -\frac{1}{11} y = \frac{1}{5} z$$

[**Ans.** $16yz - 33zx - 25xy = 0$]

4. Show that the lines drawn through the point (α, β, γ) whose direction cosines satisfy $al^2 + bm^2 + cn^2 = 0$ generate the cone $a(x-\alpha)^2 + b(y-\beta)^2 + c(z-\gamma)^2 = 0$.

5.2 CONDITION THAT THE GENERAL EQUATION OF THE SECOND DEGREE SHOULD REPRESENT A CONE, CO-ORDINATES OF THE VERTEX

We have seen that the equation of a cone with its vertex at the origin is necessarily homogeneous and conversely. Thus, any given equation of the second degree will represent a cone if, and only if there exists a point such on transfering the origin to the same the equation becomes homogeneous.

Let $f(x, y, z) = ax^2 + by^2 + cz^2 + 2fyz + 2hxy + 2ux + 2wz + d = 0$ ...(1)

represent a cone having its vertex at (x', y', z').

Shift the origin to the vertex (x', y', z') so that we change

x to $x + x'$, y to $y + y'$ and z to $z + z'$

The transformed equation is

$ax^2 + by^2 + cz^2 + 2fyz + 2gzx + 2hxy + 2[x(ax' + by' + gz' + u)$
$+ y(hx' + by' + fz' + v) + z(gx' + fy' + cz' + w)] + f(x', y', z') = 0$ ...(2)

The equation (2) represents a cone with its vertex at the origin and must, therefore, be homogenoeus. This gives

$$ax' + by' + gz' + u = 0 \quad \text{...(i)}$$
$$hx' + by' + fz' + v = 0 \quad \text{...(ii)}$$
$$gx' + fy' + cz' + w = 0 \quad \text{...(iii)}$$
$$f(x', y', z') = 0 \quad \text{....(iv)}$$

Also, $f(x', y', z') \equiv x'(ax' + by' + gz' + u) + y'(hx' + by' + fz' + v)$
$+ z'(gx' + fy' + cz' + w) + (ux' + vy' + wz' + d)$

Thus, with the help of (i), (ii) and (iii), we see that (iv) is equivalent to

$$ux' + vy' + wz' + d = 0 \quad \text{...(v)}$$

The system of equations (i), (ii), (iii), (iv) is equivalent to the system (i), (ii), (iii), (iv).

Thus, if the given equation represent a cone, there exist (x', y', z') satisfying the equations (i), (ii), (iii), (v) implying that these four equations are consistent. The condition of consistency of the system (i), (ii), (iii) and (v) of four linear equation is

$$\begin{vmatrix} a & h & g & u \\ h & b & f & v \\ g & f & c & w \\ u & v & w & d \end{vmatrix} = 0$$

This is the condition for the equation (1) of the second degree to represent a cone.

If the condition is satisfied, the co-ordinates (x', y', z') of the vertex are obtained by solving simultaneously the three linear equations (i), (ii) and (iii).

The point (x', y', z') is such that if we shift the origin to this point, the new equation will be homogeneous and as such will represent a cone.

Cor. If $F(x, y, z) \equiv ax^2 + by^2 + cz^2 + 2fyz + 2gzx + 2hxy + 2ux + 2vy + 2wz + d = 0$ represents a cone, the co-ordinates of its vertex satisfy the equations
$$F_x = 0, \ F_y = 0, \ F_z = 0, \ F_t = 0$$
where 't' is used to make $F(x, y, z)$ homogeneous and is put equal to unity after differentiations.

Making $F(x, y, z)$ homogeneous, we write
$$F(x, y, z, t) = ax^2 + by^2 + cz^2 + 2fyz + 2gzx + 2hxy + 2uxt + 2vyt + 2wzt + dt^2$$
We have
$$F_x = 2(ax + hy + gz + ut), \ F_y = 2(hx + by + fz + vt)$$
$$F_z = 2(gx + fy + cz + wt), \ F_t = 2(ux + vy + wz + dt)$$

Putting $t = 1$, we see from (i), (ii), (iii) and (iv) that the vertex (x_1, y_1, z_1) satisfies the four linear equations.
$$F_x = 0, \ F_y = 0, \ F_z = 0, \ F_t = 0$$

Note : The student should note that the coefficients of second degree term in the transformed equation (2) are the same as those in the original equation (1).

Note : The equation $F(x, y, z) = 0$ represents a cone if, and only, if the four linear equations $F_x = 0, F_y = 0, F_z = 0, F_t = 0$ are consistent. In the case of consistency the vertex is given any three of these.

In case we have
$$\begin{vmatrix} a & h & g & u \\ h & b & f & v \\ g & f & c & w \\ u & v & w & d \end{vmatrix} = 0 \text{ as well as } \begin{vmatrix} a & h & g \\ h & b & f \\ g & f & c \end{vmatrix} = 0$$
the equation will represent a pair of planes.

EXAMPLE

1. *Prove that the equation*
$$ax^2 + by^2 + cz^2 + 2ux + 2vy + 2wz + d = 0$$
represents a cone if
$$\frac{u^2}{a} + \frac{v^2}{b} + \frac{w^2}{c} = d$$

Solution. Let
$$f(x, y, z, t) \equiv ax^2 + by^2 + cz^2 + 2uxt + 2vyt + 2vyt + 2wzt + dt^2 = 0$$
$$\therefore \quad \frac{\partial F}{\partial x} = 0 \text{ for } t = 1 \text{ gives}$$
$$2ax + 2u = 0 \text{ or } x = -\frac{u}{a} \quad \dots(1)$$

Similarly, $\quad \dfrac{\partial F}{\partial y} = 0 \text{ for } t = 1 \text{ gives } y = -\dfrac{y}{b} \quad \dots(2)$

$\dfrac{\partial F}{\partial y} = 0 \text{ for } t = 1 \text{ gives } z = -\dfrac{w}{c} \quad \dots(3)$

Cones, Cylinders

and $\dfrac{\partial F}{\partial y}=0$ for $t=1$ gives $ux+vy+wz+d=0$ ...(4)

Substituting the values of x, y, z from (1), (2), (3) in (4), we get the required condition as

$$u\left(-\dfrac{u}{a}\right)+v\left(-\dfrac{v}{b}\right)+w\left(-\dfrac{w}{c}\right)+d=0$$

$$\Rightarrow \quad \dfrac{u^2}{a}+\dfrac{v^2}{b}+\dfrac{w^2}{c}=d.$$

EXERCISES

1. Prove that
$$2x^2+2y^2+7x^2-10yz-10zx+2x+2y+26z-17=0$$
represents a cone with vertex at $(2, 2, 1)$.

2. Show that the equation
$$x^2-2y^2+3z^2-4xy+5yz-6zx+8x-19y-2z-20=0$$
represents a cone with vertex $(1, -2, 3)$.

3. Show that the equation
$$2y^2-8yz-4zx-8xy+6x-4y-2z+5=0$$
represents a cone whose vertex is $\left(-\dfrac{7}{6}, \dfrac{1}{3}, \dfrac{5}{6}\right)$.

EXAMPLE

1. *Find the equations to the lines in which the plane*
$$2x+y-z=0,$$
cuts the cone
$$4x^2-y^2+3z^2=0$$

Solution. Let
$$\dfrac{x}{l}=\dfrac{y}{m}=\dfrac{z}{n}$$
be the equations of any one of the two lines in which the given plane meets the given cone so that we have
$$2l+m-n=0, \quad 4l^2-m^2+3n^2=0$$

These two equations are now to be solved for l, m, n. Eliminating n, we have
$$4l^2-m^2+3(2l+m)^2=0$$
$$8l^2+6lm+m^2=0$$

$$\Rightarrow \quad \dfrac{l}{m}=\dfrac{-6\pm\sqrt{36-32}}{16}=-\dfrac{1}{4} \text{ or } -\dfrac{1}{2}$$

We also have
$$2\dfrac{l}{m}+1-\dfrac{n}{m}=0$$

$$\dfrac{l}{m}=-\dfrac{1}{4} \quad \Rightarrow \quad \dfrac{n}{m}=\dfrac{1}{2}$$

and $\dfrac{l}{m}=-\dfrac{1}{2} \quad \Rightarrow \quad \dfrac{n}{m}=0$

Now $\dfrac{l}{m}=-\dfrac{1}{4}, \dfrac{n}{m}=\dfrac{1}{2} \quad \Rightarrow \quad \dfrac{l}{-1}=\dfrac{m}{4}=\dfrac{0}{2}$

and $\dfrac{l}{m}=-\dfrac{1}{2}, \dfrac{n}{m}=0 \quad \Rightarrow \quad \dfrac{l}{-1}=\dfrac{m}{2}; n=0$

Thus, the two required lines are
$$\frac{x}{-1} = \frac{y}{4} = \frac{z}{2}; \frac{x}{-1} = \frac{y}{2}; z = 0.$$

EXERCISES

1. Find the equation of the lines of intersection of the following planes and cones :
   (i) $x + 3y - 2z = 0$, $x^2 + 9y^2 - 4z^2 = 0$.
   (ii) $3x + 4y + z = 0$, $15x^2 - 32y^2 - 7z^2 = 0$
   (iii) $x + 7y - 5z = 0$, $3yz + 14zx - 30xy = 0$

   [**Ans.** (i) $x = 2z, y = 0, 3y = 2z, x = 0$ (ii) $\frac{x}{-3} = \frac{y}{2} = \frac{z}{1}, \frac{x}{2} = \frac{y}{-1} = \frac{z}{-2}$,
   (iii) $\frac{x}{1} = \frac{y}{2} = \frac{z}{3}, \frac{x}{3} = \frac{y}{1} = \frac{z}{2}$]

2. Show that the equation of the quadric cone which contains the three co-ordinate axes and the lines in which the plane $x - 5y - 3z = 0$ cuts the cone
   $7x^2 + 5y^2 - 3z^2 = 0$ is $yz + 10zx - 18xy = 0$

3. Find the angles between the lines of intersection of
   (i) $x - 3y + z = 0$ and $x^2 - 5y^2 + z^2 = 0$
   (ii) $10x + 7y - 6z = 0$ and $20x^2 + 7y^2 - 108z^2 = 0$
   (iii) $4x - y - 5z = 0$ and $8yz + 3zx - 5xy = 0$
   (iv) $x + y + z = 0$ and $6xy + 3yz - 2zx = 0$
   (v) $x + y + z = 0$ and $x^2 - yz + xy - 3z^2 = 0$

   [**Ans.** (i) $\cos^{-1}(5/6)$, (ii) $\cos^{-1}(16/21)$, (iii) $\pi/2$, (iv) $\pi/3$, (v) $\pi/6$]

5.3 CONE AND A PLANE THROUGH ITS VERTEX

To find the angle between the lines of intersection of the plane
$$ux + vy + wz = 0$$
and
$$f(x,y,z) \equiv ax^2 + by^2 + cz^2 + 2fyz + 2gzx + 2hxy = 0 \qquad ...(1)$$
The plane
$$ux + vy + wz = 0 \qquad ...(2)$$
will cut the cone (1) in two lines passing through the origin.
Let one of these lines be
$$\frac{x}{l} = \frac{y}{m} = \frac{z}{n} \qquad ...(3)$$
Thus, line (3) in plane (2), therefore,
$$ul + vm + wn = 0 \qquad ...(4)$$
Also, line (3) lies on (1) hence it is generator of the cone, *i.e.*, its d.c.'s satisfy the equation of the cone, hence
$$al^2 + bm^2 + cn^2 + 2fmn + 2gnl + 2hlm = 0 \qquad ...(5)$$
Putting $n = -\dfrac{ul + vm}{w}$ from (4) in (5), we have
$$al^2 + bm^2 + c\left(-\frac{ul + vm}{w}\right)^2 + (2fm + 2gl)\left(-\frac{ul + vm}{w}\right) + 2hlm = 0$$
i.e., $l^2(aw^2 + cu^2 - 2gwu) + 2lm(cuv - fwu - gvw + hw^2) + m^2(bw^2 + cv^2 - 2fvw) = 0$

Cones, Cylinders

i.e., $\dfrac{l^2}{m^2}(aw^2 + cu^2 - 2gwu) + 2\dfrac{l}{m}(cuv - fwu - gvw + hw^2)$

$$+ (bw^2 + cv^2 + 2fvw) = 0 \qquad \ldots(6)$$

Now (6) is quadratic equation in l/m and shows that plane (2) cuts cone in two lines. If their direction ratios are l_1, m_1, n_1 and l_2, m_2, n_2 then we have

$$\dfrac{l_1}{m_1} \cdot \dfrac{l_2}{m_2} = \dfrac{bw^2 + cv^2 - 2fvw}{aw^2 + cu^2 - 2gwu}$$

i.e., $\dfrac{l_1 l_2}{bw^2 + cv^2 - 2fvw} = \dfrac{m_1 m_2}{aw^2 + cu^2 - 2gwu} = \dfrac{n_1 n_2}{bu^2 + av^2 - 2huv}$ (similarly)

$$= \dfrac{l_1 l_2 + m_1 m_2 + n_1 n_2}{(b+c)u^2 + (c+a)v^2 + (a+b)w^2 - 2fvw - 2gwu - 2huv}$$

$$= \dfrac{l_1 l_2 + m_1 m_2 + n_1 n_2}{(a+b+c)(u^2 + v^2 + w^2) - f(u,v,w)}$$

Also, sum of the roots of (6) gives

$$\dfrac{l_1}{m_1} + \dfrac{l_2}{m_2} = -\dfrac{2(cuv - fwu - gvw + hw^2)}{aw^2 + cu^2 - 2gwu}$$

i.e., $\dfrac{l_1 m_2 + l_2 m_1}{-2(cuv - fwu - gvw - hw^2)} = \dfrac{m_1 m_2}{aw^2 + cu^2 - 2gwu}$

$$= \dfrac{l_1 l_2}{bw^2 + cv^2 - 2fvw} = \dfrac{n_1 n_2}{av^2 + bu^2 - 2huv}$$

$$= \dfrac{[(l_1 m_2 + l_2 m_1)^2 - 4l_1 l_2 m_1 m_2]^{1/2}}{[4(cuv - fwu - gvw + hw^2)^2 - 4(bw^2 + cv^2 - 2fvw)(aw^2 + cu^2 - 2gwu)]^{1/2}}$$

$$= \dfrac{l_1 m_2 - l_2 m_2}{\pm 2wP} \quad \text{where } P^2 = \begin{vmatrix} a & h & g & u \\ h & b & f & v \\ g & f & c & w \\ u & v & w & 0 \end{vmatrix}$$

$$= \dfrac{m_1 n_2 - m_2 n_1}{\pm 2uP} = \dfrac{n_1 l_2 - n_2 l_1}{\pm 2vP}$$

$$= \dfrac{[\Sigma (m_1 n_2 - m_2 n_1)^2]^{1/2}}{\pm 2P(u^2 + v^2 + w^2)^{1/2}}$$

If θ be the angle between the lines, then

$$\tan \theta = \dfrac{[\Sigma(m_1 n_2 - m_2 n_1)^2]^{1/2}}{l_1 l_2 + m_1 m_2 + n_1 n_2}$$

or $\qquad \tan \theta = \dfrac{2P(u^2 + v^2 + w^2)^{1/2}}{(a+b+c)(u^2 + v^2 + w^2) - f(u,v,w)}$

Cor. Condition of Perpendicularity

To find the condition, so that lines in which plane $ux + vy + wz = 0$ cuts a cone

$$f(x, y, z) \equiv ax^2 + by^2 + cz^2 + 2fyz + 2gzx + 2hxy = 0$$

may be at right angles.

The angle θ between the two lines is given by

$$\tan\theta = \pm\frac{2P(u^2+v^2+w^2)^{1/2}}{(a+b+c)(u^2+v^2+w^2)-f(u,v,w)}$$

If $\theta = 90°$, $\tan\theta = \infty$

i.e., $(a+b+c)(u^2+v^2+w^2) - f(u,v,w) = 0$

This is the required condition.

5.3.1 Mutually Perpendicular Generators of a Cone

To show that the cone

$$ax^2 + by^2 + cz^2 + 2fyz + 2gzx + 2hxy = 0 \qquad \text{...(iii)}$$

Equation of the plane through the origin perpendicular to the line (ii) is

$$\lambda x + \mu y + vz = 0 \qquad \text{(iv)}$$

If (l, m, n) be the direction cosines of any one of the two generators in which the plane cuts the given cone, we have

$$al^2 + bm^2 + cn^2 + 2fmn + 2gnl + 2hlm = 0 \qquad \text{...(v)}$$

and $l\lambda + m\mu + nv = 0$ ...(vi)

Eliminating n between (v) and (vi), we obtain

$$l^2(av^2 + c\lambda^2 - 2g\lambda v) + 2lm(c\lambda\mu + hv^2 - g\mu v + f\lambda v) + m^2(bv^2 + c\mu^2 - 2f\mu v) = 0$$

which, being a quadratic in $l : m$, we see that the plane (iv) cuts the given cone in two generators.

Hence, if (l_1, m_1, n_1), (l_2, m_2, n_2) be the direction cosines of these two generators, we have

$$\frac{l_1 l_2}{m_1 m_2} = \frac{bv^2 + c\mu^2 - 2f\mu v}{av^2 + c\lambda^2 - 2g\lambda v}$$

$$\Rightarrow \quad \frac{l_1 l_2}{bv^2 + c\mu^2 - 2f\mu v} = \frac{m_1 m_2}{av^2 + c\lambda^2 - 2g\lambda v}$$

From symmetry, each of these is further

$$= \frac{n_1 n_2}{a\mu^2 + b\lambda^2 - 2h\lambda\mu} = k, \text{ (say)}$$

Thus, we have

$$l_1 l_2 + m_1 m_2 + n_1 n_2 = k[a(\mu^2+v^2) + b(v^2+\lambda^2) + c(\lambda^2+\mu^2) - 2f\mu v + 2gv\lambda - 2h\lambda\mu]$$

$$= k(a+b+c)+(\lambda^2+\mu^2+v^2) \qquad \text{...(viii)}$$

with the help of (iii).

The two generators in which the plane (iv) intersects the curve (ii) will be at right angles if and only if

$$l_1 l_2 + m_1 m_2 + n_1 n_2 = 0$$

i.e., if and only if

$$a+b+c=0$$

We note that

$$\frac{x}{\lambda} = \frac{y}{\mu} = \frac{z}{v}$$

is an arbitrary generator of the cone and the condition that the planes through the vertex and perpendicular to the generators meet the cone in two perpendicular generators is independent of λ, μ, v.

Cones, Cylinders of Conicoids

Also we see that the two generators will themselves be perpendicular to the first generator so that the three generators will be perpendicular in pairs.

It follows that the cone (i) *admits of an infinite number of sets of three mutually perpendicular genaerators if and only if*
$$a + b + c = 0$$

In fact if this condition is satisfied, then the plane perpendicular to *any* generator OP of the cone cuts the same in two perpendicular generators OQ, OR so that OP, OQ, OR is a set of three mutually perpendicular generators.

Note : If the general equation
$$ax^2 + by^2 + cz^2 + 2fyz + 2gzx + 2hxy + 2ux + 2vy + 2wz + d = 0$$
represents a cone having sets of three mutually perpendicular generators, then also
$$a + b + c = 0$$
for, on shifting the origin to its vertex, the coefficients of the second degree term remain unaffected.

EXAMPLES

1. *Prove that the plane* $ax + by + cz = 0$ *cuts the cone* $yz + zx + xy = 0$ *in perpendicular to lines if* $\dfrac{1}{a} + \dfrac{1}{b} + \dfrac{1}{c} = 0.$

Solution. Let one of the lines of intersection be
$$\frac{x}{l} = \frac{y}{m} = \frac{z}{n}$$

The line lies on given cone and plane, hence

$$mn + nl + lm = 0 \qquad \ldots(i)$$
and
$$al + bm + cn = 0 \qquad \ldots(ii)$$

Putting the values of n from (ii) in (i), we get

$$(m + l)\left(-\frac{al + bm}{c}\right) + lm = 0$$

$$\Rightarrow \quad al^2 + (a + b - c) lm + bm^2 = 0$$

$$\Rightarrow \quad a\left(\frac{l}{m}\right)^2 + (a + b - c)\frac{l}{m} + b = 0$$

Let $\dfrac{l_1}{m_1}, \dfrac{l_2}{m_2}$ be the two roots, then

$$\frac{l_1}{m_1} \cdot \frac{l_2}{m_2} = \frac{b}{a}$$

$$\Rightarrow \quad \frac{l_1 l_2}{1/a} = \frac{m_1 m_2}{1/b} = \frac{n_1 n_2}{1/c} \quad \text{(by symmetry)}$$

The angle between the lines will be a right angle if

$$l_1 l_2 + m_1 m_2 + n_1 n_2 = 0 \quad \Rightarrow \quad \frac{1}{a} + \frac{1}{b} + \frac{1}{c} = 0$$

2. *Prove that the angle between the lines given by* $x + y + z = 0$, $ayz + bzx + cxy = 0$ *is* $\pi/2$ *if* $a + b + c = 0$ *and* $\pi/3$ *if* $\dfrac{1}{a} + \dfrac{1}{b} + \dfrac{1}{c} = 0.$

Solution. Let the plane $x + y + z = 0$ cuts the cone $ayz + bzx + cxy = 0$ in a line

$$\frac{x}{l} = \frac{y}{m} = \frac{z}{n}$$

Then $l + m + n = 0$ and $amn + bnl + clm = 0$

Eliminating n between these relations, we get

$$(am + bl)(-l - m) + clm = 0 \Rightarrow bl^2 + (a+b-c) + lm + am^2 = 0$$

$$\Rightarrow b\left(\frac{l}{m}\right)^2 + (a+b-c)\left(\frac{l}{m}\right) + a = 0. \qquad \ldots(i)$$

If the roots of this equation are $\dfrac{l_1}{m_1}$ and $\dfrac{l_2}{m_2}$, then

$$l_1 l_2 + m_1 m_2 + n_1 n_2 = 0 \Rightarrow a+b+c = 0$$

Again, from (i), we get

$$\frac{l_1}{m_1} + \frac{l_2}{m_2} = \frac{c-b-a}{b}$$

$$\Rightarrow \frac{l_1 m_2 + l_2 m_1}{m_1 m_2} = \frac{c-b-a}{b} \Rightarrow \frac{l_1 m_2 + l_2 m_1}{c-b-a} = \frac{m_1 m_2}{b} = k.$$

Now, $(l_1 m_2 - l_2 m_1)^2 = (l_1 m_2 + l_2 m_1)^2 - 4 l_1 l_2 \cdot m_1 m_2$

$$= k^2(c-b-a)^2 - 4ak \cdot bk = k^2[(c-b-a)^2 - 4ab]$$

$$= k^2(a^2 + b^2 + c^2 - 2ab - 2bc - 2ca)$$

Now, $\tan\theta = \dfrac{\sqrt{\Sigma(l_1 m_2 - l_2 m_1)^2}}{l_1 l_2 + m_1 m_2 + n_1 n_2} = \dfrac{\sqrt{k^2(3(a^2+b^2+c^2-2bc-2ca-2ab))}}{k(a+b+c)}$

If $\theta = \pi/3$, then

$$\tan^2 \frac{\pi}{3} = \frac{3(a^2+b^2+c^2-2bc-2ca-2ab)}{(a+b+c)^2}$$

$$\Rightarrow 3(a+b+c)^2 = 3(a^2+b^2+c^2-2bc-2ca-2ab) \qquad [\because \tan(\pi/3) = \sqrt{3}]$$

$$\Rightarrow 4(bc+ca+ab) = 0 \Rightarrow \frac{1}{a} + \frac{1}{b} + \frac{1}{c} = 0.$$

EXERCISES

1. Prove that the plane $lx + my + nz = 0$ cuts the cone

$$(b-c)x^2 + (c-a)y^2 + (a-b)z^2 + 2fyz + 2gzx + 2hxy = 0$$

in perpendicular lines if

$$(b-c)l^2 + (c-a)m^2 + (a-b)n^2 + 2fmn + 2gnl + 2hlm = 0$$

2. If $\quad x = \dfrac{1}{2}y = z$

represents one of a set of three mutually perpendicular generators of the cone

$$11yz + 6zx - 14xy = 0$$

find the equations of the other two. $\left[\text{Ans. } \dfrac{x}{2} = \dfrac{y}{-3} = \dfrac{z}{4}; \dfrac{x}{-11} = \dfrac{y}{2} = \dfrac{z}{7}\right]$

3. Find the angle between the lines given by

$$x + y + z = 0, \quad \frac{yz}{b-c} + \frac{zx}{c-a} + \frac{xy}{a-b} = 0$$

Cones, Cylinders

4. If the plane $2x - y + cz = 0$ cuts the cone $yz + zx + xy = 0$ in perpendicular lines, find the value of c. **[Ans. $c = 2$]**

5. Find the equations of the lines in which the plane $2x + y - z = 0$ cuts the cone
$$4x^2 - y^2 + 3z^2 = 0$$
$$\left[\text{Ans. } \frac{x}{-1} = \frac{y}{4} = \frac{z}{2}; \frac{x}{-1} = \frac{y}{2} = \frac{z}{0}\right]$$

6. If $\frac{x}{1} = \frac{y}{1} = \frac{z}{2}$ be one of a set of three mutually perpendicular generators of the cone $3yz - 2zx - 2xy = 0$, find the equations of other two generators.
$$\left[\text{Ans. } \frac{x}{2} = \frac{y}{(-4)} = \frac{z}{1}; \frac{x}{3} = \frac{y}{1} = \frac{z}{(-2)}\right]$$

7. Show that the cone whose vertex is the origin and which passes through the curve of intersection of the surface $2x^2 - y^2 + 2z^2 = 3d^2$ and any plane at a distance d, from the origin has three mutually perpendicular generators.

8. Find the locus of a point from which three mutually perpendicular lines can be drawn to intersect the central conic
$$ax^2 + by^2 = 1; z = 0 \qquad \text{[Ans. } a(x^2 + z^2) + b(y^2 + z^2) = 1]$$

9. Show that the mutually perpendicular tangent lines can be drawn to the sphere
$$x^2 + y^2 + z^2 = r^2$$
from any point on the surface
$$2(x^2 + y^2 + z^2) = 3r^2$$

10. Three points P, Q, R are taken on the ellipsoid
$$\frac{x^2}{a^2} + \frac{y^2}{b^2} + \frac{z^2}{c^2} = 1$$
so that the lines joining P, Q, R to the origin the mutually perpendicular. Prove that the plane PQR touches a fixed sphere.

5.4 INTERSECTION OF A LINE WITH A CONE

To find the points of intersection of the line
$$\frac{x - \alpha}{l} = \frac{y - \beta}{m} = \frac{z - \gamma}{n} \qquad \ldots(i)$$
and the cone.
$$f(x, y, z) \equiv ax^2 + by^2 + cz^2 + 2fz + 2gzx + 2hxy = 0 \qquad \ldots(ii)$$

The point $(lr + \alpha, mr + \beta, nr + \gamma)$ which lies on the line (i) for all values of r will lie on the cone (ii) for values of r given by the equation
$$a(lr + \alpha)^2 + b(mr + \beta)^2 + c(nr + \gamma)^2 + 2f(mr + \beta)(nr + \gamma) + 2g(lr + \alpha)(nr + \gamma)$$
$$+ 2h(lr + \alpha)(mr + \beta) = 0,$$
$$\Leftrightarrow r^2(al^2 + bm^2 + cn^2 + 2fmn + 2gnl + 2hlm) + 2r[l(a\alpha + h\beta + g\gamma) + m(h\alpha + b\beta + f\gamma)$$
$$+ n(g\alpha + f\beta + c\gamma)] + f(\alpha, \beta, \gamma) = 0 \qquad \ldots(A)$$

Let r_1, r_2 be the roots of this quadratic equation in r. The two points of intersection are
$$(lr_1 + \alpha, mr_1 + \beta, nr_1 + \gamma), (lr_2 + \alpha, mr_2 + \beta, nr_2 + \gamma)$$

Cor. A plane section of a quadratic cone is a conic, as every line in the plane meets the curve of intersection in two points.

Note : The equation (A) gives the distances of the points of intersection P and Q from the points (α, β, γ); if (l, m, n) are direction cosines.

EXERCISES

1. Show that the locus of mid-points of chords of the cone
$$ax^2 + by^2 + cz^2 + 2fyz + 2gx + 2hxy = 0$$
drawn parallell to the line
$$\frac{x}{l} = \frac{y}{m} = \frac{z}{n}$$
is the plane
$$x(al + hm + gn) + y(hl + bm + fn) + z(gl + fm + cn) = 0$$

[**Hint :** If (α, β, γ) be the middle pionts of any such chords
$$\frac{x-\alpha}{l} = \frac{y-\beta}{m} = \frac{z-\gamma}{n},$$
the two roots of the equation (A) are equal and opposite and as such their sum is zero.]

2. Find the locus of the chords of a cone which are bisected at a fixed point.

5.4.1 The tangent Lines and Tangent Plane at a Point

Let
$$\frac{x-\alpha}{l} = \frac{y-\beta}{m} = \frac{z-\gamma}{m} \qquad \text{...(i)}$$
be a line through a point (α, β, γ) of the cone
$$ax^2 + by^2 + cz^2 + 2fyz + 2gzx + 2hxy = 0 \qquad \text{...(ii)}$$
so that
$$a\alpha^2 + b\beta^2 + c\gamma^2 + 2f\beta\gamma + 2g\gamma\alpha + 2h\alpha\beta = 0$$

Thus, one of the values of r given by the equation (A) of 5.4 is zero and as such one of the two points of intersection coincides with (α, β, γ). The second point of intersection will also coincide with (α, β, γ) if the second root of the same equation is also zero. This requires
$$l(a\alpha + h\beta + g\gamma) + m(h\alpha + b\beta + f\gamma) + n(g\alpha + f\beta + c\gamma) = 0 \qquad \text{...(iii)}$$

The line (i) corresponding to the set of values of l, m, n satisfying the relation (iii) is *an tangent line* at (α, β, γ) to the cone (ii).

Eliminating l, m, n between (i) and (ii), we obtain the locus of all the tangent lines through (α, β, γ), viz.,
$$(x-\alpha)(a\alpha + h\beta + g\gamma) + (y-\beta)(h\alpha + b\beta + f\gamma) + (z-\gamma)(g\alpha + f\beta + c\gamma) = 0$$
$$\Leftrightarrow \quad x(a\alpha + h\beta + g\gamma) + y(h\alpha + b\beta + f\gamma) + z(g\alpha + f\beta + c\gamma)$$
$$= a\alpha^2 + f\beta^2 + c\gamma^2 + 2f\beta\gamma + 2g\gamma\alpha + 2h\alpha\beta = 0$$

which is a plane known as the **tangent plane**.

Clearly, the tangent plane at any point of a cone passes through its vertex.

Cor. The tangent plane at *any* point $(k\alpha, k\beta, k\gamma)$ on the generator through the point (α, β, γ) is the same as the tangent plane at (α, β, γ).

Thus, we see that the *tangent plane at any point on a cone touches the cone at all points of the generator through that point* and we say that the plane touches the cone along the generator.

EXAMPLE

1. Show that
$$\frac{x}{l} = \frac{y}{m} = \frac{z}{n}$$
the line of intersection of the tangent planes to the cone
$$ax^2 + by^2 + cz^2 + 2fyz + 2gzx + 2hxy = 0$$

Cones, Cylinders

along the lines in which it is cut by the plane
$$x(al + hm + gn) + y(hl + bm + fn) + z(gl + fm + cn) = 0$$

Sol. The tangent plane at any point (α, β, γ) of the given cone is
$$x(a\alpha + h\beta + g\gamma) + y(h\alpha + b\beta + f\gamma) + z(g\alpha + f\beta + c\gamma) = 0$$

It will contain the line
$$\frac{x}{l} = \frac{y}{m} = \frac{z}{n}$$

$$l(a\alpha + h\beta + g\gamma) + m(h\alpha + b\beta + f\gamma) + n(g\alpha + f\beta + c\gamma) = 0$$
$\Leftrightarrow \quad \alpha(al + hm + gn) + \beta(hl + bm + fn) + \gamma(gl + fm + cn) = 0$

Thus, the point (α, β, γ) lies on the plane
$$x(al + hm + gn) + y(hl + bm + fn) + z(gl + fm + cn) = 0$$

Hence, the result.

5.4.2 Condition for Tangency

To find the condition that the plane
$$lx + my + nz = 0, \qquad \ldots(1)$$
should touch the cone
$$ax^2 + by^2 + cz^2 + 2fyz + 2gzx + 2hxy = 0 \qquad \ldots(2)$$

If (α, β, γ) be the point of contact, the tangent plane
$$x(a\alpha + h\beta + g\gamma) + y(h\alpha + b\beta + f\gamma) + z(g\alpha + f\beta + c\gamma) = 0$$
thereat should be the same as the plane (1).

$$\therefore \quad \frac{a\alpha + h\beta + g\gamma}{l} = \frac{h\alpha + b\beta + f\gamma}{m} = \frac{g\alpha + f\beta + c\gamma}{n} = k, \text{ (say)}$$

Hence,
$$a\alpha + h\beta + g\gamma - lk = 0 \qquad \ldots(i)$$
$$h\alpha + b\beta + f\gamma - mk = 0 \qquad \ldots(ii)$$
$$g\alpha + f\beta + c\gamma - nk = 0 \qquad \ldots(iii)$$

Also, since (α, β, γ) lies on the plane (1), we have
$$l\alpha + m\beta + n\gamma = 0 \qquad \ldots(iv)$$

Eliminating α, β, γ, k between (i), (ii), (iii), (iv), we obtain
$$\begin{vmatrix} a & h & g & l \\ h & b & f & m \\ g & f & c & n \\ l & m & n & 0 \end{vmatrix} = 0,$$

as the required condition.
The determinant on expansion, gives
$$Al^2 + Bm^2 + Cn^2 + 2Fmn + 2Gnl + 2Hlm = 0$$
where A, B, C, F, G, H, are the co-factors of a, b, c, f, g, h respectively in the determinant
$$\begin{vmatrix} a & h & g \\ h & b & f \\ g & f & c \end{vmatrix}$$

We may see that
$$A = bc - f^2, \; B = ca - g^2, \; C = ab - h^2,$$
$$F = gh - af, \; G = hf - bg, \; H = fg - ch$$

5.4.3 Reciprocal Cones

To find the locus of lines through the vertex of the cone

$$ax^2 + by^2 + cz^2 + 2fyz + 2gzx + 2hxy = 0 \qquad \ldots(1)$$

perpendicular to its tangent planes,
Let
$$lx + my + nz = 0 \qquad \ldots(2)$$
be a tangent plane to the cone (1) so that we have
$$Al^2 + Bm^2 + Cn^2 + 2Fmn + 2Gnl + 2Hlm = 0 \qquad \ldots(3)$$
The line through the vertex perpendicular to the tangent palne (2) is
$$\frac{x}{l} = \frac{y}{m} = \frac{z}{n} \qquad \ldots(4)$$
Eliminating l, m, n between (3) and (4), we get
$$Ax^2 + By^2 + Cz^2 + 2Fyz + 2Gzx + 2Hxy = 0 \qquad \ldots(5)$$
as the required locus which is again a quadric cone with its vertex at the origin.

If we now find the locus of lines through the origin perpendicular to the tangent planes to the cone (5), we have to substitute for A, B, C, F, G, H in its equation the corresponding co-factors in the determinant

$$\begin{vmatrix} A & H & G \\ H & B & F \\ G & F & C \end{vmatrix}$$

Since, we have, by actual manipulation,
$$BC - F^2 = aD, \quad CA - G^2 = bD, \quad AB - H^2 = cD;$$
$$GH - AF = fD, \quad HF - BG = gD, \quad FG - CH = hD;$$

where
$$D \equiv abc + 2fgh + af^2 - bg^2 - ch^2$$

It follows that the required locus for the cone (5) is
$$ax^2 + by^2 + cz^2 + 2fyz + 2gzx + 2hxy = 0$$
which is the same as (1).

The two cones (1) and (5) are, therefore, such that each is the locus of the normals drawn through the origin to the tangent planes to the other and they are, on this account, called *reciprocal cones*.

We have supposed that $D \neq 0$ implying that the equation (i) does not represent a pair of planes (Refer 2.8).

Cor. *The condition for the cone*
$$ax^2 + by^2 + cz^2 + 2fyz + 2gzx + 2hxy = 0 \qquad \ldots(i)$$
passes through mutually perpendicular tangent planes is
$$A + B + C = 0$$

The cone (i) will clearly passes through three mutually perpendicular tangent planes, if its reciprocal cone
$$Ax^2 + By^2 + Cz^2 + 2Fyz + 2Gzx + 2Hxy = 0$$
has three mutually perpendicular generators and this will be so if
$$A + B + C = 0 \iff bc + ca + ab = f^2 + g^2 + h^2.$$

EXAMPLES

1. *Show that the general equation of a cone which touches the three co-ordinate plane is*
$$\sqrt{fx} \pm \sqrt{gy} \pm \sqrt{hz} = 0,$$
f, g, h being parameters.

Cones, Cylinders

Solution. The reciprocal of a cone touching the three co-ordinate planes is a cone with three co-ordinate axes as three of its generators. Now, the general equation of a cone through the three axes is
$$fyz + gzx + hxy = 0$$

Its reciprocal cone is

$$-f^2x^2 - g^2y^2 - h^2z^2 + 2ghyz + 2hfzx + 2fgxy = 0$$

$\Leftrightarrow \quad (fx + gy - hz)^2 = 4fgxy$

$\Leftrightarrow \quad fx + gy - hz = \pm 2\sqrt{fgxy}$

$\Leftrightarrow \quad fx + gy \pm 2\sqrt{fgxy} = hz$

$\Leftrightarrow \quad (\sqrt{fx} \pm \sqrt{gh})^2 = hz,$

$\Leftrightarrow \quad \sqrt{fx} \pm \sqrt{gy} \pm \sqrt{hz} = 0$

2. *Prove that the cones* $ax^2 + by^2 + cz^2 = 0$ *and* $\dfrac{x^2}{a} + \dfrac{y^2}{b} + \dfrac{z^2}{c} = 0$ *are reciprocal.*

Solution. The reciprocal cone of
$$ax^2 + by^2 + cz^2 = 0$$
$$Ax^2 + By^2 + Cz^2 + 2Fyz + 2Gzx + 2Hxy = 0 \qquad \ldots(1)$$

where
$$\Delta = \begin{vmatrix} a & 0 & 0 \\ 0 & b & 0 \\ 0 & 0 & c \end{vmatrix} = abc$$

and
$$A = \frac{\partial x}{\partial a} = bc, \; B = \frac{\partial \Delta}{\Delta b} = ac, \; C = \frac{\partial \Delta}{\partial c} = ab,$$
$$F = \frac{1}{2}\frac{\partial \Delta}{\partial f} = 0, \; G = \frac{1}{2}\frac{\partial \Delta}{\partial g} = 0, \; H = \frac{1}{2}\frac{\partial \Delta}{\partial h} = 0.$$

By putting these values, (1) becomes
$$bcx^2 + cay^2 + abz^2 = 0$$

or
$$\frac{x^2}{a} + \frac{y^2}{b} + \frac{z^2}{c} = 0 \quad \frac{x^2}{a} + \frac{y^2}{b} + \frac{z^2}{c} = 0$$

3. *Prove that the equation* $\sqrt{fx} \pm \sqrt{gy} \pm \sqrt{hz} = 0$ *represents a cone that touches the co-ordinate planes; and that the equation to the reciprocal cone is* $fyz + gzx + hxy = 0$.

Solution. The given equation can be written as
$$\sqrt{fx} \pm \sqrt{gy} = \mp\sqrt{hz}$$

$\Rightarrow \quad fx + gy \pm 2\sqrt{fgxy} = hz$

$\Rightarrow \quad (fx + gy - hz)^2 = 2fgxy$

$\Rightarrow \quad f^2x^2 + g^2y^2 + h^2z^2 - 2ghyz - 2hfzx - 2fgxy = 0 \qquad \ldots(i)$

The equation is a homogeneous equation of second degree, hence it represents a quadratic cone.

The co-ordinate plane $x = 0$ meets (i) where
$$g^2y^2 + h^2z^2 - 2ghyz = 0 \;\Rightarrow\; (gy - hz)^2 = 0$$

which being a perfect square it follows that the plane $x = 0$ touches it. Similarly, we can show that $y = 0$, $z = 0$ also touch the cone (i).

Again for the cone (i), we have

$$`a' = f^2, `b' = g^2, `c' = h^2, `f' = -gh, `g' = -hf, `h' = -fg$$

∴ $\quad A = bc - f^2 = g^2h^2 - (-gh)^2 = 0$

Similarly, $B = C = 0$,

$$F = gh - af = (-hf)(-fg) - f^2(-gh) = 2f^2 gh$$

Similarly, $G = 2g^2 hf$, $H = 2h^2 fg$

∴ The required equation of the cone reciprocal to (i) is

$$Ax^2 + By^2 + Cz^2 + 2Fyz + 2Gzx + 2Hxy = 0$$

⇒ $\quad 2f^2 ghyz + 2g^2 hfzx + 2h^2 fgxy = 0$

⇒ $\quad fyz + gzx + hxy = 0$

EXERCISES

1. Find the plane which touches the cone
$$x^2 + 2y^2 - 3z^2 + 2yz - 5zx + 3xy = 0$$
along the generator whose direction ratios are 1, 1, 1.

2. Prove that the perpendiculars drawn from the origin to be tangent planes to the cone $ax^2 + by^2 + cz^2 = 0$ lie on the cone $\dfrac{x^2}{a} + \dfrac{y^2}{b} + \dfrac{c^2}{z} = 0.$

3. Prove that tangent planes to the cone
$$x^2 - y^2 + 2z^2 - 3yz + 4zx - 5xy = 0$$
are perpendicular to the generators of the cone
$$17x^2 + 8y^2 + 29z^2 + 28yz - 46zx - 16xy = 0.$$

4. Prove that the cones
$$ayz + bzx + cxy = 0, \quad (ax)^{1/2} + (by)^{1/2} + (cz)^{1/2} = 0$$
are reciprocal.

5. Prove that the cones $fyz + gzx + hxy = 0$; $\sqrt{fx} + \sqrt{gy} + \sqrt{hz} = 0$ are reciprocal.

6. Find the condition that the plane $ux + vy + wz = 0$ may touch the cone.
$$ax^2 + by^2 + cz^2 = 0 \qquad \left[\text{Ans. } \dfrac{u^2}{a} + \dfrac{v^2}{b} + \dfrac{w^2}{c} = 0\right]$$

7. Show that a quadric cone can be found to touch any five planes which meet at a point provided no three of them intersect in a line.
Find the equation of the cone which touches the three co-ordinate planes and the planes
$$x + 2y + 3z = 0, \quad 2x + 3y + 4z = 0$$
$$[\text{Ans. } (x)^{1/2} + (-6y)^{1/2} + (6z)^{1/2} = 0]$$

5.5 INTERSECTION OF TWO CONES WITH A COMMON VERTEX

Sections of two cones, having a common vertex, by any plane are two coplanar conics which, general, intersect in four points.

The four lines joining the common vertex to the four points of intersection of these two coplanar conics are the four common generators of the two cones.

Therefore, *two cones with a common vertex have, in general, four generators in common.* In case two cones with the same vertex have five common generators, they coincide.

If $\qquad S = 0, S' = 0$

be the equations of two cones with origin as the common vertex, then
$$S + kS' = 0$$

Cones, Cylinders

Clearly, the genreal equation of a cone whose vertex is at the origin and which passes through the four common generators of the cones

$$S = 0, \ S' = 0$$

If k be so chosen that $S + kS' = 0$ becomes the product of two linear factors, then the corresponding equations obtained by putting the linear factors equal to zero represent a pair of planes through the common generators.

Such values of k are the roots of the k-cubic equation

$$\begin{vmatrix} a + ka' & h + kh' & g + kg' \\ h + kh' & b + kb' & f + kf' \\ g + kg' & f + kf' & c + kc' \end{vmatrix} = 0$$

The three values of k give the three pairs of planes through the four common generators.

EXERCISES

1. Find the equation of the cone which passes through the common generators of the cones
$$-2x^2 + 4y^2 + z^2 = 0 \text{ and } 10xy - 2yz + 5zx = 0$$
and the line with direction cosines proportional to 1, 2, 3.
[**Ans.** $2x^2 - 4y^2 - z^2 + 10xy - yz + 5zx = 0$]

2. Show that the equation of the cone through the intersection of the cones
$$x^2 - 2y^2 + 3z^2 - 4yz + 5zx - 6xy = 0 \text{ and } 2x^2 - 3y^2 + 4z^2 - 5yz + 6zx - 10xy = 0$$
and the line with direction cosines proportional to 1, 1, 1 is
$$y^2 - 2z^2 + 3yz - 4zx + 2xy = 0$$

3. Show that the plane $3x + 2y - 4z = 0$ passes through a pair of common generators of the cones $27x^2 + 20y^2 - 32z^2 = 0$ and $2yz + zx - 4xy = 0.$

4. Show that the plane $3x - 2y - z = 0$ cuts the cones
$$3yz - 2zx + 2xy = 0 \text{ and } 21x^2 - 4y^2 - 5z^2 = 0$$
in the same pair of perpendicular lines.
Also show that the plane $7x + 2y + 5z = 0$ contains the remaining two common generators.

5.6 THE RIGHT CIRCULAR CONE

5.6.1 Definition

A right circular cone is a surface generated by a line which passes through a fixed point, and makes a constant angle with a fixed line through the fixed point.

The fixed point is called the *vertex*, the fixed line the *axis* and the fixed angle the *semi-vertical angle* of the cone.

The justification for the name right circular cone is contained in the result obtained below.

Every section of a right circular cone by a plane perpendicular to its axis is a circle.

Let a plane perpendicular to the axis ON of the right circular cone with semi-vertical angle, α, meet it at N.

Let P be any point of the section. Since ON is perpendicular to the plane which contains the line NP, we have

$$ON \perp NP$$

$\Rightarrow \quad \dfrac{PN}{ON} = \tan \angle NOP = \tan \alpha$

$\Rightarrow \quad PN = ON \tan \alpha$

so that NP is constant for every position of the point P of the section. Hence, the section is a circle with N as its centre.

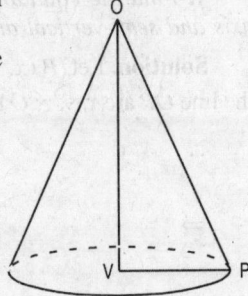

5.6.2 Equation of a Right Circular Cone

To find the equation of the right circular cone whose vertex is the poitn (α, β, γ) and whose axis is the line
$$\frac{x-\alpha}{l} = \frac{y-\beta}{m} = \frac{z-\gamma}{n}$$
and semi-vertical angle θ.

Let O be the vertex, and, OA the axis of the cone.

The required equation is to be obtained by using the condition that the line joining any point (x, y, z) on the curve to the vertex $O(\alpha, \beta, \gamma)$ makes an angle θ with the axis OA.

Direction cosines of the line OP, being proportional to
$$x-\alpha, \, y-\beta, \, z-\gamma$$

We have
$$\cos\theta = \frac{l(x-\alpha) + m(y-\beta) + n(z-\gamma)}{\sqrt{l^2+m^2+n^2}\sqrt{(x-\alpha)^2+(y-\beta)^2+(c-\gamma)^2}}$$

The required equation of the cone, therefore, is
$$[l(x-\alpha)+m(y-\beta)+n(z-\gamma)^2] = (l^2+m^2+n^2)$$
$$[(x-\alpha)^2+(y-\beta)^2+(z-\gamma)^2]\cos^2\theta$$

Cor. 1. If the vertex be the origin, the equation of the cone becomes
$$(lx+my+nz)^2 = (l^2+m^2+n^2)(x^2+y^2+z^2)\cos^2\theta$$

Cor. 2. If the vertex be the origin and axis of the cone be the Z-axis, then taking
$$z^2 = (x^2+y^2+z^2)\cos^2\theta \Leftrightarrow x^2+y^2 = z^2\tan^2\theta \qquad \text{...(1)}$$

Cor. 3. The semi-vertical angle of a right circular cone admitting sets of three mutually perpendicular generators is
$$\tan^{-1}\sqrt{2}$$
for, the sum of the coefficients of x^2, y^2, z^2 in the equation of such a cone must be zero and this means that
$$1+1-\tan^2\theta = 0, \, i.e., \, \theta = \tan^{-1}\sqrt{2} \qquad \text{[Refer (1), Cor. 2]}$$

Cor. 4. The semi-vertical angle of a right circular cone having sets of three mutually perpendicular tangent planes is
$$\tan^{-1}\sqrt{\frac{1}{2}},$$
for by Cor. to 5.4.3, this will be so if [Refer (1), Cor. 2]
$$1-\tan^2\theta - \tan^2\theta = 0 \Rightarrow \theta = \tan^{-1}\sqrt{\frac{1}{2}}.$$

EXAMPLES

1. Find the equation to the right circular cone whose vertex is at origin, the axis along x-axis and semi-vertical angle α.

Solution. Let $P(x, y, z)$ be any point on the surface of the cone, so that the direction ratios of the line OP are x, y, z; O being the origin. The direction cosines of x-axis are $1, 0, 0$.

$\therefore \qquad \cos\alpha = \dfrac{x.1 + y.0 + z.0}{\sqrt{x^2+y^2+z^2}}$

$\Rightarrow \qquad (x^2+y^2+z^2)\cos^2\alpha = x^2$

$\Rightarrow \qquad y^2+z^2 = x^2\tan^2\alpha.$

Cones, Cylinders

2. *Lines are drawn through the origin with direction cosines proportional to (1, 2, 3), (2, 3, 6), (3, 4, 12). Show that the axis of the right circular cone through them has direction cosines*

$$-\frac{1}{\sqrt{3}}, \frac{1}{\sqrt{3}}, \frac{1}{\sqrt{3}}$$

and that the semi-vertical angle of the cone is $\cos^{-1}\left(\frac{1}{\sqrt{3}}\right)$.

Obtain the equation of the cone also and show that it passes through the co-ordinate axes.

Solution. Let (l, m, n) be the direction cosines of the axes of the right circular cone. Let O be the origin and P, Q, R be the points, so that d.r.'s of OP, OQ, OR are $(1, 2, 2), (2, 3, 6), (3, 4, 12)$ respectively.

Therefore, d.c.'s of OP are $\frac{1}{3}, \frac{2}{3}, \frac{2}{3}$, those of OQ are $\frac{2}{7}, \frac{3}{7}, \frac{6}{7}$ and those of OR are $\frac{3}{13}, \frac{4}{13}, \frac{12}{13}$.

Let α be the semi-vertical angle of the cone, then

$$\cos\theta = \frac{1}{3}l + \frac{2}{3}m + \frac{2}{3}n = \frac{2}{7}l + \frac{3}{7}m + \frac{6}{7}n = \frac{3}{13}l + \frac{4}{13}m + \frac{12}{13}n$$

From first two relations

$$l + 5m - 4n = 0 \qquad \qquad \ldots(i)$$

and from first and last, we get

$$2l + 7m - 5n = 0 \qquad \qquad \ldots(ii)$$

Solving, we obtain

$$\frac{l}{-1} = \frac{m}{1} = \frac{n}{1} = \pm\frac{\sqrt{l^2 + m^2 + n^2}}{\sqrt{1+1+1}} = \pm\frac{1}{\sqrt{3}}$$

Therefore, direction cosines of the axis are

$$-\frac{1}{\sqrt{3}}, \frac{1}{\sqrt{3}}, \frac{1}{\sqrt{3}}.$$

Therefore, semi-vertical angle of the cone will be

$$\cos\alpha = \frac{1}{3}\left(-\frac{1}{\sqrt{3}}\right) + \frac{2}{3}\cdot\frac{1}{\sqrt{3}} + \frac{2}{3}\cdot\frac{1}{\sqrt{3}} = \frac{1}{\sqrt{3}} \Rightarrow \alpha = \cos^{-1}\left(\frac{1}{\sqrt{3}}\right)$$

If (x, y, z) be any point on the cone, then its equation will be

$$\cos\alpha = \frac{-1.x + 1.y + 1.z}{\sqrt{3}\sqrt{x^2 + y^2 + z^2}} = \frac{1}{\sqrt{3}}$$

$$\Rightarrow \qquad yz - zx - xy = 0$$

EXERCISES

1. Find the equation of the right circualr cone with its vertex at the origin, axis along Z-axis and semi-vertical angle α.
2. Show that the equation of the right circular cone with vertex (2, 3, 1), axis parallel to the line $-x = \frac{y}{2} = z$ and one of its generators having direction cosines proportional to (1, –1, 1) is
$$x^2 + 8y^2 + z^2 + 12yz - 12yz + 6zx + 46x + 36y + 22z - 19 = 0.$$
3. Find the equation of the circular cone which passes through the point (1, 1, 2) and has its vertex at the origin and axis the line $\frac{x}{2} = -\frac{y}{4} = \frac{z}{3}$.

[**Ans.** $4x^2 + 40y^2 + 19z^2 - 48xy - 72yz + 36xz = 0$]

4. Find the equation of the right circular cone whose vertex is origin, axis of the line $x = t, y = 2t, z = 3t$, and whose semi-vertical angle is $60°$.

[**Ans.** $38x^2 + 26y^2 + 6z^2 - 16xy - 48yz - 24zx = 0$]

5. Find the equation of the right circular cone whose vertex is $(1, -2, 1)$, axis the line
$$\frac{x-1}{3} = \frac{y+2}{4} = \frac{z+1}{5}$$
and semi-vertical angle $60°$.

[**Ans.** $7x^2 - 7y^2 - 25z^2 + 48xy + 80yz - 60zx + 22x + 4y + 17x + 78 = 0$]

6. Find the equation of the right circular cone whose vertex is $(3, 2, 1)$ axis the line
$$\frac{x-3}{4} = \frac{y-2}{1} = \frac{z-1}{3}$$
and semi-vertical angle $30°$.

7. Find the equation of right circular cone which passes through $(1, 1, 1)$, whose vertex is $(1, 1, 1)$ and axis of cone makes equal angle with co-ordinate axes.

[**Ans.** $xy + yz + zx - x - 2y - z + 1 = 0$]

8. Find the equation of the cone generated by rotating the line
$$\frac{x}{l} = \frac{y}{m} = \frac{z}{n}$$
about the line $\dfrac{x}{a} = \dfrac{y}{b} = \dfrac{z}{c}$ as axis.

[**Ans.** $(al + bm + cn)^2 (x^2 + y^2 + z^2) = (ax + by + cz)^2 (l^2 + m^2 + n^2)$]

5.7 THE CYLINDER

Def. *A **cylinder** is a surface generated by a straight line which is always parallel to a fixed line and is subject to one more condition; for instance, it may intersect a given curve or touch a given surface.*

The given curve is called the *Guiding curve.*

5.7.1 Equation of a Cylinder

To find the equation of the cylinder whose generators intersect the conic
$$ax^2 + 2hxy + by^2 + 2gx + 2fy + c = 0, z = 0 \qquad \ldots(i)$$
and are parallel to the line
$$\frac{x}{l} = \frac{y}{m} = \frac{z}{n} \qquad \ldots(ii)$$

Let (α, β, γ) be *any* point on the cylinder so that the equations of the generator through the point are
$$\frac{x-\alpha}{l} = \frac{y-\beta}{m} = \frac{z-\gamma}{n} \qquad \ldots(iii)$$

As in 5.1.2, the line (iii) will intersect the conic (i), if

$$\left(\alpha - \frac{l\gamma}{n}\right)^2 + 2h\left(\alpha - \frac{l\gamma}{n}\right)\left(\beta - \frac{m\gamma}{n}\right) + b\left(\beta - \frac{m\gamma}{n}\right)^2 + 2g\left(\alpha - \frac{l\gamma}{n}\right) + 2f\left(\beta - \frac{m\gamma}{n}\right) + c = 0$$

But this is the condition that the point (α, β, γ) should lie on the surface

$$a\left(x - \frac{lz}{n}\right)^2 + 2h\left(x - \frac{lz}{n}\right)\left(y - \frac{mz}{n}\right) + b\left(y - \frac{mz}{n}\right)^2 + 2g\left(x - \frac{lz}{n}\right) + 2f\left(y - \frac{mz}{n}\right) + c = 0$$

$$\Rightarrow a(nx - lz)^2 + 2h(nx - lz)(ny - mz) + b(ny - nz)^2 + 2gn(nx - lz)$$
$$+ 2fn(ny - nz) + cn^2 = 0$$

which is, therefore, the required equation of the cylinder.

Cones, Cylinders

Cor. If the generators be parallel to Z-axis so that
$$l = 0 = m \text{ and } n = 1$$
the equation of the cylinder becomes
$$ax^2 + 2hxy + by^2 + 2gx + 2fy + c = 0$$
as is already known to the reader.

EXAMPLES

1. *Find the equation of the cylinder whose generators are parallel to the line*
$$\frac{x}{1} = \frac{y}{-2} = \frac{z}{3}$$
and whose guiding curve is the ellipse $x^2 + 2y^2 = 1, z = 0$.

Solution. Let (α, β, γ) be any point on the cylinder, then equations of a generator through (α, β, γ) are
$$\frac{x-\alpha}{1} = \frac{y-\beta}{-2} = \frac{z-\gamma}{3}$$
This meets the plane $z = 0$ at the point given by
$$\frac{x-\alpha}{1} = \frac{y-\beta}{-2} = -\frac{\gamma}{3}$$
i.e., at $\left(\alpha - \frac{\gamma}{3}, \beta + \frac{2\gamma}{3}, 0\right)$

Therefore, the generator intersects the given curve if
$$\left(\alpha - \frac{\gamma}{3}\right)^2 + 2\left(\beta + \frac{2\gamma}{3}\right)^2 = 1$$
Hence, locus of (α, β, γ) is
$$\left(x - \frac{z}{3}\right)^2 + 2\left(y + \frac{2z}{3}\right)^2 = 1$$
$$3x^2 + 6y^2 + 3z^2 - 2zx + 8yz - 3 = 0$$

2. *Find the equation of the cylinder whose generators are parallel to*
$$\frac{x}{1} = \frac{y}{-2} = \frac{z}{3}$$
and whose guiding curve is the ellipse $x^2 + 2y^2 = 1, z = 3$.

Solution. Let (α, β, γ) be any point on the surface of the cylinder so that the equations of its generators through this point are
$$\frac{x-\alpha}{1} = \frac{y-\beta}{-2} = \frac{z-\gamma}{3}$$
This line meets the plane $z = 3$ at the point given by
$$\frac{x-\alpha}{1} = \frac{y-\beta}{-2} = \frac{3-\gamma}{3},$$
i.e., $\left(\alpha + \frac{3-\gamma}{3}, \beta + \frac{2\gamma - 6}{3}, 3\right)$

This point will lie on the surface
$$x^2 + 2y^2 = 1,$$

if
$$\left(\alpha + \frac{3-\gamma}{3}\right)^2 + 2\left(\beta + \frac{2\gamma - 6}{3}\right)^2 = 1$$

or
$$(3\alpha - \gamma + 3)^2 + 2(3\beta + 2\gamma - 6)^2 = 9$$

Hence, locus of the point (α, β, γ) will be

$$(3x - z + 3)^2 + 2(3y + 2z - 6)^2 = 9$$

or
$$3x^2 + 6y^2 + 3z^2 + 8yz - 2zx + 6x - 24y - 18z + 24 = 0$$

This is the required equation of the cylinder.

3. *Find the equation of the quadratic cylinder with generators parallel to x-axis and passing through the curve* $ax^2 + by^2 + cz^2 = 1, lx + my + nz = p$.

Solution. The equation of the required cylinder is obtained by eliminating x between the equations

$$ax^2 + by^2 + cz^2 = 1 \text{ and } lx + my + nz = p$$

For this, substituting the value of $x = \dfrac{p - my - nz}{l}$ in the other equation, we get

$$a\left(\frac{p - my - nz}{l}\right)^2 + by^2 + cz^2 = 1$$

or
$$a(p - my - nz)^2 + bl^2 y^2 + cl^2 z^2 = l^2$$

or
$$(bl^2 + am^2) y^2 + (cl^2 - am^2) z^2 + 2amyz - 2ampy - 2apnz + ap^2 - l^2 = 0$$

This is the equation of required cylinder.

EXERCISES

1. Find the equation of the cylinder whose generators intersect the curve $ax^2 + by^2 = 2z$, $lx + my + nz = p$ and are parallel to the Z-axis.
   [**Hint:** Eliminate z from the equations.]
2. Find the equation of the cylinder whose generators are parallel to
$$\frac{x}{1} = \frac{y}{2} = \frac{z}{3}$$
   and guiding curve is $x^2 + y^2 = 16, z = 0$.
3. Find the equation of the cylinder whose generators are parallel to z-axis and guiding curve is given by $ax^2 + by^2 + cz^2 = 1, lx + my + nz = p$.
   [**Ans.** $(an^2 + cl^2) x^2 + (bn^2 + cm^2) y^2 + 2lcmxy - 2cplx - 2cpmy + (cp^2 - n^2) = 0$]
4. Find the equation of cylinder whose generators is parallel to $y = mx, z = nx$ and which intersect the conic $\dfrac{x^2}{a^2} + \dfrac{y^2}{b^2} = 1, z = 0$. [**Ans.** $b^2(nx - z)^2 + a^2(ny - nz)^2 = n^2 a^2 b^2$]

5.7.2 Enveloping Cylinder

To find the equation of the cylinder whose generators touch the sphere

$$x^2 + y^2 + z^2 = a^2 \qquad \ldots\text{(i)}$$

and are parallel to the line

$$\frac{x}{l} = \frac{y}{m} = \frac{z}{n} \qquad \ldots\text{(ii)}$$

Cones, Cylinders

Let (α, β, γ) be any point on the cylinder so that the equations of the generator through it are

$$\frac{x-\alpha}{l} = \frac{y-\beta}{m} = \frac{z-\gamma}{n} \qquad \text{...(iii)}$$

The line (iii) will touch the sphere (i), if

$$(l\alpha + m\beta + n\gamma)^2 = (l^2 + m^2 + n^2)(\alpha^2 + \beta^2 + \gamma^2 - a^2)$$

But this is the condition that the point (α, β, γ) should lie on the surface

$$(lx + my + nz)^2 = (l^2 + m^2 + n^2)(x^2 + y^2 + z^2 - a^2)$$

which is, therefore, the required equation of the cylinder and is known as *Enveloping cylinder of the sphere (i)*.

EXAMPLE

1. *Find the enveloping cylinder of the sphere*

$$x^2 + y^2 + z^2 - 2x + 4y = 1$$

having its generators parallel to $x = y = z$. Also find its guiding curve.

Solution. Let (α, β, γ) be any point on the surface of the cylinder so that the equations of its generator through this point are

$$\frac{x-\alpha}{1} = \frac{y-\beta}{1} = \frac{z-\gamma}{1} = r \text{ (say)} \qquad \text{...(i)}$$

Any point on this line is

$$(\alpha + r, \beta + r, \gamma + r)$$

This point will lie on the sphere

$$x^2 + y^2 + z^2 - 2x + 4y = 1$$

if $\quad (\alpha + r)^2 + (\beta + r)^2 + (\gamma + r)^2 - 2(\alpha + r) + 4(\beta + r) = 1$

or $\quad 3r^2 + 2r(\alpha + \beta + \gamma + 1) + (\alpha^2 + \beta^2 + \gamma^2 - 2\alpha + 4\beta - 1) = 0$

Since the generators (1) touches (2), the roots of this quadratic equation in r must be identical, for which

$$4(\alpha + \beta + \gamma + 1)^2 = 12(\alpha^2 + \beta^2 + \gamma^2 - 2\alpha + 4\beta - 1)$$

or $\quad \alpha^2 + \beta^2 + \gamma^2 - \beta\gamma - \gamma\alpha - \alpha\beta - 4\alpha + 5\beta - \gamma - 2 = 0$

Therefore, the locus of (α, β, γ) is

$$x^2 + y^2 + z^2 - yz - zx - xy - 4x + 5y - z - 2 = 0$$

This is the required equation of the enveloping cylinder.

Now, equation to the plane passing through the centre $(1, -2, 0)$ of the sphere (2) and perpendicular to the generators of the cylinder whose direction cosines are proportional to $(1, 1, 1)$ is

$$1.(x-1) + 1.(y+2) + 1.(z-0) = 0$$

or $\quad x + y + z + 1 = 0 \qquad \text{...(3)}$

Clearly, the guiding curve is the curve of intersection of the sphere (2) and the plane (3), *i.e.*, the equations of the guiding curve are

$$x^2 + y^2 + z^2 - 2x + 4y - 1 = 0,$$
$$x + y + z + 1 = 0.$$

EXERCISES

1. Find the equation of the enveloping cylinder of the conicoid
$$\frac{x^2}{a^2} + \frac{y^2}{b^2} + \frac{z^2}{c^2} = 1$$
whose generators are parallel to the line (i) $\frac{x}{l} = \frac{y}{m} = \frac{z}{n}$; (ii) $x = y = z$.

$$\left[\text{Ans. (i)} \left(\frac{lx}{a^2} + \frac{my}{b^2} + \frac{nz}{c^2}\right)^2 = \left(\frac{l^2}{a^2} + \frac{m^2}{b^2} + \frac{n^2}{c^2}\right)\left(\frac{x^2}{a^2} + \frac{y^2}{b^2} + \frac{z^2}{c^2} - 1\right)\right.$$

$$\left.\text{(ii)} \left(\frac{x}{a^2} + \frac{y}{b^2} + \frac{z}{c^2}\right)^2 = \left(\frac{1}{a^2} + \frac{1}{b^2} + \frac{1}{c^2}\right)\left(\frac{x^2}{a^2} + \frac{y^2}{b^2} + \frac{z^2}{c^2} - 1\right)\right]$$

2. Obtain the equation of a cylinder whose generators touch the sphere
$$x^2 + y^2 + z^2 + 2ux + 2vy + 2wz + d = 0$$
whose generators are parallel to the line $\frac{x}{l} = \frac{y}{m} = \frac{z}{n}$.

[**Ans.** $\{l(x+u) + m(y+v) + n(z+w)\}^2$
$= (l^2 + m^2 + n^2)(x^2 + y^2 + z^2 + 2ux + 2vy + 2wz + d)$]

3. Find the equation of a right circular cylinder which envelopes a sphere with centre (a, b, c) and radius r and has the generators parallel to the direction cosines (l, m, n).

[**Ans.** $\{l(a-x) + m(b-y) + n(c-z)\}^2$
$= (l^2 + m^2 + n^2)\{(a-x)^2 + (b-y)^2 + (c-z)^2 - r^2\}$]

4. Prove that the enveloping cylinders of ellipsoid $\frac{x^2}{a^2} + \frac{y^2}{b^2} + \frac{z^2}{c^2} = 1$, whose generators are parallel to the line $\frac{x}{0} = \frac{y}{\pm\sqrt{a^2 - b^2}} = \frac{z}{c}$ meet the plane $z = 0$ in circles.

5.8 THE RIGHT CIRCULAR CYLINDER

5.8.1 Definition

A right circular cylinder is a surface generated by a line which intersects a fixed circle, called the guiding circle, and is perpendicular to its plane.

The normal to the plane of the guiding circle through its centre is called the *Axis* of the cylinder.

Section of a right circular cylinder by any plane perpendicular to its axis is called a *Normal Section*.

Clearly all the normal sections of a right circular cylinder are circles having the same radius which is also called the radius of the cylinder.

The length of the perpendicular from any point on a right circular cylinder to its axis is equal to its radius.

5.8.2 Equation of a Right Circular Cylinder

To find the equation of the right circular cylinder whose axis is the line
$$\frac{x - \alpha}{l} = \frac{y - \beta}{m} = \frac{z - \gamma}{n}$$
and whose radius is r.

Let (x, y, z) be a point on the cylinder. Equating the perpendicular distance of the point from the axis of the radius r, we get

Cones, Cylinders

$$(x-\alpha)^2 + (y-\beta)^2 + (z-\gamma)^2 - \frac{[l(x-\alpha) + m(y-\beta) + n(z-\gamma)]^2}{l^2 + m^2 + n^2} = r^2$$

which is the required equation of the cylinder.

EXAMPLES

1. Find the equation of a circular cylinder whose guiding curve is $x^2 + y^2 + z^2 = 9$, $x - y + z = 3$.

Solution. We know that the radius of a right circular cylinder is equal to the radius of the guiding curve and the axis of the cylinder is a line passing through the centre of the circle and hence of the sphere and perpendicular to the plane of the circle.

Here, radius of the sphere - 3.

Length of the perpendicular from the centre $O(0, 0, 0)$ to the given plane

$$= \frac{-3}{\sqrt{1+1+1}} = -\sqrt{3}$$

∴ Radius of the circle $= \sqrt{3^2 - 3} = \sqrt{6}$.

The axis of the cylinder passes through $(0, 0, 0)$ and is perpendicular to the plane $x - y + z = 3$.
Hence its equations are

$$\frac{x}{1} = \frac{y}{-1} = \frac{z}{1}.$$

Therefore, the equation of the circular cylinder is

$$\left(\frac{1}{\sqrt{3}}\right)^2 \left\{ \begin{vmatrix} y & z \\ -1 & 1 \end{vmatrix}^2 + \begin{vmatrix} z & x \\ 1 & 1 \end{vmatrix}^2 + \begin{vmatrix} x & y \\ 1 & -1 \end{vmatrix}^2 \right\} = (\sqrt{6})^2$$

or $(y+z)^2 + (z-x)^2 + (-x-y)^2 = 18$

or $2x^2 + 2y^2 + 2z^2 + 2yz - 2zx + 2xy = 18$

or $x^2 + y^2 + z^2 + xy + yz - zx - 9 = 0$

2. Find the right circular cylinder whose radius is 2 and axis is the line

$$\frac{x-1}{2} = \frac{y-2}{1} = \frac{z-3}{2}$$

Solution. Let $P(x_1, y_1, z_1)$ be any point on the cylinder. The length of the perpendicular from $P(x_1, y_1, z_1)$ to the given line must be equal to the radius.

∴ $2^2 (2^2 + 1^2 + 2^2) = \{2(y_1 - 2) - 1(z_1 - 3)\}^2 + \{2(z_1 - 3) - 2(x - 1)\}^2$
$$+ \{1(x_1 - 1) - 2(y_1 - 2)\}^2$$

⇒ $36 = (2y_1 - z_1 - 1)^2 + (2z_1 - 2x_1 - 4)^2 + (x_1 - 2y_1 + 3)^2$

∴ The required equation of the locus of $P(x_1, y_1, z_1)$ is

$$(2y - z - 1)^2 + (2z - 2x - 6)^2 + (x - 2y + 3)^2 = 36$$

⇒ $5x^2 + 8y^2 + 5z^2 - 4xy - 4yz - 8zx + 22x - 16y - 14z - 10 = 0$

EXERCISES

1. Find the equation of the right circular cylinder of radius 2 whose axis is the line

$$\frac{x-1}{2} = \frac{y-2}{2} = \frac{z-2}{2}$$

[**Ans.** $5x^2 + 8y^2 + 5z^2 - 4xy - 4yz - 8zx + 22x - 16y - 14z - 10 = 0$]

2. The axis of a right circular cylinder of radius 2 is
$$\frac{x-1}{2} = \frac{y}{3} = \frac{z-3}{1}$$
show that its equation is $10x^2 + 5y^2 + 13z^2 - 12xy - 6yz - 4zx - 8x + 30y - 74x + 59 = 0$.

3. Find the equation of the right circular cylinder of radius 3 and having for its axis the line
$$\frac{x-1}{2} = \frac{y-3}{2} = \frac{5-z}{7}$$
[**Ans.** $5x^2 + 5y^2 + 8z^2 - 4yz + 4zx - 6x - 42y - 96z + 225 = 0$]

4. Find the equation of the right circular cylinder whose axis is
$$\frac{x-2}{2} = \frac{y-1}{1} = \frac{z}{3}$$
and pass through (0, 0, 3).
[**Ans.** $10x^2 + 13y^2 + 5z^2 - 4xy - 6yz - 12zx - 36x - 18y + 24z + 18 = 0$]

5. Find the equation of the right circular cylinder of radius 5 and having for its axis the line
$$\frac{1}{2}x = \frac{1}{3}y = \frac{1}{6}z.$$
[**Ans.** $45x^2 + 40y^2 + 13z^2 - 12xy - 36yz - 24zx - 1225 = 0$]

6
Conicoid

6.1 THE GENERAL EQUATION OF THE SECOND DEGREE

The locus of the general equation
$$ax^2 + by^2 + cz^2 + 2fyz + 2gzx + 2hxy + 2ux + 2vy + 2wz + d = 0,$$
of the second degree in x, y, z is called a **Conicoid** or **Quadric**.

It is easy to show that every straight line meets a surface whose equation is of the second degree in two points and consequently every plane section of such a surface is a conic. This property justifies the name "Conicoid" as applied to such a surface.

The general equation of second degree contains *nine* effective constants and, therefore, a conicoid can be determined to satisfy nine conditions each of which gives rise to one relation between the constant, *e.g.*, a conicoid can be determined so as to pass through *nine* given points no four of which are coplanar.

The general equation of the second degree can, by transformation of co-ordinate axes, be reduced to any one of the following forms; the actual reduction being given in Chapter 9. (The name of the particular surface which is the locus of the equation is written along with it).

1. $x^2/a^2 + y^2/b^2 + z^2/c^2 = 1$, Ellipsoid.
2. $x^2/a^2 + y^2/b^2 + z^2/c^2 = -1$, Imaginary ellipsoid.
3. $x^2/a^2 + y^2/b^2 - z^2/c^2 = 1$, Hyperboloid of one sheet.
4. $x^2/a^2 + y^2/b^2 - z^2/c^2 = -1$, Hyperboloid of two sheets.
5. $x^2/a^2 + y^2/b^2 + z^2/c^2 = 0$, Imaginary cone.
6. $x^2/a^2 + y^2/b^2 - z^2/c^2 = 0$, Cone.
7. $x^2/a^2 + y^2/b^2 = 2z/c$, Elliptic paraboloid.
8. $x^2/a^2 - y^2/b^2 = 2z/c$, Hyperbolic paraboloid.
9. $x^2/a^2 + y^2/b^2 = 1$, Elliptic cylinder.
10. $x^2/a^2 - y^2/b^2 = 1$, Hyperbolic cylinder.
11. $x^2/a^2 + y^2/b^2 = -1$, Imaginary cylinder.
12. $x^2/a^2 - y^2/b^2 = 0$, Pair of intersecting planes.
13. $x^2/a^2 + y^2/b^2 = 0$, Pair of imaginary planes.
14. $y^2 = 4ax$, Parabolic cylinder.
15. $y^2 = a^2$, Two real parallel plans.
16. $y^2 = -a^2$, Two imaginary planes.
17. $y^2 = 0$, Two coincident planes.

The equations representing cones and cylinders have already been considered and the reader is familiar with the nature of the surface represented by them.

In this chapter we propose to discuss the nature and some of the important geometrical properties of the surfaces represented by the equation, 1, 2, 3, 4, 7, 8.

6.2 SHAPES OF SOME SURFACES

6.2.1 The Ellipsoid $\dfrac{x^2}{a^2} + \dfrac{y^2}{b^2} + \dfrac{z^2}{c^2} = 1$

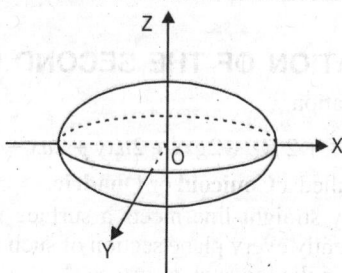

The *following facts* enable us to have an idea of the shape of the surface represented by this equation

$$\frac{x^2}{a^2} + \frac{y^2}{b^2} + \frac{z^2}{c^2} = 1.$$

(i) We have

(x, y, z) satisfies the equation $\Leftrightarrow$ $(-x, -y, -z)$ satisfies the equation.

The points (x, y, z) $(-x, -y, -z)$ lying on a straight line through the origin and equidistant from the origin, it follows that the origin bisects every chord which passes through it and is, on this account, called the *Centre* of the surface.

(ii) We have

(x, y, z) satisfies the equation $\Leftrightarrow$ $(x, y, -z)$ satisfies the equation.

The line joining the points $(x, y, z), (x, y, -z)$ is bisected at right angle by the *XOY* plane. It follows that the *XOY* plane bisects every chord perpendicular to it and the surface is symmetrical with respect to this plane.

Similarly the surface is symmetrical with respect to the *YOZ* and the *ZOX* planes.

These three planes are called **principal planes** in as much as they bisect all chords perpendicular to them. The three lines of intersection of the three principal planes taken in pairs are called **principal axes.** Co-ordinates axes are the principal axes in the present case.

(iii) x cannot take a value which is numerically greater than a, for otherwise y^2 or z^2 would be negative. Thus, we have $-a \leq x \leq a$ for every point (x, y, z) on the surface. Similarly, y and z cannot be numerically greater than b and c respectively so that we have for every point (x, y, z) on the surface

$$-a \leq x \leq a, \ -b \leq y \leq b, -c \leq z \leq c$$

Hence, the surface lies between the planes

$$x = a, \ x = -a; \ y = b, \ y = -b; \ z = c, \ z = -c$$

and so is a *closed* surface.

(iv) The *X*-axis meets the surface in the two points $(a, 0, 0)$ and $(-a, 0, 0)$, so that the surface intercepts a length $2a$ on *X*-axis. Similarly, the length intercepted on the *Y*-axis and *Z*-axis are $2b$ and $2c$ respectively.

The lengths $2a, 2b, 2c$ intercepted on the principal axes are called the lengths of the axes of the ellipsoid.

The Conicoid

(v) The sections of the surface by the planes $z = k$ which are parallel to the XOY plane are similar ellipses having equations

$$\frac{x^2}{a^2} + \frac{y^2}{b^2} = 1 - \frac{k^2}{c^2}, \quad z = k; \ -c \le k \le c. \qquad \ldots(1)$$

These ellipses have their centres on the Z-axis and diminish in size as k varies from 0 to c. The ellipsoid may, therefore, be generated by the variable ellipse (1) as k varies from $-c$ to c.

It may similarly be shown that the sections by planes parallel to the other co-ordinate planes are also ellipses and the ellipsoid may be supposed to be generated by them.

6.2.2. The Hyperboloid of One Sheet $\dfrac{x^2}{a^2} + \dfrac{y^2}{b^2} - \dfrac{z^2}{c^2} = 1$.

(i) The origin bisects all chords through it and is, therefore, the centre of the surface.

(ii) The co-ordinate planes bisect all chords perpendicular to them and are, therefore, the plane of symmetry or the **principal planes** of the surfaces. The co-ordinate axes are its **principal axes.**

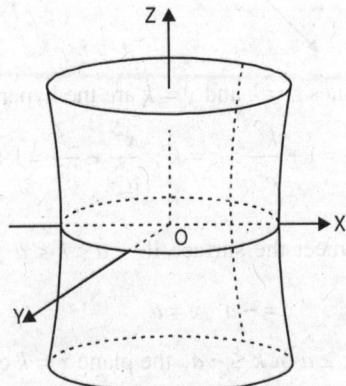

(iii) The X-axis meets the surface in the points $(a, 0, 0)$, $(-a, 0, 0)$ so that the surface intercepts length $2a$ on X-axis. Similarly the length intercepted on Y-axis is $2b$. The Z-axis does not meet the surface.

The sections by the planes $z = k$ which are parallel to the XOY plane are the similar ellipses

$$\frac{x^2}{a^2} + \frac{y^2}{b^2} = 1 + \frac{k^2}{c^2}, \ z = k \qquad \ldots(1)$$

whose centres lie on Z-axis and which increase in size as k increases. There is no limit to the increase of k. The surface may, therefore, be generated by the variable ellipse (1) where k varies from $-\infty$ to $+\infty$.

Again, the sections by the planes $x = k$ and $y = k$ are hyperbolas

$$\frac{y^2}{b^2} - \frac{z^2}{c^2} = 1 - \frac{k^2}{a^2}, \ x = k; \quad \frac{x^2}{a^2} - \frac{z^2}{c^2} = 1 - \frac{k^2}{b^2}, \ y = k$$

respectively.

Ex. Trace the surfaces

(i) $\dfrac{x^2}{a^2} + \dfrac{y^2}{b^2} + \dfrac{z^2}{c^2} = 1.$ (ii) $-\dfrac{x^2}{a^2} + \dfrac{y^2}{b^2} + \dfrac{z^2}{c^2} = 1.$

6.2.3. The Hyperboloid of Two Sheets $\dfrac{x^2}{a^2} - \dfrac{y^2}{b^2} - \dfrac{z^2}{c^2} = 1$.

(i) Origin is the **centre**; co-ordinate planes are the **principal planes**; and co-ordinate axes the **principal axes** of the surface.

(ii) X-axis meets the surface in the points $(a, 0, 0)$ and $(-a, 0, 0)$ whereas the Y and Z-axis do not meet the surface.

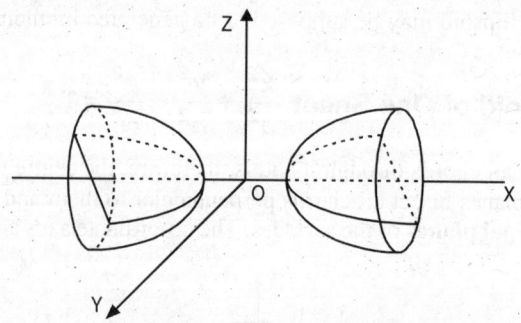

(iii) The sections by the planes $z = k$ and $y = k$ are the hyperbolas

$$\frac{x^2}{a^2} - \frac{y^2}{b^2} = 1 + \frac{k^2}{c^2}, \ z = k \ ; \ \frac{x^2}{a^2} - \frac{z^2}{c^2} = 1 + \frac{k^2}{b^2} \ ; \ y = k$$

respectively.

The plane $x = k$ does not meet the surface if $-a < k < a$ so that there is no portion of the surface between the planes

$$x = -a, \ x = a$$

when $k^2 > a^2$, i.e., when $k \geq a$ or $k \leq -a$, the plane $x = k$ cuts the surface in the ellipse

$$\frac{y^2}{b^2} + \frac{z^2}{c^2} = \frac{k^2}{a^2} - 1, \ x = k.$$

These ellipses increase in size as k^2 increases.

Ex. Trace the surfaces

(i) $-\dfrac{x^2}{a^2} + \dfrac{y^2}{b^2} - \dfrac{z^2}{c^2} = 1.$ (ii) $-\dfrac{x^2}{a^2} - \dfrac{y^2}{b^2} + \dfrac{z^2}{c^2} = 1.$

6.2.4. Central Conicoids

The equations considered above are of the form

$$ax^2 + by^2 + cz^2 = 1. \qquad \ldots(1)$$

If a, b, c are all positive, the surface is an ellipsoid; if two are positive and one negative, hyperboloid of one sheet, and finally if two are negative and one positive, hyperboloid of two sheets.

No point (x, y, z) satisfies the equation

$$ax^2 + by^2 + cz^2 = 1$$

if a, b, c are all negative.

All *these surfaces have a centre and three principal planes* and are as such known as *central conicoids*.

On the basis of the preceding discussion, the reader would do well to give precise definitions of

(i) *Centre* (ii) *Principal planes* (iii) *Principal axes* of a central conicoid.

The Conicoid

In what follows, we shall consider the equation (1) and the geometrical results deducible from it will, therefore, hold in the case of all central conicoids.

Ex. Show that the surface represented by the equation
$$ax^2 + by^2 + cz^2 + 2fyz + 2gzx + 2hxy = d$$
is a central conicoid; origin being the centre.

Note. Cone is also a central conicoid, vertex being its centre; this fact is clear from the general equation of a cone with its vertex at the origin.

6.3 INTERSECTION OF A LINE WITH A CONICOID

To find the points of intersection of the line
$$\frac{x-\alpha}{l} = \frac{y-\beta}{m} = \frac{z-\gamma}{n} \qquad \ldots(i)$$
with the central conicoid
$$ax^2 + by^2 + cz^2 = 1 \qquad \ldots(ii)$$

A point
$$(lr + \alpha, mr + \beta, nr + \gamma)$$
on the line (i) shall also lie on the surface (ii), if and only if,
$$a(lz+\alpha)^2 + b(mr+\beta)^2 + c(nr+\gamma)^2 = 1$$
$$\Leftrightarrow r^2(a^2 + bm^2 + cn^2) + 2r(al\alpha + bm\beta + cn\gamma) + (a\alpha^2 + b\beta^2 + c\gamma^2 - 1) = 0 \qquad \ldots(A)$$

Let r_1, r_2 be the two roots of (A), which we suppose to be real. Then
$$(lr_1 + \alpha, mr_1 + \beta, nr_1 + \gamma), (lr_2 + \alpha, mr_2 + \beta, nr_2 + \gamma)$$
are the two points of intersection.

Hence, *every line meets a central conicoid in two points.*

We also see that *a plane section of a central conicoid is a conic for every line in the plane meets the curve of intersection in two points only.*

The two values r_1 and r_2 of r obtained from equation (A) are the measures of the distances of the points of intersection P and Q from the point (α, β, γ) if (l, m, n) are the direction cosines of the line.

Note. The equation (A) of this article will frequently be used in what follows.

Ex. 1. Find the points of intersection of the line
$$-\frac{1}{3}(x+5) = (y-4) = \frac{1}{7}(z-11)$$
with the conicoid
$$12x^2 - 17y^2 + 7z^2 = 7. \qquad \text{[Ans. } (1, 2, -3), (-2, 3, 4)\text{]}$$

Ex. 2. Prove that the sum of the squares of the reciprocals of any three mutually perpendicular semi-diameters of a central conicoid is constant.

Ex. 3. Any three mutually orthogonal lines drawn through a fixed point C meets the quadric
$$ax^2 + by^2 + cz^2 = 1$$
in points $P_1, P_2; Q_1, Q_2; R_1, R_2$ respectively; prove that
$$\frac{P_1P_2^2}{CP_1^2 \cdot CP_2^2} + \frac{Q_1Q_2^2}{CQ_1^2 \cdot CQ_2^2} + \frac{R_1R_2^2}{CR_1^2 \cdot CR_2^2} \text{ and } \frac{1}{CP_1 \cdot CP_2} + \frac{1}{CQ_1 \cdot CQ_2} + \frac{1}{CR_1 \cdot CR_2}$$
are constants.

6.3.1. Tangent Lines and Tangent Plane at a Point

Let
$$\frac{x-\alpha}{l} = \frac{y-\beta}{m} = \frac{z-\gamma}{n} \qquad \ldots(i)$$

be a line through the point (α, β, γ) of the surface
$$ax^2 + by^2 + cz^2 = 1 \qquad \ldots(ii)$$

Thus, we have
$$a\alpha^2 + b\beta^2 + c\gamma^2 = 1 \qquad \ldots(iii)$$

One root of the equation (A) is § 6.3 is, therefore, zero.

The line (i) will touch the conicoid (ii) at the point (α, β, γ) if both the values of r given by the equation (A) in § 6.3 are zero.

The second value will also be zero, if
$$al\alpha + bm\beta + cn\gamma = 0 \qquad \ldots(iv)$$

which is thus the condition for the line (i) to be a tangent line to the surface (ii) at (α, β, γ).

The locus of the tangent line to the surface, at (α, β, γ) obtained by eliminating l, m, n between (i) and (ii), is
$$a\alpha(x-\alpha) + b\beta(y-\beta) + c\gamma(z-\gamma) = 0$$
$$\Rightarrow \quad a\alpha x + b\beta y + c\gamma z = a\alpha^2 + b\beta^2 + c\gamma^2 = 1$$

which is a plane.

Hence, the tangent lines to the surface (ii) at the point (α, β, γ) lie in the plane
$$a\alpha x + b\beta y + c\gamma z = 1$$

which is, therefore, the *tangent plane* at (α, β, γ) to the conicoid
$$ax^2 + by^2 + cz^2 = 1$$

Note. A tangent line at any point is a line which meets the surface in two coincident points and the tangent plane at a point is the locus of tangent lines at the point.

6.3.2. Condition of Tangency

To find the condition that the plane
$$lx + my + nz = p, \qquad \ldots(i)$$

should touch the central conicoid
$$ax^2 + by^2 + cz^2 = 1 \qquad \ldots(ii)$$

If (α, β, γ) be the point of contact, the tangent plane
$$a\alpha x + b\beta y + c\gamma z = 1 \qquad \ldots(iii)$$

there at should be the same as the plane (i).

Comparing the two equations (i) and (iii), we get
$$\alpha = \frac{l}{ap}, \quad \beta = \frac{m}{bp}, \quad \gamma = \frac{n}{cp}$$

and since
$$a\alpha^2 + b\beta^2 + c\gamma^2 = 1$$

we obtain the required condition
$$\frac{l^2}{a} + \frac{m^2}{b} + \frac{n^2}{c} = p^2.$$

The Conicoid

Also the point of contact then, is,
$$\left(\frac{l}{ap}, \frac{m}{bp}, \frac{n}{cp}\right)$$

Thus, we deduce that the planes
$$lx + my + nz = \pm\sqrt{l^2/a + m^2/b + n^2/c}$$
touch the conicoid (ii) for all values of l, m, n.

Cor. There are two tangent planes to a central conicoid parallel to a given plane.

6.3.3. Director Sphere

To show that the locus of the point of intersection of three mutually perpendicular tangent planes to a central conicoid is a sphere concentric with the conicoid.

Let
$$l_1 x + m_1 y + n_1 z = \left(\frac{l_1^2}{a} + \frac{m_1^2}{b} + \frac{n_1^2}{c}\right)^{1/2} \qquad \ldots(i)$$

$$l_2 x + m_2 y + n_2 z = \left(\frac{l_2^2}{a} + \frac{m_2^2}{b} + \frac{n_2^2}{c}\right)^{1/2} \qquad \ldots(ii)$$

$$l_3 x + m_3 y + n_3 z = \left(\frac{l_3^2}{a} + \frac{m_3^2}{b} + \frac{n_3^2}{c}\right)^{1/2} \qquad \ldots(iii)$$

be three mutually perpendicular tangent planes so that
$$\Sigma l_1 m_1 = \Sigma m_1 n_1 = \Sigma n_1 l_1 = 0$$
$$\Sigma l_1^2 = \Sigma m_1^2 = \Sigma n_1^2 = 1 \qquad \ldots(iv)$$

The co-ordinates of the point of intersection satisfy the three equations and its locus is, therefore, obtained by the elimination of $l_1, m_1, n_1; l_2, m_2, n_2; l_3, m_3, n_3$.

This is easily done by squaring and adding the three equations and using the relations (iv), so that we obtain
$$x^2 + y^2 + z^2 = 1/a + 1/b + 1/c$$
as the required locus which is a concentric sphere called the *Director sphere* of the given quadric. Its centre is the same as that of the central conicoid.

EXAMPLES

1. *Obtain the tangent planes to the ellipsoid*
$$\frac{x^2}{a^2} + \frac{y^2}{b^2} + \frac{z^2}{c^2} = 1,$$
which are parallel to the plane
$$lx + my + nz = 0.$$
If $2r$ is the distance between two parallel tangent planes to the ellipsoid, prove that the line through the origin and perpendicular to the planes lies on the cone
$$x^2(a^2 - r^2) + y^2(b^2 - r^2) + z^2(c^2 - r^2) = 0$$

Sol. The two tangent planes parallel to the plane $\Sigma lx = 0$, are
$$\Sigma lx = \pm\sqrt{\Sigma a^2 l^2} \qquad \ldots(1)$$

The distance between these parallel planes which is twice the distance of either from the origin is
$$2\sqrt{\Sigma a^2 l^2} / \sqrt{\Sigma l^2}$$

Thus, we have

$$\frac{2\sqrt{\Sigma a^2 l^2}}{\sqrt{\Sigma l^2}} = 2r \implies \Sigma(a^2 - r^2)l^2 = 0$$

Hence, the locus of the line

$$x/l = y/m = z/n$$

which is perpendicular to the plane (1), is

$$\Sigma(a^2 - r^2)x^2 = 0.$$

2. *A tangent plane to the conicoid $ax^2 + by^2 + cz^2 = 1$, meets the co-ordinate axes in P, Q and R. Find the locus of the centroid of the triangle PQR.*

Sol. Any tangent plane to the given conicoid is

$$lx + my + nz = \sqrt{\frac{l^2}{a} + \frac{m^2}{b} + \frac{n^2}{c}} \qquad \text{...(i)}$$

Hence,

$$P \equiv \left[\frac{1}{l}\sqrt{\frac{l^2}{a} + \frac{m^2}{b} + \frac{n^2}{c}}, 0, 0\right]$$

$$Q \equiv \left[0, \frac{1}{m}\sqrt{\frac{l^2}{a} + \frac{m^2}{b} + \frac{n^2}{c}}, 0\right]$$

$$R \equiv \left[0, 0, \frac{1}{n}\sqrt{\frac{l^2}{a} + \frac{m^2}{b} + \frac{n^2}{c}}\right]$$

If (x_1, y_1, z_1) be the centroid of ΔPQR, then

$$x_1 = \frac{1}{3l}\sqrt{\frac{l^2}{a} + \frac{m^2}{b} + \frac{n^2}{c}}, \quad y_1 = \frac{1}{3m}\sqrt{\frac{l^2}{a} + \frac{m^2}{b} + \frac{n^2}{c}}$$

and

$$z_1 = \frac{1}{3n}\sqrt{\frac{l^2}{a} + \frac{m^2}{b} + \frac{n^2}{c}}$$

$$(3lx_1)^2 = \frac{l^2}{a^2} + \frac{m^2}{b^2} + \frac{n^2}{c^2} = (3my_1)^2 = (3nz_1)^2$$

or

$$\frac{9l^2}{a} + \frac{9m^2}{b} + \frac{9n^2}{c} = \left(\frac{l^2}{a} + \frac{m^2}{b} + \frac{n^2}{c}\right)\left(\frac{1}{ax_1^2} + \frac{1}{by_1^2} + \frac{1}{cz_1^2}\right)$$

$$\implies \frac{1}{ax_1^2} + \frac{1}{by_1^2} + \frac{1}{cz_1^2} = 9$$

Hence, required locus is

$$\frac{1}{ax^2} + \frac{1}{by^2} + \frac{1}{cz^2} = 9.$$

EXERCISES

1. Show that the tangent planes at the extremities of any diameter of a central conicoid are parallel.

2. Show that the plane $3x + 12y - 6z - 17 = 0$ touches the conicoid $3x^2 - 6y^2 + 9z^2 + 17 = 0$, and find the point of contact. **[Ans. (– 1, 2, 2/3)]**

The Conicoid

3. Find the equation of the tangent planes to the curve $x^2 - 2y^2 + 3z^2 = 2$ and parallel to the plane $x - 2y + 3z = 0$. [Ans. $x - 2y + 3z = \pm 2$]

4. Find the equations to the tangent planes to the surface $4x^2 - 5y^2 + 7z^2 + 13 = 0$, parallel to the plane $4x + 20y - 21z = 0$. Find their points of contact also.
[Ans. $4x + 20y - 21z \neq 13 = 0; (\pm 1, \neq 4, \neq 3)$]

5. Find the equations to the two planes which contain the line given by $7x + 10y - 30 = 0$, and touch the ellipsoid $7x^2 + 5y^2 + 3z^2 = 60$.
[Ans. $7x + 5y + 3z - 30 = 0, 14x + 5y + 9z - 60 = 0$]

6. Find the equation to the tangent planes to $2x^2 - 6y^2 + 3z^2 = 5$ which pass through the line $x + 9y - 3z = 0 = 3x - 3y + 6z - 5$. [Ans. $2x - 12y + 9z = 5, 4x + 6y + 3z = 5$]

7. P, Q are any two points on a central conicoid. Show that the plane through the centre and the line of intersection of the tangent planes at P, Q will bisect PQ. Also show that if the planes through the centre parallel to the tangent planes at P, Q cut the chord in P', Q', then $PP' = QQ'$.

8. Prove that the locus of the foot of the central perpendicular on varying tangent planes of the ellipsoid, $\dfrac{x^2}{a^2} + \dfrac{y^2}{b^2} + \dfrac{z^2}{c^2} = 1$, is the surface
$$(x^2 + y^2 + z^2)^2 = a^2 x^2 + b^2 y^2 + c^2 z^2.$$

6.3.4. Normal

Def. *The normal at any point of a quadric is the line through the point perpendicular to the tangent plane there at.*

The equation of the tangent plane at (α, β, γ) to the surface
$$ax^2 + by^2 + cz^2 = 1 \qquad \ldots(i)$$
is
$$a\alpha x + b\beta y + c\gamma z = 1 \qquad \ldots(ii)$$

The equations to the normal at (α, β, γ), therefore, are
$$\frac{x - \alpha}{a\alpha} = \frac{y - \beta}{b\beta} = \frac{z - \gamma}{c\gamma} \qquad \ldots(ii)$$

so that $a\alpha, b\beta, c\gamma$ are the direction ratios of the normal.

If p, is the length of the perpendicular from the origin to the tangent plane (ii), we have
$$\frac{1}{a^2\alpha^2 + b^2\beta^2 + c^2\gamma^2} = p^2 \Leftrightarrow (a\alpha p)^2 + (b\beta p)^2 + (c\gamma p)^2 = 1.$$

It follows that $a\alpha p, b\beta p, c\gamma p$ are the actual direction consines of the normal at (α, β, γ).

6.3.5. Number of Normals From a Given Point

We shall now show that *through any given point six normals can be drawn to a central conicoid.*

If the normal (iii) at a point (α, β, γ) passes through a *given point* (f, g, h), we have
$$\frac{f - \alpha}{a\alpha} = \frac{g - \beta}{b\beta} = \frac{h - \gamma}{c\gamma} = r, \text{ (say)}$$

$\Leftrightarrow \quad \alpha = \dfrac{f}{1 + ar}, \beta = \dfrac{g}{1 + br}, \gamma = \dfrac{h}{1 + cr} \qquad \ldots(iv)$

Since (α, β, γ) lies on the conicoid (i), we have the relation

$$\frac{af^2}{(1+ar)^2} + \frac{bg^2}{(1+br)^2} + \frac{ch^2}{(1+cr)^2} = 1, \qquad \text{...(v)}$$

which, being an equation of the *sixth* degree, gives six values of r, to each of which there corresponds a point (α, β, γ), as obtained from (iv).

Therefore, there are *six* points on a central quadric the normal at which pass through a given point, *i.e.*, *through a given point, six normals, in general, can be drawn to a central quadric.*

6.3.6 Cubic Curve Through the Feet of Normals

The feet of the six normals from a given point to a central quadric are the intersections of the quadric with a certain cubic curve.

Consider the curve whose parametric equations are

$$x = \frac{f}{1+ar}, \quad y = \frac{g}{1+br}, \quad z = \frac{h}{1+cr} \qquad \text{...(vi)}$$

r being the *parameter*.

The points (x, y, z) on this curve, arising from those of the values of r which are the roots of the equation (v), are the six feet of the normals from the point (f, g, h).

Again, the points of intersection of this curve with any plane

$$Ax + By + Cz + D = 0$$

are given by

$$\frac{Af}{1+ar} + \frac{Bg}{1+br} + \frac{Ch}{1+cr} + D = 0$$

which determines three values of r. Hence, the curve (vi) cuts *any* plane in three points and is, as such, a *cubic* curve.

Therefore, the six feet of the normals from (f, g, h) are the intersections of the conicoid and the cubic curve (vi).

6.3.7. Quadric Cone Through Six Concurrent Normals

The six normals drawn from any point to a central quadric are the generators of a quadric cone.

We first prove that the lines drawn from (f, g, h) to intersect the cubic curve (vi) generate a quadric cone.

If *any* line

$$\frac{x-f}{l} = \frac{y-g}{m} = \frac{z-h}{n} \qquad \text{...(vii)}$$

through (f, g, h) intersects the cubic curve, we have

$$\frac{\frac{f}{1+ar} - f}{l} = \frac{\frac{g}{1+br} - g}{m} = \frac{\frac{h}{1+cr} - h}{n}$$

$$\Rightarrow \quad \frac{af/l}{1+ar} = \frac{bg/m}{1+br} = \frac{ch/m}{1+cr}$$

whence eliminating r, we get

$$\frac{af}{l}(b-c) + \frac{bg}{m}(c-a) + \frac{ch}{n}(a-b) = 0$$

which is the condition for the line (vii) to intersect the cubic curve (vi).

Eliminating l, m, n between the equations of the line and this condition, we get

The Conicoid

$$\frac{af(b-c)}{x-f} + \frac{bg(c-a)}{y-g} + \frac{ch(a-b)}{z-h} = 0$$

which represents a cone of the second degree generated by lines drawn from (f, g, h) to intersect the cubic curve.

As the six feet of the normals drawn from a point (f, g, h) to the quadric lie on the cubic curve, the normal are, in particular, the generators of this cone of the second degree.

Note. The importance of this result lies in the fact that while five given concurrent lines determine a unique quadric cone, the six normals through a point lie on a quadric cone, *i.e., the quadric cone through any of the five normals through a point also contains the six normals through the point.*

6.3.8. The General Equation of the Conicoid Through the Six Feet of the Normals

The co-ordinates (α, β, γ) of the foot of any of the six normals from (f, g, h) satisfy the relations

$$\frac{\alpha - f}{a\alpha} = \frac{\beta - g}{b\beta} = \frac{\gamma - h}{c\gamma}$$

so that we see that the feet of the normals lie on three cylinders

$$ax(y-g) = by(x-f) \Leftrightarrow (a-b)xy - agx + bfy = 0$$
$$by(z-h) = cz(y-g) \Leftrightarrow (b-c)yz - bhy + cgz = 0$$
$$cz(x-f) = ax(z-h) \Leftrightarrow (c-a)zx - cfz + ahx = 0$$

The six feet of the normals are the common points of the three cylinders and the conicoid

$$ax^2 + by^2 + cz^2 = 1$$

The equation

$$ax^2 + by^2 + cz^2 - 1 + k_1[xy(a-b) - agx + bfy] + k_2[yz(b-c) - bhy + cgz]$$
$$+ k_3[zx(c-a) - cfz + ahx] = 0$$

is satisfied by the *six* feet of the normals and contains *three* arbitrary parameters k_1, k_2, k_3. Therefore, it represents the general equation of the conicoid through them.

EXAMPLES

1. *The normal at any point P of a central conicoid meets the three principal planes at G_1, G_2, G_3 show that PG_1, PG_2, PG_3 are in a constant ratio.*

Sol. The equations of the normal at (α, β, γ) are

$$\frac{x - \alpha}{a\alpha p} = \frac{y - \beta}{b\beta p} = \frac{z - \gamma}{c\gamma p}.$$

Now since $a\alpha p, b\beta p, c\gamma p$ are the direction cosines, each of these fractions represents the distance between the points (α, β, γ) and (x, y, z).

Thus, the distance PG_1, of the point $P(\alpha, \beta, \gamma)$ from the point G_1 where the normal meets the co-ordinate plane $x = 0$ is $-1/cp$.

Similarly $\qquad PG_2 = -1/bp, \ PG_3 = -1/cp.$

Thus, we have $\qquad PG_1 : PG_2 : PG_3 :: a^{-1} : b^{-1} : c^{-1}.$

2. *Show that the lines drawn from the origin parallel to the normals to the central conicoid*

$$ax^2 + by^2 + cz^2 = 1,$$

at its points of intersection with the planes

$$lx + my + nz = p,$$

generate the cone

$$p^2\left(\frac{x^2}{a}+\frac{y^2}{b}+\frac{z^2}{c}\right)=\left(\frac{lx}{a}+\frac{my}{b}+\frac{nz}{c}\right)^2.$$

Sol. Let (f, g, h) be any point, one the curve of intersection of

$$ax^2 + by^2 + cz^2 = 1 \quad \text{and} \quad lx + my + nz = p \qquad \ldots(1)$$

The normal to the quadric at (f, g, h) is

$$\frac{x-f}{af} = \frac{y-g}{bg} = \frac{z-h}{ch} \qquad \ldots(2)$$

The line through the origin parallel to this normal is

$$\frac{x}{af} = \frac{y}{bg} = \frac{z}{ch}.$$

Also (f, g, h) satisfies the two equations (1), so that we have

$$af^2 + bg^2 + ch^2 = 1, \ lf + mg + nh = p \qquad \ldots(3)$$

The required locus is obtained by eliminating f, g, h between (2) and (3). The equations (3) give

$$af^2 + bg^2 + ch^2 = \left(\frac{lf + mg + nh}{p}\right)^2 \qquad \ldots(4)$$

which is a second degree homogeneous expression in f, g, h. From (2) and (4), we can easily obtain the required locus.

3. *If P, Q, R and P', Q', R' are the feet of the six normals from a point to the ellipsoid $\frac{x^2}{a^2}+\frac{y^2}{b^2}+\frac{z^2}{c^2}=1$, and the plane PQR is given by $lx + my + nz = p$, then the plane $P'Q'R'$ is given by*

$$\frac{x}{a^2 l}+\frac{y}{b^2 m}+\frac{z}{c^2 n}+\frac{1}{p}=0.$$

Sol. Let plate $P'Q'R'$ be

$$l'x + m'y + n'z = p'.$$

Then the feet of the six normals from any given point (x', y', z') lie on the locus given by the equation

$$(lx + my + nz - p)(l'x + m'y + n'z - p') = 0 \qquad \ldots(i)$$

Let (α, β, γ) be a foot of the normal from (x', y', z').

$$\therefore \quad (l\alpha + m\beta + n\gamma - p)(l'\alpha + m'\beta + n'\gamma - p') = 0 \qquad \ldots(ii)$$

Also, the normal at (α, β, γ) passes through (x', y', z').

$$\therefore \quad \frac{x'-\alpha}{\alpha/a^2} = \frac{y'-\beta}{\beta/b^2} = \frac{z'-\gamma}{\gamma/c^2} = \lambda,$$

$$\therefore \quad \alpha = \frac{a^2 x'}{a^2 + \lambda}, \ \beta = \frac{b^2 y'}{b^2 + \lambda}, \ \gamma = \frac{c^2 z'}{c^2 + \lambda}.$$

But (α, β, γ) lies on the given ellipsoid

$$\therefore \quad \frac{a^2 x'^2}{(a^2+\lambda)^2}+\frac{b^2 y'^2}{(b^2+\lambda)^2}+\frac{c^2 z'^2}{(c^2+\lambda)^2}=1 \qquad \ldots(iii)$$

The Conicoid

This equation being a sixth degree equation in λ, gives six values of λ corresponding to the six feet of the normals.

Also putting the values of α, β, γ in (ii), we get

$$\left(\frac{a^2 x' l}{a^2 + \lambda} + \frac{b^2 y' m}{b^2 + \lambda} + \frac{c^2 z' n}{c^2 + \lambda} - p\right)\left(\frac{a^2 x' l'}{a^2 + \lambda} + \frac{b^2 y' m'}{b^2 + \lambda} + \frac{c^2 z' n'}{c^2 + \lambda} - p'\right) = 0 \quad \ldots\text{(iv)}$$

This is also sixth degree equation in λ, gives six values of λ corresponding to the six feet of the normals. Hence, equation (iii) and (iv) are identical.

Comparing coefficients of like terms in (iii) and (iv), we get

$$\frac{a^4 l l'}{a^2} = \frac{b^4 m m'}{b^2} = \frac{c^4 n n'}{c^2} = \frac{-pp'}{1}$$

or
$$l' = \frac{-pp'}{a^2 l}, \quad m' = \frac{-pp'}{b^2 m}, \quad n' = \frac{-pp'}{c^2 n}$$

Substituting these values, the equation of plane $P'Q'R'$ may be found.

EXERCISES

1. If a point G be taken on the normal at any point P of the ellipsoid
$$\frac{x^2}{a^2} + \frac{y^2}{b^2} + \frac{z^2}{c^2} = 1,$$
such that $3PG = PG_1 + PG_2 + PG_3$,
show that the locus of G is
$$\frac{a^2 x^2}{(2a^2 - b^2 - c^2)^2} + \frac{b^2 y^2}{(2b^2 - c^2 - a^2)^2} + \frac{c^2 z^2}{(2c^2 - a^2 - b^2)^2} = \frac{1}{9}.$$

2. If a length PQ be taken on the normal at any point P of the ellipsoid
$$\frac{x^2}{a^2} + \frac{y^2}{b^2} + \frac{z^2}{c^2} = 1$$
such that $PQ = k^2 / p$, where k is a constant and p is the length of the perpendicular from the origin to the tangent planes at p, the locus of Q is
$$\frac{a^2 x^2}{(a^2 + k^2)^2} + \frac{b^2 y^2}{(b^2 + k^2)^2} + \frac{c^2 z^2}{(c^2 + k^2)^2} = 1.$$

3. Show that, in general, the normals to the ellipsoid $\frac{x^2}{a^2} + \frac{y^2}{b^2} + \frac{z^2}{c^2} = 1$ lie in a given plane. Determine the co-ordinates of the two points on the ellipsoid the normal at which lie in the plane
$$by - cz = \frac{1}{2}(b^2 - c^2). \qquad \left[\text{Ans.}\left(\pm\sqrt{\frac{1}{2}}\,a, \frac{1}{2}b, \frac{1}{2}c\right)\right]$$

4. If the feet of the three normals from P to the ellipsoid $\frac{x^2}{a^2} + \frac{y^2}{b^2} + \frac{z^2}{c^2} = 1$ lie on the plane
$$\frac{x}{a} + \frac{y}{b} + \frac{z}{c} = 1,$$
prove that the feet of the other three lie on the plane
$$\frac{x}{a} + \frac{y}{b} + \frac{z}{c} + 1 = 0$$
and P lies on the line
$$a(b^2 - c^2)x = b(c^2 - a^2)y = c(a^2 - c^2)z.$$

5. Prove that through any point (α, β, γ) six normals can be drawn to the ellipsoid $\dfrac{x^2}{a^2} + \dfrac{y^2}{b^2} + \dfrac{z^2}{c^2} = 1$ and that the feet of the normals lie on the curve of intersection of the ellipsoid and the cone
$$\frac{a^2(b^2-c^2)\alpha}{x} + \frac{b^2(c^2-a^2)\beta}{y} + \frac{c^2(a^2-b^2)\gamma}{z} = 0.$$

6. Show that the locus of points on a centre quadric, the normals at which intersect a given diameter is the curve of intersection with a cone having the principal axes of the quadric as generators.

6.4 PLANE OF CONTACT

The tangent plane
$$axx' + byy' + czz' = 1,$$
at the point (x', y', z') to the quadric $ax^2 + by^2 + cz^2 = 1$, passes through the point (α, β, γ), if
$$a\alpha x' + b\beta y' + c\gamma z' = 1.$$

This shows that the points on the quadric the tangent planes at which pass through the point (α, β, γ) lie on the plane
$$a\alpha x + b\beta y + c\gamma z = 1$$
which is called the *Plane of contact* for the point (α, β, γ).

6.5. THE POLAR PLANE OF A POINT

If a secant APQ through a given point A meets a conicoid in points P and Q and a point R be taken on this line such that points A and R divide the segment PQ internally and externally in the same ratio, then the locus of the point R is a plane called the polar plane of A.

It may be easily seen that if the points A and R divide the segment PQ internally and externally in the same ratio, then the points P, Q divide the segment AR also internally and externally in the same ratio.

Let A, be a point (α, β, γ) and let (x, y, z) be the co-ordinates of R.

The co-ordinates of the point which divides AR in the ratio $\lambda : 1$ are
$$\left(\frac{\lambda x + \alpha}{\lambda + 1}, \frac{\lambda y + \beta}{\lambda + 1}, \frac{\lambda z + \gamma}{\lambda + 1} \right)$$

This point will lie on the conicoid
$$ax^2 + by^2 + cz^2 = 1$$
for those of the values of λ which are the roots of the equation
$$a\left(\frac{\lambda x + \alpha}{\lambda + 1}\right)^2 + b\left(\frac{\lambda x + \beta}{\lambda + 1}\right)^2 + c\left(\frac{\lambda x + \gamma}{\lambda + 1}\right)^2 = 1$$
$$\lambda^2(ax^2 + by^2 + cz^2 - 1) + 2\lambda(a\alpha x + b\beta y + c\gamma z - 1) + (a\alpha^2 + b\beta^2 + c\gamma^2 - 1) = 0 \quad \ldots(1)$$

The two roots λ_1, λ_2 of this equation are the ratios in which the points P, Q divide the segment AR. Since P, Q divide the segment AR internally and externally in the same ratio, we have
$$\lambda_1 + \lambda_2 = 0$$
so that, from (1)
$$a\alpha x + b\beta y + c\gamma z - 1 = 0 \quad \ldots(2)$$

Now the equation (2) of the first degree being a relation between the co-ordinates (x, y, z) of the point gives a plane as the locus of the point R.

Thus, *the polar plane of the point (α, β, γ) with respect to the conicoid*
$$ax^2 + by^2 + cz^2 = 1$$

The Conicoid

is the plane
$$a\alpha x + b\beta y + c\gamma z = 0.$$

Any point is called the pole of its polar plane.

Note. The reader acquainted with cross-ratios and, in particular, harmonic cross-ratios, would know that the fact that the points P, Q divide AR internally and externally in the same ratio is also expressed by the statement
$$(AR, PQ) = -1$$
This is further equivalent to the relation
$$\frac{2}{AR} = \frac{1}{AP} + \frac{1}{AQ}.$$

Cor. The polar plane of a point on a conicoid coincides with the tangent plane there at and that of a point outside it coincides with the plane of contact for that point.

Ex. 1. Show that the point of intersection of the tangent planes at three points on a quadric is the pole of the plane formed by their points of contact.

Ex. 2. Find the pole of the plane $lx + my + nz = p$ with respect to the quadric
$$ax^2 + by^2 + cz^2 = 1.$$
[**Ans.** $l/ap,\ m/bp,\ n/cp$]

6.5.1. Conjugate Points and Conjugate Planes

It is easy to show that if the polar plane of a point P passes through point Q, then the polar plane of Q passes through P.

Two such points are called *Conjugate Points*.

Also it can be shown that if the pole of a plane α lies on plane β, then the pole of the plane β lies on the plane α.

Two such planes are called *Conjugate Planes*.

6.5.2. Polar Lines

Consider a line
$$\frac{x - \alpha}{l} = \frac{y - \beta}{m} = \frac{z - \gamma}{n}.$$

The polar plane of any point $(lr + \alpha, mr + \beta, nr + \gamma)$ on this line is
$$a(lr + \alpha)x + b(mr + \beta)y + c(nr - \gamma)z = 1$$
$$\Rightarrow \qquad a\alpha x + b\beta y + c\gamma z - 1 + r(alx\ bmy + cnz) = 0$$
which clearly passes through the line of intersection of the planes
$$a\alpha x + b\beta y + c\gamma z - 1 = 0,\ alx\ bmy + cnz = 0$$
for all values of r.

Thus, *the polar planes of all the points on a line l pass through another line l'*.

Now, since the polar planes of an arbitrary point P on a line l pass through every point of l', therefore, the polar planes of any point of l' will pass through the point P on l and as P is arbitrary, it passes through every point on l, i.e., passes through l.

It follows that *if the polar plane of any point on a line l passes through the line l', then the polar plane of any point on l' passes through l.*

Two such lines are said to be *Polar Lines* with respect to the conicoid.

To find the polar line of any given line, we have only to find the line of intersection of the polar planes of any two points on it.

6.5.3. Conjugate Lines

Let l, m be any two lines and l', m' their polar lines. We suppose that the line m *intersects* the line l.

We shall now show that the line l' also intersects the line m.

Let P be the point where the lines m' and l intersect.

As P lies on m' and also on l, its polar plane contains the polar lines m and l' of m' and l respectively, *i.e.*, the lines m and l' are coplanar and hence they intersect.

It follows that *if a line l intersects the polar of a line m, then the line m intersects the polar of the line l.*

Two such lines l and m are called *Conjugate Lines*.

EXAMPLE

Find the locus of straight lines drawn through a fixed point (α, β, γ) at right angles to their polars with respect to the central conicoid $ax^2 + by^2 + cz^2 = 1$.

Sol. Let
$$\frac{x-\alpha}{l} = \frac{y-\beta}{m} = \frac{z-\gamma}{n} \qquad \ldots(1)$$
be a line perpendicular to its polar line. Now the polar line of (1) is the intersection of the planes
$$a\alpha x + b\beta y + c\gamma z = 1, \quad alx + bmy + cnz = 0$$
If λ, μ, ν be the direction ratios of this line, we have
$$a\alpha\lambda + b\beta\mu + c\gamma\nu = 0, \quad al\lambda + bm\mu + cn\nu = 0$$
$$\Rightarrow \quad \frac{a\lambda}{n\beta - m\gamma} = \frac{b\mu}{l\gamma - n\alpha} = \frac{c\nu}{m\alpha - \beta l}$$

The perpendicularity of the line (1) to its polar lines implies
$$l\lambda + m\mu + n\nu = 0$$
$$\therefore \quad \frac{l(n\beta - m\gamma)}{a} + \frac{m(l\gamma - n\alpha)}{b} + \frac{n(m\alpha - l\beta)}{c} = 0$$
$$\Rightarrow \quad \alpha mn\left(\frac{1}{b} - \frac{1}{c}\right) + \beta nl\left(\frac{1}{c} - \frac{1}{a}\right) + \gamma lm\left(\frac{1}{a} - \frac{1}{b}\right) = 0$$
$$\Rightarrow \quad \frac{\alpha}{l}\left(\frac{1}{b} - \frac{1}{c}\right) + \frac{\beta}{m}\left(\frac{1}{c} - \frac{1}{a}\right) + \frac{\gamma}{n}\left(\frac{1}{a} - \frac{1}{b}\right) = 0 \qquad \ldots(2)$$

Eliminating l, m, n between (1) and (2), we see that the required locus is
$$\frac{\alpha}{x-\alpha}\left(\frac{1}{b} - \frac{1}{c}\right) + \frac{\beta}{y-\beta}\left(\frac{1}{c} - \frac{1}{a}\right) + \frac{\gamma}{z-\gamma}\left(\frac{1}{a} - \frac{1}{b}\right) = 0.$$

EXERCISES

1. Prove that the locus of the poles of the tangent planes of the conicoid $ax^2 + by^2 + cz^2 = 1$ with respect to the conicoid $\alpha x^2 + \beta y^2 + \gamma z^2 = 1$ is the conicoid
$$\frac{\alpha^2 x^2}{a} + \frac{\beta^2 y^2}{b} + \frac{\gamma^2 z^2}{c} = 1.$$

2. Show that the locus of the poles of the plane
$$lx + my + nz = p,$$
with respect to the system of conicoids
$$\frac{x^2}{a^2+\lambda} + \frac{y^2}{b^2+\lambda} + \frac{z^2}{c^2+\lambda} = 1,$$
where λ is the parameter, is a straight line perpendicular to the given plane.

3. Find the locus of straight line drawn through a fixed point (f, g, h) such that its polar lines with respect to the quadrics

The Conicoid

$$ax^2 + by^2 + cz^2 = 1 \text{ and } \alpha x^2 + \beta y^2 + \gamma z^2 = 1$$

as coplanar.

$$\left[\text{Ans. } \sum \frac{(\alpha - a)(b\gamma - c\beta)f}{x - f} = 0\right]$$

4. Find the conditions that the lines
$$\frac{x - \alpha}{l} = \frac{y - \beta}{m} = \frac{z - \gamma}{n}, \quad \frac{x - \alpha'}{l'} = \frac{y - \beta'}{m'} = \frac{z - \gamma'}{n'}$$
should be (i) polar, (ii) conjugate with respect to the conicoid
$$ax^2 + by^2 + cz^2 = 1.$$

[**Ans.** (i) $\Sigma a\alpha\alpha' = 1, \Sigma a\alpha'l = 0, \Sigma a\alpha l' = 0, \Sigma all' = 0$
(ii) $(\Sigma a\alpha l')(\Sigma a\alpha' l) = (\Sigma all')(\Sigma a\alpha\alpha' - 1)]$

6.6.1. The Enveloping Cone

Def. *The locus of tangent lines to a quadric through any points is called an enveloping cone.*
To find the enveloping cone of the conicoid
$$ax^2 + by^2 + cz^2 = 1,$$
with its vertex at (α, β, γ).

Any line
$$\frac{x - \alpha}{l} = \frac{y - \beta}{m} = \frac{z - \gamma}{n} \qquad \ldots(i)$$

through the point (α, β, γ) will meet the surface in two coincident points if the equation (A) of § 6.3 has equal roots, *i.e.*, if

$$(al\alpha + bm\beta + cn\gamma)^2 = (al^2 + bm^2 + cn^2)(a\alpha^2 + b\beta^2 + c\gamma^2 - 1) \qquad \ldots(ii)$$

Eliminating l, m, n between (i) and (ii), we obtain

$$[a\alpha(x - \alpha) + b\beta(y - \beta) + c\gamma(z - \gamma)]^2$$
$$= [a(x - \alpha)^2 + b(y - \beta)^2 + c(z - \gamma)^2](a\alpha^2 + b\beta^2 + c\gamma^2 - 1)$$

which is the required equation of the **enveloping cone**.

If we write
$$S \equiv ax^2 + by^2 + cz^2 - 1, \quad S_1 \equiv a\alpha^2 + b\beta^2 + c\gamma^2 - 1, \quad T_1 \equiv a\alpha x + b\beta y + c\gamma z - 1$$
we see that the equation of the enveloping cone can briefly be written as
$$(T_1 - S_1)^2 = (S - 2T_1 + S_1)S_1 \Leftrightarrow SS_1 = T_1^2$$
$$\Leftrightarrow (ax^2 + by^2 + cz^2 - 1)(a\alpha^2 + b\beta^2 + c\gamma^2 - 1) = (a\alpha x + b\beta y + c\gamma z - 1)^2.$$

Note. Obviously the enveloping cone passes through the points common to the conicoid and the polar plane $a\alpha x + b\beta y + c\gamma z = 1$ of the vertex (α, β, γ).

Thus, the enveloping cone may be regarded as a cone whose vertex is the given point and guiding curve the section of the conicoid by its polar plane.

EXERCISES

1. A point P moves so that the section of the enveloping cone of the ellipsoid
$$\frac{x^2}{a^2} + \frac{y^2}{b^2} + \frac{z^2}{c^2} = 1$$
with P as vertex by the plane $z = 0$ is a circle, show that P lies on one of the conics
$$\frac{y^2}{b^2 - a^2} + \frac{z^2}{c^2} = 1, \, x = 0; \quad \frac{x^2}{a^2 - b^2} + \frac{z^2}{c^2} = 1, \, y = 0.$$

2. If the section of the enveloping cone of the ellipsoid
$$\frac{x^2}{a^2} + \frac{y^2}{b^2} + \frac{z^2}{c^2} = 1,$$
whose vertex is P by the plane $z = 0$ is a rectangular hyperbola, show that the locus of P is
$$\frac{x^2 + y^2}{a^2 + b^2} + \frac{z^2}{c^2} = 1.$$

3. Find the locus of the points from which three mutually perpendicular tangent lines can be drawn to the conicoid $ax^2 + by^2 + cz^2 = 1$.

[**Ans.** $a(b+c)x^2 + b(c+a)y^2 + c(a+b)z^2 = a+b+c$]

4. A pair of perpendicular tangent planes to the ellipsoid
$$\frac{x^2}{a^2} + \frac{y^2}{b^2} + \frac{z^2}{c^2} = 1$$
passes through the fixed point $(0, 0, k)$. Show that their line of intersection lies on the cone
$$x^2(b^2 + c^2 - k^2) + y^2(c^2 + a^2 - k^2) + (z - k)^2(a^2 + b^2) = 0.$$

[**Hint** : The required locus is the locus of the line of intersection of perpendicular tangent planes to the enveloping cone of the given ellipsoid with vertex at $(0, 0, k)$.]

6.6.2. Enveloping Cylinder

Def. *The locus of tangent lines to a quadric parallel to any given line is called an Enveloping cylinder of the quadric.*

To find the enveloping cylinder of the conicoid
$$ax^2 + by^2 + cz^2 = 1$$
with its generators parallel to the line
$$\frac{x}{l} = \frac{y}{m} = \frac{z}{n}.$$

Let (α, β, γ) be a point on the *enveloping* cylinder, so that the equations of the generator through it are
$$\frac{x - \alpha}{l} = \frac{y - \beta}{m} = \frac{z - \gamma}{n} \qquad \ldots(i)$$

As in § 6.6.1, the line (i) will touch the conicoid, if,
$$(al\alpha + bm\beta + cn\gamma)^2 = (al^2 + bm^2 + cn^2)(a\alpha^2 + b\beta^2 + c\gamma^2 - 1)$$

Thus, the locus of (α, β, γ) is the surface
$$(ax^2 + by^2 + cz^2 - 1)(al^2 + bm^2 + cn^2) = (alx + bmy + cnz)^2$$
which is the required equation of the **Enveloping cylinder.**

Note. *Equation of enveloping cylinder deduced from that of enveloping cone. Use of elements at infinity.* Since each of the lines parallel to the line
$$\frac{x}{l} = \frac{y}{m} = \frac{z}{n}$$
passes through the point $(l, m, n, 0)$ which is, in fact, the point at infinity on each member of this system of parallel lines, we see that the enveloping cylinder is the enveloping cone with vertex $(l, m, n, 0)$.

The homogeneous equation of the surface being
$$ax^2 + by^2 + cz^2 - t^2 = 0$$

The Conicoid

the equation of the enveloping cylinder is

$$(ax^2 + by^2 + cz^2 - t^2)(al^2 + bm^2 + cn^2 - 0) = (alx + bmy + cnz - t.0)^2; (SS_1 = T^2)$$

so that in terms of ordinary cartesian co-ordinates, this equation is

$$(ax^2 + by^2 + cz^2 - 1)(al^2 + bm^2 + cn^2) = (alx + bmy + cnz)^2.$$

Note. Clearly the generators of the enveloping cylinder touch the quadric at points where it is met by the plane $alx + bmy + cnz = 0$ which is known as the plane of contact.

EXERCISES

1. Show that the enveloping cylinders of the ellipsoid

$$ax^2 + by^2 + cz^2 = 1,$$

   with generators perpendicular to Z-axis meet the plane $z = 0$ in parabolas.

2. Enveloping cylinders of the quadric $ax^2 + by^2 + cz^2 = 1$ meet the plane $z = 0$ in rectangular hyperbola; show that the central perpendiculars to their planes of contact generate the cone $b^2cx^2 + a^2cy^2 + ab(a+b)z^2 = 0$.

3. Prove that the enveloping cylinders of the ellipsoid

$$\frac{x^2}{a^2} + \frac{y^2}{b^2} + \frac{z^2}{c^2} = 1$$

   whose generators are parallel to the lines

$$x = 0, \ \pm \frac{y}{\sqrt{a^2 - b^2}} = \frac{z}{c}$$

   meet the plane $z = 0$ in circles.

6.7.1. Locus of Chords Bisected at a Given Point. Section With a Given Centre

Let the given point be (α, β, γ).
If a chord

$$\frac{x - \alpha}{l} = \frac{y - \beta}{m} = \frac{z - \gamma}{n} \qquad \ldots(1)$$

of the quadric $ax^2 + by^2 + cz^2 = 1$ is bisected at (α, β, γ), the two roots r_1 and r_2 of the equation (A) of § 6.3 are equal and opposite so that $r_1 + r_2 = 0$, implying

$$al\alpha + bm\beta + cn\gamma = 0 \qquad \ldots(2)$$

Therefore, the required locus, obtained by eliminating l, m, n between (1) and (2), is

$$a\alpha(x - \alpha) + b\beta(y - \beta) + c\gamma(z - \gamma) = 0$$

which is a plane and can briefly be written as

$$T_1 = S_1.$$

The section of the quadric by this plane is a conic whose centre is (α, β, γ); for this point bisects all chords of the conic through it.

Cor. *The plane which cuts the quadric* $ax^2 + by^2 + cz^2 = 1$, *in a conic whose centre is* (α, β, γ) *is*

$$\Sigma a\alpha x = \Sigma a\alpha^2$$

EXAMPLES

1. Show that the locus of the centres of sections of a central conicoid which pass through a given line is a conic.

Sol. Let the central conicoid be
$$ax^2 + by^2 + cz^2 = 1 \qquad \ldots(i)$$
The section of this conicoid whose centre is the point (α, β, γ) is given by
$$a\alpha(x - \alpha) + b\beta(y - \beta) + c\gamma(z - \gamma) = 0$$
This passes through the given line
$$\frac{x - x'}{l} = \frac{y - y'}{m} = \frac{z - z'}{n}$$
if
$$a\alpha(x' - \alpha) + b\beta(y' - \beta) + c\gamma(z' - \gamma) = 0$$
and
$$a\alpha l + b\beta m + c\gamma n = 0$$
Hence, the locus of centres is given by the equations
$$ax(x' - x) + by(y' - y) + cz(z' - z) = 0$$
and
$$alx + bmy + cnz = 0$$
or
$$ax^2 + by^2 + cz^2 = axx' + byy' + czz' \qquad \ldots(ii)$$
and
$$alx + bmy + cnz = 0 \qquad \ldots(iii)$$
These two equations determine a conic.

2. *Triads of tangent planes at right angles are drawn to the ellipsoid* $\dfrac{x^2}{a^2} + \dfrac{y^2}{b^2} + \dfrac{z^2}{c^2} = 1.$ *Show that the locus of the centre of section of the surface by the plane through their points of contact is*
$$x^2 + y^2 + z^2 = \left(\frac{x^2}{a^2} + \frac{y^2}{b^2} + \frac{z^2}{c^2}\right)(a^2 + b^2 + c^2).$$

Sol. Suppose that (α, β, γ) is the centre of section of the surface by a plane through the points of contact of a triad of mutually perpendicular tangent planes. The pole of this section must thus be a point of the director sphere
$$x^2 + y^2 + z^2 = a^2 + b^2 + c^2$$
The equation of the section is $T_1 = S_1$, i.e.,
$$\frac{\alpha x}{a^2} + \frac{\beta y}{b^2} + \frac{\gamma z}{c^2} = \frac{\alpha^2}{a^2} + \frac{\beta^2}{b^2} + \frac{\gamma^2}{c^2} \qquad \ldots(i)$$
If (f, g, h) be its pole, the equation (i) must be the same as
$$\frac{fx}{a^2} + \frac{gy}{b^2} + \frac{hz}{c^2} = 1 \qquad \ldots(ii)$$
Comparing (i) and (ii), we have
$$f = \frac{\alpha}{\Sigma(\alpha^2/a^2)}, \quad g = \frac{\beta}{\Sigma(\alpha^2/a^2)}, \quad h = \frac{\gamma}{\Sigma(\alpha^2/a^2)}$$
Since
$$f^2 + g^2 + h^2 = a^2 + b^2 + c^2$$
we have
$$\alpha^2 + \beta^2 + \gamma^2 = [(\Sigma\alpha^2/a^2)]^2(a^2 + b^2 + c^2).$$
Replacing α, β, γ by x, y, z respectively, we have the required result.

EXERCISES

1. Find the equation of the plane which cuts the surface
$$x^2 - 2y^2 + 3z^2 = 4$$
in a conic whose centre is at the point $(5, 7, 6)$. **[Ans.** $5x - 14y + 18z = 35]$

2. Find the centres of the conics
   (i) $4x + 9y + 4z = -15$, $2x^2 - 3y^2 + 4z^2 = 1$;
   (ii) $2x - 2y - 5z + 5 = 0$, $3x^2 + 2y^2 - 15z^2 = 4$. [**Ans.** (i) $(2, -3, 1)$, (ii) $(-2, 3, -1)$]

3. Prove that the plane through the three extremities of the different axes of a central conicoid cuts it in a conic whose centre coincides with the centroid of the triangle formed by those extremities.

4. Show that the centre of the conic
$$lx + my + nz = p, \quad ax^2 + by^2 + cz^2 = 1$$
is the point
$$\left(\frac{lp}{ap_0^2}, \frac{mp}{bp_0^2}, \frac{np}{cp_0^2} \right)$$
where $l^2 + m^2 + n^2 = 1$ and $p_0 = \sqrt{\Sigma l^2 / a}$.

5. A variable plane makes intercepts on the axes of a central conicoid whose sum is zero. Show that the locus of the centre of the section determined by it is a cone which has the axes of the conicoid as its generators.

6.7.2. Locus of Mid-Points of a System of Parallel Chords

Let l, m, n be proportional to the direction cosines of a given system of parallel chords and let (α, β, γ) be the mid-point of one of them.

As the chord
$$\frac{x - \alpha}{l} = \frac{y - \beta}{m} = \frac{z - \gamma}{n}$$
of the quadric is bisected at (α, β, γ), we have, as in § 6.7.1.
$$a\alpha l + b m \beta + c n \gamma = 0$$

Now, l, m, n being fixed, the locus of the mid-points (α, β, γ) of the parallel chords is the plane
$$alx + bmy + cnz = 0,$$
which clearly passes through the centre of the quadric and is known as the *Diametral plane conjugate to the direction l, m, n*.

Conversely, a plane $Ax + By + Cz = 0$ through the centre is the *diametral plane* conjugate to the direction l, m, n given by
$$\frac{al}{A} = \frac{bm}{B} = \frac{cn}{C}$$

Thus, *every central plane is a diametral plane conjugate to some direction*.

Note. If P be a point on the conicoid, then the plane bisecting chords parallel to the line OP is called the *diametral plane of OP*.

Note. Another method. Use of elements of infinity. We know that the mid-point of a line AB is the harmonic conjugate of the point at infinity on the line w.r.t. A and B. Thus, the *locus of the mid-points of a system of parallel chords is the polar plane of the point at infinity common to the chords of the system*.

We know that $(l, m, n, 0)$ is the point at infinity lying on a line with direction ratios, l, m, n. Its polar plane w.r.t. the conicoid
$$ax^2 + by^2 + cz^2 - w^2 = 0$$
expressed in cartesian homogeneous co-ordinates, is
$$alx + bmy + cnz + w.0 = 0$$
$\Leftrightarrow \quad alx + bmy + cnz = 0.$

EXERCISES

1. $P(1, 3, 2)$ is a point on the conicoid
$$x^2 - 2y^2 + 3z^2 + 5 = 0.$$
Find the locus of the mid-points of chords drawn parallel to OP. [**Ans.** $x - 6y + 6z = 0$]

2. Find the equation of the chord of the quadric $4x^2 - 5y^2 + 6z^2 = 7$ which passes through the point $(2, 3, 4)$ and is bisected by the plane $2x - 5y + 3z = 0$.

$$\left[\text{Ans. } (x - 2) = \frac{1}{2}(y - 3) = (z - 4) \right]$$

6.8 CONJUGATE DIAMETERS AND DIAMETRAL PLANES

In what follows, we shall confine our attention to the ellipsoid only.

Let $P(x_1, y_1, z_1)$ be a point on the ellipsoid
$$\frac{x^2}{a^2} + \frac{y^2}{b^2} + \frac{z^2}{c^2} = 1$$
The equation of the diametral plane bisecting chords parallel to the line OP is
$$\frac{xx_1}{a^2} + \frac{yy_1}{b^2} + \frac{zz_1}{c^2} = 1$$
Let $Q(x_2, y_2, z_2)$ be a point on the section of the ellipsoid by this plane so that we have
$$\frac{x_1 x_2}{a^2} + \frac{y_1 y_2}{b^2} + \frac{z_2 z_2}{c^2} = 0$$
which is the condition that the diametral plane of OP should pass through Q and, by symmetry, it is also the condition that the diametral plane of OQ should pass through P.

Thus, *if the diametral plane of OP passes through Q, then the diametral plane of OQ also passes through P.*

Let $R(x_3, y_3, z_3)$ be one of the two points where the line of intersection of diametral planes of OP and OQ meets the conicoid.

Since the point R is on the diametral planes of OP and OQ, the diametral plane
$$\frac{xx_3}{a^2} + \frac{yy_3}{b^2} + \frac{zz_3}{c^2} = 0$$
of OR passes through the points P and Q.

Thus, we obtain the following two sets of relations:

$$\left. \begin{array}{l} \dfrac{x_1^2}{a^2} + \dfrac{y_1^2}{b^2} + \dfrac{z_1^2}{c^2} = 1, \\[6pt] \dfrac{x_2^2}{a^2} + \dfrac{y_2^2}{b^2} + \dfrac{z_2^2}{c^2} = 1, \\[6pt] \dfrac{x_3^2}{a^2} + \dfrac{y_3^2}{b^2} + \dfrac{z_3^2}{c^2} = 1. \end{array} \right\} \dots(A) \qquad \left. \begin{array}{l} \dfrac{x_2 x_3}{a^2} + \dfrac{y_2 y_3}{b^2} + \dfrac{z_2 z_3}{c^2} = 0, \\[6pt] \dfrac{x_3 x_1}{a^2} + \dfrac{y_3 y_1}{b^2} + \dfrac{z_3 z_1}{c^2} = 0, \\[6pt] \dfrac{x_1 x_2}{a^2} + \dfrac{y_1 y_2}{b^2} + \dfrac{z_1 z_2}{c^2} = 0. \end{array} \right\} \dots(B)$$

The three semi-diameters OP, OQ, OR are called **conjugate semi-diameters** *if the plane containing any two of them is the diametral plane of the third.*

The co-ordinates of the extremities of the conjugate semi-diameters are connected by the relations (A) and (B) above.

Conjugate Planes. *The three diametral planes POQ, QOR, ROP are called conjugate planes if each is the diameteral plane of the line of intersection of the other two.*

We shall now obtain two more sets of relations (C), (D), equivalent to the relations (A), (B).

The Conicoid

By virtue of the relations (A), we see that

$$\frac{x_1}{a}, \frac{y_1}{b}, \frac{z_1}{c}; \frac{x_2}{a}, \frac{y_2}{b}, \frac{z_2}{c}; \frac{x_3}{a}, \frac{y_3}{b}, \frac{z_3}{c}$$

can be considered as the direction cosines of some three straight lines and the relations (B) show that these straight lines are also mutually perpendicular.

Hence, we have

$$\frac{x_1}{a}, \frac{x_2}{a}, \frac{x_3}{a}; \frac{y_1}{b}, \frac{y_2}{b}, \frac{y_3}{b}; \frac{z_1}{c}, \frac{z_2}{c}, \frac{z_3}{c}$$

are also the direction cosines of three mutually perpendicular straight lines. Therefore, we have

$$\left.\begin{aligned} x_1^2 + x_2^2 + x_3^2 &= a^2, \\ y_1^2 + y_2^2 + y_3^2 &= b^2, \\ z_1^2 + z_2^2 + z_3^2 &= c^2, \end{aligned}\right\} \quad \text{...(C)} \qquad \left.\begin{aligned} y_1 z_1 + y_2 z_2 + y_3 z_3 &= 0, \\ z_1 x_1 + z_2 x_2 + z_3 x_3 &= 0, \\ x_1 y_1 + x_2 y_2 + x_3 y_3 &= 0. \end{aligned}\right\} \quad \text{...(D)}$$

Properties of Conjugate Semi-Diameters

6.8.1. *The sum of the squares of three conjugate semi-diameters is constant.*

Adding the relations (C), we get

$$OP^2 + OQ^2 + R^2 = a^2 + b^2 + c^2$$

which is constant.

6.8.2. *The volume of the parallelopiped formed by three conjugate semi-diameters as coterminous edges is constant.*

The results (B) give

$$\frac{x_1/a}{\dfrac{y_2 z_3 - y_3 z_2}{bc}} = \frac{y_1/b}{\dfrac{z_2 x_3 - z_3 x_2}{ca}} = \frac{z_1/c}{\dfrac{x_2 y_3 - x_3 y_2}{ab}} = \frac{\sqrt{\Sigma x_1^2/a^2}}{\sqrt{\Sigma\left(\dfrac{y_2 z_3 - y_3 z_2}{bc}\right)}} = \pm 1$$

since $\Sigma\left(\dfrac{y_2 z_3 - y_3 z_2}{bc}\right)^2$ is the sine of the angle between two perpendicular lines with direction cosines

$$\frac{x_2}{a}, \frac{y_2}{b}, \frac{z_2}{c} \quad \text{and} \quad \frac{x_3}{a}, \frac{y_3}{b}, \frac{z_3}{c}$$

We have $\dfrac{x_1}{a} = \pm \dfrac{y_2 z_3 - y_3 z_2}{bc}, \quad \dfrac{y_1}{b} = \pm \dfrac{z_2 x_3 - z_3 x_2}{ca}, \quad \dfrac{z_1}{c} = \pm \dfrac{x_2 y_3 - x_3 y_2}{ab}$

Now the volume of the parallelopiped whose coterminous edges are OP, OQ, OR

$$= 6 \times \text{volume of the tetrahedron } OPQR$$

$$= \begin{vmatrix} 0 & 0 & 0 & 0 \\ x_1 & y_1 & z_1 & 1 \\ x_2 & y_2 & z_2 & 1 \\ x_3 & y_3 & z_3 & 1 \end{vmatrix} = \begin{vmatrix} x_1 & y_1 & z_1 \\ x_2 & y_2 & z_2 \\ x_3 & y_3 & z_3 \end{vmatrix}$$

$$= x_1(y_2 z_3 - y_3 z_2) + y_1(z_2 x_3 - z_3 x_2) + z_1(x_2 y_3 - x_3 y_2)$$

$$= \pm \frac{bc x_1^2}{a} \pm \frac{ca y_1^2}{b} \pm \frac{ab z_1^2}{c}$$

$$= \pm abc \, \Sigma \frac{x_1^2}{a^2} = \pm abc, \text{ which is a constant.}$$

The same result can also be proved in the following manner :

$$\begin{vmatrix} x_1 & y_1 & z_1 \\ x_2 & y_2 & z_2 \\ x_3 & y_3 & z_3 \end{vmatrix} \times \begin{vmatrix} x_1 & y_1 & z_1 \\ x_2 & y_2 & z_2 \\ x_3 & y_3 & z_3 \end{vmatrix} = \begin{vmatrix} \Sigma x_1^2 & \Sigma x_1 y_1 & \Sigma x_1 z_1 \\ \Sigma x_1 y_1 & \Sigma y_1^2 & \Sigma y_1 z_1 \\ \Sigma x_1 z_1 & \Sigma y_1 z_1 & \Sigma z_1^2 \end{vmatrix}$$

(By the rule of multiplication of determinatants)

6.8.3. *The sum of the squares of the areas of the faces of the parallelopiped formed with any three conjugate semi-diameters as coterminous edges is constant.*

Let A_1, A_2, A_3 be the areas of the triangles OQR, ORP, OPQ, and let l_i, m_i, n_i $(i = 1, 2, 3)$ be the direction cosines of the normals to the planes respectively.

Now, the projection of the triangle OQR on the YZ plane is the triangle with vertices $(0, 0, 0)$, $(0, y_2, z_2), (0, y_3, z_3)$ whose area is $\frac{1}{2}(y_2 z_3 - y_3 z_2)$. Also this is $A_1 l_1$. Thsu, we have

$$A_1 l_1 = \frac{1}{2}(y_2 z_3 - y_3 z_2) = \pm \frac{bcx_1}{2a}$$

Similarly
$$A_1 m_1 = \pm \frac{cay_1}{2b}, \quad A_1 n_1 = \pm \frac{abz_1}{2c}$$

Squaring and adding, we obatain

$$A_1^2 = \frac{b^2 c^2 x_1^2}{4a^2} + \frac{c^2 a^2 y_1^2}{4b^2} + \frac{a^2 b^2 z_1^2}{4c^2}$$

Similarly projecting the areas ORP and OPQ in the co-ordinate planes, we get

$$A_2^2 = \frac{b^2 c^2 x_2^2}{4a^2} + \frac{c^2 a^2 y_2^2}{4b^2} + \frac{a^2 b^2 z_2^2}{4c^2}$$

$$A_3^2 = \frac{b^2 c^2 x_3^2}{4a^2} + \frac{c^2 a^2 y_3^2}{4b^2} + \frac{a^2 b^2 z_3^2}{4c^2}$$

Adding, we get

$$A_1^2 + A_2^2 + A_3^2 = \frac{1}{4}(b^2 c^2 + c^2 a^2 + a^2 b^2)$$

which is a constant.

6.8.4. *The sum of the projections of three semi-conjugate diameters on any line or plane is constant.*

Let l, m, n be the direction cosines of any given line so that the sum of the squares of the projections of three semi-conjugate diameters OP, OQ, OR on the line is

$$= (lx_1 + my_1 + nz_1)^2 + (lx_2 + my_2 + nz_2)^2 + (lx_3 + my_3 + nz_3)^2$$
$$= l^2 \Sigma x_1^2 + m^2 \Sigma y_1^2 + n^2 \Sigma z_1^2 + 2lm \Sigma x_1 y_1 + 2mn \Sigma y_1 z_1 + 2nl \Sigma z_1 x_1$$
$$= a^2 l^2 + b^2 m^2 + c^2 n^2$$

which is a constant.

Again, let l, m, n be the direction cosines of normal of any given plane so that the sum of the squares of the projections of OP, OQ, OR on this plane is

$$= OP^2 - (lx_1 + my_1 + nz_1)^2 + OQ^2 - (lx_2 + my_2 + nz_2)^2 + OR^2 - (lx_3 + my_3 + nz_3)^2$$
$$= a^2 + b^2 + c^2 - a^2 l^2 - b^2 m^2 - c^2 n^2$$
$$= a^2 (m^2 + n^2) + b^2 (n^2 + l^2) + c^2 (l^2 + m^2)$$

which is a constant.

The Conicoid

EXAMPLES

1. *Show that the equation of the plane through the extremities*
$$(x_k, y_k, z_k), k = 1, 2, 3.$$
of the conjugate semi-diameters of the ellipsoid.
$$x^2/a^2 + y^2/b^2 + z^2/c^2 = 1,$$

is
$$\frac{x(x_1 + x_2 + x_3)}{a^2} + \frac{y(y_1 + y_2 + y_3)}{b^2} + \frac{z(z_1 + z_2 + z_3)}{c^2} = 1$$

Sol. Let
$$lx + my + nz = p$$
be the plane through the three extreminties of the given conjugate semi-diameter, so that we have
$$lx_1 + my_1 + nz_1 = p$$
$$lx_2 + my_2 + nz_2 = p$$
$$lx_3 + my_3 + nz_3 = p$$

Multiplying these by x_1, x_2, x_3 respectively and adding we obtain
$$la^2 = p \Sigma x$$

Similarly $\quad mb^2 = p \Sigma y_1$ and $nc^2 = p \Sigma z_1$

Hence, the required equation.

2. *Find the locus of the equal conjugate diameters of the ellipsoid*
$$\frac{x^2}{a^2} + \frac{y^2}{b^2} + \frac{z^2}{c^2} = 1.$$

Sol. Let OP, OQ, OR be three equal conjugate semi-diameters. We have
$$\begin{cases} OP^2 + OQ^2 + OR^2 = a^2 + b^2 + c^2; \\ OP^2 = OQ^2 = OR^2 \end{cases}$$

$\Rightarrow \qquad OP^2 = \frac{1}{3}(a^2 + b^2 + c^2)$

Let P be the point (x_1, y_1, z_1). We require the locus of the line
$$\frac{x}{x_1} = \frac{y}{y_1} = \frac{z}{z_1} \qquad \ldots(1)$$

where $\qquad x_1^2 + y_1^2 + z_1^2 = \frac{1}{3}(a^2 + b^2 + c^2) \qquad \ldots(2)$

and $\qquad \frac{x_1^2}{a^2} + \frac{y_1^2}{b^2} + \frac{z_1^2}{c^2} = 1 \qquad \ldots(3)$

From (2) and (3), we obtain the homogeneous relation
$$\frac{x_1^2}{a^2} + \frac{y_1^2}{b^2} + \frac{z_1^2}{c^2} = \frac{3(x_1^2 + y_1^2 + z_1^2)}{(a^2 + b^2 + c^2)} \qquad \ldots(4)$$

Eliminating x_1, y_1, z_1 from (1) and (4), we obtain the required locus, viz.,
$$\frac{x^2}{a^2} + \frac{y^2}{b^2} + \frac{z^2}{c^2} = \frac{3(x^2 + y^2 + z^2)}{(a^2 + b^2 + c^2)}.$$

3. *Show that the locus of the foot of the perpendicular from the centre to the plane through the extremities of three conjugate semi-diameters of the ellipsoid*
$$\frac{x^2}{a^2} + \frac{y^2}{b^2} + \frac{z^2}{c^2} = 1$$

is
$$a^2x^2 + b^2y^2 + c^2z^2 = 3(x^2 + y^2 + z^2)^2$$

Sol. Let $P(x_1, y_1, z_1), Q(x_2, y_2, z_2), R(x_3, y_3, z_3)$ be the extremities of the three conjugate semi-diameters of the given ellipsoid with centre O.

Let (α, β, γ) be the foot of the perpendicular from the centre O of the ellipsoid on the plane PQR.

Equations of the perpendicular are
$$\frac{x}{\alpha} = \frac{y}{\beta} = \frac{z}{\gamma}.$$

Therefore equation of the plane PQR is
$$\alpha(x - \alpha) + \beta(y - \beta) + \gamma(z - \gamma) = 0$$
$$\alpha x + \beta y + \gamma z = \alpha^2 + \beta^2 + \gamma^2$$

Therefore equation of the plane PQR is
$$\alpha(x - \alpha) + \beta(y - \beta) + \gamma(z - \gamma) = 0$$
or
$$\alpha x + \beta y + \gamma z = \alpha^2 + \beta^2 + \gamma^2$$

P, Q, R lie on this plane. Therefore
$$\alpha x_1 + \beta y_1 + \gamma z_1 = \alpha^2 + \beta^2 + \gamma^2, \qquad ...(i)$$
$$\alpha x_2 + \beta y_2 + \gamma z_2 = \alpha^2 + \beta^2 + \gamma^2, \qquad ...(ii)$$
and
$$\alpha x_3 + \beta y_3 + \gamma z_3 = \alpha^2 + \beta^2 + \gamma^2 \qquad ...(iii)$$

Multiplying (i) by x_1, (ii) by x_2 and (iii) by x_3, and adding, we get
$$\alpha(x_1^2 + x_2^2 + x_3^2) + \beta(x_1 y_1 + x_2 y_2 + x_3 y_3) + \gamma(z_1 x_1 + z_2 x_2 + z_3 x_3)$$
$$= (x_1 + x_2 + x_3)(\alpha^2 + \beta^2 + \gamma^2)$$

Using relations (C) and (D) of § 6.8, we have
$$\alpha a^2 = (x_1 + x_2 + x_3)(\alpha^2 + \beta^2 + \gamma^2)$$
or
$$\alpha a = \left(\frac{x_1 + x_2 + x_3}{a}\right)(\alpha^2 + \beta^2 + \gamma^2) \qquad ...(iv)$$

Similarly
$$\beta b = \left(\frac{y_1 + y_2 + y_3}{b}\right)(\alpha^2 + \beta^2 + \gamma^2) \qquad ...(v)$$
and
$$\gamma c = \left(\frac{z_1 + z_2 + z_3}{c}\right)(\alpha^2 + \beta^2 + \gamma^2) \qquad ...(vi)$$

Now squaring and adding these equations, we get
$$\alpha^2 a^2 + \beta^2 b^2 + \gamma^2 c^2 = (\alpha^2 + \beta^2 + \gamma^2)^2 \left[\left(\frac{x_1 + x_2 + x_3}{a}\right)^2 + \left(\frac{y_1 + y_2 + y_3}{b}\right)^2\right.$$
$$\left. + \left(\frac{z_1 + z_2 + z_3}{c}\right)^2\right]$$

$$= (\alpha^2 + \beta^2 + \gamma^2)^2 \left[\frac{x_1^2 + x_2^2 + x_3^2}{a^2} + \frac{y_1^2 + y_2^2 + y_3^2}{b^2} + \frac{z_1^2 + z_2^2 + z_3^2}{c^2}\right.$$
$$\left. + 2\left(\frac{x_1 x_2}{a^2} + \frac{y_1 y_2}{b^2} + \frac{z_1 z_2}{c^2}\right) + 2\left(\frac{x_3 x_1}{a^2} + \frac{y_3 y_1}{b^2} + \frac{z_3 z_1}{c^2}\right) + 2\left(\frac{x_2 x_3}{a^2} + \frac{y_2 y_3}{b^2} + \frac{z_2 z_3}{c^2}\right)\right]$$

Using relations (B) and (C) of § 6.8, we obtain

The Conicoid

$$\alpha^2 a^2 + \beta^2 b^2 + \gamma^2 c^2 = (\alpha^2 + \beta^2 + \gamma^2)^2 (1+1+1)$$

or
$$a^2\alpha^2 + b^2\beta^2 + c^2\gamma^2 = 3(\alpha^2 + \beta^2 + \gamma^2)^2$$

Hence the locus of the foot of perpendicular is
$$a^2 x^2 + b^2 y^2 + c^2 z^2 = 3(x^2 + y^2 + z^2)^2.$$

4. *Prove that the pole of the plane PQR lies on the ellipsoid*
$$\frac{x^2}{a^2} + \frac{y^2}{b^2} + \frac{z^2}{c^2} = 3,$$
where OP, OQ, OR are the conjugate semi-diameters of the ellipsoid
$$\frac{x^2}{a^2} + \frac{y^2}{b^2} + \frac{z^2}{c^2} = 1.$$

Sol. Equation of the plane PQR through the extremities
$$P(x_1, y_1, z_1), Q(x_2, y_2, z_2), R(x_3, y_3, z_3)$$

is
$$\frac{x}{a^2}(x_1 + x_2 + x_3) + \frac{y}{b^2}(y_1 + y_2 + y_3) + \frac{z}{c^2}(z_1 + z_2 + z_3) = 1 \quad \ldots(i)$$

If (α, β, γ) be the pole of the plane PQR, then its equation is
$$\frac{\alpha x}{x^2} + \frac{\beta y}{b^2} + \frac{\gamma z}{c^2} = 1 \quad \ldots(ii)$$

(i) and (ii) represent the same plane, therefore,
$$\alpha = x_1 + x_2 + x_3,$$
$$\beta = y_1 + y_2 + y_3,$$
$$\gamma = z_1 + z_2 + z_3.$$

Now multiplying these relations by $1/a, 1/b, 1/c$ respectively; squaring and adding, we get
$$\frac{\alpha^2}{a^2} + \frac{\beta^2}{b^2} + \frac{\gamma^2}{c^2} = \left(\frac{x_1 + x_2 + x_3}{a}\right)^2 + \left(\frac{y_1 + y_2 + y_3}{b}\right)^2 + \left(\frac{z_1 + z_2 + z_3}{c}\right)^2$$

On using relations (B) and (C) of § 6.8, this simplifies to
$$\frac{\alpha^2}{a^2} + \frac{\beta^2}{b^2} + \frac{\gamma^2}{c^2} = 3$$

Hence the locus of (α, β, γ) is the ellipsoid
$$\frac{x^2}{a^2} + \frac{y^2}{b^2} + \frac{z^2}{c^2} = 3.$$

EXERCISES

1. Show that the lines
$$\frac{x}{1} = \frac{y}{4} = \frac{z}{3}, \frac{x}{4} = \frac{y}{1} = \frac{z}{-9}, \frac{x}{26} = \frac{y}{-28} = \frac{z}{45},$$
are three mutually conjugate diameters of the ellipsoid
$$\frac{x^2}{2} + \frac{y^2}{4} + \frac{z^2}{9} = 1.$$

2. Find the equations of the diameters in the plane $x + y + z = 0$, conjugate to
$$x = -\frac{1}{2}y = \frac{1}{3}z$$

with respect to the conicoid $3x^2+ y^2- 2z^2 =1$. What are the equations of the third conjugate diameter ?

$$\left[\text{Ans.} \ \frac{x}{4}=\frac{y}{-9}=\frac{z}{5}, \frac{x}{34}=\frac{y}{42}=\frac{z}{3}\right]$$

3. Show that for the ellipsoid $x^2 + 4y^2 + 5z^2 = 1$ the two diameters $\frac{1}{3}x = -\frac{1}{2}y = \frac{1}{3}z$ and $x = 0, 2y = 5z$ are conjugate. Obtain the equations of the third conjuate diameter.

[Ans. $x/16 = y = - z/2$]

4. If $p_1, p_2, p_3; \pi_1, \pi_2, \pi_3$ be the projections of the three conjugate diameters on any two given lines, then $p_1\pi_1, p_2\pi_2\ p_3\pi_3$ is constant.

5. If three conjugate diameters vary so that OP, OQ lie respectively in the fixed planes

$$\frac{\alpha_1 x}{a^2}+\frac{\beta_1 y}{b^2}+\frac{\gamma_1 z}{c^2} = 0, \frac{\alpha_2 x}{a^2}+\frac{\beta_2 y}{b^2}+\frac{\gamma_2 z}{c^2}=0$$

show that the locus of OR is the cone

$$\Sigma\, \alpha^2\, (\beta_1 z - \gamma_1 y)\, (\beta_2 z - \gamma_2 y) = 0.$$

[**Hint.** The required locus of OR is obtained from the fact that the lines of intersection of the diametral plane of OR with the given planes are conjugate lines.]

6. From a fixed point H perpendiculars HA, HB, HC are drawn to the conjugate diameters OP, OQ, OR respectively; show that

$$OP^2.HA^2 + OQ^2.HB^2 + OR^2.HC^2$$

is constant.

7. OP, OQ, OR are conjugate diameters of an ellipsoid

$$x^2/a^2 + y^2/b^2 + z^2/c^2 = 1.$$

At Q and R tangent lines are drawn parallel to OP and p_1, p_2 are their distance from O. The perpendicular from O to the tangent planes at right angles to OP is p. Prove that

$$p^2 + p_1^2 + p_2^2 = a^2 + b^2 + c^2.$$

8. Show that the plane $lx + my + nz = p$ will pass through the extremities of conjugate semi-diameters if $a^2 l^2 + b^2 m^2 + c^2 n^2 = 3p^2$.

PARABOLOIDS

6.9. Having discussed the nature and geometrical properties of central conicoids, we now proceed to the consideration of *paraboloids*.

6.9.1. The Elliptic Paraboloids $x^2/a^2 + y^2/b^2 = 2z/c$

We have the following particulars about this surface :

(i) The co-ordinate planes $x = 0$ and $y = 0$ bisect chords perpendicular to then and are, therefore, its two planes of symmetry or **Principal Planes.**

(ii) z cannot be negative, and hence there is no part of the surface on the negative side of the plane $z = 0$. We have taken c positive.

(iii) The sections by the planes $z = k, (k > 0)$, parallel to the XY plane, are similar ellipses

$$\frac{x^2}{a^2}+\frac{y^2}{b^2}=\frac{2k}{c}, z= k \qquad \text{...(i)}$$

whose centres lie on Z-axis and which increase in size as k increases; there being no limit to the increase of k. The surface may thus be supposed to be generated by the variable ellipse (*i*).

The Conicoid

Hence, the surface is entirely on the positive side of the plane $z=0$, and extends to infinity.

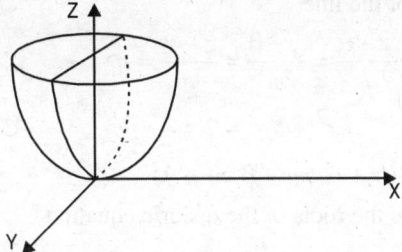

(*iv*) The section of the surface by planes parallel to the YZ and ZX planes are clearly parabolas. Figure show the nature of the surface.

Ex. Trace the surface $x^2/a^2 + y^2/b^2 = -2z/c$, $(c>0)$.

6.9.2. The Hyperbolic Paraboloid $x^2/a^2 - y^2/b^2 = 2z/c$

(i) The co-ordinate planes $x=0$ and $y=0$ are the two **Principal Planes**.

(ii) The sections by the planes $z=k$ are the similar hyperbolas

$$\frac{x^2}{a^2} - \frac{y^2}{b^2} = \frac{2k}{c}, z = k,$$

with their centres on Z-axis.

If k be positive, the real axis of the hyperbola is parallel to X-axis, and if k be negative, the real axis is parallel to Y-axis.

The section by the plane $z=0$ is the pair of lines

$$\frac{x}{a} = \frac{y}{b}, z = 0 \text{ and } \frac{x}{a} = -\frac{y}{b}, z = 0$$

(iii) The section by the planes parallel to YZ and YX planes are parabolas.

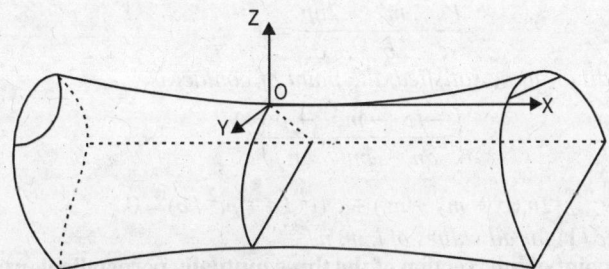

Figure shows the nature of the surface.

Note. The two equations considered in the last two articles are clearly both included in the form

$$ax^2 + by^2 = 2cz$$

This equation represents an elliptic paraboloid if a and b are both positive or both negative, and a hyperbolic paraboloid if one is positive and the other negative.

Hence, for an elliptic paraboloid, ab is positive but, for hyperbolic paraboloid, ab is negative.

The geometrical result deducible from the equation $ax^2 + by^2 = 2cz$ will hold for both the types of paraboloids.

Note. The reader would do well to give precise definitions of (i) vertex, (ii) principal planes, (iii) axis of a paraboloid.

6.9.3 Intersection of a Line With a Paraboloid

The points of intersection of the line
$$\frac{x-\alpha}{l} = \frac{y-\beta}{m} = \frac{z-\gamma}{n} = r$$

with the paraboloid
$$ax^2 + by^2 = 2cz,$$

are
$$(lr + \alpha, mr + \beta, nr + \gamma)$$

for the *two* values of r which are the roots of the quadric equation
$$r^2(al^2 + bm^2) + 2r(al\alpha + bm\beta - cn) + (a\alpha^2 + b\beta^2 - 2c\gamma) = 0 \qquad ...(A)$$

We thus see that every line meets a paraboloid in two points.
It follows from this that the *plane sections of paraboloids are conics*.

Also, if $l = m = 0$, one value of r is infinite and hence any line parallel to Z-axis meets the paraboloid in one point at an infinite distance from (α, β, γ) and so meets it in one finite point only. Such lines are called **Diameters of the paraboloid.**

In particular, Z-axis meets the surface at the origin only.

6.9.4. From the equation (A), § 6.3 above, we deduce certain results similar to those obtained for central conicoids. The proofs of some of them are left as an exercise to the student.

1. The *tangent plane to* $ax^2 + by^2 = 2cz$ at any point (α, β, γ) on the surface is
$$a\alpha x + b\beta y = c(z + \gamma).$$

In particular, $z = 0$ is the tangent plane at the origin and Z-axis is the normal thereat.
The origin O is called the vertex of the paraboloid and Z-axis, the **axis** of the paraboloid.

2. Condition of Tangency

The plane
$$lx + my + nz = p$$

will touch the paraboloid
$$ax^2 + by^2 = 2cz \qquad ...(1)$$

if
$$\frac{l^2}{a} + \frac{m^2}{b} + \frac{2np}{c} = 0$$

and assuming the condition to be satisfied, the point of contact is
$$\left(\frac{-lc}{an}, \frac{-mc}{bn}, \frac{-p}{n} \right)$$

Thus, the plane $\quad 2n(lx + my + nz) + c(l^2/a + m^2/b) = 0$
touches the surface (1) *for all values of l, m, n.*

3. Locus of the point of intersection of the three mutually perpendicular tangent planes

$$2n_r(l_r x + m_r y + n_r z) + c\left(\frac{l r_r^2}{a} + \frac{m r_r^2}{b} \right) = 0, \ (r = 1, 2, 3)$$

be three mutually perpendicular tangent planes, the locus of their point of intersection is obtained by eliminating l_r, m_r, n_r, which is done by adding the three equations and is, therefore,

$$2z + c\left(\frac{1}{a} + \frac{1}{b} \right) = 0,$$

and is a plane at right angles to the Z-axis; the axis of the paraboloid.

4. Equations of the *normal* at (α, β, γ) are
$$\frac{x-\alpha}{a\alpha} = \frac{y-\beta}{b\beta} = \frac{z-\gamma}{-c}$$

The Conicoid

5. The *polar plane* of the point (α, β, γ) is
$$a\alpha x + b\beta y = c(\gamma + z).$$

6. The *polar plane* of the *enveloping cone* with (α, β, γ) as its vertex is
$$SS_1 = T_1^2$$
$$\Leftrightarrow \quad (ax^2 + by^2 - 2cz)(a\alpha^2 + b\beta^2 - 2c\gamma) = (a\alpha x + b\beta y - cz - c\gamma)^2$$

Its *plane of contact* with the paraboloid is the polar plane
$$a\alpha x + b\beta y - cz - c\gamma = 0$$
of the vertex (α, β, γ).

7. The equation of the *enveloping cylinder* having its generagtors parallel to the line
$$\frac{x}{l} = \frac{y}{m} = \frac{z}{n}$$
is
$$(ax^2 + by^2 - 2cz)(al^2 + bm^2) = (alx + bmy - cn)^2$$

Its *plane of contact* is the plane
$$alx + bmy - cn = 0.$$

8. The *locus of chords bisected at a point* (α, β, γ) is the plane
$$T_1 = S_1$$
$$\Leftrightarrow \quad a\alpha(x - \alpha) + b\beta(y - \beta) - c\gamma(z - \gamma) = 0$$

This plane will meet the paraboloid in a conic whose centre is at (α, β, γ).

9. The *locus of mid-point of a system of parallel chords*, with direction ratios, l, m, n is the plane
$$alx + bmy - cn = 0$$
which is parallel to Z-axis, the axis of the paraboloid. The plane is called the *Diametral plane* conjugate to the given direction.

A plane $Ax + By + D = 0$ parallel to the axis of the paraboloid is easily seen, by comparison, to be the diametral plane for the system of parallel chords with direction ratios
$$A/a, B/b, -D/c$$

A plane parallel to the axis of a paraboloid is, thus, a diameteral plane.

EXAMPLES

1. *Two perpendicular tangent planes to the paraboloid $x^2/a + y^2/b = 2z$ intersect in a straight line lying in the plane $x = 0$. Show that the line touches the parabola*
$$x = 0, \quad y^2 = (a+b)(2z+a).$$

Sol. Equations of any line in the plane $x = 0$ is
$$x = 0, \, my + nz = p \qquad \ldots(i)$$

Any plane through this line is
$$\lambda x + my + nz - p = 0 \qquad \ldots(ii)$$

If this is a tangent plane to the paraboloid, then
$$a\lambda^2 + bm^2 + 2np = 0 \qquad \ldots(iii)$$

This is quadratic in λ and, therefore, gives the values λ_1, λ_2 and thus two tangent planes through the line (*i*) which are
$$\lambda_1 x + my + nz - p = 0$$
and
$$\lambda_2 x + my + nz - p = 0$$

These planes of perpendicular if
$$\lambda_1 \lambda_2 + m^2 + n^2 = 0$$

$$\Rightarrow \qquad \frac{bm^2 + 2np}{a} + m^2 + n^2 = 0$$

$$\Rightarrow \qquad (a+b)m^2 + 2np + an^2 = 0 \qquad \text{...(iv)}$$

Required parabola is the envelope of line (i) subject to the condition (iv). Eliminating p between (i) and (iv), we get

$$(a+b)m^2 + 2n(my + nz) + an^2 = 0, x = 0$$

or $\qquad (a+b)\left(\dfrac{m}{n}\right)^2 + 2y\left(\dfrac{m}{n}\right) + (a+2z) = 0, x = 0$

Therefore, the envelope of the line is given by

$$(2y)^2 - 4(a+b)(a+2z) = 0, x = 0$$

$$\Rightarrow \qquad y^2 = (a+b)(a+2x), x = 0.$$

2. Find the locus of the point of intersection of three mutually perpendicular tangent planes to the paraboloid $ax^2 + by^2 = 2cz$.

Sol. The plane $lx + my + nz = p$ touches the paraboloid $ax^2 + by^2 = 2cz$, when

$$\frac{l^2}{a} + \frac{m^2}{b} + \frac{2np}{c} = 0 \quad \text{or} \quad p = -\frac{c}{2n}\left(\frac{l^2}{a} + \frac{m^2}{b}\right)$$

By putting the value of p, we have

$$lx + my + nz + \frac{c}{2n}(l^2/a + m^2/b) = 0$$

or $\qquad 2n(lx + my + nz) + c(l^2/a + m^2/b) = 0$

This plane always touches the given paraboloid.
Let three mutually perpendicular tangent planes be

$$2n_r(l_r x + m_r y + n_r z) + c(l_r^2/a + m_r^2/b) = 0, r = 1, 2, 3 \qquad \text{...(i)}$$

Since the three planes and hence their normals are mutually perpendicular, hence

$$l_1^2 + m_1^2 + n_1^2 = 1, l_1^2 + l_2^2 + l_3^2 = 1 \text{ etc.}$$

$$l_1 l_2 + m_1 m_2 + n_1 n_2 = 0, l_1 m_1 + l_2 m_2 + l_3 m_3 = 0 \text{ etc.}$$

The locus of the point of intersection is obtained by eliminating l_r, m_r, n_r from the eqns. (i).

For, we add these planes obtained by putting $r = 1, 2, 3$ and use the above relations. Thus, we get

$$2(n_1^2 + n_2^2 + n_3^2)z + c\left(\frac{l_1^2 + l_2^2 + l_3^2}{a} + \frac{m_1^2 + m_2^2 + m_3^2}{b}\right) = 0$$

$$\Rightarrow \qquad 2z + c\left(\frac{1}{a} + \frac{1}{b}\right) = 0.$$

This is the required plane.

EXERCISES

1. Show that the plane $8x - 6y - z = 5$ touches the paraboloid $x^2/2 - y^2/3 = z$; and find the co-ordinates of the points of contact.
2. Show that the equation to the two tangent planes to the surface

$$ax^2 + by^2 = 2z,$$

which passes through the line

$$u \equiv lx + my + nz - p = 0, u' \equiv l'x + m'y + n'z - p' = 0$$

The Conicoid

is $\quad u^2\left(\dfrac{l'^2}{a}+\dfrac{m'^2}{b}-2n'p'\right)-2uu'\,(ll'+mm'-n'p-n'p)+u'^2\left(\dfrac{l^2}{a}+\dfrac{m^2}{b}-2nb\right)=0$.

3. Tangent planes at two points P and Q of a paraboloid meet in the line RS; show that the plane through RS and the middle point of PQ is parallel to the axis of the paraboloid.

4. Find the equation of the plane which cuts the paraboloid
$$x^2-\dfrac{1}{2}y^2=z$$
in a conic with its centre at the point (2, 3, 4). **[Ans.** $4x-3y-z+5=0$**]**

5. Show that the locus of the centres of a system of parallel plane sections of a paraboloid is a diameter.

6. Show that the centre of the conic
$$ax^2+by^2=2z,\ lx+my+nz=p$$
is the point
$$\left(\dfrac{l}{an},\dfrac{m}{bn},\dfrac{k^2}{n^2}\right)$$
where
$$k^2=\dfrac{l^2}{a}+\dfrac{m^2}{b}+np.$$

6.9.5. Number of Normals From a Given Point

If the normal at (α,β,γ) passes through a given piont (f,g,h), then
$$\dfrac{f-\alpha}{a\alpha}=\dfrac{g-\beta}{b\beta}=\dfrac{h-\gamma}{-c}=r,\text{ (say)}$$

$\Leftrightarrow \quad \alpha=\dfrac{f}{1+ar},\ \beta=\dfrac{g}{1+br},\ \gamma=h+cr$ ...(i)

Since (α,β,γ) lies on the paraboloid, we have the relation
$$a\dfrac{f^2}{(1+ar)^2}+b\dfrac{g^2}{(1+br)^2}=2c\,(h+cr) \qquad \ldots(ii)$$

which, being an equation of the fifth degree in r, gives five values of r, to each of which there corresponds a point (α,β,γ), from (i).

Therefore, there are five points, on a paraboloid the normals at which pass through a given point i.e., *through a given point five normals, in general, can be drawn to a paraboloid.*

Cor. I Cubic curve through the feet of the normals

If the normal at (α,β,γ) to the paraboloid
$$ax^2+by^2=2cz$$
passes through a given point (x',y',z'), we have as above
$$\alpha=\dfrac{x'}{1+a\lambda},\ \beta=\dfrac{y'}{1+b\lambda},\ \gamma=z'+c\lambda.$$

Thus, the feet of the normals lie on the curve, defined by the parametric equations, is given by
$$x=\dfrac{x'}{1+a\lambda},\ y=\dfrac{y'}{1+b\lambda},\ z=z'+c\lambda, \qquad \ldots(1)$$
where λ is the parameter.

The points where this curve meets any given plane, say,
$$ux+vy+wz+d=0 \qquad \ldots(2)$$

are given by
$$\frac{ux'}{1+a\lambda} + \frac{vy'}{1+b\lambda} + w(z'+c\lambda) + d = 0.$$

This is a cubic in λ, giving three values of λ.

Therefore the plane (2) meets the curve (1) in three points, and hence it follows that the curve is a cubic curve.

Cor. II. Cone through the five normals

If the normal at (α, β, γ) to the paraboloid $ax^2 + by^2 = 2cz$ passes through (x', y', z'), then
$$\alpha = \frac{x'}{1+a\lambda}, \beta = \frac{y'}{1+b\lambda}, \gamma = z' + c\lambda$$

Also the direction cosines of the normal at (α, β, γ) are proportional to $a\alpha, b\beta, -c$.

If the line
$$\frac{x-x'}{l} = \frac{y-y'}{m} = \frac{z-z'}{n} \qquad ...(1)$$

is a normal at (α, β, γ), then
$$\frac{l}{a\alpha} = \frac{m}{b\beta} = \frac{n}{-c}$$

$\Rightarrow \quad \dfrac{l(1+a\lambda)}{ax'} = \dfrac{m(1+b\lambda)}{by'} = \dfrac{n}{-c}$

$\Rightarrow \quad \dfrac{l\left(\dfrac{1}{a}+\lambda\right)}{x'} = \dfrac{m\left(\dfrac{1}{b}+\lambda\right)}{y'} = \dfrac{n}{-c}$

$\Rightarrow \quad \dfrac{\dfrac{1}{a}+\lambda}{x'/l} = \dfrac{\dfrac{1}{b}+\lambda}{y'/m} = \dfrac{n}{-c} = \dfrac{\dfrac{1}{a}-\dfrac{1}{b}}{\dfrac{x'}{l}-\dfrac{y'}{m}}$

Hence
$$n\left(\frac{x'}{l} - \frac{y'}{m}\right) = -c\left(\frac{1}{a} - \frac{1}{b}\right)$$

or
$$\frac{x'}{l} - \frac{y'}{m} + \frac{c}{n}\left(\frac{1}{a} - \frac{1}{b}\right) = 0$$

Therefore the locus of the normal (1) is
$$\frac{x'}{x-x'} - \frac{y'}{y-y'} + \frac{c}{z-z'}\left(\frac{b-a}{ab}\right) = 0$$

which is the equation of a cone.

Hence the five normal from (x', y', z') to the paraboloid are generators of this cone.

6.9.6. Conjugate Diametral Planes

Consider any two *diametral planes*
$$lx + my + p = 0 \qquad ...(i)$$
and
$$l'x + m'y + p' = 0 \qquad ...(ii)$$

The plane (i) bisects chords parallel to the line
$$\frac{x}{l/a} = \frac{y}{m/b} = \frac{z}{-p/c} \qquad ...(iii)$$

which will be parallel to the plane (ii), if

The Conicoid

$$\frac{ll'}{a} + \frac{mm'}{b} = 0. \qquad \ldots\text{(vi)}$$

The symmetry of the result shows that the plane (i) is also parallel to the chords bisected by the plane (ii).

Thus, if α and β be two diametral planes, such that *the plane α is parallel to the chords bisected by the plane β, then the plane β is parallel to the chords bisected by the plane α.*

Two such planes are called *conjugate diametral planes*.

Equation (iv) is the condition for the diameteral planes (i) (ii) to be conjugate.

EXAMPLES

1. *Show that the planes*

$$x + 3y = 3 \text{ and } 2x - y = 1$$

are conjugate diameter planes of the paraboloid

$$2x^2 + 3y^2 = 4z.$$

Sol. Equation of any diametral plane with respect to the paraboloid $ax^2 + by^2 = 2cz$ is

$$alm + bmy - cn = 0$$

In this case $a = 2, b = 3, c = 2$.

Therefore the equation of the diametral plane is

$$2lx + 3my - 2n = 0 \qquad \ldots(1)$$

If this plane and $x + 3y = 3$ are the same, then comparing the coefficients, we have

$$\frac{2l}{1} = \frac{3m}{3} = \frac{-2m}{-3}$$

or $\quad \dfrac{l}{1/2} = \dfrac{m}{1} = \dfrac{n}{3/2}$, i.e., $\dfrac{l}{1} = \dfrac{m}{2} = \dfrac{n}{3}$,

i.e., the direction cosines of the chords which are bisected by the plane $x + 3y = 3$ are proportional to 1, 2, 3.

This shows that these chords are parallel to the plane $2x - y = 1$.

Hence the given planes are conjugate diametral planes.

2. *Prove that any diametral plane of a paraboloid cuts it in a parabola, and that parallel diametral planes cut it in equal parabolas.*

Sol. Let the equation of the parabola be

$$ax^2 + by^2 = 2cz \qquad \ldots(1)$$

Therefore the diametral plane which bisects chords parallel to the line

$$\frac{x}{l} = \frac{y}{m} = \frac{z}{n}$$

is $\qquad alx + bmy - cm = 0 \qquad \ldots(2)$

Now to prove that the section of the paraboloid (1) by the plane (2) is a parabola, it is sufficient to prove that the projection of the section on a co-ordinate plane is a parabola, for the projection of a conic is a conic of the same species. So taking the projection of the section of (1) by the diametral plane (2) on YOZ plane, we have

$$a\left(\frac{cn - bmy}{al}\right)^2 + by^2 = 2cz, \ x = 0$$

$\Rightarrow \qquad (cn - bmy)^2 + al^2 (by^2 - 2cz) = 0, \ x = 0$

$\Rightarrow \qquad b(al^2 + bm^2)y^2 - 2bcmny + n^2 - 2cal^2 z = 0, \ x = 0$

$$\Rightarrow \quad b(al^2 + bm^2)y^2 - 2bcmny + n^2 = 2cal^2 z, \ x = 0$$

This is obviously a parabola whose latus rectum is

$$\frac{2cal^2}{b(al^2 + bm^2)},$$

which is independent of n.

Hence it also follows that sections by parallel diametral planes are equal parabolas.

EXERCISES

1. Prove that the diametral planes $2x + 3y = 4$ and $3x - 4y = 7$ are conjugate diametral planes for the paraboloid $x^2 + 2y^2 = 4z$.

2. The plane $3x + 4y = 1$ is a diametral plane of the paraboloid $5x^2 + 6y^2 = 2z$. Find the equation to the chord through $(3, 4, 5)$ which it bisects. $\left[\text{Ans. } \dfrac{x-3}{9} = \dfrac{y-4}{10} = \dfrac{z-5}{15}\right]$

3. Show that in general three normals can be drawn from a given point to the paraboloid of revolution $x^2 + y^2 = 2az$ but if the point lies on the surface $27a(x^2 + y^2) + 8(a - z)^2 = 0$ two of them coincide.

4. Show that the centre of the circle through the feet of the three normals from the point (α, β, γ) to the paraboloid $x^2 + y^2 = 2az$ is

$$\left(\frac{\alpha}{4}, \frac{\beta}{4}, \frac{\gamma + \alpha}{2}\right).$$

❑❑❑

7
Plane Sections of Conicoids

7.1 INTRODUCTION

Having seen that all plane sections of a conicoid are conics, we shall now proceed to determine the nature, the lengths, and the direction ratios of the axes of a plane section of a given conicoid.

We shall first consider the section of central conicoids, and then the paraboloids.

While determining the nature of plane sections of conicoids, we shall *assume* that the orthogonal projection of a parabola is another parabola, of a hyperbola is another hyperbola and of an ellipse is another ellipse or in some cases a circle.

7.2 NATURE OF THE PLANE SECTION OF A CENTRAL CONICOID

To determine the nature of the section of the central conicoid

$$ax^2 + by^2 + cz^2 = 1 \qquad \ldots(1)$$

by the plane

$$lx + my + nz = p \qquad \ldots(2)$$

The equation to the cylinder passing through the section and having its generators parallel to Z-axis, obtained by eliminating z from (1) and (2), is

$$x^2(an^2 + cl^2) + 2clmxy + y^2(bn^2 + cm^2) - 2clpx - 2cpmy + (cp^2 - n^2) = 0$$

The plane $z = 0$ which is perpendicular to the generating lines of the cylinder cuts it in the conic with equations

$$z = 0$$
$$x^2(an^2 + cl^2) + 2clmxy + y^2(bn^2 + cm^2) - 2clpx - 2cpmy + (cp^2 - n^2) = 0$$

This conic is the projection of the given section on the plane

$$z = 0$$

The projection and, therefore, also the given section is a parabola, hyperbola or ellipse according

as $\quad c^2 l^2 m^2 \begin{cases} = \\ > \\ < \end{cases} (an^2 \; cl^2)(bn^2 + cm^2) \Leftrightarrow bcl^2 + cam^2 + abn^2 \begin{cases} = \\ < \\ > \end{cases} 0$

Thus, we find that the section is

$$\left. \begin{array}{l} \text{a parabola} \\ \text{a hyperbola} \\ \text{an ellipse} \end{array} \right\} \text{according as } bcl^2 + cam^2 + abn^2 \begin{cases} = \\ < \\ > \end{cases} 0$$

7.2.1 Axes of a Central Plane Section

To determine the lengths and direction cosines of the section of the central conicoid

$$ax^2 + by^2 + cz^2 = 1 \qquad \ldots(1)$$

by the central plane

$$lx + my + nz = 0 \qquad \ldots(2)$$

Take a concentric sphere

$$x^2 + y^2 + z^2 = r^2 \qquad \ldots(3)$$

The extremities of all the semi-diameters of length r of the conicoid lie on the curve of intersection of the conicoid and the sphere.

The lines joining the origin to the points on this curve form a cone whose equation, obtained by making (1) and (3) homogeneous, is

$$(ar^2 - 1) x^2 + (br^2 - 1) y^2 + (cr^2 - 1) z^2 = 0 \qquad ...(4)$$

The plane (2) cuts this cone in two generators which determine the directions of two equal diameters of length $2r$ of the section and which are, therefore, equally inclined to the axes of the section.

In case $2r$ is the length of either axis of the section, the generators coincide and as such the plane touches the cone, the generator of contact being one of the axes.

Now, the condition for the plane (2) to touch the cone (4) is

$$\frac{l^2}{ar^2 - 1} + \frac{m^2}{br^2 - 1} + \frac{n^2}{cr^2 - 1} = 0$$

$$\Leftrightarrow (bcl^2 + cam^2 + abn^2) r^4 - [(b+c) l^2 + (c+a) m^2 + (a+b) n^2] r^2$$

$$+ (l^2 + m^2 + n^2) = 0 \qquad ...(5)$$

which is a quadratic equation in r^2 and has two roots r_1^2, r_2^2 which are the squares of the semi-axes of the section.

If λ, μ, ν be the direction ratios of the axis of length $2r$, the plane (2) touches the cone (4) along the line.

$$\frac{x}{\lambda} = \frac{y}{\mu} = \frac{z}{\nu},$$

and is, therefore, identical with the plane.

$$(ar^2 - 1) \lambda x + (br^2 - 1) \mu y + (cr^2 - 1) \nu z = 0 \qquad ...(6)$$

The equations (2) and (6) representing the same plane, we have

$$\frac{\lambda (ar^2 - 1)}{l} = \frac{\mu (br^2 - 1)}{m} = \frac{\nu (cr^2 - 1)}{n} \qquad ...(7)$$

The equations (7) determine the direction ratios of the axis of length $2r$; r being given by the equation (5).

7.2.2. Area of Plane Sections

Assuming that the section is an ellipse, its area

$$= \pi r_1 r_2 = \pi \cdot \frac{\sqrt{l^2 + m^2 + n^2}}{\sqrt{bcl^2 + cam^2 + abn^2}}$$

If p, be the length of the perpendicular from the origin to the tangent plane.

$$lx + my + nz = \left(\frac{l^2}{a} + \frac{m^2}{b} + \frac{n^2}{c} \right)^{1/2}$$

parallel to the given plane $lx + mx + nz = 0$, we have

$$p = \frac{\left(\frac{l^2}{a} + \frac{m^2}{b} + \frac{n^2}{c} \right)^{1/2}}{\sqrt{l^2 + m^2 + n^2}} = \frac{\sqrt{bcl^2 + cam^2 + abn^2}}{\sqrt{l^2 + m^2 + n^2}} \sqrt{\frac{1}{abc}}$$

$$\Rightarrow \qquad \text{area} = \frac{\pi}{p \sqrt{abc}}$$

Plane Section of Conicoids

7.2.3. Axes and Area of Any Central Plane Section of an Ellipsoid

$\dfrac{x^2}{a^2} + \dfrac{y^2}{b^2} + \dfrac{z^2}{c^2} = 1$ by the plane $lx + my + nz = 0$

From eqn. (5) of § 7.2.1, we have the equation quadratic in r^2 as

$$r^4(a^2l^2 + b^2m^2 + c^2n^2) - r^2[a^2(b^2+c^2)l^2 + b^2(c^2+a^2)m^2 + c^2(a^2+b^2)n^2]$$
$$+ a^2b^2c^2(l^2 + m^2 + n^2) = 0 \quad \ldots(i)$$

The direction ratios λ, μ, ν of the axis of length $2r$ are given by

$$\dfrac{\lambda(r^2 - a^2)}{a^2 l} = \dfrac{\mu(r^2 - b^2)}{b^2 m} = \dfrac{\nu(r^2 - c^2)}{c^2 n} \quad \ldots(ii)$$

And the area of the section

$$= \pi \left[\dfrac{a^2 b^2 c^2 (l^2 + m^2 + n^2)}{a^2 l^2 + b^2 m^2 + c^2 n^2} \right]^{1/2} \quad \ldots(iii)$$

$$= \dfrac{\pi abc}{p} \quad \ldots(iv)$$

where p is the length of the perpendicular distance from the centre to the parallel tangent plane

$$lx + my + nz = \sqrt{a^2 l^2 + b^2 m^2 + c^2 n^2}.$$

All the above results can also be deduced from § 7.2.1 and § 7.2.2 putting $\dfrac{1}{a^2}$ for a, $\dfrac{1}{b^2}$ for b and $\dfrac{1}{c^2}$ for c in the corresponding results.

7.2.4. Axes and Area of a Central Plane Section of the Surface

$f(x, y, z) \equiv ax^2 + by^2 + cz^2 + 2fyz + 2gzx + 2hxy = 1$

By the Plane $lx + my + nz = 0$ and to Show that the Axes are the Lines in Which the Plane Cuts a Certain Cone

Let any concentric sphere be

$$x^2 + y^2 + z^2 = r^2 \quad \ldots(i)$$

The semi-diameters of length r are generating lines of the cone given by

$$x^2\left(a - \dfrac{1}{r^2}\right) + y^2\left(b - \dfrac{1}{r^2}\right) + z^2\left(c - \dfrac{1}{r^2}\right) + 2fyz + 2gzx + 2hxy = 0 = F(x, y, z) \text{ (say)} \quad \ldots(ii)$$

Equation (ii) is obtained by making the equation of the surface homogeneous with the help of (i).

The given plane will be a tangent plane to (ii) at the extremities of the semi-axis of length r, if r equals either semi-axes of the section.

By the condition of tangency, we get

$$\begin{vmatrix} a - (1/r^2) & h & g & l \\ h & b - (1/r^2) & f & m \\ g & f & c - (1/r^2) & n \\ l & m & n & 0 \end{vmatrix} = 0 \quad \ldots(iii)$$

where r is the length of either semi-axes of the section.

$\Rightarrow \quad \sum\left[\left(b-\dfrac{1}{r^2}\right)\left(c-\dfrac{1}{r^2}\right)-f^2\right]l^2 + 2\sum\left[gh-\left(a-\dfrac{1}{r^2}\right)f\right]mn = 0$

$\Rightarrow \quad r^4\left[\sum(bc - f^2)l^2 + 2\sum(gh - af)mn\right] - r^2\left[\sum(b+c)l^2 - \sum 2fmn\right]$
$$+ l^2 + m^2 + n^2 = 0 \qquad \text{...(iv)}$$

Again, tangent plane to the cone (ii) containing the axes along

$$\dfrac{x}{\lambda} = \dfrac{y}{\mu} = \dfrac{z}{\nu}$$

is
$$x\dfrac{\partial F}{\partial \lambda} + y\dfrac{\partial F}{\partial \mu} + z\dfrac{\partial F}{\partial \nu} = 0 \qquad \text{...(v)}$$

If this is the same plane
$$lx + my + nz = 0 \qquad \text{...(vi)}$$

then comparing (v) and (vi), we get
$$\dfrac{\partial F/\partial \lambda}{l} = \dfrac{\partial F/\partial \mu}{m} = \dfrac{\partial F/\partial \nu}{n} = t \text{ (say)} \qquad \text{...(vii)}$$

which gives d.c.'s λ, μ, ν of the axes.

$\therefore \quad \dfrac{\partial F}{\partial \lambda} = lt \quad$ or $\quad \dfrac{\partial F}{\partial \lambda} - lt = 0$ etc.

$\therefore \quad \left(a - \dfrac{1}{r^2}\right)\lambda + g\nu + h\nu - lt = 0$ etc.

$\therefore \quad \left.\begin{array}{l} \dfrac{\partial f}{\partial \lambda} - \dfrac{1}{r^2}\lambda - lt = 0 \\ \dfrac{\partial f}{\partial \mu} - \dfrac{1}{r^2}\mu - mt = 0 \\ \dfrac{\partial f}{\partial \nu} - \dfrac{1}{r^2}\nu - nt = 0 \end{array}\right\} \qquad \text{...(viii)}$

and

Hence, we get
$$\begin{vmatrix} \partial f/\partial \lambda & \lambda & l \\ \partial f/\partial \mu & \mu & m \\ \partial f/\partial \nu & \nu & n \end{vmatrix} = 0 \qquad \text{...(ix)}$$

i.e., the axes lie on the cone
$$\sum (mz - 1)\dfrac{\partial f}{\partial x} = 0$$

which is obtained by eliminating λ, μ, ν between (ix) and
$$\dfrac{x}{\lambda} = \dfrac{y}{\mu} = \dfrac{z}{\nu}$$

Also the area of the section $= \pi r_1 r_2 = \pi\left[\dfrac{(l^2 + m^2 + n^2)}{\sum(bc - f^2)l^2 + 2\sum(gh - af)mn}\right]$ [from (iv)]

7.2.5. Condition for the Section to be a Rectangular Hyperbola

For a rectangular hyperbola, we have
$$r_1^2 + r_2^2 = 0$$
$\Leftrightarrow \quad (b+c)l^2 + (c+a)m^2 + (a+b)n^2 = 0.$

Plane Section of Conicoids

Ex. Obtain the condition for the section of the conicoid
$$ax^2 + by^2 + cz^2 = 1$$
by the plane $lx + my + nz = p$ to be a parabola, an ellipse, a hyperbola or a circle from the eqn. (5) of § 7.2.1.

(For a circle, $r_1^2 = r_2^2$)

[**Ans.** The conditions for a circle are
$$l = 0, m^2 (c - a) = n^2 (a - b); \text{ or } m = 0, n^2 (a - b) = l^2 (b - a)$$
or $\quad n = 0, l^2 (b - c) = m^2 (c - a)$]

7.2.6. To Find the Condition for the Two Lines

$$\frac{x}{l_1} = \frac{y}{m_1} = \frac{z}{n_1}, \quad \frac{x}{l_2} = \frac{y}{m_2} = \frac{z}{n_2} \qquad \ldots(i)$$

to be the Axes of the Section by the Plane Through Them

The quadric is
$$ax^2 + by^2 + cz^2 = 1$$

As each of the two lines in (i) will bisect chords of the section parallel to the other, we see that each of them must belong to the diametral plane conjugate to the other.

Now the diameteral plane
$$al_1 x + am_1 y + cn_1 z = 0$$
conjugate to the line
$$\frac{x}{l_1} = \frac{y}{m_1} = \frac{z}{n_1}$$
will contain the second line if
$$al_1 l_2 + bm_1 m_2 + cn_1 n_2 = 0 \qquad \ldots(ii)$$

The condition (ii) is the one sought.

In addition to (ii), we also have the condition.
$$l_1 l_2 + m_1 m_2 + n_1 n_2 = 0$$
for the axes are necessarily perpendicular.

EXAMPLES

1. *Planes are drawn through the origin so as to cut the quadric*
$$ax^2 + by^2 + cz^2 = 1,$$
in rectangular hyperbolas. Prove that the normals to the planes through the origin lie on a quadric cone.

Sol. Consider a plane $\qquad lx + my + nz = 0 \qquad \ldots(1)$
through the origin. The condition for this plane to cut the given quadric in a rectangular hyperbola is
$$(b + c) l^2 + (c + a) m^2 + (a + b) n^2 = 0 \qquad \ldots(2)$$

The normal to the plane (1) through the origin is
$$\frac{x}{l} = \frac{y}{m} = \frac{z}{n} \qquad \ldots(3)$$

Eliminating l, m, n between (2) and (3), we see that the normals, in question, lie on the surface
$$(b + c) x^2 + (c + a) y^2 + (a + b) z^2 = 0$$
which is a quadric cone.

2. *Prove that the axes of the section of the conicoid $ax^2 - by^2 - cz^2 = 1$ by the plane $lx + my + nz = 0$ lie on the cone*

$$(b-c)\frac{1}{x} + (c-a)\frac{m}{y} + (a-b)\frac{n}{v} = 0.$$

Also prove that this can pass through the normal to the plane of the section and the diameter to which the plane of section is diameteral plane.

Sol. Let λ, μ, ν be the direction ratios of the axis of the section.
Then we have

$$\frac{\lambda(ar^2-1)}{l} = \frac{\mu(br^2-1)}{m} = \frac{\nu(cr^2-1)}{n} = k \text{ (say)} \qquad ...(1)$$

$$\therefore \quad \frac{lk}{\lambda} = ar^2 - 1, \quad \frac{mk}{\mu} = br^2 - 1 \text{ and } \frac{nk}{\nu} = cr^2 - 1 \qquad ...(2)$$

Multiplying the relations in (2) by $(b-c), (c-a)$ and $(a-b)$ respectively and adding, we get

$$\frac{l(b-c)}{\lambda} + \frac{m(c-a)}{\mu} + \frac{n(a-b)}{\nu} = \frac{1}{k}[(ar^2-1)(b-c) + (br^2-1)(c-a)$$

$$+ (a-b)(cr^2-1)] = 0 ...(3)$$

Hence the axis $\dfrac{x}{\lambda} = \dfrac{y}{\mu} = \dfrac{z}{\nu}$

lies on the cone $\dfrac{l(b-c)}{x} + \dfrac{m(c-a)}{y} + \dfrac{n(a-b)}{z} = 0 \qquad ...(4)$

Now, equations of the normal to the plane

$$lx + my + nz = 0 \qquad ...(5)$$

are $\dfrac{x}{l} = \dfrac{y}{m} = \dfrac{z}{n} \qquad ...(6)$

Also, the cone whose generators are the axes of the section of conicoid

$$ax^2 + by^2 + cz^2 = 1$$

is given by equation (4).
This cone will contain the line (6) if

$$(b-c) + (c-a) + (a-b) = 0$$

which is obviously true.
Again, equations to the diameter to which (5) is the diameteral plane are

$$\frac{x}{l/a} = \frac{y}{m/b} = \frac{z}{n/c} \qquad ...(7)$$

$\therefore$ The cone (4) will pass through the above line if

$$(b-c)\frac{l}{l/a} + (c-a)\frac{m}{m/b} + (a-b)\frac{n}{n/c} = 0$$

or $\qquad a(b-c) + b(c-a) + c(a-b) = 0$

which holds identically.

3. *One axis of a central section of the conicoid*

$$ax^2 + by^2 + cz^2 = 1$$

line in the plane $\qquad ux + vy + wz = 0.$
Show that the other lies on the cone

$$(b-c) uyz + (c-a) vzx + (a-b) wxy = 0.$$

Sol. Let $\dfrac{x}{l_1} = \dfrac{y}{m_1} = \dfrac{z}{n_1}, \dfrac{x}{l_2} = \dfrac{y}{m_2} = \dfrac{z}{n_2}$

be the two axes of a central section such that the second lies in the given plane for which we have the condition

Plane Section of Conicoids

$$ul_2 + vm_2 + wn_2 = 0 \qquad \text{...(i)}$$

Also, as in § 7.2.6,

$$l_1 l_2 + m_1 m_2 + n_1 n_2 = 0 \qquad \text{...(ii)}$$

$$al_1 l_2 + bm_1 m_2 + cn_1 n_2 = 0 \qquad \text{...(iii)}$$

Eliminating l_2, m_2, n_2 from (i), (ii) and (iii), we have

$$\begin{vmatrix} u & v & w \\ l_1 & m_2 & n_1 \\ al_1 & bm_1 & cn_1 \end{vmatrix} = 0$$

$$\Leftrightarrow \qquad um_1 n_1 (b-c) + vn_1 l_1 (c-a) + wl_1 m_1 (a-b) = 0$$

With the help of this condition we see that the locus of the axis

$$x/l_1 = y/m_1 = z/n_1$$

is the cone

$$(b-c) uyz + (c-a) vzx + (a-b) wxy = 0.$$

EXERCISES

1. Show that the section of the ellipsoid $9x^2 + 6y^2 + 14z^2 = 3$, by the plane $x + y + z = 0$, is an ellipse with semi-axes $1/2$ and $\sqrt{9/22}$. Also obtain their equations.

   [**Ans.** $x/2 = y = -(1/3) z$; $x/4 = -y/5 = z$]

2. Show that the curve
   $$x^2 + 7y^2 - 10z^2 + 9 = 0, \quad x + 2y + 3z = 0$$
   is a hyperbola whose transverse axis is 6 and the direction cosines of whose axes are proportional to $(6, 3, -4)$ and $(17, -22, 9)$.

3. A_1, A_2, A_3 are the areas of three mutually perpendicular central sections of an ellipsoid; show that $A_1^{-2} + A_2^{-2} A_3^{-2}$ is constant.

4. Show that all plane sections of $ax^2 + by^2 + cz^2 = 1$ which are rectangular hyperbolas and which pass through the point (α, β, γ) touch the cone
   $$\frac{(x-\alpha)^2}{b+c} + \frac{(y-\beta)^2}{c+a} + \frac{(z-\gamma)^2}{a+b} = 0.$$

5. Show that any plane whose normal lies on the cone $bcx^2 + cay^2 + abz^2 = 0$, cuts the surface $ax^2 + by^2 + cz^2 = 1$, in a parabola.

6. The director circle of a plane central section of the elipsoid
   $$x^2/a^2 + y^2/b^2 + z^2/c^2 = 1$$
   has a radius of constant length r. Show that the plane section touches the cone
   $$\frac{x^2}{a^2 (b^2 + c^2 - r^2)} + \frac{y^2}{b^2 (c^2 + a^2 - r^2)} + \frac{z^2}{c^2 (a^2 + b^2 - r^2)} = 0.$$

7.3 AXES OF NON-CENTRAL PLANE SECTIONS

To determine the lengths and direction ratios of the section of the central conicoid

$$ax^2 + by^2 + cz^2 = 1 \qquad \text{...(1)}$$

by the plane

$$lx + my + nz = p \qquad \text{...(2)}$$

Centre of the plane section, now, is not the origin. If (α, β, γ) is the centre of the section, the plane (2) is also represented by the equation

$$a\alpha x + b\beta y + c\gamma z = a\alpha^2 + b\beta^2 + c\gamma^2$$

so that
$$\frac{a\alpha}{l} = \frac{b\beta}{m} = \frac{c\gamma}{n} = \frac{a\alpha^2 + b\beta^2 + c\gamma^2}{p} = k, \text{ (say)}$$

$\Leftrightarrow \quad \alpha = \frac{lk}{a}, \beta = \frac{mk}{b}, \gamma = \frac{nk}{c}$

$\Leftrightarrow \quad k = \frac{a\alpha^2 + b\beta^2 + c\gamma^2}{p} = \frac{k^2}{p}\left(\frac{l^2}{a} + \frac{m^2}{b} + \frac{n^2}{c}\right)$

$\Leftrightarrow \quad k = \dfrac{p}{l^2/a + m^2/b + n^2/c}$

If we write $\quad p_0^2 = \dfrac{l^2}{a} + \dfrac{m^2}{b} + \dfrac{n^2}{c}$

we get $\quad \left(\dfrac{lp}{ap_0^2}, \dfrac{mp}{bp_0^2}, \dfrac{np}{cp_0^2}\right)$

as the co-ordinates of the centre of the section. The equation of the conicoid referred to this point as origin is

$$a\left(x + \frac{lp}{ap_0^2}\right)^2 + b\left(y + \frac{mp}{bp_0^2}\right)^2 + c\left(z + \frac{np}{cp_0^2}\right)^2 = 1$$

$\Leftrightarrow \quad ax^2 + by^2 + cz^2 + \dfrac{2p}{p_0^2}(lx + my + nz) + \dfrac{p^2}{p_0^2} = 1$ ...(3)

Also, the equation of the plane (2) becomes
$$lx + my + nz = 0 \qquad \text{...(4)}$$

Now the conic
$$ax^2 + by^2 + cz^2 + (2p/p_0^2)(lx + my + nz) = 1 - (p^2/p_0^2), \; lx + my + nz = 0 \qquad \text{...(5)}$$
is the same as the conic
$$ax^2 + by^2 + cz^2 = 1 - (p^2/p_0^2), \; lx + my + nz = 0 \qquad \text{...(6)}$$

For, points whose co-ordinates satisfy the equations (5) also satisfy the equations (6) and *vice-versa*.

Putting $\quad 1 - (p^2/p_0^2) = d^2$

and replacing a, b, c by $a/d^2, b/d^2, c/d^2$ respectively in the equations (5) and (6) of the previous article 7.2.1, we get

$$\frac{l^2}{ar^2 d^{-2} - 1} + \frac{m^2}{br^2 d^{-2} - 1} + \frac{n^2}{cr^2 d^{-2} - 1} = 0$$

$$\frac{\lambda(ar^2 d^{-2} - 1)}{l} = \frac{\mu(br^2 d^{-2} - 1)}{m} = \frac{\nu(cr^2 d^{-2} - 1)}{n}$$

which give the *lengths* r_1, r_2 and the *direction ratios* l, m, n respectively at the corresponding semi-axes of the section.

7.3.1. Area of the Plane Section

Assuming that the setion is an ellipse, its area

$$= \pi r_1 r_2 = \pi d^2 \left(\frac{l^2 + m^2 + n^2}{bcl^2 + cam^2 + abn^2}\right)^{1/2}$$

$$= \pi \left(1 - \frac{p^2}{l^2/a + m^2/b + n^2/c}\right)\left(\frac{l^2 + m^2 + n^2}{bcl^2 + cam^2 + abn^2}\right)^{1/2}.$$

7.3.2. Parallel Plane Sections

Comparing the equations (7) and (8) with the equations (5) and (6) of the previous article, we see that if α, β be the lengths of the semi-axis of the section by the central plane

$$lx + my + nz = 0 \qquad \ldots(9)$$

are $\qquad d\alpha$ and $d\beta$

$$\Leftrightarrow \qquad \alpha\left(1 - \frac{p^2}{p_0^2}\right)^{1/2} \text{ and } \left(1 - \frac{p^2}{p_0^2}\right)^{1/2}$$

and the corresponding axes are parallel.

Thus, we see that *parallel plane sections of a central conicoid are similar and similarly situated conics.*

Again, if A_0 and A are the area of the sections by the planes (9) and (10), we have

$$A_0 = \pi\alpha\beta, \quad A = \pi d^2 \alpha\beta = A_0\left(1 - \frac{p^2}{p_0^2}\right)$$

It follows that $\qquad \dfrac{A}{A_0} = \left(1 - \dfrac{p^2}{\Sigma l^2/a}\right).$

Note : p/p_0 can easily be seen to be the ratio of the lengths of the perpendicular from the centre to the given plane and to the parallel tangent plane.

7.3.3. Comparison of Lengths of the Axes of Any Section and Those of Parallel Central Section of the Conicoid, $ax^2 + by^2 + cz^2 = 1$

We have already shown that the lengths of the axes of the conic $ax^2 + by^2 + cz^2 = 1$, $lx + my + nz = 1$ are given by

$$\frac{l^2}{ar^2 - 1} + \frac{m^2}{br^2 - 1} + \frac{n^2}{cr^2 - 1} = 0 \qquad \ldots(1) \text{ [By § 7.2.1]}$$

And those of the conic $ax^2 + by^2 + cz^2 = 1, lx + my + nz = p$ are given by

$$\frac{l^2}{a\dfrac{r^2}{k^2} - 1} + \frac{m^2}{b\dfrac{r^2}{k^2} - 1} + \frac{n^2}{c\dfrac{r^2}{k^2} - 1} = 0 \qquad \ldots(2) \text{ [By § 7.3]}$$

Hence if r_1^2 and r_2^2 are the roots of (1) and R_1^2 and R_2^2 those of (2), we have by comparing (1) and (2),

$$\frac{R_1^2}{k^2} = r_1^2 \quad \text{and} \quad \frac{R_2^2}{k^2} = r_2^2$$

i.e., $\qquad R_1 = kr_1 \qquad$ and $\qquad R_2 = kr_2.$

∴ The area A_0 of the section by the prallel central plane $lx + my + nz = 0$ is given by $A = \pi r_1 r_2$.

And the area A of the section by the non-central plane $lx + my + nz = p$ is given by

$$A = \pi R_1 R_2 = \pi k^2 r_1 r_2 = k^2 A_0 (1 - p^2/p_0^2).$$

EXAMPLES

1. Prove that the tangent planes to

$$\frac{x^2}{a^2} + \frac{y^2}{b^2} + \frac{z^2}{c^2} + 1 = 0$$

which cut

$$\frac{x^2}{a^2} + \frac{y^2}{b^2} + \frac{z^2}{c^2} - 1 = 0$$

in ellipse of constant area πk^2 have their points of contact on the surface

$$\frac{x^2}{a^4} + \frac{y^2}{b^4} + \frac{z^2}{c^4} = \frac{k^4}{4a^2 b^2 c^2}.$$

Sol. Equation of tangent plane to

$$\frac{x^2}{a^2} + \frac{y^2}{b^2} - \frac{z^2}{c^2} + 1 = 0$$

at any point (x_1, y_1, z_1) is

$$\frac{xx_1}{a^2} + \frac{yy_1}{b^2} + \frac{zz_1}{c^2} = -1 \qquad \ldots(1)$$

where

$$\frac{x_1^2}{a^2} + \frac{y_1^2}{b^2} + \frac{z_1^2}{c^2} = -1 \qquad \ldots(2)$$

Now, if A_0 be the area of the corresponding central section $\dfrac{xx_1}{a^2} + \dfrac{yy_1}{b^2} - \dfrac{zz_1}{c^2} = 0$ of the second conicoid

$$\frac{x^2}{a^2} + \frac{y^2}{b^2} - \frac{z^2}{c^2} = 1$$

then

$$A = \left(1 - \frac{p^2}{p_0^2}\right) A_0 \qquad \ldots(3)$$

Now, $A = \pi k^2$ given and $p^2 = (-1)^2 = 1$,

and

$$p_0^2 = a^2 \frac{x_1^2}{a^4} + b^2 \frac{y_1^2}{b^4} - c^2 \frac{z_1^2}{c^4} = -1, \qquad \text{[from (2)]}$$

and

$$A_0 = \frac{\pi ab \sqrt{-c^2} \sqrt{l^2 + m^2 + n^2}}{\sqrt{a^2 l^2 + b^2 m^2 - c^2 n^2}}$$

In the usual formula c^2 is put equal to $-c^2$,

$$= \frac{\pi ab \sqrt{-c^2} \sqrt{\dfrac{x_1^2}{a^4} + \dfrac{y_1^2}{b^4} + \dfrac{z_1^2}{c^4}}}{\sqrt{-1}} \qquad \text{[from (2)]}$$

Squaring relation (3) and putting the values, we get

$$\pi^2 k^4 = \left(1 - \frac{1}{-1}\right)^2 \frac{\pi^2 a^2 b^2 (-c^2)}{-1} \left(\frac{x_1^2}{a^4} + \frac{y_1^2}{b^4} + \frac{z_1^2}{c^4}\right)$$

or

$$\frac{k^4}{4a^2 b^2 c^2} = \frac{x_1^2}{a^4} + \frac{y_1^2}{b^4} + \frac{z_1^2}{c^4}$$

Hence the locus of (x_1, y_1, z_1), we have

$$\frac{x_1^2}{a^4}+\frac{y_1^2}{b^4}+\frac{z_1^2}{c^4}=\frac{k^4}{4a^2b^2c^2}.$$

2. *Through a given point (α, β, γ) planes are drawn parallel to three conjugate diametral planes of the ellipsoid*

$$\frac{x^2}{a^2}+\frac{y^2}{b^2}+\frac{z^2}{c^2}=1.$$

Show that the sum of the ratios of the area of the section by these planes to the areas of the parallel diametral planes is

$$3-\frac{\alpha^2}{a^2}+\frac{\beta^2}{b^2}+\frac{\gamma^2}{c^2}.$$

Sol. Equations of the conjugate dimentral planes are

$$\frac{xx_1}{a^2}+\frac{yy_1}{b^2}+\frac{zz_1}{c^2}=0 \text{ etc.} \qquad \ldots(1)$$

and that of a parallel plane through (α, β, γ) is

$$(x-\alpha)\frac{x_1}{a^2}+(y-\beta)\frac{y_1}{b^2}+(z-\gamma)\frac{z_1}{c^2}=0$$

or

$$\frac{xx_1}{a^2}+\frac{yy_1}{b^2}+\frac{zz_1}{c^2}=\frac{\alpha x_1}{a^2}+\frac{\beta y_1}{b^2}+\frac{\gamma z_1}{c^2}=p_1 \text{ (say)} \qquad \ldots(2)$$

Let A_1 be the area of the section of the ellipsoid

$$\frac{x^2}{a^2}+\frac{y^2}{b^2}+\frac{z^2}{c^2}=1$$

by (1) and A_1' be the corresponding area of the section by (2), then we know that

$$A_1' = \left(1-\frac{p_1^2}{p_0^2}\right) A_1$$

$$\therefore \quad \frac{A_1'}{A_1}=1-\frac{p_1^2}{p_0^2}$$

where

$$p_0^2 = \frac{l^2}{a}+\frac{m^2}{b}+\frac{n^2}{c}$$

when the conicoid is $ax^2+by^2+cz^2=1$ and the plane is $lx+my+nz=p$. Hence p_0^2 in this case will be

$$a^2 \cdot \frac{x_1^2}{a^4}+b^2\cdot\frac{y_1^2}{b^4}+c^2\cdot\frac{z_1^2}{c^4}=\frac{x_1^2}{a^2}+\frac{y_1^2}{b^2}+\frac{z_1^2}{c^2}=1.$$

$$\therefore \quad \frac{A_1'}{A_1}=1-p_1^2=1-\left(\frac{\alpha x_1}{a^2}+\frac{\beta y_1}{b^2}+\frac{\gamma z_1}{c^2}\right)^2 \qquad \text{[from (2)]}$$

$$\therefore \quad \Sigma \frac{A_1'}{A_1}=\left[1-\left(\frac{\alpha x_1}{a^2}+\frac{\beta y_1}{b^2}+\frac{\gamma z_1}{c^2}\right)^2\right]+\left[1-\left(\frac{\alpha x_2}{a^2}+\frac{\beta y_2}{b^2}+\frac{\gamma z_2}{c^2}\right)^2\right]$$

$$+\left[1-\left(\frac{\alpha x_3}{a^2}+\frac{\beta y_3}{b^2}+\frac{\gamma z_3}{c^2}\right)^2\right]$$

$$= 3 - \left[\frac{\alpha^2}{a^4} \Sigma x_1^2 + \frac{\beta^2}{b^4} \Sigma y_1^2 + \frac{\gamma^2}{c^4} \Sigma z_1^2 \right] \qquad \text{[other terms vanish]}$$

$$= 3 - \frac{\alpha^2}{a^2} + \frac{\beta^2}{b^2} + \frac{\gamma^2}{c^2}.$$

3. Prove that if $l_1, m_1, n_1; l_2, m_2, n_2$ are the direction cosines of the axes of any plane section of the ellipsoid

$$\frac{x^2}{a^2} + \frac{y^2}{b^2} + \frac{z^2}{c^2} = 1,$$

then $\quad \dfrac{l_1 l_2}{a^2 (b^2 - c^2)} = \dfrac{m_1 m_2}{b^2 (c^2 - a^2)} = \dfrac{n_1 n_2}{c^2 (a^2 - b^2)}.$

Sol. Let l, m, n be the d.c.'s of either axis of the section of

$$\frac{x^2}{a^2} + \frac{y^2}{b^2} + \frac{z^2}{c^2} = 1 \qquad \ldots(1)$$

by the plane $\quad Lx + My + Nz = p \qquad \ldots(2)$

then $\quad \dfrac{r^2 - a^2 k^2}{L a^2 k^2 / l} = \dfrac{r^2 - b^2 k^2}{M b^2 k^2 / m} = \dfrac{r^2 - c^2 k^2}{N c^2 k^2 / n} = \lambda \text{ (say)}$

$\therefore \quad \dfrac{L a^2 k^2}{l} (b^2 - c^2) + \dfrac{M b^2 k^2}{m} (c^2 - a^2) + \dfrac{N c^2 k^2}{n} (a^2 - b^2)$

$$= \left(\frac{1}{\lambda} \right) [(r^2 - a^2 k^2)(b^2 - c^2) + (r^2 - b^2 k^2)(c^2 - a^2)$$
$$+ (r^2 - c^2 k^2)(a^2 - b^2)] = 0$$

or $\quad \Sigma \dfrac{L a^2}{l} (b^2 - c^2) = 0$

Again, since the axes lie in the plane (2), hence

$$lL + mM + nN = 0 \qquad \ldots(4)$$

Eliminating n between (3) and (4), we get

$$\frac{L a^2 (b^2 - c^2)}{l} + \frac{M b^2 (c^2 - a^2)}{m} - \frac{N c^2 (a^2 - b^2) N}{(Ll + Mm)} = 0$$

or $\quad LM b^2 l^2 (c^2 - a^2) + lm [L^2 a^2 (b^2 - c^2) + M^2 b^2 (c^2 - a^2) - N^2 c^2 (a^2 - b^2)]$
$$+ LM a^2 m^2 (b^2 - c^2) = 0 \qquad \ldots(5)$$

Now, the d.c.'s of axes satisfy (3) and (4), and hence satisfy (5).

Thus if l_1, m_1, n_1 and l_2, m_2, n_2 are the d.c.'s of the axes, we have

$$\frac{l_1 l_2}{LM a^2 (b^2 - c^2)} = \frac{m_1 m_2}{LM b^2 (c^2 - a^2)}$$

or $\quad \dfrac{l_1 l_2}{a^2 (b^2 - c^2)} = \dfrac{m_1 m_2}{b^2 (c^2 - a^2)} = \dfrac{n_1 n_2}{c^2 (a^2 - b^2)}.$

EXERCISES

1. Find the length and directions of the axes of the section of the ellipsoid
$$9x^2 + 6y^2 + 14z^2 = 3$$
by the plane $x + y + z = 1.$ [**Ans.** $3/22, 44\sqrt{22}, (4, -5, 1), (2, 1, -3)$]

Plane Section of Conicoids

2. Show that the plane $x + y + z = 1$ cuts the quadric $11x^2 - 13y^2 - 4z^2 = 5$ in a hyperbola and find the direction ratios of its axes.   [**Ans.** $-3, 1, 2; 1, -5, 4$]

3. Show that the plane $x + 2y + 3z = 4$ cuts the conicoid $2x^2 + y^2 - 2z^2 = 1$ in a parabola, the direction cosines of whose axis are proportional to $1, 4, -3$.

4. The ellipsoid $x^2 + 2y^2 + 3z^2 = 1$ is cut by parallel planes
$$2x + 3y + 4z = 2, \quad 2x + 3y + 4z = 3;$$
show that the areas of the sections made by the planes are in the ratio $59 : 29$.

5. Find the locus of the centres of the section of the ellipsoid
$$x^2/a^2 + y^2/b^2 + z^2/c^2 = 1$$
which are of constant area πk^2.

$$\left[\textbf{Ans. } a^2 b^2 c^2 \left(\frac{x^2}{a^4} + \frac{y^2}{b^4} + \frac{z^2}{c^4}\right)\left(1 - \frac{x^2}{a^2} - \frac{y^2}{b^2} - \frac{z^2}{c^2}\right) = k^2 \left(\frac{x^2}{a^2} + \frac{y^2}{b^2} + \frac{z^2}{c^2}\right)\right]$$

6. Prove that the area of the section of the ellipsoid
$$\frac{x^2}{a^2} + \frac{y^2}{b^2} + \frac{z^2}{c^2} = 1$$
by the plane $lx + my + nz = p$, is given by
$$\frac{\pi abc\, (l^2 + m^2 + n^2)^{1/2}}{(a^2 l^2 + b^2 m^2 + c^2 n^2)^{1/2}} \left[1 - \frac{p^2}{a^2 l^2 + b^2 m^2 + c^2 n^2}\right].$$

7.4 CIRCULAR SECTIONS

To determine the circular sections of the ellipsoid
$$\frac{x^2}{a^2} + \frac{y^2}{b^2} + \frac{z^2}{c^2} = 1$$

We have to find the equations of the planes which cut the ellipsoid in circles. We suppose that $a^2 > b^2 > c^2$.

Writing the equation of the ellipsoid in the form
$$\frac{1}{b^2}(x^2 + y^2 + z^2 - b^2) + x^2\left(\frac{1}{a^2} - \frac{1}{b^2}\right) + z^2\left(\frac{1}{c^2} - \frac{1}{b^2}\right) = 0$$

$\Leftrightarrow \quad (x^2 + y^2 + z^2 - b^2) = \dfrac{a^2 - b^2}{a^2} x^2 - \dfrac{b^2 - c^2}{c^2} z^2$

we see that the two planes
$$\frac{a^2 - b^2}{a^2} x^2 - \frac{b^2 - c^2}{c^2} z^2 = 0$$
meet the ellipsoid where they meet the sphere
$$x^2 + y^2 + z^2 = b^2$$
but as a plane necessarily cuts a sphere in a circle, we find that the planes (2) cut the ellipsoid (1) in circles.

Since parallel sections are similar, *the two systems of planes*
$$\frac{x}{a}\sqrt{a^2 - b^2} + \frac{z}{c}\sqrt{b^2 - c^2} = \lambda,$$

and
$$\frac{x}{a}\sqrt{a^2 - b^2} - \frac{z}{c}\sqrt{b^2 - c^2} = \mu$$

which are parallel to those given by the equations (5) cut the ellipsoid in circles for all values of λ and μ.

Note : If we rewrite the equation of the ellipsoid in the form

$$\frac{1}{a^2}(x^2+y^2+z^2-a^2)+y^2\left(\frac{1}{b^2}-\frac{1}{a^2}\right)+z^2\left(\frac{1}{c^2}-\frac{1}{a^2}\right)=0 \qquad ...(3)$$

$$\frac{1}{c^2}(x^2+y^2+z^2-c^2)+x^2\left(\frac{1}{a^2}-\frac{1}{c^2}\right)+y^2\left(\frac{1}{b^2}-\frac{1}{c^2}\right)=0 \qquad ...(4)$$

we find that the pairs of planes

$$y^2\left(\frac{1}{b^2}-\frac{1}{a^2}\right)+z^2\left(\frac{1}{c^2}-\frac{1}{a^2}\right)=0$$

$$x^2\left(\frac{1}{a^2}-\frac{1}{c^2}\right)+y^2\left(\frac{1}{b^2}-\frac{1}{c^2}\right)=0$$

also cut the ellipsoid in circles. It may be seen, however, that these pairs of planes are not real in that there are no points which satisfy these equations.

7.4.1. Any Two Circular Sections of an Ellipsoid of Opposite Systems Lie on a Sphere

Let
$$\frac{x}{a}\sqrt{a^2-b^2}+\frac{z}{c}\sqrt{b^2-c^2}=\lambda,$$

and
$$\frac{x}{a}\sqrt{a^2-b^2}-\frac{z}{c}\sqrt{b^2-c^2}=\mu$$

be the equations of the planes of any two circular sections of opposite systems.
The conicoid

$$\frac{x^2}{a^2}+\frac{y^2}{b^2}+\frac{z^2}{c^2}-1+k\left[\frac{x}{a}\sqrt{a^2-b^2}+\frac{z}{c}\sqrt{b^2-c^2}-\lambda\right]$$

$$+\left[\frac{x}{a}\sqrt{a^2-b^2}-\frac{z}{c}\sqrt{b^2-c^2}-\mu\right]=0 \qquad ...(1)$$

which passes through the two circular sections for all values of k, will represent a sphere, if k satisfies the equations

$$\frac{1}{a^2}+\frac{k(a^2-b^2)}{a^2}=\frac{1}{b^2}=\frac{1}{c^2}-\frac{k(b^2-c^2)}{c^2}$$

Now, $k=1/b^2$ clearly satisfies these two equations.
Substituting this value of k in (1), we get the equation

$$x^2+y^2+z^2-\frac{(\lambda+\mu)\sqrt{a^2-b^2}}{a}x+\frac{(\lambda-\mu)\sqrt{b^2-c^2}}{c}z+\lambda\mu-b^2=0$$

representing the sphere through the two circular sections.
Hence, the proposition is proved.

7.4.2. Circular Sections of Any Central Conicoid

$$f(x, y, z) \equiv ax^2 + by^2 + cz^2 + 2fyz + 2gzx + 2hxy - 1 = 0$$

The given equation can be written in the following form

$$(ax^2+by^2+cz^2+2fyz+2gzx+2hxy)-\lambda(x^2+y^2+z^2)$$

$$+\lambda\left(x^2+y^2+z^2-\frac{1}{\lambda}\right)=0 \qquad ...(1)$$

Hence if $\quad (ax^2+by^2+cz^2+2fyz+2gzx+2hxy)-\lambda(x^2+y^2+z^2)=0 \qquad ...(2)$

Plane Section of Conicoids

represents a pair of planes, they will cut the given conicoid in circles.
Obviously, (2) will represent a pair of planes, if

$$\begin{vmatrix} a-\lambda & h & g \\ h & b-\lambda & f \\ g & f & c-\lambda \end{vmatrix} = 0 \qquad \ldots(3)$$

The equation (3) being a cubic in λ gives three values of λ which are all real. But only one value of λ gives real sections.

EXAMPLES

1. *Prove that the sections of hyperboloid* $\dfrac{x^2}{a^2} - \dfrac{y^2}{b^2} - \dfrac{z^2}{c^2} = 1$ *by the plane*

$$\frac{x}{a}\sqrt{a^2+b^2} + \frac{z}{c}\sqrt{b^2-c^2} = \lambda$$

is real if $\lambda^2 > a^2 + c^2$.

Sol. Equation to a sphere through the section of given hyperboloid by the given plane is

$$b^2 \left(-\frac{x^2}{a^2} + \frac{y^2}{b^2} + \frac{z^2}{c^2} + 1 \right) + \left[\frac{x}{a}\sqrt{a^2+b^2} - \frac{z}{c}\sqrt{b^2-c^2} \right]$$

$$\times \left[\frac{x}{a}\sqrt{a^2+b^2} + \frac{z}{c}\sqrt{b^2-c^2} - \lambda \right] = 0$$

The centre C of this sphere is

$$\left[\frac{\lambda\sqrt{a^2+b^2}}{2a}, 0, \frac{-\lambda\sqrt{b^2-c^2}}{2c} \right]$$

and radius is given by

$$R^2 = \frac{\lambda^2(a^2+b^2)}{4a^2} + \frac{\lambda^2(b^2-c^2)}{4c^2} - b^2$$

$$\Rightarrow \quad R^2 = \frac{\lambda^2 b^2 (c^2+a^2)}{4a^2 c^2} - b^2$$

If p be the distance of the section from the centre of the sphere, then
p = perpendicular from the centre C on the plane

$$= \frac{\dfrac{\lambda(a^2+b^2)}{2a^2} - \dfrac{\lambda(b^2-c^2)}{2c^2} - \lambda}{\left(\dfrac{a^2+b^2}{a^2} + \dfrac{b^2-c^2}{c^2} \right)^{1/2}} = \frac{\lambda b^2(c^2-a^2)}{2ca\sqrt{b^2(a^2+c^2)}}$$

Radius of the circular section, r, is given

$$r^2 = R^2 - p^2$$

It will be positive if $R^2 > p^2$

$$\Rightarrow \quad \frac{\lambda^2 b^2 (a^2+c^2)}{4a^2 c^2} - b^2 > \frac{\lambda^2 b^2 (c^2-a^2)}{4c^2 a^2 (a^2-c^2)}$$

$$\Rightarrow \quad \frac{\lambda^2 b^2}{(a^2+c^2)} > b^2$$

$$\Rightarrow \qquad \lambda^2 > (a^2 + c^2), \text{ since } b \neq 0.$$

2. *Show that the circular sections of the ellipsoid* $\dfrac{x^2}{a^2} + \dfrac{y^2}{b^2} + \dfrac{z^2}{c^2} = 1$ *passing through one extremity of x-axis are both of radius r, where*

$$\frac{r^2}{b^2} = \frac{(b^2 - c^2)}{(a^2 - c^2)}.$$

Sol. Real circular sections of the ellipsoid are

$$\frac{x}{a}\sqrt{a^2 - b^2} + \frac{z}{c}\sqrt{b^2 - c^2} = \lambda_1$$

and $$\frac{x}{a}\sqrt{a^2 - b^2} - \frac{z}{c}\sqrt{b^2 - c^2} = \lambda_2$$

The radius r of the circle in which (i) cuts the ellipsoid is given by

$$r = b\left(1 - \frac{\lambda_1^2}{a^2 - c^2}\right)^{1/2} \qquad \text{...(iii)}$$

Similarly, the radius of the circle in which (ii) cuts the ellipsoid is

$$b\left(1 - \frac{\lambda_1^2}{a^2 - c^2}\right)^{1/2}$$

As plane (i) passes through $(a, 0, 0)$

$$\therefore \qquad \lambda_1 = \sqrt{a^2 - b^2}$$

$$\Rightarrow \qquad r^2 = b^2\left(1 - \frac{a^2 - b^2}{a^2 - c^2}\right)$$

$$\Rightarrow \qquad \frac{r^2}{b^2} = \frac{b^2 - c^2}{a^2 - c^2}.$$

Similarly, we can show it for plane (ii) also.

3. *If* p_1, p_2, p_3 *be the lengths of the perpendiuclars from the extremities* P_1, P_2, P_3 *of conjugate semi-diameters on one of the planes of central circular section of the ellipsoid* $\dfrac{x^2}{a^2} + \dfrac{y^2}{b^2} + \dfrac{z^2}{c^2} = 1$, *then prove that* $p_1^2 + p_2^2 + p_3^2 = a^2c^2/b^2$.

Sol. Let $(x_1, y_1, z_1), (x_2, y_2, z_2)$ and (x_3, y_3, z_3) be the extremities of conjugate semi-diameters. One of the real central circular sections of

$$\frac{x^2}{a^2} + \frac{y^2}{b^2} + \frac{z^2}{c^2} = 1 \qquad \text{...(i)}$$

is $$\frac{x}{a}\sqrt{a^2 - b^2} + \frac{z}{c}\sqrt{b^2 - c^2} = 0, a > b > c \qquad \text{...(ii)}$$

p_1 = perpendicular from (x_1, y_1, z_1) on (ii)

$$= \frac{(x_1/a)\sqrt{a^2 - b^2} + (z_1/c)\sqrt{b^2 - c^2}}{\sqrt{(a^2 - b^2)/a^2 + (b^2 - c^2)/c^2}}$$

Plane Section of Conicoids

$$\therefore \quad p_1^2 = \frac{\left(\dfrac{x_1^2}{a^2}\right)(a^2 - b^2) + \left(\dfrac{z_1^2}{c^2}\right)(b^2 - c^2) + \left(\dfrac{2x_1 z_1}{ac}\right)\sqrt{(a^2-b^2)(b^2-c^2)}}{\dfrac{b^2}{c^2} - \dfrac{b^2}{a^2}}$$

$$= \frac{a^2 c^2}{b^2(a^2-c^2)}\left[\left(\frac{a^2-b^2}{a^2}\right)x_1^2 + \left(\frac{b^2-c^2}{c^2}\right)z_1^2 + \frac{2x_1 z_1}{ac}\sqrt{(a^2-b^2)(b^2-c^2)}\right]$$

$$\therefore \quad p_1^2 + p_2^2 + p_3^2 = \frac{a^2 c^2}{b^2(a^2-c^2)}\left[\left(\frac{a^2-b^2}{a^2}\right)(\Sigma x_1^2) + \left(\frac{b^2-c^2}{c^2}\right)(\Sigma z_1^2)\right.$$
$$\left.+ \frac{2\sqrt{(a^2-b^2)(b^2-c^2)}}{ac}\times \Sigma(x_1 z_1)\right]$$

$$= \frac{a^2 c^2}{b^2(a^2-c^2)}\left[\left(\frac{a^2-b^2}{a^2}\right)a^2 + \left(\frac{b^2-c^2}{c^2}\right)c^2\right]$$

[since $\Sigma x_1^2 = a^2$, $\Sigma z_1^2 = c^2$ and $\Sigma x_1 x_2 = 0$]

$$= a^2 c^2 / b^2.$$

EXERCISES

1. Find the locus of the centres of spheres of constant radius k which cut ellipsoid $\dfrac{x^2}{a^2} + \dfrac{y^2}{b^2} + \dfrac{z^2}{c^2} = 1$ in a pair of circles. $\left[\text{Ans. } \dfrac{x^2}{a^2-b^2} - \dfrac{z^2}{b^2-c^2} = 1 - \dfrac{k^2}{b^2},\ y = 0\right]$

2. Find the condition that the section of the ellipsoid $\dfrac{x^2}{a^2} + \dfrac{y^2}{b^2} + \dfrac{z^2}{c^2} = 1$ by the plane $lx + my + nz = p$ may be a circle. $\left[\text{Ans. } \dfrac{al}{\sqrt{a^2-b^2}} = \dfrac{m}{0} = \dfrac{cn}{\pm\sqrt{b^2-c^2}}\right]$

3. Show that the central circular sections of the hyperboloid
$$\frac{x^2}{a^2} + \frac{y^2}{b^2} - \frac{z^2}{c^2} = 1$$
are given by the planes
$$\frac{y}{b}\sqrt{a^2-b^2} \pm \frac{z}{c}\sqrt{a^2+c^2} = 0.$$
Also show that any two circular sections of opposite systems in the case of either hyperboloid lie on a sphere.

4. Find the circular sections of the following conicoids:
   (i) $2x^2 + 11y^2 + z^2 = 1$; (ii) $10x^2 - 2y^2 + z^2 + 2 = 0$; (iii) $15x^2 + y^2 - 10z^2 + 4 = 0$.
   [Ans. (i) $3y + z = \lambda$, $3y - z = \mu$; (ii) $\sqrt{3}x + y = \lambda$, $\sqrt{3}x - y = \mu$;
   (iii) $4x + 3z = \lambda$, $4x - 3z = \mu$]

5. Find the equation of the sphere which contains the two circular sections for the ellipsoid $x^2 - 3y^2 + 2z = 4$ through the point $(1, 2, 3)$. [Ans. $x^2 + y^2 + z^2 - 16y + 6z + 7 = 0$]

6. Find the radius of the circle in which the plane

$$\frac{x}{a}\sqrt{a^2-b^2} + \frac{z}{a}\sqrt{b^2-c^2} = \lambda$$

cuts the ellipsoid $\dfrac{x^2}{a^2} + \dfrac{y^2}{b^2} + \dfrac{z^2}{c^2} = 1.$ [**Ans.** $b\sqrt{\{1 - \lambda^2/(a^2-c^2)\}}$]

7.4.2. Umbilics

Def. *A point on a quadric such that the planes parallel to the tangent plane at the point determine circular sections is called an* **umbilic.**

Clearly, an *umbilic* is a point-circle lying on a quadric.

The umbilics are the extremities of the diameters which pass through the centres of the system of circular sections.

To determine the umbilics of the ellipsoid,

$$\frac{x^2}{a^2} + \frac{y^2}{b^2} + \frac{z^2}{c^2} = 1.$$

If f, g, h be an umbilic, the tangent plane

$$\frac{fx}{a^2} + \frac{gy}{b^2} + \frac{hz}{c^2} = 1 \qquad \ldots(1)$$

at the point is parallel to either of the central circular sections

$$\frac{x}{a}\sqrt{a^2-b^2} \pm \frac{z}{c}\sqrt{b^2-c^2} = 0 \qquad \ldots(2)$$

From (1) and (2), we see that

$$g = 0 \text{ and } \frac{f}{a\sqrt{a^2-b^2}} = \pm \frac{h}{c\sqrt{b^2-c^2}}$$

Also

$$\frac{f^2}{a^2} + \frac{g^2}{b^2} + \frac{h^2}{c^2} = 1.$$

Hence, $\quad f = \pm \dfrac{a\sqrt{a^2-b^2}}{\sqrt{a^2-c^2}}, \quad g = 0, \quad h = \pm \dfrac{c\sqrt{b^2-c^2}}{\sqrt{a^2-c^2}}.$

These are the co-ordinates of the four real umbilics.

EXAMPLES

1. *Prove that the perpendicular distance from the centre of the tangent planes at an umbilic of the ellipsoid* $\dfrac{x^2}{a^2} + \dfrac{y^2}{b^2} + \dfrac{z^2}{c^2} = 1$ *is* $\dfrac{ac}{b}$.

Sol. Let $P(\alpha, \beta, \gamma)$ be an umbilic. Tangent plane at P is

$$\alpha x/a^2 + \beta y/b^2 + \gamma z/c^2 - 1 = 0.$$

Its distance p from centre $(0, 0, 0)$ is given by

$$p = \frac{1}{\sqrt{(\alpha^2/a^4 + \beta^2/b^4 + \gamma^2/c^4)}} \qquad \ldots(i)$$

But $\quad \alpha = \pm \dfrac{a\sqrt{a^2-b^2}}{\sqrt{(a^2-c^2)}}, \quad \beta = 0, \quad \gamma = \pm \dfrac{c\sqrt{(b^2-c^2)}}{\sqrt{(a^2-c^2)}}$

Put the values in (i) and get the result.

Plane Section of Conicoids

2. *Prove that the umbilics of the conicoid* $\dfrac{x^2}{(a+b)} + \dfrac{y^2}{a} + \dfrac{z^2}{(a-b)} = 1$ *are the extremities of the equal conjugate diameters of the ellipse*

$$y = 0, \quad \frac{x^2}{a+b} + \frac{z^2}{a-b} = 1.$$

Sol. Let $P(\alpha, \beta, \gamma)$ be an umbilic. Then

$$\alpha = \pm \frac{\sqrt{(a+b)[(a+b)-a]}}{\sqrt{(a-b)-(a-b)}} = \pm\sqrt{(a+b)/2},\, \beta = 0 \text{ and } \gamma = \pm\sqrt{(a-b)/2}$$

Let the extremities of equal conjugate diameters of the given ellipse be

$$\{\sqrt{a+b}\cos\phi, 0, \sqrt{a-b}\sin\phi\} \text{ and } \{-\sqrt{a+b}\sin\phi, 0, \sqrt{a-b}\cos\phi\}$$

These diameters are equal if

$$(a+b)\cos^2\phi + (a-b)\sin^2\phi = (a+b)\sin^2\phi + (a-b)\cos^2\phi$$

$$\Rightarrow \qquad \tan^2\phi = 1 \text{ or } \phi = \pm\pi/4$$

∴ The extremities of equal conjugate diameters are

$$[\pm\sqrt{(a+b)/2}, 0, \pm\sqrt{(a-b)/2}].$$

EXERCISES

1. Show that the hyperboloid of one sheet has no real umbilics.
2. Find the real umbilics of the hyperboloid $\dfrac{x^2}{a^2} - \dfrac{y^2}{b^2} - \dfrac{z^2}{c^2} = 1$.

$$\left[\text{Ans. } \frac{\pm a\sqrt{(a^2+b^2)}}{\sqrt{(a^2+c^2)}}, 0, \frac{\pm c\sqrt{(b^2+c^2)}}{\sqrt{(a^2+c^2)}}\right]$$

3. Find the umbilics of the ellipsoid $2x^2 + 3y^2 + 6z^2 = 6$. $\left[\text{Ans. } \left(\pm\dfrac{1}{2}\sqrt{6}, 0, \pm\dfrac{1}{2}\sqrt{2}\right)\right]$
4. Show that the four real umbilics of an ellipsoid lie upon a circle.

7.5 SECTIONS OF PARABOLOIDS

To determine the nature of the sections of the paraboloid

$$ax^2 + by^2 = 2cz,$$

by the plane $\qquad lx + my + nz = p.$

Let $l \neq 0$ so that the plane is not perpendicular to the YZ plane which is a plane of symmetry of the surfaces. As in § 7.3, the equation of the projection of the section of the YZ plane are

$$x = 0,$$

$$(am^2 + bl^2)y^2 + 2amnyz + an^2z^2 - 2apmy - 2(apn - cl^2)z + ap^2 = 0$$

The projection and, therefore, also the section is an ellipse, parabola or hyperbola according as

$$a^2m^2n^2 - an^2(am^2 + bl^2) \begin{cases} < \\ = 0 \\ > \end{cases} \Leftrightarrow abn^2l^2 \begin{cases} > \\ = 0. \\ < \end{cases}$$

Thus, *for a parabola $n = 0$*. If $n \neq 0$, the section will be an ellipse or hyperbola according as ab is positive or negative, that is, according as the paraboloid is elliptic or hyperbolic.

If $l = 0$ and $m \neq 0$ then, by projecting on the XZ plane, we get a similar result.

If $l = m = 0$ then n cannot be equal to zero and the section is then clearly an ellipse or hyperbola according as ab is positive or negative.

Thus, we have proved the following :

All the sections of a paraboloid (elliptic or hyperbolic) which are parallel to the axis of the surface are parabolas; all other sections of an elliptic paraboloid are ellipses and of an hyperbolic paraboloid are hyperbolas.

7.5.1. Axes of Plane Sections of Paraboloids

To determine the lengths and the direction ratios of the section of the paraboloid

$$ax^2 + by^2 = 2cz \qquad \ldots(1)$$

by the plane
$$lx + my + nz = p$$

Let (α, β, γ) be the centre of the section so that the plane (2) is also represented by the equation

$$a\alpha x + b\beta y - cz = a\alpha^2 + b\beta^2 - c\gamma$$

Comparison gives

$$\frac{a\alpha}{l} = \frac{b\beta}{m} = \frac{-c}{n} = \frac{a\alpha^2 + b\beta^2 + c\gamma^2}{p}$$

$$\Rightarrow \quad \begin{cases} \alpha = -\dfrac{lc}{an}, \; \beta = -\dfrac{mc}{bn} \\ \\ c\gamma = a\alpha^2 + b\beta^2 + \dfrac{pc}{n} = \dfrac{c^2}{n^2}\left(\dfrac{l^2}{a} + \dfrac{m^2}{b} + \dfrac{np}{c}\right) \end{cases}$$

If we write
$$k = \frac{l^2}{a} + \frac{l^2}{b} + \frac{np}{c}$$

we find that the centre of the section is

$$\left(-\frac{lc}{an}, -\frac{mc}{bn}, \frac{kc}{n^2}\right)$$

The equation of the paraboloid referred to this point as the origin is

$$a\left(x - \frac{lc}{an}\right)^2 + b\left(y - \frac{mc}{bn}\right)^2 = 2c\left(z + \frac{kc}{n}\right)$$

$$\Leftrightarrow \quad ax^2 + by^2 - \frac{2c}{n}(lx + my + nz) - \frac{c(kc + np)}{n^2} = 0$$

Also, the equation of the plane (2) now becomes
$$lx + my + nz = 0$$

Now the conic

$$\left.\begin{array}{r} ax^2 + by^2 - \dfrac{2c}{n}(lx + my + nz) - \dfrac{c(kc + np)}{n^2} = 0 \\ lx + my + nz = 0 \end{array}\right\} \qquad \ldots(3)$$

is the same as the conic

$$ax^2 + by^2 = \frac{c(kc + np)}{n^2}, \; lx + my + nz = 0$$

Let us write
$$p_0^2 = c(kc + np) = c\left(\frac{l^2 c}{a} + \frac{m^2 c}{b} + 2np\right)$$

The semi-diameters of length r of the conicoid

Plane Section of Conicoids

$$ax^2 + by^2 = \frac{p_0^2}{n^2}$$

are the generators of the cone

$$ax^2 + by^2 = \frac{p_0^2}{n^2} \cdot \frac{x^2 + y^2 + z^2}{r^2}$$

$\Leftrightarrow \quad x^2(an^2r^2 - p_0^2) + y^2(bn^2r^2 - p_0^2) - z^2 p_0^2 = 0 \qquad \ldots(4)$

The plane

$$lx + my + nz = 0$$

will touch the cone if

$$\frac{l^2}{an^2r^2 - p_0^2} + \frac{m^2}{bn^2r^2 - p_0^2} - \frac{n^2}{p_0^2} = 0 \qquad \ldots(5)$$

$\Leftrightarrow \quad abn^6 r^4 - n^2 r^2 p_0^2 [(a+b)n^2 + am^2 + bl^2] + p_0^4(l^2 + m^2 + n^2) = 0$

which is a quadratic equation in r^2 and has two roots r_1^2, r_2^2, which are the squares of the semi-axes of the section.

Also, if λ, μ, ν be the direction ratios of the axis of length $2r$, the plane (2) touches the cone (4) along the line

$$\frac{x}{\lambda} = \frac{y}{\mu} = \frac{z}{\nu}$$

and is, therefore, identical with

$$(an^2r^2 - p_0^2)\lambda x + (bn^2r^2 - p_0^2)\mu y - \nu p_0^2 z = 0$$

so that we have

$$\frac{(an^2r^2 - p_0^2)\lambda}{l} = \frac{(bn^2r^2 - p_0^2)\mu}{m} = \frac{-p_0^2 \nu}{n} \qquad \ldots(6)$$

Thus determining the direction ratios of the axis of length $2r$; r being given from the equation (5).

7.5.2. The section will be a rectangular hyperbola, if

$$r_1^2 + r_2^2 = 0 \Rightarrow (a+b)n^2 + am^2 + bl^2 = 0.$$

Ex. Obtain the conclusion of § 7.5 with the help of the equation (5) of this article.

7.5.3. Area of the Section

If the section be elliptic, its area $= \pi r_1 r_2 = \dfrac{\pi p_0^2}{n^3}\left[\dfrac{l^2 + m^2 + n^2}{ab}\right]^{1/2}$

$= \dfrac{\pi c}{n^3}\left[\dfrac{l^2 c}{a} + \dfrac{m^2 c}{b} + 2np\right]\left[\dfrac{l^2 + m^2 + n^2}{ab}\right]^{1/2}$

7.5.4. If θ be the angle between the asymptotes of the section, then

$$\tan^2\theta = \frac{-4r_1^2 r_2^2}{(r_1^2 + r_2^2)^2} = \frac{-4abn^2(l^2 + m^2 + n^2)}{[(a+b)n^2 + am^2 + bl^2]^2}$$

which being independent of p, we deduce that the angle between the asymptotes of parallel plane sections is the same.

Thus, we see *that parallel plane sections of a paraboloid are similar.*

EXAMPLES

1. Find the locus of the centres of sections of the paraboloid $\dfrac{x^2}{a^2} + \dfrac{y^2}{b^2} = 2z$ which are of constant area πk^2.

Sol. Let (α, β, γ) be the centre of the section of the paraboloid by the plane
$$lx + my + nz = p$$
By § 7.5.3, the area of the section is given by
$$\pi k^2 = \frac{\pi ab}{(-1)^2}\left[a^2\left(\frac{\alpha}{a^2}\right)^2 + b^2\left(\frac{\beta}{b}\right)^2 - 2\left(\frac{\alpha^2}{a^2} + \frac{\beta^2}{b^2} - \gamma\right)\right]\left(\frac{\alpha^2}{a^4} + \frac{\beta^2}{b^4} + 1\right)^{1/2}$$
$$\Rightarrow \quad k^4 = a^2 b^2 \left(-\frac{\alpha^2}{a^2} - \frac{\beta^2}{b^2} + 2\gamma\right)^2 \cdot \left(\frac{\alpha^2}{a^4} + \frac{\beta^2}{b^4} + 1\right)$$

Hence, the locus is
$$a^2 b^2 \left(\frac{x^2}{a^4} + \frac{y^2}{b^4} + 1\right)\left(\frac{x^2}{a^4} + \frac{y^2}{b^4} - 2z\right)^2 = k^4.$$

2. Planes are drawn through a fixed point (α, β, γ) so that their sections of the paraboloid $ax^2 + by^2 = 2z$ are rectangular hyperbolas. Prove that they touch the cone
$$\frac{(x-\alpha)^2}{b} + \frac{(y-\beta)^2}{a} + \frac{(z-\gamma)^2}{a+b} = 0.$$

Sol. Any plane through (α, β, γ) is
$$l(x-\alpha) + m(y-\beta) + n(z-\gamma) = 0 \qquad \text{....(i)}$$
If r be the length of either semi-axis of the section of $ax^2 + by^2 = 2z$ by (i), then
$$abn^6 r^4 - n^2 p_0^2 r^2 [(a-b)n^2 + am^2 + bl^2] + p_0^4 (l^2 + m^2 + n^2) = 0 \qquad \text{...(ii)}$$
where
$$p_0^2 = \frac{l^2}{a} + \frac{m^2}{b} + 2\alpha(l\alpha + m\beta + n\gamma)$$

If r_1 and r_2 are the lengths of the semi-axes of the section, then the section will be a rectangular hyperbola, if
$$r_1^2 + r_2^2 = 0$$
$\Rightarrow$ if
$$(a+b)n^2 + am^2 + bl^2 = 0 \qquad \text{...(iii)}$$
Normal to (i) through (α, β, γ) is
$$\frac{x-\alpha}{l} = \frac{y-\beta}{m} = \frac{z-\gamma}{n}$$
The normal will generate the cone
$$(a+b)(z-\gamma)^2 + a(y-\beta)^2 + b(x-\alpha)^2 = 0 \qquad \text{...(iv)}$$
∴ The plane (i) will touch the reciprocal cone
$$\frac{(x-\alpha)^2}{b} + \frac{(y-\beta)^2}{a} + \frac{(z-\gamma)^2}{a+b} = 0.$$

Plane Section of Conicoids

EXERCISES

1. Show that the section of the paraboloid $ax^2 + by^2 = 2cz$, by a tangent plane to the cone
$$\frac{x^2}{b} + \frac{y^2}{a} + \frac{z^2}{a+b} = 0.$$
is a rectangular hyperbola.

2. Prove that the axis of the section of the conicoid $ax^2 + by^2 = 2z$ by the plane
$$lx + my + nz = 0$$
lie on the cone $\dfrac{bl}{x} - \dfrac{am}{y} + \dfrac{(a-b)n}{z} = 0.$

3. If the area of the section of $ax^2 + by^2 = 2cz$, be constant and equal to πk^2, the locus of the centre is $(a^2x^2 + b^2y^2 + c^2)(ax^2 - by^2 - 2cz)^2 = abc^2 k^4$.

7.6 CIRCULAR SECTIONS OF PARABOLOIDS

To determine the circular sections of the paraboloid
$$ax^2 + by^2 = 2cz \qquad \ldots(1)$$
The equation (1) can be written in the forms :
$$a\left(x^2 + y^2 + z^2 - \frac{2cz}{a}\right) + y^2(b-a) - az^2 = 0$$
$$b\left(x^2 + y^2 + z^2 - \frac{2cz}{b}\right) + x^2(a-b) - bz^2 = 0$$
Therefore, as before, the two pairs of planes
$$y^2(b-a) - az^2 = 0 \text{ and } x^2(a-b) - bz^2 = 0 \qquad \ldots(2)$$
determine circular sections through the origin.

If a or b is negative and the other positive, neither of the equations (2) give real planes.

Hence, *hyperbolic paraboloids have no real circular sections*.

Of the two pairs of planes (2), one will be real if a and b are of the same sign.

In case $a > b > 0$,
$$x^2(a-b) - bz^2 = 0,$$
gives real circular sections through the origin and the two real systems of circular sections are given by
$$x\sqrt{(a-b)} + \sqrt{b}z = \lambda, \; x\sqrt{(a-b)} - \sqrt{b}z = \mu.$$

EXERCISES

1. Show that any two circular sections of opposite systems of an elliptic paraboloid lie on a sphere.
2. Find the real circular sections of the paraboloid :
   (i) $13y^2 + 4z^2 = 2x$;   (ii) $x^2 + 5z^2 + 5y = 0$.   [**Ans.** (i) $2x \pm 3y = \lambda$; (ii) $y \pm 2z = \lambda$]

7.6.1 Umbilics of a Paraboloid

To determine the umbilics of the paraboloid
$$ax^2 + by^2 = 2cz; \; a > b > 0.$$
Circular sections are determined by the planes
$$x\sqrt{(a-b)} + \sqrt{b}z = \lambda, \; x\sqrt{(a-b)} - \sqrt{b}z = \mu.$$
If f, g, h be an umbilic, the tangent plane

$$afx + bgy - c(z+h) = 0$$

there at is parallel to either of the circular sections.

It follows that $\quad g = 0$ and $f = \pm \dfrac{c}{a}\left(\dfrac{a-b}{b}\right)^{1/2}$

Also, $\quad\quad\quad\quad af^2 + bg^2 = 2ch$

Therefore, $\quad\quad\quad h = \dfrac{(a-b)c}{2ab}$

Hence, $\quad\left[\pm \dfrac{c}{a}\left(\dfrac{a-b}{b}\right)^{1/2}, 0, \dfrac{(a-b)c}{2ab}\right]$

are two real umbilics of the paraboloid.

EXERCISES

1. Find the umbilics of the paraboloids
   (i) $4x^2 + 5y^2 = 40z$; (ii) $25x^2 + 16y^2 = 2z$.

   [**Ans.** (i) $(0, \pm 2, 1/2)$; (ii) $(\pm 3/100, 0, 9/800)$]

2. Prove that the umbilics of the paraboloid $\dfrac{x^2}{a^2} + \dfrac{y^2}{b^2} = 2cz$, $a > b$ are

   $[0, \pm \sqrt{a^2 - b^2}, c(a^2 - b^2)/2]$

❏❏❏

8
Generating Lines of Conicoids

8.1 RULED SURFACES

The surfaces which are generated by a moving straight line are called *ruled surfaces*. For example, cones, cylinders, the hyperboloids of one sheet and hyperbolic paraboloids are ruled surfaces. A ruled surface can also be defined as one through every point of which a straight line can be drawn so as to lie completely on it. The lines which lie on the surfaces are called its *generating lines*.

The ruled surfaces may be divided into two categories : *(i) developable surfaces, (ii) skew surfaces*. A developable surface is one on which the consecutive generators intersect while on a skew surface, the consecutive generating lines do not intersect. The cone is a developable surface as all the generators pass through a common vertex and the cylinder is also a developable surface as consecutive generators touch all along their length. The hyperboloid of one sheet and the hyperbolic paraboloid are skew surfaces.

8.1.1. Condition for a Line to be a Generator of the Conicoid

Let the equation of the line be

$$\frac{x-\alpha}{l} = \frac{y-\beta}{m} = \frac{z-\gamma}{n} = r \text{ (say)} \qquad \ldots(i)$$

and that on the conicoid be

$$ax^2 + by^2 + cz^2 = 1 \qquad \ldots(ii)$$

Any point on the line (i) is $(lr + \alpha, mr + \beta, nr + \gamma)$.

If it lies on the conicoid (ii), we have

$$r^2(al^2 + bm^2 + cn^2) + 2r(al\alpha + bm\beta + cn\gamma) + (a\alpha^2 + b\beta^2 + c\gamma^2 - 1) = 0 \qquad \ldots(iii)$$

If the line (i) is a generator, then it lies wholly on the conicoid, the conditions for which are

$$al^2 + bm^2 + cn^2 = 0 \qquad \ldots(iv)$$

$$al\alpha + bm\beta + cn\gamma = 0 \qquad \ldots(v)$$

$$a\alpha^2 + b\beta^2 + c\gamma^2 = 1 \qquad \ldots(vi)$$

Now, condition (iv) shows that the lines through the centre of the conicoid, *i.e.*, (0, 0, 0) and parallel to the generating lines, *i.e.*, the lines $\frac{x}{l} = \frac{y}{m} = \frac{z}{n}$ are generators of the cone

$$ax^2 + by^2 + cz^2 = 0,$$

which is called *asymptotic cone*.

The condition (v) shows that the generating lines whose direction cosines are l, m, n should lie on the plane,

$$a\alpha x + b\beta y + c\gamma z = 1$$

which is the equation of the tangent plane of the conicoid at the point (α, β, γ).

Equation (iv) and (v) give the *direction ratios of generating lines*.

8.2 GENERATING LINES OF A HYPERBOLOID OF ONE SHEET

It is interesting to see that a hyperboloid of one sheet is a ruled surface in as much it can be thought of as generated by straight lines. The consideration of this aspect of the surface will be taken up in this chapter.

We write the equation

$$\frac{x^2}{a^2} + \frac{y^2}{b^2} - \frac{z^2}{c^2} = 1, \qquad \ldots(1)$$

of a hyperboloid of one sheet in the form

$$\frac{x^2}{a^2} - \frac{z^2}{c^2} = 1 - \frac{y^2}{b^2},$$

$$\Leftrightarrow \left(\frac{x}{a} - \frac{z}{c}\right)\left(\frac{x}{a} + \frac{z}{c}\right) = \left(1 - \frac{y}{b}\right)\left(1 + \frac{y}{b}\right)$$

This may again be written in either of the two forms

$$\frac{\frac{x}{a} - \frac{z}{c}}{1 - \frac{y}{b}} = \frac{1 + \frac{y}{b}}{\frac{x}{a} + \frac{z}{c}} \qquad \ldots(2)$$

and

$$\frac{\frac{x}{a} - \frac{z}{c}}{1 - \frac{y}{b}} = \frac{1 - \frac{y}{b}}{\frac{x}{a} + \frac{z}{c}} \qquad \ldots(3)$$

We consider, now, the two *families* of lines obtained by putting the equal fractions (2) and (3) equal to arbitrary constants λ and μ respectively.

$$\frac{x}{a} - \frac{z}{c} = \lambda\left(1 - \frac{y}{b}\right), \quad 1 + \frac{y}{b} = \lambda\left(\frac{x}{a} + \frac{z}{c}\right) \qquad \ldots(A)$$

$$\frac{x}{a} - \frac{z}{c} = \mu\left(1 + \frac{y}{b}\right), \quad 1 - \frac{y}{b} = \mu\left(\frac{x}{a} + \frac{z}{c}\right) \qquad \ldots(B)$$

To each value of the constant λ, corresponds a member of the family of lines (A) and to each value of the constant μ, corresponds a member of the family of lines (B).

Now it will be shown that *every point of each of the lines* (A) *and* (B) *lies on the hyperboloid* (1).

If (x_0, y_0, z_0) be a point of a member of the family (A) obtained for some value λ_0 of λ, we have

$$\frac{x_0}{a} - \frac{z_0}{c} = \lambda_0\left(1 - \frac{y_0}{b}\right), \quad 1 + \frac{y_0}{b} = \lambda_0\left(\frac{x_0}{a} + \frac{z_0}{c}\right).$$

On eliminating λ_0 from these, we obtain

$$\frac{x_0^2}{a^2} - \frac{z_0^2}{c^2} = 1 - \frac{y_0^2}{b^2} \Leftrightarrow \frac{x_0^2}{a^2} + \frac{y_0^2}{b^2} - \frac{z_0^2}{c^2} = 1$$

which shows that (x_0, y_0, z_0) is a point of the hyperboloid (1).

A similar proof holds for the family of lines (B).

Thus, as λ and μ vary, we get two families of lines (A) and (B) each member of which lies wholly on the hyperboloid. These two families of lines are called *two systems of generating lines (or generators) of the hyperboloid*.

We shall now proceed to discussed some properties of these systems of generating lines.

8.2.1. *Through every point of the hyperboloid there passes one generator of each system.*

Let (x_0, y_0, z_0) be a point of the hyperboloid so that we have

$$\frac{x_0^2}{a^2} + \frac{y_0^2}{b^2} - \frac{z_0^2}{c^2} = 1 \qquad \ldots(4)$$

Generating Lines of Conicoids

Now the generator

$$\frac{x}{a} - \frac{z}{c} = \lambda\left(1 - \frac{y}{b}\right), \quad 1 + \frac{y}{b} = \lambda\left(\frac{x}{a} + \frac{z}{c}\right)$$

will pass through the point (x_0, y_0, z_0) if and only if λ has a value equal to each of the two fractions

$$\left(\frac{x_0}{a} - \frac{z_0}{c}\right) \div \left(1 - \frac{y_0}{b}\right), \quad \left(1 + \frac{y_0}{b}\right) \div \left(\frac{x_0}{a} + \frac{z_0}{c}\right) \qquad ...(5)$$

Also by virtue of the relation (4), these two fractions are equal.

Thus, the member of the system (A) corresponding to either of the equal valuers (5) of λ will pass through the given point (x_0, y_0, z_0). Similarly it can be shown that the member of the system (B) corresponding to either of the equal values

$$\left(\frac{x_0}{a} - \frac{z_0}{c}\right) \div \left(1 + \frac{y_0}{b}\right), \quad \left(1 - \frac{y_0}{b}\right) \div \left(\frac{x_0}{a} + \frac{z_0}{c}\right)$$

of μ passes through the given point (x_0, y_0, z_0).

8.2.2. *No two generators of the same system intersect.*

Let

I. (i) $\quad \dfrac{x}{a} - \dfrac{z}{c} = \lambda_1\left(1 - \dfrac{y}{b}\right),$ (ii) $\quad 1 + \dfrac{y}{b} = \lambda_1\left(\dfrac{x}{a} + \dfrac{z}{c}\right)$

II. (iii) $\quad \dfrac{x}{a} - \dfrac{z}{c} = \lambda_2\left(1 - \dfrac{y}{b}\right),$ (iv) $\quad 1 + \dfrac{y}{b} = \lambda_2\left(\dfrac{x}{a} + \dfrac{z}{c}\right)$

be any two different generators of the λ system.

It will be shown that these four equations in x, y, z are not consistent.

Subtracting (iii) from (i), we obtain

$$(\lambda_1 - \lambda_2)\left(1 - \frac{y}{b}\right) = 0 \Rightarrow y = b, \text{ for } \lambda_1 \neq \lambda_2$$

Again, from (ii) and (iv), we obtain

$$\left(\frac{1}{\lambda_1} - \frac{1}{\lambda_2}\right)\left(1 + \frac{y}{b}\right) = 0 \Rightarrow y = -b, \text{ for } \lambda_1 \neq \lambda_2$$

Thus, we see that these four equations are inconsistent and accordingly the two lines do not intersect.

8.2.3. *Any two generators belonging to different systems intersect.*

Let

I. (i) $\quad \dfrac{x}{a} - \dfrac{z}{c} = \lambda\left(1 - \dfrac{y}{b}\right),$ (ii) $\quad 1 + \dfrac{y}{b} = \lambda\left(\dfrac{x}{a} + \dfrac{z}{c}\right)$

II. (iii) $\quad \dfrac{x}{a} - \dfrac{z}{c} = \mu\left(1 + \dfrac{y}{b}\right),$ (iv) $\quad 1 - \dfrac{y}{b} = \mu\left(\dfrac{x}{a} + \dfrac{z}{c}\right)$

be two generators, one of each system.

It will be shown that these four equations in x, y, z are not consistent. Firstly, we solve simultaneously the equations (i), (ii) and (iii). Now, (i) and (iii) give

$$\lambda\left(1 - \frac{y}{b}\right) = \mu\left(1 + \frac{y}{b}\right) \Rightarrow y = b\frac{\lambda - \mu}{\lambda + \mu}$$

Substituting this value of y in (i) and (ii), we obtain

$$\frac{x}{a} - \frac{z}{c} = \frac{2\lambda\mu}{\lambda + \mu}, \quad \frac{x}{a} + \frac{z}{c} = \frac{2}{\lambda + \mu}$$

These given, on adding and subtracting,
$$x = a\frac{1+\lambda\mu}{\lambda+\mu}, \quad z = c\frac{1-\lambda\mu}{\lambda+\mu}$$

Now, as may easily be seen, these values of x, y, z satisfy (iv) also. Thus, the two lines intersect and the point of intersection is

$$\left(a\frac{1+\lambda\mu}{\lambda+\mu}, \; b\frac{\lambda-\mu}{\lambda+\mu}, \; c\frac{1-\lambda\mu}{\lambda+\mu}\right) \quad \ldots(6)$$

Another method. The planes

$$\frac{x}{a} - \frac{z}{c} - \lambda\left(1 - \frac{y}{b}\right) - k\left[1 + \frac{y}{b} - \lambda\left(\frac{x}{a} + \frac{z}{c}\right)\right] = 0$$

$$\frac{x}{a} - \frac{z}{c} - \mu\left(1 + \frac{y}{b}\right) - k'\left[1 - \frac{y}{b} - \mu\left(\frac{x}{a} + \frac{z}{c}\right)\right] = 0$$

pass through the two lines respectively for all values of k and k'.

Now, obviously these equations becomes identical for $k = \mu$ and $k' = \lambda$.

Thus, the two lines are coplanar and as such they intersect. Also the plane through the two lines, obtained by putting $k = \mu$ or $k' = \lambda$ is

$$\frac{1+\lambda\mu}{\lambda+\mu} \cdot \frac{x}{a} + \frac{\lambda-\mu}{\lambda+\mu} \cdot \frac{y}{b} - \frac{1-\lambda\mu}{\lambda+\mu} \cdot \frac{z}{c} = 1 \quad \ldots(7)$$

Cor. 1. The plane (7) through two generators of the opposite systems is the tangent plane to the hyperboloid (1) at the point of intersection (6) of the two generators. Since also through every point of the hyperboloid there pass two generators, one of each system, we see that *the tangent plane at a point of hyperboloid meets the hyperboloid in the two generators through the point.*

Cor. 2. *A plane through a generating line is the tangent plane at some point of the generator.* Now like every plane section, the section of the hyperboloid by a plane through a generator is a conic of which the given generator is a part. Thus, the conic is degenerate and the residue must also be a line. At the point of intersection of the lines constituting this degenerate plane section the plane will touch the hyperboloid.

Ex. Prove this result analytical also.

Cor. 3. Parametric equations of the hyperboloid. The co-ordinates (6) show that

$$x = a\frac{1+\lambda\mu}{\lambda+\mu}, \quad y = b\frac{\lambda-\mu}{\lambda+\mu}, \quad z = c\frac{1-\lambda\mu}{\lambda+\mu}$$

are the parametric equations of the hyperboloid; λ, μ being the two parameters. These co-ordinates satisfy the equation of the hyperboloid for all values of the parameters λ and μ.

EXAMPLES

1. *Find the equation to the generating lines of the hyperboloid*

$$\frac{x^2}{4} + \frac{y^2}{9} - \frac{z^2}{16} = 1.$$

which pass through the point $(2, 3, -4)$.

Sol. Any line through $(2, 3, -4)$ is

$$\frac{x-2}{l} = \frac{y-3}{m} = \frac{z+4}{n} = r \; (\text{say}) \quad \ldots(i)$$

Any point on this line is $(lr + 2, mr + 3, nr - 4)$ and it lies on the given hyperboloid if

$$\frac{(lr+2)^2}{4} + \frac{(mr+3)^2}{9} - \frac{(nr-4)^2}{16} = 1$$

Generating Lines of Conicoids

$$\Rightarrow \quad r^2\left[\frac{l^2}{4}+\frac{m^2}{9}-\frac{n^2}{16}\right]+2r\left[\frac{2l}{4}+\frac{3m}{9}+\frac{4m}{16}\right]=0 \qquad \text{...(ii)}$$

If the line (i) is a generator of the given hyperboloid, then (i) lies wholly on the hyperboloid. The conditions for this are

$$\frac{l^2}{4}+\frac{m^2}{9}-\frac{n^2}{16}=0 \text{ and } \frac{l}{2}+\frac{m}{3}+\frac{n}{4}=0.$$

Eliminating n, we get

$$\frac{l^2}{4}+\frac{m^2}{9}-\left(\frac{l}{2}+\frac{m}{3}\right)^2=0$$

$$\Rightarrow \quad -\frac{1}{3}lm=0 \Rightarrow \text{ either } l=0 \text{ or } m=0.$$

When $l=0$, $\dfrac{m}{3}=-\dfrac{n}{4}$. When $m=0$, $\dfrac{l}{2}=-\dfrac{n}{4}$.

Hence equations of the required generator are

$$\frac{x-2}{0}=\frac{y-3}{3}=\frac{z+4}{-4} \text{ and } \frac{x-2}{1}=\frac{y-3}{0}=\frac{z+4}{-2}.$$

2. Find the equations to the generators of the hyperboloid

$$\frac{x^2}{a^2}+\frac{y^2}{b^2}-\frac{z^2}{c^2}=1$$

which pass through the point $(a\cos\alpha, b\sin\alpha, 0)$.

Sol. Any line through $(a\cos\alpha, b\sin\alpha, 0)$ is given by

$$\frac{x-a\cos\alpha}{l}=\frac{y-b\sin\alpha}{m}=\frac{z}{n}=r \text{ (say)} \qquad \text{...(i)}$$

(i) will meet the hyperboloid

$$\frac{x^2}{a^2}+\frac{y^2}{b^2}-\frac{z^2}{c^2}=1 \qquad \text{...(ii)}$$

if

$$\frac{(a\cos\alpha+lr)^2}{a^2}+\frac{(b\sin\alpha+mr)^2}{b^2}-\frac{(nr)^2}{c^2}=1$$

$$\Rightarrow \quad r^2\left(\frac{l^2}{a^2}+\frac{m^2}{b^2}-\frac{n^2}{c^2}\right)+2r\left(\frac{l\cos\alpha}{a}+\frac{m\sin\alpha}{b}\right)=0 \qquad \text{...(iii)}$$

(ii) will be a generating line if (iii) is an identity, i.e., if

$$\frac{l^2}{a^2}+\frac{m^2}{b^2}-\frac{n^2}{c^2}=0 \qquad \text{...(iv)}$$

and

$$\frac{l\cos\alpha}{a}+\frac{m\sin\alpha}{b}=0 \qquad \text{...(v)}$$

From (v),

$$\frac{l\cos\alpha}{a}=-\frac{m\sin\alpha}{b}$$

or

$$\frac{l/a}{\sin\alpha}=\frac{m/b}{-\cos\alpha}=\frac{\pm\sqrt{\dfrac{l^2}{a^2}+\dfrac{m^2}{b^2}}}{\sqrt{\sin^2\alpha+\cos^2\alpha}}$$

$$\therefore \qquad \frac{l/a}{\sin \alpha} = \frac{m/b}{-\cos \alpha} = \pm \frac{n/c}{1} \quad \text{[from (iv)]} \qquad \qquad \ldots\text{(vi)}$$

∴ Equations of the required generators are given by

$$\frac{x - a \cos \alpha}{a \sin \alpha} = \frac{y - b \sin \alpha}{-b \sin \alpha} = \frac{z}{\pm c}.$$

3. CP, CQ are any conjugate diameters of the ellipse $\dfrac{x^2}{a^2} + \dfrac{y^2}{b^2} = 1$, $z = c$; $C'P'$, $C'Q'$ are the conjugate diameters of the ellipse $\dfrac{x^2}{a^2} + \dfrac{y^2}{b^2} = 1$, $z = -c$ drawn in the same direction as CP and CQ. Prove that hyperboloid $\dfrac{2x^2}{a^2} + \dfrac{2y^2}{b^2} - \dfrac{z^2}{c^2} = 1$ is generated by either PQ' or P'Q.

Sol. Let the co-ordinates of P, Q, P' and Q' are

$P(a \cos \theta, b \sin \theta, c)$, $Q(-a \sin \theta, b \cos \theta, -c)$,
$P'(a \cos \theta, b \sin \theta, -c)$ and $Q'(-a \sin \theta, b \cos \theta, -c)$

∴ Equation to PQ' is

$$\frac{x - a \cos \theta}{-a \sin \theta - a \cos \theta} = \frac{y - b \sin \theta}{b \cos \theta - b \sin \theta} = \frac{z - c}{-c - c} = r \quad \text{(say)}$$

∴ $\dfrac{x}{a} = \cos \theta - r(\sin \theta + \cos \theta)$

$\dfrac{y}{b} = \sin \theta + r(\cos \theta - \sin \theta)$

and $\dfrac{z}{c} = -2r + 1$.

Eliminating r, we get

$$\frac{x^2}{a^2} + \frac{y^2}{b^2} = r(1+1) + 1 - 2r(\sin \theta \cos \theta + \cos^2 \theta - \sin \theta \cos \theta + \sin^2 \theta)$$

$$= 2r^2 + 1 - 2r$$

$\Rightarrow \qquad \dfrac{2x^2}{a^2} + \dfrac{2y^2}{b^2} = 4r^2 - 4r + 1 + 1 = (1 - 2r)^2 + 1 = \dfrac{z^2}{c^2} + 1$

$\Rightarrow \qquad \dfrac{2x^2}{a^2} + \dfrac{2y^2}{b^2} - \dfrac{z^2}{c^2} = 1.$

EXERCISES

1. Write down the equations of the systems of generating lines of the following hyperboloids and determine the pairs of lines of the systems which pass through the given point.

(i) $x^2 + 9y^2 - z^2 = 9$, $(3, 1/3, -1)$ \qquad (ii) $x^2/9 - y^2/16 + z^2/4 = 1$, $(-1, 4/3, 2)$

[**Ans.** (i) $x + 3\mu y - z = 3\lambda$, $\lambda x - 3y + \lambda z = 3$; $x + 6y - z = 6$, $2x - 3y + 2z = 3$
$x - 3\mu y - z = 3\lambda$, $\mu x + 3y + \mu z = 3$; $x - 3y - z = 3$, $x + 3y + z = 3$
(ii) $4x - 3y + 6\lambda z = 12\lambda$, $4\lambda x + 3\lambda y - 6z = 12$; $z = 2$, $4x + 3y = 0$
$4x - 3y - 6\mu z = 12\mu$, $4\mu x + 3\mu y - 6z = 12$;
$4x - 3y + 2z + 4 = 0$, $4x + 3y - 8z + 36 = 0$]

Generating Lines of Conicoids

2. Find the equations to the generating lines of the hyperboloid $yz + 2zx - 3xy + 6 = 0$ which pass through the point $(-1, 0, 3)$. $\left[\text{Ans. } \dfrac{x+1}{0} = \dfrac{y-0}{1} = \dfrac{z-3}{0} \text{ and } \dfrac{x+1}{1} = \dfrac{y-0}{-1} = \dfrac{z-3}{3}\right]$

3. Find equations to the generating lines of hyperboloid $(x + y + z)(2x + y + z) = 6z$, which pass through the point $(1, 1, 1)$. $\left[\text{Ans. } \dfrac{x-1}{4} = \dfrac{y-1}{-5} = \dfrac{z-1}{1} \text{ and } \dfrac{x-1}{1} = \dfrac{y-1}{-3} = \dfrac{z-1}{-1}\right]$

4. A point 'm' on the parabola $y = 0, cx^2 = 2a^2 y$ is $(2am, 0, 2cm^2)$ and a point 'n' on the parabola $x = 0, cy^2 = -2b^2 z$ is $(0, 2bn, 2cn^2)$. Find the locus of the line joining the points for which (i) $m = n$, (ii) $m = -n$. $\left[\text{Ans. } \dfrac{x^2}{a^2} - \dfrac{y^2}{b^2} = \dfrac{2z}{c}\right]$

8.3. *To find the equations of the two generating lines through any point $(a \cos \theta, b \sin \theta, 0)$, of the principal elliptic section*

$$\dfrac{x^2}{a^2} + \dfrac{y^2}{b^2} = 1, \ z = 0,$$

of the hyperboloid by the plane $z = 0$.

Let
$$\dfrac{x - a \cos \theta}{l} = \dfrac{y - b \sin \theta}{m} = \dfrac{z - 0}{n}$$

be a generator through the point $(a \cos \theta, b \sin \theta, 0)$.

The point
$$(lr + a \cos \theta, \ mr + b \sin \theta, \ nr)$$
on the generator is a point of the hyperboloid for all values of r so that the equation

$$\dfrac{(lr + a \cos \theta)^2}{a^2} + \dfrac{(mr + b \sin \theta)^2}{b^2} - \dfrac{n^2 r^2}{c^2} = 1$$

$\Leftrightarrow \left[\dfrac{l^2}{a^2} + \dfrac{m^2}{b^2} - \dfrac{n^2}{c^2}\right] r^2 + 2r \left[\dfrac{l \cos \theta}{a} + \dfrac{m \sin \theta}{b}\right] = 0$

is true for all values of r. This will be so if

$$\dfrac{l^2}{a^2} + \dfrac{m^2}{b^2} - \dfrac{n^2}{c^2} = 0 \text{ and } \dfrac{l \cos \theta}{a} + \dfrac{m \sin \theta}{b} = 0$$

These give
$$\dfrac{l}{a \sin \theta} = \dfrac{m}{-b \cos \theta} = \dfrac{n}{\pm c}$$

Thus, we obtain
$$\dfrac{x - a \cos \theta}{a \sin \theta} = \dfrac{y - b \sin \theta}{-b \cos \theta} = \dfrac{z}{\pm c} \qquad \ldots(C)$$

as the two required generators.

Note : Since every generator of either system meets the plane $z = 0$ at a point of the principal elliptic section, we see that the two systems of lines obtained from (C) as θ varies from 0 to 2π are the two systems of generators of the hyperboloid. The form (C) of the equations of two systems of generators is often found more useful than the forms (A) and (B) obtained in § 8.2.

Ex. Show that the equations (A) and (B) are equivalent to the equations (C) for

$$\lambda = \tan\left(\dfrac{1}{4}\pi - \dfrac{1}{2}\theta\right), \ \mu = \cot\left(\dfrac{1}{4}\pi - \dfrac{1}{2}\theta\right).$$

8.4. To show that the projections of the generators of a hyperboloid on any principal plane are tangents to the section of the hyperboloid by the principal plane.

Consider a generator

$$\frac{x - a\cos\theta}{a\sin\theta} = \frac{y - b\sin\theta}{-b\sin\theta} = \frac{z}{c}$$

The equation

$$\frac{x - a\cos\theta}{a\sin\theta} = \frac{y - b\sin\theta}{-b\sin\theta}$$

represents the plane through the generator perpendicular to the *XOY* plane so that the projection of the generator on the *XOY* plane is

$$\frac{x - a\cos\theta}{a\sin\theta} = \frac{y - b\sin\theta}{-b\sin\theta}, \; z = 0 \iff \frac{x\cos\theta}{a} + \frac{y\sin\theta}{b} = 1, \; z = 0$$

which is clearly the tangent line to the section

$$\frac{x^2}{a^2} + \frac{y^2}{b^2} = 1, \; z = 0$$

of the hyperboloid by the principal plane $z = 0$ at the point $(a\cos\theta, b\sin\theta, 0)$.

Again

$$\frac{x - a\cos\theta}{a\sin\theta} = \frac{z}{c}$$

is the plane through the generator perpendicular to the *XOZ* plane so that the projection of the generator on the *XOZ* plane is

$$\frac{x - a\cos\theta}{a\sin\theta} = \frac{z}{c}, \; y = 0 \iff \frac{x\sec\theta}{a} - \frac{z}{c}\tan\theta = 1, \; y = 0$$

which is clearly the tangent to the section

$$\frac{x^2}{a^2} - \frac{z^2}{c^2} = 1, \; y = 0$$

of the hyperboloid by the principal plane $y = 0$ at the point $(a\sec\theta, c\tan\theta)$.

Similarly we may show that the projections of the generators on the principal plane $x = 0$ are tangents to the corresponding section.

EXAMPLES

1. *Prove that the equations to the generating lines through the hyperboloid of one sheet are*

$$\frac{x - a\cos\theta\sec\phi}{a\sin(\theta\pm\phi)} = \frac{y - b\sin\theta\sec\phi}{-b\cos(\theta\pm\phi)} = \frac{z - c\tan\theta}{\pm c}.$$

Sol. Let $P(\text{``}\theta, \phi\text{''})$ be any point on the hyperboloid

$$\frac{x^2}{a^2} + \frac{y^2}{b^2} - \frac{z^2}{c^2} = 1. \quad \text{...(i)}$$

Hence co-ordinates of *P* would be $(a\cos\theta\sec\phi, b\sin\theta\sec\phi, c\tan\phi)$.

Let equation of tangent plane at this point be

$$\frac{x}{a}\cos\theta\sec\phi + \frac{y}{b}\sin\theta\sec\phi - \frac{z}{c}\tan\phi = 1 \quad \text{...(ii)}$$

This plane meets the plane $z = 0$ in the line given by

$$\frac{x}{a}\cos\theta + \frac{y}{b}\sin\theta = \cos\phi, \; z = 0 \quad \text{...(iii)}$$

If this line meets the section of the surface by $z = 0$ in points *A* and *B* whose eccentric angles are α and β respectively, then

Generating Lines of Conicoids

$$\frac{a\cos\alpha\cos\theta}{a} + \frac{b\sin\alpha\sin\theta}{b} = \cos\phi$$

or $\cos(\theta - \alpha) = \cos\phi$

and $\frac{a\cos\beta\cos\theta}{a} + \frac{b\sin\beta\sin\theta}{b} = \cos\phi$

or $\cos(\theta - \beta) = \cos\phi$

$\therefore \quad \theta - \alpha = -\phi$ and $\theta - \beta = \phi$

so that $\theta = \frac{\alpha+\beta}{2}$ and $\phi = \frac{\alpha-\beta}{2}$,

i.e., $\alpha = \theta + \phi$ and $\beta = \theta - \phi$ ...(iv)

Since the two generating lines through P are the lines of intersection of the surface and the tangent plane P, AP and BP will be the generators through P such that $\theta + \phi = \alpha$, a constant for all points on the generator AP and $\theta - \phi = \beta$, a constant for all points on the generator BP.

Also, the direction cosines of AP are proportional to

$a(\cos\alpha - \cos\theta\sec\phi), \ b(\sin\alpha - \sin\theta\sec\phi), \ -c\tan\phi$

$\Rightarrow \quad \dfrac{a(a\cos\alpha\cos\phi - \cos\theta)}{\sin\phi}, \ \dfrac{b(\sin\alpha\cos\phi - \sin\theta)}{\sin\phi}, \ -c$

$\Rightarrow \quad a\left\{\dfrac{\cos(\theta+\phi)\cos\phi - \cos\theta}{\sin\phi}\right\}, \ b\left\{\dfrac{\sin(\theta+\phi)\cos\phi - \sin\theta}{\sin\phi}\right\}, \ -c$

$\Rightarrow \quad a\left\{\dfrac{\cos(\theta+\phi)\cos\phi - \sin(\theta+\phi-\phi)}{\sin\phi}\right\}, \ b\left\{\dfrac{\sin(\theta+\phi)\cos\phi - \sin(\theta+\phi-\phi)}{\sin\phi}\right\}, \ -c$

$\Rightarrow \quad a\sin(\theta+\phi), \ -b\cos(\theta+\phi), \ c$

$\therefore$ Equation to the generator AP are

$$\frac{x - a\cos\theta\sec\phi}{a\sin(\theta+\phi)} = \frac{y - b\sin\theta\sec\phi}{-b\cos(\theta+\phi)} = \frac{z - c\tan\phi}{c} \qquad \text{...(v)}$$

Similarly the generator BP will be

$$\frac{x - a\cos\theta\sec\phi}{a\sin(\theta-\phi)} = \frac{y - b\sin\theta\sec\phi}{-b\cos(\theta+\phi)} = \frac{z - c\tan\phi}{c} \qquad \text{...(vi)}$$

2. *The normals to $\dfrac{x^2}{a^2} + \dfrac{y^2}{b^2} - \dfrac{z^2}{c^2} = 1$ at points of a generator meet the plane $z = 0$ at points lying on a straight line, and for different generators of the same system this line touches a fixed conic.*

Sol. Any generator through

$(a\cos\theta\sec\phi, \ b\sin\theta\sec\phi, \ c\tan\phi)$

is $\dfrac{x - a\cos\theta\sec\phi}{a\sin(\theta+\phi)} = \dfrac{y - b\sin\theta\sec\phi}{-b\cos(\theta+\phi)} = \dfrac{z - c\tan\phi}{c}$ ...(i)

which shows that $(\theta + \phi) = \alpha$ ...(ii)

a constant for a given generator.

Now, the tangent plane at the point "θ, ϕ" is

$$\frac{x}{a}\cos\theta\sec\phi + \frac{y}{b}\sin\theta\sec\phi - \frac{z}{c}\tan\phi = 1$$

$$\Rightarrow \quad \frac{x}{a}\cos\theta + \frac{y}{b}\sin\theta - \frac{z}{c}\sin\phi = \cos\phi \qquad \text{...(iii)}$$

And the normal at "θ, ϕ" is

$$\frac{x - a\cos\theta\sec\phi}{\dfrac{\cos\theta}{a}} = \frac{y - b\sin\theta\sec\phi}{\dfrac{\sin\theta}{b}} = \frac{z - c\tan\phi}{-\dfrac{\sin\phi}{c}} \qquad \text{...(iv)}$$

which meets the plane $z = 0$ in the point.

$$z = 0, \quad x = a\cos\theta\sec\phi + \frac{c^2}{a}\sec\phi\cos\theta = \frac{\cos\theta\sec\phi}{a}(c^2 + a^2)$$

$$y = b\sin\theta\sec\phi + \frac{c^2}{b}\sec\phi\sin\theta = \frac{\sin\theta\sin\phi}{b}(b^2 + c^2)$$

$$\Rightarrow \quad x = \left(\frac{a^2 + c^2}{a}\right)\frac{\cos(\alpha - \phi)}{\cos\phi} = \left(\frac{a^2 + c^2}{a}\right)(\cos\alpha + \sin\alpha\tan\phi)$$

[since $\theta + \phi = \alpha$, a constant, by (ii)]

$$y = \left(\frac{b^2 + c^2}{b}\right)\frac{\sin(\alpha + \phi)}{\cos\phi} = \left(\frac{b^2 + c^2}{b}\right)(\sin\alpha - \cos\alpha\tan\phi),$$

$$z = 0$$

$$\Rightarrow \quad \frac{x}{a^2 + c^2} - \cos\alpha = \sin\alpha\tan\phi$$

$$\frac{by}{b^2 + c^2} - \sin\alpha = -\cos\alpha\tan\phi \qquad \text{...(v)}$$

Eliminating ϕ, we get

$$\frac{ax\cos\alpha}{a^2 + c^2} - \cos^2\alpha + \frac{by\sin\alpha}{b^2 + c^2} - \sin^2\alpha = 0, \quad z = 0$$

$$\Rightarrow \quad \frac{ax\cos\alpha}{a^2 + c^2} + \frac{by\sin\alpha}{b^2 + c^2} = 1, \quad z = 0 \qquad \text{...(vi)}$$

which is a fixed straight line space α is constant.
Again for different generators of the same system, α varies.
∴ Differentiating (vi) w.r.t. α, we get

$$\frac{-ax\sin\alpha}{a^2 + c^2} + \frac{by\cos\alpha}{b^2 + c^2} = 0, \quad z = 0 \qquad \text{...(vii)}$$

Squaring and adding (vi) and (vii), we get the envelope of (vi) as

$$\frac{a^2 x^2}{(a^2 + c^2)^2} + \frac{b^2 y^2}{(b^2 + c^2)^2} = 1, \quad z = 0,$$

which is a fixed conic.

3. Show that the generators through points on the principal elliptic section of

$$\frac{x^2}{a^2} + \frac{y^2}{b^2} - \frac{z^2}{c^2} = 1$$

such that the eccentric angle of one is double the eccentric angle of the other intersect on the curve given by

Generating Lines of Conicoids

$$x = \frac{a(1-3t^2)}{1+t^2}, \quad y = \frac{bt(3-t^2)}{1+t^2}, \quad z = \pm ct.$$

Sol. Let $A(a\cos\theta, b\sin\theta, 0)$ and $B(a\cos\phi, b\sin\phi, 0)$ be the two points on the principal elliptic section by the plane $z = 0$. The points of intersection P and ϕ of the generators of opposite system through them are given by

$$\frac{x}{a} = \frac{\cos\left(\frac{\theta+\phi}{2}\right)}{\cos\left(\frac{\theta-\phi}{2}\right)}, \quad \frac{y}{b} = -\frac{\sin\left(\frac{\theta+\phi}{2}\right)}{\cos\left(\frac{\theta-\phi}{2}\right)}, \quad \frac{z}{c} = \pm\frac{\sin\left(\frac{\theta-\phi}{2}\right)}{\cos\left(\frac{\theta-\phi}{2}\right)}$$

Now, we are given that $\phi = 2\theta$.

$$\therefore \quad \frac{z}{c} = \pm\tan\frac{\theta}{2} \quad \text{or} \quad z = \pm ct.$$

If we take $t = \tan\frac{\theta}{2}$.

$$x = a \cdot \frac{\cos 3\theta/2}{\cos\theta/2} = a \cdot \frac{4\cos^3\theta/2 - 3\cos\theta/2}{\cos\theta/2}$$

$$= a(4\cos^2\theta/2 - 3) = a \cdot \frac{4 - 3\sec^2\theta/2}{\sec^2\theta/2}$$

$$= a \cdot \frac{4 - 3(1 + \tan^2\theta/2)}{1 + \tan^2\theta/2} = a \cdot \frac{1 - 3t^2}{1+t^2}$$

$$y = b \cdot \frac{\sin 3\theta/2}{\cos\theta/2} = b \cdot \frac{3\sin\theta/2 - 4\sin^3\theta/2}{\cos\theta/2}$$

$$= b(3\tan\theta/2 - 4 \cdot \sin^2\theta/2 \tan\theta/2)$$

$$= b\tan\theta/2 \left(\frac{3\sec^2\theta/2 - 4\sin^2\theta/2 \sec^2\theta/2}{1 + \tan^2\theta/2}\right)$$

$$= b\tan\theta/2 \left[\frac{3(1 + \tan^2\theta/2) - 4\tan^2\theta/2}{1 + \tan^2\theta/2}\right] = \frac{bt(1-t^2)}{1+t^2}$$

Hence the generators intersect on the curve

$$x = \frac{a(1-3t^2)}{1+t^2}, \quad y = \frac{bt(3-t^2)}{1+t^2}, \quad z = \pm ct.$$

EXERCISES

1. R, S are the points of intersection of generators of opposite systems drawn at the extremities P, Q of semi-conjugate diameters of the principal elliptic section; show that

   (i) the locus of the points R, S are the ellipses $\frac{x^2}{a^2} + \frac{y^2}{b^2} = 2$, $z = \pm c$;

   (ii) the perimeter of the skew-quadrilateral $PSQR$ taken in order, is constant and equal to $2(a^2 + b^2 + 2c^2)$;

   (iii) $\cot^2\alpha + \cot^2\beta = (a^2 + b^2)/c^2$, where $\angle RPS = 2\alpha$ and $\angle RQS = 2\beta$;

   (iv) the volume of the tetrahedron $PSQR$ is constant and equal to $\frac{1}{3}abc$.

2. The generators through a point P on the hyperboloid $\frac{x^2}{a^2} + \frac{y^2}{b^2} - \frac{z^2}{c^2} = 1$ meet the principal elliptic section in points whose eccentric angles differ by a constant 2α; show that the locus of P is the curve of intersection of the hyperboloid with the cone

$$\frac{x^2}{a^2} + \frac{y^2}{b^2} = \frac{z^2}{c^2} \cos^2 \alpha.$$

3. If the generators through a point P on the hyperboloid $\frac{x^2}{a^2} + \frac{y^2}{b^2} - \frac{z^2}{c^2} = 1$ meet the principal elliptic section in two points such that eccentric angle of one is three times that of the other. Prove that P lies on the curve of intersection of the hyperboloid with the cylinder

$$y^2(z^2 + c^2) = 4b^2 z^2.$$

4. Show that the generators through any one of the ends of an equi-conjugate diameter of the principal elliptic section of the hyperboloid $\frac{x^2}{a^2} + \frac{y^2}{b^2} - \frac{z^2}{c^2} = 1$ are inclined to each other at an angle $60°$ if $a^2 + b^2 = 6c^2$. Find also the condition for the generators to be perpendicular to each other.

[**Ans.** $a^2 + b^2 = 2c^2$]

5. A variable generator of the hyperboloid $\frac{x^2}{a^2} + \frac{y^2}{b^2} - \frac{z^2}{c^2} = 1$ intersects generators of the same system through the extremities of a diameter of the principal elliptic section in points P and P'; show that

$$\frac{x_P x_{P'}}{a^2} = \frac{y_P y_{P'}}{b^2 z_P z_{P'}} = -c^2.$$

6. Show that the shortest distance between generators of the same system drawn at one end of each of the major and minor axes of the principal elliptic section of the hyperboloid $\frac{x^2}{a^2} + \frac{y^2}{b^2} - \frac{z^2}{c^2} = 1$ is $\frac{2abc}{\sqrt{a^2 b^2 + b^2 c^2 + c^2 a^2}}.$

7. Show that the shortest distance between the generators of the same system drawn at the extremities of the diameters of the principal elliptic section of the hyperboloid $\frac{x^2}{a^2} + \frac{y^2}{b^2} - \frac{z^2}{c^2} = 1$, are parallel to the XOY plane and lie on the surface

$$abz(x^2 + y^2) = \pm (a^2 - b^2) cxy.$$

8.5. To find the locus of the points of intersection of perpendicular generators of the hyperboloid

$$\frac{x^2}{a^2} + \frac{y^2}{b^2} - \frac{z^2}{c^2} = 1 \qquad \ldots(1)$$

Let (x_1, y_1, z_1) be a point the generators through which are perpendicular.
The generators are the lines in which the tangent plane

$$\frac{xx_1}{a^2} + \frac{yy_1}{b^2} + \frac{zz_1}{c^2} = 1 \qquad \ldots(2)$$

at the point meets the surface. On making (1) homogeneous with the help of (2), we obtain the equation

$$\frac{x^2}{a^2} + \frac{y^2}{b^2} - \frac{z^2}{c^2} = \left(\frac{xx_1}{a^2} + \frac{yy_1}{b^2} + \frac{zz_1}{c^2} \right)^2 \qquad \ldots(3)$$

Generating Lines of Conicoids

The curve of intersection of (1) and (2) being a pair of lines, the cone with its vertex at the origin and with the curve of intersection of (1) and (2), as the guiding curve, represented by the equation (3), reduces to a pair of planes.

If l, m, n be the direction ratios of either of the two generators, we have, since they lie on the planes (2) and (3),

$$\frac{lx_1}{a^2} + \frac{my_1}{b^2} + \frac{nz_1}{c^2} = 0 \qquad \ldots(4)$$

and

$$\frac{l^2}{a^2} + \frac{m^2}{b^2} - \frac{n^2}{c^2} = \left(\frac{lx_1}{a^2} + \frac{my_1}{b^2} + \frac{nz_1}{c^2}\right)^2 \qquad \ldots(5)$$

Now the equation (5) with the help of the equation (4) reduces to

$$\frac{l^2}{a^2} + \frac{m^2}{b^2} - \frac{n^2}{c^2} = 0 \qquad \ldots(6)$$

Eliminating n from (4) and (5), we obtain

$$\frac{l^2}{a^4}(a^2 z_1^2 - c^2 x_1^2) - \frac{2lmc^2 x_1 y_1}{a^2 b^2} + \frac{m^2}{b^4}(b^2 z_1^4 - c^2 y_1^2) = 0$$

If $l_1, m_1, n_1; l_1, m_2, n_2$ be the direction ratios of the two generators, this gives

$$\frac{l_1}{m_1} \cdot \frac{l_2}{m_2} = \frac{b^2 z_1^2 - c^2 y_1^2}{b^4} \cdot \frac{a^4}{a^2 z_1^4 - c^2 x_1^2}$$

$$\Leftrightarrow \quad \frac{l_1 l_2}{a^4 (b^2 z_1^2 - c^2 y_1^2)} = \frac{m_1 m_2}{b^4 (a^2 z_1^2 - c^2 x_1^2)} = \frac{n_1 n_2}{c^4 (a^2 y_1^2 + c^2 x_1^2)}$$

Since $l_1 l_2 + m_1 m_2 + n_1 n_2 = 0$, we obtain

$$a^4 (b^2 z_1^2 - c^2 y_1^2) + b^4 (a^2 z_1^2 - c^2 x_1^2) + c^4 (a^2 y_1^2 + b^2 x_1^2) = 0$$

$$\Rightarrow \quad b^2 c^2 x_1^2 (c^2 - b^2) + a^2 c^2 y_1^2 (c^2 - a^2) + a^2 b^2 z_1^2 (a^2 + b^2) = 0$$

$$\Leftrightarrow \quad (b^2 - c^2)\frac{x_1^2}{a^2} + (a^2 - c^2)\frac{y_1^2}{b^2} - (a^2 + b^2)\frac{z_1^2}{c^2} = 0$$

We rewrite it as

$$(a^2 + b^2 - c^2)\frac{x_1^2}{a^2} + (a^2 + b^2 - c^2)\frac{y_1^2}{b^2} - (a^2 + b^2 - c^2)\frac{z_1^2}{c^2} = x_1^2 + y_1^2 + z_1^2$$

$$\Leftrightarrow \quad (a^2 + b^2 - c^2)\left(\frac{x_1^2}{a^2} + \frac{y_1^2}{b^2} - \frac{z_1^2}{c^2}\right) = x_1^2 + y_1^2 + z_1^2$$

Since now the point (x_1, y_1, z_1) lies on the hyperboloid, this reduces to

$$x_1^2 + y_1^2 + z_1^2 = a^2 + b^2 - c^2$$

Thus, we see that the point of intersection of pairs of perpendicular generators lies on the curve of intersection of the hyperboloid and the director sphere

$$x^2 + y^2 + z^2 = a^2 + b^2 - c^2.$$

Another method : Let PA, PB be two perpendicular generators through P and PC be the normal at P so that it is perpendicular to the tangent plane determined by PA and PB. The lines PA, PB, PC are mutually perpendicular and as such the three planes CPA, APB, BPC determined by them, taken in pairs, are also mutually perpendicular.

The plane CPA through the generator PA is the tangent plane at some point of PA and the plane CPB through the generator PB is the tangent plane at some point of PB. Also the plane APB is the tangent plane at P.

Thus, the three planes CPA, APB and BPC are mutually perpendicular tangent planes and as such their point of intersection P lies on the director sphere. It follows that the locus of P is the curve of intersection of the hyperboloid with its director sphere.

EXAMPLE

Show that the angle θ between the generators through any point P of the hyperboloid is given by

$$\cot \theta = \frac{p\,(r^2 - a^2 - b^2 + c^2)}{2abc}$$

where p is the perpendicular from the centre to the tangent plane at P and r is the distance of P from the centre.

Sol. The tangent plane at $P(x_1, y_1, z_1)$ is

$$\frac{xx_1}{a^2} + \frac{yy_1}{b^2} + \frac{zz_1}{c^2} = 1 \qquad \ldots(1)$$

As in § 8.5 it can be shown that the direction ratios l, m, n of the two generators through this point are given by the equations

$$\frac{lx_1}{a^2} + \frac{my_1}{b^2} + \frac{nz_1}{c^2} = 0, \quad \frac{l^2}{a^2} + \frac{m^2}{b^2} + \frac{n^2}{c^2} = 0.$$

Proceeding as in Example 1, we can show that angle θ between the lines is given by

$$\tan \theta = \frac{\left[-4\left(\frac{x_1^2}{a^4} + \frac{y_1^2}{b^4} + \frac{z_1^2}{c^4}\right)\left(\frac{x_1^2}{a^4 b^2 c^2} - \frac{y_1^2}{b^4 c^2 a^2} + \frac{z_1^2}{c^4 a^2 b^2}\right)\right]^{1/2}}{\frac{1}{a^2}\left(\frac{y_1^2}{b^4} + \frac{z_1^2}{c^4}\right) + \frac{1}{b^2}\left(\frac{z_1^2}{c^4} + \frac{x_1^2}{a^4}\right) - \frac{1}{c^2}\left(\frac{x_1^2}{a^4} + \frac{y_1^2}{b^4}\right)}$$

Now, p, the length of perpendicular from the centre to the tangent plane (1) at (x_1, y_1, z_1), is given by

$$p = \frac{1}{\left[\sum \frac{x_1^2}{a^4}\right]^{1/2}} \Rightarrow \frac{1}{p^2} = \sum \frac{x_1^2}{a^4}$$

Also the denominator of the expression for $\tan \theta$

$$= \frac{1}{a^2 b^2 c^2}\left[\frac{x_1^2}{a^2}(c^2 - b^2) + \frac{y_1^2}{b^2}(c^2 - a^2) + \frac{z_1^2}{c^2}(a^2 + b^2)\right]$$

$$= \frac{1}{a^2 b^2 c^2}\left[\frac{x_1^2}{a^2}(c^2 - b^2 - a^2) + \frac{y_1^2}{b^2}(c^2 - a^2 - b^2) + \frac{z_1^2}{c^2}(a^2 + b^2 - c^2)\right.$$

$$\left. + (x_1^2 + y_1^2 + z_1^2)\right]$$

$$= \frac{1}{a^2 b^2 c^2}\left[r^2 - (a^2 + b^2 - c^2)\left(\frac{x_1^2}{a^2} + \frac{y_1^2}{b^2} - \frac{z_1^2}{c^2}\right)\right]$$

$$= \frac{1}{a^2 b^2 c^2}(r^2 - a^2 - b^2 + c^2)$$

$$\therefore \quad \tan\theta = \frac{\left[-\dfrac{4}{p^2}\left(-\dfrac{1}{a^2b^2c^2}\right)\left(\dfrac{x_1^2}{a^2}+\dfrac{y_1^2}{b^2}-\dfrac{z_1^2}{c^2}\right)\right]^{1/2}}{\dfrac{(r^2-a^2-b^2+c^2)}{a^2b^2c^2}} = \frac{2abc}{p\,(r^2-a^2-b^2+c^2)}.$$

8.6. CENTRAL POINT. LINE OF STRICTION. PARAMETER OF DISTRIBUTION OF A GENERATOR

Def. 1. *The **central point** of a given generator, l, is the limiting position of its point of intersection with the line of shortest distance between it and another generator, m, of the same system; the limit being taken when, m, tends to coincide wtih l.*

With some sacrifie of precision, one may say that the central point of a given generator is the point of intersection of the generator and the line of shortest distance between the generator and a consecutive generator of the system.

Def. 2. *The locus of the central points of generators of a hyperboloid is called its **line of striction**.*

Def. 3. *The **parameter of distribution** of a generator, l is*

$$\lim\left(\frac{\Delta s}{\Delta \psi}\right)$$

where, Δs, is the shortest distance and, $\Delta\psi$, the angle between l, and another generator m of the same system, the limit being taken when the generator m tends to coincide with the generator l.

8.6.1. To determine the central point of a generator.

We consider generators of the system

$$\frac{x-a\cos\theta}{a\sin\theta} = \frac{y-b\sin\theta}{-b\cos\theta} = \frac{z}{c}$$

Let any generator, *l*, of the system be

$$\frac{x-a\cos\varphi}{a\sin\varphi} = \frac{y-b\sin\varphi}{-b\cos\varphi} = \frac{z}{c} \qquad ...(1)$$

We, now, consider any other generator, *m*

$$\frac{x-a\cos\varphi'}{a\sin\varphi'} = \frac{y-b\sin\varphi'}{-b\cos\varphi'} = \frac{z}{c} \qquad ...(2)$$

of the same system.

Let the shortest distance between these generators meet them in P and Q respectively so that we have to find the limiting position of the point P on the generator *l* when $\varphi' \to \varphi$. Let C be the limit of P.

Since PQ is a chord of the hyperboloid, its limit will be a tangent line CD at the point C. Let *l, m, n* be the direction ratios of the shortest distance PQ and l_0, m_0, n_0 those of its limit. We have

$$\begin{cases} al\sin\varphi - bm\cos\varphi + cn = 0. \\ al\sin\varphi' - bm\cos\varphi' + cn = 0. \end{cases}$$

$$\Rightarrow \quad \frac{al}{\cos\varphi'-\cos\varphi} = \frac{bm}{\sin\varphi'-\sin\varphi} = \frac{cn}{\sin(\varphi'-\varphi)}$$

$$\Rightarrow \quad \frac{al}{-\sin\frac{1}{2}(\varphi'+\varphi)} = \frac{bm}{\cos\frac{1}{2}(\varphi'+\varphi)} = \frac{cn}{\cos\frac{1}{2}(\varphi'+\varphi)}$$

Let $\varphi' \to \varphi$.

Thus, we obtain
$$\frac{al_0}{-\sin\varphi} = \frac{bm_0}{\cos\varphi} = \frac{cn_0}{1}$$

Let
$$[a(r\sin\varphi + \cos\varphi), b(\sin\varphi - r\cos\varphi), cr] \qquad ...(3)$$
be the central point C on the generator (1). The equation of the tangent plance at C is
$$\frac{x(r\sin\varphi + \cos\varphi)}{a} + \frac{y(\sin\varphi - r\cos\varphi)}{b} - \frac{zr}{c} = 1$$

Since the line CD with direction ratios l_0, m_0, n_0, lies on this tangent plane, we have
$$-\frac{\sin\varphi(r\sin\varphi + \cos\varphi)}{a^2} + \frac{\cos\varphi(\sin\varphi - r\cos\varphi)}{b^2} - \frac{r}{c^2} = 0$$

$$\Rightarrow \quad r\left(\frac{\sin^2\varphi}{a^2} + \frac{\cos^2\varphi}{b^2} + \frac{1}{c^2}\right) = \left(\frac{1}{b^2} - \frac{1}{c^2}\right)\sin\varphi\cos\varphi$$

$$\Rightarrow \quad r = \frac{c^2(a^2 - b^2)\sin\varphi\cos\varphi}{(a^2b^2 + a^2c^2\cos^2\varphi + b^2c^2\sin^2\varphi)}$$

so that we have obtained r.

Substituting this value of r in (3), we see that the co-ordinates of the central point $C(x, y, z)$ are given by
$$x = \frac{a^3(b^2 + c^2)\cos\varphi}{k}, \quad y = \frac{b^3(c^2 + a^2)\sin\varphi}{k}, \quad z = \frac{c^3(a^2 - b^2)\sin\varphi\cos\varphi}{k}$$
where
$$k = a^2b^2 + a^2c^2\cos^2\varphi + b^2c^2\sin^2\varphi$$

Eliminating φ, we see that the *line of striction* is the curve of intersection of the hyperboloid with the cone
$$\frac{a^6(b^2 + c^2)^2}{x^2} + \frac{b^6(c^2 + a^2)^2}{y^2} - \frac{c^6(b^2 - a^2)^2}{z^2} = 0$$

Ex. Find the central point for a generator of the second system and show that the line of striction is the same for either system.

8.6.2. *To determine the parameter of distribution of the generator, l.*

If $\Delta\psi$ be the angle between the generators (1) and (2) of § 8.5.1, we have
$$\tan\Delta\psi = \frac{\sqrt{[b^2c^2(\cos\varphi' - \cos\varphi)^2 + c^2a^2(\sin\varphi' - \sin\varphi)^2 + a^2b^2\sin^2(\varphi' - \varphi)]}}{a^2\sin\varphi\sin\varphi' + b^2\cos\varphi\cos\varphi' + c^2}$$

$$= 2\sin\frac{1}{2}(\varphi' - \varphi) \frac{\sqrt{\left[b^2c^2\sin^2\frac{1}{2}(\varphi' + \varphi) + c^2a^2\cos^2\frac{1}{2}(\varphi' + \varphi) + a^2b^2\cos^2\frac{1}{2}(\varphi' - \varphi)\right]}}{a^2\sin\varphi\sin\varphi' + b^2\cos\varphi\cos\varphi' + c^2}$$

We write $\varphi' = \varphi + \Delta\varphi$ so that $\Delta\varphi \to 0$ as $\varphi' \to \varphi$. Then, from above, we obtain
$$\frac{d\psi}{d\varphi} = \frac{\sqrt{b^2c^2\sin^2\varphi + a^2c^2\cos^2\varphi + a^2b^2}}{a^2\sin^2\varphi + b^2\cos^2\varphi + c^2}$$

Again we shall now find the S.D., Δs between the two generators. Now the equation of the plane through (1) parallel to (2) is
$$\begin{vmatrix} x - a\cos\varphi & y - b\sin\varphi & z \\ a\sin\varphi & -b\cos\varphi & c \\ a\sin\varphi' & -b\cos\varphi' & c \end{vmatrix} = 0$$

so that cancelling a common factor $\sin \frac{1}{2}(\varphi' - \varphi)$, we obtain

$$-bcx \sin \frac{1}{2}(\varphi' + \varphi) + cay \cos \frac{1}{2}(\varphi' + \varphi) + abz \cos \frac{1}{2}(\varphi' - \varphi) + abc \sin \frac{1}{2}(\varphi' - \varphi) = 0$$

The S.D., Δs, which is the distance of the point $(a \cos \varphi', b \sin \varphi', 0)$ from this plane is given by

$$\Delta s = \frac{2abc \sin \frac{1}{2}(\varphi' - \varphi)}{\sqrt{b^2 c^2 \sin^2 \frac{1}{2}(\varphi' - \varphi) + c^2 a^2 \cos^2 \frac{1}{2}(\varphi' + \varphi) + a^2 b^2 \cos^2 \frac{1}{2}(\varphi' - \varphi)}}$$

Again putting $\varphi' = \varphi + \Delta\varphi$, we obtain

$$\frac{ds}{d\varphi} = \frac{abc}{\sqrt{b^2 c^2 \sin^2 \varphi + c^2 a^2 \cos^2 \varphi + a^2 b^2}}$$

$$\Rightarrow \quad \frac{ds}{d\psi} = \frac{ds/d\varphi}{d\psi/d\varphi} = \frac{abc \,(a^2 \sin^2 \varphi + b^2 \cos^2 \varphi + c^2)}{b^2 c^2 \sin^2 \varphi + c^2 a^2 \cos^2 \varphi + a^2 b^2}$$

8.7 HYPERBOLIC PARABOLOID

We rewrite the equation

$$\frac{x^2}{a^2} - \frac{y^2}{b^2} = \frac{2z}{c} \qquad \ldots(1)$$

of a hyperbolic paraboloid in the form

$$\left[\frac{x}{a} - \frac{y}{b}\right]\left[\frac{x}{a} + \frac{y}{b}\right] = \frac{2z}{c}$$

which may again be rewritten in either of the two forms

$$\frac{\frac{x}{a} - \frac{y}{b}}{\frac{z}{c}} = \frac{2}{\frac{x}{a} + \frac{y}{b}}, \quad \frac{\frac{x}{a} - \frac{y}{b}}{2} = \frac{\frac{z}{c}}{\frac{x}{a} + \frac{y}{b}}$$

Now, as in § 8.2 it can be shown that as λ and μ vary; each member of each of the systems of lines

$$\frac{x}{a} - \frac{y}{b} = \frac{\lambda z}{c}, \quad 2 = \lambda\left[\frac{x}{a} + \frac{y}{b}\right] \qquad \ldots(A)$$

$$\frac{x}{a} - \frac{y}{b} = 2\mu, \quad \frac{z}{c} = \mu\left[\frac{x}{a} - \frac{y}{b}\right] \qquad \ldots(B)$$

lies wholly on the hyperbolic paraboloid (1).

Thus, we see that a *hyperbolic paraboloid also admits of two systems of generating lines.*

As in the case of hyperboloid of one sheet, it can be shown that the following results hold good for the two systems of generating lines of a hyperbolic paraboloid also.

1. *Through every point of a hyperbolic paraboloid, there passes a member of each system.*
2. *No two members of the same system intersect.*
3. *Any two generators belonging to the two different systems intersect and the plane through them is the tangent plane at their point of intersection.*
4. *The tangent plane at a point meets the paraboloid in two generators through the point.*
5. *The locus of the point of intersection of perpendicular generator is the curve of intersection of the paraboloid with the plane* $2cz + a^2 - b^2 = 0$.

An Important Note : Since the generator

lies in the plane
$$2 = \lambda\left[\frac{x}{a} + \frac{y}{b}\right]$$

which is parallel to the plane
$$\frac{x}{a} + \frac{y}{b} = 0$$

whatever value λ may have, we deduce that all the generators belonging to one system of the hyperbolic paraboloid

$$\frac{x^2}{a^2} - \frac{y^2}{b^2} = \frac{2z}{c}$$

are parallel to the plane

$$\frac{x}{a} + \frac{y}{b} = 0$$

It may similarly be seen that the generators of the second system are also parallel to a plane, viz.,

$$\frac{x}{a} - \frac{y}{b} = 0.$$

8.7.1. *Tangent plane at any point meets the paraboloid in two generators through the point.*

The planes passing through the two generators of different systems λ and μ of a hyperbolic paraboloid may be given as

$$\left(\frac{x}{a} - \frac{y}{b} - 2\lambda\right) + k\left(\frac{x}{a} + \frac{y}{b} - \frac{z}{c\lambda}\right) = 0 \qquad \text{...(i)}$$

and
$$\left(\frac{x}{a} + \frac{y}{b} - 2\lambda\right) + k'\left(\frac{x}{a} - \frac{y}{b} - \frac{z}{c\mu}\right) = 0 \qquad \text{...(ii)}$$

for all values of k and k'.

These planes become identical if $k = \dfrac{1}{k'} = \dfrac{\lambda}{\mu}$.

This shows that the two generators, one of each system, are coplanar and such they intersect. The plane through them being given as [putting $k = 1/k' = \lambda/\mu$ in (i) and (ii)]

$$\mu\left(\frac{x}{a} - \frac{y}{b} - 2\lambda\right) + \lambda\left(\frac{x}{a} + \frac{y}{b} - \frac{z}{c\lambda}\right) = 0$$

$$\Rightarrow \quad \frac{x}{a}(\mu + \lambda) - \frac{y}{b}(\mu - \lambda) = \frac{1}{c}(z + 2c\lambda\mu) \qquad \text{...(iii)}$$

which is a tangent plane to the hyperbolic paraboloid at the point of intersection of the two generators. Thus, the tangent plane at a point of hyperbolic paraboloid meets it in two generators through the point, the two generators being of different systems λ and μ and the point, their common of intersection.

So we have shown that any plane through a generating line of a hyperbolic paraboloid is a tangent plane at some point of the generator.

8.7.2. *Direction cosines of the generators of the two systems given by*

$$\frac{x}{a} - \frac{y}{b} = 2\lambda, \quad \frac{x}{a} + \frac{y}{b} = \frac{z}{c\lambda} \qquad \text{...(i)}$$

$$\frac{x}{a} + \frac{y}{b} = 2\mu, \quad \frac{x}{a} - \frac{y}{b} = \frac{z}{c\mu} \qquad \text{...(ii)}$$

If l_1, m_1, n_1 and l_2, m_2, n_2 are the d.c.'s of (i) and (ii), then we have

Generating Lines of Conicoids

$$\frac{l_1}{\frac{1}{bc\lambda}} = \frac{m_1}{\frac{1}{ac\lambda}} = \frac{n_1}{\frac{2}{ab\lambda}},$$

since this generator is the line of intersection of the planes

$$\frac{x}{a} - \frac{y}{b} + 0.z = 2\lambda \quad \text{and} \quad \frac{x}{a} + \frac{y}{b} - \frac{z}{c\lambda} = 0$$

$$\Rightarrow \qquad \frac{l_1}{a} = \frac{m_1}{b} = \frac{n_1}{2c\lambda} \qquad \ldots\text{(iii)}$$

Similarly,
$$\frac{-l_2}{\frac{1}{bc\mu}} = \frac{m_2}{\frac{1}{ca\mu}} = \frac{n_2}{\frac{-2}{ab}}$$

$$\Rightarrow \qquad \frac{l_2}{a} = \frac{m_2}{-b} = \frac{n_2}{2c\mu} \qquad \ldots\text{(iv)}$$

EXAMPLES

1. *Planes are drawn through the origin O and the generators through any point P of the paraboloid given by $x^2 - y^2 = az$. prove that the angle between them is $\tan^{-1}(2r/a)$, where r is the length of OP.*

Sol. The two systems of generators of the paraboloid

$$x^2 - y^2 = az \qquad \ldots\text{(i)}$$

are given as
$$x - y = a\lambda, \; x + y = z/\lambda \qquad \ldots\text{(ii)}$$
and
$$x + y = a\mu, \; x - y = z/\mu \qquad \ldots\text{(iii)}$$

Plane through the λ-generator and the origin is
$$x + y = z/\lambda \qquad \ldots\text{(iv)}$$

Plane through the origin and the μ-generator is
$$x - y = z/\mu \qquad \ldots\text{(v)}$$

If θ is the angle between these planes (iv) and (v), then

$$\cos\theta = \frac{1 - 1 + 1/(\lambda\mu)}{\sqrt{1+1+1/\lambda^2}\sqrt{1+1+1/\mu^2}}$$

$$\Rightarrow \qquad \sec^2\theta = (2\lambda^2 + 1)(2\mu^2 + 1)$$

$$\Rightarrow \qquad \tan\theta = \sqrt{2\lambda^2 + 2\mu^2 + 4\lambda^2\mu^2} \qquad \ldots\text{(vi)}$$

Also P, the point of intersection of the generators (ii) and (iii) is,

$$\left[a\left(\frac{\lambda+\mu}{2}\right), \; a\left(\frac{\mu-\lambda}{2}\right), \; a\lambda\mu\right]$$

Then
$$OP = r = \frac{1}{2}a\sqrt{(\lambda+\mu)^2 + (\mu-\lambda)^2 + 4\lambda^2\mu^2}$$

$$= \frac{1}{2}a\sqrt{2\lambda^2 + 2\mu^2 + 4\lambda^2\mu^2} = \frac{1}{2}a\tan\theta, \qquad \text{[from (vi)]}$$

$\therefore \qquad \tan\theta = \frac{2r}{a} \Rightarrow \theta = \tan^{-1}(2r/a).$

2. *Prove that the equation $2x = ae^{2\phi}$, $y = be^{\phi}\cosh\theta$, $z = ce^{\phi}\sinh\theta$ determines a hyperbolic paraboloid and that $(\theta + \phi)$ is constant for points of a given generator of one system and $(\theta - \phi)$*

is constant for a given generator of the other.

Sol. The parameteric equations of the surface are

$$2x = ae^{2\phi}, \quad y = be^{\phi} \cosh\theta, \quad z = ce^{\phi} \cosh\theta \qquad \ldots(i)$$

$$\therefore \quad \left(\frac{y}{b}\right)^2 - \left(\frac{z}{c}\right)^2 = e^{2\phi} = \frac{2x}{a}$$

$$\Rightarrow \quad \frac{y^2}{b^2} - \frac{z^2}{c^2} = \frac{2x}{a}, \qquad \ldots(ii)$$

which is a hyperbolic paraboloid.

The different systems of generators of (ii) are given as

$$\frac{y}{b} + \frac{z}{c} = 2\lambda, \quad \frac{y}{b} - \frac{z}{c} = \frac{x}{a\lambda} \qquad \ldots(iii)$$

$$\frac{y}{b} - \frac{z}{c} = 2\mu, \quad \frac{y}{b} + \frac{z}{c} = \frac{x}{a\mu} \qquad \ldots(iv)$$

Since both the generators pass through the given point, we have

$$e^{\phi}(\cosh\theta + \sinh\theta) = 2\lambda, \quad e^{\phi}(\cosh\theta - \sinh\theta) = \frac{e^{2\phi}}{2\lambda} \qquad \ldots(v)$$

From the second of the relation in (v), we have

$$\cosh\theta - \sinh\theta = \frac{e^{\phi}}{2\lambda} \Rightarrow e^{-\theta} = \frac{e^{\phi}}{2\lambda}$$

$$\therefore \quad \lambda = \frac{e^{(\theta + \phi)}}{2} \qquad \ldots(vi)$$

Also from (v),

$$e^{\phi}(\cosh\theta + \sinh\theta) = \frac{e^{2\phi}}{2\mu}$$

$$\Rightarrow \quad \mu = \frac{e^{(\phi - \theta)}}{2} \qquad \ldots(vii)$$

Relations (vi) and (vii) show that $(\theta + \phi)$ is constant for points of a given generator of the λ-system (since λ was constant for a given generator of the λ-system) and $(\theta - \phi)$ is constant for a given generator of the μ-system.

3. *Show that the angle between the generating lines of* $\dfrac{x^2}{a^2} - \dfrac{y^2}{b^2} = 2z$ *through* (x, y, z) *is given by*

$$\tan\theta = ab\left(1 + \frac{x^2}{a^4} + \frac{y^2}{b^4}\right)^{1/2}\left(z + \frac{a^2 - b^2}{2}\right)^{-1}$$

Sol. The d.r.'s of different systems of generators are $a, b, 2c\lambda$ and $a, -b, 2c\mu$ respectively. Hence,

$$\tan\theta = \frac{[(2b\mu + 2b\lambda)^2 + (2\lambda a - 2\mu a)^2 + (-2ab)^2]^{1/2}}{a^2 - b^2 + 4\lambda\mu}$$

$$= \frac{[4b^2(\mu + \lambda)^2 + 4a^2(\mu - \lambda)^2 + 4a^2b^2]^{1/2}}{a^2 - b^2 + 4\lambda\mu}$$

Generating Lines of Conicoids

$$= \frac{\left[4b^2\left(\dfrac{x}{a}\right)^2 + 4a^2\left(\dfrac{y}{b}\right)^2 + 4a^2b^2\right]^{1/2}}{a^2 - b^2 + 2z} = ab\left(1 + \frac{z^2}{a^4} + \frac{y^2}{b^4}\right)^{1/2}\left(z + \frac{a^2 - b^2}{2}\right)^{-1}$$

EXERCISES

1. Obtain equations for the two systems of generating lines on the hyperbolic parabolid $\dfrac{x^2}{a^2} - \dfrac{y^2}{b^2} = 4z$, and hence express the co-ordinates of a point on the surface as functions of two parameters. Find the direction cosines of the generators through $(\alpha, 0, \gamma)$ and show that the cosines of the angle between them is $\dfrac{(a^2 - b^2 + \gamma)}{(a^2 + b^2 + \gamma)}$.

2. Show that the projections of the generators of a hyperbolic paraboloid on any principal plane are tangents to the section by the plane.

3. Find the locus of the perpendiculars from the vertex at the paraboloid
$$\frac{x^2}{a^2} - \frac{y^2}{b^2} = \frac{2z}{c}$$
to the generators of one system. **[Ans.** $x^2 + y^2 + 2z^2 \neq (a^2 + b^2)xy/ab = 0$**]**

4. Show that the points of intersection of generators $xy = az$ which are inclined at a constant angle α lie on the curve of intersection of the paraboloid and the hyperboloid
$$x^2 + y^2 - z^2 \tan^2 \alpha + a^2 = 0.$$

5. Through a variable generator $x - y = \lambda$, $x + y = \dfrac{2z}{\lambda}$ of the paraboloid $x^2 - y^2 = 2z$ a plane is drawn making a constant angle α with the plane $x = y$. Find the locus of the point at which it touches the paraboloid.

[Ans. Curve of intersection of the above surface and paraboloid is $x^2 - y^2 = 2z$.**]**

8.8 CENTRAL POINT. LINE OF STRICTION. PARAMETER OF DISTRIBUTION

8.8.1. To determine the central point of any generator of the system of generators.

$$\frac{x}{a} - \frac{y}{b} = \frac{\lambda z}{c}, \quad 2 = \lambda\left(\frac{x}{a} - \frac{y}{b}\right).$$

Let a generator, l, of this system be

$$\frac{x}{a} - \frac{y}{b} = \frac{pz}{c}, \quad 2 = p\left(\frac{x}{a} + \frac{y}{b}\right) \qquad ...(1)$$

We, now, consider a generator, m, of the same system

$$\frac{x}{a} - \frac{y}{b} = \frac{p'z}{c}, \quad 2 = p'\left(\frac{x}{a} + \frac{y}{b}\right) \qquad ...(2)$$

The direction ratios of these generators are $a, -b, 2c/p$; $a, -b, 2c/p'$.
If l, m, n be the direction ratios of the line of S.D., between (1) and (2), we have
$$al - bm + 2cn/p = 0, \quad al - bm + 2cn/p' = 0$$
These give $1/a, 1/b, 0$ as the direction ratios of the line of S.D., being independent of p and p', we see that the line of S.D., is parallel to a fixed line.

Let (x_1, y_1, z_1) be the central point of the generator (1). As in § 8.5.1, the limiting position of the line of S.D., is a line contained in the tangent plane

$$\frac{xx_1}{a^2} - \frac{yy_1}{b^2} = \frac{1}{c}(z + z_1)$$

at (x_1, y_1, z_1).
Thus, we have

$$\frac{x_1}{a^3} - \frac{y_1}{b^3} = 0 \qquad \ldots(3)$$

Also, since (x_1, y_1, z_1) lies on (1), we have

$$\frac{x_1}{a} - \frac{y_1}{b} = \frac{pz_1}{c}, \quad 2 = p\left(\frac{x_1}{a} + \frac{y_1}{b}\right) \qquad \ldots(4)$$

Solving (3) and (4), we obtain

$$x_1 = \frac{2a^3}{p(a^2+b^2)}, \quad y_1 = \frac{2b^3}{p(a^2+b^2)}, \quad z_1 = \frac{2c(a^2-b^2)}{p(a^2+b^2)}.$$

Eliminating p, we see that the line of striction is the curve of intersection of the surface with the plane

$$\frac{x}{a^3} + \frac{y}{b^3} = 0.$$

Ex. Find the central point of a generator of the second system and show that the corresponding line of striction is the curve of intersection of the surface with the plane

$$\frac{x}{a^3} + \frac{y}{b^3} = 0.$$

8.8.2. *To determine the parameter of distribution.*

Let $\Delta\psi$ and Δs be the angle of S.D., respectively between the generators (1) and (2). We have

$$\tan \Delta\psi = \frac{2c\sqrt{(a^2+b^2)}\,(p'-p)}{pp'(a^2+b^2)+4c^2}$$

Let $p' = p + \Delta p$ so that $\Delta p \to 0$ as $p' \to p$. We have

$$\frac{d\psi}{dp} = \frac{2c\sqrt{a^2+b^2}}{p^2(a^2+b^2)+4c^2} \qquad \ldots(5)$$

Now the plane through the generator (1) and parallel to the generator (2) is

$$\frac{x}{a} + \frac{y}{b} = \frac{2}{p}.$$

Also taking $z = 0$, we see that $(a/p', b/p', 0)$ is a point on the generator (2).

$$\therefore \quad \Delta s = \frac{\dfrac{2}{p} - \dfrac{2}{p'}}{\left(\dfrac{1}{a^2} + \dfrac{1}{b^2}\right)^{1/2}} = \frac{2(p'-p)\,ab}{pp'\sqrt{a^2+b^2}} \qquad \ldots(6)$$

Thus, as before

$$\frac{ds}{dp} = \frac{2ab}{p^2\sqrt{a^2+b^2}},$$

$$\Rightarrow \quad \frac{ds}{d\psi} = \frac{ab\,[p^2(a^2+b^2)+4c^2]}{cp^2(a^2+c^2)}$$

which is the parameter of distribution.

Generating Lines of Conicoids

Ex. For the generator of the paraboloid $\dfrac{x^2}{a^2} - \dfrac{y^2}{b^2} = 2z$ given by

$$\frac{x}{a} - \frac{y}{b} = 2\lambda, \quad \frac{x}{a} + \frac{y}{b} = \frac{z}{\lambda},$$

prove that the parameter of distribution is

$$\frac{ab(a^2 + b^2 + 4\lambda^2)}{(a^2 + b^2)}$$

and the central point is

$$\left[\frac{2a^3\lambda}{a^2 + b^2}, \frac{-2b^3\lambda}{a^2 + b^2}, \frac{2(a^2 - b^2)\lambda^2}{a^2 + b^2} \right].$$

Prove that the centtral points of the systems of generators lie on the planes

$$\frac{x}{a^3} \pm \frac{y}{b^3} = 0.$$

8.9. GENERAL CONSIDERATION

We have seen that hyperboloid of one sheet and hyperbolic paraboloid each admit of two systems of generators such that through each point of the surface there passes one member of each system and that two members of opposite systems intersect but no two members of the same system intersect. Also we know that through each point of a cone or a cylinder there passes one generator. Thus, hyperboloids of one sheet, hyperboloid paraboloids, cones and cylinders are *ruled surfaces* in as much as they can be generated by straight lines.

We now proceed to examine the case of the general quadric in relation to the existence of generaotrs.

8.9.1. Condition for a Line to be a Generator

A straight line will be a generator of a quadric if three points of the line lie on the quadric.

Let the quadric be

$$ax^2 + by^2 + cz^2 + 2fyz + 2gzx + 2hxy + 2ux + 2vy + 2wz + d = 0 \qquad ...(1)$$

The line

$$\frac{x - \alpha}{l} = \frac{y - \beta}{m} = \frac{z - \gamma}{n}$$

will be a generator of the quadric, if the point $(lr + \lambda, mr + \beta, nr + \gamma)$ on the line lies on the quadric for all values of r, i.e., the equation obtained on substituted these co-ordinates for x, y, z in (1) is an identity. As this equation is a qudric in r, it will be an identity if it is satisfied for three values of r, i.e., if three points of the line lie on the quadric.

Cor. 1. The quadric equation in r obtained above will be an identity if the coefficients of r^2, r and the constant term are separately zero. This gives

$$al^2 + bm^2 + cn^2 + 2fmn + 2gnl + 2hlm = 0 \qquad ...(2)$$

$$l(a\alpha + h\beta + g\gamma) + m(h\alpha + b\beta + f\gamma) + n(g\alpha + f\beta + c\gamma) = 0 \qquad ...(3)$$

$$a\alpha^2 + b\beta^2 + c\gamma^2 + 2f\beta\gamma + 2g\gamma\alpha + 2h\alpha\beta + 2u\alpha + 3v\beta + 2w\gamma + d = 0 \qquad ...(4)$$

The condition (4) simply means that the point (α, β, γ) lies on the quadric.

Since (2) is a homogeneous quadric equation and (3) is a homogeneous linear equation in l, m, n these two equations will determine *two* sets of values of l, m, n. Thus, we deduce that *through every point on a quadric there pass two lines, real, coincident or imaginary lying wholly on the quadric.*

Cor. 2. *A quadric can be drawn so as to contain* **three** *mutually skew lines as generators,* for the quadric determined by nine points, three on each line, will contain the three lines as generators.

8.10. QUADRICS WITH REAL AND DISTINCT PAIRS OF GENERATING LINES

8.10.1. *Of all real central quadrics, hyperboloid of one sheet only possesses two real and distinct generators through a point.*

Let
$$ax^2 + by^2 + cz^2 = 1$$
be any central quadric.

The direction ratios, l, m, n of any generator
$$\frac{x-\alpha}{l} = \frac{y-\beta}{m} = \frac{z-\gamma}{n}$$
of the quadric through the point (α, β, γ) are given by the equations
$$al^2 + bm^2 + cn^2 = 0, \quad al\alpha + bm\beta + cn\gamma = 0$$

Eliminating n from these, we obtain
$$a(a\alpha^2 + c\gamma^2)l^2 + 2ab\alpha\beta lm + b(b\beta^2 + c\gamma^2)m^2 = 0$$

Its roots will be real and distinct if, and only if
$$4a^2b^2\alpha^2\beta^2 - 4ab(a\alpha^2 + c\gamma^2)(b\beta^2 + c\gamma^2) > 0$$
$$\Leftrightarrow \quad -4abc\gamma^2(a\alpha^2 + b\beta^2 + c\gamma^2) > 0$$

Since $a\alpha^2 + b\beta^2 + c\gamma^2 = 1$, we see that the roots will be real and distinct, if and only, if abc is negative.

Now this will be the case if, a, b, c are all negative or one negative and two positive. In the former case the quadric itself is imaginary and in the latter it is a hyperboloid of one sheet

8.10.2. *Of the two paraboloids, hyperbolic paraboloid only possesses two real and distinct generators through a point.*

In the case of the paraboloid
$$ax^2 + by^2 = 2cz$$
the direction ratios, l, m, n of the generating lines through a point (α, β, γ) of the surface are given by
$$al^2 + bm^2 = 0 \qquad \ldots(1)$$
$$al\alpha + bm\beta = 0 \qquad \ldots(2)$$

The equation (1) shows that for real values of l and m, we must have a and b with opposite signs, *i.e.,* the paraboloid must be hyperbolic.

8.11. LINES INTERSECTING THREE LINES

An infinite number of lines can be drawn meeting three given mutually skew lines. For the quadric through the three given mutually skew lines a, b, c, the three lines will be generators of one system and all the other generators of the other system will intersect a, b and c.

In fact the quadric through three given mutually skew lines can be determined as the locus of lines which intersect the three given lines.

Thus, the locus is really the equation of the quadric through the three lines
$$u_r = 0 = v_r; \quad r = 1, 2, 3.$$

EXAMPLES

1. *Find the equations of the quadric containing the three lines*
$$y = b, \; z = -c; \quad z = c, \; x = -a; \quad x = a, \; y = -b.$$
Also obtain the equations of its two systems of generators.

Generating Lines of Conicoids

Sol. Any line which intersects the first two lines is given by
$$y - b + \lambda_1 (z + c) = 0, \\ z - c + \lambda_2 (x + a) = 0, \quad \ldots(i)$$

for all values of λ_1, λ_2.

This will intersect the third line $x = a, y = -b$ if
$$z = c - 2a\lambda_2 = \frac{2b}{\lambda_1} - c$$

$$\Rightarrow \quad c = \frac{b}{\lambda_1} + a\lambda_2, \quad \ldots(ii)$$

which is $f(\lambda_1, \lambda_2) = 0$.

Eliminating λ_1, λ_2 from (i) and (ii), we get
$$c = -\frac{b(z+c)}{y-b} - \frac{a(z-c)}{x+a}.$$

$$\Rightarrow \quad c(xy - bx + ay - ab) + b(xz + cx + az + ca) + a(yz - cy - bz + bc) = 0$$

$$\Rightarrow \quad ayz + bzx + cxy + abc = 0 \quad \ldots(iii)$$

which is the required quadric containing the three given lines.

To get the two systems of generators of (iii), rewrite (iii) as
$$y(ax + cx) + b(zx + ca) = 0$$

$$\Rightarrow \quad y(az + cx) + b[(x+a)(z+c) - (az + cx)] = 0$$

$$\Rightarrow \quad (y - b)(az + cx) + b(x+a)(z+c) = 0 \quad \ldots(iv)$$

which may again be written in either of the two forms.

$$\frac{b(x+a)}{(az+cx)} = \frac{-(y-b)}{(z+c)} = \lambda \text{ (say)} \quad \ldots(v)$$

and
$$\frac{-(y-b)}{b(x+a)} = \frac{z+c}{(az+cx)} = \mu \text{ (say)} \quad \ldots(vi)$$

where λ and μ are arbitrary constants.

Relations (v) and (vi) give system of generators as
$$y - b + \lambda(z+c) = 0, \quad b(x+a) - \lambda(az+cx) = 0,$$
$$\mu(az+cx) - (z+c) = 0, \quad \mu b(x+a) + (y-b) = 0.$$

2. *Find the locus of the perpendicular from a point on a hyperboloid to the generators of one system.*

Sol. Let the given point O be taken as origin and a generator through O as OX, the x-axis.

Let OZ, the normal at O be taken as the z-axis. Then XOY is the tangent plane at Q, OY being the y-axis.

Then the equation of the hyperboloid of which the x-axis is a generator and the z-axis is the normal, is of the form
$$by^2 + cz^2 + 2fyz + 2gzx + 2hxy + 2wz = 0$$

$$\Rightarrow \quad y(by + 2hx) + z(cz + 2gx + 2fy + 2w) = 0 \quad \ldots(i)$$

The system of generators of (i) are given by
$$\lambda y = z, \quad (by + 2hx) + \lambda(cz + 2gx + 2fy + 2w) = 0 \quad \ldots(ii)$$
$$y = \mu, \quad (cz + 2gx + 2fy + 2w)z + \mu(by + 2hx) = 0 \quad \ldots(iii)$$

Again let any line through origin O and perpendicular to the λ-generator be

$$\frac{x}{L} = \frac{y}{M} = \frac{z}{N} = k. \qquad \text{...(iv)}$$

Also the λ-generator is the line of intersection of the plane
$$\lambda y - z = 0$$
and $\qquad 2(h + g\lambda)x + (b + 2\lambda f)y + c\lambda z = -2w$

∴ The d.c.'s of the λ-generator are proportional to
$$(c\lambda^2 + b + 2\lambda f), \; -2(h + g\lambda), \; -2\lambda(h + g\lambda).$$

Since (iv) is perpendicular to the generator (ii), we have
$$L(c\lambda^2 + b + 2\lambda f) - 2M(h + g\lambda) - 2\lambda N(h + g\lambda) = 0 \qquad \text{...(v)}$$

Eliminating L, M, N and λ between (ii), (iv) and (v), we get
$$\frac{x}{k}\left[c\left(\frac{z}{y}\right)^2 + b - 2\frac{z}{y}f\right] - 2\frac{y}{k}\left(h + g\frac{z}{y}\right) - 2\frac{z}{k}\cdot\frac{z}{y}\left(h + g\cdot\frac{z}{y}\right) = 0$$

$\Rightarrow \quad x(cz^2 + 2fyz + by^2) - 2y^2(hy + gz) - 2z^2(hy + gz) = 0$

$\Rightarrow \quad x(cz^2 + 2fyz + by^2) - 2(y^2 + z^2)(hy + gz) = 0,$

which is the required locus.

EXERCISES

1. Find the equations of the hyperboloid through the three lines
   $$y - z = 1, \; x = 0; \; z - x = 1, \; y = 0; \; x - y = 1, \; z = 0.$$
   Also obtain the equations of its two systems of generators.

   [**Ans.** $x^2 + y^2 + z^2 - 2xy - 2yz - 2zx = 1$; $x - y - 1 = \lambda z$, $\lambda(x - y + 1) = 2x + 2y - z$,
   $x - y - 1 = \lambda(2x + 2y - z)$, $\lambda(x - y + 1) = z$]

2. The generators of one system of a hyperbolic paraboloid are parallel to the plane
   $$lx + my + nz = 0$$
   and the lines $\quad ax + by = 0 = z + c; \; ax - by = 0 = z - c$
   are two members of the same system.
   Show that the equation of the paraboloid is
   $$abc(lx + my + nz) = c(a^2nx + b^2ly + abcn).$$

3. Show that two straight lines can be drawn intersecting four given mutually skew lines.

□□□

9
General Equation of the Second Degree

9.1 REDUATION TO CANONICAL FORMS AND CLASSIFICATION

A quadric has been defined as the locus of a point satisfying an equation of the second degree. Thus, a quadric is the locus of a point satisfying an equation of the type
$$F(x, y, z) \equiv ax^2 + by^2 + cz^2 + 2fyz + 2hxy + 2ux + 2vy + 2wz + d = 0$$
which we may rewrite as
$$\Sigma(ax^2 + 2fyz) + 2\Sigma ux + d = 0 \qquad \ldots(1)$$
splitting the set of all terms into three homogeneous subsets.

We have considered so far special forms of the equations of the second degree in order to discuss *geometrical* properties of the various types of quadrics. In this chapter we shall see how the general equation of a second degree by means of an appropriate change of co-ordinate system can be reduced to simpler forms and also thus classify the types of quadrics.

Equations of various loci connected with a given quadric : We proceed to determine the equations of various loci associated with a quadric given by a general second degree equation. In this connection, we shall start obtaining a quadric in r, which will play a very important role in connection with the determination of the equations of these loci.

Consider a point (α, β, γ) and a line through the same with direction cosines (l, m, n). The co-ordinates of the point on this line at a distance r from (α, β, γ) are
$$(lr + \alpha, mr + \beta, nr + \gamma)$$
This point will lie on the quadric
$$F(x, y, z) \equiv \Sigma(ax^2 + 2fyz) + 2\Sigma ux + d = 0$$
for values of r satisfying the equation
$$\Sigma[a(lr+\alpha)^2 + 2f(mr+\beta)(nr+\gamma)] + 2\Sigma u(lr+\alpha) + d = 0$$
$$\Leftrightarrow r^2\Sigma(al^2 + 2fmn) + 2r[l(a\alpha + h\beta + g\gamma + u) + m(h\alpha + b\beta + f\gamma + v)]$$
$$+ n(g\alpha + f\beta + c\gamma + w)] + F(\alpha, \beta, \gamma) = 0 \ldots(2)$$

which is a quadric in r. Thus, if r_1, r_2 be the roots of this quadric, the two points of intersection of the line with the quadric are
$$(lr_1 + \alpha, mr_1 + \beta, nr_1 + \gamma), (lr_2 + \alpha, mr_2 + \beta, nr_2 + \gamma)$$

Note : It may be noted that the equation (2) can be rewritten as
$$r^2\Sigma(al^2 + 2fmn) + r\left(l\frac{\partial F}{\partial \alpha} + \frac{\partial F}{\partial \beta} + \frac{\partial F}{\partial \gamma}\right) + F(\alpha, \beta, \gamma) = 0$$

where $\dfrac{\partial F}{\partial \alpha} + \dfrac{\partial F}{\partial \beta} + \dfrac{\partial F}{\partial \gamma}$ denote the the values of the partial derivatives of F, w.r.t. x, y, z respectively at the point (α, β, γ).

9.1.1 The tangent plane at a point

Suppose that the point (α, β, γ) lies on the quadric so that we have
$$F(\alpha, \beta, \gamma) = 0$$

and one root of the quadric equation (2) is zero. The vanishing of value of r is also a simple consequence of the fact that one of the two points of intersection of the quadric which every line through a point of the quadric coincides with the point in question.

A line through the point (α, β, γ) on the quadric with direction cosines (l, m, n) will be a tangent line if the second point of intersection also coincides with (α, β, γ), *i.e*, if the second value of r, as given by (2) is also zero. This will be so if the coefficient of r is also zero, *i.e.*, if

$$l(a\alpha + h\beta + g\gamma + u) + m(h\alpha + b\beta + f\gamma + v) + n(g\alpha + f\beta + c\gamma + w) = 0 \quad \ldots(3)$$

which is thus the condition for the line

$$\frac{x-\alpha}{l} = \frac{y-\beta}{m} = \frac{z-\gamma}{n} \quad \ldots(4)$$

to be a tangent line at the point (α, β, γ). The locus of the tangent lines through (α, β, γ), obtained on eliminating l, m, n between (3) and (4) is

$$\Sigma (x - a)(a\alpha + h\beta + g\gamma + u) = 0$$

$$\Leftrightarrow \quad \Sigma x (a\alpha + h\beta + g\gamma + u) = \Sigma \alpha (a\alpha + h\beta + g\gamma + u)$$

Adding $u\alpha + v\beta + w\gamma + d$ to both sides, we get

$$\Sigma x (a\alpha + h\beta + g\gamma + u) + (u\alpha + v\beta + g\gamma + d) = F(\alpha, \beta, \gamma) = 0$$

Thus, the locus of the tangent line (α, β, γ) is

$$\Sigma x (a\alpha + b\beta + g\gamma + u) + (u\alpha + v\beta + w\gamma + d) = 0$$

which is a plane called the *tangent plane* at (α, β, γ).

9.1.2 The normal at a Point

The line through (α, β, γ), perpendicular to the tangent plane thereat, viz.,

$$\frac{x-\alpha}{a\alpha + h\beta + g\gamma + u} = \frac{y-\beta}{h\alpha + b\beta + f\gamma + v} = \frac{z-\gamma}{g\alpha + f\beta + c\gamma + w}$$

is the normal at the point (α, β, γ).

9.1.3 Enveloping cone from a Point

Suppose now that (α, β, γ) is a point not necessarily on the quadric. Then any line through (α, β, γ) with direction cosines (l, m, n) will touch the quadric, *i.e.*, meet the same in two coincident points, if the two roots of the quadric equation in r, are equal. The condition for this is

$$[\Sigma l (a\alpha + h\beta + g\gamma + u)]^2 = \Sigma (al^2 + 2fmn) F(\alpha, \beta, \gamma) \quad \ldots(5)$$

The locus of the line

$$\frac{x-\alpha}{l} = \frac{y-\beta}{m} = \frac{z-\gamma}{n} \quad \ldots(6)$$

through (α, β, γ) touching the quadric, obtained on eliminating l, m, n between (5) and (6), is

$$[\Sigma (x - \alpha)(a\alpha + h\beta + g\gamma + u)]^2 = [\Sigma a (x - \alpha)^2 + 2f (y - \beta)(z - \gamma)] F(\alpha, g, \gamma) \quad \ldots(7)$$

To put this equation in a convenient form, we write

$$S = F(x, y, z), S_1 = F(\alpha, \beta, \gamma), T = \Sigma x (a\alpha + h\beta + g\gamma + u) + (u\alpha + v\beta + w\gamma + d)$$

Then (7) can be written as

$$(T - S_1)^2 = S_1 (S + S_1 - 2T)$$

$$\Leftrightarrow \quad SS_1 = T^2$$

which is the *equation of the Enveloping cone of the quadric $S = 0$ with the point (α, β, γ) as its vertex.*

General Equation of the Second Degree

9.1.4 Enveloping Cylinder

Suppose now that (l, m, n) are given and we require the locus of tangent lines with direction cosines (l, m, n). If (α, β, γ) be a point on any such tangent line, we have the condition

$$[\Sigma l\,(a\alpha + h\beta + g\gamma + u)]^2 = [\Sigma\,(al^2 + 2fmn)]\,F(\alpha, \beta, \gamma)$$

as obtained in 9.1.3 above. Thus, the required locus is

$$[\Sigma l\,(ax + hy + gz + u)]^2 = \Sigma(al^2 + 2fmn)\,F(x, y, z)$$

known as Enveloping Cylinder.

This is the equation of the envelopign cylinder of the quadric $F(x, y, z) = 0$ with generators parallel to the line with direction cosines (l, m, n).

9.1.5 Section with a given centre

Suppose now that (α, β, γ) is a given point. Then a chord with direction cosines (l, m, n) through the point (α, β, γ) will be bisected thereat if the sum of the two roots of the r-quadratic (2) is zero. This will be so, if and only if

$$\Sigma l\,(a\alpha + h\beta + g\gamma + u) = 0 \qquad \ldots(8)$$

so that the locus of the chord

$$\frac{x - \alpha}{l} = \frac{y - \beta}{m} = \frac{z - \gamma}{n} \qquad \ldots(9)$$

through the point (α, β, γ) and bisected thereat, obtained on eliminating l, m, n from the relations (8) and (9) is

$$\Sigma\,(x - \alpha)\,(a\alpha + h\beta + g\gamma + u) = 0$$

which, we may rewrite as

$$T = S_1.$$

The *plane* $T = S_1$, *meets the quadric in a conic with its centre at* (α, β, γ).

9.2 POLAR PLANE OF A POINT

If a line through a point $A\,(\alpha, \beta, \gamma)$ meets the quadric in points Q, R and a point P is taken on the line such that the points A and P divide the segment QR internally and externally in the same ratio, then the locus of P for different lines through A is a plane called the *Polar plane* of the point A with respect to the quadric. It is easily seen that if the points A and P divide the segment QR internally and externally in the same ratio, then the points Q and R also divide the segment AP internally and externally in the same ratio.

Consider a line through the point $A\,(\alpha, \beta, \gamma)$ and let P be the point (x, y, z). The point dividing the segment AP in the ratio $\lambda : 1$ is

$$\left(\frac{\lambda x + \alpha}{\lambda + 1}, \frac{\lambda y + \beta}{\lambda + 1}, \frac{\lambda z + \gamma}{\lambda + 1}\right).$$

This point will lie on the quadric.

$$\Sigma\,(ax^2 + 2fyz) + 2\Sigma ux + d = 0$$

if

$$\Sigma\left[a\left(\frac{\lambda x + \alpha}{\lambda + 1}\right)^2 + 2f\left(\frac{\lambda y + \beta}{\lambda + 1}\right)\left(\frac{\lambda z + \gamma}{\lambda + 1}\right)\right] + 2\Sigma u\left(\frac{\lambda x + \alpha}{\lambda + 1}\right) + d = 0$$

$$\Leftrightarrow \quad \lambda^2 F\,(x, y, z) + 2\lambda\,[x\,(a\alpha + h\beta + g\gamma + u) + y\,(h\alpha + b\beta + f\gamma + v)$$
$$+ z\,(g\alpha + f\beta + c\gamma + w) + (u\alpha + v\beta + w\gamma + d) + F(\alpha, \beta, \gamma) = 0$$

The two value of λ give the two ratios in which the points Q and R divide the segment AP. In order that the points Q and R may divide the segment AP internally and externally in the same ratio, the sum of the two values of λ should be zero, *i.e.*,

$$x(a\alpha + h\beta + g\gamma + u) + y(h\alpha + b\beta + f\gamma + v) + z(g\alpha + f\beta + c\gamma + w)$$
$$+ (u\alpha + v\beta + w\gamma + d) = 0 \qquad ...(10)$$

which is the required locus of the point $P(x, y, z)$.

Thus, (10) is the required equation of the polar plane.

Note : The notations of *Conjugate points, Conjugate planes, Conjugate lines* and *Polar lines* can be introduced as in the case of particular forms of equations in the preceding chapters.

9.3 DIAMETRAL PLANE CONJUGATE TO A GIVEN DIRECTION

We know that if (l, m, n) be the direction cosines of a chord and (x, y, z) the mid-points of the same, then we have

$$l\frac{\partial F}{\partial x} + m\frac{\partial F}{\partial y} + n\frac{\partial F}{\partial z} = 0 \qquad ...(1)$$

Thus, if l, m, n be supposed to be given, then the equation of the locus of the mid-point (x, y, z) of parallel chords with direction cosines (l, m, n) is given by (1) above. This locus is a plane called the *Diameter plane conjugate to the direction cosines (l, m, n)*. We can rewrite *the equation (1) of the diametral plane conjugate to l, m, n* as

$$x(al + hm + gn) + y(hl + bm + fn) + z(gl + fm + cn) + (ul + vm + wn) = 0 \qquad ...(2)$$

NOTE : In this connection we should remember that there does not necessarily correspond a diametral plane conjugate to *every* given direction. Thus, we see from above that there is no diametral plane conjugate to the direction cosines (l, m, n) if l, m, n are such that the coefficients of x, y, z in the equation (2) are all zero. Thus there will be no diametral plane corresponding to a direction whose direction cosines (l, m, n) satisfy the three relations

$$al + hm + gn = 0$$
$$hl + bm + fn = 0$$
$$gl + fm + cn = 0$$

These three homogeneous linear equations in l, m, n will have a non-zero solution, if and only if

$$D = \begin{vmatrix} a & h & g \\ h & b & f \\ g & f & c \end{vmatrix} = 0$$

We also denote by A, B, C the co-factors of a, b, c in this determinant so that we have $A = bc - f^2, B = ca - g^2, C = ab - h^2$.

9.4 PRINCIPAL DIRECTIONS AND PRINCIPAL PLANES

A direction l, m, n is said to be a *Principal direction*, if it is perpendicular to the diametral plane conjugate to the same. Also then the corresponding conjugate diametral plane is called a *Principal plane* in that the chords perpendicular to itself are bisected by it.

Thus, l, m, n will be a principal direction if and only if the direction ratios

$$al + hm + gn, \ hl + bm + fn, \ gl + fm + cn$$

of the normal to the corresponding conjugate diametral plane are proportional to l, m, n i.e., if and only if there exists a number λ such that

$$al + hm + gn = l\lambda$$
$$hl + bm + fn = m\lambda$$
$$gl + fm + cn = n\lambda$$

We rewrite these as

$$(a - \lambda)l + hm + gn = 0 \qquad ...(1)$$
$$hl + (b - \lambda)m + fn = 0 \qquad ...(2)$$
$$gl + fm + (c - \lambda)n = 0 \qquad ...(3)$$

General Equation of the Second Degree

These three linear homogeneosu equations in l, m, n will posses a non-zero solution in l, m, n if and only if

$$\begin{vmatrix} a-\lambda & h & g \\ h & b-\lambda & f \\ g & f & c-\lambda \end{vmatrix} = 0$$

On expanding this determinant, we see that λ must be a root of the cubic

$$\lambda^3 - \lambda^2(a+b+c) + \lambda(A+B+C) - D = 0 \qquad \text{...(4)}$$

This cubic is known as the **Discriminating cubic** and each root of the same is called a **Characteristic root.**

The equation (4) has three roots which may not all be real or distinct. Also to each real root of (4) corresponds at least one principal direction l, m, n obtained on solving any two of the equation (1), (2) and (3).

Note : If l, m, n be a principal direction corresponding to a real root λ of the discriminating cubic, then we may easily see that the equation of the corresponding principal plane takes the form

$$\lambda(lx + my + nz) + (ul + vm + wn) = 0$$

This equation shows that we shall have no principal plane corresponding to $\lambda = 0$ if $\lambda = 0$ is a root of the discriminating cubic. In spite of this, however, we shall find it useful to say that l, m, n is a principal direction corresponding to $\lambda = 0$. Thus, every direction l, m, n satsifying the equations (1), (2), (3) corresponding to a root λ of the discriminating cubic (4) will be called a *Principal direction*.

Note 2. In the following, we shall prove some important results concerning the nature of the roots of the discriminating cubic and the **existence** of principal directions and principal planes.

Before taking up this consideration, we give a few preliminary results of algebraic character in the following section.

9.5 SOME PRELIMINARIES TO REDUCTION AND CLASSIFICATION

In this section we shall state some points which will prove useful in relation to the problem of reduction and classification.

In the following discussion, the determinant

$$\begin{vmatrix} a & h & g \\ h & b & f \\ g & f & c \end{vmatrix}$$

to be denoted by D will play an important part.

We may verify that

$$D = abc + 2fgh - af^2 - bg^2 - ch^2$$

As usual, A, B, C, F, G, H will denote the co-factors of a, b, c, f, g, h respectively in the determinant D, so that we have

$$A = bc - f^2, \; B = ca - g^2, \; C = ab - h^2;$$
$$F = gh - af, \; G = hf - bg, \; H = fg - ch$$

It can be easily verified that

$$\left.\begin{matrix} BC - F^2 = aD, \; CA - G^2 = bD, \; AB - H^2 = cD; \\ GH - AF = fD, \; HF - BG = gD, \; FD - CH = hD \end{matrix}\right\} \qquad \text{...(i)}$$

Also we have

$aA + bH + gG = 0$, $hA + bH + fG = 0$, $gA + fH + cG = 0;$

$aH + hB + gF = 0$, $hH + bB + fF = 0$, $gH + fB + cF = 0;$

$aG + hF + gC = 0$, $hG + bF + fC = 0$, $gG + fF + cC = D.$

9.5.1

If $D = 0$, then from (i), we have
$$BC = F^2, \qquad CA = G^2, \qquad AB = H^2$$
$$GH = AF, \qquad HF = BG, \qquad FG = CH.$$

Ex. Show that

(i) $D = 0$ and $A = 0 \Rightarrow H = 0, G = 0$,

(ii) $D = 0$ and $H = 0 \Rightarrow A = 0, H = 0, C = 0$ or $A = 0, B = 0, F = 0$.

Further prove that if $D = 0, A = 0, B = 0$, then F, G, H must all be zero but G may or may not be zero.

9.5.2

If $D = 0$ and $A + B + C = 0$, then
$$A, B, C, F, G, H$$
are all zero.

Now $\qquad D = 0 \Rightarrow BC = F^2, CA = G^2, AB = H^2$

$\Rightarrow \qquad A, B, C$ are all of the same sign.

Now A, B, C being all of the same sign.
$$A + B + C = 0 \Rightarrow A = 0, B = 0, C = 0$$

Further, A, B, C being zero
$$F^2 = BC, B = 0, C = 0$$
$$\Rightarrow \qquad F = 0$$
Similarly, $\qquad G = 0, H = 0$.

Note : Three homogeneous linear equations
$$a_1 x + b_1 y + c_1 z = 0, \; a_2 x + b_2 y + c_2 z = 0, \; a_3 x + b_3 y + c_3 z = 0$$
will possess a non-zero solution, *i.e.*, a solution wherefore x, y, z are not all zero, if and only if
$$\begin{vmatrix} a_1 & b_1 & c_1 \\ a_2 & b_2 & c_2 \\ a_3 & b_3 & c_3 \end{vmatrix} = 0$$

9.6 THEOREM (I)

The roots of the discriminating cubic are all real.

We, of course, suppose that the coefficients of the equation $F(x, y, z) = 0$ are all real.

Suppose that λ is a root of the discriminating cubic (4), § 9.4, and l, m, n is any non-zero set of values satisfying the corresponding equations (1), (2), (3) § 9.4.

Here it should be remembered that we cannot regard l, m, n as real, for λ is not yet proved to be real.

In the following, the complex conjugate of any number will be expressed by putting a bar over the same. Thus, $\bar{l}, \bar{m}, \bar{n}$ will denote the complex conjugate of the numbers l, m, n respectively.

Now, we have
$$al + hm + gn = l\lambda, \; hl + bm + fn = n\lambda, \; gl + fm + cn = n\lambda$$

Multiplying these by $\bar{l}, \bar{m}, \bar{n}$ respectively and adding, we obtain
$$\Sigma a l \bar{l} + \Sigma f (\overline{mn} + m\bar{n}) = \lambda \Sigma l \bar{l} \qquad \ldots(1)$$

Now, a, b, c, f, g, h are real. Also,
$$l\bar{l}, m\bar{m}, n\bar{n}$$
being the products of pairs of conjugate complex numbers, are real.

Also we notice that $\overline{mn}$ is the conjugate complex of $\overline{m}n$ so that
$$m\overline{n} + \overline{m}n$$
is real.

Similarly, $$n\overline{l} + \overline{n}l, \quad l\overline{m} + \overline{l}n$$
are real.

Finally, $\Sigma l\overline{l}$ is a non-zero real number.

Thus, λ, being the ratio of two real numbers from (1), is necessarily a real number.

Hence, the ratio of the discriminating cubic are all real. Also, therefore, the numbers l, m, n corresponding to each λ, are real.

9.6.1 Theorem (II)

The two principal directions corresponding to any two distinct roots of the discriminating cubic are perpendicular.

Suppose that λ_1, λ_2 are two distinct roots of the discriminating cubic, and
$$l_1, m_1, n_1; \quad l_2, m_2, n_2$$
are the two corresponding principal directions.

We then have

(2) $al_1 + hm_1 + gn_1 = \lambda_1 l_1$,   (5) $al_2 + hm_2 + gn_2 = \lambda_2 l_2$,

(3) $hl_1 + bm_1 + fn_1 = \lambda_1 m_1$,   (6) $hl_2 + bm_2 + fn_2 = \lambda_2 m_2$,

(4) $gl_1 + fm_1 + cn_1 = \lambda_1 n_1$,   (7) $gl_2 + fm_2 + cn_2 = \lambda_2 n_2$.

Multiplying (2), (3), (4) by l_2, m_2, n_2 respectively and adding, we obtain
$$\Sigma a l_1 l_2 + \Sigma f (m_1 n_2 + m_2 n_1) = \lambda_1 \Sigma l_1 l_2 \qquad ...(8)$$

Also multiplying (5), (6), (7) by l_1, m_1, n_1 respectively and adding, we obtain
$$\Sigma a l_1 l_2 + \Sigma f (m_1 n_2 + m_2 n_1) = \lambda_1 \Sigma l_1 l_2 \qquad ...(9)$$

From (8) and (9), we obtain
$$\lambda_1 \Sigma l_1 l_2 = \lambda_2 \Sigma l_2 l_2$$
$$\Rightarrow \qquad (\lambda_1 - \lambda_2) \Sigma l_1 l_2 = 0$$
$$\Rightarrow \qquad \Sigma l_1 l_2 = 0 \text{ for } \lambda_1 - \lambda_2 \neq 0.$$

Thus, the two directions are perpendicular. Hence, the theroem.

9.6.2 Theorem (III)

For every quadric, there exists at least one set of three mutually perpendicular principal directions.

We have to consider the following three cases :

(A) The roots of the discriminating cubic are all distinct.

(B) Two of the roots are equal and third is different from these.

(C) The three roots are all equal.

These three cases will be considered one by one.

(A) Case of three distinct roots : The roots being distinct, there will correspond a principal direction l, m, n satisfying the equation (1), (2), (3) on page 233 to each of these.

Also by Theorem II, these three directions will be mutually perpendicular. The three principal directions are unique in this case.

Thus, there exist three principal directions in this case. Moreover, these directions are as well unique in this case.

(B) Case of two equal roots : Let the discriminating cubic have two equal roots and let the third root be different from the same.

Suppose λ is a root of the D-cubic repeated twice so that λ satisfies the equation
$$\lambda^3 - \lambda^2 (a + b + c) + \lambda (A + B + C) - D = 0 \qquad ...(10)$$

and the equation
$$3\lambda^2 - 2\lambda(a+b+c) + (A+B+C) = 0 \qquad ...(11)$$
obtained on differentiating the cubic (10) with respect to λ. We write these as
$$(a-\lambda)(b-\lambda)(c-\lambda) + 2fgh - (a-\lambda)f^2 - (b-\lambda)g^2 - (c-\lambda)h^2 = 0$$
$$[(b-\lambda)(c-\lambda) - f^2] + [(c-\lambda)(a-\lambda) - g^2][(a+\lambda)(b-\lambda) - h^2] = 0$$
Here we have two relations corresponding to
$$D = 0, A + B + C = 0$$
obtained on replacing a, b, c by $a - \lambda, b - \lambda, c - \lambda$.
respectively.

Thus, we concldue that (Refer 9.5.2)
$$\left.\begin{array}{lll}(b-\lambda)(c-\lambda) = f^2 & (c-\lambda)(a-\lambda) = g^2, & (a-\lambda)(b-\lambda) = h^2 \\ (a-\lambda)f - gh, & (b-\lambda)g = hf, & (c-\lambda) = fg\end{array}\right\} \qquad ...(12)$$

corresponding to $A = 0, B = 0, C = 0; F = 0, G = 0, H = 0$. These relations show that the equation
$$(a - \lambda)l + hm + gn = 0$$
$$hl + (b - \lambda)m + fn = 0$$
$$gl + fm + (c - \lambda)n = 0$$
for the determination of l, m, n are all equivalent.

Thus, we see that if λ is a twice repeated root, the direction cosines (l, m, n) satisfy the single relation
$$(a - \lambda)l + hm + gn = 0 \qquad ...(13)$$
Suppose now that l_1, m_1, n_1 is any direction satisfying equation (13). Further we determine a direction l_2, m_2, n_2 sastifying equation (13) and perpendicular to the direction l_1, m_1, n_1. Thus, l_2, m_2, n_2 are determiend from
$$(a - \lambda)l_2 + hm_2 + gn_2 = 0$$
$$l_1 l_2 + m_1 m_2 + n_1 n_2 = 0$$
The principal direction corresponding to the third root of the discriminating cubic will, of course, be perpendicular to each of the two principal directions
$$l_1, m_1, n_1; \ l_2, m_2, n_2.$$

This, in this case also we have a set of three mutually perpendicular principal directions. Of course, they are not unique in this case.

(C) Case of all three roots equal : Suppose now that all the three roots are equal to λ. The root λ satisfies the thre equations
$$\lambda^3 - \lambda^2(a+b+c)\lambda(A+B+C) - D = 0 \qquad \text{[by equation (10)]}$$
$$3\lambda^2 - 2\lambda(a+b+c) + (A+B+C) = 0 \qquad \text{[by equation (11)]}$$
$$3\lambda - (a+b+c) = 0 \qquad ...(14)$$
In this case also the relation (12), as deduced from (10) and (11) are true.

We rewrite (14) as
$$(a - \lambda) + (b - \lambda) + (c - \lambda) = 0 \qquad ...(15)$$
Also, we have
$$(b-\lambda)(c-\lambda) = f^2, \ (c-\lambda)(a-\lambda) = g^2, \ (a-\lambda)(b-\lambda) = h^2 \qquad ...(16)$$
From (16), see that
$$a - \lambda, b - \lambda, c - \lambda$$

---
*If desired l_1, m_1, n_1 may be selected further so as to satisfy some additional suitable condition.

General Equation of the Second Degree

must all have the same sign so that with the help of (15), it follows that
$$a - \lambda = 0, \ b - \lambda = 0, \ c - \lambda = 0$$
$$\Rightarrow \quad \lambda = a = b = c$$
Also, then it follows that $f = 0, g = 0, h = 0$.
We now see that in this case the equations.
$$(a - \lambda) l + hm + gn = 0, \quad hl + (b - \lambda) m + fn = 0, \quad gl + fm + (c - \lambda) = 0$$
for the determination of the principal directions are identicalldy satisfies, *i.e.*, they are true for arbitrary values of l, m, n so that *every direction is a principal direction* in this case.'

Thus, in this case also a quadratic has a set of three mutually perpendicular principal directions. In fact, any set of three mutually perpendicular directions is a set of three mutually perpendicular principal directions in this case.

The reader may observe that the quadric is a sphere in the last case.

EXAMPLE

Find a set of three mutually perpendicular principal directions for the following conicoids:
$$3x^2 + 5y^2 + 3z^2 - 2yz + 2zx - 2xy + 2z = 0$$

Solution. We have $a = 3, b = 5, c = 3, f = -1, g = 1, h = -1$.
Therefore, the discriminating cubic is
$$\begin{vmatrix} 3-\lambda & -1 & 1 \\ -1 & 5-\lambda & -1 \\ 1 & -1 & 3-\lambda \end{vmatrix} = 0$$
$$\Leftrightarrow \quad -\lambda^3 + 11\lambda^2 - 36\lambda + 36 = 0$$
Its root are $\lambda = 2, 3, 6$
so that the characteristics roots are all different.
The principal directions corresponding to $\lambda = 2$ is given by
$$l - m + n = 0$$
$$-l + 3m - n = 0$$
$$l - m + n = 0$$
$$\Rightarrow \quad l : m : n = 1 : 0 : 1$$
Thus, the principal direction corresponding to $\lambda = 2$ is given by
$$1/\sqrt{2}, \ 0, \ -1/\sqrt{2}$$
Again the principal direction corresponding to $\lambda = 3$ is given by
$$0.l - m + n = 0,$$
$$-l + 2m - n = 0,$$
$$l - m + 0.n = 0$$
$$\Rightarrow \quad l : m : n = 1 : 1 : 1$$
and we have the corresponding principal direction
$$\frac{1}{\sqrt{3}}, \frac{1}{\sqrt{3}}, \frac{1}{\sqrt{3}}.$$
Finally the principal direction corresponding to $\lambda = 6$ is given by
$$-3l - m + n = 0$$
$$-l - m - n = 0$$
$$l - m - 3n = 0$$

whereform we may see that this principal direction is
$$\frac{1}{\sqrt{6}}, \frac{-2}{\sqrt{6}}, \frac{1}{\sqrt{6}}$$
The principal planes corresponding to the characteristics root λ being
$$\lambda(lx + my + nz) + (ul + vm + wn) = 0$$
We see that the three principal planes are
$$2x - 2z - 1 = 0, \quad 3x + 3y + 3z + 1 = 0, \quad 6x - 12y + 6z + 1 = 0$$

EXERCISES

Examine the following quadrics for principal directions and principal planes :

1. $4x^2 - y^2 - z^2 + 2yz - 8x - 4y + 8z = 0$
2. $x^2 + 2yz - 4x + 6y + 2z = 0$
3. $4y^2 - 4yz - 4zx - 4xy - 2x + 2y - 1 = 0$

ANSWERS

1. Principal directions : $1, 0, 0; 0, 1/\sqrt{2}, -1/\sqrt{2}; 0, 1/\sqrt{2}, 1/\sqrt{2}$.
   Principal planes : $x = 1, y - z + 3 = 0$.
2. Principal directions : $0, 1/\sqrt{2}, -1/\sqrt{2}$ and every direction perpendicular to it.
   Principal planes : $y - z - 2 = 0$ and every where through the line, $y + z + 4 = 0, x = 2$.
3. Principal directions : $1/\sqrt{3}, 1/\sqrt{3}, 1/\sqrt{3}; 1/\sqrt{6}, -2/\sqrt{6}, 1/\sqrt{6}; 1/\sqrt{2}, 0, -1/\sqrt{2}$.
   Principal planes : Any plane at right angle to $x = y = z - 1/2, 2(x - 2y + z) = 1$, $2(x - z) + 1 = 0$.

9.6.3 Centre

We know that if a point (x, y, z) is the mid-point of a chord with direction cosines (l, m, n) of a quadric
$$F(x, y, z) = 0$$
then we have
$$l\frac{\partial F}{\partial x} + m\frac{\partial F}{\partial y} + n\frac{\partial F}{\partial z} = 0 \qquad ...(1)$$
This shows that if (x, y, z) is such that
$$\frac{\partial F}{\partial x} = 0, \frac{\partial F}{\partial y} = 0, \frac{\partial F}{\partial z} = 0$$
then the condition (1) is satisfied, whatever values l, m, n may have, implying that every chord through (x, y, z) is bisected thereat. Such a point is known as a **Centre** of the quadric. We can rewrite these equations as
$$ax + hv + gz + u = 0 \qquad ...(2)$$
$$hx + by + fz + v = 0 \qquad ...(3)$$
$$gx + fy + cz + w = 0 \qquad ...(4)$$

It should be remembered that a quadric may or may not have a centre; also it may have more than one centre – a *line of centres* or a *plane of centres*, depending upon the nature of the solutions of the three equations (2), (3), (4).

In the following, we shall consider the different cases regarding the possible solutions of these equations. This discussion will be facilitated a good deal, if regarding x, y, z as variables, we consider the three planes represented by these equations. We have thus to eaxmine the nature of the points of intersection, if any, of these three planes to be called *Central planes*.

Before we proceed to consider the problem of the existence of the centre of a quadric, we state a preliminary result.

General Equation of the Second Degree

9.6.4

The two planes
$$p_1 x + q_1 y + r_1 z + s_1 = 0$$
$$p_2 x + q_2 y + r_2 z + s_2 = 0$$
will be
(i) same if
$$\begin{vmatrix} p_1 & q_1 \\ p_2 & q_2 \end{vmatrix} = 0, \quad \begin{vmatrix} q_1 & r_1 \\ q_2 & r_2 \end{vmatrix} = 0, \quad \begin{vmatrix} r_1 & s_1 \\ r_2 & s_2 \end{vmatrix} = 0$$

(ii) parallel but not same if
$$\begin{vmatrix} p_1 & q_1 \\ p_2 & q_2 \end{vmatrix} = 0, \quad \begin{vmatrix} q_1 & r_1 \\ q_2 & r_2 \end{vmatrix} = 0, \quad \begin{vmatrix} r_1 & s_1 \\ r_2 & s_2 \end{vmatrix} \neq 0$$

(iii) neither parallel nor same, *i.e.*, will intersect in a straight line if
$$\begin{vmatrix} p_1 & q_1 \\ p_2 & q_2 \end{vmatrix} \neq 0 \quad or \quad \begin{vmatrix} q_1 & r_1 \\ q_2 & r_2 \end{vmatrix} \neq 0$$

9.7 CASE OF A UNIQUE CENTRE

Multiplying the equations (2), (3), (4) by A, H, G respectively and adding, we obtain
$$Dx + (Au + Hv + Gw) = 0 \qquad \text{(Refer 9.5.2)}$$
Again, on multiplying (2), (3), (4) by H, B, F and by G, F, C and adding separately, we obtain
$$Dy + (Hu + Bv + Fw) = 0$$
$$Dz + (Gu + Fv + Cw) = 0$$
If $D \neq 0$, we obtain from these
$$x = -(Au + Hv + Gw)/D, \quad y = -(Hu + Bv + Fw)/D, \quad z = -(Gu + Fv + Cw)/D$$
Substituting these in (2), (3), (4) we may easily verify that the same are satisfied.

Thus, if $D \neq 0$, the quadratic has a unique centre (x, y, z) where (x, y, z) have the values given above.

9.7.1 Case of No Centre

Now suppose that $D = 0$. Then, we have
$$A(ax + hy + gz + u) + H(hx + by + fz + v) + G(gx + fy + cz + w) \equiv Au + Hv + Gw$$
(Refer 9.5.1)

This shows that the three equations cannot have a common solution, *i.e*, the quadric will not have a centre if
$$Ax + Hv + Gw \neq 0$$

Considering H, B, F and G, F, C as sets of multipliers instead of A, H, G we may similarly see that the quadric will not have a centre if
$$Hu + Bv + Fw \neq 0 \quad \text{or if} \quad Gu + Fv + Cw \neq 0$$

Thus, we see that *the quadric will* **not** *have a centre if $D = 0$ and any one of*
$$Au + Hv + Gw, Hu + Bv + Fw, Gu + Fv + Cw$$
is not zero.

9.7.2 Case of a Line of Centres

We now suppose that $D = 0$ as well as $Au + Hv + Gw = 0$.
Then we have
$$A(ax + hy + gz + u) + H(hx + by + fz + v) + G(gx + fy + cz + w) = 0$$
(i) Thus, if $A \neq 0$, we have
$$ax + hy + gz + u = -\frac{H}{A}(hx + by + fz + v) - \frac{G}{A}(gx + fy + cz + w)$$

(ii) Also, if $A \neq 0$, the two planes
$$hx + by + fz + v = 0$$
$$gx + fy + cz + w = 0$$
are neither the same nor parallel so that they intersect in a line. This is because
$$\begin{vmatrix} b & f \\ f & c \end{vmatrix} = A \neq 0$$
[Refer 9.6.4]

From (i) and (ii), we deduce that the plane
$$ax + hy + gz + u = 0$$
passes through the line of intersection of the two intersecting planes
$$hx + by + fz + v = 0, \quad gx + fy + cz + w = 0$$
Thus, in case
$$D = 0, \; Au + Hv + Gw = 0, \; A \neq 0$$
the three central planes all pass through one line so that we have a line of centres.
We may similarly see that the quadric will have a line of centres if
$$D = 0, Hu + Bv + Fw = 0, B \neq 0$$
or if
$$D = 0, Gu + Fv + Cw = 0, C \neq 0$$

Note 1. We can show that if $D = 0$, and $A \neq 0$ and $Au + Hv + Gw = 0$, then we must also simultaneously have
$$Hu + Bv + Fw = 0, \; Gu + Fv + Cw = 0$$
In fact we have
$$A(Hu + Bv + Fw) \equiv H(Au + Hv + Gw)$$
and
$$A(Gu + Fv + Cw) \equiv G(Au + Hv + Gw)$$
the equalities holding for all values of u, v and w. Thus, if $A \neq 0$, we have
$$Hu + Bv + Fw = \frac{H}{A}(Au + Hv + Gw)$$
$$Gu + Fv + Cw = \frac{G}{A}(Au + Hv + Cw)$$
The result stated now follows.

It may be remembered that if $A = 0$ then also $H = 0, G = 0$, so that $Au + Hv + Gw \equiv 0$. In this case when $A = 0, H = 0, G = 0$, we may not have
$$Hu + Bv + Fw = 0 \text{ or } Gu + Fv + Cw = 0$$
For example, consider
$$x^2 + 2y^2 + 2xy + 2x + y + 2z + 3 = 0$$
Here $a = 1, b = 2, c = 0, f = 0, g = 0, h = 1, u = 1, v = 1/2, w = 1$,
so that $\quad A = 0, B = 0, C = 1, F = 0, G = 0, H = 0, D = 0.$
Thus, we have $Au + Hv + Gw = 0$ but $Gu + Fv + Cw \neq 0$.

Note 2. The cases treated above cover the cases when $D = 0$ and one at least of A, B, C is not zero.

If we suppose that A, B, C are all zero, then it follows that F, G, H are also all zero, for
$$F^2 = BC, \; G^2 = CA, \; H^2 = AB$$
In the next sub-section we consider the case when A, B, C, F, G, H are all zero. The vanishing of D then follows from the vanishing of these co-factors in as much as we have
$$D = Aa + Hh + Gg,$$
so that $D = 0$ even if A, H, G only are known to be zero.

General Equation of the Second Degree

9.7.3 Case of no Centre

Suppose now that A, B, C, F, G, H are all zero, so that D = 0 also.
We have in this case,

(1) $\begin{cases} f(ax + hy + gz + u) - g(hx + by + fz + v) = fu - gz \\ f(ax + hy + gz + u) - h(gx + fy + cz + w) = fu - hw \end{cases}$

These show that if
$$fu - gv \neq 0 \quad \text{or} \quad fu - hw \neq 0,$$
then the quadric canot have a centre.

9.7.4 Case of a plane of centres

Suppose now that
$$fu - gv = 0 \quad \text{and} \quad fu - hw = 0$$
$\Leftrightarrow \quad fu = gv = hw.$

Then if $g \neq 0, h \neq 0$, we have from (1) above in 9.7.3, that

$$hx + by + fz + v = \frac{f}{g}(ax + hy + gz + u)$$

$$gx + fy + cz + w = \frac{f}{h}(ax + hy + gz + u)$$

so that every point of the plane
$$ax + hy + gz + u = 0$$
is also a point of the other two central planes. Thus, we have a plane of centres in this case. Similarly, we may show that if
$$fu = gv = hw$$
and some two of f, g, h are not zero, then *the quadric has a plane of centres.*

Note : If can be easily seen that if A, B, C, F, G, H are zero and one of f, g, h is known to be zero, then one more of f, g, h must also be zero. For instance, suppose that $f = 0$. Then, because
$$0 = F = gh - af$$
it follows that either g or h must also be zero. Thus, the case treated here can be stated as follows.

If A, B, C, F, G, H are all zero, none of f, g, h is zero and $fu = gv = hw$, then the quadric has a line of centres.

The case where one and, therefore, two of f, g, h are zero is treated below here.

9.7.5

Now *suppose that two of f, g, h are zero in addition to A, B, C, F, G, H being all zero and $fu = gv = hw$.*

Let $g = 0 = h$ and $f \neq 0$. In this case we see from (1) above, 9.7.3. that
$$ax + hy + gz + u = 0$$
so that $a = 0, h = 0, u = 0$.
The vanishing of u also follows from the fact that
$$fu = gv = hw \quad \text{and} \quad g = 0, h = 0, f \neq 0.$$
Consider now the two central planes
$$hx + by + fz + v = 0$$
$$gx + fy + cz + w = 0$$
the coefficients of the third central plane being all zero. As h and g are both zero, we can rewrite these as
$$by + fz + v = 0$$
$$fy + cz + w = 0$$

Here $\begin{vmatrix} b & f \\ f & c \end{vmatrix} = bc - f^2 = A = 0,$ $\begin{vmatrix} f & v \\ c & w \end{vmatrix} = fw - cv$

Thus, if $fw - cv \neq 0$, the quadric has no centre aand if $fw - cv = 0$, the quadric has a plane of centres.

We can obtain similar conditions, when
$$f = 0 = h, \ g \neq 0$$
or when
$$f = 0 = g, h \neq 0.$$

9.7.6

Now suppose that f, g, h are all zero in addition to the vanishing of A, B, C, F, G, H.

In this case two of a, b, c must be zero. Suppose that $b = c = 0$ and $a \neq 0$. Then the first of the three central planes is
$$ax + u = 0$$
and the other two are
$$0x + 0y + 0z + v = 0$$
$$0x + 0y + 0z + w = 0$$

Thus, if $v \neq 0$ or $w \neq 0$ the quadric has no centre and if $v = 0 = w$, the quadric has a plane of centres.

SUMMARY OF THE VARIOUS CASES

1. $D \neq 0$. Unique centre.
2. $\begin{cases} D = 0, \ Au + Hv + Gw \neq 0. \text{ No centre.} \\ D = 0, Hu + Bv + Fw \neq 0. \text{ No centre.} \\ D = 0, Gu + Fv + Cw \neq 0. \text{ No centre.} \end{cases}$
3. $\begin{cases} D = 0, \ Au + Hv + Fw = 0, \ A \neq 0. \text{ No centre.} \\ D = 0, Hu + Bv + Gw = 0, \ B \neq 0. \text{ No centre.} \\ D = 0, Gu + Fv + Cw = 0, \ C \neq 0. \text{ No centre.} \end{cases}$
4. A, B, C, F, G, H all zero, $fu \neq gv$ or $gv \neq hw$. No centre.
5. A, B, C, F, G, H all zero, $fu = gv = hw, \ f \neq 0, g \neq 0, h \neq 0$. Planes of centres.
6. A, B, C, F, G, H all zero, $fu = gv = hw, \ g = 0, h = 0, f \neq 0, fw - cv \neq 0$. No centres.
7. A, B, C, F, G, H all zero, $fu = gv = hw, \ g = 0, h = 0, f \neq 0, fw - cv = 0$. Planes of centres.

   We may have results similar to (6) and (7), when $f = 0, g = 0, h \neq 0$ or when $h = 0, f = 0, g \neq 0$.
8. A, B, C, F, G, H all zero, f, g, h all zero. Then two of a, b, c must be zero and one non-zero. Then we have no centre if
$$a \neq 0, v \neq 0 \ \text{or} \ w \neq 0.$$
and a plane of centres if
$$a \neq 0, v = 0 = w$$

We have similar results when $b \neq 0$ or $c \neq 0$.

Note : The results given above need not be committed to memory.

EXERCISE

Examine the following quadrics for centres :

1. $z^2 - yz + zx + xy - 2y + 2z + 2 = 0$. [**Ans.** Unique centre; $(1, 1, -1)$]

2. $2z^2 - 2yz - 2zx + 2xy + 3x - y - 2z + 1 = 0$ $\left[\textbf{Ans.} \text{ Line of centres}; \dfrac{x}{1} = \dfrac{y+2}{1} = \dfrac{2z+1}{2}\right]$

General Equation of the Second Degree

3. $4x^2 + 9y^2 + 4z^2 + 12xy + 12yz + 8zx + 3x + 4y + z = 0$ [**Ans.** No centre]

4. $x^2 + y^2 + z^2 - 2xy - 2yz + x - y + z = 0$ [**Ans.** Plane of centres; $2x - 2y + 2z + 1 = 0$]

5. $4x^2 - 2y^2 - 2z^2 + 5yz + 2zx + 2xy - x + 2y + 2z - 1 = 0$. [**Ans.** No centre]

6. $2x^2 + 2y^2 + 5z^2 - 2yz - 2zx - 4xy - 14x - 14y + 16z + 6 = 0$.
   [**Ans.** Line of centres; $x = 3 - y, z + 1 = 0$]

7. $18x^2 + 2y^2 + 20z^2 - 12zx + 12yz + x - 22y - 6z + 1 = 0$. [**Ans.** No centre]

8. $4x^2 - y^2 + 2z^2 + 2xy - 3yz + 12x - 11y + 6z + 4 = 0$ [**Ans.** Unique centre; $(-1, -2, 3)$]

9.8 TRANSFORMATION OF CO-ORDINATES

Before we take up the problem of actual reduction and classification, we shall consider two important cases of transformation of co-ordinates.

9.8.1 The Form of the Equation of a Quadratic Referred to a Centre as Origin

We suppose that the given quadric has a centre.

Let (α, β, γ) be a centre of the quadric with equation

$$F(x, y, z) \equiv \Sigma (ax^2 + 2fyz) + 2\Sigma ux + d = 0$$

Consider now a new system of co-ordinate axes parallel to the given system and with its origin at (α, β, γ). The equation of the quadric, w.r.t. the new system, obtained on replacing x, y, z by $x + \alpha, y + \beta, z + \gamma$ respectively is

$$\Sigma [a(x+\alpha)^2 + 2f(y+\beta)(z+\gamma)] + 2\Sigma u(x+\alpha) + d = 0$$

$$\Leftrightarrow \Sigma(ax^2 + 2fyz) + 2x(a\alpha + h\beta + g\gamma + u) + 2y(h\alpha + b\beta + f\gamma + v)$$
$$+ 2z(g\alpha + f\beta + c\gamma + w) + F(\alpha, \beta, \gamma) = 0$$

As (α, β, γ) is a centre, we have

$$a\alpha + h\beta + g\gamma + u = 0, \; h\alpha + b\beta + f\gamma + v = 0, \; g\alpha + f\beta + c\gamma + w = 0$$

Further, as may be easily seen

$$F(\alpha, \beta, \gamma) = \alpha(a\alpha + h\beta + g\gamma + u) + \beta(h\alpha + b\beta + f\gamma + v) + \gamma(g\alpha + f\beta + c\gamma + w)$$
$$+ (u\alpha + v\beta + w\gamma + d)$$

$$= u\alpha + v\beta + w\gamma + d$$

Thus, the required new equation is

$$\Sigma(ax^2 + 2fyz) + (u\alpha + v\beta + w\gamma + d) = 0$$

It will be seen that the second degree homogeneous part of the equation has remained unchanged and the first degree terms have disappeared.

Note 1. The discussion above is applicable whether the quadric has one centre, a line of centres or a plane of centres. In case the quadric has more than one centre (α, β, γ) may denote any one of them.

Note 2. The co-ordinates, w.r.t. the old as well as the new system of axes have both been denoted by the same symbols x, y, z.

9.8.2 The Form of the Equation of a Quadric, When the Co-ordinate Axes are Parallel to a Set of Three Mutually Perpendicular Principal Directions

Suppose that

$$(l_1, m_1, n_1); (l_2, m_2, n_2); (l_3, m_3, n_3) \qquad \ldots(1)$$

are the direction cosines of three mutually perpendicular principal directions corresponding to the three roots

$$\lambda_1, \lambda_2, \lambda_3$$

of the discriminating cubic. These roots may be all different.

We take now a new co-ordinate system through the same origin such that the axes of the new system are parellel to the directions given by (1) above.

The equation referred to the new system of axes is obtained on replacing

$$x, y, z$$

by
$$l_1 x + l_2 y + l_3 z, \; m_1 x + m_2 y + m_3 z, \; n_1 x + n_2 y + n_3 z$$
respectively.

As homogeneous linear expression are to be substituted for x, y, z we may note that a homogeneous expression for any degree will be transformed to a homogeneous expression of the same degree.

Thus, we may separately consider the transforms to the homogeneous parts

$$\Sigma (ax^2 + 2fyz) \text{ and } 2\Sigma ux$$

We shall show that the transform of the second degree homogeneous part

$$\Sigma (ax^2 + 2fyz) \qquad \text{...(2)}$$

is
$$\lambda_1 x^2 + \lambda_2 y^2 + \lambda_3 z^2$$

On direct substitution, we may see that the coefficient of x^2 in the transform of (2) is

$$al_1^2 + bm_1^2 + cn_1^2 + 2fm_1 n_1 + 2gn_1 l_1 + 2hl_1 m_1$$
$$= l_1 (al_1 + hm_1 + gn_1) + m_1 (hl_1 + bm_1 + fn_1) + n_1 (gl_1 + fm_1 + cn_1)$$
$$= l_1 (\lambda_1 l_1) + m_1 (\lambda_1 m_1) + n_1 (\lambda_1 n_1) = \lambda_1 (l_1^2 + m_1^2 + n_1^2) = \lambda_1.$$

Similarly the coefficients of y^2 and z^2 in the transform can be shown to be

$$\lambda_2 \text{ and } \lambda_3$$

respectively.

Again the coefficient of $2yz$ in the transform of (1) is

$$= al_2 l_3 + bm_2 m_3 + cn_2 n_3 + f(m_2 n_3 + m_3 n_2) + g(n_2 l_3 + n_3 l_2) + h(l_2 m_3 + l_3 m_2)$$
$$= l_2 (al_3 + hm_3 + gn_3) + m_2 (hl_3 + bm_3 + fn_3) + n_2 (gl_3 + fm_3 + cn_3)$$
$$= \lambda_3 (l_2 l_3 + m_2 m_3 + n_2 n_3) = 0.$$

Similarly, the coefficients of zx and xy in the transform can be seen to be zero.

Thus, the transform of

$$\Sigma (ax^2 + 2fyz)$$

is
$$\lambda_1 x^2 + \lambda_2 y^2 + \lambda_3 z^2$$

Finaly, we seen that the transform of

$$\Sigma (ax^2 + 2fyz) + 2\Sigma ux + d$$

is
$$\lambda_1 x^2 + \lambda_2 y^2 + \lambda_3 z^2 + 2u(l_1 u + l_2 y + l_3 z) + 2y(m_1 x + m_2 y + m_3 z)$$
$$+ 2w(n_1 x + n_2 y + n_3 z) + d$$

$$= \lambda_1 x^2 + \lambda_2 y^2 + \lambda_3 z^2 + 2x(ul_1 + vm_1 + wn_1) + 2y(ul_2 + vm_2 + wn_2)$$
$$+ 2z(ul_3 + vm_3 + wn_3) + d$$

9.9 REDUCTION TO CANONICAL FORMS AND CLASSIFICATION

We shall now consider the several cases one by one.

General Equation of the Second Degree

9.9.1 Case I.

When $D \neq 0$. In this case the quadric has a unique centre and no root of the discriminating cubic is zero.

Shifting the origin to the centre (α, β, γ) the equation takes the form

$$\Sigma(ax^2 + 2fyz) + (u\alpha + v\beta + w\gamma + d) = 0$$

Now rotating the axes so that the axes of the new system are parallel to the set of three mutually perpendicular principal directions, we see that the equation becomes

$$\lambda_1 x^2 + \lambda_2 y^2 + \lambda_3 z^2 + (u\alpha + v\beta + w\gamma + d) = 0$$

which is the required canonical form.

Below we find an elegant form for the constant term.

We have

$$a\alpha + h\beta + g\gamma + u = 0 \qquad \ldots(1)$$
$$h\alpha + b\beta + f\gamma + v = 0 \qquad \ldots(2)$$
$$g\alpha + f\beta + c\gamma + w = 0 \qquad \ldots(3)$$

Also we write

$$u\alpha + v\beta + w\gamma + d = k$$

$$\Leftrightarrow \quad u\alpha + v\beta + w\gamma + (d - k) = 0 \qquad \ldots(4)$$

Eliminating α, β, γ from (1), (2), (3) and (4), we obtain

$$\begin{vmatrix} a & h & g & u \\ h & b & f & v \\ g & f & c & w \\ u & v & w & (d-k) \end{vmatrix} = 0$$

$$\Leftrightarrow \quad \begin{vmatrix} a & h & g & u \\ h & b & f & v \\ g & f & c & w \\ u & v & w & d \end{vmatrix} - \begin{vmatrix} a & h & g \\ h & b & f \\ g & f & c \end{vmatrix} k = 0$$

$$\Leftrightarrow \quad k = \frac{\Delta}{D}, \quad (D \neq 0)$$

where we have represented the fourth order determinant on the left by Δ.

We thus see that the new equation assumes the form

$$\lambda_1 x^2 + \lambda_2 y^2 + \lambda_3 z^2 + \Delta/D = 0$$

The equation represents various types of surfaces as shown in the following table. It may be remembered that the word 'roots' refers to the characteristic roots:

	Given	Conclusion
$\Delta = 0$	Roots all > 0 or < 0	Imaginary cone.
$\Delta = 0$	Two roots > 0 and one < 0	Real cone.
$\Delta = 0$	Two roots < 0 and one > 0	Real cone.
$\Delta/D > 0$	Roots all > 0	Imaginary ellipsoid.
$\Delta/D > 0$	Roots all < 0	Real ellipsoid
$\Delta/D > 0$	Two roots > 0 and one < 0	Hyperboloid of two sheets.
$\Delta/D > 0$	Two roots < 0 and one > 0	Hyperboloid of one sheet.
$\Delta/D < 0$	Roots all > 0	Real ellipsoid
$\Delta/D < 0$	Roots all < 0	Imaginary ellipsoid
$\Delta/D < 0$	Two roots > 0 and one < 0	Hyperboloid of one sheet.
$\Delta/D < 0$	Two roots < 0 and one > 0	Hyperboloid of two sheets.

9.9.2 Case II

When $D = 0$, $Au + Hv + Gw \neq 0$. In this case the quadric has no centre and the discriminating cubic has one zero root and two non-zero roots.

We denote the non-zero roots by λ_1, λ_2. The third root λ_3 is 0.

We rotate the co-ordinate axes through the same origin so that new axes aer parallel to the set of three mutually perpendicular principal directions.

The new equation takes the form

$$\lambda_1 x^2 + \lambda_2 y^2 + 2x(ul_1 + vm_1 + wn_1) + 2y(ul_2 + vm_2 + wn_2) + 2z(ul_3 + vm_3 + wn_3) + d = 0$$
...(1)

where l_3, m_3, n_3 correspond to $\lambda_3 = 0$.

Here we notice that $\qquad ul_3 + vm_3 + wn_3 \neq 0$

If possible, let $\qquad ul_3 + vm_3 + wn_3 = 0$ ...(2)

We also have

$$hl_3 + bm_3 + fn_3 = 0 \qquad ...(3)$$
$$gl_3 + fm_3 + cn_3 = 0$$

As l_3, m_3, n_3 are not all zero, we have from (2), (3), (4)

$$\begin{vmatrix} u & v & w \\ h & b & f \\ g & f & c \end{vmatrix} = 0$$

$\Leftrightarrow \qquad Au + Hv + Gw = 0$

which is contradictory to the given condition.

Denoting the coefficients of x, y, z by p, q, r we rewrite (1) as

$$\lambda_1 x^2 + \lambda_2 y^2 + 2px + 2qy + 2rz + d = 0, \qquad \text{where } r \neq 0.$$

$\Leftrightarrow \qquad \lambda_1 \left(x + \dfrac{p}{\lambda_1}\right)^2 + \lambda_2 \left(y + \dfrac{q}{\lambda_2}\right)^2 + 2r\left[z + \dfrac{1}{2r}\left(d - \dfrac{p^2}{\lambda_1} - \dfrac{q^2}{\lambda_2}\right)\right] = 0$

so that shifting the origin to the point.

$$\left[-\dfrac{p}{\lambda_1}, -\dfrac{q}{\lambda_2}, -\dfrac{1}{2r}\left(d - \dfrac{p^2}{\lambda_1} - \dfrac{q^2}{\lambda_2}\right)\right],$$

we see that the equation takes the form

$$\lambda_1 x^2 + \lambda_2 y^2 + 2rz = 0$$

where $r = ul_3 + vm_3 + wn_3 \neq 0$.

This is the required canonical form in the present case.

This equation represents an elliptic or hyperbolic paraboloid according as λ_1, λ_2 are of the same or opposite signs.

Cor. Axis and vertex of the paraboloid : It is known that Z-axis is the axis and (0, 0, 0) is the vertex of the paraboloid

$$\lambda_1 x^2 + \lambda_2 y^2 + 2rz = 0$$

Also the principal directions of the paraboloid are those of the co-ordinate axes; the principal direction corresponding to the characteristic root zero being that of Z-axis and the principal direction corresponding to the non-zero roots λ_1, λ_2 being those of X-axis and Y-axis respectively. Further, it can be easily seen that the principal planes corresponding to the non-zero characteristic roots are the planes $x = 0, y = 0$ whose intersection Z-axis is the axis of the paraboloid. Thus, we have the following important and useful result :

General Equation of the Second Degree

The line of intersection of the principal planes corresponding to the non-zero characteristic roots is the axis and the point where the axis meets the paraboloid is the vertex. Also the axis is the line through the vertex parallel to the principal direction corresponding to the characteristic root zero.

9.9.3 Case III

When $D = 0$, $Au + Hv + Gw = 0$, $A \neq 0$. In this case the quadric has a line of centres and the discriminating cubic has one zero and two non-zero roots.

We may see that $A + B + C \neq 0$, for if it were so, then we would have A, B, C all zero and the condition $A \neq 0$, would be contradicted. Since $D = 0$ and $A + B + C \neq 0$, the discriminating cubic would have only one zero root.

Let (α, β, γ) be a centre. Shifting the origin to (α, β, γ) and rotating the axes so that the new axes are parallel to the set of mutually perpendicular principal directions, we see that the equation becomes

$$\lambda_1 x^2 + \lambda_2 y^2 + (u\alpha + v\beta + w\gamma + d) = 0$$

which is the required canonical form.

We may, as follows, obtain an expression for the constant term in a form free from α, β, γ.
In this case the central planes all pass through one line.
We select the following two equations of the centre giving

$$hx + by + fz + v = 0$$
$$gx + fy + cz + w = 0$$

so that they are different.

Now (α, β, γ) is a point satisfying these two equations. Taking $\alpha = 0$, we have

$$b\beta + f\gamma + v = 0$$
$$f\beta + c\gamma + w = 0$$

Also we write
$$v\beta + w\gamma + (d - k) = 0.$$

These give

$$\begin{vmatrix} b & f & v \\ f & c & w \\ v & w & (d-k) \end{vmatrix} = 0 \Rightarrow k = \frac{1}{A} \begin{vmatrix} b & f & v \\ f & c & w \\ v & w & d \end{vmatrix}$$

Thus, the required canonical form is

$$\lambda_1 x^2 + \lambda_2 y^2 + k = 0.$$

The equation represents various types of surfaces as shown in the following table:

	Given	Conclusions
$k = 0$	Roots both > 0 or < 0	Imaginary pair of planes.
$k = 0$	One root > 0 and other < 0	Pair of intersecting planes.
$k > 0$	Roots both > 0	Imaginary cylinder
$k > 0$	Roots both < 0	Elliptic cylinder
$k < 0$	One root > 0 and other < 0	Hyperbolic cylinder
$k < 0$	Roots both > 0	Elliptic cylinder
$k < 0$	Roots both < 0	Imaginary cylinder.
$k < 0$	One root > 0 and other < 0	Hyperbolic cylinder

Cor. 2. Axis of the Cylinder

The Z-axis is known to be the axis of the cylinder.

$$\lambda_1 x^2 + \lambda_2 y^2 + k = 0, k \neq 0$$

As in the case of the paraboloid, we have the following result regarding the axis of the cylinder.

The axis of the cylinder is the line of intersection of the principal planes corresponding to the non-zero characteristic roots. Also, it is parallel to the principal direction corresponding to the characteristic root zero. The axis is also the line of centres.

Cor. 3. Planes bisecting the angles between the planes. It may be seen that planes bisecting the angles between the two planes

$$\lambda_1 x^2 + \lambda_2 y^2 = 0$$

are
$$x = 0, y = 0$$

Thus, we see that *the two principal planes corresponding to the two non-zero characteristics roots are the two bisecting planes.*

Cor. 4. The homogeneous second degree equation

$$\Sigma (ax^2 + 2fyz) = 0$$

will represent a pair of planes if $D = 0$.

9.9.4 Case IV. When A, B, C, F, G, H are all zero and $fu \neq gv$.

In this case the quadric has no centre and two roots of the discriminating cubic are zero and one non-zero.

We rotate the axes so that the new axes are parallel to the three mutually perpendicular principal directions. The new equation takes the form

$$\lambda_1 x^2 + 2x(ul_1 + vm_1 + wn_1) + 2y(ul_2 + vm_2 + wn_2) + 2z(ul_3 + vm_3 + wn_3) + d = 0$$

As the roots λ_2, λ_3 are equal, both being zero, we know that l_2, m_2, n_2 is any direction satisfying

$$al + hm + gn = 0 \qquad \ldots(1)$$

We suppose that l_2, m_2, n_2 are so chosen that these satisfy (1) and

$$ul_2 + vm_2 + wn_2 = 0 \qquad \ldots(2)$$

Then l_3, m_3, n_3 aer chosen so as to satisfy (1) and

$$l_3 l_2 + m_3 m_2 + n_3 n_2 = 0$$

Denoting the coefficients of x and z by p, r we rewrite the equation as

$$\lambda_1 x^2 + 2px + 2rz + d = 0 \qquad \ldots(3)$$

the coefficients of y being zero by (2).

Again we rewrite (3) as

$$\lambda_1 \left(x + \frac{p}{\lambda_1}\right)^2 + 2rz + \left(d - \frac{p^2}{\lambda_1}\right) = 0 \qquad \ldots(4)$$

Also we may see that $r \neq 0$, for otherwise the quadric will have a centre. Again, we rewrite (4) as

$$\lambda_1 \left(x + \frac{p}{\lambda_1}\right)^2 + 2r\left[z + \frac{1}{2r}\left(d - \frac{p^2}{\lambda_1}\right)\right] = 0$$

Shfiting the origin to

$$\left[-\frac{p}{\lambda_1}, 0, -\frac{1}{2r}\left(d - \frac{p^2}{\lambda_1}\right)\right]$$

we see that the equation becomes

$$\lambda_1 x^2 + 2rz = 0$$

which is the required canonical form.

The equation represents a parabolic cylinder in this case.

General Equation of the Second Degree

9.9.5 Case V.

When A, B, C, F, G, H are all zero, fu = gv = hw, and no one of f, g, h is zero.

In this case the quadric has a plane of centres and the discriminating cubic has two zero and one non-zero root.

Let (α, β, γ) be a centre. Shifting the origin to (α, β, γ) and rotating the axes so that the axes of the new system are parallel to a set of three mutually perpendicular principal directions, we see that the equation becomes

$$\lambda_1 x^2 + (u\alpha + v\beta + w\gamma + d) = 0$$

The equation represents a pair of parallel or coincident planes.

Note : The case when any two or all of f, g, h are zero can be easily considered and it can be shown that we shall have a parabolic cylinder in case the quadric does not have a centre and a pair of parallel planes if the quadric has a plane of centres.

9.10 QUADRICS OF REVOLUTION

Firstly, we shall prove a lemma concerning surfaces of revolution obtained on revolving a plane curve about an axis of co-ordinates.

Lemma : *The equation of a surface of revolution obtained on revolving a plane curve about X-axis is of the form*

$$\sqrt{y^2 + z^2} = f(x)$$

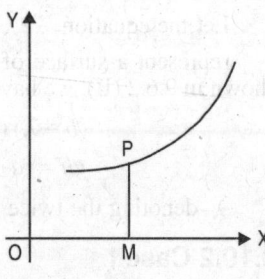

Consider the surface of revolution on revolving a curve about X-axis. Let the equations of the section of this surface by the plane $z = 0$ be

$$y = f(x), \quad z = 0 \qquad \ldots(1)$$

If P be a point on the curve and M and foot of the perpendicular from P on X-axis, we have

$$OM = x, \quad MP = y$$

so that we can rewrite.

$$y = f(x) \text{ as } MP = f(OM)$$

Now this relation remains unchanged as the curve revolves about the X-axis so that the point P describes a circle with M as its centre.

In terms of the co-ordinates (x, y, z) of the point P in any position, we have

$$MP = \sqrt{y^2 + z^2}, \quad OM = x$$

so that we can rewrite (2) as

$$\sqrt{y^2 + z^2} = f(x)$$

Hence, the result.

Similarly, the equations of the surfaces of revolution obtained on revolving plane curves about Y-axis and Z-axis are of the form

$$\sqrt{z^2 + x^2} = \phi(y), \quad \sqrt{x^2 + y^2} = \psi(z)$$

respectively.

Cor. *A quadric is a surface of revolution, if and only, if it has equal non-zero characteristic roots.* To see the truth of this result, we examine the various canonical forms which we have obtained. These are as follows :

Case I	$\lambda_1 x^2 + \lambda_2 y^2 + \lambda_3 z^2 + \Delta/D = 0$	...(1)
Case II	$\lambda_1 x^2 + \lambda_2 y^2 + 2rz = 0$	...(2)
Case III	$\lambda_1 x^2 + \lambda_2 y^2 + k = 0$	...(3)

Case IV	$\lambda_1 x^2 + 2rz = 0$	...(4)
Case V	$\lambda_1 x^2 + k = 0$	...(5)

On comparison with the equations of the surfaces of revolution we see that for the surface (1) to be that of revolution we must have two of $\lambda_1, \lambda_2, \lambda_3$ equal and for the surfaces (2) and (3) to be of revolution we must have $\lambda_1 = \lambda_2$. The equations (4) and (5) cannot be surfaces of revolution.

It will be seen that a necessary condition for a quadric to be a surface of revolution is that two of the characteristics roots are equal.

Clearly the equations (1) will represent a sphere if the characteristic roots $\lambda_1, \lambda_2, \lambda_3$ are all equal.

Hence, the result.

9.10.1 Condition for the General Equation of the Second Degree to Represent a Quadric of Revolution

Let the equation $\quad \Sigma(ax^2 + 2fyz) + 2\Sigma ux + d = 0$

represent a surface of revolution so that two of the characteristics roots are equal, so that as shown in 9.6.2 (B), we have the following two sets of necessary condition :

$$(b-\lambda)(c-\lambda) = f^2, \ (c-\lambda)(a-\lambda) = g^2, \ (a-\lambda)(b-\lambda) = h^2 \qquad ...(I)$$
$$gh = (a-\lambda) f, \ hf = (b-\lambda) g, \ fg = (c-\lambda) h \qquad ...(II)$$

λ denoting the twice repeated root.

9.10.2 Case I

Firstly, suppose that none of f, g, h is zero.

We show that in this case the set of condition I and deducible from the Set II so that the Set I is not an indepedent set of conditions and can as such be ignored. Let us assume the Set II.
Now

$$gh = (a-\lambda) f, \ hf = (b-\lambda) g$$

$\Rightarrow \qquad fgh^2 = (a-\lambda^2)(b-\lambda) fg$

$\Rightarrow \qquad (a-\lambda)(b-\lambda) = h^2;$ fg being not equal to zero.

We may similarly deduce the other two conditions of the Set I from the set of conditions II.

Thus, if the given equation represents a surface of revolution and none of f, g, h is zero, we have

$$\lambda = a - \frac{gh}{f} = b - \frac{hf}{g} = c - \frac{fg}{h}$$

$\Rightarrow \qquad a - \dfrac{gh}{f} = b - \dfrac{hf}{g} = c - \dfrac{fg}{h}$

$\Rightarrow \qquad \dfrac{F}{f} = \dfrac{G}{g} = \dfrac{H}{h} \qquad\qquad\qquad ...(III)$

Now we suppose that the conditions III are satisfied and show that the quadric is a surface of revolution.

Let $\qquad \dfrac{F}{f} = \dfrac{G}{g} = \dfrac{H}{h} = k$ (say)

$\Leftrightarrow \qquad \dfrac{gh - af}{f} = \dfrac{hf - bg}{g} = \dfrac{fg - ch}{h} = k$

$\Leftrightarrow \qquad a = \dfrac{gh}{f} - k, \ b = \dfrac{hf}{g} - k, \ c = \dfrac{fg}{h} - k$

General Equation of the Second Degree

Replacing a, b, c by $\dfrac{gh}{f} - k, \dfrac{hf}{g} - k, \dfrac{fg}{h} - k$ we get

$$F(x, y, z) = -k(x^2 + y^2 + z^2) + fgh\left(\dfrac{x}{f} + \dfrac{y}{g} + \dfrac{z}{h}\right) + 2ux + 2vy + 2wz + d$$

$$= -k(x^2 + y^2 + z^2) + 2ux + 2wz + d + fgh\left(\dfrac{x}{f} + \dfrac{y}{g} + \dfrac{z}{h}\right)^2$$

This form of the equation shows that every plane parallel to the plane

$$\dfrac{x}{f} + \dfrac{y}{g} + \dfrac{z}{h} = 0$$

cuts the surfaces in a circle. Thus, the equation represents a surface of revolution and the axis of revolution being the locus of the centres of the circular sections, is the line through the centre of the sphere

$$-k(x^2 + y^2 + z^2) + 2ux + 2vy + 2wz + d = 0$$

perpendicular to the plane (1). Thus, the axis of revolution is

$$\dfrac{x + \dfrac{u}{\lambda}}{1/f} = \dfrac{y + \dfrac{v}{\lambda}}{1/g} = \dfrac{z + \dfrac{w}{\lambda}}{1/h}$$

We have thus shown that *if no one of f, g, h is zero the necessary and sufficient condition for the equation*

$$\Sigma(ax^2 + 2fyz) + 2\Sigma ux + d = 0$$

to be a surface of revolution is

$$\dfrac{F}{f} = \dfrac{G}{g} = \dfrac{H}{h}$$

9.10.3 Case II

When any one of f, g, h is zero and not each one of f, g, h is zero.

Suppose that $\quad f = 0.$

Now $\quad gh = (a - \lambda) f$ and $f = 0$

$\Rightarrow \quad gh = 0 \Rightarrow g = 0$ or $h = 0$

Since we have supposed that not each one of f, g, h is 0, both of g, h are not zero.

Let $g = 0$ and $h \neq 0$

Now $\quad f = 0, g = 0, h \neq 0$ and $fg = (c - \lambda) h$

$\Rightarrow \quad \lambda = c.$

Also $\quad (a - \lambda)(b - \lambda) = h^2$ and $\lambda = c$

$\Rightarrow \quad (a - c)(b - c) = h^2$

We have thus shown that if the given quadric is a surface of revolution and

$$f = 0, g = 0, h = 0$$

then we necessarily have the relation

$$(a - c)(b - c) = h^2$$

We now show that this condition is also sufficient.

Since $\quad (a - c)(b - c) = h^2$

$a - c$ and $b - c$ must both be of the same sign. Suppose that they are both positive.

We have in this case
$$ax^2 + by^2 + cz^2 + 2fyz + 2hxy = (a-c)x^2 + (b-c)y^2$$
$$+ c(x^2 + y^2 + z^2) \pm \sqrt{2(a-c)(b-c)}\, xy$$
$$= (\sqrt{a-c}\, x \pm \sqrt{b-c}\, y)^2 + c(x^2 + y^2 + z^2)$$

where actually we have no ambiguity of sign in that the sign is positive or negative according as h is positive or negative.

Thus, we see that planes parallel to the plane
$$\sqrt{(a-c)}\, x \pm \sqrt{(b-c)}\, y = 0$$

cut the surface in circular sections. Thus, the quadric is a surface of revolution, the axis of revolution being the line through the centre.
$$\left(-\frac{u}{c}, -\frac{v}{c}, -\frac{w}{c}\right)$$

of the sphere
$$c(x^2 + y^2 + z^2) + 2ux + 2vy + 2wz + d = 0$$

perpendicular to the plane (4); viz., the line
$$\frac{x + u/c}{\sqrt{(a-c)}} = \pm \frac{y + v/c}{\sqrt{(b-c)}},\ z + \frac{w}{c} = 0$$

We have thus shown *that if*
$$f = 0, g = 0, h \neq 0$$

the necessary and sufficient condition for the equation
$$\Sigma(ax^2 + 2fyz) + 2\Sigma ux + d = 0$$

to represent a surface of revolution is
$$(a-c)(b-c) = h^2$$

We may similarly consider the cases $g = 0, h = 0, f \neq 0$; and $f = 0, h = 0, g \neq 0$.

9.10.4 Case III

We have in this case from Set I,
$$(b-\lambda)(c-\lambda) = 0,\ (c-\lambda)(a-\lambda) = 0,\ (a-\lambda)(b-\lambda) = 0$$

These relations imply that we necessarily have
$$a = b\ \text{or}\ b = c\ \text{or}\ c = a$$

This can also be seen as follows. The three conditions imply that the three equations
$$\lambda^2 - \lambda(b+c) + bc = 0$$
$$\lambda^2 - \lambda(c+a) + ca = 0$$
$$\lambda^2 - \lambda(a+b) + ab = 0$$

are consistent. This means that we have the relations
$$\begin{vmatrix} 1 & b+c & bc \\ 1 & c+a & ca \\ 1 & a+b & ab \end{vmatrix} = 0$$

$\Rightarrow$ $\qquad (a-b)(b-c)(c-a) = 0$

$\Rightarrow$ $\qquad a = b\ \text{or}\ b = c\ \text{or}\ c = a$

Thus, we see that if $f = g = h = 0$ and the given equation represents a surface of revolution, we necessarily have
$$a = b\ \text{or}\ b = c\ \text{or}\ c = a$$

General Equation of the Second Degree

We now show that the condition is as well sufficient.

Let $a = b$.

We have in this case

$$\Sigma (ax^2 + 2fyz) + 2\Sigma ux + d = 0$$

$$\Rightarrow [a(x^2 + y^2 + z^2) + 2\Sigma ux + d] + (c - a) z^2 = 0$$

so that planes parallel to $z = 0$ cut the surfaces in circles implying that the given quadric is a surface of revolution.

The equations of the axis of the revolution may now be easily obtained.

The cases $b = c, c = a$ may now be similarly discussed.

We have the shown that if f, g, h are all zero, a necessary and sufficient condition for the quadric to be a surface of revolution is that

$$a = b \text{ or } b = c \text{ or } c = a$$

9.11 REDUCTION OF EQUATIONS WITH NUMERICAL COEFFICIENTS

The manner in which we actually proceed in any given case depends upon whether the second degree terms do or do not form a perfect square.

Case I. *Suppose that the second degree terms form a perfect square.*

Thus, the given equation is of the form

$$(px + qy + rz)^2 + 2(ux + vy + wz) + d = 0 \quad ....(1)$$

We rewrite it as

$$(px + qy + rz + t)^2 + 2(u - pt)x + 2(v - gt)y + 2(w - rt)z + (d - t^2) = 0 \quad ...(2)$$

t, being any number whatsover. Consider now the two planes

$$px + qy + rz = 0$$

$$(u - pt)x + (v - pt)y + (w - rt)z = 0$$

We so choose t that these planes are perpendicular to each other. Thus, t is given by

$$p(u - pt) + q(v - qt) + r(w - rt) = 0$$

$$\Rightarrow (pu + qv + rw) = (p^2 + q^2 + r^2)t$$

$$\Rightarrow t = (pu + qv + rw)/(p^2 + q^2 + r^2)$$

Having thus chosen t, we rewrite (2) as

$$\left(\frac{px + qy + rz + t}{\sqrt{p^2 + q^2 + r^2}}\right)^2 = k \frac{2(u - pt)x + 2(v - qt)y + 2(w - rt)z + (d - t^2)}{2\sqrt{(u - pt)^2 + (v - qt)^2 + (w - rt)^2}}$$

where $k = -\dfrac{2\sqrt{(u - pt)^2 + (v - qt)^2 + (w - rt)^2}}{\sqrt{p^2 + q^2 + r^2}}$

Taking

$$\frac{px + qy + rz + t}{\sqrt{p^2 + q^2 + r^2}} = Y$$

$$\frac{2(u - pt)x + 2(v - qt)y + 2(w - rt)z + (d - t^2)}{2\sqrt{(u - pt)^2 + (v - qt)^2 + (w - rt)^2}} = X$$

we see that the given equation takes the form

$$Y^2 = kX$$

so that the surface is a parabolic cylinder.

EX. Show that the second degree terms form a perfect square if A, B, C, F, G, H are all zero.

Case II. The following procedure is suggested for the reduction of numerical equations when the second degree terms do not form a perfect square.

1. Find the discriminating cubic and solve the same.
2. If no characteristic root is zero, then put down the centre-giving equations and solve them.

If (α, β, γ) is a centre and $\lambda_1, \lambda_2, \lambda_3$ are the characteristics roots, then the reduced equation is

$$\lambda_1 x^2 + \lambda_2 y^2 + \lambda_3 z^2 + (u\alpha + v\beta + w\gamma + d) = 0$$

3. If one characteristic root is zero, find the principal direction l, m, n corresponding to the zero characteristic root by solving two of the three equations :

$$al + hm + gn = 0, \quad hl + bm + fn = 0, \quad gl + fm + cn = 0$$

Then find $ul + vm + wn$. If this is not zero, the reduced equation is

$$\lambda_1 x^2 + \lambda_2 y^2 + 2(ul + vm + wn)z = 0$$

is the required reduced equation.

Note : If one characteristic root is zero and two non-zero, then the line of intersection of the two principal planes corresponding to the non-zero roots is the axis, if the quadric is a parabolic or an elliptic or hyperbolic cylinder and the line of intersection of the planes, if the quadric is a pair of intersecting planes.

In the case of the elliptic and hyperbolic cylinders, and a pair of intersecting planes, the line of centres is also the axis.

EXAMPLES

1. *Reduce the equation* $2x^2 + 7y^2 - 10yz - 8zx - 10xy + 6x + 12y - 6z + 2 = 0$ *to a canonical form.*

Solution. The discriminating cubic is $\lambda^3 + 3\lambda^2 - 90\lambda + 216 = 0$.

This shows that $D = -126 \neq 0$. The roots of the discriminating cubic are $3, 6, -12$.

Again the centre-giving equations are

$$2x - 5y - 4z + 3 = 0, \quad 5x + 7y + 5z - 6 = 0, \quad 4x + 5y - 2z + 3 = 0$$

Solving these we see that the centre is

$$\left(\frac{1}{3}, -\frac{1}{3}, \frac{4}{3}\right)$$

Denoting this by (α, β, γ) we have

$$u\alpha + v\beta + w\gamma + d = -3$$

Thus, the canonical form of the equation is

$$3x^2 + 6y^2 - 12z^2 - 3 = 0 \iff x^2 + 2y^2 - 4z^2 - 1 = 0 \qquad \ldots(1)$$

which shows that the given quadric is a hyperbolic of one sheet.

The equation (1) represents the given quadric when the origin of co-ordinates is its centre and the co-ordinate axes are parallel to the principal directions, *i.e*, (1) is an equation referred to principal axes as co-ordinate axes.

2. *Reduce to canonical form the equation* $x^2 - y^2 + 4yz - 4xz - 3 = 0$ *of a quadric.*

Solution. The discriminating cubic is

$$\lambda^3 - 9\lambda = 0$$

so that the characteristic roots are

$$0, 3, -3$$

Thus, $D = 0$.

The direction cosines (l, m, n) of the principal direction corresponding to $\lambda = 0$ are given by

General Equation of the Second Degree

$$2l + 4n = 0, \quad -2m + 4n = 0, \quad 4l + 4m = 0$$

These gives
$$l : m : n = 2 : -2 : -1$$

Thus, in this case we have
$$ul + vm + wn = 0$$

so that we proceed to find the centre-giving equations. These are
$$2x + 4z = 0, \quad -2y + 4z = 0, \quad 4y + 4x = 0$$

These three planes meet in the line. Clearly $(0, 0, 0)$ is a point on it. Denoting this by (α, β, γ), we have
$$u\alpha + v\beta + w\gamma + d = -3$$

Thus, the required canonical form of the equation is
$$3x^2 - 3y^2 = 0 \implies x^2 - y^2 = 0$$

The given equation, therefore, represents a pair of intersecting planes.

Note : The fact that the given equation is free from first degree terms also shows that $(0, 0, 0)$ is a centre of the given quadric.

3. *Prove that* $5x^2 + 5y^2 + 8z^2 + 8yz + 8zx - 2xy + 12x - 12y + 6 = 0$ *represents a cylinder whose cross-section is an ellipse of eccentricity* $1/\sqrt{2}$.

Find also the equations of the axis of the cylinder.

Solution. The discrimating cubic is
$$\lambda^3 - 18\lambda^2 + 72\lambda = 0$$

so that the values of λ are
$$0, 6, 12$$

The direction cosines (l, m, n) of the principal direction corresponding to $\lambda = 0$ are given by
$$l - 5m - 4n = 0$$
$$5l - m + 4n = 0$$
$$\Leftrightarrow \quad l = 1/\sqrt{3}, m = 1/\sqrt{3}, n = -1/\sqrt{3}$$

Thus,
$$ul + vm + wn = 6/\sqrt{3} - 6/\sqrt{3} - 0/\sqrt{3} = 0$$

We have, therefore, to proceed to put down the centre-giving equations. These are
$$10x - 2y + 8z + 12 = 0 \qquad \ldots(1)$$
$$-2x + 10y + 8z - 12 = 0 \qquad \ldots(2)$$
$$8x + 8y + 16z = 0 \qquad \ldots(3)$$

Clearly (3) can be obtained on adding (1) and (2) so that as expected, these three equations are equivalent to only two. Putting $z = 0$ in (1) and (2), we obtain
$$x = -1, y = 1, z = 0$$

so that $(-1, 1, 0)$ is a centre. Thus,
$$u\alpha + v\beta + w\gamma + d = -6 - 6 + 6 = -6$$

Hence, the reduced equation is
$$12x^2 + 6y^2 - 6 = 0 \implies 2x^2 + y^2 = 1$$

The cross-section is $2x^2 + y^2 = 1, z = 0$.

Its ecentricity is now easily seen to be $1/\sqrt{2}$.

The line of centresis the axis of the cylinder so that the equations of the axis are
$$5x - y + 4z + 6 = 0, \quad x + y + 2z = 0$$

4. *Prove that* $x^2 + y^2 + z^2 - yz - zx - zy - 3x - 6y - 9z + 21 = 0$ *represents a paraboloid of revolution and find the co-orddinates of its focus.*

Solution. The discriminating cubic is
$$-4\lambda^3 + 12\lambda^2 - 9\lambda = 0$$
so that the characteristics roots are
$$0, 3/2, 3/2$$
Two values of λ being equal, the given quadric is a surface of revolution.

The direction cosines (l, m, n) of the principal direction corresponding to $\lambda = 0$ are given by any two of the three equations

$$l - \frac{1}{2}m - \frac{1}{2}n = 0$$

$$-\frac{1}{2}l + m - \frac{1}{2}n = 0$$

$$-\frac{1}{2}l - \frac{1}{2}m + n = 0$$

These give $\quad l : m : n = 1 : 1 : 1$

$\therefore \quad l = \dfrac{1}{\sqrt{3}}, m = \dfrac{1}{\sqrt{3}}, n = \dfrac{1}{\sqrt{3}}$

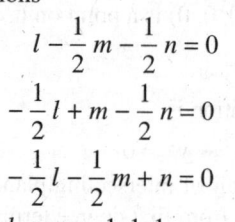

Now we have

$$ul + vm + wn = -\frac{3}{2} \cdot \frac{1}{\sqrt{3}} - 3 \cdot \frac{1}{\sqrt{3}} - \frac{9}{2} \cdot \frac{1}{\sqrt{3}} = -\frac{9}{\sqrt{3}} \neq 0$$

Thus, the quadric is a paraboloid of revolution and the reduced equation is

$$\frac{3}{2}x^2 + \frac{3}{2}y^2 - 2 \cdot \frac{9}{\sqrt{3}}z = 0 \quad \Leftrightarrow \quad x^2 + y^2 = 4\sqrt{3}z.$$

This form of the equation shows that the latus rectum of the generating parabola is $4\sqrt{3}$.

With respect to the given system of co-ordinate axes, the direction ratios of the axis of the paraboloid which is also the axis of revolution are 1, 1, 1.

We rewrite the given equations in the form

$$x^2 + y^2 + z^2 - \frac{1}{2}[(x + y + z)^2 - (x^2 + y^2 + z^2)] - 3x - 6y - 9z + 2 = 0$$

$$\Leftrightarrow \quad \frac{3}{2}(x^2 + y^2 + z^2) - 3x - 6y - 9z + 21 - \frac{1}{2}(x + y + z)^2 = 0$$

$$\Leftrightarrow \quad x^2 + y^2 + z^2 - 2x - 4y - 6z + 14 - \frac{1}{3}(x + y + z)^2 = 0$$

Thus, the axis of revolution, being the line through the centre of the sphere

$$x^2 + y^2 + z^2 - 2x - 4y - 6z + 14 = 0$$

and perpendicular to the plane

$$x + y + z = 0$$

is $\quad \dfrac{x-1}{1} = \dfrac{y-2}{1} = \dfrac{z-3}{1}$ ...(1)

which is the axis of the parabolid.

The vertex is the point where this axis meets the paraboloid. It can be shown that any point

$$(r + 1, r + 2, r + 3)$$

on the axis will be on the paraboloid if $r = -1$.

Thus, $(0, 1, 2)$ is the vertex of the paraboloid.

The required focus is the point on the axis (1) at a distance $\sqrt{3}$ from $(0, 1, 2)$. Rewriting the equation of the axis in the form

General Equation of the Second Degree

$$\frac{x-0}{1/\sqrt{3}} = \frac{y-1}{1/\sqrt{3}} = \frac{z-2}{1/\sqrt{3}}$$

we see that the point on the axis at a distance $\sqrt{3}$ from (0, 1, 2) is

(1, 2, 3)

Thus, (1, 2, 3) is the required focus.

EXERCISES

1. Show that $4x^2 - y^2 - z^2 + 2yz - 8x - 4y + 8z - 2 = 0$ represents a paraboloid. Find the reduced equation and the co-ordinates of the vertex.

   $\left[\text{Ans. } 2x^2 - y^2 + \sqrt{2}z = 0, \left(1, -\frac{9}{4}, \frac{3}{4}\right)\right]$

2. Reduce to its principal axes $2y^2 - 2yz + 2zx - 2xy - x - 2y + 3z - 2 = 0$ and state the nature of the surface represented by the equation.

   $\left[\text{Ans. } 3x^2 - y^2 = \frac{1}{2}. \text{ Hyperbolic cylinder}\right]$

3. Find the nature of the surface represented by the equation

   $$x^2 + 2y^2 - 3z^2 - 4yz + 8zx - 12xy + 1 = 0$$

   [Ans. $3x^2 + 6y^2 - 9z^2 + 1 = 0$. Hyperboloid of two sheets.]

4. Find the reduced equation of

   (i) $x^2 + 2yz - 4x + 6y + 2z = 0$

   (ii) $x^2 - y^2 + 2yz - 2xz - x - y + z = 0$

   (iii) $yz + zx + xy - 7x - 6y - 5z - 25 = 0$

   (iv) $4y^2 - 4yz + 4zx - 4xy - 2x + 2y - 1 = 0$

   (v) $2x^2 + 2y^2 + z^2 + 2yz - 2zx - 4xy + x + y + z = 0$

   (vi) $(x \cos \alpha - y \sin \alpha)^2 + (y \cos \alpha + z \sin \alpha)^2 + 2y = 1$

   (vii) $3x^2 + 6yz - y^2 - z^2 - 6x + 6y - 2z - 2 = 0$

   (viii) $4x^2 + y^2 + z^2 - 4xy - 2yz + 4zx - 12x + 6y - 6z + 8 = 0$

   (ix) $x^2 + y^2 + z^2 - 2xy - 2yz + x - 4y + z + 1 = 0$

   [Ans. (i) $x^2 + y^2 - z^2 = 10$ (ii) $3x^2 - 3y^2 = z$ (iii) $2x^2 - y^2 - z^2 = 102$

   (iv) $6x^2 - 2y^2 = 1$ (v) $\dfrac{5+\sqrt{17}}{2} x^2 + \dfrac{5-\sqrt{17}}{2} y^2 + \sqrt{2}z = 0$.

   (vi) $(1 + \sin \alpha \cos \alpha) x^2 + (1 - \sin \alpha \cos \alpha) y^2 + z \sin 2\alpha / \sqrt{1 - \sin^2 \alpha \cos^2 \alpha} = 0$

   if $\sin \alpha \neq 0$, $\cos \alpha \neq 0$; $x^2 + y^2 = 2$ if $\sin \alpha = 0$ and $y^2 + z^2 = 2$ if $\cos \alpha = 0$.

   (vii) $2x^2 + 3y^2 - 4z^2 = 4$ (viii) $3x^2 - 3\sqrt{6} x + 4 = 0$ (ix) $3y^2 = \sqrt{6}x$]

5. Show that the equation

   $$a(z-x)(x-y) + b(x-y)(y-z) + c(y-z)(z-x) = 0$$

   represents two planes whose line of intersection is equally inclined to the three co-ordinate axes.

6. Prove that the principal axes of the conicoid $ax^2 + by^2 + cz^2 + 2fyz + 2gzx + 2hxy = 1$ are given by the equations
$$x(f\lambda_r + F) = y(g\lambda_r + G) = z(h\lambda_r + H), \quad (r = 1, 2, 3)$$
where $\lambda_1, \lambda_2, \lambda_3$ are the roots of the equation
$$\begin{vmatrix} a-\lambda & h & g \\ h & b-\lambda & f \\ g & f & c-\lambda \end{vmatrix} = 0$$
and $F = gh - af$, $G = hf - bg$, $H = fg - ch$.

Also show that the cone which touches the co-ordinate planes and the principal planes of the above coincoid is
$$\sqrt{[(gH - hG)x]} + \sqrt{[(hF - fH)y]} + \sqrt{[(fG - gF)z]} = 0$$

7. If the feet of the six normals from P to the ellipsoid $\dfrac{x^2}{a^2} + \dfrac{y^2}{b^2} + \dfrac{z^2}{c^2} = 1$ lie upon a concentric conicoid of revolution, prove that the locus of P is the cone
$$\frac{y^2 z^2}{a^2(b^2 - c^2)} + \frac{z^2 x^2}{b^2(c^2 - a^2)} + \frac{x^2 y^2}{c^2(a^2 - b^2)} = 0$$
and that the axes of symmetry of the conicoids lie on the cone
$$a^2(b^2 - c^2)x^2 + b^2(c^2 - a^2)y^2 + c^2(a^2 - b^2)z^2 = 0 \quad [\textbf{Ans. } x^2 = y^2 + z^2]$$

8. Prove taht the equation $ax^2 + by^2 + cz^2 + 2fyz + 2gzx + 2hxy = 0$ will represent a right circular cone with vertical angle θ provided that
$$\frac{af - gh}{f} = \frac{bg - hf}{g} = \frac{ch - fg}{h} = \frac{(a+b+c)(1+\cos\theta)}{(1 + 3\cos\theta)}$$

9. Given the ellipsoid of revolution $\dfrac{x^2}{a^2} + \dfrac{y^2 + z^2}{b^2} = 1$, $(a^2 < b^2)$, show that cone whose vertex is one of the foci of the ellipse $z = 0$, $\dfrac{x^2}{a^2} + \dfrac{y^2}{b^2} = 1$ and whose base is any plane section of the ellipsoid is a surface of revolution.

10. Prove that if $F(x, y, z) \equiv \Sigma(ax^2 + 2fyz) + 2\Sigma ux + d = 0$ represents a paraboloid of revolution, we have
$$agh + f(g^2 + h^2) = bhf + g(h^2 + f^2) = cfg + h(f^2 + g^2) = 0$$
and that if it represents a right circular cylinder, we have also
$$\frac{u}{f} + \frac{v}{g} + \frac{w}{h} = 0$$

10
Confocal Conicoids

10.1 DEFINITIONS

The conicoids whose principal sections are confocal conics, i.e., the conics having same foci, are called **confocal conicoids**. Thus,

$$\frac{x^2}{a^2-\lambda}+\frac{y^2}{b^2-\lambda}+\frac{z^2}{c^2-\lambda}=1 \qquad \text{...(i)}$$

represents, for all values of λ, the general equation of a system of conicoids confocal with the ellipsoid

$$\frac{x^2}{a^2}+\frac{y^2}{b^2}+\frac{z^2}{c^2}=1$$

λ is known as the *parameter* of the confocal.

10.1.1 Principal Sections

The principal sections of (i) are confocal conics. Suppose $a>b>c$ and λ varies from $-\infty$ to $+\infty$.

(a) When λ is negative, the surface (i) is an ellipsoid. As λ increases, i.e., as $\lambda \to \infty$, the principal axes of the surface increase and their ratio tends to unity. Thus, a sphere of infinite radius is a limiting case of the confocals.

(b) When λ is positive and less than c^2, the surface is an ellipsoid, but the ellipsoid becomes more and more flat as λ, approaches c^2. Thus, as $z \to c^2$, the ellipsoid tends to coincide with the ellipse

$$z=0, \quad \frac{x^2}{(a^2-c^2)}+\frac{y^2}{(b^2-c^2)}=1 \qquad \text{...(ii)}$$

on the *xy*-plane.

(c) When λ lies between c^2 and b^2, the surface is a hyperboloid of one sheet. As $\lambda \to c^2$ from the right, the hyperboloid tends to coincide with the ellipse (i), and as $\lambda \to b^2$ from the left, the hyperboloid tends to coincide with the hyperbola.

$$y=0, \quad \frac{x^2}{(a^2-b^2)}-\frac{z^2}{(b^2-c^2)}=1 \qquad \text{...(iii)}$$

in the *zx*-plane

(d) When λ lies between b^2 and a^2, the surface is a hyperboloid of two sheets. As $\lambda \to b^2$ from the right, the hyperboloid tends to coincide with the hyperbola (iii), and as $\lambda \to a^2$ from left, the surface tends to reduce to the imaginary ellipse

$$x=0, \quad \frac{y^2}{(a^2-b^2)}+\frac{z^2}{(a^2-c^2)}=-1$$

When $\lambda > a^2$, the surface is always imaginary.

The conics (ii) and (iii) which are the boundaries of limiting cases of confocal conicoids, are called *called conics,* the conic (ii) is known as the *focal ellipse* and the conics (iii) as the *focal.*

In the same way, since the principal section of the paraboloids

$$\frac{x^2}{a^2 - \lambda} + \frac{y^2}{b^2 - \lambda} = 2z - \lambda \qquad \text{...(iv)}$$

and

$$\frac{x^2}{a^2} + \frac{y^2}{b^2} = 2z$$

by the planes $x = 0$ and $y = 0$ are confocal parabolas, the eqn. (iv) represents a system of confocal parabolas.

10.1.2 Confocals Through a Given Point

To prove that three conicoids confocal with a given central conicoid will pass through a given point; and one of the three is an ellipsoid, one a hyperboloid of one sheet and one a hyperboloid of two sheets.

Let the equation of the given conicoid be

$$\frac{x^2}{a^2} + \frac{y^2}{b^2} + \frac{z^2}{c^2} = 1 \, (a > b > c) \qquad \text{...(i)}$$

and the equation of a conicoid confocal to it be

$$\frac{x^2}{(a^2 - \lambda)} + \frac{y^2}{(b^2 - \lambda)} + \frac{z^2}{(c^2 - \lambda)} = 1 \qquad \text{...(ii)}$$

If it passes through the given point (α, β, γ), then

$$\frac{\alpha^2}{(a^2 - \lambda)} + \frac{\beta^2}{(b^2 - \lambda)} + \frac{\gamma^2}{(c^2 - \lambda)} = 1$$

$$\Rightarrow \quad f(\lambda) \equiv \alpha^2 (b^2 - \lambda)(c^2 - \lambda) + \beta^2 (c^2 - \lambda)(a^2 - \lambda) + \gamma^2 (a^2 - \lambda)(b^2 - \lambda)$$
$$- (a^2 - \lambda)(b^2 - \lambda)(c^2 - \lambda) = 0. \qquad \text{...(iii)}$$

This being cubic in λ, gives the parameters of three confocals which pass through the given point. By giving different values of λ, we have

λ	=	∞,	a^2,	b^2,	c^2,
$f(x)$	is	+,	+,	−,	+,

Hence, $f(\lambda) = 0$ has three real roots, such that

$$a^2 > \lambda_1 > b^2 > \lambda_2 > c^2 > \lambda_3.$$

When $\lambda = \lambda_3$, the surface is an ellipsoid, when $\lambda = \lambda_2$, it is a hyperboloid of one sheet, and when $\lambda = \lambda_1$, it is a hyperboloid of two sheets.

10.2 CONFOCALS TOUCHING A GIVEN PLANE

To prove that the conicoids confocal with a given conicoid touch a given plane.

The given plane is $\qquad lx + my + nz = p \qquad$ ...(i)

The condition that the plane touches the conicoid

$$\frac{x^2}{(a^2 + \lambda)} + \frac{y^2}{(b^2 + \lambda)} + \frac{z^2}{(c^2 + \lambda)} = 1 \qquad \text{...(ii)}$$

is $\qquad l^2 (a^2 + \lambda) + m^2 (b^2 + \lambda) + n^2 (c^2 + \lambda) = p^2 \qquad$ ...(iii)

This is a linear equation in λ and hence gives one and only one value of λ. Hence, it follows that one conicoid of a given confocal system touches any given plane.

10.2.1 Confocals Touching a Given Line

To prove that two conicoids confocal with a given conicoid touch a given straight line.
Let the given line be
$$lx + my + nz + p = 0 = l'x + m'y + n'z + p'$$
A plane through this line is
$$(lx + my + nz + p) + k(l'x + m'y + n'z + p') = 0$$
$$\Rightarrow \quad (l + kl')x + (m + km')y + (n + kn')z + (p + kp') = 0$$
This plane touches the conicoid
$$\frac{x^2}{(a^2 + \lambda)} + \frac{y^2}{(b^2 + \lambda)} + \frac{z^2}{(c^2 + \lambda)} = 1$$
if $\quad (a^2 + \lambda)(l + kl')^2 + (b^2 + \lambda)(m + km')^2 + (c^2 + \lambda)(n + kn')^2 = (p + kp')^2$

If given line is a tangent line of the conicoid, the two tangent planes through it coincide. Hence, roots of the above equation in k are equal.
$$\Rightarrow \quad [(a^2 + \lambda)l^2 + (b^2 + \lambda) + (c^2 + \lambda)n^2 - p^2]$$
$$\times [(a^2 + \lambda^2)l'^2 + (b^2 + \lambda^2)m'^2 + (c^2 + \lambda^2)n'^2 - p'^2]$$
$$= [(a^2 + \lambda)ll' + (b^2 + \lambda)mm' + (c^2 + \lambda)nn' - pp']^2$$

This is quadratic in λ. Hence, gives two confocals which touch the given line.

10.2.2 Confocals Cut at Right Angles

To prove that the confocal conicoids cut one another at right angles at all their common points, i.e., the tangent planes at any common point are at right angles.

Let (x_1, y_1, z_1) be a common point of confocals.

$$\frac{x^2}{a^2 + \lambda_1} + \frac{y^2}{b^2 + \lambda_1} + \frac{z^2}{c^2 + \lambda_1} = 1 \qquad \text{...(i)}$$

and $\quad \dfrac{x^2}{a^2 + \lambda_2} + \dfrac{y^2}{b^2 + \lambda_2} + \dfrac{z^2}{c^2 + \lambda_2} = 1 \qquad \text{...(ii)}$

Hence, $\quad \dfrac{x_1^2}{a^2 + \lambda_1} + \dfrac{y_1^2}{b^2 + \lambda_1} + \dfrac{z_1^2}{c^2 + \lambda_1} = 1 \qquad \text{...(iii)}$

and $\quad \dfrac{x_1^2}{a^2 + \lambda_2} + \dfrac{y_1^2}{b^2 + \lambda_2} + \dfrac{z_1^2}{c^2 + \lambda_2} = 1 \qquad \text{...(iv)}$

Equations of tangent planes at (x_1, y_1, z_1) to (i) and (ii), are

$$\frac{xx_1}{a^2 + \lambda_1} + \frac{yy_1}{b^2 + \lambda_1} + \frac{zz_1}{c^2 + \lambda_1} = 1 \qquad \text{...(v)}$$

and $\quad \dfrac{xx_1}{a^2 + \lambda_2} + \dfrac{yy_1}{b^2 + \lambda_2} + \dfrac{zz_1}{c^2 + \lambda_2} = 1 \qquad \text{...(vi)}$

Subtracting (iv) from (iii), we get

$$\frac{x_1^2}{(a^2 + \lambda_1)(a^2 + \lambda_2)} + \frac{y_1^2}{(b^2 + \lambda_1)(b^2 + \lambda_2)} + \frac{z_1^2}{(z^2 + \lambda_1)(z^2 + \lambda_2)} = 0$$

Which is the condition that the tangent planes (v) and (vi) are at right angles.

10.3 ELLIPTIC CO-ORDINATES

From equation (iii), 10.1.2, we have
$$f(\lambda) \equiv \alpha^2(b^2 - \lambda) + \beta^2(a^2 - \lambda)(c^2 - \lambda) + \gamma^2(a^2 - \lambda)(b^2 - \lambda)$$
$$- (a^2 - \lambda)(b^2 - \lambda)(c^2 - \lambda) = 0 \qquad \text{...(i)}$$

If the roots of this equation are $\lambda_1, \lambda_2, \lambda_3$, then
$$f(\lambda) = (\lambda - \lambda_1)(\lambda - \lambda_2)(\lambda - \lambda_3) \qquad \text{...(ii)}$$

Dividing (i) by $(a^2 - \lambda)(b^2 - \lambda)(c^2 - \lambda)$, we have

$$1 - \frac{\alpha^2}{a^2 - \lambda} - \frac{\beta^2}{b^2 - \lambda} - \frac{\gamma^2}{c^2 - \lambda} = \frac{-f(\lambda)}{(a^2 - \lambda)(b^2 - \lambda)(c^2 - \lambda)}$$

$$\Rightarrow \quad 1 - \frac{\alpha^2}{a^2 - \lambda} - \frac{\beta^2}{b^2 - \lambda} - \frac{\gamma^2}{c^2 - \lambda} = \frac{-(\lambda - \lambda_1)(\lambda - \lambda_2)(\lambda - \lambda_3)}{(a^2 - \lambda)(b^2 - \lambda)(c^2 - \lambda)} \qquad \text{...(iii)}$$

$$\Rightarrow \quad 1 - \frac{\alpha^2}{a^2 - \lambda} - \frac{\beta^2}{b^2 - \lambda} - \frac{\gamma^2}{c^2 - \lambda} - \frac{(a^2 - \lambda_1)(a^2 - \lambda_2)(a^2 - \lambda_3)}{(b^2 - a^2)(c^2 - a^2)(a^2 - \lambda)}$$
$$- \frac{(b^2 - \lambda_1)(b^2 - \lambda_2)(b^2 - \lambda_3)}{(c^2 - b^2)(a^2 - b^2)(b^2 - \lambda)} - \frac{(c^2 - \lambda_1)(c^2 - \lambda_2)(c^2 - \lambda_3)}{(a^2 - c^2)(b^2 - c^2)(c^2 - \lambda)}$$

Comparing the coefficients of $\dfrac{1}{(a^2 - \lambda)}$, $\dfrac{1}{(b^2 - \lambda)}$ and $\dfrac{1}{(c^2 - \lambda)}$ on either side, we have

$$\alpha^2 = \frac{(a^2 - \lambda_1)(a^2 - \lambda_2)(a^2 - \lambda_3)}{(b^2 - a^2)(c^2 - a^2)}, \quad \beta^2 = \frac{(b^2 - \lambda_1)(b^2 - \lambda_2)(b^2 - \lambda_3)}{(c^2 - b^2)(a^2 - b^2)}$$

$$\gamma^2 = \frac{(c^2 - \lambda_1)(c^2 - \lambda_2)(c^2 - \lambda_3)}{(a^2 - c^2)(b^2 - c^2)}.$$

These express the co-ordinates of the point $P(\alpha, \beta, \gamma)$ in terms of the parameters of the three confocal conicoids through the point P, i.e., if the parameters $\lambda_1, \lambda_2, \lambda_3$ are given, the position of the point P can be uniquely determined. Co-ordinates $\lambda_1, \lambda_2, \lambda_3$ are called the *elliptic co-ordinates* of the point (α, β, γ) with respect to the conicoid.

EXAMPLES

1. *Prove that the equation to the confocal which has a system of circular sections parallel to the plane $x = y$ is*

$$\frac{x^2}{(c^2 - a^2)(a^2 - b^2)} + \frac{y^2}{(b^2 - c^2)(a^2 - b^2)} - \frac{z^2}{2(b^2 - c^2)(c^2 - a^2)} = \frac{1}{2c^2 - (a^2 + b^2)}$$

Solution. Let the equation of confocal be

$$\frac{x^2}{(a^2 - \lambda)} + \frac{y^2}{(b^2 - \lambda)} + \frac{z^2}{(c^2 - \lambda)} = 1 \qquad \text{...(i)}$$

If can be rewritten as

$$\frac{1}{(c^2 - \lambda)} + [x^2 + y^2 + z^2 - (c^2 - \lambda)] + x^2 \left(\frac{1}{a^2 - \lambda} - \frac{1}{c^2 - \lambda}\right) + y^2 \left(\frac{1}{b^2 - \lambda} - \frac{1}{c^2 - \lambda}\right) = 0 \quad \text{...(ii)}$$

If will represent a sphere if

$$x^2 \left(\frac{1}{a^2 - \lambda} - \frac{1}{c^2 - \lambda}\right) + y^2 \left(\frac{1}{b^2 - \lambda} - \frac{1}{c^2 - \lambda}\right) = 0$$

$$\Rightarrow \quad x^2 (b^2 - \lambda)(c^2 - a^2) = y^2 (b^2 - c^2)(a^2 - \lambda) \qquad \text{...(iii)}$$

Equation (ii) represents a pair of planes $(a^2 > b^2 > c^2)$, and hence the central circular sections are given by (iii). The central circular section (ii) will be parallel to the plane $x = y$ if

$$(b^2 - \lambda)(c^2 - a^2) = (b^2 - c^2)(a^2 - \lambda)$$

Confocal Conicoids

$$\Rightarrow \qquad \lambda = \frac{2a^2b^2 - b^2c^2 - c^2a^2}{a^2 + b^2 - 2c^2}$$

Putting this value of λ in (i), the required confocal is

$$\frac{x^2}{a^2 - \frac{2a^2b^2 - b^2c^2 - c^2a^2}{a^2 + b^2 - 2c^2}} + \frac{y^2}{b^2 - \frac{2a^2b^2 - b^2c^2 - c^2a^2}{a^2 + b^2 - 2c^2}} + \frac{2z}{c^2 - \frac{2a^2b^2 - b^2c^2 - c^2a^2}{a^2 + b^2 + c^2}} = 1.$$

$$\Rightarrow \quad \frac{x^2}{(b^2 - a^2)(a^2 - b^2)} + \frac{y^2}{(b^2 - c^2)(a^2 - b^2)} + \frac{z^2}{2(b^2 - c^2)(c^2 - a^2)} = \frac{1}{2c^2 - a^2 - b^2}$$

2. *Prove that the locus of umbilics of a system of confocal ellipsoids is the focal hyperbola.*

Solution. Equation of the confocal ellipsoid is

$$\frac{x^2}{(a^2 - \lambda)} + \frac{y^2}{(b^2 - \lambda)} + \frac{z^2}{(c^2 - \lambda)} = 1 \qquad \ldots(i)$$

The umbilics of (i) are (α, β, γ), then

$$\alpha = \pm \frac{\sqrt{a^2 - \gamma}\sqrt{a^2 - b^2}}{\sqrt{a^2 - c^2}}, \quad \beta = 0, \quad \gamma = \pm \frac{\sqrt{c^2 - \lambda}\sqrt{b^2 - c^2}}{\sqrt{a^2 - c^2}}$$

$$\therefore \quad a^2 - \lambda = \frac{a^2(a^2 - c^2)}{(a^2 - b^2)}, \quad c^2 - \lambda = \frac{\gamma^2(a^2 - c^2)}{(b^2 - c^2)}, \quad \beta = 0$$

$$\Rightarrow \quad \frac{\alpha^2}{(a^2 - b^2)} - \frac{\gamma^2}{(b^2 - c^2)} = 1, \quad \beta = 0$$

Hence, the locus of umbilitics (α, β, γ) is

$$\frac{x^2}{(a^2 - b^2)} - \frac{z^2}{(b^2 - c^2)} = 1, \quad y = 0$$

which represents a focal hyperbola.

EXERCISES

1. Prove that the equation to the confocal through the point of the focal ellipse whose eccentric angle is α is

$$\frac{x^2}{(a^2 - b^2)\cos^2\alpha} - \frac{y^2}{(a^2 - b^2)\sin^2\alpha} + \frac{z^2}{(c^2 - a^2\sin^2\alpha - b^2\cos^2\alpha)} = 1$$

2. A given plane and parallel tangent plane to a conicoid are at a distance p and p_0 from the centre. Prove that the parameter of the confocal conicoid which touches the plane is $(p_0^2 - p^2)$.

3. Show that the locus of the point of intersection of three planes mutually at right angles, each of which touches one of three given confocals, is a sphere.

4. Show that the two confocal paraboloids cut everywhere at right angles.

5. Prove that the perpendiculars from the origin to the tangent planes to the ellipsoid which touch it along its curve of intersection with the confocal whose parameter is λ, lie on the cone

$$\frac{a^2 x^2}{(a^2 - \lambda)} + \frac{b^2 y^2}{(b^2 - \lambda)} + \frac{c^2 z^2}{(c^2 - \lambda)} = 0$$

6. If a straight line touches two conicoids confocal with a given conicoid then show that the tangent planes at the points of contact will be at right angles.

10.4 CONFOCALS THROUGH A POINT ON A CONICOID

To prove that the parameters of the two confocals through any point P of a conicoid are equal to the sphere of the axes of the central section of the conicoid which is parallel to the tangent plane at P, and the normals at P to the confocals are parallel to the axes of that section.

Let $P(x_1, y_1, z_1)$ be any point on the coincoid

$$\frac{x^2}{a^2} + \frac{y^2}{b^2} + \frac{z^2}{c^2} = 1 \qquad \ldots(i)$$

Let a confocal conicoid be

$$\frac{x^2}{(a^2 - \lambda)} + \frac{y^2}{(b^2 - \lambda)} + \frac{z^2}{(c^2 - \lambda)} = 1 \qquad \ldots(ii)$$

If P lies on (ii), then

$$\frac{x_1^2}{(a^2 - \lambda)} + \frac{y_1^2}{(b^2 - \lambda)} + \frac{z_1^2}{(c^2 - \lambda)} = 1 \qquad \ldots(iii)$$

Also,

$$\frac{x_1^2}{a^2} + \frac{y_1^2}{b^2} + \frac{z_1^2}{c^2} = 1 \qquad \ldots(iv)$$

From (iii) and (iv), we get

$$\frac{x_1^2}{(a^2 - \lambda)} + \frac{y_1^2}{(b^2 - \lambda)} + \frac{z_1^2}{(c^2 - \lambda)} = \frac{x_1^2}{a^2} + \frac{y_1^2}{b^2} + \frac{z_1^2}{c^2}$$

$$\Rightarrow \quad \frac{x_1^2}{a^2 (a^2 - \lambda)} + \frac{y_1^2}{b^2 (b^2 - \lambda)} + \frac{z_1^2}{c^2 (c^2 - \lambda)} = 0 \qquad \ldots(v)$$

Equation of the central section of the given conicoid parallel to the tangent plane at P is

$$\frac{xx_1}{a^2} + \frac{yy_1}{b^2} + \frac{zz_1}{c^2} = 0 \qquad \ldots(vi)$$

and the squares of the semi-axes of this section are given by

$$\frac{x_1^2}{a^2 (a^2 - r^2)} + \frac{y_1^2}{b^2 (b^2 - r^2)} + \frac{z_1^2}{c^2 (c^2 - r^2)} = 0 \qquad \ldots(vii)$$

Hence, the values of λ are the squares of the semi-axes of this section.
Again the direction cosines (l, m, n) of the semi-axis of length r are given by

$$\frac{l}{x_1/(a^2 - r^2)} = \frac{m}{y_1/(b^2 - r^2)} = \frac{n}{z_1/(c^2 - r^2)}$$

so that the axis is parallel to the normal at (x_1, y_1, z_1) to the confocal conicoids. Hence, the theorem.

10.4.1 Locus of Poles of a Plane with Respect to Confocals

To prove that the locus of the poles of a given plane with respect to a system of confocal conicoids is the normal to the plane at the point of contact with the confocal.

Let given plane be

$$lx + my + nz = 1 \qquad \ldots(i)$$

and the equation of confocal be

$$\frac{x^2}{(a^2 - \lambda)} + \frac{y^2}{(b^2 - \lambda)} + \frac{z^2}{(c^2 - \lambda)} = 1 \qquad \ldots(ii)$$

The polar plane of the point (x', y', z') is

$$\frac{xx'}{(a^2 - \lambda)} + \frac{yy'}{(b^2 - \lambda)} + \frac{zz'}{(c^2 - \lambda)} = 1 \qquad \ldots(iii)$$

On comparing it with (i), we have

Confocal Conicoids

$$\frac{x'}{(a^2-\lambda)}=l,\ \frac{y'}{(b^2-\lambda)}=m,\ \frac{z'}{(c^2-\lambda)}=n \qquad ...(iv)$$

$$\Rightarrow \frac{x'}{l}-a^2=\frac{y'}{m}-b^2=\frac{z'}{n}-c^2=-\lambda$$

and the locus of the pole is the straight line whose equation are

$$\frac{x-a^2l}{l}=\frac{y-b^2m}{m}=\frac{z-c^2n}{n}$$

These lines are at right angles to the given plane. Further, the pole with respect to that confocal which touches the plane also lies on this line. But this pole is the point of contact of the plane and the conicoid. Hence, the proposition.

10.4.2 Normals to the Confocals Through a Point

Let the confocals of the conicoid

$$\frac{x^2}{a^2}+\frac{y^2}{b^2}+\frac{z^2}{c^2}=1 \qquad ...(i)$$

which pass through $P(\alpha,\beta,\gamma)$ have parameters $\lambda_1,\lambda_2,\lambda_3$ and let p_1,p_2,p_3 be the perpendiculars from $(0,0,0)$, the centre of conicoid to the tangent planes at P to the confocals.

The equations of normal at $P(\alpha,\beta,\gamma)$ to the confocal of parameters λ_1 are

$$\frac{x-\alpha}{p_1\alpha/(a^2-\lambda_1)}=\frac{y-\beta}{p_1\beta/(b^2-\lambda_1)}=\frac{z-\gamma}{p_1\gamma/(c^2-\lambda_1)} \qquad ...(ii)$$

The co-ordinates of any point Q on it $(PQ=r)$ are

$$\left[\alpha\left(1+\frac{p_1 r}{a^2-\lambda_1}\right),\ \beta\left(1+\frac{p_1 r}{b^2-\lambda_1}\right),\ \gamma\left(1+\frac{p_1 r}{c^2-\lambda_1}\right)\right]$$

The polar plane of P, w.r.t. the given conicoid is

$$\frac{\alpha x}{a^2}+\frac{\beta y}{b^2}+\frac{\gamma z}{c^2}=1 \qquad ...(iii)$$

If Q lies in (iii), then

$$\frac{\alpha^2}{a^2}\left(1+\frac{p_1 r}{a^2-\lambda_1}\right)+\frac{\beta^2}{b^2}\left(1+\frac{p_1 r}{b^2-\lambda_1}\right)+\frac{\gamma^2}{c^2}\left(1+\frac{p_1 r}{c^2-\lambda_1}\right)=1 \qquad ...(iv)$$

Also

$$\frac{\alpha^2}{a^2-\lambda_1}+\frac{\beta^2}{b^2-\lambda_1}+\frac{\gamma^2}{c^2-\lambda_1}=1 \qquad ...(v)$$

Subtracting (iv) from (iii), we get

$$\left[\frac{\alpha^2}{a^2(a^2-\lambda_1)}+\frac{\beta^2}{b^2(b^2-\lambda_1)}+\frac{\gamma^2}{c^2(c^2-\lambda_1)}\right](p_1 r-\gamma_1)=0$$

Hence, $\quad p_1 r-\lambda_1=0 \ \Rightarrow\ r=PQ=\lambda_1/p_1.$

Similarly, we can show that if the normals at P to the other two confocals (parameters λ_2,λ_3) meet the polar plane of P in R and S, then

$$PR=\frac{\lambda_1}{p_2} \quad \text{and} \quad PS=\frac{\lambda_3}{p_3}.$$

Corollary : Putting $p_1 r=\lambda_1$, the co-ordinates of Q becomes :

$$\alpha\left(1+\frac{\lambda_1}{a^2-\lambda_1}\right),\ \beta\left(1+\frac{\lambda_1}{b^2-\lambda_1}\right),\ \gamma\left(1+\frac{\lambda_1}{c^2-\lambda_1}\right)$$

$$\Rightarrow \qquad \left(\frac{a^2\alpha}{a^2-\lambda_1},\frac{b^2\beta}{b^2-\lambda_1},\frac{c^2\gamma}{c^2-\lambda_1}\right)$$

Hence, the equation of the polar plane of Q with regard to the conicoid is

$$\frac{\alpha x}{a^2-\lambda_1}+\frac{\beta y}{b^2-\lambda_1}+\frac{\gamma z}{c^2-\lambda_1}=1$$

which is the same as the tangent plane to the confocal at the point $P(\alpha,\beta,\gamma)$. This tangent plane contains the normals PR and PS. Hence, the polar plane of Q is the plane PRS. In a similar manner, it can be proved that the polar plane of R in the plane PQS, the polar plane of S in the plane PQR and the polar plane of P is the plane QRS.

Hence, *tetrahedron PQRS is self-polar with respect to the given conicoid.*

Remark : We have seen that the tetrahedron $PQRS$ is self-polar with respect to the conicoid. It follows from it that the triangle QRS is self-polar with regard to the common section of the conicoid (i) and the enveloping cone by the plane QRS. Hence, the normals PQ, PR and PS are mutually orthogonal, they are the principal axes of the enveloping cone with the vertex at the point P.

10.4.3 Enveloping Cone

Let us take P as origin, the tangent plane at P to the three confocals as the co-ordinate planes, and the normals PQ, QR and PS as the co-ordinate axes. Then the equation to the enveloping cone is of the form

$$Ax^2+By^2+Cz^2=0 \qquad \ldots(i)$$

The centre C of the conicoid is $(-p_1,-p_2,-p_3)$ and therefore, the equations to the line PC are

$$\frac{x}{p_1}=\frac{y}{p_2}=\frac{z}{p_3} \qquad \ldots(ii)$$

As the centre of the section of the cone or the conicoid by the plane QRS lies on PC, its co-ordinates must be (kp_1,kp_2,kp_3).

Then the equation of QRS will be

$$Ap_1(x-kp_1)+Bp_2(y-kp_2)+Cp_3(z-kp_3)=0 \qquad \ldots(iii)$$

Again the normals being the axes of therefore, the plane QRS makes through $\lambda_1/p_2,\lambda_2/p_3$, λ_3/p_3 on them (10.3.2). Hence, the equation is also given by

$$\frac{x}{\lambda_1/p_1}+\frac{y}{\lambda_2/p_2}+\frac{z}{\lambda_3/p_3}=1$$

$$\Rightarrow \qquad \frac{p_1 x}{\lambda_1}+\frac{p_2 y}{\lambda_2}+\frac{p_3 z}{\lambda_3}=1 \qquad \ldots(iv)$$

Hence,
$$\frac{A}{1/\lambda_1}=\frac{B}{1/\lambda_2}=\frac{C}{1/\lambda_1}$$

Therefore, the equation of enveloping cone becomes

$$\frac{x^2}{\lambda_1}+\frac{y^2}{\lambda_2}+\frac{z^2}{\lambda_3}=0 \qquad \ldots(v)$$

10.4.4 Corresponding Points

Def. Two points (x_1,y_1,z_1) and (ξ_1,η_1,ζ_1) one on each of the co-axial conicoids whose equations are

$$\frac{x^2}{a^2}+\frac{y^2}{b^2}+\frac{z^2}{c^2}=1 \qquad \ldots(i)$$

and
$$\frac{x^2}{\alpha^2}+\frac{y^2}{\beta^2}+\frac{z^2}{\gamma^2}=1 \qquad \ldots(ii)$$

Confocal Conicoids

are said to correspond when
$$\frac{x_1}{a} = \frac{\xi_1}{\alpha}, \frac{y_1}{b} = \frac{\eta_1}{\beta}, \frac{z_1}{c} = \frac{\zeta_1}{\gamma} \qquad \text{...(iii)}$$

In order that real points on one conicoid may correspond to real points on the other, the two surfaces must be of the same nature and must be similarly placed.

Let $P(x_1, y_1, z_1)$ and $Q(x_2, y_2, z_2)$ be the two points on th first conicoid and $P'(\xi_1, \eta_1, \zeta_1)$ and $Q'(\xi_2, \eta_2, \zeta_2)$ be the corresponding points on the other conicoid which is confocal to the first, we have

$$\frac{x_1}{a} = \frac{\xi_1}{\alpha}, \frac{y_1}{b} = \frac{\eta_1}{\beta}, \frac{z_1}{c} = \frac{\zeta_1}{\gamma}$$

$$\frac{x_2}{a} = \frac{\xi_2}{\alpha}, \frac{y_2}{b} = \frac{\eta_2}{\beta}, \frac{z_2}{c} = \frac{\zeta_2}{\gamma}$$

and $\qquad a^2 - \alpha^2 = b^2 - \beta^2 = c^2 - \gamma^2 = \lambda$

Now, $\qquad PQ'^2 = (x_1 - \xi_2)^2 + (y_1 - \eta_2)^2 + (z_1 - \zeta_2)^2$

$$= \left(\frac{a}{\alpha}\xi_1 - \frac{\alpha}{a}x_2\right)^2 + \left(\frac{b}{\beta}\eta_1 - \frac{\beta}{b}y_2\right)^2 + \left(\frac{c}{\gamma}\zeta_1 - \frac{\gamma}{c}z_2\right)^2$$

and $\qquad PQ'^2 = (x_2 - \xi_1)^2 + (y_2 - \eta_1)^2 + (z_2 - \zeta_1)^2$

$\Rightarrow \qquad PQ'^2 = P'Q^2 = \Sigma\left[\left(\frac{a}{\alpha}\xi_1 - \frac{\alpha}{a}x\right)^2 - (x_2 - \zeta_1)^2\right]$

$$= (a^2 - \alpha^2)\left(\frac{\xi_1^2}{\alpha^2} - \frac{x_2^2}{a^2}\right) + (b^2 - \beta^2)\left(\frac{\eta_1^2}{\beta^2} - \frac{y_2^2}{b^2}\right)$$

$$+ (c^2 - \gamma^2)\left(\frac{\zeta_1^2}{\gamma^2} - \frac{z_2^2}{c^2}\right)$$

$$= \lambda\left[\left(\frac{\xi_1^2}{\alpha^2} + \frac{\eta_1^2}{\beta^2} + \frac{\zeta_1^2}{\gamma^2}\right) - \left(\frac{x_2^2}{a^2} + \frac{y_2^2}{b^2} + \frac{z_2^2}{c^2}\right)\right] = \lambda(1-1) = 0$$

$\therefore \qquad PQ' = P'Q$

Hence, the distance between two points one on each of the two confocal ellipsoids is equal to the distance between the two corresponding points.

10.5 EQUATION TO CONICOID REFERRED TO THE NORMALS AS AXES

Let three conicoids confocal with a given conicoid $\dfrac{x^2}{a^2} + \dfrac{y^2}{b^2} + \dfrac{z^2}{c^2} = 1$ pass through a given point P and PQ, PR, PS the normals at P to the confocals, meet the polar plane of P with respect to the given conicoid in Q, R, S. Now we are to find the equation of the conicoid with reference to the normals PQ, QR and PS as co-ordinate axes.

The conicoid will be having contact with the cone along the section of the cone and the plane QRS and hence the equation of the conicoid will be of the form

$$\left(\frac{x^2}{\lambda_1} + \frac{y^2}{\lambda_2} + \frac{z^2}{\lambda_3}\right) = k\left(\frac{p_1 x}{\lambda_1} + \frac{p_2 y}{\lambda_2} + \frac{p_3 z}{\lambda_3} - 1\right)^2 \qquad \text{...(i)}$$

The centre $O(-p_1, -p_2, -p_3)$ of the conicoid bisects all chords passing through it. The equations of the chord parallel to x-axis i.e., PQ are given by

$$\frac{x+p_1}{1} = \frac{y+p_2}{0} = \frac{z+p_3}{0} = r \text{ (say)} \qquad \ldots\text{(ii)}$$

Let the chord (ii) meets the conicoid (i) in the point $(r - p_1, -p_2, -p_3)$. Then

$$\frac{(r-p_1)^2}{\lambda_1} + \frac{p_2^2}{\lambda_2} + \frac{p_3^2}{\lambda_3} = k\left[\frac{p_1(r-p_1)}{\lambda_1} - \frac{p_2^2}{\lambda_2} - \frac{p_3^2}{\lambda_3} - 1\right]^2 \qquad \ldots\text{(iii)}$$

The equation (iii) being a quadratic in r shows that the chord (ii) meets the conicoid (i) in two points. Since the centre is the middle point of the chord, the two values of r should be equal in magnitude but opposite in signs. Hence, if r_1 and r_2 be the roots of (iii), then $r_1 + r_2 = 0$.

$\Rightarrow$ coeff. of $r = 0$

$$\Rightarrow \frac{2p_1}{\lambda_1} + \frac{2kp_1}{\lambda_1}\left(\frac{p_1^2}{\lambda_1} + \frac{p_2^2}{\lambda_2} + \frac{p_3^2}{\lambda_3} + 1\right) = 0$$

$$\Rightarrow \frac{1}{k} = \frac{p_1^2}{\lambda_1} + \frac{p_2^2}{\lambda_2} + \frac{p_3^2}{\lambda_3} + 1$$

Putting the value of k in (i), the equation of the conicoid is

$$\left(\frac{x^2}{\lambda_1} + \frac{y^2}{\lambda_2} + \frac{z^2}{\lambda_3}\right)\left(\frac{p_1^2}{\lambda_1} + \frac{p_2^2}{\lambda_2} + \frac{p_3^2}{\lambda_3} + 1\right) = \left(\frac{p_1 x}{\lambda_1} + \frac{p_2 y}{\lambda_2} + \frac{p_3 z}{\lambda_3} - 1\right)^2$$

EXAMPLES

1. *Three conicoids confocal with a given conicoid $\dfrac{x^2}{a^2} + \dfrac{y^2}{b^2} + \dfrac{z^2}{c^2} = 1$ pass through a given point P and PQ, PR, PS the normals at P to the confocals, meet the polar plane of P with respect to the given conicoid in Q, R, S. Prove that the tetrahedron PQRS is self-polar with respect to the given conicoid.*

Solution. The co-ordinates of the point Q (10.3.2) are

$$\left[\alpha\left(1 + \frac{p_1 r}{a^2 - \lambda_1}\right), \beta\left(1 + \frac{p_1 r}{b^2 - \lambda_1}\right), \gamma\left(1 + \frac{p_1 r}{c^2 - \lambda_1}\right)\right]$$

where $p_1 r = \lambda_1$. Hence, the co-ordinates reduce to

$$\left(\frac{a^2\alpha}{a^2 - \lambda_1}, \frac{b^2\beta}{b^2 - \lambda_1}, \frac{c^2\gamma}{c^2 - \lambda_1}\right)$$

Polar plane of Q with respect to the given conicoid is

$$\frac{\alpha x}{a^2 - \lambda_1} + \frac{\beta y}{b^2 - \lambda_1} + \frac{\gamma z}{c^2 - \lambda_1} = 1$$

which is also the tangent plane at P to the confocal whose parameter is λ_1, *i.e.*, the plane perpendicular to PQ.

But the normals to the three confocals through P, *viz.*, PQ, PR, PS are mutually perpendicular. Therefore, the tangent plane at P to the first confocal is the plane PRS.

It follows that the polar plane of Q is PRS, and similarly, the polar planes of R and S are PSQ and PRQ, while the polar plane of P is QRS. Hence, the tetrahedron $PQRS$ is self-polar with respect to the conicoid.

2. *Prove that the difference of the squares of the perpendiculars from the centre on any two parallel tangent planes to two given confocal conicoids is constant.*

Solution. Let the two confocal conicoids of

$$\frac{x^2}{a^2} + \frac{y^2}{b^2} + \frac{z^2}{c^2} = 1$$

be
$$\frac{x^2}{a^2-\lambda_1}+\frac{y^2}{b^2-\lambda_2}+\frac{z^2}{c^2-\lambda_1}=1 \qquad ...(i)$$

and
$$\frac{x^2}{a^2-\lambda_2}+\frac{y^2}{b^2-\lambda_2}+\frac{z^2}{c^2-\lambda_2}=1 \qquad ...(ii)$$

where λ_1 and λ_2 are constants.

Now, if p_1 and p_2 be lengths of perpendiculars from the centre on the two parallel planes

$$lx + my + nz = p_1 \qquad ...(iii)$$

and
$$lx + my + nz = p_2 \qquad ...(iv)$$

where (l, m, n) are direction cosines of the normals to these planes.

∴ If (iii) is a tangent plane of (i), we have

$$l^2(a^2-\lambda_1)+m^2(b^2-\lambda_1)+n^2(c^2-\lambda_1)=p_1^2$$

Similarly,
$$l^2(a^2-\lambda_2)+m^2(b^2-\lambda_2)+n^2(c^2-\lambda_2)=p_2^2$$

On subtraction, we have

$$(p_2^2-p_1^2)=l^2(\lambda_2-\lambda_1)+m^2(\lambda_2-\lambda_1)+n^2(\lambda_2-\lambda_1)$$

$$=(l^2+m^2+n^2)(\lambda_2-\lambda_1)=\lambda_2-\lambda_1=\text{Constant.}$$

EXERCISES

1. If λ and μ are the parameters of the confocal hyperboloids through a point P on the ellipsoid $\dfrac{x^2}{a^2}+\dfrac{y^2}{b^2}+\dfrac{z^2}{c^2}=1$, then prove that the perpendicular from the centre to the tangent plane at P to the ellipsoid is $abc/\sqrt{\lambda\mu}$. Prove that the perpendicular to the tangent planes to the hyerboloids are

$$\left[\frac{(a^2-\lambda)(b^2-\lambda)(c^2-\lambda)}{\lambda(\lambda-\mu)}\right]^{1/2} \quad \text{and} \quad \left[\frac{(a^2-\mu)(b^2-\mu)(c^2-\mu)}{\mu(\mu-\lambda)}\right]^{1/2}$$

2. If $\lambda_1, \lambda_2, \lambda_3$ are the parameters of three confocals of $\dfrac{x^2}{a^2}+\dfrac{y^2}{b^2}+\dfrac{z^2}{c^2}=1$ that pass through P, prove that the perpendiculars from the centre of the tangent plane at P are

$$\left[\frac{(a^2-\lambda_1)(b^2-\lambda_1)(c^2-\lambda_1)}{(\lambda_3-\lambda_1)(\lambda_3-\lambda_1)}\right]^{1/2} \text{ etc.}$$

3. Show that two confocal paraboloids cut everywhere at right angles.

4. If P is a point on an ellipsoid and P' is the corresponding point on a confocal whose parameter is λ, then prove that $OP^2 - OP'^2 = \lambda$, where O is the centre.

5. Find the equation to the conicoid $\dfrac{x^2}{a^2}+\dfrac{y^2}{b^2}+\dfrac{z^2}{c^2}=1$ referred to the normals to its confocals through any point P as co-ordinates axes.

$$\left[\text{Ans. } \left(\frac{x^2}{\lambda_1}+\frac{y^2}{\lambda_2}+\frac{z^2}{\lambda_3}\right)\left(\frac{p_1^2}{\lambda_1}+\frac{p_2^2}{\lambda_2}+\frac{p_3^2}{\lambda_3}+1\right)=\left(\frac{p_1 x}{\lambda_1}+\frac{p_2 y}{\lambda_2}+\frac{p_3 z}{\lambda_3}-1\right)^2\right]$$

6. If $a_1, b_1, c_1; a_2, b_2, c_2; a_3, b_3, c_3$ are the axes of the confocals of $\dfrac{x^2}{\alpha^2} + \dfrac{y^2}{\beta^2} + \dfrac{z^2}{\gamma^2} = 1$ which pass through a point (ξ, η, ζ) and p_1, p_2, p_3 are the perpendiculars from the centre to the tangent planes to the confocals at the point, prove that

(i) $\xi^2 + \eta^2 + \zeta^2 = a_1^2 + b_2^2 + c_3^2$

(ii) $\dfrac{p_1^2}{a_1^2} + \dfrac{p_2^2}{a_2^2} + \dfrac{p_3^2}{a_3^2} = 1$

(iii) $\dfrac{p_1^2}{a_1^2 - \alpha^2} + \dfrac{p_2^2}{a_2^2 - \alpha^2} + \dfrac{p_3^2}{a_3^2 - \alpha^2} - 1 = \dfrac{\alpha^2 \beta^2 \gamma^2}{(a_1^2 - \alpha^2)(a_2^2 - \alpha^2)(a_3^2 - \alpha^2)}$

10.6 FOCUS AND DIRECTRIX

There are two definitions of a conicoid which correspond to the focus and directrix definition of a conic.

1. The definition according to *Maccullagh* is as follows:

The conicoid is the locus of a point which moves so that its distance from a fixed point is in a constant ratio to its distance from a given straight line measured parallel to a given plane.

The fixed point is called the *focus* and the given line the *directrix*.

Let the given plane be $z = 0$ and the point of intersection of the given line and given plane be the origin. Choosing rectangular axes, let the fixed point be (α, β, γ) and the given line be

$$\frac{x}{l} = \frac{y}{m} = \frac{z}{n} \qquad \ldots(i)$$

Let (ξ, η, ζ) be any point on the locus. Then the plane through it parallel to $z = 0$ will meet the given line in the point $\left(\dfrac{l\zeta}{n}, \dfrac{m\zeta}{n}, \zeta\right)$.

The distance of (ξ, η, ζ) from the line measured parallel to the given plane is, therefore,

$$\sqrt{\{(\xi - l\zeta/n)^2 + (\eta - m\zeta/n)\}^2}$$

Hence, according to the definition, the locus will be

$$(x - \alpha)^2 + (y - \beta)^2 + (z - \gamma)^2 = k^2 \{(x - lz/n)^2 + (y - mz/n)^2\} \qquad \ldots(ii)$$

This is the equation of conicoid.

This conicoid is of the form

$$\lambda \varphi - (u^2 + v^2) = 0 \qquad \ldots(iii)$$

where $\varphi = (x - \alpha)^2 + (y - \beta)^2 + (z - \gamma)^2$ and $u = 0, v = 0$ represent planes.

II. The other definition is due to *Salmon* and is as follows:

The conicoid is the locus of a point the square of whose distance from a fixed point varies as the product of its distancse from two given planes.

The fixed point is called the *focus* and the line of intersection of the given planes the *directrix*.

Clearly, the equation of the locus is of the form

$$(x - \alpha)^2 + (y - \beta)^2 = k^2 (lx + my + nz + p)(l'x + m'y + n'z + p')$$

which is the equation of the conicoid.

This is of the form $\lambda \varphi - uv = 0$ where φ, u, v have the same meaning as before.

Now in either case if $S = 0$ represents the locus of the point, then

$$S - \lambda \varphi = 0$$

represents a pair of planes, which are imaginary in Case I and real in Case II. But their line of intersection, viz., $u = 0 = v$ is real in both cases.

Confocal Conicoids

Thus, we have the following rule for finding the focus and directrix of a conicoid :

If $S = 0$ is the equation to a conicoid and $\lambda, \alpha, \beta, \gamma$ are constants such that the equation $S - \lambda \varphi = 0$ represents two planes (real or imaginary), then (α, β, γ) is a focus and the line of intersection of the two planes is the corresponding directrix.

10.6.1 Foci of $ax^2 + by^2 + cz^2 = 1$

From 10.4, if (α, β, γ) is the focus, then

$$ax^2 + by^2 + cz^2 - 1 - \lambda[(x-\alpha)^2 + (y-\beta)^2 + (z-\gamma)^2] \qquad \text{...(i)}$$

must be the product of two linear factors.
Hence, λ must be equal to a, or b, or c.

(i) When $\lambda = a$, (i) becomes

$$(b-a)y^2 + (c-a)z^2 + 2a\alpha x + 2a\beta y + 2a\gamma z - a(\alpha^2 + \beta^2 + \gamma^2) - 1$$

$$\Rightarrow (b-a)\left(y + \frac{a\beta}{b-a}\right)^2 + (c-a)\left(z + \frac{a\gamma}{c-a}\right)^2 - \frac{ab\beta^2}{b-a} - \frac{ac\gamma^2}{c-a} - 1 + 2a\alpha x - a\alpha^2$$

In order that this may be resolved into two linear factors, we must have $a = 0$, and

$$\frac{ab\beta^2}{b-a} + \frac{ac\gamma^2}{c-a} + 1 = 0 \Rightarrow \frac{\beta^2}{\left(\frac{1}{b} - \frac{1}{a}\right)} + \frac{\gamma^2}{\left(\frac{1}{c} - \frac{1}{a}\right)} = 1$$

(ii) Similarly, when $\lambda = b$, we have

$$\beta = 0 \text{ and } \frac{\alpha^2}{\left(\frac{1}{c} - \frac{1}{c}\right)} + \frac{\beta^2}{\left(\frac{1}{b} - \frac{1}{c}\right)} = 1$$

and
(iii) When $\lambda = c$, we have

$$\gamma = 0 \text{ and } \frac{\alpha^2}{\left(\frac{1}{a} - \frac{1}{c}\right)} + \frac{\beta^2}{\left(\frac{1}{b} - \frac{1}{c}\right)} = 1$$

There are, therefore, three conics, one in each principal plane on which the principal foci lie.
If the surface is the ellipsoid

$$\frac{x^2}{a^2} + \frac{y^2}{b^2} + \frac{z^2}{c^2} = 1 \; (a > b > c)$$

the conics on which the foci lie are

$$x = 0, \quad \frac{y^2}{(b^2 - a^2)} + \frac{z^2}{(c^2 - a^2)} = 1$$

$$y = 0, \quad \frac{z^2}{(c^2 - b^2)} + \frac{x^2}{(a^2 - b^2)} = 1$$

and $\quad z = 0, \quad \dfrac{x^2}{(a^2 - c^2)} + \dfrac{y^2}{(b^2 - c^2)} = 1$

Of these, the first is imaginary, while the other two are real.

These are known as *focal conics*, the one being an ellipse and other a hyperbola. They are also the boundaries of limiting case of confocal conicoids as shown in 10.1.

It is obvious that confocal conicoids have the same focal conics.

EXAMPLES

1. *Show that the focal conics of a paraboloids are two parabolas.*

Solution. Let the equation of the paraboloid be
$$ax^2 + by^2 = 2z \qquad \ldots(i)$$

Let (α, β, γ) be a focus. Then
$$ax^2 - by^2 - 2z - \lambda\{(x-\alpha)^2 + (y-\beta)^2 + (z-\gamma)^2\} = 0 \qquad \ldots(ii)$$
represents a pair of planes.

(i) When $\lambda = a$, then from (ii)
$$2a\alpha x - a\alpha^2 + (b-a)y^2 + 2a\beta y - a\beta^2 - az^2 + 2(a\gamma - 1)z - a\gamma^2 = 0$$

$$\Rightarrow \quad 2a\alpha x - a\alpha^2 + (b-a)\left(y + \frac{a\beta}{b-a}\right)^2 - a\left(z - \frac{a\gamma - 1}{a}\right)^2 - \frac{ab\beta^2}{b-a} - \frac{2a\gamma - 1}{a} = 0$$

This will represent two planes, if
$$\alpha = 0 \text{ and } \frac{ab\beta^2}{b-a} + \frac{2a\gamma - 1}{a} = 0 \qquad \ldots(iii)$$

(ii) When $\lambda = b$, then
$$(a-b)x^2 + 2b\alpha x - b\alpha^2 - 2b\beta y - b\beta^2 - bz^2 - 2(b\gamma - 1)z - b\gamma^2 = 0$$

$$\Rightarrow \quad (a-b)\left(x + \frac{b\alpha}{a-b}\right)^2 + 2b\beta y - b\beta^2 - b\left(z - \frac{b\gamma - 1}{b}\right)^2 - \frac{abz^2}{a-b} - \frac{2b\gamma - 1}{b} = 0$$

and in order that this may represent two planes we must have
$$\beta = 0 \text{ and } \frac{ab\alpha^2}{a-b} + \frac{2b\gamma - 1}{b} = 0 \qquad \ldots(iv)$$

From equations (iii) and (iv), we have that the foci of the paraboloid lie on the conics
$$x = 0, \quad \frac{y^2}{\frac{1}{a} - \frac{1}{b}} + 2z - \frac{1}{a} = 0$$

and
$$y = 0, \quad \frac{x^2}{\frac{1}{b} - \frac{1}{a}} + 2z - \frac{1}{b} = 0$$

These are two parabolas and are known as the *focal parabolas* of the paraboloid.

2. *Through a straight line in one of the principal planes, tangent planes are drawn to a system of confocals. Prove that the points of contact lie on a plane and that the normals at these points pass through a fixed point in the principal plane.*

Solution. The tangent to a confocal
$$\frac{x^2}{(a^2 - \lambda)} + \frac{y^2}{(b^2 - \lambda)} + \frac{z^2}{(c^2 - \lambda)} = 1 \qquad \ldots(i)$$

at the point (x', y', z') is
$$\frac{xx'}{(a^2 - \lambda)} + \frac{yy'}{(b^2 - \lambda)} + \frac{zz'}{(c^2 - \lambda)} = 1 \qquad \ldots(ii)$$

This passes through the given line
$$\left.\begin{array}{r} lx + my = 1 \\ z = 0 \end{array}\right\} \qquad \ldots(iii)$$

in the xy-plane if $\quad \dfrac{x'}{(a^2 - \lambda)} = l \text{ and } \dfrac{y'}{(b^2 - \lambda)} = m$

Confocal Conicoids

Eliminating λ, we have
$$\frac{x'}{l} - \frac{y'}{m} = a^2 - b^2$$

Therefore, the points of contact of tangent planes through the given line in the xy-plane, lie on the plane

$$\frac{x}{l} - \frac{y}{m} = a^2 - b^2 \qquad ....(iv)$$

The normal at (x', y', z') is

$$\frac{x-x'}{x'/(a^2-\lambda)} = \frac{y-y'}{y'/(b^2-\lambda)} = \frac{z-z'}{z'/(c^2-\lambda)} \qquad ...(v)$$

This meets the plane $z = 0$, where

$$\frac{x-x'}{l} = \frac{y-y'}{m} = -(c^2-\lambda)$$

$$\Rightarrow \quad \frac{x}{l} = \frac{x'}{l} - c^2 + a^2 - (a^2-\lambda) = a^2 - c^2$$

Similarly, $\quad \dfrac{y}{m} = b^2 - c^2$

Hence, the normal at the point of contact passes through a fixed point in the xy-plane.

EXERCISES

1. Find the equations to the focal conics of the hyperboloid $x^2 + yz - 2 = 0$.

    [Ans. $x = 0$, $y^2 + z^2 + 4yz = 0$ and $y = z$, $2x^2 + 3y^2 = 12$]

2. Find the focal conics of the cone
$$ax^2 + by^2 + cz^2 = 0$$

$$\left[\text{Ans. } y = 0, \frac{z^2}{\frac{1}{c}-\frac{1}{b}} + \frac{x^2}{\frac{1}{a}-\frac{1}{b}} = 0 \text{ and } z = 0, \frac{x^2}{\frac{1}{a}-\frac{1}{c}} + \frac{y^2}{\frac{1}{b}-\frac{1}{c}} = 0\right]$$

Notes

Notes

Notes